W0105771

Molecular Biology of Hematopoiesis 5

Molecular Biology of Hematopoiesis 5

Edited by

Nader G. Abraham
The Rockefeller University
New York, New York

Shigetaka Asano
University of Tokyo
Tokyo, Japan

Günther Brittinger
Universität-Gesamthoschschule Essen
Essen, Germany

Georges J. M. Maestroni
Istituto Cantonale di Patologia
Locarno, Switzerland

and

Richard K. Shadduck
Western Pennsylvania Cancer Institute
Pittsburgh, Pennsylvania

Plenum Press • New York and London

Library of Congress Cataloging-in-Publication Data

Molecular biology of hematopoiesis 5 / edited by Nader G. Abraham ... [et al.].
p. cm.
"Proceedings of the Ninth Symposium on the Molecular Biology of Hematopoiesis, held June 23-27, 1995, in Genoa, Italy"--T.p. verso.
Includes bibliographical references and index.

1. Myeloproliferative disorders--Molecular aspects--Congresses.
2. Hematopoietic stem cell disorders--Molecular aspects--Congresses.
3. Hematopoiesis--Congresses. 4. Bone marrow--Transplantation--Molecular aspects--Congresses. I. Abraham, Nader G.
II. Symposium on the Molecular Biology of Hematopoiesis (9th : 1995 : Genoa, Italy)
[DNLM: 1. Hematopoiesis--congresses. WH 140 M7177 1996]
RC645.75.M65 1996
616.4'107--dc20
DNLM/DLC
for Library of Congress 96-27564
CIP

Proceedings of the Ninth Symposium on the Molecular Biology of Hematopoiesis, held June 23 – 27, 1995, in Genoa, Italy

ISBN-13: 978-1-4613-8031-3 e-ISBN-13: 978-1-4613-0391-6

DOI:10.1007/978-1-4613-0391-6

Softcover reprint of the hardcover 1st edition 1996

A Division of Plenum Publishing Corporation
233 Spring Street, New York, N. Y. 10013

10 9 8 7 6 5 4 3 2 1

Volumes 1 – 4 were published by Intercept, Ltd., Andover, United Kingdom

PREFACE

This volume of *Molecular Biology of Hematopoiesis* is dedicated to John W. Adamson, M.D., Tadamitsu Kishimoto, M.D., Robert C. Gallo, M.D., Arthur W. Nienhuis, M.D., and Franco Mandelli, M.D., for their contributions in developing an overall view of the state-of-the-art knowledge in the field of hematopoiesis. Richard Champlin, among other renowned clinicians, presented updated information on stem cells and T-cell depletion for bone marrow transplant. A clinical update on thrombopoietin was presented by Pamela Hunt of Amgen and by Kenneth Kaushansky. Arthur Nienhuis' and Katherine Turner's contributions to our current knowledge and advances in the fields of growth factors and gene transfer were also recognized during the 9th Symposium on Molecular Biology of Hematopoiesis in Genoa. The chapters cover such diverse areas as preclinical and clinical updates on growth factors and positive and negative regulatory molecules. "Advances in Leukemia: Mechanism and Treatment by Interferon" was presented by Professor Sante Tura. Readers will find presentation of exciting advances that have occurred in the area of hematopoiesis. The elucidation of gene structures of key growth factor proteins such as IL-12 and IL-11 will lead to new insights and new approaches in understanding the regulation of hematopoiesis, as well as application of new growth factors.

Retrovirus gene transfer techniques and genetic manipulation of murine and human cells have added new information on the regulatory steps involved in cell development and genetic disorders. Transfection of eukaryotic cells, specifically bone marrow cells and, subsequently, *in vivo* transplantation, further suggest that hematopoietic reconstitution is possible. Studies of gene transfer continue to reveal the mechanism behind various biological events in which normal and abnormal cell proliferation and differentiation will be achieved. Further, origination and isolation of stem cells, as well as the use of cord blood for bone marrow transplant, were presented. Enthusiasm for the discovery of new growth factor(s) and interleukin(s) remains strong as novel factors are introduced into the field and presented at the meeting. Also, exciting data are being generated in linking growth factors and neuroendocrines to proto-oncogenes and cell proliferation and differentiation. More data will continue to accrue, in particular, with respect to the cell surface receptors, the mechanisms of shut-off of local production of these factors, and possible clinical side effects of their continued administration. We are convinced that this volume will be a useful reference for clinicians, scientists, and investigators who are interested in the field of hematopoiesis.

Nader G. Abraham

ACKNOWLEDGMENTS

The Organizing Committee wishes to acknowledge the Menarini International Foundation and the following companies, whose support made this meeting possible:

- Amgen-Roche, SpA
- Chugai Rhône-Poulenc
- Ortho Biotech
- CellPro Inc., Europe N.V./S.A.
- Schering Plough SpA
- Wellcome Italia SpA
- Glaxo Group Research Limited
- Baxter Healthcare Corporation
- Miltenyi Biotec
- Schering Plough International
- Blackwell Science
- Le Petit (Marion-Merrell Dow)
- Impress
- Stem Cell
- SyStemix
- Ortho Biotech Inc.
- Beckman
- Sandoz
- Immunokontact
- Vestar
- Liposome
- Stockton
- Pasteur Merieux
- Fondaz. Int. Menarini
- OPA
- Intercept
- Plenum

The Organizing Committee gratefully acknowledges the patience and support, including editorial assistance, in preparation of this conference of Joyce Eshet, Symposium Coordinator.

For further information, please write to:

N.G. Abraham
New York University School of Medicine
Brooklyn Hospital Campus
Brooklyn, NY 11201
or
The Rockefeller University
1230 York Avenue
New York, NY 10021

EDITORS'/PUBLISHER'S DISCLAIMER

CONTENTS

Part I. Advances in Bone Marrow Transplant: Stem Cells and T-Cell Depletion

Part II. Clinical Update on Malignant Lymphoma and Myeloma: Mechanism/Therapy

Part III. Update on Advancement in Leukemia

Part V. Molecular Regulation of Hematopoietic Genes and Gene Transfer

Part VI. Cytokine Action and Signal Transduction

Part VII. Thrombopoiesis and Erythropoiesis: An Update

Part VIII. Cytokines and Myeloproliferative Disorders

Part IX. New Directions in Iron Chelating Therapy

T-CELLS, GRAFT-VERSUS-HOST DISEASE, AND GRAFT-VERSUS-LEUKEMIA

New Approaches for Blood and Marrow Transplantation

Richard Champlin, Sergio Giralt, and James Gajewski

Section of Bone Marrow Transplantation
Department of Hematology
Division of Medicine
University of Texas M.D. Anderson Cancer Center
1515 Holcombe Boulevard, Box 65
Houston, Texas 77030

ABSTRACT

Allogeneic bone marrow transplantation is most commonly used as treatment for leukemia. Allotransplantation is associated with several interrelated immunologic processes, graft rejection, graft-vs-host disease and graft-vs-leukemia. Graft rejection can be overcome by intensive pretransplant immunosuppressive therapy. Following transplantation immunoreconstitution must occur from donor derived progenitors and graft-vs-host disease may occur from reactivity of donor derived immunocompetant cells against host tissues. In a related, but distinct process, donor immunocompetant cells may react against the recipient leukemia and recent data confirm that this GVL effect plays a critical role to prevent relapse post transplant. This report summarizes present data regarding the mechanism of these processes. A major challange is to separate the beneficial GVL effect from GVHD, the major complication of allogeneic marrow transplantation. In this report we summarize data regarding innovative approaches to modify the composition of the transplanted marrow to optimize clinical outcome as well as use of donor lymphocyte infusions as a means to induce graft-vs-leukemia post transplant.

The biologic effects of allogeneic bone marrow transplantation depend upon the composition of the graft. The goal is to achieve engraftment and hematopoietic recovery and avoid the development of graft-versus-host disease (GVHD). For patients transplanted for leukemia, another important component is the immune mediated graft-versus-leukemia effect, important to prevent relapse.

Molecular Biology of Hematopoiesis 5, edited by Abraham et al.
Plenum Press, New York, 1996

GRAFT VERSUS HOST DISEASE

GVHD results from reactivity of immunocompetent donor cells present in the transplanted bone marrow against recipient (host) tissues. The pathophysiology of GVHD involves initial afferent phase involving activation of alloreactive T-cells followed by an efferent phase involving cell mediated cytotoxicity, as well as production of inflammatory cytokines.[1,2] CD4 and CD8-positive T-cells are involved and novel T-cell phenotypes and natural killer cells have been described within affected tissues.[3-5] In the efferent phase, macrophages and natural killer cells are recruited into affected sites. The thymus may also be a target tissue and thymic function defect may contribute to the pathophysiology of GVHD. Host cells also participate in the pathogenesis. Cytokines produced by the infiltrating cells enhance the afferent and efferent phase of the graft-versus-host response, as well as contributing directly to tissue injury. Inflammatory cytokines including interleukin-1, tumor necrosis factor and interferon-gamma appear to be important in this process.[2,6]

Depletion of T-cells from the marrow graft is effective to prevent GVHD across both major and minor histocompatibility differences.[7] When engraftment occurs, hematopoietic recovery is prompt and immune reconstitution is derived from undifferentiated progenitors or pre-T cells present within the graft. The kinetics of immune reconstitution are similar as with unmodified transplants.[8] Pan-t cell depletion effectively prevents GVHD across major or minor histocompatibility differences.[3] Techniques which deplete CD3 positive T-cells alone without targeting natural killer cells appear sufficient to prevent GVHD.[7] Selective depletion of natural killer cells has also been reported to reduce GVHD in animals, but this has not ben tested in man.[9]

Several techniques have been proposed for *ex-vivo* depletion of T-lymphocytes.[7,10] The most commonly used methods include physical depletion involving soybean lectin agglutination or elutriation, or immunologic methods using monoclonal anti-t cell antibodies either with complement, magnetic beads, or as immunotoxins to eliminate the antibody bound cells. Recently techniques for positive selection of CD34 positive progenitor cells have been developed;[11,12] this approach excludes T-cells and will prevent GVHD in animals; it has been limited, however, by problems achieving engraftment in allogeneic recipients as have other highly effective T-cell depletion techniques. Optimal physical or immunological techniques are capable of a 3 to 4 log reduction (99.9 to 99.99%) of T-lymphocytes.

GRAFT FAILURE

The risk of graft rejection is increased in T-cell depleted transplants.[13] High doses of total body irradiation do not completely ablate host immunity; viable T- and NK cells persist after the preparative regimen which can mediate graft rejection.[14,15] In man, the risk of graft failure is approximately 2% in recipients transplanted for leukemia using unmodified bone marrow from an HLA-identical sibling donor. In contrast, 10-20% of patients receiving T-lymphocyte depleted bone marrow transplants have had graft failure.[16-19] Graft failure has occurred in two clinical patterns; some patients fail to have any evidence of engraftment, while others have hematologic recovery only to experience graft failure leading to marrow aplasia. Most cases of graft failure following T-cell depleted transplants are due to immunologic rejection. In several well studied cases, host T-lymphocytes have been described that react against donor major or minor histocompatibility antigens. Less commonly, natural killer cells have been implicated.[17,20,21] Recipients of T-cell depleted transplants are more likely to be mixed chimeras than those receiving unmodified marrow.[22] The conditions leading to rejection versus stable mixed chimerism are poorly understood.

Rejection can be overcome by higher hematopoietic cell doses or intensification of the immunosuppressive conditioning regimen.[23] Engraftment is enhanced with higher doses of radiation, higher radiation dose rates, single dose or large fractions of radiotherapy or addition of splenic radiation.[19,24] Addition of busulfan or thiotepa also enhance engraftment of T-cell depleted transplants;[23,25] these agents provide little immunosuppression, but produce more complete ablation of host hematopoiesis. There is an interaction between the intensity of conditioning, the transplanted cell dose and the T-cell content of the marrow.[26] Lower cell doses are sufficient to engraft in the presence of T-cells or after intensive conditioning. Higher cell doses are required if the marrow is thoroughly depleted of t-lymphocytes.

Engraftment requires not only sufficient immunosuppressive therapy as well as the presence of facilitating accessory cells in the graft; transplantation of purified stem cells in the mouse generally fail to durably engraft across a major histocompatibility barrier. These facilitating cells are CD3+ and CD8+.[27,28] It is uncertain whether these facilitating cells are classical T-cells and their mechanism to enhance engraftment is unknown. Growth factors or lymphokines produced by T-cells may mediate some or all of their effect on engraftment, although this effect cannot be completely replaced with any known cytokine.

GRAFT-VERSUS-LEUKEMIA

Recent clinical and experimental data indicate that an important graft-versus-leukemia (GVL) effect is mediated by the transplanted bone marrow.[29,30] The mechanism of GVL and the effector cell populations are incompletely understood . The cell populations putatively mediating graft-versus-leukemia overlap with those associated with GVHD;[31,32] CD4 and CD8 positive T-cells have both been reported to participate. CD4+ and CD8+T-cell clones reactive with human leukemia cells have been described, and LAK and natural killer cells may also contribute. GVL plays a critical role in preventing recurrence of leukemia following allogeneic bone marrow transplantation. Recipients who develop graft-versus-host disease have a lower incidence of leukemia relapse than patients without GVHD.[29,30]

Cell Populations Mediating GVHD and GVL

GVHD	GVL
T-cells	T-cells
CD4+	CD4+
CD8+	CD8+
Novel t cell subsets	
Natural Killer Cells	Natural Killer Cells

The need for the GVL effect is particularly evident in patients receiving allogeneic bone marrow transplantation for treatment of CML, although acute leukemias are also affected. Following transplantation, most patients with CML have residual leukemia cells that can be detected by cytogenetics or polymerase chain reaction analysis for bcr-abl gene recombination. Leukemia cells are most frequently seen early post transplant and less frequently over the first several years post transplant. Patients receiving an unmodified generally remain in hematologic remission; their relapse rate at 3 years is approximately 12%. The residual leukemic cells are viable, however, and go on to produce overt relapse in >50% of patients receiving an identical twin or a T-cell depleted transplant.[29,30] These data indicate that GVL requires both an allogeneic immunologic target which is lacking in syngeneic transplants and T-cells in the transplanted marrow, presumably as the effectors of this process.

GVL can been induced for therapeutic purposes; patients with CML who relapse into chronic phase may achieve a second remission after infusion of donor peripheral blood lymphocytes without cytotoxic therapy.[33-36] This approach is less effective for patients in blast crisis or acute leukemias.

EFFECTS OF T-CELL DEPLETION ON LEUKEMIA FREE SURVIVAL

Techniques that effectively deplete T-lymphocytes have uniformly reduced the incidence and severity of graft-versus-host disease in recipients of HLA-identical transplants. Despite this benefit, it has not been possible to conclusively demonstrate improved overall survival following allogeneic marrow transplantation from a matched sibling donor. In patients with leukemia, the beneficial effects on GVHD have been offset by an increased risk of graft failure and leukemia relapse, abrogating improvement in leukemia free survival in controlled trials and large registry analyses.[7,19]

For transplants from unrelated HLA closely matched donors, the rate of grade ≥ 2 acute GVHD exceeds 60% despite combination posttransplant immunosuppressive therapy.[37,38] In these patients, use of t-cell depletion markedly reduces the incidence of acute GVHD, but graft rejection is an even greater problem, occurring in 20-50% of cases, unless an intensified preparative regimen is utilized. It is notable, however, that some successful transplants have been performed from fully haploidentical donors, a setting in which transplantation of unmodified marrow is rarely successful. Children with severe combined immunodeficiency disease have received successful transplants using T-cell depleted bone marrow from HLA-haploidentical parents with improvement of immunity, and successful transplants have been performed for patients with leukemia, particularly in children.[37,39-42] The risk of recurrent leukemia has not been as markedly increased with T-cell depleted transplants in this setting; GVL may be mediated by a smaller number of T-cells in this setting of greater genetic disparity between the donor and recipient. Controlled studies are required to examine the efficacy of T-cell depletion versus other strategies to prevent GVHD in these patients.

INNOVATIVE APPROACHES USING T-CELL DEPLETION

A number of novel approaches to ex vivo treatment have been proposed to improve results of transplants for leukemia as summarized in the following table.

Innovative Approaches for ex Vivo Treatment of Allogeneic Marrow Transplants

1. Intensified conditioning to enhance engraftment and eradication of malignancy
2. Optimize T-Cell Dose
   - Critical dose needed for engraftment and GVL, but below threshold for GVHD
   - Combine subtotal T-cell depletion with post transplant immunosuppression
3. Maximize hematopoietic stem cell dose
4. Add or retain graft facilitating cells while depleting GVHD effectors
5. Selective Depletion of T-cell subsets
   - CD8-positive cells
6. Retain suppressor/regulatory cells
7. Add back of immunocompetent cells to mediate GVL
8. Post transplant treatment with immunostimulatory agents

As indicated, the major problems which need to be addressed are graft failure, GVHD and recurrent leukemia. One approach is to employ an intensified more effective pretransplant preparative regimen in hope of overcoming resistance to engraftment and providing a greater antileukemic activity. Preliminary data suggests that administration of a higher dose of total body irradiation (13.5-15.75 Gy), single dose rather than fractionated radiation, or addition of total lymphoid irradiation may be associated with a lower risk of graft failure.[7,18,43] Intensification of conditioning therapy is limited by nonhematopoietic toxicity, and although many centers are actively evaluating this approach, no intensified regimen has been documented to improve overall results. Preliminary data suggest that addition of antithymocyte globulin or thiotepa to the preparative regimen may reduce graft failure.[22]

Conceivably, it may be possible to include a critical number of T-cells in the graft which would be below the threshold necessary to produce GVHD, but sufficient to allow engraftment and mediate GVL. It is likely that the dose of T-cells required for each of these processes will vary among patients depending on the degree of genetic disparity and the major and minor histocompatibility differences present. Thus, it seems unlikely that an optimal T-cell dose for all patients can be defined in the absence of a more thorough understanding of the pathogenesis of GVHD, engraftment and GVL. Subtotal depletion of T-cells might be beneficial if combined with post transplant immunosuppressive therapy. Preliminary studies suggest favorable results with a 1.5 to 2 log T-cell depletion followed by systemic therapy with cyclosporine or anti-CD5 ricin immunotoxin.[38,44,45]

Increasing the hematopoietic cell dose enhances engraftment and is likely to be beneficial for immune reconstitution. It is not possible to substantially increase the marrow cell dose without risk to the donor, but it is possible to markedly increase the cell dose by combining marrow with G-CSF mobilized peripheral blood cells collected by apheresis. Mobilized peripheral blood contains pluripotent stem cells in vitro and have been effective to restore hematopoiesis after allogeneic transplantation.[46-48] Combining mobilized blood with marrow allows collection of a 3-10 fold greater number of progenitors and may result in more rapid and complete hematopoietic recovery. Encouraging preliminary results using this approach for haploidentical transplants have recently been reported.[49]

Graft facilitator cells have recently been described which markedly enhance engraftment of purified stem cell allotransplants in mice. These cells are CD3+ and CD8+, but are reported not to express the alpha/beta or gamma/delta T-cell receptor.[28] Inclusion of these graft facilitator cells with a purified stem cell allograft enhances engraftment in mice. CD8+ cells are critical for engraftment of human allogeneic transplants[50] and whether these cells can be separated from GVHD effectors requires confirmation in clinical trials.

It is conceivable that the cellular subsets responsible for GVHD and graft-versus-leukemia may not completely overlap. CD4-positive T-cells recognize antigens presented in association with class II MHC antigens while CD8-positive cells recognize antigens presented in the context of class I loci. In murine models, CD4-positive cells primarily mediate acute GVHD in MHC class II disparate recipients, while CD8-positive cells are responsible for GVHD in class I disparate transplant mice. Depletion of the CD8-positive cytotoxic/suppressor subset of T-lymphocytes is sufficient to reduce or prevent graft-versus-host disease in most, but not all, donor-recipient murine strain combinations that are MHC compatible but mismatched for minor histocompatibility loci.[51,52] Selective depletion of CD8-positive cells in combination with post transplant cyclosporine has recently been evaluated in man; this approach resulted in a reduced rate of acute GVHD in HLA matched sibling grafts without an increase in leukemia relapse in a recent randomized study.[53] Notably, the rate of leukemia relapse in patients with CML has not been improved suggesting that this approach may separate GVHD and GVL at least in this disease.[54,55] Improved methods to selectively enhance GVL following T-cell or T-subset depleted transplants may be possible. T-cell clones or LAK/NK cell populations capable of mediating an antileukemic effect raises the possibility of infusing cloned allogeneic effector cells post transplant to enhance immune antileukemia activity.

The bone marrow contains regulatory T-cells which appear important in development of tolerance. Natural suppressor cells have been identified in mice which inhibit development of GVHD; these cells have the phenotype of CD4- CD8- T-cells or appear as null cells.[56] T-cell depletion procedures which retain this cell population may be preferable. Helper T-cell TH1 and TH2 subpopulations may also be important. TH2 cells produce cytokines inhibitory to cell mediated immune and inflammatory responses; enrichment for these cells or infusion of these antiinflammatory cytokines may reduce GVHD. Other accessory cells may also be important.

Post transplant immunostimulatory therapy with interleukin-2 or other agents may be possible with a goal of enhancing immunoreconstitution and graft-vs-leukemia. Interleukin-2 treatment increases GVHD after unmodified bone marrow transplantation, but it can be safely administered after effective T-cell depletion; preliminary data indicate that leukemia relapse rates may be reduced.[57]

DONOR LYMPHOCYTE INFUSION FOR TREATMENT OF RECURRENT LEUKEMIA FOLLOWING BMT

The recent observation that simple infusion of donor peripheral blood lymphocytes from the original transplant donor can reinduce remission in patients with CML who have relapsed after allogeneic transplantation is of fundamental importance.[58] Patients with acute leukemia are less likely to respond. This is the most direct and compelling evidence of the graft-vs. leukemia effect in man. This approach has also been effective in treatment of post-transplant Epstein-Barr virus related lymphoproliferative disease and possibly other infections as well.[59]

Approximately 70% of patients with CML relapsing into chronic phase have achieved durable cytogenetic complete remissions and many are negative for residual disease by polymerase chain reaction analysis for bcr-abl rearrangement.[34,36,60] This approach does have risks however. GVHD occurs in 80% of the patients and marrow aplasia in 56%; these have been important causes of treatment failure accounting for the death in up to 20% of responding patients. Aplasia typically develops coincident with clinical response. Following ablation of leukemic hematopoiesis, hematopoiesis must be restored by donor derived cells. If insufficient numbers of donor myeloid progenitors are present, prolonged aplasia will occur. Aplasia can be effectively treated by additional marrow infusion from the donor. Aplasia occurs more frequently in patients infused during hematologic relapse than in patients with isolated cytogenetic relapse.[61]

The efficacy of donor lymphocyte infusions in reinducing remissions encourage exploration of this strategy as a method for relapse prevention. Recent studies have shown that it is possible to safely infuse increasing number of donor lymphocytes in animals receiving T-cell depleted transplants by delaying the time period between the initial transplant and the donor lymphocyte infusion.[62] Thus, it may be possible to perform an initial T-cell depleted transplant to prevent GVHD, and later infuse donor lymphocytes to provide GVL activity.[63]

The optimal cell dose for donor lymphocyte infusion to reinduce remission are uncertain. Recently a strategy using progressively escalating doses in patients with CML, starting with a low doses unlikely to produce GVHD has been proposed.[64] Antileukemia effects are generally not seen with lymphocyte cell doses $<10^7$ cells/kg.

Since bone marrow transplants using selective depletion of CD8 postive cells results in reduced GVHD without an increase in the relapse rate in chronic myelogenous leukemia,[54] this suggests that CD8 depleted lymphocytes are capable of inducing GVL, although with a reduced potential to produce GVHD. We recently reported that CD8 depleted donor lymphocyte infusion was effective to reinduce remission in CML patients relapsing after allogeneic BMT.[65] Only 20% of patients had mild GVHD, suggesting that this approach may improve the therapeutic index.

Recently genetic manipulation transducing the donor lymphocyte infusion with Herpes virus thymidine kinase has been proposed to improve the safety of the procedure. This renders the transduced cells sensitive to ganciclovir which can be administered to abrogate GVHD if it occurs[66]. This approach has been reported successful in initial studies,[67] but further evaluation is necessary to determine the optimal means to control GVHD without also abrogating the GVL effect.

CONCLUSION

In conclusion, the cellular composition of the marrow determines the biological effects of the transplant. Depletion of T-cells is the most effective method to prevent acute and chronic graft-versus-host disease. This success has been largely offset by adverse effects on engraftment and abrogation of graft-vs. leukemia. It is likely that the composition of the graft can be modified to improve therapeutic outcomes. Results may be enhanced by further innovations in preparative regimens, doses of T-cell subsets, accessory cells and stem cells, and novel post transplant immunosuppressive and biologic therapies.

REFERENCES

1. Antin JH, Ferrara JLM: Cytokine dysregulation and acute graft-versus-host disease. Blood 80:2964, 1992
2. Ferrara JLM, Deeg HJ: Mechanisms of disease: Graft-versus-host disease. N Engl J Med 324:667, 1991
3. Korngold R, Sprent J: T cell subsets and graft-versus-host disease.. Transplantation 44:335, 1987
4. Ferrara JLM, Guillen FJ, vanDijken PJ, Marion A, Murphy GF, Burakoff SJ: Evidence that large granular lymphocytes of donor origin mediate acute graft-versus-host disease.. Transplant 47:50, 1989
5. Ferrara JLM: Advances in GVHD: Novel lymphocyte subsets and cytokine dysregulation. Bone Marrow Transplant 10 Suppl. 1:10, 1992
6. Ferrara JLM: Cytokine dysregulation as a mechanism of graft versus host disease. Curr Opin Immunol 5:794, 1993
7. Champlin RE: T-cell depletion for allogeneic bone marrow transplantation: impact on graft-versus-host disease, engraftment, and graft-versus-leukemia.. J Hematotherapy 2:27, 1993
8. Keever CA, Small TN, Flomenberg N, Heller G, Pekle K, Black P, Pecora A, Gillio A, Kernan NA, O'Reilly RJ: Immune reconstitution following bone marrow transplantation: Comparison of recipients of T-cell depleted marrow with recipients of conventional marrow grafts. Blood 73:1340, 1989
9. Johnson BD, Truitt RL: A decrease in graft-vs.-host disease without loss of graft-vs.-leukemia reactivity after MHC-matched bone marrow transplantation by selective depletion of donor NK cells in vivo. Transplantation 54:104, 1992
10. Hale G, Waldmann H: Control of graft-versus-host disease and graft rejection by T cell depletion of donor and recipient with Campath-1 antibodies. Results of matched sibling transplants for malignant diseases. Bone Marrow Transplant 13:597, 1994
11. Berenson RJ, Bensinger WI, Hill RS, Andrews RG, Garcia-Lopez J, Kalamasz DF, Still BJ, Spitzer G, Buckner CD, Bernstein ID, Thomas ED: Engraftment after infusion of $CD34^+$ marrow cells in patients with breast cancer or neuroblastoma. Blood 77:1717, 1991
12. Cottler-Fox M, Cipolone K, Yu M, Berenson R, O'Shaughnessy J, Dunbar C: Positive selection of $CD34^+$ hematopoietic cells using an immunoaffinity column results in T cell-depletion equivalent to elutriation. Exp Hematol 23:320, 1995

13. Mitsuyasu R, Champlin RE, Gale RP, Ho WG, Lanarsky C, Winston D, Selch M, Elashoff R, Giorgi JV, Wells J, Terasaki P, Billing R, Feig S: Depletion of T-lymphocytes from donor bone marrow for the prevention of graft-versus-host disease following bone marrow transplantation. Ann Intern Med 105:20, 1986
14. Reisner Y, Ben-Bassat B, Douer D, Kaploon A, Schwartz E, Ramot B: Demonstration of clonable alloreactive host T cells in a primate model for bone marrow transplantation. Proc.Natl Acad 83:4012, 1986
15. Butturini A, Seeger RC, Gale RP: Recipient immune-competent T lymphocytes can survive intensive conditioning for bone marrow transplantation.. Blood 68:954, 1986
16. Martin PJ, Hansen JA, Torok-Storb B, Durnam D, Przepiorka D, O'Quigley J, Sanders J, Sullivan KM, Witherspoon H, Deeg J, Appelbaum FR, Stewart P, Weiden P, Doney K, Buckner CD, Clift R, Storb R, Thomas ED: Graft failure in patients receiving T cell-depleted HLA-identical allogeneic marrow transplants.. Bone Marrow Transplantation 3:445, 1989
17. Kernan NA, Bordignon C, Heller G, Cunningham I, Castro-Malaspina H, Shank B, Flomenberg N, Burns J, Yang SY, Black P, Collins NH, O'Reilly RJ: Graft failure after T-cell-depleted human leukocyte antigen identical marrow transplants for leukemia: I. Analysis of risk factors and results of secondary transplants. Blood 74:2227, 1989
18. Vallera DA, Blazar BR: T cell depletion for graft-versus-host-disease prophylaxis: A perspective on engraftment in mice and humans. Transplantation 47:751, 1989
19. Marmont AM, Horowitz MM, Gale RP, Sobocinski K, Ash RC, van Bekkum DW, Champlin RE, Dicke KA, Goldman JM, Good RA, Herzig RH, Hong R, Masaoka T, Rimm AA, Ringdén O, Speck B, Weiner RS, Bortin MM: T-cell depletion of HLA-identical transplants in leukemia. Blood 78:2120, 1991
20. Kernan NA, Flomenberg N, Dupont B, O'Reilly RJ: Graft rejection in recipients of T-cell-depleted HLA-nonidentical marrow transplants for leukemia.. Transplantation 43:842, 1987
21. Bordignon C, Keever CA, Small TN, Flomenberg N, Dupont B, O'Reilly RJ, Kernan NA: Graft failure after T-cell-depleted human leukocyte antigen identical marrow transplants for leukemia: II. In vitro analyses of host effector mechanisms. Blood 74:2237, 1989
22. Mackinnon S, Barnett L, Bourhis JH, Black P, Heller G, O'Reilly RJ: Myeloid and lymphoid chimerism after T-cell-depleted bone marrow transplantation: Evaluation of conditioning regimens using the polymerase chain reaction to amplify human minisatellite regions of genomic DNA. Blood 80:3235, 1992
23. Lapidot T, Terenzi A, Singer TS, Salomon O, Reisner Y: Enhancement by dimethyl myleran of donor type chimerism in murine recipients of bone marrow allografts. Blood 73:2025, 1989
24. Lapidot T, Singer TS, Salomon O, Terenzi A, Schwartz E, Reisner Y: Booster irradiation to the spleen following total body irradiation: A new immunosuppressive approach for allogeneic bone marrow transplantation. J Immunol 141:2619, 1988
25. Terenzi A, Lubin I, Lapidot T, Salomon O, Faktorowich Y, Rabi I, Martelli MF, Reisner Y: Enhancement of T cell-depleted bone marrow allografts in mice by thiotepa. Transplantation 50:717, 1990
26. Uharek L, Gassmann W, Glass B, Steinmann J, Loeffler H, Mueller-Ruchholtz W: Influence of cell dose and graft-versus-host reactivity on rejection rates after allogeneic bone marrow transplantation. Blood 79:1612, 1992
27. Martin PJ: Donor CD8 cells prevent allogeneic marrow graft rejection in mice: Potential implications for marrow transplantation in humans. J Exp Med 178:703, 1993
28. Kaufman C, Colson YL, Wren SM, Watkins S, Simmons RL, Ildstad ST: Characterization of a novel cell that facilitates engratment of purified stem cells in an allogeneic MHC-disparate environment.. Blood 82(Suppl 1):456a, 1993
29. Champlin R: Graft-versus-leukemia without graft-versus-host disease: An elusive goal of bone marrow transplantation. Semin Hematol 29 Suppl. 2:46, 1992
30. Sullivan KM, Storb R, Buckner CD, Fefer A, Fisher L, Weiden PL, Witherspoon RP, Appelbaum FR, Banaji M, Hansen J, Martin P, Sanders JE, Singer J, Thomas ED: Graft-versus-host disease as adoptive immunotherapy in patients with advanced hematologic neoplasms. N Engl J Med 320:828, 1989
31. Faber LM, Van Luxemburg-Heijs SAP, Willemze R, Falkenburg JHF: Generation of leukemia-reactive cytotoxic T lymphocyte clones from the HLA-identical bone marrow donor of a patient with leukemia. J Exp Med 176:1283, 1992
32. Sosman JA, Oettel KR, Hank JA, Fisch P, Sondel PM: Specific recognition of human leukemic cells by allogeneic T cell lines. Transplantation 48:486, 1989
33. Kolb HJ, Mittermüller J, Clemm C, Holler E, Ledderose G, Brehm G, Heim M, Wilmanns W: Donor leukocyte transfusions for treatment of recurrent chronic myelogenous leukemia in marrow transplant patients. Blood 76:2462, 1990

34. Drobyski WR, Keever CA, Roth MS, Koethe S, Hanson G, McFadden P, Gottschall JL, Ash RC, Van Tuinen P, Horowitz MM, Flomenberg N: Salvage immunotherapy using donor leukocyte infusions as treatment for relapsed chronic myelogenous leukemia after allogeneic bone marrow transplantation: Efficacy and toxicity of a defined T-cell dose. Blood 82:2310, 1993
35. Hertenstein B, Wiesneth M, Novotny J, Bunjes D, Stefanic M, Heinze B, Hübner G, Heimpel H, Arnold R: Interferon-α and donor buffy coat transfusions for treatment of relapsed chronic myeloid leukemia after allogeneic bone marrow transplantation. Transplantation 56:1114, 1993
36. Antin JH: Graft-versus-leukemia: No longer an epiphenomenon. Blood 82:2273, 1993
37. Kernan NA, Bartsch G, Ash RC, Beatty PG, Champlin R, Filipovich A, Gajewski J, Hansen JA, Henslee-Downey J, McCullough J, McGlave P, Perkins HA, Phillips GL, Sanders J, Stroncek D, Thomas ED, Blume KG: Analysis of 462 transplantations from unrelated donors facilitated by the National Marrow Donor Program. N Engl J Med 328:593, 1993
38. Ash RC, Casper JT, Chitambar CR, Hansen R, Bunin N, Truitt RL, Lawton C, Murray K, Hunter J, Baxter-Lowe LA, Gottschall JL, Oldham K, Anderson T, Camitta B, Menitove J: Successful allogeneic transplantation of T-cell-depleted bone marrow from closely HLA-matched unrelated donors. N Engl J Med 322:485, 1990
39. Bozdech MJ, Sondel PM, Trigg ME, et al.: Transplantation of HLA haploidentical T cell depleted marrow for leukemia: addition of cytosine arabinoside to the transplant conditioning prevents graft rejection.. Exp Hematol. 13:1201, 1985
40. O'Reilly RJ, Brochstein J, Collins N, Keever C, Kapoor N, Kirkpatrick D, Kernan N, Dupont B, Burns J, Reisner Y: Evaluation of HLA-haplotype disparate parental marrow grafts depleted of T lympohcytes by differential agglutination with a soybean lectin and E-rosette depletion for the treatment of severe combined immunodeficiency.. Vox Sang 51:81, 1986
41. Reisner Y, Kapoor N, Kirkpatrick D, Pollack MS, Cunningham-Rundles S, Dupont B, Hodes MZ, Good RA, O'Reilly BJ: Transplantation for severe combined immunodeficiency with HLA-A,B,D, Dr incompatable parental marrow cells fractionated by soybean agglutinin and sheep red blood cells.. Blood 61:341, 1983
42. McGlave P, Bartsch G, Anasetti C, Ash R, Beatty P, Gajewski J, Kernan NA: Unrelated donor marrow transplantation therapy for chronic myelogenous leukemia: Initial experience of the National Marrow Donor Program. Blood 81:543, 1993
43. Uharek L, Glass B, Gassmann W, Eckstein V, Steinmann J, Loeffler H, Mueller-Ruchholtz W: Engraftment of allogeneic bone marrow cells: Experimental investigations on the role of cell dose, graft-versus-host reactive T cells and pretransplant immunosuppression. Transplant Proc 24:3023, 1992
44. Przepiorka D, Huh YO, Khouri I, Reading C, Hester J, Marshall M, Champlin RE: Graft failure and graft-versus-host disease after subtotal t-cell depleted marrow transplantation: correlations with marrow hematopoietic and lymphoid subsets.. Prog Clin Biol Res 389:557, 1994
45. Koehler M, Hurwitz CA, Krance RA, Coustan-Smith E, Williams LL, Santana V, Ribeiro RC, Brenner MK, Heslop HE: XomaZyme-CD5 immunotoxin in conjunction with partial T cell depletion for prevention of graft rejection and graft-versus-host disease after bone marrow transplantation from matched unrelated donors. Bone Marrow Transplant 13:571, 1994
46. Bensinger WI, Weaver CH, Appelbaum FR, Rowley S, Demirer T, Sanders J, Storb R, Buckner CD: Transplantation of allogeneic peripheral blood stem cells mobilized by recombinent human granulocyte colony-stimulating factor.. Blood 85:1655, 1995
47. Schmitz N, Dreger P, Suttorp M, Rohwedder EB, Haferlach T, Loffler H, Hunter A, Russell NH: Primary transplantation of allogeneic peripheral blood progenitor cells mobilized by filgrastim (granulocyte colony-stimulating factor). Blood 85:1666, 1995
48. Korbling M, Przepiorka D, Huh YO, Engel H, Van Besien K, Giralt S, Andersson B, Kleine HD, Seong D, Deisseroth AB, Andreeff M, Champlin R: Allogeneic blood stem cell transplantation for refractory leukemia and lymphoma: Potential advantage of blood over marrow allografts. Blood 85:1659, 1995
49. Aversa F, Tabilio A, Terenzi A, Velardi A, Falzetti F, Giannoni C, Iacucci R, Zei T, Martelli MP, Gambelunghe C, Rossetti M, Caputo P, Latini P, Aristei C, Raymondi C, Reisner Y, Martelli MF: Successful engraftment of T-cell-depleted haploidentical "three-loci" incompatible transplants in leukemia patients by addition of recombinant human granulocyte colony-stimulating factor-mobilized peripheral blood progenitor cells to bone marrow inoculum. Blood 84:3948, 1994
50. Champlin R, Ho W, Gajewski J, Feig S, Burnison M, Holley G, Greenberg P, Lee K, Schmid I, Giorgi J, Yam P, Petz L, Winston D, Warner N, Reichert T: Selective depletion of $CD8^+$ T lymphocytes for prevention of graft-versus-host disease after allogeneic bone marrow transplantation. Blood 76:418, 1990
51. Sprent J, Schaefer M, Lo d, Korngold R: Properties of purified T cell subsets. II. In vivo responses to class I vs. class II H-2 differences.. J Exp Med 163:998, 1986

52. Korngold R: Pathophysiology of graft-versus-host disease directed to minor histocompatibility antigens. Bone Marrow Transplant 7 Suppl. 1:38, 1991
53. Nimer SD, Giorgi J, Gajewski JL, Ku N, Schiller GJ, Lee K, Territo M, Ho W, Feig S, Selch M, Isacescu V, Reichert TA, Champlin RE: Selective depletion of CD8+ cells for prevention of graft-versus-host disease after bone marrow transplantation: A randomized controlled trial. Transplantation 57:82, 1994
54. Champlin RE, Jansen J, Ho W, et al.: Retention of graft-versus-leukemia using selective depletion of CD8 positive T lymphocytes for prevention of graft versus host disease following bone marrow transplantation for chronic myelogenous leukemia. Transplantation Proc 23:1695, 1991
55. Jansen J, Hanks S, Akard L, Martin M, Thompson J, Chang Q, Ash R, Garrett P, Figg F, English D: Selective T cell depletion with CD8-conjugated magnetic beads in the prevention of graft-versus-host disease after allogeneic bone marrow transplantation. Bone Marrow Transplant 15:271, 1995
56. Schmidt-Wolf IGH, Dejbakhsh-Jones S, Ginzton N, Greenberg P, Strober S: T-cell subsets and suppressor cells in human bone marrow. Blood 80:3242, 1992
57. Soiffer RJ, Murray C, Gonin R, Ritz J: Effect of low-dose interleukin-2 on disease relapse after T-cell-depleted allogeneic bone marrow transplantation. Blood 84:964, 1994
58. Kolb HJ, Mittermüller J, Günther W, Bartram C, Thalmaier K, Schumm M, Holler E, Lederose G: Adoptive immunotherapy in human and canine chimeras. Bone Marrow Transplant 12 Suppl. 3:S61, 1993
59. Papadopoulos EB, Ladanyi M, Emanuel D, Mackinnon S, Boulad F, Carabasi MH, Castro-Malaspina H, Childs BH, Gillio AP, Small TN, Young JW, Kernan NA, O'Reilly RJ: Infusions of donor leukocytes to treat Epstein-Barr virus-associated lymphoproliferative disorders after allogeneic bone marrow transplantation. N Engl J Med 330:1185, 1994
60. Giralt SA, Champlin RE: Leukemia relapse after allogeneic bone marrow transplantation: A review. Blood 84:3603, 1994
61. Van Rhee F, Lin F, Cullis JO, Spencer A, Cross NCP, Chase A, Garicochea B, Bungey J, Barrett J, Goldman JM: Relapse of chronic myeloid leukemia after allogeneic bone marrow transplant: The case for giving donor leukocyte transfusions before the onset of hematologic relapse. Blood 83:3377, 1994
62. Johnson BD, Truitt RL: Delayed infusion of immunocompetent donor cells after bone marrow transplantation breaks graft-host tolerance and allows for persistent antileukemic reactivity without severe graft-versus-host disease. Blood 85:3302, 1995
63. Naparstek E, OR R, Nagler A, Cividalli G, Engelhard D, Aker M, Gimon Z, Manny N, Sacks T, Tochner Z, Weiss L, Samuel S, Brautbar C, Hale G, Waldmann H, Steinberg SM, Slavin S: T-cell-depleted allogeneic bone marrow transplantation for acute leukaemia using Campath-1 antibodies and post-transplant administration of donor's peripheral blood lymphocytes for prevention of relapse. Br J Haematol 89:506, 1995
64. Mackinnon S, Papadopoulos EB, Carabasi MH, Reich L, Collins NH, Boulad F, Castro-Malaspina H, Childs B, Gillio A, Kernan NA, Small T, Young J, O'Reilly RJ: Adoptive immunotherapy using escalating doses of donor leukeocytes for relapsed chronic myeloid leukemia following allogeneic bone marrow transplantation. Blood 86: 1261, 1995.
65. Giralt S, Hester J, Huh Y, Hirsch-Ginsberg C, Rondon G, Guo J, Lee M, Gajewski J, Talpaz M, Kantarjian H, Fischer H, Deisseroth A, Champlin R: CD8+ depleted donor lymphocyte infusion as treatment for relapsed chronic myelogenous leukemia after allogeneic bone marrow transplantation: graft vs leukemia without graft vs. host disease. Blood 86:4337, 1995.
66. Tiberghien P, Reynolds CW, Keller J, Spence S, Deschaseaux M, Certoux J-M, Contassot E, Murphy WJ, Lyons R, Chiang Y, Hervé P, Longo DL, Ruscetti FW: Ganciclovir treatment of herpes simplex thymidine kinase-transduced primary T lymphocytes: An approach for specific in vivo donor T-cell depletion after bone marrow transplantation. Blood 84:1333, 1994.
67. Bonini C, Verzeletti S, Servida P, Rossini S, Traversari C, Ferrari G, Nobili N, Mavilio F, Bordignon C: Transfer of the HSV-TK gene into donor peripheral blood lymphocytes for in vivo immunomodulation of donor anti-tumor immunity after allo-BMT. Blood 84:110a, 1994.

2

COMBINED USE OF GROWTH FACTORS TO STIMULATE THE PROLIFERATION OF HEMATOPOIETIC PROGENITOR CELLS AFTER AUTOLOGOUS BONE MARROW TRANSPLANTATION (ABMT) FOR LYMPHOMA PATIENTS

Roberto M. Lemoli,* Alessandra Fortuna, Miriam Fogli, Gianantonio Rosti, Filippo Gherlinzoni, Giuseppe Visani, Lucia Catani, Alessandro Gozzetti, and Sante Tura

Institute of Hematology "Seràgnoli"
University of Bologna
Via Massarenti 9
40138 Bologna, Italia

ABSTRACT

We studied the kinetic response and concentration of bone marrow (BM) progenitor cells of patients with lymphoid malignancies submitted to ABMT, treated with G-CSF/IL-3 combination. The results were compared with those of lymphoma patients receiving the same pretransplant conditioning regimen followed by G-CSF alone. RhG-CSF was administered as a single subcutaneous (sc) injection at the dose of 5μg/Kg/day from day +1 after reinfusion of autologous stem cells while rhIL-3 was added from day +6 at the dose of 10μg /Kg/day sc (overlapping schedule). In both groups (i.e. G-CSF- and G-CSF/IL-3-treated patients) cytokine administration was discontinued when the absolute neutrophil count (ANC) was > 0.5×10^9/L of peripheral blood (PB) for 3 consecutive days.

Following treatment with colony-stimulating factor (CSF) combination, the percentage of marrow granulocyte-macrophage colony forming unit (CFU-GM) and erythroid progenitors (BFU-E) in S-phase of cell cycle increased from 9.3 ± 2% to 33.3 ± 12% and from 14.6 ± 3% to 35 ± 6%, respectively ($p < 0.05$). Similarly, we observed an increased number of actively cycling megakaryocyte progenitors (CFU-MK and BFU-MK). Conversely, G-CSF augmented the proliferative rate of CFU-GM (22.6 ±.6% compared to a

* To whom correspondence should be addressed. Tel: 051/390413; Fax: 051/398973

Molecular Biology of Hematopoiesis 5, edited by Abraham et al.
Plenum Press, New York, 1996

baseline value of 11.5 ± 3%; $p < 0.05$) but not of BFU-E, CFU-MK or BFU-MK, and the increase of S-phase CFU-GM was significantly lower than that observed in the posttreatment samples of patients receiving IL-3 in addition to G-CSF. A significant augmentation of the absolute number of both CFU-GM and BFU-E/ml of BM was reported after treatment with G-CSF/IL-3 but not G-CSF alone. Similarly, administration of the cytokine combination resulted in a higher number of CD 34+ cells and their concentration was significantly greater than that observed in the posttreatment samples of G-CSF patients. We also investigated the responsiveness to CSFs, in-vitro, of highly enriched CD 34+ cells, collected after priming with G-CSF in-vivo (i.e. after 5 days of G-CSF administration). Our results demonstrated that pretreatment with G-CSF modified the response of BM cells to subsequent stimulation with additional CSFs.

When the hematological reconstitution of patients treated with G-CSF/IL-3 was compared to that of individuals receiving G-CSF alone, the addition of IL-3 resulted in a significant improvement of granulocyte and platelet recovery, a lower transfusion requirement and shorter hospitalization. In conclusion, our results indicate that in-vivo administration of 2 cytokines increases the proliferative rate and concentration of BM progenitor cells to a greater degree than G-CSF alone and support the role of growth factor combinations for accelerating hematopoietic recovery after high dose chemotherapy.

INTRODUCTION

Several CSFs are currently under clinical investigation for accelerating BM recovery following myeloablative therapy and reinfusion of autologous stem cells. However, following treatment with G-CSF, GM-CSF or IL-3, there is still a delay of 8-12 days post graft for neutrophil recovery whereas platelet production has not been affected (1-3). Many studies in-vitro have shown the ability of various cytokine combinations to induce the proliferation and maturation of hematopoietic progenitor cells to a higher degree than single CSF (4, 5). Therefore, despite the complexity of the regulation of hematopoiesis, these findings are consistent with a regulatory model involving combination of cytokines for proliferation and maturation of early BM cells. IL-3 and G-CSF are 2 cytokines that individually enhance hemopoiesis in-vitro and in-vivo (2, 3). IL-3 has shown a broad spectrum of activity by stimulating the growth of committed progenitors (including platelet precursors BFU-MK and CFU-MK), as well as the functional activity of mature cells (6). As opposed to acting preferentially on committed progenitor cells, G-CSF appears to exert its effect on both late myeloid cells differentiating along the granulocytic pathway and very primitive hematopoietic cells (7). In-vitro studies suggest that G-CSF enhances synergistically the responsiveness of early BM progenitors to additional cytokines by shifting stem cells from G0 to the G1 stage of the cell cycle (7) and perhaps via cytokine receptors upregulation (8). Because IL-3 and G-CSF seem to affect hemopoiesis at different levels within the hematopoietic hierarchy, we hypothesized that the use of these cytokines in combination may be more effective than the same agents used individually in enhancing BM reconstitution after ABMT. Thus, we wished to investigate whether the in-vivo effects of G-CSF/IL-3 might be due to a direct activity on the number and kinetic status of BM progenitor cells. A comparison was also made with lymphoma patients who received the same pretransplant conditioning regimen followed by G-CSF alone. Our results indicate that the multilineage response to G-CSF/IL-3 in-vivo may be due to the increased proliferation of multipotential and lineage-restricted BM progenitor cells. Therefore, the combined administration of IL-3 and G-CSF may be more effective than single cytokines for hastening hematopoietic reconstitution after ABMT.

MATERIALS AND METHODS

Study Design and Patient Characteristics

Twenty two patients submitted to ABMT were studied in-vitro. Elegibility criteria included histological diagnosis of NHL or HD, no evidence of BM involvement, Karnofsky status greater than 70% and no history of previous cardiac, renal or hepatic disorder. All the patients had received 1 or more lines of treatment prior to ABMT. The 2 groups of patients (i.e. G-CSF = 5; G-CSF/ IL-3 = 17) were comparable as for age, lines of therapy prior to ABMT, NHL/HD ratio, number of nucleated cells and hematopoietic progenitors reinfused. After giving written informed consent, patients were treated according to BAVC (NHL patients) or BEAM (HD patients) myeloablative regimen and autologous BM cells were reinfused on day 0.

Recombinant human (rh) G-CSF (Neupogen, Dompè Biotec, Milano, Italy; Granulokine, Roche) was administered daily as a single sc injection at the dose of 5μ /Kg/day from day +1. As part of a phase I / II study, 17 patients also received, in addition to G-CSF, IL-3 (Sandoz, Basel, Switzerland) at the dose of 10 μg /Kg/ day sc from day +6. In both groups (i.e. G-CSF- and G-CSF/IL-3-treated patients), cytokine administration was discontinued when the ANC was >0.5 x 10^9/L for 3 consecutive days at a median time of 16 days (range 11-26) and 14 days (range 11-32) after ABMT, respectively.

BM Samples and Cell Preparation

Bone marrow was aspirated from the posterior iliac crest of study patients shortly before the administration of myeloablative chemotherapy and between 1 and 3 days after cytokine discontinuation. To minimize peripheral blood (PB) cells contamination, no more than 2-3 ml of BM were collected from each aspiration. BM smears were evaluated after staining with May-Grunwald-Giemsa stain. BM biopsy was not routinely performed.

The mononuclear cell (MNC) fraction was collected after Ficoll- Hypaque gradient (1077 g/cm^3) and either used directly in progenitor cell assay (see below) or further processed to obtain highly purified CD 34+ cells. Enriched CD34+ cells were recovered by using the Ceprate LC Kit (Cell Pro Inc, Bothell, WA, USA) as already reported (9).

Colony Assay

BM cells were cultured in semisolid medium as previously described (9). To determine the number of BM progenitor cells per unit volume before and after treatment with CSFs, the number of CFU-C/ 10^5 MNC was multiply by the MNC count in the same sample.

For late (CFU-MK) and early (BFU-MK) megakaryocyte progenitors, mononuclear non-adherent BM cells (MNAC) were obtained after overnight incubation at 37° C, in 5% CO_2 atmosphere in plastic Petri dishes, at the concentration of 10^6 cells/ml of medium (IMDM supplemented with 5% FCS) and plasma clot cultures were set up (10). After 12 and 19 days of incubation, plasma clots were fixed with methanol-acetone (1:3) for 20 min, washed with PBS and then air-dried. Fixed dishes were stored at -20°C until immunofluorescence staining was performed. CFU-MK colonies were recorded as aggregates of >3 cells strongly positive after staining with the monoclonal antibody directed to the glycoprotein-complex IIb-IIIa (CD41, Dako). BFU-MK colonies consisted of aggregates of 2-7 fluorescent foci, containing >100 cells. Binding was revealed by FITC-goat antimouse IgG (Ortho).

Cytosine Arabinoside (Ara-C) Suicide Test

To evaluate the proportion of hematopoietic progenitors in S-phase, the Ara-C suicide test was performed by incubating 1 x 10^6 cells/ ml with 2 x 10^{-6} Ara-C (Aracytin, Upjohn, Kalamazoo, MI) for 1 h at 37° C as previously reported (11). After 2 washes, cells were plated and the difference in the number of colonies between treated and untreated cultures represented the proportion of BM precursors undergoing DNA synthesis.

Flow Cytometry

The content of CD 34+ and CD 34+ CD 33- cells, before and after CSFs treatment, was determined by flow cytometry by staining BM cells with mouse-derived anti-CD34 and CD33 antibodies: HPCA-2, IgG2a-FITC; Leu-M9, IgG1-PE. The 2 conjugated monoclonal antibodies (MoAbs) were purchased from Becton-Dickinson (Mountain View, CA). Briefly, low-density cells were incubated with the MoAbs for 20 min at 4° C and respective normal IgG isotypes were used as controls. After 2 washes flourescence analysis was performed on a gated population set on scatter properties using a FACScan equipment (Becton Dickinson). A minimum of 10,000 events were collected in list mode on FACScan software.

Data Analysis

The results are expressed as the mean ± the mean of the standard error (SEM) value for each group of patients. The Student-T test (two-tailed test) for paired data was used to make comparisons between pretreatment and posttreatment results of both cytokine-treated groups. In addition, the unpaired, nonparametric Wilcoxon rank sum test was used to compare posttreatment results of G-CSF and G-CSF, IL-3 patients.

RESULTS

BM Cellularity

The enhancement of marrow cellularity by G-CSF and IL-3 (Table 1) resulted mainly from the stimulation of myelopoiesis with a shift to the left. The median myeloid to erythroid ratio was 1.5: 1 before and 4: 1 after treatment ($p < 0.03$). We did not observe any significant increase in lymphocytes, plasma cells, basophils and eosinophils. The MNC fraction was augmented from 3 ± 0.7 x 10^6 cells/ ml of BM to 4.3 ± 1.8 x 10^6 cells/ ml.($p < 0.05$) (Table 1).

Frequency and Cycling Rate of Myeloid Progenitor Cells

The frequency and the number of BM hematopoietic progenitor cells per unit volume is reported in Table 1. No significant differences were noted in the clonogenic efficiency (number of colonies formed per number of cells plated) of marrow precursors of patients receiving G-CSF/IL-3. However, given the increase in marrow cellularity , a significant augmentation in the concentration of both CFU-GM (3605 ± 712/ ml of BM vs 2213 ± 580/ml of BM; $p < 0.05$) and BFU-E (4373 ± 608 /ml vs 3027 ±516/ ml; $p < 0.05$) was noted after CSFs treatment. In contrast, G-CSF alone induced a neglegible increase in the number of mononuclear cells /ml of BM and the incidence and the absolute number of CFU-GM remained unchanged while BFU-E were slightly reduced (Table 1). The posttreatment values of CFU-GM and BFU-E were significantly higher in G-CSF/IL-3 patients as compared to patients receiving G-CSF alone (Table 1). A similar pattern of results was observed when

Table 1. Frequency of BM progenitors of patients with lymphoid malignancies administered with G-CSF or G-CSF/IL3 after ABMT

Cells	Pretreatment		Posttreatment	
	G-CSF	G-CSF/IL-3	G-CSF	G-CSF/IL-3
Total MNC(x 10^6/ml)	2.8 ± 0.5	3 ± 0.7	3 ± 1	4.3 ± 1.8[2,3]
CFU-GM(x10^5 cells)	60 ± 15	71.5 ± 20	53 ± 12	56 ± 26
CFU-GM(/ml of BM)	1906 ± 212	2213 ± 580	2018 ± 406	3605 ± 712[3,2]
BFU-E(x 10^5cells)	91.6 ± 20	83.8 ± 60	74 ± 15	65 ± 10
BFU-E(/ml of BM)	2975 ± 116	3027 ± 516	2663 ± 241	4373 ± 608[3,2]
CD34+(%)	1.8 ± 0.9	1.6 ± 1	0.5 ± 0.1[2]	1 ± 0.7[3]
CD34+(/ml of BM)	54120 ± 2000	59045 ± 16000	48640 ± 15300	131731 ± 25000[2,3]
CD34+ CD33-(%)	0.4 ± 0.05	0.6 ± 0.3	ND[1]	0.2 ± 0.01
CD34+ CD33-(/ml of BM)	18016 ± 1714	25114 ± 2500	NE[1]	16753 ± 590

BM MNC cells were collected after Ficoll sedimentation and plated in presence of PHA-LCM and EPO before and after CSF treatment.
[1]ND; not detectable: NE; not evaluable
[2]Statisticaly significant ($p < 0.05$) compared to pretreatment value.
[3]Statistically significant ($p < 0.05$) compared to G-CSF treated patients.

BM specimens were analyzed for the content of CD 34+ cells (Table 1). However, neither group of patients showed the increase of more immature, myeloid lineage negative CD 34+ CD 33- cells.

The percentage of marrow CFU-GM and BFU-E in S-phase of cell cycle increased from 9.3 ± 2% to 33.3 ± 12% and from 14.6 ± 3% to 35± 6%, respectively, in patients administered CSF combination (Table 2). Also G-CSF alone augmented the proliferative rate of CFU-GM (22 ± 6%; $p < 0.05$ compared to baseline value) but not of BFU-E. However, this increase was significantly lower than that observed in the posttreatment samples of patients receiving IL-3 in addition to G-CSF (Table 2). Four patients (G-CSF/IL-3 = 2; G-CSF = 2) were also assessed for the proliferative activity of growth factor(s) on late and early megakaryocyte progenitor cells and the results presented in Table 2 confirm that G-CSF/IL-3 combination exerts its stimulatory activity on a broad spectrum of BM precursors.

Table 2. Effect of CSFs on the percentage of BM precursors in cell cycle

Hematopoietic progenitors	Pretreatment (%)		Posttreatment (%)	
	G-CSF	G-CSF/IL-3	G-CSF	G-CSF/IL-3
CFU-GM	11.5 ± 3	9.3 ± 2	22.6 ± 6*	33.3 ± 12 *[a]
BFU-E	13.8 ± 4	14.6 ± 3	11.9 ± 3	35 ± 6 *[a]
CFU-MK	37.3 ± 10	17 ± 8	33 ± 10	77 ± 15
BFU-MK	45 ± 16	11.5 ± 8	50 ± 20	66.6 ± 18

S-phase progenitors were evaluated before and after CSFs treatment by the ARA-C suicide test, as described in Materials and Methods. CFU-MK and BFU-MK results derive from 4 patients (G-CSF/IL-3 = 2; G-CSF = 2).
*Statistically significant ($p < 0.05$) compared to pretreatment values.
[a]Statistically significant ($p < 0.05$) compared to G-CSF group.
Reprinted from Ref. 11

Table 3. Priming effect of G-CSF

	G-CSF	GM-CSF	IL-3	IL-3/G-CSF	GM-CSF/G-CSF
Clonogenic efficiency	2 ± 1	3.2 ± 2	1.2 ± 0.6	2.3 ± 1	1.3 ± 1
Total progenitors	8 ± 5	10 ± 3	6.7 ± 1	6.3 ± 3	5.5 ± 2

BM CD34+ cells were purified before and after 5 days of G-CSF treatment and cultured in presence of single CSFs or their combination. The results are presented as the mean ± SEM fold-increase of the clonogenic efficiency and the number of total progenitors/ml of BM observed in response to G-CSF priming. The plating efficiency of BM cells in presence of PHA-LCM was 41.3 ± 30 SEM and 73.7 ± 28 SEM colonies/dish before and after G-CSF administration, respectively.
Reprinted from Ref. 11.

Proliferative Response of Early Bm Cells after Priming with G-CSF *in Vivo*

A set of experiments was designed to test the responsiveness in-vitro of early BM progenitors to CSFs after priming with G-CSF in-vivo. To this end, we studied highly purified CD34+ cells collected from the BM of multiple myeloma patients (=3) and normal stem cells donors (=2), before and after treatment with G-CSF for 5 days to mobilize peripheral blood stem cells (PBSC). The results presented in Table 3 show that priming with G-CSF increased the number of CFU-C stimulated by G-CSF itself and G-CSF/IL-3 (2.3 ± 1 fold-increase). A more pronounced effect was seen on GM-CSF-responsive CFU-C (3.2 ± 2 fold-increase). Due to the enhanced BM cellularity, the total number of progenitors sharply increased after G-CSF treatment (Table 3). Taken together, these results suggest that pretreatment with G-CSF may modify the proliferative response of BM cells to CSFs. We also asked whether the G-CSF/IL-3 overlapping schedule might be more effective than the "pure" sequential G-CSF/IL-3 schedule (i.e. G-CSF treatment for 5 days followed by IL-3 alone) or the continuous exposure to G-CSF alone, for the proliferation of hematopoietic progenitor cells. For this purpose, a BM aspiration

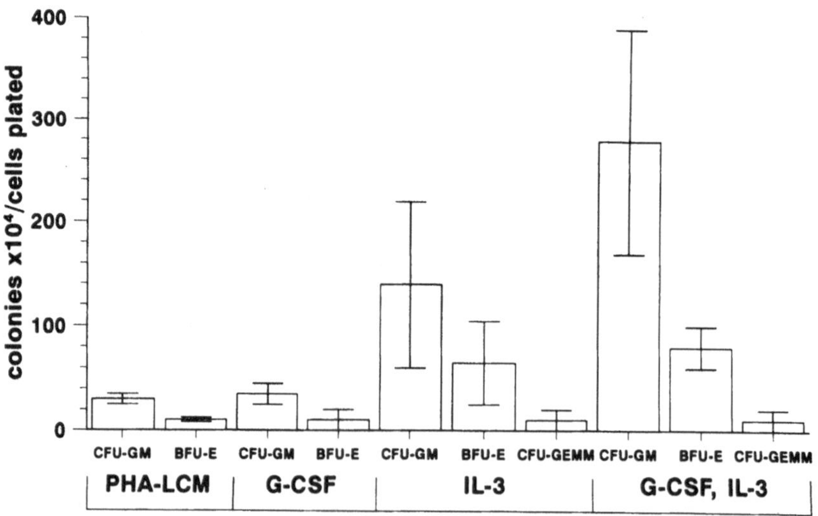

Figure 1. BM was harvested after 5 days of G-CSF administration, before the initiation of IL-3 treatment. Highly enriched CD 34+ cells were seeded in the presence of rhCSFs or PHA-LCM. The experiment was designed to compare the proliferation of myeloid progenitor cells in response to G-CSF, IL-3 or the combination of the 2 growth factors after exposure to G-CSF in-vivo. Reprinted from Ref. 11.

was performed on day 6 after reinfusion of autologous stem cells, before the initiation of IL-3 administration. The results shown in Fig. 1 demonstrate that the CSF combination induced a much higher number of CFU-GM as compared to the same growth factors used alone and a significant increase of BFU-E and CFU-GEMM in comparison with G-CSF (p always < 0.05).

BM Reconstitution of Lymphoma Patients after ABMT

The increased proliferative rate of hematopoietic progenitor cells following CSFs combination treatment was reflected by an accelerated BM recovery after ABMT as compared to patients receiving G-CSF alone. Median time to 0.5 x 10^9 ANC and 20 x 10^9 PLT/L was 11 and 15 days, respectively, versus 13.5 and 20 days ($p < 0.05$). Median time of absolute neutropenia and PLT count less than 20 x 10^9/L was 5 and 8 days versus 7.5 and 15 days, respectively ($p < 0.04$) (Fig. 2).This data resulted in a lower PLT and RBC transfusion requirement and a shorter hospitalization time (15 days versus 21 days; $p < 0.05$).

DISCUSSION

Despite the increasing number of reports on the simultaneous or sequential administration to humans of cytokine combinations, there is little data showing the effects of

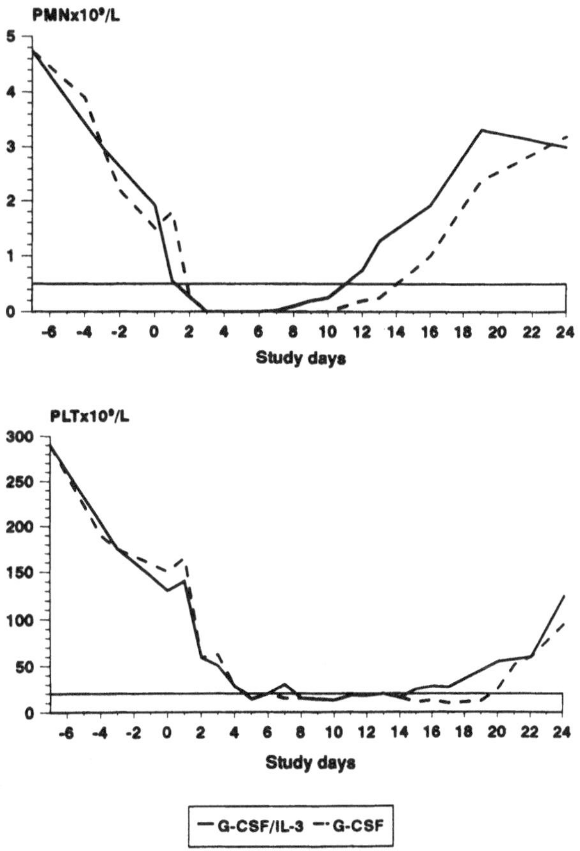

Figure 2. Hematological reconstitution of lymphoma patients administered G-CSF or G-CSF/IL-3 following ABMT.

multiple CSF treatment on hematopoietic progenitor cells. In the present study, we asked whether the multilineage response (including platelet recovery) observed in lymphoma patients undergoing myeloablative chemotherapy followed by reinfusion of autologous stem cells and G-CSF/IL-3 treatment, might be due to a direct effect of the cytokines on the number and kinetic status of early BM cells. In fact, mechanisms for growth factors positive interaction include induction of CSF receptor expression (8) and the stimulation of proliferation of BM cells by of a group of " permissive" cytokines capable of recruiting very primitive hematopoietic progenitors in cell cycle which are then able to respond to later acting growth factors. In this view, it has been shown that G-CSF, which also acts as a permissive cytokine, upmodulates the IL-3 receptor on murine BM cells (8) and enhances the IL-3-dependent proliferation of multipotential and lineage-committed hematopoietic precursors in-vitro (12, 13).

The results presented here provide evidence that the administration of G-CSF/IL-3 induces a pronounced increase in the DNA synthesis rate of CFU-GM, BFU-E and megakaryocyte precursors whereas G-CSF alone augmented the proliferative rate of CFU-GM to a lower degree than the CSF combination while it did not affect BFU-E or BFU-MK growth. Analogous to earlier studies of recombinant G-CSF and IL-3 (14, 15), the frequency of BM precursors was unchanged in G-CSF/IL-3-treated patients. However, the marked increase in marrow MNC fraction accounted for the higher number of both CFU-GM and BFU-E /ml of BM as compared to individuals administered G-CSF alone. It is also interesting to note that despite a comparable enhancement of erythroid and granulocyte-macrophage progenitor cells cycling rate, administration of G-CSF/IL-3 resulted in an increased myeloid:erythroid ratio whereas we did not observe an increase of eosinophils and basophils. This is consistent with what has been described in a rat model (16) and it may explain the extremely low incidence of allergic reactions to IL-3 reported in the same group of patients (data not shown).

A double immunofluorescence staining was used to assess the content of CD 34+ cells and more immature CD 34+ CD 33- cells before and after CSFs treatment. Similar to the clonogenic assay results, G-CSF/IL-3 combination induced an increase of the number of CD 34+ cells which was significantly higher than that of patients receiving G-CSF alone. This result derived from both the increased cellularity and the higher frequency of CD 34+ cells. Although we failed to detect a numerical increase of CD 34+ CD 33- cells in both group of patients, it should be pointed out that only in BM samples studied after G-CSF treatment the very primitive cell population was below the lower limit of detection of our immunophenotypic analysis.

Taken together, our results indicate that the administration of G-CSF after ABMT is followed by an increase in the proliferative activity of BM CFU-GM, but not BFU-E, without the amplification of the progenitor cell compartment whereas we observed a marked expansion of the mature cell pool. This data supports the well-established capacity of G-CSF to stimulate selectively neutrophil recovery in-vivo (2). Under the same clinical circumstances, G-CSF/IL-3 combination appeared more effective than G-CSF alone in inducing active cycling of BM progenitors coupled with the expansion of the number of immature precursors (i.e. clonogenic cells and CD 34+ cell fraction) as well as late myeloid cells. Based on the activity of the 2 cytokines on multiple hematopoietic lineages in-vitro (12, 13) and on initial results from animal studies (16, 17), it may have been anticipated that the CSF combination would be a broad acting stimulator of hematopoiesis in humans. In addition, the same preclinical investigations have suggested that G-CSF may have a "permissive" activity on early BM cells thus enhancing the clinical response to subsequent doses of IL-3 (17). To address this issue, we tested the responsiveness of hematopoietic progenitors to CSFs in-vitro after 5 days of G-CSF administration before the initiation of IL-3 treatment. We asked whether the priming with G-CSF induced recruitment of primitive progenitor cells which may be more effectively stimulated by the combination of G-CSF and IL-3 in comparison with the same cytokines used alone. The results presented in Fig. 1 demonstrate the synergistic activity of G-CSF and IL-3 on CFU-GM growth, the stimulation of a higher number of CFU-GM by IL-3 than G-CSF and the growth of pluripotent progenitors in presence of IL-3, alone or combined with G-CSF. By comparison, the same experiments performed prior to ABMT showed a less than additive effect of G-CSF and IL-3 on CFU-GM colony formation and the more efficient stimulation of CFU-GM by G-CSF compared to IL-3 (data not shown). These results indicate that: a) priming with G-CSF modifies the response of BM cells to subsequent treatment with additional CSFs; b) the multilineage response to G-CSF/IL-3 in-vivo after myeloablative therapy may be due to the increased proliferation of early BM cells as compared to single growth factors. The priming effect of G-CSF was also supported by the experiments showed in Table 3 where the proliferative response of BM progenitor cells in-vitro to GM-CSF, G-CSF, alone and combined with IL-3, was enhanced after treatment with G-CSF in-vivo.

In summary, this report suggests that the combination of early-and intermediate-acting CSFs shows the capacity of enhancing hematopoiesis in humans by stimulating cycling of BM progenitor cells to a greater extent than a single cytokine. The preliminary clinical data summarized in the Results section and Fig. 2 show that the increased proliferative rate of BM cells results in a faster hematopoietic recovery after ABMT. These results may help clinical investigators to design more effective growth factor treatments to hasten hematologic recovery after high dose therapy.

ACKNOWLEDGMENTS

The research was supported by: MURST 40% and MURST 60%.

REFERENCES

1. Nemunaitis J, Rabinowe S, Singer JW, Bierman PJ, Vose JM, Freedman AS, Onetto N, Gillis S, Oette D, Gold M, Buckner D, Hansen J, Ritz J, Appelbaum FR, Armitage JO, Nadler LM: Recombinant granulocyte-macrophage colony-stimulating factor after autologous bone marrow transplantation for lymphoid cancer. N Engl J Med 324: 1773, 1991.
2. Linch DC, Scarffe H, Proctor S, Chopra R, Taylor PRA, Morgestern G, Cunnhingam D, Burnett AK, Cawley JC, Franklin IM, Bell AJ, Lister TA, Marcus RE, Newland AC, Parker AC, Yver A: Randomised vehicle-controlled dose-finding study of glycosilated recombinant human granulocyte colony-stimulating factor after bone marrow transplantation. Bone Marrow Transplantation 11: 307, 1993.
3. Nemunaitis J, Appelbaum FR, Singer JW, Lilleby K, Wolff S, Greer JP, Bierman P, Resta D, Campion M, Levitt D, Zeigler Z, Rosenfeld C, Shadduck RK, Buckner CD: Phase I trial with recombinant human interleukin-3 in patients with lymphoma undergoing autologous bone marrow transplantation. Blood 82: 3273, 1993.
4. Moore MAS: Clinical implications of positive and negative hematopoietic stem cell regulators. Blood 78: 1, 1991.
5. Metcalf D: Hematopoietic regulators: Redundancy or sublety ? Blood 82: 3515, 1993.
6. Oster W, Schultz G: Interleukin-3: Biological and clinical effects. Intern J Cell Cloning 9: 5, 1991.
7. Leary AG, Zeng HG, Clark SC, Ogawa M: Growth factor requirements for survival in G0 and entry into the cell cycle of primitive human hemopoietic progenitors. Proc Natl Acad Sci USA 89: 4013, 1992.
8. Jacobsen SEW, Ruscetti FW, Dubois CM, Wine J, Keller JR: Induction of colony-stimulating factor receptor expression on hematopoietic progenitor cells: Proposed mechanism for growth factor synergism. Blood 80: 678, 1992.
9. Lemoli RM, Fogli M, Fortuna A, Motta MR, Rizzi S, Benini C, Tura S: Interleukin-11 stimulates the proliferation of human hematopoietic CD 34+ and CD34+ CD33- DR- cells and synergizes with stem

cell factor, interleukin-3 and granulocyte-macriphage colony-stimulating factor. Exp Hematol 21: 1668, 1993.

10. Gugliotta L, Bagnara GP, Catani L, Gaggioli L, Guarini A, Zauli G, Mattioli Belmonte M, Lauria F, Macchi S, Tura S: In vivo and in vitro inhibitory effect of α-interferon on megakaryocyte colony growth in essential thrombocythaemia. Br J Haematol 71: 177, 1989.
11. Lemoli RM, Fortuna A, Fogli A, Gherlinzoni F, Rosti G, Catani L, Gozzetti A, Miggiano MC, Tura S: Proliferative response of human marrow myeloid progenitor cells to in-vivo treatment with granulocyte colony-stimulating factor (G-CSF) and G-CSF in combination with Interleukin-3 (IL-3) after autologous bone marrow transplantation (ABMT). Exp Hematol, in press.
12. McNiece IK, McGrath EH, Quesenberry PJ: Granulocyte colony-stimulating factor augments in vitro megakaryocyte colony formation by interleukin-3. Exp Hemato 16: 807, 1988.
13. Ikebuchi K, Clark SC, Ihle JN, Souza LM, Ogawa M: Granulocyte colony-stimulating factor enhances interleukin-3-dependent proliferation of multipotential hematopoietic progenitors. Proc Natl Acad Sci USA 85: 3445, 1988.
14. Broxmeyer HE, Benninger L, Patel RS, Benjamin RS, Vadhan-Raj S: Kinetic response of human marrow myeloid progenitor cells to in vivo treatment of patients with granulocyte colony-stimulating factor is different from the response to treatment with granulocyte-macrophage colony-stimulating factor. Exp Hematol 22: 100, 1994.
15. Ottmann OG, Ganser A, Seipelt G, Eder M, Schlz G, Hoelzer D: Effects of recombinant human interleukin-3 on human hematopoietic progenitor and precursor cells in vivo. Blood 76: 1494, 1990.
16. Ulich TR, del Castillo J, McNiece IK, Yin S, Irwin B, Busser K, Guo K: Acute and subacute hematologic effects of multi-colony stimulating factor in combination with granulocyte colony-stimulating factor in vivo. Blood 75: 48, 1990.
17. Seiler FR, Krieter H, Mac Vittie T, Nothdurf W, Brueckner UB, Messmer K, Krumwieh D: Preclinical studies on the efficacy of CSFs in dogs and subhuman primates. In: Murphy MJ (ed) Blood cell growth factors: Their present and future use in hematology and oncology, Dayton (OH) AlphaMed Press,p.40, 1991.

3

LONG-TERM MULTILINEAGE DEVELOPMENT FROM HUMAN UMBILICAL CORD BLOOD STEM CELLS IN A NOVEL SCID-HU GRAFT

Christopher C. Fraser, Hideto Kaneshima, Gun Hansteen, Madison Kilpatrick, Ronald Hoffman, and Benjamin P. Chen*

Experimental Cell Therapy Group, SyStemix Inc.
3155 Porter Drive, Palo Alto, California 94304

ABSTRACT

Although human hematopoietic stem cells can be isolated based on cell surface antigen expression, their proliferative and developmental capacity must still be defined by their ability to produce multiple hematopoietic cell lineages for prolonged times in vivo. In order to derive a long-term in vivo multilineage SCID-hu graft we transplanted a human fetal bone and spleen adjacent to an HLA class I mismatched fetal thymus fragment in immunodeficient SCID mice (SCID-hu BTS). Grafts were analyzed at various times post transplant for cells expressing specific lineage markers. The bone marrow of SCID-hu BTS grafts maintained B cells and myeloid cells for at least 36 weeks post transplant. Analysis for progenitor content within grafts revealed that CD34+ cells, CFU-G/M, and CFU-GEMM were maintained in 94%(109/116),100%(66/66) and 79%(52/66) up to 28 weeks. HSC ($CD34^{hi}Thy\text{-}1^{+}Lin^{-}$) sorted from 20 week old SCID-hu BTS grafts demonstrated potent secondary multilineage reconstituting potential when injected into HLA mismatched SCID-hu bone grafts. In addition both immature and mature T-cells, derived from progenitors within the fetal bone, were found in 87%(101/116) of grafts analyzed up to 36 weeks in vivo. Injection of irradiated SCID-hu BTS grafts with $CD34^{+}Thy\text{-}1^{+}Lin^{-}$ umbilical cord blood cells produced B-cells, myeloid cells, T-cells and $CD34^{+}$ cells in individual grafts when analyzed 8 weeks post reconstitution, further demonstrating the multipotential nature of these HSC populations. This model is currently being used to define soluble factors or cells that are capable of enhancing human HSC engraftment.

* To whom correspondence should be addressed: B. P. Chen: Phone: (415) 813-6523; FAX: (415) 856-4919.

INTRODUCTION

The ability to assess phenotypically defined subsets of human stem cells for functional capacity will require assays that allow long term stem cell maintenance and multipotential differentiation. Our laboratory has developed the SCID-hu model as a human HSC assay (1-3). A number of models have already been established in which human fetal hematopoietic tissue is transplanted in SCID mice. Transplantation of human fetal liver and thymus together under the kidney capsule (SCID-hu Thy/Liv) derives a fusion graft in which fetal liver cells differentiate within the thymic microenvironment and produce mature human T-cells for periods of up to 1 year (2). Human myeloid progenitors, macrophages, erythroid progenitors, and B-cells derived from the fetal liver are also detectable at very low levels during this time (3,4). Fetal bone fragments implanted under the mammary fat pad of SCID mice (SCID-hu bone) maintain both B-cells and myelo-erythropoiesis for at least 20 weeks(5). Donor-derived CD19+ B-cells, CD33+ myeloid cells as well as CD34+ progenitor cells are produced when SCID-hu bone grafts are microinjected with enriched fetal and adult hematopoietic stem cell populations.

In order to develope a SCID-hu graft in which HSC in the bone marrow could differentiate into progenitors, B-cells and myeloid cells, undergo self-renewal, as well as migrate to and differentiate into T-cells within the thymic microenvironment, SCID mice were transplanted subcutaneously with fetal bone, thymus and spleen fragments (SCID-hu BTS). We found that the bone marrow of SCID-hu BTS grafts maintained B cells, myeloid cells, CD34+ cells, multilineage progenitors (CFU-MIX) and cells with a stem cell phenotype (CD34+Thy-1+Lin-) which have potent secondary reconstituting. In addition, early progenitors within the bone marrow could migrate to the thymus and differentiate into mature T-cells for at least 36 weeks in vivo. Enriched umbilical cord blood (UCB) stem cells were injected into irradiated grafts. Donor derived multilineage hematopoiesis including B-cells, myeloid cells, immature and mature T-cells and CD34+ cells was observed in individual grafts, suggesting SCID-hu BTS grafts can support multilineage hematopoiesis from a common hematopoietic stem cell pool.

MATERIALS AND METHODS

SCID-hu BTS Mice

SCID-hu BTS mice were constructed using small (2mm) human fetal spleen and thymus fragments and approximately 5 X 3 X 10 mm fetal bone fragments from femurs and tibias of 19 to 23 week gestational fetuses from elective aborteses. Fetal spleen and bone fragments were placed adjacent to an HLA mismatched fetal thymus fragment subcutaneously in the mammary fat pad of 6 to 8 week old anesthetized C.B-17 *scid/scid* mice. Fetal tissues were obtained with informed consent from procurement agencies according to federal and state regulations. For injection of sorted stem cell populations, HLA-mismatched SCID-hu BTS mice were used 12-16 weeks post implantation as recipients. Immediately prior to injection of sorted cells, mice were given a single whole body irradiation dose (400 R from a ^{137}Cs source; Gamma Cell 40, J.L. Shepard & Associates, San Fernando, CA). A 10μL volume containing sorted cells was then injected directly into grafts using a Hamilton syringe. Mice were then killed 8 weeks later and grafts analyzed for donor reconstitution.

Flow Cytometry

SCID-hu BTS mice were sacrificed 9 - 36 weeks post transplant and human tissue grafts removed. Cells were washed and analyzed after staining with directly conjugated antibodies on a FACScan fluorescence analyzer (Becton Dickinson). Antibodies included those against human CD14, CD15, CD33 and CD8 (Becton Dickinson, Moutainview, CA) and CD4, CD19, and CD45 (Caltag, South San Francisco, CA). HLA immunophenotype was determined using FITC- conjugated MAb anti-human HLA class I antibodies MA2.1, BB7.1, BB7.2 and GAP-A3, derived from hybridomas obtained from ATCC (Rockville, MD) and fluorochrome conjugated at SyStemix (kindly provided by Dr. C. Ahlem). Irrelevant Ig negative controls were directly FITC or PE-conjugated IgG1 (Becton Dickinson, MountainView, CA) and TC-conjugated IgG2A (Caltag, South San Francisco, CA).

Hematopoietic Progenitor Assay

Single cell suspensions from homogenized grafts were plated in Methocult (StemCell Technologies Inc, Vancouver, Canada) to which the following human recombinant cytokines were added: erythropoietin (1.2 U/mL) and c-kit ligand (10 ng/mL) (R and D Systems, Minneapolis, MN), IL-3 (10 ng/mL) (Sandoz Pharma, Basel, Switzerland), G-CSF (25 ng/mL) and GM-CSF (25 ng/mL) (Amgen, Thousand Oaks, CA), and cultures maintained at 37^{0}C, 5% CO_2 for 14 days before analysis.

Tissue Processing and Flow Cytometry Sorting

Human umbilical cord blood was harvested in Tucson, Arizona, and shipped overnight at room temperature to SyStemix (Palo Alto, CA). Informed consent was obtained from mothers prior to delivery. Aliquots were removed prior to freezing for HLA typing. For sorting, frozen aliquots were thawed and washed in sort buffer, then resuspended in 1 mg/mL heat inactivated gamma-globulin (Miles Inc., Elkhart, IN). Cells were incubated with lineage specific antibodies that were FITC conjugated and included CD2, CD14, CD15, CD16, CD19 (Becton Dickinson, Mountainview, CA) and glycophorin A (Amac, Westbrook, ME), then depleted by incubation with sheep anti-mouse Ig-coated magnetic beads (Dynal, Oslo, Norway). Cells were then incubated with sulphorhodamine-conjugated anti-CD34 Mab (Tuk-3) (or isotype IgG3 control mAb). Umbilical cord blood derived cells were also incubated with a PE-conjugated Fab-fragment of anti-Thy-1 mAb (GM201) or IgG1 isotype control. Cells were sorted on FACStar Plus or Vantage cell sorters (Becton Dickinson, Mountainview, CA) using electronic gates set for CD34+Thy-1+Lin-.

RESULTS

Generation of Human Fetal Bone/Thymus Fusion Grafts Capable of Long-Term T-Cell, B-Cell, Myeloid Cell, and Progenitor Production in SCID Mice

SCID mice were implanted sub-cutaneously with fetal bone (with or without a fetal spleen fragment), and an HLA-mismatched thymus fragment. Grafts were removed 19-20 weeks post transplantation and analyzed for histology as shown in Figure 1. Histological evaluation showed that the grafts are fused but maintain distinct boundaries dividing the bone, thymic and splenic tissues. The bone marrow shows foci of active hematopoiesis which contains blast cells, immature and mature granulocytes, monocytes, megakaryocytes, and lymphocytes (Fig. 1).

Grafts were analyzed by flow cytometry from 9 to 36 weeks post transplantation for human hematopoietic cells derived from the fetal bone/spleen (donor) using HLA class I

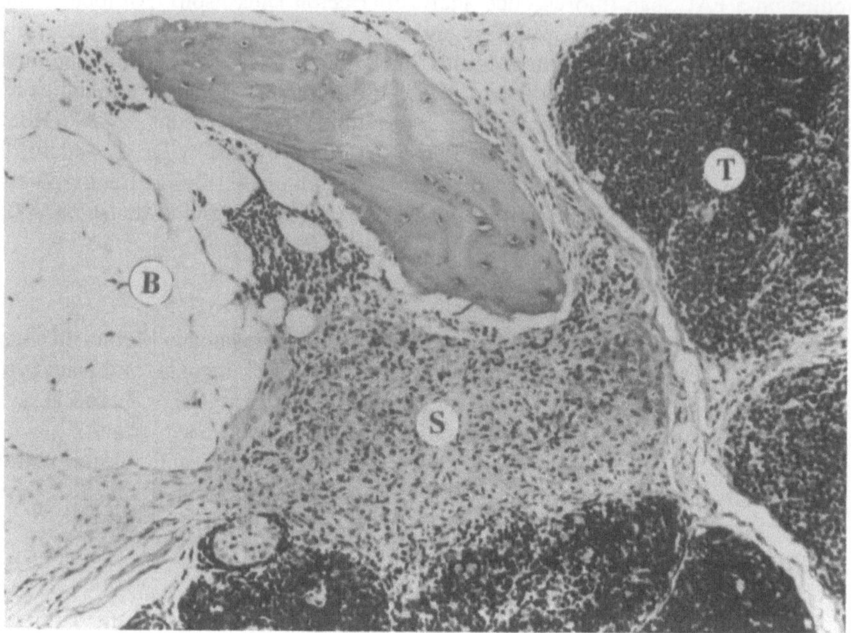

Figure 1. Histological features of 20 week old SCID-hu BTS graft. Histology of a SCID-hu BTS graft showing cellular content of the fused graft with a hypocellular splenic fragment (S) flanked by a hematopoietically active bone marrow (B), and a thymus (T) with distinct cortical and medullary areas filled with thymocytes.

allele specific antibodies. Cell suspensions were stained with combinations of monoclonal antibodies (MAb) to Class I HLA of donor cells (specific to the bone-spleen fragments and not the thymus fragment), in combination with T-cell, B-cell, myeloid cell and progenitor cell specific MAb. A typical analysis of the cellular content of one representative bone/thymus graft (SCID-hu BT) demonstrating multilineage hematopoiesis, analyzed 20 weeks post transplant is shown in Figure 2. A high proportion of CD4 and CD8 positive T-cells stain for anti-HLA specific to cells derived from the bone and spleen graft (Figure 2 A-C). Immature CD4+CD8+ double positive (DP) T-cells are low in Class I HLA expression, and acquire a higher Class I HLA expression at final stages of maturation (reviewed in (6)). Cells being processed through the thymic fragment in the SCID-hu BTS grafts should therefor have variable fluorescence intensities when stained with a donor Class I HLA MAb. Histogram analysis of single positive CD4+and CD8+ cells for donor Class I HLA expression demonstrates that mature T-cells in the SCID-hu BTS grafts express high levels of donor Class I HLA, immature DP T-cells however, express low to negative levels (Figure 2 and data not shown). Similar distributions were observed in 71 SCID-hu BTS grafts where DP T-cells were observed between 9-36 weeks post transplantation. In addition to T-cells, the same graft in Figure 2 shows donor-derived CD19 positive cells (B- cells) CD33 positive cells (myeloid cells) and CD34+ cells which are also easily detectable (Figure 2 D-F). Figure 2 indicates that early T-cell progenitors or stem cells from a common progenitor source in the bone also have the capacity to migrate to and reconstitute a fetal thymic fragment allowing T-cell production concurrent with myeloid and B-cell production in the same graft for periods of at least 20 weeks.

Analysis of 136 grafts between 9 and 36 weeks post transplant revealed that 92 (68 %) contained B-cells, T-cells and myeloid cells (Table 1). The proportion of grafts containing

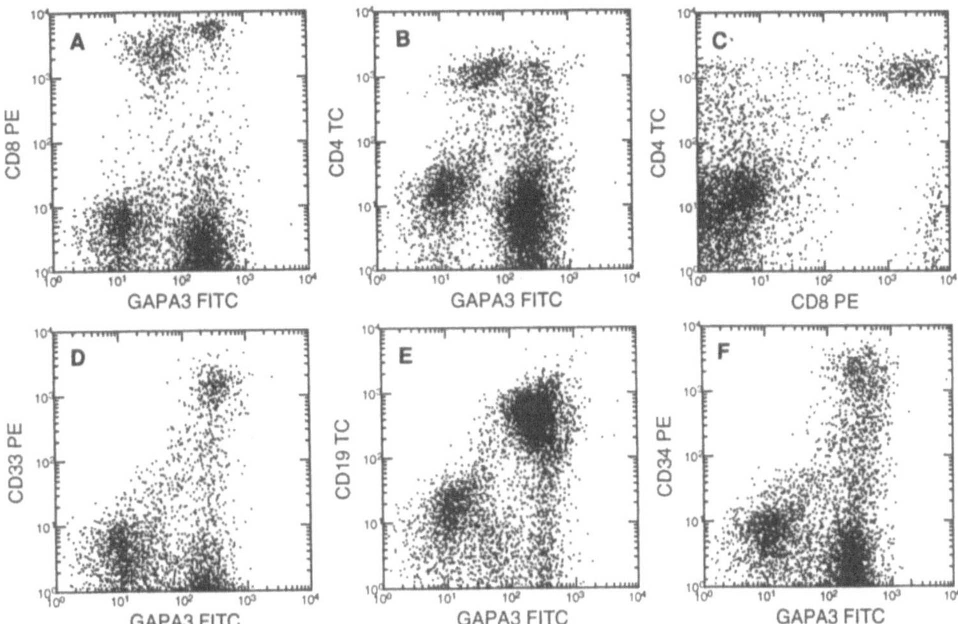

Figure 2. B-cell, T-cell and myeloid cell production in a SCID-hu BTS graft. SCID mice were transplanted subcutaneously with fetal bone (GAP-A3 +) and HLA mismatched thymus (GAP-A3 -) and sacrificed 20 weeks post transplant and graft removed. (A-C) Cell suspensions were stained with TC conjugated CD4, PE conjugated CD8 and FITC- conjugated HLA class I marker GAP-A3 (donor class I FITC), or TC conjugated CD19, PE-conjugated CD33 or CD34 and FITC conjugated GAP-A3 (D-F). Stained samples were then analysed on a FACScan fluorescent cell analyzer.

cells of all three lineages declined over time. At an early time post transplantation (9-10 weeks), a high proportion of grafts (93%, 27/29) were found to contain T-cell, B-cell and myeloid cells. Not only did the number of grafts maintaining T-cells decline with time but also the proportion of T-cells within grafts decline from 46.9 % ± 6.3% at 9-10 weeks post

Table 1. Maintenance of multiple hematopoietic cell lineages in SCID-hu BTS and SCID-hu BT grafts

	Percent of grafts positive for			
Week	T + B + M	B + M	B + T	CD4/8 DP
9-10	93% (27/29)	7% (2/29)	0% (0/29)	83% (24/29)
19-20	57% (17/30)	33% (10/30)	10% (3/30)	53% (16/30)
26-28	59% (20/34)	18% (6/34)	18% (6/34)	76% (26/34)
32-36	40% (6/15)	47% (7/15)	0% (0/15)	33% (5/15)
26-28 (BT)	79% (22/28)	21% (6/28)	0% (0/28)	75% (21/28)

SCID mice implanted with fetal bone, spleen and HLA mismatched thymus subcutaneously, or fetal bone and thymus (BT) were sacrificed, grafts removed and single cell suspensions made. Cells were stained with HLA class I specific antibodies plus either CD8-PE and CD4-TC (T, T-cells) or CD33-PE (M, myeloid cells) and CD19-TC (B, B-cells) as described in Materials and Methods. Grafts containing CD8+, CD19+ and CD33+ were considered T, B and myeloid cell positive. CD4/8 DP refers to the percentage (number) of grafts that were positive for cells co-expressing CD4 and CD8. Percentages ≤ 0.5% staining with any of the lineage antibodies were considered not detectable.

Table 2. Maintenance of CD34+ cells and multilineage hematopoietic progenitors in SCID-hu BTS and SCID-hu BT grafts

	Percent of grafts positive for			
Week	CD34	CFU-MIX	CFU-G/M	CFU-E
9-10	100% (29/29)	90% (18/20)	100% (20/20)	100% (20/20)
19-20	90% (27/30)	85% (23/27)	100% (27/27)	100% (27/27)
26-28	86% (25/29)	58% (11/19)	100% (19/19)	95% (18/19)
26-28 (BT)	100% (28/28)	ND	ND	ND

Grafts from SCID mice implanted with human fetal bone, spleen and HLA-mismatched thymus, or bone and thymus alone (BT) were removed and single cell suspensions made. Aliquots of cells were directly analysed for co-expression of donor (bone) HLA and CD34 by flow cytometry, or plated in methyl cellulose containing erythropoietin (1.2 U/mL), c-kit ligand (10 ng/mL), IL-3 (10 ng/mL), G-CSF (25 ng/mL) and GM-CSF (25 ng/mL) as described in Materials and Methods. Plates were analysed 14 days later and scored for erythroid colonies (BFU-E), myeloid colonies containing granulocytes and/or macrophages (CFU-G/M), and colonies containing all myeloid and erythroid cell populations (CFU-MIX).

transplant to 21.6% ± 8.0% at 32-36 weeks post transplant. The proportion of B-cells and myeloid cells remain relatively constant over this period (35.4 % + 6.5 % and 4.1% + 1.3%) for B and myeloid cells respectively at 32-36 weeks post transplant. Lack of a spleen fragment in grafts (SCID-hu BT) was not detrimental to the maintenance of multiple hematopoietic lineages, with 79% (22/28) of grafts containing B, myeloid and both immature and mature T-cells 26-28 weeks post transplantation (Table 1).

All grafts analyzed (66 total) contained assayable hematopoietic progenitor cells (Table 2). Both myeloid and erythroid progenitor cells, as well as CD34+ cells are maintained in a high proportion of grafts for at least 26-28 weeks (Table 2). CFU-GM were assayable in 100% of grafts analyzed up to 26-28 weeks post transplant. The proportion of grafts with detectable CFU-MIX decreased from 90% at week 9-10 to 58% at week 26-28 suggesting that active hematopoiesis may be declining with time. The number of progenitors also decreased over time from 34.0 ± 7.6 SEM CFU-GM per 1 X 10^5 cells at week 9-10 to 16.5 ± 3.4 SEM at week 26-28. The decrease in the number of grafts maintaining CD34+ cells is not as striking, with a decline from 100% of grafts positive at weeks 9-10 to 86% at weeks

Table 3. Secondary reconstitution of SCID-hu bone grafts with CD34+Thy+Lin- cells derived from 20 week old SCID-hu BTS grafts

Experiment	Injected cell number	Graft number	% donor CD34	% donor CD19	% donor CD33
1	3000	1	4.4	21.7	2.9
2	6000	1	NDt	NDt	NDt
		2	5.0	64.1	5.8
3	16000	1	6.3	54.0	11.8
		2	8.7	43.5	5.3

SCID mice were transplanted with human fetal bone, spleen and HLA mismatched thymus, and sacrificed 20 weeks later. Cells from grafts derived from each donor tissue set (18,10, and 6 grafts for experiments 1-3 respectively) were pooled, stained and sorted for CD34+Thy-1+Lin- as described in Materials and Methods. Sorted cells were pelleted and injected into fetal bone fragments that had been transplanted 8 weeks previously into the mammary fat pad of SCID mice (SCID-hu bone). Mice were irradiated (400R) immediately prior to injection. Mice were sacrificed and grafts removed 8 weeks later.

26-28 (Table 2). Interestingly, 100% of grafts composed of fetal bone and thymus only contained detectable levels of CD34+ cells for 26-28 weeks.

These data indicate that multilineage hematopoiesis at 19-20 weeks post transplantation was the result of active hematopoietic contributions from a pool of early progenitors present in the fetal bone marrow. It seems likely that hematopoietic stem cells remain present within the grafts during this period. In order to determine if stem cells were maintained, CD34+Thy-1+Lin- cells were sorted from 20 week post transplant SCID-hu BTS grafts and injected into SCID-hu bone fragments that were HLA mismatched from that of the fetal bone of the donor SCID-hu BTS graft. In 3/3 experiments from different SCID-hu BTS tissue combinations CD34+Thy-1+Lin- cells could be recovered and give rise to multiple lineages in a secondary reconstitution with as few as 3000 cells injected (Table 3).

B-Cell, T-Cell, and Myeloid Cell Production from a Common Umbilical Cord Blood Stem Cell Population in SCID-hu BTS Grafts

In order to test the SCID-hu BTS graft for both maintenance and multilineage differentiation from stem cells transplanted into the grafts, enriched stem cell populations from human UCB were injected and analyzed for their reconstituting capacity. CD34+Thy-1+Lin- cells from human UCB were isolated (Figure 3) and injected into irradiated 10-16 week old HLA mismatched SCID-hu BTS grafts. Grafts were analyzed 8 weeks later for class I HLA of the injected cells, T-cell markers CD4 and CD8, B-cell marker CD19, myeloid marker CD33 and progenitor marker CD34. Donor reconstitution was observed in 7/8 experiments (Table 4). Cell doses ranged from 4800 cells to 29000 cells per graft, with an overall donor engraftment success of 62% (24/38) of all grafts analyzed. The lineage distribution of reconstituting cells are shown in Table 4. Of 24 grafts with successful reconstitution, 7 (29 %) had reconstitution of B-cells, T-cells and myeloid cells. Representative analysis of 1 graft reconstituted with 20000 CD34+Thy-1+Lin- UCB cells as well

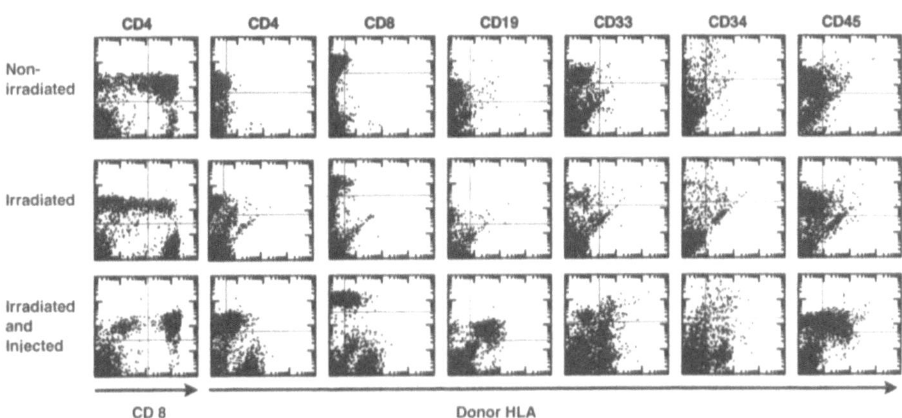

Figure 3. Umbilical cord blood CD34+Thy-1+Lin- cells can reconstitute B-cells, T-cells, myeloid cells and CD34+ cells in SCID-hu BTS grafts. Sorted cells from UCB were sorted as described, and injected into 12-16 week old SCID-hu BTS grafts. SCID-hu BTS mice were given 400R total body irradiation immediately prior to injection except for non-irradiated controls. Grafts were analysed 8 weeks later by flow cytometry for the presence of donor HLA class I positive cells plus CD4, CD8, CD19, CD33, and CD34 by three color analysis. Representative grafts that were non-irradiated, irradiated and not injected with donor cells, and irradiated and injected with donor cells from the same BTS recipient tissue combinations and analysed at the same time point are shown.

Table 4. Lineage distribution of donor cells in SCID-hu BTS grafts reconstituted with sorted UCB stem cells

Experiment	# cells transplanted	# grafts	# donor positive	Lineage distribution #B-M-T	#B-M	#B	#T
1	4800	4	0				
2	5700	1	1		1		
3	6000	8	3	1	1		1
4	8600	7	6	3	1		2
5	10000	6	4		3	1	
6	13000	5	5	2	3		
7	20000	4	4	1	3		
8	29000	3	1		1		
Total		39	24 (62%)	7	13	1	3

CD34+Thy-1+Lin- cells were sorted from umbilical cord blood cells as described in Materials and Methods. Cells were pelleted, resuspended and injected in 10μL volume into 8-12 week old HLA mismatched SCID-hu BTS grafts. SCID-hu BTS mice were irradiated (400 R) immediately prior to injection. Mice were sacrificed 8 weeks later, and grafts analysed for combinations of donor HLA with: CD4, CD8, CD19, CD33, CD34 and CD45 in a 3 color analysis by flow cytometry. 5 grafts with less than 5% human cells were not included. Grafts were scored as donor positive if > 1% of human cells was of the HLA type of the injected population. B, M, and T refer to B-cells, myeloid cells, and T-cells respectively.

as non-irradiated non-injected and irradiated non-injected controls are shown in Figure 5. Control non-irradiated and irradiated grafts show maintenance of host (donor HLA negative) cells in multiple lineages. In contrast, the injected graft has donor HLA bright cells including CD4+, CD8+, CD19+, CD33+ and CD34+ fractions. The grafts also contain donor HLA low to negative cells that are the CD4/CD8 DP thymocytes described in Figure 2. The majority of the grafts (13/24, 54%) that were reconstituted with donor cells as shown in Table 4, were reconstituted with B and myeloid cells only.

DISCUSSION

In this report we have demonstrated that transplantation of human fetal bone, spleen and thymus together at the same site subcutaneously in SCID mice (SCID-hu BTS) results in a multilineage fusion graft that can produce hematopoietic cells for at least 36 weeks post transplantation. Cells within the fetal bone marrow of SCID-hu BTS grafts produce B-cells and myeloid cells, all classes of myeloid progenitors, and hematopoietic stem cells with a CD34+Thy-1+Lin- phenotype and secondary reconstituting potential. In addition primitive cells within the fetal bone marrow migrate to and repopulate the fetal thymus fragment, where they undergo differentiation through immature CD4+CD8+ DP T-cells into mature SP T-cells. Injection of phenotypically defined umbilical cord blood stem cell populations into irradiated SCID-hu BTS grafts can result in reconstitution of all lineages assayed with donor derived immature and mature lympho-myelopoietic cells. Although human and mouse hematopoietic stem cells have been characterized with respect to expression of specific antigens (7), they must still be operationally defined by their ability to regenerate and sustain long-term myeloid and lymphoid blood cell production in vivo. It is relevant that models be developed which allow long term multilineage production from phenotypically defined subsets of hematopoietic cells.

Recently, advances have been made by transplanting human cells or tissues into SCID mice and observing human hematopoiesis for prolonged periods (10-22). Models have already been developed which allow in vivo differentiation of purified human hematopoietic stem cells into either T-cells, or B-cells and myeloid cells(8-11). The human fetal thymus provides a microenvironment for long term (greater than 1 year) T-cell production when provided a stem cell source from a human fetal liver fragment (2). We found that many grafts in SCID-hu BTS and SCID-hu BT mice which contain B-cells and myeloid cells also contained a significant proportion of T-cells derived from the progenitors in the fetal bone. Early T-cell progenitors, or stem cells present within the bone portion of the graft therefor have the capacity to seed an allogeneic fetal thymic fragment and give rise to long term immature and mature T-cell subsets. Our laboratory has previously shown that implantation of human fetal bone fragments (SCID-hu bone) can maintain progenitors for up to 12 weeks, however by 20 weeks 16/20 grafts lacked detectable human cells (5). In contrast the SCID-hu BTS grafts maintained a high proportion of CD34+ cells, and contain assayable hematopoietic progenitors in 100% of grafts analyzed 26-28 weeks post transplant. We also observed multiple hematopoietic lineages in 40% of grafts analyzed at 32-36 weeks post transplant, and in grafts that were maintained for up to 1 year (data not shown). In addition, cells with a stem cell phenotype expressing CD34+Thy-1+Lin- were isolated from 20 week old SCID-hu BTS grafts and shown to maintain secondary multilineage reconstituting potential when injected into SCID-hu bone grafts. These data suggest that the SCID-hu BTS graft may be a suitable site for engraftment and multipotential differentiation of human hematopoietic stem cell populations.

Human umbilical cord blood has also been shown to be a rich source of progenitors with in vitro multilineage and self renewal potential (12,13). Cells isolated for expression of CD34, and low levels of CD45RA and CD71 are enriched for both high proliferative potential colony-forming cells (HPP-CFC), and for high proliferative potential in liquid culture compared to Thy-1^- cells, suggesting stem cell activity lies within the Thy-1^+ fraction (14). The fact that human stem cells exist in UCB is perhaps best demonstrated by the successful hematopoietic engraftment following transplantation (15,16). Human UCB CD34+Thy-1+Lin- cells were injected into SCID-hu BTS grafts, first to determine if the grafts could support multipotential differentiation from a common stem cell pool, and second to assess the differentiation capacity of this population in vivo. Injection of cells from human UCB CD34+Thy-1+Lin- cells into irradiated SCID-hu BTS grafts can lead to production of B-cells, immature and mature T-cells, myeloid cells, and CD34+ cells when analyzed 8 weeks post injection. These observations suggest that SCID-hu BTS grafts may be an ideal model to study human hematopoietic cellular development.

SCID-hu BTS grafts have the capacity to generate concomitant T-cell, B-cell and myeloid cells for up to 9 months in vivo from a common progenitor or stem cell pool, thus providing a unique system to study endogenous or adoptively transferred defined progenitor or stem cell populations. Whether or not multiple lineages are derived from a single stem cell has not yet been determined, and may be approached with genetic marking of human HSC.

ACKNOWLEDGMENTS

We are indebted to the members of the Experimental Cellular Therapy Group for helpful discussions and technical assistance, the Comparative Medicine Group for constructing the SCID-hu mice and Dr. Schumacher of the University of Arizona for providing umbilical cord blood samples. We would also like to thank Anne Galy, Joseph M. McCune, and Irving Weissman for their helpful comments and discussions.

REFERENCES

1. McCune JM, Namikawa R, KaneshimaH, Shultz LD, Lieberman M, Weissman IL. The SCID-hu mouse: Murine model for the analysis of human hematolymphoid differentiation and function. Science 241:1632, 1988
2. Namikawa R, Weilbaecher KN, Kaneshima H, Yee EJ, McCune JM. Long-term hematopoiesis in the SCID-hu mouse. J Exp Med 172: 1055, 1990
3. Kaneshima H, Namikawa R, McCune JM. Human hematolymphoid cells in SCID mice. Curr Opp Immunol 6: 327, 1994
4. Vandekerckhove BAE, Jones D, Punnonen J, Schols D, Lin HC, Duncan B, Bacchetta R, de Vries JE, Roncarolo MG. Human Ig production and isotype switching on severe combined immunodeficient-human mice. J Immunol 151:128, 1993
5. Kyoizumi S, Baum CM, Kaneshima H, McCune JM, Yee EJ, Namikawa R. Implantation and maintenance of functional human bone marrow within SCID-hu mice. Blood 79:1704, 1992
6. Adkins B, Mueller C, Okada CY, Reichert RA, Weissman IL, Spangrude GS. Early events in T-cell maturation. Ann Rev Immunol 5: 325, 1987
7. Tsukamoto A, Weissman I, Chen B, DiGiusto D, Baum C, Hoffman R, Uchida N: Phenotypic and functional analysis of hematopoietic stem cells in mouse and man, in Levitt D, Mertelsmann R (eds): Hematopoietic stem cells: biology and therapeutic applications.New York, Marcel Dekker Inc, 1994, p85
8. Chen BP, Galy A, Kyoizumi S, Namikawa R, Scarborough J, Webb S, Ford B, Cen DZ, Chen S. Engraftment of human hematopoietic precursor cells with secondary transfer potential in SCID-hu mice. Blood 84:2497, 1994
9. DiGiusto D, Chen S, Combs J, Webb S, Namikawa R, Tsukamoto A, Chen BP, Galy AHM. Human fetal bone marrow progenitors for T, B and myeloid cells are found exclusively in the population expressing high levels of CD34. Blood 84:421, 1994
10. Baum CM, Weissman IL, Tsukamoto A, Buckle AM, Peault B. Isolation of a candidate human hematopoietic stem-cell population. Proc Natl Acad Sci USA 89:2804, 1992
11. Craig W, Kay R, Cutler RL, Lansdorp PM. Expression of Thy-1 on human hematopoietic progenitor cells. J Exp Med 177:1331, 1993
12. Lu L, Xiao M, Shen RN, Grisby S, Broxmeyer HE. Enrichment, characterization and responsiveness of single primitive CD34++ human umbilical cord blood hematopoietic progenitors with high proliferative and replating potential. Blood 81:41, 1993
13. Carow CE, Hangoc G, Broxmeyer HE. Human multipotential progenitor cells (CFU- GEMM) have extensive replating capacity for secondary CFU-GEMM: an effect enhanced by cord blood plasma. Blood 81:942, 1993
14. Mayani H, Lansdorp PM. Thy-1 expression is linked to functional properties of primitive hematopoietic progenitor cells from human umbilical cord blood. Blood 83:2410, 1994
15. Wagner JE, Broxmeyer HE, Byrd RL, Zehnbauer B, Schmeckpeper B, Shah N, Griffin C, Emanuel PD, Zuckerman KS, Cooper S, Carow C, Bias W, Santos GW. Transplantation of umbilical cord blood after myeloablative therapy: analysis of engraftment. Blood 79:1874, 1992
16. Browning MJ, Krausa P, Rowan P, Bicknell DC, Bodmer JG, Bodmer WF. Tissue typing the HLA-A locus from genomic DNA by sequence-specific PCR: Comparison of HLA genotype and surface expression on colorectal tumor cell lines. Proc Natl Acad Sci USA 90:2842, 1993

4

PERIPHERAL BLOOD STEM CELLS MOBILIZATION WITH G-CSF (FILGRASTIM) ALONE FOR AUTOLOGOUS TRANSPLANT IN MYELOID AND LYMPHOID MALIGNANCIES*

E. Archimbaud,[1] M. Michallet,[1] I. Philip,[2] C. Sebban,[1] P. Tremisi,[3]
G. Clapisson,[2] A. Belhabri,[1] and D. Fière[1]

[1] Hôpital Edouard Herriot
Lyon, France
[2] Centre Léon Bérard
Lyon, France
[3] Centre de Transfusion Sanguine
Lyon, France

ABSTRACT

In order to determine whether granulocyte colony-stimulating factor (G-CSF) alone was capable of mobilizing peripheral blood stem cells (PBSC) in myeloid as well in lymphoid malignancies, G-CSF (filgrastim), 5 μg/kg/day, SC, was administered to 48 patients with myeloid or lymphoid malignancies at various stages, after hematopoietic recovery from last chemotherapy. PBSC were harvested using daily aphereses from day 5 of G-CSF therapy. An adequate PBSC graft could be harvested in 43 patients (90%) after 2 to 5 (median 3) aphereses. Predictive factors for a good PBSC yield in a multivariate analysis included a high WBC count on day 5 of G-CSF therapy, and a diagnosis of myeloid malignancy. Thirty patients were transplanted after various conditioning regimens. Median times to neutrophil $> 0.5 \times 10^9$/l and platelet $> 50 \times 10^9$/l were 15 and 32 days respectively. The harvest of $> 10 \times 10^4$ CFU-GM/kg and the absence of busulfan in the conditioning regimen were predictive of early platelet recovery. We conclude that G-CSF therapy initiated during stable hematopoiesis is efficient to allow PBSC harvest for autografting in myeloid as well as lymphoid malignancies.

* Correspondence to: Eric Archimbaud, M.D., Service d'Hématologie, Hôpital Edouard Herriot, 69437, Lyon Cedex 03, France: Phone (33) 72.11.73.97; Fax (33) 72.11.73.10

Molecular Biology of Hematopoiesis 5, edited by Abraham et al.
Plenum Press, New York, 1996

Peripheral blood stem cells (PBSC) allow earlier hematologic recovery after autografting when compared to bone marrow stem cells (1,2,3). The optimal mobilization method in terms of PBSC yield probably includes the use of a priming chemotherapy followed by the administration of growth factors during hematopoietic recovery (4). However, granulocyte colony-stimulating factor (G-CSF) alone at dosages ranging between 5 and 24 μg/kg/day in the absence of priming chemotherapy has been shown to mobilize enough PBSC to allow their collection using a median of 3 aphereses and provide hematopoietic recovery after myeloablative therapy in solid tumors and lymphoid malignancies (1,4-8).

This study was designed to confirm these results using the lowest previously tested dosage of G-CSF (4,7) and to determine whether this approach could also be used in myeloid malignancies such as acute myeloid leukemia (AML), for which cytoxan, one of the drugs recognized as optimal for priming before mobilization of PBSC, is generally not part of baseline chemotherapy.

PATIENTS AND METHODS

Patient Selection Criteria and Characteristics

All patients with an hematopoietic malignancy and in whom an intensification of therapy followed by stem cell rescue was planed were eligible for this study provided the therapeutic protocol according to which the patient was treated allowed this form of stem cell harvest before autografting.

Forty-eight patients were included and 30 have currently been grafted. Characteristics of these 2 groups of patients are indicated in Table 1. Median age of the patients was 55 years with 12 patients aged more than 60 years. Twenty-four patients had AML while 24 had various lymphoid malignancies. Thirty patients were in complete or partial response after their first line therapy while 18 had advanced disease, 2 of them having relapsed after previous bone marrow transplantation.

Mobilization, Harvest and Cryopreservation of Peripheral Blood Stem Cells

Patients were treated with G-CSF (filgrastim), 5 μg/kg/day, SC, initiated while blood counts were within normal limits after hematopoietic recovery of previous course of

Table 1. Patient characteristics

	All patients (n = 48)	Patients grafted (n = 30)
Age (years)	55 (17 - 74)*	55 (17 - 74)
Diagnosis		
AML	24#	17#
ALL	1	1
NHL	8	7
Hodgkin	2	1
M. Myeloma	13	4
Pathology stage		
1st CR/PR/plateau phase	30	18
advanced	18§	12§

AML: acute myeloid leukemia; ALL: acute lymphoblastic leukemia; NHL: non Hodgkin lymphoma

*Median (range), 12 patients were aged >60 years; #3 cases with secondary AML; §2 with previous transplantation.

chemotherapy. Stem cell harvest was performed from day 5 of G-CSF therapy (4,7) through daily standard 2 to 3-blood volume aphereses using a Cobe Spectra (Cobe, Rungis, France) or Fenwal CS-3000 (Baxter, Maurepas, France) blood separator. The target cell yield was > 6 x 10^8 mononuclear cells (MNC)/kg of body weight of the patient, containing > 4 x 10^6/kg CD34+ cells or > 10 x 10^4/kg CFU-GM. Cells were cryopreserved in the presence of 10% DMSO and 4% serum albumin using a controlled-rate freezer and stored in liquid nitrogen.

Transplantation Procedure

At this time 30 patients have been transplanted using their harvested PBSC. Conditioning for the graft included TBI containing regimen in 9 patients, polychemotherapy regimen in 8, BCNU, 800 mg/m^2 alone in 5, and busulfan 16mg/kg alone in 8.

Statistical Analysis

Analysis of single factors related to cell yield was performed using non-parametric statistics. Multivariate analysis of factors related to the achievement of an above or below the median cell yield was performed by multiple logistic regression. The probability of neutrophil and platelet recovery after grafting was calculated according to the Kaplan and Meier method. Factors predictive for hematopoietic recovery post-transplant were analyzed using the log-rank test.

RESULTS

Efficacy of Mobilization and Harvest

Median WBC on day 5 of G-CSF therapy was 26.9 x 10^9/l (range 5.7 to 106 x 10^9/l). Therapy was well tolerated with only one case of severe (WHO grade 3) bone pain. Patients underwent a median of 3 (range 2 to 5) aphereses. The cell yields of successive apheresis are shown in Table 2. The yields of successive aphereses in the same patients where highly correlated ($p < 0.0001$), however, there was a significant decrease in the MNC and CFU-GM yields between aphereses 1 and 3 ($p = 0.001$ and $p = 0.03$ respectively). We also found a good correlation between MNC and CFU-GM yields in the same apheresis ($p < 0.0001$), while neither MNC nor CFU-GM yields were significantly correlated to CD34+ cell yield. The proportion of patients reaching global target cell yields for MNC, CD34+ cells and CFU-GM after 1, 2, or 3 aphereses is indicated in Table 3, showing that 14/29 patients tested and 20/48 patients reached the target yields for CD34+ cells and CFU-GM respectively after a single apheresis, although all patients had at least 2 aphereses. PBSC harvest was considered insufficient for a graft in 5 patients (10%). Three of them had advanced disease, one having previously received a bone marrow transplantation, while the 2 others had AML in first CR, but

Table 2. Cell yield of successive aphereses

End point	Apheresis # 1 (n = 48)	Apheresis # 2 (n = 48)	Apheresis # 3 (n = 42)	Total series (n = 48)
MNC (x 10^8/kg)	3.0 (0.9 - 9.5)*	2.7 (1.2 - 7.2)	2.3 (1.0 - 6.8)	8.0 (3.0 - 19)
CD34+ cells (x 10^6/kg)	3.4 (0 - 52)	2.9 (0 - 59)	1.6 (0 - 37)	7.8 (0 - 127)
CFU-GM (x 10^4/kg)	7.4 (0 - 180)	9.4 (0 - 110)	4.5 (0 - 107)	22 (0 - 397)

*Median (range)

Table 3. Number of patients reaching target yield

End point	Apheresis # 1	Aphereses # 1 + 2	Apheresis # 1 + 2 + 3
MNC > 6 x10^8/kg	7 / 48	22 / 48	39 / 48
CD34+ cells > 10 x10^6/kg	14 / 29	17 / 29	23 / 29
CFU-GM > 4 x10^4/kg	20 / 48	33 / 48	35 / 48

relapsed within 2 months after insufficient harvest. Predictive factors for the MNC, CD34+ cells and CFU-GM yields in univariate and multivariate analyses are indicated in Table 4: while a high WBC count on day 5 of G-CSF therapy, an underlying myeloid (as opposed to lymphoid) malignancy, and early stage of the disease at the time of collection were all predictive for high yields in the univariate analysis, only the WBC count on day 5 of G-CSF and a diagnosis of myeloid malignancy remained significant in the multivariate analysis.

Engraftment of Harvested PBMC

Patients received harvests containing a median of 8.2 (range: 5.2 to 15.3) x 10^8 MNC/kg, 7.5 (0 to 127) x 10^6 CD34+ cells/kg and 27 (4.4 to 156) CFU-GM/kg. Thirteen patients, all of them with AML in first CR conditioned with BCNU or busulfan alone, received G-CSF, 5 µg/kg IV or SC, from the day when neutrophil count dropped below 0.1 x 10^9/l after the graft until neutrophil recovery. Median time to neutrophil recovery > 0.5 x 10^9/l was 15 days (range: 9 to 23 days), with one patient not achieving neutropenia below this level. Median time to platelet recovery > 50 x 10^9/l was 32 days (range 9 to 333+ days). All but 2 patients who had early relapse became independent of platelet transfusions (platelet > 20 x 10^9/l) following transplant.

Age of the patient, the lineage (myeloid or lymphoid) and stage of underlying malignancy, the number of MNC, CD34+ cells or CFU-GM harvested for the graft and the type of conditioning regimen were not predictive for the duration of neutropenia post-transplant. The harvest of less than 10 x 10^4 CFU-GM/kg for the graft ($p = 0.03$) and the use of a busulfan containing conditioning regimen ($p = 0.04$) were associated with a longer time to recovery of platelet count above 50 x 10^9/l.

DISCUSSION

We confirm that G-CSF therapy at a dose of 5 µg/kg/day initiated during stable hematopoiesis is easy and efficient to mobilize PBSC to be harvested from day 5 of CSF administration in myeloid as well as lymphoid malignancies at various stages. The observation that cell yield decreased significantly between the first and third aphereses shows that earlier initiation of aphereses might be preferable. Apheresis have been initiated by others

Table 4. Predictive factors for good apheresis yield

	Endpoint		
Factor	MNC /kg (n = 34)	CD34+ cells/kg (n = 34)	CFU-GM /kg (n = 29)
WBC on day 5 of G-CSF	<0.0001 (*<0.001*)*	NS (*NS*)	<0.0001 (*<0.001*)
Myeloid malignancy	0.005 (*0.04*)	0.02 (*0.05*)	0.0001 (*0.005*)
Early stage of disease	0.03 (*NS*)	0.04 (*NS*)	0.007 (*NS*)

*p value in univariate (*multivariate*) analysis.

on day 4 of G-CSF therapy, however, higher dosages of G-CSF were used in these studies (1,2,5).

The observation that an underlying myeloid malignancy was also predictive for a good apheresis yield indicates that this technique is applicable to AML patients and that higher doses of CSF are perhaps not necessary in these patients, unlike found by others in lymphoid malignancies and solid tumors (6,8). On the opposite, the use of higher doses of CSF could facilitate stimulation of residual leukemic cells in AML patients (9). The use of PBSC mobilization during steady state hematopoiesis in AML might facilitate PBSC autografting in this disease, since cytoxan, the most used drug for the mobilization of PBSC in mobilizing regimen associating chemotherapy and growth factors, is of no antileukemic benefit for AML chemotherapy. Anthracyclin and cytarabine-based induction and intensive consolidation regimens generally used in these patients induce long-term aplasia and the harvest of PBSC during growth factor-induced recovery of aplasia might be hampered by persisting infections or thrombocytopenia or other complications at this time, particularly in patients in the upper age range to receive an autograft. Using this method, we have been able to autograft approximately 50% of patients aged 60 to 75 years who achieved CR1 in our center since the beginning of this study (10).

REFERENCES

1. Chao NJ, Schriber JR, Grimes K, Long GD, Negrin RS, Raimondi CM, Horning SJ, Brown SL, Miller L, Blume KG. Granulocyte colony-stimulating factor "mobilized" peripheral blood progenitor cells accelerate granulocyte and platelet recovery after high-dose chemotherapy. Blood 81:2031-2035, 1993.
2. Faucher C, le Corroller AG, Blaise D, Novakowitch G, Manonni P, Moatti JP, Maraninchi D. Comparison of G-CSF-primed peripheral blood progenitor cells and bone marrow auto transplantation: clinical assessment and cost-effectiveness. Bone Marrow Transplant 14:895-901, 1994.
3. Reiffers J, Stoppa AM, Attal M, Michallet M. Is there a place for blood stem cell transplantation for the younger adult patient with acute myelogenous leukemia? J Clin Oncol 12:1100, 1994.
4. Möhle R, Pförsich M, Fruehauf S, Witt B, Krämer A, Haas R. Filgrastim post-chemotherapy mobilizes more CD34+ cells with a different antigenic profile compared with use during steady-state hematopoiesis. Bone Marrow Transplant 14:827-832, 1994.
5. Bensinger W, Singer J, Appelbaum F, Lilleby K, Longin K, Rowley S, Clarke E, Clift R, Hansen J, Shields T, Storb R, Weaver C, Weiden P, Buckner CD. Autologous transplantation with peripheral blood mononuclear cells collected after administration of recombinant granulocyte stimulating factor. Blood 81:3158-3163, 1993.
6. Peters WP, Rosner G, Ross M, Vredenburgh J, Meisenberg B, Gilbert C, Kurtzberg J. Comparative effect of granulocyte-macrophage colony-stimulating factor (GM-CSF) and granulocyte colony-stimulating factor (G-CSF) on priming peripheral blood progenitor cells for use with autologous bone marrow after high-dose chemotherapy. Blood 81:1709-1719, 1993.
7. Bolwell BJ, Fishleder A, Andresen SW, Lichtin AE, Koo A, Yanssens T, Burwell R, Baucco P, Green R. G-CSF primed peripheral blood progenitor cells in autologous bone marrow transplantation: parameters affecting bone marrow engraftment. Bone Marrow Transplant 12:609-614, 1993.
8. Sheridan WP, Begley CG, To LB, Grigg A, Szer J, Maher D, Green MD, Rowlings PA, McGrath KM, Cebon J, Dyson P, Watson D, Bayly J, de Luca E, Tomita D, Hoffman E, Morstyn G, Juttner CA. Phase II study of autologous filgrastim (G-CSF)-mobilized peripheral blood progenitor cells to restore hemopoiesis after high-dose chemotherapy for lymphoid malignancies. Bone Marrow Transplant 14:105-111, 1994.
9. Mehta J, Powles R, Singhal S, Treleaven J. Peripheral blood stem cell transplantation may result in increased relapse of acute myeloid leukaemia due to reinfusion of a higher number of malignant cells. Bone Marrow Transplant 15:652-653, 1995.
10. Archimbaud E, Jehn U, De Cataldo F, Martin C, Thomas X. Idarubicin or mitoxantrone, VP-16 and cytarabine for induction/consolidation therapy followed by autologous stem cell transplantation in elderly patients with acute myeloid leukemia (AML) : a feasibility study. Acute Leukemias VI, Prognostic Factors and Treatment Strategies. Ann Hematol 70(suppl 2):A136, 1995.

DENSITY SEPARATION AND CRYOPRESERVATION OF UMBILICAL CORD BLOOD CELLS*

Evaluation of Recovery by Means of Short- and Long-Term Culture

Camillo Almici,[1] Carmelo Carlo-Stella,[1] John E. Wagner,[2] and Vittorio Rizzoli[1]

[1] Department of Hematology, BMT Unit
University of Parma
I-Parma, Italy
[2] Department of Pediatrics
University of Minnesota, Minnesota

ABSTRACT

The clonogenic capacity of human umbilical cord blood (UCB) has been evaluated in several studies, showing high numbers of primitive hematopoietic progenitor cells. Recently, UCB progenitor cells have been demonstrated to possess significant advantages over bone marrow (BM), in terms of proliferative capacity and immunologic reactivity. Therefore, UCB has been considered an attractive source of hematopoietic stem cells for both research and clinical applications. Previous reports have documented a significant loss of progenitor cells by any manipulation other than cryopreservation. We have evaluated the feasibility of fractionating and cryopreserving UCB samples with minimal loss of progenitor cells. We compared separation over three different densities of Percoll (1069 g/ml, 1077 g/ml, 1084 g/ml), sedimentation over poligeline (Emagel), and sedimentation over poligeline followed by separation over Ficoll/Hypaque (F/H). Separated samples (n=25) were analysed for recovery of $CD34^+$ cells and progenitor cells (CFU-GEMM, BFU-E, CFU-GM). Separation by sedimentation over poligeline followed by F/H allowed the highest depletion of RBC (hematocrit of the final cellular suspension 0.4±0.1%), while maintaining high recovery of $CD34^+$ cells (85.3%) and total recovery for CFU-GEMM, BFU-E and CFU-GM. After cryopreservation, recovery of clonogenic progenitors was 82% for CFU-GEMM, 94%

* Send correspondence to: Camillo Almici, Cattedra di Ematologia, Università di Parma, Via Gramsci 14, I-Parma, Italy. Phone: (Italy) 521-290787; Fax: (Italy) 521-292765

for BFU-E, 82% for CFU-GM and 90% for colony-forming unit after five weeks of long-term culture (LTC). Moreover, the presence of SCF significantly increased CFU-GEMM (14±4 vs. 49±5, p≤0.0005) and CFU-GM (112±18 vs. 178±19, p≤0.025), but not BFU-E (42±7 vs. 53±7, p≤0.375) growth. In conclusion, RBC depletion of UCB can be accomplished with minimal loss of committed and primitive hematopoietic progenitors. This procedure may have important implication in the large scale banking of UCB as well as ex vivo expansion/gene therapy protocols.

INTRODUCTION

The structural and functional integrity of the hematopoietic system is maintained by a relatively small population of stem cells that undergo self-renewal or differentiation into lineage-restricted progenitors (1). Bone marrow (BM) is the primary site of hematopoietic stem cells in adults (1), but those cells have been documented in peripheral blood (PB) (2) and more recently have been extensively studied in umbilical cord blood (UCB) (3). Ontologically, hematopoiesis during embryonic and fetal development is represented as a migratory phenomenon (4). Fetal blood immediately prior to delivery has been shown to contain hematopoietic progenitor cells at similar or higher frequency than those in BM (3, 5-7). Therefore, UCB, which is normally discarded, has been evaluated as a source of stem/progenitor cells for both research and clinical applications (8). A major concern for wider transplant application has been related to the low total number of progenitor cells in UCB that could be obtained in a single collection. While several studies suggest that UCB contains a higher proportion of primitive hematopoietic progenitor cells as compared to BM (4-6), it has been reported that significant numbers of UCB hematopoietic progenitor cells are lost by any manipulation prior to cryopreservation (3). In fact, gradient separation techniques, currently used for BM separation, have produced poor results, both in terms of red blood cell (RBC) depletion and progenitor cell recovery, when applied to UCB (9).

It was therefore the aim of our study to compare the efficacy of separation techniques based on different density gradient, in order to identify the separation procedure able to provide the more successful depletion of RBC and at the same time the highest recovery of progenitor cells.

MATERIAL AND METHODS

Separation Procedures

Human UCB samples, obtained with institutional review board approval from healthy full-term neonates immediately after delivery, were divided in five equal aliquots and separated using separation techniques based on different density gradients (10). We have compared separation over different densities of Percoll (Biochrom KG, Berlin, Germany) (d=1.069 g/ml; d=1.077 g/ml; and d=1.084 g/ml), sedimentation over poligeline (Emagel, Behringwerke, Marburg, Germany), sedimentation over poligeline followed by separation over Ficoll/Hypaque (F/H d=1.077g/ml, Sigma Chemical Co., St Louis, MO, USA).

Immunofluorescence Analysis

Surface antigen phenotype was determined by immunofluorescence analysis. Phenotypic analysis was performed with a FACSort flowcytometer (Becton Dickinson). Data were processed with a Hewlett-Packard (Forth Collins, CO) 340 computer using Lysis II (Becton Dickinson) software.

Cryopreservation and Thawing

UCB mononuclear (MNC) cells obtained after separation over poligeline and afterwards over F/H were cryopreserved in cryotubes with 10% (vol/vol) DMSO (Tera Pharmaceuticals, CA, USA) and 30% FCS (Hyclone, CA, USA). Rapid thawing was performed in a water bath at 37°C followed by slow stepwise dilution over 10 minutes with 10 times the volume of IMDM supplemented with 10% FCS. Cells were then washed, counted and the number of progenitor cells determined in short- and long-term culture.

Short-Term Culture Assay

The assay for CFU-GEMM, BFU-E and CFU-GM was performed as described in detail elsewhere (11). Cultures were stimulated with a mixture of human recombinant colony-stimulating factors (CSFs), including interleukin-3 (IL-3; 10 ng/ml), granulocyte-CSF (G-CSF; 10 ng/ml), granulocyte-macrophage-CSF (GM-CSF; 10 ng/ml), erythropoietin (Epo; 1 U/ml), and with or without the addition of stem cell factor (SCF; 50 ng/ml).

Long-Term Culture (LTC) Assay

LTC assay was performed according to the methods previously described (12) with minor modifications. As feeder layer we have utilised the murine stromal cell line M210B4 transfected with the IL-3 and G-CSF genes (gently provided by Dr Connie Eaves, Vancouver, BC, Canada) (13). At initiation, 2 x 10^6 MNC cells, obtained after sequential separation over poligeline and F/H, were seeded per flask. At the fifth week of culture, both non-adherent and adherent cells were harvested, counted and assayed in methylcellulose for colony growth, as indicated above.

Statistical Analysis

Statistical analysis was performed with the statistical package Statview (BrainPower Inc., Calabasas, CA, USA) run on a Macintosh II (Apple Computer, Cupertino, CA, USA) personal computer. The Student's t-test for paired data was used to test for significance of changes in the comparison of data involving counts.

RESULTS

Cell Separation Procedures

Several methods of enriching nucleated cells by removal of RBC were compared. Cell viability was > 95%, with all different separation procedures. The combination of poligeline sedimentation followed by separation over F/H (Em/FH) was able to achieve the more complete RBC depletion, being the hematocrit 0.4±0.1%. The nucleated cell recovery was ranging between 11% and 17% for the different Percoll densities, 78% for poligeline, and 23% for Em/FH (Table 1).

Table 1. Characteristics (%) of cord blood samples after separation over different density gradients

			Percoll (g/ml)		
	Emagel	Em/FH	1.069	1.077	1.084
Cell Recovery (%)	78	23	11	17	13
$CD34^+$ (%)	100	85.3	14	17	18.4
$CD33^+$ (%)	99.4	87	30.5	48	21.3
CFU-GEMM (%)	130.5	125.6	22.6	10.3	4
BFU-E (%)	67.7	28.8	10.2	7.6	5.2
CFU-GM (%)	100	120.6	73.5	35.8	43.9

Immunofluorescence Analysis

As shown in table I the gradient separation over Percoll determined an important loss in the total number of $CD34^+$ cells; on the contrary, the sedimentation over poligeline results in no loss in $CD34^+$ cells while after separation over F/H 85.3% of these cells were recovered. Similar results were also obtained for the recovery of $CD33^+$ cells (table I).

Progenitor Colony Forming Assay

Progenitor assays were performed on unseparated UCB, as well as after each separation procedure in order to calculate the recovery of hemopoietic progenitor cells. Results are reported in table I as percentage of unseparated samples. Separation over Percoll determined a major loss of progenitor cells, probably due to the high number of RBC that interfere in the separation by tracking MNC cells and therefore lowering the recovery of progenitor cells. In contrast superior results were obtained with poligeline sedimentation either alone or as a first step for RBC depletion prior to separation over F/H. Both these procedures permitted total recovery of CFU-GEMM and CFU-GM, with 67.7% and 28.8% of BFU-E recovery, respectively. Moreover, MNC cells, obtained after sequential separation over poligeline and F/H, were plated in methylcellulose stimulated with a standard combination of cytokines in the presence or absence of SCF (50 ng/ml) showing a statistically significant increase in CFU-GEMM and CFU-GM, but not BFU-E numbers (Table 2).

Cryopreservation

As shown in Table 3, cryopreservation of UCB samples, obtained after sequential separation over poligeline and F/H, was associated with minimal loss of progenitor cells.

Table 2. Effects of SCF on in vitro growth of hematopoietic progenitor cells from cord blood (n=13)

SCF	CFU-GEMM	BFU-E	CFU-GM
Absent	14±4	42±7	112±18
Present	49±5*	53±7°	178±19**

CFU-GEMM, BFU-E and CFU-GM are expressed as number of colonies per 5 x 10^4 MNC cells plated. Cultures in methylcellulose were stimulated with IL-3 (10 ng/ml), G-CSF (10 ng/ml), GM-CSF (10 ng/ml), Epo (1 U/ml), and with or without SCF (50 ng/ml). *$p \leq 0.0005$; **$p \leq 0.025$; °NS.

Table 3. Clonogenic progenitors recovery (mean±SD) after cryopreservation of cord blood mononuclear cells (n=13), separated sequentially over Emagel and Ficoll/Hypaque

Cryopreservation	Cells recovery (%)	CFU-GEMM (/5 x 10^4)	BFU-E (/5 x 10^4)	CFU-GM (/5 x 10^4)	CFU Week 5th* (/2 x 10^6)
Pre	100	49±5	53±7	178±19	204±12
Post	73±11	40±3	50±11	146±13	184±16

Cultures in methylcellulose were stimulated with IL-3 (10 ng/ml), G-CSF (10 ng/ml), GM-CSF (10 ng/ml), Epo (1 U/ml), SCF (50 ng/ml).
*Colony-Forming Unit after 5 weeks in long-term culture; assay was performed in 4 different experiments.

Cell recovery after freezing was 73±11%, with a viability > 97%. Progenitor cell clonogenic capacity was determined, demonstrating a recovery of 82%, 94% and 82% for CFU-GEMM, BFU-E and CFU-GM, respectively. Recovery of primitive progenitor cells after cryopreservation was 90±8% (n=4), as tested in long-term culture.

DISCUSSION

It has recently been demonstrated that UCB can be used as a source of transplantable stem cells (8). UCB collected at delivery has been shown to contain hemopoietic progenitor cells at similar or higher frequency than those in BM (3, 7, 10). UCB derived progenitor cells may possess significant advantages in terms of proliferative capacity and immunologic reactivity (14). UCB, which is normally discarded, can be easily collected without any danger or inconvenience to the donor (7). Therefore, UCB is an attractive source of transplantable cells that can be used for the treatment of diseases potentially curable by BMT (8).

Since UCB banks, generally, consist of unseparated samples, the advantage of an efficient separation technique would permit a reduction in the volume of samples to be cryopreserved lowering the costs of banking and the need of spaces for storage. The MNC cells' separation would allow the storage of large numbers of UCB samples with minimal space requirements, without the need for freezing unseparated blood bags. However, the major problem in separating UCB is the high contamination of RBC, that interfere in the separation by tracking mononuclear cells and lowering the recovery of progenitor cells. Poor results have been reported when procedures, derived from BM separation, are directly applied to UCB (9). Interesting results to date have been reported using modified F/H and 3% gelatine separation procedure (15, 16); however RBC contamination and progenitor cells loss have remained important problems. Therefore we investigated other possible methods of separation in the attempt to ameliorate RBC depletion and maximise the progenitor cells recovery, showing, in contrast to earlier reports (3), that the isolation of low-density UCB cells was not accompanied by massive losses in numbers of progenitor cells. We think that a sedimentation over a poligeline layer would deplete the majority of RBC. And coupling the sedimentation over poligeline with the separation over F/H would significantly reduce the final volume without affecting the recovery of progenitor cells. We report a recovery of CFU-GEMM and CFU-GM higher than 100%, caused probably by an underestimation of these progenitors cells in unseparated UCB, due to the high contamination of RBC that interfere with the scoring of smaller colonies. Our results of separation over poligeline are similar to those obtained by separation over 3% gelatine (16); however the use of poligeline seems much easier and safer, since poligeline solution (Emagel, Behringwerke, Marburg, Germany) is commercially available and in contrast gelatine comes in powder form and therefore need to be house-prepared and autoclaved for sterilisation before use.

It was reported that cryopreservation of UCB samples as whole or separated blood has minimal effects on cell recovery and viability (3, 10, 15, 16). As we report, UCB samples can be separated into MNC cell populations and frozen without any functional defects. Cryopreserved MNC cells were > 97% viable and the in vitro colony assays demonstrated that these cells were also completely functional. We have reported a 82% recovery for CFU-GEMM and CFU-GM and 94% recovery for BFU-E, and 90% recovery of more immature progenitors in long-term culture. UCB progenitor cells have been reported to possess a greater proliferative potential in comparison to BM. This finding is even more evident by the addition of SCF to CSFs mixture. SCF has been reported to influence early hematopoietic progenitors and to induce self-renewal (3, 6, 10). Results reported here confirm that SCF in combination with other CSFs significantly increases the growth of progenitor cells (CFU-GEMM and CFU-GM) (3).

In conclusion, UCB has proved to be an important source of hematopoietic stem cells suitable for clinical transplantation. Our data provide a rationale for cord blood banking as an alternative to volunteer marrow registries. In view of adult transplantation, immunoselection of $CD34^+$ cells from UCB (3, 6, 14, 17) might provide an ideal material for studies of ex vivo stem cell expansion.

ACKNOWLEDGEMENTS

This work was supported in part by grants from Consiglio Nazionale delle Ricerche (Progetto Finalizzato A.C.R.O.) and Associazione Italiana per la Ricerca sul Cancro (AIRC).

REFERENCES

1. Ogawa M, Porter P, Nakahata T. Renewal and commitment to differentiation of hemopoietic stem cells (an interpretative review). Blood 1983; 61: 823-829.
2. Siena S, Bregni M, Brando B, Ravagnani F, Bonadonna G, Gianni AM. Circulation of $CD34^+$ hematopoietic stem cells in the peripheral blood of high-dose cyclophospamide-treated patients: enhancement by intravenous recombinant human granulocyte-macrophage colony-stimulating factor. Blood 1989; 74: 1905-1914.
3. Broxmeyer HE, Douglas GW, Hangoc G, Cooper S, Bard J, English D, Arny M, Thomas L, Boyse EA. Human umbilical cord blood as a potential source of transplantable hematopoietic stem/progenitor cells. Proc Natl Acad Sci 1989; 86: 3828-3832.
4. Broxmeyer HE, Hangoc G, Cooper S, Ribeiro RC, Graves V, Yoder M, Wagner JE, Vadhan-Raj S, Rubinstein P, Broun ER. Growth characteristics and expansion of human umbilical cord blood and estimation of its potential for transplantation of adults. Proc Natl Acad Sci 1992; 89: 4109-
5. Tavassoli M. Embryonic and fetal hemopoiesis: an overview. Blood Cells 1991; 1: 269-281.
6. Hows JM, Bradley BA, Marsh JCV, Luft T, Coutinho L, Testa NG, Dexter TM. Growth of human umbilical cord blood in longterm hemopoietic cultures. Lancet 1992; 340: 73-76.
7. Pettengel R, Luft T, Henschler R, Hows JM, Dexter M, Ryder D, Testa NG. Direct comparison by limiting dilution analysis of long-term culture-initiating cells in human bone marrow, umbilical cord blood, and blood stem cells. Blood 1994; 84: 3653-3659.
8. Rubinstein P, Rosenfield RE, Adamson JW, Stevens CE. Stored placental blood for unrelated bone marrow reconstitution. Blood 1993; 81: 1679-1690.
9. Wagner JE, Kernan NA, Steinbuch M, Broxmeyer HE, Gluckman E. Allogeneic sibling umbilical cord blood transplantation in forty-four children with malignant and non-malignant disease. Lancet 1995 in press.
10. Newton I, Charbord P, Schaal JP, Herve P. Toward cord blood banking: density separation and cryopreservation of cord blood progenitors. Exp Hematol 1993; 21: 671-674.
11. Almici C, Carlo-Stella C, Mangoni L, Garau D, Cottafavi L, Ventura A, Armanetti M, Wagner JE, Rizzoli V. Density separation of umbilical cord blood and recovery of hemopoietic progenitor cells: implications for cord blood banking. Stem Cells 1995; in press.

12. Carlo-Stella C, Cazzola M, Ganser A, Bergamaschi G, Pedrazzoli P, Hoelzer D, Ascari E. Synergistic antiproliferative effect of recombinant interferon gamma with recombinant interferon alpha on chronic myelogenous leukemia hematopoietic progenitor cells (CFU-GEMM, BFU-E, CFU-GM). Blood 1988; 72: 1293-1299.
13. Sutherland HJ, Eaves CJ, Eaves AJ, Dragowskas W, Lansdorp PM. Characterization and partial purification of human marrow cells capable of iniating long-term hematopoiesis in vitro. Blood 1989; 74: 1563-1570.
14. Lansdorp PM, Dragowska W, Mayani H. Ontogeny related changes in proliferative potential of human hematopoietic cells. J Exp Med 1993; 178: 787-791.
15. Harris DT, Schumacher MJ, Rychlik S, Booth A, Acevedo A, Rubinstein P, Bard J, Boyse EA. Collection, separation and cryopreservation of umbilical cord blood for use in transplantation. Bone Marrow Transpl 1994; 13: 135-143.
16. Bertolini F, Lazzari L, Lauri E, Corsini C, Castelli C, Gorini F, Sirchia G. A comparative study of different procedures for the collection and banking of umbilical cord blood. J Hematother 1995; in press.
17. Lansdorp PM. Telomere lenght and proliferation potential of hematopoietic stem cells. J Cell Science 1995; 108: 1-6.

ECONOMIC IMPLICATIONS OF HIGH DOSE CHEMOTHERAPY PROGRAMS WITH AUTOLOGOUS STEM CELL SUPPORT

Charles L. Bennett,[1] Michael R. Bishop,[2] and Subhash C. Gulati[3]

[1] Lakeside VAMC Chicago
Northwestern University in Chicago
[2] University of Nebraska Medical Center
Omaha Nebraska
[3] The New York Hospital
Cornell University, New York, New York

INTRODUCTION

In the United States, high dose chemotherapy with stem cell support (bone marrow or peripheral blood) is rapidly becoming accepted therapy for many patients with Hodgkin's disease, non-Hodgkin's lymphoma, and breast cancer, despite concerns by insurers and patients over high costs and the potential of significant economic burden on the health care system. Issues such as comparative costs and efficacy of supportive care modalities, hematopoietic growth factors, and adjunctive measures such as purging of stem cells are important. Most often these studies have not evaluated the costs and cost-effectiveness of these pharmaceuticals and technologies. Physicians are now under increasing pressures to accept capitated payments for high dose chemotherapy patients. These capitated rates are often significantly lower than the amount that had been charged for the procedure in the previous year. Furthermore, hospital pharmacies are facing cost-containment pressures and need cost-effectiveness data on pharmaceuticals that are used in high-dose chemotherapy programs. In this paper, we address the economic implications of high dose chemotherapy programs.

1. PERIPHERAL BLOOD STEM CELL TRANSPLANTATION (PBSCT) VERSUS AUTOLOGOUS BONE MARROW TRANSPLANTATION (ABMT)

PBSCT provides one approach to high dose chemotherapy for patients who would otherwise be ineligible for autologous bone marrow due to the inability to harvest bone

Molecular Biology of Hematopoiesis 5, edited by Abraham et al.
Plenum Press, New York, 1996

marrow.[1] Bone marrow collection may not be possible if tumor has infiltrated the marrow, severe hypocellularity exists, or general anesthesia is contraindicated. Peripheral blood has increasingly become the preferred hematopoietic rescue source following high dose chemotherapy due to perceived clinical and economic advantages. Due to increased progenitor content following mobilization, transplantation of PBSC has been reported to result in more rapid hematopoietic recovery as compared to autologous bone marrow transplantation.[2,3] The reduced period of absolute neutropenia decreases the probability of a life threatening infection. In combination with increased physician experience with high dose therapy, improved supportive care measure, newer antibiotics, and the introduction of hematopoietic growth factors, PBSCT has contributed significantly to the decline in early mortality associated with high dose therapy from approximately 25% in the mid 1980s to approximately 5% in 1995.[4] Long-term survival is likely to be related to the effectiveness of the chemotherapeutic regimen, dosages, and schedule.

There are several theoretical reasons associated with economic advantages with PBSCT (Table 1). More rapid hematopoietic reconstitution may occur with shorter periods of hospitalization and lower hospital charges for room and board, blood products, and antibiotics.[5] In some clinical situations, the entire transplantation may be performed in the outpatient setting.[6,7] PBSCT does require the necessary expertise, equipment, and facilities, and therefore costs may decrease over time. Initial PBSCT trials collected PBSC in hematopoietic steady state and required multiple apheresis procedures in order to obtain a sufficient quantity of cells for transplantation. [1,8] Each apheresis product had to be processed and cryopreserved resulting in a significant cost, as compared to the processing of bone marrow harvest.[9] Mobilization of stem cells into peripheral blood markedly reduced the number of collections and therefore cost. [10,11] Total charges for stem cell collections are often significantly higher than for bone marrow collection, which may offset cost savings associated with shorter periods of neutropenia with PBSCTs. Peripheral blood stem cell collection costs include : (1) catheter placement charges which includes charges for an operating room, anesthesiologist, and surgeon; (2) cost of treatment of frequent catheter related complications; (3) blood collection facility fees for equipment and personnel; (4) additional cost of stem cell storage for multiple phereses; (5) costs for blood tests to determine the quality of the stem cell harvest (Complete blood counts (CBCs), CD-34 tumor markers etc); and (6) loss of wages (i.e. indirect medical costs) associated with multiple clinic visits for catheter placement, mobilization of stem cells with chemotherapy and/or hematopoietic growth factors.

Recently, several abstracts and papers have addressed the economic costs of PSCT versus ABMT. (Table 2). A retrospective analysis of bone marrow versus mobilized PBSCT from the University of Nebraska Medical Center found no difference in hematopoietic recovery for patients with malignant lymphoma who were matched for disease and preparative regimen.[12] Economic analyses have found that in most cases, costs of ABMTs were

Table 1. Comparison of economic implications of PSCT versus ABMT

Site	PSCT Outpatient	ABMT Inpatient/outpatient
Recovery time (WBCs)	+	
Recovery time (platelets)	+	
Collection costs		+
In-hospital costs	+	
Antibiotic costs	+	
Blood product costs	+	

+ indicates more favorable outcome.

Table 2. Costs of care for patients undergoing ABMT versus PSCT

ABMT	PSCT	Site	Reference
$69,870	$73,360	University of Nebraska	4
$82,520	$102,560	University of Nebraska	4
$41,046	$35,379	France	14
$58,281	$39,960	Germany, Belgium, UK	13
$46,694	$33,368	France	15

similar to those for PBSCTs for persons with either Hodgkin's disease or non-Hodgkin's lymphoma.[4] Among persons with non-Hodgkin's lymphoma who received total body irradiation, median costs were $82,520 for ABMTs and $102,560 for PBSCTs.

However, most studies have suggested that PSCT costs are lower than for ABMT. Smith et al reported preliminary findings from an European trial in which 58 patients with recurrent Hodgkin's disease or non-Hodgkin's lymphoma were randomized to ABMT versus PSCT. Estimated costs were $58,281 for ABMT versus $39,960 for PSCT.[13] A retrospective analysis of 20 patients with high grade non-Hodgkin's lymphoma from France reported similar findings- with estimated costs of $41,046 for 10 ABMT patients and $35,379 for 10 PSCT patients.[14] While costs of stem cell harvesting were higher for the PSCT patients ($4,745 for ABMT versus $9,030 for PSCT), lower costs were observed for hospitalization ($9,393 for ABMT vs $15,844 for PSCT), antibiotics ($5,552 for ABMT vs $3,623 for PSCT), blood products ($8,402 for ABMT versus $4,458 for PSCT), and laboratory tests/ cultures ($2,828 for ABMT vs $2,310 for PSCT). A second study from France compared costs of patients with lymphoid malignancies undergoing PSCT (n=10) with a historical control group who received ABMT (n=23).[15] Average costs were $46,694 for the ABMT patients vs $33,368 for the PSCT patients, with significant benefits associated with shorter period of platelet-transfusion dependency for the PSCT patients. However, clinical and economic results from the randomized trials of PBSCT versus ABMT will be important. While it is possible that the PBSCT methodology may be universally adopted in the near future, the long term relevance may not be all that large as rapid technologic improvements in pre-harvest treatment will markedly alter the post-transplant duration of neutropenia and thrombocytopenia.

2. NEW TECHNOLOGIES, ORGANIZATIONAL IMPROVEMENTS, AND THE LEARNING CURVE EFFECT

There is considerable debate about how rapid costs of care decrease and outcomes improve as hospitals gain experience with ABMTs and PBSCTs. While many smaller hospitals have started to perform these procedures in order to provide their patients with good access to the new technologies, private health insurers often restrict reimbursement to regional "centers of excellence" who have moved up the learning curve. Previous studies have shown better survival rates or other clinical outcomes when as surgeons gained experience with coronary artery bypass grafts, intestinal operations, total hip replacements, abdominal aneurysm repairs, prostatectomies, hysterectomies, laparoscopic cholecystectomies, and vascular repairs.[16,17,18] Similarly, there is strong evidence for a learning curve for ABMTs. [12] At the University of Nebraska Medical Center, mortality rates for Hodgkin's disease decreased from 20% in 1987 to 0% in 1991 and, for non-Hodgkin's lymphoma, from 29% in 1987 to 4% in 1991 (Figure 1). Costs decreased between 8% and 10% per year (Figure 2).

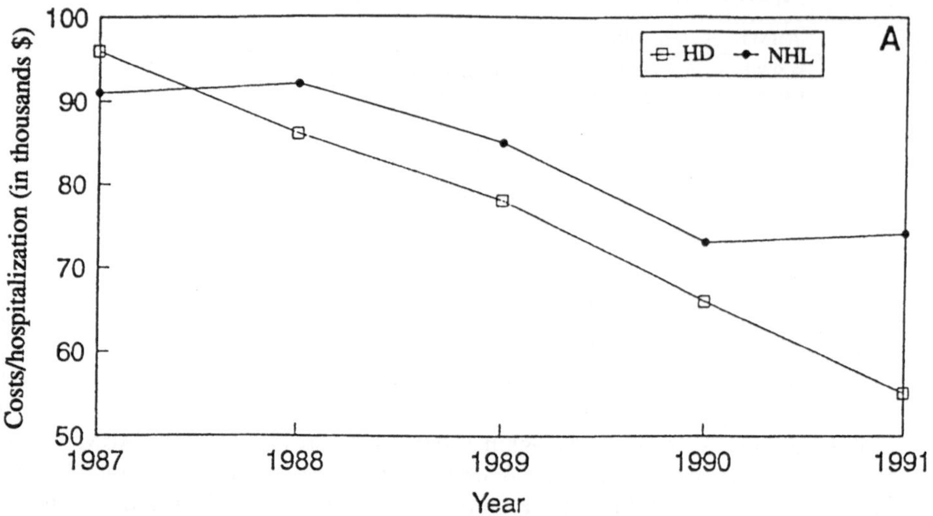

Figure 1. Average hospitalization costs for patients undergoing autologous bone marrow transplantation at the University of Nebraska Medical center (1987-1991).

Hematopoietic growth factors such as G-CSF and GM-CSF are routine parts of high dose chemotherapy programs because they decrease the duration of neutropenia and associated morbidity while increasing the safety of intensive chemotherapy programs. Prospective studies evaluating hematopoietic growth factors for autologous bone marrow transplantation are shown in Table 3. The two phase III trials for G-CSF were open label, and therefore can not compare a G-CSF group directly with a placebo group.[19,20] However, the G-CSF group had 8 fewer days of hospitalization than the control group in the study by Linch et al and 3 fewer days of hospitalization in the study by Stahel et al. G-CSF patients had significantly fewer days with fever and febrile neutropenia than non-treated patients. It appears likely that G-CSF and GM-CSF therapy decreases costs of hospitalization for persons undergoing high dose chemotherapy with autologous stem cell transplantation. The costs of the drugs are more than offset by savings due to shorter hospitalizations (with estimated savings of $14,000 in the single site study from Seattle or $18,000 in the single site study from New York.[21,22]

Table 3. Duration of neutropenia and hospitalization for ABMTs (cytokine vs placebo)

Number of patients	Cytokine	ANC500* cytokine/ placebo	Length of stay (Days) cytokine/placebo	Reference
40	GM-CSF	19/26	27/33	26, 31
24	GM- CSF	12/16	32/40	27
198	GM-CSF	18/24	32/35	34
91	GM-CSF	14/21	23/29	32
79	GM-CSF	15/28	30/31	33
58	GM-CSF	14/20	24/25	34
43	G-CSF	10/18	18/21	19
121	G-CSF	15/19	28/36	20

*number of days until absolute neutrophil count of 500 cells/mm^3 reached (cytokine/placebo)

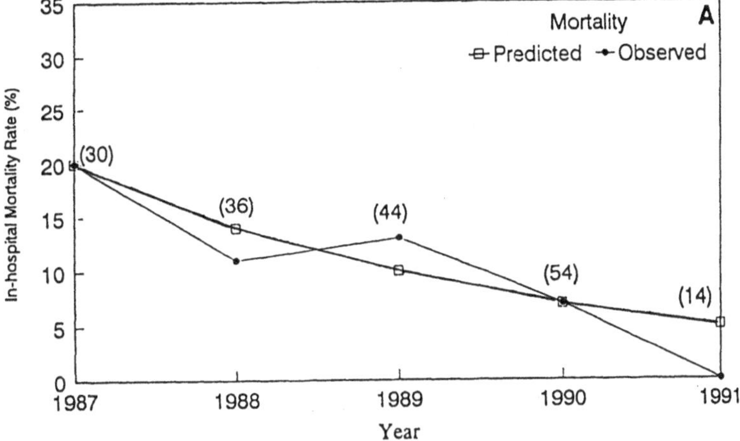

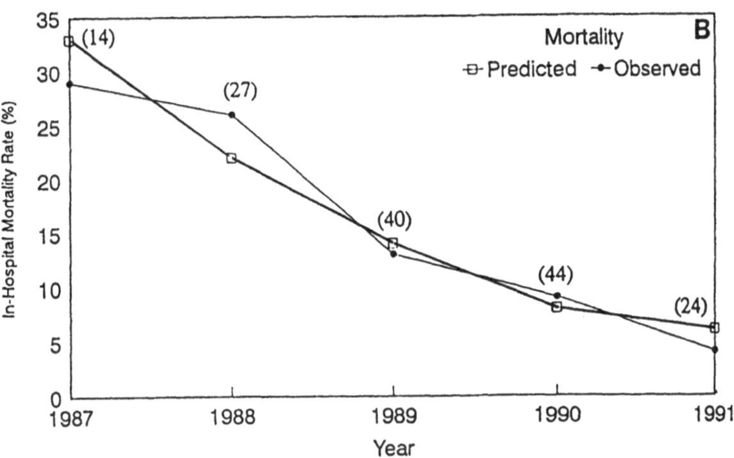

Figure 2. Predicted in-hospital mortality rates for (A) Hodgkin's disease and (B) non-Hodgkin's lymphoma patients undergoing Autologous Bone Marrow transplantation at the University of Nebraska Medical Center (1987-1991). Number of autologous bone marrow transplants performed in each year are in parenthesed.

Platelet transfusion costs account for as much as 5% of the total costs of autologous stem cell transplantation.[27] There is extensive debate about the costs, effectiveness, and cost-effectiveness of alternative platelet sources. Randomly selected donors often result in decreasing responsiveness to platlet transfusions due to alloimmunization.[23] Consequently, single donor platelets have been used to minimize alloimmunization. Recently, additional concerns for autologous stem cell transplantation programs have arisen with respect to splitting of apheresis single donor platelet products because of concern that the platelet content of the transfused product is lower than the more uniform dose obtained in the pooled concentrates. The usual platelet pheresis unit is expected to have at least 4×10^{11} platelets, while split products often have less.[24] Preliminary data from a single site study suggest that outcomes are better for patients undergoing autologous stem cell transplantation who receive high dose platelet products (4.5 - 6.1 x 10^{11} platelets/transfusion) rather than low dose products (2.5 - 3.5 x 1011 platelets/transfusion).[25] High dose products products may provide more value for the money for autologous stem cell transplantation by decreasing the duration of platelet dependency, allowing for earlier discharge from hospitals, and/or decreasing the need to return to outpatient clinics for prophylactic platelet transfusions.

3. LONG-TERM COSTS AND OUTCOMES CONCERNS

There are several factors that are likely to have an effect on long term costs of care and outcomes. For example, stem cell priming appears to result in rapid hematopoieitic engraftment in many clinical situations.[26] Long-term benefits and toxicities have not been studied extensively, but may be of significant importance with respect to clinical outcomes and total costs of care. Use of certain growth factors for priming may have a higher incidence of myelodysplastic syndrome or leukemia, whereas other growth factors may have an anti-cancer effect.[27,28] A second factor that may affect long-term outcomes is detection of minimal cancer cell contamination. Significant improvements in polymerase chain reaction (PCR) technology allows for the detection of one cancer cell in one million to ten million normal cells when the specific genetic mutation has been identified.[29] A third factor is improvements in purging technologies. Recent investigations using genetic markers clearly show that cancer cells that contaminate bone marrow transplantation can cause cancer recurrence.[30] Any method which lowers the cancer burden will eventually improve the long-term success rate of ABMTs or PBSCTs. Chemotherapy administration prior to the bone marrow harvest (in vivo purge) is one area of research, but will require more studies of up-front chemotherapy as well as the timing and quantity of the stem cell harvest. A final factor that may influence long-term costs and outcomes is post-transplant immunomodulation. Relapse after PBSCT or ABMT is the most likely cause of death. A small number, often drug-resistant, cancer cells are the primary source of the recurrence. A great deal of innovative research is addressing this problem with studies of drugs to prevent chemo-resistance, immunomodulation attempts with interleukin-2 or alpha-interferon, and post-transplant radiation or maintenance chemotherapy.

CONCLUSIONS

Technologic advances for high dose chemotherapy programs with stem cell support have resulted in marked improvements in clinical outcomes and decreases in costs of care for ABMTs and PBSCTs. While these advances are unparalleled with respect to a new technology, they have important clinical and economic implications. Factors that need to be considered include the source of stem cells, source of platelets, the adjunctive use of hematopoieitic growth factors, the learning curve effect, and technologies that address long-term benefits. The rapid proliferation of managed care programs raises concern that high technology cancer such as ABMT or PBSCTs may be at risk. However, the costs of high dose chemotherapy programs are likely to continue to decrease over time, and it is hoped that managed care programs and medical insurers will therefore continue to support the dissemination of this important technology.

REFERENCES

1. Kessinger A, Armitage JO, Smith DM et al. High-dose therapy and autologous peripheral blood stem cell transplantation for patients with lymphoma. Blood 1989; 74: 1260-5.

2. Elias AD, Ayash L, Anderson KC et al. Mobilization of peripheral blood progenitor cells by chemotherapy and granulocyte macrophage colony stimulating factor for hematologic support after high dose intensification for breast cancer. Blood 1992; 79: 3036- 44.
3. Jones HM, Jones SA, Watts MJ et al. Development of a simplified single-apheresis approach for peripheral blood progenitor cells. JCO 1994; 12: 1693-1702.
4. Bennett CL, Armitage J, Armitage G, Vose J, Bierman , Armitage JO, Anderson J. Costs and outcome for autologous bone marrow transplantation at the University of Nebraska Medical Center: 1987- 1991. J Clin Onc 1995; 13: 969-973.
5. Henon PR, Liang H, Beck-Wirth G et al. Comparison of hematopoietic and immune recovery after autologous bone marrow or blood stem cell transplants. Bone Marrow Transplant 1992; 9:285- 291.
6. Peters WP, Ross M, Vrdedenburgh JJ et al. The use of intensive clinic support to permit outpatient support to permit outpatient autologous bone marrow transplantation for breast cancer. Seminars in Oncology 1994; 21(4 suppl 7): 25-31.
7. Gilbert C, Meisenberg B, Vredenburgh J. Sequential prophylactic oral and empiric once-daily parenteral antibiotics for neurtopenia and fever after high-dose chemotherapy and autologous bone marrow support. J Clin Oncol 1994; 12:1005-11.
8. Korbling M, Martin H. Transplantation of hemapheresis-derived hematopoietic stem cells. A new concept in the treatment of patients with malignant lymphohematopoietic disorders. Plasma Ther Transfus Technol 1988; 9: 119- 123.
9. Jackson JD, Kloster T, Welniak L et al. Peripheral blood derived stem cells can be successfully cryopreserved without using controlled-rate freezing. Advances in Bone Marrow Purging and Processing. Worthington-White DA, Gee AP, Gross S (eds) Wiley-Liss: New York, 1992 pp. 367- 371.
10. To LB, Shepperd KM, Kimber RJ et al. Single high doses of cyclophosphamide enable the collection of high numbers of hemopoietic stem cells from the peripheral blood. Exp Hematol 1990; 18: 442- 7.
11. Passos-Coelho JL, Ross AA, Moss TJ et al, Abscence of breast cancer cells in a single-day peripheral blood progenitor cell collection after priming with cyclophosphamide and granulocyte-macrophage colony stimulating factor. Blood 1995; 85: 1138- 43.
12. Vose JM, Anderson JR, Kessinger A et al. High dose chemotherapy and autologous hematopoietic stem cell transplantation for aggressive non-Hodgkin's lymphoma. J Clin Oncol 1993; 11: 1846- 51.
13. Smith TJ, Hillner BE, Shmitz N et al. Economic analysis of a randomized clinical trial comparing peripheral blood progenitor cells or autologous bone marrow after high dose chemotherapy for recurrent Hodgkin's disease or lymphoma. Proceedings of ASCO 1995.
14. Woronoff-Lemsi MC, Limat S, Deconick E, Arveux P, Cahn JY. Cost comparative study of peripheral blood progenitor cells and autologous bone marrow transplantation for lyphmomas patients. Proceedings of the American Society of Hematology 1995, 822a.
15. Faucher C, Fortanier C, Protiere C et al. Comparison of peripheral blood progenitor cells and bone marrow allogeneic transplantation: clinical and cost-effectiveness study. Proceedings of the American Society of Hematology. 1995, 1528a.
16. Bennett CL, Garfinkle JB, Greenfield S et al. The relation between hospital experience and in hospital mortality for patients with AIDS-related PCP. JAMA 1989; 261: 2975- 9.
17. Bennett RL, Gilman SC, George L, Guze A, Bennett CL. Improved Outcomes in Intensive Care Units for AIDS-Related *Pneumocystis carinii* Pneumonia: 1987- 1991. JAIDS 1993; 6: 1319-1328.
18. Banta D, Bos M. The relation between quantity and quality with coronary artery bypass graft surgery. Health Policy 1991; 18: 1-10.
19. Stahel RA, Jost LM, Cerny T et al. Randomized study of recombinant human granulocyte colony stimulating factor after high dose chemotherapy and autologous bone marrow transplantation for high risk malignancies. J Clin Oncol 1994; 12: 1931- 8.
20. Linch DC, Scarffe H, Proctor V et al. Randomized vehicle controlled dose-finding study of glycosylated recombinant human granulocyte colony stimulating factor after marrow transplantation. BMT 1993; 11: 307- 11.
21. Luce BR, Singer JW, Wechsler JM et al. Recombinant human granulocyte macrophage colony stimulating factor after autologous bone marrow transplantation for lymphoid cancer: An economic analysis of a randomized double-blind placebo controlled trial. Pharmacoecon 1994; 6: 42- 8.
22. Gulati SC, Bennett CL. Granulocyte macrophage colony stimulating factor as adjunct therapy in relapsed Hodgkin's disease. Ann Int Med 1992; 116: 177- 82.
23. Gmur J, Random single donor platelet transfusions: Pros and Cons. Plasma Ther Transfus Technol 1986; 7: 463- 468
24. Practice parameter for the use of fresh frozen plasma cryoprecipitate, and platelets. JAMA 1994; 271: 777- 781

25. Herman JH, Klumpp TR, Christman RA, Goldberg SL, Mangan KF. The effect of platelet dose on the outcome of prophylactic platelet transfusion. American Association of Blood Banking presentation, November 1995, New Orleans, Louisiana.
26. Ganser A, Lindemann A, Ottman OG et al. Sequential in vivo treatment with two recombinant human hematopoietic growth factors (interleukin-3 and granulocyte macrophage colony stimulating factor) as a new therapeutic modality to stimulate hematopoiesis: Results of a phase I study. Blood 1992; 79: 2583-2591.
27. Richard C, Alsar MJ, Calavia AJ et al. Recombinant human GM-CSF enhances T-cell mediated cytotoxic function after ABMT for hematologic malignancies. Bone Marrow Trans 1993; 11: 473-8.
28. Stewart-Akers AM, Cairns JS, Tweardy DJ, McCarthy SA. Granulcoyte macrophage colony stimulating factor augmentation of T-cell dependent and t-cell receptor-independent thymocyte proliferation. Blood 1994; 83: 713-23.
29. Gulati SC. Purging in Bone Marrow Transplantation. R.G. Landes Company: Georgetown TX, 1993.
30. Brenner MK, Rill DR, Moen RC et al. Gene marking to trace origin of relapse after autologous bone marrow transplantation. Lancet 1993; 341: 85-86.
31. Bennett CL, Armitage JL, LeSage S, Gulati S, Armitage JO, and Gorin C. Economic Analyses of Clinical Trials in Cancer: Are They Helpful to Policy Makers? Stem Cells 1994; 12: 424-29.
32. Gorin NC, Coiffer B, Hayat M et al. Recombinant granulocyte macrophage colony stimulating factor after high dose chemotherapy and autologous bone marrow transplantation with unpurged and purged marrow in non-Hodgkin's lymphoma: A double-blind placebo-controlled trial. Blood 1992; 80: 1- 10.
33. Link H, Brogaerts MA, Carella AM et al. A controlled trial of recombinant human granulocyte macrophage colony stimulating factor after total body irradiation, high dose chemotherapy, and autologous bone marrow transplantation for acute lymphoblastic leukemia or malignant lymphoma. Blood 1992; 80: 188-94.
34. Khwaja A, Linch DC, Goldstone AH et al. Recombinant human granulocyte macrophage colony stimulating factor after autologous bone marrow transplantation for malignant lymphoma: A British National Lymphoma Investigation double-blind placebo controlled trial. B J of Hematol 1992; 82: 317-23

CFU-GM GROWTH FROM PERIPHERAL BLOOD STEM CELLS (PBSC) BEFORE AND AFTER CRYOPRESERVATION

Comparison with Bone Marrow Cells

M. Bonfichi, C. Brera, A. Balduini, E. P. Alessandrino, P. Bernasconi, D. Troletti, M. Boni, C. Castagnola, E. Brusamolino, G. Pagnucco, C. Perotti, L. Salvaneschi, and C. Bernasconi

Istituto di Ematologia
Univ. di Pavia Policlinico S. Matteo IRCCS Pavia
Serv. Immunoemat. e Trasf. Policlinico S. Matteo IRCCS Pavia Italy

ABSTRACT

We have compared the CFU-GM growth observed in 29 PBSC samples (8 HL, 21 NHL) to that obtained with 30 bone marrow (BM) "buffy coats" harvested for an autologous BMT program in pts without BM involvement (25 AML, 3 HL, 2 NHL). The CFU-GM tests were performed before cryopreservation, made without noncontrolled heat fusion, and after thawing. The collections of PBSC were made with a CS3000 Plus (Baxter). The apheresis were performed after mobilisation with HD CTX 4 g/m^2 and administration of G-CSF (6-15 days). The cultures of CFU-GM were made in methylcellulose for PBSC and in agar monolayer for BM cells, using a conditioned medium from 5637 cell line as source of CSF. Before freezing a mean the CFU-GM growth of from BM cells were higher than in PBSC (35.5 vs 26 : $P<0.05$), on the contrary, after thawing the proliferative activity was more elevated for PBSC (21 CFU-GM vs 18). The recovery after thawing, was better for PBSC: 21 CFU-GM 82%(range 50-100%) than for BM cells 51% (range 21-100%). Our data confirm that PBSC have growth activities comparable BM cells. The better viability demonstrated after thawing could help the faster recovery from cytopenia post chemotherapy signalled from many authors.

INTRODUCTION

PBSC are increasingly used instead of bone marrow (BM) cells for hematopoietic support after ablative chemotherapy for bone marrow transplantation (BMT). Both the procedures involve the cryopresrvation and the subsequent thawing of the products, but

Molecular Biology of Hematopoiesis 5, edited by Abraham et al.
Plenum Press, New York, 1996

Table 1. Main clinical characteristics of patients submitted to collection of PBSC or bone marrow cells

	PBSC	Bone Marrow Cells
N	29	30
Median Age: yrs (range)	35 (15-55)	46 (15-60)
Male/Female	15/14	16/14
NHL	21	2
HL	8	3
ANLL		25 (1 CR: 21)

PBSC have many advantages compared with BM cells. Their harvest does not require exposure to general anaesthesia, moreover the period of cytopenia after myeloablative chemotherapy with PBSC is shorter than with BM cells (1-2): otherwise the collection of PBSC require, after a course of high dose chemotherapy and/or growth factors therapy, one or more leukapheresis using continuous flow separators and in the procedure the anticoagulant acid citrate dextrose A (ACD-A) is used to prevent clotting of the blood. With the aim to further characterise the proliferative activity of PBSC we have compared the CFU-GM growth observed in 29 PBSC to that obtained with 30 BM "buffy coats" harvested for an autologous BMT program in patients without bone marrow involvement.

MATERIALS AND METHODS

Patients

All patients were observed in the Institute of Haematology of Policlinico S. Matteo IRCCS of Pavia from 1/1/94 to 31/12/94. Their main clinical characteristics are reported in Table 1. The procedures utilized for bone marrow harvesting and PBSC collection and their manipulations are those generally known.

Bone Marrow Harvest. Marrow cells were obtained by multiple aspirations from posterior iliac crest under general anaesthesia in the operating theatre. The marrow was concentrated to its buffy coat on a blood cell processor .

Collections of PBSC. The collections of PBSC were made with a CS3000 Plus (Baxter). The apheresis were done after mobilisation with HD CTX 4 g/m2 and following administration of after recombinant granulocyte colony stimulating factor (rh G-CSF)(6-15 days).

Cryopreservation. The cryopreservation, was made in 10% dimethylsulphoxide till - 150 C without compensation of the heat of fusion. The freezing programme was performed by an electronic programmer Planer R203 according to the following scheme: -5 C/min to +4 C; -1 C/min to -60 C; -5 C/min to -150 C. The Cells were then stored at -196 C

Thawing. Rapid warming at 37 C was used to avoid the growth of small ice crystals by the process of recrystallization.

CFU-GM Assays. The CFU-GM tests were performed before, and after thawing. The cultures of CFU-GM were made in methylcellulose for PBSC and in agar monolayer for BM

Table 2. CFU-GM growth obtained from 29 samples of PBSC and 30 BM cells before and after freezing

	PBSC		BM Cells	
	CFU-GM*	SD	CFU-GM*	SD
Before cryopreservation	26	12	35,5	12
After thawing	21,1	8,4	18,4	8

*=x 10^5 mononuclear cells; SD= standard deviation

cells, using a conditioned medium from 5637 cell line as source of CSF. BM cells: $1x10^5$/ml cells from buffy coat before freezing and $2x10^5$/cells after thawing were plated in 35 mm Petri dishes (NUNC) in triplicate at a final volume of 1 ml per plate, in IMDM medium (Gibco) 20% FCS, 0.3% agar (DIFCO), 10% conditioned medium from 5637.PBSC: 1.0 and 2.0 x 10^5 cells/ml respectively before and after freezing were plated in 35 mm Petri dishes in triplicate at final volume of 1 ml per plate, in IMDM, methilcellulose (0.8% final concentration), FCS 20%, 10-2 M/L 2 merchaptoethanol , 10-2 M/L L-Glutamine and 10% conditioned medium from 5637Colonies (aggregates of more than 40 cells) were scored to an inverted microscope after 14 days incubation in humidified incubator with 5% CO2.

The statistical analysis (T Student or CHI Square) was made with a personal computer using a statistical software (Statistica for Windows - Statsoft Inc., 1993).

RESULTS

The CFU-GM growth obtained from 29 samples of PBSC and 30 BM cells before and after freezing is reported in Table 2.

Before freezing the mean growth of BM cells was statistically significant higher than PBSC ($P<0.05$). This fact was probably due to the premature collection of some PBSC samples. The mean recovery after thawing was better for PBSC: 82%(range 50-100%) than for BM cells 51% (range 21-100%). The characteristics of CFU-GM were the same in both categories and generally the colonies from PBSC samples (fig. 1-2) were larger.

DISCUSSION

A major clinical issue in programs involving the BMT with cryopreserved hematopoietic cells from bone marrow or PBSC is quantitate the number and the kind of cells responsible for a fast reconstitution. If recently the CD34 measurement seems to provide a good indicator of the hematopoietic reconstitutive capacity, the CFU-GM assay maintains a biologic unquestionable significance to test the quality of the stem cell to infuse. Although many studies are performed about PBSC, biologic information on this cells are still limited. We have analyzed the CFU-GM growth of samples from BM harvest and PBSC collection to compare and further characterize the biological aspect of these cells.

In our experiments the controls performed before freezing have shown higher colonies numbers with BM cells. This fact is probably due to the premature collection of some PBSC, as confirmed by the fact that in other tests, performed subsequently and not here reported, the mean CFU-GM growth was more elevated.

Instead the growth activity of CFU-GM obtained after thawing was significantly better for PBSC.

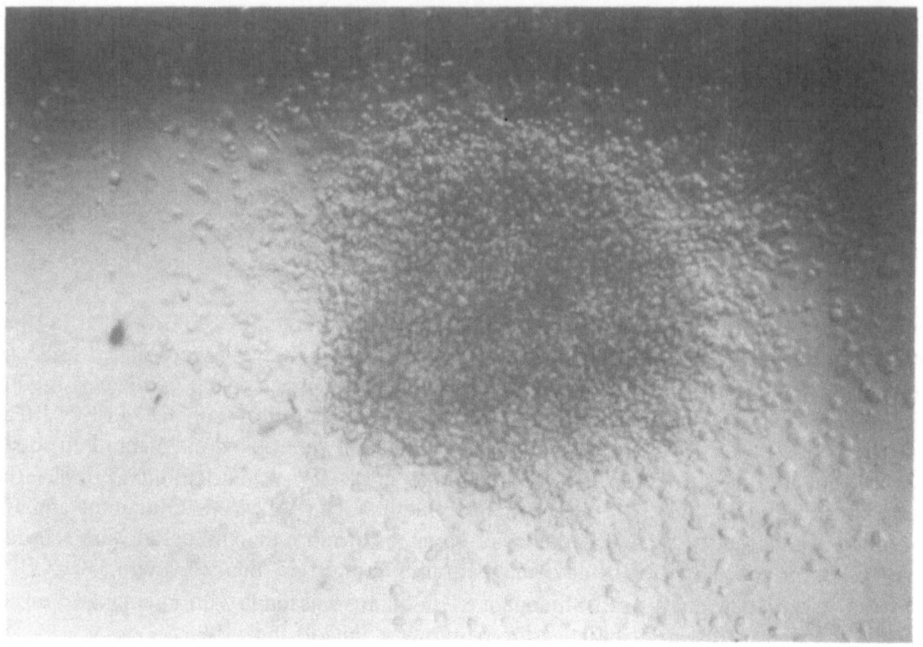

Figure 1. CFU-GM from PBSC (magnification: 3.2 x 10).

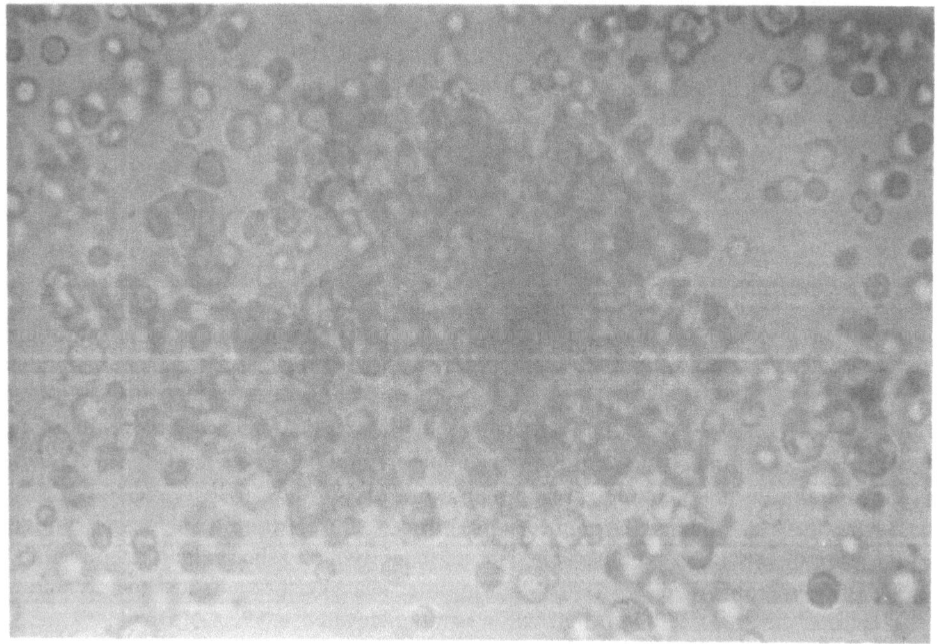

Figure 2. The same CFU-GM at 10x10 magnifications.

Therefore our data confirm that PBSC have growth activity comparable to BM cells. The better viability showed after thawing could explain the faster recovery from cytopenia post chemotherapy signalled from many authors (1-2).

REFERENCES

1. Gianni A.M., Siena S., Bregni M., Tarella C., Stern C.A., Pileri A., Bonadonna G.: Granulocyte-Macrophage Colony Stimulating Factor to harvest circulating Hemopoietic Stem Cells for autotransplantation. The Lancet 1980, 8663, 580-5
2. Besinger W.I, Longin K, Appelbaum S., Rowley S., Weaver C.H., Lilleby K., Gooley T, Lynch M., Higano T., Klarnet J., Chauncey T., Storb R., Bucker C.D.: Peripheral blood stem cells (PBSCs) collected after recombinant granulocyte colony stimulating factor (rh G-CSF) an analysis of factors correlating with the tempo of engraftment after transplantation. Br. J. Haematol 1994, 87, 825.

IS THERE AN EMBRYOLOGICAL BASIS FOR THE ASSOCIATION OF MEDIASTINAL GERM CELL TUMORS AND HEMATOLOGIC CANCERS?

A Review

William J. Larsen

Department of Cell Biology, Neurobiology, and Anatomy
University of Cincinnati College of Medicine

ABSTRACT

The unexpectedly frequent association of germ cell tumors and hematologic cancers raises a question about the developmental relationship between primordial germ cells (PGCs) and hematopoietic stem cells (HSCs). It has been known for many years that the primordial germ cells of the primitive gonad arise from the yolk sac and that the yolk sac is also the first site of blood formation. It has also become clear that the development of these two apparently disparate cell types is dependent on some of the same factors; c-kit ligand and c-kit receptor are among the most intriguing.

Even more relevant, perhaps, are recent studies investigating the very early development of the PGCs and HSCs in mammals. This new information raises alluring questions concerning the relationship between the ultimate progenitors of these cells which may also be pertinent to the unusual clinical association of germ cell and hematologic malignancies in humans.

INTRODUCTION

While primary germ cell tumors are rare,[1] patients diagnosed with apparent mediastinal nonseminomatous germ cell tumors (MNGCT), particularly MNGCTs possessing yolk sac elements, may also commonly develop hematologic cancers as well.[1-4] The most common of these hematologic malignancies (HMs) are acute megakaryoblastic leukemia and malignant histiocytosis (reviewed by Woodruff).[4]

Initial concern that the hematologic malignancy arose in patients with MNGCTs as a consequence of cis-platinum-based induction chemotherapy was probably unfounded since

Molecular Biology of Hematopoiesis 5, edited by Abraham et al.
Plenum Press, New York, 1996

HMs are also diagnosed in patients with MNGCT who had not received treatment.[1] Moreover, a growing body of cytogenetic data now supports the view that the common association of MNGCTs and HMs within such patients can be explained by the origin of the two cancers from a single progenitor cell.[1-4]

While the specific mechanism which explains this phenomenon remains unclear, it has been suggested that the clonal identity of MNGCTs and HMs could occur as a consequence of anomalous migration of a primordial germ cell from the yolk sac to an inappropriate mediastinal site during embryonic development, followed by the conversion of descendants which give rise to either the germ cell tumor or the hematologic cancer (reviewed in Nichols;[1] Woodruff[4]). Clearly, the authenticity of this hypothesis requires that the developmental interconversions it requires are probable, or at the very least possible. The purpose of this essay, therefore, is to examine pertinent studies exploring the embryological ancestries of PGCs and HSCs and to determine whether or not the present literature provides an embryological or developmental basis for the clinical association of MNGCTs and HMs.

RESULTS AND DISCUSSION

The Intraembryonic Germ Cells and Blood Cells of Humans May Both Arise from an Extraembryonic Source

Historical studies provide circumstantial evidence that neither primordial germ cells (PGCs) of the embryonic gonad[5] nor hematopoietic stem cells (HSCs) of the embryonic liver,[6] arise directly from precursors within these organs, but rather migrate to these sites from the wall of the definitive yolk sac. This evidence also supports the view that this migratory phase of germ cell and blood cell development in humans occurs during the mid-embryonic period.

Human PGCs Migrate from the Yolk Sac to the Urogenital Ridges (Presumptive Gonads) between the 4th and 6th Weeks. Evidence that PGCs of the urogenital ridge of the embryo arise from the yolk sac is convincing. While precursors of the human germ line were first observed within the endoderm of the yolk sac by Fuss as early as 1911,[7] their migration from this site to lower thoracic and upper lumbar regions of the posterior body wall was first elegantly described in a series of human embryos by Witschi in 1948.[5] The success of this cytological study depended largely upon the fact that primordial germ cells are easily recognized as large clear ovoid cells often possessing a single dense cytoplasmic inclusion and irregular borders.[5]

In early stages of development (13 somites; about 23 days) PGCs are typically localized within the endodermal layer of the yolk sac but by the 16-somite stage (about 24 days) they may also be found within the splanchnopleuric mesoderm of the yolk sac and hindgut where they mingle with hematopoietic stem cells (HSCs).

By the 20-somite stage (about 25 days) most of the PGCs are localized within both endoderm and surrounding mesenchyme of the hindgut. Subsequently (at the 30-somite stage; about 28 days) the PGCs escape into the dorsal mesentery of the gut. By 32 or 33 days they reach the regions of the future gonads and by the end of the fifth week or beginning of the sixth week they become invested by the sex cord cells of the primitive gonads. Witschi also noted that the PGCs respected few anatomical boundaries as they penetrated basement membranes and invaded foreign tissues during the course of their migration. Moreover, he noted that migrating PGCs proliferated along their migratory pathway.

However, Witschi's description of PGC migration was controversial largely because the cytologic basis for germ cell identification was not considered to be definitive (reviewed in Chiquione[8]). In 1954, however, Chiquoine[8] demonstrated that PGCs could be distinguished from somatic cells upon the basis of their unique staining with reagents that localize the activity of alkaline phosphatase and in this seminal study of alkaline phosphatase-positive PGCs in mice, was able to verify the conclusions of Witschi's cytological observations in human embryos. During the past forty years, these results have been repeated and verified by many additional studies[9-12] and the alkaline phosphatase isoform specific to PGCs has been identified.[13] Moreover, additional support for Witschi's original cytological findings has been provided by comprehensive ultrastructural studies of PGC migration.[14,15]

While extensive alkaline phosphatase, cytochemical and ultrastructural studies helped to clarify the identity and early behavior of the migrating PGCs, they also answered another important question. On the basis of their initial localization within the endoderm, Witschi and others had interpreted the PGCs to be of endodermal origin (reviewed in Witschi).[5] In contrast, other studies identified the HSCs as cells of mesodermal origin,[6] thus suggesting a relatively distant relationship between these two yolk sac cell types. However, more recent studies utilizing the alkaline phosphatase reaction to identify the PGCs of mouse embryos have traced their origin back to the extraembryonic mesoderm at the caudal end of the primitive streak in mid primitive streak stage embryos (the primitive streak is formed in human embryos on about day 16 and disappears by day 22).[10,12] Since this is the same layer of tissue which forms the first HSCs of the yolk sac (see below), these data provide support for the possibility that both PGCs and embryonic HSCs arise within the extraembryonic mesoderm of the yolk sac and that neither arises within the endoderm.

Blood Islands Containing Embryonic HSCs also Appear Within the Extraembryonic Mesoderm of the Human Definitive Yolk Sac on Day 17. At about the same time that PGCs are recognized within the extraembryonic mesoderm at the caudal end of the primitive streak,[12] HSCs also begin to form within this component of the yolk sac.[16] Moreover, yolk sacs of early mouse embryos[16] and human embryos[17] are both able to produce colony forming units (CFUs) which give rise to a range of blood cell types including granulocytic, erythrocytic, macrophage-macrocytic, and megakaryocytic cells (CUF-GEMM).[16]

Historical Evidence Has Supported the Hypothesis that Hematopoietic Cells of the Yolk Sac Seed the Liver Between the Fourth and Sixth Weeks to Form the Fetal Hematopoietic Stem Cells. In a study of human hematopoiesis, the numbers of erythroid burst forming units (BFU-E), erythroid colony-forming units (CFU-E) and granulo-macrophage progenitors (CFU-GM) within the yolk sac decline between the fourth and fifth weeks as they reciprocally increase within the liver during this same period.[17] These reciprocal relationships are closely matched by the switching of embryonic to fetal hemoglobin in human embryos.[18]

Further evidence that yolk sac HSCs seed the liver is provided by organ culture studies. It has been shown, for example, that hematopoiesis fails to occur within liver primordia explanted from mice prior to the 28 somite stage (comparable to the late-fourth week of human development) while hematopoiesis does occur within cultured liver primordia explanted after the 28 somite stage (presumably after colonization of the liver by the HSCs.[19] In addition, in a classic organ culture study using filters of different sizes, Cudennec et al.[19] provided support for the view that yolk sac HSCs were able to populate HSC-free liver explants in culture and that once these cells populated the liver, they began to produce adult hemoglobins. Their studies showed that adult hemoglobins could be extracted from liver and yolk sac explants obtained from 25 somite embryos which were cocultured for seven days on different sides of a filter containing 5 micrometer pores. In contrast, adult hemoglobin isoforms could be extracted only from the explanted yolk sac when the filter was too small to allow the transfer of cells from one explant to the other (0.1 micrometer). In this latter case, the appearance of adult hemoglobin isoforms within the cultured yolk sac suggested that the liver was capable of producing induction factors which could diffuse across the filter to induce adult hemoglobin synthesis within the HSCs within the yolk sac. Since this early study, several such HSC growth factors have been identified.

One such factor apparently required for development and survival of HSCs was called stem cell factor (SCF). But recently, SCF has been identified as the product of the steel locus, c-kit ligand, which is thought to be a trophic factor required for proliferation and survival of HSCs, PGCs, and neural crest cells (see below). Studies of both the Steel mutation (Sl) and the White dominant spotting mutation (W), a locus which encodes the receptor for c-kit ligand, also provide strong complementary support for the hypothesis that HSCs and PGCs arise from closely related progenitors.

Mice Homozygous for Sl or W Exhibit Deficiencies of Germ Cells and Blood Cells as Well as Descendants of the Neural Crest Cells

A range of weak to strong alleles of the Sl and the W locus in mice interfere with the development of blood cells, germ cells and neural crest cells (reviewed by Motro et al.[20] and Besmer et al.[21] Neural crest defects include anomalies of pigmentation[21] and development of pacemaker activity within the gut.[22] Mutants homozygous for Sl typically die before birth and exhibit pronounced anemia and lack of or reduction in the number of germ cells, while heterozygotes are viable but significantly affected.[23] Homozygotes of the original mutation of the W allele exhibit severe phenotypes similar to those observed in Sl mutants but heterozygotes of this mutation typically exhibit few defects. Other mutations of W, however, have been shown to exhibit more severe effects in heterozygotes.[21]

The Sl and W Locus Respectively Encode a Trophic Factor, C-Kit Ligand and its Receptor, C-Kit Receptor. Gene products of the steel and white dominant spotting loci have recently been cloned (reviewed respectively in Williams et al.[24] and Besmer et al.).[21] Sl encodes the c-kit ligand which specifically binds to the product of the W locus, the c-kit receptor. As noted, the steel factor has also been identified in the past as stem cell factor (SCF)[25] and also as mast cell growth factor (MGF)[26] and as kit ligand (KL).[25]

Examination of the expression pattern of c-kit ligand and c-kit receptor supports the hypothesis that the interaction of c-kit ligand with its receptor is required for the growth of oocytes and the proliferation of spermatogonia in mice.[27] It appears that c-kit ligand must be expressed within cells of the microenvironment of the migratory pathways and homing sites of PGCs, HSCs and neural crest cells during early embryogenesis while the PGCs, HSCs and neural crest cells themselves must express the receptor to ensure their survival.[20,24,28]

More direct evidence that c-kit ligand acts as a trophic factor is derived from studies which demonstrate that c-kit ligand is required for the survival of cultured PGCs removed from embryonic mice[29] and the growth of oocytes cultured in an oocyte culture system.[30] Thus, it seems likely that c-kit ligand and receptor play critical roles in the early developmental migratory behavior of these stem cells.

The common and critical role of c-kit ligand and receptor in the migration, proliferation, and survival of PGCs and HSCs would seem to support the hypothesis that PGCs and HSCs share a close developmental relationship. Alternatively, however, it could be argued that their common response to defects in c-kit ligand and receptor expression indicate only that these unrelated cells utilize similar developmental mechanisms. Indeed, this latter caveat of the hypothesis is supported by the fact, as noted above, that the migration and development of neural crest cells, an apparently unrelated line, are also affected. (Neural crest cells originate from the primary ectoderm at the lateral edges of the neural plate).[31] This caveat is further supported by the fact that the expression of c-kit ligand and receptor may play a role in an even more dissimilar cell type, namely in the process of axonal pathfinding by neurons within the brain.[20]

However, Several Recent Studies Provide Compelling Evidence that PGCs and HSCs Arise from Closely Related if Not Overlapping Populations of Primary Ectodermal Cells at the Primitive Streak Stage of Development

Ablation and Explantation Experiments and Analysis of Chimeras Generated by Microsurgical Grafting Trace the PGCs Back to the Primary Ectoderm. The primitive streak of mammalian embryos is a structure through which cells of the primary ectoderm escape to form the definitive endoderm, the intraembryonic mesoderm, and the extraembryonic mesoderm (reviewed in Larsen).[31] It thus appears that virtually the entire embryo as well as some extraembryonic tissue arises from this single layer of primary ectoderm or epiblast.

It is therefore, perhaps, not so surprising that Ozdzenski found acid phosphatase-positive cells, identified as PGCs, at the caudal end of the primitive streak as long ago as 1967[10] and that the elegant ablation and explantation experiments carried out by Snow[32] pointed so strongly to the origin of PGCs within this region of the primary ectoderm. Indeed these studies, as well as a series of microsurgical grafting studies[33] provide compelling, if somewhat indirect evidence, for the origin of PGC precursors from cells located within the primary ectoderm at the caudal end of the primitive streak. However, more direct evidence for the origin of both PGCs and HSCs from closely related cells of the primary ectoderm has been generated just within the past year.[34-36]

Cell Lineage Studies of Early Primitive Streak Stage Embryos Show That PGCs and HSCs Arise from Very Closely Related Precursors. In meticulous and difficult studies, Lawson and her colleagues[34-36] injected single cells of early primitive streak mouse embryos with either of the cell-lineage tracers, horseradish peroxidase or rhodamine-labeled dextran beads. The embryos were then cultured for a short period of time, allowed to develop and then tracer-positive descendants were identified along with characteristics that distinguished the type of cell containing the marker. For example, PGCs were identified with the alkaline phosphatase reaction[36] and blood islands were identified by cytological techniques.[35]

A striking finding of these studies is that the fate maps constructed with this data suggest that the precursors of PGCs and yolk sac HSCs are found in closely related regions of the primary ectoderm. The precursors of blood islands, however, are more closely grouped near the posterior end of the primitive streak, although the population of PGC precursors overlaps the blood island domain[36] of the primary ectoderm. Moreover, Lawson's studies revealed that some of the primary ectodermal cells injected at the earlier primitive streak stage gave rise to PGCs and HSCs along with other cell types. Even more interesting, however was the finding that when cell lineage was examined in slightly older embryos, cell fate was more restricted and in these experiments only blood islands and related yolk sac mesoderm arose from some of the cells which also formed PGCs.[36] The data also supported the conclusion that at the time of allocation of the germ line (about 7 da. 4 hr.) about 45 cells are established as founder cells for PGC development.

However, Several Recent Studies Argue for an Independent Intraembryonic Source of Definitive HSCs of the Adult

It has been suggested for some time that definitive HSCs of avian embryos arise from intraembryonic sites near the dorsal aorta.[37] More recently, however, such intraembryonic sites containing spleen colony-forming units (CFU-S) have been shown to exist within cultured explants of embryonic tissue microdissected from the "aorta, gonad, mesonephros" ("AGM") region of mice.[38] In fact the development of CFU-S within explants of tissue from this intraembryonic location was shown to occur more rapidly and in greater abundance than in explanted yolk sac. B lymphoid cells have also been shown to arise relatively early within the AGM.[39] In these latter studies, splanchnopleure from the region of the aorta or yolk sac tissue from 10 to 18 somite mouse embryos was grafted into the kidney capsule of adult immunodeficient SCID mice. Three to six months after engraftment, donor-derived serum IgM, normal numbers of IgM-secreting plasma cells and the B1a cell subset arose within mice into which this splanchnopleuric region was grafted. In contrast, the grafted yolk sac was unable to provide the host mouse with donor IgM-secreting plasma cells or the B1a cell subset. Interestingly, these cells were provided by grafts of the splanchnopleure removed as long as twelve hours prior to the time when hematopoiesis is first observed in the liver. This finding has recently been reproduced by another laboratory whose studies provide evidence that the AGM region may actually possess CFU-S activity even before the yolk sac.[40]

It should be pointed out, however, that the apparent inability of the yolk sac to repopulate adult hematopoietic organs with blood cell precursors in these studies is in some conflict with other studies.[41,42] In a study by Toles and colleagues,[41] for example, it was shown that yolk sac suspensions could repopulate adult hematopoietic organs with progenitors of the erythroid lineage when yolk sac suspensions were injected into W mutant fetuses. In addition, Weissman et al.[42] also demonstrated that yolk sac could also repopulate adult hematopoietic organs with cells of the lymphoid lineage.

The most comprehensive comparison of the presence of HSCs in embryonic yolk sac, liver and the AGM, however, has been carried out by Muller et al.[43] The operational criterion used for possession of pluripotential HSC activity was the ability of an explant to provide long-term repopulation of adult bone marrow after transplantation by intravenous injection into lethally irradiated adult mice. They found that 10 dpc (day post coitum) mouse embryo AGM, but not yolk sac or liver, was capable of long-term repopulation of bone marrow HSCs. Yolk sac and liver were not able to provide precursors capable of long term repopulation of the adult bone marrow unless they were explanted from 11 dpc mouse embryos.

It cannot be ignored, however, that HSCs and embryonic blood cells are first observed within the yolk sac at 7.5 dpc. The authors have thus suggested that the embryonic liver may be populated by different waves of HSCs; first from the yolk sac with committed multilineage progenitors, colony-forming unit-culture (CFU-C); then from the AGM with CFU-S containing myeloid and erythroid progenitors; and finally from the AGM region with long-term repopulating HSCs.[43]

However, the authors also provide another intriguing interpretation of their results, namely that embryonic HSCs cannot functionally repopulate adult recipients because the frequency of stem cell progenitors in these tissues is very low early in development. Indeed, long-term repopulation occurred in only three percent of recipients of 10 dpc mouse embryo AGM, while long term repopulation of HSCs occurred in 73 percent, 59 percent, and 40 percent of 11 dpc AGM, yolk sac and liver engrafted recipients respectively. The authors speculate that the long-term repopulating capacity of 10 dpc AGM may occur as a consequence of the presence of only one HSC within the injected suspension, thereby raising the question of the role of differences in HSC density in results obtained in these studies. They suggest that the initial function of the embryonic yolk sac and liver may be to provide the embryo and fetus with differentiated hematopoietic cells and that it is only later in development that these organs provide the definitive HSCs required by the adult—at a time when the appropriate microenvironments of the adult hematopoietic organs are formed. This speculation, of course, raises the question of the relevance of some of the experimental criteria used to establish embryological relationships between cells within embryos. For example, in many of these experiments, the tissue under study is traumatically extirpated from the embryonic microenvironment and then subjected to the foreign environment of the adult circulation, the kidney capsule or the tissue culture dish. The operational criteria used in these studies to establish HSC activity, therefore, may be too stringent (or alternatively, too conducive) to allow conclusions to be drawn about embryonic relationships from negative experimental data. Many of the strictly descriptive studies are not so subject to this criticism.

Could PGCs within the AGM Give Rise to HSCs in Explant and Transplantation Experiments?

Many Questions Are Raised by the Apparently Intimate Embryological Relationship between PGC and HSC Precursors. Whatever the potency of the AGM in producing adult HSCs, it is likely that AGM, yolk sac, and liver HSCs ultimately arise from the same progenitors of the primary ectoderm—progenitors which bear a close embryological ancestry with primordial germ cells. Indeed in the study by Medvinsky et al.,[38] it was found that during the period the AGM contained significant CFU-S activity, HSCs could not be discriminated; yet PGCs were quite evident in this region of the posterior body wall. Is it possible that these embryonic PGCs can differentiate into HSCs when cultured or when transplanted to an appropriate ectopic site? Could the production of HSCs from AGM occur in response to pertinent microenvironmental manipulations? Is it possible that yolk sac grafts in experiments where negative results were obtained did not contain the region near the allantois which would also be populated with PGCs? Indeed, ample evidence now exists to suggest that PGCs may give rise to pluripotent embryonic stem cells; ES cells which, under the right conditions can give rise to all of the hematopoietic lineages (see below).

PCGs Can Give Rise to ES Cells Which in Turn Can Form Blood Cells. It has recently been demonstrated that PGCs can give rise to pluripotent ES cells capable of forming a wide variety of cells in addition to blood cells. These relationships support the possibility that the mechanism underlying the association of germ cell tumors and hematologic malignancies (HMs) might involve an intermediary conversion of the misplaced germ cell to a pluripotent ES-like cell which then gives rise to the HM.

It has been shown, for example, that when the genital ridge (containing PGCs; 11.5 to 12.5 dpc) of male mouse embryos is grafted to an ectopic site such as testis or kidney capsule, teratocarcinomas may develop, suggesting that environmental factors may have reprogrammed the PGCs of the donor tissue resulting in development of pluripotent cells.[44-46]

More conclusive evidence for such reprogramming of PGCs has been obtained in culture studies.[47] While the addition of c-kit ligand and leukemia inhibitory factor (LIF) to culture medium can prolong the life of cultured mouse PGCs, their proliferation ultimately ceases at a point which is related to their proliferative potential within the animal. However, when basic fibroblast growth factor (BFGF) is added to the medium along with these other two additives, PGCs continue to proliferate and ultimately give rise to colonies of cells which resemble ES cells.[47] These cells can form a variety of cell types when injected into nude mice or when cultured and can produce blastocyst injection chimeras. However, when injected into neonatal SCID, and W^v/W^v and lethally irradiated mice, these cells were unable to form the myeloid lineage but were capable of forming only lymphoid cell lineages and then only at an extremely low level[48]. Again, however, it must be pointed out that microenvironment undoubtedly means everything in these kinds of studies. There is, for example, no obvious reason that one should expect to find an hospitable environment within adult animals for the conversion of ES cells to HSCs. Clearly, these negative data do not refute the possibility that the mechanism by which primordial germ cell tumors give rise to hematologic cancers involves the reprogramming of the PGC to a pluripotent ES-like cell. Nonetheless, if this mechanism can occur, the microenviromental conditions required for the conversion of PGCs in vivo to acute megakaryoblastic leukemias are presently poorly understood.

CONCLUSIONS

Whatever the final outcome of these studies, the close embryological relationship between PGCs and HSCs remains an intriguing observation and may be pertinent to an hypothesis that explains the apparent preference for conversion of germ cell tumors to hematologic malignancies. Further refinement of our understanding regarding the roles of growth factors and other variables of the microenvironment, particularly within the migratory routes of HSCs and PGCs will undoubtedly improve our understanding of an intriguing question raised by a fascinating clinical mystery.

ACKNOWLEDGEMENTS

I am grateful for discussions with Drs. Tom Doetschman, Bruce Aronow, Steve Potter, Craig Nichols, Mr. Robert Powers and Robert Flick; and for the support of Drs. H.-J. Schmoll and Nader Abraham and the Secretariat of the IXth International Symposium of the Molecular Biology of Hematopoesis.

REFERENCES

1. Nichols CR, Roth BJ, Heerema N, Griep J, Tricot G: Hematologic neoplasia associated with primary mediastinal germ-cell tumors. N Engl. J. Med. 322:1425-1429, 1990
2. Larsen M, Evans WK, Shepard FA, Phillips MJ, Bailey D, Messner H: Acute lymphoblastic leukemia: Possible origin from a mediastinal germ cell tumor. Cancer 53:441-444, 1984
3. Nichols CR, Hoffman R, Einhorn LH, Williams SD, Wheeler LA, Garnick MB: Hematologic malignancies associated with primary mediastinal germ-cell tumors. Ann. Intern. Med. 102:603-609, 1985
4. Woodruff K, Wang, N, May W, Adrone, E, Denny, C. Feig SA: The clonal nature of mediastinal germ cell tumors and acute myelogenous leukemia. Cancer Genet. Cytogenet. 79:25-31, 1995
5. Witschi E: Migration of the germ cells of human embryos from the yolk sac to the primitive gonadal folds. Contr. Embryol. Carnegie Inst., 32:67-80, 1948
6. Moore MAS, Metcalf D: Ontogeny of the haematopoietic system: Yolk sac origin of in vivo and in vitro colony forming cells in the developing mouse embryo. Br. J. Haematol. 18:279-296, 1970
7. Fuss A: Ueber extraregionare Geschlectszellen bei einem menschlichen Embryo von 4 Wochen. Anat. Anz. 39:407-409, 1911
8. Chiquoine A D: The identification, origin, and migration of the primordial germ cells of the mouse embryo. Anat. Rec. 118:135-145, 1954
9. Mintz B, Russel ES: Gene-induced embryological modifications of primordial germ cells in the mouse. J. Exp. Zool. 134:207-237, 1957

10. Ozdzenski W: Observations on the origin of the primordial germ cells in the mouse. Zool. Polo. 17:367-379, 1967
11. Tam PPL, Snow MHL: Proliferation and migration of primordial germ cells during compensatory growth in mouse embryos J.E.E.M. 64:133-147, 1981
12. Ginsburg M, Snow MHL, McLaren A: Primordial germ cells in the mouse embryo during gastrulation. Develop. 110:521-528, 1990
13. Hahnel AC, Rappollee DA, Millan JL, Manes T, Ziomek CA, Theodosiou NG, Werb Z, Pedersen RA, Schultz GA: Two alkaline phosphatase genes are expressed during early development in the mouse embryo. Develop. 110:555-564, 1990
14. Spiegelman M, Bennett D: A light and electron microscope study of the primordial germ cells in the early mouse embryo. J.E.E.M. 30:97-118, 1973
15. Clark JM, Eddy LM: Fine structural observations on the origin and association of primordial germ cells of the mouse. Devel. Biol. 45:136-155, 1975
16. Johnson GR, Moore MAS: Role of stem cell migration in initiation of mouse foetal liver haemopoiesis. Nature 258:726-728, 1975
17. Migliaccio G, Migliaccio AR, Petti S, Mavilio F, Russo G, Lazzaro D, Testa U, Marinucci M, Peschle C: Human embryonic hemopoiesis: Kinetics of progenitors and precursors underlying the yolk sac-liver transition. J. Clin Invest. 78:51-60, 1986
18. Peschle C, Mavilio F, Care A, Migliaccio G, Miglioccio AR, Salvo G. Sammoggia P, Petti S, Guerriero R, Marinucci M, Lazzaro D., Russo G, Mastroberardino G: Haemoglobin switching in human embryos. Nature 313:235-237, 1985
19. Cudennec CA, Thiery J-P, Le Dourarin NM: In vitro induction of adult erythropoiesis in early mouse yolk sac. Proc. Natl. Acad. Sci. 78:2410-2416, 1981
20. Motro B, Van der Kooy D, Rossant J, Reith A, Bernstein A: Contiguous patterns of c-kit and steel expression: analysis of mutations of W and Sl loci. Develop. 113:1207-1221, 1991
21. Besmer P, Manova, K, Duttiger R, Huang J, Packer A, Gyssler C, Bachvarova RF: The kit-ligand (steel factor) and its receptor c-kit/W: pleiotropic roles in gametogenesis and melanogenesis. Develop. Supplement 125-137, 1993
22. Maeda H, Yamagat A, Nishikawa S, Yoshinaga K, Kobayashi S, Nishi K, Nishikawa S-I: Requirement of c-kit for development of intestinal pacemaker system. Develop. 116:369-375, 1992
23. Bennett D: Developmental analysis of a mutation with pleiotropic effects in the mouse. J Morph. 98:199-229, 1956
24. Williams DE, de Vries P, Namen A, Widmer MB, Lyman SD: The steel factor. Dev. Biol. 151:368-376, 1992
25. Huang E, Nocka K, Beier DR, Chu T-Y, Buck J, Lahm H-W, Wellner D, Leder P, Besmer P: The hematopoietic growth factor KL is encoded by the Sl locus and is the ligand of the c-kit receptor, the gene product of the W locus. Cell 63:225-233, 1990
26. Copeland NG, Gilbert GJ, Cho BC, Donovan PJ, Jenkins NA, Cosman D, Anderson D, Lyman SD, Williams DE: Mast cell growth factor maps near the Sl locus and is structurally altered in a number of steel alleles. Cell 63:175-183, 1990
27. Manova K, Huang EJ, Angeles M, de Leon V, Sanchez S, Pronovost SM, Besmer P, Bachvarova RF: The expression pattern of the c-kit ligand in gonads of mice supports a role for the c-kit receptor in oocyte growth and in proliferation of spermatogonia Devel. Biol. 157:85-99, 1993
28. Matsui Y, Zsebo KM, Hogan BLM: Embryonic expression of a haematopoietic growth factor encoded by the Sl locus and the ligand for c-kit. Nature 347:667-669, 1990
29. Godin I, Deed R., Cooke J, Zsebo K, Dexter M, Wylie CC: Effects of the steel gene product on mouse primordial germ cells in culture. Nature 352:807-809, 1991
30. Packer AI, Hsu YC, Besmer P, Bachvarova RF: The ligand of the c-kit receptor promotes oocyte growth. Dev. Biol. 161:194-205, 1994
31. Larsen WJ: Human Embryology. Churchill Livingstone, 1993
32. Snow MHL: Autonomous development of parts isolated from primitive streak-stage mouse embryos. Is development clonal?. J.E.E.M. 65:269-287, 1981
33. Copp AJ, Roberts HM, Polani PE: Chimaerism of primordial germ cells in the early postimplantation mouse embryo following microsurgical grafting of posterior primitive streak cells in vitro. J.E.E.M. 95:95-115, 1986
34. Lawson KA, Meneses JJ, Pedersen R: Clonal analysis of epiblast fate during germ layer formation in the mouse embryo: Develop. 113:891-911, 1991

35. Lawson KA, Pedersen R: Clonal analysis of cell fate during gastrulation and early neurulation in the mouse. CIBA Symp. 165. Postimplantation Development in the Mouse. (Chadwick DJ, Marsh J. ed.) John Wiley and Sons, 1992
36. Lawson KA, Hage WJ: Clonal analysis of the origin of primordial germ cells in the mouse. CIBA Symp. 182. Germline Development. John Wiley and Sons, 1994
37. Dieterlen-Lievre F: On the origin of haematopoietic stem cells in the avian embryo. J.E.E.M 33:607-619, 1975
38. Medvinsky AL, Samoylina NL, Muller A, Dzierzak EA: An early pre-liver intraembryonic source of CFU-S in the developing mouse. Nature 364:64-67, 1993
39. Godin IE, Garcia-Porrero JA, Coutinho A, Dieterlen-Lievre F, Marcos, MAR: Para-aortic splanchnopleure from early mouse embryos contains B1a cell precursors. Nature 364:67-70, 1993
40. Muller A, Medvinsky A, Strouboulis J, Grosveld F, Dzierzak E.: Development of hematopoietic stem cell activity in the mouse embryo. Immunity 1:291-301, 1994
41. Toles JF, Chui DHK, Belbeck LW, Starr E, Barker JE: Hematopoietic stem cells in murine embryonic yolk sac and peripheral blood. Proc. Natl. Acad. Sci. 86:7456-7459, 1989
42. Weissman IL, Pappaloannou V, Gardner R: Fetal hematopoietic origins of the adult hematolymphoid system, in Differentiation of Normal and neoplastic Hematopoietic Cells (Cold Spring Harbor, New York: Cold Spring Harbor Laboratory Press). pp.33-47, 1978
43. Muller AM, Medvinsky A, Strouboulis J, Grosveld F, Dzierzak E: Development of hematopoietic stem cell activity in the mouse embryo. Immunity 1:291-301, 1994
44. Stevens LC: The origin and development of testicular, ovarian, and embryo-derived teratomas. in Cold Spring Harbor Conferences on Cell Proliferation, 10:23-36, 1986
45. Stevens LC, Makensen JA: Genetic and environmental influences on teratogenesis in mice. J Natl. Cancer Inst. 27:443-453, 1961
46. Nogouchi T, Stevens LC: Primordial germ cell proliferation in fetal testes in mouse strains with high and low incidences of congenital testicular teratomas. J. Natl Cancer Inst. 69:907-913, 1982
47. Matsui Y, Zsebo K, Hogan BLM: Derivation of pluripotential embryonic stem cells from murine primordial germ cells in culture. Cell 70:841-847, 1992
48. Muller A, Dzierzak EA: ES cells have only limited lymphopoietic potential after adoptive transfer into mouse recipients. Development 118:1343-1351, 1993

9

AUTOLOGOUS BMT FOR TREATMENT OF EXPERIMENTAL AUTOIMMUNE DISEASES

D. W. van Bekkum* and M. van Gelder

IntroGene B.V.
P.O. Box 3271
2280 GG Rijswijk
The Netherlands

INTRODUCTION

For many experimental autoimmune diseases (AID) the predisposition to the disease was shown to reside in the bone marrow (BM) as recently reviewed by Van Bekkum (1) and Marmont (2). Like in individual humans, genetically different inbred rodent strains vary in their susceptibility to experimentally induced AID. The permanent engraftment of BM from susceptible animals in lethally irradiated resistant animals rendered the recipients susceptible, and vice versa. The finding that the transplantation of purified hemopoietic stem cells from spontanously autoimmune-prone NZB and (NZW x BXSB) F1 mice into normal mice led to the development of AID in the recipients, suggested that genetic defects residing in the hemopoietic stem cell determine the development of AID in the case of the hereditary forms (3,4). Factors like sex and environment (5,6,7) appear to act only as modulating factors.

Only a few investigators have studied the therapeutic potential of transplantation of allogeneic bone marrow in fully developed AID (table 1). Our group showed that treatment with high dose total body irradiation (TBI) and allogeneic bone marrow transplantation (BMT) from a resistant donor strain induced complete and lasting remission in rats suffering from severe adjuvant arthritis (AA) (8) and in rats with experimental allergic encephalomyelitis (EAE) (23,24). Ikehara et al. reported complete correction of 'murine SLE' by allogeneic BMT; even full-blown glomerulonephritis resolved (9,10). In mice, treatment of type II collagen-induced arthritis with allogeneic BMT did not induce remission, but progression of the disease was completely prevented (11).

Interestingly, a few patients suffering from leukemia or aplastic anemia with coëxisting AID have been reported, who were treated with allogeneic BMT from HLA-identical sibling donors. The treatment cured the primary disease as well as their severe rheumatoid arthritis (four cases 12,13,14), psoriasis (two cases 15,16) or ulcerative colitis (one case 16).

*Tel. +31.15842662; Fax. +31.15843980

Molecular Biology of Hematopoiesis 5, edited by Abraham et al.
Plenum Press, New York, 1996

Table 1. Treatment of fully developed experimental AID with allogeneic BMT

AI strain (species)	Donor strain (conditioning)	Effect	Ref.
A. HEREDITARY AID			
B/W,BXSB (mouse)	BALB/c nu/nu (TBI)	Lasting reduction of glomerular damage and deposits of IgG and C, as well as reduction in circulating immune complexes	9
MRL/lpr (mouse)	C57BL/6 (TBI)	Complete and lasting amelioration of glomerulo-nephritis, arthritis and correction of immunological abnormalities	22
B. INDUCED AID			
AA(BUF) (rat)	WAG (TBI)	Complete remission of severe arthritis	8
EAE (BUF) (rat)	WAG, BN.1B (TBI or CY+Bu)	Remission-induction, prevention of recurrent disease (spontaneously or after reinduction) in 90% of the animals	23 24
CIA (DBA/1J) (mouse)	BALB/c (TBI)	No remission-induction, but complete prevention of progression	11

abbreviations: AA, adjuvant-induced arthritis; Bu, busulfan; CIA, collagen-induced arthritis; CY, cyclophosphamide; EAE, experimental autoimmune encephalomyelitis;TBI, total body irradiation.

The complete remissions persisted for the duration of the observation time, which was 1-7 years at the time of publication.

A completely different approach was explored by us and two other groups of investigators, i.e. the treatment of experimental AID with high dose cytoreductive agents and rescue with autologous or syngeneic BM (table 2). We showed that treatment of AA in rats with TBI and syngeneic BMT was equally effective in inducing complete and lasting remission as allogeneic BMT (8). Even the engraftment of autologous BM, obtained from the arthritic rat prior to TBI and grafted thereafter, resulted in 100% cures (17). The treatment eliminated the imflammatory component of the arthritis, but did not resolve bone deformation when performed at later stages of the disease. The need for intensive cytoreductive treatment in AA is illustrated by the observation that treatment with 'sublethal' doses of TBI was less effective. Treatment with cyclosporin A resulted in some regression of inflammation during the treatment, but arthritic symptoms fully reappeared after cessation of the treatment.

In mice suffering from collagen arthritis, treatment with TBI and syngeneic bone marrow was less effective according to Kamaiya et al.(11). The arthritis did not regress, but the disease was halted in its progression. It is still being debated whether adjuvant arthritis or collagen arthritis is the better model for human rheumatoid arthritis. Our choice to employ AA in rats was based on the close histopathological similarity between this affection and the human inflammatory joint disease, as well as on the consideration that there is no evidence that rheumatoid arthritis is induced by an autoimmune reaction against collagen.

Pestronk et al. treated rats with TBI and pseudoautologous BMT -a substitute of autologous BMT obtained from syngeneic donors with active disease-, which suffered from experimental autoimmune myasthenia gravis (EAMG) (18). EAMG is an AID mediated by anti-acetylcholine receptor (a-AChR) autoantibodies but this antibody formation appears to be T cell dependent. Treatment with high dose CY and TBI and rescue with pseudoautologous BM strongly decreased the a-AChR titer and eliminated the memory auto-antibody response. The cure of a patient with MG was reported after treatment for non-Hodgkin's lymphoma with TBI and CY and autologous stem cell rescue (19). Although very promising, a firm conclusion about a definitive lasting cure in this patient can not be made, since the observation time was only a few months.

Table 2. Treatment of fully developed experimental induced AID with syngeneic or (pseudo-) autologous BMT

AID(strain) (species, strain)	BM origin (conditioning)	Effect	Ref.
EAMG (rat: Lewis)	pseudoautologous[1] (CY + TBI)	Reduction of a-AChR titer by CY, elimination of memory response by TBI; reimmunization induces a new primary response, unless autoreactive cells were grafted	18
AA (rat:BUF)	autologous or syngeneic (TBI)	Complete and lasting remission; no regression of excessive bone formation when treated at late stage of the disease	8 17
CIA (mouse:DBA/1J)	syngeneic (TBI)	No remission-induction, but complete prevention of progression	11
EAE (mouse:SJL/J)	syngeneic (CY)	Remission-induction, prevention of spontaneous relapses in 73%; in 47% recurrent disease after reinduction	25
EAE (rat:BUF)	syngeneic pseudoautologous[1] (TBI,CY/ALS+TBI)	Remission-induction, prevention of spontaneous relapses in >75%(conditioning-dose dependant); in 50-75% recurrent disease after reinduction	21 26

Abbreviations: AA, adjuvant-induced arthritis; a-AChR, anti-acetyl choline recptor antibodies; Bu, ALS, anti-lymphocyte-serum; busulfan CIA, collagen-induced arthrits; CY, cyclophosphamide; EAE, experimental autoimmune encephalomyelitis; EAMG, experimental autoimmune myasthenia gravis; TBI, total body irradiation, [1] Pseudoautologous BM is harvested from sick syngeneic donors, during the same stage of the disease as the recipients at the time of BMT.

The third experimental AID in which the curative effects of syngeneic and autologous BMT have been studied is experimental allergic encephalomyelitis (EAE). The disease can be induced in susceptible strains of mice and rats by immunization with whole spinal cord or purified myelin protein, in combination with adjuvant (Mycobacterium tuberculosis). In certain rodent strains this results in successive attacks of paresis and paralysis, the so called chronic relapsing form of the disease. Karussis et al (25) treated mice during the first attack with high dose cyclophosphamide (CY) and syngeneic BMT which resulted in complete remission and prevention of spontanous relapses in the majority of the animals. After reimmunization of the cured animals about 50 % developed relapses.

Our experiments were carried out with Buffalo (BUF) rats in which a similar pattern of chronic-relapsing EAE can be induced. The animals were treated during the acute phase of the disease that is at 20 days after immunization with high dose TBI or other myeloablative conditioning regimens and rescued with BMT. This treatment results in a rapid regression of the neurological symptoms, the majority having recovered within 10 days.

In contrast, the untreated controls take on the average 30 days to recover from their first attack and some never remit. The incidence of spontanous and of induced relapses was determined by the nature and constitution of the bone marrow graft as will be discussed below.

In this paper we shall focus on the results obtained with autologous or PSA BM in the two AID, AA and EAE because it is generally felt that the first clinical trials with this treatment can be done best with autologous bone marrow. The main arguments for this strategy are the lower risks of transplantation associated toxicity and the universal availability of autologous marrow. Another drawback of matched allogeneic bone marrow grafts is the possibility of development of graft versus host disease (GVHD) which especially in its chronic form is difficult to distingish from certain autoimmine reactions.

The methods employed have been published extensively for both the AA (8,17) and the EAE model (21).

TREATMENT OF ADJUVANT ARTHRITIS IN BUF RATS

All published results have been obtained by conditioning with high lethal dose TBI (9 or 9.5 Gy) which causes complete and lasting cures, irrespective of the source of the BM graft. Autologous and syngeneic BM are as effective as allogeneic BM from a strain of rats that is resistant to induction of arthritis. After having demonstrated the efficacy of autologous bone marrow grafting we have continued to use pseudo-autologous bone marrow instead of autologous BM because it is a more humane procedure. As the donors of the pseudoautologous BM are taken (and sacrificed) from among the group of syngeneic arthritic rats to be treated, there should be no difference in the constitution of this marrow with that of truely autologous marrow.

The cured animals do not relapse either spontanously or after reimmunization (Figure 1). However, the reimmunization experiments were performed in rats grafted with syngeneic BM and have to be repeated with pseudo-autologous BM, because spontanous relapses may well depend on the presence of activated T lymphocytes in the graft.

Lower doses of TBI which do not require rescue with BMT are less effective, as is treatment with partial body irradiation be it on the affected joints or on the total lymphoid tissues.

It appears that the use of high dose TBI for conditioning of patients with severe AID is not favoured by many clinicians, mainly because of the increases risk of development of secondary tumors. High dose CY has been used successfully in the treatment of aplastic anemia (another autoimmune disease) in preparation for allogeneic BMT. Accordingly, we

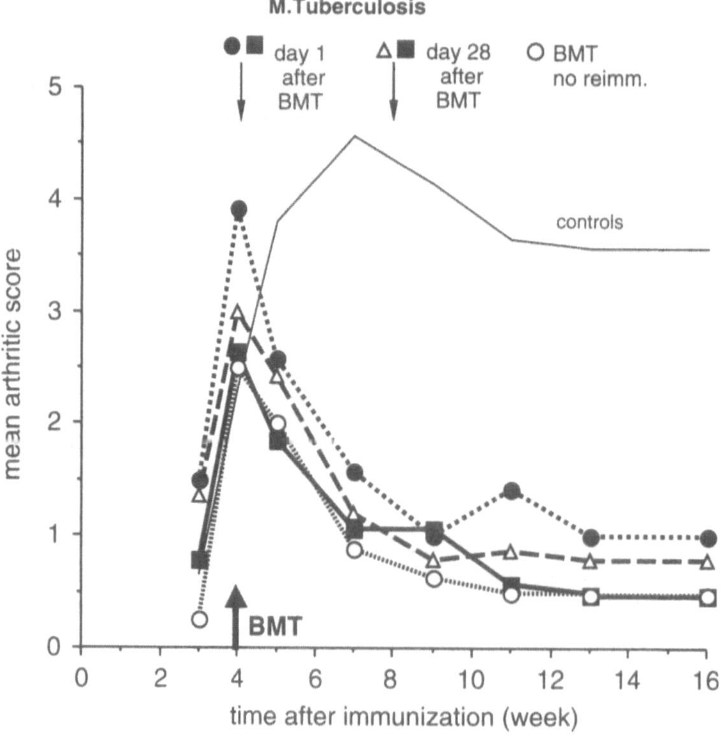

Figure 1. Effect of reimmunization with M. Tuberculosis after syngeneic bone marrow transplantation on the development of arthritis.

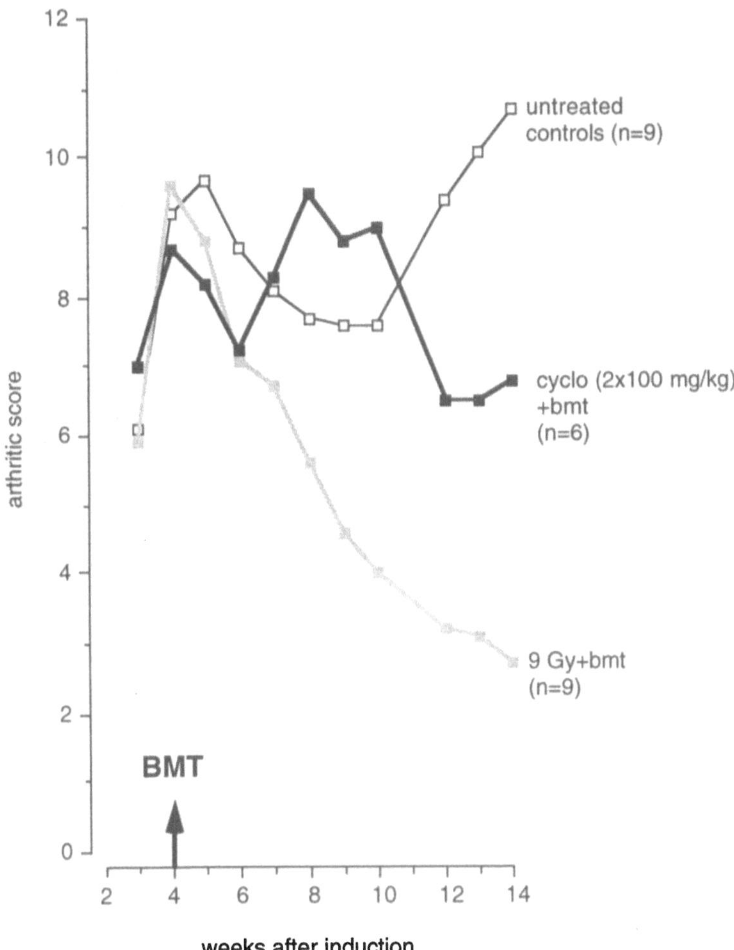

Figure 2. Pseudoautologous BMT-Adjuvant Arthritis-Rats (effect of conditioning).

have recently evaluated the efficacy of CY in the treatment of arthritic rats in combination with pseudoautologous BM. The dose of CY (2 daily doses of 150 mg/kg) was the maximal tolerated dose in combination with BM rescue. As shown in figure 2, the arthritis in the Cy group regressed at the same rate as the groups that were conditioned with TBI for two weeks and thereafter relapsed in all animals.

The cause of these relapses still has to be unraveled and it seems worth while to investigate whether the addition of other lymphotoxic agents to the CY conditioning can improve the results. For the time being, it seems advisable to employ a conditioning regimen that consists of high dose TBI, or of a combination of high dose CY with a lower dose of TBI e.g. 4 or 5 Gy.

TREATMENT OF EAE IN BUF RATS

Our results (21,23,24,26) may be briefly summarized as follows :Induction of remission occurs in all animals irrespective of the source of BM, provided conditioning is sufficiently intensive such as 10 Gy TBI, Cy (2 daily doses of 60mg/kg i.p., followed by Mesma) plus busulfan (Bu : 30 mg/kg i.p.), or 7 Gy TBI plus 2 daily doses of CY as above. BM from mismatched donors (T cell depleted to prevent GvHD) of a resistant rat strain is most effective in preventing spontanous and induced relapses (Table 3). This was attributed to the additional inactivation of residual T cells in the recipient by a subclinical GvH reaction of the allogeneic marrow and to the fact that the grafted cells are derived from a EAE resistant rat strain. With MHC matched bone marrow from a resistant strain the results were equally favourable. With PSA BM the incidence of spontanous relapses was on the average 25%, similar to that after syngeneic BMT. This suggests that the spontanous relapses are not caused by sensitized T (memory) cells from the graft, because such cells cannot be present in syngeneic marrow from normal donors. If that is correct, the spontanous relapses are caused by naive T ly from a susceptible animal source either provided by the graft or surviving the conditioning in the recipient. Obviously, the difference with allogeneic bm transplants could also be due to surviving memory cells in the recipients, which are inactivated by allogeneic lymphocytes. This would make it attractive to intensify the conditioning further when autologous BMT is envisaged. Candidate agents appear to be anti-lymphocyte globulin (ALG) and 2-chlorodeoxy-adenosine (cladribine), both of which could not be evaluated in our rat model because the available A:LG samples were toxic to hemopoietic stem cells and because the pharmacokinetics of cladribine is widely different in rats from that in humans.

The induced relapse incidence was in the order of 70% in the animals treated with PSABM, significantly higher than in those treated with syngeneic BM (40%). This difference has to be attributed to higher numbers of activated T ly in the PSABM and therefore, in the clinical situation it is mandatory to employ autologous BM only if it is ex vivo T cell depleted to the extent of 2-3 log T cell reduction.

RECOMMENDATIONS FOR THE TREATMENT OF CLINICAL AID

In view of the striking results obtained in several animal models of AID with high dose cytoreductive treatment and BMT it seems justified to investigate this approach in selected patients suffering from progressive severe disease with a poor prognosis. The experience in both the AA and the EAE models strongly suggests that this treatment can only be expected to be effective if applied during the stage of active inflammatory disease. It is not expected that end stages in which the symptomatology is mainly determined by scarring (exostoses in arthritis and demyelination in MS) will respond. As discussed earlier it is generally felt that the first attempts should be performed with autologous BMT. T cell

Table 3. Treatment of EAE in BUF rats with BMT

Conditioning	Spontanous relapses	Induced relases
Mismatched resistant donors: WAG		
10 Gy TBI	3/48	5/47
Cy + Bu	0/13	3/12
Matched resistant donors: BN.1B		
10 Gy TBI	0/18	0/18
Cy + Bu	2/29	4/29
Pseudo autologous BM		
10 Gy TBI	11/37	26/36
Cy + Bu	9/14	10/14
Cy + 7 Gy TBI	6/16	9/15

depletion of the graft should be mandatory and as complete as feasible. The conditioning regimen should contain a substantial element of TBI, except for the time being in MS patients. The problem is that in our rat model TBI, even doses as low as 1.5 Gy , invariably caused an exacerbation of short duration of the neurological symptoms. This reaction to irradiation of the inflamed CNS has not been observed in mice with EAE (our own results) nor in MS patients with chronic progressive disease (20). However, in one patient TBI was followed by a fatal aggrovation of the neurological disease (R.E.Champlin, personal communication). Untill the safety of TBI has been established in patients with active MS, it is recommended to employ the high dose CY plus BU in conjunction with ALG.

So far, in human AID the causes of relapses remain unknown and we can not even speculate whether the induced relapses as studied in our models have any relevance for the clinical situation. New information on this issue can only be provided by the results of the treatment with BMT of suitable patients.

ACKNOWLEDGEMENT

Financial support of our EAE studies was provided by the Foundation "Vrienden MS Research" The Netherlands.

REFERENCES

1. Van Bekkum DW. BMT in experimental autoimmune diseases (review) Bone Marrow Transplant. 1993, 11: 183-187
2. Marmont AM. Immune ablation followed by stem cell rescue: a new radical approach to the treatment of severe autoimmune diseases. Forum Trends Exp. Clin. Med. 1995, 5:4-15.
3. Sardina EE, Sugiura K, Ikehara S, Good RA. Transplantation of wheat germ agglutinin-positive hematopoietic cells to prevent or induce systemic autoimmune disease. Proc. Natl.Acad. Sci. USA 1991, 83: 3218-3222.
4. Ikehara S, Kawamura M, Takao F et al. Organ-specific and systemic autoimmune diseases originate from defects in hematopoietic stem cells. Proc.Natl.Acad.Sci. USA 1990, 87: 8341-8344.
5. Leiter EH. The role of environmental factors in modulating insulin dependant diabetes. The role of micro-organisms in non-infectious disease. 39-55, 1990.
6. Steinberg AD, Raveché ES, Laskin CA et al. Systemic lupus erythematosus: insights from animal models. Ann Intern. Med. 1984, 100: 714-727.
7. Denman AM. Sex hormones, autoimmune diseases and immune responses. Br. Med. J. 1991, 303: 2-3.
8. Van Bekkum DW, Bohre EPM, Houben PFJ, Knaan S. Regression of adjuvant-induced arthritis in rats following bone marrow transplantation. Proc.Natl.Acad.Sci. USA 1989, 86: 10090-10094.
9. Ikehara S, Good RA, Nakamura T et al. Rationale for bone marrow tranplantation in the treatment of autoimmune diseases. Proc. Natl. Acad. Sci. USA 1985, 82: 2483-2487.
10. Ikehara S, Yasumizu R, Inaba M et al.Longterm observations of autoimmune-prone mice treated for autoimmune disease by allogeneic bone marrow transplantation. Proc. Natl. Acad. Sci. USA 1989, 86: 3306-3310.
11. Kamiya M, Sohen S, Yamane T, Tanaka S. Effective treatment of mice with type II collagen induced arthritis with lethal irradiation and bone marrow tranplantation. J. Rheumatol. 1993, 20: 225-230.
12. Baldwin JL, Storb R, Thomas ED, Mannik M. Bone marrow transplantation in patients with gold-induced marrow aplasia. Arthritis Rheum. 1977, 20: 1043-1048.
13. Jacobs P. Vincent MD, Martell RW. Prolonged remission of severe refractory rheumatoid arthritis following allogeneic bone marrow tranplantation for drug-induced aplastic anaemia. Bone marrow transpl. 1986, 1: 237-239.
14. Lowenthal RM, Cohen ML, Atkinson K, Biggs J. Apparent cure of rheumatoid arthritis by bone marrow transplantation. J. Rheumatol. 1993, 20: 137-140.
15. Eedy DJ, Burrows D, Bridges JM, Jones FGC. Clearance of severe psoriasis after allogeneic bone marrow tranplantation. Br. Med. J. 1990, 300: 908.

16. Liu Yin JA, Jowill SN. Resolution of immune-mediated diseases following allogeneic bone marrow transplantation for leukemia. Bone marrow transplant. 1992, 9: 31-33.
17. Knaan-Shanzer S, Houben P, Kinwel-Bohré EPM, Van Bekkum DW. Remission induction of adjuvant arthritis in rats by total body irradiation and autologous bone marrow transplantation. Bone marrow transpl. 1991, 8: 33-338.
18. Pestronk A, Drachman DB, Teoh R, Adams RN. Combined short-term immunotherapy for experimental autoimmune myasthenia gravis. Ann Neurol. 1983, 14: 235-241.
19. Salzman D, Tami J, Jackson C etal. Clinical remission of myasthenia gravis (MG) in a patient (PT) after high dose therapy and autologous transplantation with CD34+ stem cells. Blood 1994, 84 suppl.1 206a.
20. Tourtelotte WW, Potvin AR, Baumhefner RW et al. Multiple sclerosis de novo CNS IgG synthesis: effect of CNS irradiation. Arch. Neurol. 1980, 37: 620-624.
21. Van Gelder M, Kinwel-Bohré EPM, Van Bekkum DW. Treatment of experimental allergic encephalomyelitis in rats with total body irradiation and syngeneic BMT. Bone marrow transpl. 1993, 11: 233-241.
23. Van Gelder M. Treatment of relapsing experimental autoimmune encephalomyelitis with partially MHC-matched allogeneic bone marrow transplantation. Thesis Leiden 1995.
24. Van Gelder M, Van Bekkum DW. Treatment of relapsing experimental autoimmune encephalomyelitis in rats with allogeneic bone marrow transplantation from a resistant strain. In press Bone Marrow Transpl.
25. Karussis DM, Slavin S, Ben-Nun A. Chronic-relapsing experimental autoimmune encephalomyelitis (CR-EAE): treatment and induction of tolerance, with high dose cyclophosphamide followed by syngeneic bone marrow transplantation. J. Neuroimmunol. 1992, 39: 201-210.
26. Van Gelder M, Mulder M, Van Bekkum DW. The feasibility of pseudoautologous bone marrow transplantation for treatment of relapsing experimental autoimmune encephalomyelitis in rats. submitted.

PERIPHERAL BLOOD PRECURSOR CELL TRANSPLANTS ACROSS A MAJOR HISTOCOMPATIBILITY BARRIER IN RABBITS*

Positive Effect of a Higher Number of Precursor Cells?

A. Gratwohl,† H. Baldomero, L. John, A. Tichelli, A. Filipowicz, C. Nissen, and B. Speck

Division of Hematology
Department of Research
Kantonsspital Basel, Switzerland

ABSTRACT

Peripheral blood precursor cells (PBPC) are used with increasing frequency as a source for hematopoietic transplants and have mainly replaced autologous bone marrow transplants. First clinical and experimental reports document the feasibility of PBPC as a source for allogeneic transplants. Few data exist on optimal procedure and the ideal number of cells for the transplant. We have previously shown in rabbits that PBPC can be used for transplants even across a major histocompatibility barrier. We used this model to test whether the number of transplanted precursor cells would influence outcome of the graft. Adult outbred Red Burgundy rabbits were used as donors, New Zealand white rabbits of the opposite sex as recipients. One individual donor was taken for one individual recipient. Conditioning consisted of single dose total body irradiation of 10 Gy followed by a short course of cyclosporine to enhance engraftment. Donor animals were treated with recombinant human granulocyte colony stimulating factor 10 μg/kg s.c. daily from day -2 until day +10. PBPC were obtained from the artery of the donor animal by repetitive centrifugation of 3 x 40 ml heparinized blood on each day of donation, i.e. days 0, +2, +3, +6, +8 and +10 and infused without further manipulation. Eight animals were transplanted. Seven of 8 engrafted, and 6 died of GvHD and pneumonia between days 12 and 55 (median survival of

* Supported in part by a a grant from the Swiss National Research Foundation, No. 32-42431.94.

† Address for correspondence: Prof. Dr. A. Gratwohl, Division of Hematology, Department of Internal Medicine, Kantonsspital Basel, CH-4031 Basel, Switzerland

Molecular Biology of Hematopoiesis 5, edited by Abraham et al.
Plenum Press, New York, 1996

all animals: 34 days). One animal is alive > 120 days. Transplanted nucleated cells varied from 7.3 to 15.4 x 10^8/kg (median 4 x 10^8/kg) and of CFU-GM from 12.3 to 176 x 10^4/kg (median 42 x 10^4/kg). There is a trend for increased survival with raised numbers of CFU-GM, (r) = 0.716, p = 0.0704. These data confirm that allogeneic PBPCT can engraft across a major histocompatiblity barrier and suggest that a higher number of CFU-GM might be advantageous.

INTRODUCTION

Transplantation of hematopoietic precursor cells is an established therapy today for a variety of congenital or acquired disorders of the bone marrow (1). The traditional source for such precursor cells, the bone marrow, has been replaced in recent years by peripheral blood precursor cells for autologous transplantation (2,3). The ease of mobilisation with hematopoietic growth factors, the faster recovery of hemopoiesis and omittance of general anesthesia, which was required for bone marrow harvesting, are the main factors leading to this development.

Theoretical considerations and practical aspects initially hindered application of PBPC for allogeneic transplants. Doubts about longterm repopulating capacity as well as fear of excessive graft-versus-host disease (GvHD) induced by the accompaning peripheral blood T lymphocytes prevented earlier studies. Recently, three single center studies and several case reports have documented clinical feasibility and renewed interest in allogeneic PBPC transplants (4-7). However, many questions are still unanswered. We have previously shown in rabbits that allogeneic PBPC can engraft even across a major histocompatibility barrier (8). We utilized this experimental animal model to test whether a high number of transplanted PBPC influences outcome. Preliminary data from these experiments form the basis of this report.

ANIMALS AND METHODS

Animals

Outbred adult Red Burgundy rabbits were taken as donors, outbred adult New Zealand White rabbits of the opposite sex as recipients. These two strains are strongly histoincompatible (8,9). They were purchased from Biological Research Laboratories, Füllinsdorf, Switzerland, a supervised animal care facility center and housed in the animal care facility center at the Kantonsspital Basel in accordance with local regulations. They were kept in single cages on a pellet diet and water ad libidum. The protocol was reviewed by the Ethical Committee for animal research of the Kanton of Basel-Stadt.

Transplant Protocol

The transplant protocol has been previously published in detail (8). In brief, total body irradiation from a cobalt 60 source of 10 Gy midline tissue dose at a dose rate of 20 cGy/min was used for conditioning followed by a brief course of cyclosporine 15 mg/kg s.c. daily from day -1 to day + 7 and then twice weekly 20 mg/kg s.c. until engraftment.

All animals were inspected daily, weight and alterations in excretion were recorded. Animals were given Bactrim 1 ml p.o. daily during the period of pancytopenia and normal saline s.c. as indicated. No platelet or red cell transfusions were given and no intravenous antibiotics were administered. Blood counts, electrolytes, renal and liver function tests were determined at regular intervals, as previously described. A full post-mortem was carried out in all animals that died post day 4, and histology was applied to define engraftment and GvHD, as previously described (8,10). Chromosomal analysis and donor skin graft was performed in long term survivors.

PBPC Mobilisation and Harvesting

Donor animals were treated with recombinant human granulocyte colony stimulating factor (G-CSF, Neupogen®, kindly provided by Roche Pharma Schweiz Ltd, Reinach) at a dose of 10 μg/kg s.c. daily beginning day -2 until day +10 (12 days total). PBPC were collected, as previously described, from the artery of the ear of donor animals on days 0, +2, +3, +5, +8 and +10, coresponding to days +2, +4, +5, +8, +10, +12 of G-CSF application, processing 3 x 40 ml of heparinized blood on each day of donation. Nucleated cells were obtained by centrifugation at 1200-1400 rpm for 20 min. Buffy coat was infused without further manipulation.

Blood counts were taken before each donation and after each step of processing and the total number of nucleated cells (TNC) and mononuclear cells (MNC) counted. The number of hematopoietic precursor cells (CFU-GM) was estimated using a semi-solid assay, as previously described.

Statistical Analysis

Means, medians and ranges of numerical variables were calculated with the Graph Instat statistical program. The correlation between transplant product and outcome was tested by a two-tailed linear regression analysis.

RESULTS

Transplant Product

The amount of blood processed and the number of cells harvested, processed and transplanted is illustrated in Table 1. A minimum of 7.3 x 10^8 nucleated cells/kg were obtained from each individual donor with a maximum of 15.4 x 10^8/kg. This corresponds approximately to a cell dose of one log more than routinely obtained in the same model with

Table 1. Number of cells processed and transplanted

		Buffy coat processing				PBPCT			
Animal		Volume ml	Processed N cells x 10^8	Harvested N cells x 10^8	% [(a)]	TNC x 10^8	TNC x 10^8/kg	CFU-GM x 10^4	CFU-GM x 10^4/kg
PBSC	18	418	57	32	55	31.0	8.2	110.3	18.6
"	19	514	54	40	75	34.9	13.2	175.3	62.4
"	20	432	49	32	66	27.1	9.6	208.8	73.7
"	21	499	61	47	77	39.2	15.4	450.8	176.8
"	22	492	40	30	75	29.1	8.8	127.4	38.5
"	23	459	41	29	70	24.8	7.3	42.5	12.3
"	24	460	59	40	67	35.8	11.4	54.3	17.2
"	25	437	37	28	75	25.1	8.0	142.9	45.6
median		460	51	32	72	30.1	9.2	133.2	42.0

[(a)] = Percentage of harvested compared to processed cells
TNC = total nucleated cell count

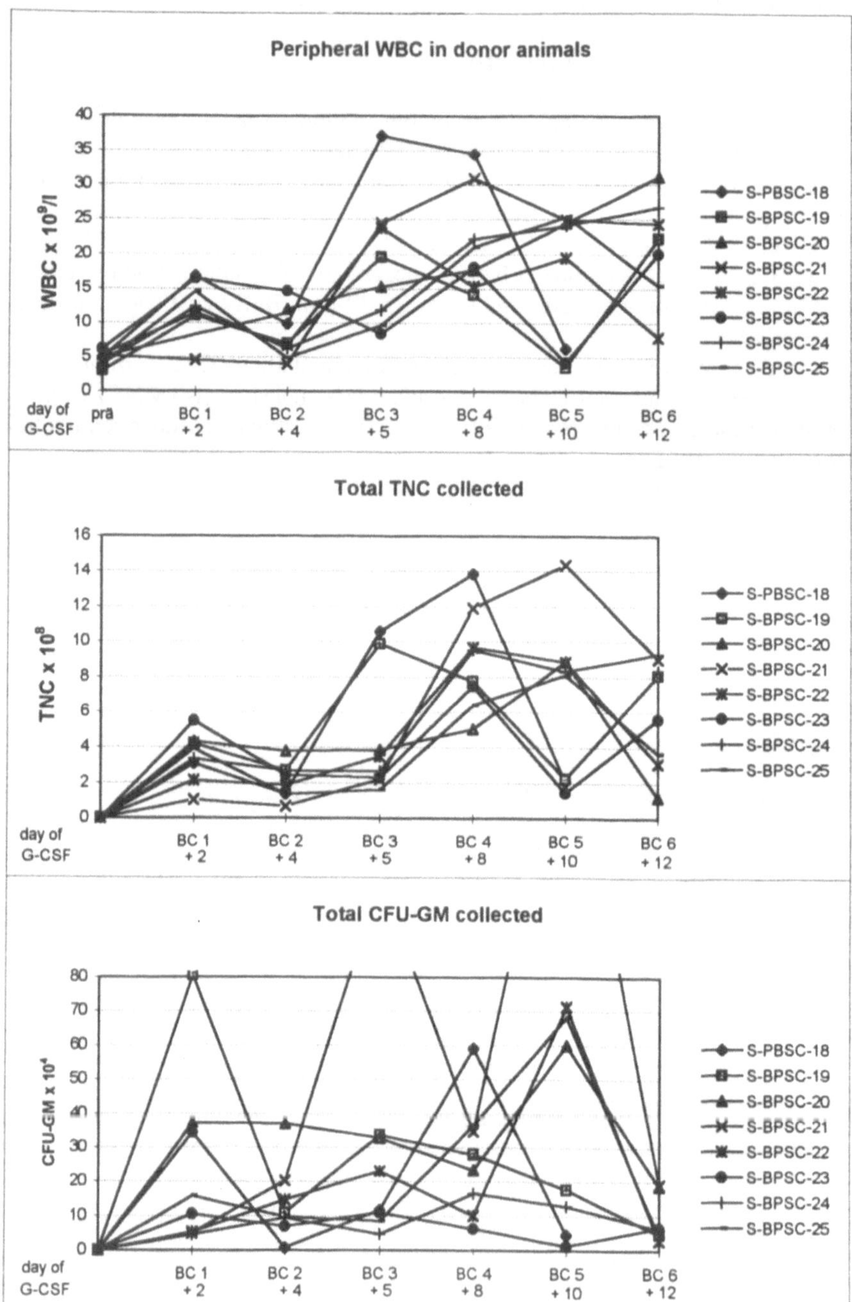

Figure 1. Peripheral blood white blood cell count in donor animals (top), total number of nucleated cells (TNC) (center) and CFU-GM (bottom) collected at days of G-CSF application. Symbols represent individual animals.

Table 2. Outcome of PBPCT

Animal	Survival (days)	Nadir of WBC x 10^9/l	Take (WBC >1x10^9/l)	Outcome
18	12	n.a.	n.a.	Bleeding
19	23	0.3	15	GvHD
20	38	> 1.0	n.a.	GvHD
21	55	0.4	17	GvHD / pneumonia
22	30	0.96	10	GvHD
23	> 120	> 1	n.a.	longterm survivors
24	39	0.82	13	GvHD / pneumonia
25	19	0.52	13	GvHD
median	34	0.82	13	

Nadir WBC: lowest WBC count observed during aplasia
Take:first day of WBC >1 x 10^9/l post transplant

bone marrow transplants. The number of transplanted CFU-GM showed a wider range from 12.3 to 176 x 10^4/kg.

All donor animals showed a rise in total white blood cell count upon stimulation with G-CSF. However, cell counts did fluctuate and were highest between days +4 and +8 of G-CSF stimulation. There was a intra-individual as well as inter-individual difference in PBPCT products harvested from day 0 to day 10. There was no correlation between peripheral white blood cell counts and total number of cells or total number of CFU-GM collected (Figure 1).

Transplant Outcome

Outcome of the individual animals is illustrated in Table 2. One animal died of bleeding on day 12. All others engrafted as evidenced by bone marrow histology, the development of clinical GvHD and the histological examination at autopsy. One animal is a longterm survivor, documentation of long term chimerism is pending. Due to the repetitive infusion of Buffy coat on days 0 through to 10, the median nadir of WBC was only 0.82 x 10^9/l. Engraftment defined as the first day of WBC > 1 x 10^9/l was rapid with a median at day 13 (Table 2).

Three animals died within 25 days, 5 animals beyond 25 days. The median number of CFU-GM/kg transplanted in the former group is 45 x 10^4/kg, 63 x 10^4/kg in the latter ($p < 0.05$, chi square test). The linear regression analysis on the number of transplanted CFU-GM compared with day of survival, (animal 23, long term survivor, excluded), shows the same trend, i.e. that higher numbers of CFU-GM might improve survival. However, in view of the very small number, the correlation is not significant with a coefficient r = 0 .716 and p value of 0.0716 (two-tailed).

DISCUSSION

The present data are compatible with our previous data and the preliminary clinical results (8,11). They confirm that allogeneic G-CSF mobilised PBPC can routinely engraft even across a major histocompatibility barrier. They illustrate the many open questions associated with PBPCT.

Mobilisation of precursor cells with G-CSF can be performend efficiently in all animals in order to obtain sufficient quantities for engraftment. However, despite standardized G-CSF application, mobilisation shows wide inter and intraindividual variations with a maximum of nucleated cells and CFU- GM between day 4 and day 10 of G-CSF mobilisation and no correlation between peripheral white blood cell count and circulating CFU- GM. In addition, mobilisation appears to show a biphasic mode. Unfortunately, the rabbit model is hampered by the fact that no antibody is available to quantitate the CD 34 equivalent in rabbits. The total number of precursor cells transplanted becomes therefore available only after transplant. The optimal mobilisation scheme and optimal day for harvesting still needs to be defined.

Fear of excessive graft-versus-host disease, caused by a high number of T-cells in the transplanted product, has initially prevented the use of PBPC for allogeneic transplants in clinics (7). Our present and previous data do not substantiate these qualms. High numbers of transplanted precursor cells might even have a protective effect by hitherto unknown mechanisms. This is illustrate by the fact that median survival of the present group, 34 days, corresponds to the median survival of a group of animals transplanted between the same two strains with T- cell depleted bone marrow (12).

Preliminary reports from single centers are compatible with our present preclinical findings (4-6). PBPC lead to regular and prompt engraftment in HLA-identical sibling transplants, and acute GvHD appears in a proportion equivalent to HLA-identical bone marrow transplantation. Prospective randomized studies comparing allogeneic BMT and allogeneic PBPCT in HLA-identical sibling transplants are currently being conducted. If they confirm present trends, PBPCT might soon replace clinical bone marrow transplantation.

REFERENCES

1. Armitage, J.O.: Bone marrow transplantation. N. Engl. J. Med. 330: 827, 1994.
2. Eaves, C.J.: Peripheral blood stem cells reach new heights. Blood 82: 1957, 1993.
3. Gratwohl, A., Hermans, J., and Baldomero, H.: Hematopoietic precursor cell transplants in Europe: Activity in 1994. Bone Marrow Transplant 17: 137, 1996.
4. Korbling, M., Przepiorka, D., Huh, Y.O., Engel, H., Van Besien, K., Giralt, S., Andersson, B., Kleine, H.D., Seong, D., Deisseroth, A.B., Andreeff, M., and Champlin, R.: Allogeneic blood stem cell transplantation for refractory leukemia and lymphoma: potential advantage of blood over marrow allografts. Blood 85: 1659, 1995.
5. Schmitz, N., Dreger, P., Suttorp, M., Rohwedder, E.B., Haferlach, T., Löffler, H., Hunter, A., and Russell, N.H.: Primary transplantation of allogeneic peripheral blood progenitor cells mobilized by filgrastim (granulocyte colony-stimulating factor). Blood 85: 1666, 1995.
6. Bensinger, W.I., Weaver, C.H., Appelbaum, F.R., Rowley, S., Demirer, T., Sanders, J., Storb, R., and Buckner, C.D.: Transplantation of allogeneic peripheral blood stem cells moblized by recombinant human granulocyte colony stimulating factor. Blood 85: 1655, 1995.
7. Goldman, J.: Peripheral blood stem cells for allografting. Blood 85: 1413, 1995.
8. Gratwohl, A., Baldomero, H., John, L., Gimmi, C., Pless, M., Tichelli, A., Nissen, C., Filipowicz, A., and Speck, B.: Transplantation of G-CSF mobilized allogeneic peripheral blood stem cells in rabbits. Bone Marrow Transplant 16: 63-68, 1995.
9. Corn, P., Gratwohl, A., and Schmid, E.: A simple technique for testing the in vitro response of rabbit lymphocytes to PHA and allogeneic cells. Experientia 35: 281, 1979.
10. Porter, K.A., and Murray, J.E.: Long-term study of x-irradiated rabbits with bone marrow homotransplants. J. Nat. Canc. Inst. 20: 189, 1958.
11. Gratwohl, A., Tichelli, A., Orth, B., John, L., Levak, A., Bargetzi, M., Gudat, H., Uhr, M., Franscini, L., Jeannet, M., Nissen, C., and Speck, B.: Transplantation von allogenen peripheren hämatopoietischen Vorläuferzellen anstelle von Knochenmark. Schweiz med. Wschr. 126:357, 1996.
12. Gratwohl, A., Forster, I., and Speck, B.: Histoincompatible skin and marrow grafts in rabbits on cyclosporin A. Transplantation 33: 361, 1982.

11

RETICULOCYTE PARAMETERS AS EARLY INDICATORS OF HEMATOPOIETIC RECOVERY AFTER BONE MARROW TRANSPLANTATION

M. Tommasi, G. d'Onofrio, G. Zini, P. Salutari, S. Sica, and G. Leone

Research Center for the Development and Clinical Evaluation of Automated Methods in Hematology
Hematology Service
Università Cattolica del Sacro Cuore-Rome, Italy

ABSTRACT

The evaluation of hematopoietic recovery after myeloablative chemotherapy in bone marrow transplant patients is based on sequential peripheral blood cell counting; particularly the absolute neutrophil count (ANC) >0,5x10^9/l is the classical indicator of BM engraftment. The development of automated flow-cytometric analyzers has made available precise reticulocyte counts (RET) and the proportion of young, highly fluorescent reticulocytes (HFR) as accurate and early indices of haemopoietic recovery. We have studied 12 patients who received autologous PBSCT; peripheral blood mononuclear cells were collected after chemotherapy and rhG-CSF administration by repeated leukapheresis until a minimum of 4x10^8 mononuclear cells. RET and percentage of HFR were obtained with the reticulocyte counter Sysmex-Toa R-1000. In our patients the three most useful indicators of BM recovery were HFR >5%, RET >20x10^9/l and ANC >0,5x10^9/l. Recovery of HFR >5% was obtained after a median time of 9 days, one day before recovery of RET and 4,5 days before recovery of ANC >0,5x10^9/l; the rise of HFR fraction was the earliest sign of engraftment in 91,7% of our patients. In conclusion, these three parameters show an excellent predictive value close to 100% for monitoring of BMT engraftment in PBSCT.

Hematopoietic regeneration after myeloablative radiochemotherapy for bone marrow transplantation (BMT) is monitored by sequential peripheral blood cell counting. Until recently the earliest practical measure of the onset of recovery was absolute neutrophil count (ANC)[1] Platelet count (PLT) is unreliable because of the many platelet transfusions given to most patients. Recent reports suggest that reticulocyte count (RET) performed by automated flow-cytometric methods is an early and accurate index of engraftment, especially if new indices which quantify the presence of young, highly fluorescent reticulocytes (HFR) are used.[4,5,8,10,14] A great heterogeneity of threshold values exists in the literature as far as RET

Molecular Biology of Hematopoiesis 5, edited by Abraham et al.
Plenum Press, New York, 1996

and reticulocyte subfractions are concerned, which makes the comparison of results obtained by different authors extremely difficult. Moreover, the transplant of stem cells collected by leukapheresis from peripheral blood (PBSCT) is characterized by very fast ANC and PLT recovery[9,13,16], but RET and HFR have never been studied in PBSCT patients.

The main objectives of our study were: 1) to define and compare the duration and characteristics of the hematopoietic suppression in PBSCT, in terms of peripheral blood cytopenia and supportive care requirements, in comparison with allo- and auto- BMT; 2) to determine the most suitable peripheral blood variables for the study of hematopoietic reconstitution; 3) to establish the clinical value of the selected variables in the monitoring of PBSCT patients.

MATERIALS AND METHODS

Patient Characteristics

We have studied 12 patients who received autologous PBSCT. Peripheral blood mononuclear cells were collected after MICMA chemotherapy and rhG-CSF administration by repeated leukapheresis performed on consecutive days until a minimum of 4 x 10^8/kg mononuclear cells were obtained. All patients received the same conditioning regimen with busulfan and cyclophosphamide (BuCy2). Results of the hematologic follow-up were compared with those obtained in two groups of patients undergoing auto- and allo-BMT; these patients had been monitored during a cooperative study for the European Reticulocyte Study Group.[7] Clinical characteristics of all patients are reported in Table 1.

Analytical Methods

Hematological tests were carried out on peripheral blood samples anticoagulated with K3 or K2-EDTA within 4 hours from phlebotomy. Total WBC and ANC were measured with

Table 1. Clinical characteristics of patients and type of bone marrow transplantation

	PBSCT	Allo-BMT	Auto-BMT
Number	12	19	12
Sex (M/F)	6/6	12/7	7/5
Age median (range)	54.5	31	47
	(23-62)	(18-58)	(18-52)
Days of follow-up (median and range)	30.5	37	38
	(20-54)	(35-85)	(23-47)
Diagnosis			
acute leukaemia	0	8	7
chronic granulocytic leukaemia	0	8	0
non Hodgkin's lymphomas	8	0	5
multiple myeloma	0	2	0
chronic lymphocytic leukaemia	0	1	0
solid tumors	2	0	0
Conditioning regimen			
irradiation + chemotherapy	0	14	8
chemotherapy alone	12	5	4
Nucleated cells x10^8/l (median and range)	9,2	3.5	2,3
	(5,2-16,2)	(1,1-5,2)	0,8-6,7)
Patients treated with rhG-CSF	2	2	6

the electronic blood cell counter Bayer H*1 with microscope check. RET and percentage of reticulocyte subfractions with different fluorescence intensity were obtained with the reticulocyte counter Sysmex-Toa R-1000 (TOA Medical Electronics, Hamburg, Germany), using an argon laser as a light source. The R-1000 determines cell sizes by forward scatter and RNA content by measurement of fluorescence after automated staining with the fluorochrome auramine-O[3,15] Platelets and leukocytes are excluded from the measurements by means of electronic thresholds; reticulocytes are separated from mature red blood cells by their fluorescence and subdivided into three subpopulations with low fluorescence (LFR), middle fluorescence (MFR) and high fluorescence (HFR). The degree of fluorescence is a function of reticulocyte age. The percentage of HFR, in particular, is normally lower than 2% and its increase reflects the premature delivery in circulating blood of macroreticulocytes with high RNA content.[11] The analytical performances of R-1000 were checked daily by analysis of control material Retcheck (TOA Medical Electronics, Hamburg, Germany).

Analysis of Data

Study of Hematopoietic Suppression. We have calculated the median duration in days of the periods in which WBC and ANC were less than 0.1 $x10^9/l$, untranfused PLT less than 30 $x10^9/l$, RET less than 10 $x10^9/l$, HFR equal or less than 0.1%. The Student's t test for unpaired data was used for the evaluation of differences.

Definition of the Indicators of Engraftment. For each variable one or more threshold values were chosen: WBC above 1.0 $x10^9/l$, ANC above 0.5 $x10^9/l$, PLT above 30, 40 or 50 $x10^9/l$ without transfusion, RET above 15, 20 or 30 $x10^9/l$, HFR above 1%, 2%, 3%, 4% or 5% and the sum of HFR and MFR above 5% or 10%. We used the two-tailed Student's t test for paired data to demonstrate the presence or absence of statistically significant differences. We also calculated the number of false positives (FP=results above the threshold value during the period of bone marrow hypoplasia and not followed by a steady increase), true negatives (TN=results below the threshold value obtained during the period of hypoplasia) and the predictive value [PV=TN/(FP+TN)] for each variable.

Efficiency of the Selected Indicators of Recovery. In the PBSCT patients, we calculated median day of recovery, PV and the percentage of cases in which each variable was the first indicator of bone marrow recovery. Statistical differences were analysed using the two-tailed Student's t test for paired data.

RESULTS

Haemopoietic Suppression in PBSCT

In patients undergoing PBSCT WBC and ANC remained below 0.1 $x10^9/l$ for median times of 1 and 6 days respectively, a much shorter period than in both allo-and auto-BMT. In PBSCT PLT below 30 $x10^9/l$ were observed for a median time of 2.5 days, compared with 12 days in the allo-BMT group; in the auto-BMT group 8 of 12 patients had not recovered a stable untransfused PLT 30 days after bone marrow infusion. The median number of days in which platelet or red cell transfusions were given was also much smaller in PBSCT than in allo- and auto-BMT (Figure 1, Table 2).

The fall of RET was earlier than granulocyte and thrombocyte nadir (median day +4 in PBSCT). The median duration of RET depression was 5 days, compared with 14 days in

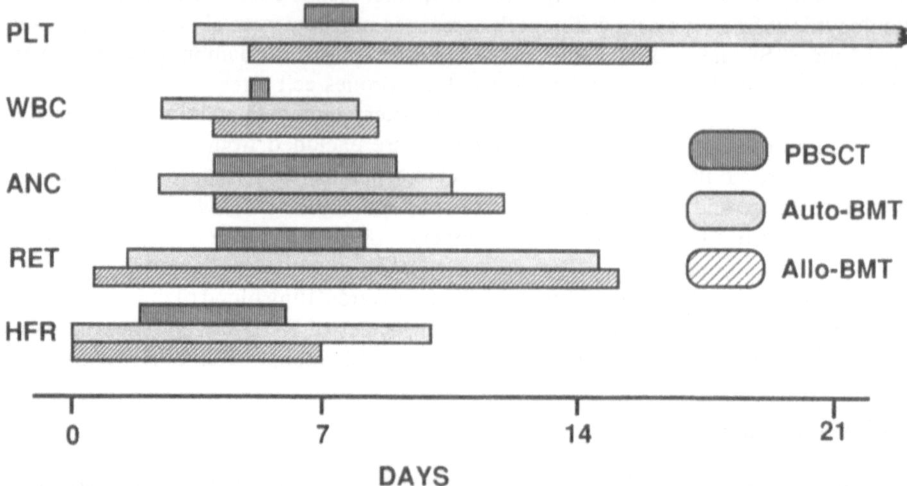

Figure 1. Graphic representation of the duration of bone marrow depression following conditioning regimens in the different types of bone marrow transplantation, as indicated by an untransfused platelet count below 30 x10^9/l (PLT), a white blood cell count equal or lower than 0.1 x10^9/l (WBC), an absolute neutrophil count equal or lower than 0.1 x10^9/l (ANC), a reticulocyte count below 10 x10^9/l (RET) and a proportion of highly fluorescent reticulocytes equal or lower than 0.1% (HFR). Day 0 is the day of bone marrow or peripheral blood stem cell infusion.

allo-BMT and auto-BMT groups. Disappearance of the HFR fraction was even earlier than the fall of RET: the nadir of HFR was reached on the median day +2 and was maintained for a median time of 5 days. The MFR fraction fell to zero a little later and for shorter periods than HFR. Statistical analysis showed that all differences between PBSCT group and the two other groups were highly significant, while differences between allo- and auto-BMT groups were not.

Table 1. Duration of the bone marrow depression induced by conditioning regimens in the different types of bone marrow transplantation

	Median duration (range)			Statistican significance (p)		
	PBSCT	Allo-BMT	Auto-BMT	Allo-BMT vs PBSCT	Auto-BMT vs PBSC	Allo vs auto-BMT
RET < 10 x10^9/l	5 (1-13)	14 (3-22)	14 (9-23)	<0.0005	<0.0001	NS
HFR < 0.1%	5 (3-12)	8 (2-21)	11 (7-17)	<0.05	<0.0001	NS
MFR < 0.1%	1.5 (1-6)	6 (2-13)	4 (1-11)	<0.005	NS	NS
WBC < 1.0 x10^9/l	1 (1-8)	5 (1-12)	7 (1-15)	NS	<0.05	NS
NEUT < 0.5 10^9/l	6 (1-8)	9 (3-17)	9 (3-20)	<0.005	<0.05	NS
PLT < 30 x10^9/l	2,5 (1-24)	12 (1-25)	>30 (15-NA)	<0.01	<0.0001	<0.0001
Days of PLT transfusion	1 (0-13)	3.5 (0-12)	8 (0-25)	NS	<0.005	<0.01
Days of RBC transfusion	0 (0-5)	7 (0-13)	7,5 (0-19)	<0.0001	<0.0005	NS

NS: not significant (p>0.05).

Table 3. Median day of recovery and predictive value for the variables selected as potential indicators of engraftment

	Median day of recovery (range)			Predictive value of selected recovery criteria		
	PBSC	Allo	Auto	PBSC	Allo	Auto
HFR > 1.0%	8 (5-11)	11 (6-20)	14 (9-22)	83.6	89.5	85.9
HFR > 3.0%	8.5 (7-11)	13 (6-20)	16.5 (9-28)	91.0	98.4	94.8
HFR > 5.0%	9 (7-14)	14 (8-33)	16.5 (9-30)	97.0	100.0	100.0
RET > 15 x10^9/l	10 (8-14)	17.5 (12-23)	20 (15-29)	95.5	93.5	99.3
RET > 20 x10^9/l	10 (9-14)	19 (13-23)	24 (20-34)	98.5	98.4	100.0
RET > 30 x10^9/l	11 (9-14)	20.5 (14-26)	24 (20-42)	100.0	100.0	100.0
WBC > 1.0 x10^9/l	12 (8-18)	18 (14-34)	20 (9-37)	97.0	100.0	100.0
NEUT > 0.5 x10^9/l	13.5 (8-18)	20 (14-34)	24 (9-45)	97.0	100.0	100.0
PLT > 30 x10^9/l	9 (7-12)	21 (10-30)	24* (12-31)	91.0	60.0	NE**

*only in 5/12 patients PLT count showed a stable recovery before day +30.

Selection of the Variables to Be Used as Predictors of Bone Marrow Recovery

Among the potential indicators of recovery, we chose HFR>5%, because of its excellent predictive value; RET >20 x10^9/l, because of its precocity and excellent predictive value; and ANC >0.5 x10^9/l , because of its clinical relevance and wide diffusion. PLT were eliminated because of the very low predictive value, due to the interference of the large number of platelet transfusions (Table 3).

Monitoring of Bone Marrow Engraftment

Table 4 includes detailed clinical data for all patients, with the days of recovery for the three indicators of engraftment (Figure 2). Specificity was excellent for all the three variables, with predictive values higher or equal to 97%. The earliest indicator of hemopoietic recovery is the proportion of HFR. Recovery of HFR above 5% was in fact obtained after a median time of 9 days (Figure 3), that is 1 day before recovery of RET (Figure 4) and 4.5 days before recovery of ANC above 0.5 x10^9/l. The rise of the HFR fraction was the earliest sign of engraftment in 11 of the 12 patients (91.7%).

DISCUSSION

Our study confirms that the duration of suppression of hematopoiesis that follows myeloablative conditioning regimens for BMT is shorter after PBSCT than after allo and auto-BMT. These results are in good agreement with recent results which show that a very fast engraftment follows PBSCT, with more rapid recovery of all hemopoietic lineages than in auto- and in allo-BMT.[9,13,16]

Table 4. Peripheral blood stem cell transplantation: patients characteristics and days of haematological recovery

Case no.	Diagnosis	State of the disease	Conditioning regimen	G-CSF	Nucleated cells ($x10^8/l$)	Day of engraftment according to HFR > 5%	RET > 20 $x10^9/l$	ANC > 0.5 $x10^9/l$
1	NHL, high grade	PR	CHT	no	10.0	10	10	12
2	NHL, high grade	PR	CHT	no	7.0	9	11	13
3	Multiple myeloma	PR	CHT	no	8.6	7	10	13
4	NHL, immunoblastic	CR	CHT	no	6.7	9	11	14
5	Ewing's sarcoma	PR	CHT	no	5.2	8	10	11
6	Ovarian cancer	St. III	CHT	yes	10.0	7	9	8
7	NHL, high grade	CR	CHT	no	8.7	8	10	14
8	NHL, low grade	PR	CHT	yes	9.7	12	9	12
9	NHL, high grade	Active	CHT	no	5.9	11	11	16
10	NHL, low grade	PR	CHT	no	10.5	8	9	14
11	NHL, high grade	Active	CHT	no	16.2	11	13	17
12	Multiple myeloma	PR	CHT	no	11.6	14	14	18
					Median	9	10	13.5
					Range	7-14	9-14	8-18
					Predictive value	97%	98.5%	97%

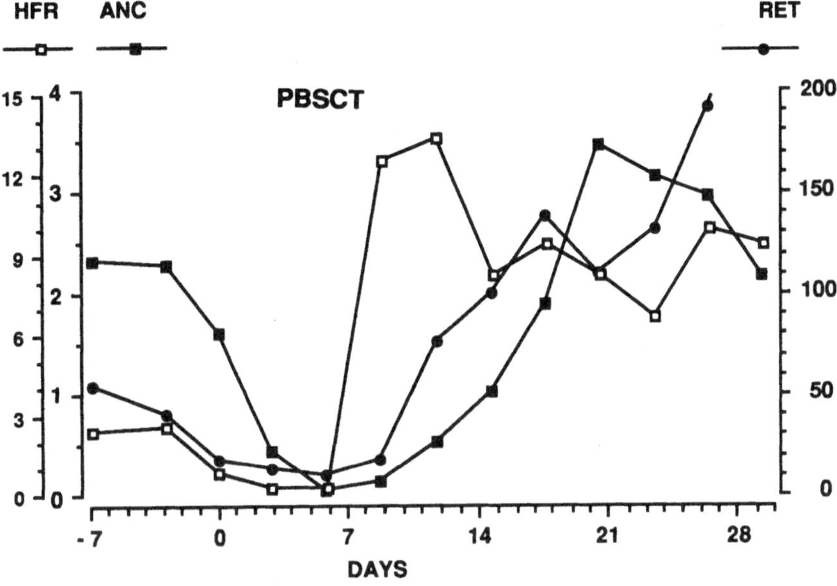

Figure 2. Median values of absolute neutrophil count (ANC), reticulocyte count (RET) and HFR percentage during the follow-up of PBSCT group. HFR percentage is the first measurement to show a prompt and sustained rise after the aplastic phase.

The suppression of erythropoietic activity was demonstrated in our patients by the fall of RET below 10 $x10^9/l$. A zero value of reticulocyte count was never seen, even if no reticulocytes at all could be seen on the peripheral blood smears during the period of aplasia. The impossibility of obtaining zero reticulocytes with the R-1000 has already been described and probably results from autofluorescence of mature erythrocytes.[15] Reticulocytopenia was

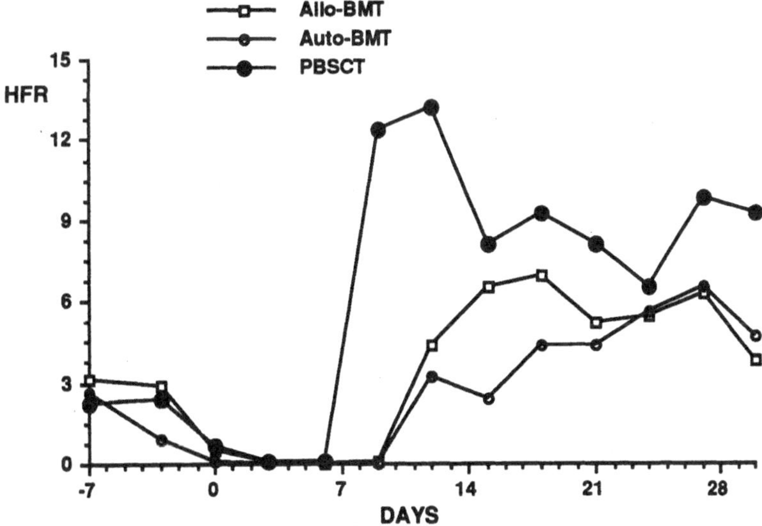

Figure 3. Median values of HFR percentage in PBSCT group compared with two other types of BMT.

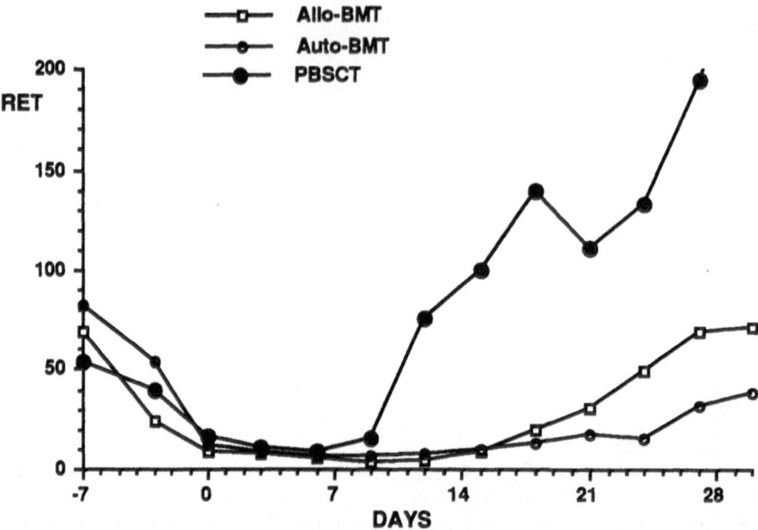

Figure 4. Median values of reticulocyte count (RET) in PBSCT group compared with two other types of BMT.

early and consistent after PBSCT, although its duration was much shorter than in the allo-BMT and auto-BMT groups (5 days instead of two weeks). Reticulocytopenia was preceded by total disappearance of the HFR fraction: this was the earliest sign of the block of haemopoietic function after conditioning treatment.

The second question adressed in our study concerned the selection of the variables to be used as predictors of engraftment. We found that the most useful and specific variables for clinical monitoring of these patients are HFR percentage with a cut-off point of 5%, RET with a cut-off point of 20 $x10^9/l$ and ANC at a cut-off point of 0.5 $x10^9/l$. These parameters showed in fact a good statistical difference from each other, together with an early median time of recovery and an excellent predictive value close to 100%.

The third objective of our study was to define which hematological parameter is the most sensitive and specific measure of haematopoietic reconstitution. Our data show that recovery of HFR is faster than recovery of both RET and ANC in 91.7% of patients and that the median day of recovery is significantly earlier for HFR than for both RET and ANC. Its specificity was also excellent at the selected cut-off point, with less than 1% false positive results and predictive value identical to that of ANC or RET. These results are in agreement with those reported by of many authors in patients undergoing allo- and auto-BMT; they have demonstrated that the rise of HFR precedes neutrophil recovery as well as the onset of a significant rise in reticulocyte count.[4,8,10,14]

REFERENCES

1. Arnold R., Schmeiser T., Heith W. Frickhofen N., Pabst G. (1986) Hematopoietic reconstitution after bone marrow transplantation. Exp. Hematol. 14:271-277
2. Beguin Y., Oris R., Fillet G. (1993) Dynamics of erythropoietic recovery following bone marrow transplantation: role of marrow proliferative capacity and erythropoietin production in autologous versus allogeneic transplants. Bone Marrow Transplant 11, 285-292
3. Bowen D., Bentley N., Hoy T., Cavill I. (1991) Comparison of a modified thiazole orange technique with a fully automated analyser for reticulocyte counting. J. Clin. Pathol. 44, 130-133

4. Davies S.V., Cavill I., Bentley N., Fegan C.D., Poynton C.H., Whittaker J.A. (1992) Evaluation of erythropoiesis after bone marrow transplantation: quantitative reticulocyte counting. Br. J. Haematol. 81, 12-17.
5. Davis B.H., Bigelow N., Ball E.D., Mills L., Gibbons G.C. III (1989) Utility of flow cytometric reticulocyte quantification as a predictor of engraftment in autologous bone marrow transplantation. Am. J. Hematol. 32, 81-87.
6. Davis B.H., DiCorato M., Bigelow N.C., Langweiler M.H. (1993) Proposal for standardization of flow cytometric reticulocyte maturity index (RMI) measurement. Cytometry 14, 318-36
7. d'Onofrio G., Tichelli A., Foures C., Theodorsen L., Cavill I., Corberand J.X., Rowan M. and the European Reticulocyte Study Group. Indicators of haematopoietic recovery after bone marrow transplantation: the role of reticulocyte parameters. Clin Lab Haematol 1995 (in press)
8. Greinix H.T., Linkesch W., Keil F. et al. (1994) Early detection of hematopoietic engraftment after bone marrow and peripheral blood stem cell transplantation by highly fluorescent reticulocyte counts. Bone Marrow Transplant 14, 307-313
9. Henon P.R., Liang H., Beck-Wirth G. et al. (1992). Comparison of hematopoietic and immune recovery after autologous bone marrow or blood stem cell transplants. Bone Marrow Transplant. 9, 285-291
10. Kanold J., Bezou M.J., Coulet M. et al. (1993) Evaluation of erythropoietic reconstitution after BMT by highly fluorescent reticulocyte counts compares favorably with traditional peripheral blood cell counting. Bone Marrow Transplant. 11, 313-318
11. Kuse R. (1993) The appearance of reticulocytes with medium or high RNA content is a sensitive indicator of beginning granulocyte recovery after aplasiogenic cytostatic drug therapy in patients with AML. Ann. Hematol. 66, 213-214
12. Lazarus H.M., Chahine A., Lacerna K. et al. (1992) Kinetics of erythrogenesis after bone marrow transplantation. Am. J. Clin. Pathol. 97, 574-583
13. Pierelli L., Iacone A., Quaglietta A.M. et al. (1994) Haemopoietic reconstitution after autologous blood stem cell transplantation in patients with malignancies: a multicentre retrospective study. Br. J. Haematol. 86, 70-75
14. Spanish Multicentric Study Group for Hematopoietic Recovery (1994). Fow cytometric reticulocyte quantification in the evaluation of hematologic recovery. Eur. J. Haematol. 53, 293-297
15. Tichelli A., Gratwohl A., Driessens A. et al. (1990) Evaluation of the Sysmex R-1000. An automated reticulocyte analyzer. Am. J. Clin. Pathol. 93, 70-78
16. To L.B., Roberts M.M., Haylock D.N. et al. (1992) Comparison of haematological recovery times and supportive care requirements for autologous recovery phase peripheral blood stem cell transplants, autologous bone marrow transplants and allogeneic bone marrow transplant. Bone Marrow Transplant. 9, 277-284

12

PRIMORDIAL GERM CELL-DERIVED HEMOPOIESIS

A New Concept for the Initiation of Hemopoiesis

Ivan N. Rich* and Frank Zimmermann

Department of Transfusion Medicine
University of Ulm, Germany

SUMMARY

A new concept for the initiation of hemopoiesis is presented in which the primordial germ cell (PGC) is viewed as the hemopoietic-initiating cell. Evidence from developmental biology and hematology is presented indicating an intimate link between the PGC population and hemopoiesis. Results are described showing that PGCs can be cultured in vitro and induced into cells of the hemopoietic system. These including erythropoietic cells as well as multipotential stem cells and the most primitive hemopoietic stem cell detected in vitro, the cobblestone area-forming cell. Finally, a pathway for the initiation of hemopoiesis from the primordial germ cell is presented.

In 1971, Metcalf and Moore[1] wrote "It is unfortunate that the role of the mammalian yolk sac has been underestimated, relegated to the status of an atavistic organ of transitory importance in mammalian development and considered a classic example of ontogeny recapitulation phylogeny. In two particular aspects, in the development of the haemopoietic system and in the production of primordial germ cells, such pessimism is hardly warranted, and there are excellent grounds for believing that the yolk sac has a unique role to play in the development of these two systems. In both instances this may be due to the early development of the yolk sac in an extraembryonic situation removed from the pressures of early embryonic differentiation. It would be singularly disastrous for the development of the embryo and the survival of the species if the first hemopoietic stem cells and primordial germ cells were to rapidly and completely succumb to differentiation pressures, such as exist within the early embryonic environment". These authors go on to state that "... haemopoietic potential is determined at the very early pre-primitive streak blastoderm stage, and is relatively uninfluenced by the course of the subsequent movement of the tissue to its

*Correspondence: Ivan N. Rich, Ph.D., Professor and Director, Stem Cell Research, Center for Cancer Treatment and Research, Department of Medicine, Richland Memorial Hospital, 7 Richland Medical Park, Columbia, South Carolina, 29212. Tel: (803) 434-4734; Fax: (803) 434-4950.

Molecular Biology of Hematopoiesis 5, edited by Abraham et al.
Plenum Press, New York, 1996

Table 1. Summary of information leading to a link between primordial germ cells and hemopoiesis

Day of Gestation	
3.5	Small population of cells is set aside in the epiblast which possess the 4C9 antigen[2]. This antigen is present on primordial germ cells.
6	4C9-labelled cells migrate from the ectoderm into the primitive streak region[2]
	Single cell labelling, using lysinated rhodamine dextran, used to follow the fate of PGCs clones from the ectoderm. These clones produce the allantois, posterior streak, yolk sac mesoderm, amnion ectoderm and mesoderm, proximal mesoderm, blood islands and embryonic mesoderm[21].
6.5	Cells from the embryo cylinder express the Brachyury (T-gene), a marker for mesoderm, the activin receptor gene, the c-myb oncogene, and genes for erythropoietin and its receptor, GATA-1, embryonic globin, and spectrin[22].
	Multipotential hemopoietic stem cells (BFUmix) and erythropoietic progenitor cells (BFU-E, CFU-E) present[18].
Late day 6	PGCs detected as an alkaline phosphatase-(AP) positive population seen at the base of the allantois in the extraembryonic mesoderm[3]. At the same location, cells also found to be positive for the 4C9 antigen[2].
7.5	Similar numbers of in vitro multipotential stem cells (BFUmix) found in the yolk sac and embryo proper[18].
	PGC colonies grown in vitro on 2A embryonic fibroblast feeder layers are both AP^{+} and stage specific embryonic antigen (SSEA-1) [9,18,23] positive, and produce maximum colony number in the presence of leukemia inhibitory factor (LIF), stem cell factor (SCF) and Interleukin-3 (IL-3)[18].
8	4C9+ cells migrate from the extraembryonic mesoderm, back into the embryo[2].
	AP+ cells migrate from the extraembryonic mesoderm back into the embryo[3].
8.5	AP+ and 4C9+ cells migrate into the hind gut wall.
	PGCs can be grown in culture on embryonic fibroblast feeder layers in the presence of LIF and/or SCF[9,23].
	In vitro culture of PGCs cells on mitomycin-C - treated embryonic fibroblast feeder layers with LIF, SCF and/or basic fibroblast growth factor (bFGF), produce AP+ and SSEA-1+ colonies[9,18,23] which can develop into different cells types and also give rise to embryonic stem cell (ES) lines[9].
	Individual PGC colonies produced on feeder layers give rise to secondary and tertiary AP+ colonies in methyl cellulose. Primary PGC colonies differentiate into erythropoietic colonies in the presence of erythropoietin (EPO) and other cytokines[18].
	Similar numbers of BFUmix colonies in visceral yolk sac and embryo proper, but absolute number increase 17.5 fold above that seen at 7.5 days[18].
	PGCs grown in the absence of feeder layers can be expanded and can produce cobblestone area forming cells (CAFC), BFUmix and

Table 1 *(continued)*

	erythropoietic progenitor cells[18].
9	AP^+ cells migrate along the mesentery and pass through the aorta/mesonephros/gonad (AMG) region or para-aortic splanchnopleura. SCF required for migration.
	No in vivo hemopoietic stem cells detectable in the yolk sac[24].
	In vivo hemopoietic stem cells (CFU-Sday8 and CFU-Sday12) and B lymphocytes detectable in AMG/para-aortic splanchnopleura region[16,17]
10.5	AP+ and 4C9+ PGCs enter the genital ridge and begin to produce germ cells. Their numbers have increased to approx. 10,000 from about 200-300 on day 8-8.5[25].

Additional Information leading to a Link between Primordial Germ Cells and Hemopoiesis

In *White Spotting* (W) mutant mice, hemopoietic stem cells lack c-kit (stem cell factor receptor) and are defective. In *Steel* (Sl) mutant mice, the hemopoietic microenvironment does not produce the *c-kit* ligand (SCF), so that stem cells cannot grow.	In *White Spotting* (W) mutant mice, primordial germ cells lack c-kit (stem cell factor receptor) and are defective. In *Steel* (Sl) mutant mice, the *c-kit* ligand (SCF), is not produced along the primordial germ cell migration pathway, so that few, if any, cells reach the genital ridges.
Both strains demonstrate a severe macrocytic anemia. The recessive homozygous trait is lethal.	Both strains can be infertile.

definitive site. It is generally accepted that the mesodermal cells that ultimately form blood islands originate from epiblast, invaginate through the primitive streak and produce, by lateral migration, the middle germ layer."

Since that time, new techniques, both at the cellular and molecular levels, have shown that much of what Metcalf and Moore postulated have been shown to be true. The mere fact that these authors considered both the hemopoietic system and the production of primordial germ cells in one breath, so to speak, implies perhaps unstated ideas. Similarly, that the two systems should not succumb completely to differentiation pressures which exist in the early embryonic environment indicates that these two systems possess something in common. Indeed, the full potentiality of both systems is not realized until much later in development. This does not mean that the proliferative and differentiation potential inherent in these two systems is not present. In many cases, these potentials are either not expressed due, for example, to the surrounding microenvironment, or they are expressed in only a very small number of cells.

Although, as stated by Metcalf and Moore, that the blood islands originate from the epiblast from cells invaginating through the primitive streak region to produce a third germ layer, the mesoderm, the cells responsible for this phenomenon and their potentialities were unknown. It is now known that a small population of cells in the epiblast carry and antigen, 4C9, to a carbohydrate moiety which appears to be important in cell adhesion[2]. The presence of this antigen later appears on cells which are alkaline phosphatase positive (AP) and present at the base of the allantois in the extraembryonic mesoderm.[3] This small population of AP^+ cells are the primordial germ cells (PGC) which will eventually give rise to the germ cells in the gonads.

That the production of PGCs and the origin of the hemopoietic system are connected is summarized in Table 1, which also provides information on some of the in vivo and in vitro characteristics of PGCs. There were three significant factors which led us to consider a more intimate connection between PGCs and the hemopoietic system, and which should be emphasized at this point.

The first is that many embryonic stem cells (ES) lines can usually differentiate spontaneously or in the presence of hemopoietic growth factors into hemopoietic cells.[4-7] Embryonic stem cells are derived from the inner cell mass of the blastocyst[8] where 4C9-positive cells have been observed.[2] Furthermore, PGCs have been shown to give rise to ES cell lines.[9] In fact, PGCs have also been shown to give rise to several different cell types.[9] This not only implies that the PGC is probably more primitive than the ES cell, but that a pathway may exist from the PGC, through an ES-like cell, to the hemopoietic stem cell. It is interesting to note, however, that to date, repopulation of the hemopoietic system by ES cells has not been demonstrated. This correlates with the absence in day 1 differentiating ES cells of both the mouse equivalent of the CD34[10] and flk2/flt3[11] genes (results to be published elsewhere). Whether these genes are present only for a very short period after removal of the ES cells from leukemia inhibitory factor (LIF) medium is not known.

The second factor is that the mouse mutants *White Spotting* (W) and *Steel* (Sl) affect both the hemopoietic system and the production and migration of PGCs.[12-14] The common denominator here is *the c-kit/c-kit ligand* (stem cell factor and its receptor) system.[15] Both mouse mutants display a severe macrocytic anemia. However, mutations in the W and Sl alleles affect hemopoiesis at the stem cell level.[13] For PGCs, defective *c-kit* (SCF receptor) or *c-kit ligand* production means that the PGCs cannot migrate to the genital ridges, thereby resulting in infertility.

The third factor was the independent reports by two groups showing that hemopoietic stem cells could be transiently detected at about day 9 in the area which comprises the aorta, mesonephric kidney and genital ridges (the so-called AMG region).[16,17] This region is similar in avian embryos and is called the para-aortic splanchnopleura.[16] The transient presence of hemopoietic stem cells in this region demonstrates that the fetal liver is not the first site of definitive hemopoiesis. However, even more important, is the fact that at the same time hemopoietic stem cells are present in this region, PGCs are also transversing this area on their way to the genital ridges. It is our view that this coincidence is not a chance act.

Taking these factors into consideration, we asked whether primordial germ cells could be responsible for the initiation of hemopoiesis in the embryo. More specifically, whether primordial germ cells are responsible for both embryonic and definitive hemopoiesis.

In a recent report, we have demonstrated that PGCs can, in fact, produce not only hemopoietic stem cells, but can be induced into the erythropoietic lineage when erythropoietin (EPO) is present.[18] Furthermore, PGCs can be expanded using sequential stationary cultures over a period of more than three weeks. At this time, the AP^+/SSEA-1^+ (stage-specific embryonic antigen-1) PGC population has increase more than 13 fold. However induction into the hemopoietic system by culturing PGCs on irradiated adult bone marrow stromal layers, produced so-called cobblestone area-forming cells (CAFC).[19,20] These are considered to constitute the most primitive in vitro hemopoietic stem cell population capable of long-term repopulation of the hemopoietic system.[19,20] In addition, less primitive multipotential hemopoietic stem cells were also produced.[18]

Molecular characterization using reverse transcription and the polymerase chain reaction (RT/PCR) of individual PGC colonies grown in the absence of mitomycin-C embryonic fibroblast feeder cell layers has shown that the cells possess c-kit and the receptors for LIF, basic fibroblast growth factor (bFGF) and Interleukin-3 (IL-3). In fact, these are the growth factors that are added to primary PGC cultures allowing them to produce colonies in vitro.[6] Although many other genes are surely expressed in PGCs, one gene, in particular, is not expressed. This is the homeobox gene HOX B6. Although this transcription factor gene is expressed in several organs and tissues during development and in the adult, recent work in this laboratory has shown that the HOX B6 gene appears to be a specific marker for commitment of hemopoietic stem cells into the erythropoietic lineage, since expression increases from the early (BFU-E) to the late erythropoietic progenitor cell (CFU-E).[21] HOX B6 is not expressed in hemopoietic stem cells, undifferentiated ES cells or PGCs.[21] When undifferentiated ES cells and PGCs are induced into the erythropoietic lineage, HOX B6 is expressed (submitted for publication).

Cellular characterization of the PGC population has been more accessible by omitting the embryonic fibroblast feeder layer. Although the growth factors LIF, SCF, IL-3 and/or bFGF are still added, the PGCs present in the allantois/primitive streak (APS) region can be grown on collagen/fibronectin-coated hydrophilic Teflon foils (Petriperm dishes) under low oxygen tension. Under these conditions, the PGC population from the APS region of the embryo can be expanded.[6] Recent studies have shown that this can be taken a few steps further by using Matrigel instead of collagen and fibronectin as the extracellular matrix and by culturing the cells in perfusion chambers (POC Chambers) (results to be published elsewhere). It is interesting that when expanded PGCs cultured under these conditions are compared with undifferentiated ES cells using flow cytometry, similar profiles are observed.

All this information have led us to postulate a new concept for the initiation of hemopoiesis. This concept is shown in Fig. 1. Primordial germ cells are present in the embryonic epiblast and delineate from the ectoderm into the primitive streak area. As shown by Lawson and Hage,[22] PGC allocation occurs producing clones of cells which give rise to the allantois, posterior streak and proximal mesoderm which are later incorporated into the extraembryonic mesoderm, the yolk sac mesoderm and the blood islands. In addition, embryonic mesoderm is also produced. Thus, the PGCs are the precursors of both mesoderm and primitive (embryonic) hemopoiesis in the visceral yolk sac. We consider this arm of the PGC-hemopoietic pathway to be transient and determined so that PGCs in this region are not thought to return into the embryo. The PGCs present at the base of the allantois in the extraembryonic mesoderm probably contribute to those that return back into the embryo to begin their migration to the genital ridges. During this migration, the cells enter and pass through the AMG/para-aortic splanchnopleura region. It is here that we consider the second arm of the PGC-hemopoietic pathway to occur, in which probably the majority of the PGCs continue to the genital ridges to give rise to the germ cells of the gonads. Some of the remaining cells are then responsible for the transient and first definitive hemopoiesis reported in this area. A proportion of undetermined PGCs may then continue their migration in other directions, but in particular, to the hepatic anlage to seed and begin the hepatic stage of hemopoiesis in the fetus. Indeed, SSEA-1^+ cells can be detected in the fetal liver (unpublished results). Whether a small PGC population is then responsible for the initiation of hemopoiesis in the fetal spleen and finally the bone marrow is unknown. If this is the case, it would also raise the possibility of a finite number of stem cells rather than a self-renewing population. Nevertheless, in this concept, both primitive and definitive hemopoiesis have their origin in the embryonic ectoderm PGC-derived population.

Finally, it is obvious that the PGC population represents one of the most primitive stem cell populations known. As such, it is theoretically the ideal population with which to perform transplantations, since it should have the most long-term repopulating potential of all stem cells populations when induced into the hemopoietic system.

ACKNOWLEDGMENTS

This work was partially supported by the DFG (SFB 322/Project A9) and the German Red Cross. The authors wish to thank Mrs. Irmgard Brackmann for her excellent technical assistance.

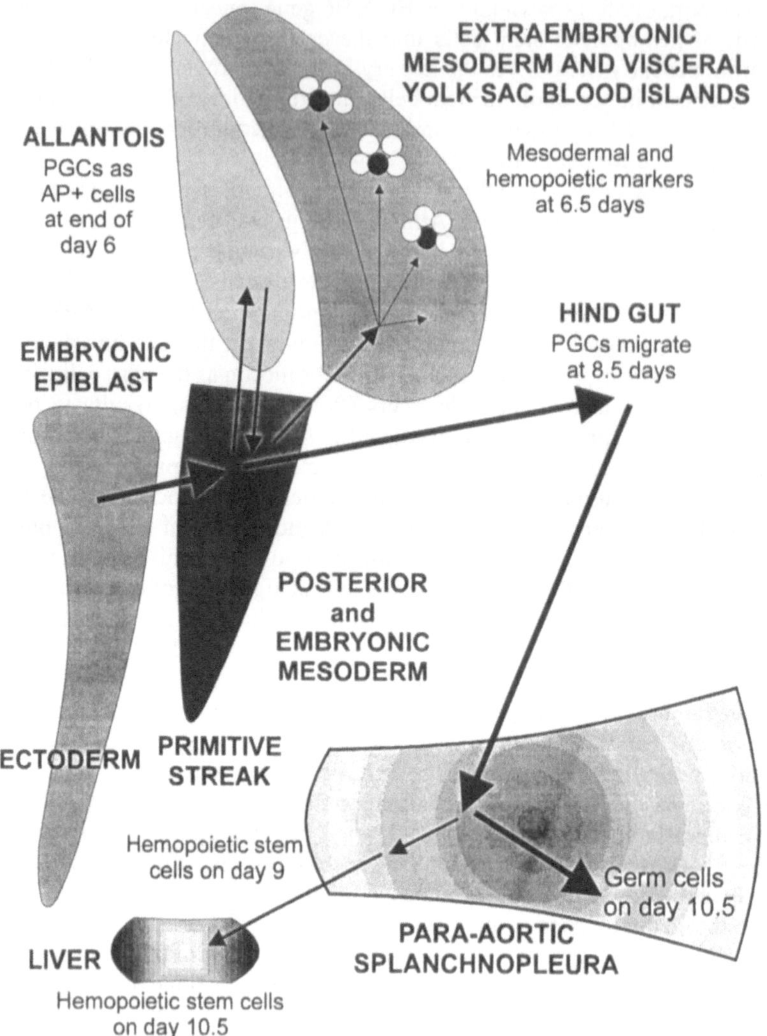

Figure 1. The primordial germ cell pathway and the initiation of hemopoiesis in the mouse embryo.

REFERENCES

1. Metcalf D, Moore M: Haemopoietic Cells. Amsterdam, North-Holland Publishing Company, pp 172. 1971.
2. Yoshinaga K, Muramatsu H, Muramatsu T: Immunohistochemical localization of the carbohydrate antigen 4C9 in the mouse embryo: a reliable marker of mouse primordial germ cells. Differentiation 48:75, 1991.
3. Ginsburg M, Snow MHL, McLaren A: Primordial germ cells in the mouse embryo during gastrulation. Development 110:21, 1990.
4. Cudennec CA, Nicolas J-F: Blood formation in a clonal cell line of mouse teratocarcinoma. J Embryol exp Morph 38:203, 19977.
5. Doetschman TC, Eistetter H, Katz M, Schmidt W, Kemler R: The in vitro development of blastocyst-derived embryonic stem cell lines: formation of visceral yolk sac, blood islands and myocardium. J Embryol exp Morph 87:27, 1985.

6. Wiles MV, Keller G. Multiple hematopoietic lineages develop from embryonic stem (ES) cells in culture. Development 111:259, 1991.
7. Keller G, Kennedy M, Papayannopoulou T, Wiles MV. Hematopoietic commitment during embryonic stem cell differentiation in culture. Mol Cell Biol 13:473, 1993.
8. Robertson E. Embryo-derived stem cell lines, in Robertson E (ed): Teratocarcinomas and Embryonic Stem Cells. A Practical Approach, Oxford, IRL Press, pp71, 1987
9. Matsui Y, Zsebo K, Hogan BLM: Derivation of pluripotential embryonic stem cells from murine primordial germ cells in culture. Cell 70:841, 1992.
10. Krause DS, Ito T, Fackler MJ, Smith OM, Collector MI, Sharkis SJ, May WS: Characterization of murine CD34, a marker for hematopoietic progenitor and stem cells. Blood 84:691, 1994.
11. Mathews W, Jordan CT, Wiegand GW, Pardoll D, Lemischka IR: A receptor tyrosine kinase specific to hematopoietic stem cell and progenitor cell-enriched populations. Cell 65:1143, 1993.
12. Mintz B, Russell ES: Gene-induced embryological modifications of primordial germ cells in the mouse. J exp Zool 134:207, 1957.
13. Russell ES: Hereditary anemias of the mouse: A review for geneticists. Adv Gen 20:357, 1971
14. Orr-Urtreger A, Avivi A, Zimmer Y, Givol D, Yarden Y, Lonai P: Developmental expression of c-kit, a proto-oncogene encoded by the W locus. Development 109:911, 1990.
15. Williams DE, de Vries P, Namen AE, Widmer MB, Lyman SD: The Steel factor. Devl Biol 151:368, 1992
16. Godin I, Garcia-Porrero JA, Coutinho A, Dieterlen-Lievre F, Marcos MAR: Para-aortic splanchnopleura from early mouse embryos contains B1a cell progenitors. Nature 364:67, 1993.
17. Medvinsky AL, Samoylina NL, MuellerAM, Dzierzak EA: An early pre-liver intra-embryonic source of CFU-S in the developing mouse. Nature 364,64, 1993.
18. Rich IN: Primordial germ cells are capable of producing cells of the hematopoietic system in vitro. Blood 86:463, 1995.
19. Ploemacher RE, Brons RHC: Separation of CFU-S from primitive cells responsible for reconstitution of the bone marrow hemopoietic stem cell compartment following irradiation: evidence for a pre-CFU-S cell. Exp Hemat 17:263, 1989.
20. van der Sluijs JP, de Jong JP, Brons NHC, Ploemacher RE: Marrow repopulating cells, but not CFU-S, establish long-term in vitro hemopoiesis on a marrow-derived stromal layer. Exp hemat 18:893, 1990.
21. Rich IN, Riedel W, Zimmermann F, Vogt C, Noé G: The initiation of the hemopoietic system. The response of embryonic cells to growth factors and expression of erythropoietin and erythroid-relevant genes during murine development, in Rich IN, Lappin TRJ (eds): Molecular, Cellular and Developmental Biology of Erythropoietin and Erythropoiesis, New York, New York Academy of Sciences, pp147, 1994.
22. Lawson KA, Hage WJ: Clonal analysis of the origin of primordial germ cells in the mouse, in Ciba Foundation Symposium on Germline Development, Chichester, UK John Wiley & Sons, pp68, 1994.

13

MONITORING OF SERUM HLA CLASS I ANTIGEN LEVELS IN ALLOGENEIC BONE MARROW TRANSPLANTATION

Francesco Puppo,[1*] Sabrina Brenci,[1] Massimo Ghio,[1] Donatella Bignardi,[1] Paola Contini,[1] Andrea Bacigalupo,[2] Maria T. Van Lint,[2] Gilberto Filaci,[1] Marco Scudeletti,[1] Soldano Ferrone,[3] and Francesco Indiveri[1]

[1] Department of Internal Medicine
University of Genova
[2] Bone Marrow Transplantation Unit
San Martino Hospital, 16132 Genova, Italy
[3] Department of Microbiology and Immunology
New York Medical College, Valhalla, New York 10595

ABSTRACT

The levels of serum HLA Class I antigens were determined at weekly intervals up to five weeks in thirty-eight patients who had undergone allogeneic BMT. In patients with GVHD grade I serum HLA Class I antigen levels did not change during the observation period. In patients with GVHD grade II - IV serum HLA Class I antigen level increased in the week before the onset of GVHD and was maximal during the GVHD episode. The mean ± SEM concentrations of serum HLA Class I antigens during GVHD grade II - IV episodes (6.3 ± 1.1 μg/ml) is significantly higher than in the first week after BMT (2.5 ± 0.3 μg/ml) and in the week preceding the GVHD episode (3.5 ± 0.4 μg/ml) ($P = 0.0002$ and $P = 0.01$, respectively). The results of the present investigation suggest that measure of serum HLA Class I antigen level may be a useful marker to detect an acute GVHD following BMT.

INTRODUCTION

Besides being expressed on the membrane of most nucleated cells (1), HLA Class I antigens are present in serum (2-6). An increase in the serum HLA Class I antigen level has been reported in acute rejection episodes following heart, liver and kidney transplants

* Corresponding author: F. Puppo, M.D., DI.M.I., Viale Benedetto XV, n.6, 16132 Genova, Italy; phone: 39-10-3537988; Fax: 39-10-3538994

(10-16). Furthermore, it has been reported that serum HLA Class I antigen level significantly increases in patients suffering from acute graft versus host disease (GVHD) episodes following allogeneic bone marrow transplantation (BMT) whereas it does not change in patients without GVHD (17). However, in this retrospective study the influence of GVHD grade on the serum HLA Class I antigen level was not investigated. In the present paper we describe the results obtained in a prospective study planned to measure HLA Class I antigen levels in sequential sera obtained from patients following BMT and to correlate changes in the serum HLA Class I antigen level with the onset and grade of acute GVHD episodes.

PATIENTS AND METHODS

Patients

Thirty-eight patients (22 men and 16 women, age 15 - 47) who had undergone allogeneic BMT were investigated. All patients received a bone marrow graft from their HLA-A, -B identical and MLR-negative siblings. The conditioning regimen and transplantation procedures have been performed as described (18). Prophylactic immunosuppression with cyclosporine or cyclosporine and methotrexate was started on day 1 before BMT as described (19).

Sera were obtained at weekly intervals up to five weeks following BMT and stored at -30° C until when used. Clinical parameters were recorded every week in order to diagnose GVHD. Acute GVHD episodes were diagnosed according to Thomas et al. (20).

Normal Controls

Serum samples from 60 healthy volunteers served as normal controls.

Monoclonal Antibodies (mAb) and Conventional Antisera

Anti-HLA Class I mAb W6/32 (21) and horseradish-peroxidase (HRP)-conjugated rabbit anti-human β_2-microglobulin (β_2-μ) antibodies were purchased from Serotec, Oxford, UK and from Dako Spa, Milan, Italy, respectively.

Double Determinant Immunoassay (DDIA) to Measure Serum HLA Class I Antigen Level

The DDIA to measure serum HLA Class I antigen level was performed as previously described (22) with minor modifications, utilizing a solid phase immuno-enzymatic assay (BEIA HLA Class I Quant, Bouty Spa, Milan, Italy). The concentrations of serum HLA Class I antigens are reported as μg/ml (mean ± SEM).

Statistical Analysis

The differences among mean serum HLA Class I antigen concentrations in various clinical conditions were calculated utilizing the Mann-Whitney U test. Two-sided *P* values less than 0.05 were accepted as significant.

RESULTS

Ten patients suffered from acute GVHD grade I and twenty-eight patients suffered from acute GVHD grade II - IV following BMT. In all patients the onset of acute GVHD episode was between the second and the third week after BMT. In patients with GVHD grade I, the mean concentrations of serum HLA Class I antigens were 2.1 ± 0.5 μg/ml, 2.6 ± 0.6 μg/ml and 2.4 ± 0.6 μg/ml in the 1st week after BMT, in the week preceding the GVHD episode and during the acute GVHD episode, respectively. The three values are not significantly different from each other (Figure 1). In patients with GVHD grade II - IV the mean concentration of serum HLA Class I antigens was 6.3 ± 1.1 μg/ml during the acute GVHD

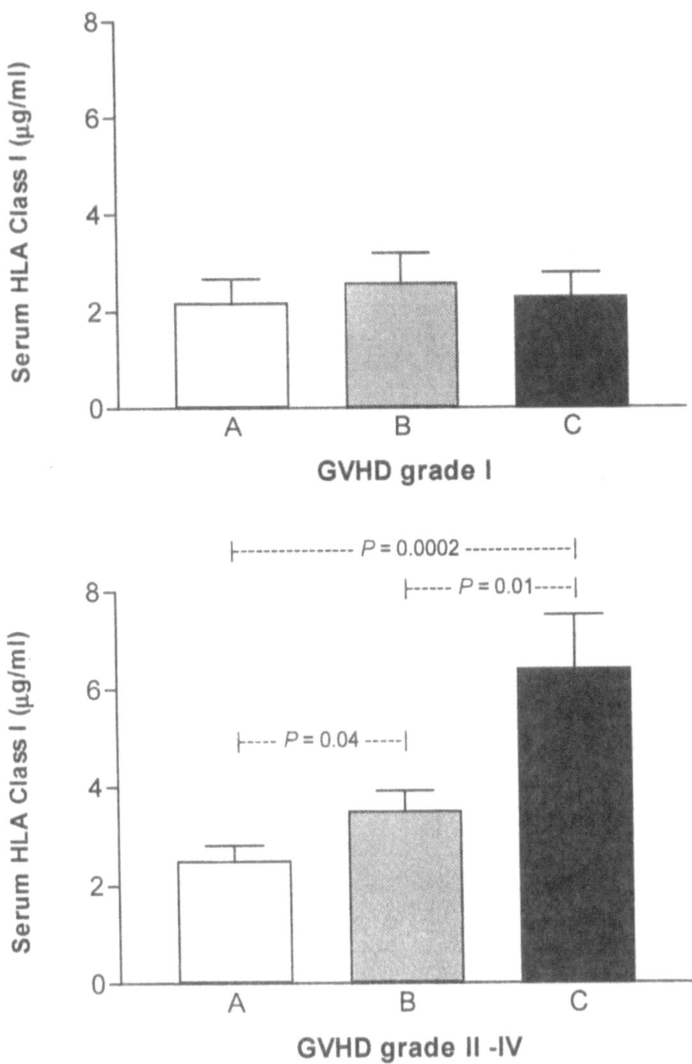

Figure 1. Serum HLA Class I concentrations in BMT recipients with GVHD. Serum HLA Class I antigen levels in the 1st week after BMT (A), in the week preceding the onset of acute GVHD episode (B) and during the GVHD episode (C) in patients suffering from GVHD grade I or GVHD grade II - IV following allogeneic BMT.

episode. This value is significantly higher than the mean serum HLA Class I antigen concentration in the first week after BMT (2.5 ± 0.3 μg/ml; $P = 0.0002$) and in the week preceding the GVHD episode (3.5 ± 0.4 μg/ml; $P = 0.01$). The latter two values significantly differ from each other ($P = 0.04$) (Figure 1). Furthermore, the serum HLA Class I antigen concentrations during acute GVHD grade II - IV episodes as well as those found in the week preceding GVHD II - IV episodes are both significantly ($P = 0.001$ and $P = 0.01$, respectively) higher than that found in sixty healthy donors. The latter has a mean ± SEM value of 1.4 ± 0.1 μg/ml. In contrast, the serum HLA Class I antigen concentration in the first week following BMT is not significantly different from that in sixty healthy donors.

DISCUSSION

The present study has shown that serum HLA Class I antigen levels in BMT recipients significantly increase in the week preceding the onset of acute GVHD (grade II - IV) and further increase during the GVHD episode. In contrast, serum HLA Class I antigen levels did not significantly change in BMT recipients who suffered from GVHD grade I. The increase in serum HLA Class I antigen levels may reflect their secretion from activated immune cells of donor origin (23,24) and their release from recipient's damaged tissues, as suggested in the course of rejection episodes in recipients of solid organ allografts (10-16).

The discrimination between donor or recipient origin of serum HLA Class I antigens cannot be done in the present study, since we used an assay which measures the total amount of serum HLA Class I antigens. It could be tested utilizing assays which measure the level of HLA Class I allospecificities (25) in HLA mismatched BMT donor-recipient combinations.

The development of GVHD grade I is associated with little morbidiy and almost no mortality, whereas GVHD grade II - IV episodes are associated with decreased survival and increased mortality rate (23). The results of the present investigation suggest that measure of serum HLA Class I antigen level may be a useful marker to monitor patients following BMT and in particular to reveal a severe acute GVHD.

ACKNOWLEDGMENT

This work was supported by grants from Ministero della Sanità - Istituto Superiore di Sanità (VII Progetto di Ricerche sull'AIDS - 1994) and from AIRC - Milano (grant to A.B.).

REFERENCES

1. Natali PG, Bigotti A, Nicotra MR, Viora M, Manfredi D, Ferrone S: Distribution of human Class I (HLA-A,B,C) histocompatibility antigens in normal and malignant tissues of nonlymphoid origin. Cancer Res 44: 4679, 1984.
2. Van Rood JJ, van Leeuwen A, van Santen MCT: Anti HL-A2 inhibitor in normal human serum. Nature 226: 366, 1970.
3. Charlton RK, Zmijewski CM: Soluble HL-A7 antigen: localization in the β-lipoprotein fraction of human serum. Science 170: 636,1970.
4. Pellegrino MA, Ferrone S, Pellegrino AG, Oh SK, Reisfeld RA: Evaluation of two sources of soluble HL-A antigens: platelets and serum. Eur J Immunol 4: 246,1974.
5. Ferrone S, Pellegrino MA, Billing R, Terasaki PI, Reisfeld RA: Production of anti-W24 xenoantisera in rabbits. Tissue Antigens 5: 41,1975.

6. Pellegrino MA, Indiveri F, Fagiolo U, Antonello A, Ferrone S: Immunogenicity of serum HLA antigens in allogeneic combinations. Transplantation 33: 530,1982.
7. Krensky AM, Weiss A, Crabtree G, Davis MM, Parham P: T-lymphocyte-antigen interactions in transplant rejection. N Engl J Med 322: 510,1990.
8. Puppo F, Brenci S, Lanza L, Bosco O, Imro MA, Scudeletti M, Indiveri F, Ferrone S: Increased level of serum HLA Class I antigens in HIV infection. Correlation with disease progression. Hum Immunol 40: 259,1994.
9. Saririan K, Wali A, Almeida RP, Russo C: Increased serum HLA class I molecule levels in elderly human responders to influenza vacciantion. Tissue Antigens 42: 9,1993.
10. Davies HffS, Pollard SG, Calne RY: Soluble HLA antigens in the circulation of liver graft recipients. Transplantation 47: 524,1989.
11. Pollard SG, Davies HffS, Calne RY: Peroperative appearance of serum class I antigen during liver transplantation. Transplantation 49: 659,1990.
12. Tilg H, Westhoff U, Vogel W, Aulitzky WE, Herold M, Margreiter R, Huber C, Grosse-Wilde H: Soluble HLA class I serum concentrations increase with transplant-related complications after liver transplantation. J Hepatol 14: 417,1992.
13. Rhynes VK, McDonald JC, Gelder FB, Aultman DF, Hayes JM, McMillan RW, Mancini MC: Soluble HLA Class I in the serum of transplant recipients. Ann Surg 217: 485,1993.
14. Puppo F, Pellicci R, Brenci S, Nocera A, Morelli N, Dardano G, Bertocchi M, Antonucci A, Ghio M, Scudeletti M, Barocci S, Valente U, Indiveri F: HLA Class-I-soluble antigen serum levels in liver transplantation. A predictor marker of acute rejection. Hum Immunol 40: 166,1994.
15. Zavazava N, Böttcher H, Müller-Ruchholz W: Soluble MHC class I antigens (sHLA) and anti-HLA antibodies in heart and kidney allograft recipients. Tissue Antigens 42: 20,1993.
16. DeVito-Haynes LD, Jankowska-Gan E, Sollinger HW, Knechtle SJ, Burlingham WJ: Monitoring of kidney and simultaneous pancreas-kideny transplantation rejection by release of donor-specific, soluble HLA Class I. Hum Immunol 40: 191,1994.
17. Westhoff U, Doxiadis I, Beelen DW, Schaefer UW, Grosse-Wilde H: Soluble HLA Class I concentrations and GVHD after allogeneic marrow transplantation. Transplantation 48: 890,1989.
18. Van Lint MT, Bacigalupo A, Frassoni F, Repetto M, Piaggio G, Congiu M, Vitale V, Scarpati D, Franzone P, Corvò R: Bone marrow transplantation (BMT) for acute lymphoblastic leukemia (ALL) in remission. Haematologica 71: 135,1986.
19. Bacigalupo A, Van Lint MT, Occhini D, Gualandi F, Lamparelli T, Sogno G, Tedone E, Frassoni F, Tong J, Marmont AM: Increased risk of leukemia relapse with high dose cyclosporine A after allogeneic marrow transplantation for acute leukemia. Blood 77: 1423,1991.
20. Thomas ED, Storb R, Clift RA, Fefer A, Johnson FL, Neiman PE, Lerner KG, Glucksberg H, Buckner CD: Bone-marrow transplantation. N Engl J Med 292: 832,1975.
21. Barnstable CJ, Bodmer WF, Brown G, Galfre G, Milstein C, Williams AF, Ziegler AZ: Production of monoclonal antibodies to group A erythrocytes, HLA and other human cell surface antigens: new tools for genetic analysis. Cell 14: 9,1978.
22. Comuzio S, Puppo F, Ruzzenenti R, Orlandini A, Grillo F, Brenci S, Lanza L, Scudeletti M, Indiveri F: Simple ELISA method for the evaluation of soluble HLA class I antigens in human serum. J Clin Lab Anal 5: 278,1991.
23. Ferrara JLM, Deeg HJ. Graft-versus-host disease. N Engl J Med 324: 667,1991
24. Vogelsang G, Hess AD: Graft-versus-host-disease: new directions for a persistent problem. Blood 84: 2061,1994.
25. Russo C, Fotino M, Carbonara A, Ferrone S: A double determinant immunoassay for HLA Class I typing using serum as an antigen source. Hum Immunol 19: 69,1987.

6. Pellegrino M.A., Indiveri F., [illegible] A., Ferrone S. [illegible] in allograft [illegible] Transplantation 35: 530, 1983.
7. [illegible] A.M., Weiss A., [illegible] G., Davis M., Crowell N. [illegible] in transplant rejection. N Engl J Med 322: 710, 1990.
8. [illegible] S., [illegible] Ferrone S. [illegible] increased level of serum HLA Class I antigen [illegible] HIV infection [illegible] 1989.
9. [illegible] Adams [illegible] Chalkeley [illegible] analysis of [illegible] Tissue Antigens 32: [illegible] 1988.
10. Davies H.S., [illegible] W.F. [illegible] soluble HLA antigens in the [illegible]
11. [illegible] transplant [illegible] 1988.
12. [illegible] H., [illegible] HLA [illegible] Hum Immunol [illegible] 1988.
13. [illegible] HLA Class I [illegible] Ann Surg [illegible]
14. [illegible] Puppo F., [illegible] Ferrone S. [illegible] HLA Class I soluble antigen serum levels in liver transplantation: a predictor marker of acute rejection. [illegible] 1994.
15. Zavazava N., [illegible] Class I [illegible] in [illegible]

14

CORRELATION OF ELISA-MEASURED ANTI-HLA CLASS I IGG AND IGG1 ANTIBODIES AND FIRST-YEAR REJECTION EPISODES

Jeffrey Regan,[1] Francisco Monteiro,[2] Daniel Speiser,[3] Jorge Kalil,[2] Ronald Kerman,[4] Philippe Pouletty,[1] and Roland Buelow[1]

[1] SangStat Medical Corporation
Menlo Park, California
[2] Heart Institute
School of Medicine
University of Sao Paulo, Brazil
[3] Swiss National Reference Laboratory for Histocompatibility
Division d'Immunologie et d'Allergologie
Hopital Cantonal Universitaire De Geneve, Switzerland
[4] Department of Surgery
University of Texas Medical School
6431 Fannin, Houston, Texas 77030

ABSTRACT

Pretransplant screening of potential recipient patient sera to determine the level of allosensitization to class I HLA antigens has traditionally been done as an aid in the management of transplant waiting lists. A soluble HLA (sHLA) ELISA was developed as an alternative to lymphocytotoxicity for anti-HLA antibody screening. This ELISA was used for the detection of anti-HLA class I IgG, IgG1, IgG2, IgG3, IgG4, and IgM antibodies and used to analyze the correlation of pretransplant allosensitization and post transplant rejection episodes in renal allograft recipients. ELISA plates were coated with 46 different sHLA preparations representing 40 different HLA class I antigens. After incubation with a serum specimen, bound antibodies were detected with a peroxidase conjugated antibody. Serum specimens from 158 microlymphocytotoxicity positive patients were analyzed. Eighty-four patients had experienced one or more rejection episodes within twelve months post transplantation. One hundred two patients were tested positive by ELISA (total IgG %PRA > 10%). A strong correlation between first year rejection and ELISA detected anti-HLA class

Molecular Biology of Hematopoiesis 5, edited by Abraham et al.
Plenum Press, New York, 1996

I IgG ($p=0.0004$) and IgG1 ($p< 0.0001$) was observed. The predictive value for IgG was 64%, and for IgG1, 73% demonstrating that ELISA results allow the identification of patients at high risk to reject the transplanted kidney. Presence of anti-HLA class I IgG2, IgG3, IgG4 or IgM was not predictive of first year rejection episodes.

INTRODUCTION

Testing to determine the level of anti-HLA class I sensitization in transplant candidates has most commonly been conducted by microlymphocytotoxicity. While this method has proven useful for the management of transplant waiting lists, there have been few studies demonstrating a correlation between microlymphocytotoxicity determined Panel Reactive Antibody % (PRA%) and clinical outcomes (1-3, 9-11). Still, the availability of a procedure to identify patients at high risk to reject a transplanted organ would be of great value to guide the choice of immunosuppressive regimens (e.g. the use of biological induction therapy) and the intensity of post-transplant monitoring.

In many ways, the lack of correlation between pretransplant anti-HLA antibodies determined by microlymphocytotoxicity and transplant outcome is not surprising. Before a transplant is conducted, the presence of anti-donor HLA antibodies is ruled out by performing a microlymphocytotoxicity based crossmatch assay. Therefore, with the presence of anti-donor antibodies ruled out, one might not expect the overall degree of sensitization of a patient to the general population of HLA class I antigens to correlate to post-transplant outcome. Additionally, any subtle correlation between pretransplant sensitization status and transplant outcome that may exist could well be obscured by the technique of microlymphocytotoxicity itself; the technique has a number of inherent limitations. PRA-testing HLA laboratories generate trays consisting of a panel of human lymphocytes from HLA-phenotyped blood donors. Cells from different donors differ in their sensitivity to complement. In addition, each laboratory uses a different source of complement. The readout of the microlymphocytotoxicity assay is often performed subjectively by microscopy. This lack of assay standardization leads to a large intra- and interlaboratory variation and low reproducibility (4-8). In addition, it is difficult to differentiate between HLA and non-HLA antibodies, IgG, IgA or IgM and autoantibodies.

To more reliably determine if a correlation between pretransplant sensitization and transplant outcome exists, a more reproducible and specific technique than microlymphocytotoxicity is desirable. We have chosen to exploit ELISA technology as one method, and have developed a soluble HLA (sHLA) ELISA that utilizes a panel of 46 different sHLA preparations for the detection of panel reactive anti-HLA class I IgG antibodies (4). An ELISA has a number of benefits over the traditional cell panel based microlymphocytotoxicity method; it does not require a viable cell preparation and it provides objective reading. Additionally, standardization of an ELISA assay is possible and intra- and interlaboratory reproducibility is high (12-15). PRA values (%) determined by ELISA in five different laboratories were highly reproducible. In contrast, %PRA values determined by lymphocytotoxicity varied significantly (4). In addition, ELISA technology allows isotype specific detection of antibodies.

Here we describe the analysis of pretransplant sera from 158 patients with known graft outcome for anti-HLA class I IgG, IgG1, IgG2, IgG3, IgG4, and IgM antibodies by ELISA. The results suggest that detection of anti-HLA antibodies using a standardized ELISA is clinically more informative than lymphocytotoxicity test results and allows the identification of patients at risk to reject a transplanted organ.

MATERIAL AND METHODS

Patients

Patient sera from three centers were analyzed: Hôpital Cantonal Universitaire de Geneve, Hospital das Clinicas Universidade de Sao Paulo and The University of Texas Medical School. Testing was conducted at the centers and at SangStat Medical Corporation. Only patients that had been sensitized (i.e. their antisera contained antibodies reactive with at least 5% of cells when tested by microlymphocytotoxicity) were included in the study. Patients were selected randomly from two groups: Group 1 did not experience any rejection episode during the first twelve months post-transplantation; group 2 experienced one or more rejection episodes during the first twelve months post-transplantation.

The patients had been transplanted between 1985 and 1994 with cadaveric donor kidneys. Immunosuppressive rejection prophylaxis consisted of cyclosporin A, azathioprine and steroids. All donors and recipients were ABO-compatible. All recipients were transplanted across a negative T cell crossmatch. Rejection episodes were clinically diagnosed based upon a biopsy in Sao Paulo and Geneva. In Houston, rejection episodes were clinically diagnosed based either upon a biopsy, or when a biopsy was not obtained, a response to therapy.

Microlymphocytotoxicity Testing

In Sao Paulo and Houston, class I and class II typing were performed serologically (1-3). In Geneva HLA A,B and C were serologically determined (1-3) and oligotyping for DRB1 was performed by hybridization with sequence specific oligonucleotide probes (SSO) after polymerase chain reaction amplification (16).

Patient sera were tested against a panel of peripheral blood lymphocytes from healthy phenotyped donors. HLA frequencies reflected those of the local population. Panel T lymphocytes were purified using nylon wool. Patient sera were tested using the NIH protocol (Sao Paulo and Geneva) or the AHG or NIH protocol (Houston) (1-3). Crossmatch testing was performed on fresh donor T lymphocytes purified from spleen using nylon wool or from peripheral blood using magnetic beads. In Geneva the crossmatch test was performed using the NIH protocol, in Sao Paulo and Houston both NIH and AHG crossmatch procedures were used.

ELISA Testing

ELISA %PRA were determined using the PRA-STAT kit (SangStat Medical Corporation, Menlo Park, CA). Ninety-six well microtiter plates were coated with monoclonal antibody TP25. This antibody reacts with the α3-domain of all known HLA-A,B,C antigens. Subsequently, sHLA antigens from the culture supernatants of 46 different EBV transformed human B cells were captured onto the antibody coated plates. After incubation of the sHLA-coated plates with patient sera, bound antibodies were detected using either a peroxidase-conjugated anti-human IgG or a peroxidase-conjugated isotype specific antibody. Bound conjugate was detected using O-phenylenediamine dihydrochloride and absorbance was read at OD495-600 using an ELISA plate reader. Corrections for non-specific binding were made for each specimen. Identification of positive reactions was accomplished by comparison to a positive reference. Assay internal cut-off value were determined using sera from non-sensitized individuals. These cut-off values were then used to identify positive reactions in each individual well. For all assays, a PRA% of greater than 10% was considered positive. For samples from Texas, only specimens with a positive total IgG PRA% were tested for IgG subclasses. Correlations of test results with post-transplant rejection episodes were analyzed using 2x2 contingency tables and a two-tailed Fisher's exact test.

RESULTS

Detection of anti-HLA Class I Ig Isotypes, and IgM by ELISA

The detection of anti-HLA class I total IgG antibodies by a soluble-HLA ELISA has recently been described (4). This ELISA was adapted to measure IgG subclasses, and IgM human anti-HLA class I antibodies by substitution of the anti-total IgG-HRP conjugate with class or subclass specific conjugates. For each subclass, a cutoff was determined by testing between 30 and 70 normal sera (depending upon the subclass). A representative assay result using an IgG1 specific secondary antibody is shown in figure 1. Serum from patient 1 reacted with 11% of the wells, serum from patient 3 and 4 with 87%. Serum from patients 2,5,6,7, and 8 were negative.

Correlation of ELISA test results with rejection episodes

The level of IgG, IgG1, IgG2, IgG3, IgG4, and IgM allosensitization was determined by ELISA in the pretransplant sera of 158 microlymphocytotoxicity positive transplant

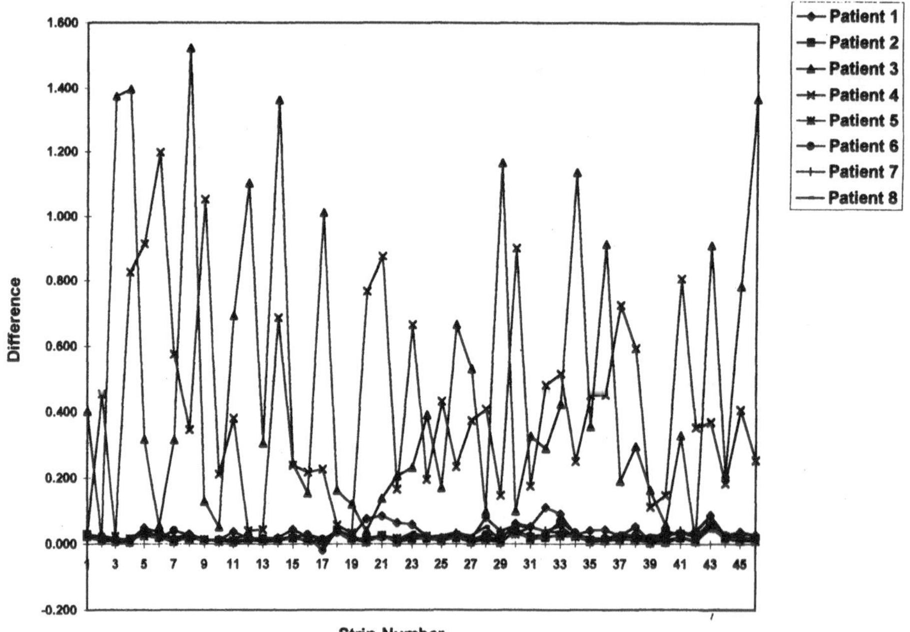

Figure 1. Detection of anti-HLA class I IgG1 by ELISA. Eight patient specimens were tested for anti-HLA class I IgG1 by ELISA using 46 different sHLA preparations. For each serum specimen, the difference between the OD of a given well which contained antigen and the OD without antigen was calculated. Patients 1, 3 and 4 were found to contain anti-HLA class I IgG1, patients 2,5,6,7 and 8 tested negative.

Table 1. Correlation of rejection and anti-HLA Class I Antibodies detected by ELISA

ELISA total IgG (n=158)

	NO REJECTION	REJECTION
POSITIVE	37	65
NEGATIVE	37	19

P = 0.0004

ELISA IgG1 (n=157)

	NO REJECTION	REJECTION
POSITIVE	20	54
NEGATIVE	53	30

P <0.0001

ELISA IgG2 (n=102)

	NO REJECTION	REJECTION
POSITIVE	16	19
NEGATIVE	33	41

P = 1.0

ELISA IgG3 (n=157)

	NO REJECTION	REJECTION
POSITIVE	10	20
NEGATIVE	61	66

P = 0.1594

ELISA IgG4 (n=157)

	NO REJECTION	REJECTION
POSITIVE	22	35
NEGATIVE	51	49

P = 0.1393

ELISA IgM (n = 84)

	NO REJECTION	REJECTION
POSITIVE	7	7
NEGATIVE	29	41

P = 0.5692

PRA% greater than 10% considered positive. P values calculated by two-tailed Fisher's exact test

recipients. All transplants were done across a negative T-cell crossmatch. In this group of patients, 84 experienced 1 or more first year rejection episodes.

In all three centers, a strong correlation between rejection episodes and anti-class I IgG1 was observed individually and combined ($p<0.0001$; Table 1 shows the combined data). Among patients that experienced one or more rejection episodes, 64% had ELISA detected anti-class I IgG1 antibodies; conversely, only 27% of the patients were IgG1 positive in the non-rejection group. The positive prediction value for IgG1 and first year rejection (number of patients that experienced rejection(s) and were tested positive by ELISA divided by number of all patients that had been tested positive by ELISA) was 73%. This demonstrates that detection of anti-HLA class I IgG1 antibodies allows the pre-transplant identification of patients at high risk to experience rejection episodes. Similarly, a correlation of anti-class I IgG with rejection was found for the Brazilian and Houston centers as well as for the combination of all three centers ($p=0.0004$). Combining the data at the 3 sites, among patients that experienced one or more rejection episodes, 77% had ELISA detected anti-class I total IgG antibodies; conversely, only 50% of the patients were IgG positive in the non-rejection group. The positive predictive value for total IgG and first year rejection was 64%. Presence of anti-HLA class I IgG2, IgG3, IgG4 or IgM did not correlate with first year rejection episodes. This indicates that patients with these kind of anti-HLA class I antibodies are not at higher risk to experience a rejection episode than patients that had been tested negative. Specificity analysis of ELISA detected anti-HLA class I antibodies did not lead to the detection of anti-donor HLA antibodies in any of the specimens that were analyzed. This confirms the lymphocytotoxicity crossmatch result that indicated the absence of anti-donor HLA antibodies.

DISCUSSION

Sera from patients on waiting lists for kidney transplantation are tested for the presence of panel reactive anti-HLA antibodies to guide the selection of serum specimens for crossmatch testing. Both panel reactive antibody and crossmatch testing are performed by microlymphocytotoxicity. However, lymphocytotoxicity methodology suffers from a number of limitations including the requirement for a viable cell preparation, variable sensitivity of cells for complement, subjective reading of results and lack of standardization. This leads to a high center to center variation of test results (4-8). In addition, in many centers lymphocytotoxicity assay results do not correlate with clinical events post transplantation (9-11).

In contrast, detection of anti-HLA IgG antibodies by ELISA has been shown to be highly reproducible. The average inter-laboratory ELISA correlation was 0.93 for 107 samples tested at 5 laboratories (4). A comparison of ELISA and microlymphocytotoxicity consensus results showed a high correlation of %PRA values (r=0.83) (4). However, due to the variability of microlymphocytotoxicity assay results comparison of %PRA values determined by ELISA with lymphocytotoxicity assay result obtained at any one center may correlate to a lower extent (4).

Dependent on the clinical relevance of assay results, higher reproducibility per se may not improve the quality of patient management. Therefore we addressed the question: Do anti-HLA class I IgG antibodies detected by ELISA correlate with the occurrence of rejection episodes within the first 12 months post transplantation? Even though all patients had been transplanted across a negative lymphocytotoxicity crossmatch, positive ELISA test results correlated with rejection. The strongest correlation was found for anti-HLA class I IgG1 antibodies. The predictive value for IgG1 and first year rejection was 73% indicating that four out of five patients that had been tested positive by ELISA experienced first year rejection episodes. Besides IgG1, a statistically significant correlation was observed with anti-HLA class I total IgG. No correlation was found with IgG2, IgG3, IgG4 and IgM. This might be due to the complement fixing capabilities of IgG1. IgG2 binds complement very poorly and IgG4 not at all; neither of these isotypes correlated with first year rejection. Both IgG1 and IgG3 bind complement efficiently. However, the concentration of IgG3 in human serum is about 10 times lower (0.8 mg/mL) compared to IgG1 (8 mg/mL). This could explain the lack of correlation of rejection episodes with anti-HLA class I IgG3 antibodies. However, this rational would not explain why IgM does not correlate with rejection, and other factors may also be at work.

All patients had been transplanted across a negative lymphocytotoxicity crossmatch. Specificity analysis of ELISA detected antibodies did not lead to the detection of anti-donor HLA antibodies confirming the negative microlymphocytotoxicity crossmatch results. These results indicate that most of the investigated kidney allograft recipients did not express detectable anti-donor HLA antibodies. The correlation of ELISA detected anti-HLA class I IgG antibodies with posttransplant rejection episodes suggests that allosensitized patients mount a stronger immune response to donor alloantigen even if no donor specific antibodies are expressed by the recipient before transplantation. This might be due to the fact that T- and B-cell epitopes are not necessarily identical. Therefore the absence of donor specific antibodies does not exclude the presence of T cells primed for certain epitopes of donor alloantigens.

Panel reactive antibody screening results obtained by lymphocytotoxicity do not always correlate with clinical events and rejection episodes. This could be due to the lack of standardization and poor reproducibility of lymphocytotoxicity assay results. Alternatively, the correlation could by hidden due to the detection of anti-HLA class I IgM antibodies, autoantibodies, or anti-class II antibodies that are known to cause positive reaction in microlymphocytotoxicity assays. These data suggest that a highly reproducible standardized ELISA procedure for the detection of anti-HLA class I IgG antibodies is clinically more informative than lymphocytotoxicity testing and may improve guidance for the treatment of high risk transplant recipients.

REFERENCES

1. Hopkins KA: Basic microlymphocytotoxicity test. In: Zachary AA, Terisi GA (eds): ASHI laboratory manual. 2d ed. Lenexa, KS: American Society of Histocompatibility and Immunogenetics, 1990.
2. Zmijewski C: Detection and Identification of HLA Antibodies. In: Zachary AA, Terisi GA (eds): ASHI laboratory manual. 2d ed. Lenexa, KS: American Society of Histocompatibility and Immunogenetics, 1990.
3. Noreen HJ: Crossmatch test. In: Zachary AA, Terisi GA (eds) ASHI laboratory manual. 2d ed. Lenexa, KS: American Society of Histocompatibility and Immunogenetics, 1990.
4. Buelow R, Mercier I, Glanville L, Regan J, Ellingson L, Janda G, Claas F, Colombe B, Gelder F, Grosse-Wilde H, Orosz C, Westhoff U, Voegeler U, Monteiro F, Pouletty P: Detection of panel reactive anti-HLA class I antibodies by ELISA or lymphocytotoxicity: Results of a blinded, controlled multicenter study. Human Immunol 44: 1, 1995
5. Hopkins KA, Maguire MG, Fink NE, Bias WB: Reproducibility of HLA-A, B, and DR typing using peripheral blood samples: results of retyping in the collaborative corneal transplantation studies. Collaborative Corneal Transplantation Studies Group, Baltimore, MD. Human Immunol 33: 122,1992
6. Marrari M, Duquesnoy R: Progress Report on the ASHI/CAP Proficiency Survey Program in Histocompatibility Testing. I. HLA-A, B, C Typing, Antibody Screening, and Lymphocytotoxicity Crossmatching. Human Immunol 39: 87, 1994
7. Albert E: Report on the tissue typers meeting at Eurotransplant on Friday, September 27, 1991. Eurotransplant Newsletter 89: 11, 1993
8. Doxiadis IN, Schreuder GMTh, Claas FHJ: ET quality control testing of tissue typing, donor retyping, crossmatching and serum screening. Eurotransplant Newsletter 98:11, 1988
9. Kerman RH, Van Buren CT Lewis RM et al.: Impact of blood transfusions and HLA on cyclosporine-treated renal transplant recipients. Transplant Proc 3(suppl 3): 264, 1988
10. Koka P, Cecka JM: Sensitization and crossmatching in renal transplants. In: Terasaki GA (ed): Clinical Transplants 1989, UCLA Tissue Typing Lab Publ : 379, 1990
11. Terasaki Pi, Cicciarelli J: Sensitization and its role in transplantation. In: Cerilli J (ed): Organ Transplantation and Replacement, Philadelphia Lippincott Publ : 197 1988
12. Doxiadis I, Westhoff U, Grosse-Wilde H: Quantification of soluble HLA class I gene products by an enzyme linked immunosorbent assay. Blut 59: 449, 1989
13. Shimizu B, Sra K, Ferrone S., Pouletty P: sHLA-STAT_ class I, an ELISA for quantification of HLA antigens in serum. Human Immunol 32: 289, 1991
14. Pouletty P, Ferrone S, Amesland F, Cohen N, Westhoff U, Charron D, Shimizy R, Grosse-Wilde H: Summary Report from the First International Workshop on Soluble HLA Antigens. Paris, August 1992. Tissue Antigens 42: 45, 1993
15. Pouletty P, Chang C, Atwood E, Kalil J, Ferrone S, Shimizu R, Howson W, Mazahari R, Del Villano B, Grumet C: Typing of serum soluble HLA-B27 by ELISA. Tissue Antigens 42: 14, 1993
16. Tierey JM, Morel C, Freidel AC, Zwahlen F, Gebuhrer L, Betuel H, Jeannet M, Mach B: Selection of unrelated donors for bone marrow transplantation is improved by HLA class II genotyping with oligonucleotide hybridization. Proc Natl Acad Sci USA 88: 7121, 1991

15

INACTIVATION OR DOWN-REGULATION OF THE OVERSTIMULATED, HLA-DR-(CD2R-, CD26-, CD38-, AND/OR CD69-) POSITIVE T CELL SUBSET ENABLES THE IMMUNOCOMPETENT CELLS TO RESTORE THE PRE-DISEASE STATE

A Novel Technique ("Microimmunosurgery"), Based on a GvHD-Free GvL/GvT- Effect, Allows a Complete (100%) Eradication of Solid Tumors by an Indirect Attack on the Tumor-Protecting CD8$^+$ HLA-DR$^+$ Cells

P. Leskovar and I. Vodenitcharov

Biochemical Research Laboratory
School of Medicine
University (TU) Munich, Germany

SUMMARY

A common element of diseases as different as cancer, autoimmune disorders and chronic (retro)viral infections, including HIV-infection, is the overstimulation of some immunocyte subsets (HLA-DR$^+$, CD38$^+$ and/or CD69$^+$-cells; β2-m, sIL-2-R/CD25, sICAM-1, sCD8, neopterin, lysozyme, cathepsinD etc.). We tested two potential approaches to down-regulate the pathologically elevated CD8$^+$ and HLA-DR$^+$ T cells: (a) In animal model, we tested the sensibility of these, disease inducing and maintaining T cell subsets to in vitro pretreated (cell death preprogrammed) semi-syngeneic and allogeneic donor T cells, both in tumor-bearing mice and in adjuvant arthritis (AA)-induced rats. (b) In the first clinical study, we used a novel combination of FDA-approved drugs which inhibits Ca^{2+}-influx and concomitantly down-regulates cytosolic cAMP in patient's overstimulated immunocompetent cells. We could achieve a 94,6-100% long-term survival in tumor-bearing mice and a 100% prevention both of the AA-induction and of a 3fold AA-rechallenge. In patients, large primary tumors and large metastases shrank by 80-85% and small metastases disappeared completely. Since in HIV-infected persons, the increase of HLA-DR$^+$ CD38$^+$T (T8) cells is

Molecular Biology of Hematopoiesis 5, edited by Abraham et al.
Plenum Press, New York, 1996

associated with a fall in CD4-level and in development of AIDS, we are looking for the elimination of these HLA-DR$^+$ targets by our novel technique in two AIDS-simulating (FIV/FeLV and SIV) animal models.

By the new technique ("microimmunosurgery"), we were able -for the first time- to show (a) a complete (100%!) tumor regression by a sole attack on tumor-protecting Ts cells (instead on tumor cells), (b) the high efficacy of the GvL- effect in solid tumors (GvT-effect), and (c) the separation of the GvL-effect from the GvH-effect.

These advantages were possible since the new approach is based on an immune reaction which is - on average - 10.000 times stronger than the reaction against a "classical" antigen (more precisely: the number of participating effector cells is 10.000 times higher than in the case of the "standard" antigen). This immunological reaction retains its efficacy even in a profoundly suppressed organism and is the only one working in vitro without a preceding in vivo antigen-priming of efffector cells.

At present, we are testing the efficacy of the novel approach in a murine leukemia model and comparing the results with those of the solid tumor model ("classical" GvL versus GvT effect).

In the planned clinical studies, the advantages of a general introduction of microimmunosurgery in the treatment of solid tumors, independently of tumor type and individuality, should be elucidated.

This report deals with a novel principle in the therapy of a series of diseases which are characterized by a pathologically hyperactivated T cell subset. This technique, which we call "microimmunosurgery," is based on a selective depletion of the pathologically overstimulated lymphocyte subpopulation by means of a combined in vivo treatment with monoclonal antibodies (Mabs) plus alloreactive donor effector cells whose lifespan was restricted ("preprogrammed") by a patented (P 4324877.2, P 4320878.9 and PCT/EP89/00403) ex vivo pretreatment.

By this special treatment, the alloreactive donor T cells are able to eliminate - to some extent selectively - the disease inducing and maintaining T cell subset, but are hindered in establishing GvHD (graft-versus-host disease),due to the preprogrammed cell death. In this way, a high GvL (graft-versus-leukemia) effect without an accompanying GvHD can be achieved. We tested this new techniqe in detail in a murine tumor model.

The B16-melanoma cells (2.10^6)which were syngeneic with respect to the C57Bl6-strain, were inoculated into (C57Bl6 x DBA2) F1-mice ("recipients").

In the first series of experiments, the primary tumor was excised 10 days after inoculation; 24 hours before this excision, 120 mg/kg cyclophosphamide (Cy) were injected i.v. and 1-3 hrs before tumor removal, the animals were injected with Mabs, directed either against mature (CD3-positive) T cells (anti-Thy 1.2.-Mabs) or against their T 8-subpopulation (anti-Lyt 2- Mabs) (Figure 1).

In the second series of experiments, the primary tumor was not removed surgically but only irradiated with 500 R four days after its inoculation (Figure 2).

The splenocytes of the C57Bl6-donor mice were preactivated in a tumor-specific or non-specific way. In addition, non-preactivated donor spleen cells were tested. In one sub-group, donor splenocytes were injected into the tumor-bearing recipients 4 days, and in the other sub-group 10 days after tumor excision.

In our system, the recipients (F1-hybrids) were tolerant to both, the tumor cells and the donor splenocytes, whereas the transfused donor spleen cells were alloreactive against the recipient but tolerant (syngeneic) with respect to tumor cells.

We compared this system with a second one, in which both, the donors and the tumor-bearing recipients belonged to the same strain (C57Bl6). The experimental conditions, including the tumor type (B16- melanoma) were identical in both systems.

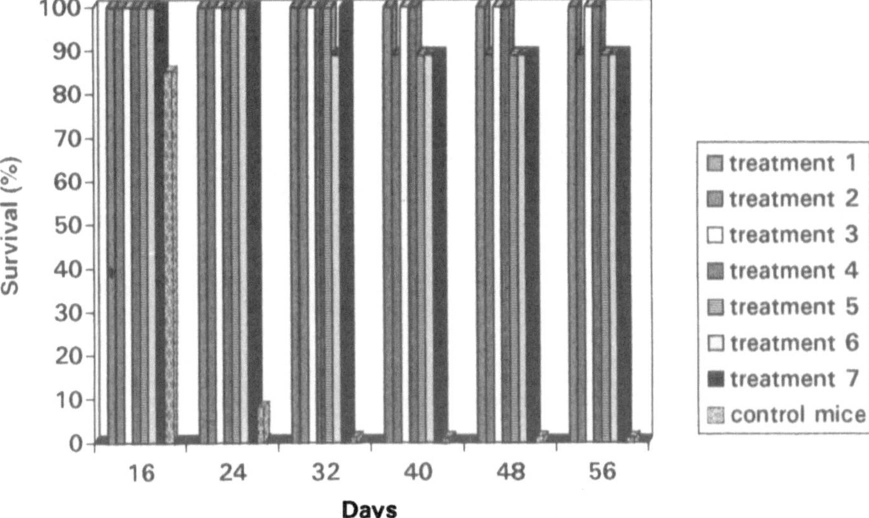

Figure 1. Permanent cure of mice with advanced tumor, treated by "microimmunosurgery," after excision of primary tumor. Recipients: (C57B16 x DBA2) F1-hybrid mice pretreatment: (a) inoculation of 2.10^6 B16 melanoma cells (belly region), (b) 10 days later: tumor excision ("day 0"), (c) 24 hrs before tumor excision: i.v.-injection cyclophosphamide (i.v., 120 mg/kg), (d) 1-3 hrs before tumor excision: i.v.-injection of pan T (CD3)-specific Mab (anti-Thy 1.2.-Mab). Donors: C57B16 strain pretreatment: see "Treatment 1-7." Transfusion of donor splenocytes into recipients: 2 days after tumor excision "Treatment 1" donor pretreatment: by B. subtilis lyophilisate (Sigma B40006) (1 mg/mouse, 30 hrs before splenectomy). "Treatment 2" donor pretreatment: inoculation of 2.10^6 living B16 tumor cells, 4 days before splenectomy. "Treatment 3" donor pretreatment: i.v.-injection of 2.10^6 inactivated (10.000R) B16 tumor cells, 4 days before splenectomy. "Treatment 4" donor pretreatment: inoculation of the mixture of 2.10^6 living B16 tumor cells plus 10 μg of B. subtilis lyophilisate (Sigma B 4006), 14 days before the harvesting of donor spleen cells. "Treatment 5" donor pretreatment: i.v.-injection of the mixture of preirradiated (10.000R) B16 tumor cells plus 100 μg of B. subtilis lyophilisate 4 days before splenectomy. "Treatment 6" donor pretreatment: 30 hours before the inoculation of 2.10^6 living B16 cells, the donor mice were preimmunized with B. subtilis (100 μg lyophilisate); donor spleen cells were harvested 4 days after tumor inoculation. "Treatment 7" donor pretreatment: identical with that of "Treatment 6," the only difference being the use of preirradiated (10.000R) instead of living B16 cells. "Control group" tumor excision (10 days after tumor inoculation); 24 hours before: cyclophosphamide (i.v., 120 mg/kg). Note: Due to a special in vitro pretreatment (preprogramming of cell death), the donor splenocytes exerted a strong GvL effect, but died before they could establish GvHD.

In this fully syngeneic system, there was a mutual tolerance between the donor, the recipient and the tumor. So, the only difference between both systems was the alloreactivity of donor splenocytes against the recipient in the first, i.e. semi-allogeneic system (Figures 3 and 4).

Since we were able, as mentioned, to prevent the establishment of the GvHD by the special pretreatment of donor effector cells, we could profit of a strong GvL-effect (graft-versus-leukemia).

One of the most exciting observations was the fact that tumor-bearing animals showed a permanent cure of cancer even in the case that donor effector cells had not been preactivated neither in a tumor-specific nor in a non-specific way. In this case the attack of donor effectors was not directed against tumor cells but exclusively against tumor-controlling recipient lymphocytes.

We tested this semi-allogeneic system both in the C57Bl6 and in the DBA2 strain; in the latter system, the donors belonged to the DBA2 strain and tumor-bearing recipients

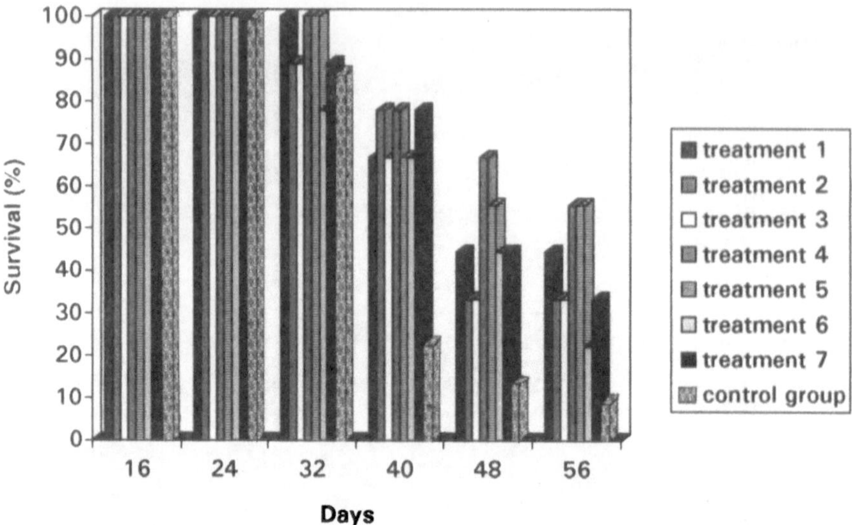

Figure 2. Tumor regression in mice, treated by "microimmunosurgery" after irradiation by 500 R. This experiment was identical with that presented in figure 1; the only difference was the irradiation of the recipients by 500 R 4 days after tumor inoculation (instead of tumor excision plus chemotherapy).

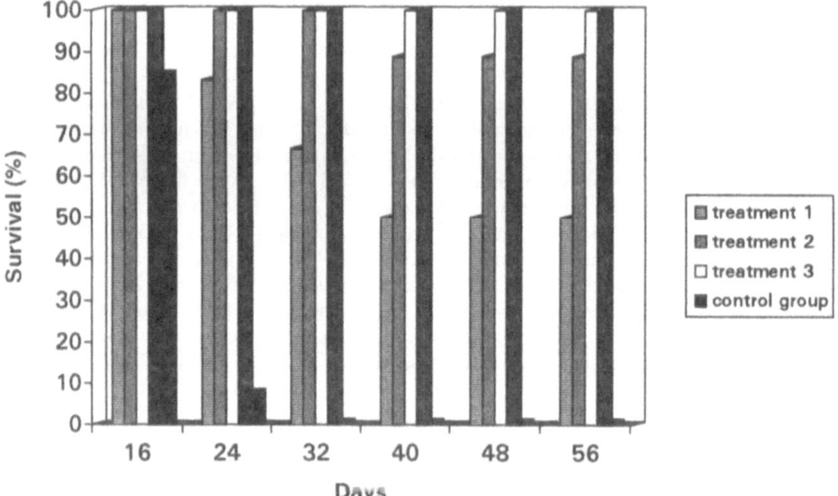

Figure 3. Comparison of the syngeneic and semi-allogeneic effector cells in tumor-excised animals. The GvL-effect seems to be even more important than the tumor-specific or non-specific preimmunization of donor mice. Recipients–pretreatment: (a) inoculation of 2.10^6 B16 melanoma cells (belly region), (b) tumor excision: 10 days later ("day 0"), (c) cyclophosphamide (120 mg/kg, i.v., 24 hrs before tumor excision), (d) i.v.-injection of pan T (CD3)-specific Mab 1-3 hrs before tumor excision. Donors–pretreatment: (a) preimmunization by B. subtilis lyophilisate (100 μg) 30 hrs before the inoculation of 2.10^6 preirradiated B16 cells, (b) harvesting of donor splenocytes 4 days after inoculation of B16 cells. Transfusion of donor splenocytes into the recipient: 2 days after tumor excision. "Treatment 1": syngeneic system (C57B16 strain); Ts-depletion in recipient by pan T-specific Mab. "Treatment 2"" semi-allogeneic system, i.e. donor: C57B16, recipient: (C57B16 x DBA2) F1; donor splenocytes alloreactive against recipient but not against tumor, recipient tolerant against tumor and against donor; Ts-depletion in recipient (in vivo) by pan T-specific Mab. "Treatment 3": like "Treatment 2", but without any preacitivation of donor mice "Control group" tumor excision by day 10 post-inoculation; 24 hrs before: cyclophosphamide (i.v., 120 mg/kg).

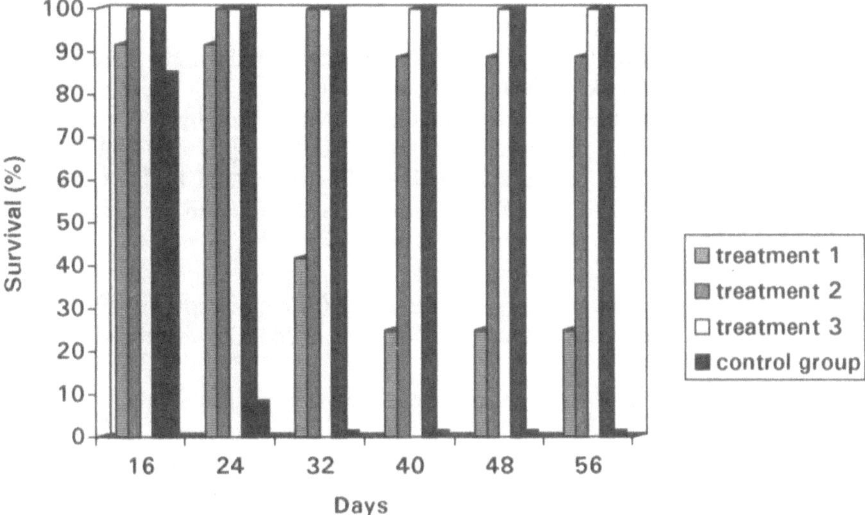

Figure 4. High efficiency of the GvLR in the 500R-irradiated tumor-bearing mice, even in the absence of a tumor-specific and non-specific preimmunization of donor spleno-cytes. Like Figure 3, the donor mice were, however, inoculated by 2.10^6 living B16 tumor cells 4 days before splenectomy.

were the (C57Bl6 x DBA2)F1-hybrids. The results of this DBA2 system (not shown here) were comparable to those, obtained in the C57Bl6 strain.

When we used tumor cells which were preadapted to the above mentioned F1-hybrid (not shown here), both in the C57Bl6 and in the DBA2 strain the percentage of permanent survivors was 100 in the case that the donor mice had been preimmunized in a tumor-specific way and nearly 100 (90-95) if non-preimmunized mice were used as the source of donor spleen cells.

When we increased the number of inoculated B16 melanoma cells from 2.10^6 to 3.10^6 and to 5.10^6, respectively (not shown here), we observed a high survival rate only if the inoculum contained 3.10^6 B16 cells and if the donors of splenocytes had been preimmunized in tumor-specific way.

When the primary tumor was excised 15 (instead of 10) days post-inoculation (data not shown here), up to 60% of animals survived permanently. If one considers that in the non-treated control group all tumor-bearing mice died within 25 days, the results of this experiment, mimicking the clinical feature of the advanced cancer, are encouraging. Experiments in which we reinoculated 1.10^5 and 2.10^6 tumor cells, respectively, 10 days after the excision of the primary tumor and 6 days after the transfusion of donor splenocytes (data not shown here), were of special interest. The reinoculated B16 tumor appeared to be completely rejected as the survival rate corresponded to that of non-reinoculated reference mice. This apparent rejection of the second tumor "worked" independently of the preimmunization of donor mice and could be confirmed in animals which were irradiated by 500R on day4 post-inoculation (instead of tumor excision).

In additional experiments, we found out that as few as 1/6 of donor spleen equivalent was sufficient to guarantee the antitumor effect in this system.

As deduced from cellular HLA-typing techniques such as primary and secondary MLR/MLC (mixed-lymphocyte-reaction/culture), CML(cell-mediated-lymphotoxycity/ lympholysis) and PLT (primed-lymphocyte-typing), the direct targets of donor effector cells appear to be the MHC II-expressing T cells (in addition to the MHC II-positive B cells and/or monocytes/ macrophages) (1)(2)(3)(4)(5).

When histoincompatible lymphocytes from person A and person B are coincubated in the one-way- MLC, cytotoxic effector cells, i.e.alloreactive CTL/Tc cells are generated only if A and B show both the class I and the class II MHC-incompatibility (6)(7). In the case that the incompatibility is restricted to the MHC I-complex, no cell proliferation (measured by the MLR-test) and no alloreactive CTLs (measured by the CML-test) are generated. If, on the other hand, the MHC-mismatch is restricted to the class II-antigens, only Th-proliferation but no CTL-generation is observed (1)(6)(7).

The critical event for the activation and clonal expansion of alloreactive CTLs seems to be the cooperation of their precursors (p-CTLs) with (primed or non-primed) Th cells on the surface (or in the microenvironment) of accessory cells (APC/macrophages) or other MHC II- coexpressing cells.

In the case that the class II-expressing cells are macrophages or other APC, both the non-primed (naive) Th cells and the primed Th cells can provide the "help" to p-CTLs. If, however, the MHC II-positive cells are activated T cells or plasma cells, the Th cells must be primed (blastogenically pretransformed).

In the normal process of T cell activation, e.g.by soluble or viral antigens, the processed antigen on the surface of the APC (macrophage) represents the competence signal (signal 1) for both the Th and CTL precursors which recognize this processed antigen by virtue of the identical CD3 (T3): Ti-receptor.

The progression signal (signal2) for Th cells is the monokine IL-1 and for the CTLs the lymphokine IL-2, secreted by the activated Th cells.

When T cells are activated by soluble or particulate antigens, the Th and Tc cells belong to the same subclone, i.e. they recognize the same antigen (epitope).

If however, T cells are activated by alloantigens, the Th and the Tc cells do not belong to the same subpopulation, expressing the same TCR (T cell receptor), since the Th cells (precursors) recognize the class II-incompatibility ("altered self- MHC II") and the CTLs (p-CTLs) the class I-incompatible structures ("altered self- MHC I") which are normally not identical (1)(2)(3)(6)(7).

It seems, that a good Th and Tc cooperation is possible even between T subpopulations with differing antigenic specificities. More important than a TCR-identity appears to be the coexpression of class II and class I MHC-antigens on the surface of (autologous or allogeneic) accessory cell, or on the surface of MHC I and MHC II double-positive (allogeneic) target cell; in other terms, the generation of both, the alloreactive Tc cells and the allospecific Th cells, requires the expression of the MHC II-complex, in addition to the MHC I-antigens on the stimulating target cells. The fact that the target cells in the CML-test have to be preactivated e.g. in an oligoclonal way by mitogenic lectins, such as ConA or PHA which are known to induce the expression of the MHC II-antigen on target cells (6)(7), supports the idea of a selective lysis of patient's pathologically (hyper) activated subclones.

Whether the lysis of target cells by alloreactive effectors is induced by the coexpression of class II and class I antigens or by unknown structures on rapidly proliferating immunoblast targets, such as the NK cell-recognized 4F2/ NKTar-structure, or by both, has not yet been elucidated.

The flow-cytometric analysis shows that diseases as different as (advanced) cancer, chronic bacterial and (retro)viral infections, including that by HIV or granulomatous inflamations (sarcoidosis) are associated with increased levels of both, the CD8-positive and the HLA-DR expressing T cells (our own unpublished data). Relapses are accompanied by an additionally increased percentage of these T cell subsets, whereas remission phases coincide with the drop of these pathologically elevated subpopulations.

An element which all these disorders seem to share is the persistence of the antigenic stimulus which leads to a local or systemic immunosuppression.

One of the markers of the persistent (chronic) immune stimulation and immune hyperactivity is β2m (β2- microglobulin) whose plasma level is increased in diseases as different as multiple myeloma (8), lymphoma (9), CLL (10)(11), further in autoimmune disorders such as SLE (12)(13) or rheumatoid arthritis (14) and viral diseases such as infection by Epstein-Barr virus (15), by hepatitis B virus (16)(17) and by HIV (18)(19).

The β2m plasma level is also increased in recipients of renal transplants (20) and in Cd-poisoned persons (21).

Neopterin, a secretory product of (hyper)activated monocytes/macrophages, shows a similarly low specificity but cannot be discussed here.

Though AIDS is characterized by a profound immune suppression, both the percentage of activated (MHC II-expressing) T cells, as well as the plasma concentration of different markers of immunocyte hyperactivation, such as neopterin/monopterin (22)(23), lysozyme (24), sCD8 (24) and acid-labile alpha-interferon (25) (in addition to the above mentioned β2m) are significantly increased in HIV-infected persons.

Resting cells are normally MHC II-negative, except a low percentage of cells with suppressive properties (26)(27). The activated state, in turn, is often associated with the coexpression of the class II-antigens (28)(29)(30).

These phenomena have been studied both in allogeneic system (31) and during T cell activation by soluble antigens (32).

During the process of cells activation, first the CD4-positive and then the CD8-positive T cells become MHC II-positive.

Cancer, chronic infections and AIDS are associated, as mentioned, with a high percentage of HLA-DR- and of CD8-positive T lymphocytes. It seems, that these T cells are active suppressor cells (" effector Ts"), since a persistent (chronic) stimulation of T cells results in a selective generation of the Ts subset.

In the case of different autoimmune disorders, the autoantigen-specific T cell subset prevails; it belongs primarily to the CD4- positive subclass and consists either of Th cells which provide "help" to the autoaggressive plasma cells or of a cytotoxic (autoaggressive) subfraction.

By a parallel attack (a) on the humoral (Mabs) and (b) on the cellular (lifespan-restricted alloreactive effectors) level, a highly efficient and to some extent selective elimination of the disease inducing or maintaining leukocyte-subsets could be achieved.

Additional details can be found in the recent patent applications P 4320878.9, P 4324877.2 and PCT/EP 89/ 00403 (89904551.2).

REFERENCES

1. Hardy,D.A., Ling,N.R., Wallin,J.M.& Aviet,T. Nature 227, 723-725 (1970)
2. Harmon,W.E., Parkman,R., Gavin,P.T., Grupe,W.E., Ingelfunger,J.R.,Yunis,E.J.& Levey,R.H. J.Immunol.129, 1573-1577 (1982)
3. Hirschberg,H., Kaakinen,H.& Thorsby, E. Tissue Antigens 7, 213-219 (1976)
4. Kristensen,T. Tissue Antigen 16, 335-367 (1980)
5. Lightbody,J.J., Bernoco,O., Miggiano, V.C.& Ceppellini,R. G.Bacteriol.Immunol. 64, 273-289 (1971)
6. Yunis,E.J. & Amos,D.B. Proc.Natl.Acad.Sci. USA 68, 3031-3035 (1971)
7. Bach,F.H. & Voynow,N.K. Science 153, 545-547 (1966)
8. Bataille,R., Grenier,J. & Sany,J. Blood 63, 468-476 (1984)
9. Amlot,P.L. & Adinolfi,M. Eur.J.Cancer 15, 791-796 (1979)
10. Simonsson,S., Wibell,L. & Nilsson,K. Scand.J.Haematol. 24, 174-180 (1980)
11. Spati,B., Child,J.A., Kermish,S. & Copper,E.H. Acta Haematol. 64, 79-86 (1980)

12. Weissel,M., Scherak,O., Fritzsche,H. & Kolarz,G. Arthritis Rheum. 19, 968-971 (1976)
13. Revillard,J.P., Vincent,C. & Rivera,S. J.Immunol. 122, 614-618 (1979)
14. Revillard,J.P., Vincent,C., Clot,J. & Sany,J. Eur.J.Rheumatol.Inflammation 5, 398-405 (1982)
15. Lamelin,J.P. Vincent,C., Fontaine-Legrand,C. & Revillard,J.P. Clin.Immunol.Immunopathol. 24,55-62 (1982)
16. Beorchia,S., Vincent,C., Revillard,J.P.& Trepo,C. Clin.Chim.Acta 109, 245-255 (1981)
17. Beorchia,S., Trepo,C., Vincent,C., Revillard,J.P. & Brette,R. Gastroenterol.Clin.Biol.6, 679-687 (1982)
18. Bhalla,R.B., Safai,B., Mertelsman,R. & Schwartz,M.K. Clin.Chem. 29, 1560-1563 (1983)
19. Crieco,M.H., Reddy,M.M., Kothari,H.B. & William,D. Clin.Immunol.Immunopathol.32, 174-184, 1984)
20. Vincent,C., Revillard,J.P. Pellet,H.& Traeger,J. Transplant.Proc.11, 438-442 (1979)
21. Kjellström,T., Evrin, P.E.& Rahnster,B. Environm.Res. 13, 303-307 (1977)
22. Abita,J.P., Milstein,S., Kaufman,S.& Saimont,G. Lancet 6, 51-52 (1985)
23. Kunze,R., Marcus,U., Jovaisas,E. & Koch,M.A. Atlanta Conference on AIDS, April 15-17 (1985)
24. Fuchs,D., Reibnegger,R., Dierich,M.P. & Wachter,H. München AIDS-Tage, Jan.19-21 (1990)
25. Abb,J. & Deinhardt,F. J.Infect.Dis. 150, 158-159 (1984)
26. Brown,M.F., Cook,R.G., Van, M. & Rich,R.R. Hum.Immunol.11, 219-223 (1984)
27. Greaves,M.F., Verbi,W. & Hayward,A. Eur.J.Immunol. 9, 356-359 (1979)
28. Ko.H.S., Fu,S.M., Yu,D.T.& Kunkel,H.G. J.Exp.Med.150, 246-249 (1979)
29. Engelman,E.G., Benike,C.J. & Charron, D.J. J.Exp.Med. 152, 114-118 (1980)
30. Pawelec,G., Blaurock,M., Schneider,E.M., Shaw,S. & Wernet,P. Eur.J.Immunol. 12, 967-969 (1982)
31. Brodsky,F.M., Parham,P. & Bodmer,W.F. Tissue Antigens 16, 30-35 (1980)
32. Fainboim,L.,Navarrete,C. & Festenstein,H. Nature 288, 391-395 (1980)

16

EUROPEAN CUP TRIAL

A Randomized Trial Comparing the Efficacy of Chemotherapy with Purged or Unpurged Autologous Transplantation in Adults with Poor Risk Relapsed Follicular NHL: An EBMT Working Party Trial

A. Porcellini,[1] H. Schouten,[2] B. Stade,[3] and R. Koll[3]

[1] Division of Hematology/BMT Center
Cremona, Italy
[2] Department of Internal Medicine
Academic Hospital
Maastricht, The Netherlands, for the EBMT Working Party
[3] Baxter Deutschland GmbH
Unterschleissheim, Germany

Follicular non- Hodgkin's lymphomas consist of the follicular centroblastic-centrocytic lymphomas (Kiel classification) and of the follicular small cleaved (FSC), follicular mixed (FM), and follicular large cell (FLC) lymphomas (International Working Formulation, IWF). The FSC and FM subtypes forming the majority of follicular lymphoma are considered to be of low-grade malignancy whereas the FLC subtype is classified as intermediate. About one third of all the patients with NHL have a follicular histology with a median age of over 50 years.

The majority of patients have generalized disease at presentation. About 10-20% of all patients will remain in stage I-II after appropriate staging. Localized follicular NHL can be cured with radiation therapy, but disseminated follicular NHL is rarely curable. Although complete remission can be achieved with intensive cytotoxic therapy, eventually almost all patients will relapse. Survival curves show no evidence of a cure plateau, although the median survival is between 4 and 10 years.

In stage III follicular lymphoma some success has been seen with radiotherapy (1,2). Approximately 75% of patients survive longer than 5 years following total nodal irradiation and 50-65% more than 10 years.

Recent evidence suggests that high dose therapy followed by autologous bone marrow transplantation in relapsed progressive diffuse NHL is effective and can result in cure (3,4). Long term disease free survival may be possible in probably more than 50% of patients depending on disease status at the time of transplantation and response to previous chemotherapy. However, only a comparatively small number of patients with relapsed follicular NHL have been treated with high dose therapy and ABMT.

Molecular Biology of Hematopoiesis 5, edited by Abraham et al.
Plenum Press, New York, 1996

EX-VIVO PURGING

Although the real efficacy of ex-vivo purging of the graft has yet to be established, neoplastic cells which remain in the harvest may represent an important problem for autologous transplantation procedures in follicular NHL since bone marrow infiltration is common at the time of diagnosis. Moreover marrow infiltration is frequently the last residual site of the disease following chemotherapy. Histology negative bone marrow may still be infiltrated by lymphoma cells as judged by molecular biology criteria.

The t(14;18) translocation can be demonstrated in approximately 85% of patients with follicular lymphomas (5). By means of the polymerase chain reaction (PCR) as few as one lymphoma cell bearing the bcl 2 translocation in 10^5 normal cells can be detected (6). In a recent study, patients in remission after conventional chemotherapy have been shown to have marrow involvement by PCR for bcl2 (7).

Several methods for eliminating clonogenic cells from autologous harvest have been developed, such as active cyclophosphamide derivatives, moabs, photodinamic dyes and immuno-magnetic procedures (8,9).

Although the real clinical efficacy of ex-vivo cleansing is still under judgment, recent results support the view that purging may be important in follicular NHL (10). This study showed that DFS of patients with no evidence of lymphoma in their marrow after immunomagnetic purging was significantly better as compared to those in which the PCR remained positive after cleansing the marrow. However the detection of bcl 2 by PCR does not necessarily reflect the presence of clonogenic tumor cells since lysed cells may release intact DNA sequences from the (14 18) breakpoint.

Although the validity of in vitro purging in reducing lymphoma cell contamination is convincing, the clinical value of this procedure in lymphoma patients is yet to be established. The fact that the majority of relapses after autologous bone marrow transplantation occur at the site of previous bulk disease, suggests that relapse is mainly due to endogenous minimal residual lymphoma cells which are resistant to high dose therapy rather than from marrow contamination. The role, therefore, of purging in ABMT/PBSCT remains to be established.

IMMUNO-MAGNETIC BEAD PURGING

Immuno-magnetic purging with anti-B cell monoclonal antibodies has been established for clinical use in ABMT of patients with B-cell lymphoma (10).

The magnetic purging has several advantages compared with purging techniques using antibody plus complement.

- Immuno-magnetic purging is 10-100 times more efficient than target cell lysis with complement activating antibodies and rabbit complement; with immuno-magnetic purging, a reduction of clonogenic cells by 4-5 orders of magnitude can be achieved.
- Immuno-magnetic purging can be more easily standardized. In addition, this method involves the least expenditure of time.
- For the analysis of the efficacy of the purging method, immuno-magnetic bead purging has some more advantage: first, there is no problem with false positive results as in the case in purging with antibody plus complement. In the latter method an underestimation of the log-kill may result from amplification of contaminating DNA carrying breakpoint sequences liberated by the lysed tumor cells. Second, false negative results can also be avoided by using immuno-mag-

netic techniques, because DNA from removed cells which are enriched in the "bead fraction" can be examined for specific breakpoint sequences by PCR.

Because of the large heterogeneity of neoplastic cells with respect to antigen-density, simultaneous use of several antibodies to the differentiation antigens (antibody cocktail) is necessary for optimal purging (11).

More recently such a technique was successfully established for peripheral blood stem cells (PBSC); in fact the first clinical experiences with immuno-magnetic purging in PBSCT proved that this is a safe procedure and patients experienced a rapid and undelayed engraftment (12).

RATIONALE FOR THE TRIAL

An international randomised phase III trial for the treatment of relapsed follicular NHL was started in Europe in 1993 (TAB I). Studies referred to the above suggest that ABMT might be an effective therapy for patients with relapsed follicular lymphoma, but it has not been proven to be superior to conventional salvage therapy. Neither is there evidence that ABMT with purged marrow is better than ABMT with unpurged marrow The high sensitivity to irradiation of follicular NHL would indicate that TBI followed by ABMT (or PBSCT) could be a curative therapy. Relatively long survival is, however, possible in patients with follicular NHL with only modest treatment. We therefore wish to establish whether patients with adverse prognostic factors, such as a short progression-free interval will benefit from the long established combination CY- TBI and ABMT/PBSCT (13). Therefore, the objectives of this protocol are:

- to compare in a randomized phase III trial the efficacy of chemotherapy versus HDT followed by either purged or unpurged stem cell support (either BM or PBSC) in patients with poor risk follicular NHL (IWF Groups: B,C and D).
- to determine the clinical usefulness of immuno-magnetic ex-vivo purging of the transplant.

TRIAL DESIGN

This is a randomized phase III trial, sponsored by the Baxter Biotech Immunotherapy Division, Munich, which involves treating adult patients, age 15-65 years, with relapsed follicular NHL. Patients who achieve either a CR or PR and who have limited bone marrow infiltration (<20% B-lymphocytes) will be randomized to either 3 further cycles of chemotherapy or high dose therapy and un-purged stem cell support or high dose therapy and purged stem cell support.

As of July 1,1995, 66 patients entered the study from 24 Euopean participating centers and of these 29 were randomly assigned to one of the 3 arm therapy 10 chemotherapy, 10 unpurged, 9 purged).

REFERENCES

1. Mc Laughlin P, Fuller LM, Velasquez WS et al: Stage III follicular lymphoma: durable remissions with a combined chemotherapy-radiotherapy regimen. J Clin Oncol 5: 867-874, 1987.
2. Mendenhall NP, Noyes WD amd Million RR: total body irradiation for stage II-IV non-Hodgkin's lymphoma: ten-year follow-up. J Clin Oncol 7:67-74, 1989.

3. Armitage JO: Bone marrow transplantation in the treatment of patients with lymphoma. Blood 73:1749-58, 1989.
4. Rohatiner AZS, Price CSA, Arnott S et al: Ablation therapy with autologous bone marrow transplantation as consolidation of remission in patients with follicular lymphoma.in Dicke K (ed): Autologous Bone Marrow Transplantation. Proc Fifth Int. Symp, 1991, pp 465-471.
5. Schouten HC, Sanger WG and Armitage JO: Chromosomal abnormalities in malignant lymphoma and Hodgkin's disease. Leukemia and Lymphoma 5:93-100, 1991.
6. Lee MS, Chang KS, Cabanillas F et al: Detection of minimal residual cells carrying the t(14;18) by DNA sequence amplification. Science 237:175-178, 1987.
7. Gribben JG, Freedman AF, Newberg D et al: Immunologic purging of marrow assessed by PCR before autologous bone marrow transplantation for B-cell lymphoma. N. Engl. J. Med. 325:1525-1533, 1991
8. Takvorian T, Canellos GP, Ritz J et al: Prolonged disease-free survival after autologous bone marrow transplantation in patients with non-Hodgkin's lymphoma with a poor prognosis. N Eng J Med 316:1499-1505 1987.
9. Kvalheim G, Sotrensen D, Fodstad O et al: Immunomagnetic removal of B-lymphoma cells from human bone marrow: a procedure for clinical use. Bone Marrow Transp 3:31-41 1988.
10. Nadler LM, Takvorian T, Botnick L et al: Anti-B1 monoclonal antibody and complement treatment in autologous bone marrow transplantation for relapsed B-cell non-Hodgkin's lymphoma Lancet 2: 427-431, 1984.
11. Kiesel S, Pezzutto A, Haas R et al: Functional evaluation of CD19 and CD22-negative variants of B-lymphoid cell lines. Immunology 64: 445-450, 1988.
12. Straka C, Drexler E, Mitterer M Langenmayer I, Pfeffernkorn L., Stade B, Koll R and Emmerich B: Autotransplantation of B-cell purged peripheral blood progenitor cells in B-cell lymphoma. The Lancet, 345:797-798, 1995.
13. Clift RA, Duckner CD, Thomas ED et al: The treatment of acute non-lymphoblastic leukemia by allogeneic marrow transplantation. Bone Marrow Transp. 2:243-248, 1987.

MECHANISMS OF CHROMOSOMAL TRANSLOCATIONS IN MALIGNANT LYMPHOMAS

Michael J. Uppenkamp,* Heinz-Gert Höffkes, Peter Meusers, and Günter Brittinger

University of Essen
Department of Medicine, Division of Hematology
45211 Essen, Hufelandstr. 55, Germany

ABSTRACT

Recurring and highly consistent chromosomal abnormalities are a feature of many hematopoietic neoplasms. They are carried throughout the malignant cells indicating that they occurred prior to clonal expansion. Over the last decade it has become clear that distinct translocations play a major role in the pathogenesis of malignant lymphomas. The main cytogenetic changes are exhibited by deletions, translocations and inversions of genetic material, resulting in loss of tumor suppressor genes, activation of proto-oncogene products or creation of tumor-specific fusion proteins.

Molecular genetic analyses of chromosomal translocations in lymphomas have demonstrated rearrangements of immunoglobulin (Ig) and T-cell receptor (TCR) genes as a consequence of chromosomal breakage. The frequent involvement of Ig and TCR genes in lymphoid chromosome translocations suggests a role of site specific recombinases in aberrant recombination. Additional mechanisms may be involved in this process: Polypurine stretches (Alu-elements) can act as targets for endonucleases and breakpoint binding proteins can stimulate homologous site-specific recombination in conjunction with nucleases. The putative molecular mechanisms leading to a new chromosomal translocation t(14;18)(q11;q23) in a T-cell line established from a patient with ataxia telangiectasia will be discussed. The characterization of these molecular events will not only provide further insight into the pathogenesis of lymphoid tumors but will enable the development of diagnostic probes for the detection of cytogenetic abnormalities.

* Corresponding author: Michael J. Uppenkamp, MD, University of Essen, Department of Medicine, Division of Hematology, 45211 Essen, Hufelandstr. 55, Germany. Tel.: 0201/723-2215; FAX: 0201/723-5934

Molecular Biology of Hematopoiesis 5, edited by Abraham et al.
Plenum Press, New York, 1996

INTRODUCTION

Chromosomal aberrations are a hallmark of many hematopoietic neoplasias. Distinct translocations in leukemias and lymphomas may lead to the activation of proto-oncogenes or creation of tumor specific fusion-proteins. With the tools of molecular genetics at hand, new chromosomal translocations have been identified that play an important role in the oncogenesis of specific malignant lymphomas. These techniques also have pinpointed the mechnisms of some of the most frequent cytogenetic changes that have been observed in some malignant lymphomas.

There are three main cytogenetic changes that may occure: deletions, translocations and inversions. Deletions often result in loss of a tumor suppressor gene as the RB-gene. Translocations or inversions can be specific or non-random for certain lymphomas or are sporadic and therefore will be seen only in the tumor from one single patient.

Many different and specific translocations have been cloned and sequenced and certain principles have evolved from these analyses (1).

The consequences of chromosomal translocation are that either the TCR- or immunglobulin-chain gene comes to lie near a proto-oncogene, thereby activating it. The chromosomal translocation breakpoints generally occur at diversity or joining region segments. The activated oncogene lies on the opposite side of the breakpoint. The transcriptional promoter of the oncogene may remain intact after the translocation. This type of translocation creates a non-fusion gene; no chimeric protein will be generated.

Another possibility is that the breaks occur within a gene on each chromosome involved, thus creating a fusion-gene encoding a chimeric protein. Exons of two different genes have become joined as a result of the translocation or inversion. Transcription of the resultant chimeric gene would run from the promoter of the gene 5' to the polyadenylation site of the joined gene 3', thus generating an mRNA encoding a fusion-protein of both genes.

In malignant lymphomas both types of chromosomal translocations can be found (Table 1). As for solid tumors, it has become evident that the main consequence of chromosomal translocations is the formation of fusion proteins.

The genes involved often encode transcription factors, indicating that altered transcription plays a major role in tumorigenesis. In addition, although Ig- or TCR-gene associated abnormalities do not generally alter the activated oncogene, there are often somatic mutations in the translocated oncogenes that may play a role in tumor progression (2).

Looking at all these different translocations and activation of oncogenes, the question arises what kind of molecular mechanisms may play an important role in the translocation process? And whether similar molecular mechanisms lead to these chromosomal breaks in different diseases?

REARRANGEMENT PROCESS

Antigen Receptor Genes

In malignant lymphomas, it became evident that almost always an immunoglobulin- or TCR-gene is involved. In order to be functionally expressed these antigen receptor genes have to undergo a rearrangement process, thereby joining contiguous sequences of either V-J or V-D-J information. The intervening DNA will be removed in this process. This rearrangement process is carried out during early T- or B-cell differentiation in a hierarchial order (3).

Table 1. Some of the chromosomal translocations in hematopoeitic neoplasias

Non-fusions

Type	Affected gene	Disease	Rearranging gene
t(8;14)(q24;q32)	c-myc (8q24)	BL, BL-ALL	IgH, IgL
t(2;14)(p12;q24)			
t(8;22)(q24;q22)			
t(1;14)(p32;q11)	tal-1 (1p32)	T-ALL	TCR-β
t(11;14)(p13;q11)	ttg-2 (11p13)	T-ALL	TCR-α/β
t(10;14)(q24;q11)	hox-11 (10q24)	T-ALL	TCR-α/β
t(3;14)(q27;q32)	bcl-6 (3q27)	NHL	IgH
t(11;14)(q13;q32)	bcl-1 (11q23)	Mantle cell	IgH
t(14;18)(q32;q21)	bcl-2 (18q21)	follicular	IgH
t(14;19)(q32;q13.1)	bcl-3 (19q13.1)	B-CLL	IgH

Gene fusions

Type	Affected gene	Protein domain	Fusion protein	Disease
inv. 14 (q11;q32)	TCR-α (14q11)	TCR-Cα	V_H-TCR-Cα	T/B-cell lymphoma
	V_H(14q32)	IgV_H		
t(9;22)(q34;q11)	c-abl (9q34)	tyrosine kinase	serine+tyrosine kinase	CML/ALL
	bcr (22q11)	serine kinase		
t(15;17)(q21;q11-22)	PML (15q21)	Zinc-finger	Zinc-finger+RAR DNA	APL
	RARA (17q21)	Retinoic acid R.-α	and ligand binding	
t(4;16)(q26;p13)	IL-2 (4q26)	IL-2	IL-2/TM	T-lymphoma
	BCM (16p13.1)	?/TM domain		

Abbreviations used: BL, Burkitt's lymphoma; ALL, acute lymphocytic leukemia; CLL, chronic lymphocytic leukemia; NHL, non-Hodgkin's lymphoma; TCR, T cell receptor; Ig, immunoglobulin (H, heavy- or L, light-chain); PML, promyelocytic leukemia; APL, acute promyelocytic leukemia; CML, chronic myelogenous leukemia; RARA, retinoic acid receptor-a; IL-2, interleukin-2; TM, TM-sequence.

Assembly of antigen receptor genes is mediated by specific enzymes, so called recombinases that recognize conserved DNA sequence elements between the V and J or D-J segments. They are composed of heptamer and nonamer regions separated from each other by a spacer region whose sequence is not conserved, but in most cases is either 12 or 23 bp in length.

Gene segments flanked by joining signals with 12 bp spacers are joined only to gene segments flanked by joining signals with 23 bp spacers. Next the recombinase cleaves at the juncture between variable and joining gene segments, deleting the intervening sequences. The V- and J-genes are then ligated and extra nucleotides may be added at the V-J or D-J junction by the well characterized enzyme terminal deoxynucleotidyl transferase (TdT) (4).

The process of recombination is very complex. Little is known about the enzymatic machinery that carries out VDJ-joinings. There are at least several genes involved in V(D)J rearrangement (5). Furthermore, there is a well defined temporal order in the rearrangement of the Ig- and TCR-genes (3,4). Yet, while different sets of genes are rearranged in developing B- and T-cells, exogenously introduced TCR-genes can be efficiently recombined in B-cells, suggesting that the B and T cell lineages use the same recombinational machinery (6).

MECHANISMS OF CHROMOSOMAL TRANSLOCATIONS

Site Specific Recombinase

Chromosomal translocation in malignant lymphomas can at least partially be attributed to mistakes in VDJ-recombination.

Croce and his coworkers showed that chromosomal translocations in T- and B-cell neoplasias may have a common mechanism (7,8). They analysed a SKW-3 cell line with a t(8;14) (q24;q11) translocation between c-myc oncogene and the TCR-α gene (8).

The breakpoint on chromosome 14 was identified in close proximity to the Jα-gene segment and a heptamer and nonamer signal sequences adjacent to the breakpoint was found, separated by a 12 bp spacer. These signal sequences correspond to a heptamer and nonamer sequence on chromosome 8. Similar to receptor gene rearrangements a site specific recombinase recognizes these conserved signals and splices right in front of the heptamer sequence and the two different chromsomes are joined. The nucleotide sequences downstream and upstream of the breakpoint of cell line SKW-3 corresponds to the germline sequences of chromosom 8 and 14. Only within the breakpoint region 6 new nucleotides were introduced by the enzyme TdT, so called N-regions.

Immunoglobulin Endonuclease

As mentioned before, another common translocation is the t(14;18) translocation in follicular lymphomas. Since the reciprocal translocation takes place during early B-cell development at the time of the D to J-rearrangement it was argued, that VDJ-recombinase is involved in generation of this breakpoint.

Korsmeyer and his group, however, showed in 1987 that chromosome 18 cleavage does not involve recognition of immunglobulin-like recombinatory signals and cleavage by immunoglobulin recombinase (9). The major breakpoint region on chromosome 18 lacks highly conserved heptamer-nonamer motifs.

The sequence analysis of the germline substrates at 14q32 and 18q21 as well as the derivative 14 and 18 reciprocal partners proposed that a pre-B-cell undergoes a faulty pairing of D_H and J_H ends on chromosome 14 with a staggered doublestrand break on chromosome 18 generated by an immunglobulin endonuclease. N-nucleotides are then added at the ends of chromosome 14, the single-stranded regions of 18 are filled in by polymerase, and the 14/18 ends are ligated to generate the reciprocal translocation. No fragile site was identified by this group within the major breakpoint region of the bcl-2 locus.

Breakpoint Binding Proteins

Jäger et al. from Vienna proposed in 1993 another mechanism of the chromosomal translocation t(14;18) in follicular lymphomas (10). This group detected a 45 Kd breakpoint binding protein which binds to a polypurine-stretch within the major breakpoint region as well as to corresponding immunoglobulin sequences.

The translocation is initiated by base pairing between polypurine tracts (CHI-like sequences) within the breakpoint region of bcl-2 and D-J gene segments of the immunoglobuline gene (Figure 1). The 45 Kd protein serves as a clamp through its capacity to bind to sequences on both chromosomes. Both breakpoint sequences are now in close contact with each other, so that a cross-over could take place. The DNA is then cleaved by the endonuclease on chromosome 18 and V(D)J-recombinase on chromosome 14 as an attempt to rearrange D to J. The chromosomes are religated after DNA repair and N-segment addition. This model can explain the selection of defined breakpoint regions for the translocation by a concerted action of sequence homology, DNA binding proteins, nucleases, and recombinases.

Other Mechanisms of Chromosomal Translocation

Other sequence elements that could generate genetic instability or facilitate homologous recombination have been identified.

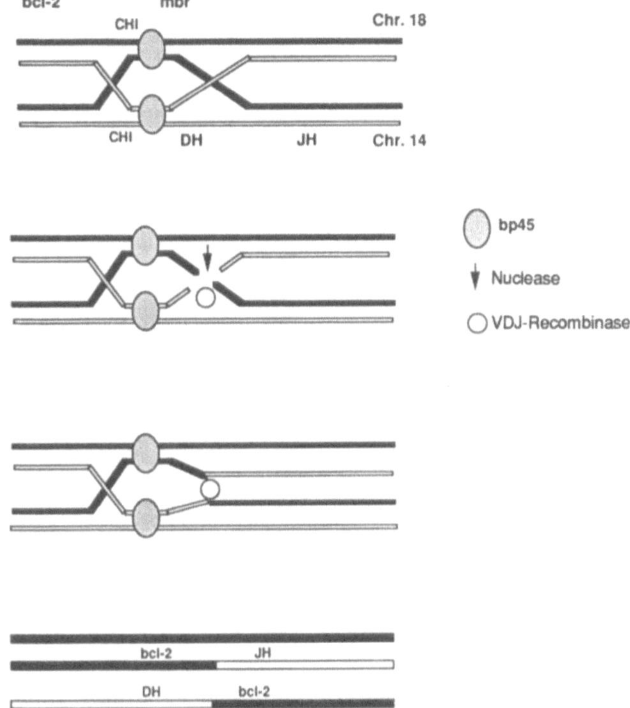

Figure 1. Proposed mechanism of t(14;18) translocation through base pairing between CHI-like sequences; a 45-Kd DNA-binding protein; cleavage by an endonuclease (chromosome 18) and a recombinase (chromosome 14); religation after DNA repair and N-segment addition.

There are hypervariable minisatellite DNA-sequences like polypurine tracts that are found around other genetic hot spots and can act as targets for endonucleases thus decreasing chromosome stability (11,12).

Similar repeats of G-rich tetranucleotides have been implicated in immunoglobulin gene class-switch (13,14).

Human *Alu*-sequences are widley dispersed, related sequences throughout the human genome. They are about 300 bp in length. *Alu*-elements have been found around many breakpoints, as for example around c-myc- and bcr-genes (15).

CHI-like sequences can activate recombinase-like mechanisms (12,16). CHI-sequence is an octamer that provides a hotspot for recombinaseA-mediated recombination in E. coli.

ATAXIA TELANGIECTASIA

A very good disorder to study chromosomal translocations is ataxia telangiectasia, which displays a chromosomal instability (17).

Ataxia telangiectasia is a rare autosomal recessive disorder characterized by progressive cerebellar ataxia, oculocutaneous telangiectases, and a primary immunodeficiency syndrome. Ataxia-patients have an elevated risk of developing malignancies, perdominantly malignant lymphomas and leukemias. Some of the malignant lymphomas and leukemias are associated with specific chromosomal translocations, involving TCR- and Ig-genes (18).

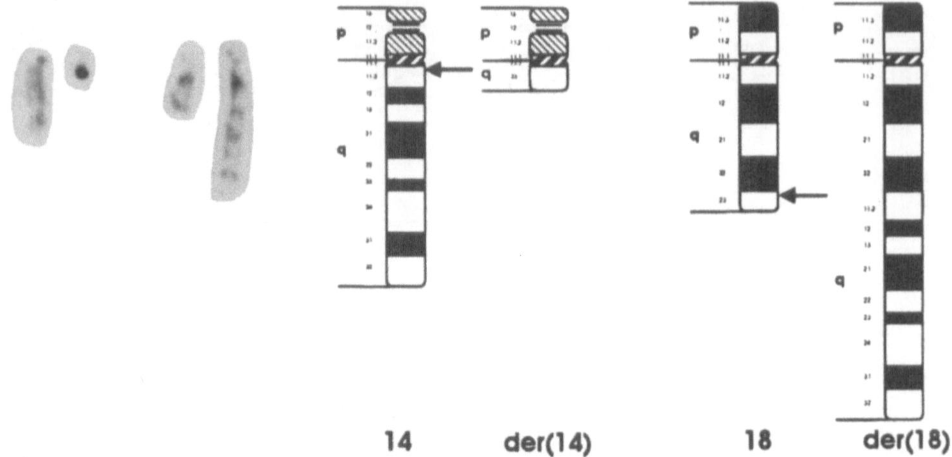

Figure 2. Reciprocal translocation t(14;18)(q11;q23) in a T-cell line of a patient with ataxia telangiectasia and chronic lymphocytic leukemia of T-cell origin.

Translocation T(14;18)(q11;q23)

We have established a T-cell line derived from a young patient with AT and a chronic lymphocytic leukemia of T-cell origin (19). These cells harbor a reciprocal translocation t(14;18)(q11;q23) (Figure 2) (20).

To chromosome 14q11 material from chromosome 18 has been added and nearly the complete long arm of chromosome 14 has been translocated to chromosome 18q23.

Moleculargenetic analyses have identified the breakpoint on chromosome 14q11 within the TCR-δ gene (Figure 3a and 3b). The restriction map of the TCR-δ gene with the V, D, J and C-region had been established earlier (21).

Using a p5DHR7 and Jδ1 probe, we could localize the breakpoint in front of the Dδ3-region gene as shown (Figure 3b). As for other rearranging genes we have identify highly conserved heptamer-nonamer signals that may have triggered the VDJ-joining.

On chromosome 18q23 we did not find such signal elements. However, we found a 300 bp *Alu*-sequence that lies 339 bp 3' to the breakpoint (Figure 3c). We speculate that this sequence plays a role in the translocation. Nevertheless, further sequencing is needed to understand the translocation process and to identify a putative oncogene on 18q23.

CONCLUSION

Consistent and specific translocations and inversions are involved in lymphoma aetilogy. Chromosomal translocations in malignant lymphomas involve Ig- and TCR-genes, which are normally rearranged in early lymphoid development. The generation of fusion-genes may also be seen in malignant lymphomas, however, this is more frequent in other hematopoietic tumors and solid tumors. At least some of the translocations are due to mistakes in V(D)J-joining. Other mechanisms include DNA-binding proteins as well as repetitive DNA-sequences that can be recognized by endonucleases. This process usually results in activation of oncogenes/transcription factors or the formation of tumor-specific fusion-proteins.

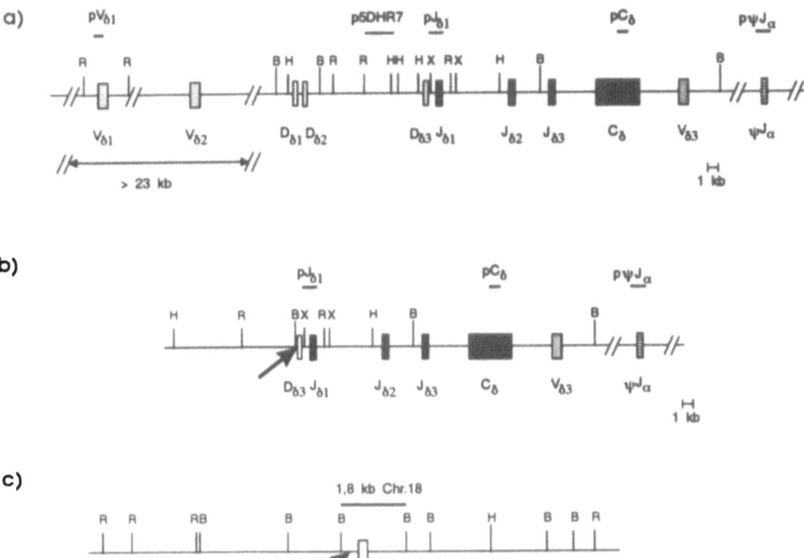

Figure 3. a) Genomic configuration of the T-cell receptor d-gene locus on chromosome 14q11 with the relevant restriction enzyme sites: E, EcoRI; B, BamHI; H, HindIII; X, XbaI. The Vd, Dd, Jd and Cd gene segements are indicated by small boxes. b) Translocation t(14;18)(q11;q23): The breakpoint is marked by an arrow. c) Restriction map of the breakpoint region on chromosome 18q23. The breakpoint is marked by an arrow; the *Alu*-sequence is shown as an open box.

Cell lines established from patients with ataxia telangiectasia who developed a malignant lymphoma are of specific interest since they may give information on the mechanisms of translocation and may also identify new genes that play a role in the pathogenesis of lymphomas. The cell line we presented here, harbors a t(14;18)(q11;q23) translocation that was also present in the original tumor cell suspension. Although the gene on chromsome 18 has not been identified yet, the mechanism of chromosomal breakage on 14q11 involves site specific recombinase, whereas an Alu-sequence near the breakpoint may play a significant role in the break on chromosome 18.

REFERENCES

1. Rabbitts TH: Chromosomal translocations in human cancer. Nature 372:143, 1994
2. Yano T, Sander CA, Clark HM, Dolezal MV, Jaffe, ES, Raffeld M: Clustered mutations in the second exon of the MYC gene in sporadic Burkitt's lymphoma. Oncogene 8:2741, 1993
3. Cossman J, Uppenkamp M, Sundeen J, Coupland R, Raffeld M: Molecular genetics and the diagnosis of lymphoma. Arch Pathol Lab Med 112: 117, 1988
4. Lewin B: Genes III; John Wiley &Sons, New York 3rd ed.:641, 1987
5. Weei-Chin Lin, Desiderio S: Regulation of V(D)j recombination activator protein RAG-2 by phosphorylation. Science 260:953, 1993
6. Schatz DG, Baltimore D: Stable expression of immunoglobulin gene V(D)J recombinase activity by gene transfer into 3T3 fibroblasts. Cell 53:107, 1988
7. Tsujimoto Y, Gorham J, Cossman J, Jaffe E, Croce CM: The t(14;18) chromosome translocations involved in B-cell neoplasms result from mistakes in VDJ joining. Science 229:1390, 1985
8. Finger LR, Harvey RC, Moore RCA, Showe LC, Croce CM: A common mechanism of chromosomal translocation in T- and B-cell neoplasia. Science 234:982, 1986

9. Bakhshi A, Wright JJ, Graninger W, Seto M, Owens J, Cossman J, Jensen JP, Goldman P, Korsmeyer SJ: Mechanism of the t(14;18) chromosomal translocation: structural analysis of both derivative 14 and 18 reciprocal partners. Proc Natl Acad Sci USA 84: 2396, 1987
10. Jaeger U, Purtscher B, Delle Karth G, Knapp S, Mannhalter C, Lechner K: Mechanism of the chromosomal translocation t(14;18) in lymphoma: detection of a 45-Kd breakpoint binding protein. Blood 81:1833, 1993
11. Wahls WP, Wallace LJ, Moore PD: Hypervariable minisatellite DNA is a hotspot for homologous recombination in human cells. Cell 60:95, 1990
12. Krowczynska AM, Rudders RA, Krontiris TG: The human minisatellite consensus at breakpoints of oncogene translocations. Nucl Acid Res 18:1121, 1990
13. Gerstein RM, Frankel WN, Chih-Lin Hsieh, Durdik JM, Rath S, Coffin JM, Nisonoff A, Selsing E: Isotype switching of an immunoglobulin heavy chain transgene occurs by DNA recombination between different chromosomes. Cell 63: 537, 1990
14. Iwasato T, Shimizu A, Honjo T, Yamagishi H: Circular DNA is excised by immunoglobulin class switch recombination. Cell 62:143, 1990
15. Chen SJ, Chen Z, D'Auriol L, LeConiat M, Grausz D, Berger R: Ph^+bcr^- acute leukemias: implication of Alu sequences in a chromosomal translocation occurring in the new cluster region within the BCR gene. Oncogene 4:195, 1989
16. Jeffreys AJ, Wilson V, Swee Lay Thein: Hypervariable 'minisatellite' regions in human DNA. Nature 314:67, 1985
17. Boder E: Ataxia-telangiectasia: an overview. In: Ataxia telangiectasia (Gatti RA & Swift M, Eds), Kroc Foundation Series. AR Liss Inc, New York 1985
18. Aurias A, Dutrillaux B: Probable involvement of immunoglobulin superfamily genes in most recurrent chromosomal rearrangements from ataxia telangiectasia. Hum Genet 72:210, 1986
19. Uppenkamp M, Gana Dresen I, Becher R, Raffeld M, Meusers P: Molecular analysis of an ataxia telangiectasia T-cell clone with a chromosomal translocation t(14;18) - evidence for a breakpoint in the T-cell receptor d-chain gene. Leukemia Res 16: 681, 1992
20. Becher R, Dührsen, U: Distinct chromosome abnormalities in ataxia telangiectasia with chronic T-cell lymphocytic leukemia. Cancer Genet Cytogenet 26:217, 1987
21. Isobe M, Russo G, Haluska F, Croce CM: Cloning of the gene encoding the d subunit of the human T-cell receptor reveals its physical organization within the a-subunit locus and its involvement in chromosomal translocations in T-cell malignancies. Proc Natl Acad Sci USA 85:3933, 1988

MOLECULAR FINDINGS AND CLASSIFICATION OF MALIGNANT LYMPHOMAS

Stefano A. Pileri,[1*] Claudio Ceccarelli,[1] Elena Sabattini,[1] Donatella Santini,[1] Ornella Leone,[1] Stefania Damiani,[1] Lorenzo Leoncini,[2] and Brunangelo Falini[3]

[1] Second Service of Pathologic Anatomy
Unit of Haematopathology
Bologna University, School of Medicine
S. Orsola Hospital, Bologna, Italy
[2] Institute of Pathologic Anatomy
Siena University, School of Medicine
Le Scotte Hospital, Siena, Italy
[3] Institute of Haematology, Haematopathology Laboratory
Perugia University, School of Medicine
Monteluce Hospital, Perugia, Italy

SUMMARY

The Authors review the problem of lymphoma classification in the light of the Revised European-American Lymphoma (R.E.A.L.) scheme, recently proposed by the International Lymphoma Study Group (ILSG). In particular, the R.E.A.L. Classification is a list of clinico-pathologic entities, all well known from the literature, which the ILSG members agreeded on. Although it contains nothing new, for the first time it organically reports all the elements - including immunophenotype and molecular data - which characterize a given lymphoma entity. This approach corresponds to the need of objective criteria, which integrate the often puzzling morphologic findings. Furthermore, a better knowledge on the molecular events which contribute to tumour development and progression, is of paramount importance for the identification of more specific and successful therapies. Some relevant molecular findings included in the Classification and additional data obtained by the ILSG members following its publication are discussed.

* Corresponding author: Professor Stefano A. Pileri, Direttore del Secondo Servizio di Anatomia Patologica - Sezione di Istologia Emolinfopatologica, Policlinico S. Orsola, Via Massarenti 9, 40138 Bologna. Tel.: +39+51+636-4674/3603/4562. Fax: +39+51+398973.

Molecular Biology of Hematopoiesis 5, edited by Abraham et al.
Plenum Press, New York, 1996

INTRODUCTION

The classification of malignant lymphomas has been a matter of debate for at least 3 decades. In the mid seventies, 6 distinct schemes[1-6] were in use, which actually made impossible any comparison among the results obtained by different clinico-pathological trials. Since 1983, following a study sponsored by the National Cancer Institute of the United States,[7] two main approaches were applied to the study and diagnosis of malignant lymphomas: the Working Formulation for Clinical Usage[7] (WF) and the Kiel Classification (KC) in its original form[5,8] and updated versions.[9,10] In particular, the WF[7] became popular in USA, while the KC[5,8-10] was extensively used in Europe. This situation did not facilitate the transatlantic communications: in fact, while the categories of the WF were tailored according to the cell size and crude survival of lymphoid tumours,[7] the KC represented a highly refined scheme, reflecting the physio-pathology of the lymphoid tissue.[5,8-10]

In 1991, 19 haematopathologists (9 from Europe, 9 from USA, and 1 from Hong-Kong) founded the International Lymphoma Study Group (ILSG) aiming to contribute to the discipline development and the solution of the existing controversies. Joint work at a multi-head microscope and two pilot-studies on mantle-cell lymphoma[11] and nodular lymphocyte predominant Hodgkin's disease[12] showed that - in spite of terminological differences - the ILSG members fully agreed on the criteria definition of most if not all lymphoid neoplams. Such a broad agreement was largely due to the regular application of immunophenotyping and molecular biology techniques to the study and diagnosis of malignant lymphomas by all the companions. On these bases, at a meeting held in Berlin in April 1993, the ILSG members figured out a proposal for a classification of malignant lymphomas, which might overcome the transatlantic variancies and represent the framework for future international studies. The proposal was published in Blood one year ago under the name "Revised European-American Lymphoma (R.E.A.L.) Classification."[13]

PRINCIPLES OF THE R.E.A.L. CLASSIFICATION

The Classification is a list of diseases (Table 1), that can be recognised with available techniques and appear to represent distinct clinico-pathologic entities.[13-15] It was produced by consensus among the ILSG members on entities already reported in the literature, but partly not included in previous classifications (e.g. mantle cell and marginal cell lymphomas).[13-15] It aims to define criteria for diagnosis, includes major clinical features, and gives synonyms in other classifications. Furthermore, it can easily be revised and updated.

Morphology is regarded as the principle basis for the identification of lymphoid neoplasms, but is recognised to be at times insufficient to distinguish between entities. Immunophenotype and genetic features are reported as an integral part of the definition of a lymphoid tumour: although they are not required for the diagnosis in every case, they are quite useful in confirming the diagnosis or in resolving the differential diagnosis problems. Normal counterparts of lymphoma cells can be postulated for some entities: however, they are unknown for many defined categories and thus cannot be the sole basis for a lymphoma classification.

The R.E.A.L. Classification incorporates all neoplasms of known lymphoid derivation, including Hodgkin's disease and multiple myeloma, whose lymphoid nature is now accepted by most authors.[16-20] In particular, three major categories are recognised: B-cell neoplasms, T-cell neoplasms, and Hodgkin's disease. Among B- and T-cell tumours, a further distinction is made between the ones derived from precursors (i.e. lymphoblastic lymphoma/leukaemia) and those originating from peripheral elements. Peripheral B- and T-cell

Table 1. List of the entities included in the R.E.A.L. Classification. * These categories are thought likely to include more than one disease entity

B-Cell Neoplasms

I. Precursor B-cell neoplasm: Precursor B-lymphoblastic leukaemia/lymphoma (B-Lb)
II. Peripheral B-cell neoplasms
  1. B-cell chronic lymphocytic leukaemia/prolymphocytic leukaemia/small lymphocytic lymphoma (B-CLL)
  2. Lymphoplasmacytoid lymphoma/immunocytoma (Ic)
  3. Mantle cell lymphoma (MCL)
  4. Follicle center lymphoma, follicular (FCL)
     Provisional cytological grades: I (small cell), II (mixed small and large cells), III (large cells)
     Provisional subtype: diffuse, predominantly small cell type
  5. Marginal zone lymphoma (MZL)
     Extranodal (MALT type ± monocytoid B cells)
     Provisional subtype: Nodal (± monocytoid B cells)
  6. Provisional entity: Splenic marginal zone lymphoma (± villous lymphocytes)
  7. Hairy cell leukaemia 8HCL9
  8. Plasmacytoma/plasma cell myeloma (PCM)
  9. Diffuse Large B-cell lymphoma* (DLBCL)
     Subtype: Primary mediastinal (thymic) B-cell lymphoma
  10. Burkitt's lymphoma
  11. Provisional entity: High-grade B-cell lymphoma, Burkitt-like* (Bt-like)

T-Cell and Putative NK-Cell Neoplasms

I. Precursor T-cell neoplasm: Precursor T-lymphoblastic leukaemia/lymphoma (T-Lb)
I. Peripheral T-cell and NK-cell neoplasms
  1. T-cell chronic lymphocytic leukaemia/prolymphocytic leukaemia (T-CLL)
  2. Large granular lymphocyte leukaemia (LGL)
     T-cell type
     NK-cell type
  3. Mycosis fungoides/Sezàry syndrome (MF/SS)
  4. Peripheral T-cell lymphomas, unspecified* (PTCL, unsp)
     Provisional cytological categories: medium-sized cell, mixed medium and large cell, large cell, lymphoepithelioid
     Provisional subtype: Hepatosplenic $\gamma\delta$ T-cell lymphoma
     Provisional subtype: Subcutaneous panniculitis T-cell lymphoma
  5. Angioimmunoblastic T-cell lymphoma (AILD-PTCL)
  6. Angiocentric lymphoma (AngC-PTCL)
  7. Intestinal T-cell lymphoma (± entreopathy associated) (Intest-PTCL)
  8. Adult T-cell lymphoma/leukaemia (ATL/L)
  9. Anaplastic large cell lymphoma, CD30+, T- and null-cell types (ALCL)
  10. Provisional antity: Anaplastic large-cell lymphoma, Hodgkin's-like (ALCL-HL)

Hodgkin's Disease
I. Lymphocyte predominance
II. Nodular sclerosis
III. Mixed Cellularity
IV. Lymphocyte depletion
V. Provisional entity: Lymphocyte-rich classical HD

lymphomas include accepted and provisional entities: the latter correspond to tumours, which have been published in sufficient details to be convincing, but are still matter of disagreement within the ILSG members or are thought to need further studies. Finally, the Classification comprises unclassifiable cases, which correspond to unusual or border-line tumours or else to cases with technical artifacts, which prevent their clear-cut categorization.

It should be underlined that — conversely to previous schemes[5,7-10] — the R.E.A.L. Classification does not introduce subdivisions into grades of malignancy. This decision is in line with the concept that each entity displays a spectrum of clinico-pathological manifestations, associated with a more or less aggressive behaviour. The course of the disease is not conditioned only by the histological type, but also by other independent factors, such as proliferative activity,[21] apoptosis,[22] multi-drug resistance,[23] cytokine production,[24] presence of specific receptors,[17] etc.

On the whole, the R.E.A.L. Classification seems to overcome most of the limitations of the previous schemes, which can be summarized as follows.

Working Formulation.[7] a) it was never intended to represent a classification as such, but merely as an attempt to make comparable the six different classifications in use in the seventies, of which only one — the Kiel Classification[5,8-10] — has actually remained in use; b) its categories were defined on the basis of about a thousand lymphomas treated with protocols in use during the sixties and seventies; c) the diagnosis of the various cases was performed only by review of slides stained with hematoxylin-eosin, completely bypassing the modern techniques of phenotyping and molecular biology; d) no distinction is made between B- and T-cell lymphomas, which seems to be relevant;[25,26] e) some categories, such as follicular center lymphomas, appear excessively fragmented, while others, such as diffuse small-cell lymphomas comprise different types of tumor both as regards their histogenesis and/or clinical profiles.

Updated Kiel Classification.[5,8-10] a) it does not provide for extranodal lymphomas, which account for more than 30% of all lymphoid tumours;[27] b) it does not include some gradings of follicular center lymphomas, which in the United States are felt to be needed;[28] c) the subclassifications of some types of lymphoma, such as the peripheral T-cell one, are scarcely reproducible either by the same observer or among different ones.[29]

MOLECULAR BIOLOGY, CYTOGENETICS, AND THE R.E.A.L. CLASSIFICATION

For the first time, the REAL Classification includes the genetic features among the characteristics of a lymphoma entity. In particular, it reports the molecular findings which can assist in diagnosing each tumour, such as specific chromosomal aberrations, integration of viruses, and clonal rearrangements of immunoglobulin (Ig) or T-cell receptor (TCR) genes and oncogenes (Tables 2 and 3).

Some of the genetic features included in the R.E.A.L. Classification[13-15] as well as recent data from the ILSG members will be discussed in the following.

Precursor B-Lymphoblastic Leukaemia/Lymphoma. Immunoglobulin (Ig) heavy chain genes are usually rearranged; light chain genes my be rearranged. Rearrangement of T-cell receptor (TCR) genes is present in a minority of the cases. Variable cytogenetic abnormalities can be seen: among these t(1;19) and t(9;22) unfavourably affect the clinical

Table 2. B-cell lymphomas: summary of main genetic features

Histotype	Ig CR	TCR CR	Main chromosomal aberration(s)	Variants	Genes Involved	Prognostic Impact
B-Lb	+	-/+	t(1;19); t(9;22)			unfavourable
			> 50 chromosomes			favourable
B-CLL	+		+12; 13q			
Ic	+		t(9;14)			
MCL	+		t(11;14)		bcl-1	
FCL	+		t(14;18)	t(2;18)	bcl-2	
				t(18;22)	bcl-2	
MZL	+		+3			
			t(11;18)			
HCL	+					
PCM	+					
DLBCL	+		3q27 aberration		bcl-6	favourable
			t(14;18)		bcl-2	unfavourable
Burkitt's	+		t(8;14)	t(2;8)	c-myc	
				t(8;22)		
Bt-like	+					

course of the disease, while the detection of more than 50 chromosomes represents a favourable prognostic indicator.

B-Cell Chronic Lymphocytic Leukaemia/Prolymphocytic Leukaemia/Small Lymphocytic Lymphoma. The genes encoding for Ig heavy and light chains are rearranged. Trisomy 12 is found in about one third of the cases, while abnormalities of 13q occur in up to 25% of the patients. In some examples of B-prolymphocytic leukaemia t(11;14) and *bcl-1* rearrangement have been reported: further studies are needed to clarify the relationship between these cases and mantle-cell lymphoma (see below).

Mantle-Cell Lymphoma. Besides Ig genes clonal rearrangements, this tumour shows a characteristic chromosomal abnormality t(11;14), which involves the Ig heavy chain locus and the *bcl-1* locus. This abnormality is observed in about 70% of the cases and seems to be almost always present in the aggressive forms of the tumour (blastoid and polymorphic).[30]

Table 3. T-cell lymphomas: summary of main genetic features

Histotype	Ig CR	TCR CR	Main chromosomal aberration(s)	Variants	Genes involved	Prognostic impact
T-Lb	-/+	+/-	14q11-14; 7q35			
			1p32-34		tal-1	
T-CLL		+	inv 14 (q11;q32)			
T-LGL		+				
NK-LGL		-				
MF		+				
PTCL, unsp		+				
AILD-PTCL		+				
AngC-PTCL	-/+	+/-				
Intest-PTCL		+				
ATL/L		+				
ALCL		+/-	t(2;5)		NPM/ALK	
ALCL,HL		?				

Thus, it has recently been proposed by Ott et al.[30] as an indicator of clonal evolution in this lymphoma. The t(11;14) results in overexpression of a gene known as PRAD1, which encodes for cyclin D1, a cell-cycle protein. The latter is not detectable in normal lymphoid cells; by contrast, it can easily be revealed in the neoplastic elements of mantle-cell lymphoma by means of specific antibodies.[31] Recent studies have shown that mantle-cell lymphoma expresses V_H genes with no or very little somatic mutations like the physiologic cells of the follicle mantle.[32]

Follicle Center Lymphoma. Ig heavy and light chain genes are clonally rearranged. In 70-95% of the cases, a specific translocation (14;18) does occur; more rarely, t(2;18) and t(18;22) can be found. All these chromosomal aberrations produce rearrangement of the *bcl-2* gene and overexpression of its product, which is a well-known anti-apoptotic agent.[22,33] It should be underlined, however, that the overexpression of the *bcl-2* product can be observed in many lymphomas other than the follicle center ones, independently of t(14;18).[22,33] This is not surprising, as the *bcl-2* product is physiologically carried by lymphoid elements, which need protection against apoptosis.[33] Noteworthly, normal follicle center cells do not express the protein in question, as they undergo clonal selection by means of apoptosis during germinal centers activation.[33] Thus, the presence and lack of the *bcl-2* protein represent useful tools for the distinction between neoplastic and hyperplastic germinal centers (*bcl-2* positive and negative, respectively).[33]

Marginal Cell Lymphoma. Clonal rearrangements of Ig genes are usually found, while *bcl-1* and *bcl-2* genes are never involved. Trisomy 3 and t(11;18)(q21;q21) have been encountered in some cases: it is still unclear whether they are associated or not with disease progression.[34] Recent studies have revealed that the neoplastic cells of extranodal marginal cell lymphomas show a high number of V_H somatic mutations — but not ongoing mutations — as usually observed in post-germinal center cells.[35]

Diffuse Large B-Cell Lymphoma. Besides the common rearrangement of the Ig genes, this is a heterogeneous group of tumours. In fact, they display in 30-40% of the cases a translocation involving the *bcl-6* gene (chromosome 3q27):[36] these probably represent *de novo* forms of the tumour, which seem to be highly sensitive to aggressive chemotherapy regimens.[37] On the other hand, about 30% of the cases show rearrangement of the *bcl-2* gene: possibly, these are neoplasms secondary to follicle center lymphomas and have a rather aggressive clinical course.[37] Rearranged *c-myc* gene is uncommon.

Burkitt's Lymphoma. Ig heavy and light chain genes are usually rearranged, as well as the *c-myc* gene. However, the type of chromosomal aberration involving the *c-myc* gene can remarkably vary: more often, the gene translocates from chromosome 8 to the Ig heavy chain region on chromosome 14 [t(8;14)], less commonly to light chain loci on 2 [t(2;8)] or 22 [t(8;22)]. In African (endemic) cases, the breakpoint on chromosome 14 involves the heavy chain joining region, while in non-endemic cases it encompasses the Ig heavy chain switch region, suggesting that the chromosomal aberration occurs at a later stage of B-cell development. EBV integration is shown in the tumour cells of most African cases (95%), is infrequent in non-endemic patients (15-20%), and has an intermediate incidence in HIV-positive subjects (30-40%).

Precursor T-Lymphoblastic Leukaemia/Lymphoma. The TCR-γ genes are usually rearranged, while the TCR-β ones may or may not. Rearrangement of the genes encoding for Ig heavy chain can sometimes be observed. The commonest chromosomal abnormalities involve 14q11-14 or 7q35. The *tal-1* gene (chromosome 1p32-34) is implicated in up to one third of the cases as a result of chromosomal translocation or microscopic deletions of its regulatory elements.

T-Cell Chronic Lymphocytic Leukaemia/Prolymphocytic Leukaemia. Besides the rearrangement of TCR genes which is virtually always present, 75% of the cases show an inversion of chromosome 14 (q11;q32).

Large Granular Lymphocyte Leukaemia, Natural Killer Cell Type (NK-LGL). TCR and Ig genes are usually germline. EBV (often in a clonal episomal form) can be demonstrated in a high percentage of cases, especially those occurring in Japan and Asians. Similar findings have recently been detected in angiocentric peripheral T-cell lymhomas of the nose in Asian patients.[38] This observation prompts additional studies in order to better define the boundaries between NK-LGL and angiocentric T-cell lymphoma.

Adult T-Cell Lymphoma/Leukaemia. On molecular grounds, this entity is characterized by rearrangement of the TCR genes and clonally integrated HTLV-1.

Anaplastic Large Cell Lymphoma (ALCL) of T- or Null-Phenotype. A proportion (12% to 50%) of the cases show t(2;5),[39] resulting in the fusion of the NPM gene (5q35) with the ALK gene (2p23):[40] the higher percentage is observed in series predominated by pediatric cases.[41-43] An increasing amount of data on this topic has become available in the last few months: in fact, besides the employment of the classical cytogenetic techniques, which can be laborious, the fusion gene can now easily be shown by a specific probe[39] or, indirectly, by the reaction between its product and a specific antibody.[44] Although some of these recent data suggest that the t(2;5) can also occur in Hodgkin's disease,[45] studies from some of the ILSG members seem to confirm its peculiar association to ALCL.[42,43,46] This strenghtens the need for the systematic search of the translocation in ALCL, Hodgkin's like, in order to solve the problem of the exact location of this provisional entity. As to what TCR genes are concerned, clonal rearrangements can be demonstrated only in a ratio of ALCLs.[47]

Hodgkin's Disease. Molecular data on Hodgkin's disease have often been conflicting. This has be attributed to the scarcity of neoplastic cells to be analysed or to the subtype of the disease investigated (for instance, nodular lymphocyte predominant Hodgkin's disease vs classical Hodgkin's disease). A typical example of the variability of the genetic findings in Hodgkin's disease is provided by the studies on the integration of EBV in the genoma of neoplastic cells: EBV is virtually absent in nodular lymphocyte predominant Hodgkin's disease (NLPHD),[12] while it can be observed in non-NLPHD cases in percentages, which remarkably differ according to the histotype (e.g. nodular sclerosing HD vs mixed cellularity HD in Western countries)[48] or geografic area (e.g. Europe vs Africa).[51] Even more contrasting are the findings on the clonality of Hodgkin and Reed-Sternberg cells (HRSC).[50-52] In this context, some interesting data have been provided by a recent study on 12 examples of HD, either of the nodular sclerosing or of the mixed cellularity type, all characterized by expression of the L26/CD20 molecule by neoplastic cells.[53] When applied to these cases, the single cell-PCR technique has shown the presence of rearrangements of the Ig genes in all HRSC: however, these turned out to be polyclonal in 6 cases, oligoclonal in 3, and monoclonal in the remaining 3. It is noteworthy that 2 of the polyclonal cases were in advanced stage (III and IV, respectively) and that HRSC could complete the mitotic cycle only in the monoclonal cases. These findings address new questions on the natural history of HD (progression from a polyclonal process to a monoclonal one through an oligoclonal phase?).

ACKNOWLEDGMENT

This paper was supported by grants from CNR (ACRO 10 project), AIRC and MURST.

REFERENCES

1. Rappaport H: Tumors of the Hematopoietic System, in Atlas of Tumor Pathology, Sect. 3, Fasc. 8. Washington, D.C.: Armed Forces Institute of Pathology 1966.
2. Lukes RJ, Collins RD: Immunologic characterization of human malignant lymphomas. Cancer 34: 1488, 1974.
3. Dorfman RF: Classification of non-Hodgkin's lymphomas. Lancet i: 1295, 1974.
4. Bennett MH, Farrer-Brown G, Henry K, Jelliffe AM: Classification of non-Hodgkin's lymphomas. Lancet ii: 405, 1974.
5. Gérard-Marchant R, Hamlin I, Lennert K, Rilke F, Stansfeld AG, van Unnik JAM: Classification of non-Hodgkin's lymphomas. Lancet ii: 406, 1974.
6. Mathé G, Rappaport H, O'Conor GT, Torloni H: Histological and cytological typing of neoplastic diseases of haematopoietic and lymphoid tissues, in International Histological Classification of Tumors, No. 14. Geneva: World Health Organization, 1976.
7. Non-Hodgkin's lymphoma pathologic classification project. National Cancer Institute sponsored study of classifications of non-Hodgkin's lymphomas: Summary and description of a Working Formulation for clinical usage. Cancer 49: 2112, 1982.
8. Lennert K: Malignant lymphomas other than Hodgkin's disease. New York, Heidelberg, Berlin, Spinger-Verlag 1978.
9. Stansfeld A, Diebold J, Kapanci Y, Kelenyi G, Lennert K, Mioduszewka O, Noel H, Rilke F, Sundstrom C, van Unnik J, Wright D: Updated Kiel classification for lymphomas. Lancet i: 292, 1988.
10. Lennert K, Feller AC: Histopathology of non-Hodgkin's lymphomas (based on the Updated Kiel Classification). Berlin, Heidelberg, New York, London, Paris, Tokyo, Hong Kong, Barcellona, Budapest, Springer-Verlag 1992.
11. Banks P, Chan J, Cleary M, Delsol G, De Wolf-Peeters C, Gatter K, Grogan T, Harris N, Isaacson PG, Jaffe E, Mason D, Pileri S, Ralfkiaer E, Stein H, Warnke R: Mantle cell lymphoma: a proposal for unification of morphologic, immunologic and molecular data. Am J Surg Pathol 16: 637, 1992.
12. Mason DY, Banks P, Chan J, Cleary M, Delsol G, De Wolf-Peeters C, Falini B, Gatter K, Grogan T, Harris N, Isaacson P, Jaffe E, Knowles D, Mller-Hermelink HK, Pileri S, Ralfkiaer E, Stein H, Warnke R: Nodular lymphocyte predominance Hodgkin's disease. A distinct clinicopathological entity. Am J Surg Pathol 18: 526, 1994.
13. Harris N, Jaffe E, Stein H, Banks PM, Chan J, Cleary M, Delsol G, De Wolf-Peeters C, Falini B, Gatter K, Grogan T, Isaacson P, Knowles D, Mason DY, Mller-Hermelink HK, Pileri S, Piris M, Ralfkiaer E, Warnke R: A Revised European-American Classification of lymphoid neoplasms: a proposal from the International Lymphoma Study Group. Blood 84: 1361, 1994.
14. Chan J, Banks PM, Cleary M, Delsol G, De Wolf-Peeters C, Falini B, Gatter K, Grogan T, Harris N, Isaacson P, Jaffe E, Knowles D, Mason DY, Mller-Hermelink HK, Pileri S, Piris M, Ralfkiaer E, Stein H, Warnke R: A proposal for classification of lymphoid neoplasms by the International Lymphoma Study Group. Histopathol 25: 517, 1994.
15. Chan J, Banks PM, Cleary M, Delsol G, De Wolf-Peeters C, Falini B, Gatter K, Grogan T, Harris N, Isaacson P, Jaffe E, Knowles D, Mason DY, Mller-Hermelink HK, Pileri SA, Piris M, Ralfkiaer E, Stein H, Warnke R: A Revised European-American Lymphoma Classification proposed by the International Lymphoma Study Group. Am J Clin Pathol 103: 543, 1995.
16. Stein H, Herbst H, Anagnostopoulos I, Niedobitek G, Dallenbach F, Kratzsch H-C: The nature of Hodgkin and Reed-Sternberg cells, their association with EBV, and their relationship to anaplastic large-cell lymphoma. Ann Oncol 2: 33, 1991.
17. Falini B, Pileri S, Pizzolo G, Durkop H, Flenghi L, Stirpe F, Martelli MF, Stein H: CD30 (Ki-1) molecule: a new cytokine receptor of the tumor necrosis factor receptor superfamily as a tool for diagnosis and immunotherapy. Blood 85: 1, 1995.
18. Chilosi M, Pizzolo G: Biopathologic features of Hodgkin's disease. Leuk Lymphoma 16: 385, 1995.
19. Corradini P, Voena C, Omede P, Astolfi M, Boccadoro M, Dalla-Favera R, Pileri A: Detection of circulating tumor cells in multiple myeloma by a PCR based method. Leuk 7: 1879, 1993.

20. Bersagel PL, Smith AM, Szozepek A, Mant MJ, Betch AR, Pilarski LM: In multiple myeloma, clonotypic B lymphocytes are detectable among CD19+ peripheral blood cells expressing CD38, CD56, and monotypic Ig light chain. Blood 85: 486, 1995.
21. Gerdes J, Stein H, Pileri S, Rivano MT, Gobbi M, Ralfkiaer E, Nielsen KM, Pallesen G, Bartels H, Palestro G, Delsol G: Prognostic relevance of tumour-cell growth fraction in malignant non-Hodgkin's lymphomas. Lancet ii: 448, 1987.
22. Leoncini L, Del Vecchio MT, Megha T, Barbieri P, Pileri S, Sabattini E, Tosi P, Kraft R, Cottier H: Correlations between apoptotic and proliferative indices in malignant non-Hodgkin's lymphomas. Am J Pathol 142: 755, 1993.
23. Pileri SA, Sabattini E, Falini B, Tazzari PL, Gherlinzoni F, Michieli MG, Damiani D, Zucchini L, Gobbi M, Tsuruo T, Baccarani M: Immunohistochemical detection of the multidrug transport protein p170 in human normal tissue and malignant lymphomas. Histopathol 19: 131, 1991.
24. Ruco LP, Pomponi D, Pigott R, Gearing AJH, Baiocchini A, Baroni CD: Cytokine production (IL-1 alpha, IL-1 beta and TNF alpha) and endothelial cell activation (ELAM-1 and HLA-DR) in reactive lymphadenitis, Hodgkin's disease, and in non-Hodgkin's lymphomas: an immunocytochemical study. Am J Pathol 140: 1337, 1992.
25. Lippman SM, Miller TP, Spier CM, Slymen DJ, Grogan TM: The prognostic significance of the immunophenotype in diffuse large-cell lymphoma: A comparative study of the T-cell and B-cell phenotype. Blood 72: 436, 1988.
26. Coiffier B, Brousse N, Peuchmaur M, Berger F, Gisselbrecht C, Bryon PA, Diebold J: Peripheral T-cell lymphomas have a worse prognosis than B-cell lymphomas: A retrospective study of 361 immunophenotyped patients treated with the LNH-84 regimen. Ann Oncol 1: 45, 1990.
27. Isaacson PG, Norton AJ: Extranodal lymphomas. Edinburgh, London, Madrid, Melbourne, New York, Tokyo, Churchill Livingstone, 1994, p 1.
28. Nathwani B, Metter G, Miller T, Burke J, Mann R, Barcos M, Kjeldsberg C, Dixon D, Winberg C, Whitcomb C, Jones S: What should be the morphologic criteria for the subdivision of follicular lymphomas? Blood 68: 837, 1986.
29. Hastrup N, Hamilton-Dutoit S, Ralfkiaer E, Pallesen G: Peripheral T-cell lymphomas: an evaluation of reproducibility of the updated Kiel classification. Histopathol 18: 99, 1991.
30. Ott MM, Ott G, Kuse R, Porowski P, Gunzer U, Feller AC, Mller-Hermelink HK: The anaplastic variant of centrocytic lymphoma is marked by frequent rearrangements of the *bcl-1* gene and high proliferation indices. Histopathol 24: 329, 1994.
31. Zukerberg LR, Yang W-I, Arnold A, Harris NL: Cyclin D1 expression in non-Hodgkin's lymphoma. Detection by immunohistochemistry. Am J Clin Pathol 103: 756, 1995.
32. Hummel M, Tamaru J, Kalvelage B, Stein H: Mantle cell (previously centrocytic) lymphomas express V_H genes with no or very little somatic mutations like the physiologic cells of the follicle mantle. Blood 84: 403, 1994.
33. Pileri S, Poggi S, Sabattini E, Santucci S, Melilli G, Falini B, Tosi P: Apoptosis as programmed cell death (PCD): cupio dissolvi in cell life. Curr Diagn Pathol 1: 48, 1994.
34. Isaacson PG: The MALT lymphoma concept updated. Ann Oncol 6: 319, 1995.
35. Stein H: Personal communication.
36. Lo Coco F, Ye BH, Lista F, Corradini P, Offit K, Knowles DM, Chaganti RSK, Dalla-Favera R: Rearrangements of bcl-6 gene in diffuse large cell non-Hodgkin's lymphoma. Blood 83: 757, 1994.
37. Offit K, Lo Coco F, Louie DC, Parsa NZ, Leung D, Portlock C, Ye BH, Lista F, Filippa DA, Rosenbaum A, Ladanyi M, Jhanwar S, Dalla-Favera R, Chaganti RSK: Rearrangement of the bcl-6 gene as a prognostic marker in diffuse large cell lymphoma. N Engl J Med 331: 74, 1994.
38. Soler J, Bordes R, Ortuno F, Montagud M, Martorell J, Pons C, Nomdedeu J, Lopez-Lopez JJ, Prat J, Rutllant M: Aggressive natural killer leukemia/lymphoma in two patients with lethal midline granuloma. Br J Haematol 86: 659, 1994.
39. Mason DY, Bastard C, Rimokh R, Dastugue N, Huret J-L, Kristoffersson U, Magaud J-P, Nezelof C, Tilly H, Vannier J-P, Hemet J, Warnke R: CD30-positive large cell lymphomas ("Ki-1 lymphoma") are associated with a chromosomal translocation involving 5q35. Br J Haematol 74: 161, 1990.
40. Morris SW, Kirstein MK, Valentine MB, Dittmer KG, Shapiro DN, Saltman DL, Look AT: Fusion of a kinase gene, ALK, to a nucleolar protein gene, NPM, in non-Hodgkin's lymphomas. Science 263: 1281, 1994.
41. Bullrich F, Morris SW, Hummel M, Pileri S, Stein H, Croce CM: Nucleophosmin (NPM) gene rearrangement in Ki-1-positive lymphomas. Cancer Res 54: 2873, 1994.

42. Lopategui JR, Sun LH, Chan JKC, Gaffey MJ, Frierson HF, Glackin C, Weiss LM: Low frequency association of the t(2;5)(p23;q35) chromosomal translocation with CD30+ lymphomas from American and Asian patients. Am J Pathol 146: 323, 1995.
43. Wellmann A, Clark HM, Otsuki T, Jaffe ES, Raffeld M: Detection of the t(2;5)(p23;q25) in classical anaplastic large cell lymphoma (ALCL) by reverse transcripatse-polymerase chain reaction. Mod Pathol 8: 123a, 1995 (abstr).
44. Shiota M, Fujimoto J, Takenaga M, Satoh H, Ichinohasama R, Abe Bae, Nakano M, Yamamoto T, Mori S: Diagnosis of t(2;5)(p23;q35)-associated Ki-1 lymphoma with immunohistochemistry. Blood 84: 3648, 1994.
45. Orscheschek K, Merz H, Hell J, Binder T, Bartels H, Feller AC: Large-cell anaplastic lymphoma-specific translocation (t[2;5] [p23;q35]) in Hodgkin's disease: indication of a common pathogenesis? Lancet i: 87, 1995.
46. Stein H, Delsol G: Data presented at the Fifth Meeting of the International Lymphoma Study Group. Hong Kong, April 24-26, 1995.
47. Herbst H, Tippelmann G, Anagnostopoulos I, Gerdes J, Schwarting R, Boehn T, Pileri S, Jones DB, Stein H: Immunoglobulin and T cell receptor gene rearrangements in Hodgkin's disease and Ki-1-positive anaplastic large cell lymphoma: Dissociation between phenotype and genotype. Leuk Res 13: 103, 1989.
48. Brousset P, Chittal S, Schlaifer D, Icart J, Payen C, Rigal-Huguet F, Voigt J-J, Delsol G: Detection of Epstein-Barr virus messanger RNA in Reed-Sternberg cells of Hodgkin's disease by in situ hybridization with biotinylated probes on specially processed modified acetone methyl bemzoate xylene (ModAMeX) sections. Blood 77: 1781, 1991.
49. Leoncini L, Pileri S, Sabattini E: Neoplastic cells of Hodgkin's disease show differences in EBV expression between Kenya and Italy. Int J Cancer 65:781, 1996.
50. Küppers R, Rajewski K, Zhao M, Simons G, Lauman R, Fischer R, Hansmann M-L: Hodgkin's disease: Hodgkin and Reed-Sternberg cells picked from histological sections show clonal immunoglobulin rearrangements and appear to derive from B cells at various stages of development. Proc Natl Acad Sci USA 91: 10962, 1994.
51. Roth J, Daus H, Trümper L, Gause A, Salamon-Looijen M, Preundschuh M: Determination of immunoglobulin heavy-chain rearrangement at the single-cell level in malignant lymphomas: No rearrangement is found in Hodgkin and Reed-Sternberg cells. Int J Cancer 57: 1, 1994.
52. Delabie J, Tierens A, Wu G, Weisenburger DD, Chan WC: Lymphocyte predominance Hodgkin's disease: Lineage and clonality determination using single-cell assay. Blood 84: 3291, 1994.
53. Hummel M, Ziemann K, Lammert H, Pileri S, Sabattini E, Stein H: Classical Hodgkin's Lymphoma of B-Immunophenotype: Frequently a Polyclonal Disease? N Engl J Med 333:901, 1995.

MANTLE CELL LYMPHOMA

Genetic Lesions and Their Role in Classification and Progression

G. Ott,[1*] M. M. Ott,[1] J. Kalla,[1] A. Helbing,[1] B. Schryen,[1] T. Katzenberger,[1] J. Bartek,[2] A. Dürr,[1] J. G. Müller,[1] H. Kreipe,[1] and H. K. Müller-Hermelink[1]

[1] Institute of Pathology
University of Würzburg, Germany
[2] Division of Cancer Biology
Danish Cancer Society
Copenhagen, Denmark

Mantle cell lymphoma, a tumor derived from CD5-positive virgin B-cells of the follicular mantle zone, accounts for 5-10% of non-Hodgkin's lymphomas in adults (Banks et al.1992, Harris et al. 1994). It was originally described in the Kiel classification (Stansfeld et al. 1974) as centrocytic lymphoma.

Beginning very early, several investigators looking at larger series of lymphomas classified as MCL or centrocytic lymphoma noticed, next to a small-cell variant of the disease, the existence of so-called blastic or anaplastic variants (Lennert 1978, Lardelli et al. 1990, Fisher et al. 1995). The significance of this finding remains controversial, as are the exact criteria and biological features to differentiate these variants.

Our present material comprises 64 cases of MCL. These cases were first classified according to morphological features and growth pattern.

The small-cell or *common* variant was separated from two types of blastic lymphomas which were termed *lymphoblastoid* and *pleomorphic* according to their cytomorphological features. In all cases included, the growth pattern, the cytologic features and the uniform immunophenotype of CD5 positive, CD10 and CD23 negative cells (Stein et al. 1984) was suggestive of mantle cell lymphoma.

Within this cytomorphological spectrum of MCL, these cases were analyzed according to proliferation indices, p53 expression, bcl-1 rearrangements at the major translocation cluster region, Cyclin D1 expression and cytogenetic features.

*Address for correspondence: Dr. G. Ott, Pathologisches Institut der Universität Würzburg, Josef-Schneider-Straße 2, D-97080 Würzburg, Germany. Tel.: 0931 2013792; Fax.: 0931 2013440.

Molecular Biology of Hematopoiesis 5, edited by Abraham et al.
Plenum Press, New York, 1996

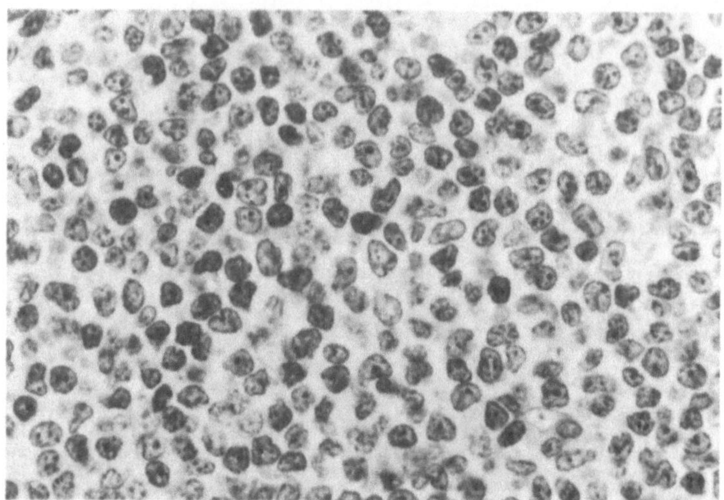

Figure 1. Small cell ("common") variant of mantle cell lymphoma with scant cytoplasma and slightly indented nuclei.

CYTOMORPHOLOGY

The *common* variant of mantle cell lymphoma (Fig.1) is composed of small-sized cells with frequently irregular nuclei, a scant cytoplasm and finely dispersed nuclear chromatin. In contrast, the *lymphoblastoid* variant (Fig.2), consisting of medium-sized blast cells, shows somewhat rounder nuclear contours with more vesicular chromatin structure. The *pleomorphic* variant (Fig.3) is either composed of large or a mixture of medium-sized and large cells with sometimes very pleomorphic or bizarre nuclei. In each subgroup,

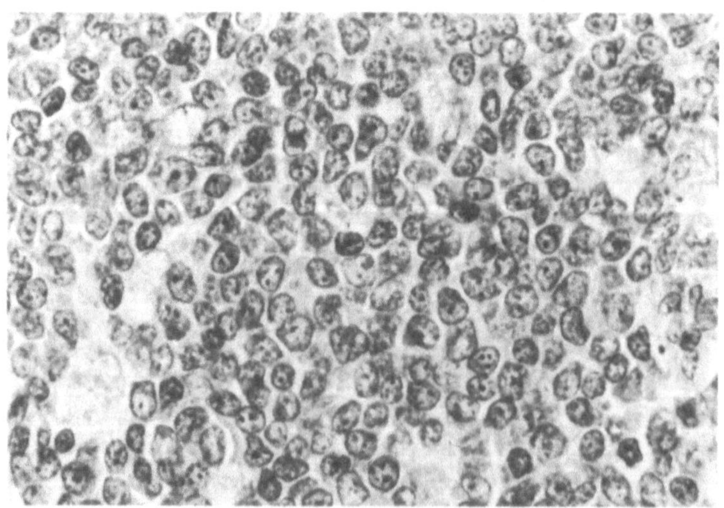

Figure 2. Lymphoblastoid variant of MCL. Note rounder nuclear contours and small, indistinct nucleoli.

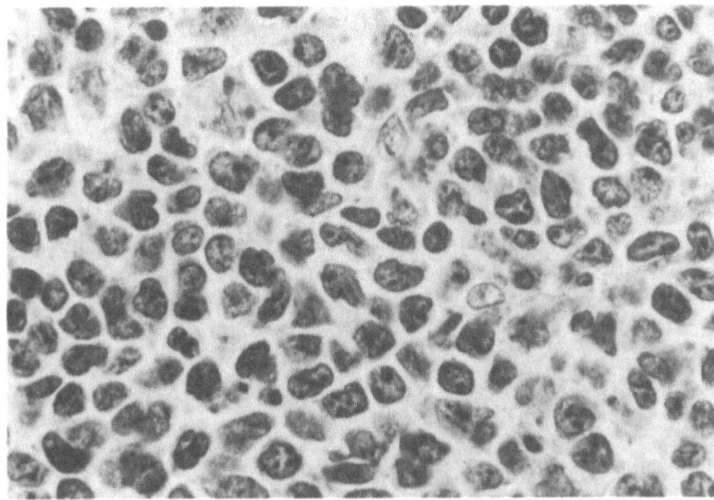

Figure 3. Pleomorphic variant of MCL. Large cells with a small rim of cytoplasm and pleomorphic, sometimes cleaved nuclei with several nucleoli.

immunophenotyping of fresh-frozen material yielded the classical immunophenotype of MCL (Stein et al. 1984).

Regarding the growth pattern, the majority of cases of the common variant displayed a perifollicular or nodular growth pattern, while both lymphoblastoid and pleomorphic MCL were predominantly characterized by a diffuse growth pattern, albeit also showing perifollicular and nodular infiltrates in a minor part of the cases.

PROLIFERATION INDICES

There was a pronounced difference in proliferation indices between the various subgroups with the median of proliferation being in the range of 25% in common variants and in the range of 50% and 60% in blastic variants, the Ki67 staining showing sometimes more than 90% of cells in the cycle.

p53 EXPRESSION

p53 expression was equally different in the cytomorphologically defined subgroups. While only 6% of common variants showed overexpression of p53 as ascertained by the DO1 antibody in paraffin material, 21% of large cell variants were positive in that assay, the lymphoblastoid variant being slightly more affected than the pleomorphic type (27% versus 14%, respectively; Ott et al. 1996b).

BCL-1 REARRANGEMENT

The molecular genetic equivalent of the t(11;14) characteristic of MCL, the rearrangement of the Cyclin D1 gene at the major translocation cluster region of the bcl-1 locus was investigated in about two thirds of our cases in Southern blotting. For the rest of the lymphomas, available only in paraffin material, we designated a seminested PCR technique allowing for the analysis of the MTC region as well (Ott et al. 1996a). There was a distinct confirmation of our previous findings (Ott et al. 1994), now extended to the analysis of 63 cases, that anaplastic variants of MCL have a preferential break at the MTC region of Bcl-1 amounting up to 57% and 64% of positivity rates in the lymphoblastoid and pleomorphic variants, respectively, as opposed to 37% in small-cell MCL. This finding may indicate that patients with breaks at this particular site of the widely scattered breakpoint region in 11q (Williams et al. 1991) may carry an increased risk of developing an high grade variant of MCL or alternatively, bcl-1 rearranged lymphomas rather tend to behave as high grade tumors.

CYCLIN D1 EXPRESSION

The consequence of this translocation, the overexpression of the Cylin D1 gene, was studied using the DCS-6 antibody (Lukas et al. 1994). Cyclin D1 was overexpressed to a detectable level in 68% of common and 85% of blastic variants, the intensity of the staining and the amount of stained nuclei also being higher in pleomorphic and lymphoblastoid variants (Ott et al. 1996a).

CYTOGENETIC FINDINGS

MCL on the cytogenetic level is characterized, next to the t(11;14), by secondary structural and numerical chromosome aberrations. These secondary changes in malignant lymphomas have been associated, for example in germinal center lymphomas, to the progression of the disease (Tilly et al. 1994). An extensive cytogenetic and interphase cytogenetic study was performed on 50 cases, which were analyzed by fluorescence in-situ hybridization (FISH) for the occurence of numerical chromosome aberrations using centromere-specific DNA probes to chromosomes 3, 7, 18, X and the Y (Ott et al. 1996b). In 14 cases, data from classical cytogenetics were available. Only a minority of cases showed defined aneusomies for single chromosomes or structural rearrangements.

In contrast, several cases were characterized by a signal distribution indicating a tetraploid chromosome number. Altogether, only 2/26 common variants, but up to 38% of lymphoblastoid and even 80% of pleomorphic variants harboured tetraploid chromosome clones, a finding that could be reproduced both by flow cytometric DNA analysis as well as by classical cytogenetics in single MCL (Ott et al. 1996b). In other types of low and high grade B-cell lymphomas, this is a very unusual finding, tetraploid chromosome clones occurring in less than approximately 10% of cases (own unpublished data).

DISCUSSION

Mantle cell lymphoma, originally described to be composed of small- to medium-sized cells and regarded as a low-grade malignant lymphoma in the Kiel classification system, could be shown to cover a broad cytomorphological spectrum ranging from predominantly small-cell to apparently blastic variants. This distinction may be of clinical importance, since anaplastic MCL variants seem to follow a more aggressive clinical course (Brittinger 1983, Fisher et al. 1995).

The morphological spectrum of MCL is distinctly reflected by different biological features separating small-cell and blastic types. These features are (1) elevated proliferation indices in anaplastic as compared to common MCL variants, (2) a higher rate of p53 expression in lymphoblastoid and pleomorphic types, (3) the preferential occurrence of breaks with involvement of the major translocation cluster region of bcl-1 in anaplastic MCL and, most notably, (4) a pronounced tendency to chromosome numbers in the tetraploid range, especially in the pleomorphic type of MCL.

It is, of course, tempting to speculate on the overexpression of cyclin D1 being a prerequisite for the development of tetraploidizations of the chromosome set in mantle cell lymphoma. Its unique mode of pathogenesis involving the t(11;14)-induced overexpression of this novel G1-cyclin, which is not normally expressed by lymphoid cells and is known to subvert the G1 phase control of the cell cycle and to drive cells into mitosis (Matsushime et al. 1991), might indeed be the reason for the occurence of two subsequent S-phases without intervening mitotic disjunction of the sister chromatids.

Mantle cell lymphoma, therefore, is not only characterized by unique morphological, immunophenotypic, cytogenetic and molecular genetic features strengthening the view of a distinct biological entity. In addition, there are distinct variants in MCL showing particular features in part related to each other such as the bcl-1 rearrangement, the overexpression of cyclin D1 and a pronounced tendency to a tetraploidization of the chromosome clones. These different features might become means to recognize prognostically important patient sub-groups and might therefore represent criteria to assess the individual risk of patients suffering from MCL.

ACKNOWLEDGMENTS

This study was supported by the Deutsche Forschungsgemeinschaft, Sonderforschungsbereich 172, Grant C8 to G.Ott and H.K.Müller-Hermelink and Grant DFG Kr 849/4-1 to H.Kreipe.

The excellent technical assistance of Mrs. Claudia Gärtner, Mrs. Karin Heintz and Mrs. Heike Brückner is gratefully acknowledged.

REFERENCES

Banks PM, Chan J, Cleary ML, Delsol G, de Wolf-Peeters C, Gatter K, Grogan TM, Harris NL, Isaacson PG, Jaffe ES, Mason D, Pileri S, Ralfkiaer E, Stein H, Warnke RA (1992). Mantle cell lymphoma: a proposal for unification of morphologic, immunologic, and molecular data. Am J Surg Pathol 16: 637-640

Brittinger G (1983) Klinik der malignen Non-Hodgkin-Lymphome, speziell der chronischen lymphatischen Leukämie. Verh-Dtsch-Ges-Pathol. 67:494-516

Fisher RI, Dahlberg S, Nathwani BN, Banks PM, Miller TP, Grogan TM (1995) A clinical analysis of two indolent lymphoma entities: mantle cell lymphoma and marginal zone lymphoma (including the mucosa-associated lymphoid tissue and monocytoid B-cell subcategories): A Southwest Oncology Group study. Blood 85: 1075-1082.

Gerard-Marchant R, Hamlin I, Lennert K, Rilke F, Stansfeld AG, van Unnik JAM (1974): Classification of non-Hodgkin's lymphomas. Lancet 2: 406-408

Harris NL, Jaffe ES, Stein H et al. (1994) A revised European-American classification of lymphoid neoplasms : A proposal from the international lymphoma study group. Blood 84: 1361-1392.

Lardelli P, Bookman MA, Sundeen J, Longo DL, Jaffe ES (1990)Lymphocytic lymphoma of intermediate differentiation. Morphologic and immunologic spectrum and clinical correlations. Am J Surg Pathol 14: 752-763

Lennert K (1978). Lymphomas of germinal-center cells. In: Lennert K (ed) Malignant lymphomas other than Hodgkin's disease. Springer, Berlin Heidelberg New York, pp 281-345

Lukas J, Pagano M, Staskova Z, Draetta G, Bartek J. (1994). Cyclin D1 protein oscillates and is essential for cell cycle progression in human tumour cell lines. Oncogene 9: 707-718

Matsushime H, Roussel MF, Ashmun RA, Sherr CJ (1991) Colony-Stimulating Factor 1 regulates novel cyclins during the G1 phase of the cell cycle. Cell 65: 701-713.

Ott MM, Ott G, Kuse R, Porowski P, Gunzer U, Feller AC, Müller-Hermelink HK (1994) The anaplastic variant of centrocytic lymphoma is marked by frequent rearrangements of the bcl-1 gene and high proliferation indices. Histopathology 24: 329-334

Ott MM, Helbing A, Ott G, Bartek J, Fischer L, Dürr A, Kreipe H, Müller-Hermelink HK. (1996a) Bcl-1 gene rearrangement and Cyclin D1 protein expression in mantle cell lymphoma. J Pathol 179;238–242.

Ott G, Kalla J, Ott MM, Schryen B, Katzenberger T, Müller JG,Müller-Hermelink HK. (1996b) Blastoid variants of mantle cell lymphoma: Frequent bcl-1 rearrangements at the MTC locus and tetraploid chromosome clones. Submitted for publication.

Stein H, Lennert K, Feller AC, Mason DY (1984) Immunohistochemical analysis of human lymphoma: Correlation of histological and immunological categories. Adv Cancer Res 42: 67-147

Tilly H, Rossi A, Stamatoullas A, Lenormand B, Bigorgne C, Kunlin A, Monconduit M, Bastard C. (1994) Prognostic value of chromosomal abnormalities in follicular lymphoma. Blood 84: 1043-1049

Williams ME, Meeker TC, Swerdlow SH (1991)Rearrangement of the chromosome 11 bcl-1 locus in centrocytic lymphoma: analysis with multiple breakpoint probes. Blood 76: 1387-1391

20

LYMPHOBLASTIC LYMPHOMA IN ADULTS

John Sweetenham*

CRC Wessex Medical Oncology Unit
University of Southampton
Southampton General Hospital
Tremona Road, Southampton, SO16 6YD, United Kingdom

INTRODUCTION

Lymphoblastic lymphoma (LBL) accounts for approximately 4% of all adult patients with non-Hodgkin's lymphoma (NHL). It is recognised as a distinct clinopathological entity in all of the recently described classifications for NHL, including the Revised European-American Lymphoma (REAL) classification. It is a neoplasm of precursor T or B lymphocytes, which is very similar to acute lymphoblastic leukaemia on the basis of morphology and phenotype. The distinction between ALL and LBL is variable between different treatment centres, and usually based on arbitrary clinical grounds, particularly the degree of bone marrow infiltration or leukaemic overspill. Because of its rarity, it has been the subject of relatively few series in the published literature, and several aspects of its management remain unclear. The results of treatment have improved in recent years, particularly with the use of intensive remission induction therapy similar to that used in acute lymphoblastic leukaemia (ALL). With intensive chemoradiotherapy, most recent series have reported remission rates of 60% to 80%, with long term disease free survival reported in 40% to 60% of patients. Therefore, although high remission rates can be achieved with conventional dose combination chemotherapy, the relapse and progression rate is high. The use of dose intensive therapy in first remission, particularly high dose therapy with stem cell transplantation, has been reported in several series, although its role remains unclear. Similarly, the selection of patients who are at high risk of relapse or progression with conventional therapy has not been reported consistently. The optimal management of patients who fail first line therapy also remains a difficult clinical problem.

PROGNOSTIC FACTORS

The identification of 'good risk' and 'poor risk' patients with LBL has been reported in several clinical series, although there are no generally accepted risk factors for this disease.

*Tel: +44-1703-796184; Fax: +44-1703-783839; E-mail - jws@soton.ac.uk.

Molecular Biology of Hematopoiesis 5, edited by Abraham et al.
Plenum Press, New York, 1996

Table 1. Adverse prognostic factors in lymphoblastic lymphoma/leukaemia. Phenotypic and genotypic characteristics

T-cell	B-cell[1,2]
CD2 negative[4-6]	t(1;19)
NK phenotype (CD16+/CD57+)[4-6]	t(9;22)
Immature phenotype[3]	11q13
	CD10-
	CD34-
	CD24-
	CD13+
	CD33+

Various phenotypic and karyotypic features have been associated with a poor outcome.[1-6] These are summarised in table 1, although it is important to emphasize that most of these have been identified in ALL rather than LBL, primarily in children.

The identification of clinical prognostic factors has been variable in different published series, reflecting small patient numbers, different treatment regimens, differences in the staging techniques used over the long periods of some retrospective studies, and inconsistencies in the distinction of LBL and ALL. The most widely accepted prognostic factors in this disease are those reported from Stanford University. In a group of 44 adult patients with LBL, they identified good risk and poor risk groups on the basis of an elevated serum lactate dehydrogenase (LDH) level, and Ann Arbor stage IV disease, with involvement of the bone marrow or central nervous system (CNS). Patients with these risk factors had a projected 5 year freedom from relapse of 19%, compared with 94% for patients without these adverse features. Other series have identified different, and often conflicting prognostic factors, which are summarised in table 2.[7-10] It is noteworthy that the presence or absence of mediastinal disease, and immunophenotype have not been shown to have prognostic significance in any of the published series.

Figure 1 shows the actuarial overall survival for 30 consecutive, unselected adult patients with LBL treated in Southampton over a 14 year period since 1980. The median age for this group was 30 years (range 15 - 81), comprising 26 male patients and 4 females. These patients have been treated with various intensive chemo-radiotherapy protocols, which have been active in our centre. Most patients have been treated with the Stanford University protocol,[7,11] a modification of the LSA_2L_2 regimen,[12] or other ALL-like induction regimens. In addition, 2 patients received high dose therapy and autologous stem cell transplanatation (ASCT) in first CR. The CR rate of 50% and 6 year actuarial overall survival (OS) of 26% is markedly inferior to most published series. This probably reflects the unselected nature of this patient group — particularly with respect to age — 8 patients (27%) in our series were over 65 years old at presentation. Univariate analysis of this small series examined the prognostic significance of age, Ann Arbor stage, bone marrow and CNS infiltration, mediastinal involvement, immunophenotype and B symptoms at presentation for response or OS. None of these factors was shown to predict for response or survival. Furthermore, all of the long term survivors in our series were 'poor risk' at presentation according to the Stanford criteria.

In summary, analysis of prognostic factors in small series of adult patients with LBL has failed to identify a poor risk group. The recently described International Prognostic System[13] has allowed the identification of risk groups in intermediate/high grade NHL, but has not been applied specifically to LBL.

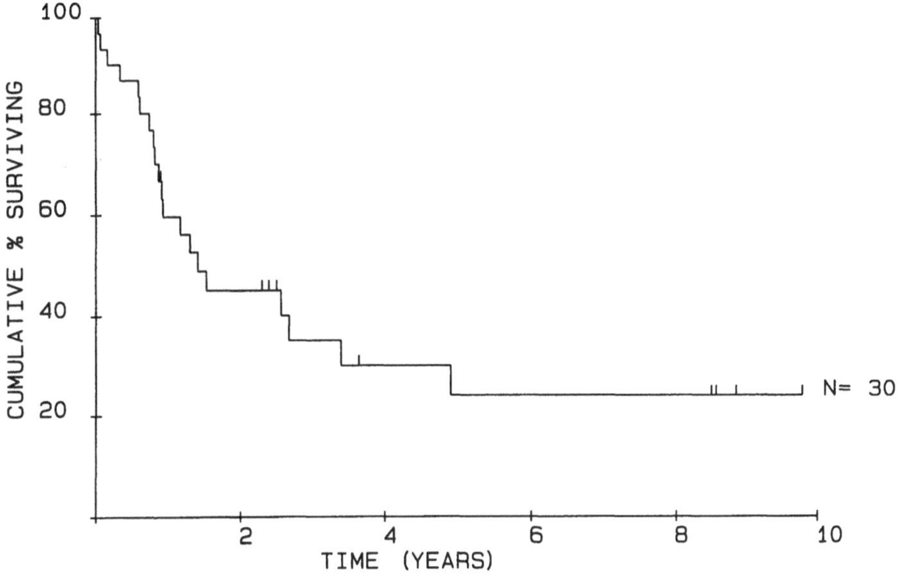

Figure 1. Lymphoblastic lymphoma overall survival.

POST REMISSION THERAPY WITH STEM CELL TRANSPLANTATION

Most previous studies have show that although high CR rates are achieved in LBL, long term DFS rates are only 40% to 60%. A high proportion of patients therefore relapse from CR, and results of salvage therapy for these patients are poor. The potential role of high dose therapy with autologous or allogeneic SCT to consolidate first remission has therefore been the subject of several single institution studies. In general, these studies have identified 'poor risk' patients in whom first remission SCT has been used, although criteria for selection of these patients have not been consistent.

Results from these single institution studies are apparently superior to those reported for 'conventional dose' strategies. Most have reported long term DFS rates of 60% to 70%.[8,14-17] The European Group for Bone Marrow Transplantation (EBMT) have previously reported 105 patients receiving high dose therapy and ASCT in first remission. The 6 year actuarial OS for this group was 63%. These results have now been updated. The EBMT lymphoma registry currently contains data on 263 patients with adult LBL undergoing ASCT in first remission. Their presenting characteristics are summarised in table 3. The actuarial OS for this group is shown in figure 2, demonstrating a projected 6 year acuarial OS of 64%.

Although all of these data demonstrate that high dose therapy can be given to patients in first remission, and may prevent relapse, they remain difficult to interpret, partly due to selection criteria, and partly because it is unclear how many patients relapsed prior to high dose therapy. The study from Genoa[14] reports 18 patients receiving high dose therapy and ASCT in first remission, with a 74% 5 year actuarial DFS. However, of the original 36 patients included in this series, only 24 reached CR after induction therapy and a further 6 patients did not undergo ASCT for various reasons. Therefore, based on intention to treat the results in this group a much poorer. In a recent study from Zurich, Jost et al have reported 17 adult patients with LBL who were treated with short term weekly chemotherapy followed

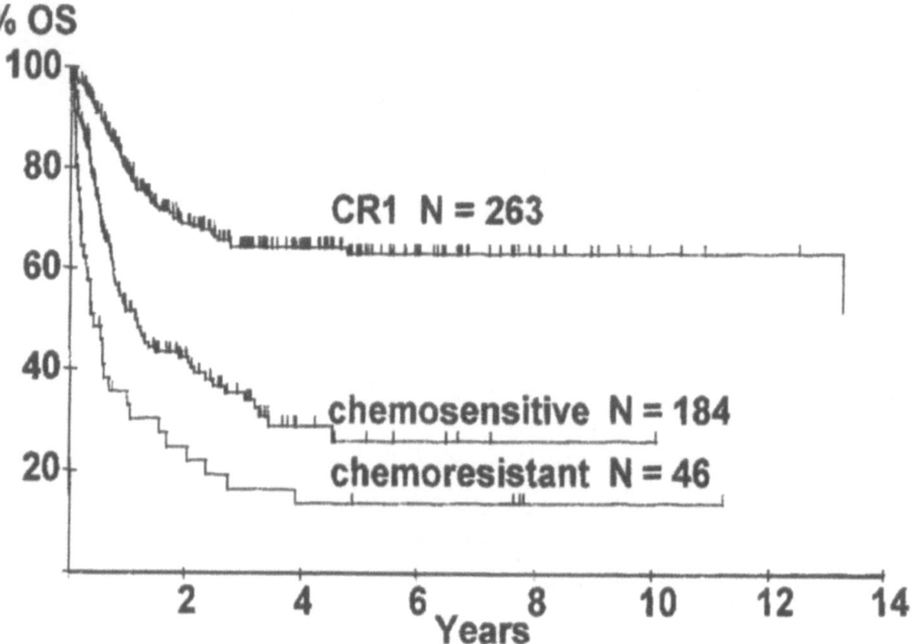

Figure 2. Overall survival – adult LBL by status at transplant.

by ASCT in first remission.[19] This study also included 7 patients with Burkitt's lymphoma. Four of the 17 LBL patients failed to achieve a CR to induction therapy and did not proceed to high dose therapy. Based on intention to treat, the actuarial 3 year overall and event free survival for this group was 48% and 31%, respectively.

The EBMT and UK Lymphoma Group are currently conducting a prospective randomised trial to assess the effect of first remission ASCT in this disease. Adult patients with LBL will receive as initial induction therapy an ALL-like regimen, after which they are randomised between conventional dose consolidation and maintenance chemotherapy, or high dose therapy with ASCT. This study is continuing to accrue patients at present.

SALVAGE THERAPY

Fewer than 10% of patients with LBL who are refractory to, or relapse after, induction chemotherapy achieve long term DFS if treated with conventional dose salvage regimens.[7-10,12] The median survival for these patients is between 6 and 9 months. High dose therapy with SCT has been used increasingly in relapsed/refractory patients, either to induce a second remission, or to consolidate second remission after conventional dose therapy. Reported series of patients treated in this way are very small. In the series reported by Morel et al.,[8] 37 of 80 patients treated for LBL required salvage chemotherapy, of whom 14 achieved a second CR. Of 7 patients treated with high dose therapy and ASCT in second CR, 3 were alive and disease free at 17+ to 70+ months.

In the series reported by the EBMT,[11] 30 patients underwent high dose therapy and ABMT in second CR. The 3 year actuarial PFS and OS for this group were 30% and 31%, respectively.

Table 2. Adverse prognostic factors in adult lymphoblastic lymphoma.

Clinical characteristics
advanced age[8]
advanced Ann Arbor stage[7,9]
bone marrow involvement[7]
central nervous system involvement[7]
serum LDH[7,8]
leukaemic overspill[10]
time to acheive complete remission[10]
presence of B symptoms[7,8]

As with other subtypes of NHL, for patients with LBL who undergo high dose therapy as a component of salvage therapy, the disease sensitivity to conventional dose therapy is predictive of subsequent outcome. The EBMT registry includes 230 patient receiving high dose therapy and ASCT as salvage therapy for LBL. As shown in figure 2, the 6 year actuarial OS for those patients with chemosensitive disease is 27%, compared with 15% for those with chemoresistant disease.

Although these results are registry based, and are therefore difficult to interpret, the results of this form of salvage therapy are apparently superior to those for conventional dose salvage. Since this is a rare disease, there are insufficient patients to conduct a meaningful randomised trial in this situation. Therefore, on the basis of these data, patients with LBL who relapse should be treated with conventional dose therapy to induce a second remission if possible. However, even in those patients with chemoresistant disease, long term DFS can be achieved in a small percentage using high dose therapy and SCT.

SOURCE OF STEM CELLS

Many centres consider allogeneic bone marrow to be the preferred source of hematopoietic stem cell for transplantation in adult LBL, although there are no comparative data to support this. Small numbers of patients in two of the published series have received allogeneic bone marrow transplants, with no apparent difference in outcome compared with autologous bone marrow recipients.[8,15] However, the patient numbers are too small to enable a meaningful comparison.

A retrospective, case controlled analysis from the EBMT compared the role of autologous and allogeneic BMT for NHL. In this study, the relapse rate for patients with LBL was lower for those undergoing allogeneic BMT compared with autologous BMT (24% versus 48%).[20] However, this difference was offset by a much higher procedure-related mortality in the allogeneic transplant patients, and allogeneic transplantation was not associated with improved survival.

Peripheral blood progenitor cells (PBPCs) are now gaining widespread use as a source of stem cells for autologous transplantation. This technique is associated with rapid and sustained multi-lineage hematopoietic recovery after high dose therapy.[21,22]

Although not specifically demonstrated in LBL, it is well documented that the protocols used to mobilise progenitor cells into the peripheral blood may also mobilise lymphoma cells.[23,24] However, as yet, this has not been shown to produce a higher relapse rate in patients receiving PBPCs.

Of the 507 patients with LBL in the EBMT database, 86% have received bone marrow, 12% peripheral blood progenitor cells, and 2% both. Since PBPC transplantation

Table 3. Presenting chracteristics of 263 patients with adult LBL undergoing first remission ASCT reported to EBMT lymphoma registry

Characteristic	%
Male	69
Female	31
Phenotype B cell	17
T cell	48
null	1
unclassified	34
Stage I	6
II	16
III	27
IV	51
B symptoms	44
CNS involvement	7
Bone marrow involvement	31
Bulky (>10cm) disease	49
Elevated LDH	67 %

is a relatively new technique, the numbers of patients having undergone this procedure is small, but at present, there is no difference in long term PFS or OS according to the source of stem cells.

CONCLUSIONS

- clinical factors do not readily identify prognostic groups
- 'biological' markers of risk are needed to identify patients suitable for high dose strategies
- the long term DFS after high dose therapy and ASCT in second CR or chemosensitive relapse is higher than with conventional dose therapy
- the long term DFS after high dose therapy and ASCT in patients with resistant relapse or refractory disease is higher than with conventional dose therapy
- the results of high dose therapy and ASCT in first remission are being assessed in a randomised clinical trial
- no meanigful comparative data exist to compare autologous and allogeneic stem cells
- autologous PBPCs are likely to become the preferred source of stem cells in this disease unless definitive evidence for a higher relapse and lower survival rate are observed.

ACKNOWLEDGMENTS

The author extends thanks to Dr. PFM Smartt and Angela Gould for the analysis of Southampton patients. Thanks are also extended to Rachel Pearce and Golnaz Taghipour for the updated analysis of the EBMT data, to Dr. AH Goldstone, Chairman of the EBMT Lymphoma Working Party, for permission to include the EBMT data, and to all of the contributing EBMT centres.

REFERENCES

1. Drexler H, Thiel E, Ludwig WD. Review of the incidence and clinical relevence of myeloid antigen-positive acute lymphoblastic leukemia. Leukemia 1991;5: 637
2. Sixth international workshop on chromosomes in leukemia: London, England, May 11- 18, 1987: Selected papers. Cancer Genet Cytogenet 1989; 40: 171
3. Weiss L, Bindl J, Picozzi V, et al. Lymphoblastic lymphoma: An immunophenotypic study of 26 cases with comparison to T cell acute lymphoblastic leukemia. Blood 1986; 67: 474
4. Swerdlow S, Habeshaw J, Richards M, et al. T lymphoblastic lymphoma with Leu-7 positive phenotype and unusual clinical course: A multiparameter study. Leuk Res 1985; 9: 167
5. Sheibani K, Winberg C, Burke J. Lymphoblastic lymphoma expressing natural killer cell-associated antigens. A clinicopathologic study of six cases. Cancer 1987; 60: 183
6. Sheibani K, Nathwani B, Winberg C. Antigenically defined subgroups of lymphoblastic lymphoma: Relationship to clinical presentation and biologic behavior. Cancer 1987; 60: 183
7. Coleman CN, Picozzi VJ, Cox RS, et al. Treatment of lymphoblastic lymphoma in adults. J Clin Oncol 1986; 4: 1626 - 1637
8. Morel P, Lepage E, Brice P, et al. Prognosis and treatment of lymphoblastic lymphoma in adults: A report on 80 patients. J Clin Oncol 1992; 10: 1078 - 1085
9. Bernasconi C, Brusamolino E, Lazzarino M, et al. Lymphoblastic lymphoma in adult patients; Clinicopathological features and response to intensive multi agent chemotherapy analogous to that used in acute lymphoblastic leukemia. Ann Oncol 1990; 1: 141 - 146
10. Slater DE, Mertelsmann R, Koriner B, et al. Lymphoblastic lymphoma in adults. J Clin Oncol 1986; 4: 57 - 67
11. Sweetenham JW, Mead GM, Whitehouse JMA. Adult lymphoblastic lymphoma: High incidence of central nervous system relapse in patients treated with the Stanford University protocol. Ann Oncol 1992; 3: 839 - 841
12. Wollner N, Burchenal JH, Lieberman PH, et al. Non-Hodgkin's lymphoma in children. A progress report on the original patients treated with the LSA_2L_2 protocol. Cancer 1979; 44: 1990 - 1999
13. The International Prognostic Factors Project. A predictive model for aggressive non-Hodgkin's lymphoma. N Engl J Med 1993; 329: 987 - 994
14. Santini G, Coser P, Chisesi T, et al. Autologous bone marrow transplantation for advanced stage adult lymphoblastic lymphoma in first complete remission. Ann Oncol 1991; 2 (suppl 2): 181 - 185
15. Milpied N, Ifrah N, Kuentz M, et al. Bone marrow transplantation for adult poor prognosis lymphoblastic lymphoma in first complete remission. Br J Haematol 1989; 73: 82 - 87
16. Verdonck LF, Dekker AW, de Gast GC, et al. Autologous bone marrow transplantation for adult poor risk lymphoblastic lymphoma in first remission. J Clin Oncol 1992; 10: 644 - 646
17. Jackson GH, Lennard AL, Taylor PRA, et al. Autologous bone marrow transplantation in poor risk high grade non-Hodgkin's lymphoma in first complete remission. Br J Cancer 1994; 70: 501 - 505
18. Sweetenham JW, Liberti G, Pearce R, et al. High dose therapy and autologous bone marrow transplantation for adult patients with lymphoblastic lymphoma: Results of the European Group for Bone Marrow Transplantation. J Clin Oncol 1994; 12: 1358 - 1365
19. Jost LM, Jacky E, Dommann-Scherrer C, et al. Short-term weekly chemotherapy followed by high-dose therapy with autologous bone marrow transplantation for lymphoblastic and Burkitt's lymphomas in adult patients. Ann Oncol 1995; 6: 445 - 451
20. Chopra R, Goldstone AH, Pearce R, et al. Autologous versus allogeneic bone marrow transplantation for non-Hodgkin's lymphoma: A case-controlled analysis of the European Bone Marrow Transplant Group registry data. J Clin Oncol 1992; 11: 1690 - 1695
21. Shpall EJ, Jones RB, Bearman SI, et al. Transplantation of enriched CD34-positive autologous bone marrow into breast cancer patients following high dose chemotherapy: influence of CD34-positive pripheral blood progenitors and growth factors on engraftment. J Clin Oncol 1994; 12: 28 - 31
22. Haas R, Witt B, Goldschmidt H, et al. Sustained long-term hematopoiesis after myeloablative therapy with peripheral blood progenitor cell support. Blood 1995; 85: 3754 - 3761
23. Sharp JC, Kessinger A, Mann S, et al. Detection and clinical significance of minimal tumor cell contamination of peripheral blood stem cell harvests. Int J Cell Cloning 1992; 10 (suppl 2): 92 - 96
24. Gribben JG, Neuberg D, Barber M, et al. Detection of residual lymphoma cells by polymerase chain reaction in peripheral blood is significantly less predictive for relapse than detection in bone marrow. Blood 1994; 83: 3800 - 3807

A NEW NON-RADIOACTIVE METHOD FOR DETECTION OF MONOCLONAL CELL POPULATIONS IN PATIENTS WITH BURKITT'S LYMPHOMA

Udo zur Stadt, Alfred Reiter, Jörg Tomeczkowski and Karl-Walter Sykora*

Department of Pediatric Hematology and Oncology
Medical School Hannover
Hannover, Germany

ABSTRACT

We developed a new non-radioactive method to investigate the clonality of biopsy material, bone marrow samples or peripheral blood mononuclear cells from children with Burkitt's lymphoma, using the polymerase chain reaction (PCR).

The variable region (VH) of the immunoglobulin heavy chain (IgH) was used as the clonal marker in B-lymphoproliferative disease.[1] For amplification of the IgH variable region, we used 7 VH region family specific primers and one joining region (JH) consensus primer in one PCR reaction.[2,3,4] The JH primer was labeled at the 5' end with digoxigenin. After PCR, the products were separated under denaturing conditions in a 4% polyacrylamide gel on a direct blotting apparatus (GATC, MWG Biotech, Ebersberg, Germany). During electrophoresis, the blotting membrane under the gel was not transported until the dye markers indicated a product length of approximately 300 bp. The membrane was then transported for 120 min. These conditions allow for wide separation of the expected products of 300-400 bp length. After blotting, the membrane was crosslinked by UV-light, blocked and incubated with an alkaline phosphatase conjugated anti-digoxigenin antibody. After washing, the specific PCR products were detected by the chemiluminescent substrate CSPD and the membrane was exposed to x-ray film. With this method we were able to differentiate between monoclonal and polyclonal material at a sensitivity of 2% clonal cells mixed into peripheral blood mononuclear cells. The specific advantage of this method is that the same membrane can be used for hybridization with a clone-specific oligonucleotide probe. This

* Corresponding author: Dr. Karl-Walter Sykora, Medical School Hannover, Department of Pediatric Hematology and Oncology IV, OE 6780, Chairman: Prof. Dr. Dr.h.c. H. Riehm, Konstanty-Gutschow-Str. 8, D-30625 Hannover, Germany. Telephone 49-511-532 ext. 9013, 9020, 3214; Fax ++49-511-532 ext. 9120.

Molecular Biology of Hematopoiesis 5, edited by Abraham et al.
Plenum Press, New York, 1996

allows characterization of the specific PCR-product by hybridization and by size fractionation at the same time, and it increases specificity and sensitivity. The method could be useful to detect malignant cells in the peripheral blood of Burkitt's lymphoma patients during therapy.

INTRODUCTION

The pediatric non-Hodgkin's lymphomas (NHL) are a heterogenous group of diseases reflecting the diverse differentiation stages of the lymphoid cells from which they originate.

In the pediatric population, the high grade lymphomas of B-cell origin (small noncleaved, mostly Burkitt's lymphoma) including B-ALL make up the largest group of 65 to 70% in the Berlin-Frankfurt-Münster (BFM) study group experience.[5] In these, the malignant cells are characterized by an L3 cytomorphology according to the FAB classification.

During B-cell development immunoglobulin heavy and light chains will be rearranged. In B-lineage cells, rearrangement of one D gene to one JH gene is followed by the addition of one of the numerous VH genes to the fused D-JH segments. If successful, the resulting VH-D-JH rearrangement is transcribed into mRNA and spliced to the constant region. When the fully assembled μ heavy chain is presented on the B-cell surface, no further rearrangement occurs (allelic exclusion).[7]

Burkitt's lymphoma cells represent a more mature stage of B-cell differentation, where immunoglobulin heavy chain (IgH) genes are already rearranged and where the enzymes involved in immunoglobulin recombination, terminal deoxynucleotide transferase (TdT) and recombination activating genes (RAG) are not expressed anymore. Therefore, after a malignant clone has been established, further clonal evolution and rearrangement would not be expected anymore, making IgH gene rearrangement a more stable marker of clonality in these more mature B-lymphoid malignancies than in B-precursor ALL.

MATERIALS AND METHODS

Primers and PCR

For immunoglobulin heavy chain amplification by PCR we used a panel of seven VH family specific primers specific for the first framework region FRI in one reaction. All six known JH regions have conserved sequences at the 3' ends which enables the use of a single primer to detect rearrangements involving any JH region. When not pointed out, primers were used as a set of seven forward VH and one reverse JH primer in one PCR reaction.[3]

About 1 μg of DNA or 0.4 μl of the reverse transcriptase (RT) reaction containing the complementary DNA (cDNA) was used in PCR in a 30 μl reaction for 28 cycles. The JH primer was 5'-labeled with digoxigenin. As positive controls for genomic DNA we used primers for the p53 gene[9] (data not shown), for cDNA we used primers for the 16S mitochondrial ribosomal RNA.[10]

Sample Preparation

Mononuclear cells from fresh biopsy material, bone marrow or peripheral blood were separated by Ficoll Paque density centrifugation (Pharmacia, Uppsala, Sweden), washed twice in phosphate buffered salt (PBS) pH 7.2 and frozen in liquid nitrogen (with RPMI 1640, 10% FCS, 10% DMSO) until use.

Total RNA and DNA were isolated using the acid guanidinium-phenol-chloroform extraction method.[8] After lysis and addition of $CHCl_3$, the suspension was stored on ice for 15 min and then centrifuged. After this step the RNA was isolated from the aqueous phase followed by additional cleaning steps. The lower part of the interphase and the organic phase containing DNA was then treated with 0.3 vol 2-propanol for DNA precipitation, stored for 2h at 4 °C and centrifuged briefly (4,500 rpm, 4 °C). The pellet was washed twice in 10% EtOH/0.1 M citric acid for at least 1h. Finally, the DNA was dissolved in 8 mM NaOH and stored at 4 °C.

cDNA was synthesized from 2 μg of total cellular RNA using 200 U of Moloney murine leukemia virus reverse transcriptase (M-MLV RT) in a 20 μl reaction.

Sequencing

PCR products were cloned into the pCR-Script SK(+) vector (Stratagene, Heidelberg, Germany). After preparation, inserts from the plasmids were amplified for 15-20 cycles using T3 and T7 promoter sequences flanking the multiple cloning site of the vector.

Fragments were purified by spin column chromatography (magic PCR-prep, Promega, Germany) and finally dissolved in 30 μl water. Half of the volume was used for non-radioactive sequencing using the PCR-cycle sequencing kit (Boehringer Mannheim, Mannheim, Germany). Sequencing products were separated on a 4% denaturing acrylamide gel, using the GATC direct blotting machine (MWG Biotech, Ebersberg, Germany).

5 to 10 clones were sequenced. The most common ones were analyzed for the clone-specific regions of the immunoglobulin heavy chain (Sequence analysis workshop program for MS-Windows, SAW, H.W. Schröder, Birmingham, AL, USA, 1993). After identification, a clone-specific oligonucleotide from the CDRIII region was synthesized with a biotin labeled 5'-end (MWG Biotech, München, Germany). The physical properties of the sequence were selected to allow both hybridization and PCR applications.

Electrophoresis

DNA and RNA from bone marrow samples and/or peripheral blood from children with non-Hodgkin lymphomas were prepared. Then, the variable region of the IgH was amplified and separated by size electrophoresis on a 2% agarose gel and on a high resolution acrylamide gel under denaturing conditions. During electrophoresis on the acrylamide gel, the products of interest were directly blotted onto a nylon membrane.

Blotting from the Agarose Gels

PCR-fragments were separated on a 2% agarose gel and stained with ethidium-bromide (EtBr). After photographing, the gel was denatured two times in 0.5 M NaOH, containing 3 M NaCl, washed twice in ddH_2O and neutralized two times in 0.5 M TRIS-HCl pH 7.5/ 0.15 M NaCl. After this procedure the DNA was blotted overnight onto a positively charged nylon membrane (Boehringer, Mannheim) by capillary blotting (20x SSC, 3 M NaCl, 0.3 M Na-citrate, pH 7.0).

Direct Blotting from Denaturing Polyacrylamide Gel

Buffers were prepared according to the manufacturers instructions (MWG Biotech, Ebersberg, Germany). PCR products were separated under denaturing conditions in a 4%

polyacrylamide gel on a direct blotting apparatus (GATC, MWG Biotech, Ebersberg, Germany). The direct blotting membrane was started after a delay of 120 min. During this delay, only products with less than 300 bp run out of the gel. The transport of the membrane was then started with a speed of 19 cm/h. During the transport, all products between approximately 300 and 400 bp were blotted onto the membrane.

Identification of Amplified Material

To detect clone-specific sequences in the membrane bound PCR products, the biotin-labeled clone-specific oligonucleotide was hybridized to the membrane and reacted with strepavidin-conjugated alkaline phosphatase (AP).

After the first detection, the membrane was heated to 85° C in 50 mM EDTA to simultaneously strip the oligonucleotide and inactivate AP. To detect all membrane bound PCR products for evaluation of length polymorphism, the digoxigenin labeled V_H IgH PCR products were reacted with an AP conjugated antibody against digoxigenin (Boehringer Mannheim, Mannheim, Germany).

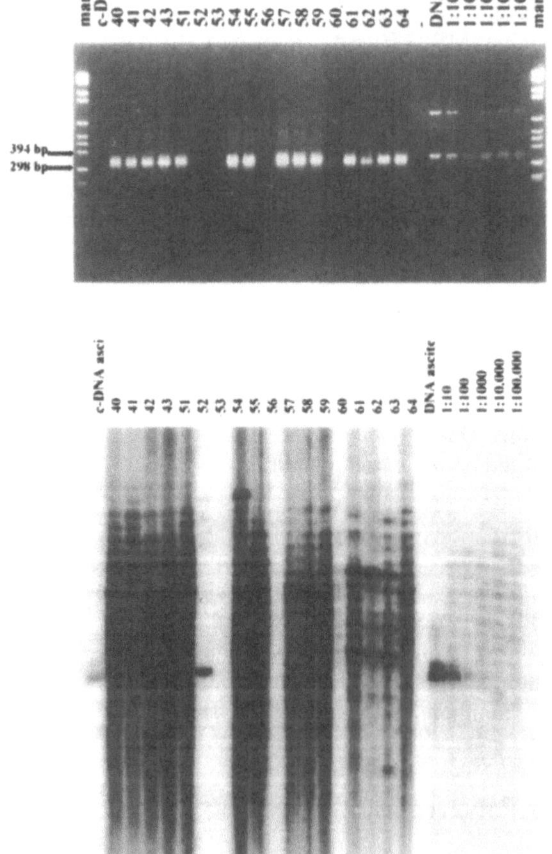

Figure 1. PCR from cDNA with a set of IgH VH amplimers. Digoxigenin-labeled PCR products were separated on a 2% agarose gel and detected by EtBr fluorescence (a) or on a high resolution denaturing acrylamide gel with direct blotting onto a nylon membrane and detected by chemiluminescence (b). Bone marrow and blood samples were obtained during therapy as follows: ascites material at the time of diagnosis; serial blood samples during therapy (# 41, 51, 52, 53, 54, 55, 59, 65); serial bone marrow samples during therapy (# 40, 42, 43,57, 58, 60, 64, 66); leukapheresis samples after chemotherapy for relapse(#61, 62, 63, 67); tumor at relapse(# 56).

Biotin-labeled oligonucleotide probes and digoxigenin-labeled PCR products were detected by chemiluminescence using CSPD (Tropix, Heidelberg, Germany) as a substrate.

RESULTS

We demonstrate the potential of this procedure in a NHL patient, who had an unusual Burkitt's type lymphoma with involvement of the bladder and the ileum and who also had malignant ascites. The patient relapsed 6 months after diagnosis while on therapy and died 2 months later. At the time of diagnosis and relapse, the patient's lymphoma cells were morphologically FAB-L3,TdT-negative, but did not express surface immunoglobulins as analyzed by flow cytometry and did not express immunoglobulin mRNA as detected by RT-PCR.

RNA and DNA were prepared from each sample, and cDNA was prepared from the mRNA by reverse transcription. These cDNAs and the genomic DNA from the same tumor or peripheral blood samples were amplified with the immunoglobulin primer set, separated on an agarose gel and stained with EtBr (Figs. 1a and 2a). A polyclonal smear was seen in the agarose gels, making it very difficult to decide whether low concentrations of clonal material were present in this smear.

In contrast to this, high resolution denaturing acrylamide gel electrophoresis with direct blotting was able to differentiate between mono- and polyclonally rearranged products (Figs. 1b and 2b) and easily allow the detection of clonal material. To determine

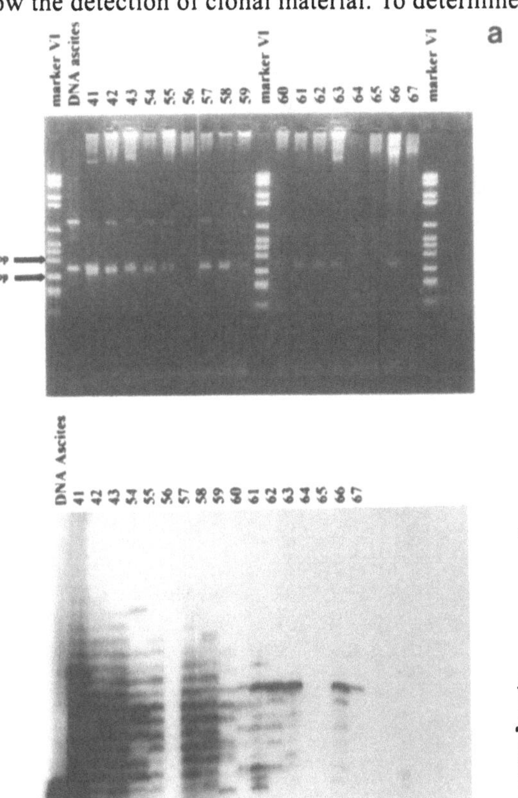

Figure 2. PCR from genomic DNA with a set of IgH VH amplimers. Digoxigenin-labeled PCR products were separated on a 2% agarose gel and detected by EtBr fluorescence (a) or on a high resolution denaturing acrylamide gel with direct blotting onto a nylon membrane and detected by chemiluminescence (b). Sample numbers are described in Fig. 1.

the sensitivity for detection of clonal material, ascites DNA was diluted into the DNA of peripheral blood mononuclear cells. The maximum dilution to detect the clonal material was about 1 to 100 in this experiment. For the amplification of cDNA, control PCRs with 16S mitochondrial ribosomal RNA primers demonstrated that comparable amounts of amplifiable DNA were present in the cDNA preparations. (Fig. 3) Using the immunoglobulin primer set on the same material showed that heavy chain mRNA was expressed in most but not in all samples. This was observed in the ascites material obtained at initial diagnosis, where no band was observed in the agarose gel (Fig 1a) and only a very weak signal in the high-resolution polyacrylamide gel (HR-gel). The tumor at relapse showed no signal, not even in the HR-gel (sample # 56, Fig. 1). This was in agreement with the flow cytometric results, where no surface immunoglobulins were detected (data not shown).

Also, when genomic DNA was used as template for amplification, the expected two PCR products of approximately 350 and 700 bp were not always detected (e.g. Fig.2, sample #56), although control amplifications of the genomic DNA with p53-specific primers

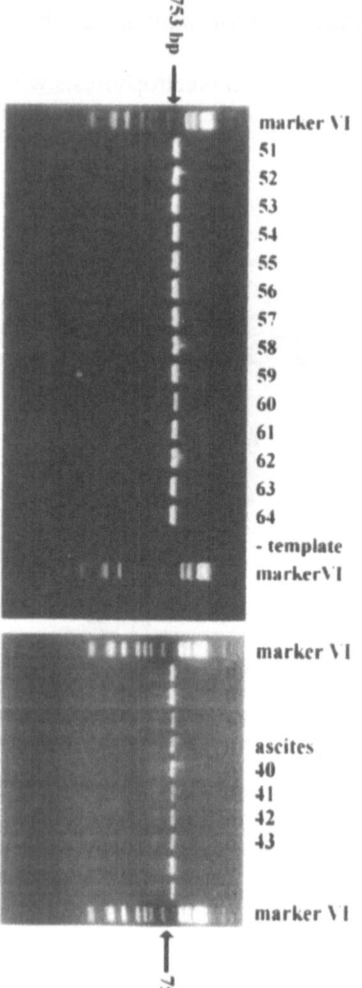

Figure 3. 16s control RT-PCR. cDNA was amplified with 16S RNA specific primers. PCR products were separated on a 2% agarose gel and detected by EtBr fluorescence. Sample numbers are as indicated in Fig 1.

demonstrated amplifiable DNA (data not shown). This indicated to us that the B-cell lymphoma itself had undergone clonal evolution in a way that made our PCR-primers ineffective, and that no contaminating normal B-cells were present in the tumor material.

A new prominent band distinct from the original ascites material appeared in the leukapheresis material (samples #61-63 and 67) harvested for autologous bone-marrow rescue after relapse chemotherapy. This could be interpreted as the evolution of a new clone, which was still detectable with the immunoglobulin primers. This band was not identical in length to the original tumor and to the tumor at relapse. Oligoclonal material was sometimes also observed in the course of the disease, and was more visible when genomic DNA (Fig. 2b) was used as the PCR template.

In order to investigate the behaviour of the specific tumor clone during therapy we generated a CDRIII specific oligonucleotide probe from the ascites material obtained at diagnosis. The immunoglobulin 350 bp product was ligated into a vector, transformed and 6 clones were sequenced. From the unique sequence, a clone-specific oligonucleotide complementary to the CDRIII region, and labeled at the 5'-end with biotin, was synthesized. This clone-specific marker was hybridized to a direct blotted membrane containing IgH PCR products from tumor material, bone marrow and from peripheral blood. The only specific hybridization was seen with the ascites material that the oligonucleotide had been prepared from (Fig. 4.a). Directly detected PCR products are shown for comparison in Fig. 4b (identical to Fig. 2b). Conventional Southern blot preparations showed the same results (data not shown).

DISCUSSION

We developed a sensitive, nonradioactive, PCR based technique for analysis of B-lymphoid clonality which relies on the detection of IgH VDJ gene rearrangement. We used the panel of VH family specific amplimers described by Deane and Norton[4] and developed a nonradioactive variant of the method described by the same authors as "immunoglobulin gene fingerprinting." The detection of length polymorphisms was optimized by using the direct blotting technique, which allows to "zoom into" the size range of interest and which resolves PCR products that differ by only one nucleotide in size. This high resolution denaturing acrylamide gel with direct blotting was used for two applications: First, specific lymphoma clones were detected by hybridization with a clone-specific oligonucleotide. Then, the oligonucleotide was removed and length polymorphisms of the membrane-bound PCR products were identified using the incorporated digoxigenin molecule.

The potential of this combined method is illustrated by the patient presented here. By using only hybridization with a clone-specific oligonucleotide in dot blots or Southern blots,[6] the evolution of a second malignant clone in this patient would not have been detected. On the other hand, a clone-specific oligonucleotide prepared against ascites did not detect clonal material at relapse, suggesting that either clonal evolution had occurred, or that the malignant lymphoma was polyclonal at presentation and that the dominant clone at relapse was not the one seen initially in ascites. It will now be possible to prepare a specific oligonucleotide from the lymphoma at relapse and determine whether this clone was detectable in the peripheral blood before clinical relapse became evident. This will be possible by probing the same membrane used for hybridization of the initial oligonucleotide. The method presented here will allow us to learn more about the in-vivo biology and clonal evolution of pediatric malignant lymphomas.

a

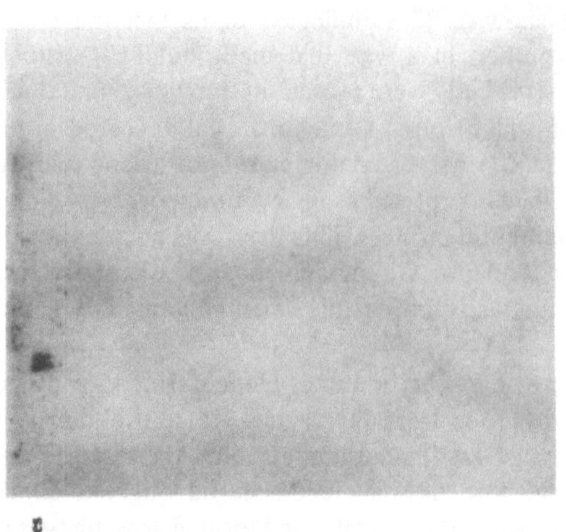

b

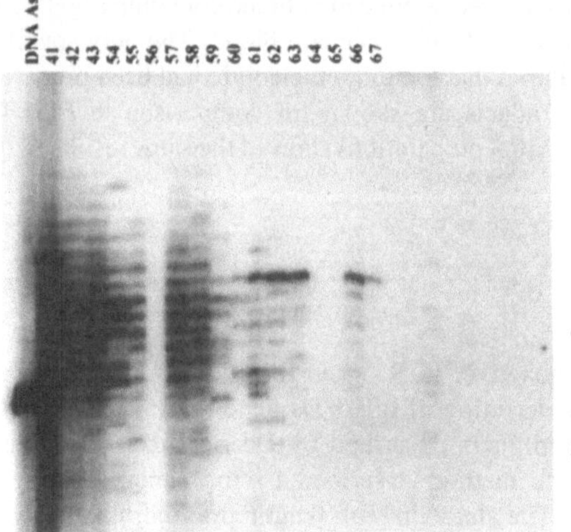

smaller products

Figure 4. Consecutive DNA-detections on a single directly blotted membrane. PCR products were blotted directly onto a nylon membrane after separation within a denaturing acrylamide gel and then first hybridized with a CDRIII specific, biotinylated oligonucleotide (a). Second, after stripping and inactivation of alkaline phosphatase, the digoxigenin-labeled membrane bound PCR products were detected by antibody reaction and chemiluminescence.

ACKNOWLEDGMENT

Supported by Deutsche Leukämieforschungshilfe and Graduiertenkolleg "Molekulare Pathophysiologie des Zellwachstums."

REFERENCES

1. Tonegawa S: Somatic generation of antibody diversity. Nature 302:575, 1983
2. Deane M, McCarthy KP, Wiedemann LM, Norton JD: An improved method for detection of B-lymphoid clonality by PCR. Leukemia 5(8): 726, 1991
3. Deane M, Norton MD: Detection of immunoglobulin gene rearrangement in B-lymphoid malignancies by PCR gene amplification. Br J Haematol 74: 251, 1990
4. Deane M, Norton MD: Immunoglobulin gene 'fingerprinting': an approach to analysis of B-lymphoid clonality in lymphoproliferative disorders. Br J Haematol 77: 274, 1991
5. Reiter A, Schrappe M, Parwaresch R, Henze G, Müller.Weirich S, Sauter S, Sykora KW, Ludwig WD, Gadner H, Riehm H: Non-Hodgkin's Lymphoma of Childhood and Aldolescence: Results of a Treatment Stratified for Biologic Subtypes and Stage- A Report of the Berlin-Frankfurt-Münster Group. J Clin Oncol 13 (2): 359, 1995

6. Yamada M, Hudson S, Tournay O, Bittenbender S, Shane SS, Lange B, Tsujimoto Y, Caton AJ, Rovera G: Detection of minimal residual disease in hematopoietic malignancies of the B-cell lineage by using third-complementarity-determining region (CDRIII)- specific probes. Proc Natl Acad Sci 86: 5123, 1989
7. Stewart AK, Schwartz RS: Immunoglobulin V regions and B cell. Blood 83 (7): 1717, 1994
8. Chomcynski P, Sacchi N: Single-step method of RNA isolation by acid guanidinium thiocyanate-phenol-chloroform extraction. Anal Biochem 162: 156, 1986
9. Serra A, Gaidano GL, Revello D, Guerrasio A, Ballerini P, Dalla Favera R, Saglio G: A new Taq1 polymorphism in the p53 gene. Nucleic Acid Research 20 (4): 928, 1992
10. Eperon JC, Anderson S, Nierlich DP: Distinctive sequence of human mitochondrial ribosomal RNA genes. Nature 286: 460, 1980

22

ONCOGENES IN THE PATHOGENESIS OF MULTIPLE MYELOMA

Paolo Corradini,* Monica Astolfi, Marco Ladetto, Silvia Campana, Mario Boccadoro, and Alessandro Pileri

Dipartimento di Medicina ed Oncologia Sperimentale
Divisione di Ematologia dell'Universita' di Torino
Azienda Ospedaliera S. Giovanni Battista, Torino, Italy

Molecular pathogenesis of multiple myeloma (MM) is largely unknown. There are several lines of evidence suggesting that malignant transformation occurs through a multistep process involving the activation of proto-oncogenes, the loss of tumor suppressor genes, and the deregulation of cytokine network. Cytogenetic studies have been difficult to perform because of low tumor mitotic activity. Karyotypic abnormalities are present in 30% of patients, no disease-specific anomaly has been detected so far. Aneuploidy is frequent, and a variety of chromosomal translocations has been reported (1).

DOMINANTLY ACTING ONCOGENES

The c-myc Oncogene

C-myc codes for a protein which acts as a transcriptional factor. It is activated through a translocation that juxtaposes it on chromosome 8 and the immunoglobulin genes on chromosome 2 or 14 or 22. Because of its frequent involvement in other human B-cell neoplasms and constant deregulation in murine plasmacytoma, c-myc was the first oncogene the involvement of which has been evaluated in MM. Alterations in the c-myc locus were only sporadically reported, by Southern blot analysis, with a probe representative of the third exon of the c-myc gene. Overexpression of c-myc RNA was observed in 25% of patients, without apparent abnormalities of c-myc RNA transcript size (2,3). It was only rarely ascribable to DNA rearrangements, and no point mutations were found in the first intron/exon putative regulatory sequences.

*Correspondence to: Paolo Corradini, M.D., Cattedra di Ematologia, Via Genova 3, 10126 Torino, Italy. Fax: 39 11 6963737; Phone 39 11 6626728 or 6625329.

Molecular Biology of Hematopoiesis 5, edited by Abraham et al.
Plenum Press, New York, 1996

The Bcl-1 Locus

The bcl-1 locus was originally identified as the breakpoint site on chromosome 11 of the t(11;14)(q13;q32) translocation, a cytogenetic lesion detectable in approximately 50% of intermediate lymphocytic lymphomas (4). The putative oncogene, called CCND1 (PRAD-1), displays homology to cyclins, a family of genes involved in the regulation of cell cycle progression. There have been sporadic reports of multiple myeloma possessing the t(11;14) translocation. In the three largest cytogenetic series, four out of 136 cases (3%) were positive. It has also been observed in two cell lines derived from patients with multiple myeloma or plasma cell leukaemia (5). Bcl-1 rearrangements have been described in five of 120 cases of multiple myeloma (4%) (6). Their incidence in multiple myeloma thus appears to be very low.

The Bcl-2 Oncogene

Bcl-2 belongs to a new class of proto-oncogenes, that apparently control cell survival by blocking programmed cell death. The bcl-2 gene was identified by molecular cloning of the t(14;18) (q32;q21) translocation, which is the most frequent chromosomal abnormality in B-cell non Hodgkin's lymphoma. The t(14;18) translocation has rarely been detected in myeloma patients. Increased expression of bcl-2 protein has been described by Durie *et al* in about 75% of cases (7,8). Such findings are of particular interest because of the potential effect of bcl-2 activation in multiple myeloma, since it is characterized by a slow proliferative activity, like follicular lymphoma. None of 60 patients with myeloma, showed rearrangement of bcl-2 gene (within major breakpoint and minor cluster regions). This finding is consistent with the rarity of t(14;18) in multiple myeloma (1).

The Family of Ras Oncogenes

This consists of three related oncogenes: H- ,K- and N-ras that encode proteins with GTPase activity involved in growth signal transduction mechanisms. Their transforming potential is acquired when point mutations occur in codons 12, 13 and 61, resulting in single amino acid substitutions. Screening for somatic point mutations has became much easier in recent years with new simple, sensitive strategies allowing the analysis of large numbers of samples. The most widely used is single-strand conformation polymorphism (SSCP) analysis, which detects mutated DNA strands because of their different electrophoretic mobility under non-denaturing running conditions. The presence of activating point mutations of N- and K-ras genes is a well-documented molecular lesion in multiple myeloma (3,9,10). Their overall incidence ranges from 9% to 31%, probably due to differences in the clinical status of the patients examined. Ras mutations are not randomly distributed. They are not present in MGUS, solitary plasmacytomas and stage I multiple myelomas, whereas they are detected in a sizable fraction of patients with advanced stage disease and adverse prognostic factors (30%). A similar incidence (30%) has been found in patients with plasma cell leukaemias. The absence of a clear difference in incidence between diagnosis and relapse rules out any mutagenic effect on the part of alkylating agents. Mutations affecting the N-ras gene at codon 61 are the most frequent finding.

In vitro transfection studies also suggest that ras oncogenes are implicated in myeloma pathogenesis, since their transfection confers both malignancy and terminal differentiated morphology to Epstein Barr virus infected B-cells (11). This biological effect seems to be peculiar to the ras gene family, since c-myc transfection induces malignant transformation without substantial changes in the differentiation phenotype of target cells, and is particularly intriguing, since plasma cell dyscrasias are typically characterized by the coexistence of a differentiated morphology and fully malignancy.

TUMOR SUPPRESSOR GENES

The role of loss or inactivation of tumor supppressor genes in the pathogenesis of various human malignancies is well estabilished. The role of these lesions in MM has been only recently investigated in MM. p53 and retinoblastoma (Rb1) genes represent the prototypes of this class of genes.

The p53 Gene

The p53 gene encodes a 53-Kd nuclear phosphoprotein that controls the normal cell cycle by regulating transcription and possibly DNA replication. Loss of this growth-inhibitor activity is usually the result of point mutations of one allele associated with the loss of the other. The majority of mutations occurs between codons 110 and 307 encompassing exon 5 to 9, and are clustered in four regions that are highly conserved among several different species.

Point mutations in exons 5 and 8 were detected in 8 of 10 human myeloma cell lines (12). The frequency of p53 mutations in bone marrow samples of patients with plasma cell dyscrasias ranges from 10% to 20% (13,14). They are typically associated with advanced and clinically aggressive forms of MM. These mutations were not found during the chronic phase in three patients, positive for p53 mutations during the terminal phases, and were more frequent in plasma cell leukemia (22%) than in aggressive myelomas (15). The presence of concomitant ras and p53 mutations has been sporadically described.

The Rb1 Gene

The Rb1 tumour suppressor gene encodes a 110-Kd phosphoprotein which accumulates in the nucleus and is associated with DNA binding activity. Rb1 gene product is expressed in human hematopoietic cells, where it is associated with the cell cycle control and the regulation of IL-6 gene. Rb1 appears to have a role in transcriptional suppression of IL-6; thus Rb1 inactivation may result in an autocrine production of IL-6 by myeloma cells. Monosomy of chromosome 13 and monoallelic loss of Rb1 have been recently shown in more than 50% of MM patients by fluorescence in situ hybridization (16). Immunohistochemical analysis has shown complete absence of the protein in 17% of MM patients, and 18% of plasma cell leukemia patients. Plasma cells normally express Rb1 protein and its absence thus leads support to the idea of a primary defect in Rb1 gene structure or expression. The molecular basis of such loss of protein expression is not clearly defined. Southern blot analysis of patients lacking protein expression with two probes spanning from exon 1 to 9 and from exon 10 to 27 showed a deletion in only one case. The absence of rearrangements or deletions in the remaining cases suggests that the lesions may be point mutations, as observed in other tumour types. It is noteworthy that patients lacking Rb1 protein presented extramedullary masses, indicating an advanced and aggressive clinical phase (13).

CONCLUSIONS

The cell from which myeloma originates is still unknown. Some progress has been made, however, by demonstrating the existence of mature B cells involved in the pathogenesis of MM (17,18). It appears that malignant plasma cell in myeloma is the result of a number of genetic lesions that can result in transformation, differentiation, and production of autocrine growth factors.

The analysis of dominant and recessive oncogenes shows an heterogeneous pattern of involvement. The frequent detection of ras, and p53 mutations, in advanced stage myeloma and plasma cell leukemias, suggests that they may play a role as tumor progression markers.

ACKNOWLEDGMENT

This work was supported by Associazione Italiana Ricerca Cancro, National Research Council, Special Project "ACRO", N. 94.01184.PF39 and Biomed Project BMHI-CT93-1407.

REFERENCES

1. Barlogie B, Epstein J, Selvanaygam P, Alexanian R. Plasma cell myeloma- New biological insights and advances in therapy. Blood 73:865-879, 1989.
2. Selvanaygam P, Blick M, Narni F, et al. Alteration and abnormal expression of the c-myc oncogene in human multiple myeloma. Blood 71:30-35, 1988.
3. Neri A, Murphy J, Cro L, et al. Ras oncogene mutation in multiple myeloma. J Exp Med 170:1715-1725, 1989.
4. Raffeld M, Jaffe ES. Bcl-1 t(11;14), and mantle cell-derived lymphomas. Blood 78:259-263, 1991.
5. Zhang XG, Gaillard JP, Robillard N, et al. Reproducible obtaining of human myeloma cell lines as a model for tumor stem cell study in human multiple myeloma. Blood 83:3654-3663, 1994.
6. Selvanaygam P, Goodacre A, Strong L, Saunders G, Barlogie B. Alterations of bcl-1 oncogene in human multiple myeloma. AACR: 76 (abs), 1987.
7. Durie BGM, Mason DY, Giles F, et al. Expression of the bcl-2 oncogene protein in multiple myeloma. Blood 76:347A, 1990.
8. Durie BGM. Cellular and molecular genetic features of myeloma and related disorders. Hematol Oncol Clin North Am 6:463-477, 1992.
9. Paquette, RL, Berenson J, Lichtenstein A, McCormick F, Koeffler HP. Oncogenes in multiple myeloma: point mutations of N-ras. Oncogene 5:1659-1663, 1990.
10. Corradini P, Ladetto M, Voena C, et al. Mutational activation of N- and K-ras oncogenes in plasma cell dyscrasias. Blood 81:2708-2713, 1993.
11. Seremetis S, Inghirami G, Ferrero D, et al. Transformation and plasmacytoid differentiation of EBV-infected human B lymphoblasts by ras oncogenes. Science 243:660-663, 1989.
12. Mazars GR, Portier M, Zhang XG, et al. Mutations of the p53 gene in human myeloma cell lines. Oncogene 7:1015-1018, 1992.
13. Corradini P, Inghirami G, Astolfi M, et al. Inactivation of tumor suppressor genes, p53 and Rb1 in plasma cell dyscrasias. Leukemia 8:758-767, 1993.
14. Portier M, Molès JP, Mazars GR, et al. p53 and ras gene mutations in multiple myeloma. Oncogene 7:2539-2543, 1992.
15. Neri A, Baldini L, Trecca D, Cro L, Polli E, Maiolo AT. p53 mutations in multiple myeloma are associated with advanced forms of malignancy. Blood 81:128-135, 1993.
16. Dao DD, Sawyer JR, Epstein J, Hoover RG, Barlogie B, Tricot G. Deletion of the retinoblastoma gene in multiple myeloma. Leukemia 8:1280-1284, 1994.
17. Corradini P, Boccadoro M, Voena C, Pileri A. Evidence for a bone marrow B cell transcribing malignant plasma cell VDJ joined to Cμ sequence in immunoglobulin (IgG)- and IgA-secreting multiple myelomas. J Exp Med 178:1091-1096, 1993.
18. Billadeau D, Ahmann G, Greipp P, van Ness B. The bone marrow of multiple myeloma patients contains B cell populations at different stages of differentiation that are clonally related to the malignant plasma cell. J Exp Med 178:1023-1031, 1993.

23

EXPRESSION AND REGULATION OF *C-KIT* RECEPTOR AND PROLIFERATIVE RESPONSE TO STEM CELL FACTOR IN CHILDHOOD LYMPHOMA AND LEUKEMIA CELLS

J. Tomeczkowski, D. Frick, B. Schwinzer, A. Reiter, K. Welte, and K. W. Sykora*

Department of Pediatric Hematology and Oncology
Medical School Hannover, Germany

ABSTRACT

The product of the protooncogene *c-kit* is a receptor tyrosine kinase for the hematopoietic cytokine stem cell factor (SCF). This cytokine is involved in the regulation of normal and neoplastic hematopoiesis, and synergizes with IL-7 to enhance the proliferation of preB-cells and thymocytes. To examine whether *c-kit* is involved in the pathogenesis of human leukemic cells, we investigated the expression and regulation of *c-kit* in childhood Burkitt's lymphoma (BL) and in T-lymphoblastic cells, as well as the mitogenic activity of recombinant human (rh) SCF on these cells. A panel of Epstein-Barr Virus (EBV) positive and negative BL cell lines, biopsy tumor cells from patients with Burkitt's lymphoma, T acute lymphoblastic leukemia (ALL) and T lymphoblastic non Hodgkin's lymphoma (LB-T NHL) were investigated. To study inducibility of *c-kit* receptor, the cells were cultured and stimulated by IL-7, PMA, and calcium ionophore A23187. c-*kit* expression was studied by RT-PCR followed by Southern blot analysis, fluorescence activated cell sorting (FACS) and by crosslinking of digoxigenin-labeled rhSCF to the cell surface. Proliferation of BL cell lines and T-lymphoblastic cells was measured under serum-free conditions in the presence and absence of rhSCF.

c-kit receptors were detected by FACS analysis in 3 out of 13 investigated LB-T-NHL, in 2 T-ALL biopsy tumor cells and in none out of the 7 biopsy BL cells investigated. Low

*Corresponding author: Karl-Walter Sykora, MD, Medical School Hannover, Department of Pediatric Hematology and Oncology IV, OE 6780, Chairman: Prof. Dr.med Dr.h.c. H. Riehm, Konstanty-Gutschow-Str. 8, D-30625 Hannover, Germany. Telephone: ++511-532 ext. 9013, 9020, 3214; Fax: ++511-532 ext. 9120.

Molecular Biology of Hematopoiesis 5, edited by Abraham et al.
Plenum Press, New York, 1996

level expression of *c-kit* mRNA was only detecable by RT-PCR followed by Southern blot analysis in 2 of 13 cultured BL cell lines. *c-kit* mRNA or receptor upregulation could be demonstrated with IL-7 or A23187 and downregulation with PMA in BL and T-ALL cells. No stimulation or inhibition by rhSCF was observed in BL and LB-T NHL cells. Two of the *c-kit* positive T-ALL cells responded to rhSCF in a dose dependent manner. Normal peripheral B and T cells showed no *c-kit* transcripts detectable by RT-PCR followed by Southern blot analysis.

These results suggest that *c-kit* receptors can be expressed at different levels in childhood B and T malignant cells and that the induction of *c-kit* by different reagents leads to a functional protein in childhood T-lymphoblastic cells but not in BL cells. We conclude that SCF can be a growth factor for T-lymphoblastic malignant cells but not BL cells.

INTRODUCTION

The pediatric non-Hodgkin's lymphomas (NHL) are a heterogenous group of malignancies of B- or T-cell origin which constitute seven to ten percent of all pediatric malignancies.[1] In Europe and in North America about 20-25% of childhood lymphomas are lymphoblastic lymphomas, one half small noncleaved cell lymphomas (mostly Burkitt's lymphoma), and the rest predominantly large cell lymphomas.[2] With current chemotherapy protocols, between 65 and 90% of all patients with BL can achieve longterm disease free survival.[3] The growth conditions of B-cell lymphomas and their dependence on growth factors are of interest, because cytokines are being added to chemotherapy protocols in patients with these conditions. The cytokines IL-2, interferon-beta,[4] granulocyte-macrophage colony-stimulating factor (GM-CSF)[5] and granulocyte colony-stimulating factor (G-CSF)[6-8] have been given to patients with malignant lymphomas in an effort to modulate the therapeutic response or to alleviate the cytopenias of chemotherapy. The direct in vivo effects on BL cells of the therapeutically used cytokines have not been studied extensively. BL cells could be shown to express IL-2 receptors[9-11] and could be activated to produce interferon gamma.[9] G-CSF receptors were absent on BL cells and could not be induced to express G-CSF receptors by different reagents and G-CSF did not influence proliferation.[12] Lymphomas of different histologies show varying degrees of expression of IL-1, IL-2, IL-4, IL-5, IL-6, GM-CSF and their receptors. The significance of this in the context of a complex cytokine network is often difficult to determine (for review see [13,14]).

The recently described hematopoietic cytokine stem cell factor (SCF), also called kit ligand (KL), mast cell growth factor (MGF) or steel factor (SF) has entered clinical testing as an adjunct in the therapy of malignant disease.[15-19] SCF is a pluripotent hematopoietic colony-stimulating factor, that binds to its specific surface receptor, a protein encoded by the *c-kit* protooncogene.[20-22] SCF acts synergistically with other growth factors, including erythropoietin (EPO), G-CSF, macrophage (M)-CSF, GM-CSF, IL-3, IL-6, IL-11 and IL-12[23-26] and plays an important role in normal and leukemic hematopoiesis. Little is known about the growth factor requirements of NHL cells in particular, although normal T- or B-cells of different stages of differentiation are known to respond to or require for their growth and differentiation the cytokines IL-1, IL-2, IL-4, IL-6, IL-7 or IL-12.[24,27] Because of its stimulatory effects on early stem cells SCF may become a candidate to support not only granulopoiesis, but also thrombopoiesis in the setting of high-dose chemotherapy. In addition, CSFs are widely used in bone marrow transplantation[6,28] and SCF appears to be a potent mobilizer of peripheral blood progenitor cells (PBSCs),[29-32] especially in conjunction with G-CSF.[6,25,33] Nothing is known so far about a direct stimulatory or inhibitory effect of SCF on childhood NHL cells.

In combination with IL-7, SCF was found to be a potent stimulus at early stages of lymphoid development.[26,27,34-36] Proliferation of murine bone marrow B-lineage cells is supported by stroma cells and IL-7; the stroma cells can be replaced by SCF for the proliferative effect but not for the further differentiation to immunoglobulin-expressing cells.[34] SCF in combination with IL-6, IL-11, or G-CSF is able to maintain the B-lymphoid differentiation potential of primary bone marrow cultures.[23] The combination of SCF and IL-7 provides a proliferative stimulus to murine B-cells only after they have differentiated to B220$^+$ cells.[34] Murine proB-cell and preB-cell clones express *c-kit*, the tyrosine kinase receptor of SCF.[26,37] The finding that *c-kit* is involved in the growth regulation of preB-cells and thymocytes makes this receptor a candidate for oncogenic involvement in the development of lymphomas. For that reason and for the beginning of clinical use of SCF, we examined the direct effect of rhSCF on pediatric BL and T-lymphoblastic cells and the expression and regulation of *c-kit* receptor to elucidate whether SCF is a cytokine which is involved in the growth regulation of these malignancies.

MATERIALS AND METHODS

Cell Lines, Fresh BL, and Fresh Lymphoblastic Cells

MHH-BL-1, MHH-BL-2, MHH-TALL-1 and MHH-TALL-2 were established in our laboratory. The cells were isolated by Ficoll-Paque centrifugation, washed and resuspended in BL culture medium and maintained at 37°C in humidified 5% CO_2 with half replacement of medium twice weekly. The BL medium consisted of RPMI 1640 (Gibco BRL, Eggenstein, Germany) with 10% fetal calf serum (FCS), supplemented with 20 nM bathocuproine disulfonic acid (Sigma, Deisenhofen, Germany), 50 μM α-thioglycerol (Sigma, Deisenhofen, Germany), 1 mM pyruvate (Gibco BRL, Eggenstein, Germany), 10 mM HEPES (Seromed Biochrom, Berlin, Germany) and 0.25 μg/ml antibiotic-antimycotic solution (Sigma, Deisenhofen, Germany). Lots of FCS were tested to support optimal growth of B-lymphoid cell lines. The BL cell lines that could be established from ascites cells were termed MHH-BL-1 (CD10$^+$, CD23$^-$, surface μ^+) and from bone marrow cells MHH-BL-2. In addition, fresh cells from MHH-BL-1 were investigated. Both T-ALL cell lines could be established from peripheral blood cells.

All other BL cell lines were initially established from EBV positive and negative BL biopsies[38-40] at the International Agency for Research on Cancer (IARC, Lyon Cedex, France) and were kindly provided by Dr. Georg W. Bornkamm (Forschungszentrum für Umwelt und Gesundheit GmbH, Munich, Germany). Two of the EBV negative BL cell lines, BL-2 and BL-41, had previously been infected in vitro by two different EBV strains[41] to study mechanisms of EBV-induced phenotypic alterations of BL cells.[42] The EBV status, karyotype and immunoglobulin expression are summarized in Table 1. To detect mycoplasma contamination, all cell lines were screened with the Gen-Probe test kit (Biermann, Bad Nauheim, Germany). The megakaryoblastic cell line MO7e[43] was used as a positive control for *c-kit.*

The fresh BL and fresh lymphoblastic cells were isolated for flow cytometry by Ficoll-Paque (Pharmacia, Uppsala, Sweden) centrifugation, washed, and stored in BL medium containing 10% dimethylsulfoxide (DMSO) and 20% FCS in liquid nitrogen until use. After thawing, the viable cells were again isolated by Ficoll-Paque centrifugation, washed and resuspended in phosphate-buffered saline (PBS). In the case of BL the percentage of L3 morphology cells was determined in each final cell preparation by Wright-Giemsa staining. Age and sex of the patient, as well as source and percentage of blast cells in the cell preparation are summarized in Table 2.

Table 1. Origin, EBV status, karyotype and immunoglobulin expression of Burkitt's lymphoma cell lines

Cell lines	Sex/Age	Ethnicity	EBV	Cytogenetics	Immunoglobulins
MHH-BL-1	F/14	Caucasian	Positive BL	not done	μ Lambda
MHH-BL-2	M/05	Turkish	Negative BL	not done	μ Lambda
IARC-BL-2	M/07	Caucasian	Negative BL	t(8:22)	μ Lambda
IARC-BL-2B95-8	M/07	Caucasian	Negative BL	t(8:22)	μ Lambda
IARC-BL-2P3HR-1	M/07	Caucasian	Negative BL	t(8:22)	μ Lambda
IARC-BL-16	F/05	Reunion	Positive BL	t(8:14)	μ Lambda
IARC-BL-18	M/03	Nth-African	Positive BL	t(8;14)	μ Lambda
IARC-BL-29	F/03	Reunion	Positive BL	t(8;14)	μ Kappa
IARC-BL-31	M/14	Caucasian	Negative BL	t(8;14)	μ Kappa
IARC-BL-41	M/08	Caucasian	Negative BL	t(8;14)	μ Kappa
IARC-BL-41B95-8	M/08	Caucasian	Negative BL	t(8;14)	μ Kappa
IARC-BL-41P3HR-1	M/08	Caucasian	Negative BL	t(8;14)	μ Kappa
IARC-BL-49	M/03	Caucasian	Negative BL	t(8;22)	μ Lambda

Fresh peripheral blood B- and T-cells were isolated by Ficoll-Paque centrifugation and subsequent separation over a nylon wool column.[44]

^{3}H-Thymidine Incorporation Assays

The mitogenic activity of rhSCF was determined under serum-free conditions in a ^{3}H-thymidine incorporation assay essentially as described.[12]

MTS Metabolism Assay

T-ALL and T-NHL cells were washed with serum-free medium. The cells were then plated in 96-well flatbottom microtiter plates in 200 μl serum-free BL-medium containing various concentrations (0, 50, 500 ng/ml) of rhSCF and 1 μl phenazine methosulfate (PMS) and 20 μl Owen's reagent (MTS) for colorimetric reaction. The absorbance of the formazan at 490 nm was measured using an ELISA plate reader after preparing the assays. The plates were incubated at 37°C and further color developments were determined the following days.

Stimulation of BL Cell Lines, T-ALL and T-NHL Cells

To assess induced expression of *c-kit* and to demonstrate *c-kit* regulation the appropriate cytokine or reagent was added to the cells (1 x 10^6 cells/ml) in 75 cm^2 tissue

Table 2. Origin of fresh Burkitt's lymphoma cells

Patient	Sex	Age	Source of cells	Percentage of L3 cells
1	F	14	ascites	45%
2	M	10	ascites	not done
3	F	11	pleural effusion	87%
4	M	7	pleural effusion	24%
5	M	10	ascites	31%
6	M	7	pleural effusion	21%
7	M	9	ascites	61%

culture flasks (Falcon 3024; Becton Dickinson, Meylan Cedex, France) containing 100 ml of BL culture medium with 10% FCS. The cells were treated with 20 ng/ml phorbol 12-myristate 13-acetate (PMA) (Sigma, Deisenhofen, Germany), with 100 ng/ml calcium ionophore A23187 (Sigma, Deisenhofen, Germany) or with 10 ng/ml IL-7 (Promega, Heidelberg, Germany).

Polymerase Chain Reaction after Reverse Transcription (RT-PCR)

Total cellular RNA from BL cell lines and fresh BL cells was extracted using the single step acid guanidinium thiocyanate-phenol-chloroform extraction described by Chomczynski and Sacchi.[45] Single-stranded cDNA was synthesized from total cellular RNA using Moloney Murine Leukemia Virus reverse transcriptase (BRL, Eggenstein, Germany).[46,47]

The primer pair used for *c-kit* amplification had the sequence 5'TGCCTGTTGTGTCTGTGTCCA 3' corresponding to position 653-678 for the upper primer and 5' TGCCTCCTTCGGTGCCTTTTA 3' to position 1166-1186 for the lower primer of the sequence published by Yarden et al.[22] Control PCR reactions were performed with specific primers for 16S mitochondrial ribosomal RNA (16S RNA) with the sequence 5' CAGATTAAAACACTGAACTGACA 3' corresponding to position 2347-2369 and the sequence 5' GGGAGGAATTTGAAGGTAGATAG 3' corresponding to position 3077-3099 of the sequence published by Eperon et al.[48] PCR reactions were started by the addition of 1 U Taq polymerase (Perkin Elmer Cetus, Überlingen, Germany) in a DNA thermal cycler (Landgraf, Hannover, Germany).

35 cycles of PCR were done as follows: Denaturation 65 seconds at 94°C, Annealing 85 seconds 50°C, Extension 70 seconds at 72°C.

Southern Blot Analysis of PCR-Products

Products from the RT-PCR were separated by electrophoresis in a 1.8% agarose gel in TBE-buffer. Gel bound DNA was denaturated, neutralized and blotted onto a positively charged nylon membrane (Boehringer, Mannheim, Germany).

C-kit-specific PCR products were detected with a 5'-Dig-labeled oligonucleotide with the sequence 5'-TTGTAAACGATGGAGAAAATGTA-3' which is complementary to nucleotide position 992 to 1014 of the sequence published by Yarden et al. [22]

Chemiluminescent detection was done according to the manufacturer's instructions (Boehringer, Mannheim, Germany) using AMPPD or CPD-star (Tropix, Heidelberg, Germany) as chemiluminescent reagent.

Labeling of SCF and Chemical Crosslinking with Digoxigenin-rhSCF (Dig-rhSCF)

1 ml rhSCF (2.2 mg/ml) in buffer pH 8.5 was labeled by incubation for 2 hours with digoxigenin-3-0-methyl-carbonyl-ε-aminocapronsäure-N-hydroxy-succinimidester (Dig) (Boehringer, Mannheim, Germany) dissolved in 150 μl ethanol at room temperature. Free Dig was removed by passage of Dig-rhSCF with PBS through a desalting column (Sephadex G-25). Dig-labeled SCF was crosslinked to cellular *c-kit* receptors, the cells were lysed and membrane proteins were separated by SDS-PAGE. Gels were blotted onto a nylon membrane and the Dig-labeled SCF-*c-kit* complex was detected by chemiluminescence as described in the manufacturer's instructions (Boehringer, Mannheim, Germany).

Flow Cytometry

Cell samples were incubated for 20 minutes on ice with saturating amounts of monoclonal antibody (mAb) in 1x PBS, 0.5% BSA, 0.1% sodium azide and were washed twice in 1x PBS, 0.1% BSA, 0.1% sodium azide. The mAb were stained by FITC-conjugated $F(ab')_2$ rabbit anti-mouse IgG (Dako, Hamburg, Germany) or goat anti mouse-PE (Dako, Hamburg, Germany) conjugated used as a second-step reagent. All samples were analyzed on a FACScan flow cytometer (Becton Dickinson, San Jose, CA, USA). Primary mAbs used were anti-CD10 (Calla) (Becton Dickinson), anti-CD117 (*c-kit*) (supernatant from clone 57A5D8B1 kindly provided by Dr. H.-J. Bühring, Tübingen, Germany) and FITC-conjugated rabbit anti-human IgG, IgA, IgM heavy and light chains (Behring, Marburg, Germany). In addition, FITC-conjugated rabbit anti-human CD3, CD10, CD19 and CD34 were used.

RESULTS

RT-PCR of 16S RNA

The quality of the generated cDNA was judged by amplification of the cDNA with 16S RNA specific primers which generate a 752 bp product. The amplification led to comparable amounts of DNA from all cells and cell lines.

RT-PCR and Southern Blot Analysis of *C-Kit* mRNA

Expression and regulation of *c-kit* receptor mRNA was investigated by RT-PCR followed by Southern blot analysis. The positive control cell line MO7e showed detectable

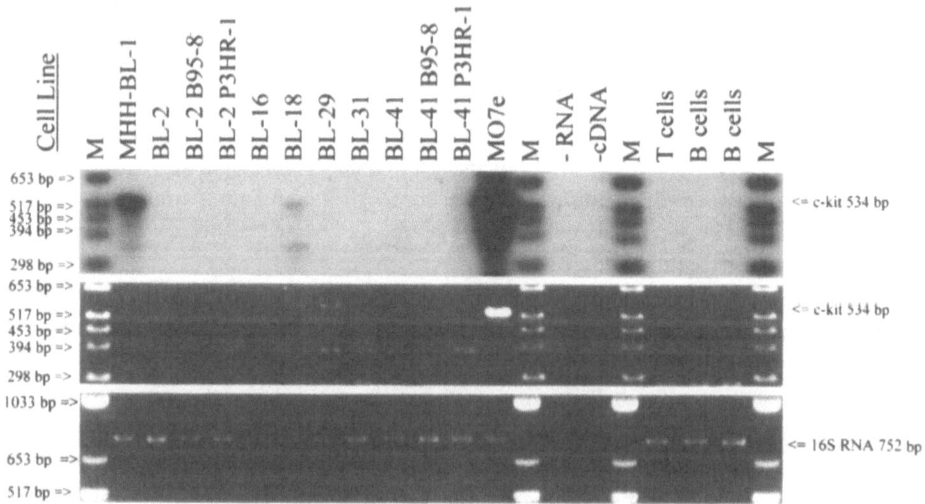

Figure 1. RT-PCR of *c-kit* transcripts. Amplified 534 bp *c-kit* transcripts were separated on a 1.8% agarose gel and stained with ethidium bromide (middle panel). The gel was blotted and hybridized with a Dig-labeled *c-kit*-specific internal oligonucleotide (upper panel). Mitochondrial 16S rRNA was amplified as positive control (lower panel). The length of size markers (M) is indicated in base pairs. Negative control reactions (-RNA and -cDNA) were performed by omitting RNA or cDNA template, respectively.

amounts of *c-kit* mRNA by RT-PCR on an ethidium bromide stained agarose gel while MHH-BL-1 and BL-18 cells showed low level expression of *c-kit* mRNA only detectable by subsequent Southern blot analysis (Fig 1). These two cell lines were always positive. On repeat PCR examination, the fresh MHH-BL-1 cells and the cell lines BL-2 P3HR-1, BL-16, BL-18, BL-41, BL-41 B95-8, BL-41 P3HR-1 were weakly positive only in some experiments indicating a very low level of expression (data not shown). *C-kit* mRNA was never detectable in the negative control cell line HL-60[20] (data not shown). No *c-kit* mRNA was detected in freshly isolated T- and B-cells and all other BL cell lines tested (Fig 1). The question, whether *c-kit* mRNA was not expressed constitutively but inducible was answered by addition of PMA or A23187 to the BL cell cultures. Only the cell line MHH-BL-1 that already expressed *c-kit* mRNA showed a gradual upregulation of *c-kit* mRNA over the time of 48 hours in the presence of A23187 (Fig 2) or IL-7 (Fig 3), while the addition of PMA led to a rapid downregulation after 1 hour (Fig 4). *C-kit* transcripts in unstimulated MHH-BL-1 cells are visible in figure 1, 3 and 4 in contrast to figure 2, due to the longer exposure time of the blot.

C-Kit Receptor Regulation and Expression on BL, T-ALL and T-NHL Cells

The SCF binding was determined by chemical crosslinking of Dig-rhSCF to its receptor. The experiment was done with the cell line MO7e used as a positive control and the cell lines MHH-BL-1 and BL-2 P3HR-1 which showed detectable *c-kit* mRNA expres-

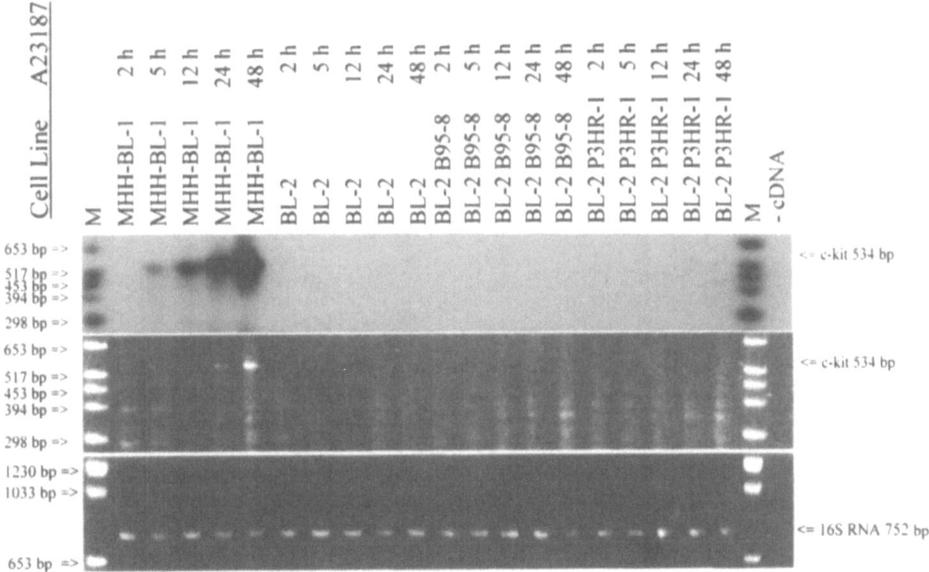

Figure 2. RT-PCR of *c-kit* transcripts in A23187 stimulated BL cells. BL cells were stimulated with A23187 for 2, 5, 12, 24, and 48 h. Amplified 534 bp *c-kit* transcripts were separated on a 1.8% agarose gel and stained with ethidium bromide (middle panel). The gel was blotted and hybridized with a labeled *c-kit*-specific internal oligonucleotide (upper panel). Mitochondrial 16S rRNA was amplified as positive control (lower panel). The length of size markers (M) is indicated in base pairs. Negative control reactions (-RNA and -cDNA) were performed by omitting RNA or cDNA template, respectively.

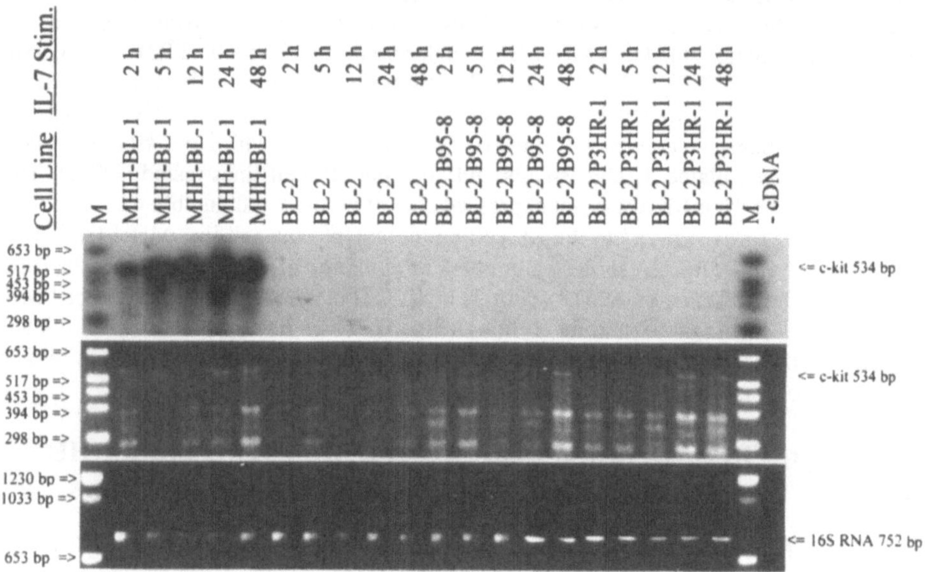

Figure 3. RT-PCR of *c-kit* transcripts in IL-7 stimulated BL cells. BL cells were stimulated with IL-7 for 2, 5, 12, 24, and 48 h. Amplified 534 bp *c-kit* transcripts were separated on a 1.8% agarose gel and stained with ethidium bromide (middle panel). The gel was blotted and hybridized with a labeled *c-kit*-specific internal oligonucleotide (upper panel). Mitochondrial 16S rRNA was amplified as positive control (lower panel). The length of size markers (M) is indicated in base pairs. Negative control reactions (-RNA and -cDNA) were performed by omitting RNA or cDNA template, respectively.

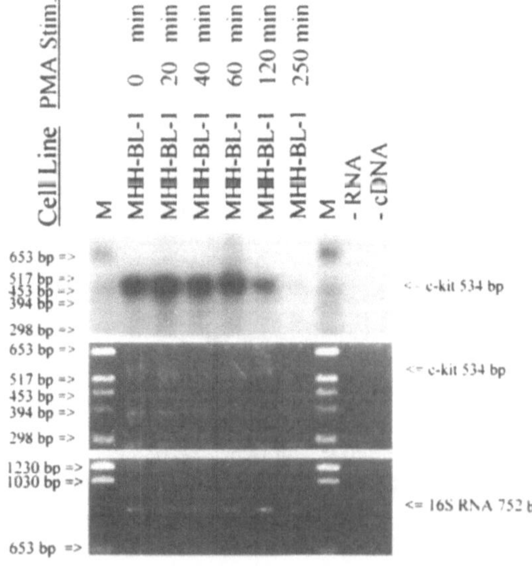

Figure 4. RT-PCR of *c-kit* transcripts in untreated and PMA treated MHH-BL-1 cells. MHH-BL-1 cells were treated 0, 20, 40, 60, 120, and 250 min with PMA. Amplified 534 bp *c-kit* transcripts were separated on a 1.8% agarose gel and stained with ethidium bromide (middle panel). The gel was blotted and hybridized with a Dig-labeled internal oligonucleotide specific for *c-kit* (upper panel). Mitochondrial 16S rRNA was amplified as positive control (lower panel). The length of size markers (M) is indicated in base pairs. Negative control reactions (-RNA and -cDNA) were performed by omitting RNA or cDNA template, respectively.

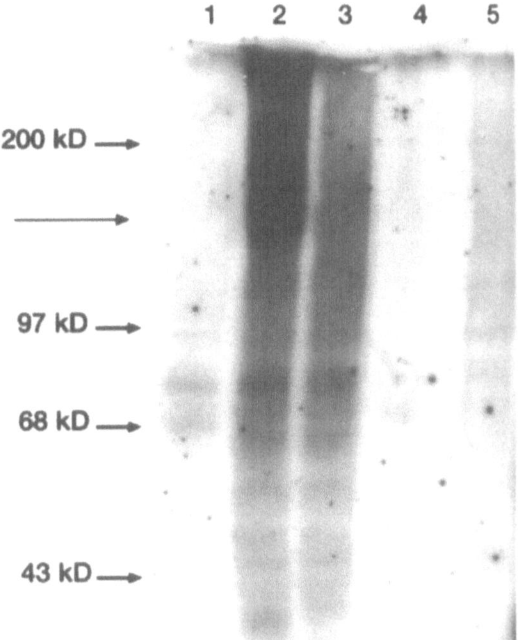

Figure 5. Chemical crosslinking of Dig-labeled SCF to its receptor on BL cell lines. Dig-labeled SCF was crosslinked to BL cells, after lysis the cellular protein was electrophoresed on a SDS-polyacrylamide gel and blotted. The complex of Dig-labeled SCF and its receptor was detected by chemiluminescence at a molecular weight of 165 to 170 Kd. Lane 1: BL-2 P3HR-1 cells; Lane 2: MO7e cells; Lane 3: MO7e cells in the presence of a 100-fold molar excess of unlabeled SCF; Lane 4: MHH-BL-1 cells, Lane 5: MHH-BL-1 cells in the presence of a 100-fold molar excess of unlabeled SCF.

sion by RT-PCR and Southern blot analysis and the cell line MHH-TALL-1. A cytokine-receptor complex of 165 to 170 Kd was found in MO7e cells (control) and MHH-TALL-1 cells (data not shown), corresponding to the expected molecular weight of the *c-kit* receptor of approximately 150 Kd. Specific binding of labeled SCF to its receptor could be demonstrated by addition of 100-fold molar excess of unlabeled SCF. BL cell lines showed no receptor-ligand complex after chemical crosslinking (Figure 5). 7 fresh BL cell preparations and the cell line MO7e were analyzed by FACS for the presence of *c-kit* receptors. All BL cells were gated for the CD10 antigen and found to express surface immunoglobulins. The CD117 antigen, the *c-kit* receptor, was found only on MO7e cells used as a positive control but on none of the BL cells (Figure 6). To determine whether A23187 treatment, which upregulates *c-kit* mRNA expression in MHH-BL-1 cells, would lead to the expression of *c-kit* receptors on the protein level, we treated MHH-BL-1 cells for 48 h with A23187 before FACS analysis. CD117 protein remained undetectable in these stimulated cells (data not shown).

FACS analysis showed *c-kit* receptors on 2 freshly isolated T-ALL cells and on 3 of 13 T-NHL cells. 1 of the T-NHL cells showed spontaneous upregulation of *c-kit* after a culture period of 48 hours and to an even higher level in the presence of A23187. In contrast, CD3 antigen was not upregulated in the presence of A23187. In one of the T-ALL cells upregulation of *c-kit* could also be demonstrated upon cultivation and to a higher extent in the presence of A23187. The CD3 antigen was downregulated upon treatment with A23187. In these cells, PMA treatment led to a rapid downregulation of the CD117 and CD3 antigen (data not shown).

Mitogenic Activity of rhSCF on BL Cell Lines, T-ALL and LB-T NHL Cells

We tested rhSCF on 13 BL cell lines, 2 T-ALL cells, 5 LB-T NHL cells and on 1 megakaryoblastic leukemia (MO7e) cell line in proliferation assays. A significant dose

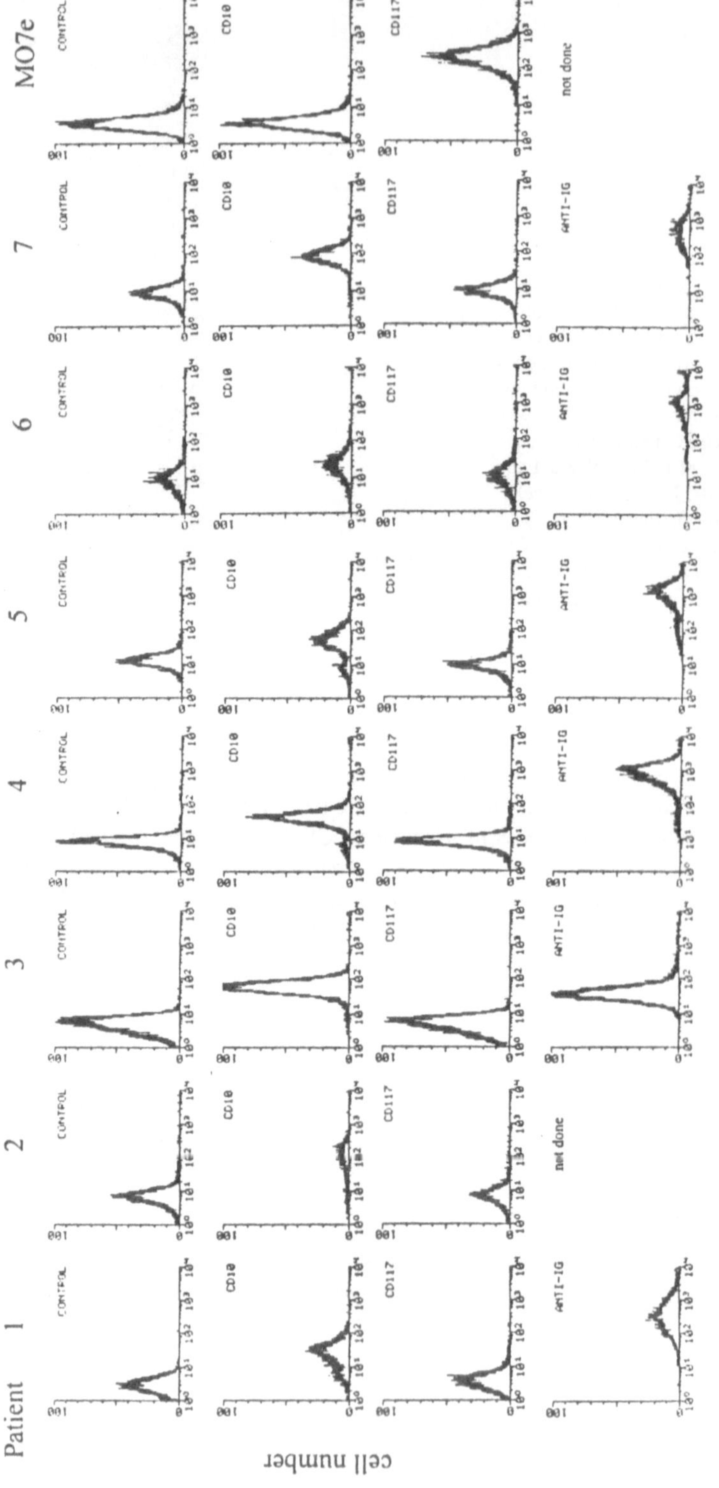

Figure 6. Surface marker analysis of freshly isolated BL cells. Purified BL cells were first stained with monoclonal antibodies against the indicated cell surface antigens and then with a FITC-conjugated F(ab')$_2$ rabbit anti-mouse IgG and analyzed by flow cytometry.

dependent proliferative response to rhSCF was detected only in the MO7e cell line used as a positive control and in 2 T-ALL cells (data not shown). Not one of the BL cell lines tested (Figure 7) and LB-T NHL cells (data not shown) showed a significant response to rhSCF.

DISCUSSION

The aim of this study was to determine the role of *c-kit* in childhood NHL cells. Because of the potential clinical use of SCF in patients undergoing autologous bone marrow harvest,[6,15-19] we investigated *c-kit* expression in childhood BL and T-lymphoblastic cell lines and fresh tumor biopsy cells.

SCF acts synergistically with other growth factors, including Erythropoietin, G-CSF, M-CSF, GM-CSF, IL-3 and IL-6 and plays an important role in normal and leukemic hematopoiesis. The *c-kit* product is expressed on various types of human cell lines derived from leukemic cells of erythroid, megakaryocytic and mast-cell lineages. It is also expressed in most cases of acute myeloblastic leukemia (AML) and in some cases of chronic myelogenous leukemia (CML) in blast crisis. By contrast, little or no expression of *c-kit* is observed in human leukemia cell lines of lymphoid lineage and in blast cells of acute lymphoblastic leukemia, although several lines of evidence indicate that SCF can influence the development of T and B cells. In combination with IL-7, SCF was found to be a potent stimulus at early stages of lymphoid development. A reason for this might be, that SCF-independent mechanisms can compensate for the lack of SCF in early stages of lymphoid development. Mature lymphocytes exhibit little responsiveness to SCF.[20]

c-kit-specific mRNA was shown by us in two BL cell lines but was only detectable by RT-PCR followed by Southern blot analysis. The possibility, that *c-kit* is not expressed spontaneously but requires induction by other cytokines or factors present in vivo was explored by in vitro stimulation experiments. BL cells in vivo are exposed to many other factors and cytokines that theoretically could serve as costimulatory factors for SCF by inducing its receptor, especially since low level expression of *c-kit* mRNA was detected in some of the BL cells.

In our experiments, the cytokines IL-1β and TNF-α did not induce the expression of *c-kit* and SCF mRNA as detected by Northern-blot analysis (data not shown). Both cytokines had been found to be highly expressed in lymph nodes from lymphoma patients.[49] These locally produced cytokines were considered most likely to be biologically relevant, since BL cells were also known to express the receptors for IL-1[50] and TNF-α.[51] In endothelial cells,[52] we previously used these cytokines to induce the expression and demonstrated regulation of SCF and *c-kit* mRNA by IL-1.

In our experiments with BL cells, the expression of *c-kit* could not be induced in BL cells treated with the IgM receptor crosslinking reagents anti IgM and SAC cells, nor by the protein kinase C activator PMA and the calcium ionophore A23187 (data not shown). To detect *c-kit* regulation on a low level, we performed RT-PCR followed by Southern blot analysis. Only MHH-BL-1 cells showed an upregulation of *c-kit* mRNA in the presence of A23187 or IL-7 and a downregulation in the presence of PMA. To further investigate the biological significance of this mRNA expression, we examined *c-kit* expression on the protein level by FACS analysis and crosslinking experiments, and the proliferative effect of SCF on BL cells by ^{3}H-thymidine incorporation. The receptor was undetectable on 7 freshly isolated BL cells by FACS analysis, and by crosslinking experiments and FACS analysis in the two BL cell lines with low level *c-kit* mRNA expression even after stimulation. RhSCF showed no mitogenic effect upon the thirteen cell lines tested.

To find out whether the absence of a proliferative response to exogenous rhSCF could be explained by the autocrine production of SCF or by binding of endogenous SCF to

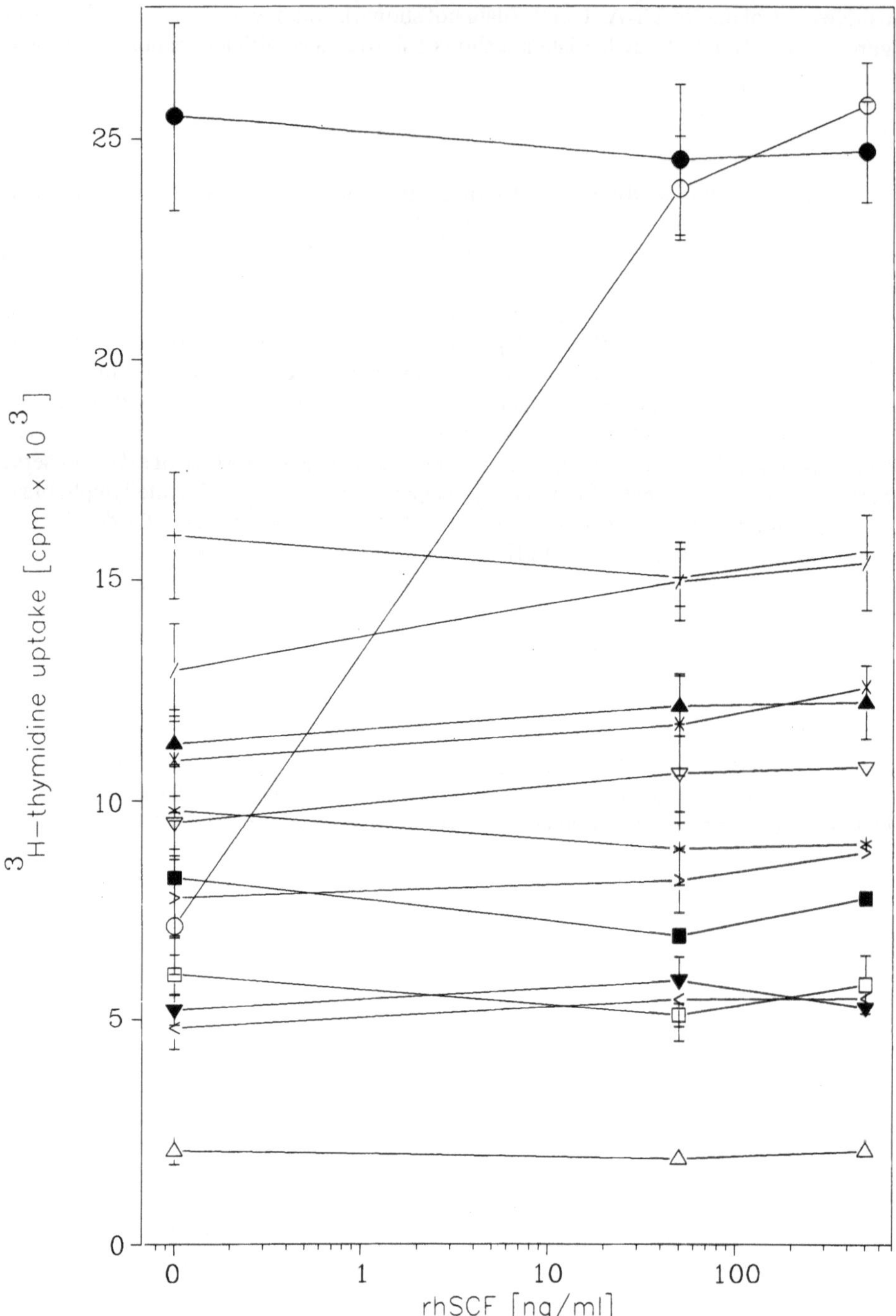

Figure 7. Effect of rhSCF on proliferation of BL cell lines. Cultures were maintained for 72 hours in serum-free medium in the presence of increasing amounts of SCF, pulsed for 4 hours with 0.5 μCi ^{3}H-thymidine and harvested for liquid scintillation counting. Data are means and standard errors of three independent samples. The megakaryoblastic leukemia cell line MO7e was used as a positive control.

intracellular *c-kit* receptors leading to its constitutive activation[53] we performed RT-PCR from unstimulated BL cells and Northern blot analysis from unstimulated and stimulated BL cells to detect SCF transcripts. Specific SCF mRNA transcripts were neither detectable in stimulated and unstimulated BL cell lines by Northern blot analysis nor by RT-PCR followed by Southern blotting (data not shown). From these results we conclude for BL cells, that the low level *c-kit* expression detected on the mRNA level has little biological significance, since *c-kit* mRNA is neither expressed nor inducible in the majority of BL cells, and because no protein expression or changes in proliferation were observed upon addition of rhSCF.

BL cells appear to be derived from more mature B-cells, that express immunoglobulins on their surface. If they behaved like their normal counterparts,[54-56] they would not be expected to express the *c-kit* receptor. Normal B-cells that had progressed in their differentiation to light-chain rearrangement were found not to express the receptor for SCF and IL-7 anymore.[54-56] Consistent with these findings in normal cells, we were not able to detect *c-kit* mRNA in freshly isolated normal B-cells by 35 cycles of RT-PCR. It is not clear, however, whether the process of malignant transformation leads to the persistence of less mature properties in BL cells. Support of this assumption is given by the recent findings showing detection of IL-7 receptor mRNA in BL cell lines.[57] Moreover, while Il-7 receptor positive BL cells are able to respond to rhIL-7, SCF did not act as a costimulatory factor in these cells (unpublished observation).

There are several observations which suggest that SCF is not mandatory for B-lineage expansion in vivo.[58] First, in normal mice treated with a neutralizing antibody against *c-kit*, almost all hemopoietic progenitor cells disappeared from the bone marrow leading to a failure to produce normal numbers of erythroid and myeloid cells while B-cell lymphopoiesis was not significantly affected.[59] Second, adult W locus anemic mice, which have functional mutations at the *c-kit* locus, maintain normal numbers of preB- and B-cells throughout postnatal life.[60] These experiments show, that B-lymphopoiesis is possible in the absence of functional SCF or *c-kit*. Nevertheless, SCF can be a costimulatory factor for IL-7 in the proliferation of a subpopulation of early B-cells.[26] Another subpopulation of B-lineage cells does not require SCF and proliferates in response to IL-7 alone.[59]

From our data we conclude, that the lack of *c-kit* expression on the protein level in BL cells is consistent with their more mature B-cell phenotype. Also, during the process of malignant transformation, the expression of *c-kit* as a marker of an immature phenotype was not maintained, as apparently happened in the case of the IL-7 receptor.[57]

The situation with respect to SCF in T-cell disease is different. T cells are derived from bone marrow stem cells that migrate to the thymus and after maturation within the thymus, mature T cells enter the peripheral circulation. Mature human peripheral blood lymphocytes express little or no *c-kit* receptors, although SCF is a potent synergistic factor with IL-2 and IL-7 in promoting the proliferation of CD4-/CD8- thymocytes and this proliferation can be completely inhibited in the presence of anti-*c-kit*.[61] These observations indicate an important role for *c-kit*/SCF interactions during early thymocyte development.[62,63] Little is known whether T-lymphoblastic leukemia or lymphoma cells can express *c-kit* receptors.

We could demonstrate *c-kit* protein by crosslinking and flow cytometry analysis on freshly isolated blast cells from 2 patients with T-ALL and on 3 of 13 patients with T-NHL (data not shown). PMA treatment led to a downregulation of *c-kit* on the protein level in T-ALL cells within 6 hours. Dubois et al. showed, that TGF-ß treatment of progenitor cell lines and murine primary progenitor cells, which constitutively expressed functional cell-surface *c-kit*, resulted in downregulation of cell-surface *c-kit* expression by a decrease of *c-kit* message stability starting after 2 hours of TGF-ß treatment.[64] Treatment of mouse mast cells and mast cell lines by phorbol myristate acetate (PMA) reduced levels of *c-kit* mRNA by 50% within 2 hours.[20] Differences in the metabolic state or cell cycle activity seems to have important effects of the *c-kit* expression. In fact, there is some evidence that *c-kit* expression may be regulated by the c-myb transcription factor, which plays an important role in regulating the cell cycle. Expression of *c-kit* was essentially abrogated by making human bone marrow mononuclear cells quiescent or by inhibiting myb gene function.[65] Consisting with these findings, we found that *c-kit* protein was upregulated after feeding of T-ALL cells with fresh medium within 20 hours and that *c-kit* protein was only dimly expressed in freshly thawed biopsy T-ALL and T-NHL cells, but expressed significantly higher levels of *c-kit* protein after a culture period of 48 hours. We showed also upregulation of *c-kit* mRNA at a low level by calcium ionophore A23187 in a Burkitt's lymphoma cell line and at the protein level in T-ALL and T-NHL cells within 48 hours. In addition, the CD3 antigen, which was also upregulated upon cultivation of these T-cells, was not upregulated in the presence of A23187. Brach et al. showed in AML cells, that TNF-treatment was associated with enhanced *c-kit* mRNA expression in all 10 samples investigated, presumably by posttranscriptional control mechanisms of *c-kit* mRNA accumulation. TNF-treated AML cells showed prolonged half-life of *c-kit* mRNA from 2 to 3 hours up to 5 to 9 hours.[66]

A restriction of *c-kit* expression to more immature, CD34+ precursors cells, is discussed. While there was a significant correlation of the *c-kit* protein expression with the CD34 antigen in myeloproliferative disorders and myelodysplastic syndromes[67] no restriction was found in AML cells. *c-kit* was not only expressed on CD34+ cells also on CD4+ CD34- precursor cells differentiating towards the monocyte lineage.[68] Kubota et al. found that in 24 of 35 myeloid leukemia cases the expression of *c-kit* and CD34 was statistically significant.[69] The immunophenotypes of *c-kit* expressing T-ALL and T-NHL cells suggest, there is no evidence for a restriction of the *c-kit* protein expression to CD34+ cells.

From our studies we conclude that *c-kit* and SCF play no role in the process of malignant transformation and in the growth of the pediatric BL cells investigated. Based on these results we do not expect a direct negative influence of rhSCF on chemotherapeutic treatment outcome in pediatric Burkitt's lymphoma. We also think that a direct stimulation of childhood BL cells, or SCF-induced direct mobilization of BL cells during an autologous peripheral stem cell harvest primed by SCF is unlikely. Secondary cytokine effects induced by the treatment of a patient with SCF, like activation of endothelial cells and stromal elements, cannot be excluded based on our results. Whether a relevant influence on growth of the lymphoma and on treatment outcome arises from the administration of this cytokine, can finally be demonstrated only in clinical treatment studies. In contrast to the situation in Burkitt's lymphoma, a direct negative influence upon treatment of children suffering from T-lymphoblastic malignancies with SCF is a possibility. *c-kit* expression may not be evident on the freshly isolated malignant cells usually used for diagnostic immunophenotyping but may only be apparent after brief periods of culture or addition of calcium-ionophore. In addition, *c-kit* positive T-lymphoblastic cells are capable of enhanced proliferation upon addition of SCF. We would consider it prudent to keep these results in mind in situations of bone marrow purging and peripheral stem cell expansion in patients with T-cell malignancy.

ACKNOWLEDGMENTS

We thank Dr. Bornkamm (GSF, Munich) for discussion about growth conditions of lymphoma cells; Dr. K. M. Zsebo (Amgen, USA) for supply of rhSCF and Dr. H.-J. Bühring (Tübingen) for supply of CD117 mAb. Supported by Deutsche Krebshilfe grant W88/91/Sy1.

REFERENCES

1. Smith SD, Rubin CM, Horvath A, Nachman J: Non-Hodgkin's lymphoma in children. Semin Oncol 17, No. 1:113, 1990
2. Magrath I: Malignant non-Hodgkin's lymphomas in children, in Pizzo PA, Poplack DG (eds): Pediatric Oncology, Philadelphia, J.B. Lippincott Company, 1993, p 537
3. Reiter A, Schrappe M, Parwaresch R, Henze G, Müller-Weihrich S, Sauter S, Sykora K-W, Ludwig W-D, Gadner H, Riehm H: Non-Hodgkin's lymphomas of childhood and adolescence: results of a treatment stratified for biological subtypes and stage. A report of the BFM group. J Clin Oncol 13:359, 1995
4. Duggan DB, Santarelli MT, Zamkoff K, Lichtman S, Ellerton J, Cooper R, Poiesz B, Anderson JR, Bloomfield CD, Peterson BA, et al. A phase II study of recombinant interleukin-2 with or without recombinant interferon-beta in non-Hodgkin's lymphoma. A study of the Cancer und Leukemia Group B. J Immunother 12:115, 1992
5. Ho AD, Haas R, Wulf G, Knauf W, Ehrhardt R, Heilig B, Korbling M, Schulz G, Hunstein W: Activation of lymphocytes induced by recombinant human granulocyte-macrophage colony-stimulating factor in patients with malignant lymphoma. Blood 75:203, 1990
6. Appelbaum FR: The use of colony stimulating factors in marrow transplantation. Cancer 72:3387, 1993
7. Pettengell R, Testa NG, Swindell R, Crowther D, Dexter TM: Transplantation potential of hematopoietic cells released into the circulation during routine chemotherapy for non-Hodgkin's lymphoma [see comments]. Blood 82:2239, 1993
8. Andreeff M, Welte K: Hematopoietic colony stimulating factors. Semin Oncol 16:211, 1989
9. Pang Y, Norihisa Y, Benjamin D, Kantor RR, Young HA: Interferon-gamma gene expression in human B-cell lines: induction by interleukin-2, protein kinase C activators, and possible effect of hypomethylation on gene regulation. Blood 80:724, 1992
10. Wagner DK, Kiwanuka J, Edwards BK, Rubin LA, Nelson DL, Magrath IT: Soluble interleukin-2 receptor levels in patients with undifferentiated and lymphoblastic lymphomas: correlation with survival. J Clin Oncol 5:1262, 1987
11. Matsuo Y, Motoori T, Irie K, Natori H, Hayashi A, Mizote H, Sagawa K, Yokoyama MM: [Establishment and characterization of an interleukin-2 receptor bearing a Burkitt's lymphoma cell line (ABL-2)]. Nippon Ketsueki Gakkai Zasshi 48:1187, 1985
12. Tomeczkowski J, Yakisan E, Wieland B, Reiter A, Welte K, Sykora K-W: Absence of G-CSF receptors and absent response to G-CSF in childhood Burkitt's lymphoma and B-ALL cells. Br J Haematol 771, 1995
13. Torcia M, Aldinucci D, Carossino AM, Imreh F, Cozzolino F: Biologic and clinical significance of cytokine production in B-cell malignancies. Eur J Haematol Suppl 43:35, 1989
14. Hsu S-M, Waldron JW, Hsu P-L, Hough AJ: Cytokines in malignant lymphomas: review and prospective evaluation. Human Pathol 24:1040, 1993
15. Kurtzberg J, Meyers F, McGuire B, Crawford J: Mobilization of peripheral blood progenitor cells in patients given recombinant methionyl human stem cell factor. Proc Am Assoc CancerRes 34:211, 1993 (abstr.)
16. Tong J, Gordon MS, Srour EF, Cooper R, McNiece I, Hoffman R: Effects of the *in vivo* administration of recombinant methionyl human stem cell factor (SCF) on human hematopietic stem cells (HSC). Proc Am Assoc CancerRes 63:216, 1993 (abstr.)
17. Demetri GD, Gordon M, Hoffman R, Hayes DF, Sledge G, Sullivan S, Edwards R, Merica E, Battiato L, Griffin JD: Effects of recombinant methionyl human stem cell factor on hematopoietic progenitor cells in vivo: preliminary results from a phase I trial. Proc Am Assoc CancerRes 34:217, 1993 (abstr.)
18. Orazi A, Gordon MS, Neiman RS, Sledge G, Battiato L, McNiece I, Hoffman R: Bone marrow effects of recombinant methionyl human stem cell factor (SCF) in patients with normal hematopiesis. Proc Am Assoc CancerRes 34:465, 1993 (abstr.)
19. Henschler R, Brugger W, Kanz L, Mertelsmann R: Haematopoietic growth factors in clinical medicine [news]. Eur J Cancer 30A:118, 1994
20. Galli SJ, Zsebo KM, Geissler EN: The kit ligand, stem cell factor. Adv Immunol 55:1, 1994
21. Zsebo KM, Wypych J, McNiece IK, Lu HS, Smith KA, Karkare SB, Sachdev RK, Yuschenkoff VN, Birkett NC, Williams LR, et al: Identification, purification, and biological characterization of hematopoietic stem cell factor from buffalo rat liver-conditioned medium. Cell 63:195, 1990
22. Yarden Y, Kuang W-J, Yang-Feng T, Coussens L, Munemitsu S, Dull TJ, Chen E, Schlessinger J, Francke U, Ullrich A: Human proto-oncogene c-kit: a new cell surface receptor tyrosine kinase for an unidentified ligand. Embo J 6:3341, 1987

23. Tsuji K, Lyman SD, Sudo T, Clark SC, Ogawa M: Enhancement of murine hematopoiesis by synergistic interactions between steel factor (ligand for c-kit), interleukin-11, and other early acting factors in culture. Blood 79:2855, 1992
24. Hirayama F, Katayama N, Neben S, Donaldson D, Nickbarg EB, Clark SC, Ogawa M: Synergistic interaction between interleukin-12 and steel factor in support of proliferation of murine lymphohematopoietic progenitors in culture. Blood 83:92, 1994
25. Bernstein ID, Andrews RG, Zsebo KM: Recombinant human stem cell factor enhances the formation of colonies by CD34+ and CD34+lin- cells, and the generation of colony-forming cell progeny from CD34+lin- cells cultured with interleukin-3, granulocyte colony-stimulating factor, or granulocyte-macrophage colony-stimulating factor. Blood 77:2316, 1991
26. McNiece IK, Langley KE, Zsebo KM: The role of recombinant stem cell factor in early B cell development. Synergistic interaction with IL-7. J Immunol 146:3785, 1991
27. Steel CM, Hutchins D: Soluble factors and cell-surface molecules involved in human B lymphocyte activation, growth and differentiation. Biochim Biophys Acta 989:133, 1989
28. Lyman SD, Williams DE: Biological activities and potential therapeutic uses of steel factor. A new growth factor active on multiple hematopoietic lineages. Am J Pediatr Hematol Oncol 14:1, 1992
29. Brandt J, Briddell RA, Srour EF, Leemhuis TB, Hoffman R: Role of c-kit ligand in the expansion of human hematopoietic progenitor cells. Blood 79:634, 1992
30. Brugger W, Mocklin W, Heimfeld S, Berenson RJ, Mertelsmann R, Kanz L: Ex vivo expansion of enriched peripheral blood CD34+ progenitor cells by stem cell factor, interleukin-1 beta (IL-1 beta), IL-6, IL-3, interferon-gamma, and erythropoietin. Blood 81:2579, 1993
31. Andrews RG, Bartelmez SH, Knitter GH, Myerson D, Bernstein ID, Appelbaum FR, Zsebo KM: A c-kit ligand, recombinant human stem cell factor, mediates reversible expansion of multiple CD34+ colony-forming cell types in blood and marrow of baboons. Blood 80:920, 1992
32. Andrews RG, Bensinger WI, Knitter GH, Bartelmez SH, Longin K, Bernstein ID, Appelbaum FR, Zsebo KM: The ligand for c-kit, stem cell factor, stimulates the circulation of cells that engraft lethally irradiated baboons. Blood 80:2715, 1992
33. Haylock DN, To LB, Dowse TL, Juttner CA, Simmons PJ: Ex vivo expansion and maturation of peripheral blood CD34+ cells into the myeloid lineage. Blood 80:1405, 1992
34. Billips LG, Petitte D, Dorshkind K, Narayanan R, Chiu CP, Landreth KS: Differential roles of stromal cells, interleukin-7, and kit-ligand in the regulation of B lymphopoiesis. Blood 79:1185, 1992
35. Funk PE, Varas A, Witte PL: Activity of stem cell factor and IL-7 in combination on normal bone marrow B lineage cells. J Immunol 150:748, 1993
36. Inui S, Sakaguchi N: Establishment of a murine pre-B cell clone dependent on interleukin-7 and stem cell factor. Immunol Lett 34:279, 1992
37. Rico Vargas SA, Weiskopf B, Nishikawa S, Osmond DG: c-kit expression by B cell precursors in mouse bone marrow. Stimulation of B cell genesis by in vivo treatment with anti-c-kit antibody. J Immunol 152:2845, 1994
38. Lenoir GM, Vuillaume M, Bonnardel C: The use of lymphomatous and lymphoblastoid cell lines in the study of Burkitt's lymphoma. IARC SCI Publ 309, 1985
39. Philip I, Philip T, Favrot M, Vuillaume M, Fontaniere B, Lenoir GM: Establishment of lymphomatous cell lines from bone marrow samples from patients with Burkitt's lymphoma. J Natl Cancer Inst 73(4):835, 1984
40. Rooney CM, Gregory CD, Rowe M, Finerty S, Edwards C, Rupani H, Rickinson AB: Endemic Burkitt's lymphoma: phenotypic analysis of tumor biopsy cells and of derived tumor cell lines. J Natl Cancer Inst 77(3):681, 1986
41. Rowe D, Heston L, Metlay J, Miller G: Identification and expression of a nuclear antigen from the genomic region of the Jijoye strain of Epstein-Barr virus that is missing in its nonimmortalizing deletion mutant, P3HR-1. Proc Natl Acad Sci USA 82:7429, 1985
42. Calender A, Billaud M, Aubry J-P, Banchereau J, Vuillaume M, Lenoir GM: Epstein-Barr virus (EBV) induces expression of B-cell activation markers on in vitro infection of EBV-negative B-lymphoma cells. Proc Natl Acad Sci USA 84:8060, 1987
43. Pietsch T, Kyas U, Steffens U, Yakisan E, Hadam MR, Ludwig WD, Zsebo K, Welte K: Effects of human stem cell factor (c-kit ligand) on proliferation of myeloid leukemia cells: heterogeneity in response and synergy with other hematopoietic growth factors. Blood 80:1199, 1992
44. Schiessl B: HLA-Bestimmung. Dreieich, Germany, Biotest AG, 1986
45. Chomczynski P, Sacchi N: Single step method of RNA isolation by acid guanidinium thiocyanate-phenol-chloroform extraction. Anal Biochem 162:156, 1987
46. PCR technology. Principles and applications for DNA amplification. New York, Stockton Press, 1989

47. PCR. A practical approach. New York, Oxford University Press, 1991
48. Eperon IC, Anderson S, Nierlich DP: Distinctive sequence of human mitochondrial ribosomal RNA genes. Nature 286:460, 1980
49. Sappino A-P, Seelentag W, Pelte M-F, Alberto P, Vassalli P: Tumor necrosis factor / cachectin and lymphotoxin gene expression in lymph nodes from lymphoma patients. Blood 75:958, 1990
50. Benjamin D, Wormsley S, Dower SK: Heterogeneity in interleukin (IL)-1 receptors expressed on human B cell lines. Differences in the molecular properties of IL-1 alpha and IL-1 beta binding sites. J Biol Chem 265:9943, 1990
51. Benjamin D, Hooker S, Miller J: Differential effects of teleocidin on TNFα receptor regulation in human B cell lines: relationship to coexpression of Il-2 and IL-1 receptors and to lymphokine secretion. Cell Immunol 125:480, 1990
52. König A, Reuter M, Tomeczkowski J, Yakisan E, Tidow N, Corbacioglu S, Welte K: Regulatory effects of bacterial pathogens and interleukin-1 on the mRNA expression of SCF and its receptor *c-kit* in human endothelial cells. Blood 82, No 10,231a, 1993 (abstr.)
53. Turner AM, Zsebo KM, Martin F, Jacobsen FW, Bennett LG, Broudy VC: Nonhematopoietic tumor cell lines express stem cell factor and display c-kit receptors. Blood 80:374, 1992
54. Rolink A, Haasner D, Nishikawa S, Melchers F: Changes in frequencies of clonable pre B cells during life in different lymphoid organs of mice. Blood 81:2290, 1993
55. Henderson AJ, Narayanan R, Collins L, Dorshkind K: Status of kappa L chain gene rearrangements and c-kit and IL-7 receptor expression in stromal cell-dependent pre-B cells. J Immunol 149:1973, 1992
56. Melchers F, Haasner D, Grawunder U, Kalberer C, Karasuyama H, Winkler T, Rolink AG: Roles of IgH and L chains and surrogate H and L chains in the development of cells of the B lymphocyte lineage, in Paul WE, Fathman CG, Metzger H (eds): Annual review of immunology, Palo Alto, California, USA, Annual Reviews, 1994, p 209
57. Benjamin D, Sharma V, Knobloch TJ, Armitage RJ, Dayton MA, Goodwin RG: B cell IL-7: human B cell lines constitutively secrete IL-7 and express IL-7 receptors. Blood 82:241a, 1993 (abstr.)
58. Ryan DH, Nuccie BL, Ritterman I, Liesveld JL, Abboud CN: Cytokine regulation of early human lymphopoiesis. J Immunol 152:5250, 1994
59. Ogawa M, Matsuzaki Y, Nishikawa S, Hayashi S-I, Kunisada T, Sudo T, Kina T, Nakauchi H, Nishikawa S-I: Expression and function of *c-kit* in hemopoietic progenitor cells. J Exp Med 174:63, 1991
60. Landreth KS, Kincade PW, Lee G, Harrison DE: B lymphocyte precursors in embryonic and adult W anemic mice. J Immunol 132:2724, 1984
61. Godfrey DI, Zlotnik A, Suda T: Phenotypic and functional characterization of c-kit expression during intrathymic T cell development. J Immunol 149:2281, 1992
62. Medlock ES, Migita RT, Trebasky LD, Housman JM, Elliott GS, Hendren RW, Deprince RB, Greiner DL: Rat stem-cell factor induces splenocytes capable of regenerating the thymus. Dev Immunol 3:35, 1992
63. Tjonnfjord GE, Veiby OP, Steen R, Egeland T: T lymphocyte differentiation in vitro from adult human prethymic CD34+ bone marrow cells. J Exp Med 1531, 1993
64. Dubois CM, Ruscetti FW, Stankova J, Keller JR: Transforming growth factor-beta regulates c-kit message stability and cell-surface protein expression in hematopoietic progenitors. Blood 83:3138, 1994
65. Ratajczak MZ, Luger SM, DeRiel K, Abrahm J, Calabretta B, Gewirtz AM: Role of the KIT protooncogene in normal and malignant human hematopoiesis. Proc Natl Acad Sci U S A 89:1710, 1992
66. Brach MA, Buhring HJ, Gruss HJ, Ashman LK, Ludwig WD, Mertelsmann RH, Herrmann F: Functional expression of c-kit by acute myelogenous leukemia blasts is enhanced by tumor necrosis factor-alpha through posttranscriptional mRNA stabilization by a labile protein. Blood 80:1224, 1992
67. Siitonen T, Savolainen ER, Koistinen P: Expression of the c-kit proto-oncogene in myeloproliferative disorders and myelodysplastic syndromes. Leukemia 8:631, 1994
68. Reuss Borst MA, Buhring HJ, Schmidt H, Muller CA: AML: immunophenotypic heterogeneity and prognostic significance of c-kit expression. Leukemia 8:258, 1994
69. Kubota A, Okamura S, Shimoda K, Harada M, Niho Y: The c-kit molecule and the surface immunophenotype of human acute leukemia. Leuk Lymphoma 14:421, 1994

24

GROWTH INHIBITION IN IL-7 RECEPTOR POSITIVE CHILDHOOD BURKITT'S LYMPHOMA CELLS IS NOT DUE TO INDUCTION OF APOPTOSIS

Karl W. Sykora,* Jörg Tomeczkowski, Konstanze Kirchhoff, Elif Yakisan, Alfred Reiter, and Karl Welte

Medizinische Hochschule Hannover
Department of Pediatrics IV
-OE 6780-
Konstanty-Gutschow-Str. 8
D-30625 Hannover, Germany

ABSTRACT

Interleukin-7 (IL-7) is a growth factor for normal precursor B and mature T-cells. We have previously shown that the IL-7 receptor was expressed to varying extent in all cells from a panel of thirteen Burkitt's lymphoma (BL) cell lines derived from nine children as detected by reverse transcriptase polymerase chain reaction (RT-PCR). The expression in two of 13 BL cells were highly positive, and also detectable by Northern blotting. We therefore investigated the functional significance of this receptor expression. Proliferation of BL cells in the presence and absence of IL-7 was measured by 3H-thymidine incorporation, and by mitochondrial dehydrogenase activity using the MTS color reagent (Celltiter 96AQ, Promega). Apoptosis was measured on cytospins of BL cells by *in-situ* incorporation of digoxigenin-labeled dUTP into DNA strand breaks using Terminal deoxynucleotidyl Transferase (TdT). Incorporated Dig-dUTP was subsequently detected using an alkaline phosphatase labeled Dig-specific antibody. Apoptosis was also estimated by forward/sidescatter FACS analysis. All proliferation assays were done in serum-free medium. IL-7 inhibited 3H-thymidine incorporation by > 50% only in the three BL cell lines with the highest IL-7 receptor expression. MTS metabolism was inhibited in the same cell lines, but only to a lesser extent of 10-30%. In-situ TdT assays showed no increase in the percentage

*Corresponding author: Dr. Karl W. Sykora, Department of Pediatrics IV -OE 6780- Medizinische Hochschule Hannover, Konstanty.Gutschow-Str. 8, D-30625 Hannover, Germany. Tel.: ++511-532 ext. 9018, 3214, 9020; Fax: ++511-532 ext. 9120.

Molecular Biology of Hematopoiesis 5, edited by Abraham et al.
Plenum Press, New York, 1996

of apoptotic cells, and the percentage of cell death was not increased on forward / sidescatter FACS analysis after IL-7 treatment. In contrast to the situation in Burkitt's lymphoma, control precursor-B-cell ALL lines showed no inhibition, but unchanged or enhanced proliferation. We conclude that IL-7 can be a growth inhibitory cytokine for a subset of BL cells and that induction of apoptosis is not the mechanism of growth inhibition in these cells.

INTRODUCTION

The pediatric non-Hodgkin's lymphomas (NHL) are a heterogenous group of diseases reflecting the diverse differentiation stages of the lymphoid cells from which they originate.[1] Burkitt's type lymphoma (BL) and B-cell acute lymphoblastic leukemia (B-ALL) account for approximately one-half of cases of NHL in childhood and adolescence.[2] In both BL and B-ALL, the malignant cells are characterized by a L3 cytomorphology according to the FAB classification, surface expression of monoclonal immunoglobulins, and the nonrandom chromosomal translocations t(8;14)(q24,q32), t(2;8)(p12,q24) and t(8;22)(q24;q11).[1,3] Both may represent different manifestations of BL rather than being different diseases.

The cytokines required for in-vitro and in-vivo growth of pediatric B-lineage NHLs have only partly been characterized. Cytokines are able to stimulate the proliferation of some B-lineage malignant cells. Among others, these include Tumor Necrosis Factor (TNF)[4] which upregulates ist own receptor and also activates normal B-Lymphocytes, and Interleukin 4 that leads to growth stimulation in 8.6% of patients with NHL but to inhibition in 60%.[5] Interleukin 6 can be be produced by B-lineage lymphoma cells and is an autocrine growth factor in a subset of these cells.[6-8]

Interleukin 7 (IL-7) is a bone marrow derived cytokine, which was initially isolated because it supported the growth of B-precursor cells in-vitro.[9] Later it became evident, that IL-7 is also a growth factor of T-cells.[10] IL-7 can be also be a growth factor for precursor B- and T- acute lymphoblastic leukemia (ALL) cells[11] and T-cell lymphomas and Szesary Cells (23,25).

Because of the special importance of BL and B-ALL in pediatric patients, we investigated the role of IL-7 in the growth and proliferation in these cells.

MATERIALS AND METHODS

Cell Lines and Fresh Cells

Fresh BL cells were obtained from one patient's ascites and one patient's bone marrow. Fresh ALL cells were obtained from three patients' peripheral blood or bone marrow. The cells were isolated by Ficoll-Paque centrifugation (Pharmacia, Uppsala, Sweden), washed and resuspended in Burkitt's lymphoma culture medium (BL medium) as described[12] and maintained at 37 °C in humidified 5% CO_2 with half replacement of medium twice weekly.

Other BL cell lines were initially established from EBV positive and negative BL biopsies by G. Lenoir,[13] International Agency for Research on Cancer (IARC) Lyon, France and kindly provided by Dr. Georg W. Bornkamm (GSF Forschungszentrum für Umwelt und Gesundheit, Munich, Germany). Two of the EBV-negative BL cell lines, BL-02 and BL-41, had been previously infected in vitro by two different EBV strains to study mechanisms of EBV-induced phenotypic alterations of B-cells.

^{3}H-Thymidine Incorporation Assays

The mitogenic activity of IL7 (Genzyme, Ismaning) was determined in a ^{3}H-thymidine incorporation assay. BL cell lines were washed three times with serumfree medium consisting of BL medium without fetal calf serum supplemented with 20 μg/ml soy bean lipids, 1 mg/ml bovine serum albumin fraction V (Cell Culture Reagents, Sigma, Deisenhofen, Germany), 21 μg/ml transferrin, 1 μg/ml insulin, 1 ng/ml sodium selenite (Iscove's Supplement, Gibco BRL, Eggenstein, Germany) and insulin-transferrin-sodium selenite supplement (Boehringer Mannheim, Germany) as described.[12]

MTS/PMS Proliferation Assay

To examine the growth and metabolic activity of BL cells in the presence and absence of IL-7, we used the MTS/PMS color reagent (Celltiter 96AQ, Promega, Madison) which measures mitochondrial dehydrogenase activity. Cells were washed twice in serum-free BL-medium, and 0,1 ml each of a 10^5 cells/ml suspension was added to 0,1 ml each of medium containing 0 or 20 ng/ml of IL-7 and the MTS/PMS reagent mixture in 96-well flatbootom plates. All tests were performed in triplicate and after incubation at 37°C in a 5% CO_2 atmosphere, plates were read at 490 nm absorbance daily for several days in a microplate ELISA reader.

Apoptosis Assay by In-Situ DNA Fragmentation

Terminal transferase quantification of apoptosis was perfomed by a variation of the assay described.[14] Lymphoma cells were centifuged onto glass slides and fixed for 20' in 3% buffered formaldehyde in phosphate buffered saline (PBS). The slides were dehydrated in an ethanol series and then treated with a proteinase K (Boehringer, Mannheim) concentration optimized for signal-strength with good preservation of morphology (10-20 μg/ml in PBS) for 30' at 37°C, again dehydrated and stored at -20°C until use. Dry slides were then incubated in TdT buffer (26 mM Tris-HCl, 200 mM potassium cacodylate, 5 mM cobalt chloride at pH 6.6) containing 0,25 U/ μl terminal deoxynucleotidyl transferase (Boehringer, Mannheim) and 0,5μM DIG-dUTP for 60' at 37 °C. After washing, the slides were incubated with anti-DIG antibody coupled to alkaline phosphatase, and positive cells were detected by the NBT-BCIP color reaction.

Flow Cytometric Determination of Size and Granularity of Burkitt's Lymphoma Cells

Cells were incubated at 37 °C for 0, 4, and 7 days in the presence and absence of 10 ng/ml IL-7. Forward- and sidescatter of the cell population was determined using a Beckton-Dickinson FACScan flow cytometer.

RESULTS

We chose a panel of thirteen Burkitt's lymphoma cell lines derived from nine children aged 3 to 14 years and two B-precursor ALL cell lines as additional controls. Most children were of caucasian origin, one child was turkish, one from North Africa and two from Reunion. Three cell lines were EBV-positive.

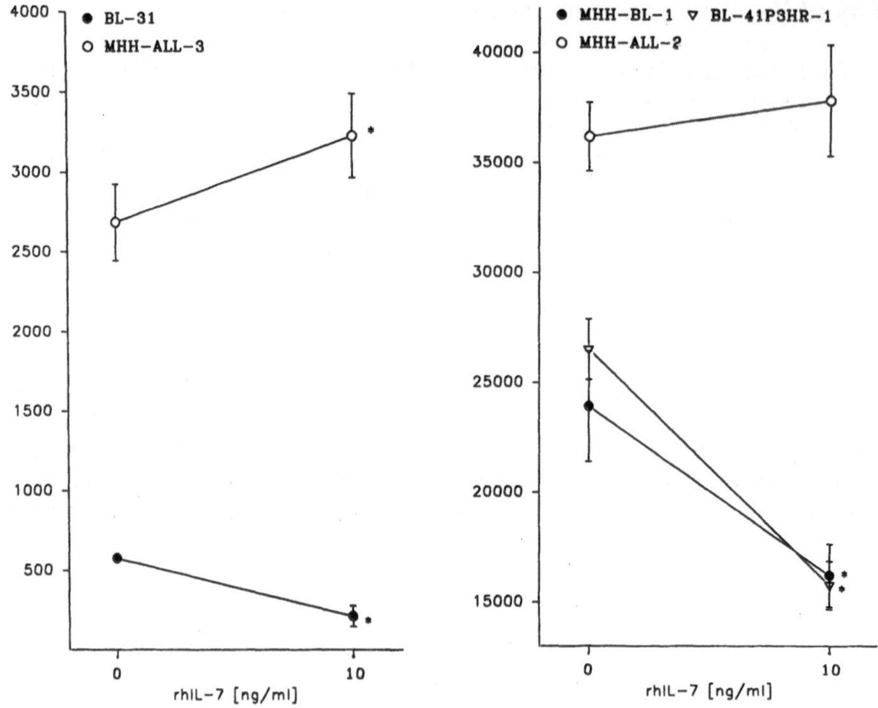

Figure 1. Inhibition of proliferation of BL cells in the presence of IL-7, 3H-Thymidine incorporation. BL cells BL-31, BL41P3HR-1 and MHH-BL-1 and control B-precursor ALL cells MHH-ALL-2 and and MHH-ALL-3 were grown for 68 hours in serum-free medium in the presence of 0 or 10 ng/ml of IL-7. ^{3}H-Thymidine was added for 4 hours, the cells were filtered on glassfiber filters, washed and incorporation was determined by liquid scintillation counting.

We and Benjamin[15,16] had previously shown, that IL-7 receptors can be expressed on Burkitt's lymphoma cell lines. Since the biological significance of this was not known so far, we examined whether IL-7 had a proliferative effect on BL cells.

A proliferative effect of IL-7 was seen in none of the BL cell lines (data not shown), quite in contrast to the B-precursor ALL cells that served as controls (Fig. 1). Three of the 13 BL cell lines (Fig. 1) reacted to IL-7 with proliferative inhibition, the other cell lines showed no reaction to IL-7 (data not shown). These were also the cell lines with the highest expression of the IL-7 receptor (data not shown). The proliferative inhibition of IL-7 shown by ^{3}H-thymidine incorporation in Fig. (1) was also confirmed by MTS metabolism. Fig. (2) shows the cell line BL-41 P3HR-1, which shows a maked proliferative inhibition during the first 6 days of culture until it reached saturation. The same is true for the MHH-BL-1 cell line shown on the next figure (Fig. 3). The more slowly growing cell line BL-31 shows only a marginal inhibition by IL-7 and BL-2 shows no effect at all. For comparison, the B-precursor ALL cell line MHH-ALL-3 shows a stimulation of proliferation (Fig. 3).

This indicated to us, that the IL-7 receptor on BL cell lines can be functional and mediate a biological effect. This effect is the opposite of the proliferative effect that is well known for B-precursor ALL cells, however.

To further delineate the mechanism of this growth inhibition, we examined whether it was due to induction of apoptosis. To detect early phases of apoptosis, the in-situ DNA fragmentation assay was used on cytospins of BL cells (Fig. 4b). This method uses

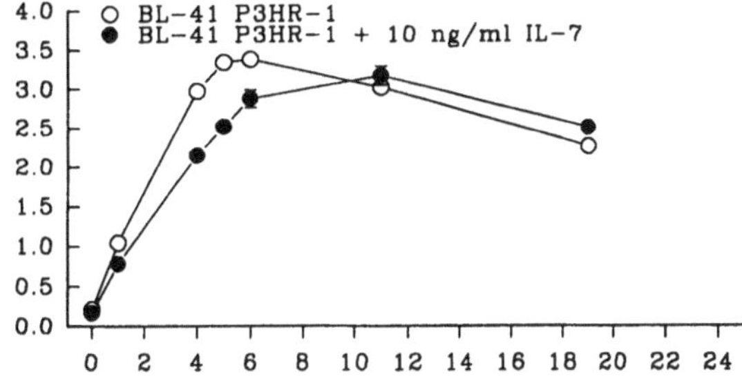

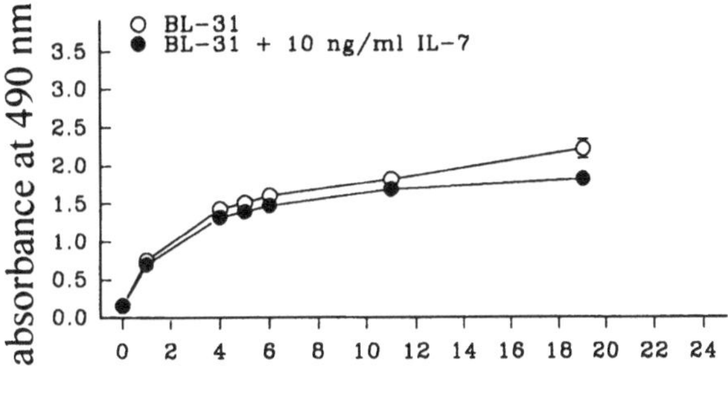

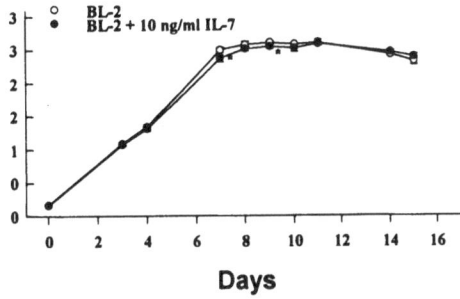

Figure 2. Inhibition of proliferation of BL cells in the presence of IL-7, MTS metabolism. 10^4 BL cells in 0,1 ml serum-free BL-medium were added to 0,1 ml each of the same medium containing 0 or 10 ng/ml of IL-7 and the MTS/PMS reagent mixture in 96-well flatbottom plates. Cells were incubated at 37 °C and absorbance was read daily at 490 nm.

terminal desoxynucleotidyl transferase (TdT) to add digoxigenin (DIG) labeled nucleotides to strand breaks in the nuclear DNA, which are then detected by anti-DIG antibodies followed by a color reaction. Negative control reactions were performed by omitting TdT in the reaction (Fig. 4a), and positive controls consisted of a cytospin that had been mildly digested with deoxyribonuclease to create DNA breaks (Fig. 4c). Cell lines were incubated in serum-free medium in the presence or absence of IL-7 for up to ten days and the percentage of remaining nonapoptotic cells was determined (Fig. 5). In this experiment, all cells that stained positive for DNA breaks were considered apoptotic. In the late stages of necrosis, however, DNA breaks can be also observed. If the characteristic morphology with nuclear condensation and fragmentation of the cell undergoing apoptosis is not also present, apoptosis cannot be formally proven. A part of the cells counted may therefore be necrotic. Fig. 5 shows that the 3 BL and one ALL cell line held in serum-free medium in the absence of IL-7 become increasingly nonviable in the course of four days. In the presence of IL-7, however, the induction of apoptosis was much delayed (BL-41 P3HR-1) and was not observed at all for ten days in the cell lines BL-31 and BL-2. The IL-7 effect on the control ALL cell line is negligible. This phenomenon was seen in repeated experiments, the extent of the protection from apoptosis is variable and appears to be

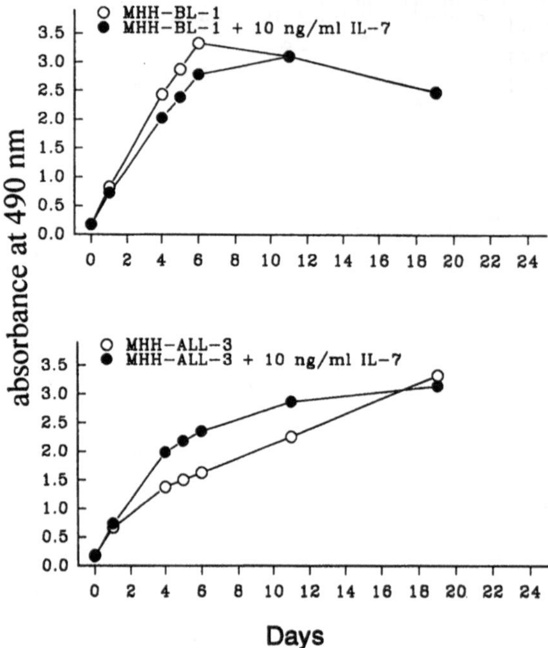

Figure 3. Stimulation of proliferation of B-precursor ALL cells in the presence of IL-7, MTS metabolism. 10^4 control B-precursor ALL (lower panel) cells in 0,1 ml serum-free BL-medium were added to 0,1 ml each of the same medium containing 0 or 10 ng/ml of IL-7 and the MTS/PMS reagent mixture in 96-well flatbottom plates. Cells were incubated at 37 °C and absorbance was read daily at 490 nm. The Burkitt's lymphoma cell line MHH-BL1 served as control (upper panel).

related to proliferative state and viability of the cells at the start of the experiment. An induction of apoptosis by IL-7 was never observed. It is also remarkable that the BL-2 cell line, which was not growth-inhibited by IL-7 (Fig. 2), was protected from apoptosis by IL-7 as well.

When looking at Wright-stained cytospins of BL-cells after serum-free culture in IL-7, we always had the impression that the BL cells looked larger, rounder and "healthier"

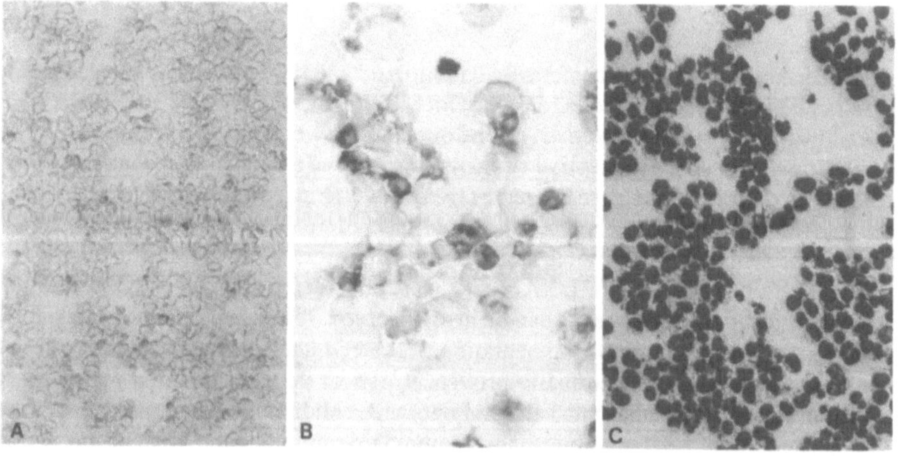

Figure 4. Apoptosis assay by in-situ DNA fragmentation. In-situ TdT quantification of DNA strand breaks was perfomed on a BL cell line (b) with positive (c) and negative (a) control reactions. For positive control reactions, DNA was fragmented in-situ by a brief controlled incubation with DNAse I. In the negative control reaction, TdT was omitted.

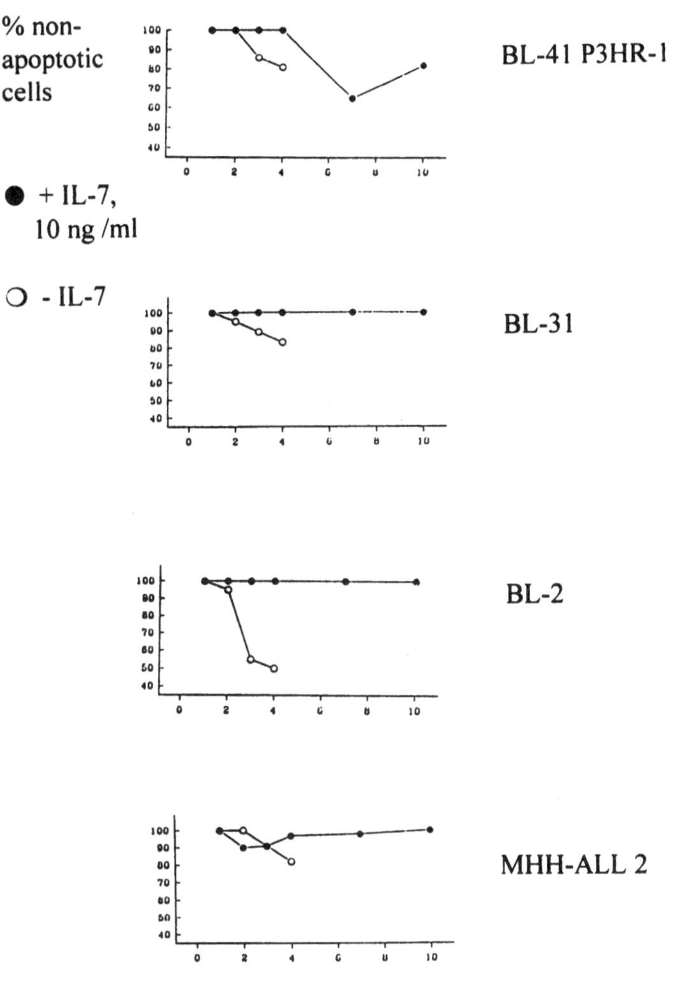

Figure 5. Apoptosis of BL cells in the presence and absence of IL-7. BL cells and one control B-precursor ALL cell (MHH-ALL-1) were incubated in serum-free medium at 37 °Cfor 10 days in the presence and absence of 10 ng/ml IL-7. Cells were centrifuged onto glass slides and the percentage of nonapoptotic cells was determined by the in-situ TdT DNA fragmentation assay.

than their non- IL-7 treated counterparts. To quantify this phenomenon, we estimated the cross-sectional diameter of the BL cell population after serum-free culture in the presence and absence of IL-7 by forward light scatter measurements in an flow cytometer (Fig. 6). After four days of culture in the presence of IL-7, BL-41 P3HR-1 cells were in average much larger in size than the cells grown without IL-7. Fig. (7) shows the same phenomenon also in the cell lines MHH-BL-1 and BL-31. The percentage of the larger cells (over channel 560) was plotted in this graph. On day 4, the average cell diameter in the IL-7 treated cultures is larger in all three BL cell lines, although the difference is smallest in the slowly growing BL-31 cell line.

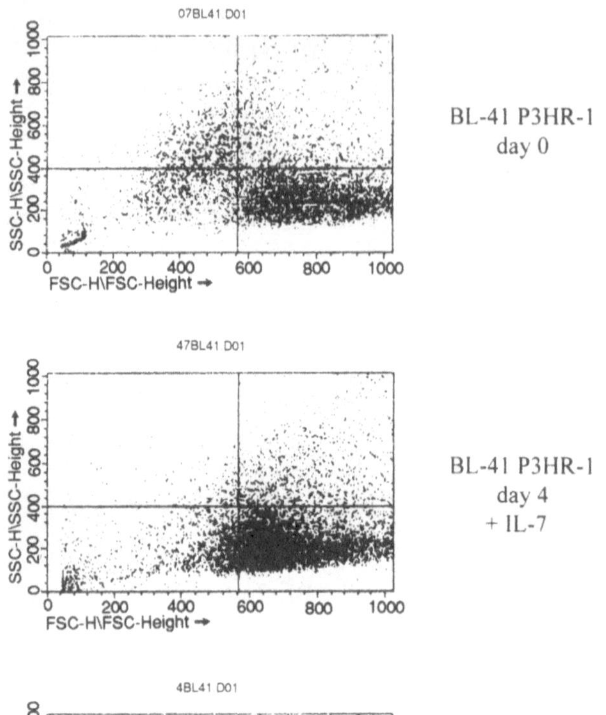

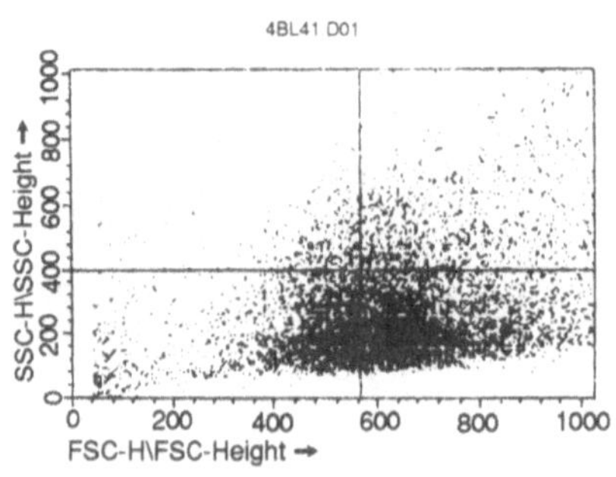

BL-41 P3HR-1
day 4
- IL-7

Figure 6. Forward / sidescatter analysis of BL-41 P3HR-1 in the presence and absence of IL-7. BL-41 P3HR-1 was grown in serum-free medium for four days in the presence or absence 10 ng/ml of IL-7. Forward/sidescatter analysis was performed using a Beckton-Dickinson FACScan flow cytometer.

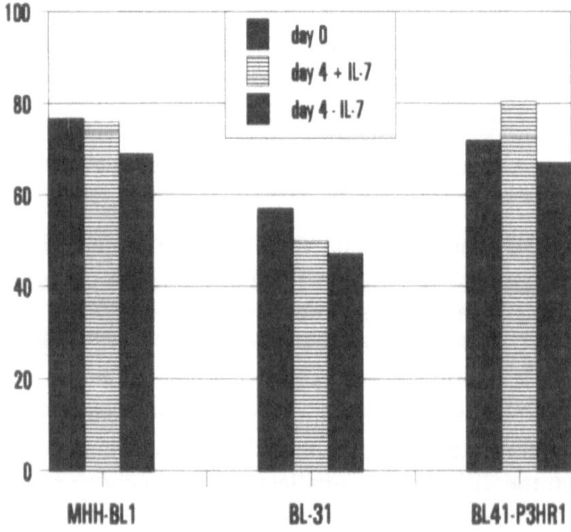

Figure 7. Cell diameter of BL cells after growth in the presence or absence of IL-7. BL cells were grown in serum-free medium for four days in the presence or absence of 10 ng/ml of IL-7. Cell size distribution was estimated by flow cytometric forward-scatter analysis. To quantify the size of the population, the percentage of cells in the gates higher than 560 was plotted.

DISCUSSION

The IL-7 receptor is known to be expressed on some Burkitt's lymphoma cell lines.[15,16] Here we demonstrate that this receptor is also capable of mediating a biological response, namely the proliferative inhibition of BL cell lines. We demonstrate, that this proliferative inhibition is not due to the induction of apoptosis in these cell lines but appears to be due to a form of differentiation.

The IL-7 treated BL cells appear to enter a dormant state where they do not proliferate and are less metabolically active, and are protected from apoptosis induced by serum starvation. Therefore, presumbly by virtue of the oncogenic process, some BL cell lines have maintained the property of pre-B cells of being responsive to IL-7. This responsiveness, however, does not result in the same biologic effect as in B-cells.

ACKNOWLEDGMENTS

We thank the chairman of our department, Prof. Dr. H. Riehm, for support, Dr. Georg W. Bornkamm (Munich) for discussion about the growth-conditions of lymphoma cells and Dr. M. Hadam for critical reading of the manuscript. Supported by Deutsche Krebshilfe grant W88/91/Sy1

REFERENCES

1. Smith SD, Rubin CM, Horvath A, Nachman J: Non-Hodgkin's lymphoma in children. Semin Oncol 17, No. 1:113, 1990
2. Reiter A, Schrappe M, Parwesch R, Henze G, Müller-Weihrich S, Sauter S, Sykora K-W, Ludwig W-D, Gadner H, Riehm H: Non-Hodgkin's lymphomas of childhood and adolescence: results of a treatment stratified for biological subtypes and stage. A report of the BFM group. J Clin Oncol 1994(in press)
3. Magrath I: Malignant non-Hodgkin's lymphomas in children, in Pizzo PA, Poplack DG (eds): Pediatric Oncology, Philadelphia, J.B. Lippincott Company, 1993, p 537
4. Heilig B, Mapara M, Bargou R, Fiehn C, Dorken B: TNF alpha therapy activates human B-lymphoma cells in vivo and may protect myelopoiesis. Leuk Res 16:769, 1992
5. Taylor CW, Grogan TM, Salmon SE: Effects of interleukin-4 on the in vitro growth of human lymphoid and plasma cell neoplasms. Blood 75:1114, 1990
6. Yee CS, Messner HA, Minden MD: Regulation of interleukin-6 expression in the lymphoma cell line OCI-LY3. J Cell Physiol 148:426, 1991
7. Emilie D, Devergne O, Raphael M, Coumbaras LJ, Galanaud P: Production of interleukin-6 in high grade B lymphomas. Curr Top Microbiol Immunol 182:349, 1992
8. Emilie D, Coumbaras J, Raphael M, Devergne O, Delecluse HJ, Gisselbrecht C, Michiels JF, Van-Damme J, Taga T, Kishimoto T, et-al: Interleukin-6 production in high-grade B lymphomas: correlation with the presence of malignant immunoblasts in acquired immunodeficiency syndrome and in human immunodeficiency virus-seronegative patients. Blood 80:498, 1992
9. Henney CS: Interleukin 7: effects on early events in lymphopoiesis. Immunol Today 10:170, 1989
10. Chazen GD, Pereira GM, LeGros G, Gillis S, Shevach EM: Interleukin 7 is a T-cell growth factor. Proc Natl Acad Sci U S A 86:5923, 1989
11. Touw I, Pouwels K, van Agthoven T, van Gurp R, Budel L, Hoogerbrugge H, Delwel R, Goodwin R, Namen A, Lowenberg B: Interleukin-7 is a growth factor of precursor B and T acute lymphoblastic leukemia. Blood 75:2097, 1990
12. Tomeczkowski J, Beilken A, Frick D, Wieland B, König A, Falk MH, Reiter A, Welte K, Sykora KW: Absence of c-kit receptor and absent proliferative response to stem cell factor in childhood Burkitt's lymphoma cells. Blood 86,1469, 1995
13. Philip I, Philip T, Favrot M, Vuillaume M, Fontaniere B, Lenoir GM: Establishment of lymphomatous cell lines from bone marrow samples from patients with Burkitt's lymphoma. JNCI 73(4):835, 1984

14. Wijsman JH, Jonker RR, Keijzer R, van de Velde CJH, Cornelisse CJ, Dierendonck JHV: A New Method to Detect Apoptosis in Paraffin sections: In Situ End-labeling of Fragmented DNA. J Histochem Cytochem 41:7, 1993
15. Sykora KW, Tomeczkowski J, Kirchhoff K, Yakisan E, Wieland B, Reiter A, Welte K: Expression of the IL-7 receptor and IL-7 induced growth inhibition in childhood Burkitt's lymphoma cells. Blood 84:159a, 1994 (abstr.)
16. Benjamin D, Sharma V, Knobloch TJ, Armitage RJ, Dayton MA, Goodwin RG: B cell IL-7: human B cell lines constitutively secrete IL-7 and express IL-7 receptors. Blood 82:241a, 1993 (abstr.)

25

EXPRESSION OF THE MULTIDRUG-RESISTANCE GENE PRODUCT, P-GLYCOPROTEIN, IN ADULT ACUTE MYELOID LEUKEMIA

Association with Stem Cell Antigen CD34 and Specific Cytogenetic Abnormalities

Elisabeth Paietta,[1*] Janet Andersen,[2] Jorge Yunis,[3] Jacob M. Rowe,[4] Peter A. Cassileth,[5] and Peter H. Wiernik,[1] for the Eastern Cooperative Oncology Group[6]

[1] Department of Oncology
Montefiore and Albert Einstein College of Medicine, Bronx, New York
[2] Division of Biostatistics
Dana Farber Cancer Institute, Boston, Massachusetts
[3] Division of Cancer Biology
Thomas Jefferson University, Philadelphia, Pennsylvania
[4] Division of Medical Oncology
University of Rochester, Rochester, New York
[5] University of Miami
Miami, Florida
[6] The Eastern Cooperative Oncology Group
Boston, Massachusetts

ABSTRACT

Response-driven retrospective analysis of 382 adult ECOG patients with *de novo* AML for prognostically significant antigen profiles revealed that expression of P-glycoprotein (Pgp) by >36.5% of blast cells predicted for failure to achieve complete remission (p=0.0026). In a univariate analysis of Pgp and the stem cell antigen CD34 in combination, response was inversely associated with the degree of expression of the two antigens in that the complete remission rate was 72% when blast cells were negative for both Pgp and CD34,

* Address of correspondence: Elisabeth Paietta, Ph.D., Montefiore Medical Center, Department of Oncology, 111 East 210th Street, Bronx, NY; Tel (718) 920-4549; FAX (718) 798-7474.

53% if one antigen was positive and 38% with double antigen positivity ($p<.0001$). In terms of associations with specific cytogenetic abnormalities, both Pgp and CD34 were characteristically absent in t(15;17) acute promyelocytic leukemia, while positivity for both antigens was significantly more likely in patients whose blast cells contained a del(5q) and/or del(7q) ($p=.0038$). These data suggest that the unfavorable prognosis of patients with del(5q) or del(7q) may be attributed to, at least in part, a high degree of immaturity (CD34 positivity) and the expression of multidrug resistance.

INTRODUCTION

Approximately 15%-30% of acute myeloid leukemia (AML) patients are primarily resistant to chemotherapy and 60-80% of patients who achieve a complete remission will inevitably relapse and succumb to their disease.[1] Classical multidrug resistance (MDR) is discussed as one major mechanism of therapy failure in AML;[2,3] it is one of the best understood mechanisms of resistance to anticancer drugs.[4,5] The classical MDR phenotype is characterized by a reduced ability of cells to accumulate drugs as compared to normal cells. The increased drug efflux is due to the activity of a 170kDa glycoprotein, the P-glycoprotein (P for permeability), a unidirectional drug-efflux pump which is encoded by the mdr1 gene.

Pgp strongly positive normal hematopoietic cells include subsets of peripheral blood cells, predominantly T-lymphocytes and natural killer cells, as well as early bone marrow progenitor cells which are characterized by high expression of the stem cell antigen CD34.[5]

We have recently shown that CD34 expression in AML correlates with classical Pgp function, as demonstrated by the decreased staining of CD34+ blast cells with the fluorescent dye Rhodamine 123 due to increased dye efflux mediated by Pgp and inhibition of Rh-efflux by known Pgp modulators such as verapamil and cyclosporine.[6] Along the same line, we found that leukemic cells from acute promyelocytic leukemia (APL), which characteristically lack CD34 expression,[7] show significantly lower Pgp and mdr1 mRNA expression levels than non-APL leukemias and lack classical Pgp function.[8] Results from a preliminary response-driven analysis of the prognostic significance of Pgp and CD34 expression in AML, showed both antigens to be associated with response. Response was inversely associated with the degree of expression of Pgp and CD34 and patients positive for both antigens had the least likelihood of achieving a complete remission.[9]

The current study aimed at evaluating the correlation of the expression of Pgp and/or CD34 with specific cytogenetic abnormalities of known prognostic significance, such as t(15;17), t(8;21), inv(16), del(5q) and/or del(7q), and a complex karyotype. The study population was a large group of patients with *de novo* adult AML, enrolled on treatment protocols of the Eastern Cooperative Oncology Group (ECOG).

MATERIALS AND METHODS

Study Population

Antigen analysis was performed in 382 patients with *de novo* AML as diagnosed by immunophenotyping and confirmed by morphology and cytochemistry.[10] The median age was 44.5 years (range 15-78). Cytogenetic data were available on 224 of those patients. Patients were recruited from four major ECOG studies, E3489, E1490, E3993 and E2491, a study of APL which explains the 17% incidence of APL patients in this analysis.

Antigen Determination and Analysis

Antigen expression was measured flow cytometrically on gated blast cells using antibody HPCA-1 (Beckton-Dickinson) for detection of the CD34 molecule and 4E3.16 (provided by Dr. R. Arceci) for detection of a cell surface peptide epitope of Pgp. The overall immunophenotype was established centrally in ECOG's immunophenotyping laboratory based on binding results with >25 antibodies directed against myeloid, lymphoid, erythroid and megakaryocytic antigens.[9] Cut-off points for defining antigen positivity of a leukemia cell population were determined retrospectively based on the antigen expression levels seen in responding versus non-responding patients using the Classification and Regression Tree (CART) analysis.[11]

Cytogenetics

Cytogenetic analyses were performed by individual ECOG institutions and, in the majority of cases, confirmed by central ECOG review. Specific cytogenetic abnormalities analyzed statistically were inv(16)(p13q22), t(15;17)(q22;q11), t(8;21)(q22;q22), del(5q) and/or del(7q) and complex karyotypes, defined as >3 numerical or structural aberrations.

Statistical Analysis

Associations between cytogenetics and antigen expression or CR rate were calculated by the Fisher's Exact Test. Trend tests for associations between cytogenetic abnormalities and CD34/Pgp expression were done by the Wilcoxon mid-rank test for ordered alternatives.

RESULTS

In 382 patients with AML, CART identified 36% and 52% as cut-off points for Pgp and CD34 expression, respectively, to give the best separation of complete response rates. Cases where more than these percent of blast cells express a given antigen are referred to as "antigen positive".[9] The likelihood of failing to respond increased with the degree of Pgp/CD34 expression in that patients negative for both antigens had a response rate of 72%, those positive for one antigen had a response rate of 53% and patients expressing both antigens above their CART cut-off points had a response rate of only 38% ($p<0.0001$).

Cytogenetic data were available in 224 of the 382 patients.

The incidence of cytogenetic abnormalities in question was inv (16) 5%, t(15;17) 17%, t(8;21) 6%, del(5q) and/or del(7q) 9% and complex karyotypes 12% of patients. Table 1 shows the CR rates for cases with and without specific cytogenetic markers. A significant impact on response to induction chemotherapy was found for inv (16) whereby 100% of patients with inv(16) achieved a CR compared to 67% of those without this cytogenetic abnormality ($p=0.019$) and for del(5q)/del(7q) in that patients with either or both of these deletions did significantly poorer than the rest of the study group ($p=0.0026$).

In terms of association between CD34 or Pgp and specific cytogenetic abnormalities, CD34 was characteristically absent in APLs with the t(15;17) ($p<0.0001$), whereas CD34 positivity was significantly more frequent in patients having a t(8;21) ($p=0.002$), a del(5q)/del(7q) ($p=0.0047$) or a complex karyotype ($p=0.007$) (Table 2). For Pgp, only a significant association between Pgp negativity (⊓ 36% of cells expressing Pgp) and t(15;17) ($p=0.013$) was found (Table 3). Because of our observation that response was inversely associated with the degree of expression of Pgp and CD34, we looked for an association between having neither CD34 nor Pgp, either CD34 or Pgp, or both antigens, and the

Table 1. The number of patients with specific cytogenetic abnormalities were: 11 patients with inv(16), 38 patients with t(15;17), 13 patients with t(8;21), 20 patients with del(5q)/del(7q) and 27 patients with complex abnormalities

CYTOGENETICS VS COMPLETE RESPONSE RATE

Cytogenetic Marker	Present CR (%)	Absent CR (%)	P-value
inv(16) (p13q22)	100	67	**0.019**
t(15;17) (q22; q11)	76	67	0.438
t(8; 21) (q22; q22)	86	68	0.235
del (5q) and / or del (7q)	38	72	**0.0026**
complex karyotype	56	71	0.125

CR, complete remission rate

cytogenetic markers. As shown in Table 4, a significant association between degree of CD34 and/or Pgp expression and cytogenetics was detected for patients with del(5q)/del(7q) in that 4% of patients who lacked both antigens fell into this category, 12% of patients who expressed either one of the two antigens and 25% of patients who expressed both antigens. Therefore, the probability of having a del(5q) and/or del(7q) increased with the expression of CD34 and Pgp by the blast cells and was highest when both antigens were present. This association was statistically highly significant (p=0.0038).

DISCUSSION

In this analysis between antigen expression, cytogenetic markers and response in *de novo* adult AML, Pgp and CD34 were dichotomized at their previously established CART cut-off points. In terms of an association between CD34 or Pgp and specific cytogenetic abnormalities, the present study confirmed the absence of both CD34 and Pgp in t(15;17) APL.[3,7,8] On the other hand, CD34 was positive to a significantly higher degree in t(8;21) when compared to all other AMLs not carrying this cytogenetic marker. Previously, increased expression of CD34 in t(8;21) AML cells compared to non-t(8;21) FAB M2-AMLs has been reported.[12] The high incidence of CD34 positivity of blast cells with del(5q) and/or del(7q) (Table 2) is in agreement with an earlier study,[13] while our data, as well as those of others,[14]

Table 2. Cytogenetics vs. CD34 expression

CYTOGENETICS VS CD34 EXPRESSION

Cytogenetic Marker	CD34 Positive	CD34 Negative	P-value
	% of Total Patients		
inv(16) (p13q22)	64	45	0.234
t(15;17) (q22; q11)	8	53	**< 0.0001**
t(8; 21) (q22; q22)	86	43	**0.002**
del (5q) and / or del (7q)	76	42	**0.0047**
complex karyotype	70	42	**0.007**

CD34 dichotomized at CART cut-off point (52.5%)

Table 3. Cytogenetics vs. P-glycoprotein expression

CYTOGENETICS VS P-GLYCOPROTEIN EXPRESSION

Cytogenetic Marker	P-glycoprotein Positive	P-glycoprotein Negative	P-value
	% of Total Patients		
inv(16) (p13q22)	18	16	0.692
t(15;17) (q22; q11)	3	19	**0.013**
t(8; 21) (q22; q22)	7	17	0.705
del (5q) and / or del (7q)	24	15	0.347
complex karyotype	15	16	1.000

P-glycoprotein dichotomozed at CART cut-off point (36.5%)

do not support an alleged association between Pgp expression and chromosome 7 abnormalities,[15] presumably involving the locus of the mdr1 gene.[16]

When analyzed in combination, we had found response to induction chemotherapy to be inversely associated with the degree of expression of Pgp and CD34 in that patients who lacked both antigens did best and those positive for both did worst.[9] Analogously, we show in this study that the probability of having a del(5q) and/or del(7q) increases significantly with expression of the two antigens by the blast cells and is highest if both antigens are positive.

In conclusion, the responsiveness of t(15;17) APL may be, in part, related to the absence of multidrug-resistance, as reflected by Pgp expression. And, the combined expression of two prognostically unfavorable antigens, Pgp and CD34, by blast cells of patients with del(5q) and/or del(7q) may contribute to the poor clinical outcome in these patients.

ACKNOWLEDGMENT

Supported by NCI, DHHS grants CA21115 (ECOG Operations Office), CA14958 (Einstein Cancer Center), CA23318 (Dana Farber Cancer Institute), CA11083 (University of Rochester), P30CA13330 (Einstein Cancer Center), the Elsa U. Pardee Foundation, the

Table 4. Progressive probability of association between cytogenetics and CD34/P-glycoprotein

PROGRESSIVE PROBABILITY OF ASSOCIATION BETWEEN CYTOGENETICS AND CD34 / P-GLYCOPROTEIN

Cytogenetic Marker Present	CD34 / P-glycoprotein Neither	Either	Both	P-value
	% of Total Patients			
inv(16) (p13q22)	3	7	6	0.27
t(15;17) (q22; q11)	33	2	6	0.02
t(8; 21) (q22; q22)	2	10	6	0.02
del (5q) and / or del (7q)	**4**	**12**	**25**	**0.0038**
complex karyotype	6	18	13	0.02

The probability of having a del(5q) and/or del(7q) increases with the expression of CD34 and P-glycoprotein by the blast cells and is highest if both antigens are positive.

Phi Beta Psi Sorority and the Chemotherapy Foundation. The contents of this work are solely the responsibility of the authors and do not necessarily represent the official views of the National Cancer Institute.

REFERENCES

1. Wiernik PH: Diagnosis and treatment of adulthood acute myeloid leukemia other than acute promyelocytic leukemia, in Wiernik PH, Canellos GP, Kyle RA, Dutcher JP (eds): Neoplastic Diseases of the Blood, 3rd ed, New York, NY, Churchill Livingston, 1995, in press
2. Marie J-P: P-glycoprotein in adult hematologic malignancies. Hematol/Oncol Clinics of North America 9:239, 1995
3. Paietta E: Immunobiology of acute leukemia, in Wiernik PH, Canellos GP, Kyle RA, Dutcher JP (eds) Neoplastic Diseases of the Blood, 3rd ed, New York, NY, Churchill Livingston, 1995, in press
4. Simon SM, Schindler M: Cell biological mechanisms of multidrug resistance in tumors. Proc Natl Acad Sci USA 91:3497, 1994
5. Licht T, Pastan I, Gottesman M, Herrmann F: P-glycoprotein-mediated multidrug resistance in normal and neoplastic hematopoietic cells. Ann Hematol 69:159, 1994
6. Paietta E, Andersen J, Racevskis J, Ashigbi M, Cassileth P, Wiernik PH: Modulation of multidrug resistance in de novo adult acute myeloid leukemia: Variable efficacy of reverting agents in vitro. Blood Reviews 9:47, 1995
7. Paietta E, Andersen J, Gallagher R, Bennett J, Yunis J, Cassileth P, Rowe J, Wiernik PH: The immunophenotype of acute promyelocytic leukemia: An ECOG study. Leukemia 8:1108, 1994.
8. Paietta E, Andersen J, Racevskis J, Gallagher R, Bennett J, Yunis J, Cassileth P, Wiernik PH: Significantly lower P-glycoprotein expression in acute promyelocytic leukemia than in other types of acute myeloid leukemia: Immunological, molecular and functional analyses. Leukemia 8:968, 1994.
9. Paietta E, Andersen J, Rowe J, Cassileth P, Wiernik PH: Myeloid blast cell maturation determines response in adult de novo acute myeloid leukemia (AML): A response-driven antigen expression analysis in 382 Eastern Cooperative Oncology Group (ECOG) patients. Am Soc Clin Oncol 14:86, #47, 1995
10. Bennett JM, Catovsky D, Daniel M-T, Flandrin G, Galton DAG, Gralnick HR, Sultan C: Proposed revised criteria for the classification of acute myeloid leukemia: A report of the French-American-British Cooperative Group. Ann Intern Med 103:620, 1985
11. Paietta E: Proposals for the immunological classification of acute leukemias. (Leukemia, in press)
12. Kita K, Nakase K, Miwa H, Masuya M, Nishii K, Morita N, Takakura N, Otsuji A, Shirakawa S, Ueda T, Nasu K, Kyo T, Dohy H, Kamada N: Phenotypical characteristics of acute myelocytic leukemia associated with the t(8;21)(q22;q22) chromosomal abnormality: Frequent expression of immature B-cell antigen CD19 together with stem cell antigen CD34. Blood 80:470, 1992
13. Geller RB, Zahurak M, Hurwitz CA, Burke PJ, Karp JE, Piantadosi S, Civin CI: Prognostic importance of immunophenotyping in adults with acute myelocytic leukemia: The significance of the stem-cell glycoprotein CD34 (My10). Br J Haematol 76:340, 1990
14. Solary E, Bidan J-M, Calvo F, Chauffert B, Caillot D, Mugneret F, Gauville C, Tsuruo T, Carli P-M, Guy H: P-glycoprotein expression and *in vitro* reversion of doxorubicin resistance by verapamil in clinical specimens from acute leukemia and myeloma. Leukemia 5:592, 1991
15. Guerci A, Merlin JL, Missoum N, Feldmann L, Marchal S, Witz F, Rose C, Guerci O: Predictive value for treatment outcome in acute myeloid leukemia of cellular daunorubicin accumulation and P-glycoprotein expression simultaneously determined by flow cytometry. Blood 85:2147, 1995
16. Chen CJ, Chin JE, Ueda K, Clark DP, Pastan I, Gottesman MM, Roninson IB: Internal duplication and homology with bacterial transport proteins in the mdr1 (P-glycoprotein) gene from multidrug-resistant human cells. Cell 47:381, 1986

26

CHARACTERIZATION OF N-RAS PROMOTER MUTATIONS IN LEUKEMIA

Harry Iland,[1*] Jacqui Thorn,[1] and Peter Molloy[2]

[1] The Kanematsu Laboratories
Royal Prince Alfred Hospital
Camperdown, NSW 2050 Australia
[2] CSIRO Division of Biomolecular Engineering
P.O.Box 184
North Ryde, NSW 2113 Australia

ABSTRACT

Many hematological malignancies are characterized by molecular abnormalities which impact upon RAS protein-mediated signal transduction pathways. In particular, mutations of N-RAS at codon 12, 13 or 61 are common in AML. Since overexpression of the normal N-RAS gene can also transform cells *in vitro*, we postulated the existence of mutations in the promoter region which had the potential to deregulate N-RAS expression, thus representing an alternative mechanism of N-RAS activation in leukemia. Single-stranded conformational polymorphism, combined with DNA sequencing and/or restriction mapping, was used to screen the N-RAS promoter in cells from 26 AML, 19 CML, 1 CLL and 13 non-leukemic samples. One AML patient had an A-to-G mutation at position 409 which abolished protein binding in the region of a c-MYB binding site. The mutation persisted during remission and relapse, implicating it as an early event in leukemogenesis. A T-to-A mutation at position 520 in a second AML patient altered DNA-protein interactions in the region of the AP-1 and c-MYC binding sites. Another T-to-A mutation lying within a protein binding site at position 378 was found in a patient with CLL. In addition, we have identified a novel CA polymorphism at position 390; loss of heterozygosity studies utilising this polymorphism have proved helpful in studying clonal evolution in a patient with CML. In conclusion, the N-RAS promoter appears to be a frequent site of mutagenic activity in human leukemia, and these mutations extend the spectrum of abnormalities which have the potential to disrupt RAS-mediated signal transduction in hematological malignancies.

* Corresponding Author: Associate Professor H. J. Iland, Kanematsu Laboratories, Royal Prince Alfred Hospital, Missenden Road, Camperdown, NSW 2050 Australia. Phone: 61-2-9515-7655; Fax: 61-2-9515-6255.

Molecular Biology of Hematopoiesis 5, edited by Abraham et al.
Plenum Press, New York, 1996

INTRODUCTION

N-RAS, K-RAS and H-RAS are members of the GTPase superfamily of signal transducer proteins. They play a pivotal role in normal cell growth and differentiation processes, and exhibit transforming properties when RAS genes acquire point mutations in codons 12, 13 or 61. These mutations result in amino acid substitutions within the guanine nucleotide binding domains of RAS that induce constitutive RAS activation, since they prevent GTPase activating proteins from enhancing the weak intrinsic GTPase activity of RAS proteins which would normally terminate signal transduction. Transforming mutations at codons 12, 13 and 61 are present in 20-25% of patients with acute myeloid leukemia (AML) [1,2]. Since increased expression of both normal and mutant RAS proteins has been identified in the majority of patients with acute leukemia [3], and since overexpression of normal N-RAS can also lead to transformation *in vitro* [4], we speculated that, at least in some patients, N-RAS activation might be due to deregulated expression secondary to mutations involving the promoter region of the N-RAS gene. We have previously characterized the 900 base pair (bp) human N-RAS promoter and have identified several transcription factor binding sites as well as a 109bp region that was critical for N-RAS expression [5]. Proteins which bind to the promoter include MYB, MYC-MAX/USF, ETS, AP-1, AP-2 and CREB. In this study, single stranded conformational polymorphism (SSCP) [6] was employed to screen the N-RAS promoter region in leukemic samples for the presence of mutations which could potentially deregulate N-RAS expression, and thereby contribute to leukemogenesis.

MATERIALS AND METHODS

DNA from Patients and Reference Plasmids

Genomic DNA was extracted [2,7] from the peripheral blood or bone marrow of patients with AML, chronic myeloid leukemia (CML), chronic lymphocytic leukemia (CLL), and from non-leukemic controls (Table 1). Reference plasmids containing 439bp sequences from the central portion of the N-RAS promoter region were also available [5,8]. pFe-2 was subcloned from a normal human fetal liver N-RAS gene. pHT-2 was subcloned from a human fibrosarcoma cell line with an N-RAS gene codon 61 mutation.

Table 1. Patient groups

Diagnosis	Number
AML	
De novo AML (at diagnosis)	23
De novo AML (at first relapse)	1
Secondary AML	2
Total	26
CML	
Chronic phase	18
Accelerated phase	1
Total	19
CLL	1
Non-leukemic controls	13
Total	59

PCR-SSCP Analysis

To maximize the detection of mutations in the N-RAS promoter, the central 439bp region containing all the known transcription factor binding sites, the 109bp region which was critical for expression, and a 150bp CpG island [5] was analyzed in two overlapping segments. The 5' end was amplified by polymerase chain reaction (PCR) for SSCP analysis using primers RPS5 (5'-GGTGAGCTCAGGGGATGTGG-3') and SSCP.1 (5'-TCCGAAC-CACGAGTCATGCGG-3') to generate a 263bp fragment (referred to as SSCP-1); the 3' end was amplified with primers SSCP.2 (5'-GCACCTAGCGCTTTCATTATTG-3') and RPH3 (5'-CTCAAGCTTCACTGCCTCTG-3') to generate a 257bp fragment (SSCP-2).

Both PCR reactions contained PCR buffer (Boehringer Mannheim, Australia), 50pmoles of each primer, 200 μM dNTP, 5 units Taq polymerase (Boehringer Mannheim), and 10pg plasmid DNA or 1 μg genomic DNA. Amplifications involved 2 mins at 94°C, 35 cycles of (30secs at 58°C, 30 secs at 75°C, 30 secs at 94°C), and 5 mins at 75°C. The PCR products were ethanol precipitated, resuspended in 5μl of formamide dye, and run on a 16cm long 0.5xMDE gel (mutation detection enhancement, AT Biochem, Gradipore, Australia) with 5% glycerol and 0.6xTBE at 4°C and 6 watts for 6 hours. The DNA bands were visualized by silver staining prior to gel drying.

DNA Sequencing and Footprinting

The promoter regions of the N-RAS reference plasmids were subcloned into M13 and sequenced by ^{32}P dideoxy Sanger reactions [5]. Patient samples which exhibited abnormal SSCP mobility were subcloned into pPCAT [5], and were subjected to cycle sequencing reactions. One μg of plasmid DNA and 4pmoles of primer were included in sequencing reactions with 9.5μl of dye deoxy terminator, performed according to Applied Biosystems protocols and visualized on an Applied Biosystems 373A DNA sequencer. DNA footprint analyses were performed as reported previously using HeLa nuclear extract [5]. Base numbering is according to the Genbank entry, ie. base number one is the first base of the 900bp promoter region [9].

RESULTS

Polymorphism at Position 390

Sequence data from the reference plasmids revealed a discrepancy at position 390; pFe-2 contained adenine (A), whereas pHT-2 and the published sequence [9] indicated cytosine (C) at this position (Figure 1). This difference was readily detectable by SSCP-1 fragment analysis, and was also demonstrable by digestion with Bsr I, since the presence of adenine results in loss of a Bsr I restriction site.

In order to clarify the significance of this variability at position 390, we performed SSCP analyses and Bsr I digestion of the SSCP-1 fragment from 13 non-leukemic controls. Both techniques confirmed the polymorphic nature of position 390, since six of 13 individuals were heterozygous for C and A (Table 2). Bsr I digestion of PCR products from AML and CML samples also showed the presence of a polymorphism at position 390 (Table 2).

Apart from the mobility differences associated with the polymorphic 390 site, SSCP analyses failed to reveal any evidence of sequence variability in either the SSCP-1 or SSCP-2 fragments of the N-RAS promoter in the non-leukemic controls.

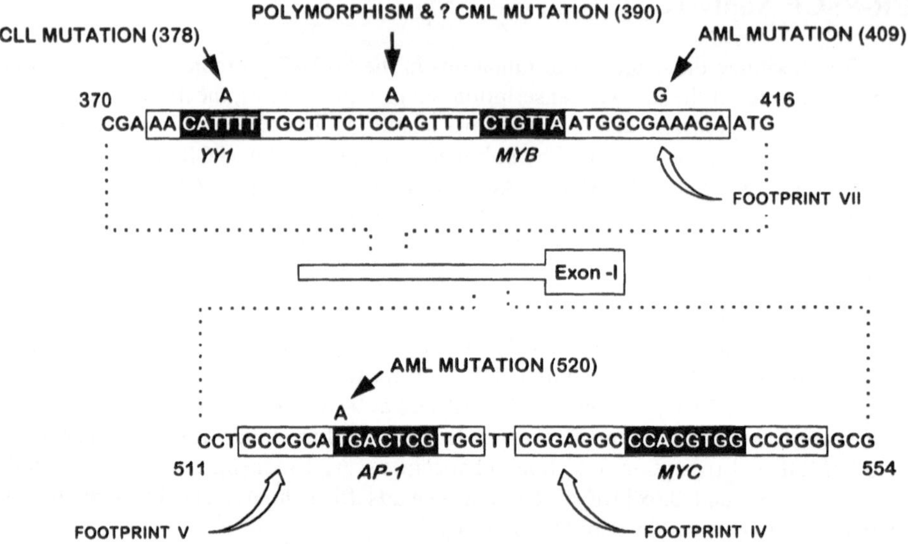

Figure 1. Sequence variability of the N-RAS promoter in leukemia. The normal DNA sequence found in footprint regions VII, V and IV of the central 439bp region of the N-RAS promoter is shown. The boundaries of the footprints are indicated by open boxes. The shaded boxes illustrate the sites in the promoter which fit the consensus sequences for binding of the YY1, MYB, AP-1 and MYC transcription factors. The positions of the polymorphism at position 390, together with the mutations found in individual patients, are indicated above the relevant sequences.

N-RAS Promoter Mutations in AML

SSCP analysis of the 26 AML patients identified two de novo AML patients with bands that were not observed in samples from the non-leukemic controls. Patient 26/0 had an abnormality in the 5' portion of the promoter (the SSCP-1 fragment), whereas patient 24/0 had an abnormality in the 3' SSCP-2 fragment. The 439bp N-RAS promoter regions from the leukemic cells of these patients were cloned into pPCAT. Clones representing both N-RAS alleles were identified by screening with SSCP; the promoter inserts were then sequenced in both orientations to determine the position and type of the mutation.

Patient 26/0. One allele contained only the polymorphic 'A' at position 390, with no other abnormalities. The other allele contained a 'C' at position 390 and an adenine-to-guanine (A-to-G) mutation at position 409, thereby altering the sequence from 5'-CGAAAG-3'

Table 2. Nucleotide at position 390

	Number of Individuals with Cytosine and/or Adenine at Position 390 of N-RAS		
Diagnosis	C only	C and A	A only
Non-leukemic controls	7	6	0
AML	22	3	1
CML	13	6	0
CLL	1	0	0
Total	43(73%)	15(25%)	1(2%)

to 5'-CGGAAG-3' (Figure 1). Position 409 is within footprint region VII of the N-RAS promoter that we had previously identified by HeLa DNase footprinting [5]. This extended footprint region binds a complex of proteins, including purified MYB protein, and the entire footprint can be abrogated by competition with a MYB-binding oligonucleotide (manuscript in preparation). DNase footprinting of the mutant sequence from patient 26/0 showed loss of protection within, and hence disruption of transcription factor binding to, region VII [7].

Five serial samples obtained from this patient after initial chemotherapy have also been analyzed by SSCP. Three samples were obtained whilst the patient was in complete remission. One sample was obtained when a deteriorating blood count suggested early relapse, and the final sample was obtained when marrow relapse had been documented. All samples showed the same SSCP-1 pattern as that observed with the initial pre-chemotherapy DNA. It is possible that the A-to-G at 409 was a constitutional variation in this patient, and not actually a leukemia-associated mutation. However, the lack of this sequence in normal controls indicates it is unlikely to be a naturally-occurring polymorphism. An alternative explanation is that this patient had persistence of leukemic hematopoiesis during apparent remission. Differentiated remissions have been reported previously using a variety of clonality markers. Similarly, continuing myelodysplasia after remission induction is not uncommon in elderly patients, and indeed some dysplastic features were apparent in our patient's remission marrows. The persistence of promoter mutations during a state of "pre-leukemia" after attainment of remission would implicate promoter mutations as early events in leukemogenesis. Unfortunately non-hematopoietic tissue was not available from this patient to conclusively discriminate between these possibilities.

Patient 24/0. Both N-RAS alleles from this patient contained a 'C' at 390 as predicted from SSCP-1 fragment analysis and Bsr I digestion. One allele contained wildtype sequence only, and is therefore a normal N-RAS allele. The other allele contained a thymine-to-adenine (T-to-A) mutation at position 520, changing the sequence from 5'-TGACTCG-3' to 5'-AGACTCG-3'. This base lies at the 5' edge of the AP-1 consensus sequence (Figure 1) within our previously identified footprint region V [5], and we have shown that AP-1 binds to the normal N-RAS sequence at this position (manuscript in preparation). DNase footprinting of the mutant sequence showed increased protection between footprint region IV, which binds MYC/USF ([5] and manuscript in preparation), and region V, although the extremities of footprint IV and V remain unaltered [7].

N-RAS Promoter Mutations in CML and CLL

No mutations were detected in samples from 18 patients with CML in chronic phase. However, we have studied one CML patient in both accelerated phase and after blast transformation (no sample from the chronic phase was available). This patient appeared to be homozygous for C at position 390 in accelerated phase, but was unexpectedly heterozygous for C and A in blast transformation (by both SSCP-1 fragment analysis and Bsr I digestion). These results suggest two possible explanations. First, the patient may have been homozygous for C; the heterozygosity seen at blast transformation being due to a point mutation in one allele at position 390. Alternatively, the patient may have been heterozygous for C and A, with loss of the A-containing allele during clonal evolution to accelerated phase. This scenario implies that a second clonal evolution took place directly from chronic phase cells rather than via the accelerated phase cells, thereby explaining the emergence of heterozygosity at the time of blast transformation.

Finally, only one patient with CLL has been studied to date, and abnormal mobility was observed in SSCP-1 fragment analysis. Sequencing of both strands confirmed the presence of a T-to-A mutation at position 378 of the N-RAS promoter, lying within a potential

YY1 binding site (Figure 1). However, the effect of this mutation on DNA-protein interactions has not yet been investigated.

DISCUSSION

In this study we have used SSCP to screen the N-RAS promoter for sequence variability in both leukemic patients and non-leukemic controls. We have identified, for the first time, a polymorphic site within the N-RAS gene at position 390, with an overall frequency of heterozygosity of 25%. This polymorphism should provide a useful marker for linker studies, and for establishing loss of heterozygosity. Indeed, we have documented changes in the proportion of the two alleles in one of 19 patients with CML during clonal evolution from chronic phase to accelerated phase and blast transformation.

In addition, we have demonstrated the existence of mutations within the N-RAS promoter region in two of 26 patients with AML, and in one patient with CLL. The effect of these mutations on DNA-protein interactions and N-RAS expression is still under investigation. We have demonstrated interference with MYB-dependent protein binding due to a point mutation at position 409, and we have also shown that a mutation at position 520 alters the interaction of proteins at MYC and AP-1 binding sites within the N-RAS promoter.

Several precedents exist in which promoter mutations play a role in the pathogenesis of human disease. In hereditary persistence of fetal hemoglobin a base substitution has been identified in the Aγ-globin gene promoter which increases promoter activity by stimulating binding of Sp1 and NF-G.C [10]. Two of 111 retinoblastoma patients, without visible gene rearrangements by Southern blotting, were shown to possess promoter mutations in the retinoblastoma gene which inhibited the binding of transcription factors to a potential Sp1 or ATF site and suppressed expression [11]. In five of seven Burkitt's lymphoma cell lines, point mutations have been identified within a 20bp protein binding region within intron 1 of the c-MYC gene [12]. Mutations in one cell line were shown to decrease binding of the nuclear protein, but no data on the effect of these mutations on expression are available. A promoter mutation in the K-RAS gene has been identified in five of eight clones from the colon carcinoma cell line, SW480 [13], and a mutation in the intron of the H-RAS gene has also been identified and shown to increase the expression of an H-RAS with a coding region mutation [14]. Thus, the spectrum of RAS mutations may extend beyond those typically found in codons 12, 13 and 61.

Data accumulated from several hematological malignancies has reinforced the pivotal role played by RAS proteins in signal transduction, particularly in myeloid leukemias [reviewed in 15]. Enhanced RAS-GTP signalling can result from: (1) mutations in codons 12, 13 or 61 of N-RAS in AML and myelodysplastic syndromes, (2) mutations in NF 1 RAS-GAP in juvenile CML, and from interactions of either (3) BCR-ABL or (4) TEL-PDGFRβ with GRB2 in Philadelphia positive CML and CMML respectively. The promoter mutations we have observed indicate that this region is a frequent site of mutagenic activity in human leukemia. If these mutations deregulate N-RAS expression, then they represent yet another mechanism of N-RAS activation.

ACKNOWLEDGMENT

This project was supported by the Australian National Health and Medical Research Council.

REFERENCES

1. Bos JL, Verlaan-de Vries M, van der Eb AJ, Janssen JWG, Delwel R, Lowenberg B, Colly LP. Mutations in N-*ras* predominate in acute myeloid leukemia. Blood 69: 1237, 1987
2. Todd AV, Radloff TJ, Ireland CM, Kronenberg H, Iland HJ. Detection of N-*ras* mutations in acute myeloid leukemia by allele specific restriction analysis (ASRA). Am J Hematol 38: 207, 1991
3. Shen WPV, Aldrich TH, Venta-Perez G, Franza BR, Furth ME. Expression of normal and mutant ras proteins in human acute leukemia. Oncogene 1: 157, 1987
4. McKay IA, Marshall CJ, Cales C, Hall A. Transformation and stimulation of DNA synthesis in NIH 3T3 cells are a titratable function of normal p21 N-ras expression. EMBO J 5: 2617, 1986
5. Thorn JT, Todd AV, Warrilow D, Watt F, Molloy PL, Iland HJ. Characterization of the human N-ras promoter region. Oncogene 6: 1843, 1991
6. Orita M, Iwahana H, Kanazawa H, Hayashi K, Sekiya T. Detection of polymorphisms of human DNA by gel electrophoresis as single-strand conformation polymorphisms. Proc Natl Acad Sci USA 86: 2766, 1989
7. Thorn JT, Molloy P, Iland H. SSCP detection of N-ras promoter mutations in AML patients. Exp Hematol 23: 1098, 1995
8. Brown R, Marshall CJ, Pennie SG, Hall A. Mechanism of activation of an N-ras gene in the human fibrosarcoma cell line HT1080. EMBO J 3: 1321, 1984
9. Hall A, Brown R. Human N-ras: cDNA cloning and gene structure. Nucleic Acids Res. 13: 5255, 1985
10. Fischer K-D, Nowock J. The T to C substitution at -198 of the Aγ-globin gene associated with the British form of HPFH generates overlapping recognition sites for two DNA-binding proteins. Nucleic Acid Res 18: 5685, 1990
11. Sakai T, Ohtani N, McGee TL, Robbins PD, Dryja TP. Oncogenic germ-line mutations in Sp1 and ATF site in the human retinoblastoma gene. Nature 353: 83, 1991
12. Zajac-Kaye M, Gelmann EP, Levens D. A point mutation in the c-myc locus of a Burkitt lymphoma abolishes binding of a nuclear protein. Science 240: 1776, 1988
13. Capon DJ, Seeberg PH, McGrath JP, Hayflick JS, Edman U, Levinson AD, Goeddel DV. Activation of Ki-ras-2 gene in human colon and lung carcinomas by two different point mutations. Nature 304: 507, 1983
14. Cohen JB, Levinson AD. A point mutation in the last intron responsible for increased expression and transforming activity of the c-H-ras oncogene. Nature 334: 119, 1988
15. Sawyers CL, Denny CT. Chronic myelomonocytic leukemia: Tel-a-kinase what ets all about. Cell 77: 171, 1994

THE ROLE OF ANTHRACYCLINES IN THE MANAGEMENT OF ADULT ACUTE LYMPHOBLASTIC LEUKAEMIA

R. Bassan,[1] *T. Lerede,[1] A. Rambaldi,[1] E. Di Bona,[2] G. Rossi,[3] E. Pogliani,[4] G. Lambertenghi-Deliliers,[5] A. Porcellini,[6] P. Coser,[7] and T. Barbui[1]

Divisione/Servizio di Ematologia, Ospedali di
[1] Bergamo
[2] Vicenza
[3] Brescia
[4] Monza
[5] Milano Università
[6] Cremona
[7] Bolzano
Italy

ABSTRACT

The role of anthracyclines (ANT) in the treatment of adult acute lymphoblastic leukaemia (ALL) is poorly defined. We reviewed ANT treatment results in adult ALL. Altogether, an early and intensive use of ANT appear to improve both initial response rate and long-term disease-free survival; the new compound idarubicin (IDR) exhibits a considerable antileukaemic activity deserving further evaluation; and the prognosis of $CD10^+$ t(9;22)/$BCR\text{-}ABL^-$ ALL can be particularly good in relation to an early dose-intensive ANT consolidation program.

INTRODUCTION

The mechanisms by which anthracyclines (ANT) induce cell death are inhibition of topoisomerase II, DNA intercalation, production of toxic free radicals, and cell membrane damage. ANT are metabolized in the liver by a ketoreductase transforming the active drug into inactive alcohol derivative. An exception occurs with idarubicin (IDR) since idarubici-

*Correspondence: Dr R. Bassan, Ematologia, Ospedali Riuniti, 24100 Bergamo, Italy. Fax: 035 269 667; Phone 035 269490.

Molecular Biology of Hematopoiesis 5, edited by Abraham et al.
Plenum Press, New York, 1996

nol is cytotoxic, exhibits a prolonged half-life of about 50-70 hours, and may cross the blood-brain barrier (1). The evidence in favour of ANT in adult ALL remains conflictual. Randomized trials from Cancer and Leukemia Group B (CALGB) demonstrated that daunorubicin (DNR) did significantly enhance the complete response (CR) rate when added to vincristine-prednisolone-asparaginase (4) but it failed to improve the durability of CR itself (5). Also, although adult ALL is an immunobiologically heterogeneous disease the effects of ANT therapy in distinct ALL subsets are unknown. These issues represent the matter of the present review, comprising a survey of the literature and a summary of Bergamo and Vicenza Hospitals (B/VH) collaborative adult ALL trials.

A LITERATURE REVIEW

Methods

Fifty-five pertinent studies published during 1976-1994 were identified. Study endpoints according to ANT therapy, type, and dose intensity were: overall CR rate, late CR rate (beyond week 4), refractoriness rate, median duration of CR and projected disease-free survival (DFS). Dose intensity (DI=total mg/m^2/week) was calculated during the first 12 weeks of treatment including the induction regimen (6).

Results

The data were previously detailed elsewhere (7). CR rates were generally better with ANT than without (Table 1). DNR, rubidazone (RDZ), and mitoxantrone were equally effective, DNR at greater than 45 mg/m^2 dose was too toxic causing worse results, and doxorubicin (DOX) given on a three-day schedule (TDS) as in the early CALGB study (4) was slightly better than the weekly schedule (WS). ANT were effective even when used sequentially after failure of ANT-free combinations. Postremission treatment results are shown in Table 2. In retrospective trials the advantage of ANT-containing combinations was statistically significant in two out of three, and the improved CR durability was found to correlate with an early DNR DI >21. The conclusions from the CALGB report showing no

Table 1. Outcome to induction therapy with/without ANT from randomized/comparative ANT studies

Group	Study	Drugs on trial	CR rate	Source
CALGB	Randomized	DNR vs none	78 vs 47 (p .003)	Blood 64: 267, 1984
CALGB	Randomized	DNR vs mitoxantrone	65 vs 63	Leukemia 5: 425, 1991
UKALL	Randomized	DNR vs MTX	88 vs 86[1]	Br J Haematol 85: 84, 1993
ECOG	Randomized	DNR vs DNR (dose)	56 vs 70[2]	Leukemia 6 (suppl 2): 178, 1992
FGTAALL	Randomized	DNR vs RDZ	78 vs 74	J Clin Oncol 11: 1990, 1993
Mexico	Randomized	DOX TDS vs WS	81 vs 70	Blood 82 (suppl 1): 56a, 1993
Bart's	Retrospective	DOX vs none	71 vs 47 (p .05)	Br Med J 1: 199, 1978
Paris	Retrospective	DNR vs none	79 vs 74	Cancer Chemother Pharmacol 1:113, 1978
Leiden	Sequential[3]	DNR	46 to 83	Blood 46: 823, 1975
CALGB	Sequential	DNR	58 to 72	Leuk Res 3: 395, 1979
Johns Hopkins	Sequential	DNR	27 to 71	Blood 82 (suppl 1): 56a, 1993

[1]shorter course to CR in DNR arm (p .04)
[2]higher CR rate with lower DNR dose (45 vs 60 mg/m^2)
[3]ANT added to nonCR patients; CR increase from baseline is indicated

Table 2. Duration of remission according to ANT therapy: results from randomized or retrospective comparative trials

Group	Study	ANT/drug on trial	5-year % DFS[1]	Source
MRC	Retrospective	DNR+DOX vs none	38 vs 18	Br J Cancer 53: 175, 1986
GATLA	Retrospective	DOX vs none	34 vs 20 (p .001)	Ann Oncol 2: 33, 1991
Verona	Retrospective	DNR vs DNR[2] (different DI)	41 vs 20 ($p<0.05$)	Leukemia 8: 376, 1994
GIMEMA	Randomized	DNR vs DNR (different dose)	22 vs 22 (4 yr)	Br J Haematol 71: 377, 1989
CALGB	Randomized	DNR vs none	29 vs 29	J Clin Oncol 9: 2002, 1991
CALGB	Randomized	DNR vs DNR+ mitoxantrone	20 vs 20 (3 yr)	Leukemia 5: 425, 1991
FGTAALL	Randomized	DNR vs RDZ	32 vs 32 (3 yr)	J Clin Oncol 11: 1990, 1993

[1]DFS, disease-free survival
[2]improved DFS with DI >21

benefit from DNR consolidation were in contrast with those from retrospective studies. Other trials are summarized in Table 3. Higher CR rates (>80%) were more common with intensive/TDS ANT treatments, whereas incidence of late response and refractory disease was higher with nonintensive/WS ANT programs and ANT-free schedules. Long-term outcome was better with ANT administered at higher DI (>20); results were proportionally inferior with lower ANT DI and without ANT.

THE B/VH GROUP AND COLLABORATIVE GROUP EXPERIENCE

Patients and Methods

Updated results from previous B/VH studies and recent collaborative trials were analysed to to assess the potential role of the new drug IDR and to evidentiate prognostic differences between ALL subtypes in relation to an early ANT DI. Between February 1979-July 1995, 328 unselected consecutive new patients were enroled into HEAV'D (DOX, high DI; n=82), OPAL-HDara-C (DOX, low DI; n=27), reinforced HEAV'D (DOX, high DI; n=39), IVAP (IDR, high DI; low DI in patients >60 years; n=118), and 07/93 (IDR, low DI; n=62) treatment protocols (8-11). An high early DI was DOX >20 or IDR

Table 3. Remission rates and duration by ANT therapy and early dose intensity (DI) (adapted from ref. 7). WS is weekly schedule, TDS is three-day schedule (induction only)

	ANT: None	WS/DI <20	TDS/DI >20
Induction of remission			
No. of evaluable studies	3	11	17
No. reporting CR >80%	0	3 (27%)	8 (47%)
No. reporting late CR >30%	3 (100%)	3/6 (50%)	1/6 (17%)
No. reporting refractory ALL >10%	1/1	7/7 (100%)	7/13 (54%)
Duration of remission			
No. of evaluable studies	7	12	5
No. reporting median CR >24 mos.	1 (14%)	4 (33%)	4 (80%)
No. reporting DFS >35%	0	4 (33%)	4 (80%)

DFS, disease-free survival

>7 (IDR is approximately three time as potent as DOX). In each program the induction schedule was with ANT-vincristine-prednisolone-L-asparaginase, and the administration of ANT was to be completed within 3 months from achievement of CR. Multi-drug intensive postremission consolidation was completed within 3-9 months, depending on study type. Afterwards, only low-dose maintenance with daily mercaptopurine and weekly methotrexate was prescribed.

Results

Experience with IDR substituting for DOX started in 1991. 180 patients commenced treatment with IDR, the largest number treated front-line with an IDR-containing regimen. The early results using a TDS (IDR 12 mg/m^2/d) were poor because of marked regimen-related toxicity (11): CR 44% (7/17). In a second step IDR was reduced to 10 mg/m^2 for two consecutive days: CR 86% (24/28, p=0.005 by the Fisher exact test). Overall CR rate with the two-day IDR regimen was 85% (119/140). Entering CR required a single 14-day course in most responders (113/119, 95%), and incidence of refractory ALL was only 6%. Although a retrospective comparison with historical DOX-based induction is difficult for several reasons, including the lack of uniform criteria to assess outcome and a younger patient age in the DOX group (median 28.5 years versus 33 years, p<0.001 by the Student *t* test), these results confirmed the remarkable activity of IDR in the initial treatment of adult ALL.

179 and 85 CR patients commenced high and low ANT DI postremission consolidation, respectively. Major diagnostic characteristics of the two groups were: age 14-64 years (median 28) and 14-73 years (median 30), FAB L2 75% and 81%, FAB L3 5% and 6%, median blast count 0-1000 x10^9/l (median 8) and 0-625 x10^9/l (median 12.7). 5-year disease-free survival (DFS) probability according to high or low ANT DI did not differ significantly in T-ALL (0.34 and 0.27, nonsignificant p value by the log-rank test) and early-B CD10$^-$ ALL (0.29 and 0.13, nonsignificant p value by the log-rank test). However, CD10$^+$ t(9;22)/BCR-ABL$^-$ early-B ALL patients treated with high ANT DI fared significantly better than any other CD10$^+$ subgroup (Figure 1), with a 5-year DFS probability of 0.54

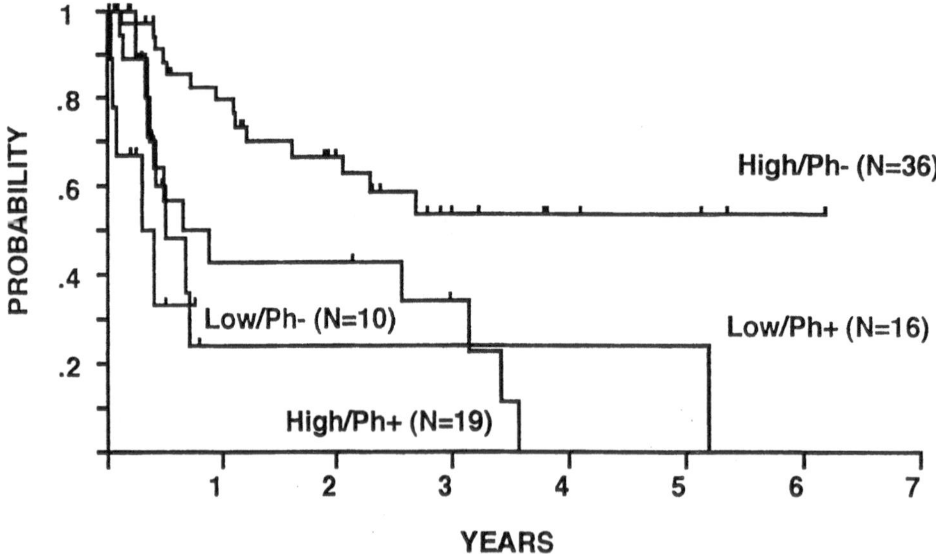

Figure 1. The role of anthracyclines in the management of adult acute lymphoblastic leukemia.

($p<0.005$ by the log-rank test). Only twelve of these cases had a recurrence (33%), 2 died of therapy-related toxicity, and one developed a secondary acute myeloid leukaemia (AML). t(9;22)/BCR-ABL$^-$ patients were younger than positive ones: median age 21 years versus 55 years ($p=0.01$ by the Student t test).

CONCLUSIONS

The contention that ANT have a role in optimizing the management of adult ALL is supported by the current survey. As regards CR induction, an aggressive regimen as the DNR TDS originally introduced by CALGB seems preferable. Because intensive ANT therapy is myelotoxic, the addition of myeloid cell growth factors should be considered in this setting. The question of whether IDR is really better than DNR or DOX is posed by the results from B/VH collaborative trials. Both the low rate of primary refractoriness and the high proportion of early responders indirectly support this assumption but at the same time this experience points to the difficulty of defining the optimal IDR dosage and schedule in ALL as opposed to AML. As far as postremission therapy is concerned, we noticed that an early intensive, brief ANT consolidation conferred a remarkably good long-term prognosis to adolescents and adults with early-B CD10$^+$ ALL lacking t(9;22)/BCR-ABL rearrangements. The data suggests that ANT contribute substantially to the eradication of the ALL clone in these cases. It remains possible that minimal residual disease following intensive ANT consolidation was cleared by standard prolonged maintenance with mercaptopurine and methotrexate. Future research should be directed at clarifying the exact clinical potential of IDR, the mechanisms of ANT resistance in specific ALL subsets, and the course of minimal residual disease during ANT treatment progression.

REFERENCES

1. Ames MM, Spreafico F. Selected pharmacologic characteristics of idarubicin and idarubicinol. Leukemia 1992; 6 (Suppl 1): 70-75.
2. Minderman H, Linssen P, van der Lely N et al. Toxicity of idarubicin and doxorubicin towards normal and leukemic bone marrow progenitors in relation to their proliferative state. Leukemia 1994; 8: 382-387.
3. Berman E, McBride M. Comparative cellular pharmacology of daunorubicin and idarubicin in human multi-drug resistant leukemia cells. Blood 1992; 79: 3267-3273.
4. Gottlieb AJ, Weinberg W, Ellison RR et al. Efficacy of daunorubicin in the therapy of adult acute lymphocytic leukemia: a prospective randomized trial from Cancer and Leukemia Group B. Blood 1984; 64: 267-274.
5. Ellison RR, Mick R, Cuttner J et al. The effects of postinduction intensification treatment with cytarabine and daunorubicin in adult acute lymphocytic leukemia: a prospective randomized clinical trial by Cancer and Leukemia Group B. J Clin Oncol 1991; 9: 2002-2015.
6. Todeschini G, Meneghini V, Pizzolo G et al. Relationship between daunorubicin dosage delivered during induction therapy and outcome in adult acute lymphoblastic leukemia. Leukemia 1994; 8: 376-381.
7. Bassan R, Lerede T, Rambaldi A, Buelli M, Viero P, Barbui T. The use of anthracyclines in adult acute lymphoblastic leukemia. Haematologica 1995; 80: 282-293.
8. Bassan R, Battista R, D'Emilio A et al. Long-term results of the HEAVD protocol for adult acute lymphoblastic leukaemia. Eur J Cancer 1991; 27: 441-447.
9. Rohatiner AZS, Bassan R, Battista R et al. High dose cytosine arabinoside in the initial treatment of adults with acute lymphoblastic leukaemia. Br J Cancer 1990; 62: 454-458.
10. Bassan R, Battista R, Montaldi A et al. Reinforced HEAV'D therapy for adult acute lymphoblastic leukemia: improved results and revised prognostic criteria. Hematol Oncol 1993; 11: 169-177.
11. Bassan R, Battista R, Corneo G et al. Idarubicin in the initial treatment of adults with acute lymphoblastic leukemia: the effect of drug schedule on outcome. Leuk Lymphoma 1993; 11: 105-110.

28

COLLECTION OF BCR/ABL NEGATIVE PERIPHERAL BLOOD PROGENITOR CELLS IN NEWLY DIAGNOSED PATIENTS WITH CHRONIC MYELOID LEUKAEMIA

N. Storey,[1*] A. L. Lennard,[1] A. M. Dickinson,[1] J. A. Irving,[1] D. Rowe,[2] D. Levett,[1] J. Dunn,[1] A. R. Cattan,[1] and S. J. Proctor[1]

[1] Department of Haematology
Royal Victoria Infirmary
Queen Victoria Road
Newcastle upon Tyne, NE1 4LP, United Kingdom
[2] Department of Human Genetics
19/20 Claremont Place
Newcastle upon Tyne, NE2 4AA, United Kingdom

ABSTRACT

Seven newly diagnosed patients with chronic myeloid leukaemia (CML), age range 22-45 years, were treated with intensive chemotherapy and G-CSF to mobilize peripheral blood progenitor cells (PBPC). Patients had previously received hydroxyurea alone to induce stable chronic phase. Cytogenetic analysis at diagnosis showed 95-100% Philadelphia (Ph) positive metaphases. PBPC were collected by leukapheresis early during white cell recovery, commencing 16-19 days after the start of chemotherapy. 2-5 daily collections were performed. CFU-GM (range 26-170x10^4/kg) considered adequate for engraftment were collected in 6 patients. Cytogenetic analysis showed Ph-negative collections of >90% in 4 patients (1 patient being 100% Ph-negative). All patients achieved >60% Ph-negative collections. Southern blot (SB) analysis for the bcr rearrangement confirmed cytogenetic results. RT-PCR in the 2 patients who were SB negative showed 1 patient to be negative after one round PCR (sensitivity 1 in 10^2-10^3) but positive after nested PCR (sensitivity 1 in10^5). The other patient was positive by one round PCR.

We have shown that it is possible to mobilize and collect Ph-negative PBPC in newly diagnosed patients with CML using chemotherapy that appears less myelosuppressive than

* Address for correspondence: Dr N. Storey, Department of Haematology, Royal Victoria Hospital, Queen Victoria Road, Newcastle upon Tyne, NE1 4LP, UK.

previously quoted. We feel this procedure is more likely to be successful in newly diagnosed patients than later in the natural history of the disease.

INTRODUCTION

In the Northern Health Region of England (population 3.08 million) 25 new cases of chronic myeloid leukaemia present annually. From a four region audit of the disease we know that 48% of patients will be under 55 years of age and potentially eligible for allogeneic bone marrow transplantation.[1] Only 20% of these patients will have a suitable donor however, and the toxicity of the procedure over the age of 35 years is substantial. Alternative forms of treatment for these patients have therefore been investigated.

It has been demonstrated that some patients with CML retain residual Philadelphia negative progenitor cells in their bone marrow. These cells may be mobilized into the peripheral blood in the early phase of marrow recovery following intensive chemotherapy and human growth factor support. If such cells can be collected by leukapheresis and cryopreserved they may be used as autologous transplant material.[2] This form of treatment does however carry risks of both morbidity and mortality.

A number of studies have shown that treatment with alpha-interferon may result in prolonged survival.[3,4] This is most marked in those patients who demonstrate a major cytogenetic response to interferon therapy, but may also include those demonstrating haematological control.[5] While treatment with interferon may produce unpleasant side-effects it does not carry the same risks as other interventional forms of treatment, and its use should be considered in all cases of CML. The duration of the interferon response remains unknown, however, and we therefore feel that the attempted collection of Philadelphia-negative progenitor cells remains a useful procedure provided it can be done with acceptable toxicity and very low risk of mortality.

This study aimed to assess the feasibility of collecting Philadelphia negative peripheral blood progenitor cells (PBPC) from patients with CML as soon after diagnosis as practically possible, and to reduce the toxicity of the procedure by using less intensive chemotherapy than previously reported. Following PBPC collection it is anticipated that all patients will proceed to a trial of interferon therapy.

MATERIALS AND METHODS

Patients

Seven newly diagnosed patients with Ph+ CML presenting in the Northern Health Region of England between April 1994 and February 1995 were entered into the study. No patient had a suitable sibling donor for allogeneic marrow transplantation. Median age of patients was 39 years (range 22-45 years), and all demonstrated 95-100% Ph+ metaphases on bone marrow examination. Sokal scores ranged from 0.37-1.34.[6] The median time from diagnosis to mobilization chemotherapy was 5 months (range2-13 months). Patients had previously received hydroxyurea alone to induce stable chronic phase.

Study Design

The plan of study is shown in Fig. 1. Chemotherapy consisted of daunorubicin 100mg/m^2 day 1, vincristine 2mg day 1, cytosine arabinoside 100mg/m^2 bd days 1-7 and prednisolone 50mg bd days1-5. G-CSF was given from the point of white cell nadir at a dose

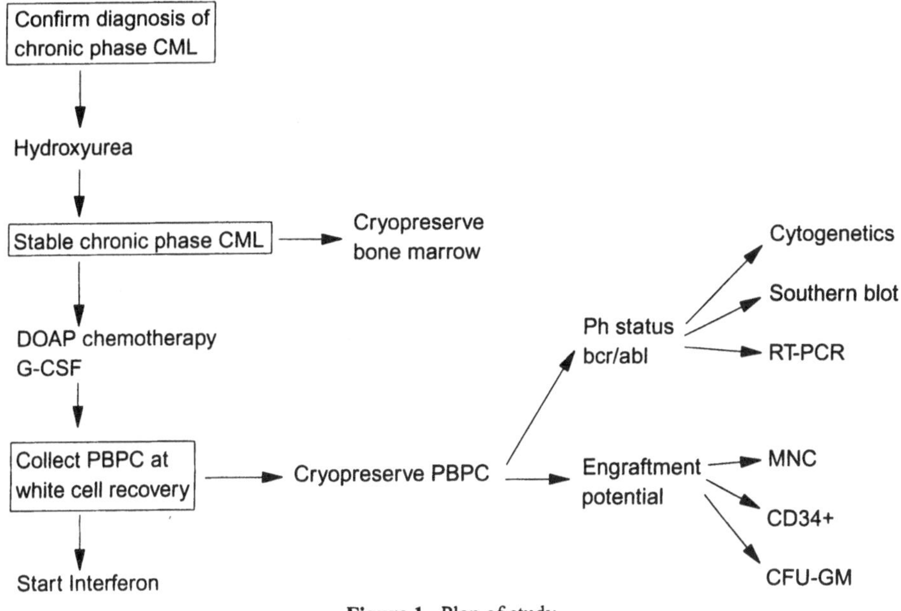

Figure 1. Plan of study.

of 300 micrograms daily (480 micrograms if patient>90kg) until completion of leukapheresis. Leukapheresis commenced the day after the absolute white count passed $0.8 \times 10^9/l$ and continued daily until $>1 \times 10^8$ mononuclear cells/kg patient weight had been collected. A bone marrow examination was performed following completion of PBPC collection, prior to start of interferon therapy, to assess Ph status.

Cytogenetic Analysis

Cytogenetic analysis was performed according to standard protocols. For each daily collection two cultures were set up in RPMI 1640 supplemented with 17% fetal calf serum and incubated for 24 hours at 37^0C. Colcemid exposure was 12 hours at a final concentration of 0.03 micrograms/ml for culture 1, and 15 minutes at a final concentration of 0.3 micrograms/ml for culture 2. Where possible, a total of 30 G-banded metaphases were examined fof Philadelphia chromosome.

Molecular Analysis

Conventional Southern blot techniques were used to probe DNA. Gene probes used were pBCR[7] or Transprobe (Oncogene Science, Inc) derived from MBCR. Each Southern Blot included a normal blood control sample, spiked with 5 or 10% Ph+ CML cells, which gave some indication of sensitivity of the method.

PBSC harvests which were Ph negative were evaluated by PCR amplification for bcr/abl transcripts using the optimized Multiplex method[8]. This method has a sensitivity of 1 in 10^2-10^3 in our hands. Samples which were negative after one round PCR were further evaluated using a nested PCR which has a sensitivity of 1 in 10.[5,9] Positive controls were normal blood spiked with K562 cells at appropriate dilutions. Precautions were taken to prevent carry over of PCR products and negative controls were never positive in either one or two round PCR.

Engraftment Potential Analysis

The engraftment potential of the PBPC harvest was assessed by performing progenitor cell colony assays, and FACS analysis for CD34+ cells.

Standard methylcellulose assays containing 100 IU/ml rG-CSF and 1.5 IU/ml erythropoietin were used.[10] Daily PBPC samples were separated over Ficoll (Nygard, Norway) and the mononuclear cell fraction used at 1-2x10^5 cells/plate. Colonies of more than 40 cells were scored at 14 days. CFU-GM and GEMM (erythroid/macrophage mixed colonies) were counted and expressed as CFU-GM/kg patient weight.

CD34+ cell yields were calculated on each daily PBPC collection using a modified flow cytometric method from Sutherland.[11] Anti-CD34PE (Anti-$HPCA_2$) and Anti-CD45FITC (Anti-HLe-1) from Becton Dickinson were used to dual label 1x10^6 cells from an EDTA sample of the unseparated PBPC collection. Using a FACScan flow cytometer 50,000 events were acquired, and the results expressed both as a CD34+ cell percentage of total white count and as CD34+ cells/kg patient weight.

RESULTS

All patients completed the planned treatment.The mobilizing chemotherapy was well tolerated and could be given as an out-patient after the first 24 hours. Periods of neutropenia and thrombocytopenia were short. The median duration of neutrophil count $<0.5\times10^9$/l was 4 days (range 3-7 days) and platelet count $<20\times10^9$/l was 2 days (range 1-3 days). Six patients required antibiotics for neutropenic fevers, but these settled quickly with appropriate antibiotic treatment. Leukapheresis started between 16 and 19 days after the start of chemotherapy in all patients, and continued daily until the target of $>1\times10^8$ mononuclear cells/kg patient weight had been collected in all but one patient. In this patient, patient 6, a delay of 2 days occured after the first PBPC collection because of recurrent fevers later thought to be due to an allergic response to antibiotics. Six patients were felt to have mobilized sufficient cells ($>20\times10^4$ CFU-GM/kg) for engraftment if transplanted. Five patients had evaluable CD34 analysis and three of these demonstrated $>1\times10^6$ CD34+ cells/kg (Table 1). Cytogenetic, DNA and RNA testing have shown that there was very good mobilization of Ph- cells in four patients (patients 1, 2, 3 and 7), with good, though less

Table 1. Engraftment potential of total PBPC collection from each patient

Patient	MNC x10^8/kg	CFU-GM x10^4/kg	CD34+ x10^6/kg
1	1.56	102	-
2	2.24	154	-
3	3.37	170	5.50
4	1.51	26	0.46
5	1.42	7	0.32
6	1.43	26	1.63
7	1.88	75	4.97

Table 2. Cytogenetic and molecular analysis of PBPC collection

Patient	d 1 %Ph+	d 2 %Ph+	d 3 %Ph	d 4 %Ph+	d 5 %Ph+
1	0 (pcr+)*	6 -	0 (pcr+)*	4 (pcr+)*	
2	0 (sb+)	3 (sb+)	7 (sb+)	3 (sb+)	
3	0 (pcr+)	0 (pcr+)			
4	17 -	27 (sb+)	20 (sb+)		
5	23	33	30	38	
6	19 (sb+)	0 (sb+)	3 (sb+)	- (sb+)	63
7	20 (sb+)	7 (sb+)	10 -		

* Required nested pcr (sensitivity 1 in 10^5cells)

sb Southern blot

marked mobilization in the remaining three (Table 2). The percentages of Ph- cells appeared to remain stable throughout the duration of the collections.

All patients have now started interferon therapy, and bone marrow examination in the four patients who have completed 6 months treatment show 100% Ph+ metaphases in all cases (Table 3).

DISCUSSION

This study has shown that Ph-negative peripheral progenitor cells may be collected in newly diagnosed patients with CML using a chemotherapy regime causing minimal morbidity. It is interesting to note that those patients with the highest percentages of Ph- cells

Table 3. Current status of study patients

Patient	Months	IFN dose	HR	%Ph+ Pre-IFN	%Ph+ At 6m
1	11	3MUx3	no	50 (11)	100
2	10	3MUx7	yes	77 (24)	100
3	8	5MUx7	yes	60 (14)	100
4	7	5MUx5	no	14 (5)	100
5	7	5MUx7	no	23 (6)	
6	5	5MUx7	no	94 (7)	
7	2	3MUx7	yes	86 (6)	

( ) Days after final PBPC harvest that sample was taken.
HR Haematological response.

also appear to have the higher CFU-GM and CD34+ numbers. This is perhaps indicative of greater residual 'normal' haematopoiesis at the time of the PBPC collection.

It has been suggested that autologous transplantation with unmanipulated chronic phase bone marrow in patients with CML might improve survival.[12] It is possible that transplant material containing a higher percentage of Ph- cells may give greater improvement. We feel that the collection of such material is easier in the newly diagnosed patient than later in the course of the disease. Further, by performing such a procedure early, a chemotherapy regime causing little morbidity may be used.

ACKNOWLEDGMENTS

Tyneside Leukaemia Research Association and Northern Region Haematology Group.

REFERENCES

1. Taylor PRA, Ross JRY, Gorst DW, Clough V, Gilbert MJ, Proctor SJP: Population based study of allogeneic marrow transplantation versus conventional treatment in chronic myeloid leukaemia, in four Regional Health Districts in the UK. Br J Haematol 86: Suppl 1,27,1994 (abstr).
2. Carella AM, Podesta M, Frassoni F, Raffo MR, Pungolino E, Vimercati R, Sessarego M, Parodi C, Rabitti C, Ferrero R, Benvenuto F, Figai O, Carlier P, Levcasic G, Valbonesi M, Vitale V, Giordano D, Pierluigi D, Nati S, Guerracio A, Rosso C, Saglio G: Collection of 'normal' blood repopulating cells during early hempoietic recovery after intensive conventioal chemotherapy in chronic myelogenous leukaemia. Bone Marrow Transplantation 12: 267, 1993.
3. Talpaz M, Kantarjian H, Kurzrock R: Interferon-alpha produces sustained cytogenetic responses in chronic myelogenous leukaemia: Ann Intern Med 114: 532, 1991.
4. The Italian Cooperative Study Group on Chronic Myeloid Leukaemia. Interferon alfa-2a as compared with conventional chemotherpy for the treatment of chronic myeloid leukaemia. N Engl J Med 330: 820, 1994.
5. Allan NC, Richards SM, Shephard PCA, on behalf of the UK Medical Research Council's Working Parties for Theraputic Trials in Adult Leukaemia. UK Medical Research Council randomised, multicentre trial of interferon-*α*n1 for chronic myeloid leukaemia: improved survival irrespective of cytogenetic response. Lancet 345: 1392, 1995.
6. Sokal JE, Cox EB, Baccarani M, Tura S, Gomez GA, Robertson JE, Tso CY, Braun TJ, Clarkson BD, Cervantes F, Rozman C, and the Italian Cooperative Study Group. Prognostic discrimination in 'good-risk' chronic granulocytic leukemia. Blood 63: 789, 1984.
7. Heisterkamp N, Stam K and Groffen J. Structural organisation of the bcr gene and its role in the Philadelphia translocation. Nature 315:758-761, 1994.
8. Cross CP, Melo VM, Feng L and Goldman JM. An optimized multiplex polymerase chain reaction (PCR) for detection of bcr/abl fusion mRNAs in haematological disorders. Leukemia 8:186-189, 1994.
9. Cross CP, Feng L, Bungey J and Goldman JM. Minimal residual disease after bone marrow transplant for chronic myeloid leukaemia detected by the polymerase chain reaction. Leukemia and lymphoma 11(S1):39-43, 1993.
10. Sutherland HJ, Eaves AC, Eaves CJ. Quantitive assays fot human hematopoietic progenitor cells, in: Bone Marrow Processing and Purging: A Practical Guide, Boca Raton CRC Press Inc, 1991;155.
11. Sutherland DR, Keating A, Nayar R, Anania S, Stewart AK. Sensitive detection and enumeration of CD34+ cells in peripheral and cord blood by flow cytometry. Exp Haem 22 : 1003, 1994.
12. McGlave PB, DeFabritiis P, Deissroth A, Goldman J, Barnett M, Reiffers J, Simonsson B, Carella A, Aeppli D. Autologous transplants for chronic myelogenous leukaemia: results from eight transplant groups. Lancet 343: 1486,1994.

CD34 QUANTITATIVE FLOW CYTOMETRY STUDY OF THE BLASTIC POPULATION IN ACUTE LEUKEMIAS

Gian Matteo Rigolin, Francesco Lanza, and Gianluigi Castoldi

Institute of Hematology
University of Ferrara
Via Savonarola, 9
44100 Ferrara, Italy

ABSTRACT

Quantitative flow cytometry analysis of membrane antigens is becoming a routinely approach in the analysis of the blastic population in acute leukemias. By flow cytometry (FACScan B.D.), we analysed CD34 quantitative expression on the blastic population of 48 patients affected by acute leukemias (39 AML, 9 ALL). The blastic population was defined by means of strategical double fluorescence combinations of monoclonal antibodies. Patients were subdivided into four groups: de novo AML (24 pts), secondary AML (15 pts), relapses of AML (8 pts) and ALL (9 pts). A significant higher intensity of expression was demonstrated in secondary AML (p=0.019) and at relapses (p=0.026). At relapse, a higher percentage of CD34 positivity and a higher intensity of CD34 expression were observed. When considering AML FAB distribution a significant higher CD34 intensity of expression was demonstrated in M0, M1 and M4 (p=0.003). When subdividing patients according to negative, weak and bright CD34 intensity of expression, the 93.3% of secondary AML had a bright expression while only a 50% of de novo AML was recorded in the bright CD34 group. Only one secondary AML had negative CD34 expression while the 33.3% of de novo AML was CD34 negative (p=0.02). Secondary AML demonstrated a lower CD33 and a higher HLA-DR intensity of expression. According this quantitative CD34 analysis and considering the models of immunological development of hemopoiesis, we may prospect that in secondary AML the neoplastic event prevalently involve a stem cell at an earlier stage of development. We think that from the combination of more antigens on a larger series of patients and considering a combined immunological and cytogenetic study we may confirm the important role of flow cytometry in the study of the heterogeneity of immunological findings in acute leukemias with biological and prognostic implications.

Molecular Biology of Hematopoiesis 5, edited by Abraham et al.
Plenum Press, New York, 1996

INTRODUCTION

Quantitative flow cytometry analysis of membrane antigens is becoming a routinely approach in the analysis of the blastic populations of acute leukemias. This is due to the stability and reproducibility of flow cytometers and to the availability of microbeads by which we can standardise the intensity of fluorescence or directly determine the antigen binding capacity (ABC) for a specific monoclonal antibody.

The study of antigen expression by the definition of the level of fluorescence and of the ABC may represent a new tool in the study of the heterogeneity of the immunological profiles of acute leukemias (AL) with possible important biological and prognostic implications [1].

In a recent work from Faharat et al. [2] TdT was quantified in AL and the highest values were found in B acute lymphoblastic leukemias (ALL) while the lowest values were recorded for acute myeloid leukemias (AML). T- ALL and biphenotypic leukemias (BAL) had intermediate levels of intensity of expression. As far as B- ALL, T-ALL, BAL and AML is concerned, no statistical difference was demonstrated as far as intensity of expression is concerned when considering cases with similar percentages of positivity.

CD10 was studied in ALL by Lavabre-Bertrand et al. [3] and an interesting association was found between CD10 level of expression and specific cytogenetic profiles, in particular it was demonstrated that a bright CD10 expression is generally associated with hyperdiploidy, while low and undetectable CD10 expressions are correlated with the t(1;19) and the t(4;11) respectively.

Aim of this study was a prospective analysis of CD34 intensity of expression in acute leukemias at diagnosis and at relapse in order to find possible biological and clinical correlations.

PATIENTS AND METHODS

Patients

48 AML patients referred and treated at the Institute of Haematology of Ferrara between 1993 and 1995 were analysed for the immunophenotypic characteristics of the blastic populations. All patients were classified in accordance to morphological and cytochemical FAB criteria [4-6]. Patients were so distributed: 24 de novo AML at first diagnosis, 15 secondary AML, 9 ALL. Eight AML patients were studied at relapse. The peripheral blood of five healthy subjects was also analysed.

Flow Cytometry (FCM) Procedures

Bone marrow or peripheral blood samples were analysed by a standard double fluorescence flow cytometry technique with a strategical panel of MoAbs [7]: CD34 FITC and CD38 PE, HLA-DR FITC and CD13 PE, CD45 FITC and CD14 PE, CD3 FITC and CD19 PE, CD7 FITC and CD33 PE (Becton Dickinson), CD10 FITC and CD19 PE (DAKO) with correspondent isotypic negative controls. Unseparated samples were incubated 30 min at 4°C and then treated for 10 minutes with FACSLysis (Becton Dickinson) solution. Subsequently they were washed twice with PBS solution and analysed by a FACScan (Becton Dickinson). Ten thousands events were analysed for each MoAb.

Instrument Calibration

Instrument calibration was performed with FITC labelled calibrating microbeads (Quantum 26 p, Flow Cytometry Standard Corporation, FCSC). The calibrating kit was run at the beginning of each batch. Within run and between run quality control assays were performed.

FCM Statistical Analysis

The analysis was performed with Lysis 2.0 software (Becton Dickinson). As far as light scattering properties is concerned [7-10], all the identifiable subpopulations were gated and then analysed in four quadrants multi-colour dot plot statistics. Gated subpopulations were subsequently analysed in histogram mode and compared to the correspondent negative control.

Statistical Analysis

MESF (Molecular Equivalents of Standard of Fluorescence) values were obtained by simple linear regression in which as independent variable we entered the assigned MESF values for each calibrating standard of fluorescence and as dependent variables the correspondent channel values of the different standards of fluorescence. CD34 MESF values were obtained by interpolation in the regression curve. A parallel CD34 analysis was performed on the peripheral leukocytes of five normal subjects in order to define the negative detection threshold of intensity for CD34.

Chi Square and Kruskall-Wallis statistics were used though-out. SPSS statistical package was used in analysing data.

RESULTS

To make results comparable with a previous study of our group, patients were subdivided into three classes of CD34 intensity of expression according to the ratio between the CD34 intensity of the blastic population and the correspondent isotypic negative control (P/N ratio). The 3 groups of patients were so defined: CD34 negative expression when the P/N ratio was < 3, weak CD34 expression when the P/N ratio was comprised between 3 and 10 and bright CD34 intensity of expression for P/N ratio > 10. Patients were considered positive when the percentage of CD34 positive cells was over 10%.

According to this subdivision of the patients, we obtained the range of MESF values for each subgroup: CD34 negative < 5000 MESF, CD34 weak 5000-25000 MESF, CD34 bright > 25000 MESF.

Definition of the 5000 MESF Cut Off

The study of CD34 expression on the leukocytes of 5 normal subjects demonstrated that CD34 intensity was in all the cases comprised in a range between 0 and 2000 MESF. The precision of our instrument was previously calculated in nearly 1300-1500 MESF. 5000 MESF represent the minimum value for a positive sample when considering the imprecision of the negative control and that of the sample (1500 MESF x 2) and the maximum value observed for a normal peripheral blood population which is by definition CD34 negative (2000 MESF).

Table 1. Distribution of the patients according to CD34 intensity od expression.

CD34	Neg.	Weak	Bright
de novo AML	8	4	12
Sec. AML	1	–	14
Relapsed AML	–	2	6
ALL	4	1	4

CD34 neg. = < 5000 MESF; CD34 weak = 5000-25000 MESF; CD34 bright = >25000 MESF

CD34 Expression in Acute Leukemias (AL)

In Table 1 and Figure 1 we reported the distribution of the patients according to CD34 intensity of expression in de novo AML, secondary AML, relapsed AML and ALL. A significant difference as far as patients distribution is concerned was found between de novo AML and secondary AML (p=0.019) and de novo, secondary and relapsed AML (p=0.026).

CD34 Intensity of Expression and FAB Classification

In Table 2 and Figure 2 we summarised the FAB distribution of all de novo and secondary AML patients according to CD34 intensity of expression. More immature FAB subtypes are correlated with a brighter CD34 intensity of expression. The only M3 case had a 15% of CD34+ cells with a weak CD34 intensity of expression. Among M4 the great part (5/6) was CD34 positive with a cut off value of 10%: only 1 case was CD34 positive with a CD34 cut off of 20%.

CD34 and AML at Relapse

Eight patients were studied at relapse. In 5 cases we studied the CD34 intensity of expression both at diagnosis and at relapse. As far as CD34 percentage of positivity is

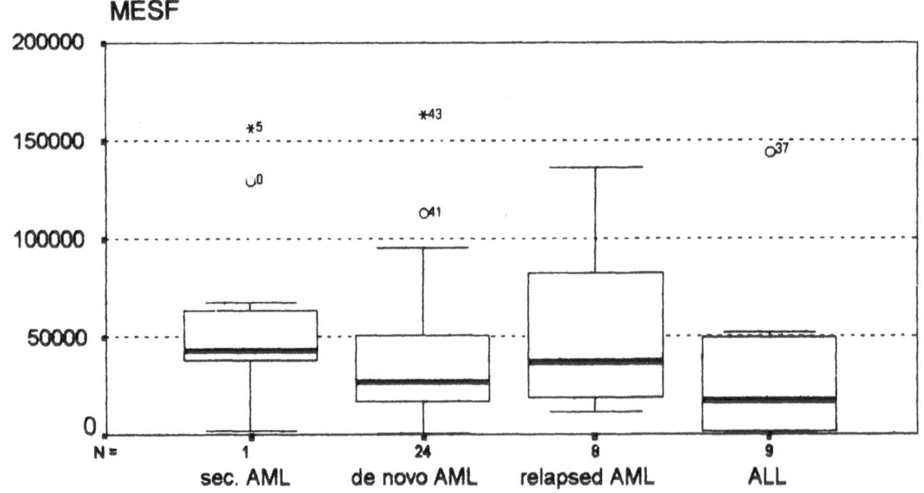

Figure 1. CD34 intensity of expression according to the different leukemic disorders. Data are expressed as MESF values.

Table 2. FAB and CD34 intensity of expression

CD34	Neg.	Weak	Bright
M0	1	–	2
M1	–	1	15
M2	2	2	2
M3	–	1	–
M4	1	–	5
M5	4	–	2
M6	1	–	–

CD34 neg. = < 5000 MESF, CD34 weak = 5000-25000 MESF, CD34 bright = >25000 MESF

concerned, 4 cases at diagnosis were negative with a cut off value of 10% (Table 3). At relapse in all cases we observed a percentage of CD34 positive cells over 10%. When comparing the intensity of expression in the CD34 positive subpopulations, all the cases demonstrated at relapse a brighter CD34 intensity of expression. Two negative CD34 cases resulted CD34 positive at relapse.

CD34 and the Other Markers

In Figure 3 we reported the mean intensity of expression (expressed as a ratio between the median intensity of expression of the MoAb and the median intensity of the correspondent negative isotypic control) of the principal markers according to the 3 classes of CD34 intensity of expression. In AML a brighter HLA-DR intensity of expression is appreciable in the CD34 bright group (p=0.03) while a weaker intensity of expression was observed as far as CD33 intensity of expression is concerned (p=0.08).

As far as ALL is concerned, CD34 bright expression was associated with a very bright HLA-DR intensity of expression.

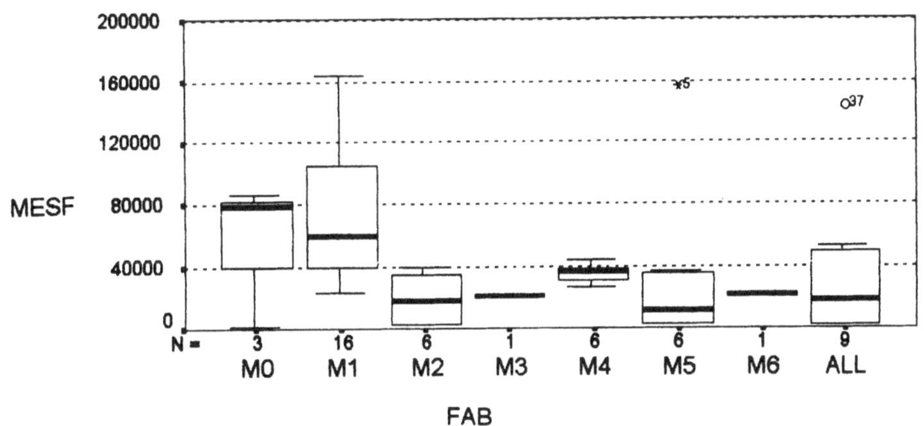

Figure 2. CD34 intensity of expression and FAB classification. Data are expressed as MESF values. CD34 populations were considered when at least a 3% of positive cells was demonstrated.

Table 3. CD34 percentages and intensities of expression at diagnosis and at relapse

Patient	CD34 intensity	CD34 percentage
A.O diagnosis	30405	17
A.O. I relapse	63244	43
A.O. II relapse	136024	39
B.E. diagnosis	0	0
B.E. I relapse	11795	28
C.S. diagnosis	0	0
C.S. I relapse	34434	14
G.V. diagnosis	25581	8
G.V. I relapse	102224	63
Z.T. diagnosis	21440	6
Z.T. I relapse	25845	11

Intensity values are expressed as MESF values

DISCUSSION

The relevance of immunophenotype analysis in the characterisation of AML is still underestimated due to the lack of a clear and reproducible prognostic impact of the different phenotypes and to the necessity of uniform diagnostic criteria and standardised technical procedures. Several recent works stressed the necessity of a cytofluorimetric analysis based on the definition of clusters of cells with similar morphological patterns and immunological properties. As far as immunological properties is concerned, quantitative flow cytometry analysis is becoming a new approach in the study of the blastic population in acute leukemias. This is due to the stability and the reproducibility of flow cytometers and to the availability of fluorescent microbeads by which we can compare results between different centres and by which we can express results as number of antigen molecules.

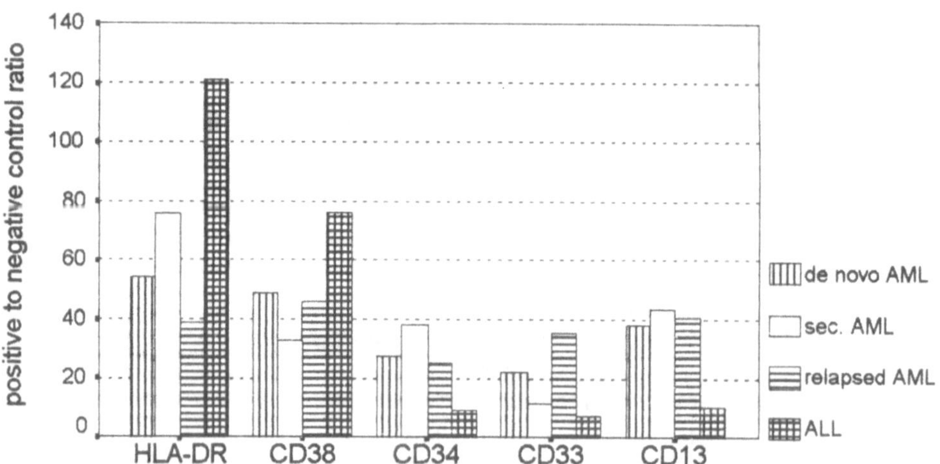

Figure 3. Intensity of expression of the principal markers in the different leukemic disorders. Data are expressed as a ratio between the median channel intensity of the blastic population and the correspondent isotypic negative control.

The usefulness of a quantitative flow cytometry analysis in AL was recently demonstrated in two different studies in which the authors evaluated CD10 and TdT intensity of expression.

CD10 expression was evaluated in acute lymphoblastic leukemias at diagnosis and at relapse [3]. In particular the authors demonstrated that CD10 over-expression may be considered a marker of leukemia and an earlier marker of relapse since CD10 expression was maintained at relapse. They also demonstrated a correlation between CD10 intensity of expression and peculiar cytogenetic findings i.e. bright CD10 and hyperdiplody, t(1;19) and weak CD10 expression, t(4;11) and undetectable levels of CD10. In a different study TdT was demonstrated to have a lower expression in AML vs. ALL and BAL in cases with comparable percentages of positive cells [2].

We decided to study CD34 intensity of expression in AL at diagnosis and at relapse in order to delineate possible biological and clinical correlations. In a previous study from our group we demonstrated a poor prognostic outcome for patients with a bright CD34 intensity of expression while patients with a weak CD34 expression had a prognostic outcome not different from that of CD34 negative patients [11].

From the analysis of CD34 intensity of expression in de novo AML and secondary AML it is evident a prevalent distribution of secondary AML in the CD34 bright group: only one case among secondary AML was CD34 negative. From the analysis of all patients with secondary AML referred at our Institute between 1991 and 1995 and considering the P/N ratio, more than the 70% of the cases was CD34 bright with a 6.7% demonstrating a weak CD34 intensity of expression. Overall secondary AML are prevalently CD34 positive and with a bright CD34 intensity of expression. In these cases CD34 bright expression is associated with a bright HLA-DR expression and with a weaker CD33 and CD38 expression. The neoplastic event in these leukemias seems to have occurred on a stem cell at an earlier stage of development according to the proposed immunological hemopoietic in vitro models [12-15].

Among de novo AML CD34 positivity was observed in the 67% of the cases; in our historical series of patients the 57.1% of cases was CD34 positive according to a cut off value of the 10%. Our results are within the range of positivity for CD34 if we considered the other series of patients (40-63%)[16-21].

The percentages of cases with a weak CD34 intensity of expression are 16 and 25.7% in this study and in our historical series respectively. The survival of these patients is similar to CD34 negative patients. The exclusion of these patients from CD34 patients results in an increased statistical significance for a worse outcome of the CD34 bright patients (data not shown).

As far as FAB distribution is concerned a higher intensity of expression was demonstrated in M0, M1 and M4. Overall these data would delineate a CD34 gradient of expression from M0 and M1 (mostly positive and with a CD34 bright expression), to M3 mainly CD34 negative but with a weak CD34 positive population in at least the 25% of cases [22]. In the middle we have the M2 FAB subtype in which we may observe an equal distribution among the CD34 negative, weak and bright groups. In this perspective we have the observation from Kita et al. that in M2 AML with the t(8;21) there is an involvement of a stem cell which is CD34 positive.

In this series of patients we observed one M3 case with a 15% of CD34 positive cells. The presence of CD34 positive cells in M3 is not a rare event as described by Paietta et al. [22]. in their study the 25% of cases had a percentage of CD34 positive cells over 38%. Significantly the CD34 intensity of expression in the M3 was weak. Interestingly most of M4 patients are CD34 positive with a cut off value of 10%. With a cut off of the 20% most of these cases would be CD34 negative. M4 CD34 positivity may be explained because of the earlier common myelomonocytic progenitor cell involved in the neoplastic event, while

M5 patients were mostly CD34 negative probably because the cell involved in the leukemic event is already committed to the monocytic lineage with a more mature phenotype.

As previously described CD34 positivity in relapsed AML is higher than at diagnosis [23]. AML tend to relapse with a more immature phenotype. Two cases in our series of patients were completely CD34 negative (CD34% of positive cell < 1%) at diagnosis, while in 3 cases we observed a CD34 percentage of positive cells comprised between 3 and 10%. All these cases resulted CD34 positive at relapse and in all the cases but one CD34 intensity of expression was higher. One patient was studied at the first and at the second relapse: CD34 intensity of expression increased in each relapse.

The phenotype of relapsed AMLs was similar to secondary AML: bright HLA-DR intensity of expression, dim CD33 expression. Immunological data would support the idea of a more immature phenotype at relapse than at the diagnosis.

The different distribution of the patients with higher intensity of expression among de novo and secondary AMLs may probably be expression of a peculiar immunophenotypic characteristic of the stem cell involved in the myelodysplastic event which may probably be correlated with a distinct cytogenetic profiles. CD34 positive AMLs were demonstrated to have a distinct cytogenetic profile since a higher percentage of abnormal kariotype was demonstrated among CD34 AML which were frequently associated with a trilineage myelodysplasia [8]. Further analysis on a larger number of patients are necessary in order to find possible correlations [25].

A flow cytometry quantitative analysis requires a careful standardisation of the methodologies and a standardisation of the analysis of fluorescence [26]. The introduction of calibrating microbeads was certainly one of the main purposes of the Fifth Workshop on Human Leukocyte Antigens held in Boston on November 1993. The usefulness of these beads is related to the correct calibration of the instrument and the standardisation of fluorescence intensity with the possibility of the comparison of the results between different centres and within the same centre. It is also possible to calculate, by means of the specific fluorescence/protein ratio, the number of surface antigen molecules.

Overall, we may prospect an important role for flow cytometry and in particular for quantitative flow cytometry analysis in the study of the heterogeneity of immunological findings of AL with biological and clinical implications.

ACKNOWLEDGMENT

Supported by 40% and 60% M.U.R.S.T. and Regional Funds.

REFERENCES

1. Terstappen LWMM, Safford M, Könemann S, Loken MR, Zurlutter K, Büchner T, Hiddemann W, Wörmann B: Flow cytometric characterisation of acute myeloid leukemia. Part II. Phenotypic heterogeneity at diagnosis. Leukemia 5: 757, 1991.
2. Farahat N, Lens D, Matutes E, Catovsky D: Differential TdT expression in acute leukemia by flow cytometry: a quantitative study. Leukemia 9: 583, 1995.
3. Lavabre-Bertrand T, Janossy G, Ivory K, Peters R, Secker-Walker L, Poerwit-MacDonald A: Leukemia-associated changes identified by quntitative flow cytometry: CD10 expression. Cytometry 18: 209, 1994.
4. Bennett JM, Catowsky D, Daniel MT, Flandrin G, Galton DAG, Gralnick HR, Sultan C: Proposed revised criteria for the classification of acute myeloid leukemia. Ann Intern Med 103: 626, 1985.
5. Castoldi GL, Cuneo A, Lanza F, Tomasi P: Diagnosis of leukemia: morphology. Leukemia 6;Suppl.4:6, 1992.

6. Castoldi GL, Liso V, Fenu S, Vegna L, Mandelli F: Reproducibility of the morphological diagnostic criteria for acute myeloid leukemia: the GIMEMA group experience. Ann Hematol 66: 171, 1993.
7. Terstappen LWMM, Könemann S, Safford M, Loken MR, Zurlutter K, Büchner T, Hiddemann W, Wörmann B: Flow cytometric characterization of acute myeloid leukemia. Part I. significance of light scattering properties. Leukemia 5: 315, 1991.
8. Verwer BJH, Terstappen LWMM: Automatic lineage assignment of acute leukemias by flow cytometry. Cytometry 14: 862, 1993.
9. Kawada H, Ichikawa Y, Watanabe S, Nagao T, Arimori S: Flow cytometry analysis of cell-surface antigen expressions on acute myeloid leukemia cell populations according to their cell-size. Leukemia Res 18: 29, 1994.
10. Rigolin GM, Lanza F, Ferrari L, Castoldi GL: CD34+/CD33+ blast cells: correlation with FAB subtypes. Leukemia Lymphoma (in press).
11. Lanza F, Rigolin GM, Moretti S, Latorraca A, Castoldi GL: Prognostic value of immunophenotypic characteristics of blast cells in acute myeloid leukemia. Leukemia Lymphoma 1994;13 suppl1:81-5.
12. Terstappen LWMM, Safford M, Unterhalt M, Könemann S, Zurlutter K, Piechotka K, Drescher M, Aul C, Büchner T, Hiddemann W, Wörmann B: Flow cytometric characterization of acute myeloid leukemia: IV. Comparison to the differentiation pathway of normal haematopoietic progenitors cells. Leukemia 6: 993, 1992.
13. Pirelli L, Teofili L, Menichella G, Rumi C, Paoloni A, Iovino S, Puggioni PL, Leoni G, Bizzi,B: Further investigations on the expression of HLA-DR, CD33 and CD13 surface antigens in purified bone marrow and peripheral blood CD34+ haematopoietic progenitor cells. Br J Haematol 84: 24, 1993.
14. Pontvert-Delucq S, Breton-Gorius J, Schmitt C, Baillou C, Guichard J, Najman A, Lemoine M: Characterization and functional analysis of adult human bone-marrow cell subsets in relation to B-lymphoid development. Blood 82: 417, 1993.
15. Terstappen LWMM, Huang S, Safford M, Lansdorp M, Loken MR: Sequential generation of hematopoietic colonies derived from single nonlineage-committed CD34+CD38- progenitors cells. Blood 77: 1218, 1991.
16. Campos L, Guyotat D, Archimbaud E, Devaux Y, Treille D, Larese A, Maupas J, Gentilhomme O, Ehrsam A, Fiere D: Surface marker expression in adult acute myeloid leukemia: correlations with initial characteristics, morphology and response to therapy. Br J Haematol 72: 161, 1989.
17. Geller R, Zahurak M, Hurwitz A, Burke PJ, Karp JE, Piantadosi S, Civin CI: Prognostic importance of immunophenotyping in adults with acute myelocitic leukemia: the significance of the stem-cell glycoprotein CD34 (My10). Br J Haematol 76: 340, 1990.
18. Borowitz MJ, Guenther KL, Shults KE, Stelzer GT: Immunophenotyping of acute leukemia by flow cytometric analysis. Am J Clin Pathol 100, 534, 1993.
19. Solary E, Casanovas O, Campos L, Béné MC, Faure G, Maingon P, Falkenrodt A, Lenhormand B, Genetet N and GEIL: Surface markers in adult myeloblastic leukemia: correlation of CD19+, CD34+ and CD14+/DR- phenotypes with shorter survival. Leukemia 6: 393, 1992.
20. Del Poeta G, Stasi R, Venditti A, Suppo G, Aronica G, Bruno A, Masi M, Tabilio A, Papa G: Prognostic value of cell marker analysis in de novo acute myeloid leukemia. Leukemia 8: 388, 1994.
21. Bradstock K, Mattews J, Benson E, Page F, Bishop J, and the Australian leukemia study group: Prognostic value of immunophenotyping in acute myeloid leukemia >Blood 84: 1220, 1994.
22. Paietta E, Andersen J, Gallagher R, Bennett J, Yunis J, Cassileth P, Rowe J, Wiernik PH: The immunophenotype of acute promyelocytic leukemia (APL): an ECOG study. Leukemia 8: 1108, 1994
23. Kita K, Nakase K, Miwa H, Masuya M, Nishii K, Morita N, Takakura N, Otsuji A, Shirakawa S, Ueda T, Nasu K, Kyo T, Dohy H, Kamada N: Phenotypical characteristics of acute myelocytic leukemia associated with the t(8;21)(q22;q22) chromosomal abnormality: frequent expression of immature B-cell antigen CD19 togheter with stem cell antigen CD34. Blood 80: 470, 1992.
24. Thomas X, Campos L, Archimbaud E, Shi ZH, Ritoutet T, Anglaret B, Fiere D: Surface marker expression in acute myeloid leukemia at first relapse. Br J Hematol 81: 40, 1992.
25. Fagioli F, Cuneo A, Carli MG, Bardi A, Piva N, Previati R, Rigolin GM, Ferrari L, Spanedda R, Castoldi GL: (1993) Chromosome aberratios in CD34-positive acute myeloid leukemia. Correlation with clinicopathologic features. Cancer Genet Cytogenet 71: 119, 1993.
26. Rigolin GM, Lanza F, Castoldi GL: Photomultiplier voltage setting: possible important source of variability in MESF calculation? Cytometry (in press).

30

CYTOKINE GENE EXPRESSION IN A CASE OF B-CELL CHRONIC LYMPHOCYTIC LEUKEMIA (B-CLL) WITH AN UNUSUAL EXPANSION OF T CELLS AT PRESENTATION

Athanasia Mouzaki,[1,2*] Vincent Kindler,[2] Nicolette Bowers,[2] Arlette Doucet,[2] Maria Melachrinou,[3] Maria-Christina Kyrtsonis,[1] and Alice Kallinikou-Maniatis[1]

[1] Hematology Laboratory
Patras University Hospital, Greece
[2] Division of Hematology
Geneva University Hospital, Switzerland
[3] Pathology Laboratory
Patras University Hospital, Greece

ABSTRACT

We present a case of a 60 year old male with lymphadenopathy, hepatosplenomegaly and leucocytosis with lymphocytosis. At presentation, his WBC was 15.1×10^9/L with 75% lymphocytes; these were mainly T-cells with an atypical morphology, and a small monoclonal population of B cells, κ type. RT-PCR of immunoglobulin genes showed a monoclonal rearrangement of the IgM type. Elf-1, bcl-2, IL-1α and β, IL-2, IL-4 and IL-6 genes were constitutively expressed to varying degrees, and superinduced upon mitogenic stimulation of the cells. Bandshifts with the lymphotropic transcription factors NF-kB, AP-1 and NF-AT, gave profiles typical of the T-cell leukemia Jurkat and not of normal (ex-vivo) T-cells or B-CLL cells, ex-vivo or stimulated. Bone marrow histology revealed a diffuse infiltration of lymphocytes, and immunohistochemistry on bone marrow sections the presence of both B and T-cells in almost equal proportions. At this stage a double (T/B) CLL was suspected. The patient was treated with alkylating agents and went into remission. Three months later, and while still on drugs, the patient's WBC was found to be 5.43×10^9/L and immunophenotyping revealed a complete change of his T/B cell ratio. RT-PCR showed that the patient's

* Editorial correspondence: Dr. A. Mouzaki, Laboratory Hematology and Transfusion Medicine, Medical School, University Hospital, Patras GR-261 10, Greece. Tel: ++30-61-999 644 and 646; FAX: ++30-61-991 991.

CD2 cells expressed none of the above genes but IL-4 when induced, and his CD19 cells IL-6 and IL-1β when induced. The expression of IgM remained monoclonal, confirming the clinical observation of B leukemia development.

INTRODUCTION

Chronic lymphocytic leukemia (CLL) is quite common in western countries (25-30% of all leukemias) with a median age of onset about 60 years. The diagnostic criteria for CLL, as defined by the National Cancer Institute Sponsored Working Group, are (i) absolute lymphocytosis in the blood (more than 5×10^9/L) for at least 4 weeks, (ii) more than 30% lymphocytosis in the bone marrow and (iii) monoclonal B- or T-cell phenotype. More than 95% of the patients have B-CLL, with T-CLL accounting for about 2% of all cases. In B-CLL the B-cells usually are $CD5^+$ with simultaneous positivity for CD20, CD19 and/or CD24 markers, and constitute the majority of all lymphocytes (up to 90%); they are also monoclonal expressing only one light chain, either κ or λ. In B-CLL, the absolute number of T-cells may be normal, decreased or even increased (23). The etiology of CLL is unknown and certain cytokines have been implicated in the pathogenesis of the disease: IL-1, IL-6 and TNFα in B-CLL (24-27) and IL-2, IL-4 and IFNγ in T-CLL (28-30). It has also been suggested that the long life of CLL cells may be due to over-expression of bcl-2 protein with its known ability to interfere with apoptosis (26, 31).

We present a case of a 60 year old male with CLL with an unusually high percentage of T-cells with atypical morphology in his peripheral blood and bone marrow. After three months of treatment the patient's lymphocytes were typical of B-CLL and his T-cells almost disappeared. We followed these events with experiments investigating gene expression of various cytokines, bcl-2 and elf-1 as well as the presence of lymphotropic transcription factors in the patient's lymphocytes.

MATERIALS AND METHODS

Case Report

A 60 year old man presented with lymphadenopathy, hepatosplenomegaly and leukocytosis with lymphocytosis. During the 3 years prior to presentation, on routine blood tests, a lymphocytosis was noted which was attributed to viral infections. One year ago, he noted enlargement of his cervical nodes but did not seek medical attention. At presentation, his WBC was 15.1×10^9/L with 75% lymphocytes. Hemoglobin and hematocrit were in the normal range but his platelet count was slightly reduced (13×10^{10}/L). The lymphocytes had an atypical morphology (Figure 1A) and immunophenotyping showed that they were mainly T-lymphocytes (Table 1) although a small population of monoclonal B-cells (κ-type) was also uncovered on repeat testing. A complete viral screening including HTLV-1 was negative. Anti-platelet antibodies were not present. Bone marrow aspiration revealed diffuse lymphocytic infiltration of the order of 70%. Bone marrow trephine biopsy showed a diffuse involvement by small lymphocytes, with a marked reduction in fat cells and normal hematopoietic cells (Figure 2A). Immunohistochemical analysis (5) showed that the small lymphocytes reacted with LCA (100%), MB1 (60-70%), MT1 (50-60%), UCHL-1 (40%), MB2 (30%), MT2 (30%) and CD3 (30%) while they did not react with L26 (Table 1 and Figure 2B & C). The patient was treated with alkylating agents. Three months later, his lymphadenopathy disappeared and his spleen size decreased significantly so that it was only palpable on inspiration. The WBC was in the normal range (5.43×10^9/L). This time, his

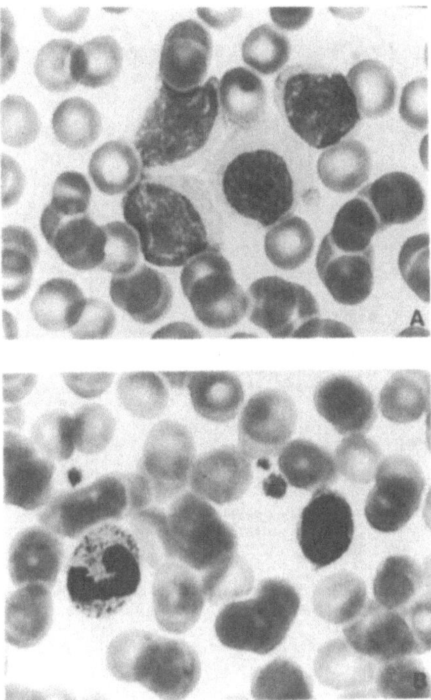

Figure 1. Peripheral blood cells of the patient before (A) and after (B) treatment. MGG x1000.

Table 1. Phenotyping of the patient's cells in peripheral blood and bone marrow

Phenotype	Healthy control (% cells)	Patient (% cells)	
		Before treatment	After/during treatment
In Peripheral Blood			
CD3	55.0	96.0	5.9
CD4	43.6	58.0	3.8
CD8	26.8	38.0	2.0
CD4/CD8	0.9	ND	0.0
CD56	30.9	ND	55.7
CD19	6.6	0.0	59.5
κ	ND	10.0	ND
λ	ND	0.0	ND
In Bone Marrow Sections			
LCA (CD45)	ND	100	ND
MB1 (CD45R)	ND	65	ND
MB2	ND	30	ND
MT1 (CD43)	ND	55	ND
UCHL-1 (CD45RO)	ND	40	ND
CD3	ND	30	ND
MT2 (CD45RA)	ND	30	ND
L26 (CD20)	ND	0.0	ND

ND=not determined

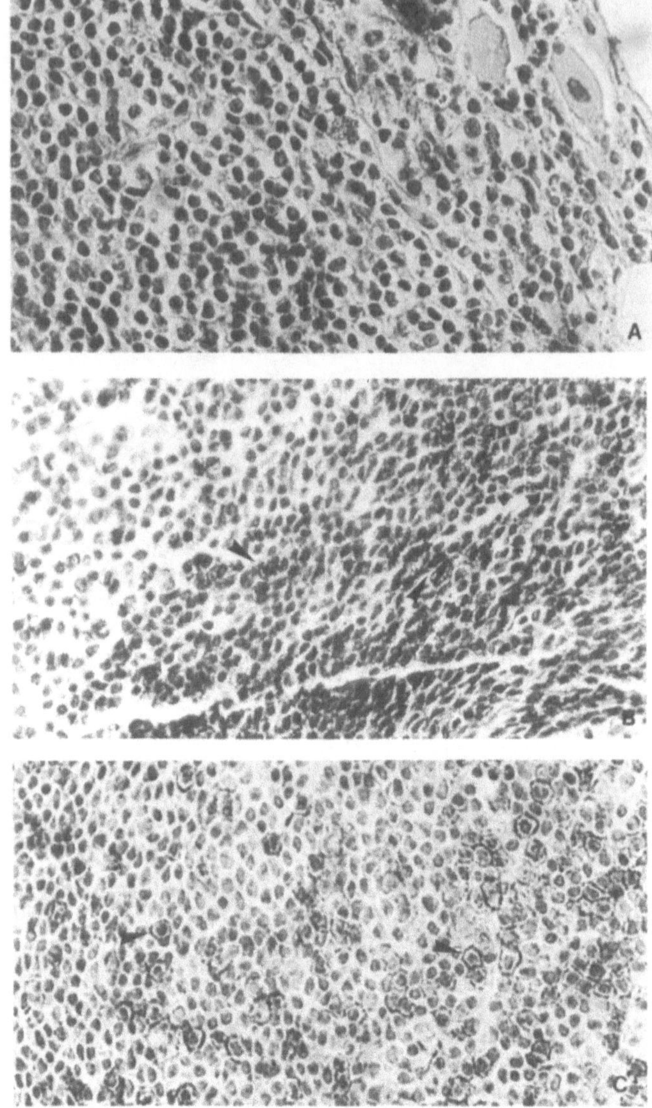

Figure 2. Bone marrow trephine biopsy and immunohistochemical analysis. (A) Diffuse bone marrow involvement by lymphocytes. HE x400. (B) MB1 (CD45R) is reactive with the majority of lymphocytes. Immunoperoxidase x400. (C) UCHL-1 (CD45RO) identifies the infiltrating T-cells. Immunoperoxidase x400.

lymphocytes had a normal morphology (see Fig. 1B). The results from the FACS analysis are shown in Table 1. During these 3 months his T-cells had diminished, overtaken by B-cells.

Reagents

Ficoll-paque was from Pharmacia, Sweden. Antibodies used in FACS analysis or immunohistochemistry were from Dako, Denmark except from LCA, MB1, MB2, MT1, MT2, anti-kappa and anti-lambda which were from Biogenex (San Ramon, CA). Pan B (anti-CD19) antibody and anti-CD2 antibody-coupled M450 magnetic beads were from Dynal, Norway. Phorbol myristate acetate (PMA) was from Sigma, St-Louis, MI. Ionomycin was from Calbiochem, San Diego, CA. AMV-reverse transcriptase, RNAsin, acetylated BSA and Taq DNA polymerase were from Promega Corp. Madison, WI. Deoxynucleotides, oligo dT primers, tRNA and all the materials used for the electrophoretic mobility shift assays (bandshifts) were from Boehringer, Germany. Oligonucleotide primers used for the polymerase chain reaction (PCR) were either synthesized at the University of Geneva, dpt. of Microbiology, Medical Centre or purchased from Genset, France. Oligonucleotides used as probes for the bandshifts were from MWG-BIOTECH, Germany.

Cells

Peripheral blood leukocytes (PBL) were obtained after Ficoll gradient. To obtain pure T and B lymphocytes, the cells were sequentially incubated with anti-CD2- and anti-CD19-coupled magnetic beads. The cells were cultured overnight in RPMI 1640 supplemented with 10% FCS culture medium as described (1), in the presence or absence of 10 ng/ml PMA and 1 μM ionomycin. For RT-PCR, the cells were lysed in 300 μl of a guanidium HCN-sarcosyl-citrate solution (2) and stored at -20°C until further processing. The B-CLL cells used have been described elsewhere (11).

RT-PCR

RNA was prepared according to Chomczinsky and Sacchi (2) and submitted to reverse transcription (RT) in a volume of 25 μl as described (3). 500 nl of the RT mixture were submitted to PCR in a volume of 10 μl in the presence of 100 μl of each deoxynucleotide, 25 U/ml Taq DNA polymerase and 500 nM of the upstream and downstream primers in a Perkin-Elmer-Cetus GeneAmp PCR system 9600 (Perkin-Elmer-Cetus, Emereyville, CA). Paraffin oil was added to avoid evaporation. Cycling conditions were 10" at 95°C for cDNA denaturation, 15" at 59-66°C for primer annealing, and 15"-25" at 75°C for processing according to the length of the amplified sequence. Half of the PCR mix was electrophoresed on a 2.5% agarose gel containing ethidium bromide and visualized under UV illumination. The primers used to detect (human) cDNA for all the genes tested but bcl-2 and elf-1, have been described elsewhere (3). The sequences (5'-3') for bcl-2 were: upstream, GCACTTC-TCCCGCCGCTACCGC and downstream, AGGCCGCATGCTGGGGCCGTAC; for elf-1 were: upstream, AGCCTAA-GAATGGCACTCCCAC and downstream, CAACACAAGTTTACTAATG-GACCG. With the exception of elf-1, all the primers used in this study were located on different exons, allowing to discriminate between genomic or cDNA amplification.

Electrophoretic Mobility Shift Assays (Bandshifts)

These were performed as described in (10 and refs. therein). Proteins were extracted as described in (12). Briefly, total cell extracts were obtained by incubating the cells on ice for 30 min. (with occasional vortexing), in 3-4 x pellet volume of a buffer containing 60 mM KCl, 15 mM NaCl, 15 mM Hepes pH 7.8, 14 mM 2-ME, 0.5 mM PMSF, 0.15 mM spermine and 0.5 mM spermidine, followed by the addition of a buffer containing 12% glycerol, 2 M KCl, 12 mM Hepes pH 7.8, 0.12 mM EDTA, 5 mM $MgCl_2$, 5 mM DTT and 0.1% Triton X-100, to give a final KCl concentration of 0.4M. The cell lysates were centrifuged at 100,000 g to sediment chromatin and the supernatant was dialyzed against KIB buffer (10). Protein concentration was determined using the Bradford assay (BioRad). The following oligonucleotide probes we used (5' to 3'): NF-kB: AGTTGAGGGATTTCACTT; NF-ATd

Table 2. RT-PCR analysis of the patient's PBL before treatment or of purified CD2 and CD19 cells after/during treatment

Gene (No. of cycles)	Healthy donor		Patient					
			Before treatment		After/during treatment			
	CM	PMA/IONO	CM	PMA/IONO	CM		PMA/IONO	
					CD2	CD19	CD2	CD19
β2-micro(25)	+++	+++	+++	+++	+++	++	+++	+++
IL-2 (25)	-	+++	+	++++	-	-	-	-
IL-4 (40)	-	+	+++	+++	-	-	+++	-
IL-6 (40)	-	-	++	+++	-	+	-	+
IL-1α (40)	-	-	++	+++	-	-	-	-
IL-1β (40)	-	-	++	+++	-	+	-	+
Elf-1 (40)	-	-	++++	++++	-	-	-	-
bcl-2 (40)	-	-	++	++	-	-	-	-
IgM-ct. (40)	++	+	+++	+++	—	+	-	+++
IgM-var.(40)	-	+	++++	++++	-	+	-	+
IgE 40)	-	-	-	-	-	-	-	-
IgG (40)	-	-	-	-	-	-	-	-
IgA 40)	-	-	-	-	-	-	-	-

(distal): AAGAAAGGAGGAA-AAACTGTTCATACAG; NF-ATp (proximal): ACAGAA-GAGGAAAACAAAGGTAATGCGTT; AP-1: GT-GACTCAGCGCG.

RESULTS AND DISCUSSION

Peripheral blood leukocytes isolated from the patient before treatment and phenotyped by FACS analysis were foung to be CD3+CD4+CD8+ and were analyzed for mRNA expression as depicted in Table 2 and Figure 3A. After overnight culture in plain medium, these cells were found to synthesize constitutively IL-1a and β, IL-2, IL-4 and IL-6 mRNA as well as elf-1 and bcl-2 mRNA. All these genes were superinduced when the cells were cultured overnight with phorbol myristate acetate (PMA) and the calcium ionophore ionomycin. Although it is possible that because of the high sensitivity of the PCR analysis genes were detected which were expressed in a few contaminating non-T-cells, nevertheless, the results are in contrast with findings in normal peripheral blood mononuclear cells which do not express these cytokine genes in their resting stage. T-cells may express IL-2, IL-3, IL-4, IL-5, IL-6, IL-10, IL-13, IFNγ and TNFβ (reviewed in 8 and 13), under specific in vivo conditions. In vitro, following a mitogenic stimulation such as PMA and ionomycin, normal T-cells usually express IL-2 and/or IL-4 and, to a certain extent, IL-6 and other cytokines, depending on the type of T-cells and on culture conditions (9). The human T-cell leukaemia Jurkat, used extensively by many researchers as a model to study aspects of T-cell activation (14, 15) also does not express any of the above genes if unstimulated. Raziuddin et al (32) have found high IL-4, IFNγ and sIL-2R serum levels in T-CLL. It has been reported elsewhere

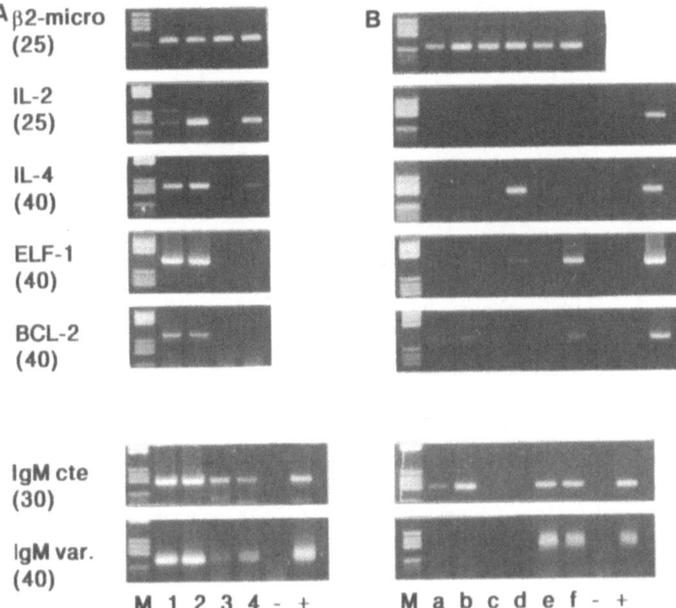

Figure 3. RT-PCR analysis of leukaemic PBL before treatment, and of purified leukaemic CD2 and CD19 cells after/during treatment. (A) Non-stimulated (1) or PMA and ionomycin (IONO)-stimulated (2) MC from the peripheral blood of the patient are compared to resting (3) or PMA/ionomycin-stimulated normal CD4 lymphocytes (4). 10^4 leukaemic cell-equivalents were loaded on the gel and compared to 10^3 normal CD4 cells. b2-micro. = b2-microglobuline; IgM ct. = transcript of the IgM constant region; IgM var. = transcript of the IgM variable region (this latter PCR amplification yields a single band when the cDNA analyzed is monoclonal for the V region and a smear, due to the variability in length of the cDNA, when it is heterogeneous (polyclonal) for V regions). Note that the sharp band in (1) and (2), suggesting monoclonality of IgM, as compared to normal MC (+) giving rise to a polyclonal smear. The few B lymphocytes contaminating the normal CD4 cells purified on Dynabeads (3) and (4), also produce a signal for IgM amplification, demonstrating the high sensitivity of the RT-PCR analysis. In brackets is the number of PCR cycles; M = molecular weight markers from Boehringer; (-) PCR negative control. (B) Purified CD19 (a) and (b) or CD2 (c) and (d), were compared to normal CD19 B lymphocyte-equivalents. (b), (d) and (f) correspond to PMA/ionomycin-induced cultures. Other symbols are as in (A).

that normal (1, 16) or B-CLL peripheral blood B-cells (11) or human EBV-transformed B clones (3, 12) can be induced in vitro and under specific culture conditions to express cytokine genes which are not normally expressed in B cells, such as IL-1α and β, IL-2 and IL-4 (in addition to TNFα, IL-6, IL-10 and TGFβ1, see ref. 1). Rambaldi et al (33) reported stable IL-1β, IL-6 and TNFα transcripts in ex-vivo B-CLL cells.

The PBL from the patient also expressed constitutively elf-1 and bcl-2: Elf-1 is the latest identified member of the ETS family of proto-oncogenes and, in T-cells, it seems to be involved in the formation of transcription factor complexes that trans-activate the IL-2, IL-2 receptor and IL-4 promoters upon cellular induction (20, 24). ETS members have been implicated in leukemogenesis (27). Bcl-2 is known to be an anti-apoptotic gene heavily implicated in neoplasia (26, 31). PCR analysis also showed an IgM monoclonality which suggested the presence of a B-cell neoplasia (see Figure 3A).

Analysis of the lymphotropic transcription factors NF-AT (=nuclear factor of activated T-cells), NF-κB and AP-1 (10, 13-15,17-19, 21, 22, 25, 28, 30) in cellular lysates of non-stimulated or mitogenicaly induced PBL from the patient by electrophoretic mobility shift assays (bandshifts), see Figure 4, gave results identical to those obtained with the T-cell

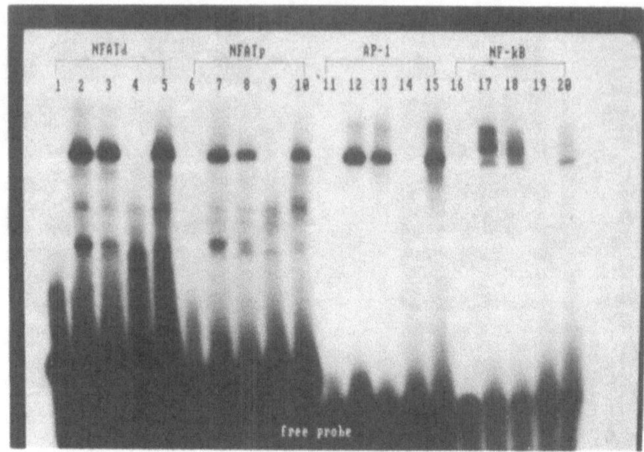

Figure 4. Key lymphotropic transcription factors in the patient's PBL as compared with B-CLL cells. Protein extraction and probes used are in materials and methods. Lanes 1, 6, 11 and 16: free probes; lanes 2, 7, 12 and 17: proteins from unstimulated B-CLL B-cells; lanes 3, 8, 13 and 18: proteins from stimulated B-CLL B-cells; lanes 4, 9, 14 and 19: proteins from unstimulated PBL of the patient; lanes 5, 10, 15 and 20: proteins from stimulated cells of the patient.

leukaemia Jurkat. In more detail, no complex formation was observed when proteins isolated from the PBL of the patient were incubated with the NF-AT probes, distal or proximal. The NF-AT complex was formed with proteins from PMA/ionomycin-stimulated cells. The same was observed with the AP-1 probe. The NF-AT complex in the T-cell leukaemia Jurkat is formed by AP-1 proteins which are synthesized de novo when the cells are activated, and a constitutive cytoplasmic protein subunit which alone does not bind to the probe (15, 17, 18, 22, 25). In normal T-cells the situation is different for NF-AT (10, 34) but the same for AP-1 (10 and refs. therein). The NF-κB complex was not formed with proteins from unstimulated cells from the patient, but it was formed from proteins extracted from stimulated cells. These results come as no surprise because in T-cells, like in most cell types including pre-B cells, the transcription factor NF-κB (p50/p65) is a ubiquitous and constitutive cytoplasmic factor complexed with its inhibitor IκB in an inactive form. Phosphorylation events involving protein kinase C inactivate IκB, resulting in the release of NF-κB which then migrates to the cell nucleus and binds to DNA. By contrast, NF-κB-like activity is constitutively present in mature B-cells and it has been suggested that in these cells there can be a cytoplasmic p50/p65, which can be translocated to the nucleus after PMA induction, and nuclear p50/c-rel at the same time (21). In Fig. 4 we also show bandshifts obtained with the same probes and proteins extracted from pure B-CLL B-cells from another patient (11). As it can be seen, all transcription factors investigated were present and exhibited binding activities in the B-CLL cells. NF-AT exists in non-T-cells also, and it is constitutively expressed in normal or immortalized B-cells (12, 19); nevertheless, it is not clear whether NF-AT in B-cells, normal or neoplastic, plays an important functional role by itself. AP-1 is also constitutively expressed in B-CLL cells (this work) and, as expected (12, 21, 29), the transcription factor NF-κB was present in both unstimulated and stimulated B-CLL cells (this work). NF-κB activity has been implicated as a putative causative agent in B-CLL (29). Our own transfections of B clones or B-CLL B-cells with recombinant plasmids containing various oligonucleotides corresponding to the NF-κB or NF-AT or AP-1 promoter elements linked to either a truncated, inactive thymidine kinase promoter or to a truncated, inactive IL-2 promoter, showed that only the NF-κB element was a strong, inducible and indispensable enhancer (11, 12).

The results from the bone marrow trephine biopsy (Fig. 2, see also case report) showed a co-expression of MT1 and MB1 by small lymphocytes, a fact that indicates their B-cell origin (6, 7). However, the relatively high percentage of UCHL-1 positive cells suggested the synchronous presence of T-cells.

The patient was given a treatment with alkylating agents (chlorambucil) and went into remission. Three months later, and while the patient was still being treated, we re-analyzed his PBL. This time, his lymphocytes had a normal morphology (see Fig. 1B). The results from the FACS analysis are shown in Table 1. During these 3 months his T-cells had diminished, overtaken by B-cells. We purified his T- and B-cells and re-analyzed gene expression by PCR in unstimulated and PMA/ionomycin-stimulated cells. The results (Table 2 and Fig. 3B) showed that his T-cells had ceased to synthesize any of the genes tested with the exception of IL-4 when induced. We offer the hypothesis that the surviving circulating T-cells were Th2-like memory cells which could confer some protection to the patient who did not present a clinical picture of immunosuppression. The patient's B-cells, expressed only IL-6 and IL-1β when PMA/ionomycin-stimulated, an occurrence which is in accordance with previously published studies showing the expression of IL-1β, IL-6 and TNFα by B-CLL cells (33). The B-cells remained monoclonal for IgM.

It is an interesting case and we shall continue to follow it up, but we are considering the following hypotheses for the development of this disease up to now: (i) It was a B-CLL from the beginning and the T-cell abnormalities regarding numbers and phenotype may reflect a strong immune response to a developing malignant B-cell clone, resulting in its inhibition; (ii) it was a T-CLL with the beginning of a B-CLL, and T-cell derived cytokines led to an uncontrolled proliferation of the malignant B clone; (iii) it was an unusual T-cell malignancy overlaying a B-CLL which (T-cell malignancy) was easily abolished by the administration of clorambucil ; (iv) dual lymphoid neoplasms may occur by chance: Harland et al (35) has reported 4 cases of CTLC with B-CLL.

REFERENCES

1. Matthes T, Werner-Favre C, Tang H, Zhang X, Kindler V, Zubler RH: Cytokine mRNA expression during in vitro response of human B lymphocytes: Kinetics of B-cell tumor necrosis factor a, interleukin (IL)-6, IL-10 and transforming growth factor b1 mRNAs. J Exp Med 178:521, 1993
2. Chomczynski P, Sacchi N: Single-step method of RNA isolation by acid guanidium thiocyanate-phenol-chlorophorm extraction. Anal Biochem 162:156, 1987
3. Tang H, Matthes T, Carballido-Perrig, Zubler RH, Kindler V: Differential induction of T-cell cytokine mRNA in Epstein-Barr virus-transformed B-cell clones: Constitutive and inducible expression of interleukin-4 mRNA. Eur J Immunol 23:899, 1993
4. Trainor K, Brisco MJ, Story CJ, Morley AA: Monoclonality in B-lymphoproliferative disorders detected at the DNA level. Blood 75:2220, 1990
5. Shi ZR, Itzkowitz SH, Kim YS: A comparison of three immunoperoxidase techniques for antigen detection in colorectal carcinoma tissues. J Histochem Cytochem 36:317, 1988
6. Poppema S, Hollema H, Visser L, Vos H: Monoclonal antibodies (MT1, MT2, MB1, MB2, MB3) reactive with leukocyte subsets in paraffin-embedded tissue sections. Am J Pathol 127:418, 1987
7. Dobson CM, Myskow MW, Krajewski AS, Carpenter FH, Horne CHW: Immunohistochemical staining of non-Hodgkin's lymphoma in paraffin sections using MB1 and MT1 monoclonal antibodies. J Pathol 153:203, 1987
8. Kuby J (ed.): Immunology, 2nd edition, W. H. Freeman and Co., New York, 1994, p 2971
9. Ehlers S, Smith KA: Differentiation of T-cell lymphokine gene expression: The in vitro acquisition of T-cell memory. J Exp Med 173:25, 1991

10. Mouzaki A, Rungger D: Properties of transcription factors regulating interleukin 2 gene transcription through the NF-AT binding site in untreated or drug-treated naive and memory T helper cells. Blood 84:2612, 1994
11. Mouzaki A, Matthes T, Miescher PA, Beris Ph: Polyclonal hypergammaglobulinaemia in a case of a B-cell Chronic Lymphocytic leukaemia: The result of IL-2 production by the proliferating monoclonal B-cells? British J Haematol (in press)
12. Mouzaki A, Serfling E, Zubler RH: Interleukin-2 promoter activity in Epstein-Barr virus-transformed B lymphocytes is controlled by nuclear factor-kB. Eur J Immunol (in press)
13. Kishimoto T (ed): Interleukins: Molecular Biology and Immunology, in: Chemical Immunology, vol. 51, Karger, Basel, 1992
14. Fujita T, Shibuya H, Ohashi T, Yamanishi K, Taniguchi T: Regulation of interleukin-2 gene: functional DNA sequences in the 5' flanking region for the gene expression in activated T lymphocytes. Cell 46: 403, 1986
15. Shaw JP, Utz PJ, Durand DB, Toole JJ, Emmel EA, Crabtree GR: Identification of a putative regulator of early T-cell activation genes. Science 241: 202, 1988
16. Kindler V, Matthes T, Jeannin P, Zubler RH: IL-2 secretion by human B lymphocytes occurs as a late event and requires additional stimulation after CD40 cross-linking. Eur J Immunol 25:1239, 1995
17. Rao A: NF-ATp: a transcription factor required for the co-ordinate induction of several cytokine genes. Immunology Today 15:274, 1994
18. Rooney JW, Hodge MR, McCaffrey PG, Rao A, Glimcher LH: A common factor regulates both Th1- and Th2-specific cytokine gene expression. EMBO J 13:625, 1994
19. Yaseen NR, Maizel AL, Wang F, Sharma S: Comparative analysis of NF-AT (nuclear factor of activated T-cells) complex in human T and B lymphocytes. J Biol Chem 268:14285, 1993
20. Thompson CB, Wang C-Y, Ho I-C, Bohjanen PR, Petryniak B, June CH, Miesfeldt S, Zhang L, Nabel GJ, Karpinski B, Leiden JM: cis-Acting sequences required for inducible interleukin-2 enhancer function bind a novel Ets-related protein, Elf-1. Mol Cel Biol 12:1043, 1992
21. Liou H-C, Baltimore D: Regulation of the NF-kB/rel transcription factor and IkB inhibitor system. Current Opinion in Cell Biol 5:477, 1993.
22. Northorp JP, Ho SN, Chen l, Thomas DJ, Timmerman LA, Nolan GP, Admon A, Crabtree GR: NF-AT components define a family of transcription factors targeted in T-cell activation. Nature 369:497, 1994
23. Rai KR, Patel DV: Chronic lymphocytic leukemia, in Hoffman R, Benz EJ, Shattil SJ, Furie B, Cohen HJ, Silberstein LE (eds): Hematology, basic principles and practice, New York, Churchill Livingstone,1995
24. John S, Reeves RB, Lin JX, Child R, Leiden JM, Thompson CB, Leonard WJ: Regulation of cell-type-specific interleukin-2 receptor alpha-chain gene expression: potential role of physical interactions between Elf-1, HMG-I(Y) and NF-kappa B family proteins. Mol Cell Biol 15:1786, 1995
25. Nolan GP: NF-AT-AP-1 and Rel-bZIP: Hybrid vigor and binding under the influence. Cell 77:795, 1994
26. Vaux DL, Cory S, Adams JM: Bcl-2 gene promotes haemopoietic cell survival and cooperates with c-myc to immortalize pre-B-cells. Nature 335:440, 1988
27. Thompson CB, Brown TA, McKnight SL: Convergence of ETS- and notch-related structural motifs in a heteromeric DNA binding complex. Science 253:762, 1991
28. Kang S-M, Beverley B, Tran A-C, Brorson K, Schwartz RH, Lenardo MJ: Transactivation by AP-1 is a molecular target of T-cell clonal anergy. Science 257:1134, 1992
29. Zaknoen SL, Christian SL, Suen R, Van Ness B, Kay NE: B-chronic lymphocytic leukemia cells contain both endogenous k immunoglobulin mRNA and critical immunoglobulin gene activation transcription factors. Leukemia 6:675, 1992
30. Baltimore D, Beg AA: A butterfly flutters by. Nature 373:287, 1995
31. Hanada M, Delia D, Aiello A, Stadtmauer E, Reed JC: bcl-2 gene hypomethylation and high-level expression in B-cell chronic lymphocytic leukemia. Blood 6:1820, 1993
32. Raziuddin S, Sheikha A, Abu-Eshy S, Al-Janadi M: Circulating levels of cytokines and soluble cytokine receptors in various T-cell malignancies. CANCER 73:2426, 1994
33. Rambaldi A, Bettoni S, Rossi V, Tini M-L, Giudici G, Rizzo V, Bassan R, Mantovani A, Barbui T, Biondi A: Transcripitonal and post-transcriptional regulation of IL-1b, IL-6 and TNF-a genes in chronic lymphocytic leukaemia. Br J Haematol 83:204, 1993
34. Mouzaki A, Rungger D: Interleukin 2 gene regulation at the Pud (NF-AT) promoter element: The complex situation in primary T lymphocytes. Molecular Biology of Hematopoiesis 3:475, 1994
35. Harland CC, Whittaker SJ, NG YL, Holden CA, Wong E, Smith NP: Coexistent cutaneous T-cell lymphoma and B-cell chronic lymphocytic leukemia. Br J Dermatol 127:519, 1992

EXPLOITATION OF FREQUENT p16 DELETION IN THE TREATMENT OF T CELL ACUTE LYMPHOBLASTIC LEUKEMIA

A. L. Yu, J. Chen, M. B. Diccianni, A. Batova, and J. Yu

University of California, San Diego Medical Center
200 West Arbor Drive
San Diego, California 92103-8447

ABSTRACT

Methylthioadenosine phosphorylase (MTAP) is essential for the salvage of both adenine and methionine. Deficiency of MTAP has been found in a variety of cancers including acute lymphoblastic leukemia (ALL). Recently, the MTAP gene has been mapped to the close proximity of tumor suppressor genes p16 and p15 which code for inhibitors of the cyclin-dependent kinases 4 and 6. We found that p16/p15 genes are frequently co-deleted in leukemic samples obtained from patients with T-cell ALL (T-ALL). Alteration of p16 gene was found in 30/49 (61%) and 22/34 (65%) of diagnosis and relapse samples, respectively. Among those samples with p16 deletion, p15 was deleted in 19 of 24 (79%) samples studied, and MTAP gene is deleted in 20/38 (53%). The finding of high frequency of MTAP deficiency in T-ALL offers an opportunity for the design of biochemically selective therapy for T-ALL. We studied the effect of methionine depletion in a T-ALL cell line, CEM, in which p16 and MTAP genes are deleted. Incubation of CEM in methionine deficient medium resulted in an initial growth inhibition followed by gradual cell death. In contrast, methionine depletion had no significant effect on the viability of normal blood mononuclear cells or proliferative response of normal T lymphocytes to PHA. In the presence or absence of methionine, the stimulation index was 90.1 ± 8.1 and 75.9 ± 7.8, respectively at 1 μg/ml PHA, and 63.7 ± 2.3 and 69.5 ± 3.9 at 2.5 μg/ml PHA. Although MTAP (-) CEM cells appear to be as sensitive as the MTAP (+) MOLT-4 cells to alanosine, an inhibitor of AMP synthesis, addition of methythioadenosine, a substrate of MTAP, protected the MTAP (+) MOLT-4 cells but not the MTAP (-) CEM cells from alanosine cytotoxicity. These findings suggest the possibility of targeting MTAP for selective therapy of T-ALL.

INTRODUCTION

Recently, a novel tumor suppressor gene, p16 ($p16^{INK4A}$, MTS1, CDKN2) has been identified by two independent groups. It encodes for inhibitors of the cell cycle regulators,

Molecular Biology of Hematopoiesis 5, edited by Abraham et al.
Plenum Press, New York, 1996

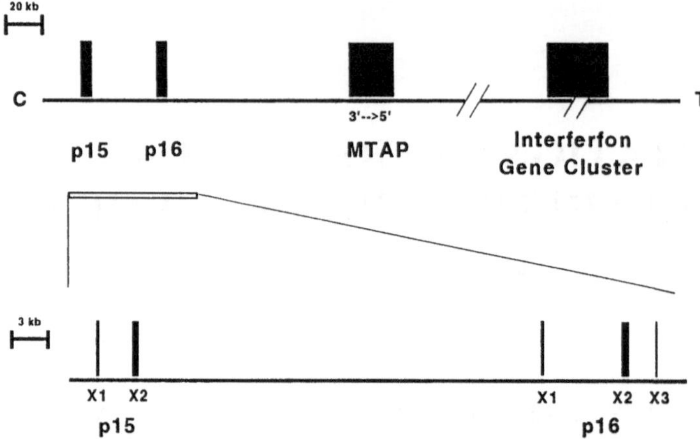

Figure 1. Genes mapped to the region of chromosome 9p21. Both the p16 gene and its homologous neighboring gene, p15, are mapped to chromosome 9p21, which is a region of frequent chromosomal alterations in ALL and other cancers. The gene encoding MTAP is located telomeric to p16.

cyclin-dependent kinases 4 and 6 (CDK4, CDK6)[1,2] and resides on chromosome 9p21 (Fig. 1), a site of frequent chromosomal abnormalities in various human tumors. Centromeric to p16 on this chromosome is a related gene, p15 ($p15^{INK4B}$, MTS2)[3]. Like p16, p15 also inhibits CDK4 and CDK6 kinases. The first 50 amino acids of p15 and p16 share 44% identity, and the next 81 amino acid residues share 97% identity, after which they diverge, with p16, but not p15, having a third exon.

A high frequency of p16 gene deletion and mutations have been found in a wide-variety of tumor cell lines.[1,2] Initially, this high frequency of p16 alterations was not found when primary tumors of lung, brain, bladder, kidney, head and neck,[4,5] and breast[5] were studied. This has led to the speculation that deletions of p16 may be an artifact of cell culture and that p16 is not a true tumor suppressor gene. However, recent studies have shown frequent p16 deletion/mutation in a variety of primary tumors,[6-11] and leukemia.[12-17] Germ-line p16 mutations have been found in some familial melanoma kindreds[18]. Furthermore, transfected p16 cDNA has been shown to inhibit the growth of glioma cell lines deleted for p16[19]. Taken together, these findings substantiate the role for p16 as an important tumor suppressor gene and suggest that abnormal regulation of CDKs may play a crucial role in tumorigenesis. The presence of p15 as a second functional member of the p16 family at chromosome 9p21 raises the possibility that loss of tumor suppression may involve inactivation of either or both genes. A few recent reports have shown that p16-deletions are often accompanied by co-deletion of the p15 gene, albeit with a lesser frequency.[12,20-21] The exact role of the p15 gene in cancer has yet to be defined. Furthermore, the prognostic significance of p16 in cancer remains to be delineated.

Methylthioadenosine Phosphorylase (MTAP) is an important salvage enzyme for both adenine and methionine. The MTAP gene has also been mapped to chromosome 9p21 in close proximity of p15 and p16 genes (Fig. 1), suggesting the possibility that the MTAP gene may also be frequently deleted in human cancers. During the synthesis of polyamines, 5'-deoxy-5'-methylthioadenosine (MTA) is generated, which is rapidly cleaved by the ubiquitous enzyme MTAP into adenine and 5-methylthioribose-1-phosphate (MTR-1-P) (Fig. 2).[22] Adenine is efficiently salvaged to form AMP by adenine phosphoribosyltrans-

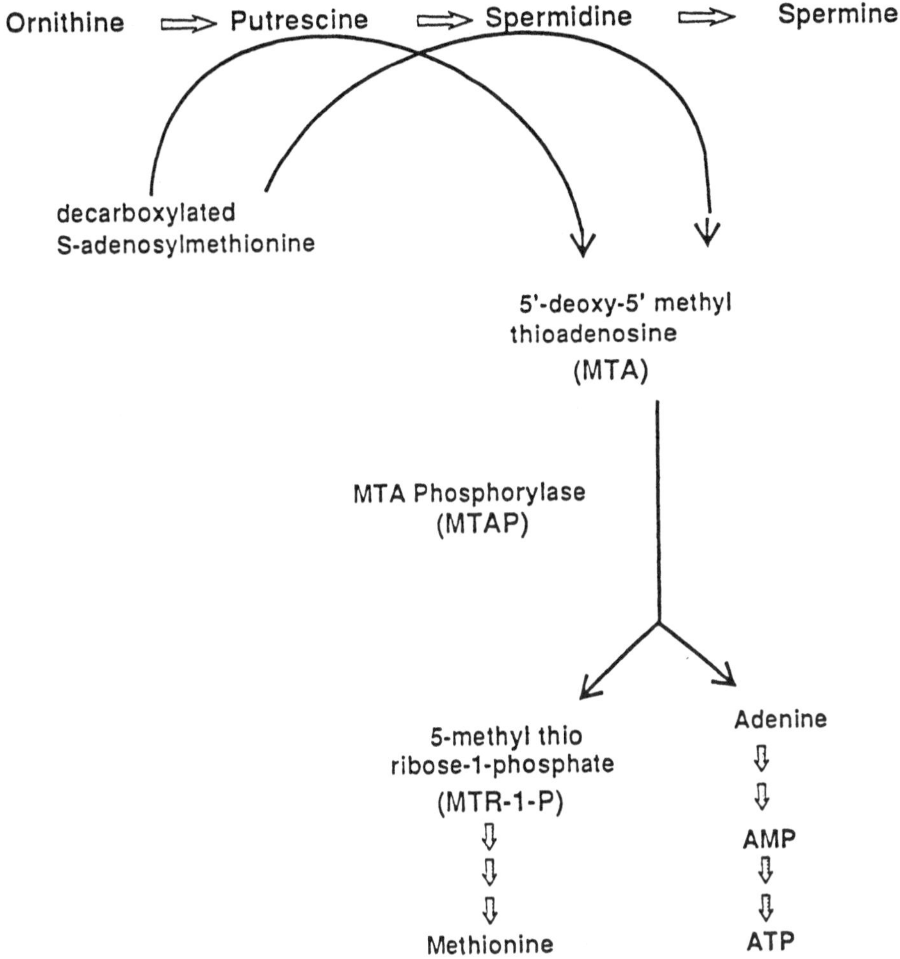

Figure 2. Role of MTAP in the salvage of adenine nucleotides from MTA.

ferase (APRT)[23] and MTR-1-P is converted to methionine by a complex oxidation via the intermediate 2-keto-4-oxo-*S*-methylthiobutyrate.[24,25]

In 1977 Toohey described four murine malignant cell lines that lacked MTAP activity, including L1210 and P388.[26] Subsequently, we reported MTAP deficiency in 2 of 20 fresh leukemia samples including one from a T-ALL patient, suggesting that the enzyme deficiency was not merely an artifact of cell culture.[27] Besides ALL, MTAP deficiency has now been documented in acute non-lymphoid leukemia,[28,29] melanoma, lung cancer, rectal adenocarcinoma,[28] among others.[30-33] In MTAP deficient cells, the salvage pathway producing methionine and adenine would be blocked, thus resulting in an increased dependency on an exogenous supply of these nutrients. Taking advantage of this metabolic difference between normal and cancer cells, we speculate that an enzyme-selective chemotherapy may be developed in which MTAP deficient cancer cells would be killed with drugs which deplete purine nucleotides and/or methionine under conditions in which MTAP (+) normal cells would be rescued with MTA as a source for purines and methionine.

In the present study, we examined the prevalence of MTAP deletions, along with p16/p15 deletions in T-ALL and determined the sensitivity of an MTAP-deficient T-ALL cell line, CEM, to purine synthesis inhibitor and methionine depletion.

METHODS

PCR

PCR amplification of p16 exon 2 was performed using 50 ng of genomic DNA in a 50 µl reaction volume containing: 10 mM Tris, pH 8.3, 25 mM KCl, 2 mM $MgCl_2$, 50 mM 4dNTP, 1% DMSO, 1 unit Taq DNA polymerase and 10 pmol each sense and antisense primer. Following an initial denaturation at 95°C for 3 minutes, amplification (quantitative for p16 exon 2) proceeded for 1 min each at 95°C, 62°C and 72°C for 30 cycles. For SSCP (single strand conformational polymorphism) analysis,[34] amplification was maximized for product yield by extending the number of cycles to 45. Sequencing was performed on agarose gel purified PCR products using the dsDNA cycle sequencing system (GIBCO BRL).

PCR amplification products were resolved on a 2% agarose gel in 1X TAE buffer, genomic DNA on a 0.7% agarose gel in 1X TAE buffer, SSCP products on a 6% non-denaturing polyacrylamide gel in 1X TBE and sequencing products on an 8% polyacrylamide/7M urea gel in 1X TBE. PCR primers and optimal conditions for amplification of exon 8 of the MTAP gene were generously provided by Dr. T. Nobori.

Primers

p16 exon 2

DGD1 (sense)	5'-GCTCTGACCATTCTGTTCTC-3'
X2A2 (antisense)	5'-GGGCTGAACTTTCTGTGCTGG-3'

Southern Blotting

10 µg of genomic DNA was digested with Eco RI and resolved on a 0.9% agarose gel in 1X TBE. The DNA was depurinated in 0.25 M HCl and denatured in 0.5M NaOH/1.5M NaCl. The gel was neutralized in 0.5M Tris-HCl, pH 7.2/1.5 M NaCl/1 mM EDTA and DNA transferred to HyBond (Amersham) by capillary blotting followed by UV crosslinking. The blot was prehybridized for 4 h at 60°C in 1M NaCl/1% SDS, then hybridized with p16 and p15 cDNA's (generous gifts from D. Beach) in 1M NaCl/1% SDS/10% Dextran Sulfate/200 mg/ml Salmon sperm DNA/30 mg/ml tRNA. p16 and p15 cDNA's were labeled by random priming to a specific activity of 1-3 X 10^9 cpm/µg with α^{32}P-dATP using a random primed kit (Boehringer Mannheim) according to the manufacturer's directions. Following an overnight hybridization at 60°C, membranes were washed twice with 2X SSC/1% SDS for 15 minutes each and then once with 0.1X SSC at 60°C for 15 minutes. p16 and p15 specific signals were visualized by phosphorimager.

RESULTS AND DISCUSSION

1. p16/p15 Deletions are Common in T-ALL at Diagnosis and Relapse

We analyzed T-ALL samples for p16/p15/MTAP deletion by Southern Blotting (Fig. 3) and PCR amplification. PCR offers the advantages of being able to analyze patient

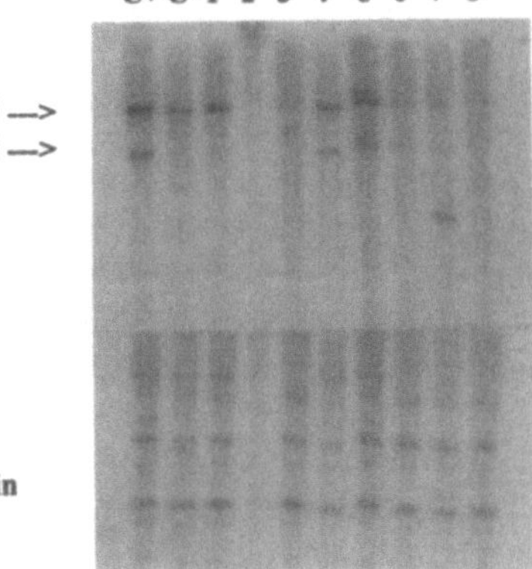

Figure 3. Southern blot analysis of p16 and p15 in relapse and diagnosis T-ALL. DNA from T-ALL cells was isolated and digested with Eco R1. Controls are as follows: C+, p16/p15 positive control (mononuclear cell DNA from a normal person); C-, p16 negative/p15 positive control (CEM cell line DNA). Patients samples are shown in lanes 1-4 (diagnosis T-ALL) and lanes 5-8 (relapse T-ALL). Actin is shown as a control. Patient ID and the percent blasts are as follows: 1) D3, 94%; 2) H2*, 90%; 3) D1, 80%; 4) C3, 99%; 5) C2, 87%; 6) J1, 86%; 7) B1, 90%; 8) G1, 76%. *Poor digestion of this sample makes the result inconclusive. However, subsequent southern blot and quantitative PCR showed this sample is deleted in p16, p15 and MTAP.

samples in which DNA quantity was insufficient for Southern blot analysis, as well as being able to verify the Southern blot results. The p16 mutations were detected by SSCP analysis,[34] followed by DNA sequencing. The representative results are shown in Table 1.

In diagnosis T-ALL, 28 of 49 (57%) patients demonstrated p16 gene deletions, one patient exhibited point mutation and one additional patient had a shifted band, suggestive of gene rearrangement. Thus, 61% (30 of 49) of diagnosis patients analyzed demonstrated p16 gene alterations. Similarly, among the relapse T-ALL, 20 of 34 patients had p16 deletion. One patient had a point mutation, and one additional patient showed gene rearrangement. Thus, 22 of 34 (65%) of the relapse patients analyzed demonstrated p16 gene alterations. P15 was also frequently deleted in T-ALL; 9 of 18 diagnosis (50%) and 11 of 18 relapse

Table 1. Comparison of p16 and p15 deletions in T-ALL at diagnosis and relapse

Diagnosis					Relapse				
		Gene status[1]					Gene status[1]		
Patient	Blast %	p16	p15	MTAP	Patient	Blast %	p16	p15	MTAP
C3	99%	+	+	+	A1	80%	– (R?)	R	+
C4	72%	–	+	+	A2	90%	+	+	+
D1	80%	–	–	–	A3	89%	–	–	+
D2	94%	–	–	+	A4	95%	+	+	+
D3	94%	–	+	+	B1	90%	R	–	+
E1	82%	–	–	–	B2	76%	+	+	+
G2	86%	–	–	+	B3	89%	–	–	–
H1	99%	R	+	+	C1	93%	+	+	+
H2	90%	–	–	–	C2	87%	+	+	+
J2	93%	–	–	+	C5	86%	–	–	–
J3	97%	+	+	+	D4	60%	–	–	+
J4	92%	+	+	+	G1	76%	–	–	–
L1	98%	–	+	+	J1	86%	–	–	–
P2	99%	+	+	+	O1	98%	–	–	–
R2	69%	–	–	–	P1	66%	+	+	+
R3	89%	+	–	+	R1	48%	–	–	–
S2	95%	–	–	+	S1	76%	–	–	–
S5	98%	+	+	+	U1	34%	–	+	+

[1]Scored for p16/p15 present (+) or absent (-).
R; shifted (rearranged) band

(61%) T-ALL patients exhibited p15 gene alterations that included deletions and rearrangements.

As presented in Table 1, deletion of the p16 gene was frequently accompanied by co-deletion of the p15 gene. In 8 of 12 (67%) diagnosis patients deleted for p16, p15 was also co-deleted. In addition, three patients had p16 deletion but intact p15 and one patient had a rearranged p16 but an intact p15. Interestingly, in one patient (R3) p15 was deleted but p16 was intact. In relapse T-ALL, 11 of 12 (92%) patients exhibiting p16 gene alterations also exhibited alterations of the p15 gene. These included 9 patients co-deleted for both genes, one patient with a rearranged p16 and a deleted p15 and another patient with p16 deletion and a p15 gene rearrangement (see Table 1). The overall higher incidence of p16 deletion over p15 is consistent with the notion that p16 rather than p15 is the target of deletion in T-ALL. However, the finding of p15 deletion sparing p16, in at least one patient, implies that p15 may also play a pathogenetic role in T-ALL.

2. MTAP Gene Deletion in T-ALL

The genes for p15 and MTAP lie on adjacent but opposite sides of the p16 gene on the chromosomal region 9p21. Thus, it is anticipated that the MTAP genes may also be frequently co-deleted with p16 in T-ALL. We examined the possible deletion of the MTAP gene, just telomeric of p16, in diagnosis and relapse T-ALL samples by quantitative PCR amplification and Southern analysis. The results are partly shown in Table 1 and summarized as follows:

	Diagnosis		Relapse	
p15	9/18	(50%)*	11/18	(61%)*
p16	30/49	(61%)*	22/34	(65%)*
MTAP	12/38	(32%)	9/25	(36%)
p15 alone	1/18	(6%)	0/18	(0%)
p16 alone	4/18	(22%)	1/18	(6%)
MTAP alone	none		none	

*includes mutation and rearrangement

Furthermore, when p16 is deleted, then:

Diagnosis		Relapse	
p15 also deleted	8/12 (67%)	p15 also deleted	11/12 (92%)
MTAP also deleted	12/24 (50%)	MTAP also deleted	8/14 (57%)

The finding of higher incidence of p16 deletion than either p15 or MTAP, again, supports the contention of p16 being the main target of deletion. Our data also reveal that a substantial portion (more than one third) of T-ALL patients had MTAP gene deletion. Since T-ALL cells are readily accessible for in vitro studies, T-ALL is an ideal model for exploring MTAP- targeted therapeutic approaches.

3. Differential Sensitivity of MTAP-Deficient T-ALL Cell Lines to Alanosine

MTAP cleaves MTA into adenine and 5'-methythioribose-1-phosphate (MTR-1-P).[22] The latter is in turn salvaged to form methionine by a complex oxidation reaction. Thus, MTAP (+) cells deprived of adenine or methionine can be rescued by MTA, whereas

Table 2. Rescue of MTAP (+) MOLT-4 cells from alanosine toxicity by MTA

	MTA µM		
	0	20	40
MOLT-4	42%	73%	81%
CEM	44%	38%	37%

*The MTAP (+) MOLT-4 and MTAP(-) CEM cells were incubated in MEM medium containing 10% dialyzed horse serum and 10 µM alanosine in the presence or absence of MTA for 72 hrs. After incubation, the number of viable cells was determined by trypan blue dye exclusion and expressed as percent of viable cells relative to control

MTAP-deficient cells cannot. This implies that one may design selective anti-cancer therapies to exploit MTAP deficiency in cancer cells. To this end, we compared the effects of the adenine synthesis inhibitor, alanosine, and methionine depletion on an MTAP-deficient T-ALL cell line, CEM, and an MTAP (+) T cell line, MOLT-4.

Alanosine inhibits the conversion of IMP to AMP by the branch point enzyme adenosylsuccinate synthetase (ASS),[35] and is therefore an ideal agent for MTAP-targeted therapy. T-ALL cells were incubated with alanosine for 72 hours and the number of viable cells was counted. Preliminary data showed that the IC_{50} of alanosine for CEM and MOLT-4 was not significantly different (approximately 6.0 µM). As MTAP can generate adenine from MTA, cells that contain MTAP should survive inhibition of purine synthesis, if MTA or a suitable substrate analogue is provided as a source for adenine salvage. As shown in Table 2, in the presence of 10 µM alanosine, addition of MTA protected the MTAP (+) MOLT-4 cells but not the MTAP-deficient CEM cells from alanosine cytotoxicity. Since all normal tissues express MTAP,[27,36] these findings suggest that alanosine may have selective therapeutic potential for MTAP-deficient cancers.

4. Methionine Dependence of T-ALL Cells

As discussed above, methionine requirement of malignant cells and their dependence on methionine salvage from MTA can be used to selectively target MTAP-deficient cells. We compared the impact of methionine depletion on the growth of malignant T-ALL cells, CEM, and on the proliferation of phytohemagglutin (PHA) stimulated normal T-lymphoblasts. As shown in Fig. 4, when CEM cells were incubated in methionine-free Dulbecco's MEM with 10% dialyzed horse serum, there was an immediate growth arrest followed by gradual decline of viable cells. In contrast, methionine depletion had little effect on the viability of normal T-lymphocytes as well as their proliferative response to PHA, other than a slight decrease in peak response and mild increase in responses to suboptimal dose of PHA (data not shown). Thus, in the presence or absence of methionine, the stimulation index was 90.1 ± 8.1 and 75.9 ± 7.8, respectively at 1 µg/ml PHA, and 63.7 ± 2.3 and 69.5 ± 3.9 at 2.5 µg/ml PHA. Similar pattern of responses were observed using lymphocytes from 3 different individuals. This is the first demonstration that mitogenic response of normal lymphocytes can proceed under methionine-free conditions. Furthermore, methionine depletion had no effect on the survival of normal un-stimulated lymphocytes. These findings support the earlier report of methionine dependence of malignant cells but not normal cells.

Next, we compared the ability of MTA to rescue MTAP-deficient and MTAP-positive cells from the deleterious effects of methionine depletion. CEM (MTAP-negative)

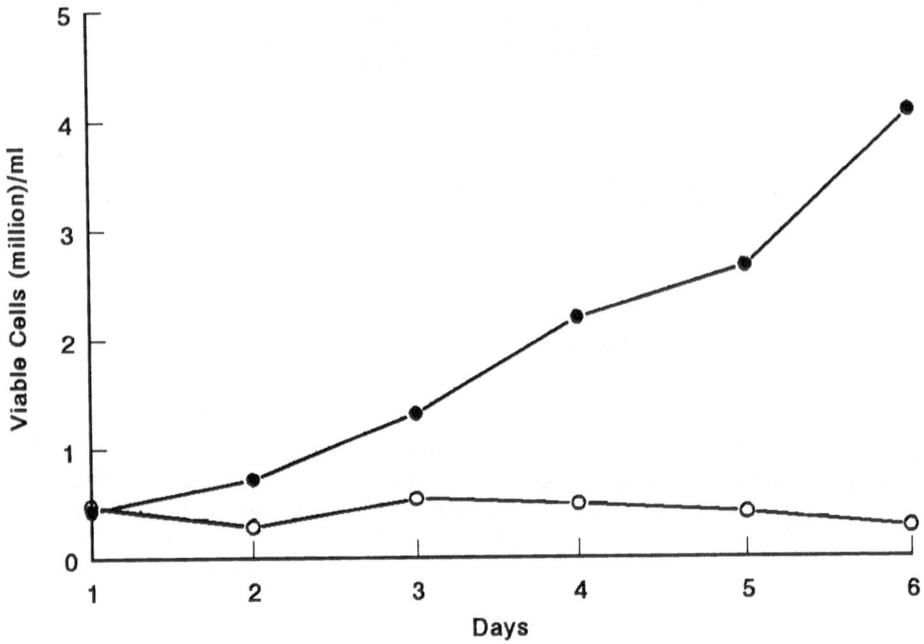

Figure 4. Effect of Methionine Depletion on Viability of CEM. Cells were plated at a density of 0.5 x 10^6 cells/ml in 24-well plates containing MEM with (●) or without (○) methionine supplemented with 10% dialyzed horse serum. Number of viable cells were determined by trypan blue exclusion on days indicated.

and MOLT-4 (MTAP-positive) cells were incubated in methionine-free medium containing 10% dialyzed horse serum and 0 to 20 μM MTA for four days, and the number of viable cells was determined. Although depletion of methionine was growth inhibitory for both CEM and MOLT-4, only the latter was partially protected by the addition of MTA. The lack of complete rescue of MOLT-4 cells is probably attributable to the undefined biochemical defect in converting MTR-1-P to methionine[37] as well as the toxicity of MTA at higher concentration, presumably due to inhibition of polyamine synthesis. These findings suggest that therapeutic strategies targeted at MTAP deficiency may indeed be very promising approach.

ACKNOWLEDGMENT

The authors would like to acknowledge the POG ALL cell bank at St. Jude Children's Research Hospital for providing some cryopreserved T-ALL samples at diagnosis. The authors would also like to thank Gregory Best for the excellent work in preparing the manuscript, and Thai Vu and Louis Bridgeman for technical assistance. Statistical analysis was supported in part by a grant from the General Clinical Research Centers M01 RR00827 of the National Center for Research Resources, National Institutes of Health. This work was supported by grants U10CA28439 and FDA 001129 (ALY) and DK 49888 (JY). MBD is supported by NIH postdoctoral training grant #T32 CA 09290. AB is supported by NIH grant for training and research on blood and blood disease #2T32-HL07107.

REFERENCES

1. Kamb A, Gruis NA, Weaver-Feldhaus J, Liu Q, Harshman K, Tavtigian SV, Stockert E, Day RS III, Johnson BE, Skolnick MH: A cell cycle regulator potentially involved in genesis of many tumor types. Science 436-440, 1994.
2. Nobori T, Miura K, Wu DJ, Lois A, Takabayashi K, Carson DA: Deletions of the cyclin-dependent kinase-4 inhibitor gene in multiple human cancers. Nature 368:753756, 1994.
3. Hannon GJ, Beach D: p15 is a potential effector of TGF-β-induced cell cycle arrest. Nature 371:257-261, 1994.
4. Zhang SY, Klein-Szanto AJP, Sauter ER, Shafarenko M, Mitsunaga S, Nobori T, Carson DA, Ridge JA, Goodrow TL: Higher frequency of alterations in the p16/CDKN2 gene in squamous cell carcinoma cell lines than in primary tumors of the head and neck. Can Res 54:5050-5053, 1994.
5. Xu L, Sgroi D, Sterner CJ, Beauchamp RL, Pinney DM, Keel S, Ueki K, Rutter JL, Buckler AJ, Louis DN, Gusella JF, Ramesh V: Mutational analysis of CDKN2 (MTS1/p16) in human breast carcinomas. Can Res 54:5262-5264, 1994.
6. Jen J, Harper JW, Bigner SH, Bigner DD, Papadopoulos N, Markowitz S, Willson JK, Kinzler KW, Vogelstein B: Deletion of p16 and p15 genes in brain tumors. Can Res 54:6353-6358, 1994.
7. Mori T, Miura K, Aoki T, Nishihira T, Mori S, Nakamura Y: Frequent somatic mutation of the MTS1/CDK41 (multiple tumor suppressor/cyclin-dependent kinase 4 inhibitor) gene in esophageal squamous cell carcinoma. Can Res 54:3396-3397, 1994.
8. Caldas C, Hahn SA, da Costa LT, Redston MS, Schutte M, Seymour AB, Weinstein CL, Hruban RH, Yeo CJ, Kern SE: Frequent somatic mutations and homozygous deletions of the p16 (MTS1) gene in pancreatic adenocarcinoma. Nature Genetics 8:27-32, 1994.
9. Washimi O, Nagatake M, Osada H, Ueda R, Koshikawa T, Seki T, Takahashi T, Takahashi T. In vivo occurrence of p16 (MTS1) and p15 (MTS2) alterations preferentially in non-small cell lung cancers. Can Res 55:514-517, 1995.
10. Okamoto A, Hussain SP, Hagiwara K, Spillare EA, Rusin MR, Demetrick DJ, Serrano M, Hannon GJ, Shiseki M, Zariwala M, et al.: Mutations in the p16INK4/MTS1/CDKN2, p15INK4B/MTS2, and p18 genes in primary and metastatic lung cancer. Can Res 55:1448-1451, 1995.
11. Cheng JQ, Jhanwar SC, Klein WM, Bell DW, Lee WC, Altomare DA, Nobori T, Olopade OI, Buckler AJ, Testa JR: p16 alterations and deletion mapping of 9p21-p22 in malignant mesothelioma. Can Res 54:5547-5551, 1994.
12. Okuda T, Shurtleff SA, Valentine MB, Raimondi SC, Head DR, Behm F, Curcio-Brint AM, Liu Q, Pui C-H, Sherr CJ, Beach D, Look AT, Downing JR: Frequent deletion of $p16^{INK4a}$/MTS1 and $p15^{INK4b}$/MTS2 in pediatric acute lymphoblastic leukemia. Blood 85:2321-2330, 1995.
13. Ogawa S, Hirano N, Sato N, Takahashi T, Hangaishi A, Tanaka K, Kurokawa M, Tanaka T, Mitani K, Yazaki Y, et al: Homozygous loss of the cyclin-dependent kinase 4-inhibitor (p16) gene in human leukemias. Blood 84:2431-2435, 1994.
14. Fizzotti M, Cimino G, Pisegna S, Alimena G, Quartatone C, Mandelli F, Pelicci PG, Lo Coco F: Detection of homozygous deletions of the cyclin-dependent kinase 4 inhibtor (p16) gene in acute lymphoblastic leukemia and association with adverse prognostic features. Blood 85:2685-2690. 1995.
15. Otsuki R, Clark HM, Wellmann A, Jaffe ES, Raffeld M: Involvement of CSKN2 ($p16^{INK4A/}$MTS1) and $p15^{INK4B}$/MTS2 om human leukemias and lymphomas. Can Res 55:1436-1440, 1995.
16. Cayuela JM, Hebert J, Sigaux: Homozygous ($p16^{INK4A}$) deletion in primary tumor cells of 163 leukemia patients. Blood 85:854, 1995.
17. Stranks G, Height SE, Mitchell P, Jadayel D, Yuille MAR, De Lord C, Clutterbuck RD, Treleaven JG, Powles RL, Nacheva E, Oscier DG, Karpas A, Lenoir GM, Smith SD, Millar JL, Catovsky D, Dyer MJS: Deletions and regarrangement of CDKN2 in lymphoid malignancy. Blood 85:893-901, 1995.
18. Hussussian CJ, Struewing JP, Glodstein AM, Higgins PAT, Ally DS, Sheahan MD, Clark WH Jr, Tucker MA, Dracopoli NC: Germline p16 mutations in familial melanoma. Nature Genetics 8:15-21, 1994.
19. Arap W, Nishikawa R, Furnari FB, Cavenee WK, Huang HJ: Replacement of the p16/CDKN2 gene suppresses human glioma cell growth. Can Res 55:1351-1354, 1995.
20. Hebert J, Cayuela JM, Berkeley J, Sigaux F: Candidate tumor-suppressor genes MTS1 ($p16^{INK4a}$) and MTS2 ($p15^{INK4B}$) display frequent homozygous deletions in primary cells from T- but not from B-cell lineage acute lymphoblastic leukemias. Blood 84:4038-4044, 1994.
21. Hatta Y, Hirama T Millere CW, Yamada Y, Tomonaga M, Koeffler HP: Homozygous deletions of the p15 (MTS2) and p16 (CDKN2/MTS1) genes in adult T-cell leukemia. Blood 85:2699-2704, 1995.
22. Pegg AE, Williams-Ashman HG: Phosphate-stimulated breakdown of 5'-methylthioadenosine by rat ventral prostate. Biochem J 115: 241-247, 1969.

23. Williams-Ashman HG, Seidenfeld J, Galletti P: Trends in the biochemical pharmacology of 5'-deoxy-5'-methylthioadenosine. Biochem Pharmacol 31: 277-288, 1982.
24. Trackman PC, Abeles RH: Methionine synthesis from 5'-S-Methylthioadenosine. Resolution of enzyme activities and identification of 1-phospho-5-S methylthioribulose. J Biol Chem 258: 6717-6720, 1983.
25. Backlund PS Jr., Smith RA: Methionine synthesis from 5'-methylthioadenosine in rat liver. J Biol Chem 256: 1533-1535, 1981.
26. Toohey JI: Methylthioadenosine nucleoside phosphorylase deficiency in methylthio-dependent cancer cells. Biophys Res Commun 83:27-35, 1978
27. Kamatani N, Yu AL, Carson DA: Deficiency of methylthioadenosine phosphorylase in human leukemic cells in vivo. Blood 60: 1387-1391, 1982.
28. Fitchen JH, Riscoe MK, Dana BW, Lawrence HJ, Ferro AJ: Methylthioadenosine phosphorylase deficiency in human leukemias and solid tumors. Can Res 46: 5409-5412, 1986.
29. Traweek ST, Riscoe MK, Ferro AJ, Braziel RM, Magenis RE, Fitchen JH: Methylthioadenosine phosphorylase deficiency in acute leukemia: pathologic, cytogenetic, and clinical features. Blood 71: 1568-1573, 1988.
30. Smaaland R, Schanche JS, Kvinnsland S, Hostmark J, Ueland PM: Methylthioadenosine phosphorylase in human breast cancer. Breast Cancer Res Treat 9: 53-59, 1987.
31. Nobori T, Karras JG, Della Ragione F, Waltz TA, Chen PP, Carson DA: Absence of methylthioadenosine phosphorylase in human gliomas. Can Res 51: 3193-3197, 1991.
32. Olopade OI, Buchhagen DL, Malik K, Sherman J, Nobori T, Bader S, Nau MM, Gazdar AF, Minna JD, Diaz MO: Homozygous loss of the interferon genes defines the critical region on 9p that is deleted in lung cancers. Can Res 53: 2410-2415, 1993.
33. Nobori T, Szinai I, Amox D, Parker B, Olopade OI, Buchhagen DL, Carson DA: Methylthioadenosine phosphorylase deficiency in human non-small cell lung cancers. Can Res 53: 1098-1101, 1993.
34. Diccianni MB, Yu J, Hsiao M, Mukerjee S, Shao LE, Yu AL: Clinical significance of p53 mutations in relapsed T-cell acute lymphoblastic leukemia. Blood 84:3105-3112, 1994.
35. Anandaraj SJ, Jayaram HN, Cooney, DA, Tyagi AK, Han N, Thomas JH, Chitnis M, Montgomery JA: Interaction of L-alanosine (NSC 153, 353) with enzymes metabolizing L-aspartic acid, L-glutamic acid and their amides. Biochem Pharmacol 29: 227-245, 1980.
36. Sahota A, Webster DR, Potter CF, Simmonds HA, Rodgers AV, Gibson T: Methylthioadenosine phosphorylase activity in human erythrocytes. Clin Chim Acta 128: 283-290, 1983.
37. Ghoda LY, Savarese TM, Dexter DL, Parks RE Jr., Trackman PC, Abeles RH: Characterization of a defect in the pathway for converting 5'-deoxy-5'-methylthioadenosine to methionine in a subline of a cultured heterogeneous human colon carcinoma. J Biol Chem 259: 6715-6719, 1984.

DETECTION OF bcr-abl m-RNA IN SINGLE PROGENITOR COLONIES BY PCR

Comparison with Cytogenetics and PCR from Uncultured Cells

Elisabeth Schulze,[1*] Rainer Krahl,[1] Karin Thalmeier,[2] and Werner Helbig[1]

[1] Department of Hematology/Oncology
University of Leipzig, Leipzig, Germany
[2] GSF
Institute of Experimental Hematology, Munich, Germany

ABSTRACT

Bone marrow and/or peripheral blood of patients with CML was investigated by the following 3 parameters: Ph' chromosome, bcr-abl expression in fresh blood and/or bone marrow and bcr-abl expression in single hemopoietic progenitor colonies generated from blood and/or bone marrow. Expression of bcr-abl was proved by a reverse "nested primer" PCR that is able to detect 1pg of hybrid m-RNA. We performed 108 investigations containing all 3 parameters: 12 on untreated patients, 7 after IFNα, 7 after low-dose Ara-C, 22 after cyclic high-dose hydroxyurea, 49 after allogeneic BMT, 5 before and 3 after stem cell mobilization and 3 after autologous stem cell transplantation (ASCT). In 53 cases (49%) cytogenetics and PCR gave identical results. In 40 cases (37%) PCR from single colonies gave additional information compared to cytogenetics (i.e. mosaic in colonies when all mitoses where positive or negative). Most interesting were the results of one patient after IFN, one patient after ASCT and 10 patients after BMT (14 investigations = 13%) showing only Ph' negative mitoses accompanied by a negative "nested primer" PCR from fresh blood/bone marrow but single positive progenitor colonies. False positive results could be widely excluded by repeated insertion of negative controls into the experiments. One explanation for these results could be that CML progenitors survive in the patients' body by being inactive and not proliferating. These cells express no or very little RNA and bcr-abl is not detectable by reverse PCR. When stimulated ex vivo in a colony assay by external growth factors, cells proliferate and produce detectable amounts of hybrid-m-RNA. The value of these observations is not clear. A follow up of the patients will show if such sleeping progenitors can be activated in vivo. Concluding our observations we can say that in special cases (therapy

* Correspondence to: Dr. Elisabeth Schulze, Miltenyi Biotec GmbH, Friedrich-Ebert-Staße 68, 51429 Bergisch Gladbach, Germany. Tel: +49-2204-8306-49; fax: +49-2204-85197.

Molecular Biology of Hematopoiesis 5, edited by Abraham et al.
Plenum Press, New York, 1996

follow up, detection of minimal residual disease) it could be useful to perform a PCR analysis of single progenitors in parallel with the routine investigations.

INTRODUCTION

Chronic myeloid leukemia (CML) is a clonal myeloproliferative disorder arising at the level of pluripotent stem cells. It is hallmarked by a specific chromosomal marker, the Philadelphia (Ph') chromosome resulting from a reciprocal translocation between chromosome 9 and chromosome 22 (1). The molecular expression of this genetic aberration is the bcr-abl hybrid gene that codes for a bcr-abl hybrid mRNA (2).

At present allogeneic bone marrow transplantation (BMT) remains the only potent curative treatment for CML patients. Several lines of evidence suggest, that in chronic phase a residual normal polyclonal hemopoiesis is present in about 50-60% of the patients at diagnosis that is suppressed by the malignant cells (3). Under certain conditions e.g. interferon (IFN) α therapy, intensive chemotherapy, autologous bone marrow transplantation (ABMT) or autologous peripheral stem cell transplantation (ASCT) this normal hemopoiesis can be reactivated leading to complete cytogenetic and even molecular remission (3-6). However, most of these remissions are not long lasting and patients usually relapse (4,7). Because allogeneic BMT is restricted to less than 10 % of all patients a potential monitoring and control of the alternative therapy regimens is necessary. Concerning IFNα therapy an accurate recording of responders and nonresponders may lead to a more relevant indication for experimental trials like ABMT, ASCT or intensive chemotherapy. This demands diagnostic methods that enable prognostification and measurement of the therapy effect. The most sensitive method for giving evidence of tumor cells in CML is the polymerase chain reaction (PCR). By using specific primers to amplify sequences from bcr-abl hybrid mRNA it is possible to detect one malignant cell within 10^5 to 10^6 normal cells (8,9).

In this context is not just the presence or absence of a malignant clone of interest but its actual amount may be critical. Several modifications of PCR have been developed allowing a quantification of the target sequences (8,9,10). In this paper we present an alternative trial for quantification of CML clone. Single progenitor cells grown from bone marrow and/or blood of patients were investigated for bcr-abl expression by using a "nested primer" PCR assay. The results were compared with that of cytogenetic examinations and PCR from uncultured cells carried out in parallel. Our aim was to get information about

1. the amount of residual normal hemopoiesis of patients in chronic phase of disease,
2. the nature of colony formation in complete cytogenetic remission,
3. persisting minimal residual disease in patients after BMT.

MATERIAL AND METHODS

Patients

A total of 68 patients was included in this study. 7 patients were newly diagnosed, 27 were included in a clinical trial,* 31 were after allogeneic BMT, 3 underwent a peripheral stem cell mobilization and one received an ASCT.

* After getting an initial therapy of cyclic high dose hydroxyurea (CHDHU) patients got either IFNa (6 pts.) or low dose (LD) AraC (4 pts.) or CHDHU was continued (17 pts.). (CML 1991 conducted by the East Germany study group).

Bone Marrow and Blood

After informed consent bone marrow (BM) and/or blood were collected. Mononuclear cells (MNC) were obtained by density separation on Lymphocyte Separation Medium (Boehringer Mannheim).

Clonogenic Progenitor Cultures

MNC from bone marrow and/or blood were cultivated in a methylcellulose assay containing 0.9 % methylcellulose in Iscoves' modified Dulbecco medium (Sigma), 30% fetal calf serum (Gibco), 1% deionized bovine serum albumin, 0.2 mg/ml iron-saturated transferrin (Sigma), 5×10^{-5} M mercaptoethanol (Ferrak), penicillin/streptomycin (Jenapharm), 10U/ml erythropoietin, 100 U/ml G-CSF (Amgen), 100 U/ml IL-3 (provided by Petra Meîßner, GSF Munich) and 50 ng/ml SCF (provided by Reinhard Mailhammer, GSF Munich). 10^5 bone marrow cells and 2×10^5 blood cells were plated into 35 mm Petri dishes in 1 ml of this mixture. After 8-9 days of culture at 37°C and 5% CO_2 single colonies containing approximately 50 - 500 cells were "picked" using a fine Pasteur-pipette and transferred into 1 ml Eppendorf cups containing 200 µl phosphate-buffered saline solution.

Cytogenetic Analysis

Samples were usually processed immediately after aspiration from nonseparated bone marrow. If the sample was obtained from an outside hospital, cells were cultured for 0.5 h and/or overnight. Chromosome preparation was performed according to the standard protocol: 0.1 µg/ml Colcemid for 30 min., hypotonic shock with 0.075 mol/l KCl for 30 min. at 37°C, followed by fixation in Carnoys' fixative (methanol : acetic acid = 3 :1). Fluorescence R-Banding was performed with chromomycin A3 and methyl green (11). A number of 50 or more analyzed metaphases was aspired. The metaphase cell preparations were photographed and analyzed in detail.

RNA Extraction, Reverse Transcription (RT) and PCR

Total RNA from uncultured MNC of patients' blood and bone marrow samples was isolated by acid phenolic precipitation according to Chomczynsky (12) using RNAzol™ B (BiotekX Laboratories). 2 and 0.2 µg of RNA preparations were reverse transcribed into cDNA in a final volume of 20 µl of RT buffer of the following composition: 1 mM dNTPs (Promega), 0.01 M DTT (Boehringer Mannheim), 0.2 µg oligo dT_{15} as a Primer for RT, 1 U/ml RNasin (Promega), 10 U/ml avian reverse transcriptase (Promega) and 4 µl of 5 x concentrated RT-reaction buffer. Reverse transcription was performed for 10 min. at room temperature, 30 min. at 42°C and finally 5 min. at 95°C to denature the RNA/cDNA hybrid strand. The first PCR amplification (40 cycles: 1 min. each 94°C, 65°C and 72°C) was performed in a final volume of 50 µl containing 10 µl of reverse transcription mixture, 5 µl of 10 x PCR-reaction buffer and 2 mM $MgCl_2$ (Gene-Amp kit, Perkin Elmer), 1 nM of a primer mixture specific for bcr-abl and 1.5 U Taq-polymerase (Ampli-taq, Perkin Elmer). Reamplification was performed under the same conditions using 2.5 µl of the first amplificate and adding another 2 mM dNTPS and primers internal of that used in the first reaction.

Cells of single colonies were lysed in 20 µl of a lysis buffer consisting of 2.5% Tween 20, 10 µM Tris/HCL; pH 8.0, 10 mM NaCl, 3 mM $MgCl_2$, 10 mM DTT and 2 U/ml RNasin by 5 min. incubation at 37°C. The cups were centrifuged to sediment nuclea and cell membranes and the lysate containing complete cytosol of cells was used as target for reverse transcription. The complete lysate of 20 µl was used for RT in a final volume of 40 µl with final concentration of 0.75 mM dNTPs, 0.01 M DTT, 0.08 µg/µl oligo dT_{15}, 1 U/ml RNasin, 10 U/ml avian reverse transcriptase and 8 µl of its specific buffer. The temperature protocol was the same as for RT from purified RNA. Also the amplification conditions were similar to that described above differing in using 20 µl of RT-mixture in 50 µl PCR reaction buffer. The conditions of the second (nested) PCR were as described. 10 µl of the PCR reaction products were analyzed in a 1.5% agarose (Pharmacia)-gel by electrophoresis, subsequently stained in 0.01 % ethidium bromide and visualized under UV light. The sensitivity of this "nested" PCR from cell lysates was determined previously. It reached a limit of 1pg purified RNA (= 0.1 - 1 cell) and 10 -100 lysed cells i.e. colonies should be detectable by this method. To exclude false positive results due to contamination we inserted negative controls (CCRF cells,* aliquots of the autoclaved double distilled water that was used as a dilution agent in RT and PCR) into every PCR experiment. False negative results were widely ruled out by running appropriate positive controls (K562 cells and RNA from K562 cells) and by amplifying β-Actin sequences simultaneously from each cDNA preparation.

RESULTS

BM and/or peripheral blood from 68 patients with CML, or after CML-indicated allogeneic BMT were investigated by the regimen shown in figure 1. BM and/or blood cells were analyzed for the presence or absence of Ph' chromosome in cytogenetics and the presence or absence of bcr-abl hybrid mRNA in a "nested primer" PCR. An aliquot of the cells was cultivated in a methylcellulose assay that gives rise to hemopoietic progenitor cell colonies. These grown colonies were also examined for the expression of bcr-abl mRNA. We were able to study these 3 parameters in parallel in 107 investigations. The patients can be divided into 3 groups. Group 1 includes patients at diagnosis and in chronic phase on therapy, group 2 are patients after allogeneic BMT and group 3 represents patients that were mobilized for a peripheral stem cell harvest.

Table 1a and b show the results obtained from patients of group 1 (Tab.1a) and 2 (Tab.1b). It is obviously that the pattern of cytogenetics, PCR from uncultured cells and PCR from individual colonies in comparison varied ranging from total correspondence to total difference. In table 2 the results of table 1a and b are summarized and compared. A high amount of the investigations in both groups showed total correspondence of all 3 parameters i.e. cytogenetics, PCR of uncultured cells and PCR of colonies were completely positive, had a comparable mosaicism in metaphases and colony PCR or all 3 parameters were negative (51% in group 1 and 47% in the group 2). In 18 cases of group 1 all metaphases and also all analyzed colonies were positive, in 7 cases metaphases and colonies as well showed a mosaicism. In a total of 20 investigations in this group the analysis of single colonies could give additional information to cytogenetics. 6 cases showed the usual problem of lacking metaphases, i.e. no information. More interesting are the residual 14 cases. In 12 of them the analysis of colonies revealed single negative progenitors while metaphases were completely positive. 2 investigations from one IFN treated patient showed a significant amount of bcr-abl-positive colonies whereas cytogenetics and also PCR from uncultivated cells were negative. In 2 cases we found a negative PCR from colonies and or purified total RNA while cytogenetic analysis could show positive metaphases.

* CCRF is a cell line arising from an acute myeloid leukemia and is proved to be bcr-abl negative.

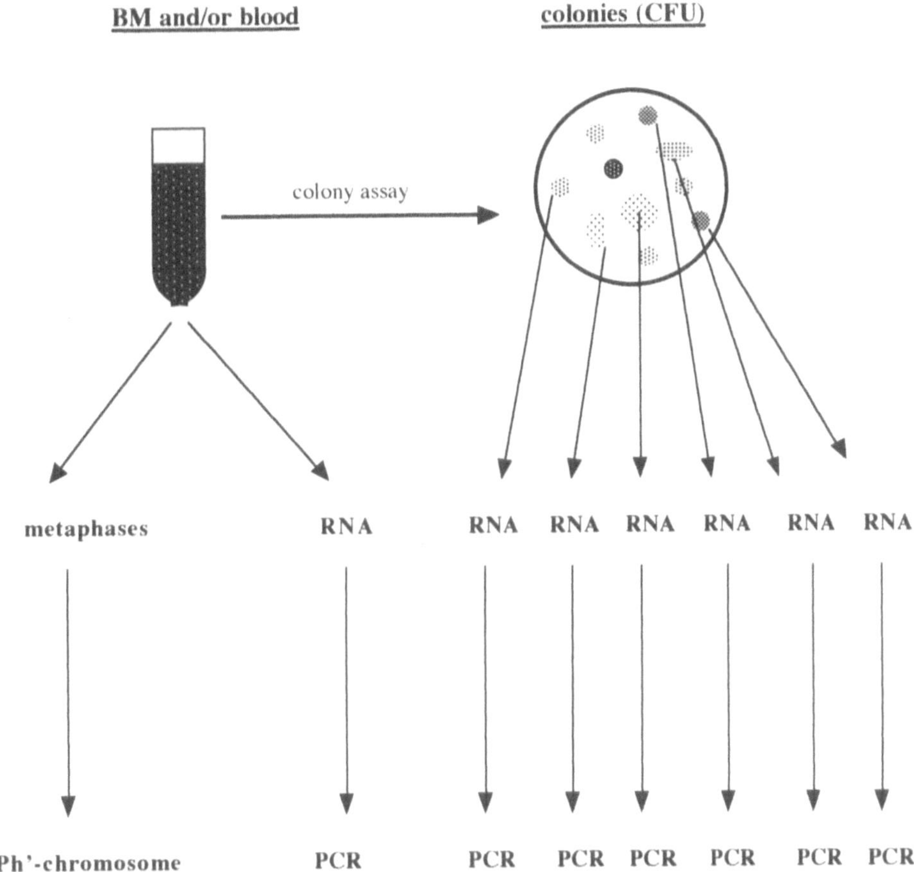

Figure 1. Schema for investigation of peripheral blood and bone marrow from patients with CML.

Table 1A. Results of cytogenetics and PCR from untreated patients and patients on therapy (group 1)

Cytogenetics	Total RNA from BM/blood	RNA from colonies	Initial	IFN	LD-Ara-C	CHD-HU
+	+	+	6	2	2	8
+	+	–/+	3 (B)	0	3	6
–/+	+	–/+	0	1	1	5
–/+	+	+	0	0	1 (B)	0
+	–	+	1	0	0	1
–/+	–	–	0	1 (B)	0	0
–	–	–/+	0	2	0	0
Ø	+ or –	+ or –	2	1	0	3

Symbols: +, – ,–/+ = positive, negative, mosaicism; B = results were obtained from peripheral blood; Ø = method failed (= no metaphases in cytogenetics, contamination of the PCR experiment)

Table 1B. Results of cytogenetics and PCR from patients after allogeneic BMT (group 2)

Cytogenetics	Total RNA from BM/blood	RNA from colonies	Results post BMT
+	+	+	4
+	+	–/+	2
–/+	+	–/+	1
–	–	–	18
–	+	+	1
–	–	–/+	12
–	+	–	4
Ø	+ or –	+ or –	2
–	–	Ø	5

Symbols: +, – ,–/+ = positive, negative, mosaicism; B = results were obtained from peripheral blood; Ø = method failed (= no metaphases in cytogenetics, contamination of the PCR experiment)

In 23 investigations of patients in group 2 cytogenetics and PCR from colonies gave identical results. The main pattern of this group was complete negativity. In 17 cases PCR from colonies could give additional information. Most interesting are 2 cases where single negative colonies indicate a residual bcr-abl negative hemopoiesis in relapse while all metaphases were positive and another 12 cases where though complete negativity of metaphases and PCR of total RNA from uncultured cells single colonies appeared bcr-abl positive. The cytogenetic and molecular results of these 12 cases (10 patients) mentioned at last after BMT were analyzed separately (Fig. 2). Except in patient 2 where bcr-abl positive colonies appeared early after BMT and relapse was diagnosed weeks later no correlation between this phenomenon and clinical outcome could be seen. There was no difference in the median interval to BMT of completely negative patients and such showing single positive colonies (117.4 weeks, range 3.9 - 444.7, for negative patients vs. 107.9 weeks, range 3.1 - 418.7, for patients with positive colonies).

Table 3 shows our very recent results of ASCT. Patients were investigated before mobilization, the stem cell harvest was analyzed and patient one was monitored after transplantation. Patient 1 was mobilized in a complete cytogenetic remission. However, PCR from colonies and total RNA of fresh bone marrow and blood was bcr-abl positive. The colony assay could also reveal positive progenitors in the transplant whereas total RNA and metaphases were negative. Initially after ASCT the patient became completely negative.

Table 2. Summary of the results from table 1a and b

Comparison of PCR and cytogenetics	Group 1	Group 2
Correspondence of cytogenetics and PCR	25 = 51%	23 = 47%
More information from colony PCR compared to cytogenetics	20 = 41 %	17 = 35%
More information from cytogenetics	1 = 2%	0
False negative results from PCR*	3 = 3%	0
False positive results in colonies**	0	5 = 10.6%
Total number of investigations	49	49

*negative PCR from uncultured cells and colonies whereas positive metaphases were found

**negative controls were positive

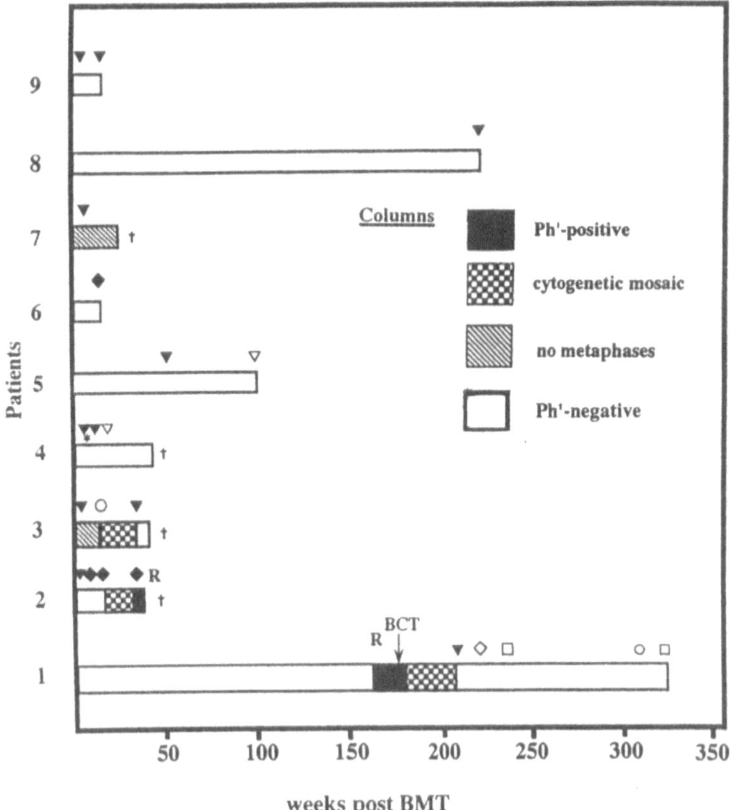

Figure 2. Monitoring after allogeneic BMT of 9 patients showing single bcr-abl positive colonies in 11 experiments whereas PCR from uncultured BM and/or blood cells was negative. (The results of one additional patient who was invesigated once 418.7 weeks after BMT showing 1/20 positive colonies is not shown in this figure.) The pattern of the columns represent the cytogenetic state. Symbols at the columns show the PCR results from single colonies and/or uncultivated cells and the outcome of the patients. Symbols: ▽ - positive colonies (total RNA not done); ▼ - positive colonies and negative total RNA; ◆ - positive colonies and positive PCR; ○ - negative colonies (total RNA not done); ● - negative colonies and negative total RNA; □ - negative PCR (colonies not done); ◇ - positive PCR (colonies not done); † - death of the patient; * - acute GvHD. Abbreviations: R - relapse; BCT - donor bufy coat transfusion.

However, 23 weeks after ASCT 1 bcr-abl positive colony was detected. The preconditions of patient 2 and 3 were worse than of patient 1. In patient 2 a mosaicism was indicated before mobilization by PCR and cytogenetics. The amount of normal hemopoiesis in the transplant seems to be increased compared to bone marrow before. Patient 3 was totally positive before mobilization. Whereas the cytogenetic analysis of the transplant showed a nearly total positivity the PCR from single colonies could reveal about 25% bcr-abl negative progenitors.

DISCUSSION

In our study we investigated groups of patients with different biological preconditions. One group (group 1) consists of untreated patients and patients in chronic phase under

Table 3. Cytogenetic and molecularbiological results from patients with CML before and after stem cell mobilization and transplantation

Patient	Date	Material	bcr/abl pattern	bcr/abl col.	bcr/abl RNA	Ph'-chrom.
1	before mobil.	BM	4 x b2/a2 and 2 x b3/a2	6/15	positive	0/9
	before mobil.	blood	b3/a2 and b2/a2	1/5	positive	–
		transplant	**b3/a2**	**4/20**	**negative**	**0/4**
	post ASCT (4W)	BM	–	0/20	negative	n.m.*
	post ASCT (10W)	BM	–	0/20	negative	0/105
	post ASCT (23W)	blood	b2/a2	1/20	negative	–
2	before mobil.	BM	b3/a2	19/20	positive	10/13
		transplant	**b3/a2 (1xb2/a2)**	**46/56**	**positive**	**40/47**
3	before mobil.	BM	b3/a2	13/13	positive	50/50
	before mobil.	blood	b3/a2	4/6	positive	–
		transplant	**b3/a2**	**26/35**	**positive**	**15/17**

* no metaphases

several treatment regimens where the malignant clone is dominating in the majority of patients. The second group (group 2) are patients after allogeneic BMT that are expected to be cured i.e. that malignant clone should be eradicated.

Interestingly regarding the summary of the results both groups are very similar. In about 50 % of the investigations in each group we found a total correspondence of cytogenetic results and results of PCR from RNA of uncultured cells and PCR from single colonies. In group 1 that means a total positivity or a comparable mosaicism in colonies and metaphases. In group 2 we found most correspondence in patients, that were Ph' and bcr-abl negative. This high and nearly identical amount of correspondence between cytogenetics and PCR in these biological different groups may indicate the validity of the PCR method for analyzing single colonies.

On the other hand in a high percentage of the experiments (40 % in group 1 and 32% in group 2) the results of PCR from single colonies and cytogenetics from uncultured cells differ. We believe that in these cases the colony assay could provide additional information. Concerning the chronic phase group most of these results 12/20 came from blood or bone marrow of patients showing only Ph-positive metaphases. PCR analysis of single colonies could detect single bcr-abl negative colonies indicating the presence a residual bcr-abl negative hemopoiesis. These findings may be interesting in regard to the prognosis of the patients and also concerning therapeutical consequences e.g. autologous transplantation of hemopoietic cells. However, these results are not very surprising. Several authors could show bcr-abl negative hemopoiesis during the last years by analyzing individual colonies (13-17). With respect to very recent reports about 70 % of newly diagnosed patients contain residual normal hemopoietic cells (18). These results were obtained by analyzing single colonies (see above) or even by more subtile methods like LTBMC (19,20,21). Nevertheless it should be pronounced that analysis of single colonies may be helpful for the evaluation of the therapy effect, particularly in view the development of new treatment regimens like autologous transplantation. The discovery of Ph' negative colonies may indicate a residual Ph' negative hemopoiesis that is able to generate Ph' negative progenitors and may provide more functional information about hemopoiesis in the patients whereas cytogenetics and conventional PCR only record the phenotype of the cells. The relation between positive and negative colonies may enable a quantification of the malignant clone. Furthermore the amplification of genes or gene expression in single cells is magnified because the first amplification occurs already when the colony is growing. Disadvantaging is the limited number of colonies that can be analyzed (< 30 / experiment).

Keating reported in 1994 on simultaneous investigation of individual progenitor colonies for bcr-abl and Ph' chromosome (22). He found within the colonies from 12 patients 23 ± 18% that did not express bcr-abl but were Ph' positive. We did not perform cytogenetics and PCR in parallel on the same individual colony. The cytogenetics in our investigations was done on fresh material. So we cannot exclude completely a Ph' positivity of our bcr-abl negative colonies

More interesting are the results of one patient on IFN (2 investigations), 9 patients after bone marrow transplantation (11 investigations) and one patient after autologous peripheral stem cell transplantation (1 investigation) where we found bcr-abl positive colonies whereas cytogenetics and also PCR from uncultured cells were negative. The fact that bcr-abl positive cells can be detected in bone marrow or peripheral blood from patients after BMT is also not new. Many authors found a temporary positivity in PCR and in cytogenetics after BMT (23-25). Recently Pichert et. al. investigated hemopoietic colonies from 5 patients after BMT for expression of bcr-abl repeatedly (26). In these patients PCR from RNA of uncultured cells has been found to be bcr-abl positive. They found an overall of 5.2 % (7/135) of positive colonies but no relationship to the outcome of the patients i.e. all 5 patients remained in stable cytogenetic remission. Concerning the individual monitoring of the 9 patients showing single positive colonies after BMT in this paper (Fig. 2) the value of these findings remains also open. One patient (pat.2) developed a relapse after showing positive colonies in the early phase after BMT. The others were 36 to 221 weeks post BMT and for making any statement a follow up of the investigations will be necessary. Interestingly in some patients (pat.1, 2 and 3) positive colonies appeared near the date when a change of the cytogenetic status was observed.

New and surprising in our study is the appearance of bcr-abl positive colonies whereas besides cytogenetics even PCR from RNA of uncultivated cells was negative. Actually one would expect the opposite. A negative PCR with a sensitivity of 1 positive cell in 10^5 - 10^6 while positive progenitors can be detected in parallel seemed to be doubtful. However, false-positive results could be widely excluded by inserting negative controls into every PCR experiment. The detection limit of our PCR assay is 1 pg of total RNA that corresponds to 0.1 - 1 cell. So it seems to be unlikely that in all these cases the PCR of RNA from fresh cells gave false-negative results. Additionally some of these negative results were confirmed in other laboratories (1 case C. Bartram, Ulm, 4 cases K. Thalmeier, Munich).

Very similar findings were published in 1993 by Bedi et al. (27) who found no or only traces of bcr-abl expression (RT-PCR) in counterflow centrifugal elutriated $CD34^{+}lin^{-}$ cells from patients with CML. On the other hand most of these cells showed the bcr-abl rearrangement in Southern blot. When these primitive cells were cultivated for 7 - 14 days in a liquid culture system containing potent stimulating cytokines like stem cell factor, Interleukin 6 (IL-6) and IL-3 the cells became bcr-abl positive in PCR analysis. His interpretation of these results was that primitive CML progenitors are possibly quiescent in vivo, and those primitive progenitors may be transcriptionally silent when in G_0, expressing bcr-abl only when they enter cell cycle.

This could also explain our results. Perhaps malignant progenitors (stem cells?) are persisting in the patients. In vivo they are inhibited by either the allogeneic graft (after BMT) or drugs (after IFN or intensive chemotherapy). When released from the in vivo conditions and stimulated by external growth factors in the colony assay they proliferate and express bcr-abl mRNA. If those "sleeping" stem cells can be reactivated in vivo remains unclear. A continuation of our investigations will be necessary to answer this question. However, our and Bedis's results may suggest that complete molecular remissions diagnosed by conventional PCR particularly of patients on alternative treatment regimens should be regarded with caution. Additional methods like analysis of individual progenitor colonies or in-situ hybridization should be used to confirm diagnosis.

ACKNOWLEDGEMENT

This study was supported in part by a grant from the "Dieter-Schlag-Stiftung". The authors are most grateful to Scarlet Musiol, Christine Günther and Christel Müller for excellent technical assistance.

REFERENCES

1. Rowley JD: A new consistent chromosomal abnormality in chronic myelogenous leukemia identified by quinacrine fluorescein and Giemsa staining. Nature 243:290, 1973
2. Stam K, Heisterkamp N, Grosfeld G, deKlein A, Verma RS, Coleman M, Dosik H, Groffen J: Evidence of a new chimeric m-RNA in patients with chronic myeloid leukemia and the Philadelphia chromosome. New Engl J Med 313:765, 1985
3. Dexter TM, Chang J: New strategies for the treatment of chronic myeloid leukemia. Blood 84:595, 1994
4. Talpaz M, Kantarjian H, Kurzrock R, Gutterman JU: Update on therapeutic options for chronic myelogenous leukemia. Semin Hematol 27:31, 1993
5. Goldman JM, Deisseroth AB: Use of autotransplants in CML. Bone Marrow Transplantation 10/S1:74, 1992
6. Dunbar CE, Steward FM: Separating wheat from the chaff: selection of benign hematopoietic cells in chronic myeloid leukemia. Blood 79:1107, 1992
7. Kantarjian HM, Deisseroth A, Kurzrock R, Estrov Z, Talpaz M: Chronic myelogenous leukemia: a concise update. Blood 82:691, 1993
8. Thompson JD, Brodsk I, Yunis JJ: Molecular quantification of residual disease in chronic myelogenous leukemia after bone marrow transplantation. Blood 79:1629, 1992
9. Cross NCP, Feng L, Chase A, Bungey J, Hughes TP, Goldman JM: Competitive Polymerase chain reaction to estimate the number of bcr-abl transcripts in chronic myeloid leukemia patients after bone marrow transplantation. Blood 82:1929, 1993
10. Cross NCP: Quantitative PCR techniques and applications. Brit J Haematol 89:693, 1995
11. Sahar E, Latt S.A.: Enhancement of banding pattern in human metaphase chromosomes by energy transfer. Proc Natl. Acad SCI USA 75:5650, 1078
12. Chomczynsky P, Sacci N: Single-step method of RNA isolation by guanidinium thiocyanate-phenol-chloroform extraction. Anal Biochem 162:156, 1987
13. Leemhuis T, Leibowitz D, Cox G, Silver R, Srour EF, Tricot G, Hoffman R: Identification of bcr-abl-negative primitive hematopoietic progenitor cells within chronic myeloid leukemia marrow. Blood 81:801, 1993
14. Leemhuis T, Leibowitz D, Cox G, Srour EF, Tricot G, Hoffman R: Selection of bcr-abl-negative progenitor cells from chronic myeloid leukemia marrow. Progr Clin Biol Res 377:231, 1992
15. Sadamura S, Umemura T, Takahira H, Hirata J, Nishimura J, Nawata H: Analysis of the clonality at the level of progenitors in chronic myelogenous leukemia using the polymerase chain reaction. Leuk Res 16:371, 1992
16. Agarwal R, Doren S, Hicks B, Dunbar CE: Long-term culture of chronic myelogenous leukemia marrow cells on stem cell factor-deficient stroma favors benign progenitors. Blood 85:1306, 1995
17. Carlo-Stella C, Mangoni L, Piovani G, Garau D, Almici C, Rizzoli V: Identification of Philadelphia-negative granulocyte-macrophage colony-forming units generated by stroma-adherent cells from chronic myelogenous leukemia patients. Blood 83:1373, 1994
18. Carella AM, Frassoni M, Podesta M, Pollicardo N, Pungolino E, Lerma E, Prencipe E, Ghio R, Vassalo F, Ferrero R, Soracco M, Benvenuto F, Giordano D, Chimiri F: Mobilization and transplantation of Ph.'-negative BPCs in CML. 9th Symposium on Molecular Biology of Hematopoiesis, Genua June 27-27, 1995
19. Eaves C, Udomsadki C, Cashman J, Barnett M, Eaves A: The biology of normal and neoplastic stem cells in CML. Leuk Lymph 11:245, 1993

20. Verfaillie C, Hurley R, Bhatia R, McCarthy JB: Role of bone marrow matrix in normal and abnormal hematopoiesis. Crit Rev Oncol Hematol 16:201, 1994
21. Podesta M, Carella AM, Frassoni F: Primitive hemopoietic cells (LTC-IC) in mobilized peripheral blood cells after chemotherapy-induced aplasia. Bone Marrow Transplant 14, S3:S42, 1994
22. Keating A, Wang X-H, Laraya P: Variable transcription of bcr-abl by Ph+ cells arising from hematopoietic progenitors in chronic myeloid leukemia. Blood 83:1744, 1994
23. Miyamura K, Tahara T, Tanimoto M, Morishita Y, Kawashima K, Morishima Y, Saito H, Tzuzuki S, Takeyama K, Kodera Y, Matsuyama K, Hirabayashi N, Yamada H, Naito K, Imai K, Sakamaki H, Asai O, Mitzutani S: Long persistent bcr-abl positive transcript detected by poymerase chain reaction after bone marrow transplant for chronic myelogenous leukemia without clinical relapse: A study of 64 patients. Blood 81:1089, 1993
24. Gaiger A, Lion T, Kahls P, Mitterbauer G, Henn T, Haas O, Födinger,M., Kier P, Forstinger C, Quehenberger P, Hinterberger W, Jäger U, Linkesch W, Mannhalter C, Lechner K: Frequent detection of bcr-abl specific m-RNA in patients with chronic myeloid leukemia (CML) following allogeneic and syngeneic bone marrow transplantation (BMT). Leuk 7:1766, 1993
25. Saglio G, Guerrasio A, Gottardi E, Parziale A, Martinelli G, Zaccaria A, Frassoni F: Significance of detection of bcr-abl transcript after allogeneic and autologous transplant. Bone Marrow Transplant 14, S3:S62, 1994
26. Pichert G, Aleya EP, Soiffer RJ, Roy D-C, Ritz J: Persistence of myeloid progenitor cells expressing BCR-ABL mRNA after allogeneic bone marrow transplantation for chronic myelogenous leukemia. Blood 84:2109, 1994
27. Bedi A, Zehnbauer BA, Collector MI, Barber JP, Zicha MS, Sharkis SJ, Jones RJ: bcr-abl gene rearrangement and expression of primitive hematopoietic progenitors in chronic myeloid leukemia. Blood 81:2898, 1993

33

QUANTITATIVE RT-PCR OF PML-RARα TRANSCRIPTS IN ACUTE PROMYELOCYTIC LEUKEMIA

Harry Iland,* Francisca Springall, Tao Zeng, and Katrina Bradfield

The Kanematsu Laboratories
Royal Prince Alfred Hospital
Camperdown, NSW 2050 Australia

ABSTRACT

Detection of PML-RARα fusion transcripts in acute promyelocytic leukemia (APL) by RT-PCR is a useful adjunct to cytogenetic analysis at the time of diagnosis, and is the method of choice for detection of minimal residual disease. Since PML-RARα transcripts are frequently detectable after successful remission induction with all-trans retinoic acid (ATRA), we have established a competitive RT-PCR protocol for quantitation of PML-RARα transcripts which could be used to assess the effects of ATRA induction and subsequent chemotherapy consolidation. In addition to quantitation of the target transcript (PML-RARα), simultaneous quantitation of a second template (transferrin receptor transcripts) provides a control for the quality of RNA extraction and for the efficiency of cDNA synthesis, since the level of PML-RARα transcripts can be expressed relative to the level of transferrin receptor transcripts. For patients with APL, we anticipate that decisions regarding the requirement for additional chemotherapy, or the optimal timing of allogeneic transplantation, will be facilitated by the ability to detect progressive increases in fusion transcripts.

INTRODUCTION

Gene rearrangements resulting from chromosomal translocations are being increasingly recognized in hematological malignancies [1]. One of the consequences of gene rearrangements is the formation of novel fusion genes which encode abnormal chimeric protein products. Acute promyelocytic leukemia (APL), the FAB-M3 subtype of acute

*Corresponding author: Associate Professor H. J. Iland, Kanematsu Laboratories, Royal Prince Alfred Hospital, Missenden Road, Camperdown, NSW 2050 Australia. Phone # 61-2-9515-7655; Fax # 61-2-9515-6255.

Molecular Biology of Hematopoiesis 5, edited by Abraham et al.
Plenum Press, New York, 1996

myeloid leukemia (AML), is uniquely associated with a t(15;17) reciprocal translocation, and the breakpoints involved have been characterized in detail [reviewed in 2,3]. The disrupted genes include the retinoic acid receptor alpha gene (RARα) at chromosome 17q21, and the PML gene at chromosome 15q22. The RARα protein is a member of the steroid/thyroid hormone receptor superfamily of nuclear transcription factors, and sequence data indicate that the PML protein is also a DNA-binding transcriptional regulator. Actively transcribed fusion genes are generated on both derivative chromosomes, but the PML-RARα fusion protein derived from the derivative chromosome 15 is more likely to be associated with leukemogenesis than its reciprocal counterpart. PML-RARα fusion proteins have the potential to disrupt normal PML- and RARα-mediated transcription, and may also interfere with the retinoid X receptor pool whose members interact with an even broader range of transcription factors.

The breakpoint on chromosome 17 consistently occurs in the 2nd intron of the RARα gene, whereas the chromosome 15 breakpoint is more variable, and is localized within one of three breakpoint cluster regions in the 6th intron (bcr-1), the 6th exon (bcr-2), or the 3rd intron (bcr-3). Thus RNA, rather than DNA, is more suitable as a template for polymerase chain reaction (PCR) detection of the rearrangement, but cDNA must first be synthesized by reverse transcriptase (RT-PCR). Although multiple PML-RARα fusion transcripts are generated by variable splicing and polyadenylation patterns, PCR primers which target the 3rd RARα exon and the 5' region of the 3rd PML exon are able to amplify fusion RNA sequences in virtually all patients.

Whilst RT-PCR is a highly sensitive technique for the detection of minimal residual leukemia, the significance of persistent RT-PCR positivity varies considerably, depending on the precise type of leukemia under consideration, and on the clinical and therapeutic context in which it has been detected. Unfortunately, PCR is inherently non-quantitative, and yet is used to monitor a clinical spectrum of residual disease. Furthermore, numerous sources of variability in RT-PCR technology complicate the interpretation of conventional RT-PCR results. Patients with chronic myeloid leukemia (CML) typically remain RT-PCR positive after interferon, regardless of the extent of cytogenetic response [4]. After allogeneic marrow transplantation, RT-PCR positivity often persists for up to 12 months, even in patients who are shown ultimately to have been cured [5]. Several reports have indicated that quantitation of RT-PCR products provides superior prognostic value in these situations. In APL, patients often remain RT-PCR positive despite successful induction with all-trans retinoic acid (ATRA) [6]. Conversion from negative to positive often heralds relapse, but it would be advantageous if quantitative data were available to facilitate decision making with respect to reintensification of therapy.

MATERIALS AND METHODS

Competitive RT-PCR

Several strategies are available to ensure that PCR is quantitative [7]. We chose competitive RT-PCR, which has become the most widely used form of quantitative RT-PCR. Quantitation of target transcripts is achieved by the inclusion of competitor templates in RT-PCR reactions; PCR products derived from competitor templates can be distinguished from products derived from endogenous templates by differences in size. In competitive PCR, serial dilutions of competitor template are added to fixed amounts of endogenous template cDNA and the mixtures are subjected to amplification. The amount of target template in the patient's sample is given by the amount of competitor added that produces equal molar amounts of target and competitor PCR products. The competitor we have

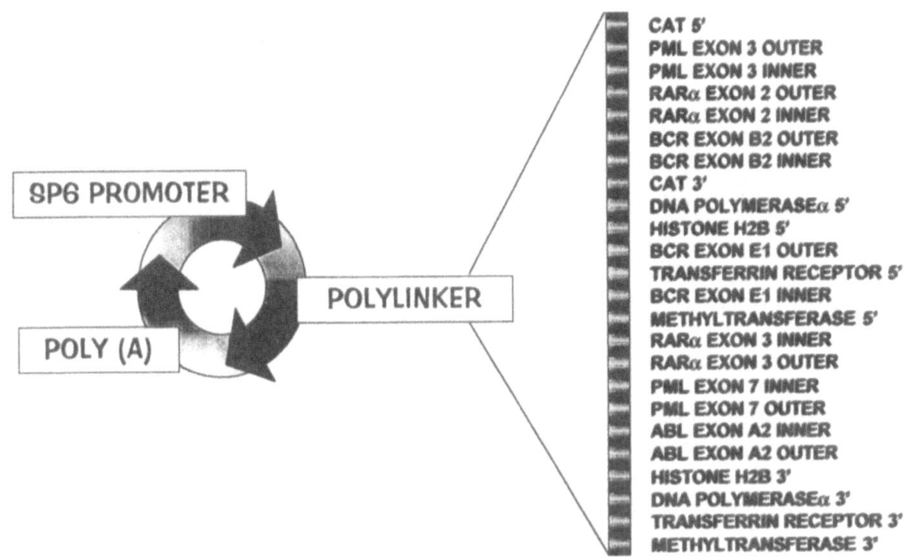

Figure 1. Structure of competitor for quantitative RT-PCR. A synthetic DNA sequence, containing the primer sites shown, was inserted into pSP64 (poly-A), downstream from the SP6 promoter site. Whole plasmid DNA was used as the competitor molecule in competitive RT-PCR reactions. The insert had been synthesized in six fragments on an Applied Biosystems 381A synthesizer, and pairs were used to make double-stranded fragments which were ligated and inserted by standard methods.

produced (Figure 1) contains sites for primers that are able to co-amplify the following endogenous templates: PML-RARα, the reciprocal RARα-PML, and BCR-ABL (for both b2a2 and b3a2 junctions in CML, and for the e1a2 junction in Philadelphia chromosome positive acute lymphoblastic leukemia). The competitor also contains primer sites for several housekeeping genes; amplification of the corresponding transcripts can be used as internal controls. They include transferrin receptor, DNA methyltransferase, DNA polymeraseα, and histone H2B. The competitor DNA has been cloned into an expression vector, pSP64 poly(A) (Promega).

Separate quantitations are performed for the target transcript (eg. PML-RARα or BCR-ABL) and for the control template (we currently use the transferrin receptor). Quantitation of transferrin receptor transcripts provides a control for the quality of RNA extraction and for the efficiency of cDNA synthesis, since the level of the target transcripts can be expressed relative to the level of transferrin receptor transcripts. To ensure accurate quantitation of PCR products, we perform standard semi-nested PCR reactions, but include fluorescent R110-labelled dUTP ([R110]dUTP). The PCR products are electrophoresed in acrylamide gels and visualized in an Applied Biosystems 373A automated sequencer running Genescan software.

RT-PCR Amplification of PML-RARα Transcripts from a Patient with APL

RNA was extracted by the method of Chomczynski and Sacchi [8] from bone marrow aspirated at the time of diagnosis. The marrow had morphological features consistent with APL, and cytogenetic analysis confirmed the diagnosis by demonstrating t(15;17). cDNA was synthesized from 10μg total RNA in a 100μl reaction containing 1x RT buffer (Boehrin-

Table 1. Primer sequences for amplification of PML-RARα and transferrin receptor transcripts

Primer target and location	Primer (5' → 3')
For amplification of PML-RARα transcripts	
5' PML exon 3 primer for 1st and 2nd rounds	GAGGCTGTGGACGCGCGGTA
3' RARα exon 3 primer for 1st round	CAGGCGCTGACCCCATAGTGG
3' RARα exon 3 primer for 2nd round	ACAAAGCAAGGCTTGTAGATGCGG
For amplification of transferrin receptor transcripts	
5' Transferrin Receptor exon 12 primer	CTGTTGTATACGCTTATTGAGAA
3' Transferrin Receptor exon 16 primer	TTCAATTCAACATCATGGGTTAG

ger Mannheim), 0.8mM dNTP, 10μM random hexamers, 69.6 units of Super AMV RT (Molecular Genetic Resources), and 140 units RNAsin in water. The reaction was incubated at 42°C for one hour, and then heated to 95°C for five minutes.

For PCR, replicates of patient cDNA (1μl) were spiked with serial dilutions of competitor DNA (4μl) and then amplified in a 50μl reaction containing 200μM dNTPs, 2.5 units Taq polymerase (Perkin Elmer), 1x PCR buffer (Boehringer Mannheim), and primers (100nM for PML-RARα, 600nM for transferrin receptor). The number of competitor molecules in each reaction varied from 10^6 down to 10 for PML-RARα, and from 10^7 down to 10^2 for transferrin receptor. Primer sequences are listed in Table 1. A second round of PCR, using an internal 3' RARα primer was performed for PML-RARα transcripts only, using 2μl of the outer reaction as template. Fluorescent [R110]dUTP was included in the transferin receptor reaction and the second round PML-RARα reaction at a 1:750 ratio of [R110]dUTP:dTTP. Amplification at each round was for 35 cycles; PML-RARα was amplified in a 2-step reaction (70°C and 94°C) in both the first and second rounds of PCR, whereas transferrin receptor was amplfied in a 3-step reaction (60°C, 72°C and 94°C). Unincorporated [R110]dUTP was removed by passage through a Centricon-30 column prior to electrophoresis.

RESULTS

RT-PCR of PML-RARα Transcripts

When a bcr-1 breakpoint is present in the PML gene, as was the case for the patient presented here, two PCR products are observed due to alternative splicing such that PML exon 5 sequences, comprising 144 bp, are not always present. With our primer system, these two products measure 854 and 710 bp respectively (after the 2nd round of PCR). The competitor molecule generates a 292 bp product after the 2nd round. Occasionally, we have also observed small amounts of PCR product of 451 bp in length, corresponding to alternate splicing of the 259 bp PML exon 6. The presence of multiple PCR products complicates quantitation and comparison with competitor PCR product, particularly when ethidium bromide stained agarose gels are used. For this purpose, we have utilized the ability of Genescan 672 software in combination with an Applied Biosystems 373A automated DNA sequencer to accurately determine PCR product length and quantity. An example of an electrophoretogram for one lane in a competitive PCR assay is shown in Figure 2. Panel A shows two major peaks, corresponding to the 710 and 854 bp products, as well as a minor peak at 451 bp, and a peak at 292 bp derived from the competitor. Panel B shows Genescan-2500 fluorescent molecular weight markers. These are run in the same lane as the

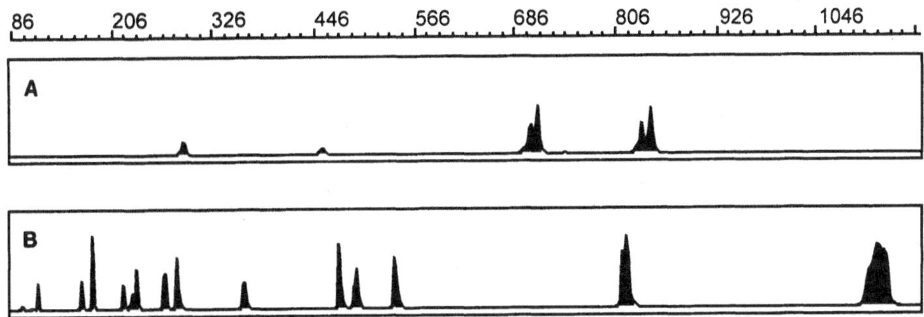

Figure 2. Sample electrophoretogram from a competitive RT-PCR reaction in a patient with APL. Panel A shows the peaks generated by PCR products which were labelled during PCR with fluorescent [R110]dUTP. The 292 bp product is derived from the competitor molecule, whereas the other products are derived from PML-RARα fusion transcripts in the patient sample, and result from alternative splicing of exons 5 and 6. Panel B contains molecular weight markers used by Genescan software to assign accurate sizes.

sample, and are distinguished by a different fluorescent colour from the sample peaks, although they have been presented separately here in a monochrome display. Genescan software utilises the markers to accurately calculate the size of the PCR products in the sample present in the same gel lane. The 710 and 854 bp products run as split peaks, since both strands of each product run separately with slight mobility differences due to differences in sequence composition.

Competitive RT-PCR of PML-RARα Transcripts

For quantitation of PML-RARα, the areas under the curves of both components of any individual PCR product peak are added together, and then corrected for differences in the number of potential [R110]dUTPs that can be incorporated in each strand. Where multiple PCR products are present, as is the case for bcr-1 breakpoints, the corrected areas under the curve are added together. The data are then plotted as the logarithm of the ratio of the sum of the sample peaks to the competitor peak (on the x-axis), versus the logarithm of the number of competitor molecules added (on the y-axis). Linear regression is performed to ensure a straight line relationship, and the y-intercept (where x=0) corresponds to the equivalence point, ie. the point at which the number of target cDNA molecules in the sample equals the number of competitor DNA molecules (Figure 3) [7]. The same analysis is

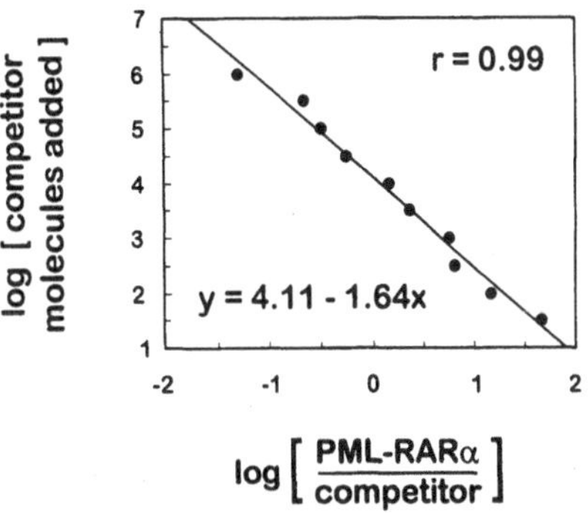

Figure 3. Quantitation of PML-RARα fusion transcripts in APL. The sum of the areas under the curves (AUC) of each PCR product derived from the fusion transcript (after correction for length differences) were divided by the AUC for the corresponding competitor PCR product in the same reaction. The logarithm of this ratio was then plotted against the logarithm of the number of competitor molecules added to each reaction, and linear regression used to plot the line of best fit. The point of equivalence (where x=0) indicates that $10^{4.11}$ (ie. 12903) molecules of competitor produced an equivalent signal to that derived from the patient's sample containing PML-RARα fusion transcripts.

Table 2. Serial quantitation of PML-RARα and transferrin receptor transcripts in a patient with APL

	PML-RARα	Transferrin Receptor	PML-RARα / Transferrin Receptor
Pre- ATRA	12903	2394970	.00539
Day 35 ATRA	529	557442	.00095
	(4.1%)	(23.3%)	(17.6%)

performed for the control gene, viz. the transferrin receptor, and the results are expressed as the ratio of PML-RARα to transferrin receptor (Table 2).

DISCUSSION

Traditional methods available for the demonstration of gene rearrangements include Southern and Northern analysis. More recently, PCR has been employed because it is more rapid, more sensitive, and requires less tissue. PCR systems have been developed for most of the gene rearrangements associated with hematological malignancies that have been characterized to date. In addition to providing a diagnostic assay, PCR and RT-PCR can also be used to detect minimal residual disease, since they are sensitive to the presence of as few as one malignant cell in a background of 10^5-10^6 normal cells [9]. In broad terms, the clinical validity of PCR-based detection of minimal residual disease has been demonstrated for CML following allogeneic bone marrow transplantation [10,11], for APL after ATRA and chemotherapy [6], for purged autografts in non-Hodgkin's lymphoma [12], and for ALL following chemotherapy-induced remission [13].

To some extent, however, the detection of minimal residual disease by highly sensitive PCR-based systems has been hindered by their non-quantitative nature. This has particularly been evident in CML and APL. After allogeneic bone marrow transplantation for CML, BCR-ABL fusion transcripts are frequently detectable by RT-PCR for periods up to 12 months [5]. Negative results indicate a very low probability of relapse, whereas positive results are of somewhat limited value since they do not always indicate that hematological relapse is imminent. Similar considerations apply for patients with APL [14-16]. PML-RARα fusion transcripts often remain positive during ATRA induction therapy, despite other evidence of leukemic cell differentiation, and some patients remain RT-PCR positive even after complete remission has been obtained. The ability to quantitate PCR products reliably, thereby allowing discrimination between declining levels of fusion transcripts and rising levels, has the potential to dramatically alter the clinical management of patients with CML and APL. For example, a progressive rise in BCR-ABL fusion transcripts following allogeneic transplantation in CML would suggest that residual leukemia was both present and increasing in activity, and hence additional post-engraftment therapy was appropriate. In contrast, a declining level of fusion transcripts would suggest that residual leukemic cells were being eliminated by a graft-versus-leukemia effect; under these circumstances further myelosuppression and/or immunosuppression would be inappropriate, and these patients could be spared the toxicity of unnecessary additional therapy.

There is now a growing body of opinion that quantitative RT-PCR is the most informative method for monitoring patients with CML [17-20]. In fact, the European Investigators on Chronic Myeloid Leukemia Group has recommended the use of quantitative PCR for monitoring minimal residual disease, and has accepted the detection of PCR relapse as a basis for therapeutic decisions [21]. While the place for quantitative RT-PCR in determining therapy in APL has not yet been clearly established, it is likely that useful information will be derived from such data when they are available. The methodology reported here represents the first description of a quantitative system for the detection of PML-RARα fusion transcripts. Our protocol effectively controls for the major sources of variability in quantitative RT-PCR (ie. quality and amount of RNA due to variability in RNA extraction efficiency, amount of cDNA template used in the PCR arising from variability in reverse transcription efficiency, and finally PCR amplification efficiency).

Our results demonstrate a marked reduction in PML-RARα transcripts following remission induction with ATRA, even though the patient was still PCR positive by conventional (non-quantitative) RT-PCR when remssion was first attained. Although the control transcript used in this study, transferrin receptor, also showed a significant reduction, the amount of fusion transcript was still reduced to 17.6% of the initial value after correction for transferrin receptor expression. Since transferrin receptor expression is cell cycle related, the fall observed in its expression is probably real, rather than due to reduced RNA extraction or cDNA synthesis. It could therefore be argued that transferrin receptor is not useful as a control in this situation, and the fall in PML-RARα transcripts to 4% of pre-treatment values may be more realistic. However, it is likely that transferrin receptor expression will be a more valid control when comparing serial values of PML-RARα expression during ongoing remission, when transferrin receptor expression is likely to be relatively constant. The clinical relevance of quantitation in APL can now be addressed in serial studies of large cohorts of patients treated with ATRA, chemotherapy, and/or stem cell transplantation.

ACKNOWLEDGMENT

This project was supported by the Leo & Jenny Leukaemia and Cancer Foundation.

REFERENCES

1. Rabbitts TH. Chromosomal translocations in human cancer. Nature 372:143, 1994.
2. Warrell RP Jr, de Thé H, Wang Z-Y, Degos L. Acute promyelocytic leukemia. New Engl J Med 329:177, 1993.
3. Zelent A. Translocation of the RARα locus to the PML or PLZF gene in acute promyelocytic leukaemia. Brit J Haematol 86:451, 1994.
4. Dhingra K, Kurzrock R, Kantarjian H, Baine R, Eastman PS, Ku S, Gutterman JU, Talpaz M. Minimal residual disease in interferon-treated chronic myelogenous leukemia: results and pitfalls of analysis based on polymerase chain reaction. Leukemia 6:754, 1992.
5. Miyamura K, Barrett J, Kodera Y, Saito H. Minimal residual disease after bone marrow transplantation for chronic myelogenous leukemia and implications for graft-versus-leukemia effect: a review of recent results. Bone Marrow Transplantation 14:201,1994.
6. Miller WH Jr, Kakizuka A, Frankel SR, Warrell RP Jr, DeBlasio A, Levine K, Evans RM, Dmitrovsky E. Reverse transcription polymerase chain reaction for the rearranged retinoic acid receptor α clarifies diagnosis and detects minimal residual disease in acute promyelocytic leukemia. Proc Natl Acad Sci (USA) 89:2694, 1992.
7. Cross NCP. Quantitative PCR techniques and applications. Brit J Haematol 89:693, 1995.
8. Chomczinski P and Sacchi N. Single-step method of RNA isolation by acid guanidinium thiocyanate-phenol-chloroform extraction. Analytical Biochem 162:156, 1987.
9. Morgan GJ, Shiach C, Potter M. The clinical value of detecting gene rearrangements in acute leukaemias. Brit J Haematol 88:459, 1994.
10. Kohler S, Galili N, Sklar JL, *et al.* Expression of bcr-abl fusion transcripts following bone marrow transplantation for Philadelphia chromosome-positive leukemia. Leukemia 4:541, 1990.
11. Hughes TP, Morgan GJ, Martiat P, *et al.* Detection of residual leukemia after bone marrow transplant for chronic myeloid leukemia: Role of polymerase chain reaction in predicting relapse. Blood 77:874, 1991.

12. Gribben JG, Freedman AS, Neuberg D, *et al.* Immunologic purging of marrow assessed by PCR before autologous bone marrow transplantation for B-cell lymphoma. N Engl J Med 325:1525, 1991.
13. Brisco MJ, Condon J, Hughes E, *et al.* Outcome prediction in childhood acute lymphoblastic leukaemia by molecular quantification of residual disease at the end of induction. Lancet 343:196, 1994.
14. LoCoco F, Diverio D, Pandolfi PP, *et al.* Molecular evaluation of residual disease as a predictor of relapse in acute promyelocytic leukaemia. Lancet 340:1437, 1992.
15. Huang W, Sun G-L, Li X-S, *et al.* Acute promyelocytic leukemia: Clinical relevance of two major PML-RARα isoforms and detection of minimal residual disease by retrotranscriptase/polymerase chain reaction to predict relapse. Blood 82:1264, 1993.
16. Miller WH Jr, Levine K, DeBlasio A, et al. Detection of minimal residual disease in acute promyelocytic leukemia by a reverse transcriptase polymerase chain reaction assay for the PML/RAR-α fusion mRNA. Blood 82:1689, 1993.
17. Thompson JD, Brodsky I & Yunis JJ. Molecular quantification of residual disease in chronic myelogenous leukemia after bone marrow transplantation. Blood 79:1629, 1992.
18. Malinge MC, Mahon FX, Delfau MH, Daheron L, Kitzis A, Guilhot F, Tanzer J, Grandchamp B. Quantitative determination of the hybrid Bcr-Abl RNA in patients with chronic myelogenous leukaemia under interferon therapy. Brit J Haematol 82:701, 1992.
19. Lion T, Henn T, Gaiger A, *et al.* Early detection of relapse after bone marrow transplantation in patients with chronic myelogenous leukaemia. Lancet 341:275, 1993.
20. Cross NC, Feng L, Chase A, *et al.* Competitive polymerase chain reaction to estimate the number of BCR-ABL transcripts in chronic myeloid leukemia patients after bone marrow transplantation. Blood 82:1929, 193.
21. Lion T. Clinical implications of qualitative and quantitative polymerase chain reaction analysis in the monitoring of patients with chronic myelogenous leukemia. The European Investigators on Chronic Myeloid Leukemia Group. Bone Marrow Transplantation 14:505, 1994.

34

AIDS-RELATED NON-HODGKIN LYMPHOMAS

Molecular Genetics, Viral Infection, and Cytokine Deregulation

Gianluca Gaidano,[1*] Cristina Pastore,[1] Annunziata Gloghini,[2] Daniela Capello,[1] Gisella Volpe,[1] Paolo Ghia,[3] Giuseppe Saglio,[1,4] and Antonino Carbone[2]

[1] Laboratorio di Medicina e Oncologia Molecolare
Dipartimento di Scienze Biomediche e Oncologia Umana
Ospedale San Luigi, Università di Torino, Torino, Italy
[2] Divisione di Anatomia Patologica
I.R.C.C.S.-Centro di Riferimento Oncologico, Aviano, Italy
[3] Basel Institute for Immunology
Basel, Switzerland
[4] CNR-CIOS
Università di Torino, Torino, Italy

ABSTRACT

AIDS-related non-Hodgkin lymphomas (AIDS-NHL) are almost invariably derived from B cells and are grouped into three distinct histologic categories, including small non cleaved cell lymphoma (SNCCL), diffuse large cell lymphoma (DLCL), and anaplastic large cell lymphoma (ALCL). In addition, AIDS-NHL presenting solely as a body cavity effusion are thought to be a peculiar clinico-pathologic entity and are defined as body-cavity-based lymphoma (BCBL). At the biologic level, AIDS-related lymphomagenesis is characterized by activation of proto-oncogenes, inactivation of tumor suppressor genes, viral infection of the tumor clone, and deregulated cytokine production. Distinct AIDS-NHL types associate with specific molecular pathways. The first pathogenetic pathway clusters with AIDS-SNCCL, and is characterized by a relatively mild degree of host immunodeficiency. AIDS-SNCCL consistently associates with c-*MYC* rearrangements and *p53* inactivation in 100% and 60% of the cases respectively, whereas infection by Epstein-Barr virus (EBV) is restricted to 30% of the cases. Production of high levels of IL-10 is an additional peculiar

* Corresponding author: Gianluca Gaidano, M.D., Ph.D., Laboratorio di Medicina e Oncologia Molecolare, Sezione 5A, Dipartimento di Scienze Biomediche e Oncologia Umana, Ospedale San Luigi, Università di Torino, Orbassano - Torino 10043, Italy. Tel: (39)(11)9026-605; Fax: (39)(11)9038-636.

Molecular Biology of Hematopoiesis 5, edited by Abraham et al.
Plenum Press, New York, 1996

feature of EBV positive AIDS-SNCCL. The second pathogenetic pathway associates with AIDS-DLCL, which is usually accompanied by a marked immunodeficiency of the host. AIDS-DLCL is characterized by EBV infection in the large majority of cases and by the mutually exclusive presence of *BCL*-6 rearrangements and c-*MYC* translocations in 40% of the cases. A third pathway characterizes AIDS-BCBL, which associates virtually in all cases with infection by EBV and with the presence of DNA sequences of the recently identified Kaposi sarcoma herpes virus (KSHV) in the apparent absence of other known genetic lesions. Finally, the pathogenetic features of AIDS-ALCL are still under investigation.

INTRODUCTION

Non-Hodgkin lymphomas (NHL) represent one of the most common malignancies associated with AIDS, and, starting in 1985, AIDS-related NHL (AIDS-NHL) are recognized as an AIDS defining illness.[1-3] Since their initial observation in 1982, the incidence of AIDS-NHL has been consistently increasing and they now represent the most frequent AIDS-related malignancy in some AIDS risk groups, namely the hemophiliacs.[1-3]

AIDS-NHL are almost invariably derived from B cells and belong to the high grade lymphomas group[1-4]. Three main histologic categories may be recognized, including small non cleaved cell lymphoma (SNCCL), diffuse large cell lymphoma (DLCL), and anaplastic large cell lymphoma (ALCL). AIDS-related DLCL (AIDS-DLCL) and AIDS-related-SNCCL (AIDS-SNCCL) account for approximately 65% and 30% of AIDS-NHL, respectively, whereas AIDS-related ALCL (AIDS-ALCL) occurs more rarely.[1-3] When compared to NHL of similar histology in the immunocompetent host, AIDS-NHL display distinctive features, including widespread extent of disease at presentation, poor prognosis, and the frequent involvement of extra-nodal sites.[1-3] Recently, it has been proposed that AIDS-NHL presenting solely as a body cavity effusion should be singled out as a specific clinico-pathologic entity and termed as body-cavity-based lymphoma (BCBL).[5,6]

Intriguingly, the distribution of the different AIDS-NHL types varies according to the patients' CD4 counts and the time elapsed since HIV infection.[7] Thus, it is now well established that AIDS-DLCL generally associates with lower CD4 counts and occurs as a later manifestation of HIV infection than AIDS-SNCCL.

Despite the accumulation of a relatively vast body of knowledge regarding the clinico-pathologic features of AIDS-NHL,[4,8] the biologic bases of these tumors is still largely unclarified. Several factors are thought to contribute to AIDS-related lymphomagenesis, including *a)* host predisposing factors; *b)* accumulation of genetic lesions within the tumor clone; *c)* tumor infection by viruses; and *d)* deregulation of cytokine loops.

HOST PREDISPOSING FACTORS

The association between an immunodeficiency state and the development of lymphoma has been long since recognized in several clinical conditions other than AIDS, including congenital and iatrogenic immunodeficiencies.[9] Although the exact immune alteration contributing to lymphomagenesis is not clear in the instance of AIDS, at least three major components are thought to be involved, including *a)* disturbed immunosurveillance; *b)* chronic antigen stimulation; and *c)* pre-existent viral infection.

The role of disturbed immunosurveillance has been substantiated by several observations. First, one of the AIDS-associated immunologic defects selectively impairs the immunosurveillance against EBV infected cells, which are present in increased numbers in the patients' blood and lymphoid organs and may be responsible for minor clonal B cell expansions that precede neoplastic transformation.[10, 11] Second, and most importantly, the risk for NHL development in AIDS patients inversely correlates with the levels of circulating CD4 positive T cells.[7]

Chronic antigen stimulation is thought to be involved in AIDS-related lymphomagenesis based on the indirect observation that AIDS patients frequently develop persistent generalized lymphadenopathy (PGL) and polyclonal hypergammaglobulinemia, presumably resulting from chronic B cell expansion and activation.[1-3] In addition, the role of chronic antigen stimulation is supported by the experimental evidence that AIDS-NHL produce self reactive antibodies and carry somatic mutations within the hypervariable regions of the immunoglobulin genes, the nature of which is consistent with antigen selection.[12] Finally, the contribution of viral infection will be discussed in a later section.

MOLECULAR GENETICS OF AIDS-NHL

The contribution of genetic lesions to AIDS-related lymphomagenesis has been clarified to a certain extent, leading to the notion that multiple genetic aberrations are necessary for the development of AIDS-NHL (Table 1).[3, 8] Analysis of the different types of AIDS-NHL has shown that c-*MYC* activation occurs in 100% of AIDS-SNCCL, as well as in a limited fraction of AIDS-DLCL and AIDS-ALCL.[13-15] Conversely, AIDS-BCBL are consistently devoid of c-*MYC* alterations.[16] As in Burkitt lymphoma of the immunocompetent host, c-*MYC* activation occurs through gene rearrangements following chromosomal translocations between 8q24, the site of the c-*MYC* proto-oncogene, and an immunoglobulin chromosomal locus.[17] In addition to gross truncations of the gene, c-*MYC* deregulation in AIDS-NHL is consistently accompanied by mutations within the c-*MYC* regulatory regions, as well as by aminoacid changes at the site of the transcriptional activation domain of gene.[17,18] The role of c-*MYC* activation in AIDS-related lymphomagenesis is substantiated by the experimental evidence that activated c-*MYC* alleles lead to the neoplastic transformation of Epstein Barr virus (EBV)-immortalized B cells.[19] Furthermore, the *in vivo* targetted expression of c-*MYC* oncogenes in the B cell lineage of transgenic animals induces the development of malignancy at a relatively high frequency.[20]

Among other dominantly acting oncogenes commonly involved in the pathogenesis of NHL of the general population, *BCL*-1 and *BCL*-2 do not seem to play a role in AIDS-related lymphomagenesis.[8] On the contrary, alterations of *BCL*-6 are a relatively frequent lesion which selectively associates with the AIDS-DLCL histologic type.[21] The mechanism of *BCL*-6 rearrangements in AIDS-DLCL is similar to that of DLCL of the immunocompetent host and includes truncation within the 5' regulatory regions of the gene.[17] Intriguingly, alterations of *BCL*-6 and c-*MYC* in AIDS-DLCL appear to be mutually exclu-

Table 1. Frequency of genetic lesions in AIDS-NHL[a, b, c]

Histology	EBV	KSHV	c-*MYC*	*p53*[c]	*BCL*-6	*RAS*	*BCL*-1	*BCL*-2
AIDS-SNCCL	30%	-,–	100%	60%	–	15%	–	–
AIDS-DLCL	70%	–	20%	–	20%	15%	–	–
AIDS-ALCL	70%	–	–	–	–	ND	–	–
AIDS-BCBL	100%	100%	–	ND	ND	–	–	–

[a] AIDS-SNCCL, AIDS-related small non cleaved cell lymphoma; AIDS-DLCL, AIDS-related diffuse large cell lymphoma; AIDS-ALCL, AIDS-related anaplastic large cell lymphoma; AIDS-BCBL, AIDS-related body-cavity-based lymphoma.

[b] –, genetic lesion not involved.

[c] ND, not done.

sive genetic alterations, suggesting the existence of (at least) two distinct molecular pathways in AIDS-DLCL development.[21] Other dominantly acting oncogenes frequently activated in human neoplasia, though not in NHL of the general population, appear to be involved in AIDS-NHL, as exemplified by the case of *RAS* mutations which are detected in a small subset of AIDS-NHL independent of histology.[13]

Finally, the role of tumor suppressor gene inactivation in AIDS-NHL pathogenesis has also become evident from a number of observations.[8] Mutations and/or losses of *p53* have been found to selectively associate with 60% of AIDS-SNCCL.[13, 14] The frequent association between c-*MYC* deregulation and *p53* inactivation in AIDS-SNCCL may underlie a putative synergistic effect of these two lesions in this specific AIDS-NHL type. Deletions of the long arm of chromosome 6 represent an additional site of a putative tumor suppressor gene relevant to AIDS-related lymphomagenesis, being present in a significant proportion of AIDS-NHL.[22]

VIRAL INFECTION IN AIDS-NHL

EBV infection of the tumor clone has been repeatedly associated with AIDS-NHL. The frequency of EBV infection is strictly dependent upon the AIDS-NHL histology.[1-4, 8, 13, 23-25] Thus, EBV sequences are detected in 70% of AIDS-DLCL and AIDS-ALCL and in 30% of AIDS-SNCCL. Furthermore, EBV sequences are also found in virtually 100% of AIDS-BCBL.[16] In addition to the well recognized ability of EBV to transform B cells in vitro,[17] several other observations support the role of EBV in AIDS-related lymphomagenesis. First, EBV infection in the context of PGL predisposes to the development of lymphoma.[26] Second, by using the heterogeneity of EBV genomic termini as a marker of clonality of the infection within the tumor population, it has been shown that EBV infection precedes the expansion of the tumor clone.[13] Finally, the consistent association between EBV infection and selected AIDS-NHL categories suggest that EBV might be a *sine qua non* in the pathogenesis of at least some AIDS-NHL types.[1-4, 8, 13, 23-25] Curiously, the pattern of expression of the EBV transforming proteins EBNA-2 and LMP-1 markedly differs throughout the histologic spectrum of AIDS-NHL.[23-25] Thus, neither EBNA-2 nor LMP-1 are expressed by EBV positive AIDS-SNCCL, whereas EBV positive AIDS-DLCL and AIDS-ALCL, as well as EBV positive AIDS-BCBL (our unpublished observation), generally express LMP-1 and, occasionaly, also EBNA-2.[23-25] The phenotypic heterogeneity of EBV antigenic expression among AIDS-NHL may be partially explained by the fact that EBV uses different cellular factors depending on the maturation stage of the host cell to regulate the synthesis of its own transformation-associated proteins.[27]

Recently, Kaposi sarcoma herpesvirus (KSHV), a novel member of the herpesvirus family, has been proposed as a relevant pathogenetic factor in AIDS-related lymphomagenesis. Although KSHV has been initially identified because of its 100% association with Kaposi sarcoma,[28] subsequent studies have also detected KSHV genomic sequences in a subset of AIDS-NHL presenting as body cavity effusions and in all cases corresponding to the clinico-pathologic definition of BCBL.[6,7,16] All other types of AIDS-NHL are consistently devoid of KSHV DNA sequences. The selectivity of KSHV infection for BCBL among AIDS-NHL suggests that KSHV may play a key role in the development of this putative AIDS-NHL entity.

CYTOKINE DEREGULATION IN AIDS-NHL

Disruption of normal cytokine networks is a key feature of HIV infection.[29] Yet, the contribution of cytokines to AIDS-related lymphomagenesis is a relatively unexplored field.

Among the many cytokines regulating B cell growth, IL-6 and IL-10 have been documented to be involved in AIDS-NHL growth.[1-3, 8] Regarding IL-6, it is well established that HIV drives its expression from monocytes and macrophages,[29] thus establishing paracrine loops between monocytes/macrophages residing in lymphoid organs and the lymphoma cells.[30] In addition, it has been shown that a substantial fraction of AIDS-SNCCL produce IL-6, putatively contributing to autocrine loops which sustain the lymphoma growth (Pastore *et al.*, in preparation).

IL-10 is a pleotropic cytokine with a striking homology to BCRF1, an EBV protein. IL-10, but not BCRF1, is a potent B cell stimulator.[31] EBV positive AIDS-SNCCL produce large quantities of IL-10, whereas EBV negative AIDS-SNCCL fail to produce the cytokine or produce it at extremely low levels (Pastore *et al.*, in preparation). Production of large quantities of IL-10 is a peculiar feature of SNCCL arising in the AIDS context, since IL-10 production is absent or very low in both sporadic and endemic Burkitt lymphoma cases.[32] Intriguingly, although BCRF1 expression is never detected in AIDS-SNCCL, production of high levels of IL-10 clusters with AIDS-SNCCL cases carrying EBV infection (Pastore *et al.*, in preparation). The role of IL-10 in AIDS-related lymphomagenesis is probably a complex one. The B cell differentiating activity of IL-10 might be responsible for enhanced immunoglobulin secretion,[31] and, consequently for the hypergammaglobulinemia typically observed in AIDS patients. On the other hand, IL-10 is also a viability factor for B cells[31] and, consequently, may facilitate AIDS-NHL cell expansion by inhibiting apoptosis.

Recently, we have investigated the expression and/or production of a large panel of cytokines by AIDS-SNCLL cell lines (Pastore *et al.*, in preparation). Our data indicate that, beside IL-6 and IL-10, AIDS-SNCCL frequently associates also with production of TNFβ and with expression of IL-13 and CD30 ligand. The production of these two latter cytokines is currently being tested. Conversely, production of IL-1α, IL-1β, IL-2, IL-3, IL-4, IL-5, IL-7, IL-8, IL-12, TNFα, γIFN, TGFβ, GM-CSF, G-CSF and SCF is consistently absent in AIDS-SNCCL.

MOLECULAR PATHWAYS IN AIDS-RELATED LYMPHOMAGENESIS

During the last few years, the concept has emerged that the repertoire of genetic lesions in AIDS-NHL differs substantially according to the histologic type of the lymphoma.[1-3, 8] As outlined in the previous pages, these lesions include activation of dominantly acting oncogenes (c-*MYC*, *BCL*-6, *RAS*), inactivation of tumor suppressor genes (*p53*, 6q deletions), and viral infection of the tumor clone (EBV, KSHV). At least three independent patterns of genetic lesions can be identified in AIDS-NHL, leading to distinct molecular pathways in AIDS-related lymphomagenesis (Fig. 1 and Table 1).

The first of these pathways associates with AIDS-SNCCL, and is characterized by relatively mild immunodeficiency of the host and by the frequent presence of a PGL phase pre-existent to the lymphoma.[1-3, 8] Genetically, AIDS-SNCCL associates consistently with c-*MYC* deregulation and, in the majority of cases, also with *p53* disruption.[13, 14] EBV infection is restricted to a fraction of AIDS-SNCCL cases, which, however, fail to express the EBV transforming antigens.[13, 23-25] The role of chronic antigen stimulation has also been experimentally substantiated in AIDS-SNCCL, and appears to involve self antigens at least in the cases examined.[12] In addition, the large production of IL-10 by EBV positive AIDS-SNCCL suggests that this cytokine may play a relevant role in the lymphomagenesis process (Pastore *et al.*, in preparation; and [32]). Overall, the most puzzling issue regarding AIDS-SNCCL is its unique association with AIDS among immunodeficiencies in man.[9]

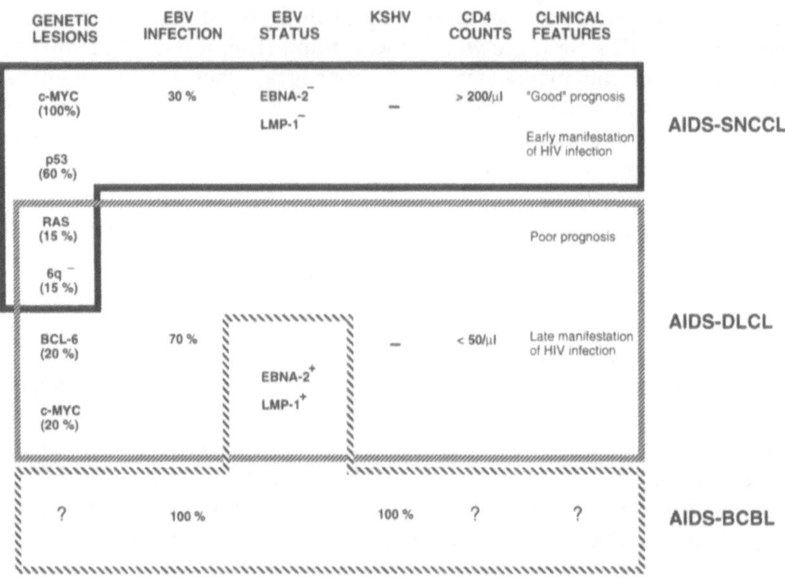

Figure 1. Clinical and molecular features of AIDS-related small non cleaved cell lymphoma (AIDS-SNCCL), AIDS-related diffuse large cell lymphoma (AIDS-DLCL), and AIDS-related body-cavity-based lymphoma (AIDS-BCBL). The molecular features of the other known type of AIDS-related NHL, namely AIDS-related anaplastic large cell lymphoma, are still under investigation.

Unravelling this discrepancy, putatively due to a causative agent closely linked to HIV and absent in other forms of immunodeficiency, may well enhance our understanding of AIDS-related lymphomagenesis.

A second molecular pathway clusters with AIDS-DLCL. These tumors are characterized by severe immunodeficiency of the host and by a very poor prognosis.[1-3, 8] The actual role of prolonged exposure to immunosuppression in the genesis of AIDS-DLCL is testified by the outburst of these neoplasms in AIDS long term survivors and by the direct correlation between decreased CD4 counts and increased risk of AIDS-DLCL development.[7] At the biologic level, AIDS-DLCL associates with EBV infection in a significant proportion of cases.[13] Since LMP-1 may be expressed in EBV positive AIDS-DLCL,[23-25] it is reasonable to think that EBV is indeed a driving force for tumor growth and expansion. In addition to EBV infection, AIDS-DLCL also associates with rearrangements of c-*MYC* or *BCL*-6 in 40% of the cases.[13, 21] Curiously, alterations of c-*MYC* and of *BCL*-6 appear to be mutually exclusive lesions in AIDS-DLCL,[21] suggesting a certain degree of heterogeneity in the pathogenesis of this tumor.

The third pathway is a distinguishing feature of AIDS-BCBL and includes EBV infection as well as the consistent presence of KSHV genomic sequences in the apparent absence of other known genetic lesions.[16] As stated above, BCBL is a novel clinicopathologic entity represented by NHL growing in liquid phase and clinically presenting as body cavity effusions.[5, 6] AIDS-BCBL preferentially express large or anaplastic cell morphotype and a null phenotype, although immunogenotypic analysis demonstrates their origin from the B cell lineage;[5, 6, 16] in addition, AIDS-BCBL generally express several activation markers such as CD30 and CD38.[5, 6, 16] Because AIDS-BCBL has been identified as a distinct NHL category only recently, further studies are needed in order to precisely clarify the clinical features and the molecular genetics of these tumors.

Finally, no specific genetic pathway has been found to associate with AIDS-ALCL, although it has been suggested that this AIDS-NHL type may be strictly related to EBV infection, as it is also the case for Hodgkin's disease occurring in patients with AIDS.[4, 8, 23-25] As for AIDS-BCBL, detailed investigations of AIDS-ALCL are urged in the near future.

ACKNOWLEDGMENT

This work has been supported by IX AIDS project, Istituto Superiore di Sanità, Rome, Italy (grants # 9404-33 and 9404-04). C.P. is being supported by a fellowship from A.I.R.C., Milan, Italy. G.V. is being supported by a fellowship from "Fondazione Piera, Pietro e Giovanni Ferrero", Alba, Italy.

REFERENCES

1. Karp JE, Broder S: Acquired immunodeficiency syndrome and non-Hodgkin's lymphomas. Cancer Res 51:4743, 1991
2. Beral V, Peterman T, Berkelman R, Jaffe H: AIDS-associated non-Hodgkin lymphomas. Lancet 337:805, 1991
3. Levine AM: Acquired immunodeficiency syndrome-related lymphoma. Blood 80:8, 1992
4. Gaidano G, Carbone A: AIDS-related lymphomas: From pathogenesis to pathology. Br J Haematol, 90:235, 1995
5. Walts AE, Shintaku IP, Said JW: Diagnosis of malignant lymphoma in effusions from patients with AIDS by gene rearrangement. Am J Clin Pathol 94:170, 1990
6. Green I, Espiritu E, Ladanyi M, Chaponda R, Wieczorek R, Gallo L, Feiner H: Primary lymphomatous effusions in AIDS: a morphologic, immunophenotypic and molecular study. Mod Pathol 8:39, 1995
7. Pluda JM, Venzon D, Tosato G, Lietzau J, Wywill K, Nelson DL, Jaffe ES, Karp JE, Broder S, Yarchoan R: Parameters affecting the development of non-Hodgkin's lymphoma in patients with severe human immunodeficiency virus infection receiving antiretroviral therapy. J Clin Oncol 11:1099, 1993
8. Gaidano G, Dalla-Favera R: Molecular pathogenesis of AIDS-related lymphomas. Adv Cancer Res, 67:113, 1995
9. Frizzera G: Atypical lymphoproliferative disorders. In Neoplastic Hematopathology. DM Knowles, ed. Baltimore: Williams & Wilkins, pp. 459-495. 1992
10. Birx DL, Redfield RR, Tosato G: Defective regulation of Epstein-Barr virus infection in patients with acquired immunodeficiency syndrome (AIDS) or AIDS-related disorders. N Engl J Med 314:874, 1986
11. Pelicci PG, Knowles DM, Arlin ZA, Wieczorek R, Luciw P, Dina D, Basilico C, Dalla-Favera R: Multiple monoclonal B cell expansions and c-myc oncogene rearrangements in acquired immune deficiency syndrome-related lymphoproliferative disorders. Implications for lymphomagenesis. J Exp Med 164:2049, 1986
12. Riboldi P, Gaidano G, Schettino EW, Knowles DM, Dalla-Favera R, Casali P: Two AIDS-associated Burkitt's lymphomas produce specific anti-i IgM cold agglutinins utilizing somatically mutated V_H4-21 segments. Blood 83:2952, 1994
13. Ballerini P, Gaidano G, Gong JZ, Tassi V, Saglio G, Knowles DM, Dalla-Favera R: Multiple genetic lesions in acquired immunodeficiency syndrome-related non-Hodgkin's lymphoma. Blood 81:166, 1993
14. Gaidano G, Parsa NZ, Tassi V, Della-Latta P, Chaganti RSK, Knowles DM, Dalla-Favera R: In vitro establishment of AIDS-related lymphoma cell lines: Phenotypic characterization, oncogene and tumor suppressor gene lesions, and heterogeneity in Epstein-Barr virus infection. Leukemia 7:1621, 1993
15. Chadburn A, Cesarman E, Jagirdar J, Subar M, Mir RN, Knowles DM: CD30 (Ki-1) positive anaplastic large cell lymphomas in individuals infected with the human immunodeficiency virus. Cancer 72:3078, 1993
16. Cesarman E, Chang Y, Moore PS, Said JW, Knowles DM: Kaposi's sarcoma-associated herpesvirus-like DNA sequences in AIDS-related body-cavity-based lymphomas. N Engl J Med 332:1186, 1995
17. Gaidano G, Dalla-Favera R: Molecular biology of lymphoid neoplasms. In The Molecular Basis of Cancer. Mendelsohn J, Howley PM, Israel MA, Liotta LA, eds. Philadelphia: WB Saunders, pp. 251-279. 1995

18. Bhatia K, Spangler G, Gaidano G, Hamdy N, Dalla-Favera R, Magrath I: Mutations in the coding region of c-myc occur frequently in acquired immunodeficiency syndrome-associated lymphomas. Blood 84:883, 1994
19. Lombardi L, Newcomb EW, Dalla-Favera R: Pathogenesis of Burkitt's lymphoma: Expression of an activated c-myc oncogene causes the tumorigenic conversion of EBV-infected human lymphoblasts. Cell 49:161, 1987
20. Adams JM, Harris AW, Pinkert CA, Corcoran LM, Alexander WS, Cory S, Palmiter RD, Brinster RL: The c-myc oncogene driven by immunoglobulin enhancers induces lymphoid malignancy in transgenic mice. Nature 318:533, 1985
21. Gaidano G, Lo Coco F, Ye BH, Shibata D, Levine AM, Knowles DM, Dalla-Favera R: Rearrangements of the BCL-6 gene in AIDS-associated non-Hodgkin's lymphoma: Association with diffuse large cell subtype. Blood 84:397, 1994
22. Gaidano G, Hauptschein RS, Parsa NZ, Offit K, Rao PH, Lenoir G, Knowles DM, Chaganti RSK, Dalla-Favera R: Deletions involving two distinct regions of 6q in B-cell non-Hodgkin lymphoma. Blood 80:1781, 1992
23. Carbone A, Tirelli U, Gloghini A, Volpe R, Boiocchi M: Human immunodeficiency-virus-associated lymphomas may be subdivided into two main groups according to Epstein-Barr viral latent gene expression. J Clin Oncol 11:1674, 1993
24. Carbone A, Gloghini A, Zanette I, Canal B, Volpe R: Demonstration of Epstein-Barr virus genomes by in situ hybridization in acquired immune deficiency syndrome-related high grade and anaplastic large cell $CD30^+$ lymphomas. Am J Clin Pathol 99:289, 1993
25. Carbone A, Gloghini A, Volpe R, Boiocchi M, Tirelli, and the Italian Cooperative Group on AIDS and Tumors: High frequency of Epstein-Barr virus latent membrane protein-1 expression in AIDS-related Ki-1 (CD30)-positive anaplastic large cell lymphomas. Am J Clin Pathol 101:768, 1994
26. Shibata M, Weiss LM, Nathwani BN, Brynes KK, Levine AM: Epstein-Barr virus in benigh lymphnode biopsies from individuals infected with the human immunodeficiency virus is associated with concurrent or subsequent development of non-Hodgkin's lymphoma. Blood 77:1527, 1991
27. Contreras-Bodin BA, Anvret M, Imreh S, Altiok E, Klein G, Masucci MG: B cell phenotype-dependent expression of the Epstein-Barr virus nuclear antigens EBNA-2 to EBNA-6: Studies with somatic cell hybrids. J Gen Virol 72:3025, 1991
28. Chang Y, Cesarman E, Pessin MS, Lee F, Culpepper J, Knowles DM, Moore PS: Identification of herpesvirus-like DNA sequences in AIDS-associated Kaposi's sarcoma. Science 266:1865, 1994
29. Fauci AS, Schnittman SM, Poli G, Koenig S, Pantaleo G: Immunopathogenetic mechanisms in human immunodeficiency virus (HIV) infection. Ann Int Med 114:678, 1991
30. Emilie D, Coumbaras J, Raphael M, Devergne O, Delecluse HJ, Gisselbrecht C, Michelis JF, Van Damme J, Taga T, Kishimoto T, Crevon MC, Galanaud P: Interleukin-6 production in high grade B lymphomas: correlation with the presence of malignant immunoblasts in acquired immunodeficiency syndrome and in human immunodeficiency virus-negative patients. Blood 80:498, 1992
31. Callard R, Gearing A: The Cytokine Factsbook. London: Academic Press, 1994
32. Benjamin D, Knobloch TJ, Dayton MA: Human B-cell interleukin-10: B-cell lines derived from patients with acquired immunodeficiency syndrome and Burkitt's lymphoma constitutively secrete large quantities of interleukin-10. Blood 80:1289, 1992

35

ANTI-HIV VIRAL INTERFERENCE INDUCED BY RETROVIRAL VECTORS EXPRESSING A NON-PRODUCER HIV-1 VARIANT

M. Federico,[1] R. Bona,[1] P. D'Aloja,[1] M. Baiocchi,[1] K. Pugliese,[1] F. Nappi,[1] C. Chelucci,[2] F. Mavilio,[3] and P. Verani[1*]

[1] Laboratory of Virology and
[2] Laboratory of Hematology and Oncology
Istituto Superiore di Sanità, Rome
[3] DIBIT-Istituto Scientifico
S. Raffaele Hospital, Milan, Italy

ABSTRACT

A Hut-78 cell clone (F12) harboring a non-producer human immunodeficiency virus (HIV-1) variant shows complete resistance to HIV-1 or HIV-2 superinfection. The F12-HIV provirus produces an altered HIV-1 protein pattern and cannot generate even immature viral particles. We demonstrated that HeLa CD4+ cells transfected with the F12-HIV genome resist HIV superinfection through a CD4-independent mechanism. As F12-HIV appears to be a useful system to induce anti-HIV intracellular immunization, we constructed various retroviral vectors containing the F12-HIV genome, modified by elimination of the F12 3' LTR and part of its *nef* gene, inserted "antisense" with respect to the Moloney murine leukemia virus 5' LTR. Here we show that recombinant retroviral particles carrying the N2/F12-HIV *nef*-(as) construct can stably transduce both CEMss human cells and primary human peripheral blood lymphocytes, inducing the expression of the F12-HIV genome. These results could open the way to an anti-AIDS gene therapy strategy based on F12-HIV-induced intracellular immunization.

INTRODUCTION

Most of the current strategies for gene therapy to cure AIDS are based on containment of the viral infection, until an approach can be found to excise the integrated HIV genome

* Corresponding author: Prof. Paola Verani, Laboratory of Virology, Istituto Superiore di Sanità, Viale Regina Elena, 299, 00161 Rome, Italy; Tel: 0039 6 4440143, Fax 0039 6 4453904.

Molecular Biology of Hematopoiesis 5, edited by Abraham et al.
Plenum Press, New York, 1996

from the host cells, thus rendering the patient truly virus-free. However, a different strategy may be represented by a gene therapy project able to render cells HIV-resistant. Engineering the CD34+ early hematopoietic cells would allow transmission of HIV-resistance to the whole hematopoietic progeny.

In our laboratory, a HUT-78 cell clone was isolated, carrying an HIV variant, named F12, whose characteristics are the inability to produce even immature viral particles and the ability to confer to the cell resistance to superinfection by a wild type HIV, a phenomenon known as viral interference (1). We have shown that transfected HeLa CD4+ cells expressing the F12 genome became resistant to HIV superinfection without any evidence of CD4 down-regulation. This viral interference was also present when the pNL4-3 infectious molecular clone was transfected, indicating that a block in a step following retrotranscription may occur (2).

As a gene therapy protocol based on induction of viral interference appears to be a promising tool in AIDS therapy, a reliable means for transducing the F12-HIV genome in human primary cells becomes necessary. Among the different methodologies developed so far, the use of a murine retroviral vector appears most suitable for clinical use, due to its safety and efficiency. For these reasons, the aim of this study has been the construction of a retroviral particle able to transduce the functionally defective and interfering F12-HIV genome in human primary cells.

MATERIALS AND METHODS

Molecular Constructs

The full-length F12-HIV proviral DNA was previously cloned from a genomic library of the HUT-78/F12 cell DNA (3). The four different N2/F12-HIV molecular constructs obtained and tested are (Fig. 1): i) N2/F12-HIV *nef-* "sense" (s), where the F12-HIV genome was inserted in the same transcription orientation as N2, after being deprived of part of the

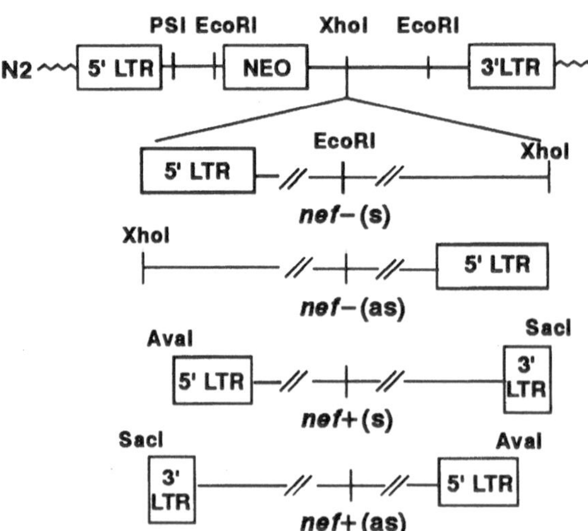

Figure 1. Schematic maps of the four N2/F12-HIV molecular constructs. XhoI and EcoRI sites are indicated as are the restriction sites where the F12-HIV LTRs are truncated in the *nef+* constructs. The N2 packaging site (psi) and mutagenized polyA signal in the 5' F12-HIV LTR are also shown.

nef gene (truncated in the XhoI site); ii) N2/F12-HIV *nef-* "antisense" (as), where the F12-HIV modified as described above was inserted in the opposite orientation with respect to the N2 vector; iii) N2/F12-HIV *nef+* "sense" (s), where an F12-HIV genome mantaining its *nef* gene but lacking both the remaining part of the 3' LTR (cut in the SacI site) and the "negative regulatory elements" (NRE) of its 5' LTR (through the AvaI cut) was inserted in the same orientation as the N2 vector; iv) N2/F12-HIV *nef+* "antisense" (as), where the F12-HIV genome, modified as for *nef+* (s) and the N2 retroviral vector are in opposite transcriptional orientations.

To avoid the premature polyadenylation of RNA transcribed from the N2 5' LTR in the "sense" constructs, the AATAAA consensus of F12-HIV 5' LTR was mutagenized to AACCAA.

Cell Cultures and Transfections

CEMss cells were maintained in RPMI 1640 supplemented with 10% fetal calf serum (FCS), HeLa, NIH 3T3, GP+E86 and PA317 were grown in Dulbecco's modified MEM supplemented with 10% FCS. Human peripheral blood lymphocytes (PBLs) obtained as in (4) were stimulated with phytohemagglutinin (5 μg/ml) for 48-72 hours and then cultivated in RPMI 1640 supplemented with 20% FCS and 50 units/ml of recombinant human interleukin (IL)-2 (Roche, Nutley, N.J., USA). For the G418 (GIBCO, Grand Island, N.Y., USA 50% of activity) selection, CEMss cells were treated with 1 mg/ml of the antibiotic, whereas both the "packaging" and the NIH 3T3 cells were selected with 0.5 mg/ml of G418. Transfections of "packaging" cell monolayers were performed by the calcium-phosphate protocol, as described in (5).

Virus Infection

Supernatants from GP+E86 and PA317 cells were used to infect subconfluent monolayers of PA317 (single cycle of infection) or CEMss cells (four cycles of infection), respectively. Co-cultivations were performed in order to infect with the amphotropic retrovirus released by PA317 "producer" cell clones either CEMss or human PBLs. 10^5 "producer" cells were seeded together with either 10^5 CEMss cells or 10^6 PBLs, in both cases in 1 ml of culture. After 24 h, PA317 cells were eliminated by removing the medium and reseeding the suspension cultures. Forty-eight h after the end of the infection cycles, the G418 selection was started. Titers of the amphotropic retrovirus preparation were assayed as colony forming units (CFU/ml) on NIH 3T3 cells as described in (6).

Molecular Characterization: Analysis of DNA and F12-HIV Expressed Proteins

DNA from amphotropic retrovirus infected cells was extracted by the phenol-chloroform method, digested with the appropriate restriction endonucleases, run in a 0.8% agarose gel and Southern blotted as described elsewhere (2). The F12-HIV genome excised from the N2/F12-HIV construct was ^{32}P radioactively labelled by the random primed method. The determination of intracellular p24 antigen in retrovirus infected cells was carried out by lysing 10^5 cells in 200μl of TNE buffer plus 0.1% Triton X-100. Supernatants were then tested in ELISAs by an antigen capture assay (Abbott, North Chicago, IL, USA).

Table 1. Amount of intracytoplasmic F12-HIV p24 protein (pg/10^5) in N2/F12-HIV-transfected cells

	48 h posttransfection	After G418 selection			
	HeLa	NIH 3T3	PA317	HeLa	NIH 3T3
N2	<10	<10	<10	<10	<10
nef- (s)	124.3±10.8	37.3	63.5	>246	184.5
nef- (as)	239.2±20	25.7	121.1	>246	130.6
nef+ (s)	160.2±18.3	40	44.7	>246	211.5
nef+ (as)	203.1±1.2	11.2	117.5	210	90.5
pUc/F12-HIV	134.5±10	88.4	67.4	>246	ND

Average values of four experiments ± SD are reported. ND = Not done.

RESULTS

To verify the functionality of the four retroviral constructs, they were transiently transfected in HeLa, NIH 3T3 and PA317 cells. The intracellular presence of the F12-HIV *gag* protein in the cells was assessed 2 days after transfection by the ELISA test. Cells carrying the N2 vector alone were used as a negative control. As shown in Table 1, neither the modifications introduced into the viral genome nor the presence of the Moloney murine leukemia virus (MoMLV) LTR affect the expression of the F12-HIV *gag* protein, when compared to the same genome inserted in the pUc19 plasmid. In fact, no significant differences in the amounts of p24 protein could be detected among cells transfected with N2/F12-HIV retroviral vectors compared to cells transfected with a full-length pUc/F12-HIV molecular clone.

To determine whether the four N2/F12-HIV retroviral constructs can generate infective recombinant particles, they were transfected in the ecotropic GP+E86 packaging cell line, and supernatants were used to infect the amphotropic PA317 packaging cell line. Furthermore, PA317 cells were also directly transfected with the same constructs, and the supernatants used to infect the human CEMss cell line.

Fig. 2 shows the Southern blotted (XhoI-digested) genomic DNA of the infected PA317 and CEMss cell lines. The N2/F12-HIV *nef-* (as) construct appears to be the most suitable for our purposes. In fact, in the CEMss population infected with the supernatants from PA317 transfected with this construct, the 9-kb DNA is present, corresponding to the unspliced F12-HIV RNA, together with a lower molecular weight signal, probably derived

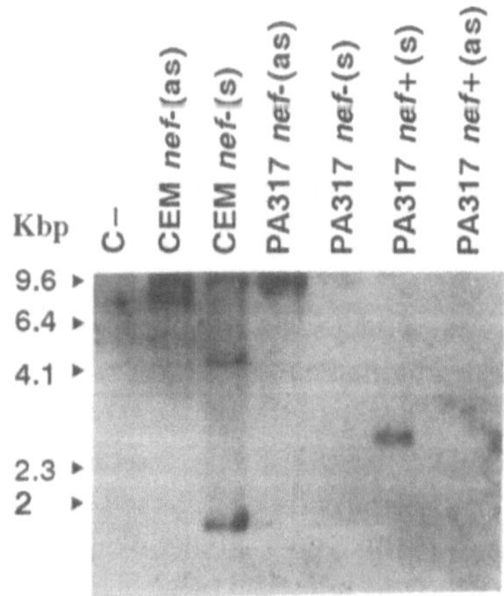

Figure 2. F12-HIV probed Southern blot of XhoI cut genomic DNA from transduced neo^r cell populations. DNA from uninfected CEMss cells was used as a negative control (c-). HindIII digested lambda phage DNA (on the left) was used as the molecular-wright marker.

Table 2. Results following infection with sense(s) or antisense(as) N2/F12-HIV *nef-* recombinant retroviruses: intracytoplasmic F12-HIV p24 protein[1] in G418 resistant cell populations (I) and number of F12-HIV expressing cell clones out of total cell clones scored (II)

	I		II	
	s	as	s	as
PA 317	<10	23.3±2.4	0/14	11/92
CEMss	11±1.2	106.3±7	0/140	6/26

[1]Values are expressed as picograms p24 HIV protein in the lysates of 10^5 G418 resistant cells.

from a genomic rearrangement. Moreover, the 9 kb band, in the absence of any other signal, is present in PA317 cells infected with the supernatants from GP+E86 cells transfected with the same construct.

CEMss cells infected with the supernatants from the N2/F12-HIV *nef-*(s) transfected PA317 cells show the integration of three different F12-HIV DNAs, probably originating from the three major forms of the F12-HIV RNA (i.e. unspliced, single- and double-spliced forms of 9, 4.8 and 2 kb, respectively), while PA317 cells infected with supernatants from GP+E86 cells carrying the same construct fail to integrate F12-HIV specific sequences.

Regarding the N2/F12-HIV *nef+* constructs, it was impossible to recover G418-resistant cells from a CEMss population after infection with PA317 transfected cell supernatants, for both the (s) and (as) construct. Furthermore, the analysis of PA317 cells, infected with the GP+E86 supernatants and selected on G418, shows no detectable signal in the case of the *nef+* (as) construct, and the presence of only a 3-kb band in the case of the *nef+* (s) virions.

To assess the functionality of the transduced N2/F12-HIV constructs, the intracytoplasmic accumulation of the F12-HIV *gag*-proteins had to be tested in the neo^r cell populations. In view of the results described above, we only analyzed *nef-* expressing cells by ELISA assay. As reported in Table 2, panel I, no positivity was detectable in the neo^r PA317 cells infected with the *nef-* (s) virions, whereas low and transient positive values were detectable in the neo^r CEMss population; they became negative a few weeks after the neo selection.

In contrast, higher levels of F12-HIV *gag* protein were detectable in both PA317 and CEMss cells integrating the *nef-* (as) genome. These values remained stable over a long period of time (> 1 year, data not shown).

To evaluate the percentage of cells stably expressing the F12-HIV genome, the neo^r PA317 and CEMss cells integrating the *nef-* genome in either (s) or (as) orientations were cloned by the limiting-dilution method. Tab. 2, panel II, shows the number of clones expressing F12-HIV *gag* protein. As expected, no F12-HIV *nef-*(s)-expressing clones were isolated either from PA317 or CEMss neo^r cell populations, even if the latter was initially positive in the ELISA test. In contrast, several PA317 cell clones expressing the *nef-*(as) genome were obtained; moreover, about 23% of CEMss clones were able to synthetize *gag* protein.

Finally, we also succeeded in transducing the N2/F12-HIV *nef-*(as) construct in fresh human PBLs. After 24 h of co-cultivation of PBLs with selected PA317 producer cell clones, the transduction efficiency was assessed by ELISA testing intracytoplasmic extracts for the presence of *gag* proteins. Positive values were obtained from 10^5 PBLs 5 days after the end of co-cultivation with either of two selected PA317 clones. Southern analysis of the XhoI

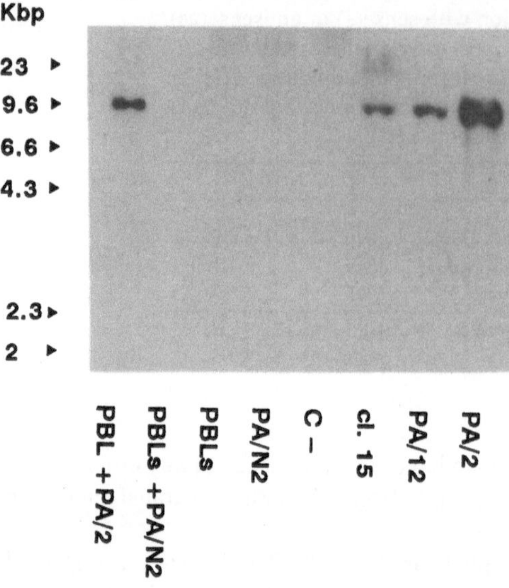

Figure 3. Southern blot analysis of genomic DNA from fresh human PBLs 5 days after the end of co-cultivation with a selected PA317 clone (clone 2). DNAs were Southern blotted after XhoI digestion and hybridized with a full-length F12-HIV molecular probe. As positive controls, DNA from a N2/F12-HIV *nef*-(as) transduced CEMss clone (clone 15) and from two selected producer PA317 clones (clones 2 and 12) was used. As negative controls, DNA from fresh human PBLs, either transduced with the N2 vector alone or untransduced, and from N2 transduced or untransduced PA317 (c-) were utilized. On the left, DNA molecular-weight markers as in Fig. 2 are reported.

digested DNA extracted from PBLs 5 days after the end of co-cultivation with a PA317 clone confirms these data (Fig. 3). An apparently intact form of the F12-HIV genome is clearly detectable in the Southern blot, where the DNA extracted from transduced PBLs is overloaded (100 versus 15 μg of DNA in the other lanes). In conclusion, we estimate by FACS analysis that about 5% of PBLs were successfully transduced by the co-cultures (data not shown).

DISCUSSION

The aim of the present work was to produce a retroviral vector able to transduce the non-producer, interfering F12-HIV variant in human primary cells, with a view of developing a gene therapy protocol for AIDS.

To reduce, as much as possible, the theoretical possibility of rearrangements between MoMLV and F12-HIV LTRs, we removed the F12-HIV LTR sequences not directly involved in the provirus promoter functions, i.e.: i) the whole 3' LTR in the *nef*- constructs, and ii) the 5' LTR NRE (shown to negatively regulate HIV expression) (7), and part of the 3' LTR encompassing both the whole U5 region and part of the R region (including the polyA signal) in the *nef*+ constructs. A Southern blot of transduced G418-resistant cell populations shows that substantially negative results were obtained with the N2/F12-HIV *nef*+ constructs. This may suggest that the presence of the complete *nef* gene and/or the lack of the 5'LTR NRE sequences could prevent either the formation of a competent recombinant retrovirus and/or the survival of the transduced cells.

Conversely, the transfection of N2/F12-HIV *nef*- constructs in packaging cells led to the formation of a recombinant retrovirus able to infect, retrotranscribe and integrate in the host cells, as demonstrated by Southern blot analysis of neo^r cells. Cells infected by the *nef*-(s) retrovirions were transduced by three major families of retrovirions, each containing the F12-HIV genome originating from unspliced, single- and double-spliced F12-HIV RNAs. No additional signals were detectable, indicating that the products of possible unexpected rearrangements are not included in infectious recombinant retroviruses and/or in cells surviving the G418 selection.

Unfortunately, we were unable to isolate CEMss clones expressing the F12-HIV genome inserted in the *nef*-(s) construct. In fact, every cell cloning experiment produced neo^r clones lacking all of the F12-HIV sequences (data not shown). This may suggest an intrinsic instability of such a molecular construct. Conversely, cells transduced by the N2/F12-HIV *nef*- (as) construct integrated stably and expressed both the F12-HIV and the G418 resistance gene. This allowed us to select producer PA317 cell clones and to succesfully transduce the F12-HIV genome in human PBLs.

The production of a MoMLV-based retroviral vector carrying the F12-HIV variant, able to stably transfect the provirus in both the human CEMss cell line and in human primary lymphocytes represents an important step towards achieving F12-HIV-mediated intracellular immunization both "in vivo" and "ex vivo". We are currently committed to demonstrating that F12-HIV-induced homologous viral interference can be reproduced in human primary lymphocytes.

ACKNOWLEDGMENTS

We are indebted to Ms. A. Lippa, D. Wool and A. Caratelli for their excellent secretarial assistance. This study was supported by grants from the AIDS Projects of the Italian Ministry of Health.

REFERENCES

1. Federico M, Titti F, Buttò S, Orecchia A, Carlini F, Taddeo B, Macchi B, Maggiano N, Verani P, Rossi GB: Biologic and molecular characterization of producer and non producer clones from HUT-78 cells infected with a patient HIV isolate. AIDS Res Hum Retrov 5: 385-396, 1989.
2. Federico M, Nappi F, Bona R, D'Aloja P, Verani P, Rossi GB: Full expression of transfected non-producer interfering HIV-1 proviral DNA abrogates susceptibility of human HeLa CD4+ cells to HIV. Virology 206: 76-84, 1995.
3. Carlini F, Federico M, Equestre M, Ricci S, Ratti G, Zibai Q, Verani P, Rossi GB: Sequence analysis of an HIV-1 proviral DNA from a non producer chronically infected HUT-78 cellular clone. J. Viral Diseases 1: 40-55, 1992.
4. Rossi GB, Verani P, Macchi B, Federico M, Orecchia A, Nicoletti L, Buttò S, Lazzarin A, Mariani G, Ippolito G, Manzari V: Recovery of HIV-related retroviruses from Italian patients with AIDS or AIDS-related complex and from asymptomatic at-risk individuals. Ann NY Acad Sci USA 511: 390-400, 1987.
5. Wigler M, Sweet R, Sim GK, Wold B, Pellicer A, Lacy E, Maniatis T, Silverstein S, Awel R: Transformation of mammalian cells with genes from procaryotes and eucaryotes. Cell 16: 758-777, 1979.
6. Federico M, Taddeo B, Carlini F, Nappi F, Verani P, Rossi GB: A recombinant retrovirus carrying a non-producer human immunodeficiency virus (HIV) type 1 variant induces resistance to superinfecting HIV. J Gen Virol 74: 2099-2110, 1993.
7. Lu Y, Touzjian N, Stensel M, Dorfman T, Sodroski JG, Haseltine WA: Identification of cis-acting repressive sequences within the negative regulatory element of human immunodeficiency virus type 1. J Virol 64: 5226-5229, 1990.

Unfortunately, we were unable to isolate CEMss clones expressing the F12-HIV genome inserted in the [illegible] construct. In fact, every cell cloning experiment produced neo[r] cells lacking all of the F12-HIV sequences (data not shown). This may suggest an intrinsic instability of such a [illegible] construct. However, [illegible] cells transduced by the [illegible] F12-HIV [illegible] construct integrated stably and expressed both the F12-HIV and the [illegible] genes. The [illegible] F12-HIV [illegible] in human PBL.

The production of a [illegible]-based retroviral vector carrying the F12-HIV variant, [illegible] human CEMss cell line and [illegible] [illegible] F12-HIV [illegible] [illegible] introduced in human [illegible]

ACKNOWLEDGMENTS

We are indebted to [illegible], E. Wool and A. Geraldo for their excellent [illegible]. This study was supported by grants from the [illegible] of the [illegible]

SUPPRESSION OF HEMATOPOIETIC SUPPORT FUNCTION IS ASSOCIATED WITH OVER-EXPRESSION OF IL-4 AND TGFβ1 IN LP-BM5 MuLV INFECTED STROMAL CELL LINES

Vincent S. Gallicchio,* Kam-Fai Tse, Jennifer Morrow, and Nedda K. Hughes

Hematology/Oncology Division
Departments of Internal Medicine, Clinical Sciences, Pathology, and Toxicology
Lucille P. Markey Cancer Center
Chandler Medical Center, University of Kentucky
and Department of Veterans Affairs, Lexington, Kentucky

ABSTRACT

Murine acquired immunodeficiency syndrome (MAIDS) induced by defective LP-BM5 MuLV murine leukemia virus is a disease with many similarities to human AIDS. Our previous studies demonstrated the depressed hematopoiesis observed in LP-BM5 MuLV infected marrow cultures was attributed to a defective hematopoietic stroma. We report now the generation of permanent stroma cell lines from non-infected and LP-BM5 MuLV infected marrow cultures. Retrovirus infection was confirmed by polymerase chain reaction for detection of viral genome expresson, p12 envelope glycoprotein. The ability of these cell lines to support *in vitro* hematopoiesis was evaluated. The results demonstrated when cocultured with normal or infected nonadherent mononuclear cells, noninfected cell lines efficiently supported the production of hematopoietic progenitors, whereas virus-infected cell lines induced suppresson of both normal and virus-infected progenitors. Expression of cytokine genes in stromal cell lines was also examined. All cell lines expressed equivalent levels of transcripts for IL-1β, IL-2, IL-3, IL-6, IL-7, IL-10, IFN, TNF-α, stem cell factor (SCF). However, infection was associated with higher expression of IL-4 and TGFβ1. These

* Corresponding author: Vincent S. Gallicchio, Ph.D., MT(ASCP), Hematology/Oncology Division, Room CC-406, Lucille P. Markey Cancer Center, Chandler Medical Center, University of Kentucky, Lexington, Kentucky, 0536-0084. Telephone: 606-323-5688; Fax: 606-257-7715

Molecular Biology of Hematopoiesis 5, edited by Abraham et al.
Plenum Press, New York, 1996

findings demonstrate that infected stromal cell lines generate a defective hematopoietic microenvironment that produces altered cytokine expression resulting in faulty hematopoiesis. Further characterization of these defective cell lines should assist in the elucidation of the mechanism(s) whereby retrovirus produce altered hematopoiesis ultimately leading to the generation of immunodeficiency.

INTRODUCTION

The hematopoietic microenvironment or stroma plays a critical role in the maintainence of normal hematopoiesis. Stromal cells of the bone marrow (BM) provide a favorable environment for self-renewal, differentiation, and the growth of hematopoietic cells. Long-term marrow cultures (LTMC) provide a valuable *in vitro* system for investigating interactions between the stromal cell compartment and developing progenitor cells. The key characteristic in LTBMC is a marrow-derived stromal cell layer comprising a variety of cell types such as reticular cells, adipocytes, macrophages, fibroblasts, and endoenthelial cells. Under appropriate *in vitro* conditions, fresh marrow-derived hematopoietic stem cells seeded onto an established stromal layer initiated under LTBMC conditions undergo proliferation and differentiation and maintain active hematopoiesis for many months. Studies using purified stromal cell cultures derived from LTBMCs demonstrated the role of specific stromal cell types, growth factors, and the extracellular matrix proteins in the maintenance of hematopoietic linages.

Susceptible mouse strains infected with the murine leukemia virus (MuLV) complex LP-BM5 develop an immunodeficiency syndrome known as murine acquired immunodeficiency syndrome (MAIDS), which has many similarities to human immunodeficiency virus type-1 (HIV-1) infection. The similar features to human AIDS are B-lymphocyte activation, hypergammaglobulinemia, lymphadenopathy, BM abnormalities, and profound immunodeficiency. However, MAIDS differs from human AIDS in that MAIDS mice die of respiratory failure caused by mediastinal lymph node enlargement and not of recurrent infections, which is commonly associated with human AIDS.

LP-BM5 MuLV is a mixture of viruses, consisting of a nonpathogenic replication competent B-trophic ecotropic virus, a mink cell focus-inducing virus, and replication-defective virus. Recent studies have identified the replication-defective virus. Studies have identified the replication-defective virus. Recent studies have identified the replication-defective virus as the etiologic agent of MAIDS. Our previous study demonstrated that LTBMC prepared from LP-BM5 MuLV infected mice exhibited a defective function for normal hematopoiesis compared with marrow cultures from non-infected mice[1]. The addition of zidovudine (AZT) did not reverse this defect in infected cultures.

The mechanism(s) inducing the stromal cell defect of viral-infected cultures is unknown. Virus infection might alter BM stromal cell functions, resulting in reduced support capacity for multilineage and committed progenitor cells in culture. Furthermore, an imbalance or a defect in production of cytokines by infected stroma may have a major effect on the development of the immune deficient state. To further characterize the defective nature of virus-infected stroma, we generated permanent stromal cell lines from LP-BM5 MuLV infected and non-infected LTBMC. Our results demonstrated the establishment of stromal cell lines of primarily fibroblastic morphology. Retrovirus infection of infected stromal cell lines has been confirmed and these cell lines, as well as the non-infected control, were then tested for their ability to support and maintain reconstituting hematopoietic cells harvested from normal and MAIDS mice. The results indicated an alteration in support and maintenance of hematopoietic progenitors from normal as well as viral-infected cells when cocultured with the infected stromal cell lines. Finally, studies were performed to analyze the expression of cytokine genes in established stromal cell lines. Infection with LP-BM5 MuLV was found to be associated with increased expression of IL-4 transcript and exclusive TGFβ1 expression in infected stromal cells.

MATERIALS AND METHODS

Mice

C57BL6 mice were obtained from Harlan (Indianapolis, IN). Animals were housed in a temperature-humidity-controlled environment and fed Purina Lab Chow (St. Louis, MO) and water *ad libitum.*

Infection with LP-BM5 MuLv Murine Immunodeficiency Virus

The LP-BM5 MuLV stock used for inoculation consisted of a mixture of mink cell focus-forming (BM5) virus, ecotropic helper (BM5e) virus, and replication defective (BM5d) virus that were harvested from chronically infected SC-1 cells (clone G6) cultures that are routinely maintained in the laboratory. Normal 5-week old C57BL6 mice were inoculated intraperitoneally with 0.2 ml of cell-free virus stock containing 10 μg total protein. This is the recommended standard infection procedure for induction of MAIDS *in vivo.*[2] Decreased progression was characterized by extensive lymphadenopathy and splenomegaly, hypergammaglobulinemia, development of B-cell lymphoma at late stages of disease, and death as the result of acute respiratory failure 16-20 weeks post-viral infection.

LTBMCs and Establishment of Permanent Stromal Cell Lines

LTBMCs were established as described previously[1]. Briefly, normal noninfected and 4-week post-infected mice were killed by cervical dislocation. The contents of a tibia and femur were flushed with growth medium containing Fischer's medium (GIBCO), supplemented with 20% horse serum (Hyclone), 17 mmol/L sodium bicarbonate (Sigma), 5×10^{-7} mol/L hydrocortisone sodium succinate (Upjohn), and 80 U/ml penicillin-streptomycin (GIBCO). Single-cell suspensions of pooled BM were seeded into 25-cm2 flasks (Corning) and incubated at 33° C in humidified 5% CO_2. The cultures were maintained for a period of 6 months by weekly removal of half of medium from the flask and replenishment with fresh growth medium. Six months after the establishment of LTBMCs, the adherent cells from representative noninfected and LP-BM5 MuLV-infected culture flasks were removed by treatment with 0.25% trypsin-EDTA (GIBCO) and were passed in growth medium supplemented with 5×10^{-7} hydrocortisone sodium succinate. Cultures were then passed weekly at a 1:2 split for a period of 5 weeks to select highly proliferative cells. After further passage and selection, two noninfected stromal cell lines, designated KLT1 and KLT2, and three viral-infected stromal cell lines, designated KLTM1, 2, and 3, respectively, were successfully established from three separate studies. All cell lines, whether infected or noninfected, isolated from these attempts were morphologically similar with an appearance of fibroblast-like cells became the dominant cell morphology after subsequent repeated passage. Cell lines T2 and M1 proliferated roughly at the same rate and exhibited similar cell densities. All cell lines grew as monolayers and were contact inhibited when they reached confluency. All cell lines were devoid of committed progenitors confirmed by the absence of detectable hematopoietic progenitors, i.e., CFU-GM, CFU-Meg and BFU-E.

Characterization of Noninfected and Infected Stromal Cell Lines

Cytochemical and histochemical stains were used to characterize established stromal cell lines. Staining for alkaline phosphatase, acid phosphatase, myeloperoxidase, naphthol AS-D chloroacetate, and α-naphthyl acetate were performed using commerically available diagnostic kits (Sigma) after collection of cells on slides by cytospin and appropriate fixation, according to the specifications recommended by the manufacturer. Smears of fixed normal mouse BM were used as positive controls according to the manufacturers methods.

Determination of Retrovirus Production in BM-Derived Stromal Cell Lines

To determine viral particle production in infected stromal cell lines from both normal and infected cell lines were removed with a rubber policeman and centrifuged (400*g*; model TJ-6R, Beckman). At least 1x 10^6 cells were pelleted and prepared for examination by electron microscopy. The pelleted cells were minced into less than 1-mm cubes, fixed in 3% glutaraldehyde in buffer, postfixed in 1% osminum tetroxide in buffer, dehydrated, and embedded in epon. After selecting appropriate blocks, thin sections were cut and stained with Reynold's lead citrate and manyl acetate and examined with a Phillips 400 transmission electron microscope.

Preparation of Nonadherent Mononuclear Cells from Normal and LP-BM5 MuLV Infected Mice

Normal noninfected mice and LP-BM5-infected mice were killed 4 weeks postinfection by cervical dislocation. Femoral marrow was flushed and single-cell suspensions of BM were prepared in RPMI (GIBCO) supplemented with 10% fetal bovine serum (Hyclone), 2 mmol/L L-glutamine (GIBCO), 100 U/mL pencillin, and 100 µg/mL streptomycin (GIBCO). After centrifugation at 400*g* for 10 minutes at 4°C, cells were resuspended in RPMI medium and then overlayered on equal volume of Ficoll-Paque (specific gravity 1.077 g/ml, Pharmacia). Density centrifugation was performed at 400g for 25 minutes at 4°C and the interface layer of low-density mononuclear cells as collected, washed, and resuspended in fresh medium before adherent cell depleton. The low-density mononuclear cells were further separated into adherent and nonadherent populations after incubation on culture flasks for 90 minutes at 37°C. The nonadherent cell fraction was then collected, centrifuged, and resuspended in medium for cell counting with a Coulter Counter (Model ZBI, Coulter Electronics). To confirm the efficacy of the separation, aliquots of this nonadherent progenitor-enriched fraction were cultured alone in tissue culture flasks and were observed for growth of adherent stromal cell layers. These nonadherent normal and viral-infected low-density mononuclear cells were used as the source of marrow progenitors in the following reconstitution studies.

Stromal Cell Lines Reconstituted with Hematopoietic Cells

To assess the ability of established stromal cell lines to support hematopoietic stromal cells from noninfected and viral-infected cell lines were simultaneously seeded at 5 x 10^5 cells per 25-cm^2 flasks. The subconfluent stromal layers were then reconstituted with 2.5 x 10^4 fresh nonadherent-enriched marrow mononuclear cells (MNCs) harvested from both virus-free and LP-BM5-infected mice as described above. In control experiments, enriched MNC were added to 25-cm^2 flasks in the absence of performed stromal cell layers to assess the survival of these progenitors in culture flasks. These control MNC cultures failed to produce detectable stromal cells after up to 3 weeks in culture and no viable MNCs were detected after 1 week. Stromal cell cultures incubated without reconstituting MNC were also established. No supernatant-derived nonadherent cells could be detected after 3-weeks in culture. The reconstituted flasks were maintained at 33°C in 5% CO_2 with weekly one-half volume media change with LTBMC medium supplemented with 5 x 10^{-7} mol/L hydrocorticosterone sodium succinate. The reconstituted MNCs attached and formed "cobblestone islands" on each stromal cell layer. At weekly intervals after reconstitution for up to 3 weeks, culture supernatants were removed from reconstituting flasks and nonadherent cells were harvested and enumerated for total nonadherent cells production. These collected nonadherent cells were used to assay for progenitor cells giving rise to CFU-GM, CFU-Meg and BFU-E in hematopoietic progenitor cell assays. The methods for these procedures were described previously.[1] The reconstitution studies reported here were performed with noninfected line T2 and infected M1 cell lines.

RNA Extraction

Total RNA was isolated from infected and noninfected stromal cell lines and from stimulated spleen and Wehi-3 cells using the method as previously described by Chomczynski and Sacchi.[3] Approximately 5 x 10^6 to 5 x 10^7 stromal cells were lysed in 1 mL of denaturing solution containing 4 mol/L guanidinium thiocyanate (Sigma), 25 mmol/L sodium citrate (pH 7.0), 0.5% N-lauroylsarcosine, and 0.1 mol/L β-mercaptoethanol. Sequentially, 0.2 mL of 2 mol/L sodium acetate (pH 4.0), 1 mL of water-saturated phenol, and 0.4 mL of chloroform was added to the homogenate. The mixture was vigorously shaken for 10 seconds and then centrifuged at 11,000*g* for 20 minutes at 4°C. After centrifugation, the acqueous phase containing the RNA was transferred to a fresh tube. An equal volume of isopropanol was added to the tube and the RNA was precipitated at -20°C for 1 hour. The precipated RNA was pelleted at 11,000*g* for 20 minutes at 4°C and washed twice with 70% ethanol. The pellet was dried and dissolved in 100 μl diethylpyrocarbonate (DEPC)-H_2O. The yield and purity of the RNA was determined by spectrophotometric analysis at 260 and 280 nm. The quanity of each individual preparation of RNA was assessed by electrophoresis on 1% agarose-formaldehyde gels and stained with ethidium bromide.

Defective Virus and Cytokine Gene Expression

Reverse transcription-polymerase chain reactions (RT-PCR) were performed to analyze the expression of the LP-BM5 MuLV defective virus genome and several cytokine genes in both infected and noninfected stromal cell lines. For first-strand cDNA synthesis, 1 μg of total RNA was reversed transcribed in a 20 μL reaction volume containing reaction buffer (50 mmol/L KCl, 10 mmol/L Tris-HCl, pH 8.3); 5 mmol/L $MgCl_2$; 1 mmol/L each of dGTP, dATP,dCTPand dTTP; 20 U RNAse inhibitor; 50 U MuLV reverse transcriptase; and 2.5 μmol/L Oligo d(T) primer (all reagents were supplied by Perkin Elmer) for 15 minutes of incubation at 42°C. The reaction was terminated by 5 minutes of incubation at 99°C. The resulting first-strand cDNA was used as template in the PCR. After first-strand synthesis, PCR was performed by adding 20 μL of the cDNA solution to a 80 μL master mix containing 4 μL of 25 mmol/L $MgCl_2$, 8 μL of 10x buffer (500 mmol/L KCl, 100 mmol/L Tris-HCl, pH 8.3), 65.5 μL DEPC-H_2O, 2.5 U AmpliTaq DNA polymerase (Perkin Elmer), and 0.15 μmol/L of each specific primer. The samples were amplified in a Model 480 thermal cycler (Perkin Elmer) programmed for the following cycle parameters: stem cell factor (SCF), tumor necrosis factor α, and p12 *gag:* 95°C for 1 minute, 60°C for 1 minute, and 72°C for 2 minutes for 30 cycles; GM-CSF, IFN∂, TGFβ1: initial denaturation at 94°C for 1 minute,

94°C for 1 minute, 60°C for 2 minutes, and 72°C for 3 minutes for 35 cycles, followed by an extension at 72°C for 7 minutes; IL-1α, IL-2, IL-3, IL-4, IL-6, IL-7, IL-10; 94°C for 5 minutes, 60°C for 1 minute, and 70°C for 1 minute and 30 seconds for 1 cycle followed by 94°C for 45 seconds, 60°C for 1 minute and 30 seconds for 39 cycles. The specific primers for the murine cytokines examined have been previously reported and are not repeated here.[4] Primers specific for murine β-actin were used as an internal control in all PCR reactions. All primer pairs produced products of the predicted size. All PCR analyses included a negative (reagent) control to rule out reagent contamination. Total RNA extracted from concanavalin A (con A) or lipopolysaccharide (LPS)-stimulated spleen and Wehi-3 cells, was reverse transcribed and PCR amplified as positive controls for the cytokines analysed. In addition, a SC-1 cell line chronically infected with LP-BM5 MuLV was also used as the positive control for the p12 *gag* primers. The resulting PCR products were analyzed by electrophoresis of 20 μL of each reaction through 2% agarose gels containing 0.5 μg/mL of ethidium bromide and visualized with UV fluorescence. The gels were photographed using Polaroid 667 film (Polaroid). The negatives were scanned using BioImage System (Millipore) and the density of the bands was determined using BioImage whole band analysis protocol for 1-D electrophoresis image.

Statistical Analysis

Student's *t*-test was used to analyze the difference between the means of the experimental and control data. A P value <0.05 was used to determine the level of significance.

RESULTS

Morphology and Characteristics of LTBMC-Derived Stromal Cell Lines

In the attempt to elucidate possible mechanism(s) responsible for the hematopoietic microenvironment defect observed in LP-BM5-infected primary cultures, permanent stromal cell lines were established from adherent cells of LP-BM5 infected and noninfected control. Two noninfected and three viral-infected stromal cell lines were isolated. These five stromal cell lines were all morphologically similar. When cultured at 33°C and examined by phase contrast microscopy, these cell lines resembled fibroblastoid cells with a pleomorphoric appearance. Phaselucent lipid vacuoles were occasionally seen in the cultures. All cell lines grew as monolayers and were contact inhibited when reaching confluency. These cell lines proliferated with a doubling time ranging from 28 ± 6 to 32 ± 5 hours and with a plateau phase cell density from 1.5×10^5 cells/cm^2 to 2.0×10^5 cells/cm^2. Histochemical and cytochemical studies showed that all the cell lines were acid phosphatase and α-naphthyl acetate esterase (nonspecific esterase) positive, findings consistent with a macrophage histiocyte cell lineage. In contrast, all cell lines were negative for alkaline phosphatase and myeloperoxidase, suggesting that stromal-associated myeloid precursors were absent. Naphthol AS-D chloroacetate esterase activity was inconclusive for all cell lines.

LP-BM5 Virus Production in Established Stromal Cell Lines

To confirm retrovirus infection, established stromal cell lines were examined for the presence of viral particles. Electron microscopy examination of noninfected and viral-infected cell lines demonstrated that only the latter expressed viral particles that were budding from the cell membrane. Retrovirus infection has been maintained in infected cell lines despite repeated passages.

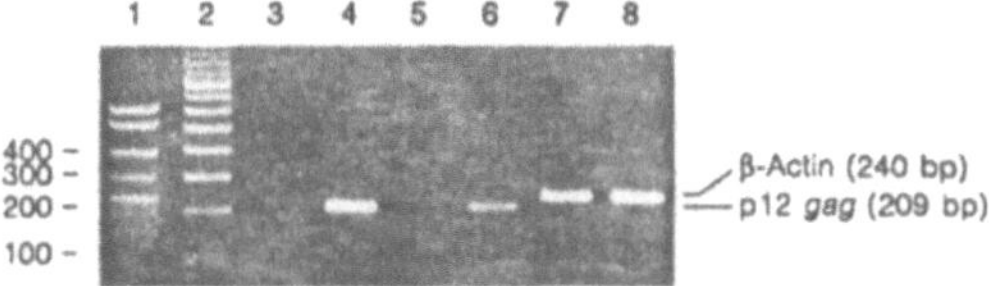

Figure 1. Detection of defective virus in established stromal cell lines by RT-PCR. Total RNA was isolated from noninfected stromal cell lines, LP-BM5 MuLV-infected stromal cell lines, and SC-1 cells chronically infected with LP-BM5 virus. Defective viral genome was amplified with p12 gag-specific primers as described in Materials and Methods. Amplified products were separated on a 2% agarose gel containing ethidium bromide. The size for amplified products was 209 bp. Lane 1, molecular weight marker (*Msp* I-digested pBR322 DNA); Lane 2, 100-bp ladder molecular weight marker; Lane 3, reagent control; SC-1 control; lanes 5 and 7, noninfected cell line (T2); lanes 6 and 8, LP-BM5-infected cell line (M1). β-Actin was used as the internal control.

Detection of Defective Virus p12 gag in Infected Stromal Cell Lines by RT-PCR

LP-BM5 MuLV is a mixture of viruses, consisting of nonpathogeneic replication-competent B-trophic ecotrophic virus, a mink cell focus-inducing virus, and an etiologic, replication-defective virus. The demonstration of viral particles production may not necessarily reflect the presence of defective virus in infected stromal cell lines. Thus, analysis for defective virus genome was performed in established stromal cell lines. Southern and Northern blot hybridization have been previously used by several research groups to detect the defective virus genome in infected mice. In the present study, a more sensitive and specific RT-PCR method was used to detect the expression of defective virus in the infected stromal cell lines. To amplify the defective viral RNA, a specific 3' primer that corresponds to the unique sequence in the p12 coding region and a 5' primer that corresponds to the p15 coding region that is not unique to the virus were used. This primer pair amplified a 209-bp fragment that is specific for the LP-BM5 MuLV defective virus. The specificity of this primer pair was shown by the amplification of total RNA extracted from SC-1 cells chronically infected with the LP-M5 MuLV mixture (Fig 1, lane 4). Using this primer pair, we were able to detect the presence of defective viral genomes in established infected stromal cell lines. In contrast, PCR analysis was unable to detect defective virus expression in all the noninfected stromal cell lines. A representative of PCR analysis is shown in Fig 1. β-actin was used as an internal control in all PCR analyses.

In Vitro Support of Reconstituted Hematopoietic Cells by Noninfected and Infected Cell Lines

Developing hematopoietic cells in the BM are closely associated with the underlying supportive stroma. Thus, the ability of established stromal cell lines to maintain and support reconstituted nonadherent hematopoietic cells was used. Representative results are reported and were obtained from the noninfected T2 and the infected M1 stromal cell pair. Subconfluent stromal cultures of noninfected and infected cell lines were simultanously reconstituted with identical inocula of enriched MNCs obtained from virus-free and MAIDS mice. To ensure that the MNCs reconstituted into the stromal cell lines not be contaminated with stromal cells, enriched MNCs were plated alone and incubated for 3 weeks. No adherent layer was detected amd MNCs in the cultures did not survive past 1 week. In addition, endogenous hematopoiesis was also tested in each cell line when cultured alone. Nonadherent hematopoietic cells were not detected in the stromal culture media throughout the study period. At week 1 after reconstitution small "hematopoietic islands" indicating the presence of hematopoietic activity became visable within the stromal cell layer derived from T2. These islands enlarged over the next two weeks and eventually became indistinguishable. In contrast, hematopoietic islands were less visable within the stromal layer derived from M1 throughout the duration of the study. Hematopoietic cells reconstituted on each stromal cell line generated nonadherent cells, which were harvested weekly by removing half of media and refeeding the cultures with an equal volume of fresh medium. T2 when reconstituted with normal MNCs, supported nonadherent cell production for more than 3 weeks in culture ($9.5 \pm 0.71 \times 10^5$ cells/flask). In contrast, the production of nonadherent cells M1 reconstitution was significantly reduced for more than 3 weeks ($6.9 + 0.21 \times 10^5$ cells/flask, $P < 0.05$). A similar pattern was observed when these stromal cell lines were cocultured with MNCs derived from MAIDS mice. Production of reconstituted nonadherent cells was significantly lower in the M1 derved underlayer ($P < 0.05$). The depressed nonadherent cell production in M1 cell line was not caused by cell growth deficiency because both T2 and M1 proliferated at approximately the same rate. By week 4, underlayers from both stromal cell lines were overgrown in all culture flasks, resulting in cell detachment and termination of the study.

Maintenance of Reconstituted Hematopoietic Progenitors by Noninfected and Viral-Infected Stromal Cell Underlayers

To determine the ability of noninfected and viral-infected cell lines to support hematopoietic progenitors, identical inocula of noninfected and viral-infected MNCs were reconstituted onto T2 and M1 stromal layers. Nonadherent cells harvested weekly from reconstituted cultures were assayed for CFU-GM, CFU-Meg and BFU-E as described. T2 positively supported CFU-GM progenitor cells derived from normal and virus-infected BM at week 3 after reconstitution. In contrast, the number of CFU-GM progenitor cells produced at week 3 by M1 was at a level of only 2% to 4% that of the control culture ($P < 0.01$). When CFU-Meg progenitor cell production was assayed, M1 supported megakaryocyte progenitor cells at only 3% to 4% compared to T2 control ($P < 0.03$). M1 also demonstrated a significant reduced capacity to support erythroid progenitors. Erythroid progenitor colonies were rarely detected from cells harvested after cocultivation with the M1 line. Only 8% to 14% of BFU-E progenitor cells were maintained compared with the control cell line ($P < 0.03$). Thus, infected cell line M1 demonstrated a significant decreased support capacity for either normal or infected hematopoietic progenitors compared with the normal control T2.

Cytokine Gene Expression in Noninfected and LP-BM5-Infected Stromal Cell Lines

BM-derived noninfected and viral-infected stromal cell lines have been analyzed for their *in vitro* hematopoiesis supporting ability. These stromal cell lines differ in their ability to support reconstituting hematopoietic cells, although they have similar growth characteristics and morphology. Altered patterns of cytokine expression and/or different cytokine gene regulation could contribute to these functional differences. To assess whether LP-BM5 MuLV expression led to a general alteration in cytokine gene expression in infected stromal cells, mRNA for all factors tested except for IL-4 and TGFβ1 were as observed from the non-infected T2 cells examined using RT-PCR (Table 1). The PCR method was chosen for its high specificity and sensitivity, which allows detection of both small amounts of mRNA and minor changes in mRNA levels. First strand cDNA templates for PCR amplication were

Table 1. Comparison of cytokine mRNA transcripts detected by RT-PCR from stromal cells, non-infected (T2) versus virus-infected (M1). See Materials and Methods for details

Cytokine	Non-Infected (T2)	Virus-Infected (M1)
IL-1 α	+	+
IL-2	+	+
IL-3	–	–
IL-4	+	+++
IL-6	+	+
IL-7	+	+
IL-10	+	+
Stem cell factor	+	+
GM-CSF	+	+
IFN∂	+	+
TNF α	+	+
TGFβ1	–	++++

synthesized using oligo-dT priming in the reverse trnascription reaction to exclude immature cytokine transcripts in the subsequent PCR amplification. Irrespective of the primers used, no PCR products were detected in the reagent controls in which no RNA was included in the first-strand reverse transcription reaction. To determine the relative levels of cytokine transcripts present in noninfected and infected cell lines, equivalent amounts of cDNA from both cell lines were used and β-actin cDNA was also amplified as an internal control in each RT-PCR analysis. Because the same concentration of total cDNA was used for each stromal cell sample, differences in the amount of amplified product produced per sample reflect differences in the expression level of the mRNA in question. However, because PCR analyses are not truly quantitative, no comparison can be made for the levels of expression among different cytokines. To test the specificity of each primer pair, total RNA from spleen cells stimulated with mitogens and from Wehi-3 cells was reverse transcribed and the resulting cDNA was amplified with specific primer pairs. In each case, the appropriate size fragment was amplified by RT-PCR and detected by electrophoresis through an agarose gel. The results from representative RT-PCR are shown in Figs 2 and 3. Based on densitometric analysis, results indicated that mRNA transcripts for stem cell factor (SCF) and tumor necrosis

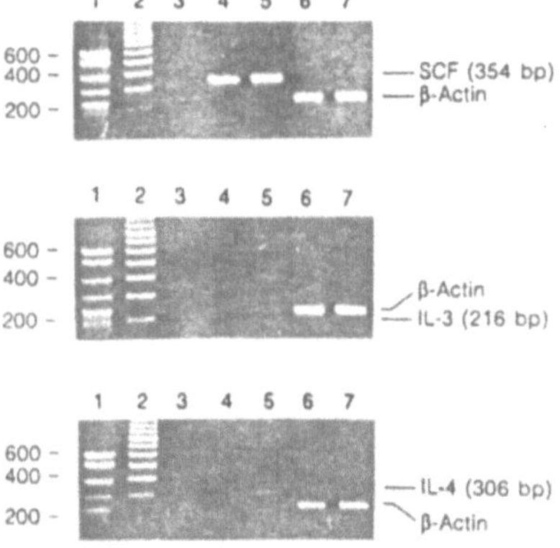

Figure 2. RT-PCR analyses of SCF, IL-3, and IL-4 mRNA expression by noninfected and LP-BM5-infected stromal cell lines. Total RNA was reverse transcribed to cDNA. Equivalent amounts of cDNA were amplified by PCR. PCR products were electrophoresed through an 2% agarose gel containing ethidium bromide. The predicted products and their sizes are indicated. Results were obtained from represenitive stromal cell lines, T2 and M1. β-Actin (240 bp) served as the internal control for each PCR analysis. (A) SCF, (B) IL-3, and (C) IL-4. Lane 1, molecular weight marker (*Msp* I-digested pBR322 DNA); Lane 2, 100-bp ladder molecular marker T2 cells; lanes 5 and 7, LP-BM5-infected M1 cells.

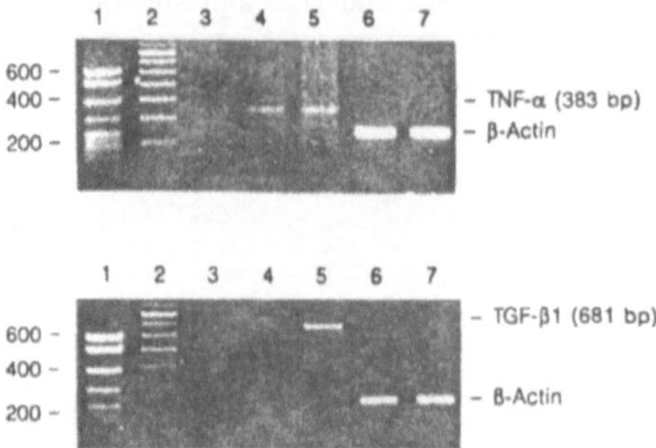

Figure 3. RT-PCR analyses of TNFα and TGFβ1 mRNA expression by noninfected and LP-BM5-infected stromal cell lines. Total RNA was reverse transcribed to cDNA. Equivalent amounts of cDNA were amplified by PCR. PCR products were electrophoresed through 2% agarose gel containing ethidium bromide. The predicted products and their sizes are indicated. Results were obtained from representative stromal cell lines, T2 and M1. β-Actin (240 bp) served as the internal control for each PCR analysis. (A) TNFα and (B) TGFβ1. lane 1, molecular weight marker (*Msp* I-digested pBR322 DNA); Lane , 100-bp ladder molecular marker; Lane 3, reagent control; Lanes 4 and 6, noninfected T2 cells; Lanes 5 and 7, LP-BM5 infected M1 cells.

factor-α (TNF-α). However, the transcripts for IL-3 could not be detected in either infected or noninfected cell lines. To ensure that the conditions for the IL-3 cDNA-PCR reactions were correct, total RNA extracted from Wehi-3 RNA was also reverse transcribed and amplified by PCR, in parallel with other samples tested. The results demonstrated that IL-3 transcripts were readily detected in Wehi-3 cells under these conditions (data not shown), further suggesting that neither noninfected nor infected stromal cell lines synthesized detectable mRNA for IL-3. Both noninfected and infected stromal cell lines expressed IL-4 transcripts. However, the densitometric analysis showed that the levels of expression were found to be five times higher in infected stromal cell lines (Fig 2). We next examined the levels of TGFβ1 in these stromal cell lines. On conversion of total mRNA to cDNA and amplification by PCR, transcripts for TGFβ1 were not detected in noninfected stromal cell lines, whereas it was exclusively in infected stromal cell lines.

DISCUSSION

The bone marrow in MAIDS has many similarities when compared with the marrow from AIDS patients, which is characterized by a decrease in cellularity and fat atrophy. The mechanism inducing such abnormalities is currently unknown. Previous studies in LTBMCs using marrow harvested from MAIDS mice suggested that there was a defect in the support and maintenance of the hematopoietic microenvironment required for normal hematopoiesis. The production of several progenitor cell lineages in MAIDS-LTBMCs was depressed compared with noninfected controls. The present studies report that permenant marrow stromal cell lines were established from MAIDS-LTBMCs and from normal noninfected cultures as controls. Cytochemical and histochemical characterizations of the established stroma cell line showed a heterogenous stromal cell population in these cultures. The viral-infected cell lines were morphologically equivalent to the noninfected lines. Spindle-shaped fibroblastic cells and macrophage-like cells that stained positively for acid phosphatase and α-naphthyl acetate esterase were the predominant cell types found in all cell lines. Production of viral particles has been demonstrated using electron microscopy and this was correlated with the detection of defective virus expression by RT-PCR only in viral-infected but not in noninfected stromal cell lines.

The ability of these cell lines to support in vitro hematopoiesis was examined. After reconstitution with virus-free or viral-infected hematopoietic progenitor cells, altered paterns of hematopoiesis were repeatedly observed when the viral-infected were used as the support underlayer compared with the reconstituted, noninfected cell lines. Production of nonadherent cells was detected at a significantly lower level in the infected cell lines grown for 3 weeks. The numbers of CFU-GM, CFU-Meg and BFU-E progenitors were also significantly reduced when cocultured with the infected cell lines compared with controls. It is unlikely that hematopoietic progenitor cell infection by virus could explain the altered patterns of growth as marrow cells derived from virally infected mice proliferated and differentiated normally when cocultured with the noninfected stromal cell lines. On the contrary, virus-free progenitor cell growth was significantly depressed when the infected cell lines were used as the supportive underlayer. These results further suggest that as the result of virus infection, the marrow microenvironment is altered such that a defective stroma is incapable of supporting normal hematopoiesis. Furthermore, these data show that progenitor cells derived from LP-BM5 MuLV infected mice are capable of responding to the systems that regulate hematopoiesis in LTBMCs because their support was not reduced when overlayed on normal, non-virus-infected stromal cells. Whereas these and other results presented in this report were generated from stromal cell lines derived from in vivo infected BM, results from subsequent studies using normal stromal cultures transfected in vitro also generated similar results, thus supporting the conclusions presented here. Therefore, the defect observed in this viral system lies within the stroma rather than at the progenitor cell level.

The mechanism inducing the infected stromal cell defect is unclear. An imbalance in production or regulation of cytokines in infected stromal cells might have a major consequence on the support and maintenence of hematopoietic progenitor cells. Cytokine genes that are constitutively expressed in marrow-derived stromal cells and mediate positive and negative regulation of hematopoiesis have been reported.[5,6] In the present study, the levels of cytokine gene expression in noninfected and viral-infected stromal cell lines were examined using RT-PCR. With the exception of mRNA for IL-3, all other four cytokine genes were expressed, although some at different levels, by the established stromal cell lines. The absence of mRNA for IL-3 in our analysis does not necessarily mean that there is no expression of IL-3 transcripts, but merely that levels of expression were below the method level of detection. Several investigators have reported low levels of IL-3 expression from stromal cells.[7,8] The constitutive expression of SCF transcripts, the ligand for the c-kit receptor, in both cell lines described here is of considerable interest because of the well-documented activity of this cytokine in supporting proliferation and differentiation of a broad spectrum of cell types in the hematopoietic progenitor and stem cell hierarchy.[9] Thus, the depressed hematopoiesis in infected stromal cell lines reported here could not be correlated with a low level of SCF gene expression. The levels of TNFα were found to be very low and were equivalent between noninfected and infected stromal cell lines. In this respect, our results correlate with the recent findings of Cheung et al[10] and Munis et al[11] who demonstrated low levels of TNFα expression in retrovirus-infected cells.

Another finding in the cytokine analysis was the elevated levels of IL-4 expression in viral-infected stromal cell lines compared with those in the noninfected lines. Increased IL-4 expression has been demonstrated and correlated with virus-specific gene expression in MAIDS mice.[12] A report by Kanagawa et al[13] using IL-4 deficient mice (IL-4$^{-/-}$) mice showed that IL-4 is critical to the development of retrovirus-induced immunodeficiency in mice. In addition, several reports showed that stimulaton with IL-4 produced growth inhibitory activity against either with IL-4 produced growth inhibitory activity against either myeloid or lymphoid lineage cell proliferation in BM stromal cells[14] and rapid disappearance of pluripotential and committed myeloid hematopoietic progenitors on addition of IL-4 to co-culture.[14,15] These data provide evidence to suggest that increased levels of IL-4 expression in infected stromal cell lines may contribute to the depressed hematopoiesis described in this retroviral model of stromal hematopoiesis.

Another significant finding is that TGFβ1 was expressed only by stromal cell lines infected with virus. TGFβ1 has been demonstrated to be a potent negative regulator of growth and differentiation of early hematopoietic cells.[16] Thus, the upregulated expression of TGFβ1 from infected cells may have contributed to the early arrest of progenitors, resulting in the impaired hematopoiesis observed in our system. Based on the cytokine analysis, it can be postulated that virus infection of stromal cells led to deregulation of two of the five cytokines examined. The change in cytokine gene expression might induce cytokine imbalance and thus induce a negative regulation of hematopoiesis by the infected stroma. As a result, both virus-free and infected reconstituting progenitors could only be supported by noninfected but not by viral-infected stromal underlayer, as observed in the coculture studies.

Infection of the nonhematopoietic microenvironment has been reported by several research groups. The report by Scadden et al[17] on HIV infection and replication in a human stromal cell line and the consequent suppression of hematopoietic support function suggests an effect of viral infection on the stromal cells in the marrow microenvironment. Such an effect of viral infection on the ability of marrow stromal cells to support hematopoiesis in vitro has been described for cytomegalovirus.[18] These studies also suggest that as the result of CMV infection, which produces an ineffective stroma, BM failure develops.

The results from these studies demonstrate that stromal cell lines generated following retroviral infection in marrow cultures retain their defective characteristics of the parental cells. Such defective characteristics perhaps relate to different levels of cytokine gene expression that would contribute to the impaired growth and differentiation of early hematopoietic progenitors. Further characterization on the effect of retroviral alteration of marrow stromal cell function should assist in the elucidation on the definitive role played by the stroma in the pathogenesis of MAIDS.

ACKNOWLEDGMENTS

The authors acknowledge the assistance of Michael Cibull, M.D., Department of Pathology, and thank Amrit Dhopper, Doreen Jezek, John May, and Hope Gaines for technical assistance. These studies were supported in part by a Merit Review Grant from the Department of Veterans Affairs, Washington, D.C.

REFERENCES

1. Tse KF, Hughes NK, Gallicchio VS. Failure to establish long-term marrow cultures from immunodeficient mice (MAIDS): Effect of zidovudine *in vitro.* J Leukoc Biol 53:658, 1993.
2. Harley JW, Frederickson TN, Yetter RA, Makino M, Morse HC III. Retrovirus-induced murine acquired immunodeficiency syndrome: natural history of infections and differing susceptibility of different mouse strains. J Virol 63:1223, 1989.
3. Chomczynski P ans Sacchi N. Single step method of RNA isolation by acid guanidinium thiocyanate-phenol-chloroform extraction. Anal Biochem 162:156,1987.
4. Tse KF, Morrow JM, Hughes NK, Gallicchio VS. Stromal cell lines derived from LP-BM5 MuLV murine leukemia virus-infected long-term bone marrow cultures impair hematopoiesis *in vitro.* Blood 84:1508, 1994.

5. Shenoy S, Chattopadhyay S, Morse H, Pitha P. Expression of defective virus and cytokine genes in murine MAIDS. J Virol 65:833, 1991.
6. Magnani N, Brandi G, Rossi L, Carnevall A, Fraternale A, Albano A. The bone marrow in murine AIDS. Br J Haematol 84:539, 1993.
7. Kittler E, McGrath H, Temeles D, Crittenden R Kister K, Quesenberry P. Biologic significance of constitutive and subliminal growth factor production by bone marrow stroma. Blood 79:3168, 1992.
8. Gutierrez-Ramos J, Olsson C, Palacios R. Interleukin (IL-1 to IL-7) gene expression in fetal liver and bone marrow stromal clones: Cytokine-mediated positive and negative regulation. Exp Hematol 20:986,1992.
9. Williams D, Uries P, Namen A, Widmer M, Lyman S. The steel factor. Dev Biol 151:368, 1992.
10. Cheung S, Chattopadhyay S, Hartley J, Morse H, Pitha P. Aberrant expression of cytokine genes in peritoneal macrophages from mice infected with LP-BM5 MuLV, a murine model of AIDS. J Immunol 146:121, 1991.
11. Munis J, Richman D, Kornbluth R. HIV infection of macrophages *in vitro* neither induces TNF/cachectin gene expression nor alter TNF/cachectin induction by lipopolysaccharide. J Clin Invest 85:591, 1990.
12. Bradley W, Ogata N, Good R, Day N. Alteration of *in vivo* cytokine gene expression in mice infected with a molecular clone of the defective MAIDS virus. J AIDS 7:1, 1994.
13. Kanagawa O, Vaupel B, Gayamaq S, Koehler G, Kepf M. Resistance of mice deficient in IL-4 to retrovirus-induced immunodeficiency syndromes (MAIDS). Science 262:240, 1993.
14. Peschel C, Green I, Paul W. Interleukin-4 induces a substance in bone marrow stromal cells that reversibly inhibits factor-dependent and factor-independent cell proliferation. Blood 73:1130, 1989.
15. Rennick D, Yang G, Muller-Sieburg C, Smith C, Arai N, Takabe Y, Germmell L. Interleukin-4(B-cell stimulatory factor 1) can enhance or antagonize the factor-dependent growth of hematopoietic progenitor cells. Proc Natl Acad Sci USA 84:6889, 1987.
16. Keller J, Mantel G, Sing L, Ellingsworth S, Ruscetti S, Ruscetti F. Transforming growth factor β, selectively regulates early murine hematopoietic progenitors and inhibits the growth of IL-3 dependent myeloid leukemia cell lines. J Exp Med 168:737, 1988.
17. Scadden DT, Zeira M, Woon A, Wang Z, Schieve L, Ikeuchi K, Lim B, Groopman JE. Human immunodeficiency virus infection of human bone marrow fibroblasts. Blood 76:317, 1990.
18. Apperley JF, Dowling C, Hibbin J, Buiter J, Matutes F, Sistsons PJ, Gordon M, Goldman JM. The effect of cytomeglovirus on hematopoiesis: *In vitro* evidence for selective infection of marrow stromal cells . Exp Hematol 17:38, 1989.

ROLE OF HIV-1 IN THE FUNCTIONAL IMPAIRMENT OF CD34$^+$ HEMATOPOIETIC PROGENITOR CELLS

Giorgio Zauli*

Institute of Human Anatomy
University of Ferrara
Via Fossato di Mortara 66, 44100 Ferrara, Italy

ABSTRACT

The role played by HIV-1 infection in the pathogenesis of peripheral blood cytopenias frequently found in HIV-1 seropositive individuals was initially elusive. However, a body of in vivo and in vitro experimental evidence suggests that HIV-1 can be directly involved in the suppression of hematopoietic progenitor cells through either direct or indirect mechanisms: (1) infection (productive or non-productive) of a subset of CD34$^+$ hematopoietic progenitor cells, co-expressing the CD4 antigen with growth defects of infected cells; (2) membrane interactions of CD34$^+$ hematopoietic progenitor cells with HIV-1 virions or immune complexes containing env gp120, which can directly lead CD34$^+$ cells to apoptotic cell death. Both the viral load and the biological characteristics of the virus play an important role in causing these suppressive effects, since different isolates displayed a differential ability to suppress hematopoiesis; (3) infection with HIV-1 and/or exposure of bone marrow accessory cells to viral proteins (gp120 and Tat) with increased production of inhibitory factors, such as TNF-α or TGF-β1.

Hematologic disorders are frequently observed in the majority (70-80%) of HIV-1 infected patients during the course of HIV-1 disease (1-2). Most peripheral blood cytopenias take place in symptomatic patients and usually their frequency increases as the disease progresses towards overt acquired immunodeficiency syndrome (AIDS). On the other hand, isolated thrombocytopenia can occur early in the natural history of HIV-1 infection as an isolated hematological manifestation (3).

It is fairly well established that the pathogenesis of hematological abnormalities in AIDS patients is multi-factorial (1-2). In particular, B cell lymphomas spread to the bone marrow, a variety of opportunistic agents and a reduced production of erythropoietin can

* Please send all correspondence to: Dr. Giorgio Zauli, Institute of Human Anatomy, University of Ferrara, Via Fossato di Mortara 66, 44100 Ferrara, Italy. Fax: +39-532-200-558; Tel: +39-532-207-310.

Molecular Biology of Hematopoiesis 5, edited by Abraham et al.
Plenum Press, New York, 1996

significantly contribute to the bone marrow suppression. Moreover, peripheral blood cytopenias in HIV-1 infected individuals can be markedly worsened by the treatment with zydovudine (4-5) or other anti-retroviral or antineoplastic agents. The main clinical relevance of peripheral blood cytopenias in HIV-1 infected patients is represented by the fact that they often entail the discontinuation or suspension of zydovudine or cytotoxic therapy.

A body of experimental evidence, however, suggests that also HIV-1 infection may be directly involved in the pathogenesis of peripheral blood cytopenias of HIV-1 seropositive individuals through different mechanisms: (i) infection of mature hematopoietic precursors; (ii) infection of marrow accessory cells; (iii) inhibitory interaction(s) of HIV-1 virions with $CD34^+$ cells.

The paradigmatic example of HIV-1 infection of mature hematopoietic cells is represented by infection of bone marrow megakaryocytes, which occurs in at least 50% of HIV-1 seropositive individuals harbouring a peripheral thrombocytopenia (6-7) and it is thought to result in a reduced production of platelets.

A productive infection of cells belonging to the bone marrow microenvironment of HIV-1 seropositive subjects (8-12), in particular of T lymphocytes (8-9) and monocytes (10), has been demonstrated. Such infection might induce disruption of the physiological control of hematopoietic stem/progenitor cells by an imbalanced cytokine production (13-14). Moreover, the experimental infection of long term bone marrow cultures (LTBMCs) pointed out that the viral burden and the kind of viral isolate used to infect LTBMCs may be critical factors in the HIV-1 mediated suppression of hematopoiesis (15-17).

Several defects in the colony forming ability of hematopoietic progenitors have been identified in HIV-1 infected individuals (20-29). The general picture emerging from all these studies is that the frequency of hematopoietic progenitor cells in the marrow and peripheral blood of HIV-1 infected individuals progressively decreases as the disease progresses. The only remarkable exception is represented by patients with isolated and persistent thrombocytopenia, who showed a selective defect of the CFU-Meg progenitor cell compartment (28).

At least two distinct mechanisms of inhibition are responsible for the impaired colony growth of hematopoietic progenitor cells in HIV-1 seropositive patients (Table 1): an indirect suppression by accessory cells present in the bone marrow and peripheral blood samples and an intrinsic defect in stem/progenitors. Evidence for an inhibitory activity of accessory cells comes from studies showing that the defective colony formation could be partially restored by either T-cell depletion (18, 24), treatment of marrow cells with anti-sense oligonucleotides to Tat or nef regulatory gene sequences (13) or addition of anti-TNF-α in culture (26). However, also purified $CD34^+$ cells from symptomatic HIV-1 seropositive individuals are defective in colony formation (8, 25, 27-29), which implicates the presence of an intrinsic defect of the stem/progenitor cells.

Several groups of investigators have attempted to evaluate whether the reduced colony forming ability of hematopoietic progenitor cells in HIV-1 seropositive individuals could be due to a direct HIV-1 infection of $CD34^+$ cells. As summarized in Table 2, most investigators reported rare infection of $CD34^+$ cells, purified from the bone marrow of HIV-1 seropositive individuals at various stages of the disease (8-9, 15, 25, 28-30). The higher percentage of low level (1 proviral DNA copy in 500 $CD34^+$ cells) infection was reported by Stanley et al. (27) in a significant subset of HIV-1 seropositive patients with advanced stage disease. Also the presence of proviral DNA in hematopoietic colonies at the end of the culture time was only occasionally reported (15, 27). Consistently, a number of studies have reported the ability of HIV-1 to infect *in vitro* $CD34^+$ cells purified from either normal bone marrow (31-33) or peripheral blood (34) and the presence of proviral DNA has been recovered in fully developed granulocyte/macrophage and erythroid colonies. The picture emerging from all these studies clearly suggests that infection of $CD34^+$ hematopoietic

Table 1. HIV-1 related mechanisms of impaired hematopoietic colony formation in HIV-1 seropositive individuals

Indirect suppression by accessory cells present in the bone marrow and peripheral blood samples	
	Maciejwski et al. J Immunol 53:4303, 1994
	Carlo Stella C et al. J Clin Invest 80:286, 1987
	Balleari et al. Ann Hematol 63:320, 1991
	Louache et al. Blood 80:2991, 1992
Intrinsic defect in $CD34^+$ stem/progenitors	
	Davis et al. J Virol 65:1985, 1991
	Zauli et al. Blood 79: 2680, 1992
	Zauli et al. J Infect Dis 166:710, 1992
	Stanley et al. J Immunol 149:689, 1992
	De Luca et al. Br J Haematol 85: 20, 1993

Table 2. Evidence for an *in vivo* infection with HIV-1 of $CD34^+$ hematopoietic progenitor cells

	N° of positive cases/total cases
Molina et al. Blood 76:2476, 1990	0/6
von Lear et al. Blood 76:1281, 1990	1/14
Davis et al. J Virol 65:1985, 1991	1/12
Zauli et al. Blood 79: 2680, 1992	2/17
Zauli et al. J Infect Dis 166:710, 1992	0/6
Stanley et al. J Immunol 149:689, 1992	22/77

progenitor cells does not seem to occur frequently *in vivo* and is not significantly related to their functional impairment *in vitro* (Table 3).

On the other hand, we have demonstrated that a brief exposure (2 hours) to either primary isolates obtained from HIV-1 symptomatic carriers showing peripheral blood cytopenias (25) or a laboratory strain (IIIB) of HIV-1 (35) resulted in an impaired survival/proliferation capacity of $CD34^+$ hematopoietic progenitor cells. Remarkably, the defects induced in CFU-GM, BFU-E, and CFU-Meg derived colony formation as well as in the ability of exposed $CD34^+$ cells to proliferate/survive in IL-3 containing liquid cultures took place in absence of productive or latent infection of the stem/progenitor cells. This further strengthen the notion that a direct infection of hematopoietic progenitor cells was not required to observe the HIV-1 mediated suppressive effect.

Table 3. Evidence against a primary role of $CD34^+$ cell infection with HIV-1 in the pathogenesis of peripheral blood cytopenias in HIV-1 seropositive individuals

Infection of $CD34^+$ cells *in vivo* mainly involves only a subset of patients in advanced stages of HIV-1 disease;
The viral load present in $CD34^+$ infected cells is not sufficient to explain their impaired colony forming ability;
The colony number is similar in patients with detectable infection in $CD34^+$ cells in comparison with those absent of HIV-1 infection in $CD34^+$ cells;
Hematopoietic colonies harboring proviral DNA do not show gross abnormalities in their size (number of cells/colony) and morphology.

Table 4. Mechanisms by which HIV-1 and/or viral proteins can directly impair the survival/proliferation of $CD34^+$ hematopoietic progenitor cells

A subset of hematopoietic progenitor cells are susceptible to HIV-1 infection both *in vitro* and *in vivo*;
HIV-1 virions and recombinant gp120 are able to induce apoptotic cell death of $CD34^+$ hematopoietic progenitor cells. This effect is greatly enhanced in the presence of anti-gp120 antibody;
The viral load and the biological characteristics of the virus play an important role in causing this suppressive effect.

The nature of the interaction of HIV-1 virions with hematopoietic progenitor cells was further analyzed using the TF-1 factor-dependent hematopoietic cell line (TF-1) as a model system (36). Besides HIV-1 virions, also cross-linked recombinant gp120 (1 µg/ml) and Leu 3a anti-CD4 monoclonal antibody were able to induce a significant increase of apoptosis in the presence of low concentrations of IL-3 (1 ng/ml or less). On the other hand, higher concentrations (2 ng/ml or more) of IL-3 showed a protective effect on TF-1 cells. These findings suggest that gp120 is an essential component of the inhibitory effect of intact virions and that this inhibitory effect was specifically mediated by the CD4 antigen. Consistently, mouse hemopoietic progenitors express CD4 marker on their surface (37-38). This hypothesis has been next verified demonstrating that both TF-1 cell line and a subset of primary BM $CD34^+$ cells express a mature form of CD4 antigen, which is able to functionally bind recombinant gp120 (39). Therefore, HIV-1 virions may trigger apoptosis of subset of hematopoietic progenitor cells through the specific interaction of envelope gp120 with the CD4 receptor, as previously demonstrated for $CD4^+$ T-lymphocytes (40-41). It is noteworthy that also defective virions and/or free glycoprotein gp120, which are produced in abundance by infected cells (42), may be equally effective than infective virions in the induction of apoptosis expecially in the presence of anti-gp120 antibody.

In conclusion, several data are compatible with a model in which HIV-1 and/or viral proteins can directly impair the survival/proliferative capacity of $CD34^+$ purified hematopoietic progenitor cells (Table 4). This suppressive effect greatly depend on the biological characteristics of the viral isolates (17) as well as by the viral load, as also suggested by the fact that cytopenias occur most frequently during times of increased viremia (e.g. not only in the AIDS stage of disease, but also during symptomatic acute primary infection) (43). This scenario offers a rational basis to the combination of anti-retroviral therapy with hematopoietic growth factors in the therapy of peripheral blood cytopenias of HIV-1 seropositive individuals (44). This therapeutic approach, counteracting the ability of both viral and cellular products to induce the apoptosis of hematopoietic progenitor cells, could increase their survival/proliferation capacity and possibly restore their ability to reconstitute a functional hematopoietic and lymphoid system in HIV seropositive individuals.

ACKNOWLEDGMENTS

This study was supported by Istituto Superiore di Sanita', Progetto AIDS and by MURS 40%.

REFERENCES

1. Zon LI, Arkin C, Groopman, JE: Haematologic manifestation of the human immune deficiency virus (HIV). Br J Haematol 66: 251, 1987
2. Scadden DT, Zon LI, Groopman JE: Pathophysiology and management of HIV associated hematologic disorders. Blood 4:1455, 1989
3. Ratner L: Human immunodeficiency virus (HIV) associated autoimmune thrombocytopenic purpura: a review. Am J med 86:194, 1989
4. Dainiak N, Worthington M, Riordan MA, Kreczo S, Goldman L: 3'-azido-3'-Deoxythymidine (AZT) inhibits proliferation in vitro of human haematopoietic progenitor cells. Br J Haematol 69: 299, 1988
5. Richman DD, Fischl MA, Grieco MH, Gottleib MS, Volberdig PA, Laskin OL, Leedom JM, Groopman JE, Mildvan D, Hirsch MS, Jackson GG, Durack DT, Nusinoff-Lehrman S, and the AZT collaborative working group: The toxicity of azidothymidine (AZT) in the treatment of patients with AIDS and AIDS-related complex. New Engl J Med 317:192, 1987
6. Zucker-Franklin D, Cao Y: Megakaryocyte of human immunodeficiency virus-infected individuals express viral RNA. Proc Natl Acad Sci USA 86:5595, 1989
7. Louache F, Bettaieb A, Henri A., Oksenhendler E, Farcet JP, Bierling P, Seligmann M, Vainchenker W: Infection of megakaryocytes by human immunodeficiency virus in seropositive patients with immune thrombocytopenic purpura. Blood 78:1697, 1991
8. Davis BR, Marx JC, Johnson CE, Berry JM, Lyding J, Zander A, Merigan TC and Schwartz J: Absent or rare HIV infection of bone marrow stem/progenitor cells in vivo. J Virol 65:1985, 1991
9. von Lear D, Hufert FT, Fenner TE, Schwander S, Dietrich M, Schmitz H, Kern P: CD34+ hematopoietic progenitor cells are not a major reservoir of the human immunodeficiency virus. Blood 76:1281, 1990
10. Canque B, Marandin A, Rosenzwajg M, Louache F, Vainchenker W, Gluckman JC: Succeptibility of human marrow stromal cells to human immunodeficiency virus. Virology 208: 779, 1995
11. Scadden DT, Zeira M, Woon A, Wang Z, Schieve L, Ikebuchi K, Lim B, Groopman JE: Human immunodeficiency virus infection of human bone marrow stromal fibroblasts. Blood 76:317, 1990
12. Steiberg HN, Anderson J, Crumpacker CS, Chatis PA: HIV infection of the BS-1 human stroma cell line: effect on murine hemopoiesis. Virology 193:524, 1993
13. Maciejwski JP, Weichold FF, Young NS: HIV-1 suppression of hematopoiesis in vitro mediated by envelope glycoprotein and TNF-α. J Immunol 53:4303, 1994
14. Zauli G, Davis BR, Re MC, Visani G, Furlini G, La Placa M: Tat protein stimulates production of transforming growth factor-beta 1 by marrow macrophages: a potential mechanism for HIV-1 induced hematopoietic suppression. Blood 80:3036, 1992
15. Kaczmarski RS, Davison F, Blair E, Sutherland S, Moxham J, McManus T, Mufti GJ: Detection of HIV in hematopoietic progenitors. Br J Haematol 82:764, 1992
16. Calenda V, Sebahoun G, Chermann JC: Modulation of normal human erythropoietic progenitor cells in long term liquid cultures after HIV-1 infection. AIDS Res Hum Retroviruses 8: 61, 1992
17. Cen D, Zauli G, Szarnicki R, Davis BR: Differential effect of human immunodeficiency virus type 1 isolates on bone marrow hematopoiesis. Br J Haematol 85:596, 1993
18. Carlo Stella C, Ganser A, Hoelzer D: Defective in vitro growth of hematopoietic progenitor cells in the acquired immunodeficiency syndrome. J Clin Invest 80:286, 1987
19. Leiderman IZ, Greenberg ML, Adelsberg BR, Siegal FP: A glycoprotein inhibitor of in vitro granulopoiesis associated with AIDS. Blood 70:1267, 1987
20. Donahue RE, Johnson MM, Zon LI, Clark SC, Groopman JE: Suppression of in vitro haematopoiesis following human immunodeficiency virus infection. Nature 326:200, 1987
21. Lunardi-Iskandar Y, Nugeyere MT, Georgoulias V, Barré-Sinoussi F, Jasmin C, Chermann JC: Replication of the human immunodeficiency virus 1 and impaired differentiation of T cells after in vitro infection of bone marrow immature T cells. J Clin Invest 83:610, 1989
22. Ganser A, Ottman OG, von Briesen H, Volkers B, Rubsamen-Waigmann H, and Hoelzer D: Changes in the haematopoietic progenitor cell compartment in the acquired immunodeficiency syndrome. Res Virol 41:185, 1990
23. Bagnara GP, Zauli G, Giovannini M, Re MC, Furlini G, La Placa M: Early loss of circulating hematopoietic progenitors in HIV-1 infected subjects. Exp Hematol 18:426, 1990
24. Balleari E, Timitilli S, Puppo F, Gaffuni L, Mussell C, Rizzo F, Indivieri F, Ghio R: Impaired in vitro growth of peripheral blood hematopoietic progenitor cells in HIV-infected patients: evidence for an inhibitory effect of autologous T lymphocytes. Ann Hematol 63:320, 1991

25. Zauli G, Re MC, Visani G, Furlini G, Mazza P, Vignoli M, La Placa M: Evidence for an Human Immunodeficiency Virus-type 1 mediated suppression of uninfected hematopoietic ($CD34^+$) progenitor cells in AIDS patients. J Infect Dis 166:710, 1992
26. Louache F, Henri A, Bettaieb A, Oksenhendler E, Raguin G, Tulliez M, Vainchenker W: Role of human immunodeficiency virus replication in defective in vitro growth of hematopoietic progenitors. Blood 80: 2991, 1992
27. Stanley SK, Kessler SW, Justement JS, Schnittman SM, Greenhouse JJ, Brown CC, Musongela L, Musey K, Kapita B, Fauci AS: $CD34^+$ bone marrow cells are infected with HIV in a subset of seropositive individuals. J Immunol 149:689, 1992
28. Zauli G, Re MC, Davis BR, Sen L, Visani G, Gugliotta L, Furlini G, La Placa M: Impaired in vitro growth of purified ($CD34^+$) hematopoietic progenitors in HIV-1 seropositive thrombocytopenic individuals. Blood 79: 2680, 1992
29. De Luca A, Teofili L, Antinori A, Iovino MS, Mencarini P, Visconti E, Tamburrini E, Leone G, Ortona L: Haemopoietic $CD34^+$ progenitor cells are not infected by HIV-1 in vivo but show impaired clonogenesis. Br J Haematol 85:20, 1993
30. Molina JM, Scadden DT, Sakaguchi M, Fuller B, Woon A and Groopman JE: Lack of evidence for infection of or effect on growth of hematopoietic progenitor cells after in vivo or in vitro exposure to human immunodeficiency virus. Blood 76:2476, 1990
31. Folks TM, Kessler SW, Orenstein JM, Justment JS, Laffe E, Fauci AS: Infection and replication of HIV-1 in purified progenitor cells of normal human bone marrow. Science 242: 919, 1988
32. Kitano K, Abboud CN, Ryan DH, Quan SG, Baldwin GC, Golde DW: Macrophage-active colony-stimulating factors enhance human immunodeficiency virus type 1 infection in bone marrow stem cells. Blood 77:1699, 1991
33. Steinberg HN, Crumpacker CS, Chatis PA: In vitro suppression of normal human bone marrow progenitor cells by human immunodeficiency virus. J Virol 65:1765, 1991
34. Chelucci C, Hassan HJ, Locardi C, Bulgarini D, Pelosi E, Mariani G, Testa U, Federico M, Valtieri M, Peschle C: In vitro human immunodeficiency virus-1 infection of purified hematopoietic progenitors in single-cell culture. Blood 85:1181, 1995
35. Zauli G, Re MC, Furlini G, Giovannini M, La Placa M: Human immunodeficiency virus type 1 envelope glycoprotein gp120-mediated killing of human hematopoietic progenitors (CD34+ cells). J Gen Virol 73:417, 1992
36. Zauli G, Vitale M, Re MC, Furlini G, Zamai L, Falcieri E, Gibellini D, Visani G, Davis BR, Capitani S, La Placa M: In vitro exposure to human immunodeficiency virus type-1 (HIV-1) induces apoptotic cell death of the factor-dependent TF-1 hematopoietic cell line. Blood 83:167, 1994
37. Zauli G, Furlini G, Vitale M, Re MC, Gibellini D, Zamai L, Visani G, Borgatti P, Capitani S, and La Placa M: A subset of human CD34+ hematopoietic progenitors express low levels of CD4, the high affinity receptor for human immunodeficiency virus-type 1. Blood 84:1896, 1994
38. Wineman JP, Gilmore GL, Gritzmacher C, Torbett BE, Muller-Sleburg CE: CD4 is expressed on murine pluripotent hematopoietic stem cells. Blood 80:1717, 1992
39. Onishi M, Nagayoshi K, Kitamura K, Hirai H, Takaku F, Nakauchi H: $CD4^{dull+}$ hematopoietic progenitor cells in murine bone marrow. Blood 81:3217, 1993
40. Newell MK, Haughn LJ, Maroun CR, Julius MH: Death of mature T cells by separate ligation of CD4 and the T-cell receptor for antigen. Nature 347:286, 1990
41. Banda NK, Bernier J, Kurahara DK, Kurrle R, Haigwood N, Sekaly DK, Finkel TH: Crosslinking CD4 by human immunodeficiency virus gp120 primes T cells for activation-induced apoptosis. J Exp Med 176:1099, 1992
42. Schneider J, Kaaden O, Copeland TD, Oroszlan S, Hunsmann G: Shedding and interspecies type sero-reactivity of the envelope glycopeptide gp120 of the human immunodeficiency virus. J Gen Virol 67:2533, 1986
43. Fauci A: Multifactorial nature of human immunodeficiency virus disease: implications for therapy. Science 262:1011, 1993
44. Miles SA: Hematopoietic growth factors as adjuncts to antiretroviral therapy. AIDS Res Hum Retro 8:1073, 1992

HUMORAL IMMUNE RESPONSE TO HUMAN CYTOMEGALOVIRUS

Diagnostic and Clinical Implications

Maria Paola Landini,* Tiziana Lazzarotto, and Paola Dal Monte

Section of Microbiology
Department of Clinical and Experimental Medicine
University of Bologna, Italy

ABSTRACT

Human Cytomegalovirus (CMV) is associated with several diseases in immunocompromised individuals. Diagnosis of CMV infection can be obtained by direct demonstration of the virus or virus components in pathological materials or indirectly through serology. Serological diagnosis gives only indirect evidence of the presence of the virus, and is problematic because of the immunological disorders occurring in most patients at risk of developing a CMV infection. Furthermore, antigenic reagents used in commercially available kits are not standardized and discordant results are often obtained. However serology is cheaper than the other diagnostic tests, requires a short execution time, is safe and can be completely automized.

Furthermore the rapid evolution triggered by the detailed study of the viral genome and its antigenic gene products has allowed a rapid progress in CMV serology. Therefore, it is worthwhile exploring the possible application fields in which the use of serology is justified now adays. This is what this review will attempt to do in general terms and in bone marrow transplant recipients in particular.

INTRODUCTION

Human cytomegalovirus (CMV) is a ubiquitous herpesvirus in man. It is rarely pathogenic in healthy adults but is associated with several diseases in immunocompromised individuals (such as HIV-infected people and transplant recipients). Bone marrow transplant (BMT) recipients are particularly vulnerable to CMV infection and the lung is the most important target organ. Morbidity and mortality rates as high as 90% for CMV pneumonitis

* Address for correspondence: Maria Paola Landini, MD, Institute of Microbiology, University of Bologna, St. Orsola General Hospital, Via Massarenti n.9, 40138 Bologna, Italy. Fax n.: 39.51.341632.

Molecular Biology of Hematopoiesis 5, edited by Abraham et al.
Plenum Press, New York, 1996

have been reported in these patients. Furthermore, CMV is the most common cause of congenital infection in humans. Intrauterine primary infections are second only to Down's syndrome as a known cause of mental retardation. Less severe complications are the result of secondary infections (for review see ref. 1 and 2). As infections are either asymptomatic or accompanied by symptoms that are not specific of CMV (such as fever and leukopenia) laboratory techniques are the sole means of diagnosing acute CMV infection. Diagnosis of CMV infection can be obtained by direct demonstration of the virus or virus components in pathological materials or indirectly through serology (for review see ref. 3). Serological diagnosis in general is less attractive than virological diagnosis because it gives only indirect evidence of the presence of the virus, and is problematic because of the immunological disorders occurring in most patients at risk of developing a CMV infection. Finally, antigenic reagents used in commercially available kits are not standardized and discordant results are often obtained. However serology is cheaper than the other diagnostic tests, requires a short execution time, is safe and can be completely automized and recombinant antigens are available to improve the specificity and sensitivity of the tests. Therefore, it is worthwhile exploring the possible application fields in which the use of serology is justified now adays.

HCMV-SPECIFIC IgG

A) When Do We Need to Search for CMV-IgG?

The detection of CMV-specific IgG has to be performed to determine whether a subject has been infected in the past. This is useful for the following reasons:

1. to determine susceptibility to primary infection in women of fertile age as primary infection is linked to a higher percentage of congenital infection with fetal damage.[1,2]
2. to determine the donor/recipient pre-transplant antibody status as an index of susceptibility to primary infection in solid organ transplant recipients. This is important because primary infections are more frequently clinically relevant.[1,2] This does not seem to be enough in BMT recipients as it has been demonstrated that neither donor nor recipient pre-transplant antibody status determines the post-transplant antibody status. Therefore decisions about the use of chronic suppressive acyclovir and CMV hyperimmune globulin after BMT should perhaps be reviewed on the basis of both pre-transplant and post-transplant immune status for CMV rather than just the pre-transplant status as is the usual practice.[4]
3. to select CMV-seronegative blood donors as it has been repeatedly shown that the incidence of post transfusion CMV infection in high risk groups is much lower when the transfused blood is seronegative (2.6%) than seropositive (16-30%).[1]
4. for epidemiological studies related to the diffusion of CMV infection in different geographical areas, socio-economical settings etc.

Furthermore anti CMV IgG detection might be useful as a parameter of risk in the following situations:

1. In solid organ transplant recipients where high viral loads and high incidences of CMV disease are related to delayed onset and low titres of specific immunity (humoral and cell-mediated).[5]
   Such a correlation was not found in BMT recipients where a similar IgG titre was found in symptomatic and asymptomatic patients with high, low or absent antigenemia (Lazzarotto and Bandini personal comunication).

2. In pregnant women, since women who delivered infants developing hearing loss have a more intense and prolongued antibody response.[6,7]

Finally, if performed in two or more consecutive blood samples, the detection of CMV-specific IgG is also useful to diagnose a primary infection on the basis of a seroconversion. The same goal can be obtained measuring IgG avidity,[8] a low avidity being associated with primary infection and high avidity with reactivations and reinfections.

The search for CMV-IgG is therefore useful in the diagnosis and prevention of CMV infection.

B) How to Search for CMV-IgG?

Numerous test methods (such as complement fixation, passive hemoagglutination, latex agglutination, radioimmunoprecipitation, radioimmunoassay, counter immunoelettrophoresis, enzyme immunoassay, immunofluorescence, immunoprecipitation, immunoblotting, dot immunoblotting etc.) are available for the determination of serum-anti-CMV IgG titre with different degrees of sensitivity. The different procedures, with the exception of immunofluorescence show an acceptable overall agreement (table 1).[9-12] The low agreement of immunofluorescence with other tests is probably caused by the visual reading of the results which gives rise to subjective interpretation. The most widely used procedure is undoubtely the enzyme-linked immunosorbent assay and different products are commercially available. The overall agreement among the different assays is acceptable, however a high cross reactivity with the other members of the *Herpesviridae* family has been reported.[13]

In order to overcome this problem and possibly lower the costs of the reagents, antigenic materials composed of single well characterised viral proteins, or portions of them, produced via molecular biology or peptide chemistry, should be used. Therefore a problem arises regarding the most promising antigens and antigenic portions able to detect HCMV-specific IgG in sera from different groups of CMV-infected (acutely and latently) individuals.

Table 1. Comparison between different serological procedures to detect CMV-specific IgG

Test	Commercial name	Sensitivity	Specificity	Agreement	Ref.
Dot immunoblotting	CUBE (Difco)	96.3	99.3		8
ELISA	CMV STAT (Wittaker)	99.3	96.2		
Dot immunoblotting	CUBE (Difco)	97.5	99.4		8
Latex agglutination	CMV SCAN (BBL)	99.4	97.5		
Immunofluorescence	(Viramed)				9
ELISA	(Behring)			56.4	
ELISA	(Biotest)			58.2	
ELISA	(Medac)			56.4	
Western blotting	home made	83.3	92.6	88.9	10
Latex agglutination	CMV SCAN (BBL)	82.3	87.5	85.7	
ELISA	home-made	78.4	98.1	90.0	
MEIA	IMX (Abbott)	100.0	84.8	90.0	
ELISA	Enzygnost α (Behring)	100	93.7	97.9	11
Complement Fixation	CMV (Virgo-Roche)				
ELISA	Behring			nd	12
ELISA	Biotest			95.7	
ELISA	Medac			96.8	
ELISA	Biomerieux			99.2	

Table 2. CMV proteins and fragments best reacting with human anti CMV IgG

Protein	aa	Ref.
ppUL32(pp150)	862-1048	15
	1005-1048	16
	495-691	17
	595-614	18
	695-854	19
	719-880	17
ppUL44(pp52)	202-434	16
	297-434	15
ppUL83(pp65)	297-458	16
	372-546	15
ppUL99(pp28)	undefined	2016
gpUL55(gB)	552-635 (AD1)	27
	28-84 (AD2)	25
	783-906 (AD3)	24
gpUL75(gH)	undefined	31,32

The analysis of the humoral immune response elicited during natural infection has repeatedly shown that the basic phosphoprotein of 150 KD encoded by UL32 (ppUL32) and localized in the viral tegument is highly immunogenic and is recognized by sera from nearly 100% of the HCMV-seropositive subjects tested (for review see ref. 14). It has been shown that only the sera from patients in the early phase of a primary CMV infection lack antibody to this protein.[15] In this molecule different portions have been described to react efficiently with human immunoglobulins. They are located either at the C' terminus[16,17] or in the middle of the molecule.[18-20] Another HCMV phosphoprotein which reacts very well with IgG is ppUL44 (for review see ref. 14) the non structural DNA-binding protein of 52KD. The C' terminus of the molecule has always been considered for serological studies[15,16] because besides its efficient IgG binding, it does not contain relevant aminoacid sequences cross-reacting with the homologous protein of other herpesviruses (homologous sequences are localized in the NH_2 half). Another structural protein highly reacting with IgG is ppUL99(p28)[21] but the mapping of the immunogenic regions has not been performed yet.

An antigen mixture has been recently proposed to replace the virus (and the infected cells) in the detection of CMV-specific IgG[22] in serological tests. The proposed mixture contains three fragments of ppUL32 (495-691, 695-854 and 862-1048) and one large portion of ppUL44 (297-433). This mixture of epitopes has been shown to be very efficient in capturing CMV-specific IgG, however the absence of any envelope glycoprotein in it indicates that neutralizing antibodies are not evaluated. In order to capture neutralizing IgG, the inclusion of the envelope glycoproteins (or portions of them) that elicit neutralizing antibodies should be considered.

NEUTRALISING ANTIBODIES

A) When Do We Need to Search for CMV-Nt Antibodies?

The virus is cleared by immune surveillance, but this is a complex process because HCMV itself may cause immunosuppression and can interact with cells of the immune system.

Neutralizing antibodies do not seem to play a role in the recovery from primary infection since the infection generally resolves before antibody levels rise. Yet administration of neutralizing antibodies protects animals from disease and this appears to occur naturally in newborns that suckle immune mothers. In humans, HCMV can be transmitted despite the presence of passively acquired antibody, although HCMV disease may be attenuated by administering of antibodies or by vaccination with HCMV vaccines. In conclusion it is generally accepted that although humoral immune response and especially neutralizing antibodies do not protect from infection, they do protect from severe diseases (for review see ref 23 and 24).

The serological determination of Nt antibodies is therefore important in judging whether a subject is naturally protected against severe HCMV disease.

Furthermore it would also be important in determining the quality of immunoglobulin preparations used prophylactically in transplant recipients.

Despite multiple treatment efforts, the mortality remains high among BMT recipients with proven CMV pneumonia. Administration of CMV immunoglobulin or plasma has been tried for a number of years for prophylactic treatment of CMV disease. The results of these studies remain controversial[23] which could, at least in part be due to the variable titre of neutralizing antibodies against CMV in immunoglobulin preparations. In fact it has been repeatedly shown that commercially available immunoglobulin preparations differ significantly in neutralization titer. Since the clinical effect of the administration of anti-CMV hyperimmune gamma globulin is believed to depend on its neutralizing activity, preparations with significantly higher neutralizing capacity should improve the effects of CMV prophylaxis.

At this regard we have been able to created IgG pools with 5-10 fold higher Nt titres than conventional IgG preparations by screening sereum reactivity against antigenic domains known to bind neutralizing antibodies (see below).[25]

B) How to Search for CMV-Nt Antibodies?

The determination of Nt antibodies is traditionally achieved by plaque reduction assay altough in the last 7-8 years many procedures have appeared in the literature describing the possibility of detecting Nt antibodies by rapid methods (for review see ref. 3) always

Table 3. CMV proteins (and fragments) best reacting with CMV-specific IgM

Protein	Epitopes (aa)	References
ppUL32(pp150)	862-1048	15
	1024-1048	16,34
	1005-1048	16,34
	495-691	17
	595-614	18
	695-854	19
ppUL44(pp52)	202-434	16
	297-434	15
ppUL83(pp65)	297-458	16
	372-546	15
ppUL80A	105-373	16
ppUL57	545-601	21,42,43
	1144-1233	
gpUL75(gH)		

based on the reduction in viral activities (such as immediate early antigens production). Recently we have set up an ELISA system with recombinant glycoproteins that can be used for capturing Nt antibodies.[25]

The most abundant glycoprotein in the envelope of CMV and the most immunogenic for the humoral immune response is gpUL55(gB). Using protein truncated at the C' terminus and expressed in CHO cells an approximately 60% positivity rate in sera from immunocompetent individuals was reported.[26] When full length gB was used, expressed from recombinant baculovirus, 100% of CMV-positive human sera were found positive (Urban and Mach, unpublished observation). Similar observations were made by others.[27] A possible explanation for the discordant results could be antibody binding to the C' terminal part of gB which occurs in approximately 35% of the sera. gB carries several epitopes recognized by neutralizing and non neutralizing antibodies as well as by convalescent human sera.

Neutralizing antibodies bind to AD1 and AD2. AD1 is the immunodominant epitope in gB, in fact 100% of the sera that react with gB contain antibodies to AD1.[28] 50% of gB-positive sera react with AD2[29] and 35-50% with AD3.[28,30] AD1 is a complex structural domain[27] capable of inducing neutralizing and competing non neutralizing antibodies (32,33). AD2 contains two independent antibody binding sites. Site I (aa 68-77) is conserved among isolates and is recognized by a human monoclonal antibody (C23) which is capable of neutralizing the virus without complement. Site II (aa50-54) is not conserved among strains and therefore induces the synthesis of strain-specific antibodies.[29]

As a statistically significant correlation was found between IgG titre to gB or to AD 1 and neutralizing capacity of convalescent sera[25, 27] AD1 is the gB portion which should be included in a serological test for capturing neutralizing antibodies. AD1 is also the fragment that can be used determine the level of Nt antibodies in different lots of IVIGs.

gH is the second most important antigen in the HCMV envelope. Data regarding antibodies to gH during natural infection differ somewhat. Rasmussen and colleagues investigated gH-specific antibodies using a truncated gH of strain AD169 expressed in mammalian cells and reported a seropositivity between 0-10% in healthy individuals.[26] In another study using procaryotically expressed fusion proteins a seropositivity of 35% was found.[34] More recent investigations using baculovirus-expressed gH indicate a higher reactivity when a full length gH is used and denaturation is avoided (Urban and Mach,

Table 4. Comparison between recombinant ELISA and conventional ELISA for IgM detection in several groups of subjects (data from ref 39 and 41)

		Rec-EIA +		Con EIA+	
Patients studied by recombinant IgM ELISA	N of patients	N	(%)	N	(%)
Individuals with CMV mononucleosis	8	8	100	6	75
CMV-infected pregnant women who transmitted the infection	8	8	100	6	75
CMV-infected pregnant women who did not transmit the infection	25	20	80	12	60
CMV-uninfected pregnant women	18	0	0	0	0
Congenitally infected newborns	6	2	33	0	0
Newborns excreting CMV during the first year of life	19	10	53	2	10
CMV-uninfected newborns	10	0	0		
Antigenemia-positive heart transplant recipients	52	43	83	26	50
CMV uninfected heart transplant recipients	21	0	0	1	5

unpublished results). A similar observation was obtained with full length gH expressed in astrocytoma cells (Reschke and Landini, unpublished data).

On gH there seems to be only one antibody binding region that is independent of conformation of the antigen. It is located at the aminoterminal part of gH (aa 34-43). This domain has sequence variation and antibody-binding is strictly strain-specific.[34] Another two antibody-binding regions were identified but they are strictly conformation-dependent.[35]

In conclusion the determination of which gH fragments can capture CMV-Nt antibodies is problematic due to strain variation and strict conformation-dependency.

CMV-SPECIFIC IgM

A) How to Search for CMV-IgM?

Before answering the question of when to search for CMV-IgM, it is necessary to develop a sensitive and specific detection test as those commercially available present several drawbacks and the agreement between the results obtained with different kits available on the market is not satisfactory.[36] As for IgG, antigenic materials composed of single well characterised viral proteins, or portions of them, produced via molecular biology or peptide chemistry, should be used in order to increase specificity and sensitivity and facilitate antigen standardization. A few proteins are considered best reacting with CMV-specific IgM. The C' terminus of ppUL32 has been shown (by synthetic peptides) to react with approximately 80% of IgM-positive sera.[37] The central portion of ppUL32 also reacts with IgM[17,22,38] and the two regions (one from the C' terminus, the other from the central part of the molecule) fused together were shown to produce a double epitope fusion protein which can replace the entire p150 molecule in its IgM-binding ability.[39] ppUL44 reacts very well with IgM and the same regions chosen for IgG are used to capture IgG.[22,40] A fairly good IgM response has also been found against ppUL83(pp65), the main reacting region being between aa 297 and 546[16] and ppUL80A (pp38) between aa 105 and 373.[17,41]

Recently the product of UL57 (a DNA binding protein of 110 Kd) was described as a major target for CMV-IgM antibodies[21,42] and two regions seem to be responsible for the binding.[42,43]

Very recently two different antigenic mixtures have been proposed to capture CMV-specific IgM.[22,44] The first mixture contains two ppUL32 fragments (aa 595-614, 1005-1048), one large portion of pUL44 (aa202-434) a large portion of ppUL82 (aa 297-510) and a large portion of ppUL80a (aa 117-373). The second mixture conteins one portion of ppUL32 (aa695-854) a portion of ppUL44 (aa 297-433) and 60 aa of ppUL57. Both mixtures gave very good results and open the possibility of evaluating the usefulness of IgM detection in several clinical conditions.

B) When to Search for CMV-IgM?

Using the first mixture we tested several sera from different groups of subjects and found encouraging results.[44] In particular we found 100% positivity in individuals with CMV-mononucleosis and in pregnant women (18-21 week of pregnancy) who transmitted the infection to their fetus. 80% of CMV-infected pregnant women who did not transmit the infection gave a positive result by recombinant IgM-ELISA

Unsatisfactory results were obtained with congenitally infected newborns (33% of positivity) and newborns excreting CMV during the first year of life (53%). Therefore in these cases although much better than conventional ELISA, recombinant IgM-ELISA is not the diagnostic procedure of choice.

In solid organ transplant recipients recombinant IgM-ELISA gave a positive result in 83% of antigenemia-positive patients and was the only procedure that detected 12 % of (asymptomatic)

CMV infections in agreement with published data.[45] In BMT recipients only 50% of infections were detected by IgM recombinant serology (Lazzarotto and Landini, unpublished data).

Furthermore the detection of CMV-IgM might be useful as parameter of risk in solid organ transplant recipients. In fact in these patients a much higher IgM titre to p52 was obtained in primary infection than in secondary infection. IgM to p52 were also much higher in symptomatic patients than in asymptomatic CMV infected subjects.[46] This result is not surprising but reflects the fact that primary infections are more frequently symptomatic and confirms previous findings indicating that a high IgM to p52 should be considered a marker of primary infection in different groups of individuals.[47] Moreover, a high immune response against p52 may represent a marker of CMV disease.[47,48] This does not seem to be true in BMT recipients (Lazzarotto and Bandini, personal communication) where the level of IgM to p52 does not correlate with clinical importance of CMV infection.

CMV-SPECIFIC IgA AND IgE

It is well appreciated that IgG can persist in the serum for many years after CMV infection, usually being associated with lifelong immunity. In contrast serum IgA responses were considered to be transient and a complete correspondence with IgM was shown.[49] On the contrary other Authors reported results supporting the hypothesis that IgA persist as long as IgG in at least 42% of CMV-infected individuals.[50]

Due to the fact that data available on CMV-IgA are scant and contraddictory, before taking into consideration an CMV-IgA test as a diagnositic test further data are needed.

Specific CMV-IgE response has been reported by some authors and was proposed as a valuable marker of CMV infection (for review see ref. 51). The presence of IgE to CMV when detected by immunoblotting was shown to be always present in cases of severe CMV clinical manifestations.[51] As for CMV-IgA, data available on the IgE response to CMV are scant and further data are needed before considering the presence of CMV-IgE a diagnostic tool.

CONCLUSIONS

In conclusion, the rapid evolution triggered by the detailed study of the viral genome and its antigenic gene products has allowed a rapid progress in CMV serology and broadened the horizons of future research. The detection of CMV-specific antibodies when performed with the appropriate antigens proved very useful in several clinical settings. However in BMT recipients the value of CMV serology is limited to the following points:

1. Determination of pre/post transplant serological status (IgG mainly reactive with tegument proteins and non structural DNA-binding proteins)
2. To select IgG preparations with significantly higher neutralizing capacity (IgG to antigenic domains known to bind neutralizing antibodies)

REFERENCES

1. Ho M Cytomegalovirus. Biology and Infection. II Edition. Plenum Medical Press, New York and London,1991.
2. Britt, W.J. Infections associated with human cytomegalovirus. Chapter 2 of "Herpesvirus Infections" R. Glaser and J.F. Jones Editors, Marcel Dekker, Inc, New York. 1994.
3. Landini M P. New approaches and perspectives in cytomegalovirus diagnosis. Prog Med Virol 1993; 4, 157-177.
4. Epstein JB, Phillips K, Sherlock CH Viral serology after bone marrow transplantation. Viral immunol 1991; 4: 133-137.

5. The TH, van der Berg AP, van Son WJ, et al. Recent advances in the early and reliable immunodiagnosis of cytomegalovirus infection in immunocompromised hosts. in Landini MP, ed: Progress in Cytomegalovirus Research, Elsevier, Amsterdam, 1991:209.
6. Alford, CA, Hayes K, Britt W. Primary cytomegalovirus infection in pregnancy: Comparison of antibody responses to virus-encoded proteins between women with and without intrauterine infection. J Inf Dis 158: 917-924, 1988.
7. Boppana, SB, Pass R, Britt W. Virus-specific antibody responses in mothers and their newborn infants with asymptomatic congenital infections. J Inf Dis 167: 72-77, 1993.
8. Blackburn NK, Besselaar TG, Schoub BD, O'Connell KF. Differentiation of primary cytomegalovirus infection from reactivation using the urea denaturation test for measuring antibody avidity. J Med Virol 33: 6-9, 1991.
9. Gleaves CA, Wendt SF, Dobbs DR, Meyers JD. Evaluation of the CMV-CUBE assay for detection of Cytomegalovirus serological status in marrow transplant patients and marrow donors. J Clin Microbiol 28: 841-842, 1990.
10. Koerner K, Kilian D, Zimmermann B, Nebel Schickel H, Horn J. Comparison of four different ELISAs and indirect immunofluorescence for screening of blood donors for antibodies to Cytomegalovirus. Biotest Bull 4: 119-123, 1990.
11. Kraat YJ, Hendrix RMG, Landini MP, Bruggeman CA. Comparison of four techniques for detection of antibodies to Cytomegalovirus. J Clin Microbiol 30: 522-524, 1992.
12. Gutierrez J, del Carmen Maroto C, Piedrola G. Evaluation of a new reagent for anti-cytomegalovirus and anti-Epstein barr Virus immunoglobulin G. J Clin Microbiol 32: 2603-2605, 1994.
13. Doerr HW and Albert S. New developments in CMV antibody screening. Biotest Bull 4: 125-130, 1990.
14. Landini, M.P. Antibody response to human cytomegalovirus proteins. Rev Med Virol 2: 63-72, 1992
15. Landini MP, Rossier E, Schmitz H. Antibodies to human cytomegalvirus structural polypeptides during primary infection. J Virol Meth 22: 309- 317,1988,
16. Vornhagen R, Plachter B, Hinderer W, The TH, Van Zanten J, Matter L, Schmidt CA, Sonneborn HH, Jahn G. Early serodiagnosis of acute cytomegalovirus infection by Enzyme-linked immunosorbent assay using recombinant antigens. J Clin Microbiol 32: 981-986, 1994.
17. Landini MP, Guan MX, Jahn G, Lindenmeier W, Mach M, Ripalti A, Necker A, Lazzarotto T, Plachter B. Large scale screening of human sera with Cytomegalovirus recombinant antigens. J Clin Microbiol 28: 1375-1379,1990
18. Scholl BC, Von Hintzestein B, Borisch B, Traupe B, Broker M, Jahn G. Procaryotic expression of immunogenic polypeptides of the large phosphoprotein (pp150) of human cytomegalovirus. J Gen Virol 69: 1195-1204, 1988
19. Novak J, Sova P , Krchnak V, Hamsikova E, Zavadova H E Mapping of serologically relevant regions of human cytomegalovirus phosphoprotein pp150 using synthetic peptides. J Gen Virol 72: 1409-1413, 1991
20. Plachter B, Wieczorek L, Scholl B-C, Ziegelmaier R and Jahn G. Detection of Cytomegalovirus antibodies by an enzyme linked immunosorbent assay using recombinant polypeptides of the large phosphorylated tegument protein pp150. J Clin Microbiol 30: 201-206, 1992
21. Meyer H, Bankier AT, Landini MP, Brown CM, Barrell BG, Ruger B, Mach M. Identification and procaryotic expression of the gene coding for the highly immunogenic 28Kilodalton st5ructural phosphoprotein (pp28) of human cytomegalovirus. J Virol 62: 2243-2250, 1988.
22. Vornhagen R, Plachter B, Hinderer W, Bein G, The TH, Matter L, Sonneborn HH, Jahn G. Serodiagnosis of acute and past HCMV-infection using recombinant autologous fusion proteins. Cytomegalovirus workshop, Stockholm 1995
23. Snydman DR. Cytomegalovirus Immunoglobulins in the prevention and treatment of Cytomegalovirus disease. Rev Inf Dis 12 (sup7): 839-848.
24. Mach M. and Britt W Immunoprophylaxix of human cytomegalovirus infections in Progress in Progress in Cytomegalovirus Research, MP Landini Ed, 1991, Elsevier, Amsterdam.
25. Utz U, Britt WJ, Vugler L, Mach M. Identification of a neutralizing epitope on glycoprotein gp58 of human cytomegalovirus. J Virol 63: 1995-2001, (1989).
26. Rasmussen L, Matkin C, Spaete R, Pachl C, merigan TC. Antibody response to human cytomegalovirus glycoprotein B gB and gH after natural infection in humans. J Inf Dis 164: 835-842, 1991.
27. Marshall GS, Rabalais GP, Stout GG , Waldeyer SL. (1992) Antibodies to recombinant-derived glycoprotein B after natural cytomegalovirus infection correlate with neutralizing activity. J Inf Dis 165: 381-384, 1992.
28. Kniess N, Mach M, Fay J Britt WJ. Distribution of linear antigenic sites on glycoprotein gp55 of human cytomegalovirus. J Virol 65: 138-146, 1991.
29. Meyer H, Sundqvist VA, Pereira L, Mach M. Glycoprotein gp116 of human cytomegalovirus contains epitopes strain common and strain-specific antibodies. J Gen Virol 73: 2375-2383, 1992.

30. Silvestri M, Sundqvist VA, Ruden U, Wharen B. Characterization of a major antigenic region on gp55 of human cytomegalovirus. J Gen Virol 72: 3017-3023, 1991.
31. Wagner B, Kropff B, Kalbacher H, Britt WJ, Sundqvist VA, Ostberg L, Mach M. A continuous sequence of more than 70 aminoacids is essential for antibody binding to the dominant antigenic site of glycoprotein gp58 of human cytomegalovirus. J Virol 66: 5290-5297, 1992a.
32. Ohlin M, Sundqvist, VA, Gilljam G, Ruden U, Gombert FO, Wharen B, Borrebaeck CA. Characterization of human monoclonal antibodies directed against the pp65 matrix antigen of human cytomegalovirus. Clin Expt Immunol 84: 508-514, 1991.
33. Kropff B, Landini MP, Mach M. An Elisa using recombinant proteins for the detection of neutralizing antibodies against human cytomegalovirus. J Med Virol 39: 187-195, 1993.
34. Urban M, Britt WJ, Mach M. The dominant linear neutralizing antibody-binding site of glycoprotein gp86 of human cytomegalovirus is strain-specific. J Virol 66: 1303-1311, 1992.
35. Simpson JA, Chow JC, Baker J, Avdalovic N, Yuan S, Au D, Co MS, Vasquez M, Britt WJ, Coelingh KL. Neutralizing monoclonal antibodies that distinguish three antigenic sites on human glycoprotein H have distintict binding sites. J Virol 67: 489-496, 1993.
36. Lazzarotto T, Dalla Casa B, Campisi B, Landini MP. Enzyme-linked immunoadsorbent assay for the detection of cytomegalovirus-IgM: Comparison between eight commercial kits, immunofluorescence and immunoblotting. J Clin Lab Anal 1992; 6: 216-218.
37. Landini MP, Ripalti A, Sra K , Pouletty P. Human cytomegalovirus structural proteins: immune raction against pp150 synthetic peptides. J Clin Microbiol 1991; 29: 1868-1872.
38. Ripalti A, Ruan Q, Boccuni MC, Campanini F, Bergamini G and Landini MP Construction of a polyepitope fusion antigens of human cytomegalovirus ppUL32: reactivity with human antibodies J Clin Microbiol 1994 ; 32: 358-363.
39. Ripalti A, Boccuni MC, Campanini F, Bergamini G, Lazzarotto T, Battista MC, Dalla Casa B, Landini MP. Construction of a polyepitope fusion antigen of human cytomegalovirus ppUL32 and detection of specific antibodies by ELISA. Microbiologica 18: 1-12, 1994
40. Ripalti A, Dal Monte P, Boccuni, MC, Campanini, F, Lazzarotto T, Campisi B, Ruan Q, Landini MP. Prokaryotic expression of a large fragment of the most antigenic cytomegalovirus DNA-binding protein (ppUL44) and its reactivity with human antibodies. J Virol Methods 1994; 46: 39-50.
41. Lindenmeier W, Necker A, Krause S, Bonewald R, Collins J. Cloning and characterization of major antigenic determinants of human cytomegalovirus AD169 seen by the human immune system. Arch Virol 1990; 113: 1-16.
42. Vornhagen, R, Hinderer, W, Sonneborg, HH, Bein, G, Matter, L, The, TH, Jahn, G, Plachter, B. The DNA-binding protein pUL57 of Human Cytomegalovirus is a major terget antigen for the immunoglobulin M antibody response during acute infection. J Clin Microbiol 33: 1927-1930, 1995.
43. Maine, GT, Lazzarotto, T, Chovan, LE, Flanders, R, and Landini, MP. The DNA-binding protein pUL57 of human Cytomegalovirus: Comparison of specific Immunoglobulin M (IgM) reactivity with IgM reactivity to other major terget antigens. Clin Diag Lab Immunol 3: 358-360, 1996.
44. Landini MP, Lazzarotto T, Maine GT, Ripalti A, Flanders R. recombinant mono and poly antigens to detect Cytomegalovirus-specific IgM in human sera by enzyme immunoassay. J Clin Microbiol 1995; submitted.
45. Weber W, Nestler U, Ernst W, Rabenau H, Braner J, Birkenbach A, Scheuermann EH, Schoeppe W, Doerr HW. Low correlation of human cytomegalovirus DNA amplification byn polymerase chain reaction with cytomegalovirus disease in organ transplant recipients. J Med Virol. 1994; 43: 187-193.
46. Ghisetti V, Barbui A, Lazzarotto T, Donegani E, Ripalti A, Dal Monte P Bobbio, di Summa M , Marchiaro G, Landini MP. Comparison between virology and serology for the follow upof cytomegalovirus infection in heart transplant recipients. Transplantation, submitted.
47. Landini MP, Lazzarotto T, Ripalti A, Guan MX, La Placa M. Antibody response to recombinant Lambda gt11 fusion proteins in Cytomegalovirus infection. J Clin Microbiol 1989; 27: 2324-2327.
48. Basson J, Tardy JC, Aymard M. Characterization of immune complexes containing Cytomegalovirus-specific IgM antibodies following a kidney graft. J Med Virol 1991; 33: 205-210.
49. Strand OA, Hoddevik GM The diagnostic significance of specific serum IgA detection in Cytomegalovirus infection. Archives of Virology 1984; 82: 173-180.
50. Morris GE, Coleman RM, Best JM, Benetato BB, Nahmias AJ Persistence of serum IgA antibodies to herpes simplex, Varicella Zoster, Cytomegalovirus and Rubella virus detected by enzyme-linked immunosorbent assays. Journal of Medical Virology 1985; 16: 343-349.
51. Vargas, MA, Bertrand, F, Mulongo, N, Squifflet, JP, Lamy, ME. Specific IgE detected by ELISA and immunoblott after human cytomegalovirus infection in renal transplant recipients. Clin Diag Virol 6: 1-9, 1996.

TARGETED GENE TRANSFER INTO CD4 POSITIVE CELLS BY HIV-BASED RETROVIRAL VECTORS

Takashi Shimada and Koichi Miyake

Department of Biochemistry and Molecular Biology
Nippon Medical School, Tokyo 113, Japan

ABSTRACT

Because CD4 is the major receptor for HIV infection, HIV based retroviral vectors are capable of targeted and efficient gene transfer into CD4 positive helper T cells. The strict T cell tropism of the HIV vector should be important for the development of gene therapy for AIDS. Another feature is that HIV can infect non-dividing cells. Therefore, the HIV vector may also be useful for gene therapy targeting slow- or non-dividing cells such as hematopoietic stem cells and neural cells. We developed a strategy to use the HIV vector for gene transfer into non-lymphoid cells. A replication defective adenovirus vector containing the human CD4 gene was constructed. Using this recombinant adenovirus vector, the CD4 gene was efficiently transferred and expressed in HeLa, K562, and Raji cells. These cells were stably transduced with an HIV vector containing the neoR gene. These results indicate that transient expression of CD4 by the adenovirus vector is sufficient to render non-T cells susceptible to gene transfer by the HIV vector. Since adenovirus can infect non-dividing cells, the combination of the adenovirus vector containing the CD4 gene and the HIV vector may be used for stable gene transfer into various types of cells arrested in the cell cycle.

INTRODUCTION

Retroviral vectors derived from Moloney murine leukemic virus (MoMLV) with the amphotropic envelope are the most commonly used vector in clinical trials of human gene therapy.[1] MoMLV is a member of the oncoretrovirus subfamily. The advantages of this vector system include broad host range, stable integration into the chromosomes, and availability of packaging cell lines. The host range of retrovirus is determined by the interaction between the virus envelope and the cell surface receptor. The receptor for amphotropic retrovirus has been recently identified as a sodium-dependent phosphate transporter.[2,3] The homologous protein is widely distributed on various tissues and species. However, the level of expression of the protein appears to vary and might limit viral entry into particular target cells such as T cells and hematopoietic stem cells.

Another disadvantage of retrovirus mediated gene transfer is inability to infect growth arrested cells. It has been reported that breakdown of the nuclear envelope at mitosis is necessary for nuclear transport of the oncoretroviral preintegration complex.[4,5] Because most of the target cells are quiescent, this requirement for cell division has limited the applicability of retroviral vectors for therapeutic gene transfer.

Human immunodeficiency virus (HIV) is a etiological agent of acquired immunodeficiency syndrome (AIDS). HIV is a human retrovirus and belongs to the lentivirus subfamily. CD4 molecules on human helper T lymphocytes have been identified as a major receptor for HIV infection.[6] Therefore, HIV infect specifically CD4 positive cells and destroy the immune system of the patients. An important feature of HIV is that, unlike oncoretroviruses, HIV and probably other lentiviruses have the ability to infect non-dividing cells.[7] This property seems to be due to the existence of a nuclear localization signal (NLS) in the viral matrix protein which enable the viral complex actively transport through the nucleopore.[8]

These properties of HIV suggest that HIV based vectors may be used for targeted gene transfer into T lymphocytes and also gene transfer into non-dividing cells. To examine these possibilities, we developed a packaging system for producing recombinant HIV vectors. Gene transfer by HIV vectors was shown to be dependent upon CD4 expression on target cells. In addition, we described a new strategy to use the HIV vector for gene transfer into non-lymphoid cells.

PRODUCTION OF HIV VECTORS

We designed packaging and vector plasmids for producing HIV vectors (Fig. 1). The first packaging plasmid, pCGPE, consists of the all HIV coding sequences, the CMV

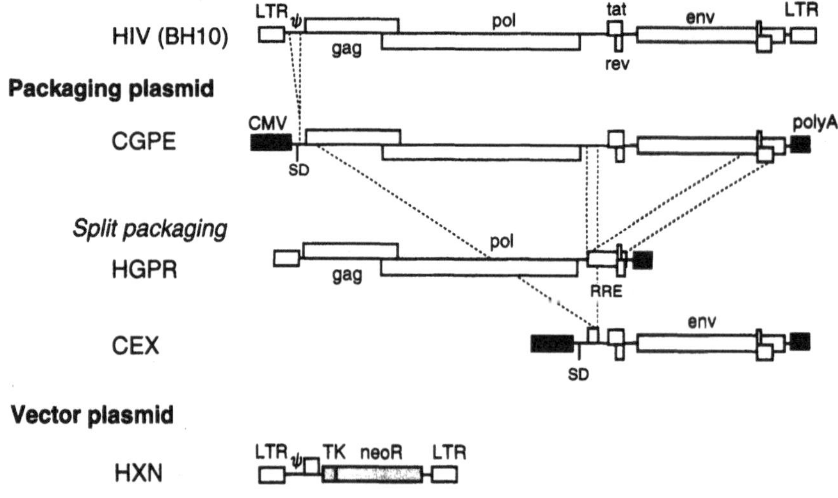

Figure 1. Structure of plasmids for production of HIV vectors. Plasmid construction of pCGPE and pHXN was described previously[9]. The HGPR plasmid was constructed by inserting the HIV-LTR between the XhoI (nt 8927[14]) and the HindIII (nt 9646) sites, the coding sequences for Gag and Pol proteins between the RsaI (nt 745) and the NdeI (nt 5155) sites, and the RRE sequence between BglII (nt 7651) and the BamHI (nt 8505) sites in the L vector. The CEX plasmid was constructed by deleting the 4.9 kb ClaI-NcoI (nt 827-5708) fragment from pCGPE.

promoter, and the β-globin processing signal in a single plasmid. The putative packaging signal was deleted from the packaging plasmid. The vector plasmid contains the thymidine kinase promoter driven neomycin resistance gene (pHXN) flanked by the two HIV LTR. To eliminate the production of replication competent virus, the homologous region between these two plasmids was restricted in the 5' end region.

Cos cells were co-transfected with both packaging and vector plasmids and after 48 hours, the culture medium was collected and characterized. High levels of expression of p24 and gp120 were detected by immunological techniques, and virus-like particles were observed by the electron microscopic examination. In addition, blot hybridization showed that recombinant RNA genomes were associated with viral particles.[9] We tested for replication competent HIV using the HTLV-1 transformed human T cell line, MT-2, that is highly sensitive to cytopathic effects of HIV. However, syncytium formation was not observed in MT-2 culture with the conditioned medium of the Cos cells transfected with pCGPE + pHXN, even the MT-2 cells were cultured for three months after incubation with Cos cell medium. These results indicate that no detectable infectious viruses were generated in this packaging system.

THE SPLIT PACKAGING SYSTEM FOR HIV VECTORS

We wanted to developed a split packaging system to further minimize the possibility of generation of replication competent virus. However, because the regulation of HIV gene expression is very complicated, it is difficult to design plasmid vectors which express each viral protein at high efficiencies. We tested various constructs, and the high level of gag/pol expression was observed with pHGPR in which the minimal coding sequences for Gag and Pol proteins and the Rev responsive element (RRE) were transcribed from the HIV-LTR promoter. We also found that pCEX expressed high levels of Env, Tat, and Rev proteins. In this vector, most part of the gag region was deleted from pCGPE, but the major splicing donor and acceptor sites were retained. Therefore, transcripts are alternatively spliced to produce distinct RNA molecules for each protein.

Recombinant HIV-neo vectors could be produced by using pHGPR and pCEX as packaging plasmids. However, the titer of the HIV vector obtained in this system was approximately one fifth of that produced with a single packaging plasmid, pCGPE. Therefore, although the split packaging system is preferable from a safety standpoint, the production of the HIV vector in this system is less efficient than that in the simple single plasmid packaging system.

TARGETED GENE TRANSFER INTO CD4 POSITIVE CELLS

We studied tissue specificity of HIV vectors. CD4 positive HeLa cells (T4H) and CD4 negative parental HeLa cells were incubated with the HXN vector and selected in the presence of G418 for two weeks (Fig. 2). Resistant colonies appeared only in CD4 positive cells. No colonies were detected in CD4 negative cells.

When 10 ug/ml soluble CD4 molecules were co-incubated with virus vectors, colony formation was completely inhibited. Soluble CD4 at this concentration did not inhibit cell growth. These results clearly showed that gene transfer by HIV vectors is mediated by CD4 on the target cell surfaces. Therefore, HIV vectors can be used for targeted gene transfer into CD4 positive T lymphocytes.

T-lymphocytes are important potential target for therapeutic gene transfer both in vitro and in vivo. The most attractive approach is to use the HIV vector to treat lymphocytes

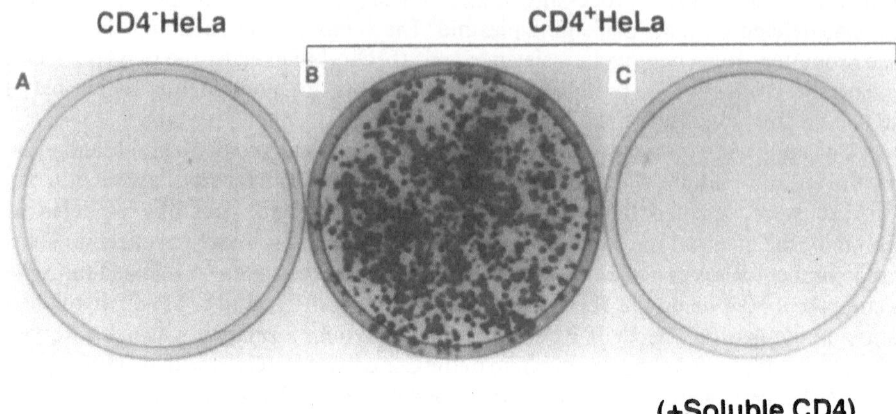

Figure 2. Targeted gene transfer into CD4 positive HeLa cells. A HeLa cell subline (T4H) was established by stable transduction with CCD4/LH. CD4- parental HeLa cells (A) or T4H cells (B and C) were incubated with HXN. Ten ug/ml soluble CD4 molecules were included in C. The cells were cultured in the presence of G418.

of AIDS patients. Various anti-HIV genes which dominantly interfere with HIV replication have been identified.[10] If we could introduce these anti-HIV genes into CD4+ lymphocytes, these cells may become resistant to HIV infection. Because human lymphocytes are relatively resistant to murine retroviral vectors, HIV vector mediated gene transfer should be useful for these therapeutic applications.

GENE TRANSFER INTO NON-LYMPHOID CELLS BY HIV VECTORS

It has been documented that terminally differentiated macrophages are infected with HIV. Lewis et al.[11] showed that HIV can integrate into irradiated cells and produce viral RNA and protein. Subsequently, Trono and colleagues[8] showed that the matrix protein which contains a nuclear localization signal (NLS) plays a key role in nuclear import of core particles in non-dividing cells. These results strongly suggest that HIV vectors are capable of stable gene transfer into non-dividing cells. Therefore, HIV vectors should be useful for gene therapy targeting non-lymphoid cells.

In order to use HIV vectors for gene transfer into various types of cells other than CD4 positive helper T cells, the CD4 gene must be expressed in these target cells prior to HIV vector mediated transduction. To express CD4 in non-T cells, we used adenovirus vector containing the human CD4 gene. Recombinant adenovirus vectors are capable of gene transfer into many types of cells at very high efficiencies. However, the transferred genome is maintained extrachromosomally and therefore, the gene expression is transient. In this strategy, the adenovirus vector is used for transient expression of the receptor molecules. Subsequent CD4 specific transduction with HIV vectors should result in stable gene transfer and expression.

A recombinant adenovirus vector (AdexCAG-CD4) that express CD4 was constructed according to Saito's method.[12] The CMV/β-actin hybrid promoter (CAG) was used to achieve high levels of CD4 expression.[13] Three CD4 negative cell lines, HeLa, K562 and Raji cells were transduced with the AdexCAG-CD4 vector at various MOI ratios, and expression of CD4 was analyzed by flow cytometry. HeLa cells were highly sensitive to

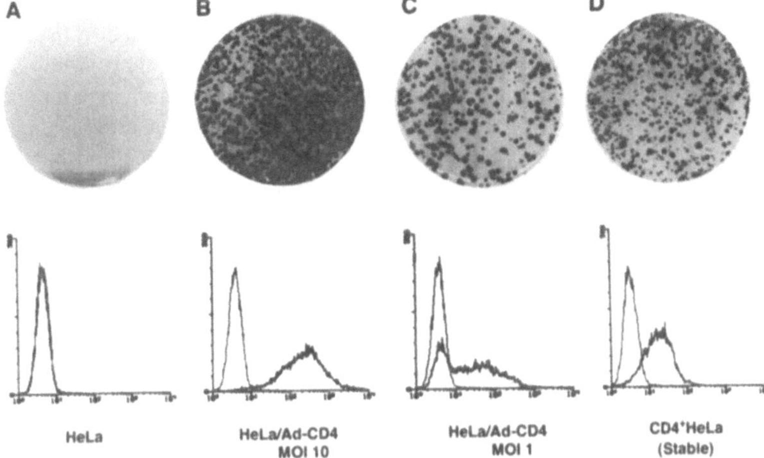

Figure 3. Transduction of HeLa cells by a combination of the adenovirus and HIV vectors. HeLa cells were first incubated with AdexCAG-CD4 at the MOI of 1 : 1 (B) or 1 : 10 (C). Non-transduced HeLa cells (A) and CD4 positive HeLa cells (CD4H) (D) were included as negative and positive controls. The cells were subsequently transduced with HXN and selected for G418 resistance.

adenovirus transduction, and 100 % cells could be transduced at an MOI ratio of 10. Floating hematologic cells were relatively resistant to adenovirus transduction, and , at an MOI of 100, about 80 % of K562 and 60 % of Raji cells were transduced with the adenovirus vector.

HeLa cells were first transduced with the AdexCAG-CD4 vector, and were subsequently transduced with the HXN vector. The cells were cultured in the medium containing 1 mg/ml G418 for two weeks. The use of these two different vector systems allowed high efficient stable gene transfer into HeLa cells (Fig. 3). Without adenovirus transduction, the HIV vector alone did not transduce HeLa cells.

This combination method was also useful for gene transfer into floating cells. K562 and Raji cells could be transduced with the HIV vector at the efficiency comparable to that for CD4 positive T cell lines (Table 1). These results clearly showed that transient expression of CD4 by the adenovirus vector is sufficient to render non-lymphoid cells susceptible to gene transfer by the HIV vector.

Table 1. Transduction efficiency of Non-T Cells by a combination of Adeno-CD4 and HIV-neo vectors

Cells	AdexCAG-CD4	HXN
CEM (CD4+T)		76
H9 (CD4+T)		96
K562 (Erythroleukemic)	-	0
	+	51
Raji (B lymphocytic)	-	0
	+	28

Cells were first incubated with or without AdexCAG-CD4 and subsequently transduced with HXN. The cells were dispensed into 96 well culture plates at 10^3 cells/well. Wells containing G418 resistant cells were counted.

The combination of the adenovirus vector with the CD4 gene and the HIV vector may be useful for stable gene transfer into various types of cells arrested in the cell cycle. Target cells for many gene therapy protocols are non-dividing or rarely dividing cells, and therefore, resistant to retroviral mediated gene transfer. Neural cells and hematopoietic stem cells are important target cells for various genetic diseases. However, previous methods of gene transfer into these cells have been extremely inefficient. The method based on two different viral vectors described here may be applicable for gene therapy targeting such cells.

CONCLUSION

We have developed a packaging system for producing replication incompetent helper free HIV vectors which are capable of targeted and efficient gene transfer into CD4 positive cells. Transient expression of CD4 by the adenovirus vector rendered non-lymphoid cells susceptible to HIV vector mediated gene transfer. Because both adenovirus and HIV can infect non-dividing cells, a combination of the adenovirus vector with the CD4 gene and the HIV vector might be used for stable gene transfer into various types of cells arrested inthe cell cycle.

ACKNOWLEDGMENT

We thank Drs. A.W. Nienhuis, H. Matsuoka, T. Tohyama, T. Igarashi, E. Shinya, and S. Suzuki for their contributions.

REFERENCES

1. Miller AD: Retrovirus packaging cells. Hum Gene Ther 1:5, 1993
2. Miller DG, Edwards RH, Miller AD: Cloning of the cellular receptor for the amphotropic murine retrovirus reveals homology to that for gibbon ape leukemia virus. Proc Natl Acad Sci U S A 91:78, 1994
3. Van Zeijl M, Johann SV, Closs E, Cunningham J, Eddy R, Shows TB, O'hara B: A human amphotropic retrovirus receptor is a second member of the gibbon ape leukemia virus receptor family. Proc Natl Acad Sci U S A 91:1615, 1994
4. Humphries EH, Temin HM: Requirment for cell division for initiation of transcription of Rous sarcoma virus RNA. J Virol 14:531, 1974
5. Miller DG, Adam MA, Miller AD: Gene transfer by retrovirus vectors occurs only in cells that are actively replicating at the time of infection. Mol Cell Biol 10:4239, 1990
6. Maddon PJ, Dalgleish AG, McDougal JS, Clapham PR, Weiss RA, Axel R: The T4 gene encodes the AIDS virus receptor and is expressed in the immune system and the brain. Cell 47:333, 1986
7. Brice Weinberg J, Matthews TJ, Cullen BR, Malim MH. Productive human immunodeficiency virus type 1 (HIV-1) infection of nonproliferating human monocytes. J Exp Med 174:1477, 1991
8. Gallay P, Swingler S, Aiken C, Trono D: HIV-1 infection of nondividing cells: C-terminal tyrosine phosphorylation of the viral protein is a key regulator. Cell 80:379, 1995
9. Shimada T, Fujii H, Mitsuya H, Nienhuis AW: Targeted and highly efficienct gene transfer into CD4+ cells by a recombinant human immunodeficiency virus retroviral vector. J Clin Invest 88:1043, 1991
10. Yu M, Poeschla E, Wong-Staal F: Progress towards gene therapy for HIV infection. Gene Ther 1:13, 1994
11. Lewis P, Hensel M, Emerman M: Human immunodeficiency virus infection of cells arrested in the cell cycle. EMBO J 11:3053, 1992
12. Nakamura Y, Wakimoto H, Abe J, Kanegae Y, Saito I, Aoyagi M, Hirakawa K, Hamada H: Adoptive immunotherapy with murine tumor-specific T lymphocytes engineered to secrete interleukin 2. Cancer Res 54:5757, 1994
13. Niwa H, Yamamura K, Miyazaki J: Efficient selction for high-expression transfectants with a novel eukaryotic vector. Gene 108:193, 1991

14. Ratner L, Haseltine W, Patarca R, Livak KJ, Starcich B, Joseph SF, Doran ER, Rafalski JA, Whitehorn EA, Baumeister K, Ivanoff L, Petteway Jr SR, Pearson ML, Lautenberger JA, Papas TS, Ghrayeb J, Chang NT, Gallo RC, Wong-Staal F: Complete nucleotide sequence of the AIDS virus HTLV-III. Nature 313:277, 1985

IN VITRO RETROVIRAL VECTOR-MEDIATED TRANSFER OF THE RAT BETA-GLUCURONIDASE cDNA INTO CANINE FETAL LIVER CELLS AND WEANLING MPS VII BONE MARROW CELLS

Margret L. Casal, Mark E. Haskins, and John H. Wolfe*

Department of Pathobiology
School of Veterinary Medicine
University of Pennsylvania
3800 Spruce Street, Philadelphia, Pennsylvania 19104

A beta-glucuronidase cDNA was transferred into normal canine fetal liver cells and bone marrow cells obtained from weanling dogs affected with mucopolysaccharidosis (MPS) type VII. The cells were transduced by direct cocultivation with a recombinant retrovirus-producing cell line, and by culturing the hematopoietic cells and vector packaging cells separated by a 0.45 μm filter in a dual-chambered cocultivation system. The weanling MPS VII dog bone marrow cells were also transduced in long-term cultures with high titer vector virus. Gene transduction was achieved in all fetal liver cells obtained from 24 and 35 day old dog fetuses, but not 55 day old fetuses. The post-natal bone marrow cells were transduced by each of these methods. The results indicate that early gestation fetal liver cells can be transduced in culture conditions without contaminating virus-producing packaging cells.

Mucopolysaccharidosis type VII (MPS VII) in humans, mice, and dogs is a well characterized progressive, genetic disease caused by a deficiency of beta-glucuronidase (GUSB) activity.[1-4] Microscopic signs of disease are present at birth,[5] thus the animal models can be used to determine if providing GUSB activity at an early age or in utero reduces the pathology associated with MPS VII.

Bone marrow transplantation of hematopoietic stem cells (HSC) provides a means of treating a variety of lysosomal storage diseases by providing a source of normal enzyme.[4,6] However, the efficacy of bone marrow transplantation is variable depending on the type of storage disease and the age at which transplantation is performed.[6-8] In addition, matched donors frequently are not available and the recipient must be conditioned either by irradiation or chemotherapy, both of which bear the risk of diseases associated with these pretransplan-

* To whom correspondence should be addressed.

Molecular Biology of Hematopoiesis 5, edited by Abraham et al.
Plenum Press, New York, 1996

tation treatments. Potential complications of bone marrow transplantation are graft rejection and graft versus host disease.[9]

Many of the side effects of bone marrow transplantation may be avoided by transferring a functional exogenous gene into autologous hematopoietic stem cells in bone marrow. Retroviral vectors efficiently transfer exogenous genes into murine hematopoietic stem cells.[10-14] However, transplantation of transduced hematopoietic stem cells into adults may not lead to reversal of all aspects of disease.[14] In addition, studies in larger animal species have shown that efficiency of transduction and expression of the transferred gene are low.[15-17] Introduction of an exogenous gene into fetal or neonatal hematopoietic stem cells may result in higher transduction rates and better expression. In utero intervention may reduce the pathology caused by the gene defect, or prevent manifestation of disease altogether. In this study, we used a retroviral vector containing GUSB cDNA that has been shown to transduce adult hematopoietic stem cells[14] and express GUSB in affected canine cells.[18-20] This study examined retroviral transduction of canine fetal liver cells of 24, 35, and 55 days gestational age, and canine MPS VII bone marrow cells obtained 42 days post-natally.

MATERIALS AND METHODS

Preparation of Canine Fetal Liver Cells

Canine fetal livers were obtained by surgically removing fetuses at 24, 35, and 55 days gestational age (E24, E35, and E55). The liver cells were prepared in IMDM containing 10% FBS, L-glutamine, and penicillin-streptomycin-fungizone. Single liver cell suspensions were obtained by repeated pipeting using successively smaller pipets after the livers had been sliced with a scalpel blade. The suspension was left to rest for five minutes at room temperature, the top two-thirds were removed, and centrifuged at 1000 rpm for 5 minutes at 4°C. The pellet was resuspended in IMDM with 10% FBS, 3% BSA, 5% dog serum, IL-1B (200U/ml), IL-6 (200 U/ml), recombinant canine stem cell factor (a gift from Amgen, Thousand Oaks, CA), and 5% phytohemagglutinin-lymphocyte-conditioned-medium.[21] Cell counts were performed by trypan blue exclusion.

Preparation of Canine Bone Marrow Cells

Bone marrow cells were obtained from 2 six-week-old pups affected with MPS VII by aspiration biopsy from the iliac crest and femur. The white blood cells and their precursors were separated from the red blood cells by Ficoll® density gradient centrifugation.[21]

Gene Transfer Protocol

The vector used in this study (NTKBGEO-A3S6) consisted of the neomycin resistance gene (neo) and rat GUSB cDNA driven by the thymidine kinase (TK) promoter.[18] The construct was converted to amphotropic viruses in GP+envAm 12 cells.[22] The following procedures were used to transduce the fetal liver cells and bone marrow cells. Immediately after preparation, fetal liver cells were plated at 5×10^5 to 1×10^6 and bone marrow cells were plated at 1×10^7 per well in 6-well plates and incubated for 48 hours at 37°C (5% CO_2 in humidified air). One day before infection, virus-producing fibroblasts were seeded either into 6-well plates or into the top well of 3.5 cm double-chambered cell culture plates (Transwell plates, Costar, Cambridge, MA) at 2.5×10^5 cells/well for fetal liver cells and 1×10^6/well for bone marrow cells and incubated (24 hours, 37°C, 5% CO_2 in humidified air). Thereafter, the fetal liver and bone marrow cells were either directly seeded on the monolayer of virus-producing cells (direct cocultivation) or were transferred into the bottom well (Transwell), which was separated from the top well by a membrane with 0.45 µm pores preventing contamination of the hematopoietic cells with virus-producing fibroblasts but allowing diffusion of infectious viral particles.[22,23] In addition, fetal liver cells were transferred into the bottom well of a Transwell plate that did not contain any packaging cells, as a negative control. After adding fresh IMDM containing growth factors and polybrene at a total concentration of 8 µg/ml, both culture systems were incubated for another 48 hours (fetal liver cells) or 72 hours (bone marrow cells) at 37°C (5% CO_2 in humidified air).

The fetal liver cells were analyzed for presence of GUSB cDNA after 48 hours of cocultivation. Bone marrow cells that were directly cocultivated with vector-virus producing cells were harvested in two fractions: After 72 hours of cocultivation, adherent cells were assayed for the presence of vector sequences separately from the non-adherent bone marrow cells.

In a longer term experiment, weanling dog bone marrow cells cultured in 6-well plates (1 x 10^7 cells/well) with IMDM containing growth factors, polybrene (8 µg/ml), and 1 ml of NTKBGEO-A3S6 viral supernatant at a titer of 2.3 x 10^6. A second 6-well plate containing bone marrow cells, IMDM, and growth factors, but no virus was also prepared as a negative control. Both plates were incubated for 7 days (37°C, 5% CO_2 in humidified air). On day 7 and 14, half of the adherent and non-adherent cells and half of the supernatant were collected for PCR and enzyme activity, and the rest of cells were left to incubate overnight. The next day (day 8 and 15), fresh medium was added to both plates and 1 ml of the high titer virus was added to the first plate. All of the cells and the supernatant were collected for enzyme assay and PCR on day 21.

Transduction Assays

Fetal liver cells and bone marrow cells were assayed by polymerase chain reaction (PCR) before and after transduction to determine the presence of the vector sequences. The GUSB cDNA was amplified by PCR using the following primers: 5'AGAATTCTGGTCATCGATGAGTGTCCC3' (forward primer) and 5'GGCAATCCTCCAGTATCTCTCTCGC3' (reverse primer). The PCR product specific for the GUSB cDNA is 682 bp long, which can easily be distinguished from the genomic GUSB, which is too large to be amplified by PCR under the conditions used.[24]

Enzyme Assays

GUSB activity in bone marrow cells and in their culture supernatant was determined using a fluorometric assay with 4-methylumbelliferyl-beta-D-glucuronide.[25]

RESULTS

Cell counts obtained at 0, 48, and 96 hours revealed a two-fold increase within the first 48 hours of incubation (Table 1). During the next 48 hours, the number of fetal liver cells derived from E24 and E35 fetuses remained constant, while the number of E55 cells decreased precipitously between 48-96 hours.

In both experiments using the Transwell system with E24 and E35 fetal liver cells, the cells were positive for the presence of the GUSB cDNA in the provirus by PCR after vector infection (Figure 1). E24 fetal liver cells also were positive for the vector provirus after direct cocultivation with vector-virus producing packaging cells, while uninfected

Table 1. Retroviral vector transduction of normal canine fetal liver cells

Age	Method	Percent of starting fetal liver cells 0 h	48 h	96 h	PCR
E24	None	100	211	220	-
	Direct	100	220	220	+
	Cocultivation	100	220	232	+
	Transwell	100	220	173	+
E35	Transwell	100	ND	243	+
E55	Transwell	100	208	1.7	-
	Transwell	100	167	1.3	-
	Transwell	100	193	3.5	-

+: PCR product present in fetal liver cell cultures; -: PCR product not present; ND: not determined.

control cells were negative after 96 hours. Vector-virus sequences were not detected in any of the E55 fetal liver cells after cocultivation in the Transwell plates.

The adherent and non-adherent post-natal MPS VII bone marrow cells that were directly cocultivated with vector-virus producing cells were positive by PCR for proviral sequences (Table 2). GUSB cDNA was present in 5/7 samples of bone marrow cells which were cocultivated with the packaging cells in the dual-chambered culture system.

Vector-encoded GUSB activity could not be directly assayed in MPS VII target cells in these experiments because the packaging cells secret normal GUSB which cross-corrects the target cells.[22] Therefore, to demonstrate that the vector could correct the enzyme deficiency in MPS VII cells, vector virus containing medium was used to infect weanling MPS VII cells in long-term bone marrow cultures.[26] On day 7, 14, and 21, all bone marrow cells in the plate incubated without virus were negative for vector sequences by PCR. GUSB activity was at background levels in the adherent and non-adherent cells, as well as in the supernatant (Table 3). In the non-adherent bone marrow cells incubated with high titer virus, the integrated provirus was detected by PCR in 5/6, 4/6, and 5/6 wells at 7, 14, and 21 days, respectively. GUSB activity measured in the transduced bone marrow cells and secreted into

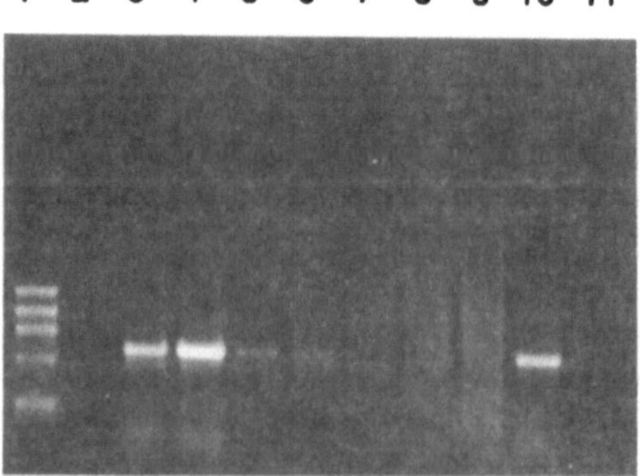

Figure 1. Agarose gel with the 682 bp amplification product after transduction of fetal dog liver cells. Lane 1: ΦX174 HAE III size marker. Lane 2-5, E24 cells. Lane 2, uninfected cells. Lane 3 and 4, direct cocultivation. Lane 5, Transwell. Lane 6, E35 cells, Transwell. Lane 7-9, E55, Transwell. Lane 10, positive control, NTKBGEO infected fibroblasts[24]. Lane 11, H_2O.

Table 2. Retroviral vector transduction of weanling (day 42) MPS VII dog bone marrow cells

			PCR Reaction		
			Direct cocultivation		Transwell
Day	Dog	Sample	Non-adherent BM	Adherent BM	BM
2	A	1	+	+	+
		2	ND	+	+
	B	1	+	+	-
		2	+	+	+
5	A	1	+	+	-
		2	+	+	+
	B	1	+	+	+
		2	+	+	ND

BM: bone marrow cells; +: PCR product present; -: PCR product not present; ND: not determined

the supernatant was present at high levels. The activity progressively increased following subsequent infections.

DISCUSSION

Fetal hematopoietic stem cell transplants have been performed in several animal species and resulted in some degree of chimerism.[27-33] Graft rejection as well as graft versus host disease are avoided, because of immunotolerance of the fetal cells.[34] In genetic disorders such as the storage diseases, only a small amount of the functional gene product (enzyme) may be necessary to correct the clinical signs.[14] However, donor cells may not be readily available. A feasible alternative is autologous transplantation of fetal hematopoietic stem cells that have been transduced in vitro with a vector-virus containing the appropriate gene. In this study, the data indicate that the amphotropic retrovirus can transduce canine fetal liver cells obtained during midgestation, but not during late gestation when hematopoeisis has shifted to the bone marrow. Others have reported being unable to transduce E15 (third trimester) murine fetal liver cells with an amphotropic virus.[35]

Positive PCR results obtained when fetal liver and non-adherent bone marrow cells were directly cocultured with the vector-virus producing packaging cells could be explained

Table 3. Transduction of, and GUSB activity in long-term culture of weanling dog MPS VII bone marrow cells

		Nonadherent bone marrow cells		Adherent bone marrow cells		Supernatant
Day	Virus	#PCR+	GUSB (±SD)*	#PCR+	GUSB (±SD)	GUSB (±SD)
7	-	0/6	6.7±1.8	ND	ND	1.2±0.5
	+	5/6	238±30	ND	ND	105±72
14	-	0/6	15±9	ND	ND	4.2±2.0
	+	4/6	322±48	ND	ND	300±253
21	-	0/6	27±7	0/6	10±3.2	4.0±1.4
	+	5/6	1360±431	4/6	1407±564	93±41

*Units of GUSB expressed as nmol substrate degraded/hour/mg protein.

by contaminating packaging cells in the fetal liver and bone marrow cells. The GUSB cDNA present in the packaging cells was probably also detected in the adherent bone marrow cells because they could not be separated. In contrast, the positive results obtained in the dual-chambered culture system are unequivocal, because virus producing packaging cells are unable to pass through the membrane separating the two cell populations.[22,23]

Cell counts from cultures of E24 and E35 fetal livers increased significantly during 96 hours of culture in medium containing appropriate growth factors, whereas E55 fetal liver cells decreased rapidly during the last 48 hours of culture. The increase of midgestational fetal liver cells but not late gestational cells corresponds to the ages when the fetal liver is a hematopoietic organ.

These studies show that the dual-chambered culture system allows transduction of canine fetal liver and bone marrow cells. Transduction of weanling dog bone marrow cells can also be achieved in long-term cultures with high titer virus. Therefore, canine cells can be transduced and safely transplanted into a recipient without contaminating packaging cells.

ACKNOWLEDGMENTS

This work was supported by a grant from the National Institute of Diabetes and Digestive and Kidney Diseases (DK-46637) and MLC was supported by an NIDDK fellowship (DK-09185).

REFERENCES

1. Sly WS, Quinton BA, McAlister WJ, Rimoin DL: Beta-glucuronidase deficiency. Report of clinical, radiologic and biochemical features of a new mucopolysaccharidosis. J Pediatr 82:249-257, 1973.
2. Birkenmeier EH, Davisson MT, Beamer WG, Ganshow RE, Vogler CA, Gwynn B, Lyford KA, Maltais LM, Wawrzyniak CJ: Murine mucopolysaccharidosis type VII. Characterization of a mouse with β-glucuronidase deficiency. J Clin Invest 83:1258-1266, 1989.
3. Haskins ME, Desnick RJ, DiFerrante N, Jezyk PF, Patterson DF: Beta-glucuronidase deficiency in a dog: A model of mucopolysaccharidosis VII. Pediatr Res 18:980-984, 1984.
4. Neufeld EF, Muenzer J In *The Metabolic Basis of Inherited Disease*; 6th ed.; C. R. Scriver, A. L. Beaudet, W. S. Sly and D. Valle, Ed.; McGraw Hill: New York, 1989; pp 1565-1587.
5. Vogler C, Birkenmeier EH, Sly WS, Levy B, Pegors C, Kyle JW, Beamer WG: A murine model of mucopolysaccharidosis type VII. Characterization of a mouse with beta-glucuronidase deficiency. Am J Pathol 136:207-217, 1990.
6. Haskins ME, Baker HJ, Birkenmeier E, Hoogerbrugge PM, Poorthuis BJHM, Sakiyama T, Shull RM, Taylor RM, Thrall MA, Walkley SU In *Treatment of Genetic Diseases*; R. J. Desnick, Ed.; Churchill Livingstone: New York, 1991; pp 183-201.
7. Sands MS, Barker JE, Vogler C, Levy B, Gwynn B, Galvin N, Sly WS, Birkenmeier E. Treatment of murine mucopolysaccharidosis type VII by syngeneic bone marrow transplantation in neonates. Lab Invest 68:676-686, 1993.
8. Taylor RM, Farrow BRJ, Stewart GJ: Improvement in the neurological signs and storage lesions of fucosidosis in dogs given marrow transplants at an early age. Trans Proc 21:3818-3819, 1989.
9. Deeg HJ: Delayed complications and long-term effects after bone marrow transplantation. Hematol Oncol Clin North Am 4:641-657, 1990.
10. Williams DA, Lemischka IR, Nathan DG, Mulligan RC: Introduction of new genetic material into pluripotent haematopoietic stem cells of the mouse. Nature 310:476, 1984.
11. Eglitis MA, Kantoff P, Gilboa E, Anderson WF: Gene expression in mice after high efficiency retroviral-mediated gene transfer. Science 230:1395-1398, 1985.
12. Belmont JW, Henkel-Tigges J, Chang SM, Wager-Smith K, Kellems RE, Dick JE, Magli MC, Phillips RA, Bernstein A, Caskey CT: Expression of human adenosine deaminase in murine hematopoietic progenitor cells following retroviral transfer. Nature 322:1986.

13. Dzierzak EA, Papayannopoulou T, Mulligan RC: Lineage-specific expression of a human beta-globin gene in murine bone marrow transplant recipients reconstituted with retrovirus-transduced stem cells. Nature 331:35, 1988.
14. Wolfe JH, Sands MS, Barker JE, Gwynn B, Rowe LB, Vogler CA, Birkenmeier EH: Reversal of pathology in murine mucopolysaccharidosis type VII by somatic cell gene transfer. Nature 360:749-753, 1992.
15. Kantoff PW, Gillio A, McLachlin J, Bordognon C, Eglitis MA, Kernan NA, Moen RC, Kohn DB, Yu S-F, Karson E, Karlsson S, Zwiebel JA, Gilboa E, Blaese RM, Nienhuis AW, O'Reilly RJ, Anderson WF: Expression of human adenosine deaminase in non-human primates after retroviral mediated gene transfer. J Exp Med 166:219, 1987.
16. Stead RB, Kwok WW, Storb R, Miller AD: Canine model for gene therapy: Inefficient gene expression in dogs reconstituted with autologous marrow infected with retroviral vectors. Blood 71:742-747, 1988.
17. Zwiebel JA, Freeman SC, Kantoff PW, Cornetta K, Ryan US, Anderson WF: High-level recombinant gene expression in rabbit endothelial cells transduced by retroviral vectors. Science 243:220, 1989.
18. Wolfe JH, Schuchman EH, Stramm LE, Concaugh EA, ME MEH, Aguirre GD, Patterson DF, Desnick RJ, Gilboa E: Restoration of normal lysosomal function in mucopolysaccharidosis type VII cells by retroviral vector-mediated gene transfer. Proc Natl Acad Sci USA 87:2877-2881, 1990.
19. Stramm LE, Wolfe JH, Schuchman EH, Haskins ME, Patterson DF, Aguirre GD: β–Glucuronidase mediated pathway essential for retinal pigment epithelial degradation of glycosaminoglycans. Disease expression and in vitro disease correction using retroviral mediated cDNA transfer. Expl Eye Res 50:521-532, 1990.
20. Smith BF, Hoffman RK, Giger U, Wolfe JH: Genes transferred by retroviral vectors into normal and mutant myoblasts in primary cultures are expressed in myotubes. Molec Cell Biol 10:3268-3271, 1990.
21. Shull RM, Suggs SV, Langley KE, Okino KH, Jacobsen FW, Martin FH: Canine stem cell factor (c-kit ligand) supports the survival of hematopoietic progenitors in long-term canine marrow culture. Exp. Hematol. 20:1118-1124, 1992.
22. Taylor RM, Wolfe JH: Cross-correction of beta-glucuronidase deficiency by retroviral vector-mediated gene transfer. Exp Cell Res 214:606-613, 1994.
23. Germeraad WTV, Asami N, Fujimoto S, Mazda O, Katsura Y: Efficient retrovirus-mediated gene transduction into murine hematopoietic stem cells and long-lasting expression uusing a transwell coculture system. Blood 84:780-788, 1994.
24. Wolfe JH, Kyle JW, Sands MS, Sly WS, Markowitz DG, Parente MK: High level expression and export of β-glucuronidase from murine mucopolysaccharidosis VII cells corrected by a double-copy retrovirus vector. Gene Therapy 2:70-78, 1995.
25. Glaser JH, Sly WS: β-glucuronidase deficiency mucopolysaccharidosis: methods for enzymatic diagnosis. J Lab Clin Med 82:969-977, 1973.
26. Carter RF, Abrams-Ogg AcG, Dick JE, Kruth SA, Valli VE, Kamel-Reid S, Dube ID: Autologous transplantation of canine long-term marrow culture cells genetically marked by retroviral vectors. Blood 79:356-364, 1992.
27. Zanjani ED, Mackintosh FR, Harrison MR: Hematopoietc chimerism in sheep and nonhuman primates by in utero transplantation of fetal hematopoietic stem cells. Blood Cells 17:349-363, 1991.
28. Pearce RD, Kiehm D, Armstrong DT, Little PB, Callahan JW, Klunder IR, Clarke JTR: Induction of hemopoietic chimerism in the caprine fetus by intraperitoneal injection of fetal liver cells. Experientia 45:307-308, 1989.
29. Kantoff PW, Flake AW, Eglitis MA, Scharf S, Bond S, Gilboa E, Erlich H, Harrison MR, Zanjani ED, Anderson WF: In utero gene transfer and expression: A sheep transplantation model. Blood 73:1066-1073, 1989.
30. Fleischman RA, Mintz B: Prevention of genetic anemias in mice by microinjection of normal hematopoietic stem cells into the fetal placenta. Proc Natl Acad Sci USA 76:5736-5740, 1979.
31. Harrison MR, Slotnick RN, Crombleholme TM, Golbus MS, Tarantal AF, Zanjani ED: In-utero transplantation of fetal liver haemopoietic stem cells in monkeys. Lancet 2:1425-1427, 1989.
32. Flake AW, Harrison MR, Adzick NS, Zanjani ED: Transplantation of fetal hematopoietic stem cells in utero: The creation of hematopoietic chimeras. Science 233:776-778, 1986.
33. Zanjani ED, Pallavicini MG, Ascensao JL, Flake AW, Reitsma M, MacKintosh FR, Langlois R, Stutes D, Harrison MR, Tavassoli M: Engraftment and long term expression of human fetal hematopoietic stem cells in sheep following transplantation in utero. J Clin Invest 89:1178, 1992.
34. Zanjani ED, Ascensao JL, Flake AW, Harrison MR, Tavassoli M: The fetus as an optimal donor and recipient of hemopoietic stem cells. Bone Marrow Transp 10 Supp:107-114, 1992.
35. Richardson C, Ward M, Podda S, Bank A: Mouse fetal liver cells lack functional amphotropic retroviral receptors. Bllod 84:433-439, 1994.

41

THE SIMULATION OF HEMATOPOIESIS

Implications for the Gene Therapy of Lysosomal Enzyme Disorders

Janis L. Abkowitz,* Monica T. Persik, Sandra N. Catlin, and Peter Guttorp

Departments of Medicine and Statistics
University of Washington, Seattle, Washington 98195

ABSTRACT

Although the hematopoietic stem cell is an attractive target cell for gene transfer, little is known about its biology *in vivo* in large animals (or man). We have studied the *in vivo* behavior of hematopoietic stem cells in glucose-6-phosphate dehydrogenase (G6PD) heterozygous (female Safari) cats, and demonstrated that clonal instability persists for up to 4.5 years after autologous marrow transplantation. This contrasts the 2-6 months of clonal disequilibrium reported in comparable murine studies. Our data also suggests that hematopoietic stem cells do not self-renew more frequently than once per three weeks. These data may impact strategies for optimizing gene therapy in large animals, and by extension in man.

INTRODUCTION

The hematopoietic stem cell is a reasonable target cell for gene therapy because of its biology. First, it is the head of a cascade which supports hematopoiesis and is able to differentiate into all lineages. Second, it has a large, although likely finite, self-renewal potential (reviewed, 1-5). Thus stem cells could provide constant and stable sources of mature blood, and blood-derived cells, such as monocytes and macrophages. Studies in mice have reinforced these concepts. It is clear that a single hematopoietic stem cell can reconstitute all lineages (including lymphoid) of hematopoiesis, and maintain contributions to hematopoiesis throughout a mouse's life span. In fact, this the operational definition of an hematopoietic stem cell (5).

* Address correspondence to: Janis L. Abkowitz, M.D., Professor of Medicine, University of Washington, Hematology, Room K136C, Box 357710, Seattle, Washington 98195.

Molecular Biology of Hematopoiesis 5, edited by Abraham et al.
Plenum Press, New York, 1996

Table 1. The hematopoietic demand of different species

Species	RBC per ml	Weight	Blood volume, ml	RBC lifespan, days	RBC per day	Life expectancy, years	RBC per life of animal
Mouse	9.0×10^9	25 g	1.8	50	3.2×10^8	2	*2.4×10^{11}*
Man	6.0×10^9	70 kg	4900	120	*2.5×10^{11}*	80	7.3×10^{15}
Cat	7.5×10^9	4 kg	280	70	*3.0×10^{10}*	15	1.6×10^{14}

A typical adult Safari female cat also weighs 4 kg. RPC = red blood cells. The italicized values demonstrate that a mouse during its lifetime makes as many red blood cells as a man does in 1 day for a cat does in 8 days.

For 2 to 6 months after transplantation, other investigators have demonstrated the clonal contributions to hematopoiesis are unstable. After this time, clonal disequilibrium ceases, and in certain circumstances, a single hematopoietic stem cell can near exclusively support all hematopoiesis (6). This has led to the concept that if one were able to correct a defective gene within one or few hematopoietic stem cells, and return the engineered cells to the host through autologous transplantation, that the patient, after 2-6 months, would be cured. This promise, however, has not become a reality in large animal or human gene transfer studies to date. There are difficulties with the expression of transferred genes, but in addition, there are several cellular constraints that impact gene therapy.

Table 1 shows the hematopoietic demand of different species, and demonstrates that a mouse in its life-time makes the same number of erythroid cells as does a man in 1 day. Similar calculations can be done for the other blood cell lineages. Therefore it is possible that the regulation of stem cells and/or the kinetics of their differentiation in a large animal or man, is different than in mice. As shown in Table 1, a cat in 8 days makes the same number of red cells as a mouse in its life-time.

RESULTS AND DISCUSSION

To study this issue, we have analyzed hematopoiesis in Safari cats. Safari cats are the offspring of matings between Geoffroy cats (a South American wildcat) and domestic cats (which are of Eurasian origin). These species have electrophoretically distinct phenotypes of the X-chromosome-linked enzyme glucose-6-phosphate dehydrogenase (G6PD). Because of X-chromosome inactivation, female Safari cats have some somatic cells which contain domestic-type G6PD, and other cells contain Geoffroy-type G6PD. Our isoelectric focusing technique is sufficiently sensitive that the G6PD phenotype of erythroid bursts and GM

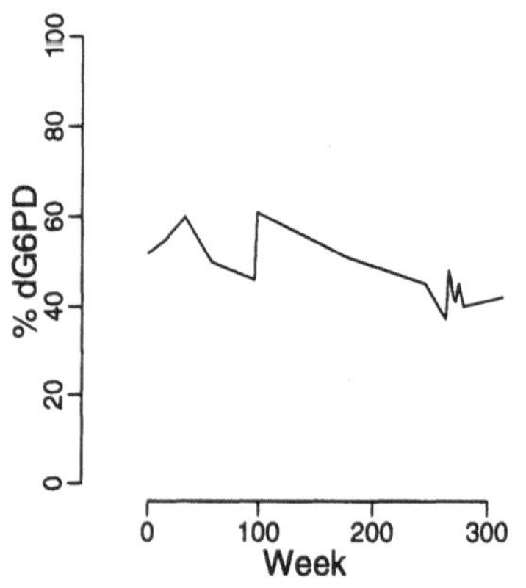

Figure 1. The G6PD phenotype of progenitor cells in a control (untransplanted) female Safari cat. The percentage of BFU-E and CFU-GM with domestic-type G6PD is plotted on the x-axis. Repeated studies were obtained for 6 years. Similarly, approximately 50% of red cells, granulocytes, and T lymphocytes from this animal expressed domestic-type G6PD throughout this time of observation (data not shown).

colonies derived in culture from single BFU-E or CFU-GM, respectively, can be determined. In individual animals, the ratio of G6PD phenotypes among BFU-E is always equivalent of that among CFU-GM. Female Safari cats are generally balanced heterozygotes, with on average, equal numbers of progenitor cells (and differentiated blood cells) of each parental phenotype (7,8).

By tracking the G6PD phenotype of progenitor cells, we have studied the contributions of hematopoietic stem cells to the progenitor cell compartment (8,9). As shown in Figure 1, the percentage of progenitors with domestic-type G6PD is relatively constant over time, and variation above the mean value is not significant with chi-square analysis. In this and subsequent Figures, the y-axis plots the percentage of progenitors (BFU-E and CFU-GM) with domestic-type G6PD ± the standard error of this determination, and the x-axis shows time in weeks. Therefore, hematopoiesis appeared polyclonal and stable during the 6 years of repeated observations in this representative normal animal.

To determine the behavior of hematopoietic stem cells in the circumstance of a limited marrow reserve, we performed autologous transplantations. These data have been reported in Proc. Natl. Acad. Sci, USA (8,9). After lethal irradiation, the animals received small numbers of nucleated marrow cells (1-2 $\times$ 10^7/kg vs. the 1-2 $\times$ 10^8/kg target cell dose for autologous human transplantation for clinical diseases). By 4-5 weeks after transplantation, peripheral blood counts were normal. By 10 weeks, the frequencies of BFU-E and CFU-GM in marrow aspirates and their cell cycle kinetics were normal, suggesting that progenitor cell compartment was entirely regenerated.

Although hematopoiesis remained normal throughout the duration of the study (6 years), the contribution of stem cell clones to the progenitor cell compartment was very different than that observed in untransplanted animals (9). For example, prior to transplantation, a mean of 35% of progenitors from cat 40005 contained domestic-type G6PD, and there was no variation about this mean. After transplantation, the percentage of progenitors with domestic-type G6PD fluctuated extensively.

Figure 2 extends these data to 6 years. During the first 78 weeks after transplantation, the percentage of progenitors of domestic-type G6PD varied between 4 and 99. From week 78 to week 240, this percentage fluctuated less widely, but the fluctuations still exceeded binomial variability. After 240 weeks, however, the variations subsided, and 94-100% of progenitors, essentially all progenitors, contained domestic-type G6PD. All granulocytes, all red cells, and 85% of T cells also contained domestic-type G6PD at this time (9). Thus,

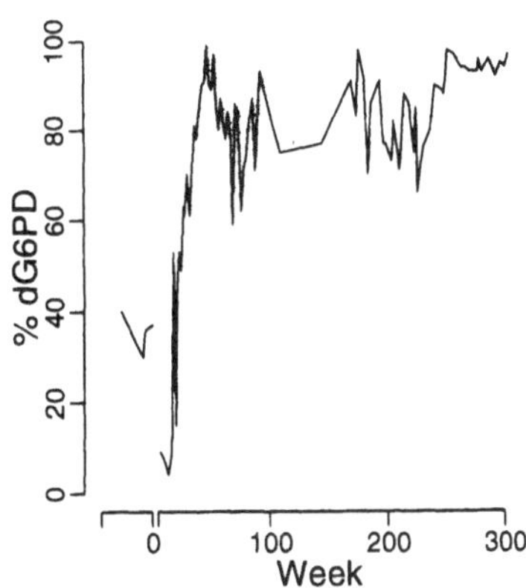

Figure 2. Studies in experimental cat 40005. At week 0, this cat received autologous marrow cells harvested prior to irradiation. See text for details.

it appeared that hematopoietic reserve had been reconstituted by the progeny of one or few stem cells, each of which contained domestic-type G6PD.

We have followed clonal contributions to hematopoiesis for over 4 years in 6 cats. Each cat has a unique pattern (9). Following transplantation, the percentage of progenitors of domestic-type G6PD always varied. The variation extended between 1 and 41/2 (see Figure 2) years, when analyzed with sequential chi-square analyses (not 2-6 months as seen in the comparable murine studies). We presume that this is the length of time that is required for the few transplanted cells to replicate sufficiently to reconstitute the large hematopoietic reserve of cats.

These data led to several questions. Could the divergent and unique outcomes which we observed be explained by stochastic differentiation? What are the implications for gene therapy?

To answer these questions we simulated hematopoiesis (10). We began with "aspirated marrow" containing quiescent stem cells and differentiating cells. Rather than transplant these cells into a cat, we placed them in a computer. We let differentiation happen randomly, and tracked outcome. For each quiescent stem cell at all times, there was a probability of replication, a probability of differentiation, and a probability of cell death or apoptosis. When a stem cell differentiated, its progeny formed a clone which contributed to hematopoiesis for a random period of time. Therefore, each differentiating stem cell clone had a probability of extinction. BFU-E and CFU-GM were sampled from this differentiating cell compartment and the G6PD phenotypes of sampled cells were recorded. We then tested different values for each probability and for the numbers of transplanted cells to determine if any set of values could lead to the spectrum of outcomes which we observed in female Safari cats.

Figure 3 shows 20 independent simulations of marrow transplantation with 30 hematopoietic stem cells. Again the y-axis contains the percentage of progenitors with domestic-type G6PD and the x-axis contains 6 years of computed time. Outcomes similar

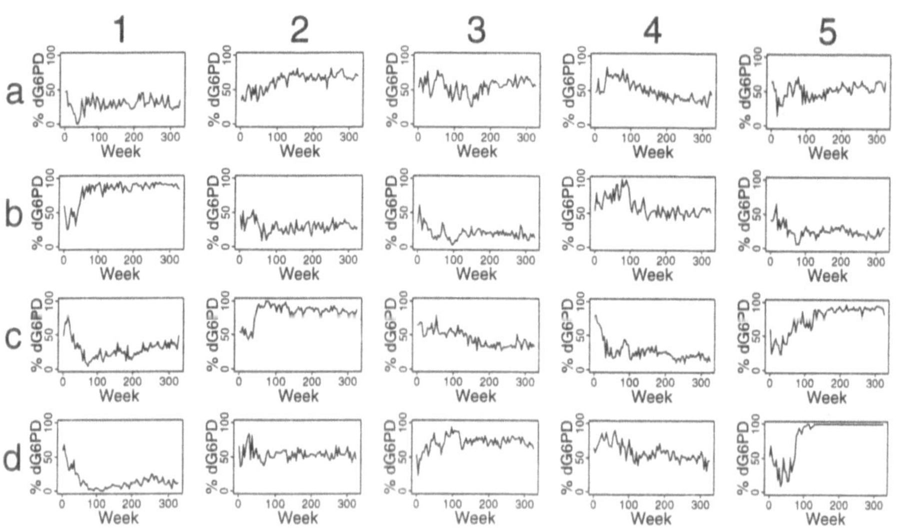

Figure 3. Simulation of hematopoiesis after autologous transplantation in female Safari cats. Each plot represents an independent simulation. For these simulations, 30 quiescent hematopoietic stem cells were transplanted. Hematopoietic stem cells self-renewed on average once each 10 weeks, and initiated a differentiation program on average once every 12.5 weeks. Differentiating clones were extinguished, on average once each 6.7 weeks. See the text and reference 10 for further details.

to cat 40005 are seen (for example b1, c5, d5). Other outcomes are similar to those observed in other animals. In fact, variation extends from 1 to 41/2 years in these simulations and then subsides. At times, the subsequent hematopoiesis is dominated by cells of a single phenotype. At times, it is more balanced, as was observed in our Safari cat transplantation studies.

If one simulates transplantation with 10, and not 30, quiescent stem cells, the initial variation about the mean value is extensive. Monoclonality develops far too frequently. Some cats run out of stem cells. In composite, the patterns do not look like those of animal experiments (10). Similarly, simulation of marrow transplantation with 100 quiescent stem cells leads to no significant variation in the percentage of progenitors with domestic-type G6PD about its mean value, and does not reproduce the observed data (10).

Using broad criteria to assess the similarity of simulations to observed data, we derived feasible values for all parameters (10). An optimal number of quiescent transplanted stem cells was 30 (see Figure 3). Acceptable outcomes, however, were seen with quiescent stem cell numbers ranging from 15 to 50. In contrast, the number of clones contributed to hematopoiesis at the time of engraftment was not an important determinant of outcome. Values between 0 and 2000 yielded equivalent results. This is consistent with the observation that long-term reconstitution is a property of quiescent stem cells and not of differentiating stem cell clones.

In the simulations of Figure 3, hematopoietic stem cells replicated on average once every 10 weeks. If the mean rate of self-renewal was more frequent than once every 3 weeks, we could not simulate the observed outcomes (10). As retroviral integration requires stem cell self-renewal, these data might also explain the difficulty with gene transfer into earliest cells in large animal models.

In further studies we simulated gene therapy experiments. If 1 of 30 cells expressed the gene of interest, and 30 quiescent stem cells were transplanted, only small, generally insignificant, contributions of marked clones to hematopoiesis were seen. Contributions persisted over 100 weeks in only 3 of the 20 simulations. When 2 of 30 cells express the gene of interest, contributions over 1%, lasting over 100 weeks, were present in 7 of the 20 outcomes. When 6 of 30 cells (20%) express the gene of interest, 15 of 20 outcomes were "successful". Also, in many simulations there was significant variation of contributions of clones over time. If one made observations at infrequent intervals, inappropriate conclusions could be drawn. For example, in one circumstance when 2 of 30 marked clones were transplanted, no marked clone contributed to hematopoiesis at 20 weeks, but all of the clones contributing to hematopoiesis at 45 weeks after transplantation were marked. This variability over time is minimized when one transplants higher numbers of hematopoietic stem cells. However, more target cells may lead to a decreased multiplicity of infection and less effective retroviral gene transfer. Other methods for gene delivery may be similarly impacted.

Taken together, these data suggest that the kinetics of hematopoietic stem cells need to be considered when designing optimal strategies for gene therapy, and that it is inappropriate to generalize from single "successful" experiments. Similarly the kinetics of the differentiated progeny of hematopoietic stem cells, such as microglial cells within the central nervous system, and Kupffer cells within the liver, may also impact gene therapy approaches.

ACKNOWLEDGMENTS

The authors thank Allan Dimaunahan for help in preparation of this manuscript. This work was supported by Grant HL 46598 from the National Heart, Lung, and Blood Institute of the National Institutes of Health. JLA is a recipient of a Faculty Research Award from the American Cancer Society.

REFERENCES

1. Lemischka IR: The haematopoietic stem cell and its clonal progeny: mechanisms regulating the hierarchy of primitive haematopoietic cells. Cancer Surveys 15:3, 1992
2. Spangrude GJ: Characteristics of the hematopoietic stem cell compartment in adult mice. Int. J. Cell Cloning 10:277, 1992
3. Uchida N, Fleming WH, Alpern EJ, Weissman IL: Heterogeneity of hematopoietic stem cells. Current Opinion in Immunol. 5:177, 1993
4. Ogawa M: Differentiation and proliferation of hematopoietic stem cells. Blood 81:2844, 1993.
5. Orlic D, Bodine DM: What defines a pluripotent hematopoietic stem cell (PHSC): will the real PHSC please stand up! Blood 84:3991, 1994
6. Jordan CT, Lemischka IR: Clonal and systemic analysis of long-term hematopoiesis in the mouse. Gen. Dev. 4:220, 1990
7. Abkowitz JL, Ott RL, Nakamura JM, Steiman L, Fialkow PJ, Adamson JW: Feline G-6-PD cellular mosaicism: application to the study of retrovirus-induced pure red cell aplasia. J. Clin. Invest. 75:133, 1985
8. Abkowitz JL, Linenberger ML, Newton MF, Ott RL, Guttorp P: Evidence for the maintenance of hematopoiesis in a large animal by the sequential activation of stem cell clones. Proc. Natl. Acad. Sci. USA 87:9062, 1990
9. Abkowitz JL, Persik MT, Shelton GH, Ott RL, Kiklevich JV, Catlin SN, Guttorp P: The behavior of hematopoietic stem cells in a large animal. Proc. Natl. Acad. Sci. USA 92:2031, 1995
10. Abkowitz JL, Catlin SN, Guttorp P: Evidence that hematopoiesis may be a stochastic process *in vivo*. Submitted.

HOMOLOGOUS GENE TARGETING FOR HUMAN GENE THERAPY

Virus-Mediated Gene Transfer, Homology-Associated Nonhomologous Recombination, and Homology-Length Dependence

Ichizo Kobayashi*

Department of Molecular Biology
Institute of Medical Science
University of Tokyo, Shiroganedai, Tokyo 108 Japan

INTRODUCTION

Gene targeting, the designed alteration of genomic information by homologous recombination, has provided a powerful means for the genetic analysis of the mammalian systems. The targeted correction of mutations is believed to provide an ideal means for the therapy of diseases caused by mutations. This would be a better alternative, in many respects, to the strategy of expressing wild type c-DNA — complementation approach — in particular, from the point of view of gene expression and the absence of potential mutagenesis.

Its application to human gene therapy is, however, hampered by its low efficiency: Only a very small fraction of the treated cells will acquire the designed change. The overall inefficiency results from the rarity of homologous recombination and from the low frequency of gene transfer. The former problem is related to the unusual characteristics of DNA recombination in mammalian somatic cells: Frequent non-homologous recombination, steeper-than-linear dependence of targeted homologous recombination on the homology length,[1] predominance of non-conservative homologous recombination,[2] and independence of the target copy number.[3]

* Corresponding author: Ichizo Kobayashi, Department of Molecular Biology, Institute of Medical Science, University of Tokyo, 4-6-1 Shiroganedai, Tokyo 108, Japan. FAX: (81) 3-5449-5422; e-mail: ikobaya@hgc.ims.u-tokyo.ac.jp.

Molecular Biology of Hematopoiesis 5, edited by Abraham et al.
Plenum Press, New York, 1996

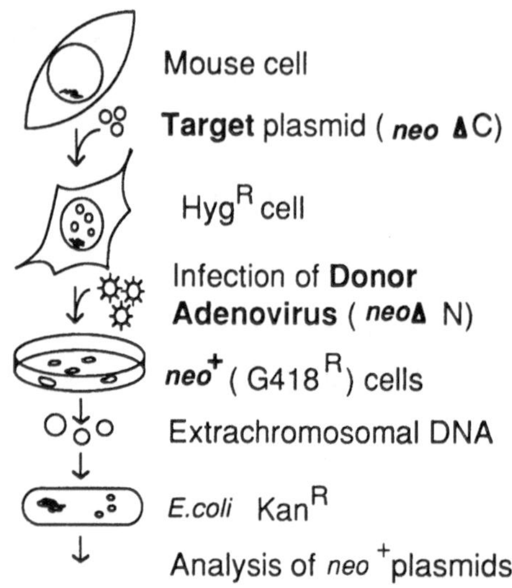

Figure 1. Homologous gene targeting mediated by a replication-defective adenovirus vector. In the first step, we established a mouse C127 cell line carrying the target plasmid (pIK423), taking advantage of the phenotype of hygromycin-resistance (Hyg^{R}) conferred by *hph* gene. In the second step, we infected this cell line with the donor adenovirus constructs. Homologous recombination between the two *neo* genes should restore a functional *neo* gene (*neo*$^{+}$) and should make the cell resistant to drug G418. In the third step, we recovered these *neo*$^{+}$ plasmid molecules in a *recA* strain of *E. coli* by selecting the kanamycin resistance conferred by the *neo*$^{+}$ gene. The structure of the recovered plasmids was analyzed by restriction mapping and sequencing.

HIGH-EFFICIENCY GENE-TARGETING WITH A REPLICATION-DEFECTIVE DNA VIRUS VECTOR

A replication-defective adenovirus vector was used for efficient delivery of donor DNA in order to bypass the problem of low efficiency.[4] Homologous recombination between a donor *neo* gene inserted in the adenovirus vector and a target mutant *neo* gene on a nuclear papillomavirus plasmid was selected for, as shown in Figure 1. These recombinant adenoviruses allowed gene transfer to 100% of the treated cells without impairing their viability.

Homologous recombinants were obtained at a frequency much higher (1000 fold or so) than with electroporation or the calcium phosphate procedure. Their structure was analyzed in detail after recovery in *E. coli*. All of the recombinants examined had experienced precise correction of the mutant *neo* gene.

Some of them had a non-homologous rearrangement as well. One type of non-homologous recombination took place at the end of the donor-target homology (Figure 2). The vector adenovirus DNA was inserted into some of the products obtained at a high multiplicity of infection. The insertion was at the end of the donor-target homology with a concomitant insertion of a filler sequence in one of them. There may be a possible relationship between these arrangements and the homologous recombination events, as discussed below.

These results indicate that this replication-defective vector is applicable to gene targeting.

NON-HOMOLOGOUS REARRANGEMENTS ASSOCIATED WITH HOMOLOGOUS INTERACTION

Non-homologous recombination of DNA underlies many of the changes that occur in genomes. It involves little or no homology between recombining DNAs and has been considered unrelated to homologous recombination, which requires longer regions of homology. In mouse cells, however, we found recombination products whose structures suggest that homologous interaction between DNAs causes non-homologous recombination with another DNA.[5]

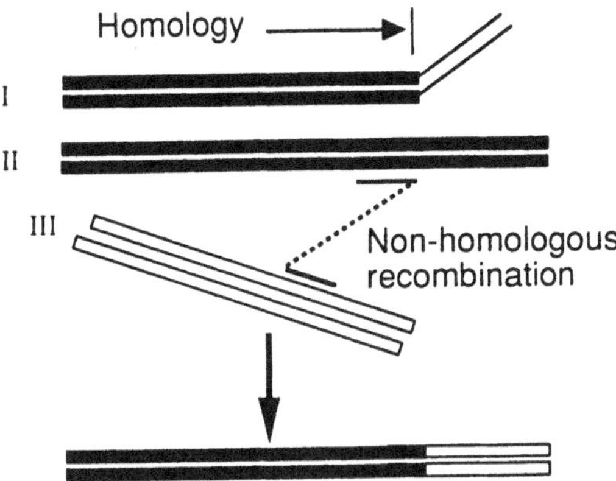

Figure 2. Non-homologous recombination at the end of the homologous region. DNA I and DNA II share long homology. At the end of the homology, DNA II experiences non-homologous recombination with a third, unrelated DNA (III).

Non-homologous recombination events appear to have occurred when intermediates of homologous recombination were trapped at various stages. In one product, the non-homologous recombination disrupted gene conversion. In another, it took place exactly at the end of a region of long homology shared between two DNAs (Figure 2). This finding explains why gene targeting needs long uninterrupted homology and why mammalian homologous recombination is often non-conservative.[2]

This mechanism could explain several gene rearrangements reported previously, and may have contributed to genome evolution. Furthermore, it is possible that this homology-driven gene destruction mechanism may have some positive roles to play, such as the elimination of insertion mutations.[6]

DEPENDENCE OF FREQUENCY OF HOMOLOGOUS RECOMBINATION ON HOMOLOGY LENGTH

The frequency of the homologous recombination is believed to be a linear function (a - bN) of the length (N) of homology between DNAs. Here, the N-intercept (a/b) is believed to be determined by a threshold length below which some physical (or structural) constraint is effective. In the mammalian gene targeting systems, however, the frequency is more than linear with respect to the length of homology.

To explain both the linear and non-linear dependencies, a reaction model was proposed where the branch point of the reaction intermediate is assumed to "walk randomly" along the homology and be destroyed if it ever reaches either of the ends of the homology (Figure 3).[7,8]

A parameter h characteristic of the system turns out to play a key role. This parameter reflects the efficiency of the intermediate processing, and therefore, the unlikelihood of destruction at the ends. The frequency is proportional to N^3 for smaller N and is a linear function of N for larger N. Where the shift from N^3-dependence to linear-dependence takes place is determined by the parameter h. The N-range showing the N^3-dependence becomes narrower as h becomes larger, and is lost at high values of h.

The non-linear behavior which is observed not only in mammalian gene targeting,[1] but also in bacteriophage T4, in *E. coli* and yeast systems, agrees well with this N^3-depend-

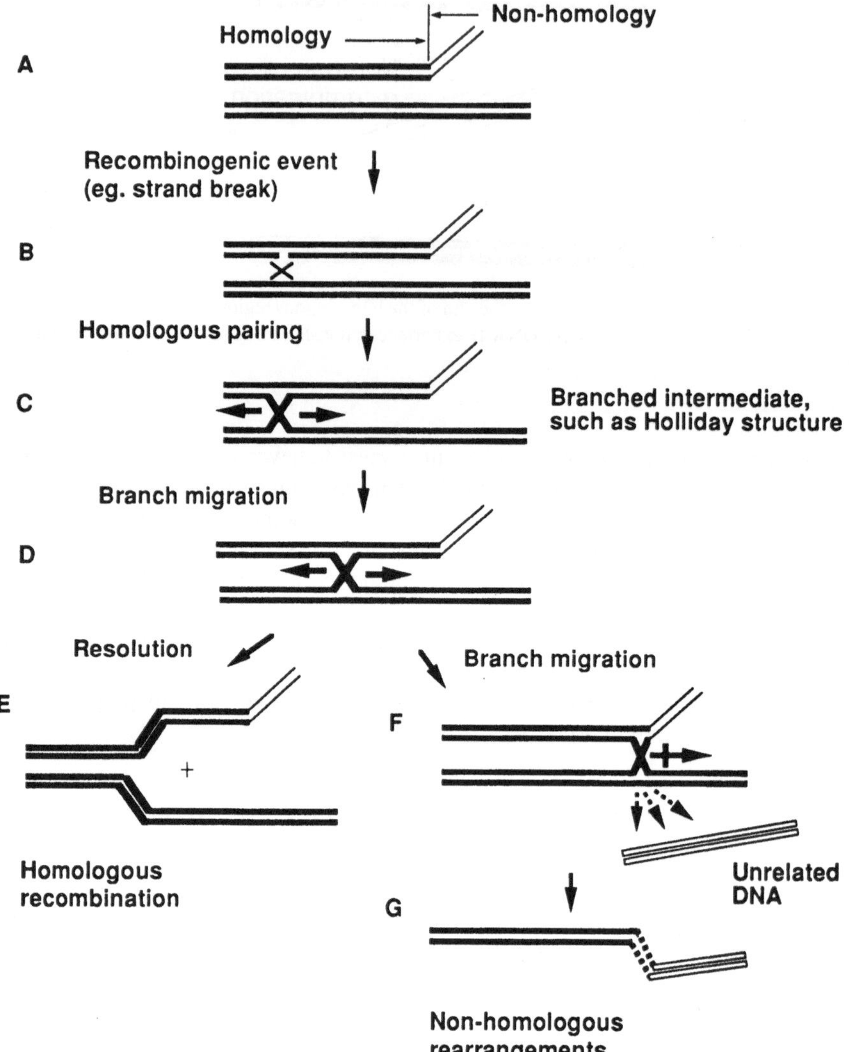

Figure 3. Plausible steps of homologous interaction in mammalian cells. (A) Two DNA's sharing homology in one region. (B) A recombinogenic event in one of them leads to their homologous pairing. The event could be a strand break, for example. (C) The resulting cross-stranded structure is called a Holliday intermediate. (D) The branch point of the intermediate moves along the homology in a process known as branch migration. (E) Resolution of the intermediate leads to completion of homologous recombination. (F) When the intermediate encounters non-homology, an unusual structure, a double-strand break, for example, is formed at the homology/non-homology boundary. (G) It causes non-homologous rearrangement with a third unrelated DNA.

ence. According to this model, the N-intercept is determined not by physical or structural constraints but by h alone.

DISCUSSION

The above results indicate that replication-defective virus vectors may provide a powerful means of gene targeting. The advantages of adenovirus vector in gene transfer have already been exploited in the complementation approach. These advantages include its broad host range, high transfer efficiency, high cell viability, non-integration and potentiality of *in vivo* use (as opposed to *ex vivo* use). The high-efficiency of gene targeting per treated cell might allow *in vivo* treatment and direct delivery of recombinant virus particles to the patient, as opposed to *ex vivo* treatment involving reimplantation of selected and expanded cell clones.

The rarity of non-homologous recombination events means safety over other *in vivo* methods of gene delivery used in gene targeting even if some of them might be intrinsic to the homologous interaction process. According to our simple model, the increase of the donor-target homology has the effect of increasing the absolute efficiency of homologous recombination and decreases the chance of this type of non-homologous rearrangement.

ACKNOWLEDGMENTS

I am grateful to my collaborators, whose names appear in the reference list. The work in our laboratory was supported by grants from Department of Education, Department of Health, Uehara Foundation, and Nissan Science Foundation.

REFERENCES

1. Deng C, Capecchi M R: Reexamination of gene targeting frequency as a function of the extent of homology between the targeting vector and the target locus. Mol. Cell. Biol. 12: 3365, 1992
2. Kitamura Y, Yoshikura H, Kobayashi I: Homologous recombination in a mammalian plasmid. Mol. Gen. Genet. 222, 18, 1990
3. Zheng H, Wilson J H: Gene targeting in normal and amplified cell lines. Nature 344: 170, 1990
4. Fujita A, Sakagami K, Kanegae Y, Saito I, Kobayashi I.: Gene targeting with a replication-defective adenovirus vector. J. Virol. 69: 6180, 1995
5. Sakagami K, Tokinaga Y, Yoshikura H, Kobayashi I: Homology-associated non-homologous recombination in mammalian gene targeting. Proc. Natl. Acad. Sci. USA 91: 8527, 1994
6. Kobayashi I, Sakagami K, Kusano K, Fujitani Y, Takahashi-Kobayashi N, Yoshikura H: Homologous interaction as a gene destruction process: Hypotheses for homology-driven non-homologous recombination and double-strand break repair, in Proceedings of Japanese-American joint meeting on modification of gene expression and non-Mendelian inheritance. (Oono, K. Takaiwa, F. eds.) 55. Japan, Tsukuba, National Institute of Agrobiological Resources, 1995
7. Fujitani Y, Yamamoto K, Kobayashi I: Dependence of frequency of homologous recombination on homology length. Genetics140: 797-809, 1995
8. Fujitani Y, Kobayashi I: Random walk model of homologous recombination. Phys. Rev. E.52:6607, 1995

TRANSFECTION OF THE HUMAN HEME OXYGENASE GENE INTO RABBIT CORONARY MICROVESSEL ENDOTHELIAL CELLS

Protective Effect against Heme and Hemoglobin Toxicity

N. G. Abraham

The Rockefeller University
1230 York Avenue
New York, New York 10021

ABSTRACT

Heme oxygenase* (HO) is a stress protein and has been suggested to participate in defense mechanisms against agents which may induce oxidative injury such as metals, endotoxin, heme-hemoglobin and various cytokines. Overexpression of HO in cells might therefore protect against oxidative stress produced by certain of these agents, specifically heme and hemoglobin, by catalyzing their degradation to bilirubin, which itself has anti-oxidant properties. We report here the successful *in vitro* transfection of rabbit coronary microvessel endothelial cells with a functioning gene encoding the human HO enzyme. A plasmid containing the cytomegalovirus promoter and the human HO cDNA complexed to cationic liposomes (Lipofectin) was used to transfect rabbit endothelial cells. Cells transfected with human HO exhibited a ≈3.0-fold increase in enzyme activity and expressed a several-fold induction of human HO mRNA as compared to endogenous rabbit HO mRNA. Transfected and non-transfected cells expressed Factor VIII antigen and exhibited similar acetylated low density lipoprotein uptake (two important features which characterize endothelial cells) with greater than 85% of cells staining positive for each marker. Moreover, cells transfected with the human HO gene acquired substantial resistance to toxicity produced by exposure to recombinant hemoglobin (rHb) and heme as compared to non-transfected cells. The protective effect of HO overexpression against heme/hemoglobin toxicity in endothelial cells shown in these studies provides direct evidence that the inductive response of human

* HO refers to the HO-1 isozyme of heme oxygenase unless otherwise specified.

Molecular Biology of Hematopoiesis 5, edited by Abraham et al.
Plenum Press, New York, 1996

HO to such injurious stimuli represents an important tissue adaptive mechanism for moderating the severity of cell damage produced by these blood components.

INTRODUCTION

A variety of oxidative stress-inducing agents, such as metals, ultra-violet (UV) light, heme and hemoglobin have been implicated in the pathogenesis of the inflammatory process. The cellular response to such agents involves the production of a number of soluble mediators including acute phase proteins, eicosanoids and various cytokines.

The rate-limiting enzyme in heme catabolism, heme oxygenase (HO), is a stress-response protein and its induction has been suggested to represent an important cellular protective response against oxidative damage produced by free heme and hemoglobin (1-5). Induction of HO may specifically decrease cellular heme (pro-oxidant) and elevate bilirubin (anti-oxidant) levels (5-9). Two HO isozymes, the products of distinct genes, have been described (10,11). HO-1, which is ubiquitously distributed in mammalian tissues, is strongly and rapidly induced by many compounds which elicit cell injury; the natural substrate of HO, heme, is itself a potent inducer of the enzyme (10,11). HO-2, which is believed to be constitutively expressed, is present in high concentrations in such tissues as the brain and testis, and is believed to be non-inducible (10).

Endotoxin, IL-1 and other stress agents cause a rapid (within 5-10 minutes) activation of the HO gene and a subsequent accumulation of HO mRNA (12-14). This process involves transcriptional activation of several regulatory sites in the HO promoter region. AP-1 and IL-6 responsive elements are found in the promoter region of this gene (13,15). A recent study from this laboratory demonstrated that the proximal promoter region of the human HO gene also contains NF-kB and AP-2 binding sequences (16). The finding of AP-2 and NF-kB binding sites on the HO promoter suggests the importance of HO in stress/injury responses, when these transcriptional factors are known to be activated (16).

Our goal in these studies was to augment HO activity by transfecting rabbit coronary microvessel endothelial cells with the human cDNA for HO so as to distinguish expression of the transfected HO from that of rabbit HO and to determine whether the transfected human HO gene could functionally protect the endothelial cells against the toxic effects of free heme and hemoglobin. We used plasmid DNA complexed to a cationic liposome preparation, Lipofectin, for transfection of the human HO gene; other investigators have also used these cationic liposomes with success in *in vivo* and *in vitro* DNA delivery systems (17-21).

Our data demonstrate that it is possible to transfect rabbit coronary endothelial cells with human HO cDNA and to achieve selective overexpression of the human HO mRNA to levels significantly higher than those of the endogenous rabbit HO mRNA. The overexpression of the human HO in the rabbit cells was not associated with phenotypic changes in the cells or alterations of endothelial cell markers such as acetylated-low density lipoprotein (Ac-LDC) uptake or Factor VIII antigen. Our study also demonstrates that augmentation of human HO activity in the transfected cells confers substantial protection against cellular toxicity produced by free heme/hemoglobin. These data thus directly demonstrate the protective effect of human HO against the cellular toxic effects of these blood components which are known to participate in inflammatory reactions at sites of hemmorhage, thrombosis and trauma.

METHODS

Rabbit Coronary Microvessel Endothelial (RCME) Cell Culture

Cells were isolated from the midportion of the rabbit myocardium by collagenase digestion, filtration, homogenization and centrifugation as described by Gerritsen et al. (22). The cells were seeded onto fibronectin-coated six-well culture plates and incubated in Dulbecco's modified Eagles medium (DMEM) containing 20% PDS (plasma derived serum), 100μg/ml ECGF (endothelial cell growth factor), and 2mM L-glutamine at 37°C in a standard humidified incubator. After 2 h non-adherent cells were removed. Endothelial cell colonies appeared in 2-5 days and were initially characterized by their morphology (i.e., closely apposed cells with a polygonal morphology). Cultures free of pericytes and smooth muscle cells were subcultured. Homogeneity of RCME cell cultures was assessed by Ac-LDL labeling followed by visual fluorescence microscopy and fluorescence-activated cell sorting. Cells used in the experiments (passages 12-25) were cultured in DMEM containing 10% fetal calf serum (FCS) supplemented with 20mM HEPES, L-glutamine (0.35mg/ml), and gentamicin (0.06 mg/ml).

Factor VIII Antigen Immunofluorescence

Fibronectin-coated Lab-Tek® tissue culture chamber slides (Fisher Inc., Springfield, NJ) were used to culture RCME cells. At confluence, the medium was removed and cells were washed with Earls' balanced salt solution (EBSS). The cells were fixed with 100% methanol (-20°C) for 3 minutes and then washed three times with EBSS. A 1:80 dilution of fluorescenated anti-human factor VIII (IncStar Corp., Still Water, MN) was added and the cells were incubated for 60 minutes, then washed with EBSS. The slides were mounted in 50% glycerol in PBS, and viewed and photographed with a phase microscope equipped with an epifluorescent light source.

Uptake of Ac-LDL

Preconfluent cultures (1 T-75 flask/sort) of cells were incubated with 10μg/ml of 1,1-dioctadecyl-3,3,3',3'-tetramethylindocarbocyanine perchlorate labelled Ac-LDL (Molecular Probes, Eugene, OR) in growth medium for 4 h at 37°C. Cells were washed several times with probe-free media and visualized by fluorescence microscopy.

Transfection of Human HO-cDNA Into Rabbit Endothelial Cells

The plasmid used was a human HO construct, pRc/CMV-human HO, containing the entire protein-coding region, the XhoI/XbaI fragment (-63/924) of pH HO, in a correct orientation under the promoter/enhancer of cytomegalovirus. The plasmid was constructed as follows: both ends of the XhoI/XbaI fragment were converted to blunt ends with Klenow enzyme, ligated to the Hind III linker, then cloned into the Hind III site of pRc/CMV (Invitrogen, San Diego, CA). The same plasmid without the HO cDNA was used as a control.

For stable transfection approximately one million cells were seeded in a 75 cm^2 flask using standard culture media (see above) at 37° C in 5% CO_2. Four hours later the growth medium was changed to DMEM with 0.5% FCS, and cells were transfected with 20μg of plasmid DNA-Lipofectin complex using the mammalian transfection kit (Gibco, BRL Life Technology, Grand Island, NY). Cells were selected for neomycin resistance in medium containing G418 (500μg/ml). Positive colonies were pooled and subcultured, and used in all subsequent studies. CYP4A1 cDNA (2.1 Kb) cloned in a pBluescript SK+/- vector was cut out by digestion with XbaI and XhoI and inserted into the XbaI and XhoI sites of the pBK-CMV vector (Stratagene, 4512bp). RCME cells were transfected with the CYP4A1-CMV vector using the lipofectamine method and subjected to selection with neomycin. Total RNA was extracted and hybridized with the CYP4A1 cDNA probe. Only transfected cells expressed CYP4A1 mRNA and CYP4A1 protein as measured by Northern and Western hybridization, respectively (data not shown).

RNA Extraction and Northern Blot Analysis

Confluent cultures of transfected and non-transfected cells were used for Northern blot analysis and probed with either human HO cDNA or rat HO cDNA, since rat HO cDNA is able to hybridize with rabbit HO mRNA. Total RNA was isolated by lysis of the cells in 4 M guanidium isothiocyanate and quantitated by spectrophotometry. Total RNA (10μg) was denatured and size-separated by electrophoresis on 1.2% agarose gels containing 2.2 M formaldehyde. To verify the integrity of the samples, gels were stained for 5 min with 0.5μg/ml ethidium bromide in diethylpyrocarbonate-treated water, destained and photographed. RNA was transferred to nylon membranes (Gene Screen Plus Membranes, NEN, Boston, MA) using 20xSSC (0.15 M NaCl, 0.015 M Na citrate) buffer as recommended by the manufacturer. The membranes were then baked for 2 hrs at 80°C. Blots were hybridized at 60°C overnight and washed as previously described (16). The hybridization mixture consisted of 1% BSA, 7% SDS and 1mM EDTA in 3xSSC, pH 7.0. Hybridized blots were washed at 60°C in a solution containing 0.5% BSA, 5% SDS and 1mM EDTA in 0.3xSSC followed by a second wash with 0.1xSSC containing 1% SDS and 1mM EDTA. Filters were exposed to autoradiography film (DuPont NEN, Boston, MA) at 70°C. All filters were reprobed with cDNA encoding glyceraldehyde-3-phosphate dehydrogenase (G3PDH) (Clontech, Palo Alto, CA) to ensure that equal amounts of RNA were loaded onto each lane.

Measurement of HO Enzyme Activity

Enzyme activity was assessed using microsomes (6xT-175 flasks) from nontransfected and transfected endothelial cells. Cells were harvested and microsomes prepared following cell homogenization and centrifugation (100,000xg). Microsomes (1mg protein) were incubated with heme (50μM), rat liver cytosol (2mg/ml), $MgCl_2$ (1mM), glucose-6-phosphate dehydrogenase (3 units), glucose-6-phosphate (G-6-P) (1mM), and $NADP^+$ (2mM) in 0.5ml potassium phosphate buffer, 0.1 M, pH 7.4, for 30 min at 37°C. The reaction was stopped by placing the tubes on dry ice and bilirubin was extracted with chloroform as described previously (14). The amount of bilirubin generated was estimated using a scanning spectrophotometer and defined as the difference between 463-520nm (14). Results are expressed as pmol/$5x10^6$ cells/30min.

Cell Viability

Cells seeded in 12-well culture plates ($1x10^5$ cells/well) were grown for 24 h in standard culture medium. Recombinant human hemoglobin (rHb) was obtained from Somatogen (Boulder, CO) and prepared in phosphate buffered saline. Heme was prepared as previously described (23). Heme (1-200μM) or rHb (1-400μM) was added to the cells for 24 h. Control cultures received the appropriate vehicles. Experiments were performed in triplicate. Cell viability was measured by trypan blue exclusion. Briefly, cell suspensions were prepared by trypsinization followed by washing and resuspension in 1 ml of Hanks' HBSS. Cell suspensions (0.2ml) were mixed with 0.5ml Trypan blue (0.5%) and 0.3ml HBSS and allowed to stand for 10min at room temperature. Cells were counted using a hemocytometer. Cell viability was calculated as the percent viable (unstained) cells of total (stained and unstained) cells. Statistical analysis was performed using a Kruskal-Wallis ANOVA followed by a Dunn's comparison.

RESULTS

Effect of HO Gene Transfection on Factor VIII Levels and Ac-LDL Uptake

Factor VIII antigen and Ac-LDL uptake are widely used as specific markers for the characterization of the endothelial cell phenotype. As seen in Figure 1A and B both cell types, transfected as well as control endothelial cells, expressed factor VIII antigen; Ac-LDL uptake, as a marker to aid in identifying possible effects of human HO on the endothelial cell phenotype, was also assessed. Both transfected and non-transfected cells exhibited a similar degree of Ac-LDL uptake (Fig. 1, C and D). In both cases, i.e., factor XIII antigen and Ac-LDL uptake, 85-90% of transfected and control cells stained positive. The presence of factor VIII antigen and Ac-LDL uptake supports the view that transfection of endothelial cells with the human HO gene did not significantly alter the cell phenotype.

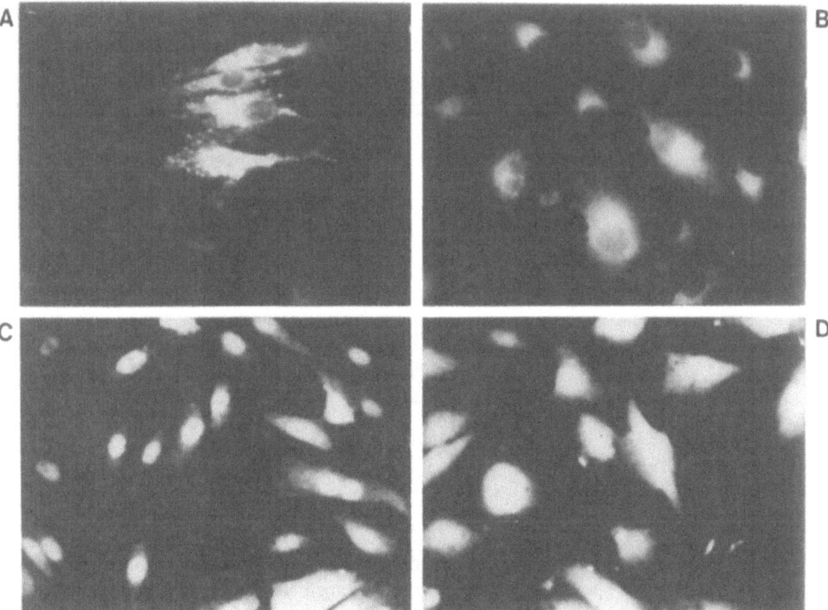

Figure 1. Effect of human HO gene transfection on endothelial cell factor VIII antigen and Ac-LDL uptake in endothelial cells. Immunofluorescent staining of control (A) and HO-transfected endothelial cells (B) with factor VIII antibodies. Fluorescent Ac-LDL uptake in the endothelial cells of control (C) and HO transfected cells (D), respectively.

Expression of Human HO mRNA in Rabbit Endothelial Cells

Transfection of endothelial cells with the human HO gene was assessed by parallel determinations of endogenous rabbit HO mRNA and the expressed human HO mRNA in control cells, cells transfected with human HO plasmid DNA, and cells transfected with control plasmid PRC/CMV (Invitrogen, San Diego, CA). As seen in Figure 2, endothelial cells transfected with the human HO cDNA-Lipofectin complex displayed significant expression of human HO mRNA (lane 3). Endothelial cells transfected with the control plasmid did not show detectable human HO mRNA (lane 2). Similarly, control endothelial cell RNA did not hybridize with human cDNA indicating the absence of a hybridizable band similar to human HO mRNA (lane 1). To examine the effect of the control plasmid on endogenous rabbit HO, we performed Northern blot analysis on total RNA from non-transfected and transfected cells and probed with the rat HO cDNA. Rat HO cDNA hybridizes to rabbit HO mRNA whereas human HO cDNA does not. The presence of a hybridizable rabbit HO mRNA is seen in all three cell types; cells transfected with control plasmid; cells transfected with the human HO plasmid; and untransfected cells (Figures 2, middle panel). As seen in Fig. 2, there were no differences in endogenous HO mRNA levels in all cell types following transfection with control plasmid. To assure the integrity and quantity of transferred RNA, nitrocellulose filters were washed and re-hybridized with radiolabelled G3PDH cDNA. As seen in Figure 2 (lower panel) signals corresponding to G3PDH confirmed the integrity and transfer of equal amounts of RNA to the filters in each lane of the paired experiments.

Enhancement of HO Activity in Endothelial Cells Transfected with Human HO

We examined HO activity in untransfected endothelial cells, cells transfected with control plasmid (pRC/CMV) and plasmid containing human HO; the results are depicted in Table I. The basal level of HO activity in endothelial cells was 160±17 pmol bilirubin formed/5x10^6 cells/30 min as compared to 482±77 pmol bilirubin formed/5x10^6 cells/30 min

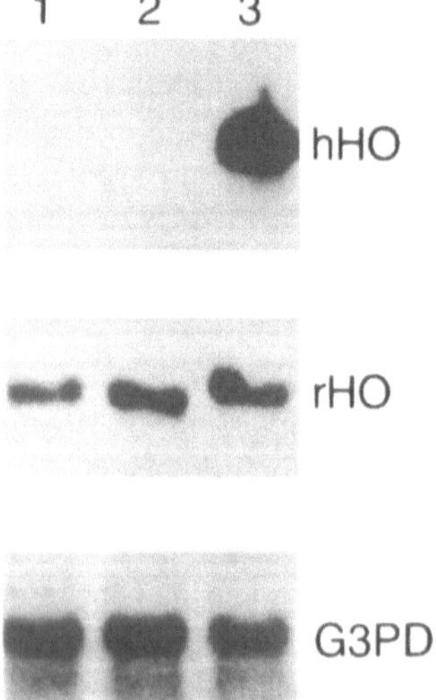

Figure 2. Northern blot analysis of RNA from control endothelial cells (lane 1), cells transfected with a plasmid control (lane 2) and cells transfected with the human HO gene (lane 3). Analysis of total RNA (10µg/lane) with a human HO cDNA (HHO, upper panel), rat HO cDNA (middle panel) and glyceraldehyde-3-phosphate dehydrogenase probe (G3PDH, lower panel). The human HO-specific probe does not detect rabbit HO mRNA transcripts and rat HO cDNA does not detect human HO mRNA.

Table 1. Effect of Sn-mesoporphyrin on control and transfected endothelial cells heme oxygenase activity

Conditions	Heme oxygenase activity (pmol/5x10^6 cells/30 min mean±SE)
Control cells	160 ± 14
Control cells + Sn-mesoporphyrin	14 ± 4
Cells transfected with control plasmid	152 ± 27
Transfected cells with human HO gene	482 ± 77
Transfected cells with human HO gene + Sn-mesoporphyrin	120 ± 12

Heme oxygenase was assayed as described in Materials and Methods. Sn-mesoporphyrin was added at a final concentration of 5μM in the incubation mixture. Results are presented as the mean±SD.

following human HO cDNA transfection ($p<.001$). In contrast, transfection of endothelial cells with control plasmid did not result in an increase in the endogenous rabbit HO activity (Table 1). To further ascertain the characteristics of the expressed HO protein, we tested the effect of Sn-mesoporphyrin, a potent inhibitor of HO activity, on microsomal preparations from both nontransfected and transfected cells. Addition of 5μM Sn-mesoporphyrin to cell preparations inhibited the enzyme activity by ≈90% and ≈75% ($p<.001$) in both control and transfected cells, respectively (Table 1), as has been demonstrated with the normal and induced mammalian HO (24-28).

Effect of Recombinant Hemoglobin and Heme on Transfected and Non-Transfected Endothelial Cells

Since released hemoglobin from circulating, damaged erythrocytes results in the generation of free radicals leading to oxidative stress which can produce damage to vascular endothelial cells (29), we examined the effect of exposure to rHb and free heme on endothelial cells transfected with the human HO gene. Both transfected and non-transfected cells were exposed to sublethal and lethal doses of rHb and heme and evaluated for cell viability after 24 hrs. Control or transfected endothelial cells, not exposed to rHb, were greater than 98% viable as measured by trypan blue exclusion. Exposure of nontransfected endothelial cells to 200μM rHb resulted in 30% cell death within 24 hrs (Fig. 3A). Endothelial cells transfected with the human HO gene were remarkably well preserved and resistant to hemoglobin toxicity with survival rates increased to 95±3.6 at 200μM rHb. Exposure of endothelial cells to free heme resulted in more toxicity than was produced by rHb (Fig. 3B). For example, heme at a concentrations of 100μM and 200μM caused 25% and 44% cell death as compared to 16% and 30% after exposure to 100μM and 200μM rHb, respectively ($p<.05$). As in the case of rHb, cells transfected with the HO gene were significantly resistant to heme toxicity as demonstrated by a 35-55% increase in cell survival over control cells (Fig. 3B). In contrast, tranfection of these cells with a plasmid containing the CYP4A1 cDNA which resulted in high expression of CYP4A1 mRNA and protein did not confer protection against hemoglobin/heme toxicity (Figures 3A and 3B). Likewise, cells transfected with the control plasmid ("sham" infected cells) did not display enhanced HO expression or a cytoprotective effect (data not shown). These results confirm that transfected endothelial cells with the HO gene have a better survival rate compared to untransfected cells after exposure to high concentrations of free heme or hemoglobin and further substantiate the specificity of HO in mediating this protective effect.

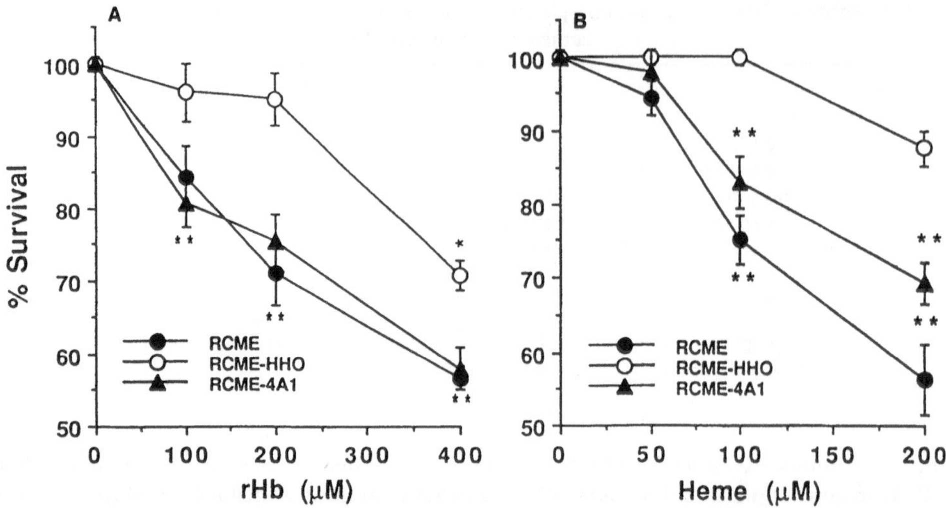

Figure 3. Effects of heme and hemoglobin (rHb) on endothelial cell viability. Control (RCME), HO gene-transfected (RCME-HHO) and CYP4A1 transfected (RCME-YA1 cells were treated with and without A) rHb (1-400μM) and B) heme (1-200μM) for 24 hrs. Cell viability was assessed by trypan blue exclusion as described in Methods. Results are from four different determinations in triplicate and are mean±SEM. *$p<0.01$ Significant from untreated cells (without rHb or heme). **$p<0.01$ Significant from RCME-HHO cells and from untreated cells (without rHb or heme).

DISCUSSION

Induction of HO has been suggested to be an adaptive response to oxidative stress agents (4,6,7,30); for example, hemoglobin released from damaged erythrocytes at tissue sites of hemorrhage or other injury, is a major pro-oxidant and a potential source of free radicals (29). Furthermore, heme moieties released from hemeproteins and hemoglobin have been shown to promote the formation of oxygen radicals, generating reactive species toxic to endothelial cells (8,29). Oxidant-mediated tissue damage may contribute significantly to the pathogenesis of diseases such as atherosclerosis and to reperfusion injury after myocardial ischemia and strokes. Thus, overexpression of HO may offer a means of cellular protection against heme/hemoglobin oxidative injury by enhancing the degradation of these pro-oxidants to bile pigments, which themselves have anti-oxidant properties (6,7).

The iron release resulting from HO activity is believed to be the cause of the increased expression of ferritin synthesis, which serves to sequester the metal, thus rendering this potential cellular oxidant inactive (30,31). Bilirubin and biliverdin both act as antioxidants *in vitro* and *in vivo* (6) and their increased local concentrations following HO induction may be beneficial in the protection of endothelial cells from injury. Recently, Neuzil and Stocker (32) have demonstrated that free and albumin-bound bilirubin efficiently inhibits lipid oxidation and that this antioxidant activity is likely due to an interaction of bilirubin with the antioxidant, α-tocopherol and lipoprotein (32). Finally, CO, a by-product of heme degradation, may mimic NO and thus also serve as a modulator of endothelial cell functions following hemorhagic/oxidative injury. CO is also a vasodilator; it may, therefore, together with other vasodilator substances, counteract the vasoconstrictive properties of hemoglobin and heme (33).

Pertinent to the role of HO in hemmorhagic injury is the study of Nath et al (7); these investigators demonstrated that *in vivo* induction of HO prevents renal failure and drastically reduces mortality following glycerol-induced rhabdomyolysis in the rat, a condition which is characterized by an increased release of myoglobin and hemoglobin into the extracellular renal space resulting in irreversable tissue injury, renal failure and death. These investigators also showed the coupling of ferritin synthesis to enhanced HO activity *in vivo* (7). Similarly, Balla and co-workers (29) demonstrated that endothelial cells *in vitro* respond to heme/hemoglobin by induction of HO and the production of large amounts of ferritin and they suggested that both HO and ferritin are major intracellular factors assisting cells in resisting oxidative damage. Skin fibroblasts also respond to oxidative injury such as that produced by UV light in a similar manner, i.e., rapid induction of HO and accumulation of ferritin (34). In these and other cell types, induction of HO prior to exposure to oxidative assaults is thought to provide cell protection and an increase in cell viability (34,35).

In the present study we were able to successfully transfect human HO cDNA into rabbit coronary endothelial cells using Lipofectin as a means of gene delivery. The transfected DNA directed increased expression of the human HO gene, without affecting expression of the endogenous rabbit HO gene, and without altering, morphologically or biochemically the endothelial cell phenotype. Transfection of the human HO gene caused a significant increase in HO activity indicating a successful and functional gene transfer. Overexpression of the human HO gene considerably enhanced resistance of the endothelial cells to oxidative injury produced by concomitant exposure to free heme or hemoglobin. The protective effect was dose-related and substantial, up to concentrations of 200μM of heme/hemoglobin. Cellular protection against the damaging consequences of H_2O_2 exposure by catalase gene transfer into endothelial cells has previously been demonstrated (36). Although catalase is important in the dissipation of H_2O_2, the enzyme does not appear to be a stress protein and in any case does not act physiologically to degrade heme moieties - a physiological function specifically reserved to HO. Thus, HO activity can assist other antioxidant systems in diminishing the overall production of reactive oxygen species and can thereby contribute to cellular resistance to such injury, as this study demonstrates. It can also be postulated that under conditions in which glutathione levels are reduced (37) or catalase and superoxide dismutase activities are compromised (38), induction of HO may be further beneficial.

The ability to transfect the human HO-gene into endothelial cells and to demonstrate its overexpression in such cells offers a wide range of experimental possibilities for studying the role of this human enzyme in tissue protective mechanisms against oxidative injury. The direct evidence provided in this study that human HO overexpression in coronary endothelial cells provides these cells with significant protection against heme/hemoglobin-induced toxicity suggests that this enzyme could play an important role in moderating tissue damage associated with many pathologic processes in humans, as for example trauma, hemorrhage, thrombosis and re-perfusion injury. Pharmacological manipulation of HO to further enhance its synthesis and activity may thus have important therapeutic potential in these clinical circumstances.

ACKNOWLEDGMENTS

This work was supported in part by grants R01 HL54138 and EY06531 from the National Institutes of Health.

REFERENCES

1. Shibahara, S., Muller, R. M. & Taguchi, H. (1987) *J. Biol. Chem.* 262, 12889-12892.
2. Mitani, K., Fujita, H., Sassa, S. & Kappas, A. (1989) *Biochem. Biophys. Res. Commun.* 165, 437-441.

3. Keyse, S. M. & Tyrrell, R. M. (1991) *Proc. Natl. Acad. Sci. USA* 86, 99-103.
4. Taketani, S., Kohno, H., Yoshinaga, T. & Tokunaga, R. (1989) *FEBS Lett.* 245, 173-176.
5. Abraham, N. G., Lin, J. H., Schwartzman, M. L., Levere, R. D. & Shibahara, S. (1988) *Int. J. Biochem.* 20, 543-558.
6. Stocker, R., Yamamoto, Y., McDonagh, A. F., Glazer, A. N. & Ames, B. N. (1987) *Science* 235, 1043-1046.
7. Nath, K. A., Balla, G., Vercellotti, G. M., Balla, J., Jacob, H. S., Levitt, M. D. & Rosenberg, M. E. (1992) *J. Clin. Invest.* 90, 267-270.
8. Paller, M. S. & Jacob, H. S. (1994) *Proc. Natl. Acad. Sci. USA* 91, 7002-7006.
9. Stocker, R. (1990) *Free Radical Research Communications* 9, 101-112.
10. McCoubrey, W. K.,Jr., Ewing, J. F. & Maines, M. D. (1992) *Arch. Biochem. Biophys.* 295, 13-20.
11. Shibahara, S., Yoshizawa, M., Suzuki, H., Takeda, K., Meguro, K. & Endo, K. (1993) *J. Biochem.* 113, 214-218.
12. Cantoni, L., Rossi, C., Rizzardini, M., Gadina, M. & Ghezzi, P. (1991) *Biochem. J.* 279, 891-894.
13. Rizzardini, M., Terao, M., Falciani, F. & Cantoni, L. (1993) *Biochem. J.* 290, 343-347.
14. Lutton, J. D., DaSilva, J-L., Moquattash, S., Brown, A. C., Levere, R. D. & Abraham, N. G. (1992) *J. Cell Biol.* 49, 1-7.
15. Alam, J. & Zhining, D. (1992) *J. Biol. Chem.* 267, 21894-21900.
16. Lavrovsky, Y., Schwartzman, M. L., Levere, R. D., Kappas, A. & Abraham, N. G. (1994) *Proc. Natl. Acad. Sci. USA* 91, 5987-5991.
17. Lim, C. S., Chapman, G. D., Gammon, R. S., Muhlestein, J. B., Bauman, R. P., Stack, R. S. & Swain, J. L. (1991) *Circulation* 83, 2007-2911.
18. Brigham, K. L. & Schreier, H. (1993) *J. Lipid Res.* 3, 31-49.
19. Mayhew, E. G. & Papahadjopoulos, D. (1983) in *Liposomes*, ed. Ostro, M. J. (Marcel Dekker, New York), pp. 289-341.
20. Brigham, K. L., Canonico, A. E., Conary, J. T. & Meyrick, B. O. (1992) in *Mediators of Sepsis*, eds. Lamy, M. & Thijs, L. G. (Springer-Verlag, New York), pp. 393-405.
21. Nabel, E. G., Plautz, G. & Nabel, G. J. (1990) *Science* 249, 1285-1288.
22. Gerritsen, M. E., Carley, W. & Milici, A. J. (1988) *Adv. Cell Cult.* 6, 35-67.
23. Brenner, D. A. & Bloomer, J. R. (1980) *N. Engl. J. Med.* 302, 765.
24. Kappas, A., Drummond, G. S., Henschke, C., Petmezaki, S. & Valaes, T. (1994) in *CURRENT PROGRESS IN PERINATAL MEDICINE. The Proceedings of the 2nd World Congress of Perinatal Medicine, Rome, Italy*, eds. Cosmi, E. V. & DiRenzo, G. C. (The Parthenon Publishing Group, London), pp. 623-629.
25. Valaes, T., Petmezaki, S., Henschke, C., Drummond, G. S & Kappas, A. (1994) *Pediatrics* 93, 1-11.
26. Drummond, G. S., Galbraith, R. A., Sardana, M. K. & Kappas, A. (1987) *Arch. Biochem. Biophys.* 255, 64-74.
27. Landaw, S. A., Drummond, G. S. & Kappas, A. (1989) *Pediatrics* 84, 1091-1096.
28. Drummond, G. S., Rosenberg, D. W., Kihlstrom Johanson, A. C. & Kappas, A. (1989) *Pharmacology* 39, 273-284.
29. Balla, J., Jacob, H. S., Balla, G., Nath, K., Eaton, J. W. & Vercelloti, J. M. (1993) *Proc. Natl. Acad. Sci. USA* 90, 9285-9290.
30. Kutty, G., Hayden, B., Osawa, Y., Wiggert, B., Chader, G. J. & Kutty, R. K. (1992) *Curr. Eye Res.* 11, 153-160.
31. Vile, G. G. & Tyrrell, R. M. (1993) *J. Biol. Chem.* 268, 14678-14681.
32. Neuzil, J. & Stocker, R. (1994) *J. Biol. Chem.* 269, 16712-16719.
33. Verma, A., Hirsch, D. J., Glatt, C. E., Ronnett, G. V. & Snyder, S. H. (1993) *Science* 259, 381-384.
34. Vile, G. F., Basu-Modak, S., Waltner, C. & Tyrrell, R. M. (1994) *Proc. Natl. Acad. Sci. USA* 91, 2607-2610.
35. Morishita, T., Peresleny, T., Noiri, E., Staudinger, R., Abraham, N. G. & Goligorsky, M. S. (1994) *J. Am. Soc. Nephrol.* 5, 904(Abstract).
36. Erzurum, S. E., Lemarchand, P., Rosenfeld, M. A., Yoo, J-H. & Crystal, R. G. (1993) *Nucleic Acids. Res.* 21, 1607-1612.
37. Jornot, L. & Junad, A. F. (1993) *Am. J. Physiol.* 264, L482-L489.
38. Llesury, S., Evelson, P., Gonzales-Flecha, B., Peralta, J., Caneras, M. C., Poderoso, J. J. & Boveris, A. (1994) *Free Radic. Biol. Med.* 16, 445-451.

44

EXPRESSION AND SIGNAL TRANSDUCTION OF THE FLT3 TYROSINE KINASE RECEPTOR

Olivier Rosnet,[1] Hans-Jorg Bühring,[4] Odile deLapeyrière,[1] Nathalie Beslu,[2] Chrystel Lavagna,[1] Sylvie Marchetto,[1] Irène Rappold,[4] Hans G. Drexler,[5] Françoise Birg,[3] Robert Rottapel,[7] Charles Hannum,[6] Patrice Dubreuil,[2] and Daniel Birnbaum[1*]

[1] Molecular Oncology Laboratory
U.119 INSERM, 27 Bd Leï Roure
13009 Marseille, France
[2] Molecular Hematology Laboratory
U.119 INSERM, Marseille, France
[3] Molecular Hematology and Cytogenetics Laboratory
U.119 INSERM, Marseille, France
[4] Second Department of Internal Medicine
University of Tübingen, Tübingen, Germany
[5] DSM-German Collection of Microorganisms and Cell Cultures
Braunschweig, Germany
[6] DNAX Research Institute of Molecular and Cellular Biology
901 California Avenue, Palo Alto, California 94304-1104
[7] Wellesley Hospital
Toronto, Ontario, Canada

The biology of hematopoietic stem cells is far from being completely understood. However, much progress has been made recently in at least two directions: the enrichment of hematopoietic stem cells populations on the basis of expression or non expression of specific markers, and the characterization of cytokines acting on the survival, self-renewal or differentiation of these cells. This has allowed ex vivo expansion of hematopoietic progenitor cells and the beginning of stem cell replacement therapy.[1,2]

Two peptide regulatory factors play central roles in the physiology of early hematopoietic cells. These factors are the steel factor, or KIT ligand, and the FLT3 ligand, hereafter designated KL and FL, respectively, and are the ligands of two tyrosine kinase receptors encoded by the *KIT* and *FLT3* genes.

* Address correspondence to: D. Birnbaum. U.119 Inserm, 27 Bd. Leï Roure, 13009 Marseille, France; Fax 33 91 26 03 64.

Molecular Biology of Hematopoiesis 5, edited by Abraham et al.
Plenum Press, New York, 1996

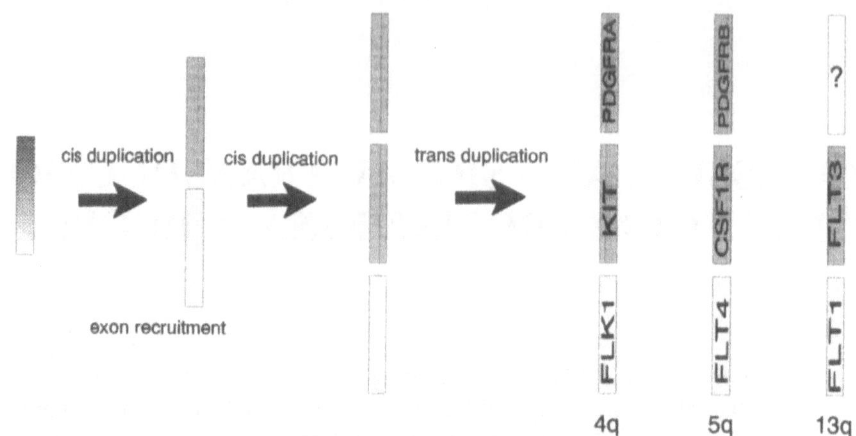

Figure 1. Putative evolutionary scheme common to genes encoding receptor tyrosine kinases with five (class III)(lightly shaded) and seven (heavily shaded) Immunoglobulin-like domains in their extracellular region.

KIT and FLT3[3] (also named FLK2[4] or STK1[5]) are two receptors endowed with an intrinsic tyrosine kinase capacity activated upon stimulation by their cognate ligands. Their extracellular region is composed of five Immunoglobulin-like domains (IgL). A third, related receptor which binds the CSF1/MCSF molecule is encoded by the *FMS* proto-oncogene. These three receptors, together with the two receptors for PDGFs, constitute the so-called class III of receptor tyrosine kinases (RTKs).

The human *FLT3* gene is located in chromosomal band 13q12, closely linked[6] with the *FLT1* gene which encodes a receptor for the Vascular Endothelial Growth Factor. In contrast to FLT3, the FLT1 receptor has seven IgL in its extracellular region but is still highly related to class III RTKs. The exon-intron organization of the *FLT3* gene is identical to those of *FMS* and *KIT*.[7] *FLT3* encodes a protein of 993 amino acid residues.[8] The mouse *Flt3* gene[3] codes for a sligthy longer molecule of 1000 residues, synthesized as two glycosylated isoforms with apparent molecular masses of 155 and 132 kDa, respectively.[9] It is located on chromosome 5 and is also linked with the mouse *Flt1* gene. The class III genes are clustered with genes encoding the VEGF receptor family (FLT1, FLK1/KDR and FLT4). This organization has led to investigations[10] and hypotheses (Fig 1) about their putative evolution. The two classes of genes are thought to have evolved from a common ancestor through a series of cis- and trans-duplications.

In contrast to the mouse *Kit* gene, which is allelic to the *W* locus, also on chromosome 5 but more centromeric, no locus for a naturally occuring mutation has been described in the *Flt3* region. In contrast to FMS and KIT, no viral form of FLT3 has yet been uncovered. Because of these two facts, little information was initially available on the possible biological effects of FLT3.

The ligand of FLT3, FL, has been cloned and characterized by two groups.[11,12,13] Purified recombinant FL was assayed in various tests to establish its range of activities of this new cytokine.[14,15,16] It was shown to have little effect by itself but to support the proliferation of primitive murine lymphohematopoietic progenitors in synergy with other cytokines including IL3, IL6, IL11 and G-CSF. FL alone or in synergy with KL or IL7 can support proliferation of B cell progenitors. It is able to promote growth of early progenitors from human fetal liver, in synergy with IL3, KL or GM-CSF. In contrast, it has no effect on erythroid progenitors, even in the presence of EPO, or on mast cells.[11-16]

In order to gain some insights in FL/FLT3 functions, the patterns of expression and the signal transduction pathway of FLT3 were investigated.

SPECTRUM OF EXPRESSION OF FLT3

Initial studies have shown that *Flt3* is expressed in the hematopoietic and nervous systems, the gonads and the placenta.[3,17] The major site of expression and function appears to be the hematopoietic system. Primitive fetal liver hematopoietic cells, defined by the presence of the AA4.1 marker and the low expression of lineage specific markers, express the *Flk2/Flt3* gene.[4] More recently, a refined analysis using monoclonal antibodies has shown that cycling stem cells from both murine fetal liver and bone marrow express FLT3.[18] In humans, initial reports described the presence of *FLT3* mRNA in most lymphohematopoietic organs,[3, 8] in $CD34^+$ cells[5] and in almost all acute leukemias samples[19] and several cell lines including, in particular, pro/pre B cell lines.[20, 21]

We first performed a detailed analysis of *Flt3* gene expression using in situ hybridization of paraffin-embedded sections of mouse fetal and adult tissues.[22] *Flt3* RNA transcripts could be detected in three major sites namely the lymphohematopoietic organs, central nervous system and placenta. *Flt3* was expressed in the medullary area of fetal and newborn thymus, in the paracortical regions of lymph nodes and in the red pulp of spleen. In placenta, labyrinthine trophoblasts expressed *Flt3*. *Flt3* mRNAs were found in cerebellar Purkinje cells and in some nuclei of adult mesencephalon. The significance of *Flt3* expression in nervous tissues is not known. In this respect, two facts must be noted. First, although we did not detect transcripts for the *Flt3* ligand gene in these tissues, in the mouse, *FL* may be expressed at low level or in some restricted areas of the nervous system. Second, the FLT3 protein was detected as two bands of 132 and 155 kDa in Western blots of lymphohematopoietic tissues and placenta, but as a single 145 kDa band in brain and cerebellum. This leaves therefore the possibility that the FLT3 molecule found in the nervous system functions differently from the hematopoietic receptor.

A second series of experiments designed to study FLT3 expression was performed in humans and at the protein level. Monoclonal antibodies (mAbs) directed against the extracellular region of human FLT3 were generated and used to study the cell surface expression of the receptor on normal bone marrow cells.[23] Expression of FLT3 was restricted to a subset of both CD34-positive and CD34-negative bone marrow cells. Co-expression with KIT was also determined; as in the mouse,[18] four subpopulations of primitive CD34+ cells could be identified: FLT3+KIT+, FLT3-KIT-, FLT3+KIT- and FLT3-KIT+. Fig 2 represents a schematic summary of our present, still preliminary, view of FLT3 expression in the lymphohematopoietic system.

In addition, FLT3 expression was studied on leukemic blasts from patients with acute leukemias. In agreement with previous findings on *FLT3* mRNA expression in almost all acute leukemias,[19] the FLT3 protein was found expressed at a high level on leukemic blasts of most acute leukemias, in particular on pre-B-ALL and AML cells. Recent studies showed that FL led to a significant proliferative response in some cases of acute myeloid leukemia, both primary leukemia samples from patients and leukemia cell lines, but not in pre-B-ALL cells despite their strong expression of the receptor.[24] Furthermore, in the model system of one FL-responsive AML cell line, co-stimulation with IL-3, GM-CSF or KL had a synergistic effect on FL-induced cell growth. TGF-β1 partially inhibited the FL-promoted proliferation but FGF2/bFGF significantly abrogated the inhibitory effects of FGF-β1.

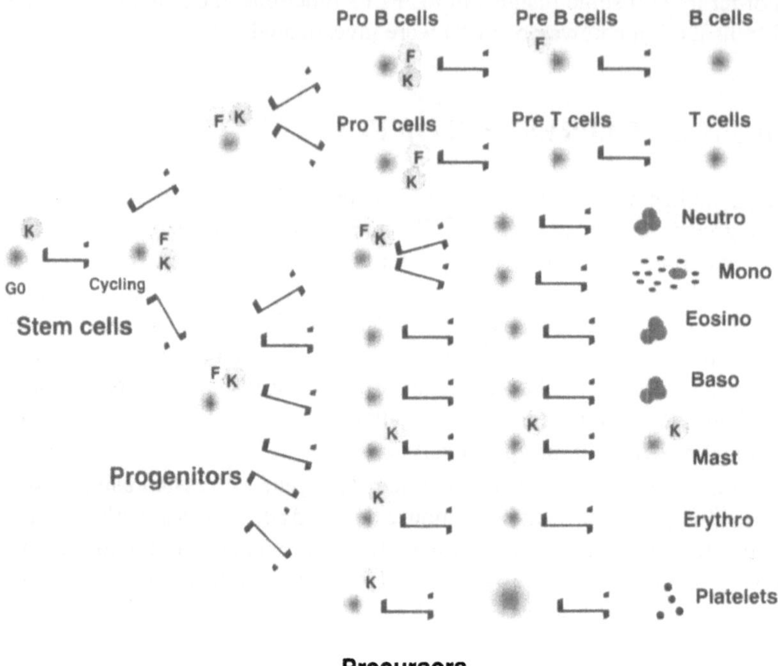

Figure 2. Tentative comprehensive summary of present available data and speculative hypotheses on FLT3/FLK2 receptor gene and/or protein expression in the lymphohematopoietic system. F: FLT3/FLK2; K: KIT.

SIGNAL TRANSDUCTION VIA THE FLT3 RECEPTOR

The FLT3 receptor transduces activation signals through the association with and/or the phosphorylation of cytoplasmic proteins, including RAS GTPase activating protein, phospholipase C gamma, VAV and SHC[25, 26] (Fig 3A). The p85 subunit of phosphatidyl inositol 3' kinase (PI3K) and the RAS pathway associated GRB2 adaptor associate with the activated receptor though their SH2 domains[26, 27] (Fig 3B). Consensus sequences for putative GRB2 binding sites are present in the C-terminus of FLT3 (Fig 3B). Phosphopeptide competition experiments have shown[26, 27] that, in contrast to the two related receptors, FMS/CSF1R and KIT, p85 associates with FLT3 at a specific docking site located in the C-terminal end of the molecule, which contains the tyrosine residue 958 (Y958). Lack of association and activity of PI3K was demonstrated for FLT3 mutants in which Y958 had been substituted by F958. Yet, the expression of these mutated receptors in fibroblast cells had no effect on ligand-induced proliferation. The 958 mutation had no effect on the internalization of the receptor, in contrast to the observation made with other class III RTKs in which the PI3K binding site had been eliminated.

CONCLUSIONS AND FUTURE DIRECTIONS

Our work has aimed at establishing the pattern of expression of the FLT3 receptor in both murine and human tissues. FLT3 expression in tissues is more restricted than that of its ligand, and such a restriction is likely to play a major role in the function of the FL/FLT3

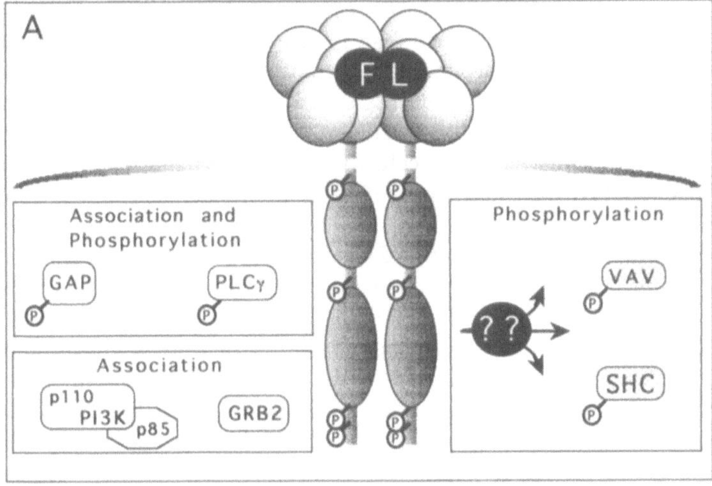

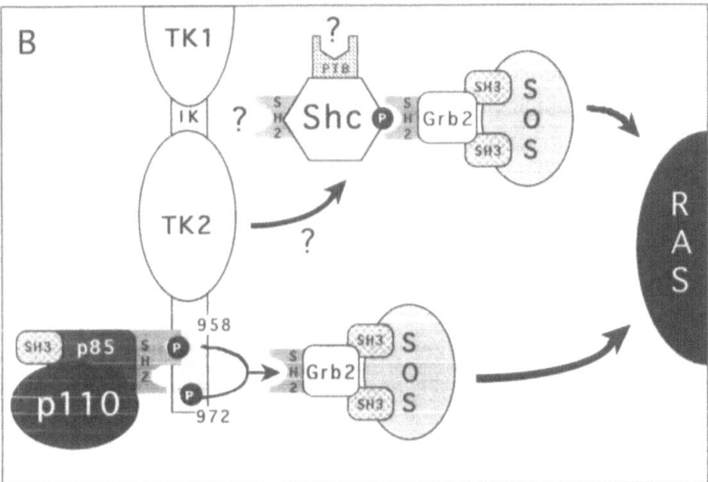

Figure 3. Schematic representation of the transduction pathway initiated by FL/FLT3 interaction. A. Summary of FLT3 autophosphorylation sites and identified substrates. B. Detailed representation of the association of p85 subunit of phosphatidylinositol 3' kinase with phosphorylated tyrosine 958. SHC is phosphorylated after FLT3 activation but does not seem to associate with FLT3. GRB2 associates with FLT3 at potential sites Y958 and Y972.

pair. It is still not completely clear which subpopulations of hematopoietic cells express FLT3. However, the established pattern of expression (schematized in Fig 2) correlates relatively well with the demonstrated properties of the FLT3 ligand. Major sites of expression/function of FLT3 are some stem and progenitor cells, in particular the cycling stem cell subpopulations and pro B cells.

The overall structure of the FLT3 ligand is similar to those of KL and CSF1. Its extracellular region is predicted[12] to fold into a structure made of four helical domains common to other cytokines.[28] This similarity between KL and FL, and the relatedness of their cognate receptors, has functional consequences. Although studies of the biological activity of the FLT3 ligand have just begun, available data suggest that it acts in a fashion very similar to that of the KIT ligand. Like KL, it has little proliferative capacity by itself, but potentializes other multilineage cytokines such as IL3, IL6 or IL7. However, some differences are noticeable. The effect of FL on primitive lymphoid cells in combination with IL7 is consistent with the pattern of expression of the FLT3 receptor. It suggests that the FL/FLT3 pair plays an important role in the growth control of cells of the lymphoid lineage. In contrast to KL, FL has no effect on cells of the erythroid and mast cell lineages. Mice devoid of a functional KL/KIT pair have defects in both mast cells and erythroid cells but not in the other lineages. There is obviously no redundancy between KIT and FLT3 in mast and red cell lineages. It is more difficult to draw firm conclusions for the other cell lineages. One of the primary targets of FL is the lymphoid lineage. Indeed, *Flt3* knock-out mouse mutants have defects in primitive B lymphoid progenitors.[29] They also present defects in T and myeloid cells in a competitive repopulation assay.[29] Thus, although KIT plays some role in the biology of lymphoid cells, defects in FL/FLT3 function are not totally compensated by the KL/KIT pair. One may envisage a major but redundant function of KL/KIT and FL/FLT3 pairs at the level of the stem cell. Available data suggest KL/KIT and FL/FLT3 have both overlapping and distinct activities. Accordingly, double knock-out mutant mice have severely reduced myelopoiesis and lymphopoiesis, showing that no other major compensatory system exists.[29]

Given the pattern of expression of its receptor, availability of recombinant FL may become most helpful in the ex vivo expansion of purified stem cells, once included in the already existing multicytokine cocktails. It may also be useful in transplantation and gene therapy protocols.

ACKNOWLEDGMENTS

This work was supported by Inserm and grants from Association pour la Recherche sur le Cancer.

REFERENCES

1. Krauss J: Hematopoietic stem cell gene replacement therapy. Biochim Biophys Acta 1114:193, 1992.
2. Whetton A, Dexter M: Influence of growth factors and substrates on differentiation of hematopoietic stem cells. Curr Op Cell Biol 5:1044, 1993.
3. Rosnet O, Marchetto S, deLapeyriere O, Birnbaum D: Murine *Flt3* , a gene encoding a novel tyrosine kinase receptor of the PDGFR/CSF1R family. Oncogene 6:1641, 1991.
4. Matthews W, Jordan C, Wiegand G, Pardoll D, Lemischla I: A receptor tyrosine kinase specific to hematopoietic stem and progenitor cell-enriched populations. Cell 65:1143, 1991.
5. Small D, Levenstein M, Kim E, Carow C, Amin S, Rockwell P, Witte L, Burrow C, Ratajczak M, Gewirtz A, Civin C: STK-1, the human homolog of Flk-2/Flt-3, is selectively expressed in CD34+ human bone marrow cells and is involved in the proliferation of early progenitors/stem cells. Proc Natl Acad Sci USA 91:459, 1994.
6. Imbert A, Rosnet O, Marchetto S, Ollendorff V, Birnbaum D, Pébusque MJ: Characterization of a yeast artificial chromosome from human chromosome band 13q12 containing the FLT1 and FLT3 receptor-type tyrosine kinase genes. Cytogenet Cell Genet 67:175, 1994.
7. Agnès F, Shamoon B, Dina C, Rosnet O, Birnbaum D, Galibert F: genomic structure of the downstream part of the human *FLT3* gene: exon/intron structure conservation among genes encoding receptor tyrosine kinase (RTK) of subclass III. Gene 145:283, 1994.
8. Rosnet O, Schiff C, Pébusque MJ, Marchetto S, Tonnelle C, Toiron Y, Birg F, Birnbaum D: Human *FLT3/FLK2* gene: cDNA cloning and expression in hematopoietic cells. Blood 82:1110, 1993.
9. Maroc N, Rottapel R, Rosnet O, Marchetto S, Lavezzi C, Mannoni P, Birnbaum D, Dubreuil P: Biochemical characterization and analysis of the transforming potential of the FLT3/FLK2 receptor tyrosine kinase. Oncogene 8:909, 1993.

10. Rousset D, Agnès F, Lachaume P, André C, Galibert F: Molecular evolution of the genes encoding receptor tyrosine kinase with immunoglobulinlike domains. J Mol Evol 41:421-429, 1995.
11. Lyman S, James L, Vanden Bos T, deVries P, Brasel K, Gliniak B, Hollingsworth L, Picha K, McKenna H, Splett R, Fletcher F, Maraskovsky E, Farrah T, Foxworthe D, Williams D, Beckmann P: Molecular cloning of a ligand for the flt3/flk2 tyrosine kinase receptor: a proliferative factor for primitive hematopoietic cells. Cell 75:1157, 1993.
12. Hannum C, Culpepper J, Campbell D, McClanahan T, Zurawski S, Bazan F, Kastelein R, Hudak S, Wagner J, Mattson J, Luh J, Duda G, Martina N, Peterson D, Menon S, Shanafelt A, Muench M, Kelner G, Namikawa R, Rennick D, Roncarolo MG, Zlotnik A, Rosnet O, Dubreuil P, Birnbaum D, Lee F: Ligand for FLT3/FLK2 receptor tyrosine kinase regulates growth of hematopoietic stem cells and in encoded by variant RNAs. Nature 368:643, 1994.
13. Lyman S, Williams D: Biology and potential clinical applications of flt3 ligand. Current Op Hematol 2:177, 1995.
14. Lyman S, James L, Johnson L, Brasel K, deVries P, Escobar S, Downey H, Splett R, Beckmann P, McKenna H: Cloning of the human homologue of the murine flt3 ligand: a growth factor for early hematopoietic progenitor cells. Blood 83:2795, 1994.
15. Muench M, Roncarolo MG, Menon S, Xu Y, Kastelein R, Zurawski S, Hannum C, Culpepper J, Lee F, Namikawa R: FLK-2/FLT-3 ligand regulates the growth of early myeloid progenitors isolated from human fetal liver. Blood 85:963, 1995.
16. Hirayama F, Lyman S, Clark S, Ogawa M: The *flt3* ligand supports proliferation of lymphohematopoietic progenitors and early B-lymphoid progenitors. Blood 85:1762, 1995.
17. Rossner M, McArthur G, Allen J, Metcalf D: Fms-like tyrosine kinase 3 catalytic domain can transduce a proliferative signal in FDCP1 cells that is qualitatively similar to the signal delivered by c-Fms. Cell Growth Differ 5:549, 1994.
18. Zeigler F, Bennett B, Jordan C, Spencer S, Baumheuter S, Carroll K, Hooley J, Bauer K, Matthews W: Cellular and molecular characterization of the role of the FLK-2/FLT-3 receptor tyrosine kinase in hematopoietic stem cells. Blood 84:2422, 1994
19. Birg F, Courcoul M, Rosnet O, Bardin F, Pébusque MJ, Marchetto S, Tabilio A, Mannoni P, Birnbaum D: Expression of the *FMS/KIT*- like gene *FLT3* in human acute leukemias of the myeloid and lymphoid lineages. Blood, 80:2584, 1992.
20. DaSilva N, Hu ZB, Ma W, Rosnet O, Birnbaum D, Drexler H: Expression of the *FLT3* gene in human leukemia-lymphoma cell lines. Leukemia 8:885, 1994.
21. Meierhoff G, Dehmel U, Gruss HJ, Rosnet O, Birnbaum D, Dirks W, Drexler H: Expression of FLT3 receptor and FLT3-ligand in human leukemia-lymphoma cell lines. Leukemia, 9:1368-1372, 1995.
22. deLapeyrière O, Naquet P, Planche J, Marchetto S, Rottapel R, Gambarelli D, Rosnet O, Birnbaum D: Expression of *Flt3* tyrosine kinase receptor gene in mouse hematopoietic and nervous tissues. Differentiation, 58:351, 1995.
23. Rosnet O, Bühring HG, Marchetto S, Rappold I, Lavagna C, Sainty D, Chabannon C, Kanz L, Hannum C, Birnbaum D: The human FLT3/FLK2 hematopoietic receptor tyrosine kinase is expressed at the surface of normal and malignant hematopoietic cells. Leukemia 10:238-248, 1996.
24. Dehmel U, Zaborski M, Meierhoff G, Rosnet O, Birnbaum D, Ludwig WD, Quentmeier H, Drexler HG: Effects of FLT3 ligand on human leukemia cells. I. Proliferative response of myeloid leukemia cells. Leukemia 10:261-270, 1996.
25. Dosil M, Wang S, Lemischka I: Mitogenic signaling and substrate specificity of the Flk2/Flt3 receptor tyrosine kinase in fibroblasts and interleukin 3-dependent hematopoietic cells. Mol Cell Biol, 13:6572, 1993.
26. Rottapel R, Turck C, Casteran N, Liu X, Birnbaum D, Pawson T, Dubreuil P: Substrate specificities and identification of a putative binding site for PI3K in the carboxy tail of the murine Flt3 receptor tyrosine kinase. Oncogene 9:1755, 1994
27. Casteran N, Rottapel R, Beslu N, Lecocq E, Birnbaum D, Dubreuil P: Analysis of the mitogenic pathway of the FLT3 receptor and characterization in its C terminal region of a specific binding site for the PI3' kinase. Cell Mol Biol 40:443, 1994.
28. Kaushansky K. and Karplus A: Hematopoietic growth factors: understanding functional diversity in structural terms. Blood, 82:3229, 1993.
29. Mackarehtschian K, Hardin J, Moore K, Boast S, Goff S, Lemischka I: Targeted disruption of the Flk2/Flt3 gene leads to deficiencies in primitive hematopoietic progenitors. Immunity, 3:147-161, 1995.

JAK2 IS CONSTITUTIVELY ASSOCIATED WITH C-KIT AND IS PHOSPHORYLATED IN RESPONSE TO STEM CELL FACTOR*

Diana Linnekin,[1†] Sarah R. Weiler,[1] Sherry Mou,[2] Candy S. DeBerry,[1] Jonathan R. Keller,[2] Francis W. Ruscetti,[1] Douglas K. Ferris,[2] and Dan L. Longo[1]

[1] Laboratory of Leukocyte Biology, Biological Response Modifiers Program
Division of Cancer Treatment, National Cancer Institute
Frederick Cancer Research and Development Center
Frederick, Maryland 21702
[2] Biological Carcinogenesis and Development Program
SAIC Frederick
National Cancer Institute-Frederick Cancer Research and Development Center
Frederick, Maryland, 21702

ABSTRACT

Stem cell factor (SCF) interacts with the receptor tyrosine kinase c-kit and has potent effects on hematopoiesis. We have examined the role of JAK2 in the SCF signal transduction pathway. JAK2 and c-kit were constitutively associated and treatment with SCF resulted in rapid and transient tyrosine phosphorylation of JAK2. Incubation of cells with JAK2 antisense oligonucleotides resulted in significant decreases in SCF-induced proliferation. These data suggest that JAK2 plays a role in SCF-induced proliferation.

---

* The content of this publication does not necessarily reflect the views or policies of the Dept. of Health and Human Services, nor does mention of trade names, commercial products or organizations imply endorsement by the U.S. Government.

† To whom correspondence should be addressed: Diana Linnekin, Ph.D., Blg. 567, Rm 226, Laboratory of Leukocyte Biology, Biological Response Modifiers Program, Division of Cancer Treatment, National Cancer Institute, Frederick Cancer Research and Development Center, Frederick, MD 21702. Telephone: 301-846-5188; FAX 301-846-6107.

INTRODUCTION

Stem cell factor (steel factor, kit ligand and mast cell growth factor) is an important component of normal hematopoiesis (reviewed in 1). The receptor for SCF, the c-kit proto-oncogene product, is a receptor tyrosine kinase and is closely related to the receptors for colony stimulating factor-1 (CSF-1) and platelet-derived growth factor (2, 3). A number of signal transduction components have been implicated in SCF/c-kit stimulus-response coupling mechanisms. Included among these are tec, PI3 kinase, MAP kinases, c-raf, SHC, vav and phospholipase C gamma (4-12). Interestingly, in contrast to the hemapoietin receptor superfamily, little is known about the role of JAK family members in relation to receptor tyrosine kinase signal transduction (13, 14).

We have examined the SCF/c-kit signal transduction pathway to determine the potential involvement of the JAK2 protein tyrosine kinase. Our results demonstrate that c-kit is constitutively associated with JAK2 and that SCF induces rapid and transient phosphorylation of JAK2. Further, JAK2 antisense oligonucleotides inhibited SCF-induced proliferation by 57%. These data suggest that JAK2 is a component of the SCF signal transduction pathway.

MATERIAL AND METHODS

Cells and Growth Factors

Mo7e cells were maintained in RPMI 1640, 10% fetal calf serum, 2 mM L-glutamine and 1% penicillin-streptomycin (cell culture media) supplemented with recombinant human GM-CSF and human interleukin 3. FDCP-1 cells were maintained in cell culture media supplemented with recombinant murine IL-3. Human GM-CSF, SCF and IL-3 as well as murine IL-3 were purchased from PeproTech (Rocky Hill, NJ).

Immunoprecipitations, Electrophoresis, and Immunoblotting

Cells were resuspended in RPMI 1640, 1% FCS, stimulated with SCF for the time(s) indicated in the text, rapidly pelleted and resuspended in lysis buffer (1% Triton X 100, 50 mM NaCl, 10 mM Tris, 5 mM EDTA, 30 mM Sodium Pyrosphosphate, 50 mM Sodium Fluoride, 100 uM Sodium Orthovanadate, 1 mM PMSF and 0.1% Bovine Serum Albumin, pH 7.6). Lysates were incubated on ice, clarified and immunoprecipitations performed on the clarified supernatant. Immunoprecipitates were washed 6 times with lysis buffer and samples were eluted from the protein A sepharose (PAS) with SDS sample buffer. Electrophoresis and immunoblotting were performed as previously described (15). The blots were visualized using the ECL Western Blotting detection system (Amersham, Arlington Heights, IL).

Proliferation Assays

FDCP-1 cells were washed four times and resuspended at $5x10^4$/ml in cell culture media. Cells were pretreated for 5 hours in the presence of sense or antisense oligonucleotides. Media or murine SCF (200 ng/ml) were then added to triplicate wells in the presence of additional oligonucleotides. The microtiter plates were incubated 72 hrs at 37° C, 5% CO_2, pulsed with ^{3}H-thymidine (6.7 Ci/mM, NEN, Boston, MA), harvested (Scatron Semiautomatic Cell Harvester) and counted in a liquid scintillation counter (Model 1216,

LKB, Piscataway, NJ). The sequence for the sense oligonucleotide was 5'-ATG GAG GCA ACC TCC ACA-3' and for the antisense oligonucleotide was 5'-TGT GGA GGT TGC CTC CAT-3'. Oligonucleotides were synthesized with an automated synthesizer as described previously (16).

RESULTS

Stimulation of Mo7e cells with SCF results in phosphorylation of a number of proteins. Shown in Figure 1 are SCF-induced phosphoproteins detected using either antiphosphotyrosine immunoprecipitates from ^{32}P-orthophosphate radiolabeled cells (panel A) or antiphosphotyrosine immunoblotting of antiphosphotyrosine immunoprecipitates (panel B). Interestingly, a 125 kDa protein was one of the phosphotyrosylproteins observed in SCF-treated cells. Because this was the approximate size of the protein tyrosine kinase JAK2, we chose to examine the effects of SCF on JAK2 phosphorylation. Shown in Fig. 2A is an antiphosphotyrosine immunoblot of JAK2 immunoprecipitates from Mo7e cells stimulated 2 minutes with SCF. Increases in JAK2 phosphorylation were clearly observed in the absence of detectable changes in JAK2 protein levels (Fig. 2A and 2B). The kinetics of SCF-induced phosphorylation of JAK2 are shown in Fig. 2C. Increases in JAK2 tyrosine phosphorylation

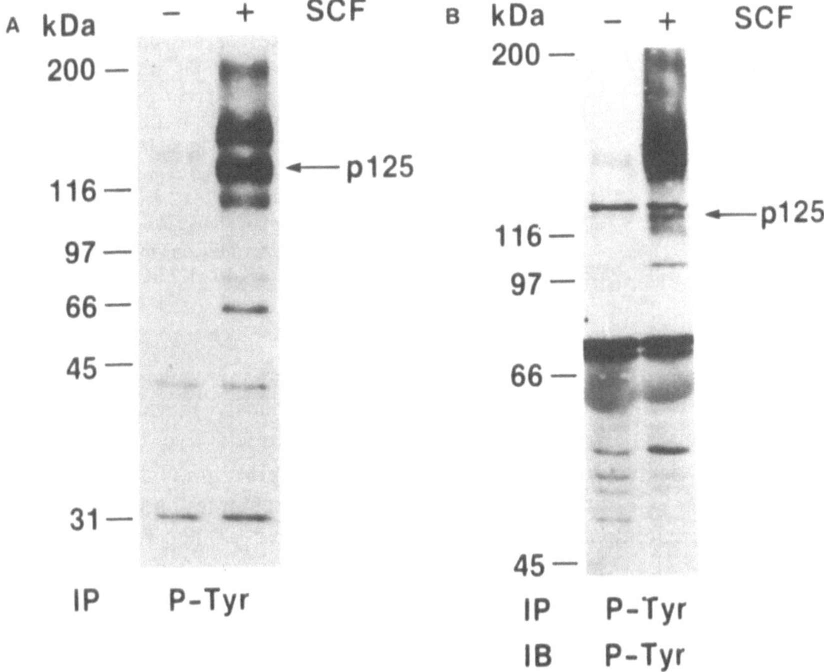

Figure 1. SCF-induced tyrosine phosphorylation of a 125 kDa protein. A. Mo7e cells were radiolabeled with ^{32}P-orthophosphate (1 mCi/ml), stimulated with 100 ng/ml SCF, lysed and phosphotyrosylproteins immunoprecipitated with PY-20. Proteins were resolved with SDS-PAGE and visualized with autoradiography. B. Mo7e cells were stimulated for 2 minutes with 100 ng/ml of human SCF, lysed, clarified and immunoprecipitated with the PY-20 antiphosphotyrosine antibody. Phosphotyrosylproteins were resolved by SDS-PAGE, transferred to Immobilon and immunoblotted with 4G10, an antibody specific for phosphotyrosine. Proteins were visualized using ECL.

A

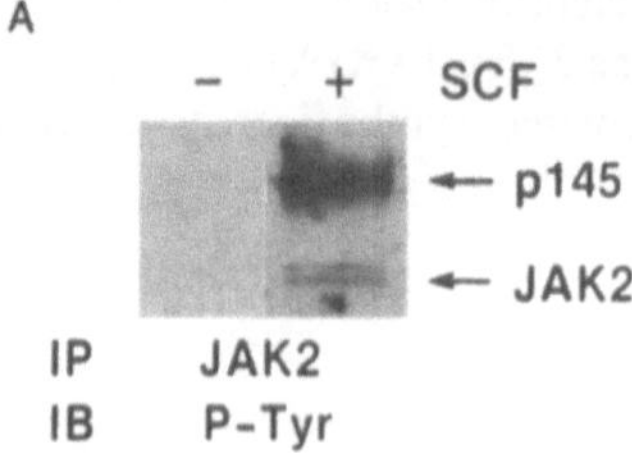

B

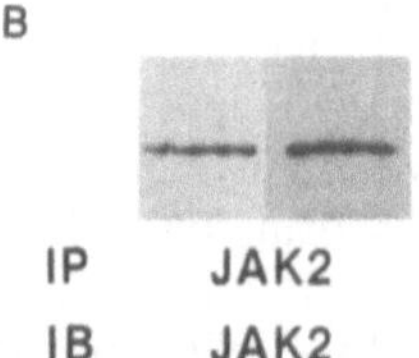

C

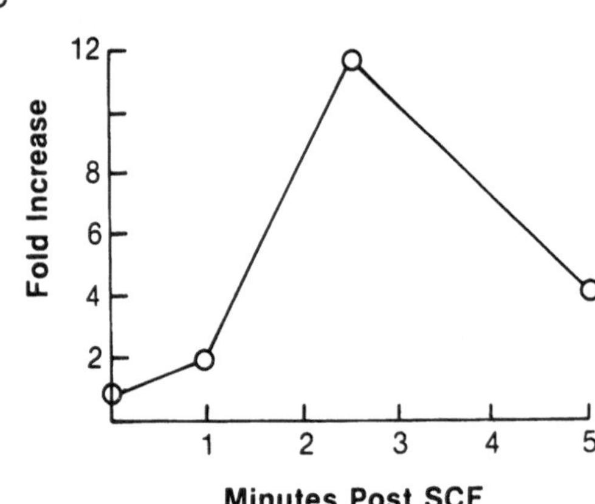

Figure 2. SCF induces rapid tyrosine phosphorylation of JAK2. **A.** SCF-induced tyrosine phosphorylation of JAK2. Mo7e cells were incubated 2 minutes with 100 ng/ml of human SCF, lysed and immunoprecipitated with JAK2 antisera. Proteins were resolved by SDS-PAGE, transferred to Immobilon and immunoblotted with 4G10. **B.** JAK2 immunoblot. The immunoblot from panel A was stripped and reprobed with antisera specific for JAK-2. **C.** Kinetics of SCF-induced JAK2 phosphorylation. Laser densitometry was performed on antiphosphotyrosine immunoblots of JAK2 immunoprecipitates.

were detectable within 1 minute, peaked by 2.5 minutes and approached baseline levels by 5 minutes. Identical results were obtained with the SCF-responsive murine cell line FDCP-1, as well as with normal human progenitor cells isolated from fetal liver (data not shown).

Fig. 2A also demonstrates that a 145 kDa phosphotyrosylprotein coimmunoprecipitates with JAK2 after SCF stimulation. Because the SCF receptor is approximately 145 kDa, we addressed the possibility that JAK2 was associated with c-kit. Shown in Fig. 3 is a c-kit immunoblot of control, c-kit and JAK2 immunoprecipitates from lystates of unstimulated Mo7e cells. These results suggest that JAK2 is constitutively associated with c-kit.

The association of JAK2 with c-kit, and, its subsequent phosphorylation after SCF treatment suggested that JAK2 may play a role in the SCF/c-kit signal transduction pathway. To address this possibility, we examined the effects of JAK2 antisense oligonucleotides on SCF-induced proliferation. These studies were performed using murine cells and oligonucleotides specific for the murine JAK2 sequence since, to date, the human JAK2 sequence has not been published (17, 18). The murine cell line FDCP-1 was incubated 5 hours with the indicated concentration of sense or antisense oligonucleotides prior to initiation of proliferation assays. The results of a representative experiment are shown in Fig. 4. Treatment with antisense oligonucleotides resulted in 12%, 32% and 57% inhibition of SCF-in-

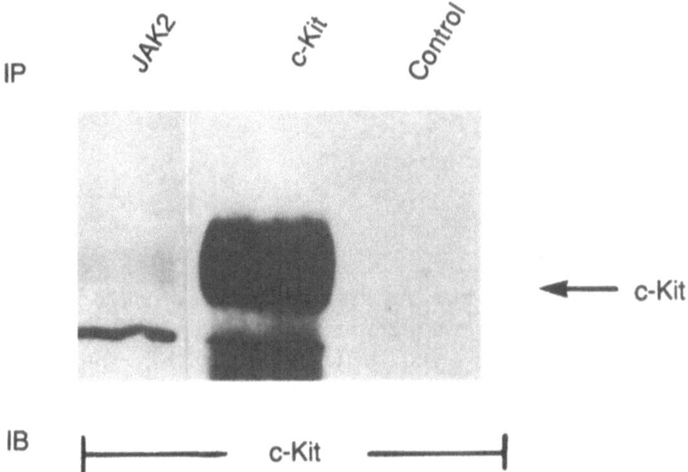

Figure 3. Association of JAK2 with c-kit. Constitutive association of JAK2 and c-kit. Mo7e cells were lysed and immunoprecipitated with either control, c-kit or JAK2 antibodies. Proteins were resolved by SDS-PAGE, transferred to Immobilon and immunoblotted with antibody specific for c-kit.

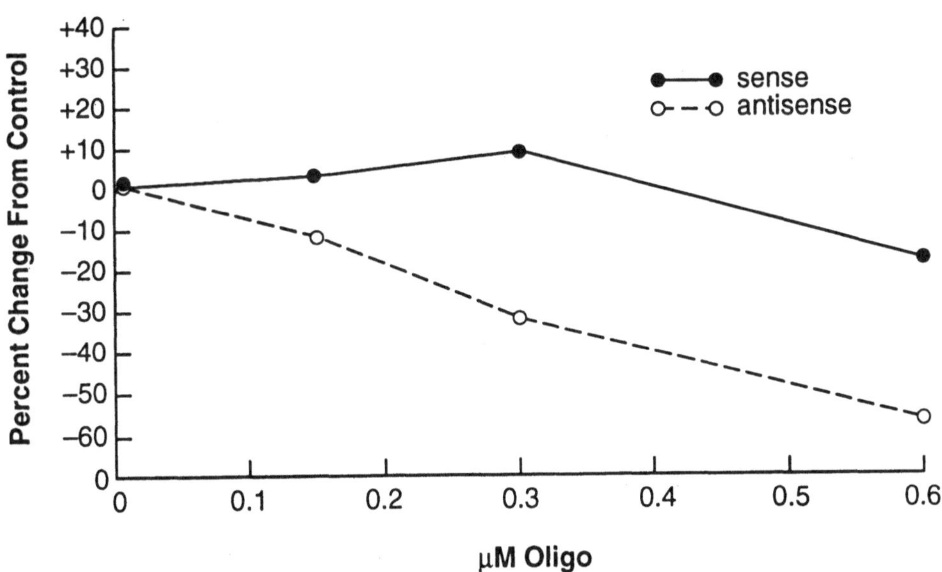

Figure 4. JAK2 partially mediates SCF-induced proliferation. Inhibition of SCF-induced proliferation by JAK2 antisense oligonucleotides. FDCP-1 cells were preincubated 5 hours with the indicated concentration of sense or antisense oligonucleotides and proliferation assays performed as described in Materials and Methods. The data are presented as CPM X 10^3 and are the average of triplicate points.

duced proliferation at concentrations of 0.15, 0.3 and 0.6 uM oligonucleotides, respectively. In contrast, treatment with comparable concentrations of sense oligonucleotides had minimal effects on SCF-induced proliferation. PCR analysis of JAK2 expression indicated that the antisense oligonucleotides specifically inhibited expression of JAK2 mRNA (Weiler et al, manuscript in preparation).

DISCUSSION

Our results demonstrate that JAK2 is a component of the SCF signal transduction pathway. JAK2 and c-kit exist in a preformed complex and SCF induces rapid tyrosine phosphorylation of JAK2 (Figs 2 and 3). Inhibition of JAK2 expression with antisense oligonucleotides results in a significant decrease in SCF-induced proliferative responses (Fig 4). It is important to note that treatment of cells with antisense oligonucleotides to JAK2 resulted in only a partial inhibition of SCF-induced proliferation. This could be attributed to several factors. First, expression of JAK2 protein may not be completely abrogated with antisense oligonucleotides over the entire course of the proliferation assay. Second, the JAK2 pathway may serve as only one component of the SCF signal transduction pathway. Third, SCF induces a relatively weak phosphorylation of JAK1 as well as JAK2 (DML, unpublished observations). Thus, elimination of both JAK1 and JAK2 may be required for complete inhibition of SCF-induced proliferation.

While our findings are in agreement with the recent report of Brizzi et al, they contrast with the results of Tang et al. who did not observe SCF-induced phosphorylation of JAK2 (5, 19). Our data suggest that the rapid and transient nature of SCF-induced JAK2 phosphorylation is a likely explanation for these conflicting observations. In contrast to the slower, more protracted phosphorylation of JAK2 in response to GM-CSF or IL-3, SCF-induced phosphorylation of JAK2 peaks 2-3 minutes after stimulation and has generally returned to basal levels after 5 minutes of stimulation.

Though recent work has focused on the role of JAK family members in the signal transduction of hemapoietin receptor superfamily members, less is known about the role of JAK family members in the signal transduction pathways of receptor tyrosine kinases (18, 20-25). Although EGF has been shown to stimulate tyrosine phosphorylation of JAK1, the role of JAK1 in the EGF signal transduction pathway is less clear (26). In addition, tyrosine phosphorylation of Stat1 has been observed in response to PDGF, EGF, and CSF-1 (27-29). Interestingly, EGF stimulation results in association of Stat1 with the EGF receptor as well as phosphorylation of Stat3 (27, 30). These findings, in conjuction with our data, suggest that the JAK family of protein tyrosine kinases may also serve as important components of receptor tyrosine kinase signal transduction pathways.

ACKNOWLEDGMENTS

The authors would like to thank Dr. Axel Ullrich for the c-kit antibody and Lin Grove for secretarial assistance.

REFERENCES

1. Witte ON: Steel locus defines new multipotent growth factor. Cell 63:5, 1990

2. Qiu F, Ray P, Brown K, Barker E, Jhanwar S, Ruddle FH, Besmer P: Primary structure of c-kit: relationship with the CSF-1/PDGF receptor kinase family - oncogenic activation of v-kit involves deletion of extracellular domain and C terminus. EMBO J 7:1003, 1988
3. Yarden Y, Kuang W-J, Yang-Feng T, Coussens L, Munemitsu S, Dull TJ, Chen E, Schlessinger J, Francke U, Ullrich A: Human proto-oncogene c-kit: a new cell surface receptor tyrosine kinase for an unidentified ligand. EMBO J 6:3341, 1988
4. Rottapel, R, Reedijk M, Williams DE, Lyman SD, Anderson DM, Pawson T, Bernstein A: The *steel/W* transduction pathway: Kit autophosphorylation and its association with a unique subset of cytoplasmic signaling proteins is induced by the Steel factor. Mol Cell Biol 11:3043, 1991
5. Tang, B, Mano H, Yi T, Ihle JN: Tec kinase associates with c-kit and is tyrosine phosphorylated and activated following stem cell factor binding. Mol Cell Biol 14:8432, 1994
6. Miyazawa K, Hendrie PC, Mantel C, Wood K, Ashman LK, Broxmeyer HE: Comparative analysis of signaling pathways between mast cell growth factor (c-*kit* ligand) and granulocyte-macrophage colony-stimulating factor in a human factor-dependent myeloid cell line involves phosphorylation of Raf-1, GTPase-activating protein and mitogen-activated protein kinase. Exp Hematol 19:1110, 1991
7. Okuda K, Sanghera JS, Pelech SL, Kanakura Y, Hallek M, Griffin JD, Druker BJ: Granulocyte-macrophage colony- stimulating factor, interleukin-3, and Steel factor induce rapid tyrosine phosphorylation of p42 and p44 MAP kinase. Blood 79:2880, 1992
8. Welham MJ, Schrader JW: *Steel* factor-induced tyrosine phosphorylation in murine mast cells. J Immunol 149:2772, 1992
9. Hallek M, Druker B, Lepisto EV, Wood KW, Ernst TJ, Griffin JD: Granulocyte-macrophage colony-stimulating factor and Steel factor induce phosphorylation of both unique and overlapping signal transduction intermediates in a human factor-dependent hematopoietic cell line. J Cell Physiol 153:176, 1992
10. Matsuguchi T, Salgia R, Hallek M, Eder M, Druker B, Ernst TJ, Griffin JD: Shc phosphorylation in myeloid cells is regulated by granulocyte macrophage colony-stimulating factor, interleukin-3, and Steel factor and is constitutively increased by $p210^{BCR/ABL}$. J Biol Chem 269:5016, 1994
11. Alai M, Mui AL-F, Culter RL, Bustelo XR, Barbacid M, Krystal G: Steel factor stimulates the tyrosine phosphorylation of the proto-oncogene product, $p95^{vav}$, in human hemopoietic cells. J Biol Chem 267:18021, 1992
12. Matsuguchi T, Inhorn RC, Carlesso N, Xu G, Druker B, Griffin JD: Tyrosine phosphorylation of $p95^{Vav}$ in myeloid cells is regulated by GM-CSF, IL-3 and Steel factor and is constitutively increased by $p210^{BCR/ABL}$. EMBO J 14:257, 1995
13. Wilks AF, Harpur AG: Cytokine signal transduction and the JAK family of protein tyrosine kianses. BioEssays 16:313, 1994
14. Ihle JN, Witthuhn BA, Quelle FW, Yamamoto K, Thierfelder WE, Kreider B, Silvennoinen O: Signaling by the cytokine receptor superfamily: JAKs and STATs. Trends Biochem Sci 19:222, 1994
15. Linnekin D, Howard Z, Park L, Farrar W, Ferris D, Longo D: HCK expression correlates with GM-CSF induced proliferation in HL-60 cells. Blood 84:94, 1994
16. Muszynski KW, Ruscetti FW, Heidecker G, Rapp U, Troppmair J, Gooya JM, Keller JR: Raf-1 protein is required for growth factor-induced proliferation of hematopoietic cells. J Exp Med, in press.
17. Harpur AG, Andres A-C, Ziemiecki A, Aston RR, Wilkes AF: JAK2, a third member of the JAK family of protein tyrosine kinases. Oncogene 7:1347, 1992
18. Silvennoinen O, Witthuhn BA, Quelle FW, Cleveland JL, Yi T, Ihle JN: Structure of the murine Jak2 protein-tyrosine kinase and its role in interleukin 3 signal transduction. Proc Natl Acad Sci USA 90:8429, 1993
19. Brizzi MF, Zini MG, Aronica MG, Blechman JM, Yarden Y, Pegoraro L: Convergence of signaling by interleukin-3, granulocyte-macrophage colony-stimulating factor, and mast cell growth factor on JAK2 tyrosine kinase. J Biol Chem 269:31680, 1994
20. Witthuhn BA, Quelle FW, Silvennoinen O, Yi T, Tang B, Miura O, Ihle JN: JAK2 associates with the erythropoietin receptor and is tyrosine phosphorylated and activated following stimulation with erythropoietin. Cell 74:227, 1993
21. Quelle FW, Sato N, Witthuhn BA, Inhorn RC, Eder M, Miyajima A, Griffin JD, Ihle JN: JAK2 associates with the β_c chain of the receptor for granulocyte-macrophage colony-stimulating factor, and its activation requires the membrane-proximal region. Mol Cell Biol 14:4335, 1994
22. Miura O, Nakamura N, Quelle FW, Witthuhn BA, Ihle NJ, Aoki N: Erythropoietin induces association of the JAK2 protein tyrosine kinase with the erythropoietin receptor in vivo. Blood 84:1501, 1994
23. He T-C, Jiang N, Zhuang H, Quelle DE, Wojchowski DM: The extended box 2 subdomain of erythropoietin receptor is nonessential for Jak2 activation yet critical for efficient mitogenesis in FDC-ER cells. J Biol Chem 269:18291, 1994

24. Barber DL, D'Andrea AD: Erythropoietin and interleukin-2 activate distinct JAK kinase family members. Mol and Cell Biol 14:6506, 1994
25. Miura O, Miura Y, Nakamura N, Quelle FW, Witthuhn BA, Ihle JN, Aoki N: Induction of tyrosine phosphorylation of VAV and expression of Pim-1 correlates with Jak2-mediated growth signaling from the erythropoietin receptor. Blood 84:4135, 1994
26. Shuai K, Ziemiecki A, Wilks AF, Harpur AG, Sadowski HB, Gilman MZ, Darnell JE: Polypeptide signalling to the nucleus through tyrosine phosphorylation of Jak and Stat proteins. Nature 366:580, 1993
27. Fu X-Y, Zhang J-J: Transcription factor p91 interacts with the epidermal growth factor receptor and mediates activation of the c-fos gene promoter. Cell 74:1135, 1993
28. Ruff-Jamison S, Chen K, Cohen S: Induction by EGF and interferon-γ of tyrosine phosphorylated DNA binding proteins in mouse liver nuclei. Science 261:1733, 1993
29. Silvennoinen O, Schnindledr C, Schlessinger J, Levy DE: Ras-independent growth factor signaling by transcription factor tyrosine phosphorylation. Science 261:1736, 1993
30. Zhong Z, Wen Z, Darnell J: Stat3: A STAT Family Member Activated by Tyrosine Phosphorylation in Response to Epidermal Growth Factor and Interleukin-6. Science 264: 95, 1994

ERYTHROID-SPECIFIC ACTIVATION OF THE DISTAL (TESTIS) PROMOTER OF *GATA1* DURING DIFFERENTIATION OF PURIFIED NORMAL MURINE HEMATOPOIETIC STEM CELLS

Anna Rita Migliaccio,[1,2*] Giovanni Migliaccio,[1,2] Eishi Ashihara,[1] Emanuela Moroni,[3] Barbara Giglioni,[3] and Sergio Ottolenghi[4]

[1] The New York Blood Center
New York, New York
[2] Istituto Superiore di Sanità
Rome, Italy
[3] Centro di Studio sulla Patologia Cellulare
CNR, Milan, Italy
[4] Dipartimento di Genetica e Biologia dei Microorganismi
University of Milan, Milan, Italy

ABSTRACT

To understand the molecular mechanisms of erythroid differentiation, we analyzed by semiquantitative RT-PCR the expression of the transcription factor *GATA1*, the erythropoietin receptor (*EpoR*), and erythroid (β-*globin*) differentiation markers in purified hematopoietic stem cells (HSCs) after in-vitro-induced differentiation. Whether *GATA1* transcription was from the proximal (with respect to the AUG, also known as erythroid) or the distal (also known as testis) promoter was analyzed as well. Low-density marrow cells which bind to wheat germ agglutinin, but not to the antibody 15.1.1, and which do or do not retain the dye rhodamine 123 (Rho-bright and Rho-dull, respectively), were purified. Rho-dull, but not Rho-bright, cells permanently reconstitute lymphomyelopoiesis in W/W^v and severe-combined-immunodeficiency mice and, therefore, contain HSCs. Both Rho-dull and Rho-bright cells give rise to progenitor and differentiated cells (peak values at days 15 and 5, respectively) in liquid culture. Multilineage, erythroid-restricted or myeloid-restricted

* Address correspondence to: Anna Rita Migliaccio, Laboratory of Hematopoietic Growth Factors, New York Blood Center, 310 East 67th Street, New York, New York 10021. Phone: (212) 570-3014; Fax: (212) 570-3195.

Molecular Biology of Hematopoiesis 5, edited by Abraham et al.
Plenum Press, New York, 1996

differentiation is observed when the cultures are stimulated with stem cell factor (SCF) + interleukin (IL)-3, SCF + IL-3 + Epo, or SCF + IL-3 + granulocyte-colony-stimulating factor, respectively. Rho-dull cells have barely detectable reconstitution potential at day 5 of culture. None of the genes examined were expressed in purified Rho-bright or Rho-dull cells. The only exception was *GATA1* which was expressed at maximal levels in Rho-bright cells at the onset of culture. Rho-dull cells did not express *GATA1* before day 3 of culture (maximal expression at days 10 - 15). Activation of *GATA1* and *EpoR* was observed in all growth factor combinations. There was a significant correlation between the amount of mRNA for the two genes expressed by the cells. In contrast, β-*globin* mRNA was detected only in the presence of Epo. The transcription of *GATA1* was exclusively from the proximal promoter in the absence of Epo but both proximal and distal transcripts were observed in its presence. Maximum transcription from the distal promoter (approximately equal to 0.2% of total *GATA1* mRNA) coincided with maximal globin mRNA levels (day 5 or day 15 for Rho-bright and Rho-dull cells, respectively). These results indicate that *GATA1* is activated at the transition point between HSC and pluripotent progenitor cells and erythroid specific *GATA1* regulation involves activation of the distal *GATA1* promoter.

INTRODUCTION

Hematopoietic stem cells (HSCs) give rise to all the differentiated elements of the blood by generating a series of cellular compartments which progressively lose proliferative potential and commit themselves to expressing genes specific for a particular differentiation lineage (1, 2). HSCs are defined by their ability to reconstitute both the lymphoid and the myeloid lineages *in vivo* for the life of the animal (> 1-2 years). The immediate progeny of the HSCs on the other hand, are the multilineage progenitor cells which are capable of myeloid and lymphoid differentiation in vivo and in vitro but do not reconstitute syngeneic hosts (3). HSCs and multilineage progenitor cells can be physically separated on the basis of their staining with the supravital dye rhodamine-123 (Rho). HSCs are purified in the Rho-dull fraction while multilineage progenitors copurify with Rho-bright cells (3).

The molecular differences between HSCs and multilineage progenitors have not been characterized as yet. Our working hypothesis is that HSCs and multilineage progenitor cells differ in their expression of genes involved in the activation of the differentiation programs, (i.e. genes activating erythroid differentiation). Therefore, to understand the differences between HSCs and multilineage progenitor cells, we have analyzed by semiquantitative RT-PCR the expression of several genes involved in erythroid differentiation in Rho-dull and Rho-bright cells purified according to published techniques (4). The cells were then induced to differentiate in liquid cultures with combinations of growth factors which would either allow multilineage [stem cell factor (SCF) + interleukin (IL)-3]- or erythroid [SCF + IL-3 + erythropoietin (Epo)]-, or myeloid [SCF + IL-3 + granulocyte-colony-stimulating-factor (G-CSF)]-restricted differentiation (5, 6). The RNA isolated from the purified cells and from the cells in culture at different time points (days 1-10) was analyzed by specific semiquantitative RT-PCR for the expression of *GATA1*, the erythropoietin receptor (*EpoR*) and β-*globin*. It has been reported that while in hematopoietic cell lines, transcription of *GATA1* starts at multiple sites immediately upstream of the first (untranslated) exon [the erythroid or proximal, with respect to the AUG, promoter (7,8)], Sertoli cells of the testis also express *GATA1* but from a promoter (the testis or distal, with respect to the AUG promoter) located about 8 kilobases upstream of the proximal promoter (9). It has been suggested that this promoter is involved in the activation of HSC differentiation. To test this hypothesis, *GATA1* cDNA was amplified either with primers internal to the gene, or with primers specific for transcripts derived from the proximal or the distal promoters.

METHODS

Cell Purification

Murine stem cells were purified from the bone marrow of normal mice (*C57BL*, Jackson Laboratories, Bar Harbor, ME, USA) according to modifications of standard procedures (4) which include density separation on a discontinuous gradient of metrizamide (density $1.050 < \rho < 1.080$; Nycodenz, Gibco, Grand Island, NY, USA) and sorting of cells double labelled with wheat germ agglutinin-allophycocyanine (WGA-APC, Becton Dickinson, San Jose, CA, USA) and 15.1.1-fluorescein isothiocyanate (generously provided by Dr. J. Visser) on a FACStar Plus (Becton Dickinson). WGA^+ and $15.1.1^-$ cells were stained with Rho (Molecular Probes, Eugene, OR, USA) and Rho-bright and Rho-dull cell fractions separated by a second sorting. The morphology of the sorted cells was evaluated by light microscopy on cytocentrifuged smears. The cells were also injected into W/W^v mice to analyze their HSC content and cultured in semisolid medium to determine their colony-forming capacity. Aliquots were also lysed in mRNA buffer for RT-PCR analysis or induced to differentiate in liquid culture (see below).

Cell Culture

Rho-dull or Rho-bright cells (10^3 - 10^4 cells/ml) were cultured for up to 15 days in liquid culture under serum-deprived conditions as described elsewhere (5). The cultures were stimulated with the following growth factor combinations: SCF + IL-3, SCF + Epo and SCF + IL-3 + Epo or SCF + G-SCF and SCF + IL-3 + G-CSF. At sequential time points the cells from each well were recovered and the numbers of HSCs, progenitor cells, and differentiated cells determined. The number of progenitor cells was measured in duplicate 1 ml semisolid cultures each containing 10^3 sorted cells or their progeny obtained in liquid culture with combinations of growth factors shown in previous studies (5, 10) to be optimal for colony growth under serum-deprived culture conditions [SCF (100 ng/ml), IL-3 (100 U/ml), G-CSF (100 ng/ml), granulocyte/macrophage-colony-stimulating factor (GM-CSF, 100 U/ml), and Epo 15 ng/ml)]. Rat SCF and human G-CSF and Epo were a gift from Amgen (Thousand Oaks, CA, USA); murine IL-3 and GM-CSF were provided by Dr. K. Kaushansky, (University of Washington, Seattle, WA, USA). Plates were incubated at 37 °C in a fully humidified incubator with 5% CO_2 in air and scored at 7-9 days for the presence of erythroid bursts, GM colonies and mixed-cell colonies. Part of the sample was also lysed and RNA extracted.

Stem Cell Determination

The repopulation ability of purified HSCs was analyzed in a genetically stem cell defective repopulation assay, the W/W^v model (11). Since the host and the donor red cells and myeloid cells differ in the electrophoretic mobility of their hemoglobin (Hb), and of the enzyme glucose phosphate isomerase (GPI), erythroid and myeloid reconstitution is readily documented by analyzing the Hb and GPI types of the circulating cells. To measure the number of HSC in a given population, increasing numbers (10^2-10^4) of cells, either Rho-bright or Rho-dull, or the progeny generated by 10^2-10^4 Rho-dull cells after 5 days in culture, are injected into the tail vein of the mice (6-10 mice/group) and the Hb and the GPI expressed by cells circulating in the blood of the transplanted animals are analyzed periodically during the life of the animals (1.5-2 years). Analysis for the presence in the red cells of the different Hb isoforms (Hb^d or Hb^s) was performed according to the method of Whitney et al. (12).

The isoforms of GPI in myeloid and lymphoid cells were determined as described by Eicher and Washburn (13).

RNA Isolation and RT-PCR Analysis

10^4-10^6 cells (purified and cultured as described above) were lysed with 4 M guanidinium isothiocyanate/25 mM sodium citrate/0.5% Sarkosyl/0.1 M β-mercaptoethanol. The RNA was isolated by phenol/$CHCl_3$ extraction (14) in the presence of 20 μg 16S rRNA as carrier (Boehringer Mannheim, Indianapolis, IN, USA). RNA extracted from approximately 10^4 cells was reverse-transcribed in 100 μl by the M-MLV reverse-transcriptase (200 U/μl Gibco-BRL) with random hexamers (Promega, Madison, WI, USA) for 1 h at 37 °C, according to standard procedures. All the cDNA was subsequently amplified with volumes of retrotranscription reaction corresponding to approximately 10^3 cells. The PCR was performed in 100 μl of 10 mM Tris/50 mM KCl/1% Triton X-100/2.5 mM $MgCl_2$/0.2 mM dNTPs/0.25 μl Taq DNA polymerase (5 u/μl Promega)/500 ng of each primer with a Perkin Elmer (Norwalk, CT, USA) thermocycler (denaturing, 94 °C 1 min; annealing, 60 °C 1 min; primer extension, 72 °C 2 min). The primer pairs and the internal oligos were described by Schmitt et al. (15).

The *GATA1* cDNA was amplified either with primers within the coding region (15) or with primers specific for the *GATA1* proximal and distal transcripts. In these cases, cDNA was amplified in the presence of a common antisense *GATA1* oligonucleotide mapping to the second exon (nucleotides 118-139) and either a distal-specific (nucleotides 1-20) (9) or proximal-specific (nucleotides 1-22) (16) first exon sense primer.

The cDNAs from each time point were amplified up to 35 cycles. At specific cycles, 10 μl of each PCR reaction were withdrawn, the amplified band separated by electrophoresis on agarose gels (1.8%), transferred to zeta-probe GT membrane (Bio-Rad, Richmond, CA, USA) and hybridized with the internal oligodeoxynucleotides (15), labelled at the 5' end by ^{32}P-γ-ATP with polynucleotide kinase. In the case of proximal and distal *GATA1* PCR, the blots were hybridized to a cloned cDNA probe (nucleotides 1-184 of the distal transcript) comprising the whole distal first exon. Membranes were exposed to Kodak X-Omat AR film (Sigma, St. Louis, MO, USA) for 5-15 min at room temperature. All the procedures were carried out according to standard protocols (17). Each experiment routinely included two negative controls: one in which no reverse transcriptase was added in the RT reaction and another one in which no template was used in the PCR reaction (not shown).

RESULTS

Numbers of Precursor, Progenitor and Stem Cells in Sorted Rho-Bright and Rho-Dull Cells at Onset (Day 0) or after 5-10 Days of Culture

Rho-bright and Rho-dull cells had predominantly (> 70%) the morphology of small blasts with limited cytoplasm. They contained similar numbers of contaminating cells recognizable by morphological (20-30%) and clonogenic (15%) criteria. However, reconstitution of 20% of W/W^v animals was observed with as few as 250 Rho-dull cells and 10^4 Rho-bright cells. Rho-dull cells were, therefore, > 50 fold more enriched than Rho-bright for cells giving rise to permanent reconstitution of W/W^v animals.

In liquid culture, 10^3 purified cells, either Rho-bright or Rho-dull, gave rise to approximately 10^5 cells at day 5 and more than 10^6 cells at day 12 in all three growth factor combinations used. The number of HSC in Rho-dull cells cultured in SCF + IL-3 decreased

1000 fold by day 5 (< 1 HSC per original 10^3 Rho-dull cells). On the other hand, the number of early progenitor cells (both erythroid and myeloid) increased proportionately (from 150 to > 4000 per culture). CFU-E, which were below the level of detection (< 1 per 2.5 x 10^3 cells) in the purified cells, were generated in high numbers (too many to be counted) but only in the presence of SCF + IL-3 + Epo. Differentiated cells with a morphologically recognizable phenotype were usually observed for the first time at day 3 in the culture of Rho-bright and at day 5 in cultures of Rho-dull cells. In the presence of SCF + IL-3, Rho-bright and Rho-dull cells generated granulocytes of all types, macrophages and mast cells (which became the predominant population by day 15-21). Erythroid cells, although detected, represented a minority of all the cell types observed. In contrast, more than 80% of the cells obtained with SCF + IL-3 + Epo were erythroid while >99% of the cells obtained with SCF + IL-3 + G-CSF were either granulocytes of the neutrophil lineage (50-60%) or monocytes (40-50%).

RT-PCR Analysis of *GATA1*, *EpoR*, β-*Globin* and *Actin* Expression in Rho-Bright and Rho-Dull Cells at Onset (Day 0) or after 5-15 Days of Liquid Culture

Analysis by semiquantitative RT-PCR of the expression of *GATA1*, *EpoR*, and β-*globin* is presented in fig. 1, 2. In purified Rho-dull cells (fig. 1), *GATA1* was undetectable or, occasionally (one purified sample out of six analyzed, not shown) barely visible after prolonged exposure. *EpoR* was never detected in these cells (fig. 1). Purified Rho-bright cells, however, expressed almost maximal levels of *GATA1* but did not express detectable levels of β-*globin*. *EpoR* mRNA, not expressed in the analysis depicted in figure 1, was occasionally detectable, although at very low levels (see table 1). Cultured Rho-dull cells were induced to express *GATA1* and *EpoR* (not shown) at day 3-5 by all of the growth factor combinations investigated (fig. 2, table 1). Maximal levels of expression were reached by day 5 (*GATA1*/*actin* = 4.0 ± 1.3, Table 1). *GATA1* was expressed at equivalent levels in cultures stimulated with either SCF + IL-3 or SCF + IL-3 + Epo. SCF + IL-3 + G-CSF, however, induced approximately a quarter of maximal gene expression (*GATA1*/*actin* = 1.5 ± 1.0/1.0 ± 0.2). There was a significant linear correlation ($p < 0.01$) between the ratio of

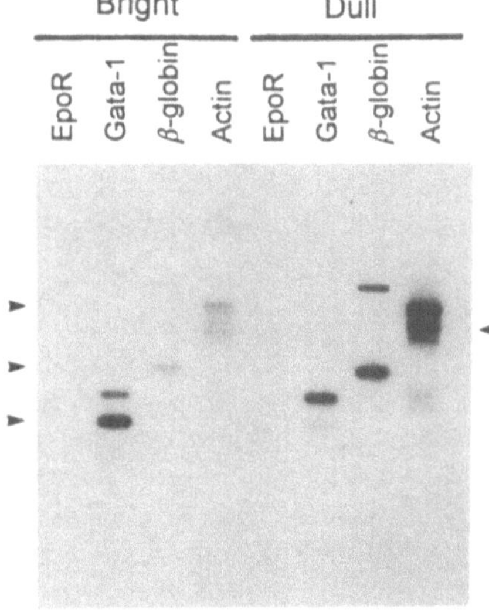

Figure 1. RT-PCR analysis of the expression of *EpoR*, *GATA1*, β-*globin* and *actin* (30 cycles each) in purified Rho-bright and Rho-dull cells. The size of the amplified bands presented corresponds to that predicted on the basis of the cDNA sequence (and is indicated by an arrow). The upper band observed in each lane corresponds to the amplification of genomic DNA which was contaminating the particular sample. Rho-bright cells at day 0 expressed detectable levels of *GATA1*. Normalization for *actin* expression indicates that these cells express maximal levels of *GATA1*. Rho-dull cells did not express significant levels of *GATA1*.

Table 1. Ratios between *GATA1*, *EpoR*, β-*globin* and *actin* cDNAs coamplified during the *in vitro* differentiation of Rho-bright and Rho-dull cells*

	Rho-bright cells			Rho-dull cells		
Days in culture	0	5	10	0	5	10
SCF + IL-3						
GATA1/actin	3.9 ± 2.1	4.0 ± 2.0	1.0 ± 0.6	0.2 ± 0.1	1.2 ± 0.7	0.6 ± 0.3
EpoR/actin	0.2 ± 0.1	0.6	0.07	0.04 ± 0.04	0.1	0.06
β-*Globin*/actin	n.d.**	0.14	0.08	n.d.	0.05	0.04
SCF + IL-3 + Epo						
GATA1/actin	3.9 ± 2.1	5.2 ± 2.1	4.1 ± 2.3	0.2 ± 0.1	4.0 ± 1.3	1.3 ± 0.3
*GATA1*t/e	0	1.5 ± 0.5	1.1 ± 0.4	0	0.8 ± 0.2	2.4 ± 0.8
EpoR/actin	0.2 ± 0.1	0.8	0.4	0.04 ± 0.04	1.2	0.3
β-*Globin*/actin	n.d.	4.5	2.7	n.d.	1.6	0.6
SCF + IL-3 + G-CSF						
GATA1/actin	3.9 ± 2.1	1.5 ± 1.0	2.1 ± 1.2	0.2 ± 0.1	1.0 ± 0.2	0.8 ± 0.5
EpoR/actin	0.2 ± 0.1	0.05	0.1	0.04 ± 0.04	0.2	0.09
β-*Globin*/actin	n.d.	0.1	0.06	n.d.	0.02	0.03

**GATA1/actin* cDNA ratios were calculated at 25-27 cycles. β-*Globin* and *EpoR/actin* cDNA ratios were evaluated at 15 and 30 cycles, respectively. The results represent the mean (± SD) of at least three independent amplifications. Where SD are not presented, the quantification was done only in one representative experiment.

**n.d.= Not detectable.

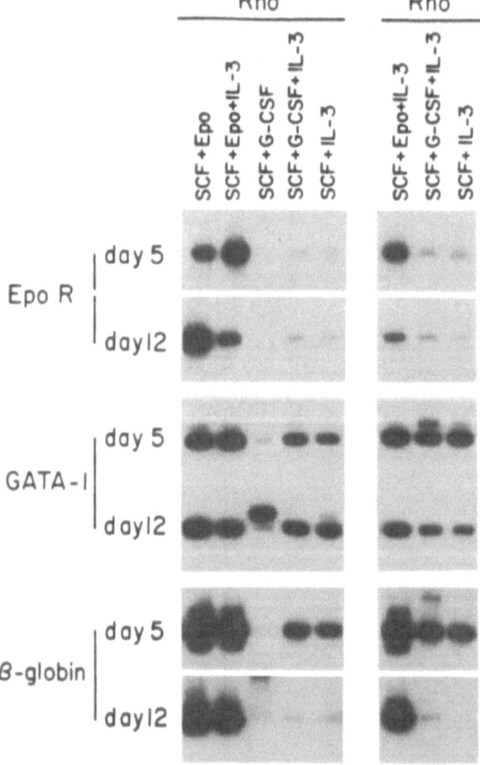

Figure 2. RT-PCR analysis of the expression of the *EpoR*, *GATA1* and β-*globin* (30 cycles each) in purified Rho-bright (Rho+) and Rho-dull (Rho-) cells after 5 and 12 days of liquid culture stimulated with multi-lineage (SCF + IL-3), erythroid (SCF + Epo or SCF + IL-3 + Epo) or myeloid (SCF + G-CSF or SCF + IL-3 + G-CSF) growth factor combinations. The size of the amplified bands presented corresponds to that predicted on the basis of the cDNA sequence. In the case of *GATA1*, an upper band is observed in some of the samples and corresponds to the genomic DNA. In culture, *GATA1* and *EpoR* expression were activated by all growth factor combinations (compare the levels of expression with those presented in figure 1). Activation of β-*globin* was observed only in the growth factor combinations which included Epo. It should be noted that the apparent expression of β-*globin* observed at 5 days in the presence of myeloid growth factor combinations is a technical artifact. In fact, the β-*globin* cDNA was amplified for 30 cycles. Therefore, the differences between amplifications from expressing and nonexpressing cells are underestimated.

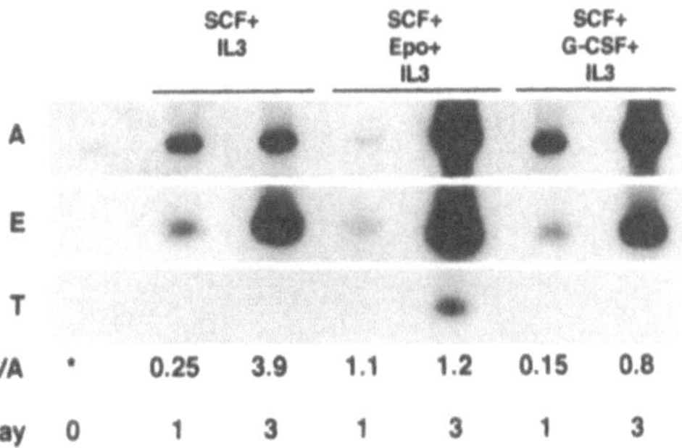

Figure 3. Expression of *GATA1* mRNA from the proximal and distal promoters in purified Rho-dull cells (day 0) or after 24 or 72h in culture stimulated with SCF + IL-3, SCF + IL-3 + Epo or SCF + IL-3 + G-CSF. Rho-dull did not express *GATA1* but its expression was induced after 72 h in culture by all growth factor combinations. The activation was from the proximal promoter (E) in all of the cases. In the presence of Epo, however, activation was both from the proximal (E) and from the distal (T) promoter.

GATA1/actin and *EpoR/actin* mRNA expressed by Rho-dull cells at each experimental point. In contrast, there was no correlation between the levels of *GATA1* and β-*globin* mRNAs which were expressed only in cultures stimulated with Epo (table 1, fig. 2).

Relative Expression of *GATA1* from the Proximal or the Distal Promoter during in Vitro-Induced Differentiation of Rho-Dull and Rho-Bright Cells

Figures 3 and 4 present the results of the specific RT-PCR, analyzing the promoter from which *GATA1* messages are transcribed during HSC differentiation. The transcription of *GATA1* was exclusively from the proximal promoter in the absence of Epo (see also table 1), while both proximal and distal transcripts were observed in the presence of Epo (fig. 3, 4). Both Rho-bright and Rho-dull cells began to express low levels of *GATA1* (approximately 1%, relative to total *GATA1*, table 1) from the distal promoter after day 3 of Epo stimulation (fig. 3 and data not shown); maximal levels of expression (2-3% of total *GATA1* mRNA) were found at day 5-10 in cultures of Rho-bright and Rho-dull cells, respectively (fig. 4 and table 1). There was no statistical correlation between the level of expression of *GATA1* from the distal promoter and the overall expression of the gene (see also table 1).

DISCUSSION

Murine HSC did not express *GATA1*. In contrast, their immediate progeny, either purified directly from the bone marrow (Rho-bright cells) or obtained *in vitro* (Rho-dull at day 3 of culture) expressed high levels of *GATA1* from the proximal promoter. Since Rho-bright cells are as capable as the Rho-dull cells of differentiating toward the myeloid and the lymphoid series (3), *GATA1* was activated *in vivo* during the transition from HSC to multilineage progenitor. In agreement with this notion, activation of *GATA1* in Rho-dull cells was detected very early in culture (by day 3), before loss of reconstitution potential was observed (by day 5).

Although Rho-bright and Rho-dull cells generated early erythroid progenitors (mixed-cell CFU and BFU-E) in cultures if stimulated with either erythroid or myeloid growth factor combinations (ref. 5 and data not shown), generation of CFU-E and erythroid precursors and expression of the globin genes (fig. 2, table 1) were observed only in the presence of growth factor combinations which included Epo. The fact that Rho-dull and Rho-bright cells expressed low (1-3%) but consistent levels of *GATA1* from the distal promoter after 3 days of Epo exposure, indicates that Epo specifically reorganized the control

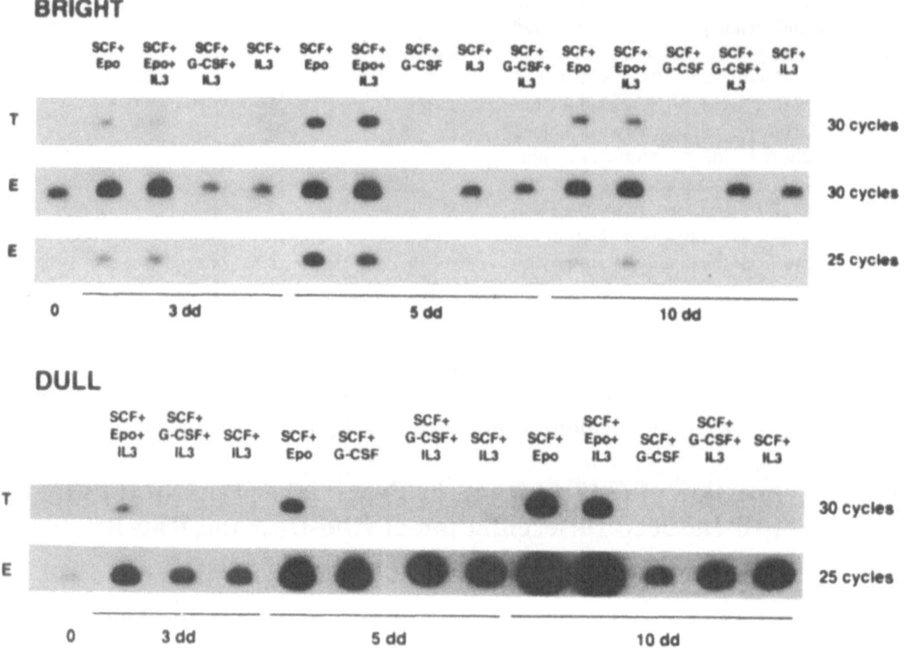

Figure 4. Expression of *GATA1* mRNA from the proximal (E) or the distal (T) promoter in purified normal murine HSCs at onset (day 0) or after 3, 5, and 10 days of liquid culture, stimulated with multilineage (SCF + IL-3), erythroid (SCF + Epo and SCF + IL-3 + Epo) or myeloid (SCF + G-CSF and SCF + IL-3 + G-CSF) growth factor combinations. Purified Rho-bright cells (top panel) expressed maximal levels of *GATA1* from the proximal promoter. Purified Rho-dull cells did not express significant levels of *GATA1* but were induced by all growth factor combinations to express *GATA1* from the proximal promoter by day 3 of culture. Both Rho-bright and Rho-dull cells in the presence of Epo activated the expression of *GATA1* from the distal promoter. This activation was first seen after 3 days of culture.

of *GATA1* expression by activating the distal promoter as part of the induction of the erythroid differentiation program.

In conclusion, our data indicate that both the distal and the proximal promoter are involved in the expression of *GATA1* in hematopoietic cells. *GATA1* activation from the proximal promoter is an early event in the process of stem cell differentiation, occurring at the transition point between HSC and multilineage progenitors. On the other hand, activation of the testis promoter is a late lineage-restricted event associated with cellular progression along the erythroid differentiation program and is specifically induced by Epo.

ACKNOWLEDGMENT

The authors thank Dr. John Adamson for support and encouragement and Mr. Harold Ralph for assisting with the sorting of the bone marrow cells. This study was supported by a Johnson & Johnson Focused Giving Grant, institutional funds of the Lindsley F. Kimball Research Institute of the New York Blood Center, Progetto Finalizzato CNR Ingegneria Genetica and Applicazioni Cliniche della Ricerca sul Cancro, by a grant Biotec from EEC and by Associazione Italiana per la Ricerca sul Cancro.

REFERENCES

1. Metcalf D. The molecular control of cell division, differentiation commitment and maturation in haemopoietic cells. Nature 339:27-30, 1989
2. Ogawa M, Porter PN, Nakahata T. Renewal and commitment to differentiation of hemopoietic stem cells (an interpretive review). Blood 61:823-829, 1983
3. Spangrude GJ, Johnson G. Resting and activated subsets of mouse multipotent hematopoietic stem cells. Proc. Natl. Acad. Sci. USA 87:7433-7437, 1990
4. Visser JWM, Bauman JGJ, Mulder AH, Eliason JF, de Leeuw AM. Isolation of murine pluripoent hemopoietic stem cells. J. Exp. Med. 59:1576-1590, 1984
5. Migliaccio G, Migliaccio AR, Valinsky J, Langley K, Zsebo K, Visser JW, Adamson JW. Stem cell factor (SCF) induces proliferation and differentiation of highly enriched murine hematopoietic cells. Proc. Natl. Acad. Sci. USA 88:7420-7424, 1991
6. Migliaccio G, Migliaccio AR, Druzin ML, Giardina PJ, Zsebo KM, Adamson JW. Long-term generation of colony-forming cells in liquid culture of CD34+ cord blood cells in the presence of recombinant human stem cell factor. Blood 79:2620-2627, 1992
7. Tsai SF, Strauss E, Orkin SH. Functional analysis and in vivo footprinting implicate the erythroid transcription factor GATA-1 as a positive regulator of its own promoter. Genes Dev. 5:919-931, 1991
8. Nicolis S, Bertini C, Ronchi A, Crotta S, Lanfranco L, Moroni E, Giglioni B, Ottolenghi S. An erythroid specific enhancer upstream to the gene encoding the cell-type specific transcription factor GATA-1. Nucleic Acids Res. 19:5285-5291, 1991
9. Ito E, Toki T, Ishihara H, Ohtani H, Gu L, Yokoyama M, Engel JD, Yamamoto M. Erythroid transcription factor GATA-1 is abundantly transcribed in mouse testis. Nature, 362:466-468, 1993
10. Migliaccio G, Migliaccio AR, Visser JWM. Synergism between erythropoietin and interleukin-3 in the induction of hemopoietic stem cell proliferation and erythroid differentiation. Blood 72:944-951, 1988
11. Barker JE, Braum J, McFarland-Starr EC: Erythrocyte replacement precedes leukocyte replacement during repopulation of W/W^v mice with limiting dilutions of +/+ donor marrow cells. Proc. Natl. Acad. Sci. USA 85:7332-7335, 1988.
12. Whitney JB. Simplified typing of mouse hemoglobin (Hbb) phenotypes using cytamine. Biochem. Genet. 16:667-672, 1978
13. Eicher EM, Washburn LL. Assignment of genes to regions of mouse chromosomes. Proc. Natl. Acad. Sci. USA 75:946-950, 1978
14. Chomczynski P, Sacchi N. Single-step method of RNA isolation by acid guanidinium thiocyanate-phenol-chloroform extraction. Anal. Biochem. 162:156-159, 1987
15. Schmitt RM, Bruyns E, Snodgrass HR. Hematopoietic development of embryonic stem cells in vitro: cytokine and receptor gene expression. Genes Dev. 5:728-740, 1991
16. Tsai SF, Martin DI, Zon LI, D'Andrea AD, Wong GG, Orkin SH. Cloning of cDNA for the major DNA-binding protein of the erythroid lineage through expression in mammalian cells. Nature, 339:446-451, 1989
17. Sambrook J, Fritsch EF, Maniatis T. (eds.) Molecular Cloning: A Laboratory Manual. Cold Spring Harbor Laboratory Press, Cold Spring Harbor, 1989

47

PROTO-ONCOGENE PRODUCTS Vav AND c-Cbl ARE INVOLVED IN THE SIGNAL TRANSDUCTION THROUGH GRB2/ASH IN HEMATOPOIETIC CELLS

Yutaka Hanazono,[1] Hideharu Odai,[1] Ko Sasaki,[1] Akihiro Iwamatsu,[2]
Yoshio Yazaki,[1] and Hisamaru Hirai[1*]

[1] Third Department of Internal Medicine, Faculty of Medicine
University of Tokyo
Hongo 7-3-1, Bunkyo-ku, Tokyo 113, Japan
[2] Kirin Brewery Co. Ltd., Central Laboratory for Key Technology
Fukuura 1-13-5, Kanazawa-ku, Yokohama, Kanagawa 236, Japan

SUMMARY

Grb2/Ash is composed of one SH2 and two SH3 domains and functions as an adapter linking tyrosine-kinase receptors and Ras in fibroblasts. We have investigated the nature of proteins interacting with Grb2/Ash in hematopoietic cells.

The product of the *vav* proto-oncogene (Vav) is exclusively expressed in hematopoietic cells and has the guanine nucleotide exchange activity. Here we report that granulocyte-macrophage colony-stimulating factor (GM-CSF), interleukin-3 (IL-3), and erythropoietin (EPO) induce rapid and transient tyrosine phosphorylation of Vav and that Vav is constitutively associated with the SH3 domain of Grb2/Ash in a human leukemia cell line UT-7. These data suggest that Vav is implicated in a signaling pathway leading to activation of Ras or Ras-related proteins in hematopoietic cells.

Furthermore, we have shown that the proto-oncogene c-*cbl* product (c-Cbl) is also tyrosine-phosphorylated by stimulation with GM-CSF or EPO and is constitutively associated with the SH3 domain of Grb2/Ash in UT-7. However, we could not find the homologous regions with guanine nucleotide exchange factors or GTPase-activating proteins in the c-*cbl* gene. Therefore, Grb2/Ash might also transduce a signal that is different from the signal leading to the small-G protein regulation.

[*] Correspondence should be addressed to: Hisamaru Hirai, M.D. Ph.D., Third Department of Internal Medicine, Faculty of Medicine, University of Tokyo, Hongo 7-3-1, Bunkyo-ku, Tokyo 113, Japan. Tel: 81-3-3815-5411 ext. 3116; Fax: 81-3-3815-8350.

Molecular Biology of Hematopoiesis 5, edited by Abraham et al.
Plenum Press, New York, 1996

INTRODUCTION

The Grb2/Ash protein is a 27-kDa protein composed of SH2 and SH3 domains in the order of SH3-SH2-SH3.[1,2] The SH2 domain of Grb2/Ash binds to tyrosine-phosphorylated proteins such as the epidermal growth factor (EGF) receptor,[3,4] Shc,[5] insulin receptor substrate-1 (IRS-1),[6] and Syp.[7] On the other hand, the SH3 domain of Grb2/Ash binds to the proline-rich domains of Sos,[3, 8] dynamin,[9,10] and C3G[11] which regulate Ras or Ras-related proteins. As the results of these association, the Grb2/Ash seems to be involved in coupling tyrosine kinases to the Ras regulators. Hematopoietic growth factors induce tyrosine phosphorylation of Shc and form the complex of Shc and Grb2/Ash, and this complex has a key role in Ras activation which is critical for proliferation.[12-16] These lines of evidence motivated us to search for signaling molecules which interact with Grb2/Ash and participate in the signal transduction of hematopoietic growth factors. We here report that two proto-oncogene products, Vav and c-Cbl, constitutively bind to the SH3 domain of Grb2 in hematopoietic cells and that they are tyrosine-phosphorylated by stimulation with granulocyte-macrophage colony-stimulating factor (GM-CSF), interleukin-3 (IL-3), or erythropoietin (EPO).

MATERIALS AND METHODS

Cell Lines and Growth Factors

UT-7 cells were maintained in RPMI 1640 medium containing 8% bovine serum and 10 ng/ml GM-CSF. Recombinant human GM-CSF and recombinant human IL-3 were supplied by Kirin Brewery Co. Ltd. (Tokyo, Japan) and recombinant human EPO was delivered by Chugai Pharmacy Co. Ltd. (Tokyo, Japan).

Antibodies

Rabbit polyclonal antibody to Grb2/Ash (c-23) purchased from Santa Cruz Biotechnology (Santa Cruz, CA) and monoclonal antibody to Grb2/Ash purchased from MBL (Nagoya, Japan) were used for immunoprecipitation and immunoblotting of Grb2/Ash, respectively. Polyclonal anti-Vav antibody was prepared from serum of a rabbit immunized against a synthetic peptide. The peptide sequence was KKDKLHRRAQDKKRNELGLP corresponding to the downstream of the nuclear localization signal sequence.[17] This antibody was a kind gift from Dr. Toshihide Mimura (University of Tokyo, Tokyo, Japan) and used for immunoprecipitation of Vav. Monoclonal antibody to Vav which was purchased from Upstate Biotechnology (Lake Placid, NY) was used for immunoblotting of Vav. Rabbit polyclonal anti-c-Cbl antibody purchased from Santa Cruz Biotechnology (Santa Cruz, CA) was used for immunoprecipitation and immunoblotting of c-Cbl. A mouse monoclonal anti-phosphotyrosine antibody (anti-Ptyr) 4G10 was used for immunoblotting of phosphotyrosine-containing proteins. 4G10 was generously provided by Dr. Deborah K Morrison (National Cancer Institute, Frederick, MD).

GST Fusion Proteins

The bacterial expression plasmids coding to the GST fusion proteins containing the full-length, the N-terminal SH3 domain, or the SH2 domain of Grb2/Ash[10] were generously provided by Dr. Tadaomi Takenawa (University of Tokyo, Tokyo, Japan). These plasmids were transformed into XL I-Blue strain of *Escherichia coli* and the resulting transformants were induced with isopropyl-1-thio-β-D-galactopyranoside to produce GST fusion proteins. The bacteria were collected by centrifugation and resuspended in the *E.coli* lysis buffer containing 40 mM Tris/HCl (pH 7.5), 5 mM ethylenediaminetetraacetic acid (EDTA), 0.1 mM phenylmethylsulfonyl fluoride (PMSF), and 1% Triton X-100. Vigorous sonication was performed before centrifugation at 25,000 x *g* for 20 min. The resulting supernatants were saved as crude extracts containing GST fusion proteins.

Preparation of Cell Lysates

UT-7 cells were incubated in RPMI 1640 medium containing 0.1% bovine serum albumin without serum or growth factors for 8-15 hr prior to stimulation with growth factors and then resuspended in RPMI 1640 medium containing 100 mM Na_3VO_4. The cells were treated with 10 ng/ml GM-CSF or 20 units/ml EPO for 5 min at 37°C and then lysed at 4°C in the lysis buffer containing 20 mM Tris/HCl (pH 8.0), 1% Nonidet P-40 (NP-40), 1 mM PMSF, 500 units/ml aprotinin, 2 mM EDTA, 50 mM NaF, and 1 mM Na_3VO_4. Unsolubilized materials were removed by centrifugation at 15,000 x *g* at 4°C for 10 min.

Binding of Cellular Proteins to GST Fusion Proteins

Lysates from $1x10^7$ cells were mixed with 40 mg of the fusion protein noncovalently coupled to glutathione-agarose beads (Sigma, St. Louis, MO) for 3 hr at 4°C. Beads were washed with the lysis buffer before resuspension in Laemmli's sample buffer.

Immunoprecipitation

For immunoprecipitation, lysates from $1x10^7$ cells were mixed with the specific antibody for 3 hr at 4°C. The immunoprecipitates were collected with protein A-Sepharose (Sigma, St. Louis, MO). All the immunoprecipitates were intensively washed with the lysis buffer before resuspension in Laemmli's sample buffer.

Immunoblotting

Samples were subjected to SDS-PAGE and electrotransferred onto polyvinylidene difluoride (PVDF) filters (Millipore, Bedford, MA). Filters were blocked with the buffer containing 10 mM Tris/HCl (pH 8.0), 150 mM NaCl, 10% skimmed milk, and 0.05% Triton X-100. For immunoblotting, filters were incubated with the specific antibody, and then with the alkaline phosphatase-conjugated antibody (Promega, Madison, WI). After each incubation, filters were washed three times in the buffer containing 10 mM Tris/HCl (pH 8.0), 150 mM NaCl, and 0.05% Triton X-100. Color reaction was performed using nitro blue tetrazolium (NBT) and 5-bromo-4-chloro-3-indolyl-phosphate (BCIP) (Promega, Madison, WI) according to the manufacturer's protocol.

RESULTS

Tyrosine Phosphorylation of Vav

UT-7 is a human leukemia cell line which is dependent for growth on GM-CSF, IL-3, or EPO.[18] We have used this cell line in subsequent experiments. Vav was precipitated from UT-7 cells with the specific antibody and blotted with the anti-phosphotyrosine antibody. Vav was shown to be tyrosine-phosphorylated by stimulation with GM-CSF,

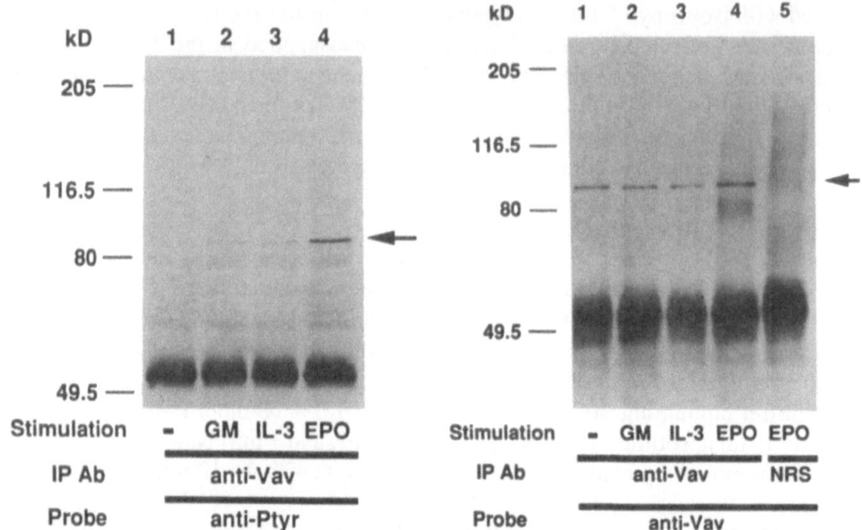

Figure 1. GM-CSF, IL-3, and EPO induce tyrosine phosphorylation of Vav in UT-7 cells. The lysates from 1 x 10^7 UT-7 cells unstimulated (lane 1) or stimulated with GM-CSF (lane 2), IL-3 (lane 3), or EPO (lanes 4 and 5) were mixed with the polyclonal anti-Vav antibody (lanes 1-4) or with normal rabbit serum (NRS, lane 5). The immunoprecipitates were collected with protein A-Sepharose, subjected to 7% SDS-PAGE, and immunoblotted with the anti-phosphotyrosine antibody 4G10 (A) and with the monoclonal anti-Vav antibody (B). The arrow indicates the position of Vav.

IL-3, or EPO (Fig. 1A). The amounts of Vav were not affected by the stimulation (Fig. 1B). Tyrosine phosphorylation of Vav is rapid and transient. Vav was phosphorylated within 1 min after the stimulation and dephosphorylated within 60 min (data not shown). Tyrosine phosphorylation of Vav was observed at the physiological concentrations of GM-CSF (data not shown).

Association of Vav with Grb2/Ash

Fig. 2 shows the association of Vav with Grb2/Ash. Grb2/Ash was immunoprecipitated from UT-7 cells and blotted with the anti-Vav antibody. Vav was revealed to be co-precipitated with Grb2/Ash regardless of the stimulation (Fig. 2B). Conversely, Vav was immunoprecipitated from UT-7 cells and blotted with the anti-Grb2/Ash antibody. Grb2/Ash was shown to be co-precipitated with Vav regardless of the stimulation (Fig. 2A). Thus, Vav is constitutively associated with Grb2/Ash in UT-7 cells. To determine which domain of Grb2/Ash binds to Vav, we have prepared the SH2 or SH3 domain of Grb2/Ash expressed as the GST fusion proteins. The lysates from UT-7 cells were mixed with the GST fusion proteins and the resulting precipitates were immunoblotted with the anti-Vav antibody. It was shown that Vav bound to the GST-SH3, but not to the GST-SH2 or to the GST alone (Fig. 3). The bands that migrated faster than Vav in Fig. 3 are non-specific. Thus, Vav binds to the SH3 domain of Grb2/Ash.

Tyrosine Phosphorylation of c-Cbl

c-Cbl was immunoprecipitated from UT-7 cells and blotted with the anti-phosphotyrosine antibody. c-Cbl was shown to be tyrosine-phosphorylated by stimulation with EPO

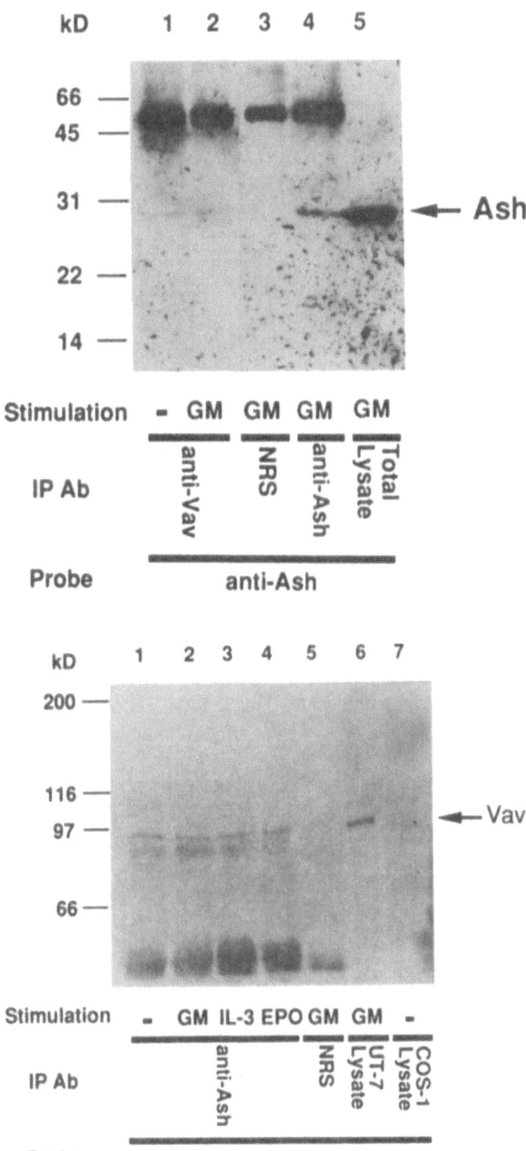

Figure 2. (A) Anti-Vav immunoprecipitates contain Grb2/Ash in UT-7. The lysates from 1 x 10^7 UT-7 cells unstimulated (lane 1) or stimulated with GM-CSF (lanes 2-4) were mixed with the polyclonal anti-Vav antibody (lanes 1 and 2), normal rabbit serum (NRS, lane 3), or the polyclonal anti-Grb2/Ash antibody (lane 4). The immunoprecipitates were collected with protein A-Sepharose, subjected to 12% SDS-PAGE, and immunoblotted with the monoclonal anti-Grb2/Ash antibody. The whole lysate (100 μg protein per lane) from UT-7 cells was also applied for reference (lane 5). The arrow indicates the position of Grb2/Ash. (B) Anti-Grb2/Ash immunoprecipitates contain Vav in UT-7. The lysates from 1 x 10^7 UT-7 cells unstimulated (lane 1) or stimulated with GM-CSF (lanes 2 and 5), IL-3 (lane 3), or EPO (lane 4) were mixed with the polyclonal anti-Grb2/Ash antibody (lanes 1-4) or with normal rabbit serum (NRS, lane 5). The immunoprecipitates were collected with protein A-Sepharose, subjected to 7% SDS-PAGE, and immunoblotted with the monoclonal anti-Vav antibody. The whole lysates (100 μg protein per lane) from UT-7 cells (lane 6) and COS-1 cells (lane 7) were also applied for reference. COS-1 cells do not express Vav and they were used for the negative control. The arrow indicates the position of Vav.

or GM-CSF (Fig. 4A). The amounts of c-Cbl were not affected by the stimulation (Fig. 4B). Tyrosine phosphorylation of c-Cbl is rapid and transient. c-Cbl was phosphorylated within 2 min after the stimulation and dephosphorylated within 30 min (data not shown).

Association of c-Cbl with Grb2/Ash

Fig. 5 shows the co-precipitation of c-Cbl with Grb2/Ash. c-Cbl was immunoprecipitated from UT-7 cells and blotted with the anti-Grb2/Ash antibody. Grb2/Ash was

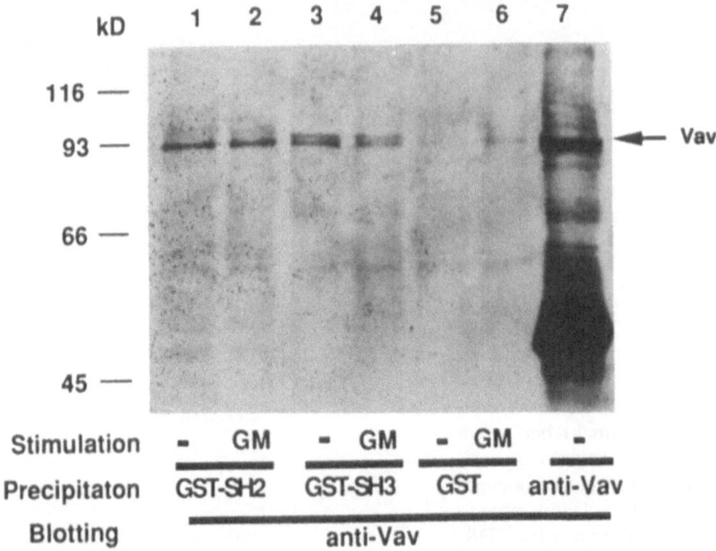

Figure 3. Vav binds to the SH3 domain of Grb2/Ash. The lysates from UT-7 cells unstimulated (lanes 1, 3, and 5) or stimulated with GM-CSF (lanes 2, 4, and 6) were mixed with the GST fusion protein containing the SH2 domain (lanes 1 and 2) or the SH3 domain (lanes 3 and 4) of Grb2/Ash, or the GST alone (lanes 5 and 6). The resulting precipitates and the anti-Vav immunoprecipitate (lane 7) were subjected to SDS-PAGE and immunoblotted with the anti-Vav antibody. The arrow indicates the position of Vav.

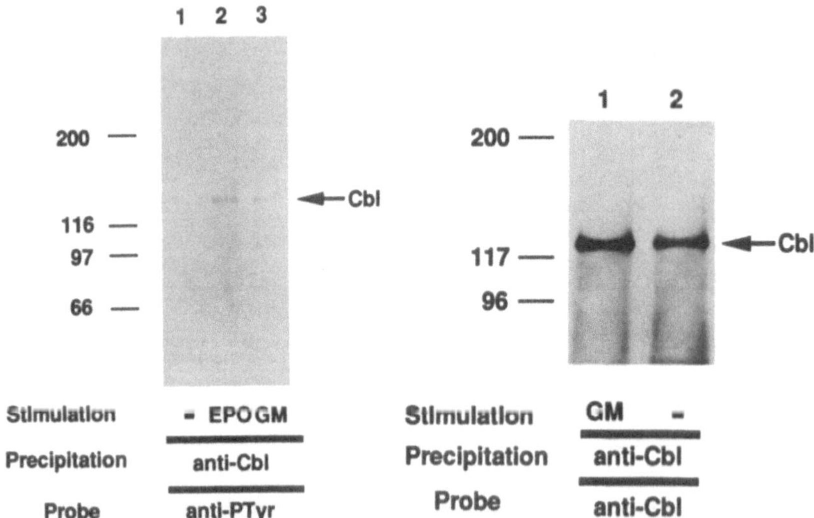

Figure 4. (A) GM-CSF and Epo induce tyrosine phosphorylation of c-Cbl. The lysates from UT-7 cells unstimulated (lane 1) or stimulated with Epo (lane 2) or GM-CSF (lane 3) were mixed with the anti-c-Cbl antibody and the immunoprecipitates were subjected to SDS-PAGE and immunoblotted with anti-Ptyr (4G10). Molecular weight markers, indicated at the left, are given in kDa. The arrow indicates the position of c-Cbl. (B) The amounts of c-Cbl are not affected by the stimulation. The lysates from UT-7 cells stimulated with GM-CSF (lane 1) or unstimulated (lane 2) were mixed with the anti-c-Cbl antibody and the precipitates were subjected to SDS-PAGE and immunoblotted with the anti-c-Cbl antibody. Molecular weight markers, indicated at the left, are given in kDa. The arrow indicates the position of c- Cbl.

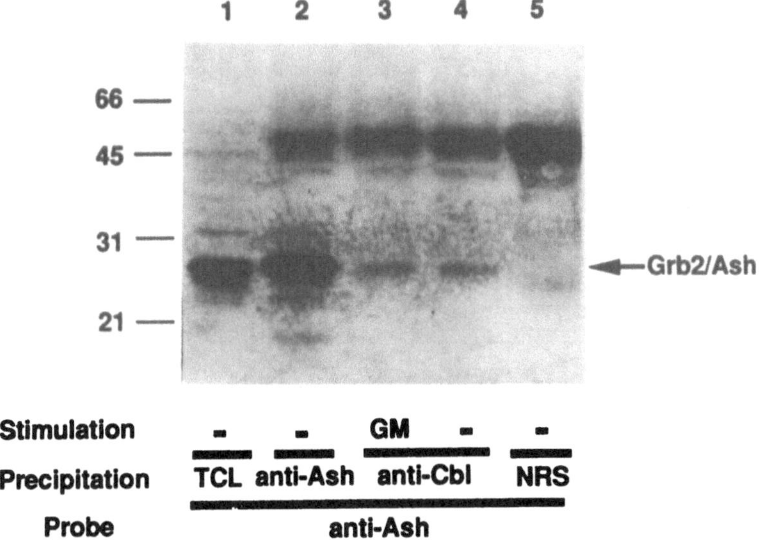

Figure 5. The association of c-Cbl with Grb2/Ash. The lysates from UT-7 cells stimulated with GM-CSF (lane 3) or unstimulated (lanes 2, 4, and 5) were mixed with the anti-Grb2/Ash antibody (lane 2), the anti-c-Cbl antibody (lanes 3 and 4), or normal rabbit serum (NRS, lane 5). The resulting precipitates were subjected to SDS-PAGE and immunoblotted with the anti-Grb2/Ash antibody. The total cell lysate of UT-7 (TCL, lane 1) was also applied for reference. Molecular weight markers, indicated at the left, are given in kDa.

revealed to be co-precipitated with c-Cbl regardless of the stimulation. Fig. 6 shows that c-Cbl binds to the SH3 domain of Grb2/Ash. The lysates from UT-7 cells were mixed with the GST fusion proteins containing the SH2 or SH3 domain of Grb2/Ash, and the resulting precipitates were immunoblotted with the anti-c-Cbl antibody. It was shown that c-Cbl only bound to the GST-SH3, but not to the GST-SH2 or to the GST alone.

DISCUSSION

Vav is the proto-oncogene product at the molecular weight of 95 kDa in humans.[17] Vav is exclusively expressed in hematopoietic cells. The oncogenic protein loses the N-terminal region. Vav has one SH2 and two SH3 domains, but no kinase activity.[19, 20] Vav has a Dbl/Cdc24-like sequence[21] and is, in fact, reported to have the guanine nucleotide exchange activity.[22,23] We have shown that GM-CSF, IL-3, and EPO induce rapid and transient tyrosine phosphorylation of Vav in a human leukemia cell line UT-7. Furthermore, we have demonstrated that Vav is constitutively associated with the SH3 domain of Grb2/Ash in UT-7. These data suggest that tyrosine kinases, the adapter Grb2/Ash, and the guanine nucleotide exchange factor Vav are members of a signaling pathway leading to Ras activation in hematopoietic cells. Another guanine nucleotide exchange factor Sos is also associated with Grb2/Ash in UT-7 (data not shown). The relative contribution of Sos and Vav to the receptor signaling to Ras in the cells remains to be investigated.

The c-*cbl* gene was cloned as the cellular homolog of the v-*cbl* oncogene which is the transforming component of a murine tumorigenic retrovirus CAS NS-1.[24] *gag*-v-*cbl* lacks the proline-rich domain and the leucine zipper but still possesses the basic region. c-Cbl has no kinase activity or no Src-homology.[25] The biological roles of c-Cbl are unknown for the

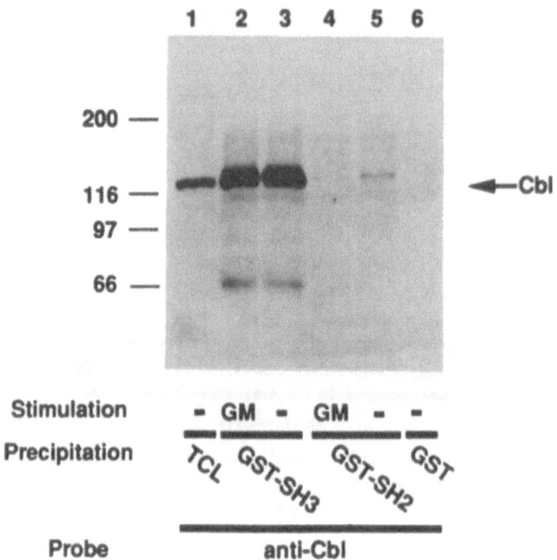

Figure 6. The association of c-Cbl with the SH3 domain of Grb2/Ash. The lysates from UT-7 cells unstimulated (lanes 3, 5, and 6) or stimulated with GM-CSF (lanes 2 and 4) were mixed with the GST fusion protein containing the SH3 domain (lanes 2 and 3) or the SH2 domain (lanes 4 and 5) of Grb2/Ash, or the GST alone (lane 6). The resulting precipitates were subjected to SDS-PAGE and immunoblotted with the anti-c-Cbl antibody. The total cell lysate of UT-7 (TCL, lane 1) was also applied for reference. Molecular weight markers, indicated at the left, are given in kDa.

present. We have shown that c-Cbl is also tyrosine-phosphorylated by stimulation with GM-CSF or EPO and is constitutively associated with the SH3 domain of Grb2/Ash in UT-7. However, we could not find any homologous regions with guanine nucleotide exchange factors or GTPase-activating proteins in the c-*cbl* gene. Therefore, Grb2/Ash might also transduce a signal that is different from the signal leading to Ras regulation.

Several proteins bind to the SH3 domain of Grb2/Ash including Sos, Vav, and c-Cbl in hematopoietic cells. The binding sites of Sos to Grb2/Ash are the proline-rich domains.[3,8] The c-Cbl has a proline-rich domain[25] and we could find in this domain two sequences which resemble the proposed consensus sequence for the Abl SH3-binding site[26] and the Grb2/Ash-binding sites on mouse Sos1.[27] Therefore, it is considered that c-Cbl binds to the SH3 domain of Grb2/Ash through these two proline-rich sequences of c-Cbl. On the other hand, other investigators have reported that the SH3 domain of Vav binds to the SH3 domain of Grb2/Ash.[28] This is unusual protein-protein interaction that is through dimerization of the SH3 domains of each protein.

Vav and c-Cbl are unique because they are phosphorylated on tyrosine residues, although other SH3-binding proteins including Sos, C3G, and dynamin have not been reported to be tyrosine-phosphorylated. It is possible that Vav and c-Cbl binds to SH2-containing proteins by GM-CSF or EPO stimulation. We are now identifying these proteins

ACKNOWLEDGMENT

This work was in part supported by Grants-in-Aids from the Ministry of Education, Science and Culture of Japan, and from the Ministry of Health and Welfare of Japan.

REFERENCES

1. Matuoka K, Shibata M, Yamakawa A, Takenawa T: Cloning of ASH, a ubiquitous protein composed of one Src homology (SH) 2 and two SH3 domains, from human and rat cDNA libraries. Proc Natl Acad Sci USA 89:9015, 1992
2. Suen KL, Bustelo XR, Pawson T, Barbacid M: Molecular cloning of the mouse *grb2* gene: differential interaction of the Grb2 adaptor protein with epidermal growth factor and nerve growth factor receptors. Mol Cell Biol 13:5500, 1993
3. Buday L, Downward J: Epidermal growth factor regulates $p21^{ras}$ through the formation of a complex of receptor, Grb2 adapter protein, and Sos nucleotide exchange factor. Cell 73:611, 1993
4. Rozakis-Adcock M, Fernley R, Wade J, Pawson T, Bowtell D: The SH2 and SH3 domains of mammalian Grb2 couple the EGF receptor to the Ras activator mSos1. Nature 363:83, 1993
5. Rozakis-Adcock M, McGlade J, Mbamalu G, Pelicci G, Daly R, Li W, Batzer A, Thomas S, Brugge J, Pelicci PG, Schlessinger J, Pawson T: Association of the Shc and Grb2/Sem5 SH2-containing proteins is implicated in activation of the Ras pathway by tyrosine kinases. Nature 360:689, 1992
6. Skolnik EY, Lee CH, Batzer A, Vicentini LM, Zhou M, Daly R, Myers Jr MJ, Backer JM, Ullrich A, White MF, Schlessinger J: The SH2/SH3 domain-containing protein GRB2 interacts with tyrosine-phosphorylated IRS1 and Shc: implications for insulin control of *ras* signalling. EMBO J 12:1929, 1993
7. Li W, Nishimura R, Kashishian A, Batzer AG, Kim WJH, Cooper JA, Schlessinger J: A new function for a phosphotyrosine phosphatase: linking GRB2-Sos to a receptor tyrosine kinase. Mol Cell Biol 14:509, 1994
8. Chardin P, Camonis JH, Gale NW, van Aelst L, Schlessinger J, Wigler MH, Bar SD: Human Sos1: a guanine nucleotide exchange factor for Ras that binds to GRB2. Science 260:1338, 1993
9. Gout I, Dhand R, Hiles ID, Fry MJ, Panayotou G, Das P, Truong O, Totty NF, Hsuan J, Booker GW, Campbell ID, Waterfield MD: The GTPase dynamin binds to and is activated by a subset of SH3 domains. Cell 75:25, 1993
10. Miki H, Miura K, Matuoka K, Nakata T, Hirokawa N, Orita S, Kaibuchi K, Takai Y, Takenawa T: Association of Ash/Grb-2 with dynamin through the Src homology 3 domain. J Biol Chem 269:5489, 1994
11. Tanaka S, Morishita T, Hashimoto Y, Hattori S, Nakamura S, Shibuya M, Matsuoka K, Takenawa T, Kurata T, Nagashima K, Matsuda M: C3G, a guanine nucleotide-releasing protein expressed ubiquitously, binds to the Src homology 3 domains of CRK and GRB2/ASH proteins. Proc Natl Acad Sci USA 91:3443, 1994
12. Cutler RL, Liu L, Damen JE, Krystal G: Multiple cytokines induce the tyrosine phosphorylation of Shc and its association with Grb2 in hemopoietic cells. J Biol Chem 268:21463, 1993
13. Damen JE, Liu L, Cutler RL, Krystal G: Erythropoietin stimulates the tyrosine phosphorylation of Shc and its association with Grb2 and a 145-Kd tyrosine phosphorylated protein. Blood 82:2296, 1993
14. Sato N, Sakamaki K, Terada N, Arai K, Miyajima A: Signal transduction by the high-affinity GM-CSF receptor: two distinct cytoplasmic regions of the common β subunit responsible for different signaling. EMBO J 12:4181, 1993
15. Lioubin MN, Myles GM, Carlberg K, Bowtell D, Rohrschneider LR: Shc, Grb2, Sos1, and a 150-kilodalton tyrosine-phosphorylated protein form complexes with Fms in hematopoietic cells. Mol Cell Biol 14:5682, 1994
16. Welham MJ, Duronio V, Leslie KB, Bowtell D, Schrader JW: Multiple hematopoietins, with the exception of interleukin-4, induce modification of Shc and mSos1, but not their translocation. J Biol Chem 269:21165, 1994
17. Katzav S, Martin-Zanca D, Barbacid M: *vav*, a novel human oncogene derived from a locus ubiquitously expressed in hematopoietic cells. EMBO J 8:2283, 1989
18. Komatsu N, Nakauchi H, Miwa A, Ishihara T, Eguchi M, Moroi M, Okada M, Sato Y, Wada H, Yawata Y, Suda T, Miura Y: Establishment and characterization of a human leukemic cell line with megakaryocytic features: dependency on granulocyte-macrophage colony-stimulating factor, interleukin 3, or erythropoietin for growth and survival. Cancer Res 51:341, 1991

19. Bustelo XR, Ledbetter JA, Barbacid M: Product of *vav* proto-oncogene defines a new class of tyrosine protein kinase substrates. Nature 356:68, 1992
20. Margolis B, Hu P, Katzav S, Li W, Oliver JM, Ullrich A, Weiss A, Schlessinger J: Tyrosine phosphorylation of *vav* proto-oncogene product containing SH2 domain and transcription factor motifs. Nature 356:71, 1992
21. Boguski MS, Bairoch A, Attwood A, Michaels GS: Proto-*vav* and gene expression. Nature 358:113, 1993
22. Gulbins E, Coggeshall KM, Baier G, Katzav S, Burn P, Altman A: Tyrosine kinase-stimulated guanine nucleotide exchange activity of Vav in T cell activation. Science 260:822, 1993
23. Gulbins E, Coggeshall KM, Langlet C, Baier G, Bonnefoy-Berard N, Burn P, Wittinghofer A, Katzav S, Altman A: Activation of Ras in vitro and in intact fibroblasts by the Vav guanine nucleotide exchange protein. Mol Cell Biol 14:906, 1994
24. Langdon WY, Hartley JW, Klinken SP, Ruscetti SK, Morse HC III: v-*cbl*, an oncogene from a dual-recombinant murine retrovirus that induces early B-lineage lymphomas. Proc Natl Acad Sci USA 86:1168, 1989
25. Blake TJ, Shapiro M, Morse HC III, Langdon WY: The sequences of the human and mouse c-*cbl* proto-oncogenes show v-*cbl* was generated by a large truncation encompassing a proline-rich domain and a leucine zipper-like motif. Oncogene 6:653, 1991
26. Ren R, Mayer BJ, Cicchetti P, Baltimore D: Identification of a ten-amino acid proline-rich SH3 binding site. Science 259:1157, 1993
27. Olivier JP, Raabe T, Henkemeyer M, Dickson B, Mbamalu G, Margolis B, Schlessinger J, Hafen E, Pawson T: A Drosophila SH2-SH3 adaptor protein implicated in coupling the sevenless tyrosine kinase to an activator of Ras guanine nucleotide exchange, Sos. Cell 73:179, 1993
28. Ye ZS, Baltimore D: Binding of Vav to Grb2 through dimerization of Src homology 3 domains. Proc Natl Acad Sci USA 91:12629, 1994

48

IL3BP1, A TRANSCRIPTION FACTOR WITH DUAL FUNCTION

Wei Zhang,[1] A. Thomas Look,[2] Toshiya Inaba,[2] and Stephen D. Nimer[1*]

[1] Laboratory of Molecular Aspects of Hematopoiesis
Division of Hematologic Oncology
Memorial Sloan-Kettering Cancer Center
New York, New York
[2] Department of Experimental Oncology
St. Jude Children's Research Hospital
Memphis, Tennessee

Transcriptional regulation of eukaryotic gene expression involves interactions between transcription factor proteins and specific DNA sequences in the enhancer/promoter regions of regulated genes. These transcription factors can be activators, which can act by increasing transcription initiation or enhancing mRNA elongation (1), or they can serve as transcriptional suppressors and down regulate transcription (2,3). There are many transcription factors that can function either as activators or suppressors, depending on the context. For example, the Drosophila transcription factor Kruppel functions as an activator at low concentration, by interacting with TFIIB as a monomer, but it functions as a suppressor at high concentration, forming a dimer and interacting with TFIIEβ(4). The human thyroid hormone receptor-β (hTRβ) converts from a repressor to an activator upon binding of thyroid hormone, and YY1 suppresses transcription when bound to the initiation region of the promoter but converts to an activator when bound by the adenovirus E1a protein (5).

We recently cloned a transcription factor (NF-IL3A) based on its ability to bind to regulatory sequences in the human interleukin-3 promoter (6). The nearly identical transcription factor E4BP4 was cloned based upon its ability to bind to the adenovirus E4 promoter region (7). This factor, which has been given the name IL3BP1 by the Human Genome Nomenclature Committee (alternately named E4BP4 and NF-IL3A), has been shown to act as a repressor of transcription in HepG2 and HeLa cells (7), act as an activator of interleukin-3 (IL-3) promoter function in T cells (6) and have minimal or a slight repressive effect on gene expression in pre-B cells (8).

* Address correspondence to: Stephen D. Nimer, M.D., Department of Medicine, Memorial Sloan-Kettering Cancer Center, 1275 York Avenue, New York, NY 10021. Phone: (212) 639-7871; Fax: (212) 794-5849.

THE CLONING OF IL3BP1(NF-IL3A)

To isolate transcription factors that regulate human IL-3 expression in T cells, we synthesized complementary oligonucleotides that contain the DNase I footprint "A" region of IL-3 promoter [from -165 to -128 relative to the transcription start site (9,10)] and used it as a radiolabeled probe to screen a λgt11 expression cDNA library made from mRNA isolated from PHA stimulated human primary T cells. We isolated a 1.9 kb cDNA, which we termed NF-IL3A (6) (now called IL3BP1) that belongs to the basic leucine zipper (bZIP) family of transcription factors. Comparing the basic region of NF-IL3A to other bZIP members we found NF-IL3A is more closely related to the PAR family of bZIP transcription factors such as DBP, TEF and HLF, than to other bZIP proteins such as AP-1 (Fig. 1). NF-IL3A is nearly identical to E4BP4, a bZIP protein cloned from a human placental cDNA library by Cowell and colleagues (7), which contains a single amino acid change, an arginine in NF-IL3A vs. a glycine in E4BP4. There are also nine nucleotide differences between NF-IL3A and E4BP4 in the 3' untranslated region. Our Southern blot analysis has shown that the human IL3BP1 gene is a single copy gene, and chromosomal localization studies, using fluorescence in situ hybridization, has shown that IL3BP1 is located on human chromosome 9q22 (11).

THE EXPRESSION OF IL3BP1

IL3BP1 (NF-IL3A) mRNA is expressed in resting MLA-144 T cells and the mRNA level increases following mitogen stimulation. This increase was not observed in the HTLV-infected S-LB-1 T cell line, which do not express IL-3 (6). IL3BP1 is also expressed in numerous B-lineage lymphoid cell lines UOC-B1 [t(17;19)], 697 [t(1;19)], RS4;11 [t(4;11)], Sup-B2 and 920, although the mRNA levels in the first three B cell lines are

```
                              Basic Region

NF-IL3A   P D E K K D A M Y W E K R R K N N E A A K R S R E K R R L N D L V
VBP/TEF   P D E Q K D E K Y W T R R K K N N V A A K R S R D A R R L K E N Q
DBP       P E E Q K D E K Y W S R R Y K N N E A A K R S R D A R R L K E N Q
HLF       P D D L K D D K Y W A R R R K N N M A A K R S R D A R R L K E N Q
c-JUN     M E S Q E R I K A E R K R M R N R I A A S K C R K R K L E R I A R
C-FOS     L S P E E E E K R R I R R E R N K M A A A K C R N R R R E L T D T

                         Leucine Zipper Region
          *                 *                 *                 *                 *
NF-IL3A   L E N K L I A L G E E N A T L K A E L L S L K L K F G L I S S T A
VBP/TEF   I T I R A A F L E K E N T A L R T E V A E L R K E V G K C K T I V
DBP       I S V R A A F L E K E N A L L R Q E V V A V R Q E L S H Y R A V L
HLF       I A I R A S F L E K E N S A L R Q E V A D L R K E L G K C K N I L
c-JUN     L E E K V K T L K A Q N S E L A S T A N M L R E Q V A Q L K Q K V
C-FOS     L Q A E T D Q L E D E K S A L Q T E I A N L L K E K E K L E F I L
```

Figure 1. Comparison of the basic and leucine zipper regions of NF-IL3A with other bZIP proteins.

significantly lower than in the latter two cell lines (8). We probed a multiple human tissue Northern blot (purchased from Clonetech) that contains poly A$^+$ RNA from colon, bladder, kidney, skeletal muscle, liver, cecum, basal cell and stomach with the IL3BP1 cDNA and observed significant expression of IL3BP1 in the bladder and a very weak signal in muscle (unpublished data), demonstrating some tissue specificity of IL3BP1 expression. The expression of IL3BP1/E4BP4 by placenta cells has also been reported (7). Using a more sensitive RT-PCR assay we have been able to show the presence of IL3BP1 mRNA in several human myeloid leukemia cell lines such as KG-1, HL-60, U937 and CMK, indicating broad expression of IL3BP1 RNA in hematopoietic cell lines and more restricted expression in other tissue types. It is not known whether IL3BP1 is similarly expressed in normal hematopoietic cells.

UV crosslinking data, using MLA-144 T cell nuclear extract and an IL-3 promoter "A" region probe, demonstrated proteins of 56kDa and 58kDa, the same molecular weight as in vitro translated IL3BP1 (12). Two weak anti IL3BP1 rabbit antisera, generated against two different portions of the recombinant IL3BP1 molecule, precipitated 56kDa and 58kDa proteins, which were also detected in UOC-B1 or Nalm-6 cell lysates (although at a much lower concentration) (8). IL3BP1 (E4BP4) protein was not detected in any cell lines by Cowell et al (7), and it is possible that production of IL3BP1 protein is predominantly controlled at the translational level, similar to the expression of DBP and TEF (13,14). Nonetheless, IL3BP1 protein is present in T cell and B cell hematopoietic cell lines.

THE DNA BINDING SEQUENCE OF IL3BP1

We used in vitro translated IL3BP1 to perform DNase I footprinting of the IL-3 promoter/enhancer region; both the coding and the antisense strand showed protection of the sequence TAATTACGTCTG. When this sequence was mutated to TAATTACacgTC, the binding affinity for IL3BP1 was reduced, but not eliminated, whereas the responsiveness of IL-3 promoter CAT constructs to IL3BP1 transactivation was eliminated by this mutation. Electrophoretic mobility shift assays, using wild type and mutant IL-3 promoter sequences, suggesting that the sequence ATTACG is the minimal DNA sequence required for binding of IL3BP1. Cowell et al (7) defined a consensus binding site for E4BP4 as (G/A)T(G/T)A(C/T)GTAA(C/T), which is similar to the consensus binding site for HLF (GTTACGTAAT), another member of the PAR family of bZIP proteins (15,16). These sequences all match the consensus DNA binding sequence required for PAR family of transcriptional factors, (A/G)TT(A/G)(C/T)(G/A)T(C/A)(A/T)(T/G) (17). IL3BP1 can bind to DNA sequences related to its consensus sequence including the adenovirus E4 promoter sequence ATGACGTAAC (7), the HLF consensus sequence GTTACGTAAT (15), the human IL-3 sequence ATTACGTCTG and the murine, rat and human γ -interferon promoter sequence ATTACGTAAT (Fig. 2).

IL3BP1 ACTIVATES THE IL-3 PROMOTER IN T CELLS

To define the regulatory effects of IL3BP1 on IL-3 promoter activity we cotransfected IL-3 promoter CAT constructs with an IL3BP1 expression plasmid (BC12-NF-IL3A) into MLA-144 T cells, and observed a 5 fold increase in reporter gene activity, which was not observed when the IL3BP1 binding sequence was mutated to ATTACacgTG. The IL-3 promoter sequences in the "A" region are sufficient for conveying IL3BP1 responsiveness; one or two copies of an oligonucleotide containing the "A" region sequence conferred IL3BP1 responsiveness on a enhancerless reporter plasmid (pTE2). In addition to the IL3BP1

Interleukin-3 promoter	AGAAAGTCATGGATGAATAATTACGTCTGTGGTTTTCT
γ-interferon promoter	AGGATTACGTATTTTCACAAGTTT
Adenovirus E4 promoter	CTTCTAAAAAATGACGTAACGGAAGCTT
HLF consensus binding site	GTTACGTAAT
E4BP4 consensus binding site	(G/A)T(G/T)A(C/T)GTAA(C/T)

Figure 2. Different promoter sequences shown to bind NF-IL3A. Underlined sequences identify the region of homology with the NF-IL3A/E4BP4 consensus binding sequence.

binding sequence, the "A" region also contains an Oct-1-like binding sequence (ATGAATAAT) at the 5' end and an AML-1 binding site (TGTGGT) at the 3' end, which suggests that other proteins might also be involved in the regulatory effects of the "A" region in T cells (9,18,19). In fact, EMSA data suggest an interaction of IL3BP1 with Oct-1 in binding to the "A" region sequences. Although the γ-interferon promoter contains an IL3BP1 consensus binding site, and can bind IL3BP1 in in vitro assays, the γ-interferon promoter sequence failed to confer IL3BP1 responsiveness to the enhancerless reporter plasmid pTE2. Similarly, the adenovirus E4 promoter sequences were not responsive to IL3BP1 in T cell lines. These results suggest that binding of other proteins may affect the activity of IL3BP1 on the "A" region in T cells.

IL3BP1 IS ANTAGONISTIC TO E2A/HLF IN B CELLS

HLF (hepatic leukemia factor) is a transcription factor that is a member of the PAR family of bZIP proteins. This family includes DBP and TEF, which are normally expressed in human liver, kidney and lung tissue, but not in normal or transformed lymphoid cells (16,20). In human B-lineage acute lymphoblastic leukemia with a t(17;19) chromosomal translocation, the E2A gene on chromosome 19 (encoding a basic helix-loop-helix transcription factor) is fused to the HLF gene on chromosome 17, leading to the expression of the E2A-HLF fusion protein. The E2A-HLF DNA binding sequence is nearly identical to the HLF consensus binding sequence GTTACGTAAT (15), which also matches the IL3BP1 (E4BP4) consensus sequence (G/A)T(G/T)A(C/T)GTAA(C/T). HLF does not act as a transcriptional activator in B cells, whereas E2A-HLF is an activator of transcription. In Nalm-6 pro-B cells, E2A-HLF activates transcription 10-15 fold of a reporter gene containing a heterologous promoter with several HLF consensus binding sites cloned upstream. In contrast, neither HLF nor IL3BP1 have transactivating potential via the HLF consensus sequence in Nalm-6 cells. When the E2A-HLF and IL3BP1 cDNAs were co-transfected into Nalm-6 cells, trans-activation of the CAT gene by E2A-HLF was repressed by IL3BP1 (8). This repression required the presence of the IL3BP1 DNA-binding domain and was largely abrogated by elimination of a C-terminal putative transcriptional repression domain (21). Although E2A-HLF readily formed dimers with DBP, it did not form heterodimers with IL3BP1. This suggests that IL3BP1 and E2A-HLF compete for binding to the HLF consensus binding site.

IL3BP1 SUPPRESSES TRANSCRIPTION IN HELA AND HepG2 CELLS

Studies performed by Cowell and colleagues (7,21) demonstrated that when HeLa and HepG2 cells were cotransfected with an IL3BP1(E4BP4) expression vector and a CAT reporter plasmid containing the IL3BP1 consensus binding sequence TTATGTAA, the CAT activity was reduced compared to the control (a co-transfected, empty expression vector). This trans-repression is mediated by the C-terminal repression domain of IL3BP1, which is located between amino acids 297 and 363.

SUMMARY

We have shown that IL3BP1 can transactivate the IL-3 promoter in T cells (via the "A" region sequences) but has no effect on the adenovirus E4 promoter binding sequences or the human γ-IFN promoter consensus binding sequences in these cells. IL3BP1 has minimal effect on the activation of the HLF consensus binding site in B cells but it can dominantly repress transactivation by E2A-HLF in these cells. Cowell and colleagues have shown that IL3BP1 can repress transcription in Hela cells. This data suggests that IL3BP1 cooperates with other DNA binding proteins to exert its function. There are several examples of transcription factors that can function as both a repressor and an inducer of gene expression (5); IL3BP1 can now be added to the list.

REFERENCES

1. Ptashne M. How transcriptional activator work. Nature 335:683, 1988
2. Jaynes JB, O'Farell PH. Active repression of transcription by the engrailed homeodomain proteins. EMBO J. 10:1609, 1991
3. Auwerx J, Sassone-Corsi P. IP-1: A dominant inhibitor of fos/jun whose activity is modulated by phosphorylation. Cell 64:983, 1991
4. Sauer F, Fondell JD, Ohkuma Y, Roeder RG, Jackle H: Control of transcription by Kruppel through interactions with TFIIB and TFIIEβ. Nature 375:162, 1995
5. Roberts SGE, Green MR: Dichotomous regulators. Nature 375:105, 1995
6. Zhang W, Zhang J, Konuc M, Kwan K, Frank R, Nimer SD: Molecular cloning and characterization of NF-IL3A; a transcriptional activator of the human IL-3 promoter. Mol. Cell. Biol. 15:6055, 1995
7. Cowell IG., Skinner A, Hurst HC: Transcriptional repression by a novel member of the bZIP family of transcription factors. Mol. Cell. Biol. 12:3070, 1992
8. Inaba T, Nimer S, Yoshihara T, Zhang W, Look AT. The mechanism of transformation by E2A-HLF may involve genes normally regulated by IL3BP1. (Submitted for publication).
9. Davies K., TePas EC, Nathan DG, Mathey-Prevot B: Interleukin-3 expression by activated T cells involves an inducible, T cell-specific factor and an octamer binding protein. Blood 81:928, 1993
10. Wolin MM, Kornuc Lau R, Shin SK, Lee F, and Nimer SD. Differential effect of HTLV infection and HTLV Tax on IL-3 expression. Oncogene 8:1905, 1993
11. Zhang W, Inaba T, Look AT, Nimer SD. Chromosomal localization and characterization of the human NF-IL3A gene. (Submitted for publication).
12. Nimer SD, Zhang W, Zhang J, Uchida H. Interacting DNA binding proteins regulate interleukin-3 gene expression. (Manuscript in preparation).
13. Mueller CR, Maire P, Schibler U. DBP, a liver enriched transcriptional activator is expressed late in ontogeny and its tissue specificity is determined posttranscriptionally. Cell 61:279, 1990
14. Drolet DW, Scully KM, Simmons DM, Wegner M, Chu K, Swanson LW, Rosenfeld MG. TEF, a transcription factor expressed specifically in the anterior pituitary during embryogenesis, defines a new class of leucine zipper proteins. Genes Dev. 5:1739, 1991

15. Inaba T, Shapiro LH, Funabiki T, Sinclair AE, Jones BG, Ashmun RA, Look AT. DNA-binding specificity and transactivating potential of the leukemia associated E2A hepatic leukemia factor fusion protein. Mol. Cell. Biol. 14:3403, 1994
16. Inaba T, Roberts WM, Shapiro LY, Jolly KW, Raimondi SC, Smith SD, Look AT. Fusion of the leucine zipper gene HLF to the E2A gene in human acute B-lineage leukemia. Science 257:531, 1992
17. Haas NB, Cantwell CA, Johnson PF, Burch JBE. DNA-binding specificity of the PAR basic leucine zipper protein VBP partially overlaps those of the C/EBP and CREB/ATF families and is influenced by domains that flank the core basic region. Mol. Cell. Bio. 15:1923, 1995
18. Mathey-Prevot B, Andrews NC, Murphy HS, Kreissman SG, Nathan DG. Positive and negative elements regulate human interleukin-3 expression. Proc. Natl. Acad. Sci. USA 87:5046, 1990
19. Shoemaker SG, Hromas R, Kaushansky K. Transcriptional regulation of interleukin-3 gene expression in T-lymphocytes. Proc. Natl. Acad. Sci. USA 87:9650, 1990
20. Hunger SP, Ohyashiki K, Toyama K, Cleary ML. HLF, a novel hepatic bZIP protein, shows altered DNA-binding properties following fusion to E2A in t(17;19) acute lymphoblastic leukemia. Genes Dev. 6:1608, 1992
21. Cowell IG, Hurst HC. Transcriptional repression by the human bZIP factor E4BP4: definition of a minimal repression domain. Nucleic Acid Res. 22:59, 1994

ADHESION OF HUMAN HEMATOPOIETIC PROGENITOR CELLS TO STROMAL CELLS IS ENHANCED BY ANTIBODIES TO CD44

Robert A. J. Oostendorp,* Elisabeth Spitzer, and Peter Dörmer

GSF-Forschungszentrum für Umwelt und Gesundheit
Institut für Experimentelle
Hämatologie, München

ABSTRACT

CD44 has been implicated to mediate adhesive interactions between hematopoietic progenitor cells and the stromal microenvironment. Ligands of CD44 include several extracellular matrix components, such as hyaluronic acid and fibronectin. Antibodies against CD44 have been shown to induce homotypic T-cell aggregation, and to stimulate T- and NK-cell activity. We hypothesized that CD44 could similarly amplify interactions of blast colony-forming cells and bone marrow stromal cells. Indeed, we have previously found that the anti-CD44 antibody NKI-P2 enhanced VLA-4 dependent interactions. Here, we studied an additional panel of 19 anti-CD44 antibodies from the 5th Workshop on Leukocyte Differentiation antigens, to find out whether amplification was associated with a particular CD44-epitope. None of these antibodies showed inhibitory activity, whereas nine significantly increased the number of blast colonies more than 2-fold. Seven of these recognized epitope 1, and two epitope 2. More than 4-fold enhancement was only observed with epitope 1 antibodies: 4.C3 (4.4-fold), 212.3 (6.3-fold), L178 (9.1-fold), and NIH44-1 (9.2-fold). Our data suggest that primarily epitope 1 is associated with enhancement of colony formation. Furthermore, the findings support the role of CD44 as an amplifier in progenitor-BMSC interactions.

INTRODUCTION

CD44 is a cell surface proteoglycan which has been implicated to play an important role in adhesive interactions between hematopoietic progenitors and bone marrow-derived

*Corresponding author: Robert A. J. Oostendorp, Ph. D., GSF-Institut für Experimentelle Hämatologie, Marchioninistrasse 25, D-81377 München, Germany. Telephone: +49 - 89 7099 213; Fax: +49 - 89 7099 225; E-mail: oostendorp@haemal14.gsf.de

Molecular Biology of Hematopoiesis 5, edited by Abraham et al.
Plenum Press, New York, 1996

stromal cells (BMSC). Ligands of CD44 include several extracellular matrix components, such as hyaluronic acid (HA) and fibronectin. In long-term culture (LTC) systems, monoclonal antibodies (mAb) against CD44 have been shown to suppress[1] or even abrogate the production of non-adherent myeloid and lymphoid progenitors.[2] In addition, a CD44-related proteoglycan has been demonstrated, like VLA-4, to be involved in the interaction between the heparin-binding domains of fibronectin and $Lin^-/CD34^+/HLA\text{-}DR^+$ BFU-E and GM-CFC.[3]

In contrast to these inhibitory activities of anti-CD44 mAb, enhancing effects of such mAb were noted in studies using more mature cells, such as T-lymphocytes,[4,5] and NK-cells[6] (Oostendorp et al, unpublished). Similarly, enhancing effects have been described in a preliminary study in which anti-CD44 mAb increased the binding of B-cell precursors to BMSC.[7] In addition, we noted that the anti-CD44 antibody NKI-P2[8] increased the number of blast colonies from stroma-adherent progenitors (blast colony-forming cells, Bl-CFC).[9]

To find out whether this enhancing activity of anti-CD44 mAb is associated with a particular CD44-epitope and inhibitory activities with another, we studied an additional panel of 19 antibodies, all of which were included in the CD44 panel of the 5th Workshop on Human Leukocyte Differentiation Antigens (5th HLDA).[10]

MATERIALS AND METHODS

Progenitor Cells

Bone marrow cells were prepared from spongiosa of patients undergoing heart or lung surgery. Mononuclear cells were separated using NycoPrep density centrifugation (d=1.077 g/ml, Life-Technologies, Eggenstein, Germany), washed three times in modified McCoy's 5a Medium (GIBCO, Life Technologies) supplemented with 5% (v/v) fetal calf serum (FCS) (McCoy's /FCS). The mononuclear cells were depleted of plastic-adherent cells by incubation (37°C, 5% CO_2) for 2 hours in culture flasks (Nunc, Wiesbaden-Biebrich, Germany). The resulting low-density, plastic non-adherent cells ($BMC_{LD/NA}$) were washed twice in McCoy's /FCS, and used as source of progenitor cells. The plastic-adherent cells were cultured to obtain the BMSC as described below.

In some experiments, $CD34^+$ cells were enriched from these $BMC_{LD/NA}$ with the use of an antibody coupled to magnetic beads (Dynabeads-CD34, Deutsche Dynal, Hamburg, Germany) as recommended by the manufacturer. After detachment from the magnetic beads, $CD34^+$-cells, which were 70-95% pure as determined by flow cytometry, were washed twice with McCoy's/FCS and used immediately.

Bone Marrow-Derived Stromal Cells (BMSC)

Stromal cell preparations were established from the plastic-adherent mononuclear cell fraction in long-term culture (LTC) medium: McCoy's 5a medium (GIBCO), supplemented with preselected batches of 12.5% FCS (Boehringer) and 12.5% horse serum (Boehringer), as well as with 1% sodium bicarbonate, 1% sodium pyruvate, 0.4% modified Eagle's medium (MEM) non-essential amino acid solution, 0.8% MEM essential amino acids, 1% vitamins, 1% antibiotics (all solutions from GIBCO) and 1.0 μM hydrocortisone (Sigma Chemie, Deisenhofen, Germany). After 3 to 4 weeks of primary culture, confluent stromal cell layers were passaged. BMSC were washed twice in phosphate-buffered saline (PBS, pH=7.2) and treated with 0.25% trypsin (GIBCO) for 10 minutes. An equal volume of FCS was added, after which detached cells were harvested and washed twice in LTC medium. Detached BMSC were either reseeded in culture flasks ($3 \times 10^4/cm^2$) and cultured for another 2 to 3 weeks before renewed passage or irradiated (15 Gy) in a ^{137}Cs gamma-source (Gamma-Cell, Ontario, Canada), and plated in 24- or 96-well plates (Nunc) at a density of 3×10^4 cells/cm^2. Passaged stromal cell cultures were allowed to adhere and spread for at least 1 day at 37°C in a humidified CO_2 atmosphere before use.

Antibodies

Monoclonal antibodies (mAb) against CD44 have been obtained from the 5th HLDA. Additional aliquots of the anti-CD44 mAb L178 were obtained from Becton Dickinson (Heidelberg Germany). The anti-CD44 mAb NKI-P2 was a generous gifts of C. Figdor (Academisch Ziekenhuis St Radboud, Nijmegen, The Netherlands). Throughout this study, the mAb MP30-1 (IgG2, against CD47), which was included in the Workshop panel, and M10 (IgG1 against dog CD8),[11] which was a generous gift of Dr. E. Kremmer (GSF-Forschungszentrum, München, Germany), have been used as isotype control mAb.

Progenitor Cell Assays

Assays for erythroid burst-forming units (BFU-E), granulocyte/ macrophage colony-forming cells (GM-CFC), and Bl-CFC have been performed as described earlier.[9] Adherent progenitors were obtained from adherent stromal layers after trypsin-treatment as described above. Trypsin-treatment did not affect the number of BFU-E or GM-CFC, although a slight decrease in size of colonies was noted.

RESULTS

Previously, we have tested a large number of antibodies against adhesion molecules for their capacity to block adhesive interactions between blast-CFC and BMSC.[9] Surprisingly, the anti-CD44 mAb NKI-P2 increased the number of blast-CFC adhering to BMSC by an average of 2.8-fold. Here, we confirm and extend these findings.

Since opposing effects of anti-CD44 antibodies have been observed in different experimental systems (see introduction), we investigated the CD44 panel of antibodies from the 5th HLDA.[10] In this Workshop, three different CD44 epitopes have been defined (table 1). To study the effect of these antibodies on Bl-CFC, we have performed experiments in which the mAb were present during the initial 2 hour adhesion phase in a 1 to 100 dilution. None of the antibodies tested, showed inhibitory activity (less or smaller blast colonies). Nine of the mAb significantly enhanced the formation of blast colonies more than two-fold. Seven of these mAb belonged to the epitope 1-binding mAb (n=9) and two to the HA-binding site (epitope 2) mAb (n=7). The mAb against epitope 3 (n=3) displayed no activity (figure 1). More than four-fold enhancement was only observed with epitope 1-binding mAb: 4.C3 (4.4-fold), 212.3 (6.3-fold), L178 (9.1-fold), and NIH-44-1 (9.2-fold).

The enhancing effect of NKI-P2, L178 and NIH44-1 was not limited to Bl-CFC. In experiments using $BMC_{LD/NA}$ immunomagnetically enriched for expression of CD34, we found more stroma-adherent BFU-E and GM-CFC after treatment with these mAb. The total number of these progenitors (adherent + non-adherent) does not change with regard to the untreated or control antibody (M10)-treated control cultures (data not shown). In figure 2, it is shown that L178 dose-dependently increases the adhesion of both BFU-E and GM-CFC to BMSC. Similar results have been obtained with NKI-P2 (data not shown). In contrast to the results in the Bl-CFC assay, more than 5-fold increase of adherent BFU-E or GM-CFC was not observed after treatment with L178 or NIH44-1.

Table 1. MAb against CD44 can be subdivided in 3 epitope groups according to their cross-blocking with other mAb. Additional arguments for the epitope group division come from experiments measuring adhesion to hyaluronic acid (HA), and reactivity with trypsin-treated cells or baboon leukocytes

Epitope group	1	2	3
mAb blocks binding of mAb			
A3D8	+	–	–
5F12	–	+	–
BRIC225	–	+	–
mAb blocks adhesion to hyaluronic acid	–[a]	+[b]	–
mAb reacts with trypsin-treated cells	–	–[c]	+
mAb reacts with baboon leukocytes	–	–	++

Data have been compiled from Ref. 10.
Remarks: a: mAb HP2/9 blocks HA-binding while at the same time cross-blocking A3D8 binding; b: only subgroup (BRIC235, 3F12, BU75, 50B4) blocks HA-binding (epitope 2a); c: epitope 2 antibody BRIC214 also reacts with trypsin-treated cells.

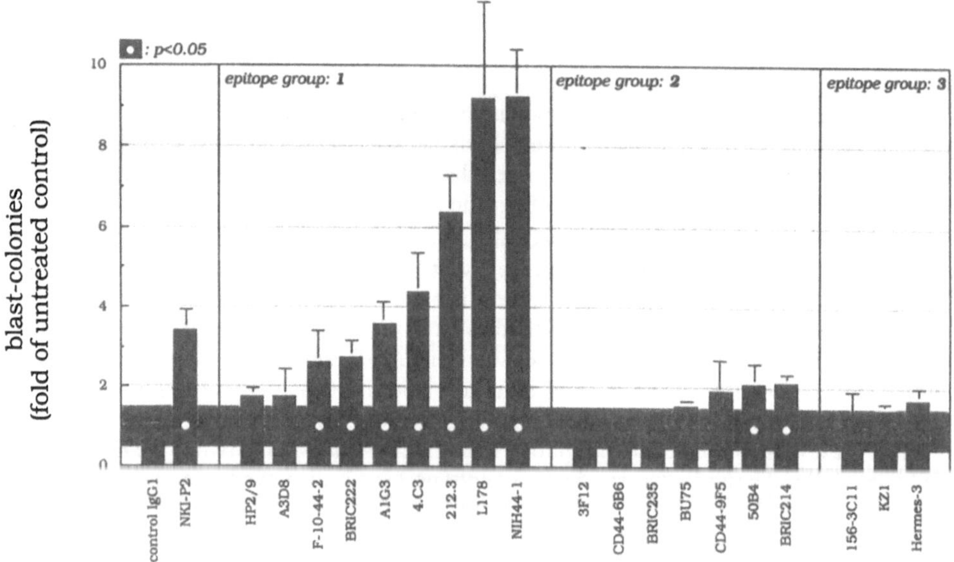

Figure 1. Some antibodies from the CD44 antibody panel from the 5th HLDA increase the number of blast colonies. All antibodies (1:100 dilution) were present only during the initial 2 hour adhesion phase. The number of colonies derived from each single donor were standardised with regard to the untreated control (fold increase: number of blast colonies after mAb treatment divided by number of colonies from untreated control). The standardized values for the treatment with control antibody (MP30-1 or M10) were then compared with the standardized value of the treatment with test antibody using the paired Student's *t*-test. P-values of less than 0.05 were considered statistically significant. The results represent the mean ± SD of 3 or 4 independent experiments per mAb tested.

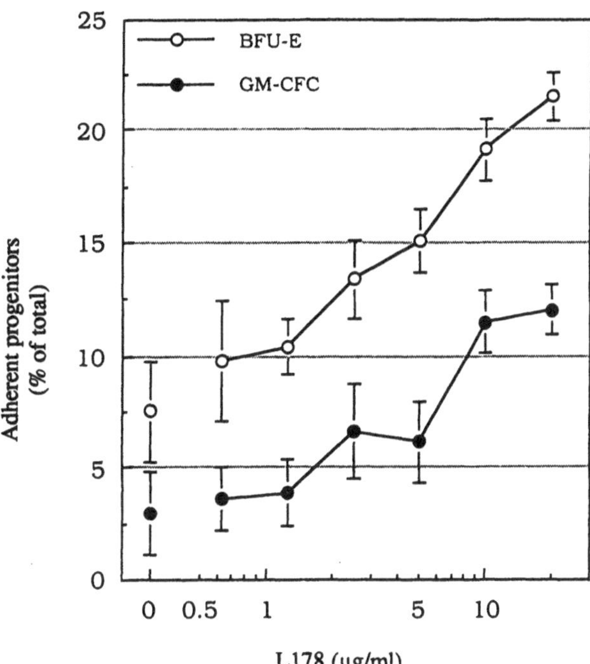

Figure 2. The anti-CD44 (epitope 1) mAb L178 dose-dependently enhances adhesion of CD34+ progenitors to bone marrow stromal cells. The antibody L178 was present during the initial 2 hour adhesion phase. Adherent progenitors were treated with trypsin. Subsequently, adherent and non-adherent progenitors were plated separately in methylcellulose-based semi-solid medium to allow for growth of BFU-E and GM-CFC. The values in the figures represent the percentage of progenitors found to adhere to BMSC (={adherent progenitors/[adherent + non-adherent progenitors]}∞100%). The total recovery of BFU-E and GM-CFC was similar (range 73-95%) in untreated control and L178-treated cultures.

DISCUSSION

In the present study we show that several anti-CD44 antibodies enhance the adhesive interactions between hematopoietic progenitors and BMSC. Furthermore, we show that this enhancement is associated with the epitope 1 of CD44; the epitope 2, which includes antibodies that inhibit CD44-HA interactions, as well as the epitope 3 mAb display neither relevant enhancing nor inhibitory activity.

The ability of anti-CD44 mAb to induce or stimulate adhesive processes have been well documented. They may induce homotypic adhesion of T-cells[4,5] and heterotypic adhesion of T cells to dermal fibroblasts[12] or endothelial cells.[13] Moreover, anti-CD44 antibodies may trigger effector functions of T cell clones[14] and NK cells[6] (R. A. J. Oostendorp: unpublished observations), and monocytes.[15] These findings suggest that CD44 may mediate outside-in signaling. Indeed, it was demonstrated that binding of HA to CD44 induces intracellular signals,[16] and postulated that anti-CD44 mAb only stimulate cells when the mAb is reactive with the hyaluronate binding-site of CD44.[13] In addition, it has been demonstrated that murine CD44 has intrinsic GTPase activity.[17] With regard to the latter observations, we were surprised to find that anti-CD44 mAb which did not react with the HA-binding site increased adhesion, whereas those that did (3F12, BU75, 50B4, BRIC235, HP2/9)[10] demonstrated no activity in our system. This would suggest that the enhancing effect we found is mediated by a mechanism different from that mediated through HA binding.

Alternative mechanisms may explain the enhanced binding of progenitors to stroma. The mAb may directly induce the above described intracellular signals which may ultimately lead to enhanced affinity of adhesion molecules by an inside-out mechanism. For instance, the mAb 212.3, which enhances blast colony output 6.3-fold (figure 1) has been demonstrated to elevate intracellular cAMP and modulate OKT3-induced Ca^{2+} levels in T cells.[18]

Whether induction of such signals require cross-linking of CD44 is unclear at present. Another possibility is that the anti-CD44 mAb cross-link the progenitors and BMSC, which both express CD44. Investigations with F(ab) fragments should resolve both issues related mechanisms involving cross-linking. A third possibility is that the mAb induce the production of adhesion-enhancing cytokines such as chemokines in either progenitors or BMSC.

The extensive studies reported in the CD44 Workshop of the 5th HLDA have identified a number of antibodies capable of enhancing CD2-induced T cell activation and IL-2 production as well as NK-cell activity.[10] However, none of these activities is limited to one specific epitope of CD44. In contrast, both precursor B-cell binding[7] to as well as Bl-CFC interactions (this study) with BMSC appear to be specifically associated with the epitope 1 of CD44. This suggests that interactions through this epitope of CD44 are important in adhesion of committed lympho/ hematopoietic progenitors. At present, it is unclear whether a similar role for CD44 can be assumed in adhesion of early human progenitors to BMSC. To date, only a single report has addressed this question. Gunji et al reported that the anti-CD44 antibody BU52 abrogated production of non-adherent colony-forming cells in human LTC[1] in a similar fashion as reported earlier by Miyake and coworkers in murine LTC.[2] In our hands, when NKI-P2 was present during the initial 2 hour adhesion phase, an increased output of non-adherent GM-CFC after four or more weeks of Dexter-type LTC of the stroma-adherent cells was observed (unpublished). The mechanism of this enhancement is, however, still under investigation, since the LTC system we initially chose[19] did not discriminate between prolonged survival of committed progenitors, enhanced adhesion of early progenitors, or enhanced growth of (a subset of) early progenitors. Our results do not contradict the results of Gunji and coworkers, since from the results shown in figure 1 it is clear that different anti-CD44 mAb, even against the same epitope of CD44, may well yield opposing results when tested in LTC.

Future investigations will focus on the mechanisms of the enhancing effects we have described here, and their relation to some of the inhibitory effects of anti-CD44 mAb described by others.[1,3] Our present findings point to a key role of CD44 in adhesive interactions between hematopoietic progenitors and the stromal microenvironment. In such a scheme, CD44-mediated interactions may affect adhesion or proliferative capacity of progenitors, or exert their effects through the stromal cells.

ACKNOWLEDGMENTS

For their generous gifts of antibodies we thank S. M. Denning (Duke University Medical Center, Durham NC: *5th HLDA antibodies*), and C. Figdor (Academisch Ziekenhuis St Radboud, Nijmegen, The Netherlands: *NKI-P2*). Furthermore, we thank Drs G. Überfuhr, and H. Dienemann (Klinikum Großhadern, Munich Germany) for their gift of bone marrow samples.

REFERENCES

1. Gunji Y, Nakamura M, Hagiwara T, Hayakawa K, Matsushita H, Osawa H, Nagayoshi K, Nakauchi H, Yanagisawa M, Miura Y, Suda T: Expression and function of adhesion molecules on human hematopoietic stem cells: $CD34^+LFA\text{-}1^-$ cells are more primitive than $CD34^+LFA\text{-}1^+$ cells. Blood 80: 429, 1991
2. Miyake K, Medina KL, Hayashi S-I, Ono S, Hamaoka T, Kincade PW: Monoclonal antibodies to Pgp-1/CD44 block lympho-hemopoiesis in long-term bone marrow cultures. J Exp Med 171: 477, 1990
3. Verfaillie CM, Benis A, Iida J, McGlave PB, McCarthy JB: Adhesion of committed human hematopoietic progenitors to synthetic peptides from the C-terminal heparin-binding domain of fibronectin: cooperation between the integrin $\alpha4\beta1$ and the CD44 adhesion receptor. Blood 84: 1802, 1994

4. Koopman G, van Kooyk Y, de Graaff M, Meijer CJLM, Figdor CG, Pals ST: Triggering of the CD44 antigen on T-lymphocytes promotes T-cell adhesion through the LFA-1 pathway. J Immunol 145: 3589, 1990
5. Belitsos PC, Hildreth JEK, August JT: Homotypic cell aggregation induced by anti-CD44(Pgp-1) monoclonal antibodies and related to CD44(Pgp-1) expression. J Immunol 144: 1661, 1991
6. Tan PHS, Santos EB, Rossbach H-C, Sandmaier BM: Enhancement of natural killer activity by an antibody to CD44. J Immunol 150: 812, 1993
7. Dittel BN, LeBien TW: Amplification of B cell precursor/bone marrow stromal cell adhesion with antibodies to CD44. Tissue Antigens 4: 271 (AS056), 1993
8. Pals ST, Hogervorst F, Keizer GD, Thepen T, Horst E, Figdor CG: Identification of a widely distributed 90 kDa glycoprotein that is homologous to the Hermes-1 human lymphocyte homing receptor. J Immunol 143: 851, 1989
9. Oostendorp RAJ, Reisbach G, Spitzer E, Thalmeier K, Dienemann H, Mergenthaler H-G, Dörmer P: VLA-4 and VCAM-1 are the principal adhesion molecules involved in the interaction between blast colony-forming cells and bone marrow stromal cells. Brit J Haematol, 97:275, 1995
10. Denning SM, Telen MJ, Hale LP, Liao HX, Haynes BF: CD44 and CD44R cluster report. Leukocyte Typing V, Oxford University Press, 1995
11. Voß C, Kremmer E, Hoffmann-Fezer G, Schumm M, Günther W, Kolb H-J, Thierfelder S: Identification and characterization of a mous monoclonal antibody (M10) against canine (dog) $CD8^+$ lymphocytes. Vet Immunol Immunopathol 38: 311, 1993
12. Bruynzeel I, Koopman G, van der Raaij LMH, Pals ST, Willemze R: CD44 antibody stimulates adhesion of peripheral blood T cells to keratinocytes through the LFA-1/ICAM-1 pathway. J Invest Dermatol 100: 424, 1993
13. Toyama-Sorimachi N, Miyake K, Miyasaka M: Activation of CD44 induces ICAM-1/LFA-1-independent, Ca^{2+} Mg^{2+}-independent adhesion pathway in lymphocyte endothelial cell interaction. Eur J Immunol 23: 439, 1993
14. Galandrini R, Albi N, Tripodi G, Zarcone D, Terenzi A, Moretta A, Grossi CE, Velardi A: Antibodies to CD44 trigger effector functions of human T cell clones. J Immunol 150: 4225, 1993
15. Gruber MF, Webb DSA, Gerrard TL: Stimulation of human monocytes via CD45, CD44, and LFA-3 triggers macrophage-colony-stimulating factor production. Synergism with Lipopolysaccharide and IL-1β. J Immunol 148: 1113, 1992
16. Galandrini R, Galluzzo E, Albi N, Grossi CE, Velardi A: Hyaluronate is costimulatory for human t cell effector functions and binds to CD44 on activated T cells. J Immunol 153: 21, 1994
17. Lokeshwar VB, Bourguignon LYW: The lymphoma transmembrane glycoprotein gp85 (CD44) is a novel guanine nucleotide-binding protein which regulates gp85 (CD44)-ankyrin interaction. J Biol Chem 267: 22073, 1992
18. Rothman BL, Kennure N, Kelley KA, Katz M, Aune TM: Elevation of intracellular cAMP in human T lymphocytes by an anti-CD44 mAb. J Immunol 151: 6036, 1993
19. Mergenthaler H-G, Brühl P, Dörmer P: Kinetics of myeloid progenitor cells in human micro long-term bone marrow cultures. Exp Hematol 16: 145, 1988

50

Fas AND FasL INTERACTIONS IN CYTOTOXICITY AND ITS ROLE IN LYMPHOPROLIFERATIVE DISEASE

Denise Hammond-McKibben,[1] Asimah Rafi,[1] Mitzi Nagarkatti,[2] and Prakash Nagarkatti[1*]

[1] Department of Biology, College of Arts and Sciences
[2] Department of Biomedical Sciences and Pathobiology
Virginia-Maryland Regional College of Veterinary Medicine
Virginia Tech, Blacksburg, Virginia 24061

ABSTRACT

The murine *lpr* gene encodes for an aberrant form of Fas (CD95) a molecule involved in apoptosis. Also, the *gld* gene leads to a nonfunctional expression of Fas ligand (FasL). Mice homozygous for *lpr* and *gld* mutations develop severe lymphoproliferative and autoimmune disease characterized by the appearance of unique CD4$^-$ CD8$^-$ (double negative, DN) T cells. The nature and functional significance of these DN T cells is not clear. In the current study, we demonstrate that *lpr* DN T cells but not *gld* DN T cells mediate spontaneous cytolytic activity as well as redirected lysis of tumor targets when activated through the TCR. This cytolytic activity was totally dependent on the interaction between Fas and FasL despite the fact that both lpr and gld DN T cells constitutively expressed perforin as well as other cytokines such as IFN-γ and TNF-α. Our studies also demonstrated that DN T cells can mediate perforin dependent lysis after Fas-FasL interactions. Together these studies demonstrate that although DN T cells are cytotoxic in nature they may not be able to exert this function *in* vivo in the natural host, due to the defect in the expression of Fas or FasL and this may permit unregulated expansion of the autoreactive B cells. Furthermore the DN T cells by virtue of expressing constitutively several cytokines may trigger an inflammatory response leading to autoimmune disease.

* Address correspondence to: Dr. Prakash Nagarkatti, Dept. of Biology, Virginia Tech, Blacksburg, VA 24060. Telephone: (540) 231-5029; Fax: (540) 231-9307.

INTRODUCTION

Mice homozygous for the lpr and gld mutation exhibit a defect in the expression of Fas and FasL respectively, leading to massive accumulation of an unusual subset of $\alpha\beta$TCR$^+$ T cells that are CD4$^-$CD8$^-$ (DN).[1] Despite extensive research, the nature and significance of the DN T cells in autoimmune disease is not clear. Studies from our lab have demonstrated that DN T cells from *lpr* mice constitutively express perforin gene and exhibit spontaneous NK like cytotoxicity.[2] These studies demonstrated that DN T cells may represent activated T cells that may participate in autoimmune disease by being able to mediate lysis of autologous cells. In the current study however, we demonstrate that DN T cells isolated from *gld* mice fail to mediate spontaneous and redirected cytotoxicity despite constitutively expressing perforin. The inability of the DN T cells to function as cytotoxic cells *in vivo* may lead to uncontrolled growth and differentiation of the B lymphocytes producing autoantibodies.

MATERIALS AND METHODS

Mice. MRL *lpr/lpr (lpr)*, MRL+/+, C57BL/6 +/+ and C57BL/6 *gld/gld (gld)* were bred in our animal facilities as described.[3]

Medium and Reagents. Tissue culture medium RPMI 1640 supplemented with additional nutrients and 10% fetal bovine serum was used as described elsewhere.[2] All antibodies used were prepared from the culture of hybridomas.[2,3] The tumor targets such as YAC-1 and P815 were passaged *in vitro*[2].

DN T Cells. The DN T cells were purified by treating the LN cells from four month old *lpr* or *gld* mice with antibodies against CD4 and CD8 plus complement as described.[2]

Cytotoxic Assay. The DN T cells were tested for their ability to mediate cytolytic activity against various tumor targets using ^{51}chromium release assay as described elsewhere.[2,4] In the redirected cytotoxic assays, antibodies against the TCR or other adhesion molecules were added.

PCR Analysis of Cytokine Gene Expression in DN T Cells. PCR method was employed to study whether *lpr* or *gld* DN T cells spontaneously expressed various cytokine genes as described previously.[2] Briefly, total RNA was prepared from the DN T cells and was reverse transcribed to cDNA. The resulting cDNA was subjected to PCR amplification using synthetic oligonucleotide primers for TNF-α, IFN-γ, perforin and β-actin as a control. The PCR product was electrophoresed on 1.5% agarose gel stained with ethidium bromide.

RESULTS

Lpr but not *Gld* DN T Cells Exhibit Cytotoxic Activity *in Vitro*

We have recently demonstrated that *lpr* DN T cells can mediate spontaneous and redirected cytotoxicity against certain tumor targets.[2] In the current study we investigated whether *gld* DN T cells would also mediate similar cytolytic activity. The data shown in Fig. 1 indicated that unlike *lpr* DN T cells, the *gld* DN T cells failed to mediate spontaneous

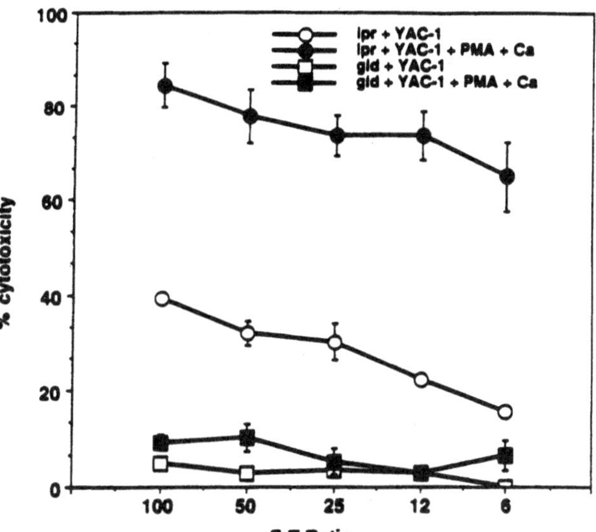

Figure 1. Double-negative (DN) T cells from *lpr* but not *gld* mice exhibit spontaneous cytolytic activity against YAC-1 targets. Freshly isolated DN T cells from *lpr* or *gld* strains were tested for cytotoxicity in the absence or presence of PMA + calcium ionophore at various effector:target (E:T) cell ratios. The cytotoxicity was studied using ^{51}Cr-release assay.

lytic activity as well as cytotoxicity in the presence of PMA + calcium ionophore, against YAC-1 tumor targets. Furthermore, the *gld* DN T cells failed to mediate redirected lytic activity in the presence of mAbs against the TCR (data not shown). These findings together suggested that the gld DN T cells may not be able to mediate cytolytic activity due to a defect in the functional expression of Fas ligand (FasL). To further confirm this, we tested the ability of the *lpr* DN T cells to mediate cytolytic activity against B cell blasts that were prepared from either +/+ (Fas$^+$ mice) or lpr/lpr (Fas$^-$ mice). In these experiments it was observed that *lpr* DN T cells could mediate cytolytic activity against Fas$^+$ B cell targets but not Fas$^-$ B cell targets (data not shown). Also, the cytolytic activity of *lpr* DN T cells was significantly inhibited in the presence of mAbs against Fas. These data indicated that both *lpr* and *gld* DN T cells were totally dependent on the interaction between Fas and FasL to mediate cytotoxic activity.

Expression of Adhesion Molecule and Cytokine Genes by *Lpr* and *Gld* DN T Cells

It was possible that the reason why *gld* DN T cells could not mediate cytotoxicity was due to the fact that *gld* DN T cells may express differential levels of various adhesion molecules involved in cytotoxicity when compared to the *lpr* DN T cells. To address this, the DN T cells from both *gld* and *lpr* mice were stained with various mAbs directed against the TCR and other adhesion molecules. These experiments revealed that *gld* DN T cells exhibited similar levels of various adhesion molecules when compared to the *lpr* DN T cells. Furthermore, the *gld* DN T cells were also found to constitutively express similar levels of perforin (Fig 2) and IFN-γ and TNF-α genes (data not shown). These data together suggested that the *gld* DN T cells are very similar to the *lpr* DN T cells both in their ability to express various adhesion molecules as well as spontaneously transcribe various cytokine genes.

DN T Cells Can Mediate Cytotoxicity of Endothelial Cells

We have previously shown[2] that DN T cells could mediate cytolytic activity when activated through homing receptors such as CD44 and gp90$^{\mathrm{Mel14}}$. Inasmuch as, endothelial

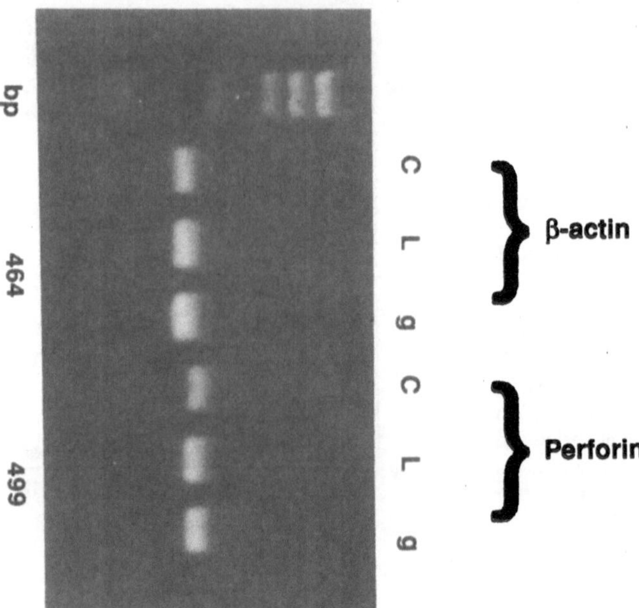

Figure 2. Constitutive expression of perforin gene by lpr and gld DN T cells. The expression of mRNA for perforin in the DN T cells was studied using PCR. Lane 1 is a molecular marker ($\phi\chi$174 HaeIII digest), C represents a positive control consisting of a cytotoxic T cell line and lanes L and g represent DN T cells from *lpr* and *gld* mice respectively. The bands at 464 bp and 499 bp represent those for β-actin and perforin respectively.

cells express the ligands for these homing receptors, it was suggested that such cytotoxic T cells may be able to mediate lysis of endothelial cells.[2,4] In the current study we addressed whether *lpr* DN T cells would mediate spontaneous cytotoxicity of a syngeneic SV40 transformed endothelial cell line designated TME-3H3. These studies indicated that *lpr* DN T cells were able to mediate efficient lysis of endothelial cells whereas similar cells from *gld* mice failed to exhibit cytolytic activity. Furthermore, the cytotoxicity of the endothelial cells was MHC-unrestricted and TCR independent. Also, this cytolytic activity was inhibited in the presence of hyaluronic acid thereby suggesting that the endothelial cell lysis caused by DN T cells was dependent on the interaction between CD44 and hyaluronic acid.

Role of Perforin in DN T Cell-Mediated Cytotoxicity

Recent studies have demonstrated that the cytotoxic activity can be attributed to two independent mechanisms, first Fas-based and secondly, perforin-based.[5] In the current study, we observed that the DN T cells constitutively expressed perforin but were dependent on Fas and FasL interactions to mediate cytotoxic activity. In order to address whether perforin plays a significant role, we added EGTA into the cytotoxicity assay to deplete extracellular calcium and study its effect on cytotoxicity because earlier studies have shown that perforin dependent cytotoxic activity is calcium dependent whereas Fas mediated cytolytic activity is calcium independent.[5] The data shown in Fig. 3 suggested that when *lpr* DN T cells were tested for cytolytic activity against P815 tumor targets in the presence of anti-CD3 mAbs, this lytic activity was inhibited in the presence of EGTA whereas it was only partially inhibited in the presence of anti-Fas mAbs. These data therefore suggested that the *lpr* DN

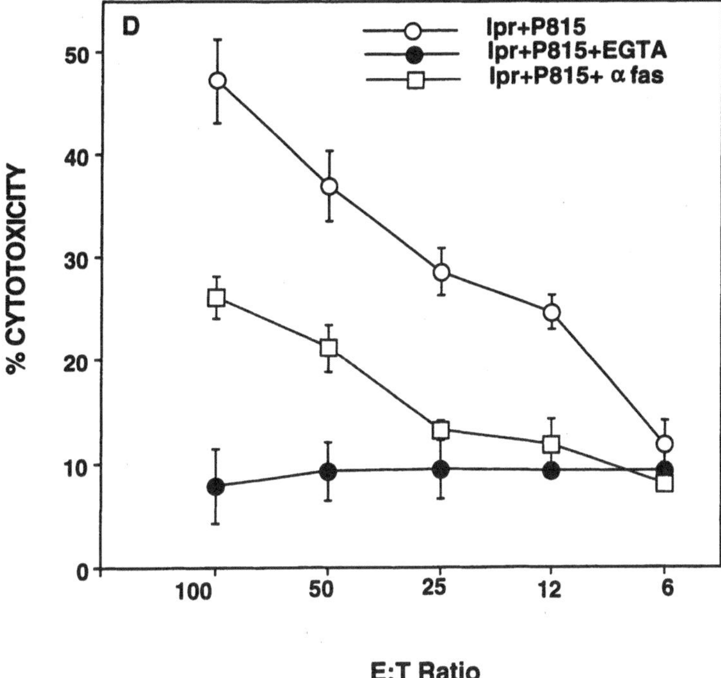

Figure 3. Role of perforin in the cytotoxic activity of lpr DN T cells. The lpr DN T cells were tested for redirected cytotoxic activity against P815 targets in the presence of anti-CD3 mAbs. The cytotoxicity was carried out in the absence or presence of EGTA (6mM) + $MgCl_2$ (3mM) or anti-Fas (Jo2) mAbs (200 ng/ml).

T cell-mediated cytoxicity when activated through the TCR is perforin-dependent and that for granule exocytosis, interaction between Fas and FasL maybe critical.

DISCUSSION

Mice homozygous for *lpr* and *gld* mutations develop a severe lymphoproliferative disease characterized by the accumulation of large numbers of DN Tcells. These cells possibly represent autoreative T cells which proliferate in response to stimulation with self antigens and are unable to undergo apoptosis and deletion *in vivo* due to lack of expression of Fas or FasL, necessary for the cells to undergo apoptosis. In the current study we demonstrated that both *lpr* and *gld* DN T cells are phenotypically very similar and express several cytokine genes such as perforin, IFN-γ and TNF-α. Interestingly however, the *gld* DN T cells failed to mediate cytolytic activity unlike the *lpr* DN T cells which participate in both spontaneous and redirected cytotoxicity of tumor targets.

Several lines of evidence suggested that the reason why *gld* DN T cells are unable to mediate cytotoxicity is due to the defect in the functional expression of FasL. Furthermore, the *lpr* DN T cells which mediate spontaneous cytolytic activity were also found to depend on the interaction between Fas and FasL to mediate cytolytic activity. These data together therefore suggested that although DN T cells are cytotoxic in nature they may not be able to exert this function *in vivo* in the natural host because in *lpr* mice the autologous target cells would be deficient in the expression of Fas and in *gld* mice, the cytolytic effector cells would be deficient in the functional expression of FasL. Although the DN T cells may fail to exhibit the cytolytic functions *in vivo* they may contribute towards the induction of autoimmunity by being unable to mediate cytolytic activity against autologous B cells which respond to self antigens. This may permit uncontrolled growth and differentiation of B lymphocytes leading to the production of autoantibodies as seen in these strains of mice. Secondly, although the DN T cells are unable to mediate cytolytic activity in the natural host, the fact that they can constitutively express several cytokine genes suggests that these cytokines that are produced *in vivo* may contribute towards the induction of an inflammatory response particularly at the site of blood vessels, thereby leading to vasculitis as well as other types of inflammatory pathology that is commonly seen in these autoimmune strains of mice.

Recent studies have characterized two distinct mechanisms of cytotoxicity, Fas-based and perforin-based, which appear to be independent based on the fact that cytotoxic cells from perforin knockout mice can lyse target cells using Fas-dependent pathway and *gld* cytotoxic cells can lyse by perforin pathway.[5] In this context, our data suggest that DN T cell are unique, inasmuch as, they are dependent on Fas pathway despite constitutively expressing perforin. Also, our data suggested that for certain forms of cytolytic activity, such as when activated through the TCR, the DN T cells could mediate perforin-based cytolytic activity and although such cytotoxic activity was also dependent on the interaction between Fas and FasL. These data suggested that in DN T cells, the perforin and Fas dependent pathways may be linked.

ACKNOWLEDGMENT

This work was supported in part by a grant from American Cancer Society.

REFERENCES

1. Cohen, BL, and Eisenberg, RA: *Lpr* and *gld*: Single gene models of Systemic Autoimmunity and Lymphoproliferative Disease. Annual Review of Immunology 9:243, 1991.
2. Hammond, D, Nagarkatti, PS, Gote, L R, Seth, A, Hassuneh, MR and Nagarkatti, M: Double negative T cells from MRL *lpr/lpr* mice mediate cytolytic activity when triggered through adhesion molecules and constitutively express perforin gene. J Exp Med 178:2225, 1993.
3. Seth, A, Pyle, RH, Nagarkatti, M and Nagarkatti, PS: Expression of the J11d marker on peripheral T lymphocytes of MRL *lpr/lpr* mice. J of Immunol 141:1120, 1988.
4. Seth, A, Gote. L, Nagarkatti, M and Nagarkatti, PS: T cell receptor independent activation of cytolytic activity of cytotoxic T lymphocytes mediated through CD44 and gp90^{Mel14}. Proc Natl Acad Sci USA 88:7877, 1991.
5. Nagata, S and Goldstein, P: The Fas death factor. Science 267:1449, 1995.

51

ANTISENSE OLIGONUCLEOTIDES TO mdr1 INCREASE ADRIAMYCIN SUSCEPTIBILITY IN BREAST CANCER CELL LINES

D. Frank Andrews III,* Edward A. Faber Jr., Holly Hawk, and Stanley J. Geyer

Western Pennsylvania Cancer Institute
and Department Of Pathology
Western Pennsylvania Hospital, Pittsburgh, Pennsylvania

ABSTRACT

Breast cancer remains the leading case of death in woman between the ages of 30 and 55 in industrialized societies. In premenopausal women, approximately 40% have estrogen receptor negative disease. While usually being resistant to hormonal manipulation, disease of this type is generally highly susceptible to anthracycline based chemotherapy. Indeed, anthracyclines constitute the single most active group of drugs in this disease. As in other malignancies, P-glycoprotein/*mdr1*-mediated multi-drug resistance to anthracycline chemotherapy emerges as a selected phenotype after breast cancer has undergone drug exposure. Multi-drug resistant breast cancer presents a therapeutic problem particularly suited to antisense oligonucleotide therapy. In the present study, *mdr*-antisense oligonucleotide therapy is successfully applied in anthracycline-resistant, p-glycoprotein-positive breast cancer cell lines in order to restore anthracycline responsiveness.

INTRODUCTION

In 1990, it was estimated that between 800,000 and 900,000 new cases of breast cancer were diagnosed worldwide. Breast cancer remains a leading cause of mortality among woman in industrialized world as well as developing countries. A number of prognostic indicators identify women at greatest risk for breast cancer recurrence following resection.[1] Adjuvant chemotherapy remains standard care for those women at greatest risk for relapse.

* Please address all correspondence to: D. Frank Andrews III M.D., F.A.C.P., Director, Bone Marrow Transplant Program, Western Pennsylvania Hospital, 4800 Friendship Ave., Suite 2303, Pittsburgh, Pennsylvania 15224.

Molecular Biology of Hematopoiesis 5, edited by Abraham et al.
Plenum Press, New York, 1996

In this population as well as in women with metastatic breast cancer, anthracyclines continue to serve as frontline chemotherapy. Studies of doxorubicin as a single agent in clinical trials have demonstrated superior overall response rates of 30 to 50%.[2,3] Several recently published adjuvant chemotherapy trials have demonstrated disease free survival and overall survival advantages for women receiving adjuvant chemotherapy regimens containing doxorubicin.[4-6]

In women with relapsed or metastatic disease, the emergence of tumor cells resistant to chemotherapy remains the major cause of death. Although the reduction in 10 year mortality by adjuvant chemotherapy is evident from clinical trials, induction of chemotherapeutic resistance by prior exposure to chemotherapy is a documented mechanism *in vitro*.[7] Indeed, drug resistant cell lines are routinely developed by selecting cells in culture under selective pressure of low dose *in vitro* chemotherapy.[8] Various mechanisms arise for drug resistance depending upon the therapeutic agent in question. The multidrug resistance gene *mdr1* and its protein product the P-glycoprotein are a major resistance mechanism developed by cells exposed to anthracyclines or vinca alkaloids. *mdr* has now been recognized in breast cancer cells *in vitro* and *in vivo*.[9-11]

The multidrug resistant phenotype was initially described by Ling *et al* in the 1970's in Chinese Hamster Ovary Cells.[12,13] The drug resistant phenotype was noted to confer resistance to structurally unrelated chemotherapeutic compounds simultaneously. Particularly this mechanism resisted naturally occurring compounds of the anthracycline class and vinca alkaloid class of agents. In 1986, Gros *et al* reported the isolation of a cDNA sequence from Chinese Hamster Ovary cells which encoded the multidrug resistant phenotype or *mdr* gene.[14] Roninson *et al* reported the isolation of human *mdr* cDNA sequences from the KB carcinoma cell line.[15] The *mdr1* gene is located at human chromosome locus 7q21.1 and encodes 1280 amino acid protein of approximately 170 Kd designated the P-glycoprotein (P for permeability).[16] This protein contains 12 transmembrane domains. Two sites within the molecule are lipophilic and appear to have ATP binding activity. P-glycoprotein shares considerable homology with a series of bacterial transport proteins (permeates) found in gram negative bacteria.[17] It appears that proteins of this type are membrane bound, pore-forming molecules with partial specificity for certain substrates. Expression of the P-glycoprotein can be found in numerous normal tissues as well as human tumor types including breast cancer.[10,18,19] Amplification and overexpression of the *mdr* gene has been described in cultured cells in association with particular cytogenetic abnormalities such as homogeneously staining regions (HSR's). Transfection of the *mdr1* encoding cDNA can confer multidrug resistance on previously sensitive cell lines.[20,21] In studies of acute myelogenous leukemia, generally less than 25% of AML's express *mdr1* prior to therapy. However, at the time of relapse, between 50% and 80% of AMLs express the P-glycoprotein. In a study by Santo *et al*, expression of P-glycoprotein in AML correlated directly with response of disease to chemotherapy.[22-26] Similarly, while normal breast cancer tissues express low levels of P-glycoprotein, Ro *et al* observed that frequent expression of P-glycoprotein is found in breast cancers following preoperative chemotherapy and that increased expression significantly correlated with poor response to further chemotherapy.[27] Verrelle *et al* found that 85% of breast cancer patients expressed P-glycoprotein in some of their tumor cells by immunohistochemical staining.[28] Thus, *mdr1* is present within cells as an inherent mechanism within the cells from which acquired drug resistance derives after exposure to antineoplastic agents.

The objective of the present study is to demonstrate that antisense therapy can be successfully employed to downregulate the expression of P-glycoprotein-positive human breast cancer cell lines. In order to accomplish this, *mdr1* expressing breast cancer cell lines were analyzed by flow cytometry for surface expression of P-glycoprotein in order to select the chemotherapy resistant cell line. Resistance to doxorubicin was further established by toxicity assays. *mdr*-antisense ODN's were then used to restore anthracycline susceptibility in a resistant breast cancer cell lines which express the *mdr1*/P-glycoprotein product in high levels.

MATERIALS AND METHODS

Breast Carcinoma Cell Lines

Breast cancer cell lines which are described in the medical literature as either anthracycline-resistant or anthracycline-sensitive are and are commercially available were obtained. (American Type Culture Collection, Rockville, MD). These include the MCF7 cell line which although adriamycin sensitive is known to harbor the P-glycoprotein.[29,30] The MDA-MB-231 cell line was selected specifically in the presence of adriamycin. Also available as a control cell line, the BT-474 line is shown to be P-glycoprotein negative, and the HS578bst cell line was derived from normal breast tissue. The above cell lines are chosen specifically to serve as either controls or as cell lines likely to be doxorubicin resistant due to prior therapy. These cell lines are all derived from different women, including both Afro-American and Caucasian females.

Analysis of Cell Lines by Flow Cytometry

The MCF-7, MDA-MB-231, and BT-474 cell lines were analyzed by flow cytometry to detect the mdr1 gene product, the P-170-glycoprotein (Pgp-170). Each cell line was cultured until 80 % confluent and collected on ice with 1 mM EDTA in PBS (pH 7.4). Cells were initially labelled with clone MRK16, a mouse anti-human anti-P-glycoprotein IgG antibody, and subsequently conjugated with goat anti-mouse IgG2A antibody labelled with FITC.[Caltag, S. San Francisco] Negative controls of the same isotype as the MRK16 antibody (IgG2A-FITC: mouse anti-human) were run simultaneously for each sample tested. Results of flow cytometry are given in Figure III. Flow cytometry was performed on a FacSCAN flow cytometer, (Becton Dickinson, San Jose, CA.) and data analysis was performed by Lysis II, Version 1.1, software.(Becton Dickinson, San Jose, CA.)

Flow cytometry results in conjunction with drug resistance assays were used to select the appropriate cell line for Pgp antisense studies.

Colony Assay of Selected Breast Cancer Cell Lines

The MCF-7, MDA-MB-231, and BT-474 cell lines were analyzed by a colony assay to determine the inhibitory concentration of 50% (IC_{50}) of adriamycin. Each cell line was grown to 80% confluence in a 150 cm^2 tissue culture flask, collected by trypsin, washed with PBS and resuspended at 500 viable cells/ml of media. A total of 1000 cells were plated in 6-well plates and allowed to attach for 24 hours. Adriamycin was administered in serial log dilutions beginning with 100 nM and ending with 0.01 nM, along with a single untreated control. These conditions were repeated in triplicate. After 24 hours, the media was removed, replaced with fresh media, and the cells were allowed to form colonies for 7 to 10 days. The media was removed and the cells washed twice with PBS. The colonies were fixed with 2 ml of methanol for 15 minutes at room temperature and again washed twice with PBS. The colonies were fixed with 2 ml 20 % Giemsa stain for 15 minutes at room temperature and washed thrice with PBS. Once the colonies were allowed to dry for 1 hour, they were counted over an inverted microscope at 40X magnification. A colony was determined to be a cell aggregation containing greater than 50 cells.

Antisense Oligonucleotide Colony Assay of MDA-MB-231 Cell Line

The MDA-MB-231 cell line was selected for antisense oligonucleotide administration since it possessed a characteristic survival curve and a high level of Pgp-170 expression. MDA-MB-231 cells were grown to 80 % confluence in a 150 cm_2 tissue culture flask, collected by trypsin, washed with PBS and resuspended at 2000 viable cells/ml of media. 1000 cells were plated in 24 wells plates and allowed to attach for 24 hours. For each oligo combination, 6 wells were established, with a total of 24 wells seeded in 4 lanes. Three oligo combinations were utilized: a) A1, which anneals to position -9 to 6; b) A2, which anneals to position 993 to 1008; and c) A1 + A2. In order to reverse anthracycline drug resistance with antisense ODN's directed against *mdr1*, these two 15-mers spanning separate positions of the *mdr1* gene were used. These sequences include 5'(CTC CAT CAC CAC CTC)3' and 5'(GAG TGA CAT AAG AAA) 3'. (Figure 6) These 15-mers were modified at their 3' ends to include phosphorothioate modified bases ("S-Oligos") in order to resist exonucleases. The former sequence (-9 to +6bp) is notable in that it includes the AUG initiation colon. The second sequence located between 993-1008 bp is a key loop forming site. Antisense directed against these sites has proven effective in former studies.[31-33]

All oligos were administered at a dose of 10 μM. In the combined experiments, the concentration (A1 + A2)was 10 μM. The oligos were administered every 24 hours for 72 hours, followed by the addition of Adriamycin for 24 hours in serial log dilutions beginning with 100 nM and ending with 0.01 nM. After 24 hours, the cells in each well were collected by trypsin, replated in 6-well plates, and allowed to form colonies for 7 to 10 days at which point they were fixed with methanol and stained with 20% Giemsa and counted. These conditions were repeated in three separate experiments.

For each condition, a colony was defined as an aggregation with greater than 50 cells. The number of colonies from each adriamycin concentration was added together and the average was calculated from experiments performed in triplicate.

RESULTS

Pgp Expression by Flow Cytometry

Pgp-170 expression by each cell line was analyzed by flow cytometry. Pgp expression detected by the MRK 16 antibody was performed as described under Methods. The results of flow cytometry demonstrated variable Pgp expression among the various cell lines (Figure 1). Pgp expression was determined by gating of cells which stained with the MRK 16 antibody (filled curve) which were not included under the curve of the isotype control antibody (open curve). Normal human peripheral blood was used as a control and expressed Pgp on 19.65% of cells. Pgp expression by BT-474, MCF-7, and MDA-MB-231 were 3.4%, 17.79%, and 21.0% respectively. For this reason, the MDA-MB-231 cell line was selected as the Pgp expressing cell line for study in anthracycline-resistance experiments.

Dose Responsiveness of Selected Breast Cancer Cell Lines to Adriamycin

Surviving cell viability after 72 hours exposure to log increased doses of adriamycin are given in Table 2 and Figure 2. The survival demonstrates that the normal breast cell line HS578bst was the most susceptible while the BT-474 cell line was less susceptible to adriamycin in this assay. These data were used to determine the proliferation curves given

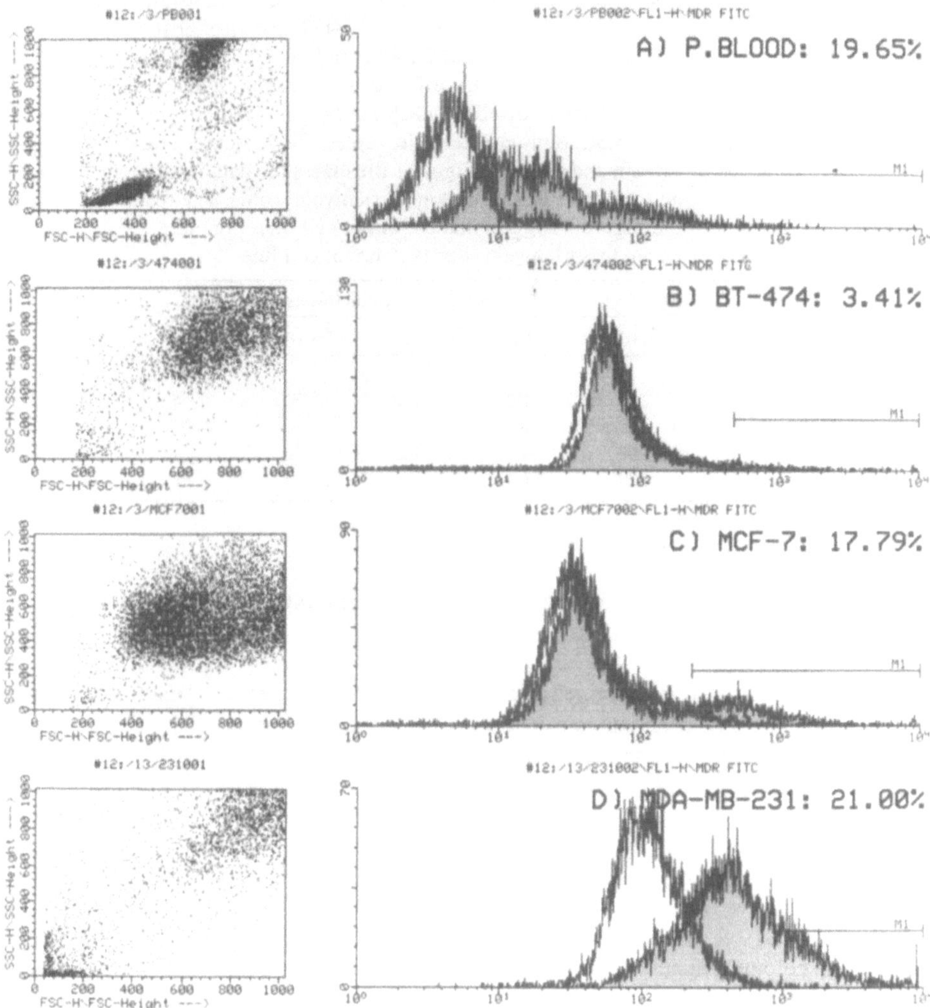

Figure 1. Light Scatter and Histogram Plots of Peripheral Blood and Breast Tumor Cell Lines. Normal peripheral blood (PB002) was used as a control. Samples were stained with a mouse anti-human MRK16, followed by conjugation to a goat anti-mouse IgG2a-FITC. Negative isotopic controls consisted of mouse anti-human IgG2a-FITC and are represented by the (color) overlays in each histogram. Percentage of the total number of cells expressing Pgp-170 is indicated in the upper right hand corner of each histogram. Negative controls of goat-antimouse isotype which is a FITC labelled antibody were employed as shown under the unfilled area under the curve. M1 denotes the gate setting. Percentage of cells expressing Pgp is shown by the area under the colored curve for each cell line. The 231 cell line shows the greatest Pgp expression while the negative 474 breast cancer cell line shows the lowest degree of expression.

in Table 3 and Figure 3. On this curve the BT-474 cell line actually proliferated in the presence of 0.01 and 0.1 nM adriamycin. Between 1 nM and 10 nM adriamycin all cell lines showed a dramatic decrease in survival and proliferation Figures 2 and 3. Thus, the BT-474 cell line while expressing low levels of Pgp demonstrated remarkable resistance to the anthracycline. The MDA-MB-231 line and the MCF-7 cell line showed somewhat lower survival and decreased proliferation curves.

Table 1. Dose response data of selected breast cancer cell lines. Data and results of treating breast cancer cell lines MCF-7, MDA-MB-231, BT-474, and Hs578bst with adriamycin. MCF-7, MDA-MB-231, and BT-474 are tumor cell lines, while Hs578bst is a normal breast cell line. Briefly, 500,000 cells were plated in a 75 cm^2 tissue culture flask and allowed to attach for 24 hours. Adriamycin was administered at the concentrations contained in Table 1 for 72 hours. The cells were collected by trypsin and counted. Figure 1 illustrates the data in a graphic format by plotting total cell number vs. log [adriamycin] (nM). The results in Table 1 and Figure 1 suggest that the tumor cell lines are more resistant to adriamycin than the normal cell line

	Number of Cells (thousands)					
[adriamycin] (nM)	0.01	0.1	1	10	100	1000
Cell Line						
MCF-7	350	282	188	20	5	0
MDA-MB-231	450	388	250	30	35	0
BT-474	740	642	442	40	15	0
Hs578bst	300	160	100	0	0	0

Effect of Increasing Doses of Adriamycin on Breast Cancer Cell Line Colony Formation

The colony assay was performed as described under *Methods* above. Again, colonies were subjected to log increases of adriamycin dose. For each of the 3 malignant cell lines, colony assay for survival was performed in triplicate at each dose level. (Table 4.) The survival curve plotted for these data is given in Figure 4. As determined from this curve, the IC_{50} was determined. IC_{50} for the MDA-MB-231 cell line, BT-474 and MCF-7 cell lines were 18 nM, 8.8 nM, and 4.1 nM respectively. Thus, by this assay the MDA-MB-231 cell line was more resistant to increasing doses of adriamycin. For this reason in conjunction with its expression of Pgp by flow cytometry, the MDA-MB-231 cell line was selected for antisense studies. The three antisense oliogonucleotides (15-mers) were used singly or in combination to treat cells in culture prior to adriamycin exposure and colony formation as described above. Again, all experiments were performed in

Table 2. Proliferation data of selected breast cancer cell lines. Data and results of treating breast cancer cell lines MCF-7, MDA-MB-231, BT-474, and Hs578bst with adriamycin. For each cell line and concentration of adriamycin, percent increase in cell number was calculated by dividing each corresponding total cell number in Table I by the original cell number, or 500,000 cells. The results in Table II and Figure II reinforce the suggestion that the tumor cell lines are more resistant to adriamycin than the normal cell line since the tumor cell lines are either proliferating more or not dying as fast as the normal cell line. From Figures I and II, it is suggested that the utility of the cell lines for the antisense oligonucleotide studies would by BT-474>MDA-MB-231>MCF-7, with BT-474 being the optimal cell line

Cell Line	Number of Cells (thousands)					
[adriamycin] (nM)	0.01	0.1	1	10	100	1000
MCF-7	350	282	188	20	5	0
MDA-MB-231	450	388	250	30	35	0
BT-474	740	642	442	40	15	0
Hs578bst	300	160	100	0	0	0

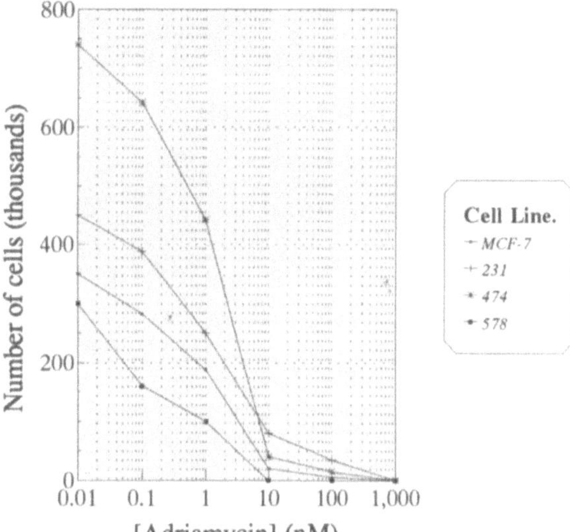

Figure 2. Chemotherapy sensitivity of each cell line was established. This graphic of a dose response curve shows the number of cells surviving at each adriamycin dose level for each cell line. The 474 cell line shows the greatest degree of survival on this simple survival graph. The Pgp expressing 231 cell line shows the second best survival. The 578 is the normal human mammary cell line.

triplicate. (Table 5). The IC_{50} was again calculated from survival curves plotted from the data given in Table 5.(Figure 5) Each individual antisense ODN increased the susceptibility of the MDA-MB-231 cell line to adriamycin. In each case, the IC_{50} on the curve was shifted to the right as compared to cells untreated with antisense prior to adriamycin exposure. Oligo A1 produced an IC_{50} of 1.8 nM while oligo A2 produced 700 pM. The two antisense ODN's in combination (A1 + A2) were synergistic in reducing cell survival with even low doses of adriamycin. The IC_{50} for the combination A1 + A2 was 90 pM as shown in Figure V. Thus, the combination of ODN's (A1 + A2) was more effective than either ODN alone in restoring adriamycin responsiveness.

DISCUSSION

mdr1 represents a potentially reversible mechanism of acquired resistance to an otherwise highly effective class of drugs, the anthracyclines, in breast cancer. Fojo *et al*

Table 3. Proliferation data of selected breast cancer cell lines. Data and results of treating breast cancer cell lines MCF-7, MDA-MB-231, BT-474, and normal breast cell line Hs578bst with adriamycin. For each cell line and concentration of adriamycin, percent increase in cell number was calculated by dividing each corresponding total cell number in Table II by the original cell number, or 500,000 cells. Figure III illustrates the data in a graphic format by plotting proliferation vs log [adriamycin] [nM]. The results in Table !II and Figure III substantiate the evidence that the tumor cell lines are more resistant to adriamycin than the normal cell line. The tumor cell lines are either proliferating faster or living longer than the normal cell line

Cell Line	Number of Cells (thousands)					
[adriamycin] (nM)	0.01	0.1	1	10	100	1000
MCF-7	70.00	56.40	37.40	4.00	1.00	0.00
MDA-MB-231	90.00	77.60	50.00	16.00	7.00	0.00
BT-474	148.00	128.4	88.40	8.00	3.00	0.00
Hs578bst	60.00	32.00	20.00	0.00	0.00	0.00

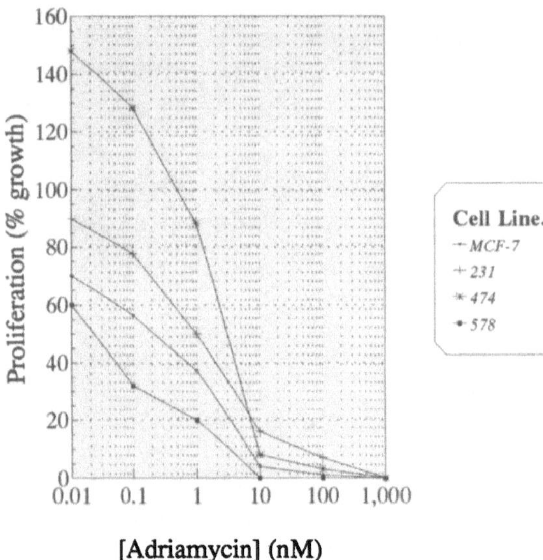

Figure 3. Percent growth of cells in various concentrations of adriamycin. In these cell cultures, 500,000 were plated in a 75cc tissue culture flasks and allowed to attach for 24 hours. adriamycin was administered at the concentrations shown for 72 hours. The cells were collected and stained. Cell survival and percent cell proliferation were determined. Note that the 231 cell line shows minimal inhibition to the normal cell line.

demonstrated that high levels of P-glycoprotein expression were associated with increased drug efflux thus preventing intracellular drug accumulation.[34] Recently, agents such as calcium channel blockers, tamoxifen and cyclosporines have been investigated as methods for modulating the activity of this efflux pump.[35-37] In most incidences, however, it appears that doses of these agents are associated with a number of side effects when adequate drug levels are achieved.[38,39]

Methods for effectively delivering ODN's to the tumor without exposing them to nuclease destruction poses an important problem. Lipofection is a recently developed modality for delivery of the new molecular therapies currently under development in cancer

Table 4. Preliminary colony data of selected breast cancer cell lines

	Number of Colonies > 50 Cells						
[adriamycin] (nM)	0.01	0.1	1	10	100	1000	Control
MCF-7							
Group I	157	118	100	63	20	0	162
Group II	140	136	114	93	33	0	200
Group III	147	117	118	74	34	0	175
Average	148	124	111	77	29	0	179
Survival	82.68	69.27	62.01	43.02	16.20	0	—
MDA-MB-231							
Group I	178	133	160	92	18	0	200
Group II	168	175	134	120	21	0	158
Group III	157	164	133	100	28	0	153
Average	166	157	142	104	22	0	170
Survival	97.65	92.35	83.53	61.18	12.94	0	—
BT-474							
Group I	103	166	101	82	26	0	188
Group II	171	144	116	96	28	0	176
Group III	176	136	152	87	32	0	179
Average	150	149	123	88	29	0	181
Survival	82.87	82.32	67.80	48.62	16.02	0.00	—

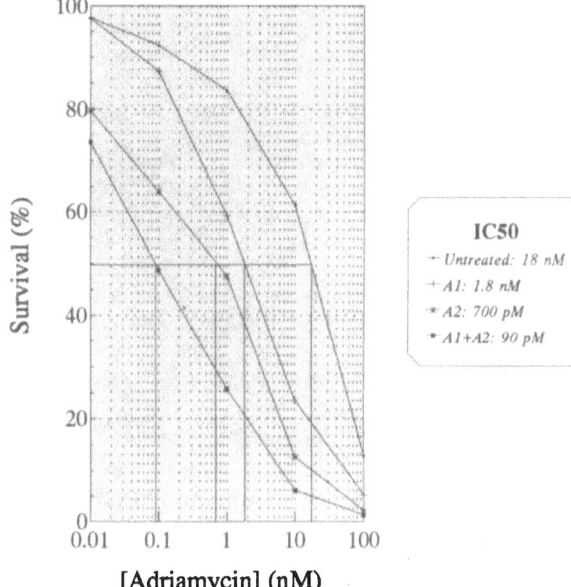

Figure 4. While the 474 cell line had previously shown superior survival in the tissue flask assay, the Pgp positive 231 cell line showed superior survival in the colony assay. These data in conjunction with the flow cytometry data, led to selection of the 231 cell line for use in antisense studies.

treatment and represents a promising strategy for the implementation of antisense therapy in breast cancer. While other modalities such as retroviral vectors have been employed, these pose particular safety risks and are difficult and time consuming to devise.[40] The recent successful application of liposomal amphotericin-B in human clinical trials attests to the safety and efficacy of liposomal vehicles.[41-43] Liposomal/antisense therapy has several practical advantages: 1) oligonucleotides or expression vectors are protected from serum

Table 5. Colony data of 231 cells treated with antisense oligonucleotides

	Number of Colonies Containing > 50 Cells					
[adriamycin] (nM)	0.01	0.1	1	10	100	Control
Oligo A1: -9...6						
Experiment I	418	447	300	87	25	448
Experiment II	459	379	280	117	22	444
Experiment III	465	373	232	121	26	482
Average	447	400	271	108	24	458
Survival	97.60	87.34	59.17	23.58	5.24	—
Oligo A2: 993...1008						
Experiment I	421	304	204	59	13	448
Experiment II	314	289	232	63	8	444
Experiment III	361	285	219	53	9	482
Average	365	293	218	58	10	458
Survival	79.69	63.97	47.60	12.66	2.18	—
Oligo A1 and A2						
Experiment I	338	253	130	30	8	448
Experiment II	292	203	126	31	6	444
Experiment III	382	213	98	22	5	482
Average	337	223	118	28	6	458
Survival	73.58	48.73	25.76	6.11	1.31	—

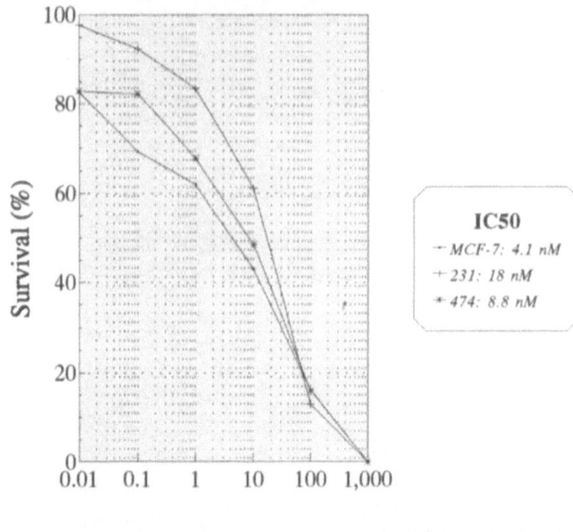

[Adriamycin] (nM)

Figure 5. Data and Results of treating breast cancer cell line MDA-MB-231 with antisense oligonucleotides and adriamycin. Table 5 depicts a culmination of three experiments. For each condition, a colony was defined as an aggregation with greater than 50 cells. The number of colonies of each condition for the three experiments was added, averaged and survival was calculated by dividing the average number of colonies for each condition by the average value of the control. Figure 5 illustrates a plot of survival vs. adriamycin concentration for each oligo combination and the data for the untreated control is found in Table 4. Inhibitory concentration of 50% (IC_{50}) was determined from the graph, as indicated by the lines extrapolated from 50%. The curves show percentage colony survival in experiments in which antisense oligos were used. Cells were exposed to increasing concentrations of adriamycin after exposure to ODN's. The inhibitory concentration of 50% as determined from the graph shows the greatest survival in 231 cells untreated with antisense oligos. Treatment with the combined antisense oligos showed the lowest IC_{50} and therefore the greatest susceptibility to adriamycin. The use of either antisense oligo in isolation showed a level of adriamycin resistance intermittent between and the combined A1 + A2 curves.

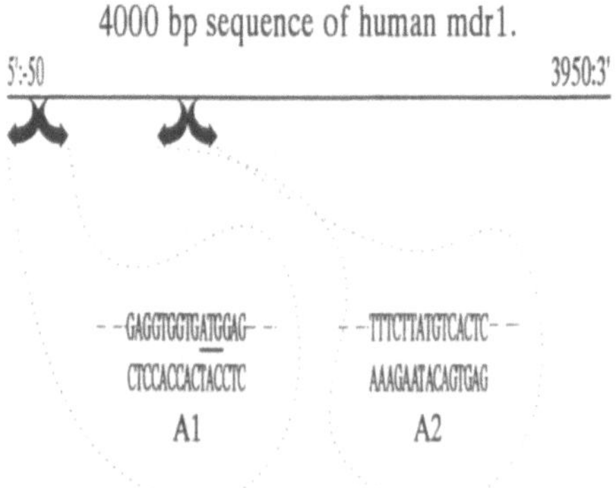

Figure 6. The antisense DNA sequences directed against human mdr1. Of note, oligo A1 binds to the extreme 5' position of the mdr1 mRNA which includes the ribosome binding sequence. Oligo A2 binds to position in mdr1 of which forms a key loop forming site in the mRNA. A2 presumably prevents the formation of a secondary structure necessary to translation of the mRNA.

nucleases when complexed with liposomal membranes; 2) lipofection can be targeted through the use of monoclonal antibodies incorporated into the lipsomal membranes thereby sparing normal tissues from excessive exposure to the antisense modality (which may have unknown side effects on normal tissues which may require on mdr 1 function); and finally, 3) with monoclonal targeted delivery, the antisense effect will be less diluted by uptake by untargeted tissues. Lipofection reagents such as DOTMA and DOPE are positively charged lipids that can effectively complex with nucleic acids.[44] When applied in cell culture, the lipid/nucleic acid complex fuses with plasma membranes and transfers the nucleic acid into the cytoplasm.[45,46] This application of cationic lipids has been used to successfully transfect numerous mammalian cell lines including human cells. More recently, lipofection has been successfully applied *in vivo* in rats and mice for the purpose of nonviral gene transfection.[44,47] Furthermore, it has been possible to selectively target specific mouse lung endothelial cells through the use of N-terminal modified poly (L-lysine)-antibody (NPLL) conjugates in order to specifically target particular cell types.[48] The association of the antibody, NPLL-cationic liposome, and DNA forms a stable ternary electrostatic complex.[49] As the monoclonal antibody specifically targets antigens found on particular cell types, the cationic liposome fuses to the specific cell membrane resulting in introduction of the DNA into the appropriate cell. Thierry *et al* have successfully employed liposomally encapsulated antisense ODN's to introduce antisense ODN's directed against the *mdr1* mRNA into doxorubicin resistant human ovarian carcinoma cells.[31] In so doing they were able to successfully reverse drug resistance to doxorubicin as demonstrated in cell toxicity assays. Antisense therapy directed against P-glycoprotein-mediated drug resistance constitutes a potential therapeutic modality with specific, clinical application in the short term. Therapeutic application of antisense oligonucleotides (ODN's) introduced via antibody-targeted liposomes can be developed initially through the employment of breast cancer cell lines and eventually the use of animal models. If these models are successful, more broadly based animal studies could rapidly justify the use of agents in phase I human trials. Indeed, liposomally encapsulated agents such as amphotericin-B have already entered clinical trials. If this modality is successful in *mdr*-mediated drug resistance, potentially this liposome-antisense method might find wider application in many tumors.

From the present study of *mdr1* mediated resistance in breast cancer cell lines, several conclusions may be drawn or inferred. It is also true that developing cell lines under selective pressure will enhance *mdr1* expression. The MDA-MB-231 cell line taken from a patient previously treated with adriamycin shows increased resistance although the cell line itself was grown without selective pressure of adriamycin. Thus, the inherent resistance of the cell line developed *in vitro* was still present *in vivo* a cell line of this type is, therefore, useful in studies without being grown *in vivo* under selective pressure. Flow cytometry proved a useful tool for detecting the presence of Pgp in the MDA-MB-231 cell line suggesting that the increased resistance to adriamycin results from the presence of *mdr1* expression. This fact along with toxicity assay was used to select the MDA-MB-231 cell line as the cell line of choice for the antisense experiment. It also served to demonstrate the usefulness of flow cytometry in detecting Pgp expression on the various breast cancer cell lines.

The BT-474 cell line on one assay showed a high level of resistance to adriamycin suggesting that this cell line possesses a mechanism of resistance other than Pgp. While *mdr1* is known to be a mechanism of chemotherapy resistance for a number of tumors, breast cancer is known to possess additional mechanisms of resistance. Such mechanisms increase protein kinase C, topoisomerase II, glutathione-S-transferase and dihydrofolate-reductase have also been demonstrated in breast cancer cell lines.[50-53] Certainly a mechanism such as topoisomerase II could account for the resistance in the BT-474 cell line.

The MDA-MB-231 cell line demonstrated increased susceptibility to adriamycin in the presence of *mdr1* antisense ODN's as compared to the same cell line untreated with

ODN's. Each ODN singly increased the cell line susceptibility to doxorubicin. It might be anticipated that the A1 ODN is more effective in restoring drug susceptibility than is the A2 ODN as A1 contains the initiation codon. The remarkable further shift to the right of the survival code and IC_{50} in the A1 + A2 experiment is of note as the total ODN concentration in the tissue culture median was the same as that with either ODN alone. This suggests a potent synergy when ODNs are used in combination. The data in the present study in combination with the findings of other investigators confirm the potential usefulness of ODN therapy in drug resistant malignancy. In particular these data show the potential usefulness of antisense therapy as a therapeutic modality in *mdr1* mediated resistant breast cancer.

REFERENCES

1. Consensus conference. Adjuvant chemotherapy for breast cancer. J A M A 265:391, 1985
2. Hoogstraten B, George SL, Samal B, et al. : Combination chemotherapy and Adriamycin in patients with advanced breast cancer. Cancer 38:13, 1976
3. Ahmann D, Bisel H, Eagan R, et al. : Controlled evaluation of Adriamycin (NSC-123127) in patients with disseminated breast cancer. Cancer Chemother Rep 58:877, 1974
4. Buzzoni R, Bonadonna G, Valagussa P, Zambetti M: Adjuvant chemotherapy with doxorubicin plus cyclophosphamide, methotrexate, and fluorouracil inthe treatment of resectable breast cancer with more than three positive axillary nodes. J Clin Oncol 9:2134, 1991
5. Misset JL, Gil-Delgado M, Chollet P, Belpomme D, Fargeot P, Fumoleau P: Ten years results of the French trial comparing adriamycin, vincristin, 5-fluorouracil, and cyclophosphamide to standard CMF as adjuvant therapy for noed positive breast cancer. Proc ASCO 11:54, 1992
6. Tormey DC, Gray R, Abeloff MD, Roseman DL, Gilchrist KW, Barylak EJ, Stott P, Falkson G: Adjuvant therapy with a doxorubicin regimen and long-term tamoxifen in premenopausal breast cancer patients: an Eastern Cooperative Oncology Group trial. J Clin Oncol 10:1848, 1992
7. McGuire WL, Clark GM: Prognostic factors and treatment decisions in axillary-node-negative breast cancer. N Engl J Med 326:1756, 1992
8. Mimnaugh EG, Fairchild CR, Fruehauf JP, Sinha BK: Biochemical and pharmacological characterization of MCF-7 drug-sensitive and AdrR multidrug-resistant human breast tumor xenografts in athymic nude mice. Biochem Pharmacol 42:391, 1991
9. Sanfilippo O, Ronchi E, De Marco C, Di Fronzo G, Silvestrini R: Expression of P-glycoprotein in breast cancer tissue and in vitro resistance to doxorubicin and vincristine. Eur J Cancer 27:155, 1991
10. Wishart GC, Plumb JA, Going JJ, McNicol AM, McArdle CS, Tsuruo T, Kaye SB: P-glycoprotein expression in primary breast cancer detected by immunocytochemistry with two monoclonal antibodies. Br J Cancer 62:758, 1990
11. Hedley DW: Flow cytometric assays of anticancer drug resistance. Ann N Y Acad Sci 677:341, 1993
12. Ling V, Thompson LH: Reduced permeability in CHO cells as a mechanism of resistance to colchicine. J Cell Physiol 83:103, 1974
13. Ling V, Kartner N, Sudo T, Siminovitch L, Riordan JR: Multidrug resistance phenotype in Chinese Hamster Ovary cells. Caner Treat Rep 67:869, 1983
14. Gros P, Croop J, Housman D: Mammalian multidrug resistance gene: complete cDNA sequence indicates strong homology to bacterial transport proteins. Cell 47:371, 1986
15. Roninson IB, Chin JE, Choi K, Gros P, Housman DE, Fojo A, Shen D-W, Gottesman MM, Pastan I: Isolation of human mdr DNA sequences amplified in multidrug-resistant KB carcinoma cells. Proc Natl Acad Sci USA 83:4538, 1986
16. Callen DF, Baker E, Simmers RN, et al. : Localization of the human multiple drug resistance gene, *mdr*1, to 7q21.1. Hum Genetics 77:142, 1987
17. Chen C, Chin JE, Ueda K, Clark DP, Pastan I, Gottesman MM, Roninson IB: Internal duplication and homology with bacterial transport proteins in the mdr1 (P-glycoprotein) gene from multidrug-resistant human cells. Cell 47:381, 1986
18. Thiebaut F, Tsuruo T, Hamada H, Gottesman MM, Pastan I, Willingham MC: Cellular localization of the multidrug-resistance gene product P-glycoprotein in normal human tissues. Proc Natl Acad Sci USA 84:7735, 1987
19. Fojo AT, Ueda K, Slamon DJ, Poplack DG, Gottesman MM, Pastan I: Expression of a multidrug-resistance gene in human tumors and tissues. Proc Natl Acad Sci USA 84:265, 1987

20. Gros P, Fallows DA, Croop JM, Housman DE: Chromosome-mediated gene transfer of multidrug resistance. Mol Cell Biol 6:3785, 1986
21. Shen D-W, Fojo A, Roninson IB, Chin JE, Soffir R, Pastan I, Gottesman MM: Multidrug resistance of DNA-mediated transformants is linked to transfer of the human mdr1 gene. Mol Cell Biol 6:4039, 1986
22. Musto P, Melillo L, Lombardi G, et al. : High risk of early resistant relapse for leukemic patients with presence of multidrug resistance associated P-glycoprotein positive cells in complete remission. Br J Haematol 77:50, 1981
23. Goldstein IJ, Galski H, Fojo A, et al. : Expression of a multidrug resistance gene in human cancers. JNCI 81:116, 1989
24. Herweijer H, Sonneveld P, Baas F: Expression of mdr1 and mdr3 multidrug-resistance genes in human leukemias. Leuk Res 82:1133, 1993
25. Sato H, Preisler H, Day R, et al. : Mdr1 transcript levels as an indication of resistant disease in acute myelogenous leukemia. Br J Haematol 75:340, 1990
26. Nooter K, Sonneveld P, Oostrum R, et al. : Overexpression of the mdr1 gene in blast cells from patients with acute myelocytic leukemia is associated with decreased anthracycline accumulation that can be restored with cyclosporine. Int J Cancer 45:263, 1990
27. Ro J, Sahin A, Ro JY, et al. : Immunohistochemical analysis of P-glycoprotein expression correlated with chemotherapy resistance in locally advanced breast cancer. Hum Pathol 21:787, 1990
28. Verrelle P, Meissonnier F, Fonck Y, et al. : Clinical relevance of immunohistochemical detection of multidrug resistance P-glycoprotein in breast carcinoma. JNCI 83:111, 1991
29. Yeh GC, Lopaczynska J, Poore CM, Phang JM: A new functional role for P-glycoprotein: efflux pump for benzo(alpha)pyrene in human breast cancer MCF-7 cells. Cancer Res 52:6692, 1992
30. Plumb JA, Milroy R, Kaye SB: The activity of verapamil as a resistance modifier in vitro in drug resistant human tumour cell lines is not stereospecific. Biochem Pharmacol 39:787, 1990
31. Thierry AR, Rahman A, Dritschilo A: Overcoming multidrug resistance in human tumor cells using free and liposomally encapsulated antisense oligodeoxynucleotides. Biochem Biophys Res Commun 190:952, 1993
32. Efferth T, Volm M: Modulation of P-glycoprotein-mediated multidrug resistance by monoclonal antibodies, immunotoxins or antisense oligodeoxynucleotides in kidney carcinoma and normal kidney cells. Oncology 50:303, 1993
33. Rivoltini L, Colombo MP, Supino R, Ballinari D, Tsuruo T, Parmiani G: Modulation of multidrug resistance by verapamil or mdr1 anti-sense oligodeoxynucleotide does not change the high susceptibility to lymphokine-activated killers in mdr-resistant human carcinoma (LoVo) line. Int J Cancer 46:727, 1990
34. Fojo A, Akiyama S-i, Gottesman MM, Pastan I: Reduced drug accumulation in multiply drug-resistant human KB carcinoma cell lines. Caner Treat Rep 45:3002, 1985
35. Kirk J, Houlbrook S, Stuart NS, Stratford IJ, Harris AL, Carmichael J: Differential modulation of doxorubicin toxicity to multidrug and intrinsically drug resistant cell lines by anti-oestrogens and their major metabolites. Br J Cancer 67:1189, 1993
36. Kirk J, Houlbrook S, Stuart NS, Stratford IJ, Harris AL, Carmichael J: Selective reversal of vinblastine resistance in multidrug-resistant cell lines by tamoxifen, toremifene and their metabolites. Eur J Cancer 29A:1152, 1993
37. Clarke R, Currier S, Kaplan O, Lovelace E, Boulay V, Gottesman MM, Dickson RB: Effect of P-glycoprotein expression on sensitivity to hormones in MCF-7 human breast cancer cells [see comments]. J Natl Cancer Inst 84:1506, 1992
38. List AF, Spier C, Greer J, Wolff S, Hutter J, Dorr R, Salmon S, Futscher B, Baier M, Dalton W: Phase I/II trial of cyclosporin as a chemotherapy-resistance modifier in acute leukemia. J Clin Oncol 11:1652, 1993
39. Pennock GD, Dalton WS, Roeske WR, et al. : Systemic toxic effects associated with high dose verapamil infusion and chemotherapy administration. JNCI 83:105, 1991
40. Temin HM: Safety considerations in somatice gene therapy of human disease with retrovirus vectors. Hum Gene Ther 1:111, 1990
41. Ringdën O, Meunier F, Tollemar J, Ricci P, Tura S, Kuse E, Viviani MA, Gorin NC, Klastersky J, Fenaux P: Efficacy of amphotericin B encapsulated in liposomes (AmBisome) in the treatment of invasive fungal infections in immunocompromised patients. J Antimicrob Chemother 28 Suppl B:73, 1991
42. Chopra R, Fielding A, Goldstone AH: Successful treatment of fungal infections in neutropenic patients with liposomal amphotericin (AmBisome)—a report on 40 cases from a single centre. Leuk Lymphoma 7 Suppl:73, 1992
43. Meunier F, Prentice HG, Ringdën O: Liposomal amphotericin B (AmBisome): safety data from a phase II/III clinical trial. J Antimicrob Chemother 28 Suppl B:83, 1991

44. Akamizu T, Ikuyama S, Saji M, Kosugi S, Kozak C, McBride OW, Kohn L: Cationic liposomes in in vivo gene transfection. Proc Natl Acad Sci USA 87:5677, 1990
45. Legendre J-Y, Szoka FC: Delivery of plasmid DNA into mammalian cell lines using pH-sensitive liposomes: comparison with cationic liposomes. Pharm Res 9:1235, 1992
46. Ellens H, Bentz J, Szoka FC: pH-induced destabilization of phophatidylehtanolamine-containing liposomes: role of bilayer contact. Biochemistry 23:1532, 1984
47. Leibiger B, Leibiger I, Sarrach D, Zuehlke H: Expression of exogenous DNA in rat liver cells after liposome-mediated transfection in vivo. Biochem Biophys Res Commun 174:1223, 1991
48. Trubetskoy VS, Torchilin VP, Kennel S, Huang L: Cationic liposomes enhance targeted delivery and expression of exogenous DNA mediated by N-terminal modified poly(L-lysine)-antibody conjugate in mouse lung endothelial cells. Biochim Biophys Acta 1131:311, 1992
49. Gershon H, Ghirlando R, Guttman SB, Minsky A: Mode of formation and structural features of DNA-cationic liposome complexes used for transfection. Biochemistry 32:7134, 1993
50. Giai M, Biglia N, Sismondi P: Chemoresistance in breast tumors. Eur J Gynaecol Oncol 12:359, 1991
51. Taylor CW, Dalton WS, Parrish PR, Gleason MC, Bellamy WT, Thompson FH, Roe DJ, Trent JM: Different mechanisms of decreased drug accumulation in doxorubicin and mitoxantrone resistant variants of the MCF7 human breast cancer cell line. Br J Cancer 63:923, 1991
52. Chen YN, Mickley LA, Schwartz AM, Acton EM, Hwang JL, Fojo AT: Characterization of adriamycin-resistant human breast cancer cells which display overexpression of a novel resistance-related membrane protein. J Biol Chem 265:10073, 1990
53. Doroshow JH, Akman S, Esworthy S, Chu FF, Burke T: Doxorubicin resistance conferred by selective enhancement of intracellular glutathione peroxidase or superoxide dismutase content in human MCF-7 breast cancer cells. Free Radic Res Commun 12-13 Pt 2:779, 1991

52

TRANSCRIPTIONAL ACTIVATION OF THE HUMAN HEME OXYGENASE-1 GENE DURING ACUTE-PHASE REACTION*

Shigeru Sassa,† Tadashi Nagai, Mayumi Nagai, Kinuko Mitani, Yoshiaki Fukuda, and Hiroyoshi Fujita

The Rockefeller University
New York, New York, 10021

ABSTRACT

The effects of human interleukin-6 (hIL-6), the major acute-phase inducer, on the level of the transcript of microsomal heme oxygenase (HO-1) were examined in human hepatoma cell lines, Hep3B, HepG2 and HepG2f. While Hep3B and HepG2 cells express well-differentiated phenotype of hepatocytes, HepG2f cells do not. mRNAs encoding HO-1 and haptoglobin (Hpt) increased following hIL-6 treatment in a time- and dose-dependent manner, in Hep3B and HepG2, but not in HepG2f cells. The hIL-6-mediated induction of HO-1 mRNA in Hep3B cells was completely abrogated by simultaneous treatment of cells with actinomycin D, but not with cycloheximide, suggesting that the induction occurs at the level of transcription.

To study the molecular mechanism of activation of the HO-1 gene by hIL-6, a chimeric plasmid was constructed by fusing a 5'-flanking region of the HO-1 gene to the firefly luciferase gene, and introduced into human HepG2 hepatoma cells. The luciferase activity in HepG2 cells transfected with the chimeric plasmid was significantly increased by treatment of cells with hIL-6. Deletion analyses revealed that the region between nucleotides -120 and -80 contains a regulatory element involved in the hIL-6-mediated induction of the gene in HepG2 cells. In this region, there is a sequence homologous to the type A IL-6 responsive element, which has been shown to be involved in the activation of certain acute-phase protein genes. Gel mobility shift assays demonstrated the existence of a nuclear factor which specifically binds to this region. These findings suggest that HO-1 is a positive acute-phase reactant in human liver-derived cell lines such as Hep3B and HepG2, and that

*This work was supported in part by grants from U.S.P.H.S. DK-39264 and the Yamanouchi Molecular Medicine Fund.

†Correspondence to: Shigeru Sassa, M.D.,Ph.D., The Rockefeller University, New York, New York 10021. Phone: (212) 327-8497; Fax: (212) 327-8872; E-mail address: sassa@rockvax.rockefeller.edu.

the nuclear factor specific to the IL6-RE may be involved in the activation of the HO-1 gene after hIL-6 treatment.

INTRODUCTION

Stimuli such as injury and infection cause a profound change in the hepatic synthesis and circulating concentrations of plasma proteins. These changes represent an acute-phase response. Acute-phase proteins are thought to play an important role in host's defense, by acting as opsonins, proteinase inhibitors, or carriers of various hormones (1). The acute-phase response can also be induced *in vitro* in human hepatoma cell lines such as Hep3B and HepG2 by stimulating cells with interleukin-6 (hIL-6) (2), which is the major inducer of liver-specific gene expression during inflammation (3). As a result of acute-phase induction, hepatic heme levels are decreased, resulting in the down-regulation of cytochrome P450 synthesis (4).

Microsomal heme oxygenase-1 (HO-1; EC 1.14.99.3) is the rate-limiting enzyme in the oxidative metabolism of heme which yields biliverdin IXα (5). HO-1 and its mRNA are increased in many cell types by treatment with hemin, the substrate for the enzyme, as well as non-heme substances (6). HO-1 mRNA is also induced by a transcriptional mechanism in human hepatoma cells by treatment with heat-shock or IL-6 (7,8). Thus, HO-1 is not only a key enzyme in heme metabolism but also a stress-inducible protein (7-9).

In order to examine the mechanism(s) of HO-1 gene activation in the acute-phase reaction, a chimeric plasmid containing a 5'-flanking region of the human HO-1 gene and the firefly luciferase gene was constructed, and introduced into human hepatoma cells. The results of our study demonstrate that the sequence between nucleotides -120 and -80 in the HO-1 gene contains a regulatory element responsible for the induction of HO-1 by hIL-6, and that there is a nuclear factor which binds specifically to this region.

MATERIALS AND METHODS

Cell Cultures and hIL-6

HepG2 cells were a generous gift from Dr. Barbara B. Knowles, Wistar Institute, Philadelphia, PA. Hep3B cells were obtained from American Type Culture Collection, Rockville, MD. Cells were cultivated as described previously (7). HepG2f cells were a variant of HepG2 cells which had been isolated in our laboratory, and were shown to have less differentiated properties than HepG2 (10). hIL-6 was a gift from either Genetics Institute, Boston, MA, or Kirin Brewery Company, Ltd., Maebashi, Japan.

Northern Blot Analysis

Northern blot analysis was performed using 15 μg total RNA as described previously (7).

Plasmid Construction

A fragment of the human HO-1 gene (11) corresponding to the sequence from -507 to +18 was amplified by a polymerase chain reaction (PCR) using a human genomic DNA library (CLONTECH Laboratories, Inc. Palo Alto, CA). The primers used for PCR were as follows:

Forward: 5'-CCCAAGCTTAAGCAGTCAGCAGAGGATTC-3'
Backward: 5'-CCCTCTAGAAGGAGGCAGGCGTTGACT-3'

The fragment was then cloned into the pGL2-basic vector (Promega, Madison, WI). This hybrid plasmid was designated as pLuc-HOa. Two deletion mutants, pLuc-HOb and pLuc-HOc were generated by digesting pLuc-HOa with PstI (-282) and SmaI (-120), respectively, in combination with HindIII. The fragments were blunt-ended and ligated to the pLuc-basic vector. pLuc-HOd mutant was generated by PCR using following primers;

Forward: 5'-CCCAAGCTTGCCAGAAAGTGGGCATCA-3'
Backward: 5'-CCCTCTAGAAGGAGGCAGGCGTTGACT-3'

The amplified fragment was cloned into the pLuc-basic vector. The authenticity of the constructs was confirmed by sequencing using the Sequenase system (United States Biochemical Corp., Cleveland, OH). pLuc-HOa* was prepared by introducing two point mutations, T $\rightarrow$ A and A $\rightarrow$ T, at position -92 and -86, respectively, using Transformer™ site-directed mutagenesis kit (CLONTECH).

Transfection

Transfection was performed according to the procedure described by Graham et al. (12). Briefly, cells were seeded at a density of 3.6×10^5 cells/60mm dish, and cultured for 24 h. The medium was replaced with fresh medium 3 h before transfection with the plasmid. The cells were incubated with 1ml of calcium phosphate solution (124mM $CaCl_2$, 25mM HEPES (pH7.1), 140mM NaCl, 0.75mM Na_2HPO_4), and DNA solution which contained 8μg of the Luc plasmid of interest, and 2μg of pSV-βGal (Promega) (as an internal control) for 16 h. Then the cells were incubated in fresh medium for 24 h, and incubated with or without hIL-6 for 24 h.

Luciferase Assay

Luciferase and β-galactosidase activities were determined using Luciferase Assay System (Promega) and β-Galactosidase Enzyme Assay System (Promega), respectively. Luciferase activity was determined using Packard Tri-carb 1900CA liquid scintillation counter, and enzyme activity was normalized on the basis of β-galactosidase activity.

Gel Mobility Shift Assay

An oligonucleotide corresponding to -120 to -79 of the human HO-1 gene (11) (GGGCCGGGCTGGGCGCGGGCCCTGCGGGTGTTGCAACGCCCG) was end-labeled with γ-^{32}P-ATP by T4 polynucleotide kinase (Promega). Nuclear extracts were prepared according to the method of Miskimins et al. (13). Nuclear protein (5μg) was incubated with ^{32}P-labeled oligonucleotide at room temperature for 15 min in 25μl of reaction mixture containing 20mM HEPES, 60mM KCl, 0.2mM EDTA, 10% glycerol, 0.5mM DTT, 6mM $MgCl_2$, and an equimolar mixture of poly(dI-dC) and poly(dA-dT) (1.5μg). DNA-protein complexes were then separated by electrophoresis in 4% polyacrylamide gel containing 0.25xTBE. The assay was performed twice using independently prepared nuclear extracts, and representative results are shown.

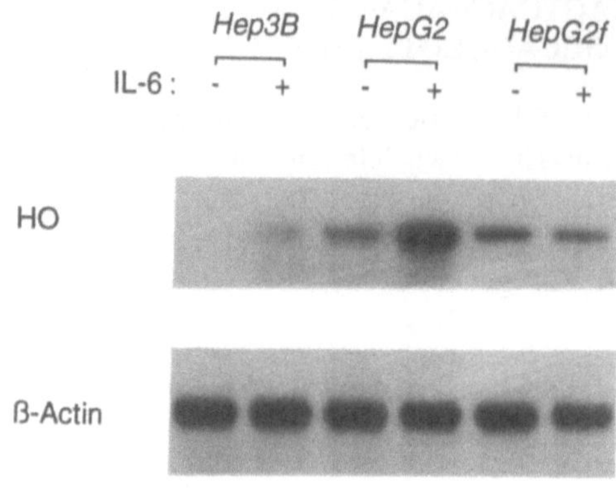

Figure 1. Effect of IL-6 on HO-1 mRNA in human hepatoma cells. Cells were cultured in the presence or absence of 100U/ml IL-6 for 48 h and total RNA was examined by Northern blot analysis using the probe for HO-1 mRNA as described in *materials and methods*.

RESULTS

Effect of hIL-6 on Human HO-1 mRNA

The effect of hIL-6 on the expression of mRNA encoding human HO-1 was examined using three hepatoma cell lines, Hep3B, HepG2 and HepG2f. As shown in Fig. 1, the level of mRNA for HO-1 was lower in untreated Hep3B cells than in untreated HepG2 and HepG2f cells. However, HO-1 mRNA levels were increased in Hep3B as well as HepG2 cells at least by 5-fold following treatment with hIL-6. These changes are significant since there was no change in the level of β-actin mRNA. In contrast to Hep3B and HepG2 cells, HepG2f cells did not show any increase in HO-1 mRNA by hIL-6 treatment, consistent with the less differentiated property of these cells than that of Hep3B or HepG2 cells (10).

Induction of the HO/Luc Hybrid Gene Expression by hIL-6 and Identification of the Regulatory Element Necessary for Induction

In order to examine the mechanisms of the induction of human HO-1 by hIL-6, a chimeric plasmid, pLuc-HOa, containing 5'-flanking sequence (-507 to +18) of the human HO-1 gene (11) fused to a reporter gene for luciferase was constructed (Fig. 2). In addition, three 5' deletion mutants, pLuc-HOb, pLuc-HOc, and pLuc-HOd (Fig. 2), were constructed. Each plasmid was transfected into HepG2 cells, and luciferase activities in the transfectants incubated in the presence or absence of hIL-6 were measured. The results were then compared with those obtained with pLuc-HOa transfection.

Luciferase activity in HepG2 cells transfected with pLuc-HOa was significantly increased by treatment with hIL-6 (Fig. 2). No significant decrease in the induction of luciferase activity was observed after deleting the region from -507 to -283 or the region from -507 to -121 (Fig. 2). Rather, the deletion of these regions increased both the basal activity (in the absence of hIL-6) and the hIL-6 induced activity, by retaining a similar inducibility, suggesting that there may be an element which suppresses HO-1 gene expression in this region. In contrast, the deletion from -507 to -80 completely abrogated luciferase induction by hIL-6. These results strongly suggest that the sequence involved in the hIL-6 dependent induction of the human HO-1 is located between -120 and -80 in the gene. This region, at nucleotide position from -93 through -85, contains AGTGANGNAA, i.e., a type A consensus sequence of IL-6 responsive element (IL-6RE) (14), which has also been identified in several human acute-phase protein genes. The mutant construct which contained two point mutations, T $\rightarrow$ A and A $\rightarrow$ T, at position -92 and -86, respectively, did not show

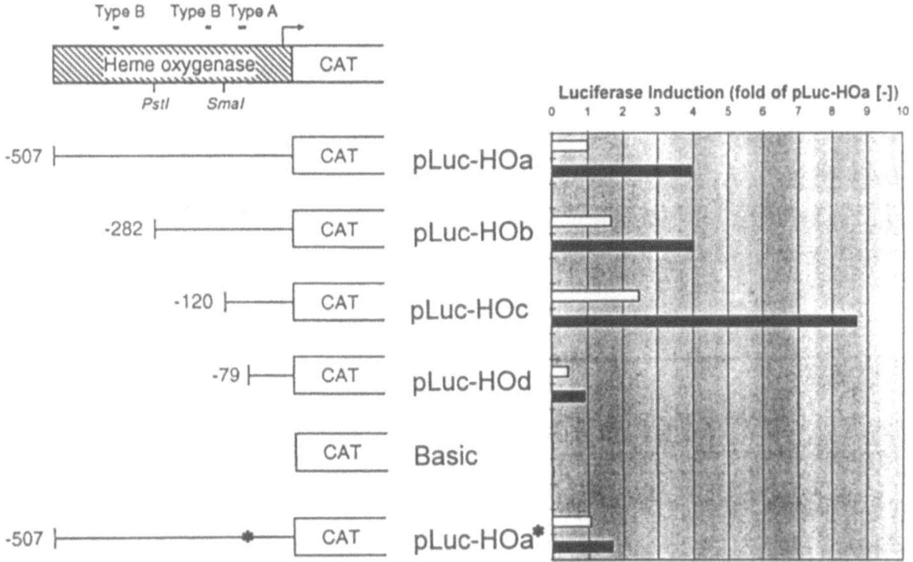

Figure 2. Structure of the human HO/Luc constructs, and the effect of IL-6 on luciferase activity in HepG2 cells transfected with the plasmids. pLuc constructs were prepared as described in *materials and methods*. HepG2 cells transfected with each plasmid were cultured in the presence (solid bar), or in the absence (open bar) of 100U/ml IL-6 for 24 h. Luc activity was determined as described in *materials and methods*. The left panel shows the constructs with the deletions or mutations. The location of three IL6-REs, a type A and two type B, are indicated. The transcription start site is indicated by the arrow. The right panel shows the relative induction of luciferase activity in the transfectants as compared with that of the pLuc-HOa control. The values are the means ± S.E. for three determinations.

any hIL-6-mediated induction (Fig. 2), corroborating with the critical nature of this region in the HO-1 gene activation.

A Nuclear Factor that Binds to the Regulatory Region of the Human HO-1 Gene

To identify a nuclear protein(s) necessary for transcriptional activation of the human HO-1 gene, *in vitro* binding studies were performed using gel mobility shift assays. As a probe, we used an end-labeled oligonucleotide which corresponds to the sequence from -120 to -79 of the human HO-1 gene which contains type A IL6-RE. As shown in Fig. 3, two retarded bands representing DNA-protein complexes were observed in an extract from untreated HepG2 cells. Treatment of cells with hIL-6 for 6 h, however, did not influence the amount of the DNA-protein complex. These results were reproduced in 2 separate experiments using independently isolated nuclear extracts. Addition (a 100-fold molar excess) of the cold oligonucleotide, but not of a nonspecific oligonucleotide, to the binding reaction inhibited the formation of the complex (Fig. 3).

DISCUSSION

The results presented in this paper demonstrate that HO-1 mRNA is induced by hIL-6 both in Hep3B and HepG2 cells, but it is not inducible in HepG2f cells (Fig. 1). In our earlier

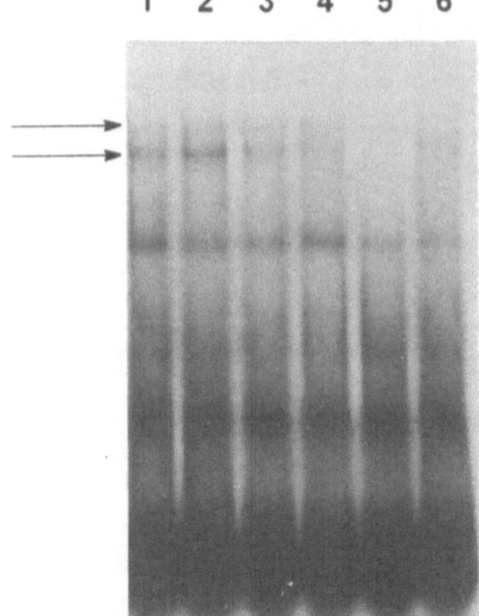

Figure 3. Gel mobility shift assay using oligonucleotides specific for the human HO-1 gene. Gel mobility shift assay was performed as described in *materials and methods*. The arrow indicates the specifically retarded band. Lane 1. untreated cells, 3 h; Lane 2. untreated cells, 6 h; Lane 3. IL-6 (100U/ml) treated cells, 3 h; Lane 4. IL-6 treated cells, 6 h; Lane 5. IL-6-treated cells, 6 h + 100-fold molar excess on the cold oligonucleotide (200μg); Lane 6. IL-6-treated cells, 6 h + 200μg of an oligonucleotide unrelated to the human HO-1 gene (AP-1 bindings site of the human erythroid-specific δ-aminolevulinate synthase promoter).

study, we demonstrated that HO-1 mRNA levels increased within 6 h of treatment with hIL-6, which continuously increased and reached a maximum of 5-fold at 48 h (8). Recent studies reported that there are two types of IL-6RE, type A (or α) and type B (or β), in various acute-phase genes (14,15). The human HO-1 gene also contains three putative IL-6REs, one type A and two type B elements. The type A element, GGTGTTGCAA, is located between nucleotide -94 and -85, while the type B elements, CTGGGACG and CAAGGGTC, are located between -371 and -364 and between -161 and -154, respectively (Fig. 2). Using transient expression experiments, we demonstrated that the critical element for HO-1 gene activation is located between nucleotides -120 and -80 in the 5'-flanking region of the HO-1 gene. A nuclear factor that specifically binds to this regulatory region has also been demonstrated using gel mobility shift assays. These results strongly suggest that the type A element in this region (nucleotides - 93 to -85) may play an important role in the expression of the human HO-1 gene during the acute-phase reaction, although it is possible that other regulatory element(s) may also be present further upstream and may participate in gene activation.

There are two isozymes of heme oxygenase, i.e., HO-1 and HO-2 (16). While HO-2 is not inducible by stress, HO-1 is inducible by various stress stimuli, including acute-phase reaction. Our findings suggest that the increase of HO-1 mRNA concentration may serve an indicator of an acute-phase reaction. Although acute-phase reactants are generally believed to be involved in the host-defence response (1,3), their exact role is not well understood. HO-1 is unique in that its function as an enzyme to catalyze the oxidative breakdown of heme is well defined. Free heme is a potent pro-oxidant, while bile pigments, the products of the HO-1 reaction, are physiologically important anti-oxidants (17) (Fig. 4). As such, induction of HO-1 can provide protection against the potentially toxic effects of the heme-mediated oxidative stress (Fig. 4). In this respect, it is of interest to note that there is, as yet, no known genetic deficiency of HO-1 described, while genetic defects of each enzyme in the heme biosynthetic pathway have been recognized. It is possible, therefore, that HO-1 may be important in fundamental cellular processes (17-19), such as the acute-phase reaction, and that a genetic deficiency of this enzyme may be deleterious to the host.

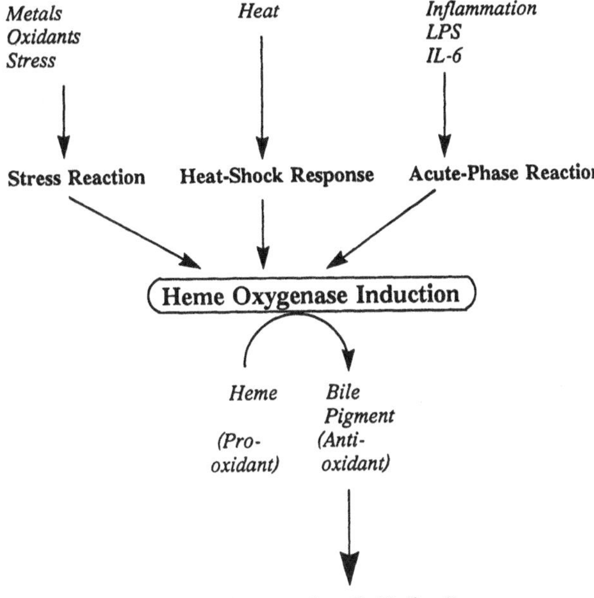

Figure 4. Heme oxygenase induction in stress reactions.

ACKNOWLEDGMENTS

We are grateful to Dr. Shigeki Shibahara for supplying HO-1 cDNA, to Dr. Barbara B. Knowles for supplying HepG2 cells, to Genetics Institute, Boston, MA, and Kirin Brewery Co., Ltd., Maebashi, Japan, for supplying recombinant hIL-6. The excellent technical assistance of Ms. Luba Garbaczewski is gratefully acknowledged.

REFERENCES

1. Koj A, Magielska-Zero D, Kurdowska A, Bereta J, Rokita H, Dubin A (1988) The acute-phase response to infection and inflammation. Adv Clin Enzymol 6:143-151
2. Iwasa F, Galbraith RA, Sassa S (1988) Effects of dimethyl sulphoxide on the synthesis of plasma proteins in the human hepatoma HepG2. Induction of an acute-phase-like reaction. Biochem J 253:927-930
3. Heinrich PC, Castell JV, Andus T (1990) Interleukin-6 and the acute phase response. Biochem J 265:621-636
4. Fukuda Y, Ishida N, Noguchi T, Kappas A, Sassa S (1992) Interleukin-6 down regulates the expression of transcripts encoding cytochrome P450 IA1, IA2 and IIIA3 in human hepatoma cells. Biochem Biophys Res Commun 184:960-965
5. Tenhunen R, Marver HS, Schmid R (1969) Microsomal heme oxygenase: Characterization of the enzyme. J Biol Chem 244:6388-6394
6. Kikuchi G, Yoshida T (1983) Function and induction of the microsomal heme oxygenase. Mol Cell Biochem 53/54:163-183
7. Mitani K, Fujita H, Sassa S, Kappas A (1989) Heat shock induction of heme oxygenase mRNA in human Hep 3B hepatoma cells. Biochem Biophys Res Commun 165:437-441
8. Mitani K, Fujita H, Kappas A, Sassa S (1992) Heme oxygenase is a positive acute-phase reactant in human Hep3B hepatoma cells. Blood 79:1255-1259
9. Shibahara S, Muller RM, Taguchi H (1987) Transcriptional control of rat heme oxygenase by heat shock. J Biol Chem 262:12889-12892

10. Iwasa F, Galbraith RA, Sassa S (1990) Phenotypic variation in human HepG2 hepatoma cells: alterations in cell growth, plasma protein synthesis and heme pathway enzymes. Int J Biochem 22:303-310
11. Shibahara S, Sato M, Muller RM, Yoshida T (1989) Structural organization of the human heme oxygenase gene and the function of its promoter. Eur J Biochem 179:557-563
12. Graham FL, Eb AJ, van der. (1973) A new technique for the assay of infectivity of human adenovirus 5 DNA. Virology 52:456-467
13. Miskimins WK, Roberts MP, McClelland A, Ruddle FH (1985) Use of a protein-blotting procedure and a specific DNA probe to identify nuclear proteins that recognize the promoter region of the transferrin receptor gene. Proc Natl Acad Sci USA 82:6741-6744
14. Akira S, Kishimoto T (1992) IL-6 and NF-IL6 in acute-phase response and viral infection. Immunol Rev 127:25-50
15. Majello B, Arcone R, Toniatti C, Ciliberto G (1990) Constitutive and IL-6-induced nuclear factors that interact with the human C-reactive protein promoter. EMBO J 9:457-465
16. McCourbrey WK Jr, Maines MD (1994) The structure, organization and differential expression of the gene encoding rat heme oxygenase-2. Gene 139:155-161
17. Stocker R (1990) Induction of haem oxygenase as a defence against oxidative stress. Free Radical Research Communications 9:101-112
18. Abraham NG, Levere RD, Lutton JD (1991) Eclectic mechanism of heme regulation of hematopoiesis. Int J Cell Cloning 9:185-210
19. Abraham NG (1991) Molecular regulation—biological role of heme in hematopoiesis. Blood Reviews 5:19-28

53

TRANSCRIPTIONAL CONTROL OF THE HUMAN HEME OXYGENASE-1 GENE BY STRESS

Shigeki Shibahara,* Kazuhisa Takeda, Shoji Okinaga, Miki Yoshizawa, Kazuhiro Takahashi, and Hiroyoshi Fujita

Department of Applied Physiology and Molecular Biology
Tohoku University School of Medicine, Sendai, Japan

ABSTRACT

Heme oxygenase-1 is an essential enzyme in heme catabolism, releasing carbon monoxide, iron, and biliverdin. By transient expression analysis of the fusion genes, containing the firefly luciferase gene as a reporter under the human heme oxygenase-1 gene promoter, we have identified a 10-bp sequence, TGCTAGATTT, that confers the cadmium-mediated induction on the fusion gene, located about 4 kb upstream from the transcriptional initiation site (J. Biol. Chem. 269: 22858-67, 1994). This cadmium-responsive element (CdRE) is different from the consensus sequence of the metal-regulatory element of the metallothionein genes. Both heme oxygenase-1 and metallothionein are considered to be involved in the defense system against metal toxicity. Here we showed that the CdRE of the human heme oxygenase-1 gene is not responsive to zinc, whereas a metal-regulatory element (MRE) of the human metallothionein IIA gene is able to respond to both cadmium and zinc. These results suggest that the metal-selective activation of each gene promoter is mediated by a separate mechanism.

INTRODUCTION

Heme oxygenase-1 (E.C.1.14.99.3) is an essential enzyme in heme catabolism, cleaving heme to release carbon monoxide, iron, and biliverdin (1, 2), the latter of which is subsequently converted to bilirubin by biliverdin reductase (3). The activity of heme oxygenase is higher in the spleen, liver, and bone marrow, where senescent erythrocytes are

* To whom correspondence should be addressed. Department of Applied Physiology and Molecular Biology, Tohoku University School of Medicine, Aoba-ku, Sendai, Miyagi 980-77, Japan. Tel. 81-22-717-8117; Fax 81-22-717-8118.

sequestrated and destroyed by resident macrophages (4). Thus, heme oxygenase plays a key role in hematopoiesis by recruiting iron to erythroblasts in the bone marrow. Furthermore, the levels of heme oxygenase-1 mRNA were reduced during chemically-induced differentiation of erythroid cells (5), in which transcription of the genes for heme synthetic enzymes were activated. Such reduced expression of heme oxygenase-1 is of physiological significance, because the large amounts of heme are required for hemoglobin synthesis during erythroid differentiation. It is also noteworthy that transcription of the heme oxygenase-1 gene is activated during differentiation of a human monocytic leukemia cell line THP-1 to macrophages (6). Heme oxygenase-1 therefore plays an important role in functional differentiation of bone marrow cells.

Heme oxygenase-1 activity is highly induced in human cell lines by its substrate heme (7-10) and by various environmental derangements (reviewed in Ref. 11), including heavy metals (9, 10, 12, 13). Recently, we have shown that cadmium increased the transcription of the human heme oxygenase-1 gene (13) and identified the cadmium-responsive element (CdRE), TGCTAGATTT, of the human heme oxygenase-1 gene that is responsible for induction by cadmium (13). However, the CdRE of the human heme oxygenase-1 gene is different from the consensus sequence of the metal-regulatory elements (MREs) of the metallothionein genes, TGCRCNC (R, purine; N, any nucleotides) (14), except for the TGC trinucleotides at its 5' end. Metallothioneins constitute a family of the cystein-rich proteins with a low molecular weight and are involved in multiple cellular processes such as metal homeostasis, adaptation to stress, and heavy metal detoxification (reviewed in Ref. 15). The expression of the metallothionein genes is controlled mainly at the transcriptional level and can be induced by a wide range of different stimuli, including exposure to heavy metals such as zinc and cadmium (14, 16, 17). Here, we compared the function of CdRE with that of an MRE of the metallothionein IIA gene promoter.

EXPERIMENTAL PROCEDURES

Transient Expression Analysis of Fusion Genes

HeLa cells were cultivated in minimum essential medium (MEM) containing 10% fetal calf serum. For transient transfection, HeLa cells, about 70% confluent in a 6-cm dish, were incubated with plasmid DNA precipitated with Ca phosphate for 16 h, refed with fresh medium, and incubated for 24 h (13). The DNA used for cotransfection was 7 μg of each promoter-luciferase fusion gene and 1 μg of β-galactosidase expression vector pCH110 (Pharmacia LKB Biotechnology Inc.), containing the simian virus 40 (SV40) early promoter, as an internal control. The fusion genes used were pSVLCd2 and pSVLMREb, which contain the synthetic oligonucleotides, containing the CdRE of the human heme oxygenase-1 gene and MREb, an MRE of the human metallothionein IIA gene, respectively, in the upstream of the enhancerless SV40 promoter of pSVLE(-) (18). Following a 24-h incubation, cells were treated for 5 h with $CdCl_2$ (5 and 50 μM) or $ZnCl_2$ (50 and 100 μM) in MEM supplemented with 10% fetal calf serum. To normalize the variability in transfection efficiency, each luciferase activity was divided by β-galactosidase activity (relative luciferase activity) as described previously (13, 18). The data shown are means ± standard deviations for at least three independent experiments.

RESULTS

As shown in Fig. 1, the CdRE of the human heme oxygenase-1 gene is located about 4 kb upstream from its transcriptional initiation site, while the human metallothionein IIA

Heme oxygenase-1 gene

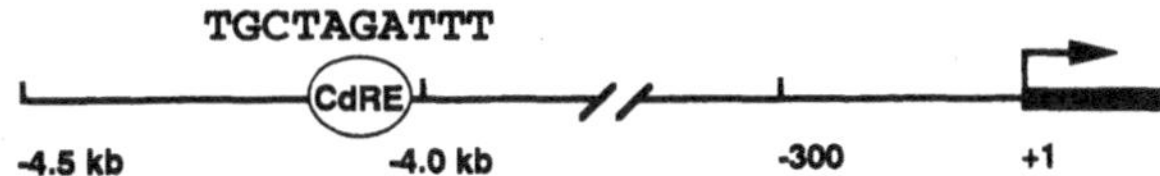

Metallothionein II A gene

TGCACTC

h g f e d c b a

-300 -200 -100 +1

Figure 1. Schematic representation of the human heme oxygenase-1 and metallothionein IIA gene promoters. Shown are the 5'-flanking regions of the human heme oxygenase-1 and metallothionein IIA genes. The *arrows* indicate the transcription initiation sites and the *closed boxes* represent the first exons. The CdRE of the heme oxygenase-1 gene (13) and the eight MREs of the metallothionein IIA gene (14) are indicated. The MREb is indicated by *box*. The *numbers* shown are the nucleotide residues from the transcription initiation sites.

gene promoter contains multiple copies of MREs in the proximal 5'-flanking region (14). We have shown that both heme oxygenase-1 and metallothionein IIA mRNAs are remarkably induced in HeLa cells by the treatment with either cadmium (5 and 50 μM) or zinc (50 and 100 μM) in a dose-dependent manner (18). To examine the effects of these metals on the human heme oxygenase-1 and metallothionein IIA promoter activities, we performed transient expression assays of the fusion genes containing either CdRE or an MRE of the human metallothionein IIA gene, MREb (positions -57/-43) (14). The luciferase activity was induced by cadmium in the cells transfected with pSVLCd2 or pSVLMREb by two- to three-fold (Fig. 2), whereas zinc increased the luciferase activity in HeLa cells transfected with pSVLMREb but not with pSVLCd2. In the case of pSVLMREb, luciferase activity was increased noticeably in a dose-dependent manner by zinc.

DISCUSSION

We show that the CdRE of the heme oxygenase-1 gene is not responsive to zinc, although MREb is responsive to both cadmium and zinc (18). The CdRE, TGCTAGATTT (13), is different from the consensus sequence of MRE, TGCRCNC (14), except for the TGC trinucleotides at its 5' end. By gel mobility shift assays, it was shown that nuclear protein(s) prepared from untreated HeLa cells did bind to CdRE but was unable to bind to MREb (18). The CdRE-binding activities were not apparently affected even if nuclear proteins were prepared from the cadmium-treated cells (13, 18). These results also support the functional difference between CdRE and MREb. A biological role of cadmium has not been defined yet, although cadmium is present in minute traces in living organisms. Because cadmium is able to bind various cellular components such as free sulfhydryl groups and nucleic acids (reviewed in Ref. 19), cadmium may simply mimic or modify a physiological mediator that is involved in a signal transduction system leading to activation of the heme oxygenase-1 gene transcription.

Heme oxygenase-1 activity is highly induced in human cell lines by various environmental derangements (reviewed in Ref. 11). However, it is not necessarily easy to evaluate whether induction of heme oxygenase-1 is beneficial or harmful, because heme breakdown yields unique products (Fig.3): biliverdin/bilirubin, iron, and carbon monoxide (CO), all of which are considered as the metabolic double-edged swords. Bilirubin, a final heme

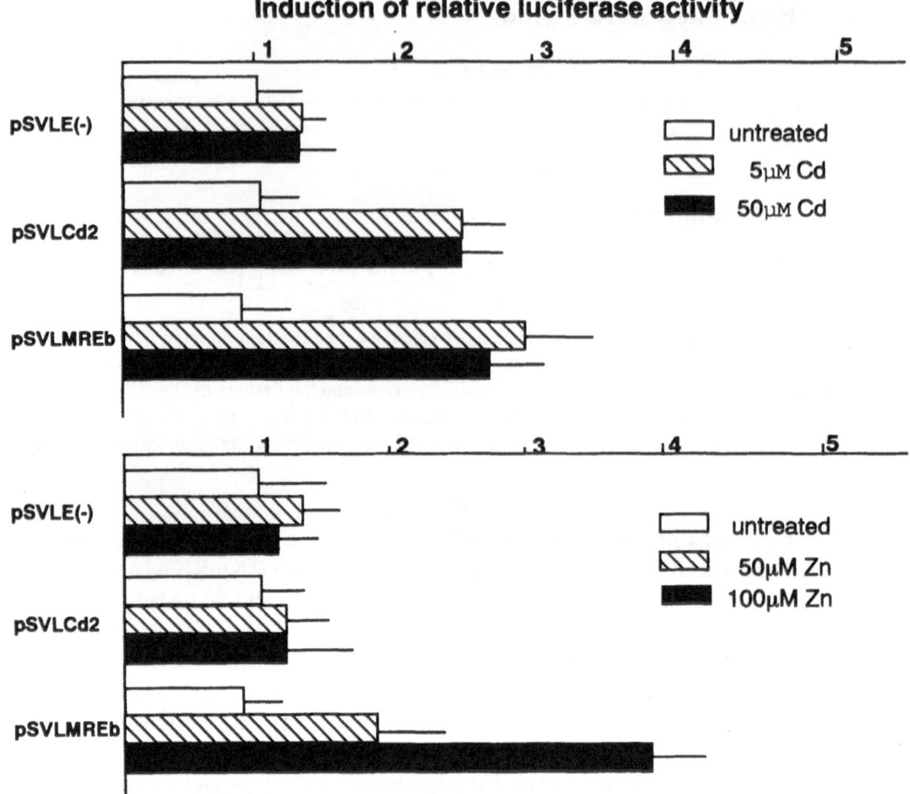

Figure 2. Functional analysis of the two types of metal-regulatory elements using a heterologous promoter. The function of CdRE and MREb was analyzed by transient expression assays. The Cd2 oligonucleotide carried by pSVLCd2 represents the sequence, AGGCGGATTTTGCTAGATTT, including CdRE (13), and the MREb oligonucleotide represents the promoter region (-62/-36) of the metallothionein IIA gene, including MREb as detailed previously (18). The data were taken from ref. 18. The magnitude of induction shown represents the ratio to the relative luciferase activity obtained with an enhancerless construct, pSVLE(-).

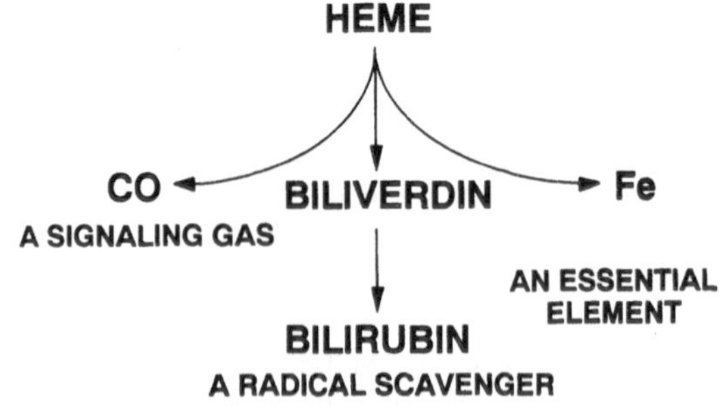

Figure 3. Potential physiological roles of heme breakdown products.

degradation product, was traditionally considered as a toxic waste product, but it was shown that biliverdin or bilirubin produced locally may work as a physiological antioxidant (20, 21). It is thus proposed that heme catabolism constitutes a member of the defense system against oxidative stress. Iron is an essential metal for living organisms, but in the presence of iron, superoxide anion radical and hydrogen peroxide can be involved in the formation of hydroxyl radical, a highly reactive species of radicals. By analogy to nitric oxide (NO), CO has been suggested to be a physiological regulator of cGMP and function as a neurotransmitter (Reviewed in Ref. 22).

Here we show that the CdRE of the human heme oxygenase-1 gene is a novel type of the *cis*-acting element that is responsive only to cadmium. It is also noteworthy that both heme oxygenase-1 and metallothioneins differ in their structures; namely, in contrast to metallothioneins, human heme oxygenase-1 contains no cysteine residues (7). It appears reasonable that living organisms have acquired the two independent systems involved in detoxification of heavy metals.

ACKNOWLEDGMENTS

This work was supported in part by a Grant-in-Aid for Scientific Research on Priority Areas from the Ministry of Education, Science, and Culture of Japan.

REFERENCES

1. Tenhunen R, Marver HS, Schmid R: The enzymatic conversion of heme to bilirubin by microsomal heme oxygenase. Proc. Natl. Acad. Sci. USA 61: 748, 1968
2. Tenhunen R, Marver HS, Schmid R: Microsomal heme oxygenase. Characterization of the enzyme. J. Biol. Chem. 244: 6388, 1969
3. Tenhunen R, Ross ME, Marver HS, Schmid R: Reduced nicotinamide-adenine dinucleotide phosphate dependent biliverdin reductase: partial purification and characterization. Biochemistry 9: 298, 1970
4. Tenhunen R, Marver HS, Schmid R: The enzymatic catabolism of hemoglobin: stimulation of microsomal heme oxygenase by hemin. J. Lab. Clin. Med. 75: 410, 1970
5. Fujita H, Sassa S: The rapid and decremental change in haem oxygenase mRNA during erythroid differentiation of murine erythroleukaemia cells. Br. J. Haematol. 73: 557, 1989
6. Muraosa Y, Shibahara S: Identification of a *cis* -regulatory element and putative *trans*-acting factors responsible for 12-*O*-tetradecanoylphorbol-13-acetate (TPA)-mediated induction of heme oxygenase expression in myelomonocytic cell lines. Mol. Cell. Biol. 13: 7881, 1993
7. Yoshida T, Biro P, Cohen T, Muller RM, Shibahara S: Human heme oxygenase cDNA and induction of its mRNA by hemin. Eur. J. Biochem. 171: 457, 1988
8. Shibahara S: Regulation of heme oxygenase gene expression. Seminars Hematol. 25: 370, 1988
9. Taketani S, Kohno H, Yoshinaga T, Tokunaga R: The human 32-kDa stress protein induced by exposure to arsenite and cadmium ions is heme oxygenase. FEBS Lett. 245: 173, 1989
10. Sato M, Ishizawa S, Yoshida T, Shibahara S: Interaction of upstream stimulatory factor with the human heme oxygenase gene promoter. Eur. J. Biochem. 188: 231, 1990
11. Shibahara S: Heme oxygenase - Regulation of and physiological implication in heme catabolism, in Fujita H (ed): Regulation of Heme Protein Synthesis. Dayton, OH, AlphaMed Press, 1994, p 103
12. Keyse SM, Tyrrell RM: Heme oxygenase is the major 32-kDa stress protein induced in human skin fibroblasts by UVA radiation, hydrogen peroxide, and sodium arsenite. Proc. Natl. Acad. Sci. USA 86: 99, 1989
13. Takeda K, Ishizawa S, Sato M, Yoshida T, Shibahara S: Identification of a *cis*-acting element that is responsible for cadmium-mediated induction of the human heme oxygenase gene. J. Biol. Chem. 269: 22858, 1994
14. Stuart GW, Searle PF, Palmiter RD: Identification of multiple metal regulatory elements in mouse metallothionein-I promoter by assaying synthetic sequences. Nature 317: 828, 1985
15. Hamer DH: Metallothionein. Annu. Rev. Biochem. 55: 913. 1986

16. Karin M, Haslinger A, Holtgreve H, Richards RI, Krauter P, Westphal HM, Beato M: Characterization of DNA sequences through which cadmium and glucocorticoid hormones induce human metallothionein-II_A gene. Nature 308: 513, 1984
17. Karin M, Haslinger A, Heguy A, Dietlin T, Cooke T: Metal-responsive elements act as positive modulators of human metallothionein-II_A enhancer activity. Mol. Cell. Biol. 7: 606, 1987
18. Takeda K, Fujita H, Shibahara S: Differential control of the metal-mediated activation of the human heme oxygenase-1 and metallothionein IIA genes. Biochem. Biophys. Res. Commun. 207: 160, 1995
19. Vallee BL, Ulmer DD: Biochemical effects of mercury, cadmium, and lead. Annu. Rev. Biochem. 41: 91, 1972
20. Stocker R, Glazer AN, Ames BN: Antioxidant activity of albumin-bound bilirubin. Proc. Natl. Acad. Sci. USA 84: 5918, 1987
21. Stocker R, Yamamoto Y, McDonagh AF, Glazer AN, Ames BN: Bilirubin is an antioxidant of possible physiological importance. Science 235: 1043, 1987
22. Marks GS, Brien JF, Nakatsu K, McLaughlin BE: Does carbon monoxide have a physiological function? Trends Pharmacol. Sci. 12: 185, 1991

MEGAKARYOCYTE GROWTH AND MATURATION FROM PURIFIED PERIPHERAL BLOOD PROGENITORS IN UNILINEAGE SERUM-FREE LIQUID CULTURE

H. J. Hassan,* R. Guerriero, U. Testa, M. Gabbianelli, G. Mattia, E. Montesoro, G. Macioce, A. Pace, B. Ziegler, and C. Peschle

Department of Hematology-Oncology
Istituto Superiore di Sanità, Rome, Italy
Thomas Jefferson University
Philadelphia, Pennsylvania
Department of Clinical Physiology and Occupational Medicine
University of Ulm, Germany

ABSTRACT

Hematopoietic progenitor cells (HPCs), 80-90% purified from peripheral blood, were induced to megakaryocytic differentiation/ maturation in serum-free liquid suspension culture treated with a hematopoietic growth factor (HGF) cocktail (IL-3, KL and IL-6) and/or recombinant mpl ligand (thrombopoietin, Tpo). Particularly, (i) the HGF cocktail induced the growth of a 40% megakaryocyte (MK) cell population, i.e., 4×10^4 cells at day 0 generated 2×10^5 MK at terminal maturation; (ii) further addition of Tpo increased the MK purity level to 80% with a final yield of 4×10^5 MKs; (iii) treatment with Tpo alone resulted in a 97-99% MK population with a mild increase of cell number (up to 1×10^5 cells). In all culture conditions, morphological evaluation indicated the presence of putative mononuclear MK precursors and then mature polynucleated MK peaking at day 5 and 12 respectively. Membrane phenotype analysis showed a gradual decline of $CD34^+$ HPCs, coupled with an inverse rise of MK-specific antigens (e.g., CD61/62/42b) starting prior to MK detection by morphology analysis. In situ hybridization showed the expression of MK-specific von Willebrand gene in both immature and mature MKs.

* Correspondence to: Dr. H.J. Hassan, Department of Hematology-Oncology, Istituto Superiore di Sanità, V.le Regina Elena, 299 - 00161 Rome, Italy.

Molecular Biology of Hematopoiesis 5, edited by Abraham et al.
Plenum Press, New York, 1996

This HPC differentiation culture system allows for growth of a relatively large number of highly purified or "pure" immature and then mature MKs, thus providing an in vitro experimental tool to dissect the cellular and molecular basis of megakaryocytopoiesis.

INTRODUCTION

Hematopoietic stem cells (HSCs) have the ability to self-renew and differentiate into progenitor cells (HPCs) (1,2), that are committed to specific lineage(s) (1). HPCs in turn differentiate into morphologically recognizable precursors that mature to terminal elements circulating in peripheral blood (PB).

Megakaryocytopoiesis involves proliferation of megakaryocytic precursors and maturation of megakaryocytes (MKs) that increase their size by polyploidization. Diverse cytokines and hematopoietic growth factors (HGFs), although not specific for megakaryocytopoiesis, stimulate MK differentiation (3). Recently, the murine and human mpl ligands (thrombopoietin, Tpo) have been cloned (4-6). Numerous data indicate that Tpo exerts in vivo and in vitro a stimulatory activity essentially restricted to the MK lineage (4-8): its function is similar to that postulated for thrombopoietin, which promotes MK maturation and possibly MK-CSF, which induces proliferation of MK progenitors (3).

In this study we describe the proliferation and gradual differentiation /maturation of a highly purified or pure MK progeny in purified PB HPC serum-free liquid culture.

MATERIALS AND METHODS

Recombinant Human Hematopoietic Growth Factors and Medium

Interleukin-3 (IL-3, specific activity, 2-4 x 10^6 U/mg) and IL-6 (2-7 x 10^6 U/mg) were supplied by Genetics Institute (Cambridge, MA), KL (1 x 10^5 U/ml) by Immunex (Seattle, WA). Tpo was kindly provided by Boehringer Mannheim, Penzberg, Germany, as serum-supplemented (10% FCS) or serum-free conditioned media from producer COS cells and by Genentech (S. Francisco, CA) as purified HGF.

HPC Liquid Suspension Culture

Adult PB was obtained from 20-40 year-old healthy male donors after informed consent. PB HPCs were purified from the buffy coat by a three-step procedure according to a slight modification (9) of the method described in (10). Purified HPCs (4x10^4 cells/ml) were grown in FCS$^-$ liquid culture (11) supplemented with various recombinant human cytokine combinations. Cells were incubated in a fully humidified atmosphere of 5% CO_2, 5% O_2, 90% N_2. The initial cell density was 4 x 10^4 cells/ml. Cell concentration was adjusted to 1x10^5 cells/ml and fresh HGFs were added, twice weekly.

MK Characterization

Cells cytocentrifuged onto glass slides were stained with May-Grünwald Giemsa.

MoAbs to CD34, CD62 (Becton-Dickinson) conjugated with phycoerythrin (PE) and MoAbs to CD34, CD42b, CD61 (Becton-Dickinson), CD41a (Serotec, Oxford, England) conjugated with fluorescein isothiocyanate (FITC) were used. Cells were treated and analyzed by FACScan (Becton-Dickinson, Lysis II program) as in (12).

In situ hybridization on MK-enriched cells was performed utilizing a 49-mer oligonucleotide as probe complementary to the 3018-3067 bp sequence of the von Willebrand (vW) factor mRNA (13), which was labeled with [$\gamma^{35}S$] dATP (New England Nuclear) at the 5' end by T4 polynucleotide kinase and hybridized as in (14).

RESULTS

Purified PB Step IIIP cells are ~ 90% CD34$^+$ with a HPC frequency of 81.5 ± 1.2% (mean ± SEM values) (15). More recently, the HPC clonogenetic assay has been optimized by adding to the formerly employed HGFs (KL, IL-3, GM-CSF, Ep) saturating dosages of G-CSF, M-CSF and FLT3 ligand: in a series of 13 consecutive purification experiments the HPC frequency and recovery rise to 90.3 ± 2.1 and 69.0 ± 5.3 respectively.

4x10^4 Step IIIP cells were cultured in liquid suspension with (i) early-acting HGFs [IL-3 (0.01 U/ml) + KL (10 ng) + IL-6 (10 ng)], or (ii) Tpo alone or (iii) Tpo combined with the HGFs, and analyzed for their ability to proliferate and differentiate to MKs. Cell growth in the culture containing HGFs is shown in Fig. 1; after 10-12 days of culture, morphological analysis indicated the presence of 40% MKs. Further addition of Tpo did not modify the growth curve, but markedly increased the MK percentage on day 12 from 40% to 80%; while Tpo alone induced only a moderate cell proliferation, but resulted in a virtually pure MK progeny (97-99%) (Fig.1).

Morphology of the IL-3 + KL + IL-6 culture showed an increase in morphologically identifiable MK population from day 7 to 12 from 5% to 39% of total cells (Fig. 2). The membrane phenotype of the cells treated with anti-platelet MoAbs (CD61, CD62, CD41a) and analyzed by flow cytometry from day 0 to day 14 of culture is shown in Fig. 2.

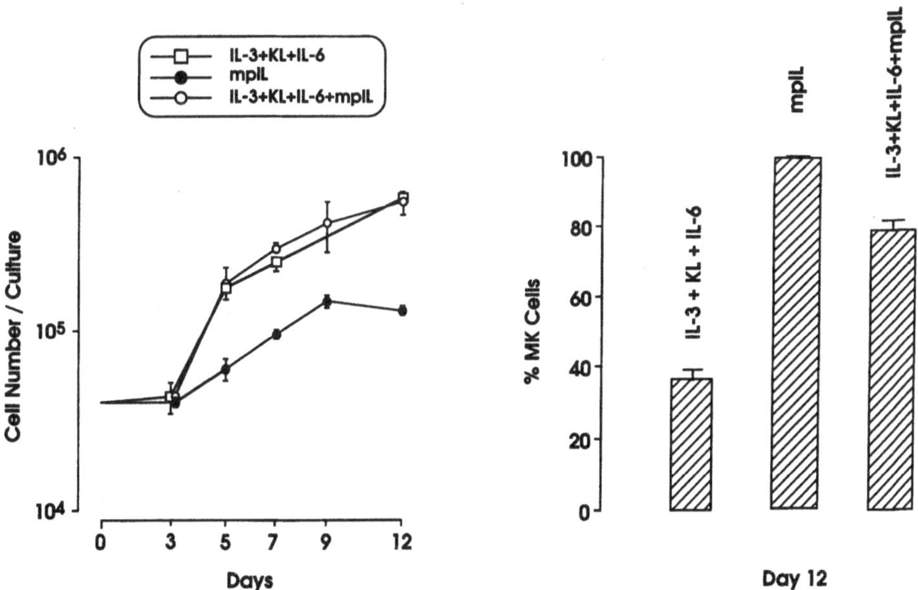

Figure 1. Left: Growth curve of purified Step IIIP HPCs grown in FCS$^-$ liquid suspension culture. Right: Percentage of MKs at day 12, as evaluated by morphological analysis (mean ± SEM values from 3 separate experiments). Reprinted from (16).

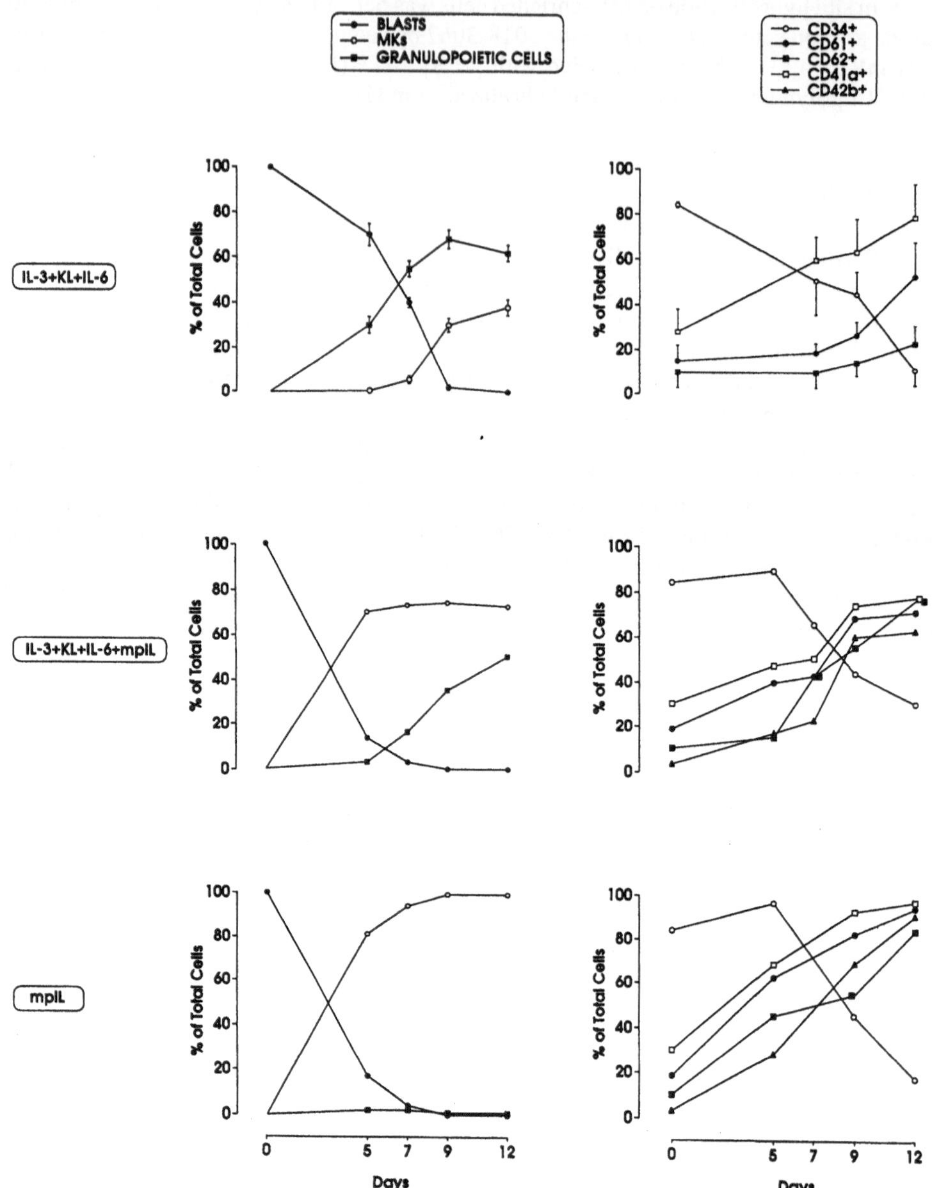

Figure 2. Comparative analysis of morphology (left) and membrane phenotype (right) in Step IIIP HPC liquid suspension culture treated with HGFs (top), HGFs + Tpo (middle), or Tpo alone (bottom). Mean ± SEM values from 3 independent experiments. Modified from (16).

Particularly, the frequency of $CD61^+$ cells showed a gradual rise, which paralleled that of MKs.

In Tpo + early HGFs and Tpo alone cultures the membrane phenotype pattern was similar, although (i) the final MK purity was lower in the Tpo + HGFs culture (80% vs 97-99%), (ii) the maturation process was slightly delayed in the Tpo + HGFs culture. In both culture conditions, phenotypical analysis indicated that CD34 was expressed at high level

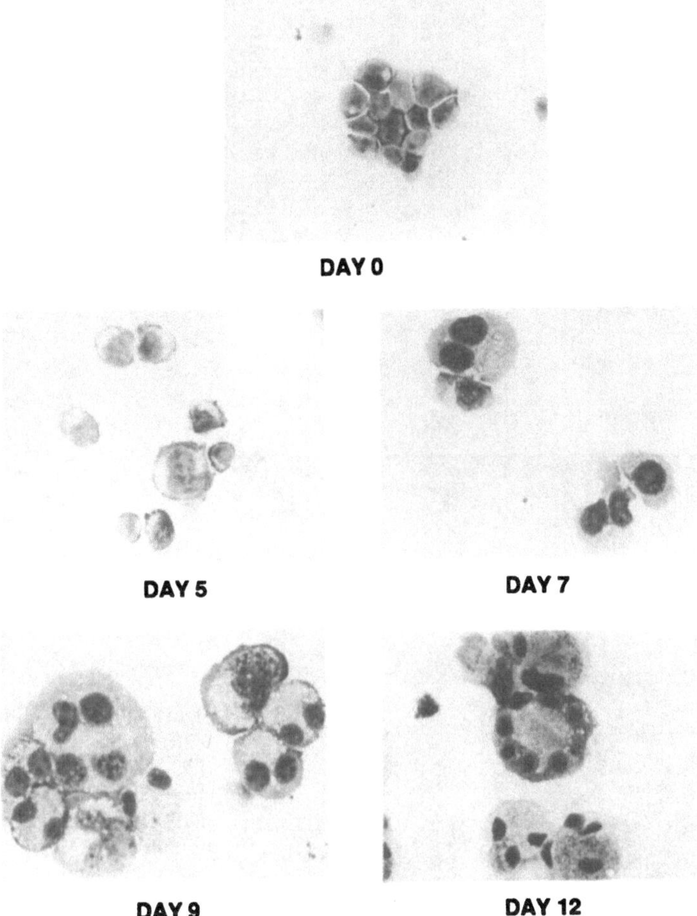

Figure 3. Cell morphology in Tpo-treated Step IIIP HPC liquid suspension culture at sequential days (magnification X 400). Representative results are presented. Modified from (16).

from day 0 to 5 and progressively declined. At day 0, 10-20% of the HPCs express CD61, but not CD42b, which recognize an early and late MK-specific marker respectively. At day 12 most of the cells were positive for both CD61 and CD42b markers. Interestingly, all the MK-specific membrane markers showed a more rapid rise after treatment with Tpo alone as compared to Tpo + early HGFs (Fig. 2).

The Tpo-supplemented culture systems gave rise to a virtually "pure" MK progeny, thus allowing identification of MK morphological features at sequential stages of differentiation/ maturation (Fig. 3). Purified Step IIIP cells are essentially composed by small undifferentiated blasts with round shape and a small thin blue cytoplasm. At day 5 most cells are large and mononuclear, while only a small percentage (~10%) of them is binuclear. At this stage the mononuclear cells represent putative MK precursors, that appear large with an oval or round shape and a large and pale cytosplasm exhibiting a "white" perinuclear area and a peripheral blue area. At day 7 most cells display two nuclei, while the chromatin is more dense. At day 9 the majority of cells reached the polyploid stage with two, four or more nuclei. Finally, at day 12 the large

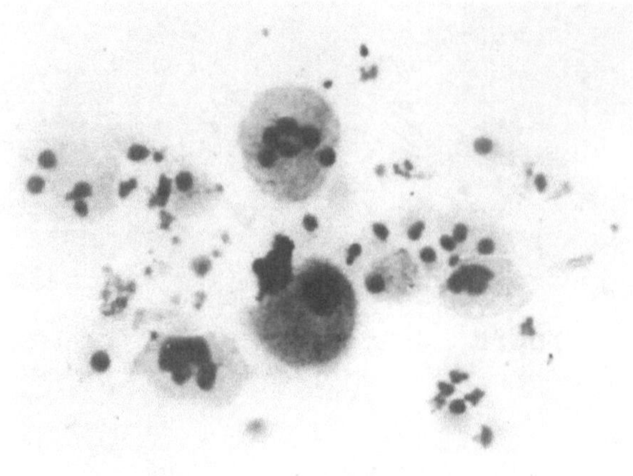

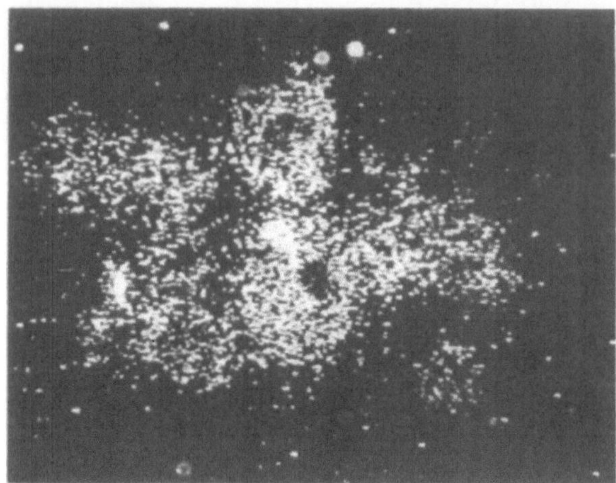

Figure 4. In situ hybridization of von Willebrand factor on day 12 Step IIIP HPC grown in liquid suspension culture in the presence of Tpo + early HGFs. Bright and dark field images are presented (May-Grünwald staining, magnification X 500). Modified from (16).

majority of MKs were highly polyploid and exhibited platelet formation (i.e., they contain innumerable granules that stain pink, magenta and blue scattered throughout the cytoplasm). At the periphery of the cytoplasm of MKs, platelet budding was sometimes observed.

Furthermore, cells, at day 12 of early HGF + Tpo culture, were analyzed for MK specific vW factor mRNA expression by in situ hybridization. Fig. 4 shows specific hybridized areas in all mature MKs, but not in contaminant cells, which were essentially neutrophils. Mononuclear cells not morphologically recognizable as polynucleated mature MKs but positive for this transcript were identified as MK precursors. The percentage of vW factor positive cells (80%) detected by in situ hybridization in the presence of Tpo + early HGFs corresponds to the results obtained by morphological analysis.

DISCUSSION

Megakaryocytes represent a very small proportion of human bone marrow cells rendering difficult their isolation and analysis.

We have developed an FCS⁻ liquid suspension culture for gradual differentiation of Step IIIP HPCs along the megakaryocytic lineage (16) as previously performed with the erythroid (17), the granulopoietic-neutrophilic (18) and the monocytic lineages (12).

Cultures supplemented with early HGFs (IL-3, KL, IL-6) caused extensive MK proliferation, which was associated with prevailing growth of granulocytic cells; the combined addition of early-acting HGFs and Tpo provides optimal MK growth, in terms of purity (80% MK cells) and yield. HPCs cultured in the presence of Tpo alone gave rise to a virtually pure (97-99%) MK progeny, through sequential stages of cell differentiation and maturation, although with limited amplification of cell number: particularly, (i) in the first week of culture HPCs apparently differentiate along the megakaryocytic lineage, as shown by the gradual decrease of $CD34^+$ cells and blasts and progressive rise of CD61 and other megakaryocytic membrane markers; (ii) in the second week of culture we observed the proliferation/differentiation of megakaryocytic precursors indicated by the gradual appearance of maturing MKs (from the binuclear to more than 8 nuclei stage). Finally we observed large polynucleated MKs, coupled with increasing expression of lineage-specific membrane (CD41a, CD42b, CD61 and CD62) and cytoplasmic (vW factor) markers and a further drop of $CD34^+$ cell frequency.

MK-specific antigens (CD41a, CD42b, CD61 and CD62) were more rapidly induced in cultures supplemented with Tpo alone than with both Tpo and early HGFs. This pattern is consistent with the observation that Tpo induces selective MK differentiation, while further addition of early HGFs also causes a marked amplification of megakaryocytic cell number. These results suggest that: (i) Tpo induces essentially the differentiation of MK-committed HPCs; (ii) IL-3, KL and IL-6 trigger extensive growth of HPCs, which are then channeled by Tpo into the megakaryocytic differentiation pathway.

Altogether, these results indicate that highly purified HPCs proliferate and differentiate through the megakaryocytic lineage giving rise to relatively large numbers of highly purified or virtually pure megakaryocytic precursors and then mature MKs. This culture system will facilitate the analysis of the cellular and molecular events underlying megakaryocytopoiesis. Particularly, murine gene targeting studies indicate a key role for the transcription factor NF-E2 in MK maturation (19): the present methodology may provide an important tool to delineate the regulatory role of NF-E2 and other transcription factors in adult human megakaryocytic differentiation and maturation.

REFERENCES

1. Metcalf D: The molecular control of cell division, differentiation, commitment and maturation in hematopoietic cells. Nature 339:27, 1989
2. Ogawa M: Differentiation and proliferation of hematopoietic stem cells. Blood 81:2844, 1993
3. Gewirtz AM: Developmental biology of megakaryocytes and platelets. Curr Opinion Hematol p. 256, 1993
4. Lok S, Kaushansky K, Holly RD, Kuijper JL, Lofton-Day CE, Oort PJ, Grant FJ, Heipel MD, Burkhead SK, Kramer JM, Bell LA, Sprecher CA, Blumberg H, Johnson R, Prunkard D, Ching AFT, Mathewes SL, Bailey MC, Forstrom JW, Buddle MM, Osborn SG, Evans SJ, Sheppard PO, Presnell SR, O'Hara PJ, Hagen FS, Roth GJ, Foster DC: Cloning and expression of murine thrombopoietin cDNA and stimulation of platelet production in vivo. Nature 369:565, 1994
5. De Sauvage FJ, Hass PE, Spencer SD, Malloy BE, Gurney AL, Spencer SA, Darbonne WC, Henzel WJ, Wong SC, Kuang WJ, Oles KJ, Hultgren B, Solberg LA,Jr., Goeddel DV, Eaton DL: Stimulation of megakaryocytopoiesis and thrombopoiesis by the c-Mpl ligand. Nature 369:533, 1994
6. Bartley TD, Bogenberger J, Hunt P, Li YS, Lu HS, Martin F, Chang MS, Samal B, Nichol JL, Swift S, Johnson MJ, Hsu RY, Parker VP, Suggs S, Skrine JD, Merewether LA, Clogston C, Hsu E, Hokom MM, Hornkohol A, Choi E, Pangelinan M, Sun Y, Mar V, McNinch J, Simonet L, Jacobsen F, Xie C, Shutter J, Chute H, Basu R, Selander L, Trollinger D, Sieu L, Padilla D, Trail G, Elliott G, Izumi R, Covey T,

Crouse J, Garcia A, Xu W, Del Castillo J, Biron J, Cole S, Hu MC-T, Pacifici R, Ponting I, Saris C, Wen D, Yung YP, Lin H, Bosselman RA: Identification and cloning of a megakaryocyte growth and development factor that is a ligand for the cytokine receptor mpl. Cell 77:1117, 1994

7. Kaushansky K, Lok S, Holly RD, Broudy VC, Lin N, Bailey MC, Forstrom JW, Buddle MM, Oort PJ, Hagen FS, Roth GJ, Papayannopoulou T, Foster DC: Promotion of megakaryocyte progenitor expansion and differentiation by the c-Mpl ligand thrombopoietin. Nature 369:568, 1994
8. Wendling F, Maraskovsky E, Debili N, Florindo C, Teepe M, Titeux M, Methia N, Breton-Gorius J, Cosman D, Vainchenker W: c-Mpl ligand is a humoral regulator of megakaryocytopoiesis. Nature 369:571, 1994
9. Labbaye C, Valtieri M, Testa U, Giampaolo A, Meccia E, Sterpetti P, Parolini I, Pelosi E, Bulgarini D, Cayre YE, Peschle C: Retinoic acid downmodulates erythroid differentiation and GATA-1 expression in purified adult progenitor culture. Blood 83:651, 1994
10. Gabbianelli M, Sargiacomo M, Pelosi E, Testa U, Isacchi G, Peschle C: "Pure" human hematopoietic progenitors: permissive action of basic fibroblast growth factor. Science 249:1561, 1990
11. Valtieri M, Gabbianelli M, Pelosi E, Bassano E, Petti S, Russo G, Testa U, Peschle C: Erythropoietin alone induces erythroid burst formation by human embryonic but not adult BFU-E in unicellular serum-free culture. Blood 74:460, 1989
12. Gabbianelli M, Pelosi E, Montesoro E, Valtieri L, Luchetti L, Samoggia P, Vitelli L, Barberi T, Testa U, Lyman S, Peschle C: Multi-level effects of flt3 ligand on human hematopoiesis: expansion of putative stem cells and proliferation of granulomonocytic progenitors/monocytic precursors. Blood 86: 1661, 1995
13. Bonthron D, Orr EC, Mitsock LM, Ginsburg D, Handin RI, Orkin SH: Nucleotide sequence of pre-pro-von Willebrand factor cDNA. Nucl Acids Res 14:7125, 1986
14. Lewis ME, Sherman TG, Burke S, Akil H, Davis LG, Arentzen R, Watson SJ: Detection of proopiomelanocortin mRNA by in situ hybridization with an oligonucleotide probe. Proc Natl Acad Sci USA 83:5419, 1986
15. Giampaolo A, Pelosi E, Valtieri M, Montesoro E, Sterpetti P, Samoggia P, Camagna A, Mastroberardino G, Gabbianelli M, Testa U, Peschle C. HOXB gene expression and function in differentiating purified hematopoietic progenitors. Stem Cells 13(Suppl 1):90, 1995
16. Guerriero R, Testa U, Gabbianelli M, Mattia G, Montesoro E, Macioce G, Pace A, Ziegler B, Hassan HJ, Peschle: Unilineage megakaryocytic proliferation and differentiation of purified hematopoietic progenitors in serum-free liquid culture. Blood (in press)
17. Sposi NM, Zon LI, Carè A, Valtieri M, Testa U, Gabbianelli M, Mariani G, Bottero L, Mather C, Orkin SH, Peschle C: Cell cycle-dependent initiation and lineage-dependent abrogation of GATA-1 expression in pure differentiating hematopoietic progenitors. Proc Natl Acad Sci USA 89:6353, 1992
18. Labbaye C, Valtieri M, Barberi T, Meccia E, Masella B, Pelosi E, Condorelli G, Testa U, Peschle C: Differential expression and functional role of GATA-2, NF-E2 and GATA-1 in normal adult hematopoiesis. J Clin Invest 95:2346, 1995
19. Shivdasani RA, Rosenblatt MF, Zucker-Franklin D, Jackson CW, Hunt P, Saris CJM, Orking SH: Transcription factor NF-E2 is required for platelet formation independent of the actions of thrombopoietin/MGDF in megakaryocyte development. Cell 81:695; 1995

55

ROLES OF JAK2 IN HUMAN GM-CSF RECEPTOR SIGNALS

Sumiko Watanabe, Tohru Itoh, and Ken-ichi Arai*

Department of Molecular and Developmental Biology
Institute of Medical Science, University of Tokyo
4-6-1, Shirokane-dai, Minato-ku, Tokyo 108, Japan

ABSTRACT

The IL-3 and GM-CSF receptors (hGMR) consist of two subunits, α and β, both of which are members of the cytokine receptor superfamily. We analyzed the mechanism of c-*fos* mRNA activation by GM-CSF using several hGMRβ subunit mutants. In addition to box1 region, a membrane distal region (a.a. 544-589) of hGMRβ is required for c-*fos* activation. Only one tyrosine residue (Tyr577) exists within the region 544-589, and substitution of Tyr577 to phenylalanine in GMRβ 589 resulted in the loss of c-*fos* activation. In contrast, the same substitution in a wild type receptor did not affect GM-CSF-induced activities such as c-*fos* mRNA induction and proliferation but abolished Shc phosphorylation. These results suggest that the activation of Shc is not essential for c-*fos* activation and several tyrosine residues co-ordinate to activate c-*fos* activation.

It is well documented that IL-3 or GM-CSF activates JAK2 in various cells. However the role of JAK2 in IL-3/GM-CSF functions is largely unknown. We examined the role of JAK2 in GM-CSF-induced signaling pathways. Dominant negative JAK2 (ΔJAK2) lacking the C-terminus kinase domain, suppressed IL-3/GM-CSF induced c-*fos* activation, c-*myc* activation and proliferation suggesting that JAK2 is involved in both signaling pathways. Several tyrosine residues are known to be phosphorylated by GM-CSF within this region and JAK2 expressed transiently in COS7 cells phosphorylated certain tyrosine residues within hGMRβ. JAK2 also phosphorylated PTP1D in COS7 cells. PTP1D and Shc are phosphorylated by IL-3/GM-CSF in BA/F3 cells, however these phosphorylation events were inhibited by expression of ΔJAK2. Taken together, these results indicate that JAK2 is a primary kinase regulating all the known activities of GM-CSF. JAK2 mediates GM-CSF induced c-*fos* activation through receptor phosphorylation and Shc/PTP1D activation.

*Corresponding author : Ken-ichi Arai. Phone number: 81-3-5449-5660; Fax number: 81-3-5449-5424.

INTRODUCTION

GM-CSF is a cytokine which regulates the differentiation and proliferation of various hemopoietic cells (1). The receptor of hGM-CSF is composed of two subunits, α and β, both of which are members of the cytokine receptor family (10, 11). In both the mice and the humans, GM-CSF, IL-3 and IL-5 receptors share a common β subunit. These receptors are composed of a common β subunit and an α subunit specific to each cytokine. In addition to the common β subunit (AIC2B), in the mouse system, an additional β subunit (AIC2A) specific to IL-3 exists. We previously showed that distinct hGMR signaling pathways are involved in c-*myc* mRNA induction /cells proliferation and in the induction of c-*fos*/c-*jun* mRNAs, which is dependent upon a membrane proximal region (a.a. 455-544) and a more cytoplasmic region (a.a. 544-589) of the β subunit respectively (24). To analyze the signals for the activation of c-*myc* and proliferation, we established protocols to monitor the activities of the c-*myc* promoter and DNA replication using a polyoma-replicon (21, 22). Experiments using these systems revealed that E2F/p107 complexes play an important role in the regulation of c-*myc* induction. Further analyses to clarify signaling molecules involved in c-*myc* activation and the initiation of DNA replication are necessary. Although GMR has no intrinsic tyrosine kinase activity, phosphorylation of tyrosine residues in GMR β and several cellular proteins are observed with GM-CSF stimulation. The involvement of JAK family kinases in cytokine signaling has been well documented. The JAK family of kinase consists of JAK1, JAK2, JAK3 and Tyk2 in mammalian species (6) but their roles in hematopoiesis remain to be determined. Interestingly, a dominant mutation of Drosophila homolog, *hop* gene (*hopscotch*$^{\text{Tumorous-lethal}}$) resulted in hematopoietic defects (9). Much attention has been directed to JAK family kinases since their roles in interferon (IFN) signals were recognized (3). Studies with IFN receptor signals revealed that JAK family kinases are involved in IFN-specific gene expression in cooperation with STAT proteins (12, 18, 25). Subsequent studies of IL-6 and Prolactin signaling revealed that the JAK-STAT system plays a role in cytokine specific gene expression (20, 26). However, it is unclear whether or not JAK family kinase is involved in activities shared by many cytokines such as the induction of cell proliferation or activation of immediate response genes. JAK2 is phosphorylated or activated by many cytokines including IL-3 and GM-CSF (14, 15, 16) and association of JAK2 with the common β subunit of IL-3R and GMR was reported (2). In the present work, we attempted to determine whether or not JAK2 is involved in hGMR signals using BA/F3 and COS7 cells. We found that JAK2, which is activated through the box1 region of hGMRβ, plays essential roles in both c-*myc*/proliferation activation and c-*fos*/c-*jun* activation signaling pathways.

THE ROLE OF TYROSINE RESIDUES IN THE hGMRβ SUBUNIT AND c-*fos* ACTIVATION

Similar to other growth factors such as EGF and PDGF, GM-CSF induces cell proliferation and transcription of immediate early genes such as c-*fos*, c-*jun* and c-*myc* (23, 24). In addition, GM-CSF induces ID1, cln and egr genes. Using a series of deletion mutants of the β subunit and tyrosine kinase inhibitors, we examined the signaling mechanism inducing these events (24). There are at least two hGMR signaling pathways (Fig.1). One pathway, involving the membrane proximal region (AA 455-517) containing box1 of receptor β subunit, is required for the activation of c-*myc* transcription and proliferation. This pathway is sensitive to genistein indicating that a tyrosine kinase sensitive to genistein plays an essential role to activate c-*myc* and proliferation. The other pathway, involving a more membrane distal region (AA 544-589) in addition to the membrane proximal region,

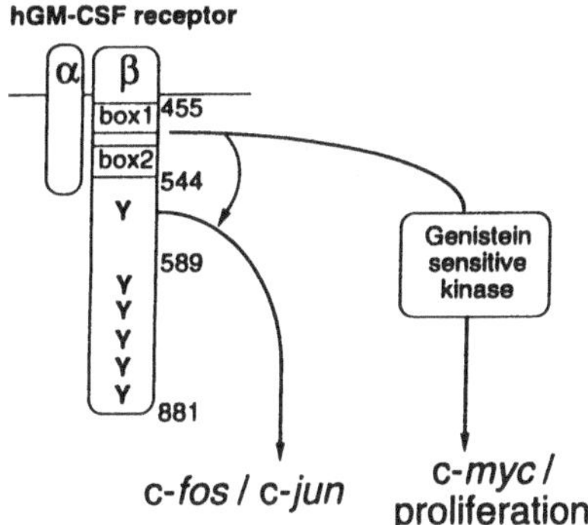

Figure 1. Two distinct signaling pathways of hGMR signals.

leads to the activation of c-*fos* and c-*jun* transcription. In contrast to the former pathway, this pathway is not suppressed by genistein, and c-*fos* activation is rather augmented by genistein.

Because there are several tyrosine residues within the C-terminus region of the β subunit, we next examined the roles of tyrosine residues in hGMRβ and the involvement of tyrosine kinase in GM-CSF signals. To analyze the requirement of tyrosine residues, we constructed several hGMRβ mutants and analyzed signals through these mutants in BA/F3 cells (Fig. 2A). Fig. 2B is a summary of various analyses with these mutants. Deletion up to 589 did not affect any of the activities. Further deletion to 544 resulted in the loss of c-*fos* activation and phosphorylation of tyrosine residues of hGMR and cellular proteins. Activation of proliferation is not affected because the box1 region is sufficient for activity. These results suggest that a region between 589 to 544 is essential for activation of c-*fos* promoter. We next examined role of tyrosine residues. There are 6 tyrosine residues in C-terminus region of hGMRβ. Because there is only one tyrosine residue, 577 in the region covering 589 to 544, we first substituted tyrosine residue 577 to phenylalanine of mutant 589. As expected, this mutant did not activate the tyrosine phosphorylation of cellular proteins or the GMRβ nor did it activate the c-*fos* promoter. On the other hand, the same substitution in the wild type hGMRβ did not affect these activities. This means that tyrosine 577 plays an essential role in the signaling of mutant 589, but in the wild type receptor, multiple tyrosine residues are involved in the c-*fos* gene activation. It is tempting to speculate that multiple signals arise from each tyrosine residue. To test this hypothesis we next determined the involvement of tyrosine 577 in Shc and PTP 1D activation. Shc (13) and PTP1D (19) are known to be phosphorylated by IL-3/GM-CSF (27) and their positive roles in the MAPK cascade have been discussed (8). We analyzed the hGM-CSF induced phosphorylation of these molecules by immunoprecipitation followed by Western blotting using anti-phosphotyrosine antibody 4G10 (α PTyr). BA/FGMR cells were depleted of mIL-3 for 5 hr, and stimulated with 5 ng/ml of hGM-CSF for 5 min. Cells were harvested and lysed with lysis buffer. Immunoprecipitations were done with either anti-Shc or PTP 1D proteins and Western blotting were performed with either α PTyr, anti-Shc or PTP 1D antibodies. As shown in Fig. 3A, Shc is phosphorylated with hGM-CSF stimulation in BA/FGMR cells. The phosphorylation of Shc is abolished with a phenylalanine substitution of tyrosine 577 in GMRβ mutant 589. For the same substitution in the wild type receptor, phosphorylation of Shc cannot be observed even though this mutant can activate c-*fos* activation. These results suggest that tyrosine 577 of hGMRβ is essential for Shc activation, and activation of Shc is not essential for activation of c-*fos* promoter. PTP 1D is also phosphorylated following hGM-CSF stimulation, and substitution of tyrosine 577 to phenylalanin in the 589 mutant abolished GM-CSF induced PTP 1D phosphorylation (Fig. 3B). In

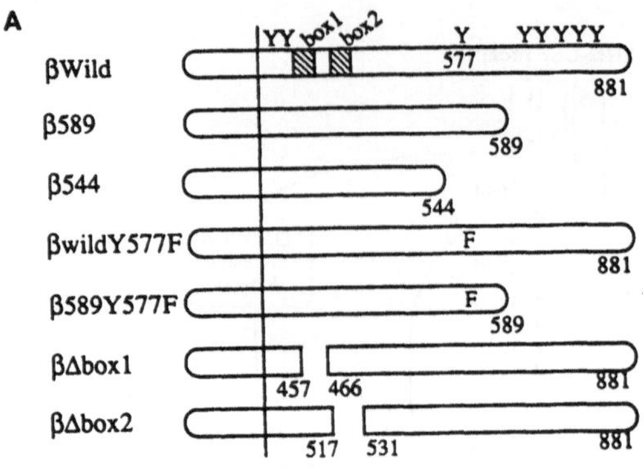

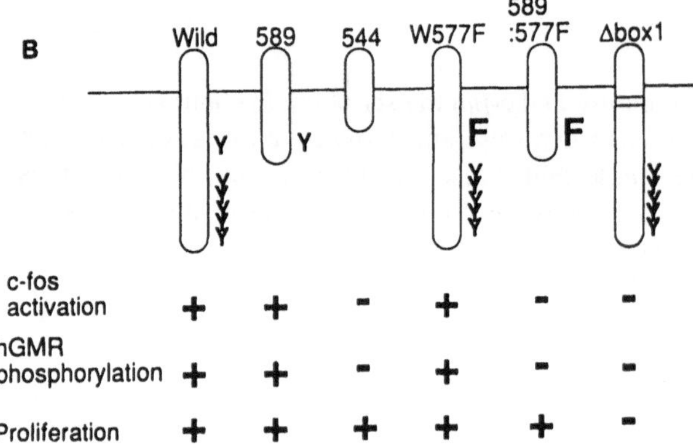

Figure 2. Analysis of requirement of tyrosine residues of hGMR on various signals. A; Schematic drawing of various hGMRβ mutants. B; Summary of several mutants of hGMRβ in c-*fos* activation, tyrosine phosphorylation, cell proliferation in BA/F3 cells.

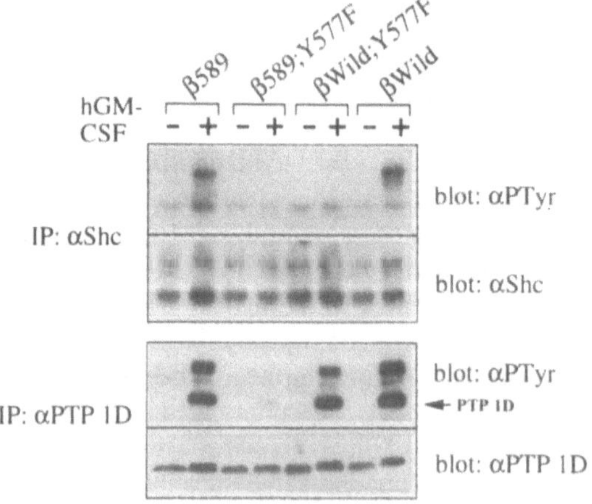

Figure 3. Effects of tyrosine residue substitution in hGMRβ on Shc and PTP 1D phosphorylation. BA/F3 cells expressing hGMRα and various mutants of hGMRβ were depleted mIL-3 for 5hr and restimulated with hGM-CSF for 5 min. Cells were lysed and proteins were immunoprecipitated and analyzed by Western blotting with either anti-Shc (A) or anti-PTP 1D (B) antibodies.

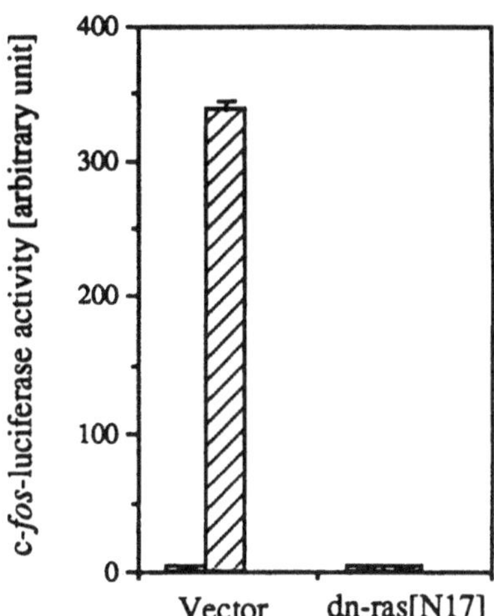

Figure 4. Effect of dominant negative Ras (N17) for GM-CSF induced c-*fos* activation. hGMRα and hGMRβ plasmids were co-transfected with either control vector or dominant negative Ras (N17) and c-*fos*-luciferase activities induced by hGM-CSF (diagonal lines) were analyzed.

contrast to Shc activation, the wild type receptor containing phenylalanine 577 can phosphorylate PTP 1D. These results suggest that tyrosine 577 is not essential for PTP 1D activation.

The RAS protein is also known to be involved in the MAP kinase cascade (7). We examined the role of Ras protein using a dominant negative type of mutant Ras, Ras N17. Co-expression of dominant negative Ras completely suppressed GM-CSF induced c-*fos* activation (Fig. 4). Taken together, we proposed a model of hGMRβ and signaling molecules as illustrated in Fig. 5. Shc may be activated through tyrosine 577 of the β subunit and PTP 1D can activate either tyrosine 577 or other C-terminal tyrosine residues. Signals transduced by Shc and PTP 1D are integrated at Ras and lead to activation of c-*fos* transcription. We next analyzed the involvement of the tyrosine kinase responsible for these events.

ROLE OF JAK2 IN GM-CSF SIGNALS

Several tyrosine kinases such as the src family tyrosine kinases or Tec kinase were reported to be activated by GM-CSF stimulation (14, 15, 17). JAK2 is activated by IL-3 or

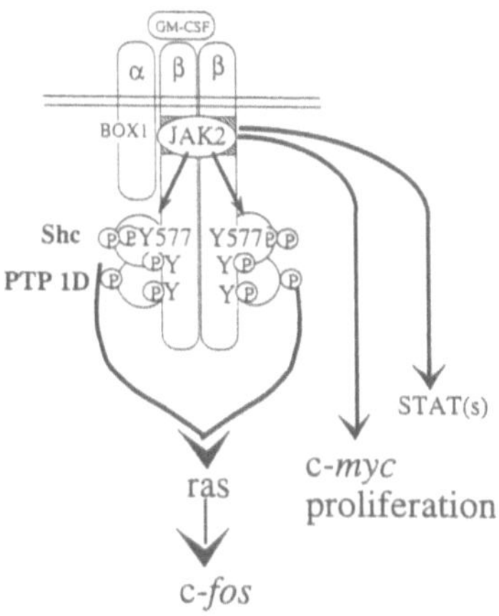

Figure 5. Schematic illustration of signals from hGMR to activate c-*fos* promoter.

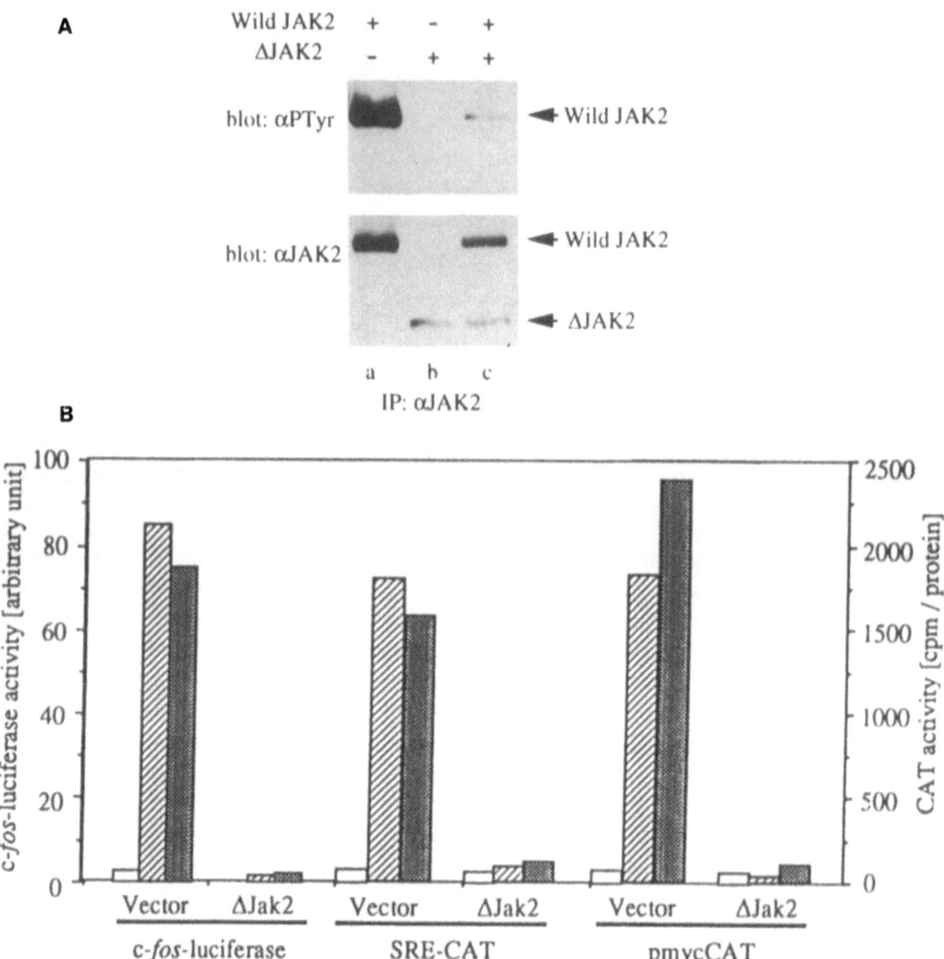

Figure 6. Effects of dominant negative JAK2 in COS7 and BA/F3 cells. A; Wild type JAK2 were co-transfected with or without dominant negative JAK2 in COS7 cells. Tyrosine phosphorylation of JAK2 was examined by Western blotting using anti-PTyr 4G10. B; Effect of dominant negative JAK2 in various promoter activities. Plasmid containing fragment of c-*fos* promoter fused to the luciferase gene, three tandem SRE sites fused to the CAT coding region, or c-*myc* promoter fragment fused to the CAT coding region were transfected with or without ΔJAK2 and mIL-3-(diagonal lines) or hGM-CSF-(screen) induced luciferase or CAT activities were analyzed by luminometer or diffusion assay, respectively.

GM-CSF and it was reported that the membrane-proximal region, box1 of hGMRβ, is required for phosphorylation of JAK2 (Watanabe et al. submitted). Because the membrane proximal region is essential for all the known activities of hGM-CSF, we analyzed the roles of JAK2 in hGMR signals. We analyzed the involvement of JAK2 in GMR signals with dominant negative JAK2. JAK2 lacking the C-terminal kinase domain was constructed. This mutant dominant negatively inhibited autophosphorylation of JAK2 (Fig. 6A) but not JAK1 or JAK3 in COS7 cells (data not shown). We first analyzed the involvement of JAK2 in the pathway to activate the c-*myc* gene and cell proliferation activation in BA/F3 cells. We previously established a c-*myc* promoter transient assay (21) using the CAT gene as a reporter gene. As shown in Fig. 6B, c-*myc* CAT activity induced by IL-3 or GM-CSF was completely suppressed by co-expression of dominant negative JAK2. DNA replication was monitored using the polyoma replication origin (22) and was also abolished by dominant negative JAK2 (data not shown). These results suggest that JAK2 plays an essential role for signals regulating c-*myc* promoter activation and cell proliferation.

The signaling pathway regulating c-*fos* activation by GM-CSF is resistant to the tyrosine kinase inhibitor genistein (24). Because genistein does not affect GM-CSF induced JAK2 phosphorylation, we tested the possibility that JAK2 is involved in c-*fos* activation. For this purpose, we used a luciferase plasmid fused with a 0.4 kb fragment of the c-*fos* promoter. As shown in Fig. 6B, dominant negative JAK2 suppressed c-*fos* promoter activation by IL-3 or GM-CSF completely in BA/F3 cells. The c-*fos* promoter carries an SIE site which contains the GAS sequence, a target site of the STAT protein (4). To examine whether JAK2 exerts its effect through SIE or SRE, we tested the effect of dominant negative JAK2 on the SRE site using tandem repeats of the SRE site fused to CAT coding region (5). Again, dominant negative JAK2 suppressed IL-3 or GM-CSF induced SRE activation suggesting that JAK2 is involved in the STAT independent signaling pathway to the c-*fos* promoter. To analyze the role of JAK2 in the activation of SRE or c-*fos*, we next examined the effect of dominant negative JAK2 on the signal transducing molecules, Shc and PTP1D which are known to be involved in the activation of the c-*fos* promoter. hGMRα and β were transiently transfected into BA/F3 cells and immunoprecipitation was done with either anti-Shc or PTP 1D antibodies. Phosphorylation of Shc or PTP 1D through transiently expressed hGMR was observed and this phosphorylation was completely abolished by co-expression of dominant negative JAK2. It appears that dominant negative JAK2 interferes with signaling event(s) upstream of Shc or PTP 1D activation, thereby indicating that JAK2 plays an essential role in the activation of both signaling molecules (data not shown).

THE hGMR β SUBUNIT AND PTP 1D ARE PHOSPHORYLATED IN COS7 CELLS BY EITHER JAK1, JAK2 OR JAK3

We next asked whether or not PTP 1D could be phosphorylated in COS7 cells by JAK2. PTP 1D, when expressed alone, was not phosphorylated whereas it was heavily phosphorylated when JAK2 and PTP 1D were co-expressed. PTP 1D is known to be activated by various mitogens such as insulin and insulin-like growth factor-1. We examined the specificity of the JAK family kinases with regard to their potential to phosphorylate PTP 1D. Interestingly, co-expression of either JAK1 or JAK3 resulted in PTP 1D phosphorylation. We then examined whether or not JAK1 or JAK3 is also capable of phosphorylating hGMRβ in COS7 cells. Indeed we found they could phosphorylate hGMRβ (Fig. 7). These results indicate that in COS7 cells there is no target sequence specificity among JAK family members. However, it should be noted that, in BA/F3 cells, mIL-3 or hGM-CSF preferentially activates JAK2 but not JAK1 or JAK3 suggesting that JAK1 and JAK3 do not play a major role in mIL-3 or hGM-CSF signaling.

CONCLUSION AND FUTURE ASPECT

In summary, we conclude that JAK2 is the primary kinase regulating all the known GM-CSF signals such as the activation of proliferation, c-*myc* promoter and c-*fos* promoter. Activation of JAK2 is dependent on the GMR box1 region essential for both signaling pathways, and JAK2 phosphorylation. JAK2 may be responsible for hGMRβ phosphorylation and activates a signaling pathway including the phosphorylation of PTP 1D leading to the expression of c-*fos* promoter. These results indicate that JAK2 is involved in multiple

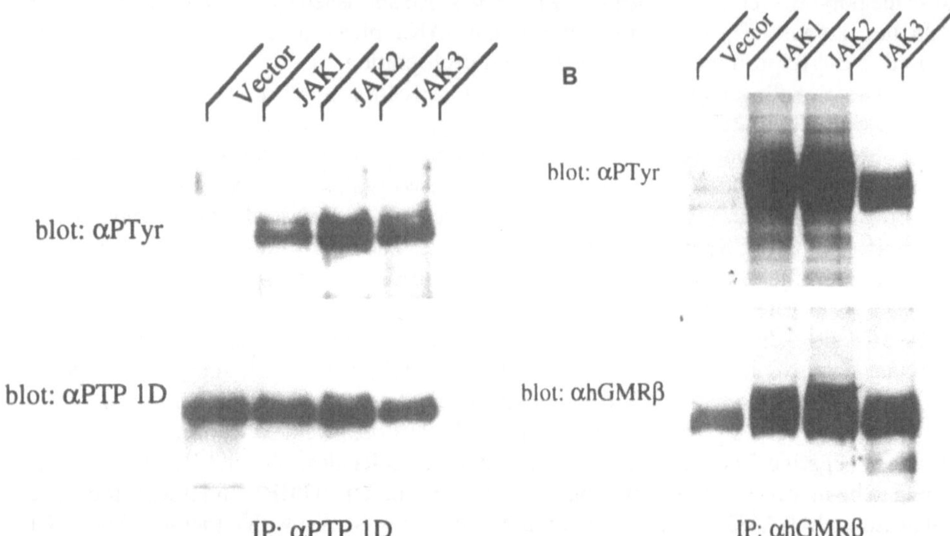

Figure 7. Phosphorylation of hGMRβ and PTP 1D by JAK family kinases in COS7 cells. A, B; Plasmid encoding hGMRβ (A) or PTP 1D (B) and either control vector, JAK1, JAK2 or JAK3 were co-transfected to COS7 cells. Immunoprecipitations were done with either anti hGMRβ (A) or PTP 1D (B) followed by Western blotting.

pathways of hGMR signals in addition to the STAT dependent pathway. How is JAK2 activated by GM-CSF and its receptor? Because it has been reported that JAK2 is constitutively bound to the hGMRβ subunit, a possible mechanism involves ligand-induced dimerization of the β subunit leading to the phosphorylation and activation of JAK2 kinase followed by phosphorylation of the β subunit. To test this possibility we examined the organization of hGMR subunits using a chemical cross linker. BA/FGMR cells were treated with the chemical cross linker, BS3, and immunoprecipitation was done with anti β subunit. We found that the β subunit dimer is formed even in the absence of hGM-CSF stimulation. And the α subunit is associated with the β subunit only when the receptor was stimulated (data not shown).

In summary, we wish to propose a model of activation of hGMR as shown in Fig. 8. JAK2 forms a homo dimer with the β subunit in the absence of hGM-CSF, and its activation was triggered by hGM-CSF stimulation. Because JAK2 is constitutively bind to hGMRβ and GMRβ forms a homodimer, the mechanism of JAK2 activation seems to involve other mechanisms than simply dimerization induced by hGM-CSF. JAK2 is involved in all of the known activities and signals were extinguished by phosphatase PTP 1C. We obtained evidence that the C-terminal region is responsible for the activation of PTP 1C. We are currently working on the mechanism of JAK2 activation and the roles of other tyrosine kinases such as JAK1, Src family kinases and Tec in GM-CSF signals.

ACKNOWLEDGMENTS

Authors thanks M. Dahl for the comments to the manuscript.

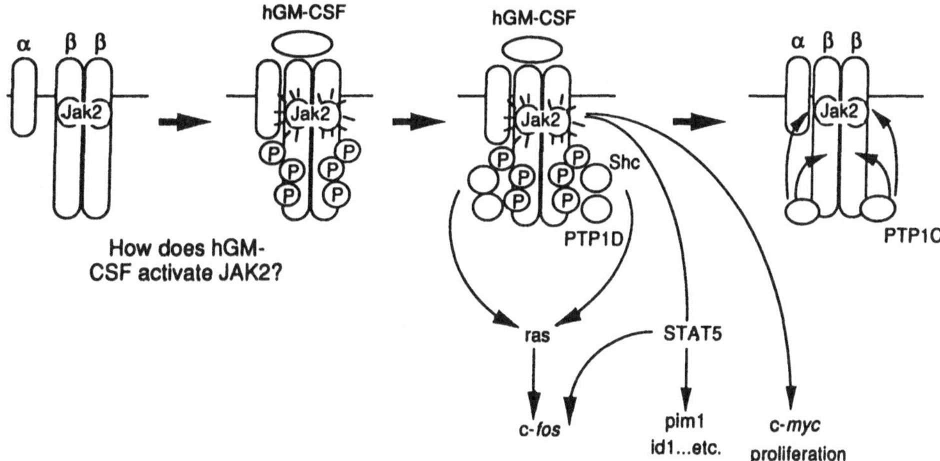

Figure 8. Schematic representation of GM-CSF receptor signals.

REFERENCES

1. Arai, K., F. Lee, A. Miyajima, S. Miyatake, N. Arai and T. Yokota 1990. Cytokines: coordinators of immune and inflammatory responses. Ann. Rev. Biochem. 59: 783-836
2. Brizzi, M. F., M. G. Zini, M. G. Aronica, J. M. Blechman, Y. Yarden and L. Pegoraro 1994. Convergence of signaling by interleukin-3, granulocyte-macrophage colony-stimulating factor, and mast cell growth factor on JAK2 tyrosine kinase. J. Biol. Chem. 269: 31680-31684
3. Darnell Jr., J. E., I. M. Kerr and G. R. Stark 1994. Jak-STAT pathways and transcriptional activation in response to IFNs and other extracellular signaling proteins. Science. 264: 1415-1421
4. Fu, X.-Y. and J.-J. Zhang 1993. Transcription factor p91 interacts with the epidermal growth factor receptor and mediates activation of the c-*fos* gene promoter. Cell. 74: 1135-1145
5. Fukumoto, Y., K. Kaibuchi, N. Oku, Y. Hori and Y. Takai 1990. Activation of the c-*fos* serum-response element by the activated c-Ha-ras protein in a manner independent of protein kinase C and cAMP-dependent protein kinase. J. Biol. Chem. 265: 774-780
6. Ihle, J. N., B. A. Witthuhn, F. W. Quelle, K. Yamamoto, W. E. Thierfelder, B. Kreider and O. Silvennoinen 1994. Signaling by the cytokine receptor superfamiy: JAKs and STATs. TIBS. 19: 222-227
7. Kaziro, Y., H. Itoh, T. Kozasa, M. Nakafuku and T. Satoh 1991. Structure and function of signal-transducing CTP-binding proteins. Annu. Rev. Biochem. 60: 349-400
8. Li, W., R. Nishimura, A. Kashishian, A. G. Batzer, W. J. H. Kim, J. A. Cooper and J. Schlessinger 1994. A new function for a phosphotyrosine phosphatase: Linking Grb2-Sos to a receptor tyrosine kinase. Mol. Cell. Biol. 14: 509-517
9. Luo, H., W. P. Hanratty and C. R. Dearolf 1995. An amino acid substitution in the Drosophila $^{hopTum\text{-}l}$ Jak kinase causes leukemia-like hematopoietic defects. EMBO J. 14: 1412-1420
10. Miyajima, A., T. Kitamura, N. Harada, T. Yokota and K. Arai 1992. Cytokine receptors and signal transduction. Ann. Rev. Immunol. 10: 295-331
11. Miyajima, A., A. L.-F. Mui, T. Ogorochi and K. Sakamaki 1993. Receptors for granulocyte-macrophage colony-stimulating factor, interleukin-3, and interleukin-5. Blood. 82: 1960-1974
12. Muller, M., J. Briscoe, C. Laxton, D. Guschin, A. Ziemiecki, O. Silvennoinen, A. G. Harpur, G. Barbieri, B. A. Witthuhn, C. Schindler, S. Pellegrini, A. F. Wilks, J. N. Ihle, G. R. Stark and I. M. Kerr 1993. The protein tyrosine kinase JAK1 complements defects in interferon-α/β and -γ signal transduction. Nature. 366: 129-135

13. Pelicci, G., L. Lanfrancone, F. Grignani, J. McGlade, F. Cavallo, G. Forni, I. Nicoletti, F. Grignani, T. Pawson and P. G. Pelicci 1992. A novel transforming protein (Shc) with an SH2 domain is implicated in mitogenic signal transduction. Cell. 70: 93-104
14. Quelle, F. W., N. Sato, B. A. Witthuhn, R. C. Inhorn, M. Eder, A. Miyajima, J. Griffin and J. N. Ihle 1994. JAK2 associates with the βc chain of the receptor for granulocyte-macrophage colony-stimulating factor, and its activation requires the membrane-proximal region. Mol. Cell. Biol. 14: 4335-4341
15. Silvennoinen, O., B. A. Witthuhn, F. W. Quelle, J. L. Cleveland, T. Yi and J. N. Ihle 1993. Structure of the murine Jak2 protein-tyrosine kinase and its role in interleukin 3 signal transduction. Proc. Natl. Acad. Sci. USA. 90: 8429-8433
16. Taniguchi, T. 1995. Cytokine signaling through nonreceptor protein tyrosine kinases. Science. 268: 251-255
17. Torigoe, T., R. O'Connor, D. Santoli and J. C. Reed 1992. Interleukin 3 regulates the activity of the Lyn protein-tyrosine kinase in myeloid-committed leukemic cell lines. Blood. 80: 617-624
18. Velazquez, L., M. Fellous, G. R. Stark and S. Pellegrini 1992. A protein tyrosine kinase in the interferon α/β signaling pathway. Cell. 70: 313-322
19. Vogel, W., R. Lammers, J. Huang and A. Ullrich 1993. Activation of a phosphotyrosine phosphatase by tyrosine phosphorylation. Science. 259: 1611-1614
20. Wakao, H., F. Gouilleux and B. Groner 1994. Mammary gland factor (MGF) is a novel member of the cytokine regulated transcription factor gene family and confers the prolactin response. EMBO J. 13: 2182-2191
21. Watanabe, S., S. Ishida, K. Koike and K. Arai 1995. Characterization of cis-regulatory elements of the c-*myc* promoter responding to human GM-CSF or mouse interleukin 3 in mouse proB cell line BA/F3 cells expressing the human GM-CSF receptor. Mol. Biol. Cell. 6: 627-636
22. Watanabe, S., Y. Ito, A. Miyajima and K. Arai 1995. Granulocyte macrophage colony stimulating factor dependent replication of Polyoma virus replicon in hematopoietic cells: Analyses of receptor signals for replication and transcription. J. Biol. Chem. 270: 9615-9621
23. Watanabe, S., A. L.-F. Mui, A. Muto, J. X. Chen, K. Hayashida, A. Miyajima and K. Arai 1993. Reconstituted human granulocyte-macrophage colony-stimulating factor receptor transduces growth-promoting signals in mouse NIH 3T3 cells: Comparison with signalling in BA/F3 pro-B cells. Mol. Cell. Biol. 13: 1440-1448
24. Watanabe, S., A. Muto, T. Yokota, A. Miyajima and K. Arai 1993. Differential regulation of early response genes and cell proliferation through the human granulocyte macrophage colony-stimulating factor receptor: Selective activation of the c-fos promoter by genistein. Mol. Biol. Cell. 4: 983-992
25. Watling, D., D. Guschin, M. Muller, O. Silvennoinen, B. A. Witthuhn, F. W. Quelle, N. C. Rogers, C. Schindler, G. R. Stark and J. N. Ihle 1993. Complementation by the protein tyrosine kinase JAK2 of a mutant cell line defective in the interferon-γ signal transduction pathway. Nature. 366: 166-170
26. Wegenka, U. M., C. Lutticken, J. Buschmann, J. Yuan, F. Lottspeich, W. Muller-Esterl, C. Schindler, E. Roeb, P. C. Heinrich and F. Horn 1994. The interleukin-6-activated acute-phase response factor is antigenically and functionally related to members of the signal transducer and activator of transcription (STAT) family. Mol. Cell. Biol. 14: 3186-3196
27. Welham, M. J., U. Dechert, K. B. Leslie, F. Jirik and J. W. Schrader 1994. Interleukin (IL)-3 and granulocyte/macrophage colony-stimulating factor, but not IL-4, induce tyrosine phosphorylation, activation, and association of SHPTP2 with grb2 and phosphatidylinositol 3'-kinase. J. Biol. Chem. 269: 23764-23768

NEURAL REGULATION OF HEMATOPOIESIS BY THE TACHYKININS

Implications for a "Fine Tuned" Hematopoietic Regulation

Pranela Rameshwar* and Pedro Gascón

UMDNJ-New Jersey Medical School
Department of Medicine
Division of Hematology, Newark, New Jersey 07103

ABSTRACT

The neuromodulators/neurotransmitters, substance P (SP) and neurokinin-A (NK-A) belong to the tachykinin family. Both peptides are widely distributed in the central and peripheral nervous systems. Although lymphoid organs such as the bone marrow (BM) are innervated by peptidergic fibers, non-neural sources for these peptides are possible. Specific cells (endothelial, macrophages and eosinophils) produce SP. In addition, we have found that cytokines induce SP-immunoreactivity (SP-IR) in BM stroma. Although both peptides interact with each of the three cloned NK- receptors (NK-1R, NK-2R, NK-3R), SP and NK-A exhibit preferences for NK-1R and NK-2R respectively. SP and NK-A interact with the G-protein coupled NK-like receptors on hematopoietic cells (CD34+, stroma) to regulate hematopoiesis. Our data indicate the presence of NK-1 and NK-2 like receptors, as well as subtypes of these receptors in BM cells. We have reported that SP stimulates *in vitro* hematopoiesis (erythroid and myeloid). Compared to SP, NK-A is less stimulatory to erythroid and inhibits myeloid colonies. The regulatory effects by the tachykinins are partly determined by the particular type of NK-Rs being stimulated. NK-1R appears to mediate a stimulatory response whereas, NK-2R inhibits hematopoiesis. Most of the effects by the tachykinins are indirect through the induction of growth factors. We have shown that SP induces IL-1, IL-3, IL-6, GM-CSF and *c-kit* ligand in BM cells and NK-A induces MIP-1α and TGF-β. Our data indicate that the stroma is involved in most of the tachykinin-mediated effects through the induction of cytokines which in turn regulate the induction, expression and binding affinities of NK-R on BM cells. These studies suggest that the tachykinins can mediate a network of regulatory interactions among BM cells and, that this can lead to "fine-tuned" hematopoiesis. The results also provide evidence that in addition to being

* Address correspondence to: Pranela Rameshwar, Ph. D. , UMDNJ- New Jersey Medical School, 185 South Orange Avenue, MSB E-585, Newark, New Jersey 07103.

Molecular Biology of Hematopoiesis 5, edited by Abraham et al.
Plenum Press, New York, 1996

neuro-transmitters/-modulators, the tackykinins may be a link of a putative neurohematopoietic axis.

INTRODUCTION

Experimental evidence indicates that the nervous and immune hematopoietic systems communicate bidirectionally through soluble factors: neuropeptides, cytokines and neurotrophic factors via receptor-specific interactions.[1,2] The possibility for a neural control of hematopoiesis can be implied from morphological studies that showed that BM innervation occurs late in fetal life prior to the onset of hematopoiesis.[3] In the adult, the nervous and the immune-hematopoietic systems share an anatomical connection through innervation of the primary (BM and thymus) and, secondary (spleen, lymph nodes and gut associated lymphoid tissue) lymphoid organs by sympathetic and peptidergic fibers.[4,5,6] The BM is an organ innervated by several types of nerve fibers including peptidergic fibers.[6] The precise anatomical distribution of these fibers within the BM is not completely known. However, electron microscopic studies have shown that nerve terminals are in close contact with the cytoplasmic processes of BM stromal cells (reticular and fibroblastoid cells).[7] Furthermore, peptidergic nerve endings have been found in the periosteum of rats.[8] Based on this information, it can be assumed that nerve fibers are in close vicinity with hematopoietic cells, particularly since nerve endings are distributed throughout the entire BM cavity.

Neuropeptides reach their target tissues as neurohormones through the circulation or as neurotransmitters when released by nerve fibers locally, within the tissue.[9] Through specific receptors, neuropeptides can function as immune and hematopoietic modulators. Immune/ hematopoietic cells express multiple neuropeptide/neurotransmitters receptors: β-adrenergic, acetylcholine, endorphins, enkephalins, somastotatin, calcitonin gene related peptide, vasoactive intestinal peptide and, members of the tachykinins.[9] Their respective ligands, derived from neural and non-neural sources are found in the lymphoid microenvironment.[10]

In our lab, we have been studying the role of tachykinins to mediate neuro-hematopoietic interactions based on the following: 1) tachykinins receptors are present on cells found in the BM, 2) The BM cavity is innervated by substance P-immunoreactive (SP-IR) fibers and, 3) BM cells can produce, at least, one of the tachykinins, SP, implying that the BM has two potential sources of SP, neural and non-neural.

TACHYKININS

The tachykinins are a family of structurally related peptides that share a common -COOH terminal sequence, Phe-X-Gly-Leu-Met-NH_2 where X is either an aromatic or branched aliphatic amino acid. Substance P (SP), neurokinin (NK)-A and NK-B are the three major mammalian tachykinins. SP and NK-A are alternately spliced products from the preprotachykinin I (PPT-I) gene (Figure 1), and are found in the central and peripheral nervous systems.[11,12,13] NK-B, although reported in a few peripheral tissues is found mostly in the brain and is derived from the PPT-II gene. Three tachykinin receptors have been cloned: neurokinin (NK)-1R, NK-2R and NK-3R.[14] Biological assays and ligand binding studies indicate subtypes of each NK-R. SP, NK-A and NK-B exhibit binding preferences for NK-1R, NK-2R and NK-3R respectively. The NK-Rs are expressed in immune and hematolopoietic cells and belong to a family of G-protein coupled receptors with seven transmembrane spanning segments.[15] There are exceptions, such as reported for macro-

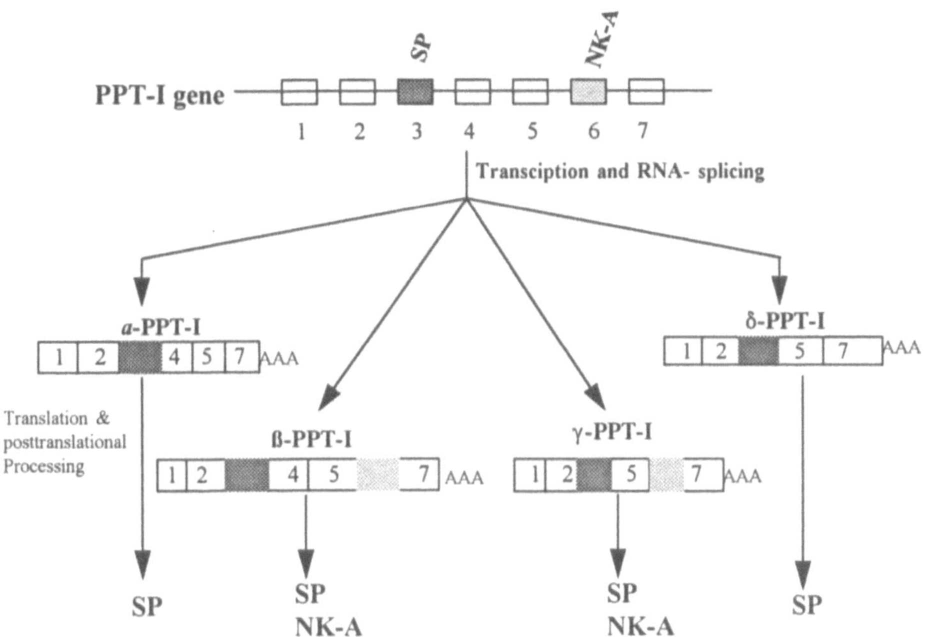

Figure 1. Diagrammatic representation of the preprotachykinin-I (PPT-I) gene processing. References 12 and 13.

phages which express the classical NK-Rs and also, a non-neurokinin receptor coupled to Gi-protein.[16]

SP has been the most studied tachykinin and, most of the reports pertain to its immune mediated functions. This neuropeptide is involved in several inflamatory responses such as stimulation of lymphocytes and regulation of tissue repair via enhanced proliferation of fibroblasts and endothelial cells[9]. SP also enhances phagocytosis, stimulates chemotaxis and the release of histamine, induces mast cell degranulation and eosinophil migration, enhances immunoglobulin production and, functions as a terminal differentiation co-factor for B-cells.[17,18] SP induces also cytokine production in immune, hematopoietic cells and BM stroma. These include IL-1, IL-2, IL-3, IL-6, TNF-α, IFN-γ, GM-CSF and *c-kit* ligand.[19-24]

SP STIMULATES *IN VITRO* HEMATOPOIESIS

SP is synthesized in the cell body of neurons from the PPT-I gene and, transported axoplasmically for storage in the nerve terminals. However, the BM has potential non-neural sources of SP from resident cells such as macrophages and BM stroma cells (unpublished data).[25] The expression of SP-receptors has been reported on T and B cells, macrophages, hematopoietic progenitors, and endothelial cells which make them potential targets for interactions with SP to regulate hematopoiesis.[9,26] We have reported that SP, at very low concentrations (10^{-11}- 10^{-9}M), exerts a significant stimulatory effect on in vitro hematopoiesis: erythroid (BFU-E and CFU-E) and myeloid progenitors (CFU-GM).[19] Most of these effects are mediated by the adherent BM-cell population (stroma) via the induction of cytokines. Our lab and others have reported that SP can induce hematopoietic growth factors (HGF) such as IL-1 and IL-6 in immune/hematopoietic cells,[19,22] IL-3 and GM-CSF in hematopoietic cells[20] and, IL-1 and *c-kit* ligand in BM stroma.[21] The effects of SP on erythroid and myeloid progenitor colonies are seen even in serum-free experimental conditions. Furthermore, SP is capable of substituting for the required exogenous HGF (IL-3 and GM-CSF) in erythroid and myeloid colonies but, not for erythropoietin in erythroid colonies.[19] These stimulatory effects by SP are abrogated by a specific NK-1R antagonist, CP-96,345-1 (Pfizer, Groton, CT) implying the involvement of the high affinity SP-R (NK-1like receptor).[19]

EFFECTS OF NK-A IN *IN VITRO* HEMATOPOIESIS

SP and NK-A are products of the same gene (PPT-I) (Figure 1) and therefore, they can potentially co-exist in the same nerve fibers. However, their hematopoietic effects differ substantially. SP exerts a potent stimulatory effect on erythroid and myeloid progenitors.[19] NK-A stimulates BFU-E and CFU-E (although to a lesser extent than SP), and is suppressive to CFU-GM (Table 1B). The NK-A effect appears to be mediated through the release of hematopoietic inhibitors (MIP-α and TGF-β).[27] Studies with specific NK-1R (CP-96,345) and NK-2R antagonists (SR 48968, SANOFI Recherche, Montpellier Cedex, France) indicate that the induction of these two hematopoietic inhibitors is mediated by an NK-2 like receptor (unpublished data). Other studies on progenitor cell proliferation by SP were performed by blocking the NK-1 and NK-2 receptors with specific antagonists. Blocking of

Table 1. SP-mediated stimulation on hematopoiesis by NK-1 like receptors is partially modulated by NK-2 like receptors. B. Relative *in vitro* hematopoietic effects induced by SP and NK-A

A

Receptor stimulation	BFU-E	CFU-E	CFU-GM
NK-1 like + NK-2 like	↑↑	↑↑	↑↑
NK-1 like	↑↑↑	↑↑↑	↑↑↑

B

Tachykinin	BFU-E	CFU-E	CFU-GM
SP	↑↑	↑↑	↑↑
NK-A	↑	↑	↓↓

the NK-2R enhanced the stimulatory effects of SP suggesting that the NK-2R may be regulating the ability of NK-1R to induce the proliferation of myeloid and erythroid progenitors (Table 1A).

REGULATION OF NK-1 RECEPTOR EXPRESSION IN BM STROMA BY CYTOKINES

In a recent report, we have shown that both IL-1 and *c-kit* ligand can regulate NK-1R induction and SP-binding sites in human BM stroma.[21] Ligand binding studies indicate that the number of SP-binding sites as well as the dissociation constants in BM cells vary according to the exposure time to IL-1 and *c-kit* ligand.[21] Twenty four hour stimulation of BM stroma cells induces the expression of high affinity SP-binding sites. Prolongation of this stimulation to seventy two hours results in a loss of the high affinity receptors and return to baseline low affinity receptors.[21] Amplification by RT-PCR of cDNA prepared from stromal cells using specific NK-1R primers demonstrates the presence of products slightly bigger than the predicted size.[21] Together, this data suggest that the BM stroma may be expressing different subtypes of neurokinin receptors.

DISCUSSION

Neural regulation of hematopoiesis is suggested by three major anatomo-physiological findings: 1) the BM is an organ well innervated containing SP-immunoreactive fibers among many other types, 2) cells present within the BM cavity express receptors for neuropeptides/ neurotransmitters and, 3) the BM contains neural (nerve endings) and non-neural (macrophages and BM stromal cells) sources of SP. Results from our lab and others strongly suggest that, at least for SP, these hematological effects are mediated by the BM stroma via the induction of cytokines through specific NK-Rs.

We have hypothesized that in the BM, cytokines and neurotrophic factors can induce PPT -I gene transcription in neural and non-neural sources (Figure 2), similar to the events described in the ganglia, where SP is induced by IL-1 and leukemia inhibitory factor (LIF).[28,29] In our lab, we have recently found that several cytokines and neurotrophic factors are capable of inducing the production of SP (unpublished data). Similar induction by these factors on the mRNA for NK-1 like receptors was also observed in BM stroma.[21]

In the case of the neuro-hematopoietic axis, the product of the PPT-I gene will be determined by the signals (cytokines/neurotrophic factors) from the hematopoietic environment. Since SP and NK-A exert different hematopoietic effects, the end product of the PPT-I gene can determine the hematopoietic outcome. Cytokines, neuropeptides and neurotrophic factors are hematopoietic regulators and have the potential to provide the signals for PPT-1 gene induction in neural tissues. These signals would be retrogradely taken by neural tissue from the BM microenvironment, via specific receptors in the axon terminal.[30,31,32] The final products of PPT-I induction (SP or NK-A) would be then determined by a particular stimulus capable of influencing the direction of alternate splicing and post-translational processing. *In vitro* studies indicate that cytokines can influence nerve plasticity suggesting an alternate mechanism for a change in neuropeptide content within the BM.[1] On the other hand, the ability of neuropeptides to induce cytokine production in BM cells suggest that neuropeptides and cytokines may be regulating the synthesis of each other.

Extrapolation of *in vitro* studies on neuropeptides/neurotransmitters by our lab and others to the BM microenvironment will lead to our proposed model of hematopoietic regulation by the

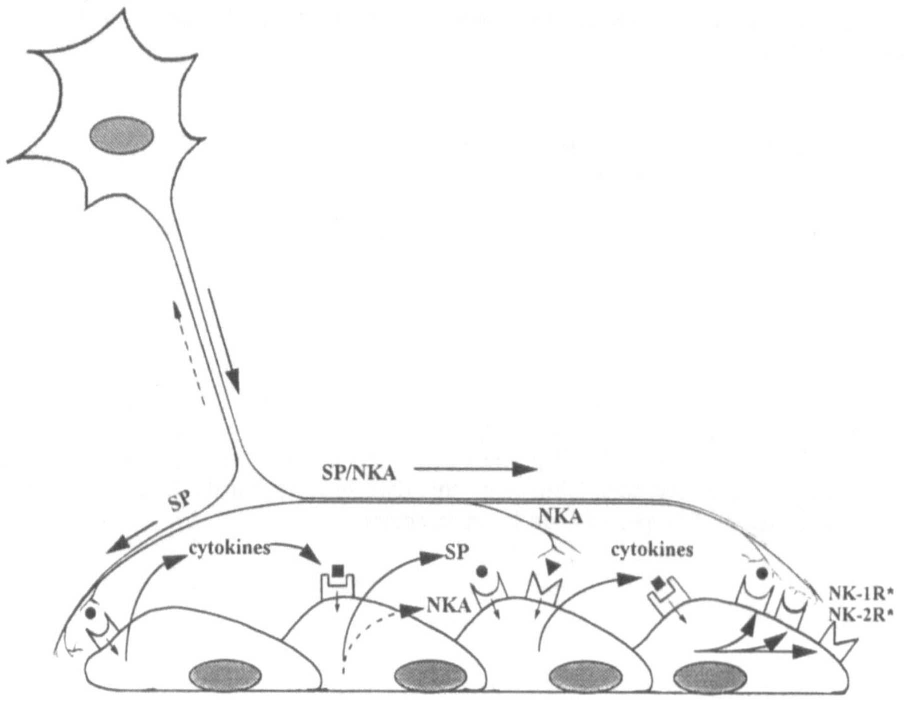

Figure 2. Hypothetical diagram of a bidirectional communication between the nervous and hematopoietic systems. Tachykinins from nerve fibers and stroma interact with specific receptors on bone marrow stroma to induce cytokines. The cytokines in turn modulate the expression and induction of tachykinins and their receptors. *NK-1 like receptor and NK-2 like receptors.

tachykinins (Figure 2). In that model it would be possible to have a self-sustain system composed of the BM stroma and nerve endings communicating among themselves through cytokines and neuromodulators. In that same model, neuromodulators will be able to induce cytokines and vice versa and, either one to regulate the expression of tachykinin receptors. Once these factors are released in the BM microenvironment they could be taken retrogradely through the axon termini to induce the PPT-I gene. Products of the induced gene would then be transported to the nerve endings to act as stroma signals. By doing so, this bidirectional communication among the stroma-nerve-stroma would result in a steady-state, "fine tuned" hematopoietic regulation.

ACKNOWLEDGMENT

Supported by Grant No. HL 54973 from the National Institutes of Health.

REFERENCES

1. Savino W, Dardenne M: Immune-neuroendocrine interactions. *Immunol Today* 16:318, 1995.
2. Blalock JE: The syntax of immune-neuroendocrine communications. *Immunol Today* 15:503, 1994.
3. Miller ML, McCuskey RS: Innervation of bone marrow in the rabbit. *Scand J Haematol* 10:17, 1973.

4. Felten LD, Felten YS, Ackerman DK, Bellinger DL, Kelly SM, Carlson SL, Livnat S: Peripheral innervation of lymphoid tissue. *in Neuroendocrine-Immune Network* (Freier S, ed), CRC, Boca Raton, pp 9, 1990.
5. Weihe E, Nohr D, Michel S, Müller S, Zentel H-J, Fink T, Krekel J: Molecular anatomy of the neuro-immune connection. *Intern J Neurosci* **59**:1, 1991.
6. Felten SY, Felten DL, Bellinger DL, Olschowka JA: Noradrenergic and peptidergic innervation of lymphoid organs. *in Neuroimmunoendocrinology* (Blalock JE, ed) Karger, Basel pp 25, 1992.
7. Yamazaki K, Allen TD: Ultrastructural morphometric study of efferent nerve terminals on murine bone marrow stromal cells, and the recognition of a novel anatomical unit: The "Neuro-Reticular Complex". *Am J Anat* **187**:261, 1990.
8. Calvo W: The innervation of the bone marrow in laboratory animals. *Am J Anat* 123:315, 1986.
9. Maggi CA, Giachetti A, Dey RD, Said SI: Neuropeptides as regulators of airway function: Vasoactive intestinal peptide and the tachykinins. *Physiol Rev* 75:277, 1995.
10. Bulloch K: Neuroanatomy of lymphoid tissue: a review. *in* Neural Modulation of Immunity. (Guillemin et al, eds.). Raven Press, NY, p 11, 1985.
11. Pernow B: Substance P. *Pharmacol Rev* 35:85, 1983.
12. Krause JE, Chirgwin JM, Carter MS, Xu ZS, Hershey D: Three rat preprotachykinin mRNAs encode the neuropeptides substance P and neurokinin A. *Proc Natl Acad Sci USA* 84:881, 1987.
13. Khan I, Collins SM: Fourth isoform of preprotachykinin messenger RNA encoding for substance P in the rat intestine. *Biochem Biophy Res Commun* 202:796, 1994.
14. Gerard NP, Bao L, Ping HX, Gerard C: Molecular aspects of the tachykinin receptors. *Regul Pept* 43:21, 1993.
15. Patacchini R, Maggi CA: Tachykinin receptors and receptor subtypes. *Arch Int Pharmacodyn* 329:161, 1995.
16. Kavelaars A, Broeke D, Jeurissen F, Kardux J, Meijer A, Franklin R, Gelfand EW, Heijnen CJ: Activation of human monocytes via a non-neurokinin substance P receptor that is coupled to Gi protein, calcium, phospholipase D, MAP kinase, and IL-6 production. *J Immunol* 153:3691, 1994.
17. Payan DG: The role of neuropeptides in inflammation. *in Inflammation:Basic principles and clinical correlates* (Gallin R et al, eds), Ravel Press, NY, pp 177, 1992.
18. Bost KL, Pascual DW: Substance P: a late-acting B lymphocyte differentiation cofactor. *Am Physiol Soc* 31:C537, 1992.
19. Rameshwar P, Ganea D, Gascón P: *In vitro* stimulatory effect of substance P on hematopoiesis. *Blood* 81:391, 1993.
20. Rameshwar P, Ganea D, Gascón P: Induction of IL-3 and granulocyte-macrophage colony-stimulating factor by substance P in bone marrow cells is partially mediated through the release of IL-1 and IL-6. *J Immunol* 152:4044, 1994.
21. Rameshwar P, Gascón P: Substance P (SP) mediates production of stem cell factor and interleukin-1 in bone marrow stroma. Potential autoregulatory role for these cytokines in SP receptor expression and induction *Blood* 86:482, 1995.
22. Lotz M, Vaughan JH, Carson DA: Effect of neuropeptides on production of inflammatory cytokines by human monocytes. *Science* **241**:1218, 1988.
23. Rameshwar P, Gascón P, Ganea D: Stimulation of interleukin 2 production in murine lymphocytes by substance P and related tachykinins. *J Immunol* **151**:2484, 1993.
24. Wagner F, Fink R, Hart R, Dancygier H: Substance P enhances interferon-g production by human peripheral blood mononuclear cells. *Regulat Peptides* **19**:355, 1987.
25. Pascual DW, Bost KL: Substance P production by P388D1 macrophages: a possible autocrine function for this neuropeptide. *Immunology* **71**:52, 1990.
26. Rameshwar P, Gascón P: Regulation of substance-P (SP)-like receptors in CD34+ cells by cytokines and neurotrophic factors. *FASEB J* 1432:A248, 1994.
27. Rameshwar P, Gascon P: Induction of negative hematopoietic regulators by neurokinin-A in bone marrow stroma. *Blood* 88:98, 1996.
28. Freidin M, Kessler JA: Cytokine regulation of substance P expression in sympathetic neurons. *Proc Natl Acad Sci USA* 88:3200, 1991.
29. Hart RP, Shadiack AM, Jonakait GM: Substance P gene expression is regulated by interleukin-1 in cultured sympathetic ganglia. *J Neurosci Res* 29:282, 1991.
30. Curtis R, Krystyna MA, Yuan Z, Harkness PJ, Lindsay RM, DeStefano PS: Retrograde axonal transport of ciliary neurotrophic factor is increased by peripheral nerve injury. *Nature* 365:253, 1993.
31. JeanJean AP, Moussaaoui SM, Maloteaux JM, Laduron PM: Interleukin-1β induces long-term increase of axonally transported opiate receptors and substance P. *Neuroscience* 68:151, 1995.

32. Patterson PH, Nawa H. 1993. Neuronal differentiation factors cytokines and synaptic plasticity. *Cell 72/Neuron 10 (Suppl)*:123.

HISTORICAL ASPECTS OF PHYSIOLOGICAL CELL DEATH PRIOR TO DESCRIPTION OF APOPTOSIS

Eugene P. Cronkite*

Medical Department
Brookhaven National Laboratory
Upton, New York 11973-5000

ABSTRACT

Cell death as a physiological process was recognized in the early 1920's. There is a large and rich literature on cell death in hematopoiesis, embryogenesis, immunology, cancer, after exposure of lymphocytes to ionizing radiation or cortisone. Cell death and its measurement has been an essential component in study of cell kinetics of all tissues in which cell turnover is evident.

The objective of this concise and incomplete review on cell death, prior to emergence of the term apoptosis, was to document that non-necrotic cell death in normal and pathologic processes has been appreciated for several decades. When it was discovered that several tissues in the adult continued to have mitoses and that with each mitosis there was a net gain of one cell it was self-evident that there had to be a compensatory loss of cells by death, a process which might appropriately be called cytocide.

Cells are programmed to die in order to fashion organs during embryogenesis, to allow the development of the immune system, to maintain homeostasis in the endocrine-dependent tissues, to limit reactive hyperplasia, or to maintain a steady state in self-renewing tissues in addition to differentiation. Cell death is also a pathophysiological mechanism by which normal and malignant cells may die in response to ionizing radiation and/or chemotherapeutic drugs and toxic chemicals. In the landmark publication in 1972 by Kerr et al. (1) the term apoptosis was coined to dignify the phenomenon of cell death. After the coining of the name apoptosis and its cytological description and subsequently the identification of its genetic controls, opened the flood gates for a torrent of publications on apoptosis. There were eight publications in 1980 and 800 in 1993. Publications continued to expand in '94 and '95.

*Communications to: Eugene P. Cronkite, M.D., Medical Department - Bldg.490, Brookhaven National Laboratory, Upton, New York 11973. Telephone: 516-282-7538; Fax: 516-282-5311.

Molecular Biology of Hematopoiesis 5, edited by Abraham et al.
Plenum Press, New York, 1996

EARLY STUDIES ON CELL DEATH

Granulopoiesis

Florence Sabin in 1925 (2) showed that 6% of blood neutrophils were dying per hour. Accordingly, all neutrophils would be replaced in about 17 hrs for a half time in the blood of about 8.5 hrs. Cartwright, Athens, and Wintrobe (3) developed a technique of labelling of granulocytes with DFP^{32} after separation of the peripheral leukocytes from the blood and autotransfusion showing that there was a random loss from the blood with a half time of about 6.9 hrs.

Fliedner, Cronkite, and Robertson in 1964 (4) observed the senescence of granulocytes in the peripheral blood of human beings describing the pyknosis of a small fraction of granulocytes and measured the kinetics of the phenomenon utilizing tritiated thymidine labelling of granulocytic precursors in the bone marrow, the emergence of labelled cells into the peripheral blood and the time that it takes for this to appear in the pyknotic granulocytes. Patt and Maloney (5,6) conceived of a model of granulopoiesis in which an "ineffective" component of intramedullar myeloid cell production occurs normally, suggesting a mechanism by which the system could respond rapidly to demand. Jamuar and Cronkite (7) searched for granulocytes in the alveolar walls of pathogen-free mice where there would be a greater probability of seeing them. Granulocytes were not seen even in mice with a granulocytosis of 100,000/mm^3 induced by a tumor producing granulocyte colony stimulating factor. Granulocytes were only seen in the spleen where a few were being phagocytosed by macrophages. This suggested that circulating granulocytes in the normal steady state are selectively removed by splenic perhaps marrow macrophages and do not to a detectable extent, migrate out of the blood into tissues at random. It would appear therefore, in the absence of an inflammatory process the major means by which steady state granulopoiesis is maintained is by phagocytosis of apoptotic blood granulocytes by splenic and other macrophages. In fact, phagocytosis of apoptotic granulocytes may stimulate production of granulopoietic molecular regulators by macrophages to maintain a steady state.

Lymphocytes

Schrek (8,9) in 1946 clearly described the development of pyknosis after exposure of lymphocytes to graded doses of x-rays demonstrating the quantitative effect. In 1962, (9) he and associates evaluated an *in vitro* test for radiosensitivity of leukemic lymphocytes utilizing the development of pyknosis. Trowell in 1952 (10) described sensitivity of lymphocytes to ionizing radiation and established dose effect relationships using the pyknotic index. He extended his studies in 1953 to show that cortisone produced pyknosis of lymphocytes *in vitro* (11). Fliedner et al. (12) described the process of cell death in germinal centers of the spleen. He labeled lymphocytes by tritiated thymidine and documented by autoradiography that the label commenced its appearance in the pyknotic nuclear fragments by 30 min after the injection of tritiated thymidine. Labeled fragments of DNA were found in macrophages. Cronkite et al. (13) compared the effect of injection of graded amounts of tritiated thymidine on the pyknosis of small lymphocytes compared to graded doses of x-ray. Twenty-five μCi/gm simulates the effect of 75 rad whole body x-ray. No effect was observed at 5 μCi/gm. Labeled pyknotic nuclei of both large and small lymphocytes were found after doses of tritiated thymidine exceeding 5 μCi/gm and phagocytic cells laden with nuclear debris were present.

Lymphocytes and in particular, thymic derived cells are essential in immunological processes. Burnet (14) in 1959 drew attention to the importance of cell death in his clonal selection theory of immunological processes.

Erythropoiesis

Ashby in 1919 (15) demonstrated that the non-nucleated erythrocyte of human beings had a mean life span of 120 days. Simple arithmetic, using red cell count, average blood volume and the life span of the red cell showed an enormous death rate of 2 x 10" erythrocytes per day. In a steady state this required an equal production rate. Although the erythrocyte is non-nucleated it was the first clearly defined portrayal of the magnitude of cell production and death rates.

Six years later granulocyte turnover was established by Sabin et al. (2) and its magnitude defined by Cartwright et al. (3) as 1 x 10" per day.

Ineffective erythropoiesis was described by Stohlman in 1959 (16). He portrayed that a significant fraction of erythrocytic precursors died in the bone marrow and suggested that a method of increasing red cell production would be the termination of this intramedullary death of red cell progenitors.

OTHER BIOLOGICAL PROCESSES IN WHICH CELL DEATH PLAYS A ROLE

Glucksman in 1951 (15) and Saunders in 1966 (18) pointed out the importance and magnitude of cell death that occurs in organogenesis during embryogenesis.

CELL DEATH IN HUMAN HEMATOPOIETIC MALIGNANT DISEASES

Killman et al. (19) in 1962 labelling myeloma cells with tritiated thymidine and autoradiography demonstrated that the generation time of labeled cells ranged from 2-6 days average of 3 days. Therefore, from a single cell 40 doublings would replace the whole marrow in 120 days and from this single cell the whole body mass with 6 more doublings for a total of 58 days. The inevitable conclusion is multiple myeloma is not a disease of uncontrolled growth. There is substantial cell death. Mauer et al. (20) in 1973 also showed that cell death is a significant factor in acute lymphoblastic leukemia of childhood. Dameshek (21) from an extensive study of the clinical progression of chronic lymphocytic leukemia concluded that it was a disease due to the accumulation of cells whose survival is greater than normal a prediction borne out by the more recent publications showing that chronic lymphocytic leukemia and many B-cell lymphomata are characterized by overexpression of the *bcl*-2 gene which delays or prevents apoptosis (22).

CELL RENEWAL SYSTEMS – MATHEMATICAL CONSIDERATIONS

The cell renewal systems in the mammal and their properties have been extensively described and discussed by Leblond and Walker in 1956 (23). Mathematical description of cell renewal systems are many and varied. Johnson described these and proposed his own (24). Cronkite (25) in 1964 presented a mathematical concept based on the notions of Johnson (24). This mathematical model states that the population (N) doubles with each generation so that

$$N = N_o \cdot 2^{\frac{t}{t_G}} = N_o\,(e^{0.693})^{\frac{t}{t_G}}$$

and the growth constant

$$\lambda = \frac{0.693}{t_G}$$

or

$$N = N_o e^{\lambda t}$$

where t_G = generation time

$$t = \text{time } t_o \text{ to } t;\ \frac{t}{t_G} = \text{number of generations (n)}$$

$dN/dt = \lambda N$ the growth rate in an expanding population. If cells are removed from the expanding population by apoptosis, by non-apoptotic death or differentiation and subsequent desquamation (as in skin or gastrointestinal tract), then the net rate of change is $dN/dt = \lambda N - \lambda$(sub 1)N.

λ_l = loss constant from all causes and in the steady state $\frac{dN}{dt} = 0$.

CONCISE DESCRIPTION OF APOPTOSIS

Apoptosis (programmed cell death) is a process in which the cell actively sets in motion a sequence of events involving the cell membrane and the nucleus. The membrane is altered in a manner making the cell subject to phagocytosis by adjoining macrophages. In parallel, specific endonucleases are activated by influx of Ca^{2+} ions that degrades DNA into pyknotic chunks visible by light microscopy and units of 185 base pairs detectable by electrophoresis. Numerous genes have been identified whose products participate in the initiation of apoptosis, degradation of cell and its phagocytosis. Two genes ced-3 and ced-4 could aptly be called cytocide genes, their products killing the cell in which they are activated. Inactivation of ced-3 or 4 the population grows in size. In addition, ced-9 when activated its product suppresses cell death. Conversely, loss of ced-9 function results in cell death. For further details see Tomei and Cope (26).

SUMMARY AND CONCLUSIONS

The phenomenon of physiological cell death in cell renewal symptoms and induction in some cell systems by ionizing radiation and corticosteroids has been known for decades and has been quantified in some systems. The elegant and detailed description by Kerr et al. (1) set the scene for elucidation of its molecular basis which is being described by a torrent of papers in the last ten years leading to exploitation of the phenomenon in study of cancer and its therapy.

ACKNOWLEDGMENT

This research was supported in part by the U.S. Department of Energy under Contract DE-AC02-76CH00016.

REFERENCES

1. Kerr JFR, Wyllie AH, Currie AR: Apoptosis: A basic biological phenomenon with wide ranging implications in tissue kinetics. Br. J. Cancer 26:239, 1972
2. Sabin FR, Cunningham RS, Doan CA, Kindawall JA: The normal rhythm of the white blood cells. Bulletin of the Johns Hopkins Hospital 37:14, 1925
3. Cartwright G, Athens JW, Wintrobe MM: The kinetics of granulopoiesis in normal man. Blood 24:780, 1964
4. Fliedner TM, Cronkite EP, Robertson JS: Granulocytopoiesis I senescence and random loss of neutrophilic granulocytes in human being. Blood 24:402, 1964
5. Patt HM, Maloney MA: A model of granulocyte kinetics. Annals of the New York Acad. of Sci. 113:515, 1964
6. Maloney MA, Patt HM, Lund JE: Granulocyte dynamics and a question of ineffective granulopoiesis. Cell Tissue Kinetics 4:201, 1971
7. Jamuar N, Cronkite EP: The fate of blood granulocytes. Exp. Hemat. 8:884, 1980
8. Schrek R: Studies *in vitro* on cellular physiology; the effect of x-rays on the survival of cells. Radiology 46:395, 1946
9. Schrek R, Leithold SL, Freedman IA, Best WR: Clinical evaluation of an *in vitro* test for radiosensitivity of leukemic lymphocytes. Blood 20:432, 1962
10. Trowell CA: Radiation, lymphocytes and pyknosis. The sensitivity of lymphocytes to ionizing radiation. J. of Path. and Bact. 64:687, 1952
11. Trowell CA: The action of cortisone on lymphocytes *in vitro*. J. of Physiology 119:274, 1953
12. Fliedner TM, Kesse M, Cronkite EP, Robertson JS: Cell proliferation in germinal centers of the rat spleen. Annals of the New York Acad. of Sci. 113:578, 1964
13. Cronkite EP, Fliedner TM, Killman SA, Rubini JR: Tritium-labeled thymidine: its somatic toxicity and use in the study of growth rates and potential in normal and malignant tissue of man and animals, in Tritium in the Physical and Biological Sciences, vol 2. International Atomic Energy Agency, Vienna, 1962, p 189
14. Burnet M: The clonal selection theory of acquired immunity. Vanderbilt University Press, Nashville, 1959, p 58
15. Ashby W: Determination of the length of life of transfused red blood cells in man. J. Exp. Med. 29:267, 1919
16. Stohlman Jr F: Observations on the Kinetics of Red Cell production, in Kinetics of Cellular Prolifration, Grune and Stratton, NY, 1959, p 318
17. Glucksmann A: Cell deaths in normal vertebrate ontogeny. Biological Reviews of the Cambridge Philosophical Society 26:59, 1951
18. Saunders Jr JPW: Death in embryonic system. Science 54:604, 1966
19. Killman SA, Cronkite EP, Fliedner TM, Bond VP: Cell proliferation in multiple myeloma studied with tritiated thymidine *in vivo*. Laboratory Investigation 11:845, 1962
20. Mauer AM, Evert CF, Lampkin BC, MacWilliams NB: Cell kinetics in human acute lymphoblastic leukemia: computer simulation with discreet modeling techniques. Blood 41:141, 1973
21. Dameshek W: Chronic lymphocytic leukemia - an accumulative disease of immunologically incompetent lymphocytes. Blood 29:566, 1967
22. Korsmeyer SJ: Bcl-2 initiates a new category of oncogenes: Regulators of cell death. Blood 80:879, 1992
23. Leblond CP, Walker DE: Renewal of cell populations. Physiological Reviews 36:235, 1956
24. Johnson HA: Some problems associated with histological study of cell proliferation kinetics. Cytologia 26:32, 1961
25. Cronkite EP: Enigmas underlying the study of hematopoietic cell proliferation. Fed. Proc. 23:649, 1964
26. Tomei L David, Frederick O. Cope Frederick O (eds): Apoptosis: The molecular basis of cell death. Current Communications in Cell and Molecular Biology. Cold Spring Harbor Laboratory Press, 1991.

THROMBOPOIETIN

Biologic Expectations, Physiologic Surprises, and Clinical Promise

Kenneth Kaushansky,[1] * Nancy Lin,[1] Angelika Grossmann,[3] Katherine Sprugel,[3] Ewa Sitnicka,[2] and Virginia Broudy[1]

[1] Division of Hematology, Box 357710
[2] Department of Pathology
University of Washington School of Medicine
Seattle, Washington 98195
[3] ZymoGenetics, Inc.
1201 Eastlake Avenue E.
Seattle, Washington 98102

Maintenance of the primary hemostatic system is an enormous task. Each day the adult human produces approximately 2×10^{11} platelets. This number can increase ten-fold in times of increased demand. Studies over the past few decades have revealed that blood cell production is dependent on cellular and extracellular components of the bone marrow stroma, hematopoietic stem and progenitor cells, and polypeptide growth factors (reviewed in[1,2]). Although at least 20 such cytokines have now been cloned and characterized, it has become clear that each blood cell lineage is supported by one or more early-acting, pleotropic cytokines, responsible for expanding primitive cells to supply adequate numbers of lineage-committed progenitors, and a late-acting growth factor, required for the maturation of these progenitors into mature blood elements. In erythropoiesis, these two functions are supported by kit ligand (KL; also termed stem cell factor, steel factor or mast cell growth factor) and erythropoietin (Epo).[3,4] In myeloid cell development, interleukin (IL)-3 or granulocyte-macrophage (GM)-colony- stimulating factor (CSF) and G-CSF fill these respective roles.[5-7] However, until recently, the cytokines which support megakaryocyte and platelet development were enigmatic. For although IL-3, GM-CSF or KL were shown able to support megakaryocyte colony formation in serum or plasma containing cultures,[8-10] the late acting maturation factor for this process, termed thrombopoietin (Tpo) some 37 years ago,[11] had not been purified or cloned. Within the last year, all of this has changed.

* To whom correspondence should be addressed.

THROMBOPOIETIN EXISTS

Based on pioneering work with the murine myeloproliferative virus MPLV, Wendling and her colleagues described an orphan cytokine receptor, mutation of which formed the basis for the pan-myeloid expansion characteristic of the disease.[12-14] As the receptor was originally cloned from a biphenotypic leukemic cell line with megakaryocytic potential,[14] we and others surmised that the receptor might be that for Tpo. Within a short time the Mpl ligand was cloned and characterized.[15-19] In initial studies, we showed that the murine homologue was a powerful megakaryocyte differentiation factor; it increased the size, ploidy and expression of platelet specific membrane glycoproteins on marrow megakaryocytes.[20] Moreover, although contrary to previous expectation, the Mpl ligand was shown to be a most potent megakaryocyte CSF.[20,21] When administered to animals, murine Mpl ligand increased platelet production approximately five-fold,[15,20] by expanding marrow megakaryocytic progenitors and mature megakaryocyte numbers and size.[20] And Wendling and coworkers demonstrated that the Mpl ligand accounted for all of the thrombopoietic activity of plasma and that its levels varied inversely with platelet count.[21]

THROMBOPOIETIN IS ESSENTIAL FOR FULL MEGAKARYOCYTIC MATURATION

Although it is now clear that the Mpl ligand is identical to Tpo, it has not been established that it is essential for megakaryocyte and platelet development. A number of investigators, including our own group, have shown that IL-3, GM-CSF and KL are capable of producing megakaryocytes in semi-solid in vitro cultures.[8-10] If Tpo is essential for megakaryocyte development, then these cultures must depend on endogenous thrombopoietin present in the culture system. To test this hypothesis, we established that marrow cells can produce Tpo mRNA. Next, in order to neutralize the Tpo in these cultures, we developed a soluble form of the Mpl receptor, on the premise that it would neutralize the biological activity of its ligand. Subsequent studies proved this to be true. Using KL in the absence or presence of IL-6 or IL-11, we showed that although megakaryopoiesis was brisk, the soluble Mpl receptor essentially obliterated megakaryocyte colony formation.[22] These results were consistent with our hypothesis, and suggest that KL is dependent on Tpo for activity. However, in contrast to these results, the soluble receptor could reduce but never eliminate megakaryocyte colony formation from cultures initiated with IL-3, with or without IL-6 or IL-11. Thus, IL-3 could support megakaryocyte formation in the absence of Tpo. Moreover, based on the number of megakaryocytes in each colony in IL-3 and in Tpo-induced cultures, the IL-3-responsive progenitors appeared to have greater proliferative potential. However, these results do not speak to the level of megakaryocyte maturation in the cultures. To address this issue we studied the size, ploidy and cytoplasmic maturation of megakaryocytes grown in the presence of IL-3 with or without IL-11, in the presence of the soluble Mpl receptor. All of the megakaryocytes in IL-3 containing cultures were small, were 2N or 4N and failed to demonstrate demarcation membranes, platelet fields or platelet-specific granules. The addition of IL-11 increased the size of the cells, and was associated with some demarcation membrane formation, but the soluble Mpl receptor appeared to block most platelet field and granule formation in IL-3 plus IL-11 stimulated cells. Thus, Tpo appears to be essential for the terminal stages of normal megakaryocyte development.

While this work was underway, Gurney and coworkers at Genentech generated a homozygous Mpl knock-out animal using the techniques of homologous recombination. The homozygous animals had greatly reduced marrow megakaryocytes and blood platelet levels.[23] These results argue that Tpo is the primary regulator of platelet production. However, they also suggest that there is a minor alternative pathway to platelet production. Our results would suggest that this is mediated by IL-3 plus either IL-6 or IL-11.

THROMBOPOIETIN IS A POLYFUNCTIONAL CYTOKINE

Although Tpo is the primary regulator of platelet production, we next sought to determine whether it might also affect other cell lineages. There were several reasons to test this hypothesis. First, the erythroid and megakaryocytic lineages display several common features.[24] For example, they share a number of transcription factors and cell surface molecules (perhaps the most important of which are the receptors for IL-3, KL, Epo and Tpo). Moreover, many if not most erythroid leukemic cell lines display or can be induced to display markers of megakaryocytic differentiation, and megakaryocytic lines express or can be induced to display erythroid markers. And the introduction of a gp IIb suicide gene eliminates not only megakaryocytic development, but early erythroid colony formation as well (G. Marguerie, personal communication). Using purified cytokines we found that Tpo augments the development of large erythroid bursts in the presence of IL-3 or KL and Epo (Figure 1).[25] Similar findings were also recently reported by Ogawa, who also demonstrated that Tpo acts directly on erythroid progenitors (ISHAGE meeting, Vancouver, Canada). In suspension culture, Tpo enhances the generation of CFU-E, again, only in the presence of Epo.[25] Given these in vitro effects, we sought to determine whether Tpo affected erythropoiesis in vivo. In normal animals, Tpo increased the number of marrow erythroid bursts, and redistributed the CFU-E from marrow to spleen. However, the reticulocyte count was only minimally elevated and no statistically significant effects were noted on red cell levels.[25] In retrospect this finding was not surprising. All of the in vitro erythropoietic effects of Tpo were dependent on the presence of Epo. Should Tpo stimulate erythropoiesis in vivo, Epo levels would fall, thereby abrogating the Tpo effect on erythropoiesis. Thus, we next

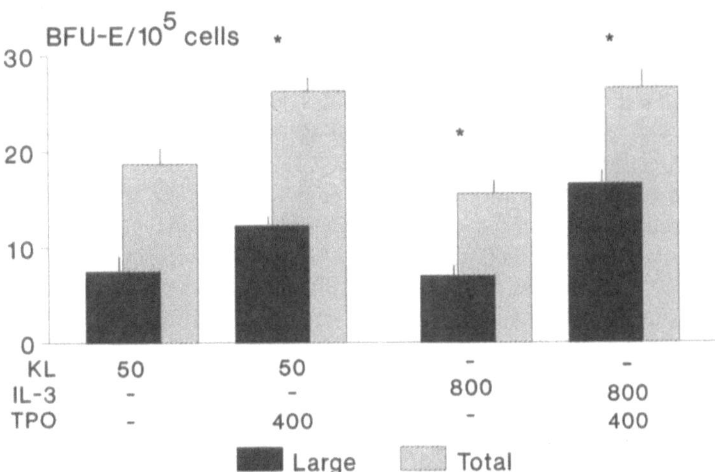

Figure 1. Tpo enhances the formation of large erythroid bursts in the presence of KL or IL-3 and Epo. Semi-solid cultures containing purified murine KL (50 ng/ml) or murine IL-3 (800 units/ml), with or without Tpo (400 U/ml) were evaluated. All cultures contained 2U/ml human Epo. The total number of erythroid bursts, as well as those greater than 10^4 cells/burst were enumerated on day 8. The results are for a representative experiment and has been repeated three times.

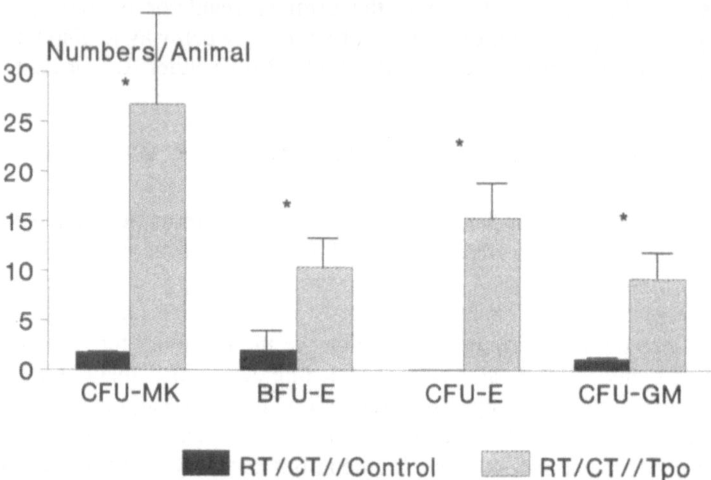

Figure 2. Tpo enhances the recovery of all hematopoietic progenitors following myelosuppressive therapy. Balb/C mice treated with 350 cGy gamma irradiation and a single dose of 1.2 mg carboplatinum were administered vehicle or recombinant murine Tpo subcutaneously twice daily for 13 days (the midpoint of hematopoietic recovery in this model) and sacrificed. Marrow and spleen cells were harvested and the number of CFU-Mk (x 10^{-3}), BFU-E (x 10^{-3}), CFU-E (x 10^{-5}) and CFU-GM (x 10^{-4}) determined in standard colony forming assays using optimal levels of cytokines (IL-3 plus Tpo for CFU-Mk and CFU-GM, PWM-SCM plus Epo for BFU-E and Epo for CFU-E).[20] The results represent the mean values of three experiments containing five animals in each group. * $p< .01$.

determined the effect of Tpo on erythropoiesis in anemic animals. Using a combined radiation/chemotherapy protocol we found that in addition to greatly accelerated platelet recovery,[26] Tpo-treated animals had far greater numbers of both marrow and splenic erythroid bursts and CFU-E midway through hematopoietic recovery, had a normal reticulocyte count two weeks before vehicle treated animals, and recovered red cell levels far quicker (Figure 2).[25] These results are more fully discussed in an accompanying paper in this volume.[27]

More recently, we have expanded these studies to myeloid progenitor populations. In the presence of KL, but surprisingly not IL-3 or GM-CSF, Tpo enhanced the production of CFU-GM derived colonies (Figure 3). Vainchenker has recently reported similar findings using purified progenitor cell populations (ISHAGE meeting, Vancouver, Canada). And like the findings for erythroid progenitors and blood counts in vivo, the administration of Tpo to normal mice increases marrow CFU-GM but not peripheral blood leukocyte counts, but does hasten neutrophil recovery when G-CSF levels are high following cytoreductive therapy.

Finally, we have begun to explore the effects of Tpo on the most primitive hematopoietic cells using murine marrow cells selected to display the physical characteristics of blasts, depleted of lineage restricted cells, and retaining very low levels of rhodamine and Hoechst dye ($lin^-/Ho^{lo}/Rh^{lo}$). This subpopulation of marrow can repopulate all of hematopoiesis with as little as ten cells.[28] We have found that Tpo increases the proliferation of these cells if added to cultures containing sub-optimal levels of cytokines, and acts to induce these cells, normally quiescent, into active cell cycle. It is noteworthy that Tpo acts like G-CSF in this regard, a relatively lineage-restricted maturational cytokine which can induce primitive, pleuripotent cells to divide.[29] These findings may make teleologic sense. In states of iatrogenic or natural marrow failure, neutropenia and thrombocytopenia lead to increases in G-CSF and Tpo plasma levels. In addition to promoting the final stages of maturation of cells restricted to each lineage, these cytokines, perhaps not surprisingly, also act to induce

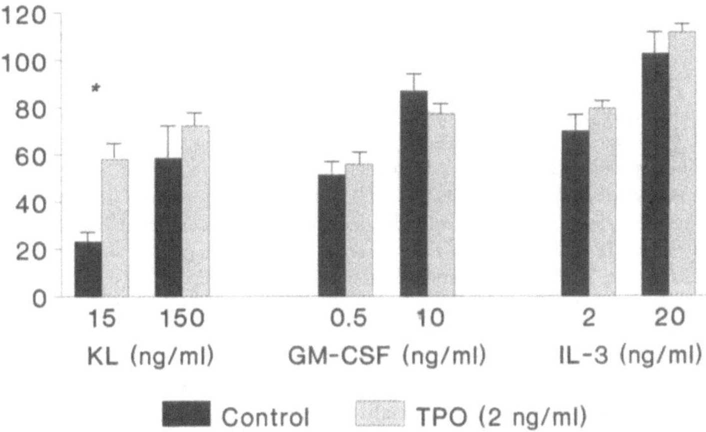

Figure 3. Tpo acts synergistically with KL to enhance granulocyte-macrophage colony formation. Semi-solid cultures containing two concentrations of murine KL, GM-CSF or IL-3 were initiated with or without 750 U/ml murine Tpo. GM colonies were enumerated on day 5 of culture. The results for KL plus Tpo are statistically different* (t-test) from those for KL alone ($p<0.02$).

immature cells to begin the process of lineage commitment (which appears to require cell division) and thereby ensure the adequate supply of progenitors on which to act.

In the span of one brief year, thrombopoietin has evolved from a hazy concept to a cloned growth promoting maturational hormone. Its biological attributes place it firmly in the class of Epo and G-CSF. Given the therapeutic successes of these agents, Tpo has been tested for its capacity to protect animals from the complications of myelosuppressive therapy. These studies have shown the agent to be a potent stimulus of not only thrombopoiesis, but also, at least in certain murine models, a potent stimulus of erythropoietic and myeloid recovery as well. Little if any toxicity has been encountered in these studies. As such, Tpo has recently entered its first clinical trials. It is hoped that it will fill an important and much needed role in promoting the recovery of hematopoiesis in states of natural and iatrogenic marrow failure.

REFERENCES

1. Metcalf D: Hematopoietic regulators: Redundancy or subtlety? Blood 82:3515, 1993
2. Ogawa M: Differentiation and proliferation of hematopoietic stem cells. Blood 81:2844, 1993
3. Dai CH, Krantz SB, Zsebo KM: Human burst-forming units-erythroid need direct interaction with stem cell factor for further development. Blood 78:2493, 1991
4. Geissler K, Stockenhuber F, Hinterberger W, Balcke B, Lechner K: Recombinant human erythropoietin: a multipotential hemopoietic growth factor *in vivo* and *in vitro*. In Schaefer RM, Heidland A, Hörl WH (eds.): Erythropoietin in the 90s. Contrib Nephrol Vol. 98, Basel, Karger, 1990, p 1
5. Emerson SG, Yang Y-C, Clark SC, Long MW: Human recombinant granulocyte-macrophage colony stimulating factor and interleukin 3 have overlapping but distinct hematopoietic activities. J Clin Invest 82:1282, 1988
6. Broxmeyer HE, Williams DE, Cooper S, Shadduck RK, Gillis S, Waheed A, Urdal DL, Bicknell DC: Comparative effects in vivo of recombinant murine interleukin 3, natural murine colony-stimulating factor-1, and recombinant murine granulocyte-macrophage colony-stimulating factor on myelopoiesis in mice. J Clin Invest 79:721, 1987

7. Kobayashi M, Van Leeuwen BH, Elsbury S, Martinson ME, Young IG, Hapel AJ: Interleukin-3 is significantly more effective than other colony-stimulating factors in long-term maintenance of human bone marrow-derived colony-forming cells in vitro. Blood 73:1836, 1989
8. Kavnoudias H, Jackson H, Ettlinger K, Bertoncello I, McNiece I, Williams N: Interleukin-3 directly stimulates both megakaryocyte progenitor cells and immature megakaryocytes. Exp Hematol 20:43, 1992
9. Kaushansky K, O'Hara PJ, Berkner K, Segal GM, Hagen FS, Adamson JW: Genomic cloning, characterization, and multilineage expression of human granulocyte-macrophage colony-stimulating factor. Proc Natl Acad Sci USA 83:3101, 1986
10. Briddell RA, Bruno E, Cooper RJ, Brandt JE, Hoffman R: Effect of c-kit ligand on in vitro human megakaryocytopoiesis. Blood 78:2854, 1991
11. Kelemen E, Cserhati I, Tanos B: Demonstration and some properties of human thrombopoietin in thrombocythemic sera. Acta Haematol (Basel) 20:350, 1958
12. Wendling F, Varlet P, Charon M, Tambourin P: A retrovirus complex inducing an acute myeloproliferative leukemia disorder in mice. Virology 149:242, 1986
13. Souyri M, Vigon I, Penciolelli J-F, Heard J-M, Tambourin P, Wendling F: A putative truncated cytokine receptor gene transduced by the myeloproliferative leukemia virus immortalizes hematopoietic progenitors. Cell 63:1137, 1990
14. Vigon I, Mornon J-P, Cocault L, Mitjavila M-T, Tambourin P, Gisselbrecht S, Souyri M: Molecular cloning and characterization of *MPL*, the human homolog of the v-*mpl* oncogene: Identification of a member of the hematopoietic growth factor receptor superfamily. Proc Natl Acad Sci USA 89:5640, 1992
15. Lok S, Kaushansky K, Holly RD, Kuijper JL, Lofton-Day CE, Oort PJ, Grant FJ, Helpel MD, Burkhead SK, Kramer JM, Bell LA, Sprecher CA, Blumberg H, Johnson R, Prunkard D, Ching AFT, Mathewes SL, Balley MC, Forstrom JW, Buddle MM, Osborn SG, Evans SJ, Sheppard PO, Presnell SR, O'Hara PJ, Hagen FS, Roth GJ, Foster DC: Cloning and expression of murine thrombopoietin cDNA and stimulation of platelet production in vivo. Nature 369:565, 1994
16. de Sauvage FJ, Hass PE, Spencer SD, Malloy BE, Gurney AL, Spencer SA, Darbonne WC, Henzel WJ, Wong SC, Kuang W-J, Oles KJ, Hultgren B, Solberg Jr LA, Goeddel DV, Eaton DL: Stimulation of megakaryocytopoiesis and thrombopoiesis by the c-Mpl ligand. Nature 369:533, 1994
17. Bartley TD, Bogenberger J, Hunt P, Li Y-S, Lu HS, Martin F, Chang M-S, Samal B, Nichol JL, Swift S, Johnson MJ, Hsu R-Y, Parker VP, Suggs S, Skrine JD, Merewether LA, Clogston C, Hsu E, Hokom MM, Hornkohl A, Choi E, Pangelinan M, Sun Y, Mar V, McNinch J, Simonet L, Jacobsen F, Xie C, Shutter J, Chute H, Basu R, Selander L, Trollinger D, Sieu L, Padilla D, Trail G, Elliott G, Izumi R, Covey T, Crouse J, Garcia A, Xu W, Del Castillo J, Biron J, Cole S, Hu MC-T, Pacifici R, Ponting I, Saris C, Wen D, Yung YP, Lin H, Bosselman RA: Identification and cloning of a megakaryocyte growth and development factor that is a ligand for the cytokine receptor Mpl. Cell 77:1117, 1994
18. Sohma Y, Akahori H, Seki N, Hori T, Ogami K, Kato T, Shimada Y, Kawamura K, Miyazaki H: Molecular cloning and chromosomal localization of the human thrombopoietin gene. FEBS Letters 353:57, 1994
19. Kuter DJ, Beeler DL, Rosenberg RD: The purification of megapoietin: a physiological regulator of megakaryocyte growth and platelet production. Proc Natl Acad Sci USA 91:11104, 1994.
20. Kaushansky K, Lok S, Holly RD, Broudy VC, Lin N, Bailey MC, Forstrom JW, Buddle M, Oort PJ, Hagen FS, Roth GJ, Papayannopoulou T, Foster DC: Promotion of megakaryocyte progenitor expansion and differentiation by the c-Mpl ligand thrombopoietin. Nature 369:568, 1994.
21. Wendling F, Maraskovsky E, Debili N, Florindo C, Teepe M, Titeux M, Methia N, Breton-Gorius J, Cosman D, Vainchenker W: c-Mpl ligand is a humoral regulator of megakaryocytopoiesis. Nature 369:571, 1994
22. Kaushansky K, Broudy, VC, Lin N, Jorgensen M, McCarty J, Fox N, Zucker-Franklin D, Lofton-Day C: Thrombopoietin, the Mpl-ligand, is essential for full megakaryocyte development. Proc Natl Acad Sci USA 92:3234, 1995
23. Gurney AL, Carver-Moore K, de Sauvage FJ, Moore MW: Thrombocytopenia in *c-mpl*-deficient mice. Science 265:1445, 1994
24. McDonald TP, Sullivan PS: Megakaryocytic and erythrocytic cell lines share a common precursor cell. Exp Hematol 21:1316, 1993
25. Kaushansky K, Lin N, Grossman A, Humes J, Sprugel K, Broudy V: Thrombopoietin expands erythroid, granulocyte-macrophage and megakaryocyte progenitor cells in normal and myelosuppressed mice. J Clin Invest *(in press)*.
26. Sprugel KH, Humes JM, Grossmann A, Ren HP, Kaushansky K: Recombinant thrombopoietin stimulates rapid platelet recovery in thrombocytopenic mice. Blood 84(Suppl 1):242a, 1994 (abstr)
27. A.Grossmann, this volume.

28. Wolf NS, Koné A, Priestley GV, Bartelmez SH: In vivo and in vitro characterization of long-term repopulating primitive hematopoietic cells isolated by sequential Hoechst 33342-rhodamine 123 FACS selection. Exp Hematol 21:614, 1993
29. Ikebuchi K, Clark SC, Ihle JN, Souza LM, Ogawa M: Granulocyte colony-stimulating factor enhances interleukin 3-dependent proliferation of multipotential hemopoietic progenitors. Proc. Natl. Acad. Sci. USA 85:3445, 1988

BIOLOGIC EFFECTS OF THROMBOPOIETIN (TPO)

A. Grossmann,[1*] J. S. Lenox,[1] J. M. Humes,[1] H. P. Ren,[1] K. Kaushansky,[2] R. A. Nash,[3] and K. H. Sprugel[1]

[1] ZymoGenetics Inc.
1201 Eastlake Avenue E
Seattle, Washington 98102
[2] Division of Hematology
Department of Medicine
University of Washington
Seattle, Washington 98115
[3] Fred Hutchinson
Cancer Research Center
1124 Columbia Street
Seattle, Washington 98104

ABSTRACT

The recent cloning of thrombopoietin (TPO) has allowed us to study its *in vivo* effect in normal and myelosuppressed mice. In normal mice, megakaryocytes were more than 50% higher in the bone marrow after four days of hTPO treatment when compared to controls. This increase was accompanied by increased platelet counts in the peripheral blood. After 7 days of treatment with hTPO, peak platelet counts were reached on day 8.

TPO treatment did not affect red or white blood cell counts in normal mice. We also studied the effect of TPO treatment in myelosuppressed mice, since the development of blood cells is highly dependent on the interactions of hematopoietic stem cells, cytokines and bone marrow stroma. Mice were irradiated with 350 cGy TBI and dosed with 1.2 mg carboplatin to damage the bone marrow. Platelet counts recovered to 50% of baseline 13 days earlier in TPO treated mice than vehicle treated mice. In addition, TPO treatment increased the recovery of white and red blood cell counts. These results suggest that TPO affects multiple cell lineages in myelosuppressed mice.

* To whom correspondence should be addressed.

Molecular Biology of Hematopoiesis 5, edited by Abraham et al.
Plenum Press, New York, 1996

INTRODUCTION

For more than 30 years it has been suggested that platelet production was controlled by specific factors.[1,2,3] However it has been difficult to study the differentiation of megakaryocytes and platelets from megakaryocytic progenitors, because this factor was neither purified nor cloned. Recently, the cDNA responsible for this activity was cloned by several groups and subsequently expressed and purified.[4-8]

TPO has been shown to be a 70 kilodalton glycoprotein which can stimulate megakaryocyte colony formation and increase the size and ploidy of megakaryocytic cells *in vitro*. When administered to normal mice, TPO dramatically increases the circulating platelet counts with concomitant increases in marrow and splenic megakaryocyte number, size and ploidy.[5,9] In addition, TPO has been shown to stimulate megakaryocyte production in stem cell populations.[10] Thus, TPO appears to be able to act on early proliferation and differentiation of megakaryocytes, as well as induce their maturation to produce platelets.

The recent cloning of TPO has enabled us to evaluate the effects of TPO on megakaryocytes in the bone marrow and peripheral blood cells in normal and myelosuppressed animals.[11,12] Kaushansky, et. al.,[13] have shown that normal mice and myelosuppressed mice treated with TPO showed significantly higher myeloid progenitors of all lineages in the bone marrow and spleen than in vehicle treated mice. Studies reported here determined the effect of TPO on peripheral blood cells and megakaryocytes in normal and myelosuppressed mice. In normal mice, TPO treatment increased platelet counts exclusively. In contrast, myelosuppressed mice treated with TPO recovered their platelet, red and white blood cell counts earlier than vehicle controls, suggesting that TPO can act on multiple blood cell lineages.

MATERIAL AND METHODS

Animals

Six week old female Balb/c mice (Jackson Laboratories, Bar Harbor, ME) were maintained in ALAAAS-certified animal care facilities with a 12 hour light-dark cycle. Water and standard laboratory rodent chow were supplied *ad libitum*. All procedures were reviewed and approved by the ZymoGenetics Animal Care and Use Committee.

Study Design

The myelosuppressive regimen was similar to that of Leonard et al (1994).[14] On day 0, mice were exposed to 350 cGy total body irradiation from a ^{137}Cs source (Gammacell 40 Irradiator, Nordian International, Inc. Kanata, Canada), immediately followed by intraperitoneal injection of 1.2 mg carboplatinum (Carboplatin, Bristol Myers Squibb, Cambridge, MA). Treatment with hTPO or vehicle started on day 1 and continued through day 8. hTPO was formulated in a vehicle of 20 mM Tris (pH 8.0)/0.9% NaCl/0.25% rabbit serum albumin (RSA, Sigma Chemical Co, St. Louis, MO A-0639) and injected subcutaneously twice daily. Each treatment group consisted of 5 mice.

Recombinant TPO

hTPO was expressed in mammalian cells and assayed as previously described.[5,15] Conditioned medium was collected and concentrated by ultrafiltration. An affinity column consisting of recombinant mouse c-mpl receptor coupled to CNBr Sepharose was used to purify hTPO. After buffer exchange to 20 mM Tris/1mM EDTA, the protein was loaded onto the affinity column and eluted with 3 M KSCN. Fractions containing material absorbing at A280 were pooled and dialyzed in 20 mM Tris (pH 8). The final material consisted of a mixture of TPO molecular weight forms with the 70 kD and 30 kD forms being most prominent. Doses are expressed as units, where 50 units are defined as the amount of hTPO stimulating a half maximal response in the BaF3/mouse mpl mitogenesis assay.

Hematological Evaluation

Peripheral blood (60 ul) was collected from the retro-orbital sinus under ether anesthesia into heparinized capillary tubes and immediately transferred to EDTA coated microtainer tubes (Becton-Dickinson, San Jose, CA). Complete blood counts were performed on 20-40 ul blood with a Baker-Serono 9010 Hematology analyzer (Serono Diagnostics, Allentown, PA) using mouse discriminator settings. Baseline values were collected for each mouse 7 days prior to irradiation and chemotherapy treatment.

Cytospin of the Bone Marrow

Bone marrow cells from each mouse were flushed from one femur with 400 ul buffer (HaemaLine II, Serono Diagnostics, Allentown, PA) with 1% RSA (Sigma Chemical Co, St. Louis, MO A-0639). 20ul of each sample were then diluted into 200 ul of buffer and cytospin slides were prepared (Shandon Cytospin 3) and stained with Wright-Giemsa (EM Science, Gibbstone, NJ). The total number of megakaryocytes in the 20 ul cytospin were counted and multipled by 20 to calculate the total number of megakaryocytes/femur.

Statistics

The effects of treatment on various parameters was analyzed using the non-parametric Mann-Whitney Test. A criteria of $p<0.05$ was considered significant.

RESULTS

Effect of TPO in Normal Mice

The activity of full length hTPO was measured in 6-8 week old normal mice by injecting hTPO subcutaneously twice daily for 7 days in doses ranging from 500 U/day to 20,000 U/day. A dose dependent increase in the number of platelets was observed with hTPO (Figure 1). Platelet counts peaked at day 8 and returned to normal levels between day 16 and 19. To determine the effect of TPO on normal bone marrow, mice were treated for 4 days with 20 kU TPO/day. On day 4, analysis of cytospin slides revealed a two fold increase in megakaryocytes in the bone marrow from TPO treated mice compared to vehicle controls.

Effect of TPO in Myelosuppressed Mice

To explore whether TPO could reduce thrombocytopenia, we used a mouse model of myelosuppression which results in a depression of circulating platelets to below 15% of baseline values. On day 0, mice received a dose of sublethal irradiation and an injection of

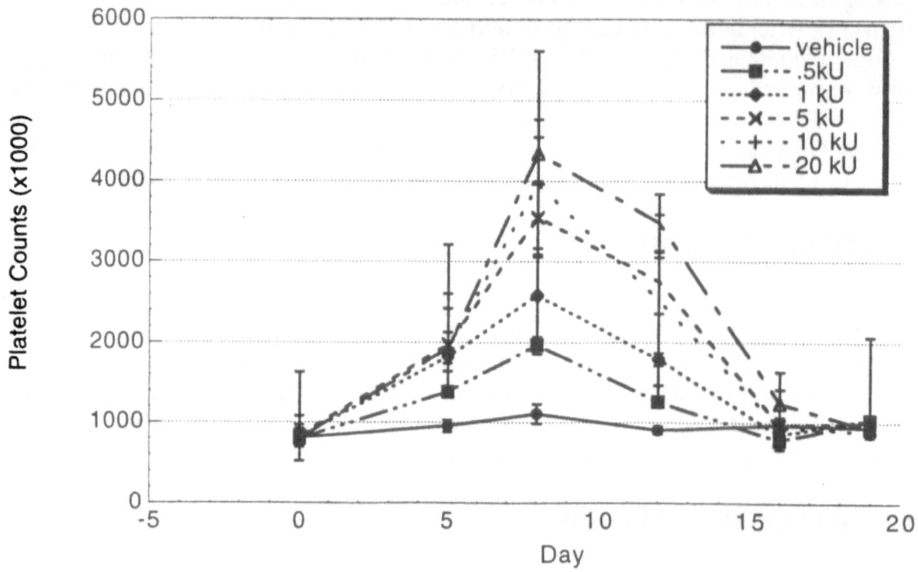

Figure 1. Dose response of human TPO in normal mice. Mice received sc injections of TPO twice daily for 7 days. Points represent mean ± SD (n = 4/point).

carboplatin. Five mice/group were treated with 10,000 units hTPO/day or vehicle twice daily by subcutaneous injection for 8 days. In the vehicle-treated group platelet counts dropped below 15% of baseline by day 9 and remained below this level through day 13. Mice treated with 10,000 U hTPO/day for 7 days showed marked improvements in platelet recovery. Recovery to 50% of baseline was significantly ($p<0.02$) shortened by TPO treatment and occurred on day 7 ± 2 in TPO treated mice and on day 20 ± 2 in vehicle controls.

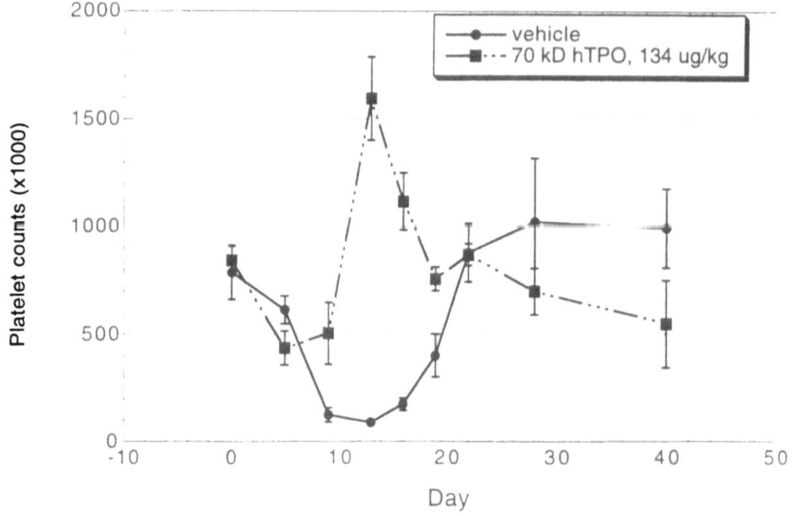

Figure 2. Effect of hTPO treatment for 8.5 days on platelet counts in myelosuppressed mice. Points represent mean ± SD (n=5/point).

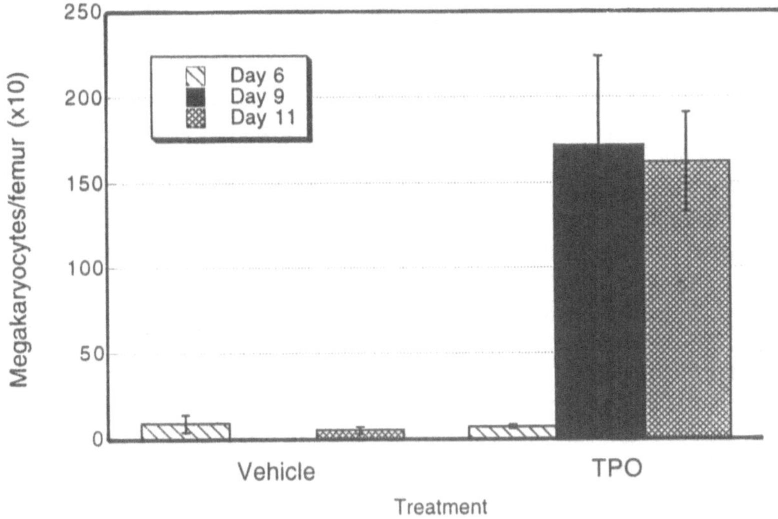

Figure 3. Effect of hTPO treatment for 8.5 days on the number of megakaryocytes/femur in myelosuppressed mice. Points represent mean ± SD (n=3/point).

Effect of TPO on Bone Marrow Recovery

The effect of TPO on bone marrow recovery was determined in myelosuppressed mice which were sacrificed on days 6, 9 and 13. Cytospin slides were made from flushed bone marrow cells. Mice treated with vehicle had low numbers of megakaryocytes in their bone marrow at all three time points. TPO treated mice also showed low numbers of megakaryocytes on day 6, however, megakaryocyte numbers were significantly increased on days 9 and 13 when compared to vehicle controls (Figure 3).

In addition to more rapid platelet recovery, myelosuppressed mice experienced a shorter period of anemia and neutropenia. In vehicle controls, red blood cells were dropped to 35% of baseline at the nadir on day 19. TPO treated mice had only a 25% reduction in red blood cell counts. Mean time to recovery (defined as 75% of baseline) was 12 days in TPO treated mice compared to 26 days in controls (p=0.02). These experiments suggest that hTPO enhances erythropoiesis in myelosuppressed mice. Indeed, the TPO related red blood cell effect was accompanied by a significantly earlier increase in the reticulocyte index starting on day 9 in the TPO treated mice as opposed to day 16 in the vehicle treated mice. These results establish that at least part of the improvement in red blood cell counts was due to stimulation of erythropoiesis. White blood cells also recovered more rapidly, reaching baseline on day 13 in TPO treated mice compared to day 19 in vehicle controls. Recent experiments suggest that the increased white blood cell counts are related to increased numbers of neutrophils (data not shown).

DISCUSSION

Treatment of mice with TPO resulted in a dose dependent increase in platelet counts (up to 6 fold) and increased numbers of megakaryocytes in the bone marrow. In addition to working in mice, hTPO is able to increase platelet counts in normal rats, dogs and monkeys, demonstrating that hTPO can act via the mpl receptor in these species.[16,17] In normal primates and dogs, as in normal treated mice, platelets are the only peripheral blood cells affected by TPO treatment (Sprugel and Nash, personal communication). These data indicate that TPO stimulates the maturation of only the megakaryocytic lineage in normal bone marrow.

TPO treatment dramatically increased the recovery of megakaryocytes and platelets in myelosuppressed mice. In addition, myelosuppressed mice showed accelerated recovery of red and white blood cells when compared to vehicle controls. This effect is not evident in normal mice. The regulation of hematopoiesis in the bone marrow is very complex and requires the interaction of many cytokines. Myelosuppression alters the cytokine environment in the bone marrow. We hypothesized that in damaged bone marrow with altered cytokine levels, TPO treatment could result in expansion of precursors other than CFU-Mk. In fact, TPO treatment of myelosuppressed and normal mice increases marrow BFU-E, CFU-GM and CFU-Mk.[13] These effects may be related to the activation of the mpl receptor which has not only been found on megakaryocytes and platelets, but also on a small number of CD34 positive cells.[18] In contrast to normal mice, the bone marrow cytokine environment in myelosuppressed mice may allow the differentiation of BFU-E, CFU-GM and CFU-Mk into mature cells. TPO is not the only cytokine which can affect multiple cell types. Other cytokines such as G-CSF, EPO or GM-CSF were also initially described to have lineage specific effects, but were later found to influence other cell types.[19-22]

Mouse models of thrombocytopenia have the disadvantage that platelet counts are still relatively high compared to myelosuppressed humans and recovery is fast. A prolonged, severe thrombocytopenia can be induced in dogs with a single dose of 200 cGy TBI.[23] Using this model, we evaluated the effect of 28 days of TPO treatment. Platelet counts were higher during the treatment period with TPO compared to vehicle controls (Nash, personal communication). This is similar to what occurs in mice. Vehicle treated dogs had, on average, 14 days of platelet counts below 10,000/cu mm, whereas TPO treated dogs had, on average, only 4 days of platelet counts below 10,000/cu mm. TPO treatment also enhanced neutrophil recovery in myelosuppressed dogs. The mechanism(s) for the accelerated recovery of granulocytes in either dogs or mice are not clear. TPO may act synergistically with endogenous G-CSF or other cytokines in the generation of early or late progenitor cells. Further studies are needed to elucidate the mechanisms.

These experiments indicate that in myelosuppressed animals TPO can hasten myeloid recovery and may therefore be useful in a clinical setting in myelosuppressed patients.

ACKNOWLEDGMENTS

We thank Minako Lee of the University of Washington Dept. of Biological Structure for access to the Gammacell 40 Irradiator, Leslie Bestow and the Tissue Culture group for production of rhTPO and John Forstrom and the Protein Chemistry group (especially Michele Buddle, Rachel Stevenson, and Deb Gilbertson) for preparation and characterization of the rhTPO used in these studies. We also want to thank Greg Price for his invaluable technical assistance, and Margo Rogers and Molly Bernard for their help in preparing the manuscript.

REFERENCES

1. Kelemen E, Cserhati I, Tanos B: Demonstration and some properties of human thrombopoietin in thrombocythemic sera. Acta Haematol (Basel) 20: 350, 1958
2. McDonald TP: Thrombopoietin: its biology, clinical aspects, and possiblities. Am J Pediatr Hematol Oncol 14: 8, 1992

3. Hoffman R: Regulation of megakaryocytopoiesis. Blood 74: 1196, 1989
4. de Sauvage FJ, Hass PE, Spencer SD, Malloy BE, Gurney AL, Spencer SA, Darbonne WC, Henzel WJ, Wong SC, Kuang WJ, Oles KJ, Hultgren B, Solberg LAJ, Goeddel DV, Eaton DL: Stimulation of megakaryocytopoiesis and thrombopoiesis by the c-Mpl ligand. Nature 369: 533, 1994
5. Lok S, Kaushansky K, Holly RD, Kuijper JL, Lofton-Day CE, Oort PJ, Grant FJ, Heipel MD, Burkhead SK, Kramer JM, Bell LA, Sprecher CA, Blumberg H, Johnson R, Prunkard D, Ching AFT, Mathewes SL, Bailey MC, Forstrom JW, Buddle MM, Osborn SG, Evans SJ, Sheppard PO, Presnell SR, O'Hara PJ, Hagen FS, Roth GJ, Foster DC: Cloning and expression of murine thrombopoietin cDNA and stimulation of platelet production in vivo. Nature 369: 565, 1994
6. Bartley TD, Bogenberger J, Hunt P, Li YS, Lu HS, Martin F, Chang MS, Samal B, Nichol JL, Swift S, Johnson MJ, Hsu R, Parker VP, Suggs S, Skrine JD, Merewether LA, Clogston C, Hsu E, Hokom MM, Hornkohl A, Choi E, Pangelinan M, Sun Y, Mar V, McNinch J, Simonet L, Jacobsen F, Xie C, Shutter J, Chute H, Basu R, Selander L, Trollinger D, Sieu L, Padilla D, Trail G, Elliott G, Izumi R, Covey T, Crouse J, Garcia A, Xu W, Del Castillo J, Biron J, Cole S, Hu M, Pacifici R, Ponting I, Saris C, Wen D, Yung YP, Lin H, Bosselman RA: Identification and cloning of a megakaryocyte growth and development factor that is a ligand for the cytokine receptor Mpl. Cell 77: 1117, 1994
7. Miyazaki H, Kato T, Ogami K, Iwamatsu A, Shimada Y, Souma Y, Akahori H, Horie K, Kokubo A, Kudo Y, Maeda E, Kawamura K, Sudo T: Isolation and cloning of a novel human thrombopoietic factor. Exp Hematol 22: 838, 1994
8. Kuter DJ, Beeler DL, Rosenberg RD: The purification of megapoietin: a physiological regulator of megakaryocyte growth and platelet production. Proc Natl Acad Sci U S A 91: 11104, 1994
9. Kaushansky K, Lok S, Holly RD, Broudy VC, Lin N, Bailey MC, Forstrom JW, Buddle MM, Oort PJ, Hagen FS, Roth GJ, Papayannopoulou T, Foster DC: Promotion of megakaryocyte progenitor expansion and differentiation by the c-Mpl ligand thrombopoietin. Nature 369: 568, 1994
10. Zeigler FC, deSauvage F, Widmer HR, Keller GA, Donahue C, Schreiber RD, Malloy B, Hass P, Eaton D, Matthews W: In vitro megakaryocytopoietic and thrombopoietic activity of c-mpl ligand (TPO) on purified murine hematopoietic stem cells. Blood 84: 4045, 1994
11. Lok S, Foster DC: The structure, biology and potential therapeutic applications of recombinant thrombopoietin. Stem Cells (Dayt) 12: 586, 1994
12. Kaushansky K: The mpl ligand: molecular and cellular biology of the critical regulator of megakaryocyte development. Stem Cells (Dayt) 12 Suppl 1: 91, 1994
13. Kaushansky K, Broudy VC, Grossmann A, Humes J, Lin N, Ping Ren H, Bailey MC, Papayannopoulou T, Forstrom J, Sprugel KH: Thrombopoietin expands erythroid progenitors, increases red cell production and enhances erythroid recovery following myelosuppressive therapy. J Clin Invest (in press)
14. Leonard JP, Quinto CM, Kozitza MK, Neben TY, Goldman SJ: Recombinant human interleukin-11 stimulates multilineage hematopoietic recovery in mice after a myelosuppressive regimen of sublethal irradiation and carboplatin. Blood 83: 1499, 1994
15. Foster DC, Sprecher CA, Grant FJ, Kramer JM, Kuijper JL, Holly RD, Whitmore TE, Heipel MD, Bell LA, Ching AF, McGrane V, Hart C, O'Hara PJ, Lok S: Human thrombopoietin: gene structure, cDNA sequence, expression, and chromosomal localization. Proc Natl Acad Sci U S A 91: 13023, 1994
16. Sprugel KH, Humes JM, Grossmann A, Ren HP, Kaushansky K: Recombinant thrombopoietin stimulates rapid platelet recovery in thrombocytopenic mice. Blood 84: 242A, 1994 (abstr)
17. Farese AM, Hunt P, Boone T, MacVittie TJ: Recombinant human megakaryocyte growth and development factor stimulates thrombocytopoiesis in normal nonhuman primates. Blood 86: 54, 1995
18. Debili N, Wendling F, Cosman D, Titeux M, Florindo C, Dusanter-Fourt I, Schooley K, Methia N, Charon M, Nador R, Bettaieb A, Vainchenker W: The Mpl receptor is expressed in the megakaryocytic lineage from late progenitors to platelets. Blood 85: 391, 1995
19. Erslev AJ: The effect of anemic anoxia on the cellular development of nucleated red cells. Blood 14: 386, 1959
20. Metcalf D, Nicola NA: Proliferative effects of purified granulocyte colony-stimulating factor (G-CSF) on normal mouse hemopoietic cells. J Cell Physiol 116: 198, 1983
21. Ikebuchi K, Clark SC, Ihle JN, Souza LM, Ogawa M: Granulocyte colony-stimulating factor enhances interleukin 3-dependent proliferation of multipotential hemopoietic progenitors. Proc Natl Acad Sci U S A 85: 3445, 1988
22. Geissler K, Stockenhuber F, Hinterberger W, Balcke B, Lechner K: Recombinant human erythropoietin: a multipotential hemopoietic growth factor in vivo and in vitro. Contrib Nephrol 87: 1, 1990
23. Nash RA, Seidel K, Storb R, Slichter S, Schuening FG, Appelbaum FR, Becker AB, Bolles L, Deeg HJ, Graham T, Hackman RC, Burstein SA: Effects of rhIL-11 on normal dogs and after sublethal radiation. Exp Hematol 23: 389, 1995

ANTISENSE OLIGONUCLEOTIDES TO BOTH HUMAN AND MURINE ACETYLCHOLINESTERASE PROMOTE MYELOID STEM CELL EXPANSION AND REDUCE ERYTHROID AND MEGAKARYOCYTE LINEAGES IN MURINE CFU-GEMM*

Deborah Patinkin†

Department of Bone Marrow Transplantation
Hadassah University Hospital
P.O. Box 12000
Jerusalem, Israel 91120

INTRODUCTION

Hematopoietic stem cell differentiation is caused by a complex interaction of cytokines, stromal elements and apoptotic agents and is frequently disturbed in malignant and toxic conditions. The presence of the neurotransmitter hydrolyzing enzyme acetylcholinesterase (AChE) in cells such as the erythrocyte and the megakaryocyte has raised questions as to its possible role in hematopoietic differentiation. Thought-provoking in this regard have been the frequent breakpoints in the 7q22 location of the AChE gene in leukemias (1,2) and the increased risk to leukemia induced by exposure to AChE-inhibitory insecticides (3). The gene is also subject to incomplete amplification in peripheral blood cells of patients with acute myeloblastic leukemia (4) and to frequent mutability (5,6). In an *in vitro* study, addition of AChE to resting differentiated erythroleukemia cells caused inhibition of growth as measured by cell mass (7). We have previously shown that antisense oligodeoxynu-

* Research Support: A Bundesministerium fur Forshchung und Technologie Biotechnology Program Grant, the Israel Cancer Research Society, the Israel Health Ministry, and the Israel Ministry of Science and Technology.

† Telephone: 972-2-776-561; Fax: 972-2-422-731.

cleotides (AS-oligos) targeted to the AChE-related gene, BChE (8), and a novel gene, CHED, homologous to the cdc kinases (9), reduced megakaryopoiesis in CFU-MK.

We first analyzed the link between AChE gene expression and the hematopoietic system by introducing AS-oligos targeted to ACHEmRNA into serum-free murine bone marrow (BM) cultures (10). We probed the effect of AS-ACHE interference with growth of CFU-MK and CFU-GEMM to evaluate their particular influence on megakaryocyte and erythrocyte development. Using RT-PCR amplification, we observed that AS-ACHE causes a 90% reduction of ACHEmRNA after two hours and a subsequent increase of 10-fold after 4 days of culture over its level in either S-ACHE or fresh bone marrow (BM) cultures (10). In a differential PCR display with other cellular transcripts, we found altered transcription patterns between the fresh BM and the oligo-treated cells (10).

Introduction of AS-ACHE into cultures of CFU-MK grown with WEHI-CM as a source of IL-3, caused a drop in colony number after four days from 5μM on. This was accompanied by a rise in cell numbers, indicating that the number of progenitors was reduced but their proliferation potential increased (10). Adding Epo to our cultures for the CFU-GEMM assay resulted in a five-fold expansion ($P<0.001$) rather than a reduction of colonies at a peak of 12μM and a two-fold increase in cell number, demonstrating a dose-dependent stem cell expansion. Measurements of DNA yields in AS-, S-ACHE-treated CFU-GEMM and fresh BM revealed five-fold higher DNA yields in AS-ACHE-treated cells when compared to those of S-ACHE-treated cells. After electrophoresis, blotting and hybridization with a ^{32}P-labeled mouse genomic DNA probe, DNA from S-ACHE and untreated cultures exhibited the extensive DNA fragmentation patterns typical of apoptosis whereas AS-ACHE-treated and fresh BM DNA lanes were almost devoid of these fragments. We therefore concluded that AS-ACHE not only induces expansion of multipotential precursors but also suppresses apoptosis (10).

A differential analysis of cell compositions of the CFU-GEMM cultures revealed a striking dose-dependent increase in macrophages and neutrophils and corresponding decreases in erythrocytes and megakaryocytes in AS-ACHE cultures (10). In CFU-MK cultures, on the other hand, 2.5 μM of AS-ACHE induced the formation of large numbers of primitive blasts that labeled positively with either anti-glycophorin or anti-GpIIb/IIIa antibodies, suggesting they represented bipotential precursors to both erythroid and megakaryocyte lineages. In line with this idea was the 10-fold rise in mRNA of the transcription factor GATA-1 measured by RT-PCR in liquid CFU-MK cultures treated with AS-ACHE but not with S-ACHE(10).

In an extension of the previous study, we attempted to evaluate further whether the AS-ACHE-treated CFU-MK blasts were truly bipotential in nature by replating them after 24 hours in methyl cellulose with erythropoietin (Epo) and fetal calf serum (FCS). To ensure that the effects obtained with the human AS-ACHE on murine BM cells were not limited to this gene, we employed an AS-oligo targeted to Exon 2 of the previously cloned murine AChE (11) gene. Finally, we delved into the mechanism of AS-ACHE action on hematopoietic stem cells by investigating the effect of cell concentration on CFU-GEMM.

MATERIALS AND METHODS

Cell Culture. Bone marrow (BM) cells from 8-12 week-old C3H/HeJ mice (Bar Harbor, ME, The Jackson Laboratories) were grown 4 days at a concentration of 1×10^5 cells/ml in serum-free methyl cellulose-IMDM medium, containing 1% BSA and 10% WEHI-CM as a source of IL-3. For CFU-GEMM, 2.8×10^{-4}M transferrin (Boehringer-Mannheim) and 2u /ml Epo (1,000u/mg protein, Stem Cell Technologies, Vancouver) were added to the above medium and incubated for 8-9 days at 37 C. and 5% CO_2. Colonies were scored with a Zeiss dark field optic fibre microscope and cytospin preparations were stained with May-Grunwald-Giemsa.

Replating Experiments. 2 X 10^5/ml BM cells were grown with 2.5 μM AS-ACHE in serum-free liquid culture with WEHI-CM as inducer. After 24 hrs., the cells were scraped loose and transferred to semi-solid methyl cellulose-IMDM medium with the addition of 2u/ml Epo and 20% FCS. There was no further addition of AS-ACHE after the first day. Incubation was then resumed for another seven days before scoring.

AS- and S-oligos were diluted to concentrations of 1-30 μM in PBS, added directly to the culture medium and retained there throughout the culture period.

M-CSF antibody was a polyclonal antibody, a kind gift of E.R. Stanley.

RT-PCR and Oligos. AS-ACHE (5'-CTGCGGGGCCTCAT-3')(12,13) and the complementary S-ACHE oligo were synthesized either in the fully phosphorothioated form (14, 15) or with three 3'-terminal phosphorothioates to reduce non-specific toxicity. Oligos were heated 15' at 65 C. prior to their addition to culture.

PCR conditions were denaturation at 94 C for 1 min (first step, 3 min), annealing at 65 C for 1 min and elongation at 72 C for 1 min. For semiquantitative comparison of mRNA levels, samples were removed every two cycles.

RESULTS

The replating of CFU-MK in semi-solid medium with Epo and FCS, resulted in control cultures devoid of intact colonies but containing platelets and disintegrating megakaryocytes. AS-ACHE-treated cultures, on the other hand, exhibited a very large increase in macrophages, up to 70% of the total cell number, as well as erythroblasts and young megakaryocytes (Figs 1 and 2). As can be seen from Fig 1, treatment first with 2.5μM AS-ACHE and subsequent addition of Epo and FCS resulted in colonies with differential

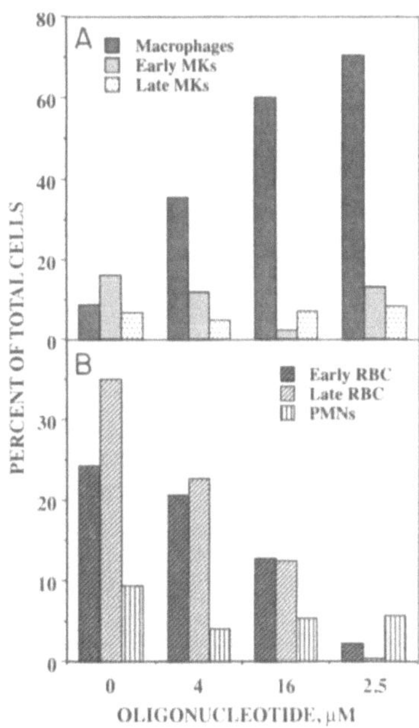

Figure 1. Effect of replating of AS-ACHE-treated CFU-MK on differential cell composition. Cells in columns O, 4 and 16 μM AS-ACHE were grown 8 days with 2u/ml Epo and 10% WEHI-CM in serum-free medium with simultaneous addition of AS-ACHE. Cells in column 2.5 μM AS-ACHE were first grown with AS-ACHE and 10% WEHI-CM in liquid culture to form blasts. After 24 hours, the cells were transferred to methylcellulose-IMDM with the addition of 2u/ml Epo and 20% FCS and incubated for another 7 days. Colonies were cytospinned and stained with May-Grunwald-Giemsa. Abbreviations: E. MK - early megakaryocyte; L. MK = late megakaryocyte: E.RBC = early erythroid cell; L. RBC = late erythroid cell; PMN = neutrophil.

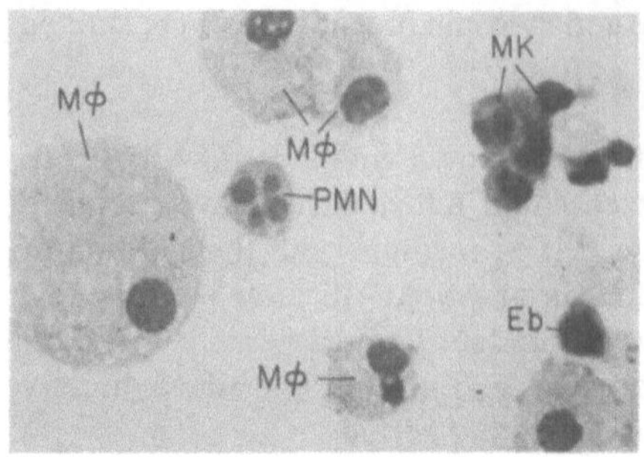

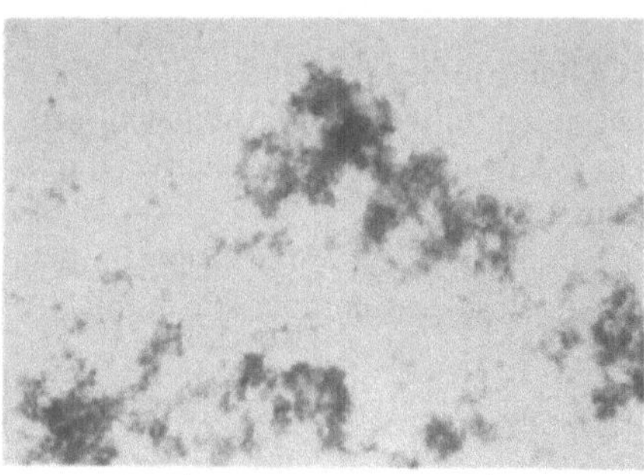

Figure 2. Typical fields of untreated and AS-ACHE-treated CFU-MK replated with Epo and FCS. (A) AS-ACHE-treated cultures with many macrophages(Mϕ), young megakaryocytes (MKs), neutrophil (PMN) and erythroblast(Eb) (B). Untreated cultures. Clumps of platelets. Colonies were cytospinned and stained with May-Grunwald-Giemsa. Original magnification X 1,000.

cell composition very similar to that obtained previously(10) with the much higher concentration of 16μM AS-ACHE in CFU-GEMM cultures (Fig 1). Typical fields of control and AS-ACHE-treated CFU-MK are displayed in Fig 2.

The cloning and sequencing of the murine AChE gene by another group (11) enabled us to examine whether murine CFU-GEMM progenitors behaved in a similar fashion to both human and murine genes. We prepared a 20-mer AS-oligo to the mRNA of Exon 2 (AS-mE2) of the murine AChE gene that was completely different in sequence from the original AS-ACHE oligo and, of particular importance, was lacking in its tetra G component(13). An inverted form of murine Exon 5 was used as the sense-oligo. From a dose-response analysis of the effect of AS-mE2 on CFU-GEMM (Fig 3), a peak of colonies was observed at 5μM

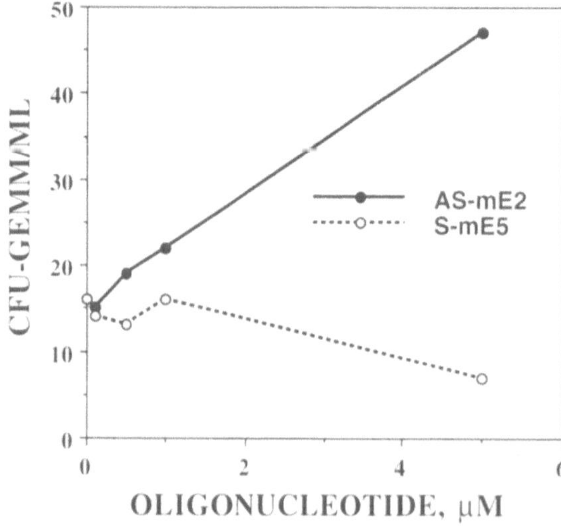

Figure 3. Effect of increasing AS-mE2 concentration of CFU-GEMM colonies. Murine BM cells, 1 X 10^5/ml, were grown in serum-free methyl cellulose-IMDM with 2 u/ml Epo and 10% WEHI-CM with the simultaneous addition of different concentrations of AS-mE2. Colonies were scored after 8 and 9 days. Peak effects are seen at 5 μM while colony number (not shown) declined at 10 μM AS-oligo.

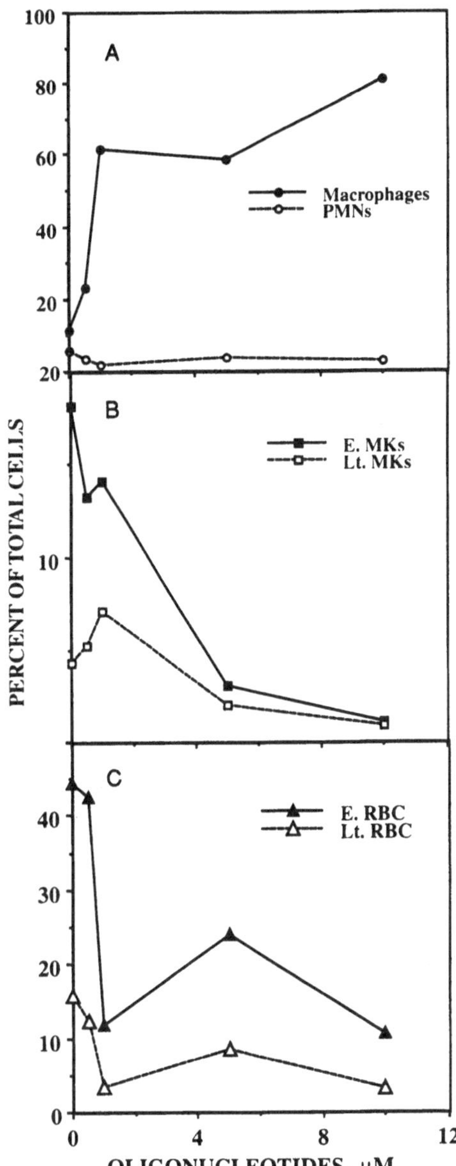

Figure 4. Effect of increasing AS-mE2 concentration on differential cell composition of murine CFU-GEMM. (A) Macrophages show a pronounced increase at low concentrations while neutrophils remain low in number. (B) Both early and late MKs decline sharply with rising concentration. (C) Early and late erythroid cells decline with rising concentration.

of AS-mE2, as opposed to a peak of 12-16 μM for the human AS-oligo. This represents a 2.5 to 3-fold decline in the requirement of the AS-oligo for optimal function as well as a 3-fold expansion of stem cells over control values. Differential cell analysis revealed similar diversion of hematopoiesis to that noted with the human gene (Fig 4) with a sharp rise in myeloid cells and a fall in erythroid and megakaryocyte cell numbers. From 5μM on, large increases in macrospocpic macrophage colonies (possible HPP-CFC) were also observed, above and beyond the striking increase in macrophage content of the CFU-GEMM themselves. These colonies exhibited only an occasional erythroblast among the tremendous clusters of macrophages and were white and club-shaped as opposed to the pink or red color of the true CFU-GEMM.

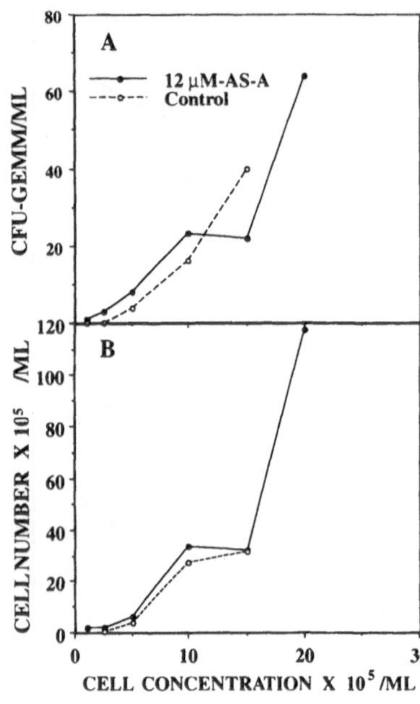

Figure 5. Effect of increasing cell concentration of human AS-ACHE-treated CFU-GEMM on colony and final cell number. Murine BM cells were grown 8 days in serum-free methyl cellulose-IMDM with 2u/ml Epo, 10% WEHI-CM and peak AS-ACHE concentrations of 12 µM. (A) CFU-GEMM colonies. (B) Final cell numbers of all CFU-GEMM/ml after 8 days. Differential cell counts were virtually impossible to determine at higher cell concentrations because of the enormous clumping of macrophages.

In recent cell concentration-dose-response experiments, a sharp increase in CFU-GEMM number was observed at the highest cell concentration of 2 X 10^5 cells, suggesting that the effect of AS-ACHE is an indirect one (Fig 5). The addition of M-CSF antibody, at 1:10,000 dilution, to AS-ACHE and untreated cultures resulted in a drop in CFU-GEMM of about 2.5-fold in both types of cultures. Thus, AS-ACHE does not exhibit any specific effect on M-CSF and the mechanism of induction of myeloidopoiesis noted in its presence requires further analysis.

DISCUSSION

AS-ACHE has been shown to cause a specific short-term destruction of ACHEmRNA in CFU-MK cultures followed by a subsequent rise of 10-fold in mRNA over control levels (10). Its role in hematopoietic cell culture is complex, exhibiting inhibition of lineage-specific CFU-MK yet increased proliferation of primitive blasts that label with both erythroid and megakaryocyte markers. Although it is generally accepted that both erythrocytes and megakaryocytes stem from a common precursor (16), further confirmation of this thesis was provided by the considerable increase in these cultures of GATA-1mRNA, a transcriptional factor important in both early erythroid and megakaryocyte development(17,18). Also in keeping with the idea of a common precursor expressing high levels of both ACHE and GATA-1mRNA is the ten-fold rise in ACHEmRNA noted previously at four days following treatment of CFU-MK with AS-ACHE(10). Final evidence of the multipotential nature of these blasts was afforded by the extension of their growth in culture to eight days with the addition of Epo and FCS resulting in mixed colony formation. These results are in line with our previous findings regarding AS-ACHE-treated CFU-GEMM, confirming the diversion of hematopoiesis to myelopoiesis and the decline in erythroid and megakaryocyte lineages. They also provide further evidence for both the multipotential stem cell expansion and increased cell survival properties of AS-ACHE when contrasted to the poor survival and lineage-specific results observed in untreated cultures.

In the presence of Epo and IL-3 during CFU-GEMM formation, AS-ACHE induces opposite results to those described for CFU-MK. Rather than an inhibition of colony formation, there is an increase in colony numbers, i.e. stem cell expansion, characterized by a pronounced surge of myeloid cell production. The depression of erythroid and megakaryocyte lineages is indicative of a negative influence of AS-ACHE on multipotential precursors of these cells and thus conversely, their positive regulation by AChE. In both lineage-specific and multipotential colonies we perceive a dominant role of the action of AS-ACHE over the growth-inductive properties of the cytokines, although it should be remembered that the concentration of the AS-oligos, 1-15 μM, is far greater than that of the cytokines, that usually range between 10^{-9} to 10^{-11}M. The indirect effect of AS-ACHE may be due to its interaction with some unknown cell-surface protein(s) such as the beta-neurexins(19), some known to have domains homologous to ACHE. Another possibility is that it may be caused by a secondary induction of cytokines such as GM-CSF and IL-6. Interesting in this regard is the recent finding that functional G-CSF receptors are produced by megakaryocytes(20) and are present on the platelet membrane(21) while G-CSF is responsible for the secondary aggregation of platelets. Whether some of the effects mentioned above can be attributed to the absence of the hydrolyzing action of ACHE on acetylcholine, as has been suggested(22), remains to be clarified.

Examination of the effects of antisense to Exon 2 of the murine AChE on murine BM has confirmed in large degree our findings with the human AS-oligos. We have observed diversion of hematopoiesis to the myeloid lineages, their stem cell expansion with concomitant depression of erythroid and megakaryocyte functions. We have then definitive biological evidence for a commonality of behavior of two oligos from diverse mammalian sources and diverse sequences; the human exhibiting a G-C-rich sequence(13) and the mouse devoid of this potential complication.

Although the exact molecular mechanism of AS-ACHE action in hematopoiesis remains to be elucidated, it should be emphasized that the very pronounced increase in myeloidogenesis and blast formation noted is a leukemia-like phenomenon. It may offer an explanation for the increased incidence of myeloid leukemia in farmers exposed to excessive amounts of ACHE-inhibitory organophosphorous insecticides(3). In addition, where defects in AChE gene expression have been observed (1), a situation may arise that is analogous to AS-ACHE inhibition, permitting reduced apoptosis of hematopoietic cells and thus terminating in leukemia.

REFERENCES

1. Ehrlich G, Viegas-Pequignot E, Ginzberg D, Sindel L, Soreq H, Zakut H: Mapping of the acetylcholinesterase gene to chromosome 7q22 by fluorescent *in situ* hybridization coupled with selective PCR amplification from a somatic hybrid cell panel and chromosome-sorted DNA libraries. Genomics 13:1192, 1992
2. Soreq H & Zakut H: Human Cholinesterases and Anticholinesterases, Academic Press, San Diego, 1993
3. Brown LM, Blair A, Gibson R, Everett GD, Kantor KP, Schuman LM, Burmeister LF, Van Lier SF, Dick F: Pesticide exposures and other agricultural risk factors for leukemia among men in Iowa and Minnesota. Cancer Res 50:6585, 1990
4. Zakut H, Lapidot-Lifson Y, Beeri R, Ballin A, Soreq H: *In vivo* gene amplification in non-cancerous cells: cholinesterase genes and oncogenes amplify in thrombocytopenia associated with lupus erythematosis. Mutat Res 276:275, 1992
5. Koury MJ: Mini Review. Programmed cell death (apoptosis) in hematopoiesis. Exp Hematol 20:391, 1992
6. Lapidot-Lifson Y, Prody CA, Ginzberg D, Meytes D, Zakut H, Soreq H: Co-amplification of human acetylcholinesterase and butyrylcholinesterase genes in blood cells: correlation with various leukemias and abnormal megakaryocytopoiesis. Proc Natl Acad Sci USA 86:4715, 1989

7. Paoletti F, Mocali A, Vannucci AM: Acetylcholinesterase in murine erythroleukemia (Friend) cells: evidence for megakaryocyte-like expression and potential growth-regulatory role of enzyme activity. Blood 79:2873, 1992
8. Patinkin D, Seidman S, Eckstein F, Benseler F, Zakut H, Soreq H: Manipulations of cholinesterase gene expression modulate murine megakaryocytopoiesis *in vitro*. Mol Cell Biol 10:6046 1990
9. Lapidot-Lifson Y, Patinkin D, Prody CA, Ehrlich G, Seidman S, Ben-Aziz R, Benseler F, Eckstein F, Zakut H, Soreq H: Cloning and antisense oligonucleotide inhibition of a human homolog of cdc2 required in hematopoiesis. Proc Natl Acad Sci, USA 89:579, 1992
10. Soreq H, Patinkin D, Lev-Lehman E, Grifman M, Ginzberg D, Eckstein F, Zakut H: Antisense oligonucleotide inhibition of acetylcholinesterase gene expression induces progenitor cell expansion and suppresses apoptosis *ex vivo*. Proc Natl Acad Sci 91:7907, 1994
11. Rachinsky TL, Camp S, Li Y, Ekstrom TJ, Newton M, Taylor P: Molecular cloning of mouse acetylcholinesterase: tissue distribution of alternatively spliced mRNA species. Neuron 5:317, 1990
12. Rakonczay Z, Brimijoin S: Biochemistry and pathophysiology of the molecular forms of cholinesterases. Subcell Biochem 12:335-378, 1988
13. Soreq H, Ben-Aziz R, Prody CA, Seidman S, Gnatt A, Neville L, Lieman-Hurwitz J, Lev-Lehman E, Ginzberg D, Lapidot-Lifson Y, Zakut H: Molecular cloning and construction of the coding region for human acetylcholinesterase reveals a G-C-rich alternating structure. Proc Natl Acad Sci, USA 87:9688, 1990
14. Eckstein F. Nucleoside phosphorothioates. Annu Rev Biochem 54:367, 1985
15. Stein CA, Cohen JS: Oligonucleotides as inhibitors of gene expression: a review. Cancer Res 48:2659, 1988
16. McLeod DL, Shreeve MM, Axelrad AA: Chromosome marker evidence for the bipotentiality of BFU-E. Blood 56:318, 1980
17. Romeo PH, Prandini MH, Joulin V, Mignotte V, Preant M, Vainchenker W, Marguerie G, Uzan G: Megakaryocytic and erythrocytic lineages share specific transcription factors, Nature 344:447, 1990
18. Orkin SH: Perspective GATA-binding transcription factors in hematopoietic cells. Blood 80:575, 1992
19. Ichtchenko K, Hata Y, Nguyen T, Ullrich B, Missler M, Moomaw C, Sudhof TC: Neuroligin 1: A splice site-specific ligand for beta-neurexins. Cell 81:435, 1995
20. Shimoda K, Okamura S, Harada N, Okamura T, Niho Y: Identification of a functional receptor for granulocyte colony stimulating factor on platelets. J Clin Invest 91:1310, 1993
21. Kanji T, Okamura T, Nagafuji K, Iwasaki H, Shimoda K, Okamura S, Niho Y: Megakaryocytes produce the receptor for granulocyte colony-stimulating factor. Blood 85:3359, 1995
22. Burstein SA, Adamson JW, Harker LA: Megakaryocytopoiesis in culture: modulation by cholinergic mechanisms. J Cell Physiol 103:201, 1980

61

GATA TRANSCRIPTION FACTORS NEGATIVELY REGULATE ERYTHROPOIETIN GENE EXPRESSION*

Shigehiko Imagawa,[1†] Masayuki Yamamoto,[2] and Yasusada Miura[1]

[1] Division of Hematology
Department of Medicine
Jichi Medical School Minamikawachi-machi
Tochigi-ken, 329-04, Japan
[2] Department of Biochemistry
Tohoku University School of Medicine
Aoba-ku, Sendai-shi, Miyagi-ken, 980, Japan

ABSTRACT

We examined regulation of the human erythropoietin (Epo) gene through the GATA sequence in the Epo promoter, and demonstrated that Hep3B and $HepG_2$ cells express human GATA-2 (hGATA-2) mRNA and protein. Nuclear extracts of QT6 cells transfected with hGATA-1, 2 or 3 transcription factors revealed specific binding to the GATA element in the human Epo gene promoter by gel mobility shift assay. Transient transfection of Hep3B cells with hGATA-1, 2 or 3 demonstrated that each of these transcription factors significantly decreased the level of expression of Epo mRNA as assessed by a competitive polymerase chain reaction (PCR). Furthermore, transient transfection of Hep3B cells with hGATA-1, 2 and 3 and an Epo-reporter gene construct showed significant inhibition of the Epo promoter. We conclude that the hGATA-1, 2 and 3 transcription factors specifically bind to the GATA element in the human Epo gene promoter and negatively regulate Epo gene expression.

---

* Supported by grants-in-aid for scientific research from the Ministry of Education, Science and Culture of Japan, the Uehara Memorial Foundation, the Yamanouchi Foundation, the Ichiro Kanehara Foundation and the Chugai Foundation.

† To whom all correspondence should be addressed: Tel: 81-285-44-2111 ext. 3455; Fax: 81-285-44-5258.

Molecular Biology of Hematopoiesis 5, edited by Abraham et al.
Plenum Press, New York, 1996

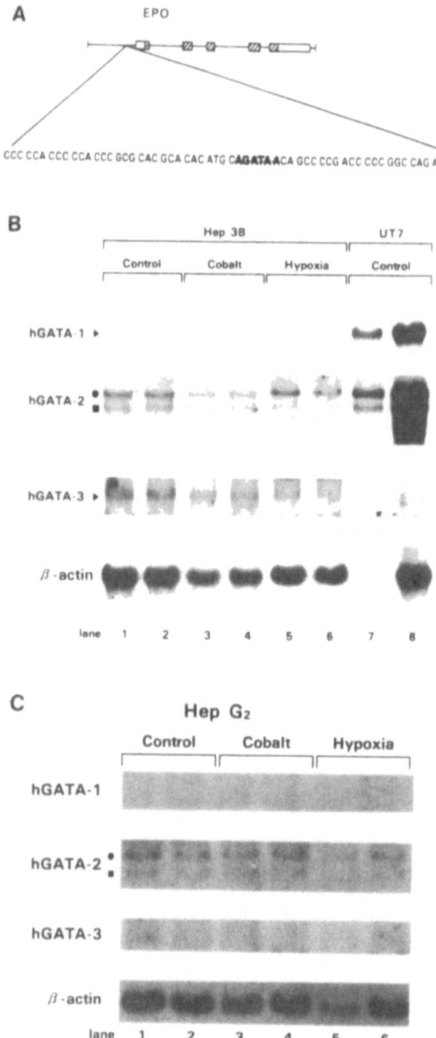

Figure 1. hGATA specific mRNA expression. A: GATA binding site in the Epo 5' promoter region. (shaded region). B: Northern blot analysis of Hep3B cells. Northern blot analysis was performed using 20 μg total RNA from control Hep3B cells (lanes 1, 2), Hep3B cells treated with 50 μM $CoCl_2$ for 24 hrs (lanes 3, 4), and Hep3B cells treated with 1% O_2 for 24 hrs (lanes 5, 6). UT7 cells (lanes 7, 8) were used as a positive control for hGATA-1 and 2. The filter was hybridized to a probe of hGATA-1, 2 and 3 cDNA and then stripped and rehybridized to a β-actin cDNA probe. C: Northern blot analysis of $HepG_2$ cells. Northern blot analysis was performed using 20 μg total RNA from control $HepG_2$ cells (lanes 1, 2), $HepG_2$ cells treated with 50 μM $CoCl_2$ for 24 hrs (lanes 3, 4), and $HepG_2$ cells treated with 1% O_2 for 24 hrs (lanes 5, 6). The filter was hybridized to a probe of hGATA-1, 2 and 3 cDNA, respectively and then stripped and rehybridized to a β-actin cDNA probe.

INTRODUCTION

Erythropoietin (Epo) is produced in the kidney and fetal liver in response to hypoxia[1] as well as $CoCl_2$.[2] However, little is understood about the intracellular pathway by which hypoxia leads to an increase in Epo expression. Goldberg *et al*[3] have demonstrated that, in the human hepatoma cell lines Hep3B and $HepG_2$, Epo protein and mRNA can be induced in response to hypoxia or $CoCl_2$. They have proposed that the cells have an oxygen-sensing mechanism in which a ligand-dependent conformational change in a heme protein in response to either hypoxia or cobalt results in an induction of Epo expression.[4]

Positive and negative regulatory elements in the Epo gene were studied using synthetic oligonucleotides. These oligonucleotides were designed to control Epo transcription by means of an antigene strategy.[5] We recently used this method to demonstrate that CACCC elements at -60 bp from the CAP site are positive regulatory sites of the Epo gene, whereas the GATA element at -30 bp is a negative regulatory element for transcription of the

Epo gene[5] (Fig. 1A). Furthermore, we demonstrated that transcription factors in nuclear extracts from Hep3B cells specifically bind to CACCC elements or the GATA element.[5]

Recently, it has been hypothesized that the TATA-binding protein of TFIID interacts with the core promoter of the GATA motif in the Epo gene, thereby competing with the GATA protein.[6] However, the significance of the GATA protein in the regulation of Epo transcription is not understood.

In the present study, we examined transcription factors which regulate the human Epo gene, especially through the GATA specific sequence. We determined that Hep3B and $HepG_2$ cells express human GATA-2 (hGATA-2) mRNA and protein by Northern blotting analysis and immunohistochemical staining, respectively. QT6 cells were transfected with hGATA-1, 2 and 3 transcription factor constructs and binding of these hGATA transcription factors to the GATA element in the human Epo gene promoter was assessed by gel mobility shift assay. The effects of hGATA transcription factors on the level of Epo mRNA in Hep3B cells was assessed by measuring Epo mRNA by competitive PCR following transient transfection of hGATA-1, 2 and 3 expression plasmids into Hep3B cells. Furthermore, the effects of hGATA transcription factors on the activity of Epo promoter were determined on an Epo-reporter gene following transient transfection of hGATA-1, 2 and 3 expression plasmids into Hep3B cells. We conclude that the hGATA-1, 2 and 3 transcription factors specifically bind to the GATA element of the human Epo gene promoter and negatively regulate Epo gene expression. Moreover, in Hep3B and $HepG_2$ cells, the hGATA-2 transcription factor is a strong negative regulator of the Epo gene.

MATERIALS AND METHODS

Cell Culture and RNA Preparation

The Hep3B, $HepG_2$ and QT6 cell lines were obtained from the American Type Culture Collection. These cells were cultured in Dulbecco's modified Eagle's medium (Life Technologies, Inc.), supplemented with penicillin (100 units/ml), streptomycin (100 μg/ml), and 10% heat-inactivated fetal bovine serum (Hyclone) in 25-cm^2 tissue culture flasks. Cells were maintained in a humidified 5% CO_2, 95% air incubator at 37 °C. One day prior to experimentation, cells were harvested and plated at a density of 3×10^6 cells/25-cm^2 flask. Cells were grown in the presence or absence of 50 μM $CoCl_2$ or hypoxia (1% oxygen), as previously described.[5] After 24 h of incubation with 50 μM $CoCl_2$ or 1% oxygen, extracts from stimulated or unstimulated cells were prepared. Total cellular RNA was harvested by conventional methods.[7]

RNA Blot Analysis

Probes were labelled with [α-^{32}P]-dCTP by random priming[8] and used in RNA blot hybridization.[8] Formaldehyde gels for RNA electrophoresis were prepared as described.[8] RNA blot hybridization was performed using 20 μg total RNA from Hep3B or $HepG_2$ cells. Filters were hybridized to probes of hGATA-1, 2 or 3 cDNA. The same filter was stripped and rehybridized to a β-actin cDNA probe to determine the level of RNA in each lane.

Immunohistochemical Analysis

Immunohistochemical staining was performed as previously described.[9] UT7/Epo, Jurkat, HD3, MSB1, Hep3B and $HepG_2$ cells were loaded on glass slides using a Cytospin (Shandon), fixed in 1% paraformaldehyde for 5 min, and then fixed in cold acetone for 5 min.

After incubation in 2% sheep serum, specimens were reacted with specific anti-hGATA-1, 2, 3 monoclonal antibodies overnight at 4°C as previously described.[9] After washing in PBS, samples were incubated with a horseradish peroxidase-conjugated F(ab')$_2$ fragment of antimouse immunoglobulin G (1:100 dilution; Amersham) overnight at 4°C. Diaminobenzidine was used as the chromogen. Nuclei were counter-stained with methylgreen for 1 hr. The chromogen diaminobenzidine stains antibody-reactive structures brown, while the counter-stain details the position of the nuclei in all cells. The fixation and photographic procedure used shows only the nuclei in these stained cells.

Transfection

Electroporation of hGATA transcription factors into Hep3B cells were performed as previously described.[10] A total of 3-10x10^6 cells were electroporated in 1 mL of 20 mmol/L HEPES buffer, pH 7.05, with 137 mmol/L NaCl, 5 mmol/L KCL, 0.7 mmol/L Na_2HPO_4, 6 mmol/L dextrose containing 20 µg of vector DNA (closed circular rather than linearized) and 500 µg carrier salmon sperm DNA at a voltage of 250V and a capacitance 960 µF (Bio-Rad). The time constant of the shock was 12~14 ms. The competitive polymerase chain reaction (PCR) for Epo mRNA was performed 24 hrs after transfection. Growth hormone (GH) was assayed in the media after a 4-day incubation, because production of GH was linear during a 4-day period, as previously described and the cell pellet was assayed for CAT.[10]

DNA Binding Assays

Nuclear extracts were prepared by published methods.[11] Protein concentrations were determined by a Bio-Rad assay with bovine serum albumin standards. Sense strand oligonucleotides were CATGCAGATAACAGCCCCGACCCCCGGCCA end-labeled with T_4 polynucleotide kinase (Toyobo) and annealed to a 4-fold excess of the corresponding unlabeled antisense oligonucleotide. A probe (2.0 ng) was used in each binding reaction. The binding buffer consisted of 10 mM Tris-HCl (pH 7.5), 1mM EDTA, 4% Ficoll, 1 mM dithiothreitol, and 75 mM KCl. An equimolar mixture of 1.5 µg poly[d(I-C)] and poly[d(A-T)] (Sigma) was used as nonspecific competitor. Binding reactions (25 µl) were incubated for 15 min at 4 °C and electrophoresed on 5% nondenaturing polyacrylamide gels in 0.25 x TBE buffer (22 mM Tris borate, 22 mM boric acid, 0.5 mM EDTA) at room temperature at 150 V for 1.5 hrs as previously described.[12] Gels were vacuum-dried and autoradiographed with intensifying screens at -80°C for 2~24 hrs.

Competitive PCR

Epo Map with Position of Oligonucleotide Primers. Epo-A: 5'-GTCGGGCA GCAGGCCGTAGAAGTCTGGCAG-3' and Epo-B: 5'-AGATGTCATTGCTGGCACTGGAGT GTCCAT-3' flanking primers were each 30 bp, had 67 and 50% G + C contents, respectively, and lacked 3' complementarity between primer pairs as previously described.[5]

Reverse Transcription of RNA into cDNA. Epo mRNA was reverse transcribed from total RNA and was co-amplified with a competitive template by competitive PCR.[5] After initial denaturation of the RNA at 65 °C to eliminate possible secondary structure, 5 µg of RNA from Hep3B cells was reverse transcribed in a reaction mixture containing 20 nmol of each dNTP (U. S. Biochemical Corp.), antisense (Epo-B) primer (40 pmol), 1 unit of RNasin (Boehringer Mannheim), and 16 units of avian myeloblastosis virus reverse transcriptase

(Life Sciences, Inc.) in a 40-μl total volume of (1x) PCR buffer for 1h at 37 °C. The transcription reaction was terminated by incubating the samples at 65 °C for 10 min. The cDNAs were immediately applied for competitive PCR. The 10-fold PCR buffer contained 500 mM Tris-HCl (pH 8.2), 15 mM $MgCl_2$, 500 mM KCl, and 0.01% (w/v) gelatin. Each reaction mixture contained dNTPs (200 μM final concentration in each), 0.2 μM each primer, and 5 units/ml of *Taq* polymerase (Cetus) in 1x PCR buffer in a final volume of 100 μl. In all experiments, the presence of possible contaminants was checked by control reactions in which amplification was carried out on samples in 1) the absence of reverse transcriptase and 2) lysis buffer alone. Samples were amplified by 60 cycles at 94 °C for 1 min, 62 °C for 2 min, and 72 °C for 3 min, containing various amounts of Epo genomic DNA competed against a fixed volume (10 μl) of Epo cDNA. An aliquot of each reaction mixture was subjected to electrophoresis on 1% agarose, 2% NuSieve gels. Gels were stained with ethidium bromide, photographed, and analyzed by densitometry (Immunomedica Co. Ltd.).

Plasmid Vectors

The GH gene was used as a reporter gene in a construct where expression of GH is driven by the mouse metallothionein-I promoter (XGH). We constructed a vector, 5AXGH, which harbors a 1192-bp HindIII-XbaI fragment, extending from 378 bp upstream of the cap site through the first exon of Epo (contains only 13 bp of coding sequence) and entire first intron, that is inserted into the polylinker upstream of XGH. To obviate the problem of false initiation of translation from the Epo ATG start codon, this site was mutated to TAG by preparation of a gapped heteroduplex and a 15 bp mutant primer. Mutant colonies were selected by Grunstein hybridization and verified by dideoxy sequencing. A second vector, 5AXGH3Ac, contains an additional a 255 bp AccI-BglII fragment that extends 67 bp upstream from the Epo termination codon and covers much of the 3' noncoding region that is homologous with the mouse Epo gene.[10] hGATA-1, 2 and 3 transcription factor expression plasmids have been previously described.[12]

Assays

The RIA for hGH was performed with a kit produced by Nichols Institute Diagnostics (San Juan Capistrano, CA). Chlorampehnicol acetyltransferase (CAT) was measured as described by Neumann *et al.*[13]

RESULTS

Hep3B and $HepG_2$ Cells Express hGATA-2 mRNA and Protein

Human hepatoma cell lines Hep3B and $HepG_2$ can be induced to produce large amounts of biologically active and immunological identifiable Epo in response to hypoxia as well as cobalt.[1,2] Furthermore, upon such stimulation, markedly increased levels of Epo mRNA have been observed. Therefore, the expression of hGATA transcription factors was examined in Hep3B and $HepG_2$ cells by Northern blotting analysis. Hep3B cells express hGATA-2 and 3 transcription factors (Fig. 1B) while $HepG_2$ cell expresses hGATA-2 (Fig. 1C). To determine whether Hep3B and $HepG_2$ cells express hGATA factor proteins, immunohistochemical staining was performed. UT7/Epo, HD3 and MSB1 cells were used as positive controls for hGATA-1, 2 and 3 transcription factors, respectively (Fig. 2 a, e, i), and Jurkat cells were used as a negative control for hGATA-1 and 2 (Fig. 2 b, f). hGATA-2 was abundantly expressed within the nucleus of Hep3B and $HepG_2$ cells (Fig. 2g, h). However,

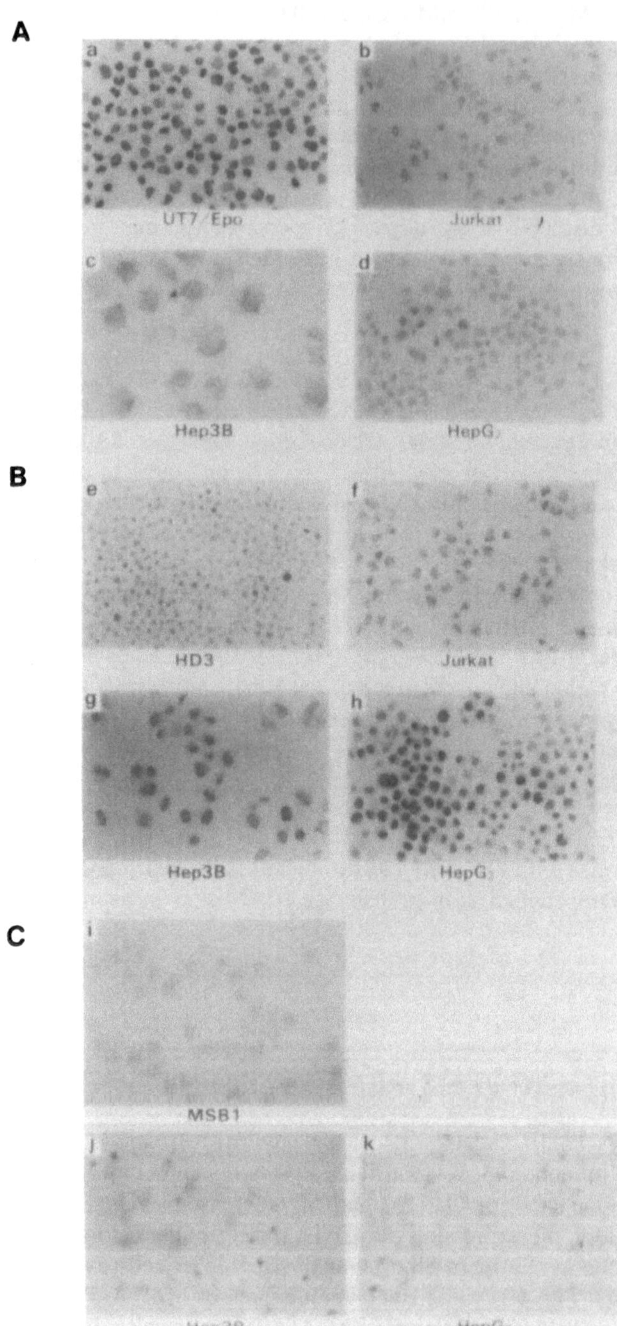

Figure 2. Expression of the hGATA proteins. A: Immunohistochemical analysis of hGATA-1 expression in four cell lines (UT7/Epo [a], Jurkat [b], Hep3B [c] and $HepG_2$ [d]). hGATA-1 was abundantly expressed within the nucleus of UT7/Epo [a], whereas expression was not detected in control cells (Jurkat [b]). Hep3B [c] and $HepG_2$ [d] showed undetectable staining. B: Immunohistochemical analysis for hGATA-2 in four cell lines (HD3 [e], Jurkat [f], Hep3B [g] and $HepG_2$ [h]). hGATA-2 was abundantly expressed within the nucleus of HD3 [e], whereas expression was not detected in control cells (Jurkat [f]). Hep3B [g] and $HepG_2$ [h] showed positive staining. C: Immunohistochemical analysis for hGATA-3 in three cell lines (MSB1 [i], Hep3B [j] and $HepG_2$ [k]). hGATA-3 was expressed within the nucleus of MSB1 [i], whereas expression was not detected in (Hep3B [j]) and $HepG_2$ [k].

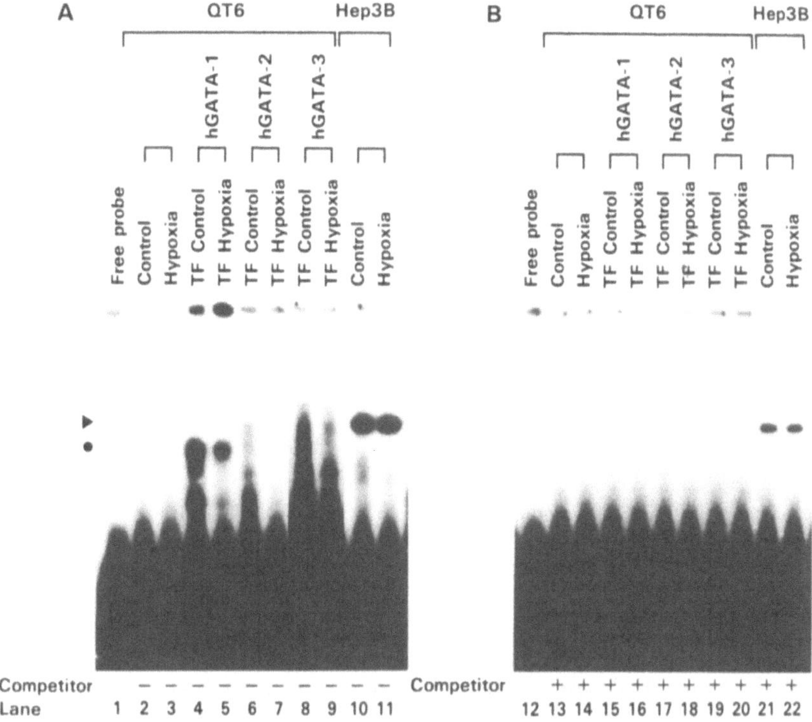

Figure 3. Gel mobility shift assay in transfected QT6 cells with GATA transcription factors. A: Gel mobility shift assays were performed using 2.5 μg of total protein from QT6 cells (lanes 2, 3), hGATA-1 transfected QT6 cells (lane 4), hGATA-1 transfected QT6 cells under 1% O_2 for 24 hrs (lane 5), hGATA-2 transfected QT6 cells (lane 6), hGATA-2 transfected QT6 cells under 1% O_2 for 24 hrs (lane 7), hGATA-3 transfected QT6 cells (lane 8), hGATA-3 transfected QT6 cells under 1% O_2 for 24 hrs (lane 9), Hep3B cells (lane 10) and Hep3B cells under 1% O_2 for 24 hrs (lane 11). The position of the complex is noted with a spot and a triangle. B: Gel mobility shift assay was repeated using a competitor for the GATA elements.

hGATA-1 and 3 were not expressed (Fig. 2 c, d, j, k), suggesting that hGATA-2 negatively regulates the Epo gene in Hep3B and $HepG_2$ cells.

hGATA-1, 2 and 3 Transcription Factors Specifically Bind to the GATA Element in the Human Epo Gene Promoter

Constructs of hGATA-1, 2 and 3 transcription factors were transfected into QT6 cells by $CaPO_4$ precipitation. Binding of proteins from nuclear extracts of QT6 cells transfected with hGATA transcription factors was assessed by a gel mobility shift assay (Fig. 3). DNA-protein interactions in nuclear extracts of cells transfected with hGATA-1, 2 and 3 and in Hep3B cells (Fig. 3A spot, triangle) were identified as retarded complexes using the GATA element as a probe. The addition of non-radiolabeled GATA element oligonucleotide demonstrated that these DNA-protein interactions were specific (spot, triangle in Fig. 3A, B). These experiments demonstrated that hGATA-1, 2 and 3 transcription factors specifically bind to the GATA element of the human Epo gene promoter.

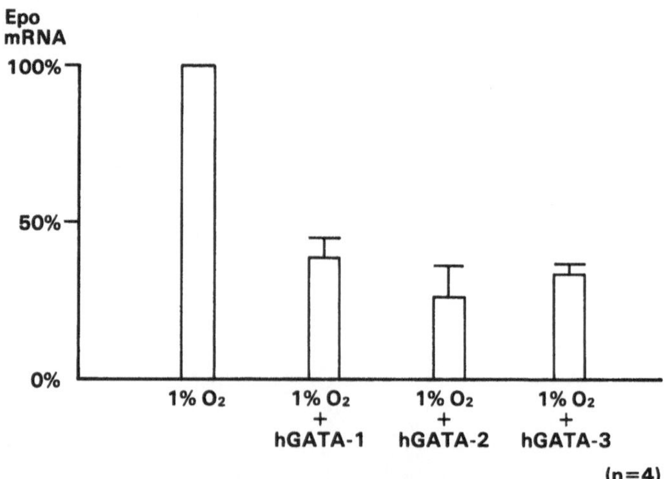

Figure 4. Effect of the GATA transcription factors on Epo mRNA expression in Hep3B cells. The level of Epo mRNA was measured by competitive PCR as described in the experimental procedures. These results were normalized to the level of Epo mRNA in Hep3B cells incubated in 1% O_2 without transfection of hGATA factors.

hGATA-1, 2 and 3 Transcription Factors Decrease Expression of Epo mRNA in Hep3B Cells

To examine the effect of hGATA transcription factors on Epo mRNA in Hep3B cells, hGATA-1, 2 or 3 expression plasmids were transfected into Hep3B cells by electroporation. The level of Epo mRNA was then measured in the transfected Hep3B cells by competitive PCR. We determined that 80 fg of Epo mRNA/µg of RNA was at the limit of dectability in our assay. The measured change in the amount of Epo mRNA following $CoCl_2$ treatment or hypoxia was a 20-fold or 100-fold increase, respectively. These results demonstrated that hGATA-1, 2 and 3 transcription factors decreased the expression level of Epo mRNA, to 38.3±6.4%, 36.1±9.8% and 33.8±2.4%, respectively, when compared to Epo mRNA levels following incubation in 1% O_2 without transfection of hGATA transcription factors (Fig. 4).

hGATA-1, 2 and 3 Transcription Factors Inhibit the Activity of the Epo Promoter

To further clarify the negative regulation of human Epo gene expression by hGATA transcription factors, cis-elements of the Epo gene fused with a reporter gene were transfected into Hep3B cells by electroporation. These constructs were made as described previously[10] (Fig. 5A, B). 5A is an 1192 bp HindIII- XbaI fragment that extends from 378 bp upstream of the Epo CAP site through the first exon and the entire first intron. To obviate the problem of false initiation from the Epo ATG start codon, this site was mutated to TAG by site directed mutagenesis. 3A is a 255 bp ACCI-BglII fragment that extends 67 bp upstream from the Epo termination codon and covers much of the 3' noncoding region and is homologous with the mouse Epo gene. We used GH as a reporter gene with the mouse metallothionein I promoter (XGH), and 5AXGH has 5A upstream of the mouse metallothionein I promoter while

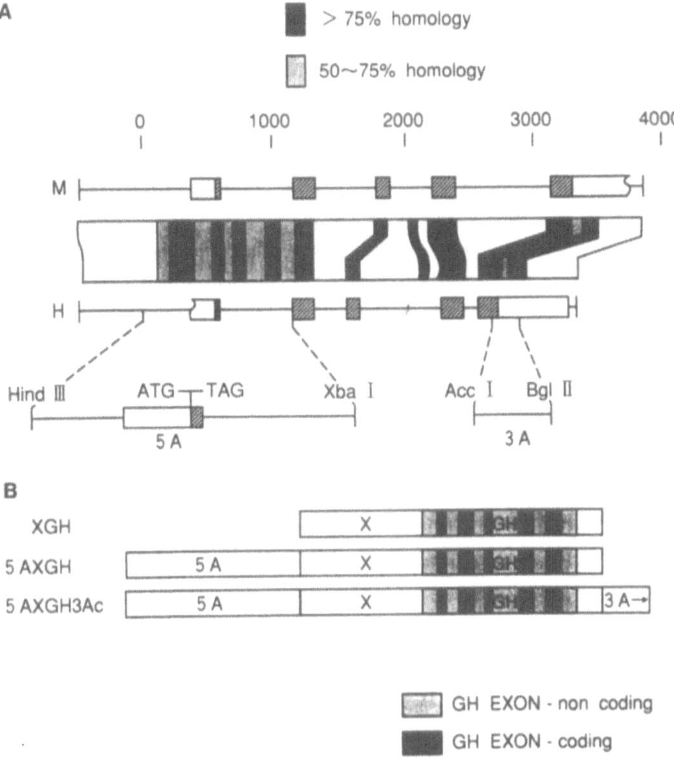

Figure 5. A: Homology between the mouse and human Epo genes. The five exons are shown as rectangles. Coding portions are hatched. The black regions between the genes represent >75% homology, while the shaded regions represent 50% to 75% homology. The two fragments used in this study: Epo 5A and 3A, are shown. B: Constructs prepared from the fragments shown in (A) and inserted into HGH-PUC12 plasmids. The coding and noncoding portions of the GH exons are shown by solid and hatched areas, respectively.

5AXGH3Ac has 3A in the correct orientation downstream of the GH gene. Hep3B cells were transfected with each of Epo GH constructs, hGATA transcription factors as well as RSVCAT as an internal standard. At the end of a 4 day incubation with 1% O_2 or 21% O_2 (control), GH was measured in the cell media, and the cell pellet was assayed for CAT. To correct for variations in transfection efficiency, these results were normalized for expression of CAT. The values of GH/CAT in a 1% O_2 incubator were normalized to the values with 21% O_2. The results in the presence of hGATA-transcription factors were then normalized to the results without hGATA-transcription factors. The effect of hGATA transcription factors on the activity of Epo promoter were expressed by normalizing the results in cells with the XGH construct alone. The results demonstrated that the hGATA-1 transcription factor significantly inhibited the activity of the Epo promoter in the presence of 5A, or 5A and 3Ac to 39.8±20.9% and 11.9±6.3%, respectively (Fig. 6A). The hGATA-2 transcription factor also inhibited the activity of the Epo promoter in the presence of 5A, or 5A and 3Ac to 23.1±3.8 % and 43.0±3.2%, respectively (Fig. 6B). Furthermore, the hGATA-3 transcription factor inhibited the activity of the Epo promoter in the presence of 5A, or 5A and 3Ac to 30.3±18.0% and 41.5±23.3%, respectively (Fig. 6C).

DISCUSSION

A comparison of the sequences of mouse and human Epo genes provides insight into candidate regulatory cis elements. An equivalent degree of homology exists in three stretches of noncoding sequence: (a) the 140-bp region upstream of the transcription start site, (b) two segments within the first intron, and (c) a fragment extending 100 to 220 bp downstream of the translation termination codon. We have previously demonstrated, using GH reporter gene constructs and Hep3B cells, the presence of promoter and enhancer elements within the 5' flanking region, the first intron, and the 3' flanking region of the human Epo gene that are responsive to both hypoxia and cobalt treatment.[10] The effects of these regions, however,

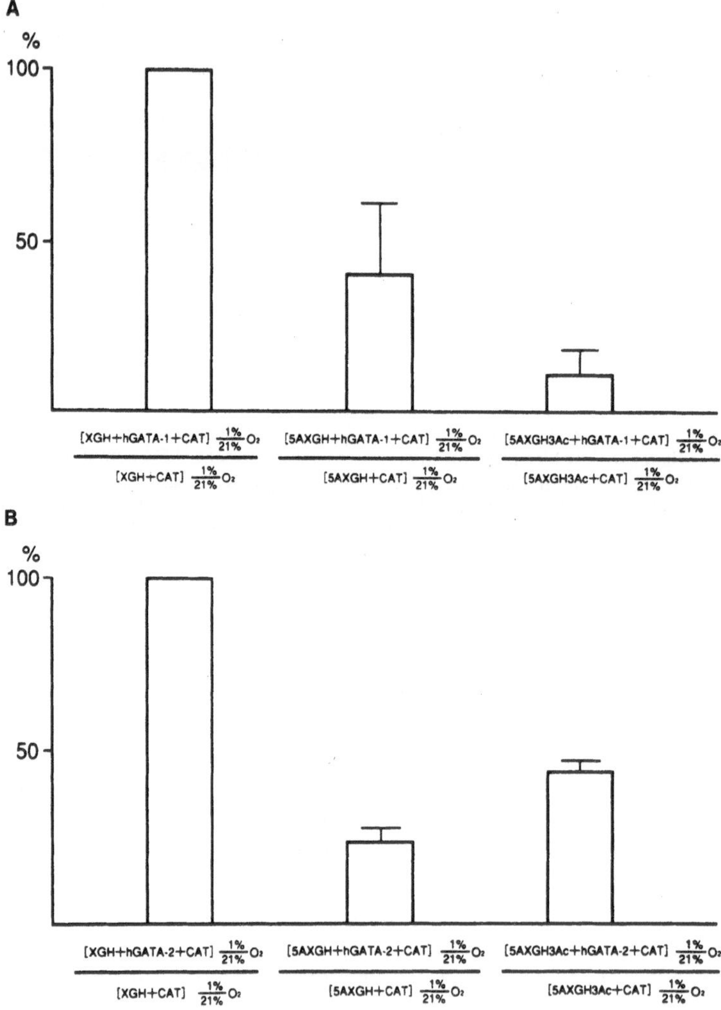

Figure 6. Effect of hGATA-1, hGATA-2 and hGATA-3 transcription factors on induction of the GH reporter in Hep3B cells. Hep3B cell were transfected with each of the Epo-GH constructs and hGATA transcription factors, hGATA-1 (A), hGATA-2 (B) and hGATA-3 (C) as well as RSVCAT as an internal standard. At the end of a 4-day incubation in either 1% O_2 (hypoxia) or 21% O_2 (control), GH was measured in the cell medium, and the cell pellet was assayed for CAT. To correct for variations in transfection efficiency, these results were normalized for expression of CAT (part C is on the following page).

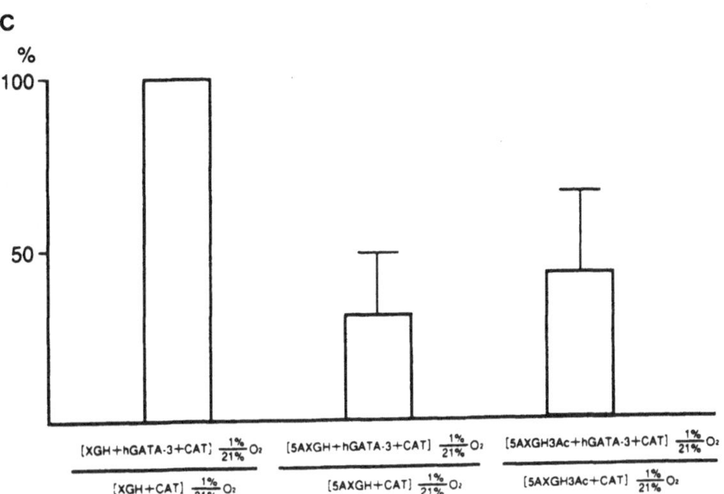

were weaker than expected, suggesting that these regulatory regions include both positive and negative regulatory elements.

In order to investigate positive and negative regulatory elements within the Epo gene in more detail, synthetic oligonucleotides were designed to control Epo transcription by means of an antigene strategy. By this method, we recently demonstrated that the CACCC elements at -60 bp from the CAP site are positive regulatory elements of the Epo gene, whereas the GATA element at -30 bp is a negative regulatory element.[5]

A hypoxia-inducible enhancer was defined in the 3'-flanking sequence of the Epo gene.[14-17] This hypoxia-inducible enhancer is functionally tripartite, with the first two sites essential for hypoxia inducibility and a third site functioning to amplify the induction signal.[16-20] A hypoxia induced DNA-binding protein (HIF-1), that binds specifically to site 1 of the hypoxia-inducible enhancer, was identified.[17] The promoter also contains an element that respond to hypoxia and cooperates with the enhancer element to faithfully reproduce the transcriptional induction seen *in vivo*.[19] The regions of the promoter and enhancer that are essential for the hypoxic induction of Epo transcription contain steroid response elements (SREs).[19]

One of the orphan nuclear receptors, hepatic nuclear factor 4 (HNF-4) and EAR3/COUP-TF1, bound specifically to the SREs in the Epo promoter and enhancer.[21] HNF-4 enhances hypoxic induction, and EAR3/COUP-TF1 negatively regulates hypoxic induction by competing with HNF-4 for binding to the Epo gene.[21] The relative levels of HNF-4 and the COUP family members may control the fine tuning of Epo expression in response to hypoxia.

Similar fine tuning of Epo production at the GATA box-containing promoter of Epo gene has also been demonstrated.[6] Aird *et al.* investigated the function of GATA box-containing promoters *in vitro* and showed that the TATA-binding protein of TFIID was required for initiation of transcription from these GATA box-containing promoters.[6] GATA-1 and GATA-2 interacted with the core promoter GATA motif of the Epo gene and inhibited generation of preinitiation complex by competing with TFIID as shown by a gel shift assay.[6] They concluded that initiation of *in vitro* transcription from a GATA box-containing core promoter might be inhibited by GATA proteins.[6] This study clearly demonstrates that over-expression of hGATA-1, 2 and 3 transcription factors significantly inhibits Epo mRNA expression in Hep3B cells. This decrease in expression may be due to inhibition of Epo promoter activity by hGATA-1, 2 and 3 transcription factors. In Hep3B and $HepG_2$ cells, which regulate Epo protein and mRNA expression in response to hypoxia and $CoCl_2$, the hGATA-2 transcription factor may bind the GATA element in the human Epo gene promoter, negatively regulating Epo gene expression. Since Hep3B cells already express hGATA-2, this additional decrease in Epo expression after transient transfection of the hGATA-2 plasmid may be due to an increase in hGATA-2 expression.

ACKNOWLEDGMENT

We thank H. Franklin Bunn, M.A. Goldberg, J. Fandrey, W. Jelkmann and D. Engel for helpful suggestions and discussions, and H. Motohashi and H. Harigae for immuno-histchemical staining. We thank Motoko Yoshida for preparing this manuscript and Kyoko Kubo for expert technical assistance.

REFERENCES

1. Graber SE, and Krantz SB: Erythropoietin: Biology and clinical use. *Hematol. Oncol. Clin. North Am.* 3: 369, 1989
2. Goldwasser E, Jacobson LO, Fried, W, Plzak, L.F: Studies on erythropoiesis. V. The effect of cobalt on the production of erythropoietin. *Blood* 13: 55, 1958
3. Goldberg MA, Glas, GA, Cunningham JM, Bunn HF: The regulated expression of erythropoietin by two human hepatoma cell lines. *Proc. Natl. Acad. Sci. U.S.A.* 84: 7972, 1987
4. Goldberg MA, Dunning SP, Bunn HF: Regulation of the erythropoietin gene: Evidence that the oxygen sensor is a heme protein. *Science* 242: 1412, 1988
5. Imagawa S, Izumi, T, Miura Y: Positive and negative regulation of the erythropoietin gene. *J. Biol. Chem.* 269: 9038, 1994
6. Aird WC, Parvin JD, Sharp PA, Rosenberg RD: The interaction of GATA-binding proteins and basal transcription factors with GATA box containing core promoters. *J. Biol. Chem.* 269: 883, 1994
7. Gilman M: Current Protocols in Molecular Biology, Wiley Interscience, New York, 1988, Vol. 1, p 4.1.4
8. Riddle RD, Yamamoto M, Engel JD: Expression of δ-aminolevulinate synthase in avian cells: Separate genes encode erythroid-specific and non specific isozymes. *Proc. Natl. Acad. Sci.* U.S.A. 86: 792, 1989
9. Yang Z, Gu L, Romeo P-H, Bories D, Motohashi H, Yamamoto M, Engel JD: Human GATA-3 trans-activation, DNA-binding, and nuclear localization activities are organized into distinct structural domains. *Mol. Cell. Biol.* 14: 2201, 1994
10. Imagawa S, Goldberg MA, Doweiko J, Bunn HF: Regulatory elements of the erythropoietin gene. *Blood* 77: 278, 1991
11. Costa-Giomi P, Caro J, Weinmann R: Enhancement by hypoxia of human erythropoietin gene transcription in vitro. *J. Biol. Chem.* 265: 10185, 1990
12. Yamamoto M, Ko LJ, Leonard MW, Beng H, Orkin SH, Engel JD: Activity and tissue-specific expression of the transcription factor NF-E1 multi gene family. *Genes & Dev.* 4: 1650, 1990
13. Neumann JR, Morency CA, Russian KO: A novel rapid assay for chloramphenicol acetyltransferase gene expression. *BioTechniques* 5, 444, 1987
14. Beck I, Ramirez S, Weinmann R, Caro J: Enhancer element at the 3'-flanking region controls transcriptional response to hypoxia in the human erythropoietin. J. Biol. Chem. 266: 15563, 1991
15. Pugh CW, Tan CC, Jones RW, Ratciffe PJ: Functional analysis of an oxygen-regulated transcriptional enhancer lying 3' to the mouse erythropoietin gene. *Proc. Natl. Acad. Sci. U.S.A.* 88: 10553, 1991
16. Semenza GL, Nejfelt MK, Chi SM, Antonarakis SE: Hypoxia-inducible nuclear factors bind to an enhancer element located 3' to the human erythropoietin gene. *Proc. Natl. Acad. Sci. U.S.A.* 88: 5680, 1991
17. Semenza GL, Wang GL: A nuclear factor induced by hypoxia via de novo protein synthesis binds to the human erythropoietin gene enhancer at a site required for transcriptional activation. *Mol. Cell. Biol.* 12: 5447, 1992

18. Beck I, Weinmann R, Caro J: Characterization of hypoxia-responsive enhancer in the human erythropoietin gene shows presence of hypoxia-inducible 120-kd nuclear DNA-binding protein in erythropoietin-producing and nonproducing cells. *Blood* 82: 704, 1993
19. Blanchard KL, Acquaviva AM, Galson DL, Bunn HF: Hypoxic induction of the human erythropoietin gene: Cooperation between the promoter and enhancer, each of which contains steroid receptor response elements. *Mol. Cell. Biol.* 12: 5373, 1992
20. Madan A, Curtin PT: A 24-base-pair sequence 3' to the human erythropoietin gene contains a hypoxia-responsive transcriptional enhancer. *Proc. Natl. Acad. Sci. U.S.A.* 90: 3928, 1993
21. Galson DL, Tsuchiya T, Tendler DS, Huang LE, Ren Y, Ogura T, Bunn HF: The orphan receptor hepatic nuclear factor 4 functions as a transcriptional activator for tissue-specific and hypoxia-specific erythropoietin gene expression and is antagonized by EAR3/COOP-TF-1. *Mol. Cell. Biol.* 15: 2135, 1995

[illegible]

THE RIBOSOMAL S6 KINASE p70^{S6k} IS INVOLVED IN THE REGULATION OF THE PROLIFERATION OF HEMATOPOIETIC CELL LINES BUT NOT IN THE INDUCTION OF ERYTHROID DIFFERENTIATION BY ERYTHROPOIETIN[*]

Robert Jaster,[1†] Thomas Bittorf,[1] S. Peter Klinken,[2] and Josef Brock[1]

[1] Institute of Medical Biochemistry
Medical Faculty of the University Rostock
Schillingallee 70, 18057 Rostock, Germany
[2] Department of Biochemistry
Royal Perth Hospital
Perth, Western Australia

ABSTRACT

The ribosomal S6 kinase p70^{S6k} is supposed to be involved in the translational control through the phosphorylation of the 40S ribosomal subunit protein S6. Here we show that the proliferative status of several cytokine-responsive hematopoietic cell lines correlates well with p70^{S6k} activity. In contrast, erythropoietin-induced erythroid maturation of J2E cells proceeds independently of p70^{S6k} function. In the erythropoietin-dependent cell line HCD-57, p70^{S6k} appears to be regulated through different erythropoietin-induced pathways.

Protein synthesis is supposed to be regulated in part by the multiple phosphorylation of the 40S ribosomal protein S6[1], which is mediated by the serine/threonine specific protein kinase p70^{S6k}.[2] In rat embryo fibroblasts, p70^{S6k} function was shown to be essential throughout the G1 phase of the cell cycle.[3] Activation of the enzyme requires multiple independent inputs[4] and is associated with phosphorylation of several clustered serine/threonine residues in a putative autoinhibitory domain of the kinase.[5] P70^{S6k} is regulated through common protein kinase C (cPKC)-dependent and independent pathways.[6] In some cell types, the PI 3-kinase was shown to mediate activation of p70^{S6k}.[7,8]

---

[*] Supported by Grants from the Deutsche Forschungsgemeinschaft and the SANDOZ-Foundation.
[†] Corresponding author. Telephone: (0049) (381) 494 5756; fax: (0049) (381) 455 383.

Several lines of evidence suggest an important role of the $p70^{S6k}$/S6 pathway in the regulation of hematopoiesis. First, a *Drosophila melanogaster* S6 knockout results in abnormal differentiation of blood cells, overgrowth of lymph glands and development of melanotic tumors.[9,10] Secondly, the T lymphocyte growth factor IL-2 stimulates the phosphorylation and activation of $p70^{S6k}$. The macrolide rapamycin blocks the IL-2 induced $p70^{S6k}$ activation and is a strong inhibitor of T-cell proliferation.[11] Thirdly, the same immunosuppressive drug was shown to block the cytokine-driven proliferation of bone marrow cells *in vitro* as well as hematopoietic recovery after myelodepression *in vivo,*[12] although the molecular basis of these phenomena remains to be elucidated.

Erythroid maturation of lineage committed progenitors is coupled with continuous replication, and both processes are mainly regulated by erythropoietin (EPO).[13,14] Despite considerable progress in the characterization of signaling cascades induced by the activated EPO-receptor (reviewed by Wojchowski and He[15]) as well as other hematopoietin-receptors the relationships between pathways stimulating replication and erythroid maturation are poorly understood.

In this study, we have investigated the correlation between $p70^{S6k}$ activity and proliferation as well as erythroid maturation of cytokine-responsive cell lines.

MATERIALS AND METHODS

Reagents

The polyclonal $p70^{S6k}$ antibody (from rabbit) as well as the synthetic peptide RRRLSSRA derived from the S6 sequence (referred to as S6 peptide) were obtained from Santa Cruz Biotechnology (Santa Cruz, CA, USA.). The recombinant human growth factors EPO, granulocyte-macrophage colony-stimulating factor (GM-CSF) and interleukin-3 (IL-3) were purchased from Boehringer Mannheim (Germany). Staurosporine (Sigma, Deisenhofen, Germany) dissolved in DMSO and rapamycin (Calbiochem, Bad Soden, Germany) dissolved in ethanol were stored as stock solutions at -20°C and diluted appropriately prior to use. Radiochemicals were supplied by Amersham (Bucks, UK).

Cell Culture

TF-1[16] and HCD-57 cells[17] were cultivated as described.[18,19] J2E cells were grown in suspension culture in Dulbecco's modified Eagle's medium supplemented with 10% fetal calf serum (FCS), penicillin 100 U/ml and streptomycin 100 µg/ml. The percentage of terminally differentiated J2E cells was determined by benzidine staining of hemoglobin.[20] Cell viability was calculated by the trypan blue exclusion test.

Cell Proliferation

J2E cells were serum-deprived for 4 hours and seeded in 24-well plates at a final concentration of 1×10^5 cells/ml. 30 min prior to the readdition of serum, cells were exposed to rapamycin as indicated. EPO was applied together with the serum, and cells were incubated for 20 hours at 37° C. Cell proliferation was assayed by determining [^{3}H] thymidine incorporation into DNA. Measurements were performed in triplicates by incubating 90 µl aliquots of the samples in 96-well plates with 0.6 µCi [^{3}H] thymidine (83 Ci/mmol [methyl ^{3}H] thymidine) in 10 µl culture medium. After an incubation period of 4 hours, incorporation of radioactivity was quantified by liquid scintillation counting.

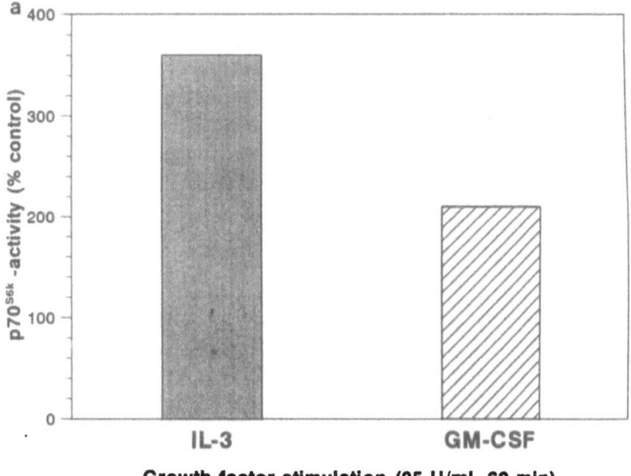

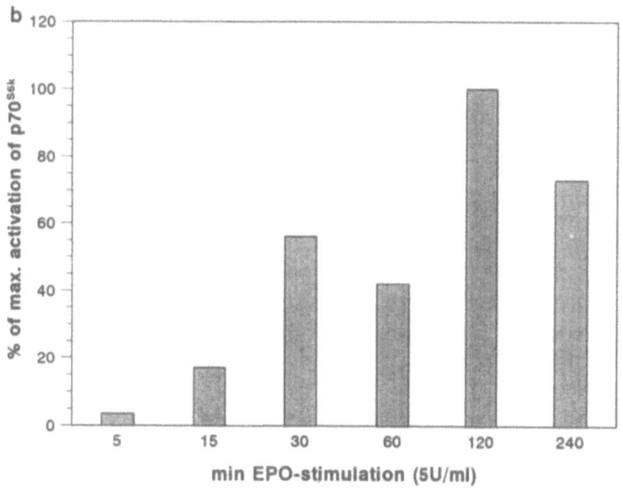

Figure 1. Effects of growth factors on p70S6k activity in hematopoietic cell lines TF-1- (a) and HCD-57 cells (b) were deprived of their growth factors GM-CSF and EPO, respectively, for 24 hours. Afterwards, cells were exposed to the indicated hematopoietin(s) followed by cell lysis, immunoprecipitation of p70S6k and assay of S6 peptide phosphotransferase activity. Data represent means of at least three independent experiments.

P70 S6 Kinase Assay

TF-1- and HCD-57 cells were pretreated and growth factor-stimulated as indicated. Aliquots of 2 x 10^6 cells were lysed for 10 min on ice in 500 μl ice-cold lysis buffer (20 mM Tris-HCl, pH 7.4, 137 mM NaCl, 0.1% SDS, 0.5% sodium deoxycholate, 1% Triton X-100, 10% glycerol, 2 mM EDTA, 1 mM phenylmethylsulphonylfluoride, 0.15 U/ml aprotinin, 1 mM sodium orthovanadate and 25 mM β-glycerophosphate). Immunoprecipitation of p70S6k and mesurement of p70S6k activity using the S6 peptide as substrate were performed essentially as described.[19]

RESULTS

To investigate the relationship between growth factor-stimulated proliferation of hematopoietic cells and activation of p70S6k, we performed assays of the enzyme activity in cytokine-responsive cell lines (Fig.1).

Table 1. Effect of staurosporine on EPO-stimulated $p70^{S6k}$ activity. HCD-57 cells were cultured for 24 hours in the absence of EPO and staurosporine was applied 1 hour before the readdition of the growth factor. Values are means of three independent experiments

Staurosporine nM	EPO - stimulation (5U/ml), time in min	$p70^{S6k}$ activity %
0	30	= 100
100	30	6
0	120	= 100
100	120	45

The human erythroleukemia cell line TF-1 proliferates in the presence of GM-CSF or IL-3.[16] Stimulation of proliferation by IL-3 or GM-CSF was accompanied by an activation of $p70^{S6k}$. The maximum enzyme activities (60 min after application of each of the two cytokines) are shown in Fig.1a. In TF-1 cells, EPO promotes short-time survival only and displays a slight growth-stimulatory effect.[16,18] In contrast to GM-CSF or IL-3, EPO did not induce a significant activation of $p70^{S6k}$ (data not shown).

HCD-57, a murine erythroid cell line, proliferates in a strictly EPO-dependent manner.[17] Induction of erythroid differentiation, however, requires the simultaneous presence of hemin.[21] EPO-stimulation of factor-deprived HCD-57 cells induced a rapid activation of $p70^{S6k}$ with two peaks of enzyme activity 30 and 120 min after EPO-application (Fig 1b). In HCD-57 cells, early phase of EPO-induced activation of $p70^{S6k}$ was much more suppressed by the potent protein kinase C inhibitor staurosporine than late phase (Table 1) suggesting different regulatory mechanisms.[19]

In contrast to HCD-57- or TF-1 cells, the murine erythroid cell line J2E proliferates in the absence of cytokines. However, in response to EPO-stimulation J2E cells undergo

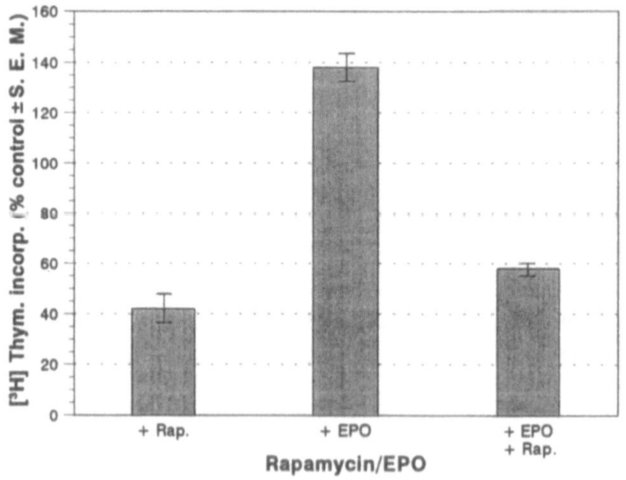

Figure 2. Influence of rapamycin (10 nM) on basal and EPO (5 U/ml)-stimulated proliferation of J2E cells determined by measuring [^{3}H] thymidine uptake. Data are expressed as percentage of the untreated control ± ±S.E.M.

Table 2. Erythroid maturation of J2E cells is not affected by rapamycin. Terminally differentiated cells were detected by benzidine staining of hemoglobin 48 hours after application of the additives

Rapamycin nM	EPO U/ml	Sodium butyrate mM	% Benzidine + cells ± S.E.M.
-	-	-	4 ± 1.2
100	-	-	2 ± 0.8
-	5	-	27 ± 0.8
100	5	-	28 ± 4.7
-	-	0.5	56 ± 5.0
100	-	0.5	59 ± 4.8

enhanced replication and erythroid differentiation.[22] In our experiments, the growth factor stimulated DNA-synthesis by approximately 40% (Fig.2) and induced maturation of almost 30% of the cells. When the chemical inducer sodium butyrate was applied, 56% differentiated cells were detected 48 hours later (Table 2).

At the non-cytotoxic concentration of 10 ng/ml, the potent p70^{S6k}-inhibitor rapamycin almost completely abolished p70^{S6k} activity in J2E cells within 15 min of application, and this strong inhibition was maintained for at least 24 hours independently of the presence or absence of EPO (data not shown). Simultaneously, both EPO-stimulated and basal DNA-synthesis of the cells were reduced by more than 50% (Fig.2). Nevertheless, the growth advantage of EPO-treated cells remained detectable (Fig.2). Although EPO-promoted growth of J2E cells was accompanied by a slight increase of p70^{S6k} activity, this stimulation was insignificant compared to the strong effect of the growth factor on the kinase activity in strictly EPO-dependent HCD-57 cells (data not shown).

Together, our data indicate a strong correlation between the proliferative status of various hematopoietic cell lines and p70^{S6k} activity.

We also investigated the effect of rapamycin on erythroid maturation of J2E cells (Table 2). Even at 100 ng/ml, rapamycin neither induced maturation itself nor displayed any significant effect on EPO- or sodium butyrate-induced terminal differentiation. Therefore, erythroid maturation of J2E cells proceeds independently of p70^{S6k} function.

DISCUSSION

Signal transduction through members of the cytokine receptor superfamily[23] has been intensively studied in the last few years, and the progressive elucidation of signaling cascades as the Ras/MAP kinase- or the JAK/STAT pathway (reviewed by Ihle et al[24]) attracts main attention to the regulation of gene expression, now. In contrast to the signaling cascades mentioned above, the p70^{S6k} pathway appears to be mainly involved in the control of protein translation (see Erikson[6] for a review concerning S6 kinases).

The data presented in this study suggest an important role of p70^{S6} in the regulation of hematopoietin-stimulated cell growth. When proliferation of factor-dependent cell lines was induced by the appropriate growth factor GM-CSF, IL-3 or EPO, p70^{S6k} activity rose, as it was described, for example, for the epidermal growth factor (EGF)[25] or IL-2.[26] On the other hand, downregulation of the enzyme activity suppressed both basal as well as EPO-stimulated proliferation of J2E cells.

In TF-1 cells, the growth-stimulatory effect of EPO is insignificant compared to IL-3 or GM-CSF,[18] and the hormone does not induce an activation of p70^{S6k}. Interestingly, in this cell line a translocation breakpoint in exon 8 of one EPO receptor gene has been described.[27] The mutated gene is highly expressed and produces an abnormal protein.[28] Our own preliminary cDNA sequence data indicate that the position of the stop codon is altered. The relationships between the EPO receptor gene mutation, missing p70^{S6k} activation and only slightly and temporary stimulated cell growth remain to be established.

Using the J2E model of terminal erythroid differentiation[22] we further show that an active p70^{S6k} pathway is not a precondition for the maturation of the cells. Thus, at least in this system, p70^{S6k} function seems to be restricted to the control of cell growth.

Regulation of p70^{S6k} function is complex and only incompletely understood. Our data suggest that in HCD-57 cells EPO-induced activation of p70^{S6k} is mediated through more than one pathway and that PKC is involved in the regulation of enzyme activity. Further investigations are necessary to identify all elements of this signaling cascade.

REFERENCES

1. Thomas G, Martin-Perez J, Siegmann M, Otto AM: The effect of serum, EGF, $PGF_{2\alpha}$ and insulin on S6 phosphorylation and the initiation of protein and DNA synthesis. Cell 30:235, 1982
2. Chung J, Kuo CJ, Crabtree GR, Blenis J: Rapamycin-FKBP specifically blocks growth-dependent activation of and signaling by the 70 kd S6 protein kinases. Cell 69:1227, 1992
3. Lane HA, Fernandez A, Lamb NJC, Thomas G: P70^{S6k} function is essential for G1 progression. Nature 363:170, 1993
4. Weng Q-P, Andrabi K, Kozlowski MT, Grove JR, Avruch J: Multiple independent inputs are required for activation of the p70 S6 kinase. Mol Cell Biol 15:2333, 1995
5. Ferrari S, Bannwarth W, Morley SJ, Totty NF, Thomas G: Activation of p70^{S6k} is associated with phosphorylation of four clustered sites displaying Ser/Thr-Pro motifs. Proc Natl Acad Sci USA 89:7282, 1992
6. Erikson RL: Structure, expression and regulation of protein kinases involved in the phosphorylation of ribosomal protein S6. J Biol Chem 266:6007, 1991
7. Chung J, Grammer TC, Lemon KP, Kaziauskas A, Blenis J: PDGF- and insulin-dependent pp70^{S6k} activation mediated by phophatidylinositol-3-OH kinase. Nature 370: 71, 1994
8. Monfar M, Lemon KP, Grammer TC, Cheatham L, Chung J, Vlahos CJ, Blenis J: Activation of pp70/85 S6 kinases in interleukin-2-responsive lymphoid cells is mediated by phosphatidyl- inositol 3-kinase and inhibited by cyclic AMP. Mol Cell Biol 15:326, 1995
9. Stewart MJ, Denell R: Mutations in the *Drosophila* gene encoding ribosomal protein S6 cause tissue overgrowth. Mol Cell Biol 13:2524, 1993
10. Watson KL, Konrad KD, Woods DF, Bryant PJ: *Drosophila* homolog of the human S6 ribosomal protein is required for tumor suppression in the hematopoietic system. Proc Natl Acad Sci USA 89:11302, 1992
11. Kuo CJ, Chung J, Fiorentino DF, Flanagan WM, Blenis J, Crabtree GR: Rapamycin selectively inhibits interleukin-2 activation of p70 S6 kinase. Nature 358:70, 1992
12. Quesniaux FJ, Wehrli S, Steiner C, Joergensen J, Schuurman H.-J, Hermann P, Schreier M.H., Schuler W.: The immunosuppressant rapamycin blocks in vitro responses to hematopoietic cytokines and inhibits recovering but not steady-state hematopoiesis in vivo. Blood 84:1543, 1994
13. Youssouffian H, Longmore G, Neumann D, Yoshimura A, Lodish HF: Structure, function and activation of the erythropoietin receptor. Blood 81:2223, 1993
14. Krantz SB: Erythropoietin. Blood 77:419, 1991
15. Wojchowski DM, He T-Ch: Signal transduction in the erythropoietin receptor-system. Stem Cells 11:381, 1993
16. Kitamura T, Tange T, Terasawa T, Chiba S, Kuwaki T, Kiyoshi M, Piao Y-F, Miayazono K, Urabe A, Takaku F: Establishment and characterization of a unique human cell line that proliferates dependently on GM-CSF, IL-3, or erythropoietin. J Cell Physiol 140: 323, 1989

17. Hankins WD, Chin K, Dons R, Szabo J: Isolation of erythropoietin-dependent cell lines suggests viability role for developmental hormones. Blood 70:161a, 1987 (abstr)
18. Bittorf Th, Jaster R, Brock J: Rapid activation of the MAP kinase pathway in hematopoietic cells by erythropoietin, granulocyte-macrophage colony-stimulating factor and interleukin-3. Cell Signal 6:305, 1994
19. Jaster R, Bittorf Th, Markewitz M, Selig G, Brock J: Erythropoietin induces biphasic activation of p70^{S6k}: evidence for a different regulation of early and late phase of activation. Cell Signal 7:325, 1995
20. Cooper MC, Levy J, Cantor LN, Marks PA, Rifkind RA: The effect of erythropoietin on colonial growth of erythroid precursor cells *in vitro*. Proc Natl Acad Sci USA 71: 1677, 1974
21. Sigounas G, Cao H, Fox H, Schecter A, Hankins WD: Hemin inducibility of erythropoietin dependent cell lines. Blood 70: 161a, 1987 (abstr)
22. Busfield SJ, Klinken SP: Erythropoietin-induced stimulation of differentiation and proliferation in J2E cells is not mimicked by chemical induction. Blood 80:412, 1992
23. Bazan JF: A novel family of growth factor receptors: A common binding domain in the growth hormone, prolactin, erythropoietin, and IL-6 receptors and the IL-2 receptor beta chain. Biochem Biophys Res Commun 164:788, 1989
24. Ihle JN, Witthuhn BA, Quelle FW, Yamamoto K, Thierfelder WE, Kreider B, Silvennoinen O: Signaling by the cytokine receptor superfamily: JAKs and STATs. TIBS 19:222, 1994
25. Susa M, Olivier AR, Fabbro D, Thomas G: EGF induces biphasic S6 kinase activation: late phase is protein kinase C-dependent and contributes to mitogenicity. Cell 57:817, 1989
26. Calvo V, Crews CM, Vik TA, Bierer BA: Interleukin 2 stimulation of p70 S6 kinase activity is inhibited by the immunosuppressant rapamycin. Proc Natl Acad Sci USA 89: 7571, 1992
27. Ward JC, Harris KW, Penny LA, Forget BG, Kitamura T, Winkelmann JC: A structurally abnormal erythropoietin receptor gene in a human erythroleukemia cell line. Exp Hematol 20:371, 1992
28. Winkelmann JC, Ward J, Mayeux P, Lacombe C, Schimmenti L, Jenkins RB: A translocated erythropoietin receptor gene in a human erythroleukemia cell line (TF-1) expresses an abnormal transcript and a truncated protein. Blood 85:179, 1995

ROLE OF HYDROGEN PEROXIDE IN HYPOXIA-INDUCED EXPRESSION OF THE ERYTHROPOIETIN AND THE VASCULAR ENDOTHELIAL GROWTH FACTOR GENES

J. Fandrey,[1] S. Frede,[1] and W. Jelkmann[2*]

[1] Institute of Physiology
University of Bonn
D - 53115 Bonn, Germany
[2] Institute of Physiology
Medical University of Lübeck
D - 23538 Lübeck, Germany

ABSTRACT

Haemoproteins of the cytochrome b family have been proposed to play a role in the control of pO_2-dependent processes such as the synthesis of erythropoietin. Because extramitochondrial cytochromes of the b type promote the formation of H_2O_2, we have studied effects of modulation of H_2O_2 abundance on hypoxia-induced erythropoietin production in cultures of the human hepatic cell line HepG2. Erythropoietin mRNA levels and erythropoietin secretion were lowered by manoeuvres and compounds that increase H_2O_2 levels (aminotriazole, phorbol ester, NADPH, menadione, exogenous H_2O_2). Erythropoietin production was restored by compounds that reduce H_2O_2 levels (catalase, dithiothreitol, tetra-methylthiourea). Additional studies revealed that H_2O_2 also inhibited the hypoxic induction of VEGF (vascular endothelial growth factor) mRNA formation in HepG2 cells. These findings support the hypothesis that H_2O_2 - and possibly other reactive O_2 species - suppress the expression of the erythropoietin and VEGF genes at high pO_2.

* Correspondence should be addressed to: Wolfgang Jelkmann, M.D., Institut fuer Physiologie, Medizinische Universitaet, Ratzeburger Allee 160, D - 23538 Luebeck, Germany. Tel: xx49-451/500-4150; Fax.: xx49-451/500-4151.

Molecular Biology of Hematopoiesis 5, edited by Abraham et al.
Plenum Press, New York, 1996

INTRODUCTION

There are several defence mechanisms which protect the organism from lack of O_2. Induced by arterial chemoreceptors, alveolar ventilation increases acutely when the arterial pO_2 is abnormally low. To improve the supply with O_2, blood flow increases in most organs in response to hypoxia, i.e., there is local arteriolar dilation. Only in the lung the vessels constrict upon lowered pO_2. Apart from the acute neuronal and muscular responses, pO_2 changes cause prolonged reactions at the gene level. The production of certain protective proteins increases at low pO_2, while overall DNA transcription and protein synthesis decline in hypoxia. The protective proteins include erythropoietin (Epo), vascular endothelial growth factor, and several glycolytic enzymes [1].

Our knowledge about the O_2 sensing mechanisms underlying the acute and chronic adaptive responses to hypoxia is only fragmentary. Recent evidence suggests that reactive O_2 species such as H_2O_2 act as signalling molecules. A role for H_2O_2 has been proposed for O_2 sensing in carotid body chemoreceptors [2], pulmonary resistance vessels [3], and Epo-producing hepatic cells [4, 5]. The major enzymic sites of the formation of reactive O_2 species are NAD(P)H oxidases, xanthine oxidase, the cytochrome P_{450} system and mitochondria.

Studies utilising the Epo-producing cell line HepG2 have shown that the capacity to generate H_2O_2 increases at high pO_2 [5]. Possibly, H_2O_2 interacts with transcription factors that control the expression of the Epo gene. Indeed, both positive and negative *trans*-acting factors have been described that bind to regulatory elements flanking the Epo gene [6].

The present report describes the effects of H_2O_2 -generating and -scavenging compounds on Epo gene expression and Epo secretion in hypoxic HepG2 cells. In addition, preliminary findings will be reported on the effects of H_2O_2 on VEGF (vascular endothelial growth factor) mRNA formation, because VEGF is produced by hepatoma cells in a pO_2-dependent manner, similar to Epo [7].

MATERIAL AND METHODS

Cell Culture

HepG2 cells from the American Type Culture Collection (ATCC no. HB8065) were maintained in RPMI 1640 medium (Flow Laboratories, Meckenheim) supplemented with 10% foetal bovine serum (Gibco, Eggenstein) and $NaHCO_3$ (2.2 g/l) in a humidified atmosphere (5% CO_2 in air) at 37°C (Heraeus Incubators, Hanau). Cell monolayers were grown to confluence in 24well polystyrene dishes (Falcon, Becton Dickinson, Heidelberg). The cultures had a density of 5×10^5 cells/cm^2 when used. The medium (0.5 ml/cm^2) was renewed 24 h before the experiment. The following compounds were added to the cultures to increase the cellular H_2O_2 concentration [8]: exogenous H_2O_2; the catalase inhibitor aminotriazole; and menadione, phorbol 12-myristate 13-acetate (PMA) and NADPH to activate NADPH oxidase. To reduce H_2O_2 levels, catalase from bovine liver, dithiothreitol, N-acetyl-L-cysteine, 4-hydroxy-3-methoxy-acetophenone (HMA) and di- (DMTU) or tetramethylthiourea (TMTU) were applied [9]. In addition, superoxide dismutase, pyrrolidine dithiocarbamate and α-tocopherol were tested for their effects on Epo production. With the exception of HMA (Roth, Karlsruhe), DMTU and TMTU (Fluka, Neu Ulm), all agents were obtained from Sigma (Deisenhofen).

For study of Epo secretion, cells were incubated with the respective agent in air - 5% CO_2 for 24 h. Under these conditions, about 10 U Epo per hour and g cellular protein are produced because the cultures are hypoxic due to the diffusion - limited O_2 supply [10]. For study of Epo mRNA levels, cells were first exposed to 95% O_2 and 5% CO_2 for 6 h to overcome the hypoxia. During this preincubation, Epo mRNA declines to baseline levels [11]. At the beginning of the experiment, fresh prewarmed medium with the respective agent was added and the cells were switched to 1% O_2, 94% N_2 and 5% CO_2 for the induction of hypoxia.

Biochemical Determinations

For determination of Epo protein the medium was collected and frozen at -20°C. Epo was measured by radioimmunoassay [10]. The cell layer was washed with PBS and lysed with SDS/NaOH (5 g/l SDS in 0.1 M NaOH). Total cellular protein was determined by a micro determination kit (Sigma Diagnostics, Taufkirchen).

Cytotoxicity was assessed by means of the [3-(4,5-dimethylthiazol-2-yl)-2,5-diphenyl]tetrazolium bromide assay (MTT) as recently described [12].

For Epo and VEGF mRNAs quantification, cells were lysed with 4 M guanidinium isothiocyanate/0.1 M 2-mercaptoethanol. Total RNA was isolated by CsCl centrifugation, redissolved in water and the concentration determined by measuring A_{260}. Competitive polymerase chain reaction (PCR) was performed as recently reported [11].

Statistics

Data are expressed as the mean ± standard deviation (SD). Student's t-test was applied for estimation of significance ($P < 0.05$) of the difference between two group means. Dunnett's test was applied to compare a control mean with several treatment means.

RESULTS

The addition of H_2O_2 to the cells resulted in a dose-dependent inhibition of the 24 h-rate of the secretion of Epo. Because the half-life of exogenous H_2O_2 is only about 9 min in HepG2 cultures [5], relatively high bolus doses of H_2O_2 were required for inhibition (Fig. 1). Accordingly, the effect of H_2O_2 was rapid as demonstrated by washout-experiments. The addition of H_2O_2 for 60 min sufficed to exert full inhibition of Epo production (Fig. 2). Based on MTT assay, cytotoxicity was not detectable up to 500 μM H_2O_2. At 1 mM H_2O_2 the capacity of the cells to reduce formazan was lowered by about 10%.

Simultaneous treatment with catalase (260 U/ml; Fig. 1) or dithiothreitol (1 mM; data not shown) completely prevented the effect of H_2O_2 on Epo production.

It is likely that H_2O_2 acted primarily on gene transcription. Treatment of the cells with exogenous H_2O_2 prevented the increase in Epo mRNA levels usually occuring following the induction of hypoxia. Likewise, the increase in VEGF mRNA was suppressed by H_2O_2 (Fig. 3).

Epo production was suppressed by all of the compounds known to lower H_2O_2 degradation or to stimulate its production, which were tested in the present study. The concentration needed to reduce Epo production by 50% was determined for aminotriazole, menadione, NADPH and PMA (Table 1). In addition to the agents given in Table 1, xanthine (200 μM) and xanthine oxidase (1 mU/ml), given in combination, lowered Epo production (from 323 ± 68 to 220 ± 31 U Epo/g cell protein in 24 h). The addition of catalase - but not of superoxide dismutase (SOD) - prevented the inhibitory effect of PMA on Epo production (Table 2). Note, that exogenous catalase did not restore Epo production in NADPH-treated cells.

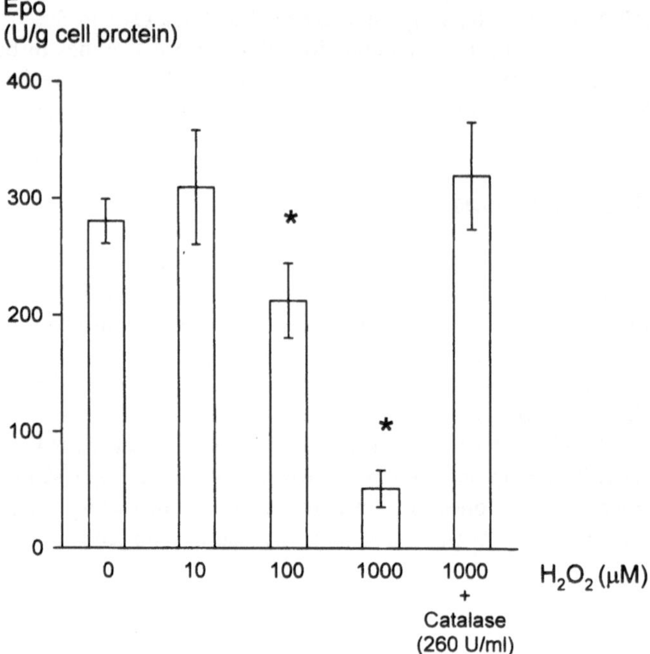

Figure 1. Dose-dependent inhibition of the production of Epo (24 h-rates) by H_2O_2, and its reversal by catalase. Mean ± SD of 4 expt. * Significantly different ($P < 0.05$) from HepG2 cultures without exogenous H_2O_2.

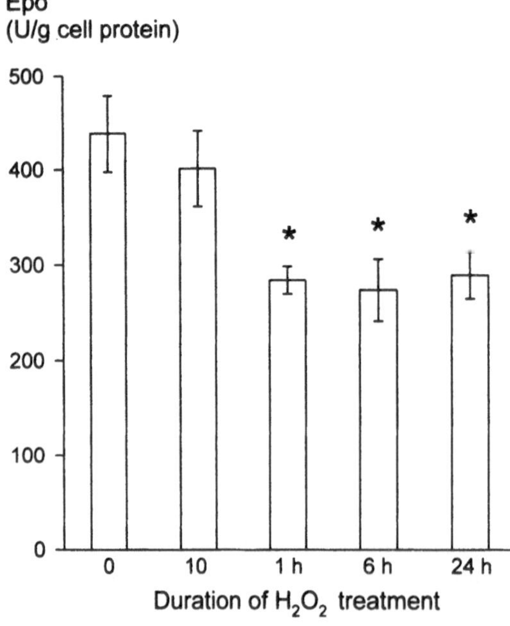

Figure 2. Epo production (24 h-rates) in HepG2 cultures treated with H_2O_2 (300 μM) for different periods followed by its removal. Mean ± SD of 4 expt. * Significantly different ($P < 0.05$) from untreated controls.

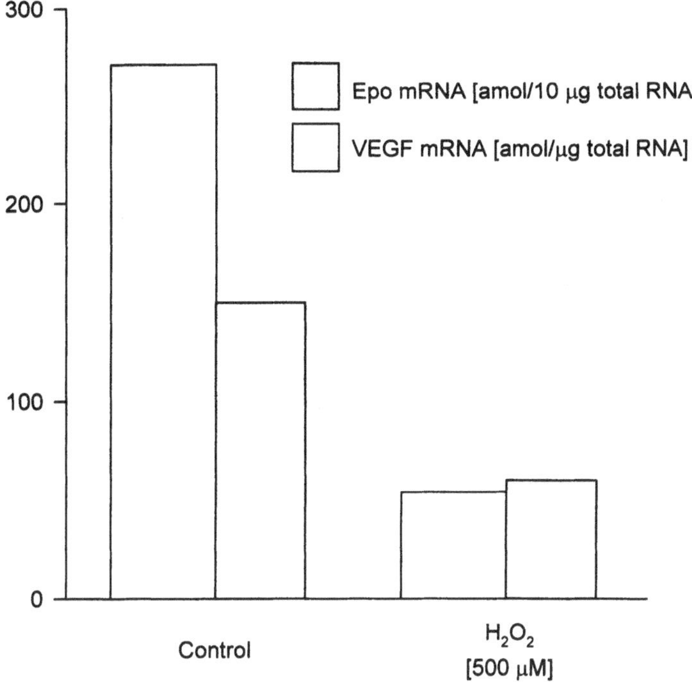

Figure 3. Effects of H_2O_2 (500 μM) on VEGF mRNA and Epo mRNA levels in hypoxic HepG2 cultures (expt. period 1 h; mean of 2 expt.).

Finally, several agents believed to act as antioxidants were tested for their effects on Epo production. Of these, only tetra-methylthiourea and α-tocopherol acted stimulatory (Table 3).

DISCUSSION

In the living organism O_2 molecules are reduced permanently to reactive intermediates. These include superoxide and hydroxyl radicals, and H_2O_2. Reactive O_2 species are

Table 1. Concentration, needed to reduce the 24 h-rate of the production of Epo by 50% (IC_{50}) in hypoxic HepG2 cultures, of agents that increase cellular H_2O_2 levels

Agent	IC_{50}	Cytotoxicity *
Aminotriazole	30 mM	no (≥ 100 mM)
Menadione	75 μM	no (≥ 100 μM)
NADPH	200 μM	no (> 500 μM)
PMA	10 nM	no (> 1 μM)

*Upper dose limit without significant cytotoxicity given in brackets.

Table 2. Effects of catalase and superoxide dismutase (SOD) on phorbol myristate acetate (PMA)-induced inhibition of Epo production in HepG2 cells during a 24 h experimental period

PMA	Catalase	SOD	Epo (U/g cell protein)
–	–	–	389 ± 36
+	–	–	80 ± 4 *
+	+	–	243 ± 13 *
+	–	+	79 ± 10 *
+	+	+	244 ± 22 *
–	+	+	366 ± 26

PMA 50 nM; Catalase 260 U/ml; SOD 50 U/ml.
Values are means ± SD of 4 parallel cultures.
*$P < 0.05$ compared to cultures without agent (-).

considered to be hazardous because oxidative stress may cause damage to oligonucleotides, proteins and lipids. During episodes of hypoxia, the generation of O_2 intermediates declines.

Recent studies have also assigned a regulatory function to O_2 intermediates, as these activate transcription factors controlling gene expression. H_2O_2 is a signalling molecule of major interest for the following reasons. First, H_2O_2 is not generally toxic because it is not reactive enough to oxidise organic molecules to a greater extent. Second, compared to O_2-derived radicals, H_2O_2 has a relatively long biological half-life. Third, H_2O_2 is capable of rapidly diffusing through hydrophobic membranes, because it is nonionized.

We have recently demonstrated that H_2O_2 production declines steeply in HepG2 cells stimulated with menadione when the pO_2 is lowered extracellularly from 70 to 30 mmHg [5]. It is tempting to correlate these pO_2-dependent changes in H_2O_2 production to Epo mRNA levels which increase in the relevant pO_2 range in hepatoma cells (Figure 4).

The present study shows that Epo production ceased when H_2O_2 was added to HepG2 cells or when endogenous H_2O_2 formation was stimulated. Although a variety of enzymic

Table 3. Effects of antioxidants on the 24 h-rate of the production of Epo in hypoxic HepG2 cultures

Agent	Change in Epo production (% versus untreated controls)
None (controls)	0(± 16)
N-acetyl-L-cysteine (10 mM)	-40(± 10)*
Pyrrolidine dithiocarbamate	
(30 µM)	-32(± 14)*
(100 µM)	-83 (± 2)*
4-Hydroxy-3-methoxy-acetophenone 100 µg/ml	+6(± 17)
N, N^1-di-methylthiourea (3 mM)	-10(± 13)
N, N, N^1, N^1-tetra-methylthiourea	
(1 mM)	+15(± 12)
(3 mM)	+40(± 14)*
α-Tocopherol(10 mg/ml)	+26(± 10)*

Values are means ± SD of 4-5 cultures.
*Significantly different from untreated control cultures ($P < 0.05$).

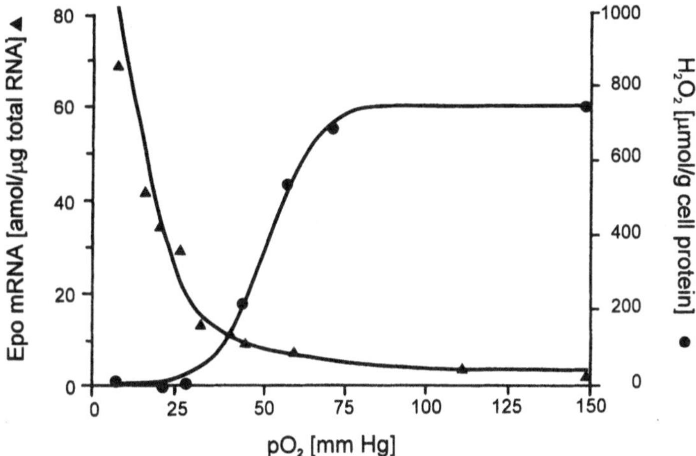

Figure 4. pO_2-dependence of Epo mRNA levels (in HepB3 cells) and H_2O_2 production rates (in HepG2 cells). Epo mRNA data are from Fandrey and Bunn [11] and refer to 6 h incubation periods. H_2O_2 production was measured in HepG2 cultures stimulated with menadione for 2 h (from [5]).

systems involving haemoproteins may generate H_2O_2 [13], it is likely that most of the compounds, which were used herein, acted on plasmalemmal NADPH oxidase. These compounds included NADPH [14], which is unable to permeate through the cell membrane, and PMA, which activates NADPH oxidase via protein kinase C translocation [15]. The inhibition of Epo production by exogenous H_2O_2 and PMA was partially prevented by catalase, which degrades H_2O_2. This observation supports the concept that H_2O_2 or metabolites derived from it, rather than superoxide or hydroperoxyl radicals, were the inhibitory derivatives. In addition, exogenously added superoxide dismutase (SOD) had no effect on the rate of Epo production. This finding was less surprising, because SOD does not enter the cell. On the other side, the superoxide radicals rarely reach the extracellular space [13].

The stimulation of Epo production by tetra-methylthiourea (TMTU) is a novel finding. Methylated thiourea compounds are potent scavengers of hydroxyl radicals [9, 16]. The precise role of hydroxyl radicals in Epo production still needs to be investigated. In general, their effects are restricted to specific metal-containing sites, because the presence of transition metals is required for the formation of hydroxyl radicals [17]. The diffusion capacity of the resulting hydroxyl radicals is extremely small [13].

A number of genes are activated by reactive O_2 species [18, 19]. Others, such as the Epo and VEGF genes, are suppressed. Epo gene expression is under the control of hypoxia-inducible factor 1, which is a heterodimer composed of HIF-1α and HIF-1β [20]. Hypoxia induces the *de novo* synthesis of HIF [21]. In turn, HIF dimers bind to the hypoxia-inducible enhancer in the 3'-flanking region of the Epo gene [6, 20]. In preliminary form, it has been reported that redox changes exert significant effects on HIF DNA-binding activity [22]. Pretreatment of cells with H_2O_2 blocked hypoxia-induced HIF DNA-binding and Epo gene expression. Oxidation of sulfhydryl groups produced a similar DNA-binding inhibition.

In summary, based on present knowledge we propose that H_2O_2 (or hydroxyl radicals derived from it) acts as a negative signalling molecule connecting O_2 sensitive haemoproteins with Epo gene transcription factors. The lack of H_2O_2 in hypoxia allows for increased HIF DNA-binding and Epo gene transcription.

ACKNOWLEDGMENTS

This work was supported by research travel grants from Cilag GmbH, Sulzbach, and Boehringer Mannheim GmbH. Thanks are also due to Mrs. Lisa Zieske for her expert secretarial help.

REFERENCES

1. Fandrey J: Hypoxia-inducible gene expression. Respir Physiol 1995; in press
2. Cross AR, Henderson L, Jones OTG, Delpiano MA, Hentschel J, Acker H: Involvement of an NAD(P)H oxidase as a pO_2 sensor protein in the rat carotid body. Biochem J 1990; 272: 743-747
3. Mohazzab-H KM, Wolin MS: Properties of a superoxide anion-generating microsomal NADH oxidoreductase, a potential pulmorary artery pO_2 sensor. Am J Physiol 1994; 267: L823-L831
4. Görlach A, Holtermann G, Jelkmann W, Hancock JT, Jones SA, Jones OTG, Acker H: Photometric characteristics of haem proteins in erythropoietin-producing hepatoma cells (HepG2). Biochem J 1993; 290: 771-776
5. Fandrey J, Frede S, Jelkmann W: Role of hydrogen peroxide in hypoxia-induced erythropoietin production. Biochem J 1994; 303: 507-510
6. Semenza GL: Regulation of erythropoietin production. Hematol Oncol Clin North America 1994; 8: 863-884
7. Goldberg MA, Schneider TJ: Similarities between the oxygen-sensing mechanisms regulating the expression of vascular endothelial growth factor and erythropoietin. J Biol Chem 1994; 269: 4355-4359
8. Kinnula VL, Whorton AR, Chang L-Y, Crapo JD: Regulation of hydrogen peroxide generation in cultured endothelial cells. Am J Respir Cell Mol Biol 1992; 6: 175-182
9. Satriano JA, Shuldiner M, Hora K, Xing Y, Shan Z, Schlondorff D: Oxygen radicals as second messengers for expression of the monocyte chemoattractant protein, JE/MCP-1, and the monocyte colony-stimulating factor, CSF-1, in response to tumor necrosis factor-α and immunoglobulin G. J Clin Invest 1993; 92: 1564-1571
10. Wolff M, Fandrey J, Jelkmann W: Microelectrode measurements of pericellular pO_2 in erythropoietin-producing human hepatoma cell cultures. Am J Physiol 1993; 265: C1266-C1270
11. Fandrey J, Bunn HF: In vivo and in vitro regulation of erythropoietin mRNA: Measurement by competitive polymerase chain reaction. Blood 1993; 81: 617-623
12. Wolff M, Jelkmann W: Effects of chemotherapeutic and immunosuppressive drugs on the production of erythropoietin in human hepatoma cultures. Ann Hematol 1993; 66: 27-31
13. Yu BP: Cellular defenses against damage from reactive oxygen species. Physiol Rev 1994; 74: 139-162
14. Meier B, Radeke HH, Selle S, Younes M, Sies H, Resch K, Habermehl GG: Human fibroblasts release reactive oxygen species in response to interleukin-1 or tumour necrosis factor-α. Biochem J 1989; 263: 539-545
15. Gennaro R, Florio C, Romeo D: Co-activation of protein kinase C and NADPH oxidase in the plasma membrane of neutrophil cytoplasts. Biochem Biophys Res Commun 1986; 134: 305-312
16. Fox RB: Prevention of granulocyte-mediated oxidant lung injury in rats by a hydroxyl radical scavenger, dimethylthiourea. J Clin Invest 1984; 74: 1456-1464
17. Makrigiorgos GM, Bump E, Huang C, Baranowska-Kortylewicz J, Kassis AI: A fluorimetric method for the detection of copper-mediated hydroxyl free radicals in the immediate proximity of DNA. Free Radical Biol Med 1995; 18: 669-678
18. Datta R, Taneja N, Sukhatme VP, Qureshi SA, Weichselbaum R. Kufe DW: Reactive oxygen intermediates target $CC(A/T)_6GG$ sequences to mediate activation of the early growth response 1 transcription factor gene by ionizing radiation. Proc Natl Acad Sci USA1993; 90: 2419-2422
19. Lu D, Maulik N, Moraru II, Kreutzer DL, Das DK: Molecular adaptation of vascular endothelial cells to oxidative stress. Am J Physiol 1993; 264: C715-C722
20. Wang GL, Semenza GL: Purification and characterization of hypoxia-inducible factor 1. J Biol Chem 1995; 270: 1230-1237
21. Wang GL, Jiang B-H, Rue EA, Semenza GL: Hypoxia-inducible factor 1 is a basic- helix-loop-helix-PAS heterodimer regulated by cellular O_2 tension. Proc Natl Acad Sci USA 1995; 92: 5510-5514
22. Huang LE, Bunn HF: Involvement of a redox mechanism in the regulation of hypoxia-inducible factor-1 DNA-binding activity and erythropoietin gene expression. Blood 1994; 84 No 10, Suppl. 1: 509a

64

REDOX AND HEAT SHOCK PROTEIN HSP70 AFFECT THE BINDING OF ERYTHROPOIETIN RNA BINDING PROTEIN TO ERYTHROPOIETIN mRNA*

Aline B. Scandurro,[1,2†] Eric McGary,[2] Isaac J. Rondon,[2] Russell B. Wilson,[3] and Barbara S. Beckman[1,2]

[1] Tulane Cancer Center
[2] Department of Pharmacology
[3] Department of Pathology and Laboratory Medicine
Tulane University Medical Center, New Orleans, Louisiana

ABSTRACT

Factors affecting complex formation between the cytosolic protein identified in our laboratory to bind to erythropoietin (EPO) mRNA (ERBP) and this mRNA have been investigated. It was observed that reducing agents, dithiothreitol and 2-mercaptoethanol, enhanced ERBP binding activity in a dose-dependent manner. Oxidizing agents such as diamide abolished complex formation. Treatment with the sulfhydryl alkylating agent N-ethylmaleimide also resulted in inhibition of complex formation, suggesting a possible role of sulfhydryl groups in the interaction between this protein and EPO mRNA.

Since heat shock proteins are induced in response to a variety of stresses including hypoxia the role of these proteins in ERBP-EPO mRNA complex formation was studied. Using an electrophoretic mobility shift assay (EMSA) in which human anti-hsc70 was added to ERBP-containing human hepatoma cell (Hep3B) lysates it was found that the ERBP-EPO mRNA complex was supershifted when compared to normal controls. However, no supershift was observed when hsp70 antibody was added to the reaction. Western blots of these EMSA gels show hsc70 localized to the ERBP-EPO mRNA complex. These findings suggest the involvement of hsc70 in ERBP-EPO mRNA complex formation and perhaps a novel role for hsps in the regulation of Epo mRNA stability. It is possible that hsc70 might be involved in the complex process that controls red blood cell production during oxygen deprivation.

* The sections containing redox effects on ERBP have been published (5).

† Address: Dept. of Pharmacology, 1430 Tulane Ave., New Orleans, Louisiana 70112. Tel: (504) 588-5444; Fax: (504) 588-5283; E-mail: sigtrans@ tmcpop.tmc.tulane.edu.

Molecular Biology of Hematopoiesis 5, edited by Abraham et al.
Plenum Press, New York, 1996

INTRODUCTION

The erythropoietin (EPO) gene is post-transcriptionally regulated. Factors believed to mediate this control include a *cis*- acting 120 bp region located 3' of the EPO mRNA and a *trans*-acting 70 and 135 kDa. protein complex, termed erythropoietin RNA binding protein (ERBP). Using a gel band shift assay this protein complex has been previously shown to specifically bind to the 120 bp region of EPO mRNA and in several murine tissues, greater binding of ERBP to EPO mRNA was observed after hypoxic treatment (1). To further characterize the nature of the interaction between ERBP and EPO mRNA we examined the effects of reducing and oxidizing agents on complex formation. A number of other mRNA binding proteins (e.g.- IRE-BP and AUBF) have been reported to be affected by reduction-oxidation (2, 3).

Since heat shock proteins are induced in response to a variety of stresses, including hypoxia, the role of these proteins in ERBP-EPO mRNA complex formation was investigated (4). Addition of a polyclonal HSP70 antibody to the gel band shift reaction mixture results in a supershift of the ERBP-EPO mRNA complex. This antibody was originally characterized as an anti-hsc70 antibody, however this antibody has now been shown to be anti-hsp70. Further, commercially prepared monoclonal antibodies (StressGen) have also been used in the present study to determine that the inducible hsp70 affects ERBP-EPO mRNA complex formation. These studies suggest a novel role for heat shock factors, regulation of EPO mRNA stability, and suggest that hsp70 might be involved in the complex process that controls red blood cell production during oxygen deprivation.

METHODS

Cell Culture

Hep3B cells obtained from ATCC (Rockville, MD) were routinely cultured in Eagle's minimal essential medium (MEM) supplemented with 10% fetal bovine serum (FBS), 0.1 mM nonessential amino acids, 1 mM sodium pyruvate, 100 U/ml penicillin G, and 100 ug/ml streptomycin in a humidified atmosphere of 5% CO_2, 95% air at 37°C. Media and all supplements were purchased from Life Technologies, Inc. (Gaithersburg, MD). Hypoxic conditions were obtained by placing cells in a hypoxic chamber (Billups-Rothenburg, Del Mar, CA) (1% O_2, 5% CO_2, 94% N_2) for 6-8 hrs.

Preparation of Lysates

After trypsinization (0.05% trypsin and 0.53 mM EDTA, GIBCO), Hep3B cells were pelleted at 500 rpm for 2.5 min. and washed twice with PBS (without calcium and magnesium). Following the second wash the cells were gently resuspended in PBS, transferred to an Eppendorf tube, pelleted at 1000 x g for 5-10 seconds, PBS removed and 25 mM Tris (pH 7.9), 0.5 mM EDTA added so that the final concentration was about 2.0×10^5 cells/µl. Cells were frozen and thawed through repetitive cycles (1). The protein concentration of each cytoplasmic lysate was determined by the method of Bradford (BioRad Laboratories, Hercules, CA) or BCA (Pierce, Perstorp Biotec Co.; Rockford, IL.).

In Vitro Transcription

EPO cDNA was kindly provided by Dr. Jerry Powell (University of California, Davis) and subcloned into the EcoRI site of pcDNAI (Invitrogen, San Diego, CA). Radiolabeled

EPO mRNA was produced by T7 RNA polymerase (Promega) from 1 μg XhoI-digested pcDNAI containing EPO cDNA at the EcoRI site and labeled with [^{32}P] UTP (1). Following phenol-chloroform extraction transcription mixtures were passed through push-trap columns (Stratagene, La Jolla, CA). Specific activity was typically 10^7 cpm/μg RNA.

Band-Shift Assay

Two μg of cytoplasmic lysate were incubated for 10 min with reducing agents, oxidizing agents or vehicles followed by 5 x 10^4 cpm of EPO mRNA in 12 mM HEPES (pH 7.9), 10% glycerol, 15 mM KCl, 0.25 mM EDTA, 5 mM $MgCl_2$ and 200 μg/ml yeast transfer RNA in a total volume of 10 μl for 10 min at 30°C. Twenty units RNase T1 were added and reaction mixtures were incubated for an additional 30 min at 37 °C prior to electrophoresis in a 7% native polyacrylamide gel with 0.25 x TBE (Tris-Borate-EDTA) running buffer. After drying, gels were exposed overnight to film (Amersham-Hyperfilm) with two intensifying screens or exposed 30 min. to 2 hrs. to a Fuji phosphorimaging board.

For competition studies unlabeled EPO mRNA was added to incubation mixtures on ice 30 min prior to addition of radiolabeled EPO mRNA and subsequently incubated for 10 min at 30°C. For reactions with antibodies, these were added to incubations on ice for 1 hr. prior to addition of radiolabeled Epo mRNA and incubation at 30°C for 10 min.

RESULTS

Comparison of Normoxic and Hypoxic Lysates Treated with Various Concentrations of DTT and 2-ME

It has been hypothesized that ERBP plays a role in EPO mRNA activity (1) and that it may mediate ribonuclease activity (5). However, it is not known how the binding activity of ERBP is regulated or if other factors recruited or bound to ERBP regulate its activity. In addition, it is known that the EPO mRNA concentration reaches a peak level after 5 to 8 hours of hypoxia (5) suggesting that the effect of ERBP on EPO mRNA should occur at this time. In order to investigate the redox status of ERBP as a possible hypoxia sensor mechanism that controls the binding activity of ERBP normoxic and hypoxic (8 hr) lysates were treated with various concentrations of DTT (0-10 mM) or 2-ME (0.01-10%) and their ERBP binding activity was assessed by band shift assay. Fig. 1 represents the dose-dependent effect of DTT in increasing ERBP-EPO mRNA binding. Results were quantitated after exposure of the dried gel band shift gel to a phosphorimaging board (Fuji Co., Japan) using the MacBas analyzing program (version 2.0) and are depicted graphically. Fig. 2 also shows that reduction by 2-ME treatment results in increased ERBP binding. Consistently treatment of lysates with 10% 2-ME resulted in a smear which we attributed to total protein denaturation. These results suggest that ERBP-EPO mRNA binding might involve prior, local reduction of ERBP's sulfhydryl groups.

Effect of Reducing and Oxidizing Agents on the Formation of ERBP-EPO mRNA Complex

To determine the effect of reducing and oxidizing agents on the formation of ERBP-EPO mRNA complex, 2 μg of normoxic Hep3B lysates were pre-incubated for 10 min at 30°C with 0.2 mM DTT, 1% 2-ME or the oxidizing agent diamide (1 or 5 mM) with or without 1% 2-ME prior to the addition of radiolabeled EPO mRNA. This treatment was

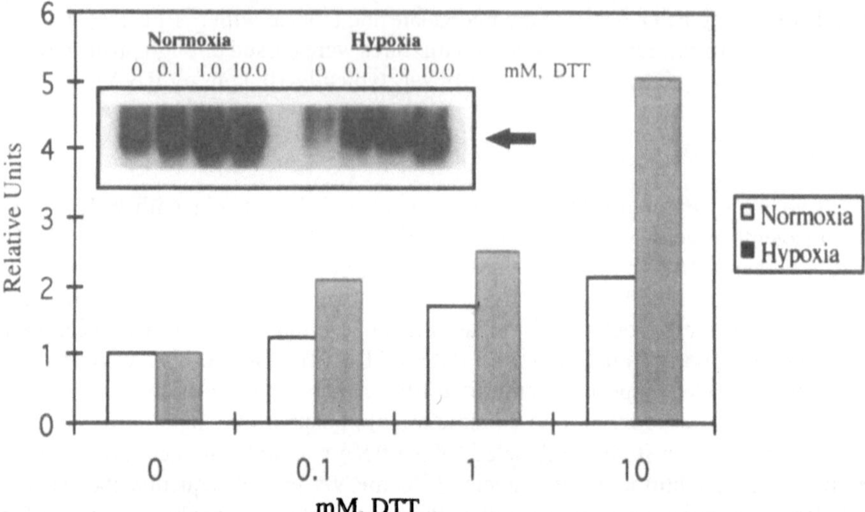

Figure 1. Comparison of normoxic and hypoxic (8 hr.) lysates treated with various concentrations of dithiothreitol. Hep3B lysates were prepared from cells harvested after 8 hr normoxic or hypoxic stimulus as described in Materials and Methods. Subsequently these lysates were pretreated for 10 min with varying concentrations of DTT (0-10 mM) prior to gel band shift assay and separation on 7% native polyacrylamide gel as described. Radioactive values for each band were obtained after exposing the dried gels to a Fuji phosphoimaging board. The data presented are the means of duplicate samples performed with lysates from three independent experiments. Standard error values for each sample are: normoxia, 0.1 mM (+/- 0.133), 1 mM (+/- 0.37), 10 mM (+/-0.82); hypoxia, 0.1 mM (+/- 0.59), 1.0 mM (+/- 0.94) and 10 mM (+/-1.38). The insert shown is a representative autoradiogram used in data analysis.

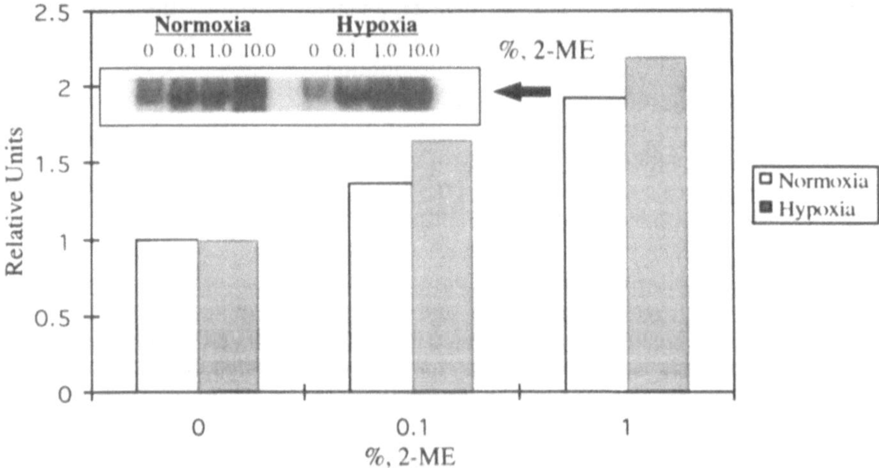

Figure 2. Comparison of normoxic and hypoxic (8 hr.) lysates treated with various concentrations of 2-ME. Reaction mixtures were as described, except for the pretreatment of the lysates with different concentrations of 2-ME for 10 min, prior to the addition of the radiolabeled EPO mRNA. The standard error values for each sample are: normoxia, 0.1 mM (+/- 0.19), 1.0 mM (+/- 0.24); hypoxia, 0.1 mM (+/- 0.42), and 1.0 mM (+/- 0.67). Consistently 10 mM 2-ME treatment of the lysates resulted in retention of large protein complexes in the native gel. For this reason this particular sample has been omitted from the densitometric data graph. The insert shown is a representative autoradiogram used in data analysis. The arrow indicates ERBP-EPO mRNA complex.

followed by RNase T1 digestion, electrophoresis and autoradiography. Treatment of lysates with DTT or 2-ME resulted in an increase in ERBP binding activity whereas treatment with diamide abolished ERBP binding activity. Binding activity could be restored after diamide treatment by a brief incubation with 0.1% 2-ME in a dose-dependent manner. A representative band-shift assay is shown in Fig. 2. These data indicate that alteration in the redox state of the cytoplasmic lysate affect ERBP binding activity.

Effect of N-Ethylmaleimide (NEM), an Irreversible Oxidizing Agent, on the Formation of ERBP-EPO mRNA Complex

Reduced sulfhydryl groups can function in different ways. We have previously demonstrated that ERBP is composed of two subunits, 70 and 130-140 kDa, which may be attached via intermolecular disulfide bonds (1). Likewise, AUBF is composed of three subunits and redox changes appear to affect its intermolecular disulfide bonds (3) while redox changes appear to affect sulfhydryl groups on IRE-BP that mediate ligand binding (2). In order to determine which of these possible interactions is operative for ERBP-EPO

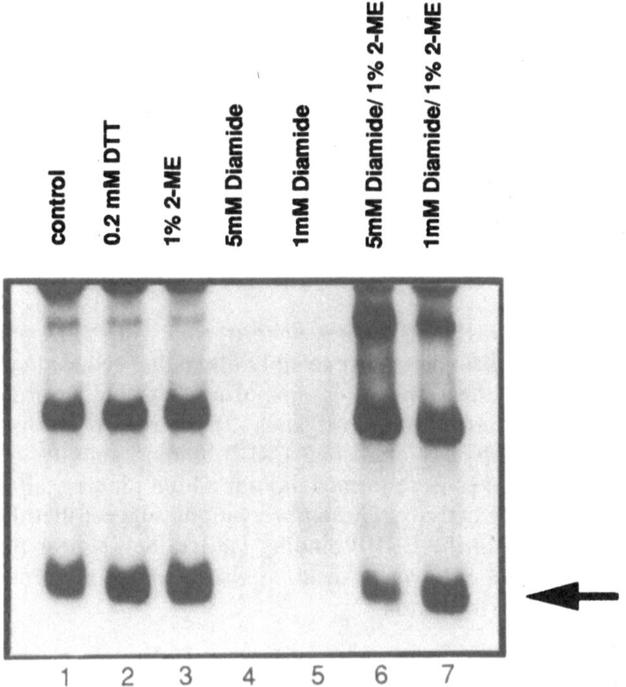

Figure 3. Effect of reducing and oxidizing agents on the formation of ERBP-EPO mRNA complex. Normoxic Hep3B cytoplasmic lysates from near confluent cultures were prepared by freeze-thaw lysis as described in "Materials and Methods". mRNA was transcribed by T7 RNA polymerase and labeled with [^{32}P]UTP from XhoI-digested pcDNAI containing EPO cDNA at the EcoRI sites. Two ug of cytoplasmic lysate were incubated for 10 min with reducing agents, oxidizing agents or vehicle followed by 5 x 10^4 cpm of mRNA in 12 mM HEPES and 200 μg/ml yeast transfer mRNA in a total volume of 10 μl for 10 min at 30°C. Twenty units RNase T1 was added and reaction mixtures were incubated for an additional 30 min at 37°C prior to electrophoresis in a 7% native polyacrylamide gel with 0.25X TBE running buffer. After drying, the gel was exposed overnight to film. Lane 1, vehicle/control; Lane 2, 0.2 mM DTT; Lane 3, 1% 2-ME; Lane 4, 5mM diamide; Lane 5, 1mM diamide; Lane 6 and 7, 5 and 1 mM diamide respectively, followed by an additional 10 min incubation with 1% 1-ME. The arrow indicates ERBP-EPO mRNA complex.

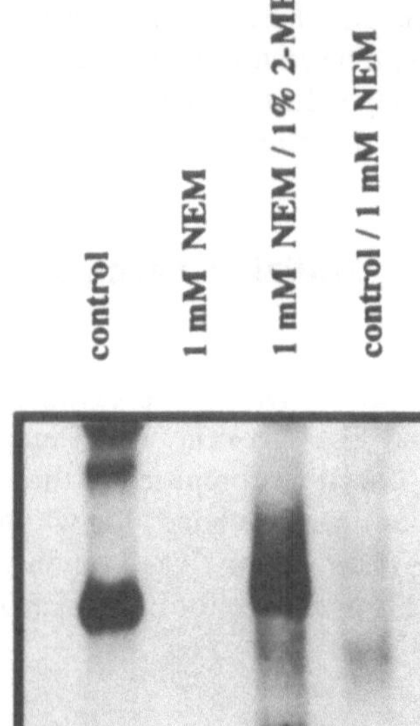

Figure 4. Effect of N-ethylmaleimide, an irreversibe oxidizing agent, on the formation of ERBP-EPO mRNA complex. Reaction mixtures were as described in Fig. 3 except for the pretreatment of the lysates. Lane 1, vehicle; Lane 2, 1 mM NEM; Lane 3, 1 mM NEM followed by 10 min incubation with 1% 2-ME; Lane 4, treatment of the mixture with 1 mM NEM after the complex has already formed. The arrow indicates ERBP-EPO mRNA complex.

mRNA complex formation, Hep3B lysates were treated with NEM prior to band shift assay. NEM differs from diamide by irreversibly alkylating reduced sulfhydryl groups. In addition, NEM has no effect on -SH groups of mRNA binding proteins involved in transient Michael adduct formation with ligands (2). As shown in Fig. 4 preincubation of lysates with NEM completely abolished ERBP binding activity. NEM added after ERBP-EPO mRNA complexes were formed did not inhibit binding, although it seems to have decreased the binding activity. These observations suggest that the -SH groups of ERBP might be involved in the ERBP binding site as they appear to be for IRE-BP. Additionally there may be -SH groups residing elsewhere on the protein whose redox state might affect complex formation.

Effect of Various Anti-Heat Shock Protein Antibodies on the Formation of ERBP-EPO mRNA Complex

Since heat shock proteins are induced in response to a number of physiologic stresses including hypoxia, the role of these proteins in ERBP-EPO mRNA complex formation was analyzed. Fig. 5 shows the result of preincubation of the Hep3B lysates with: human anti-hsp70; bovine anti-hsp70; human monoclonal anti-hsp70 (inducible); human monoclonal anti-hsp70 (cognate); chicken gammaglobulin; murine monoclonal protein kinase C; or reconstituted bovine hsp70. Only sample mixtures that were preincubated with human anti-hsp70 antibodies resulted in inhibition of the ERBP-EPO mRNA. Sample mixtures pretreated with the other reagents showed no effect.

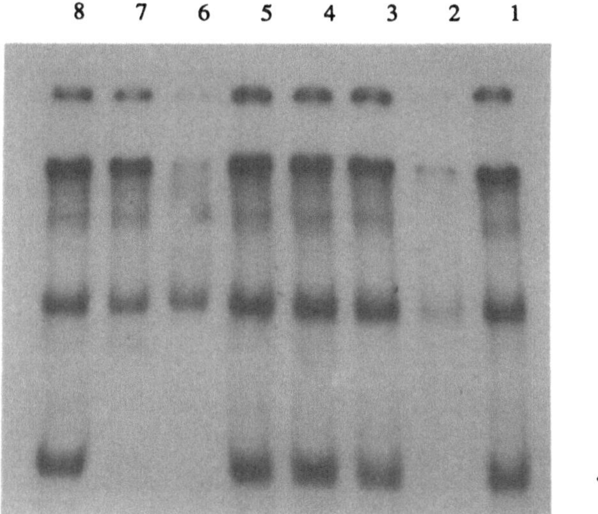

Figure 5. Effect of Various Anti-Heat Shock Protein Antibodies on the Formation of ERBP-EPO mRNA Complex. Reaction mixtures were as described except that the cytoplasmic lysates were preincubated for 1 hr. on ice with: Lane 1, control, no antibody; lane 2, human polyclonal anti-hsp70; lane 3, 1:10 diluted human polyclonal anti-hsp70; lane 4, bovine anti-hsp70; lane 5, chicken gamma globulin; lane 6, human monoclonal hsp70 (StressGen); lane 7, human monoclonal hsp70 (Affinity Bioreagents) ; lane 8, 50 ng bovine recombinant hsp70.

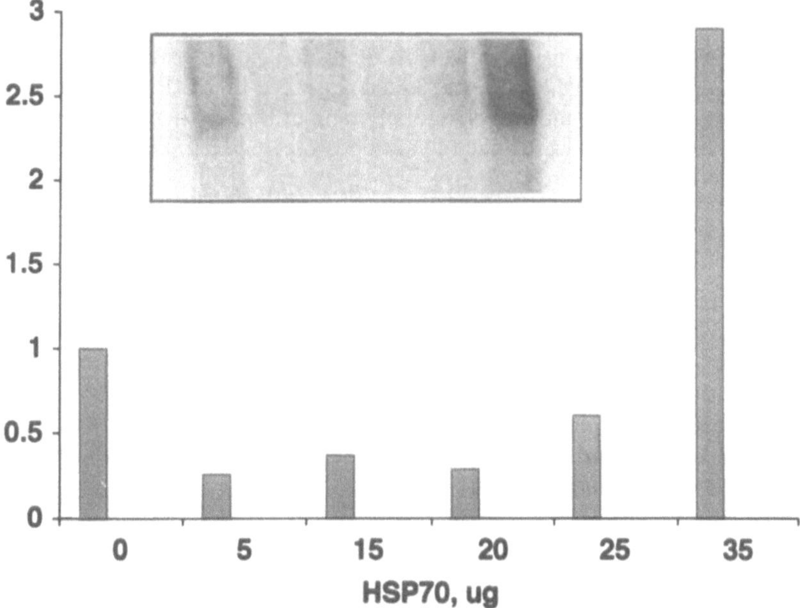

Figure 6. Rescue of ERBP-EPO mRNA Complex by Inducible Hsp70. Hep3B lysates were pretreated with increasing amounts of purified inducible HSP70 (0-35 μg) for 30 min on ice prior to incubation on ice for 1 hr. with an equal amount of monoclonal anti-hsp70 antibody (StressGen). Subsequently reaction mixtures were treated as described in Fig. 3. Insert is a sample autoradiogram used in data analysis.

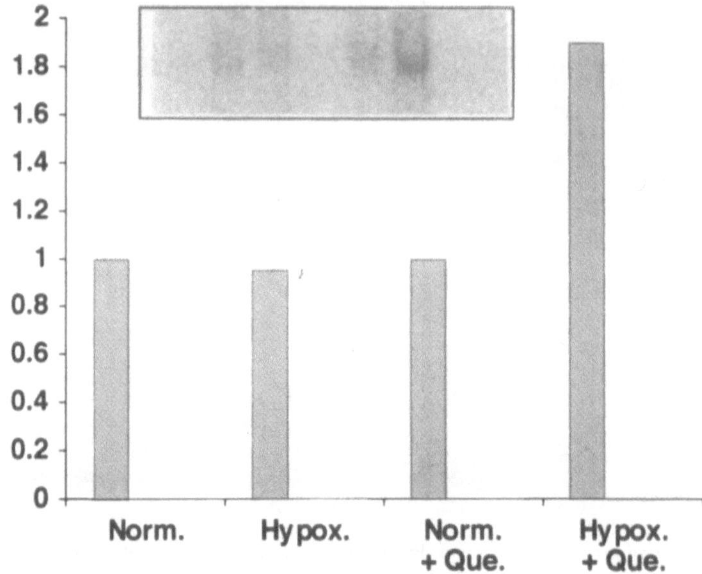

Figure 7. *In vivo* Effects of Inducible Hsp70 on ERBP Binding Activity. Hep3B cells were incubated in the presence or absence of quercetin (Que., 0.1 mM) for 4 hrs prior to incubation at normoxia or hypoxia for an additional 6 hrs. Lysates were prepared and analyzed in the gel band shift assay as described in Fig. 3. Insert is a representative autoradiogram used in data analysis.

Rescue of ERBP-EPO mRNA Complex by Purified Inducible HSP70

To show that the anti-hsp70 effect on the complex formation was specific for hsp70, homogeneous inducible hsp70 was added to rescue the ERBP-EPO mRNA complex. As depicted in the inset and corresponding graph of Fig. 6 when increasing amounts of inducible hsp70 (0-35 μg) were added to antibody-treated Hep3B lysates the complex reappeared.

In Vivo Effects of Inducible HSP70 on ERBP Binding Activity

Quercetin is a bioflavonoid known to inhibit Hsp70 induction (6). To assess whether inhibition of hsp70 induction *in vivo* by quercetin would affect ERBP binding activity, Hep3Bcells were pretreated for 4 hrs with 100 μM quercetin before incubation in normoxic or hypoxic conditions for an additional 6 hrs. Fig. 7 shows that inhibition of inducible hsp70 resulted in increased ERBP binding only when cells were treated under hypoxic conditions. Cells treated with quercetin and grown under normoxic conditions did not exhibit any change when compared with untreated samples.

DISCUSSION

Investigations were carried out to analyze the factors affecting ERBP-EPO mRNA complex formation. First, the effects of changes in reduction-oxidation of Hep3B lysates were studied as had been reported for other mRNA binding proteins. Treatment of ERBP-containing Hep3B lysates with the reducing agents, DTT and 2-ME resulted in increased

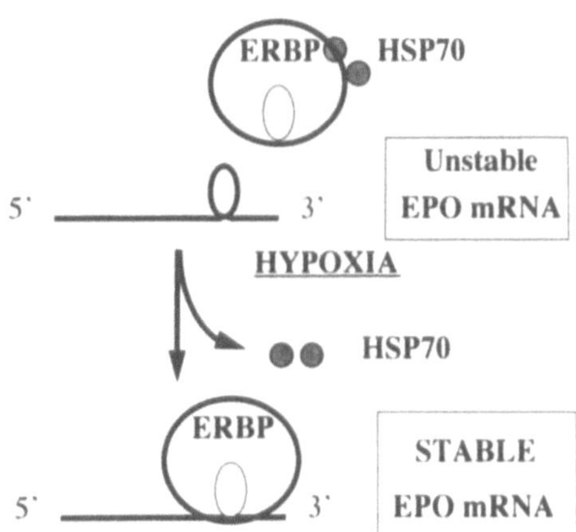

Figure 8. Model for ERBP Activity. Epo mRNA shown with ERBP-binding site contained within a 3'loop. Large circle represents the ERBP complex. Small shaded circles are hsp70 molecules. During normal physiologic conditions inducible hsp70 binds and keeps ERBP from binding and stabilizing EPO mRNA. During hypoxia hsp70 is sequestered away leaving ERBP free to bind and stabilize EPO mRNA. This interaction is most likely mediated by reduced sulfhydryl groups in the ERBP complex.

ERBP-EPO mRNA binding. Conversely, treatment with the oxidizing agent diamide inhibited complex formation. It is likely as has been reported for IRE-BP interaction with IRE that ERBP-EPO mRNA complex formation requires free sulfhydryl groups for the protein's interaction with the mRNA. This is supported by the observation that NEM did not affect already formed ERBP-EPO mRNA complexes (Fig. 4).

The involvement of the heat shock protein, hsp70 in ERBP-EPO mRNA complex formation was studied since hsps belong to a class of proteins that respond to hypoxia (7). Preincubation of Hep3B lysates with monoclonal antibodies to hsp70 inhibited complex formation. Further, addition of homogeneous inducible hsp70 to the antibody treated lysates rescued complex formation. Thus, inducible hsp70 plays a role in ERBP-EPO mRNA complex formation. Inhibition of hsp70 induction *in vivo* by quercetin resulted in increased binding in hypoxic but not normoxic treated cells. Thus, it appears that hsp70's recruitment away following hypoxia is not sufficient for increased binding but requires another hypoxic signal, possibly reduction of the sulfhydryl groups mentioned above.

We would like to propose a model (Fig. 8) by which to explain the results reported here. Under normal physiologic conditions the multimeric protein ERBP is kept from binding EPO mRNA by hsp70 sequestration. Hsp70 binding may also serve to prime ERBP by exposing the sulfhydryl groups necessary for EPO mRNA binding. Similar mechanisms have been described for heat shock proteins as for example hsp90 binding to the glucocorticoid receptor (8). During hypoxia these sulfhydryl groups are reduced either directly or mediated by an accessory protein (e.g.- *ref*-like) (9). Concomitantly, hsp70 is recruited away to the nucleus to serve its stress function. The reduced, hsp70-free ERBP complex is now able to bind to its target sequence on the EPO mRNA thereby stabilizing it. This stabilization most likely occurs by the physical obstruction of the ERBP complex bound to the mRNA of the destabilizing machinery suggested to assemble downstream of the ERBP binding site (10). Studies to prove this model are currently the focus of the laboratory.

ACKNOWLEDGMENT

This work was supported by National Institutes of Health grant #DK 40501 and a grant from the Tulane Cancer Center.

REFERENCES

1. Rondon I.J., Mac Millan L.A., Beckman B.S., Goldberg M.A., Schneider T., Bunn H.F., and Malter, J.S: Hypoxia up-regulates the activity of a novel erythropoietin mRNA binding protein. J. Biol. Chem. 266:16594, 1991.
2. Klausner R.D., Roualt T.A. and Hartford J.B.: Regulating the fate of mRNA: the control of cellular iron metabolism. Cell 72: 19, 1993.
3. Malter J.S. and Hong Y.: A redox switch and phosphorylation are involved in the post-translational up-regulation of the adenosine-uridine binding factor by phorbol ester and ionophore.
4. Morimoto RI, Tissieres A, and Georgopoulos C: The Biology of Heat Shock Proteins and Molecular Chaperones. Chp. 18, 20. Cold Spring Harbor Laboratories Press. 1994.
5. Rondon IJ, Scandurro AB, Wilson RB, and Beckman BS: Changes in redox affect the activity of erythropoietin RNA binding protein. FEBS Lett. 359:267, 1995.
6. Nagai N, Nakai A, and Nagata K: Quercetin suppresses heat shock response by down regulation of HSF1. Biochem Biophys Res Com 208:1099, 1995.
7. Morimoto RI, Tissieres A, and Georgopoulos C: The Biology of Heat Shock Proteins and Molecular Chaperones. Chp. 18, 20 and 22. Cold Spring Harbor Laboratories Press. 1994.
8. Hutchinson KA, Dittmar KD, Czar MJ, and Pratt WB: Proof that hsp70 is required for assembly of the glucocorticoid receptor into a heterocomplex with hsp90. J. Biol. Chem. 269:5043, 1994.
9. Xanthoudakis S and Curran T: Identification and characterization of Ref-1, a nuclear protein that facilitates AP-1 DNA binding activity. EMBO J. 11:653, 1992.
10. Ho V, Acquaviva A, Duh E, and Bunn HF: Use of a marked erythropoietin gene for investigation of its cis-acting elements. J. Biol. Chem. 270: 10084, 1995.

65

THE ROLE OF HEMOPOIETIC CYTOKINES IN THE MYELODYSPLASTIC SYNDROMES

Peter L. Greenberg

Hematology Division
Stanford University Medical Center
and
VA Palo Alto Health Care System
Palo Alto, California 94304

ABSTRACT

Hemopoietic cytokines have been used to attempt to improve the cytopenias and to modulate the pathogenetic mechanisms in myelodysplastic syndromes (MDS). Neutropenia has generally been alleviated in 80-90% of MDS patients with *G-CSF* and *GM-CSF* therapy in numerous short-term and several long-term trials. A recent multi-institutional Phase III randomized trial of G-CSF vs. observation in RAEB/RAEB-T subtypes of MDS demonstrated that G-CSF caused no alteration in the incidence or rate of evolution to AML. Survival in RAEB-T patients was the same in both groups; however, RAEB patients receiving G-CSF had decreased survival. This effect was apparent only in the increased number of "high-risk" patients in the G-CSF group. Several trials utilizing the combination of GM-CSF plus low dose AraC have demonstrated either good responses, stable disease or disease progression in approximately equal thirds of the patients, with a high degree of treatment-related toxicity. No increased survival was shown compared to prior studies using either agent alone. These data indicate that other approaches are needed to attempt to alter the natural history of MDS. Cytopenias have been modestly improved with *IL-3* as have thrombocytopenias in a minor proportion of patients with *IL-6* treatment, although substantial toxicity was noted at high doses. Erythropoietin (Epo) in relatively high doses has improved the anemias in only approximately 20-25% of MDS patients. Epo combined with G-CSF has demonstrated erythropoietic synergistic activity and has led to an approximate 40-45% erythroid response rate in MDS. The biologic effects of the hemopoietic cytokines provide a basis for planning current clinical trials in this problematic disease.

This paper will focus on recent advances in attempts to modulate the cytopenias and pathogenetic mechanisms of myelodysplastic syndromes (MDS) by treatment with hemopoietic cytokines. Therapeutic results will be placed in the context of newer clinical and biological approaches for stratifying MDS.

Molecular Biology of Hematopoiesis 5, edited by Abraham et al.
Plenum Press, New York, 1996

CLINICAL AND BIOLOGIC FEATURES OF MDS

Stratification of MDS Subgroups

The myelodysplastic syndromes (MDS) describe patients with clonal refractory cytopenias whose marrows show characteristic dysplastic changes in at least two of the three hemopoietic cell lines and have a propensity to undergo transformation into acute myeloid leukemia (AML). The French American and British (FAB) morphologic classification[1] has been a useful risk stratification scheme which permitted consistency in case reporting and described prognostic findings (reviewed in References 2 and 3). MDS patients having refractory anemia with excess of blasts (RAEB, 5-20% marrow blasts) and those with RAEB in transformation (RAEB-T, 20-30% marrow blasts) generally had relatively poor prognoses, with median survivals generally ranging from 5-12 months. In contrast, refractory anemia (RA, < 5% blasts) or RA with ringed sideroblasts (RARS, < 5% blasts) MDS subtypes had median survivals of approximately 3-6 years. The proportion of these individuals who transformed to AML varied similarly, as in the **high-risk** RAEB/RAEB-T patients this incidence was 40-50%, compared to 5-15% in the **low-risk** group RA/RARS group. These studies also indicated that patients could be further subdivided by considering other clinical prognostic features such as age, platelet count and percent marrow blasts.[3]

In addition to these clinical findings, biologic features such as cytogenetic abnormalities, particularly if they are complex, and defective in vitro culture clonogenicity have also defined patients with poorer prognoses.[4,5] These biologic abnormalities generally correlate with the more advanced FAB classifications. Several major studies have recently demonstrated the prognostic import of the well defined non-random marrow cytogenetic anomalies in MDS.[6,7] Due to the heterogeneity of prognoses in MDS subgroups, these patients need to be appropriately stratified when considering approaches to treatment and analysis of therapeutic outcomes.

TREATMENT WITH HEMOPOIETIC CYTOKINES

Defective proliferation of hemopoietic precursors within MDS marrow is related to decreased responsiveness to and/or decreased production of hemopoietic cytokines/growth factors (HGFs). However, as some leukemic cells have enhanced proliferative responses to the colony-stimulating factors (CSFs) in vitro[8-10] concern exists regarding the safety of using such agents in responsive neoplastic cells. Further, intrinsic abnormalities of responsiveness occur in the myeloid precursors of these patients.

A variety of treatment approaches have been used in MDS due to the elderly nature of these patients and their relatively poor responses to standard chemotherapy. Currently, transfusion and antibiotic supportive care are the mainstay of therapy. To attempt to modify the cytopenias in these disorders, and their consequent morbidities, HGFs have been utilized in MDS. The effects of such cytokine therapy will be reviewed.

Treatment with GM-CSF

Following the initial Phase I-II trials which showed the efficacy of GM-CSF for short-term improvement of neutrophil counts in the majority of MDS patients,[11-16] a multicenter Phase III study[17] provided information regarding the use of GM-CSF in relatively low risk MDS patients (i.e., RA, RAEB patients with < 10% marrow blasts), for longer periods (i.e., periods up to two months). Eighty-two patients (50 RA, 32 RAEB) received either of two different fixed daily dose levels of GM-CSF. As indicated above, these low risk MDS patients untreated would have been expected to generally have relatively good prognoses, with median survivals of several years and a low incidence of evolution to AML. Nearly all of the MDS patients treated with GM-CSF responded with increased neutrophil counts. However, only 35% of the patients completed eight weeks of treatment and the drug was discontinued in the others due to progression of disease, local infiltrates, flu-like syndromes, hyperleucocytosis or bone pain. In 25% of the patients, platelet counts decreased during GM-CSF administration to less than 50% of baseline values, whereas two patients had increases in platelets. Six patients had progressive disease, two of whom with RAEB developed acute leukemia; erythroid responses did not occur. No differences in responses were demonstrated between the two dose levels of GM-CSF used. The impact of GM-CSF on progression of disease could not be addressed because of the small number of patients and short duration of treatment. A preliminary report has indicated similar results of another multi-institutional randomized trial of GM-CSF treatment for period generally up to 6 months vs. observation in 21 patients with MDS, with crossover occurring in patients with infections.[18] A decrease in infections in the GM-CSF group was reported. These studies demonstrated some of the relative tradeoffs of potential difficulties and benefits of chronic administration of a predominantly proliferative myeloid growth factor, such as GM-CSF, in patients with MDS.

In two recent studies, low and high risk subtypes of MDS were received **GM-CSF plus low dose Ara-C** for two weeks at different schedules for several months.[19,20] Evaluation of responses indicated that approximately equal thirds of the patients demonstrated either clinical improvement, stable disease or progressive disease/toxic deaths. Adverse events were noted in over half of the patients, including major hemorrhage and infections, often leading to discontinuation of treatment, with most of the adverse events being due to the low dose Ara C. The median survival of these individuals was similar to that of prior controlled studies using low dose Ara C alone.[21]

Treatment with G-CSF

Following initial Phase I-II trials which demonstrated a high degree of efficacy for improving neutrophil counts in the vast majority of MDS patients with G-CSF therapy,[22-24] a **Phase III international randomized trial** for 102 patients with high risk MDS (RAEB/RAEB-T), comparing long-term G-CSF administration to observation, was performed to attempt to determine the impact of G-CSF on the natural history of the disease.[25] The G-CSF treatment, which improved neutrophil counts in nearly all patients, was generally well tolerated, with few patients withdrawing from study due to adverse events. No difference was noted in the incidence of or time to progression to AML for RAEB or RAEB-T patients who were in the G-CSF or observation arms of the study. Survival for RAEB-T patients was similar in both groups. However, for RAEB patients, the median survival was shorter in patients receiving G-CSF (10.4 months vs. 21.4 mo, p=0.02), with an increase in disease-related non-leukemic deaths. The RAEB patients receiving G-CSF demonstrated median survival time similar to previously reported RAEB survival data in the literature, while the RAEB patients in the observation group had prolonged survival. Although balanced for most clinical parameters, an increased proportion of RAEB patients receiving G-CSF (29% vs. 14%) were in the high prognostic risk category, based on the scoring system utilizing the proportion of marrow blasts, platelets and age.[3] Decreased survival in RAEB patients receiving G-CSF was evident only in this high-risk group, compared with that for the high-risk RAEB patients in the observation group (p=0.006). The difference of survival in RAEB patients may be related to the increased number of high-risk patients included in the G-CSF group or the unusually long survival of high-risk patients in the observation group. These disparate therapeutic responses indicate that major biologic differences in MDS are defined by these clinical risk features, stressing the importance of these parameters in the design and analysis of clinical trials. The impact of G-CSF treatment on the incidence of infections is being evaluated in these patients.

Treatment with Interleukin-3

Relatively short-term clinical trials have reported the effects of interleukin-3 (IL-3) therapy in low risk MDS patients.[26,27,28] These studies indicated modest improvements in neutrophils which, however, were not as prominent as those demonstrated with G-CSF or GM-CSF. Further, only limited responses occurred in the other cell lines. These data indicate that IL-3 will need to be combined with other hemopoietic cytokines to achieve substantial improvement in the cytopenias in MDS.

Treatment with Interleukin-6

To attempt to improve circulating platelet levels in low risk MDS, 22 such patients were treated with IL-6.[29] Platelet count responses occurred in 8 patients. However, these responses were noted in only 3 of 11 patients (27%) with <20,000 platelets/mm.[3] Moderate to severe toxicity with constitutional symptoms occurred without leukocyte improvement, and worsening anemia developed in a substantial portion of these patients. Due to the adverse events, very few patients could continue this treatment for several months.

Treatment with Erythropoietin

Serum erythropoietin (Epo) levels may be suboptimally elevated in MDS patients relative to their degree of anemia.[30] Thus, recombinant human Epo therapy has been instituted to attempt to correct the hypoproductive anemias. Numerous reports have detailed the erythroid responses of MDS patients to this form of treatment (reviewed in Reference 2). The initial studies utilizing Epo in MDS[31-37] indicated that 14 of 75 (19%) patients responded to Epo. Generally, the patients required relatively high doses of Epo (> 200 U/kg/day) for their responses. This limited in vivo responsiveness of MDS marrow cells to Epo is not totally unexpected as the defective erythroid precursors in MDS have demonstrated suboptimal in vitro responses to Epo alone, particularly for BFU-E growth.[4,38]

Hemopoietic cytokines such as G-CSF are synergistic with Epo, enhancing marrow BFU-E numbers or responsiveness to Epo in vitro in MDS,[39] suggesting their potential to provide more prominent in vivo erythroid responses in combination than either agent alone. Two studies describing effects of such **combination therapy with G-CSF and Epo** to treat the anemia of MDS[40,41,42] have substantiated this thesis. Approximately 40-45% of patients receiving this combination treatment had substantial erythroid responses (i.e., decreased transfusion requirements and increased hemoglobin levels), and all patients had neutrophil responses. Responses were more frequent in patients with less advanced pancytopenia, lower endogenous Epo levels and in those with marrow ringed sideroblasts. Patients with ringed sideroblasts, who respond poorly to Epo alone, showed a response rate of 60%. Recent extension of these trials indicated that erythroid responses persisted for many months in the majority of patients receiving both factors.[43] Upon discontinuing G-CSF, approximately half maintained their responses, whereas the remainder required both factors, consistent with the synergistic effects of these two agents for a portion of MDS patients. Studies are ongoing to attempt to further define features predictive of these clinical responses.

EVIDENCE OF CLONAL RESPONSES

Cytogenetic evaluations and investigations analyzing restriction fragment length polymorphisms (RFLP) of X-linked genes were performed in several of these studies[2,23,44-46] to determine whether selective responses to CSFs of normal vs. abnormal clones occurred in MDS. The generally persisting cytogenetic abnormalities and clonal RFLP studies after treatment with these CSFs suggested induced differentiation of the abnormal clone in MDS. Most evidence has indicated, with one exception in which polyclonal hemopoiesis developed after GM-CSF treatment,[46] that GM-CSF or G-CSF did not preferentially stimulate normal marrow stem cells to proliferate nor had the ability to eradicate the cytogenetically abnormal clone by inducing terminal differentiation.[47]

DIRECTIONS

These studies analyzing effects of G-CSF, GM-CSF and IL-3 in MDS indicate that although improved neutrophil counts occur with these interventions, other approaches are required to attempt to modulate the natural history of this problematic disorder. Further investigations are needed to better select anemic MDS patients likely to benefit from Epo alone or in combination with other cytokines. Also, other cytokines (such as the recently described thrombopoietin),[48] singly or in combination, are still required to augment platelet levels in order to modify this major cause of morbidity in MDS. Stratification of MDS patients into prognostic subgroups will be important in order to appropriately design and analyze therapeutic trials. "High intensity" (chemotherapy, BMT) vs. "low intensity" (cytokines, differentiation-inducing agents) therapies require evaluation in coordinated comparative trials to determine the potential of these treatments for altering disease natural history.

REFERENCES

1. Bennett JM, Catovsky D, Daniel MT, et al: FAB Cooperative Group: Proposal for the classification of the myelodysplastic syndromes. Br J Haematol 1982; 51:189-199.
2. Greenberg, PL: Treatment of MDS with hemopoietic growth factors. Semin Oncol 1992; 19:106-114.
3. Sanz GF, Sanz MA, Vallespi T, et al: Two regression models and a scoring system for predicting survival and planning treatment in myelodysplastic syndromes: A multivariate analysis of prognostic factors in 370 patients. Blood 1989; 74:395-408.
4. Greenberg P: In vitro hemopoietic cell culture studies in MDS. Sem Oncol 1992; 19:34-46.
5. Pierre R, Catovsky D, Mufti G, et al: Clinical cytogenetic correlations in myelodysplasia (preleukemia). Cancer Genet Cytogeniet 1990; 44:15-26.
6. Morel P, Hebbar M, Lai JL, et al: Cytogenetic analysis has strong independent prognostic value in *de novo* MDS and can be incorporated in a new scoring system: A report on 408 cases. Leukemia 1993; 7:1315.
7. Toyama K, Ohyashiki K, Yoshida Y, et al: Clinical implications of chromosomal abnormalities in 401 patients with MDS: A multicentric study in Japan. Leukemia 1993; 7:499-508.
8. Metcalf D: The molecular biology and functions of the granulocyte-macrophage colony-stimulating factors. Blood 1986; 67:257-267.
9. Miyauchi J, Kelleher CA, Yang YC, et al: The effects of three recombinant growth factors, IL3, GM-CSF and G-CSF, on the blast cells of acute myeloblastic leukemia maintained in short-term suspension culture. Blood 1987; 76:657-663.
10. Vellenga E, Young DC, Wagner K, et al: The effects of GM-CSF and G-CSF in promoting growth of clonogenic cells in acute myeloblastic leukemia. Blood 1987; 69:1771-1776.

11. Vadhan-Raj S, Keating M, LeMaistre A, et al: Effects of recombinant human granulocyte-macrophage colony-stimulating factor in patients with myelodysplastic syndromes. N Eng J Med 1987; 317:1545-1552,
12. Antin JH, Smith BR, Holmes W, et al: Phase I/II study of recombinant granulocyte-macrophage colony-stimulating factor in aplastic anemia and myelodysplastic syndrome. Blood 1988; 72:705-713.
13. Ganser A, Volkers B, Greher J, et al: Recombinant human GM-CSF in patients with MDS- A Phase I/II trial. Blood 1989; 73:31-37.
14. Herrmann F, Lindemann A, Klein H, et al: Effect of recombinant GM-CSF in patients with myelodysplastic syndrome with excess blasts. Leukemia 1989; 3:335-338.
15. Thompson JA, Lee DJ, Kidd P, et al: Subcutaneous granulocyte-macrophage colony-stimulating factor in patients with MDS: Toxicity, pharmacokinetics, and hematological effects. J Clin Oncol 1989; 7:629-637.
16. Gradishar W, Le Beau MM, O'Laughlin R, et al: Clinical and cytogenetic responses to GM-CSF in therapy-related myelodysplastic syndrome. Blood 1992; 80:2463-2470.
17. Willemze R, van der Lely N, Zwierzina H, et al: A randomized phase I/II multicenter study of recombinant human GM-CSF therapy for patients with myelodysplastic syndromes and a relatively low risk of acute leukemia. Ann Hematol 1992; 64:173-180.
18. Schuster MW, Thompson JA, Larson R, et al: Randomized trial of subcutaneous GM-CSF versus observation in patients with myelodysplastic syndrome or aplastic anemia. Proc Am Soc Clin Oncol 1990, p. A793.
19. Economopoulos T, Papageorgiou E, Stathakis N, et al: Treatment of myelodysplastic syndromes with human granulocyte-macrophage colony-stimulating factor (GM-CSF) or GM-CSF combined with low-dose cytosine arabinoside. Eur J Haematol 1992; 49:138-142.
20. Gerhartz HH, Marcus R, Delmer A, et al: A randomised Phase II study of low-dose cytosine arabinoside plus GM-CSF in MDS with a high risk of developing leukemia. Leukemia 1994; 8:16-23.
21. Miller KB, Kim K, Morrison FS, et al: Evaluation of low dose Ara C vs. supportive care in the treatment of MDS: An intergroup study by the ECOG and SWOG. Blood 1988; 72(suppl 1):215A.
22. Negrin RS, Haeuber DH, Nagler A, et al: Treatment of myelodysplastic syndromes with recombinant human granulocyte colony stimulating factor. Annals Internal Med 1992; 110:976-984.
23. Negrin RS, Nagler A, Kobayashi Y, et al: Maintenance treatment of patients with myelodysplastic syndromes using recombinant human granulocyte colony stimulating factor. Blood 1992; 78:36-43.
24. Yoshida Y, Hirashima K, Asano S, et al: A Phase II trial of recombinant human granulocyte colony-stimulating factor in the myelodysplastic syndromes. Br J Haematol 1991;78:378-384.
25. Greenberg P, Taylor K, Larson R, Koeffler P, Negrin R, et al: Phase III randomized multicenter trial of G-CSF vs. observation for MDS. Blood 1993; 82(suppl 1):196a.
26. Ganser A, Seipelt G, Lindemann A, et al: Effects of recombinant human interleukin-3 in patients with myelodysplastic syndromes. Blood 1990; 6:455
27. Kurzrock R, Talpaz M, Estrov Z, et al: Phase I study of recombinant human interleukin-3 in patients with bone marrow failure. J Clin Oncol 1991; 9:1241-1250.
28. Ganser A, Ottmann OG, Seipelt G, et al: Effect of long-term treatment with recombinant human interleukin-3 in patients with myelodysplastic syndromes. Leukemia 1993; 7:696-701.
29. Gordon MS, Nemunaitis J, Hoffman R, et al: A Phase I trial of recombinant human interleukin-6 in patients with myelodysplastic syndromes and thrombocytopenia. Blood 1995; 85:3066-3076.
30. Jacobs A, Janowska-Wieczorek A, Caro J, et al: Circulating erythropoietin in patients with MDS. Br J Haematol 1989; 73:36-39.
31. Bessho M, Jinnai I, Matsuda A: Improvement of anemia by recombinant erythropoietin in patients with myelodysplastic syndromes and aplastic anemia. Int J Cell Cloning 1990; 8:445-458.
32. Stebler C, Tichelli A, Dazzi H, et al: High-dose recombinant human erythropoietin for treatment of anemia in MDS and paroxysmal nocturnal hemoglobinuria: A pilot study. Exp Hematol 1990; 18:1204-1208.
33. Stein R, Abels R, Krantz S: Pharmacologic doses of recombinant human erythropoietin in the treatment of myelodysplastic syndromes. Blood 1991; 78:1658-1665.
34. van Kamp H, Prinsze-Postema T, Kluin PM, et al: Effect of subcutaneously administered human recombinant erythropoietin in patients with myelodysplasia. Brit J Haematol 1991; 78:488.
35. Hellstrom E, Birgegård G, Lockner D, et al: Treatment of myelodysplastic syndromes with recombinant human erythropoietin. Eur J Haematol 1991; 47:355-360.
36. Schouten HC, Vellenga E, van Rhenen D, et al: Recombinant human erythropoietin for patients with MDS. Blood 76:317, 1990
37. Bowen D, Culligan D, Jacobs AJ: The treatment of anaemia in the MDS with recombinant human erythropoietin. Brit J Haematol 1991; 77:419.

38. Merchav S, Nielsen OJ, Rosenbaum H, et al: In vitro studies of erythropoietin-dependent regulation of erythropoiesis in myelodysplastic syndromes. Leukemia 1990; 4:771-774.
39. Greenberg, PL, Negrin R, Ginzton N: G-CSF synergizes with erythropoietin for enhancing erythroid colony formation in myelodysplastic syndromes. Blood 1991; 78:38a.
40. Negrin RS, Stein R, Doherty K, et al: Treatment of the anemias of MDS using recombinant human granulocyte colony-stimulating factor in combination with erythropoietin. Blood 1993; 82:737-743.
41. Hellstrom-Lindberg E, Birgegard G, Carlsson M, et al: A combination of G-CSF and erythropoietin may synergistically improve the anaemia in patients with MDS. Leuk Lymph 1993; 11:221-228.
42. Hellstrom-Lindberg E, Negrin R, Stein R, et al: Prediction of response to G-CSF and Epo treatment for the anemia of myelodysplastic syndromes. Leuk Res 1994; 18 (Suppl):53a.
43. Negrin RS, Stein R, Doherty K, Cromwell J, Vardiman J, Krantz S, Greenberg P: G-CSF plus erythropoietin for the maintenance treatment of the anemia of myelodysplastic syndromes. Blood 1996; 87:4076-4081.
44. Janssen JWG, Buschle M, Layton M, et al: Clonal analysis of MDS: Evidence of multipotent stem cell origin. Blood 1989; 73:248-254.
45. Gilliland DG, Blachard KL, Levy J, Perrin S, Bunn H: Clonality in myeloproliferative disorders: Analysis by means of the polymerase chain reaction. PNAS 1991; 88:6848.
46. Vadhan-Raj S, Broxmeyer HE, Spitzer G, et al: Stimulation of nonclonal hematopoiesis and suppression of the neoplastic clone after treatment with recombinant human GM-CSF in a patient with therapy-related myelodysplastic syndrome. Blood 1989; 74:1491-1498.
47. Kohler S, Busque L, DeHart D, Le Beau M, Negrin R, Ganser A, Greenberg P, Gilliland DG: Clonality analysis in myelodysplasia patients treated with G-CSF. Blood 1993; 82(suppl 1):376a.
48. Kaushansky K, Lok L, Holly R, et al: Promotion of megakaryocyte progenitor expansion and differentiation by the c-mpl ligand thrombopoietin. Nature 1994; 369:568.

STEM CELL FACTOR AS A SURVIVAL AND GROWTH FACTOR IN HUMAN NORMAL AND MALIGNANT HEMATOPOIESIS

H. T. Hassan* and A. R. Zander

Bone Marrow Transplantation Center
Department of Hematology and Oncology
Hamburg University Hospital Eppendorf
Martinistraße 52, 20246 Hamburg, Germany

ABSTRACT

Stem cell factor (SCF) is an essential hematopoietic cytokine that interacts with other cytokines to preserve the viability of hematopoietic stem and progenitor cells, to influence their entry into cell cycle and to facilitate their proliferation and differentiation. SCF on its own is unable to drive noncycling hematopoietic progenitor cells into cell cycle but does prevent their apoptotic death. SCF when combined with other cytokine(s) increases the cloning efficacy of hematopoietic progenitor cells from all lineages. Also, SCF stimulates the growth of CD34 positive leukemic progenitor cells from most patients with acute myeloid leukemia (AML). The mRNA expression of the SCF receptor: *c-kit* has been shown to be significantly increased in all fresh AML blast cells as compared with normal controls (healthy volunteers), particularly in CD34 positive cells. Only two inhibitory cytokines: TGF-ßeta and IL-4 could decrease *c-kit* expression, whereas TNF-alpha increased *c-kit* expression in AML cells. However, none of the differentiation inducing agents or chemotherapeutic drugs tested showed any effect on *c-kit* mRNA expression.

Apoptosis has been shown to be directly related to high complete remission rate in AML patients following induction therapy. Since, SCF has been shown to stimulate the proliferation of mainly CD34 positive AML cells we have investigated whether the poor response of patients with CD34 positive myeloid leukemia cells to chemotherapy could be due to the resistance to apoptosis by the SCF. The effect of SCF on the apoptosis induced by each of three commonly used chemotherapeutic drugs in the treatment of AML:

* Correspondence: Dr. H. T. Hassan, MD, PhD, Bone Marrow Transplantation Center, Pavillon 23, Hamburg University Hospital Eppendorf, Martinistraße 52, D-20246 Hamburg, Germany. Tel. no.: 49 40 47176421, Fax no.: 49 40 47174871.

Molecular Biology of Hematopoiesis 5, edited by Abraham et al.
Plenum Press, New York, 1996

Cytarabine, Daunorubicin and Carboplatin was examined in the human CD34 positive myeloid leukemia cells in serum free cultures.

SCF significantly reduced the induced-apoptosis by more than 50% in all CD34 positive human leukemia cells treated by any of the three chemotherapeutic drugs. Using antibodies blocking *c-kit* reversed the significant inhibitory effect of SCF on chemotherapy-induced apoptosis, confirming the role of SCF in the resistance to chemotherapy-induced apoptosis in CD34 positive human leukemia.The present results suggest that the poor response of patients with CD34 positive leukemia cells could be at least partially due to less chemotherapy-induced apoptosis as a result of protection by SCF as an adjuvant mechanism for drug resistance in myeloid leukemia.

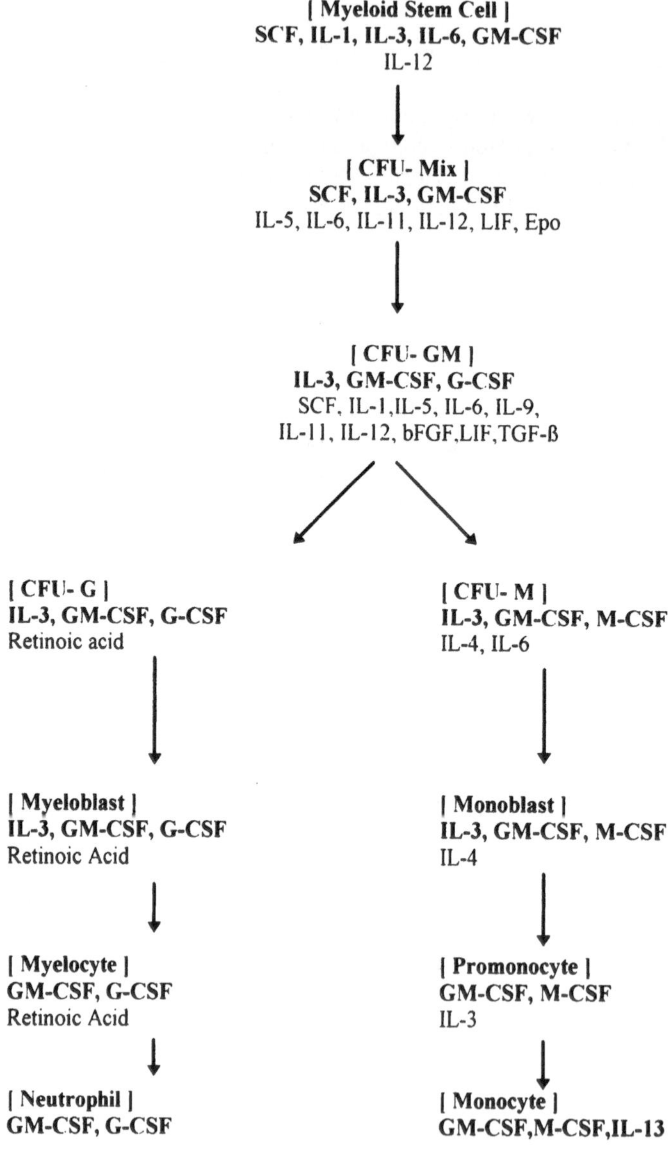

Figure 1. Cytokine regulation of normal human myelopoiesis.

We conclude that use of an antisense strategy to block *c-kit* expression in AML blast cells may be of value in decreasing the chemoresistance of AML patients. The abrogation of leukemic resistance to apoptotic death through anti-SCF/ *c-kit* expression combined with chemotherapy offers potential in the design of novel therapeutic approaches for refractory AML patients.

Human stem cell factor (SCF) is an early acting hematopoietic growth factor of 248 amino acids produced by marrow stromal cells with a growth-promoting activities on all hematopoietic cell lineages (figures 1-4). The SCF gene has a length of 70 kb, contains 21 exons and is located on the long arm of human chromosome 12q22-24[1] and its receptor, *c-kit*, is expressed in 76% of CD34 positive and all CD34+HLA-DR- marrow cells[2] and belongs

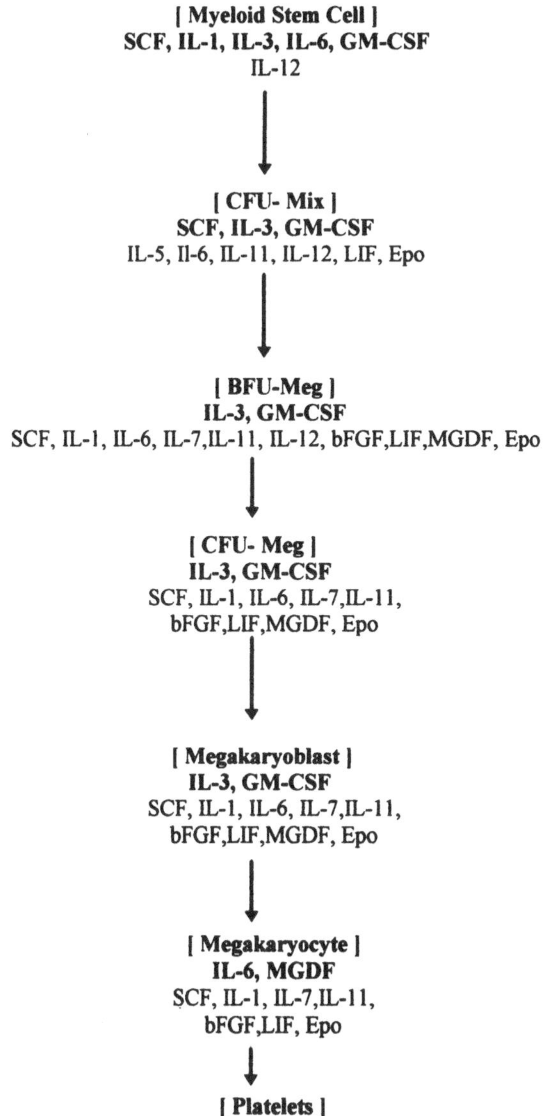

Figure 2. Cytokine regulation of normal human megakaryocytopoiesis.

to a family of receptors with tyrosine kinase activity including *c-fms*(M-CSF receptor), Flk-2/Flt3, Stk-1 and PDGF-receptor.[3, 4, 5]

Soluble SCF is normally detectable in human serum (mean 3.3 ng/ml, range 1.3 - 8.0),[6] and significantly reduced in serum from patients with myelodysplastic syndromes,[7] severe aplastic anemia[8] and Fanconi's anemia.[9]

Stem Cell Factor and Human Normal Hematopoiesis

SCF alone does not stimulate myelopoiesis but significantly enhances the GM-CSF-, G-CSF-, IL-3, IL-3/GM-CSF-fusion protein (PIXY321)- or IL-6-stimulated myelopoiesis in both short[10,11,12] and long[13,14] term bone marrow cultures (figure 1). Also, SCF expands the GM-CSF-, G-CSF-, IL-3-, PIXY321- or IL-6-stimulated myelopoiesis from both peripheral blood[15] and umbilical cord blood[16] CD34 positive cells. Moreover, SCF enhances IL-5, IL-3 and GM-CSF-stimulated eosinopoiesis from normal human bone marrow.[17]

SCF alone niether stimulate megakaryocytopoiesis nor alter megakaryocyte markers nor increase the cell size, ploidy and DNA content of magakaryocytes but significantly increases the number and size of GM-CSF-, IL-3 or IL-6-stimulated human megakaryocytic colonies[18, 19, 20] (figure 2).

SCF alone, in the presence of Erythropoietin (Epo), stimulates the formation of macroscopic human erythroid burst colonies (BFU-E) of more than 1mm from human bone marrow cells[10] (figure 3).

Also, SCF enhances IL-7- and IL-2-stimulated proliferation of early B-lymphopoiesis[21] (figure 4).

Not only SCF acts as a survival factor in vitro for both myeloid[22] and erythroid[23] progenitor cells by reducing their apoptosis (programmed cell death), but also in vivo treatment with SCF expanded human marrow haemopoietic progenitor and stem cells in cancer patients.[24]

Stem Cell Factor and Human Malignant Hematopoiesis

SCF significantly enhances the eyrthropoietin-stimulated erythroid burst colonies (BFU-E) from patients with myelodysplastic syndromes,[25] severe aplastic anemia,[26,27] Fanconi's anemia,[28] Diamond- black anemia,[29] and Bone marrow failure syndrome.[30] The potent effect of SCF on erythropoiesis in these malignant diseases is encouraging for future clinical applications in patients, particularly after selection by in vitro tests before starting therapy.

Also, SCF stimulates the growth of CD34 positive leukemic progenitor cells from most patients with myelodysplastic syndromes,[25] and acute myeloid leukemia (AML)[31,32] but not from patients with chronic myeloid leukemia.[33]

The mRNA expression of *c-kit* has been shown to be significantly increased in all fresh AML blast cells as compared with normal controls (healthy volunteers), in particular CD34 positive cells.[34] Also, in all myeloproliferative and myelodysplastic syndromes, *c-kit* mRNA expression is significantly increased, being highest in the CD34 positive cells during blastic transformation of CML and MDS.[35] A study of 164 human leukemia-lymphoma cell lines showed the *c-kit* to be expressed in all erythroid and megakaryoblastic cell lines, most of myeloblastic cell lines but not in any of the monocytic or lymphoid cell lines.[36] In human myeloid leukemia cells, only two inhibitory cytokines: transforming growth factor-ßeta[37] and interleukin-4 as well as phorbol-myristate-acetate[38] could decrease *c-kit* mRNA expression and almost completely abolished the proliferative stimulatory effect of SCF in CD34 positive cells (Table 1), whereas tumor necrosis factor-alpha increased *c-kit* mRNA expression in fresh AML[34] and GM/SO myeloid leukemia[39] cells. However, none of the differen-

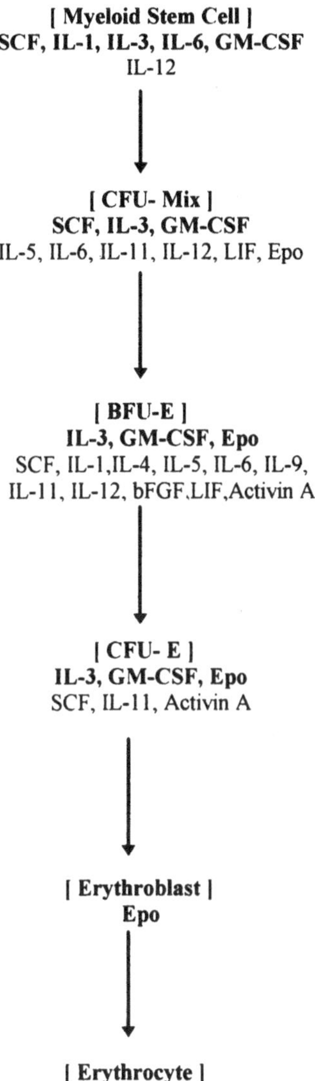

Figure 3. Cytokine regulation of normal human erythropoiesis.

tiation inducing agents or chemotherapeutic drugs tested showed any effect on *c-kit* mRNA expression (Table 1).

Apoptosis and Its Positive and Negative Triggers in Myeloid Leukemia

Several chemotherapeutic drugs that are often used in the chemotherapy of cancer patients, have been shown to induce apoptosis (programmed cell death) in both myeloid and lymphoid leukemia cells.[40] Apoptosis, determined by the presence of in situ end labeling (ISEL) of fragmented DNA positive cells, has been shown to be directly related to high complete remission rate in AML patients following induction therapy.[41] In murine myeloid leukemia cells, several hemopoietic cytokines including interleukin-1, interleukin-6, and

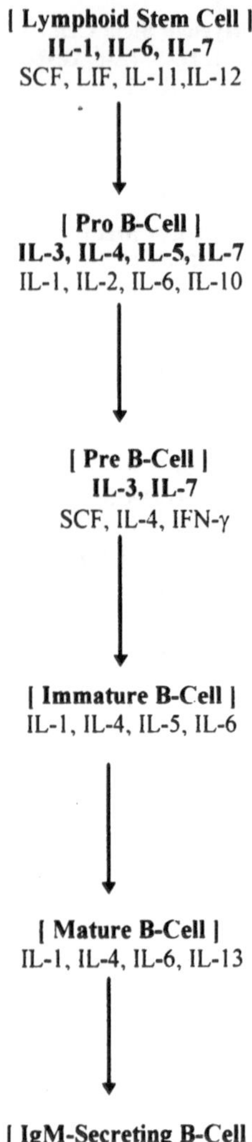

Figure 4. Cytokine regulation of normal human b-lymphopoiesis.

G-CSF could inhibit apoptosis induced by transforming growth factor-ßeta and several chemotherapeutic drugs.[42] Since, SCF has been shown to stimulate the proliferation of mainly CD34 positive AML cells[43] we have investigated whether the poor response of patients with CD34 positive myeloid leukemia cells to chemotherapy could be due to the resistance to apoptosis by the stem cell factor.

The effect of SCF on the apoptosis induced by each of three commonly used chemotherapeutic drugs in the treatment of AML: Cytarabine, Daunorubicin and Carboplatin was examined in the human CD34 positive myeloid leukemia cells in serum free cultures. The CD34 positive myeloid leukemia cells were cultured at a concentration of 100 thousand cells per ml. at day 0 in 24-well tissue culture plates at 37°C in humidified incubator for 96 hours under CO_2. SCF was added at a concentration of 200 ng/ml, which has been found to be the optimum concentration for the proliferation of CD34 positive myeloid leukemia cells. Apoptosis was determined after 48, 72 and 96 hours by both flow cytometry and immuno-

Table 1. Effect of cytokines, chemotherapeutic drugs and differentiation inducing agents on *c-kit* mRNA expression in human myeloid leukemia cells

Agent	Effect on c-kit Expression	Human Myeloid Leukemia Cells
Human Cytokines		
Interleukin-4	Decrease	MHH225 and fresh AML
TGF-beta	Decrease	MHH225 and fresh AML
Interleukin-6	No Effect	MHH225
Interferon-gamma	No Effect	HEL, M07e
Erythropoietin	No Effect	MHH225
Tumor Necrosis Factor	Increase	GM/SO and fresh AML
Differentiation Inducers		
Butyric Acid	No Effect	HEL
Retinoic Acid	No Effect	HEL
Vitamin D3	No Effect	HEL
Phorbol-myristate-acetate	Decrease	HEL, M07e, fresh AML
Chemotherapeutic Drugs		
Carboplatin	No Effect	MHH225
Cytarabine	No Effect	MHH225
Daunorubicin	No Effect	MHH225

HEL: megakaryoblastic leukemia, MHH225: CD34 positive leukemia, M07e: megakaryoblastic leukemia, TGF: transforming growth factor.

cytochemistry of DNA cell content and APO-1 (CD95) antigen in chemotherapy-treated leukemia cells as well as the total viable cell count and cell morphology in May-Grünwald-Giesma stained smears by scoring apoptotic cells that were small with a fragmented nucleus, condensed chromatin and formation of apoptotic bodies in both the presence and absence of SCF.

Table 2 shows the significant inhibitory effect on chemotherapy-induced apoptosis in the presence of SCF. SCF significantly reduced the induced-apoptosis by more than 50% in all CD34 positive human leukemia cells treated by any of the three chemotherapeutic drugs (Table 2). Using antibodies blocking *c-kit* reversed the significant inhibitory effect of SCF on chemotherapy-induced apoptosis, confirming the role of SCF in the resistance to chemotherapy-induced apoptosis in CD34 positive human leukemia.The present results suggest that the poor response of patients with CD34 positive leukemia cells could be at least partially due to less chemotherapy-induced apoptosis as a result of protection by SCF as an adjuvant mechanism for drug resistance in myeloid leukemia.

We conclude that use of an antisense strategy to block *c-kit* expression in AML blast cells may be of value in decreasing the chemoresistance of AML patients. The abrogation of leukemic resistance to apoptotic cell death through anti-SCF/ *c-kit* expression combined with cytotoxic chemotherapy offers potential in the design of novel therapeutic approaches for refractory and/or relapsed AML patients.

Table 2. Effect of stem cell factor on chemotherapy-induced apoptosis in CD34 positive human myeloid leukemia cells

Drug	Percentage of Apoptosis in Human CD34 positive myeloid leukemia		
(μg/ml)	Without SCF	With SCF	P
Cytarabine (0.5)	79.1 ± 6.0	37.4 ± 5.1	< 0.01
Daunorubicin (0.05)	86.2 ± 8.1	42.8 ± 4.6	< 0.01
Carboplatin (0.05)	93.1 ± 5.7	38.0 ± 3.2	< 0.01

Stem Cell Factor and Mobilisation of Peripheral Blood Stem and Progenitor Cells

SCF has been shown to enhance the granulocyte colony stimulating factor (G-CSF) mobilisation of peripheral blood stem and progenitor cells in several animal models.[44-48] The biological effects of SCF administration in humans are similar to those observed in other species with the main toxicities of local and distant allergic reactions due to its effect on mast cell activation which can be minimised by antihistamine prophylaxis.[49] Recently, the first results of on-going clinical studies have demonstrated that the addition of SCF to G-CSF resulted in 2-3 fold greater mobilisation as well as more rapid hemopoietic recovery after high dose chemotherapy in cancer patients.[50] Further clinical and in vitro studies are warranted to determine the therapeutic role of SCF in the mobilisation and ex vivo expansion of peripheral blood stem and progenitor cells for transplantation after high dose chemotherapy in cancer patients.

REFERENCES

1. Geissler EN, Liao M, Brook JD, Martin FH, Zsebo KM, Housman DE, Galli SJ: Stem cell factor, a novel hematopoietic growth factor and ligand for c-kit tyrosine kinase receptor, maps on human chromosome 12 between 12q14.3 and 12qter. Somat Cell Mol Genetic 17:207,1991.
2. Simmons PJ, Aylett GW, Niutta S, To LB, Juttner CA, Ashman LK: c-kit is expressed by primitive human hematopoietic cells that give rise to colony-forming cells in stroma-dependent or cytokine-supplemented culture. Exp Hematol 22:157,1994.
3. Qui F, Ray P, Brown K, Barker PE, Jhawer S, Ruddle FH, Besmer P: Primary structure of c-kit: relationship with the CSF-1/PDGF receptor kinase family: oncogenic activation of c-kit involves deletion of extracellular domain and C-terminus. EMBO J 7:1003,1988.
4. Lyman SD, James L, Vandenbos T, Dervies P, Brasel K, Gliniak B, Hollingsworth LT, Picha KS, McKenna HJ, Splett RR: Molecular cloning of a ligand for the Flt3/Flk2 tyrosine kinase receptor: a proliferative factor for primitive hematopoietic cells. Cell 75:1157,1993.
5. Small D, Levenstein M, Kim E, Carow CE, Amin S, Rockwell P, Witte L, Burrow C, Ratajczak M, Gewirtz AM, Civin CI: STK-1, the human homolog of Flk2/Flt3 is selectively expressed in CD34+ human bone marrow cells and is involved in the proliferation of early progenitor/stem cells. Proc Natl Acad Sci USA 91:459,1994.
6. Langley KE, Bennet LG, Wypych J, Yancik SA, Liu XD, Westcott KR, Chang DG, Smith KA, Zsebo KM: Soluble stem cell factor in human serum. Blood 81:656,1993.
7. Bowen D, Yancik S, Bennett L, Culligan D, Resser K: Serum stem cell factor concentration in patients with myelodysplastic syndromes. Br J Haematol 85: 63-66.
8. Wodnar-Filipowicz A, Yancik S, Moser Y, dalle-Carbonare V, Gratwohl A, Tichelli A, Speck B, Nissen C: Levels of soluble stem cell factor in serum of patients with aplastic anemia. Blood 81: 3259,1993.
9. Wunder E, Mortensen BT, Schilling F, Henon PR: Anomalous plasma concentrations and impaired secretion of growth factors in Fanconi's anemia. Stem Cells Dayt 11:suppl.2,144.
10. McNiece IK, Langley KE, Zsebo KM: Recombinant human stem cell factor synergises with GM-CSF, G-CSF, Il-3 and Epo to stimulate human progenitor cells of the myeloid and erythroid lineages. Exp Hematol 19:226,1991.
11. Carow CE, Hangoc G, Cooper SH, Williams DE, Broxmeyer HE: Mast cell growth factor (c-kit ligand) supports the growth of human multipotential progenitor cells with a high repopulating potential. Blood 78:2216,1991.
12. Bernstein ID, Andrews RG, Zsebo KM: Recombinant human stem cell factor enhances the formation of colonies by CD34+ and CD34+lin- cells and the generation of colony-forming cell progeny from CD34+lin- cells cultured with IL-3, G-CSF or GM-CSF. Blood 77: 2316,1991.
13. Firkin F, Dunlop J, Bertoncello I: Expansion of hemopoietic activity in long-term culture of human bone marrow by c-kit ligand (stem cell factor). Growth Factors 8:135,1993.
14. Liesveld JL, Broudy VC, Harbol AW, Abboud CN: Effect of stem cell factor on myelopoiesis potential in human Dexter-type culture systems. Exp Hematol 23:202,1995.

15. Sekhsaria S, Malech HL: Recombinant human stem cell factor enhances myeloid colony growth from human peripheral blood progenitors. Blood 81: 2125,1993.
16. Gabutti V, Timeus F, Ramenghi U, Crescenzio N, Marranca D, Miniero R, Cornaglia G, Bagnara GP: Expansion of cord blood progenitors and use for hemopoietic reconstitution. Stem Cells Dayt 11:suppl.2,105.
17. Effect of c-kit ligand (stem cell factor) in combination with IL-5, GM-CSF and Il-3 on eosinophil lineage. Int J Hematol 58:21, 1993.
18. Briddel RA, Bruno E, Cooper RJ, Brandt JE, Hoffman R: Effect of c-kit ligand on in vitro human megakaryocytopoiesis. Blood 78:2854,1991.
19. Avraham H, Vannier E, Cowley S, Jiang SX, Chi S, Dinarello CA, Zsebo KM, Groopman JE: Effects of stem cell factor, c-kit ligand, on human megakaryocytic cells. Blood 79:365,1992.
20. Imai T, Nakahata T: Stem cell factor promotes proliferation of human primitive megakaryocytic progenitors, but not megakaryocytic maturation. Int J Hematol 59:91, 1994.
21. McNiece IK, Langley KE, Zsebo KM: The role of recombinant stem cell factor in early B cell development: synergistic interaction with IL-7. J Immunol 146:3785, 1991.
22. Wineman JP, Nishikawa S-I, Müller-Sieburg CE: Maintenance of high levels of pluripotent hematopoietic stem cells in vitro: effect of stromal cells and c-kit ligand. Blood 81:365,1993.
23. Muta K, Krantz SB: Apoptosis of human erythroid colony forming cells is decreased by stem cell factor and insulin-like growth factor I as well as erythropoietin. J Cell Physiol 156:264, 1993.
24. Tong J, Gordon MS, Srour EF, Cooper RJ, Orazi A, McNiece I, Hoffman R: In vivo administration of recombinant methionyl human stem cell factor expands the number of human marrow hematopoietic stem cells. Blood 82:784, 1993.
25. Soligo D, Servida F, Cortelezzi A, Pedretti D, Uziel L, Morgutti M, Pogliani E, Lambertenghi-Deliliers G: Effects of recombinant human stem cell factor on colony formation and long-term bone marrow cultures in patients with myelodysplastic syndromes. Eur J Haematol 52:53,1994.
26. Amano Y, Koike K, Nakahata T: Stem cell factor enhances the growth of primitive erythroid progenitors to a greater extent than IL-3 in patients with aplastic anemia. Br J Haematol 85:663, 1993.
27. Bacigalupo A, Piaggio G, Podesta M, Raffo MR, Tedone E, Sogno G, Benvenuto F, Figari O, Grassia L, Bagnara GP: In vitro effect of stem cell factor on colony growth from acquired severe aplastic anemia. Stem Cells Dayt 11:suppl.2, 175, 1993.
28. Bagnara GP, Strippoli P, Bonsi L, Brizzi MF, Avanzi GC, Timeus F, Ramenghi U, Piaggio G, Tong J, Podesta M: Effect of stem cell factor on colony growth from acquired and constitutional Fanconi's anemia. Blood 80:382, 1992.
29. Abkowitz JL, Sabo KM, Nakamoto B, Blau CA, Martin FH, Zsebo KM, Papayannopoulou T: Diamond-Blackfan anemia: in vitro response of erythroid progenitors to the ligand for c-kit. Blood 78: 2198, 1991.
30. Alter BP, Knobloch ME, He L, Gillio AP, O'Reilly RJ, Reilly LK, Weinberg RS: Effect of stem cell factor on in vitro erythropoiesis in patients with bone marrow failure syndromes. Blood 80:3000, 1992.
31. Nara N, Tanikawa S, Nagata K, Hamaguchi H, Tomiyama J: Effects of stem cell factor (SCF) on the in vitro growth of leukemic blast progenitors in acute non-lymphocytic leukemia. Int J Hematol 58:27, 1993.
32. Ikeda H, Kanakura Y, Furitsu T, Kitayama H, Sugahara H, Nishiura T, Karasuno T, Tomiyama Y, Yamatodani A, Kanayama Y: Changes in phenotype and proliferative potential of human acute myeloblastic leukemia cells in culture with stem cell factor. Exp Hematol 21:1686, 1993.
33. Strife A, Perez A, Lambek C, Wisniewiski D, Bruno S, Darzynkiewicz Z, Clarkson B: Characterisation of lineage-negative blast subpopulations derived from normal and chronic myelogenous leukemia bone marrows and determination of their responsiveness to human c-kit ligand. Cancer Res 53: 401, 1993.
34. Brach MA, Buhring HJ, Gruss HJ: Functional expression of c-kit by AML blasts is enhanced by TNF-alpha through posttranscriptional mRNA stabilisation by a labile protein. Blood 80: 1224, 1992.
35. Siitonen T, Savolainen ER, Koistinen O: Expression of the c-kit proto-oncogene in myeloproliferative disorders and myelodysplastic syndromes. Leuk 8: 631, 1994.
36. Hu ZB, Ma W, Uphoff CC, Quentmeier H, Drexler HG: c-kit expression in human megakaryoblastic leukemia cell lines. Blood 83: 2133, 1994.
37. de Vos S, Brach MA, Asano Y: TGF-beta interferes with the proliferation inducing activity of stem cell factor in myelogenous leukemia blasts through down regulation of the c-kit proto-oncogene product. Cancer Res 53: 3638, 1993.
38. Asano Y, Brach MA, Ahlers A: Phorbol ester (TPA) down regulates expression of the c-kit proto-oncogene product. J Immunol 151: 2345, 1993.
39. Oez S, Birkmann J, Smetak M, Corbacioglu S, Hofmann-Wackersreuther G, Welte K, Gallmeier WM: Regulation of the density of the stem cell factor receptor (c-kit) by tumor necrosis factor alpha on a human myeloid cell line. Eur Cytokine Netw 4:439, 1993.

40. Green DR, Bissonnettee RP and Cotter TG: Apoptosis and Cancer. In: Important Advances in Oncology 1994, V.T. De Vita, Hellman S, Rosenberg SA, editors, pp. 37-52, 1994, Philadelphia: J.B. Lippincott Company.
41. Parchaidou A, Saikia T, Loew J: Assessment of Apoptosis in serial bone marrow biopsies of 53 patients with newly diagnosed high risk AML following induction therapy. Blood 84: 312a, 1234, 1994.
42. Sachs L, Lotem J: Control of programmed cell death in normal and leukemic cells: new implications for therapy. Blood 82: 15, 1993.
43. Pietsch T, Kyas U, Ludwig W-D and Welte K.: Mitogenic effect of human stem cell factor alone and in combination with G-CSF, GM-CSF, and IL-3 on fresh leukemic blast cells from patients with stem cell factor receptor-positive AML. In: Cytokines in Hemopoiesis, Oncology and Aids II, Freund M et al., eds., pp. 171-178, 1992, Springler Verlag.
44. Molineux D, Migdalska A, Szmitkowski M, Zsebo K, Dexter TM: The effect of hematopoiesis of recombinant stem cell factor (ligand for c-kit) administered in vivo to mice either alone or in combination with G-CSF. Blood 78: 961, 1991.
45. Briddell RA, Hartley C, Stoney G, McNiece I: SCF synergises with G-CSF in vivo to mobilise murine peripheral blood progenitor cells with enhanced engraftment potential. Exp Hematol 21:1150, 1993.
46. Andrews RG, Briddell RA, Kintter GH, Rowley SD, Appelbaum FR, McNiece I: Rapid engraftment by peripheral blood progenitor cells mobilised by recombinant human SCF and recombinant human G-CSF in nonhuman primates. Blood 85: 15, 1995.
47. Drize NJ, Chertkov JL, Zander AR: Hematopoietic stem cell mobilisation into peripheral blood of mice by combination of recombinant rat SCF and recombinant human G-CSF. Exp Hematol 23: in press, 1995.
48. Yan XQ, Hartley C, McElroy T, McNiece I: Serial transplantation of SCF plus G-CSF mobilised PBPC demonstrated long term engraftment in secondary recipient mice. Blood 84, suppl.1: 344a, 1361, 1994.
49. Andrews RG, Briddell RA, Appelbaum FR, McNiece IK: Stimulation of hematopoiesis in vivo by stem cell factor. Curr Opinion Hematol 1:187, 1994.
50. Begley CG, Basser R, Mansfield R, Maher D, To B, Juttner C, Fox R, Cebon J, Grigg A, Szer J, McGrath K, Thomson B, Sheridan W, Menchaca D, Collins J, Russell I, Green M: Randomised prospective study demonstrating a prolonged effect of SCF with G-CSF on PBPC in untreated cancer patients: early results. Blood 84, suppl.1: 25a, 90.

THE ROLE OF GM-CSF IN CHRONIC INFECTIONS

Charlotte Nath,[1*] Joginder Nath,[1] and S. C. Gulati[2]

[1] West Virginia University
Morgantown, West Virginia, 26506
[2] The New York Hospital-Cornell Medical Center
New York, New York, 10021

ABSTRACT

Significant progress has been made in understanding wound healing. Various growth factors interact in a complex manner resulting in eventual wound healing. GM-CSF, platelet derived growth factor, tumor necrosis factor, alpha and various angiogenesis factors may have a significant role and in the future, may have to be combined for the maximum benefit in wound healing.

INTRODUCTION

Chronic, non-healing cutaneous wounds are the result of inadequate tissue repair or faulty wound healing. These wounds tend to be resistant to the standard treatment of antibiotics, protective dressings, rest, revascularization and debridement, ultimately progressing to pervasive infection and subsequent amputation. Approximately 15% of people with diabetes develop cutaneous ulcers during their lifetime.[1] Forty-five percent of lower-extremity amputations (LEA) are performed on people with diabetes, at an average cost of $25,000 each.[2,3] The relative risk for LEAs among persons with diabetes is 15 times greater than that in persons without diabetes.[4]

Pecoraro, et al[5] reviewed possible causal pathways responsible for LEAs in the US Veteran Administration system. In 80 consecutive diabetic male subjects during a 3-month period at a single hospital, the sequence of minor trauma, ulceration and faulty wound healing applied to 72% of the amputations. Consideration was limited to seven potential causes (alone or in combination), based on major pathophysiological mechanisms (ischemia,

* Address correspondence to: Dr. Charlotte Nath, Department of Family Medicine, PO Box 9152, Robert C Byrd Health Sciences Center-WVU, Morgantown, West Virginia 26506-9152. Phone: (304)-293-8188; fax: (304)-293-5860.

Molecular Biology of Hematopoiesis 5, edited by Abraham et al.
Plenum Press, New York, 1996

infection, neuropathy and faulty wound healing), soft tissue complications (ulceration and gangrene), and minor trauma.

For each amputation the contributory cause(s) was determined. Faulty wound healing after a cutaneous ulceration emerged as the most prevalent (81%) pathophysiological component cause leading to amputation. As an independent causal factor, wound failure demonstrated linkages with other causes that were critical in the progression to amputation. Rarely was a single factor sufficient by itself to produce amputation. The multiplicity of interactions may explain why singular interventions have not worked particularly well in salvaging limbs.

RESULTS AND DISCUSSION

Wound healing can be divided into three phases, inflammation, granulation and reconstruction. Each phase involves different biochemical mediators and cells that function together to resolve and or repair injured tissue. The inflammation phase seals the wound and then dissolves the clot to initiate the reconstructive phase of healing. Macrophages release angiogenesis factor (AF) and fibroblast-activating factor (FAF) to attract epithelial cells, vascular endothelial cells and fibroblasts. The granulation phase includes the clean up of the wound site and the initiation of regeneration processes. The reconstructive phase is characterized by fibroblast proliferation which is followed by collagen synthesis by the fibroblasts, epithelialization, and cellular differentiation. Healing is completed with differentiation of cells, contraction of the wound, scar formation, and remodeling of the scar.

Successful healing then represents the completion of a sequence of inflammation, cell proliferation, matrix formation, remodeling, wound contraction and epithelization. Faulty wound healing may occur during any phase of the wound-healing process and may involve insufficient repair, excessive repair, or infection.

In the diabetic subject, studies have shown that polymorphonuclear leukocytes do not function normally, as demonstrated by poor responses to chemotactic stimuli, impaired migration, phagocytosis, and intracellular killing.[6,7] Erythrocyte membrane fluidity abnormalities have been demonstrated and polymorphonuclear leukocyte membrane abnormalities may be present also. The authors concluded that the decreased membrane fluidity may contribute to the impaired function of the neutrophils.

Recently there has been a cascade of research into the effects of growth factors in wound healing. These autologous peptides act as part of a biologic response triggered by the wound. A direct causal relationship between systemic growth factor function and faulty wound healing has not yet been established in the diabetic patient.[8] However, growth factors have been found to influence four general healing related functions: 1) stimulation of chemotactic activity, 2) signaling cells to proliferate, 3) altering option of the phenotypic status of the cell and 4) modulation of wound matrix synthesis.[8]

Some of the factors in wound healing are epidermal growth factor (EGF), fibroblast growth factor (FGF), platelet derived growth factor (PDGF), transforming growth factor (TGF), tumor necrosis factor (TNF), interlukin - 1 (IL-1), platelet releasate topical factor (CT-102), and granulocyte-macrophage colony stimulating factor (GM-CSF). Clinical trials to date have been less than conclusive in demonstrating the specific role of particular factors in wound healing. The blood levels of factors may have poor correlation with the levels present or needed at the wound site. The synergistic and inhibitory effects of various factors are poorly understood. Additionally, in any given wound, various phases of healing are occurring simultaneously. Perhaps, a combination of factors is needed to achieve curative results since many growth factors can involve several functions depending on cell types and wound environment.

Granulocyte-macrophage colony-stimulating factor (GM-CSF) is another refinement in the wound healing repertoire. While hematopoietic growth factors have been used for increasing white blood cell count recovery, their wound-healing effects appear to be more significant than those of the PDGFs. GM-CSF acts to enhance antimicrobial activity by activating neutrophils and monocytes/macrophages thereby increasing phagocytosis, increasing oxidative metabolism, and increasing the release of chemotactic factors. GM-CSF appears to induce macrophage accumulation in greater amounts than PDGF and TNF-alpha. GM-CSF has been reported to enhance migration of endothelial cells and proliferation of keratinocytes. Furthermore, GM-CSF stimulated alpha-SM actin expression in fibroblastic cells while the other factors did not.[9]

It has been established that during the healing of an open wound, myofibroblast differentiation is important but this process has remained mysterious. Cytokines such as IL-1, TNF-alpha and PDGF have been identified as contributors to granulation tissue formation. GM-CSF is emerging as the cytokine necessary for not only increased granulation tissue production but also for alpha-SM actin-expressing myofibroblasts. The local application of GM-CSF produces an extensive inflammatory reaction marked by polymorphonuclear (pmn) cell and macrophage accumulations. Gradually, the pmn cells decrease and the macrophages cluster. Also, GM-CSF induces neovascularization of granulation tissue as reported by Bussolino et al.[10]

Chronic wounds that involve the loss of subcutaneous tissue (decubitus ulcer), failure of re-epithelialization(venous ulcer) or necrosis and infection (diabetic ulcer) pose a clinical challenge for healing. Wound treatment is approached differently by the health care professional group and the clinical research group. The health care professional group tends to view wound treatment by examining the effects of neuropathy, pressure, infection, weight bearing, and ischemia. The clinical research group tends to look at the mechanism of cellular repair, inhibitors to growth, optimizing the wound environment for healing, nutrition, and active wound repair. A combination of all treatment modalities is needed because the healing process is multifactorial and the treatment approach should combine the expertise of each health-team member.

Anecdotal success of GM-CSF in Kaposi lesions, skin grafts, breast cancer infiltrating skin, decubitus and venous ulcers has been reported. Although a factor combination formula has been used successfully to accelerate healing of chronic diabetic neurotrophic foot ulcers,[11] the use of GM-CSF in diabetic ulcer has not been reported. Boenta, et al.[12] reported healing of a chronic Kaposi sarcoma lesion. During a 6 week course of chemotherapy (vinblastine and vincristine) lesions on the patient's thorax flattened but foot lesion did not change. During three months after chemotherapy the foot lesions continued to grow. 400 ug rhGM-CSF was injected at multiple sites around the lesion. Several days later the lesion was surrounded by erythema and signs of inflammation. After 10 days the lesion disappeared and had not recurred at 6 months out. There was no change in lesions not injected.

Pojda and Struzyna[13] applied recombinant human granulocyte macrophage colony-stimulating factor (rh GM-CSF) to non-healing ulcers and skin grafts in humans. In a study involving only 6 patients, each of the patients had two grafts applied, one pretreated with GM-CSF and one prepared in the standard manner. In 3 patients only the GM-CSF treated grafts took, in 2 patients both grafts took and in 1 patient neither graft took. No adverse effects were noted and peripheral white blood cell counts did not change during therapy.

Marques da Costa[14] reported healing of chronic leg ulcers in 3 patients (78-86 years) who received 400 ug GM-CSF by subcutaneous perilesional injection. Complete resolution of ulcer was seen in two patients after a single application. Second treatment was given to remaining patient with wound closure completed at week 5. No side effects were noted apart from itching during the first three days. Itching is commonly a sign of healing.

Given the clinical challenge of chronic wounds and the primary action of growth factors, there appears to be a role for growth factors in wound healing. The exact mechanism of action needs to be understood. Topical applications of a combination of growth factors have been successful in active wound repair of large, severe, open wounds. The combination growth factors (platelet-derived wound healing formula-PDWHF) have shown marked success in healing wounds in a shorter period of time and salvaging limbs.[15] Platelet-derived wound healing formulas are combinations of several growth factors (most commonly PDGF, TGF, and platelet derived EGF as well as platelet factor 4 (PF-4) and platelet derived angiogenesis factor (PDAF)). These growth factors actively cause the growth of granulation tissue, capillaries, and epithelial cells.

Growth factors interact in a variety of ways *in vitro* and *in vivo* systems.[16] Several combinations shown to be synergistic or inhibitory have been reported and suggest that the most effective treatment of wounds may require a mixture of growth factors. There are relatively few prospective trials and many of those lack adequate controls. Interventions involving sequential or concomitant use of growth factors will have to not only prove their benefit in wound healing but their economic and clinical benefit in the newly emerging competitive managed care environment.

Longer survival of diabetic patients allows time for progressive neuropathy and atherosclerosis to develop, resulting in an increasing number of foot wounds. Foot wounds are responsible for more inpatient hospital days than any other complication of diabetes. A study released in 1993 established a model for economic evaluation of alternative treatment methods for diabetic foot ulcers.[17] Cost effectiveness was estimated for traditional, growth factor and saline solution treatments. The growth factor treatment was more cost effective in cost per patient healed.

While the utilization of new treatment techniques to manage the chronic, cutaneous non-healing wound has shown promise, the large, double-blind, multi-center, controlled studies still need to be done. The specific role of factors and the appropriate combination of factors need to be determined. The advantages of GM-CSF demonstrated in animal, *in vivo* and *in vitro* systems offer promise for not only a more rapid wound healing process but for fewer break-downs or recurrences of the wound. Certainly there is need to define the correct use of the existing growth factors and to explore new factors that will improve and accelerate wound healing and repair.

REFERENCES

1. Levin ME, O'Neal LW, (eds): *The Diabetic Foot* 4th ed. St. Louis, Mo, CV Mosby Co, 1988, p ix
2. Palumbo PJ, Melton, LJ III: Peripheral vascular disease and diabetes, in *Diabetes in America: Diabetes Data Compiled in 1984*, Washington, DC, US Govt Printing Office, Chap XV, NIH publication, 1985, p 85
3. Levin ME, Poucher RL, Stavosky JW: Neuropathic ulcers and the diabetic foot, in *Treatment of Chronic Wounds*, Curative Technologies, Inc., Setauket, NY, 1994
4. U.S. Dept. of Health and Human Services Public Health Service: *Diabetes Surveillance, 1980-1987: Policy Program Research - 1990 Annual Report*, Centers for Disease Control, Atlanta, GA, 1990
5. Pecoraro R, Reiber G, and Burgess E: Pathways to diabetic limb amputation: basis for prevention, *Diabetes Care*,1990, p 513
6. Valerius NH, Eff C, Hansen NE, Karle H, Nerup J, Soeberg B, Sorensen SF: Neutrophil and lymphocyte function in patients with diabetes mellitus, *Acta Med Scand*, 1982, p 463
7. Goodson WH III, Hunt TK: Wound healing and the diabetic patient, *Surg Gynecol Obstet*, 1979, p 149
8. Keyser, J: Diabetic wound healing and limb salvage in an outpatient wound care program, *S Med J*, 1993, p 311
9. Vyalov S, Desmouliere A, and Gabbian, G: GM-CSF-induced granulation tissue formation: relationships between macrophage and myofibroblast accumulation, *Virchows Archiv B Cell Pathol*, 1993, p 231

10. Bussolino F, Ziche M, Wang J, Alessi D, Morbidelli L, Cremona O, Bosis A, Marchisio P, Mantovani A: In vitro and in vivo activation of endothelial cells by colony-stimulating factors, *J Clin Invest*, 1991, p 986
11. Steed DL, Goslen JB, Holloway GA, Malone JM, Hunt TJ, Webster, MW: Randomized prospective double-blind trial in healing chronic diabetic foot ulcers. *Diabetes Care*, 1992, p 1598
12. Boenta P, Sampaio C, Brandao M, Moreira E: Local peri-lesional therapy with rhGM-CSF for Kaposi's sarcoma, *The Lancet*, 1993, p 1154
13. Pojda Z and Struzyna J: Successful skin grafting in non-healing ulcerations after in vitro pre-treatment of graft material with recombinant human granulocyte-macrophage colony-stimulating factor (rhGM-CSF), *The Lancet*, 1994, p 1100
14. Marques da Costa R, Aniceto C, Jesus FM, Mendes M: Quick healing of leg ulcers after molgramostim, *The Lancet*, 1994, p 481
15. Knighton D, Ciresi K, Fiegel V, Schumerth S, Butler E, and Cerra F: Stimulation of repair in chronic, non-healing, cutaneous ulcers using platelet-derived wound healing formula, *Surg, Gynecol, & Obstet*, 1990, p 56
16. Holloway GA, Steed DL, DeMarco MJ, Masumoto T, Moosa HH, Webster MW, Bunt TJ, Polansky M: A randomized, controlled, multicenter, dose response trail of activated platelet supernatent, topical CT-102 in chronic, nonhealing, diabetic wounds, *Wounds*, 1993, p 198
17. Bentkover J, Champion A: Economic evaluation of alternative methods of treatment for diabetic foot ulcer patients: cost-effectiveness of platelet releasate and wound care clinics. *Wounds*, 1993, p 207

68

DYSREGULATION OF CYTOKINE GENE EXPRESSION AS A CAUSE OF T CELL TRANSFORMATION AND *IN VIVO* TUMORIGENICITY

Mona R. Hassuneh,[2] Prakash S. Nagarkatti,[2] and Mitzi Nagarkatti[1*]

[1] Department of Biomedical Sciences and Pathobiology
[2] Department of Biology
Virginia-Maryland College of Veterinary Medicine
Virginia Tech, Blacksburg Virginia, 24061

ABSTRACT

Neoplasia develops due to several molecular perturbations and dysregulation of autocrine growth factor production is one such factor. Recently, we reported that the *in vitro* spontaneous transformation of a T cell clone resulted exclusively due to constitutive IL-2 autocrine stimulation. However, whether the tumors which originate spontaneously *in vivo* also use autocrine growth factor stimulation as a mechanism for tumorigenesis is not clear. In the current study, we demonstrate using a murine T cell lymphoma line designated LSA that the tumor growth is dependent on IL-2 autocrine stimulation. The LSA T cell lymphoma line constitutively expressed mRNA for IL-2 and IL-2 receptor β-chain as well as for IL-4 and IL-4 receptors. This cell line was inhibited from growing *in vitro* in the presence of antibodies against the growth factors or their receptors. Furthermore, the *in vivo* originated LSA tumor cell line was found to constitutively express immunosuppressive cytokines such as interleukin 10 and TGF-β. Together our data suggest that dysregulation in autocrine growth factor production such as IL-2 and IL-4 can lead to T cell transformation *in vivo* and furthermore, constitutive production of immunosuppressive cytokines may facilitate the growth of such tumors *in vivo*.

* Address correspondance to: Dr. Mitzi Nagarkatti, Departmeny of Biomed. Sciences and Pathobiology, Virginia-Maryland College of Vet. Medicine, Blacksburg Virginia 24061. Telephone: (540) 231-5035, Fax (540) 231-7367.

Molecular Biology of Hematopoiesis 5, edited by Abraham et al.
Plenum Press, New York, 1996

INTRODUCTION

The development of tumors is triggered by several molecular perturbations and dysregulation of autocrine growth factor production may constitute one such mechanism[1] . In this process, a cell may constitutively express a growth factor and its receptor and together this may lead to autonomous growth and transformation. Such a mechanism was first described for cells transformed by infection with transforming viruses and subsequently several studies have shown that a variety of different factors such as platelet derived growth factors, hematopoietic growth factors, epidermal/ transforming growth factor-α etc. all act as autocrine growth factors contributing to cell transformation.[1]

Recent studies have also suggested that interleukin 2 (IL-2) may serve as a T cell autocrine growth factor responsible for T cell transformation. In an earlier study, we demonstrated that a T cell line which underwent spontaneous transformation *in vitro* was totally dependent on IL-2 for autonomous growth.[2] This cell line induced tumors in nude mice but not in normal mice and furthermore, the cell line could be completely inhibited from growing *in vivo* using monoclonal antibodies against IL-2 or IL-2 receptors (IL-2R). These data suggested that dysregulation in IL-2 can lead to spontaneous transformation *in vitro* and raised the possibility of whether the transformed T cell lines which originate *in vivo* would also use a similar mechanism of tumorigenesis. In the current study, we demonstrate using an *in vivo* originated T cell lymphoma line that the autonomous growth of this tumor cell line *in vitro* is dependent on IL-2 and IL-4 autocrine growth factors.

MATERIALS AND METHODS

Tumor Cell Line

LSA, a T cell lymphoma syngeneic to C57BL/6 mouse was used in the current study. This cell line has been extensively characterized and used in our previous studies.[3] Other cell lines such as P815 a mastocytoma was grown *in vitro* and used as described elsewhere.[4]

The transformed cell lines were maintained by culturing them in tissue culture flasks in RPMI 1640 complete medium with 10% fetal bovine serum[2].

Cell Proliferation

The proliferation of the LSA *in vitro* was measured by both the uptake of ^{3}H-thymidine as well as counting viable cells using trypan blue dye exclusion assay[2]. In experiments studying the effect of cyclosporin A (CsA) (kindly provided by Sandoz pharmaceutical) on cell proliferation, CsA was prepared by dissolving 1mg of CsA in 0.1 ml of ethanol and 0.02 ml of tween 80 followed by addition of 1 ml of RPMI 1460.

PCR Analysis of Cytokine and Cytokine Receptor Gene Expression

The PCR method was employed to study whether the transformed cells constitutively express IL-2, IL-4, IL-2R, IL-4R genes as described in detail elsewhere.[2,5] Briefly, total RNA was extracted from the cells at various time intervals and reverse transcribed. The cDNA samples were subjected to PCR amplification using synthetic oligonucleotides primers for various cytokine genes and their receptors, with β-actin serving as an internal control. The PCR product was electrophoresed through a 1.5% agarose gel containing ethidium bromide.

Antisense Oligonucleotides

The antisense oligonucleotide specific for IL-2 were designed to hybridize at sequences immediately down stream from the initiation codon of the IL-2 gene mRNA. Since different cell lines may have different exonuclease or endonuclease activities, the selected oligonucleotides were modified to be nuclease-resistant by making them 100% phosphorothioated. The antisense oligonucleotide sequence for IL-2 was$^{5'}$ GAGCTGCATGCTGTA.$^{3'}$ The oligonucleotides were dissolved in Tris-EDTA buffer at pH 7.4.

RESULTS

Constitutive Expression of IL-2 and IL-2R Genes in the In Vivo Transformed LSA Tumor Cell Line

LSA is a spontaneously originated tumor cell line in C57BL/6 strain of mouse, which grows and kills the syngeneic host once injected *in vivo*. In the current study, we first investigated whether this cell line would express constitutively T cell growth factor and growth factor receptor genes. To address this, the total RNA extracted from the cell line at various time intervals was reverse transcribed and the cDNA samples were subjected to PCR amplification using synthetic oligonucleotide primers for IL-2, IL-4, IL-2R and IL-4R, with β-actin serving as an internal control. The data shown in Figure 1 indicated that the LSA tumor cell line constitutively expressed IL-2 and IL-2R β-chains when tested at various time intervals. The LSA tumor cells also constitutively expressed IL-4 and IL-4R as well as IL-10 and TGF-β genes (data not shown).

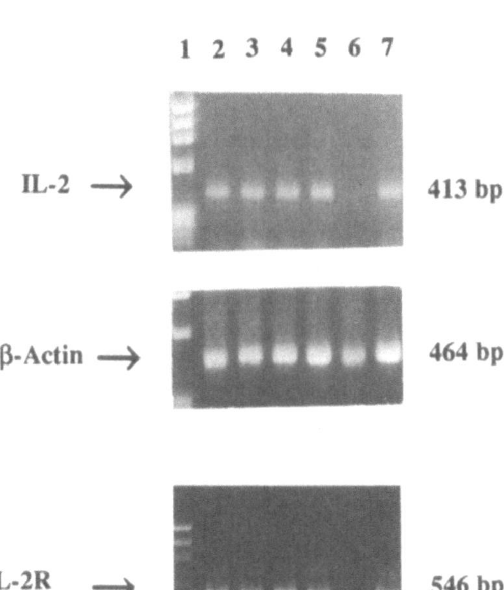

Figure 1. Demonstration of constitutive expression of IL-2 and IL-2R genes by LSA tumor cell line, using PCR. Lane 1 is a molecular marker φχ174/ Hae III, lane 2-5 represent LSA cells isolated at 12, 24, 36 and 48 hours of *in vitro* culture, lane 6 is a negative control consisting of naive spleen cells and lane 7 is a positive control consisting of spleen cells cultured with ConA.

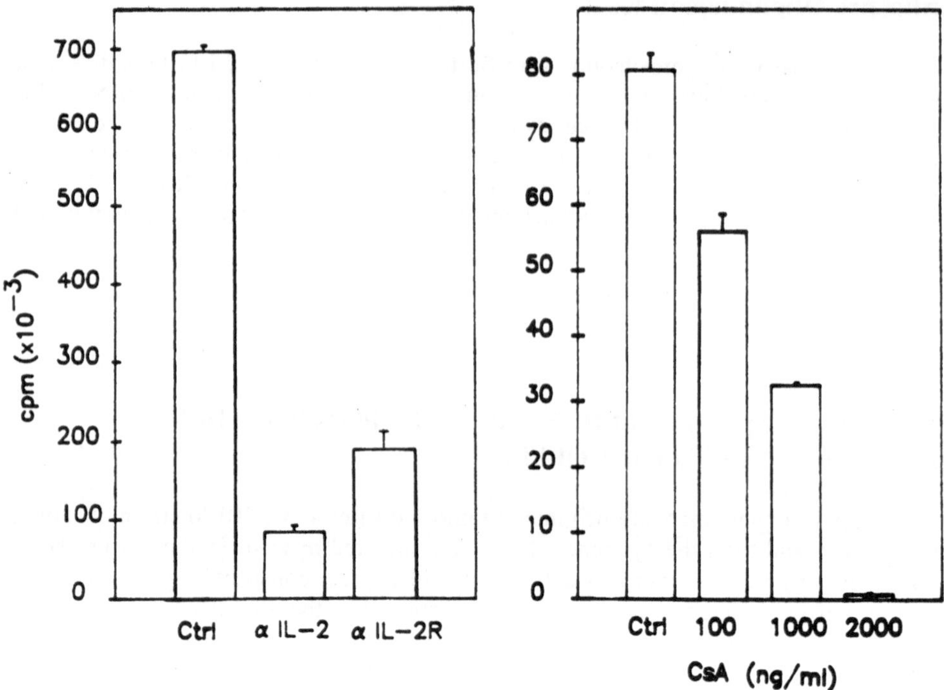

Figure 2. Inhibiting LSA tumor growth *in vitro* in the presence of mAbs against IL-2 or IL-2R or a drug, Cyclosporin A (CsA). LSA tumor cells were cultured in the presence of mAbs (left panel) or various concentrations of CsA (right panel) for 48 hours and the cell proliferation was measured by ^{3}H-thymidine uptake assay.

Inhibition of the LSA Tumor Growth In Vitro using Monoclonal Antibodies against the Growth Factors

We next addressed whether monoclonal antibodies against IL-2 and IL-2R would inhibit the autonomous growth of the LSA tumor cell line *in vitro*. The data presented in Figure 2 (left panel) demonstrated that both monoclonal antibodies against IL-2 and IL-2R were capable of significantly inhibiting the autonomous growth of the LSA tumor cell line. Furthermore, addition of CsA which is known to inhibit cytokine secretion by T cells caused a dose dependent inhibition of proliferation of the LSA tumor cell line (Fig 2, right panel). Addition of antibodies against IL-4 as well as a combination of antibodies against IL-2 and IL-4 could also significantly suppress the growth of the LSA tumor (data not shown). Further studies using antisense oligonucleotides also corroborated the above data that the IL-2 and IL-4 were acting as autocrine growth factors leading to the autonomous growth of the LSA tumor. The data shown in Figure 3 indicated that addition of antisense oligonucleotides specific for IL-2 caused a significance inhibition in the growth of LSA tumor whereas similar concentrations of antisense oligonucleotides fail to significantly inhibit the growth of P815 mastocytoma which has been shown previously in our lab to be independent of IL-2 for autonomous growth. Furthermore, addition of antisense oligonucleotides against IL-4 as well as a combination of antisense against IL-2 and IL-4 or antisense oligonucleotides against IL-2R or IL-4R was shown further to significantly suppress the growth of the LSA tumor cell line (data not shown). These data conclusively demonstrated that both IL-2 and IL-4

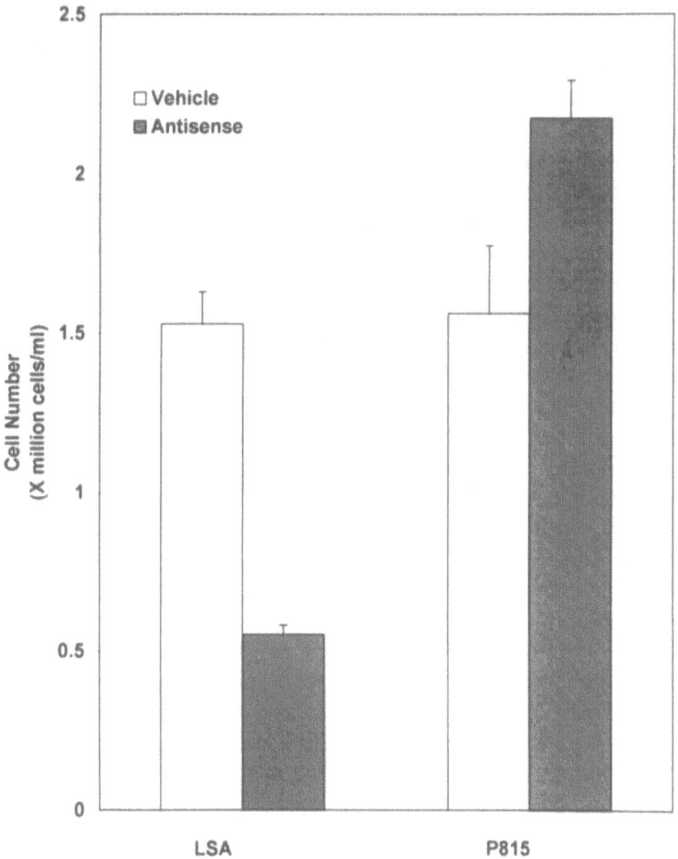

Figure 3. Antisense oligonucleotides specific for IL-2 inhibit LSA cell proliferation. LSA and P815 tumor cells were cultured in the presence of vehicle or antisense oligonucleotides for 48 hours and the cell proliferation was measured by counting the number of viable cells.

were acting as autocrine growth factors in the autonomous growth of the LSA tumor cell line.

DISCUSSION

Although, it is becoming increasingly clear that dysregulation in autocrine growth factor production may play an important role in the transformation of cells, most of these studies are based on the ability of the growth factor or their receptor antagonists to block *in vitro* cell proliferation. Thus, whether a similar mechanism operates *in vivo* leading to cell transformation is not clear. In the current study, we demonstrate for the first time that a tumor cell line which spontaneously originated in C57BL/6 strain of mouse designated LSA, was dependent on both IL-2 and IL-4 for autonomous growth *in vitro*. This was demonstrated by several observations such as the constitutive expression of IL-2 and IL-4 and their receptor genes, as well as the ability of antibodies against IL-2, IL-2R and antisense oligonucleotides specific for IL-2 , IL-4 and their receptors, to inhibit the growth of the LSA tumor *in vitro*. This combined with our earlier studies demonstrating the role of IL-2 as an autocrine growth factor in the tumorigenic transformation of an *in vitro* transformed cell line[2] suggest that T cell lymphomas or leukemias may indeed originate following dysregulation in the production of T cell growth factors such as IL-2 and IL-4. Recent studies have demonstrated in adult T cell leukemia patients as well as in T cell lines infected with HTLV that autocrine IL-2-induced self-stimulation may play an important role in T cell transformation.[1,6]

It is interesting to note that in our earlier study we observed that the *in vitro* transformed T cell line was unable to grow in normal mice but was able to cause tumor in immunodifecient, nude mice.[2] Further studies revealed that the in vitro transformed T cell line does not grow *in vivo* in normal mice probably because the cell line was able to trigger a strong immune response because of its ability to secrete IL-2. However, in the current study we noted that the LSA cell line was able to induce tumors even in normal syngeneic mice despite the fact that the cell line was constitutively expressing IL-2 and IL-4 genes. This can be explained by two mechanisms: First, the LSA cell line may not secrete the cytokines exogenously . Secondly, that the LSA tumor cell line may also expresses constitutively, immunosupressive cytokines which might suppress the immune system of the normal C57BL/6 mice. To this effect, we have shown recently that LSA tumor cell line expresses constitutively IL-10 and TGF-β which have been known to act as immunosuppressive molecules whereas the *in vitro* transformed cell line[2] namely AutoD1.4, fails to express such cytokines at significant levels.

The fact that transformed T cells as well as other tumors constitutively express IL-2R suggests that immunotherapy with IL-2 may not be effective against such tumors. In fact, the IL-2 may facilitate the growth of these tumors. Secondly our studies also demonstrate that it should be possible to treat such autocrine growth factor-triggered T cell malignancies using monoclonal antibodies or antisense oligonucleotides against the growth factors or their receptors or drugs such as CsA which would inhibit the growth factor production and its ability to support autonomous growth of transformed cells *in vivo*.

ACKNOWLEDGMENT

This work was supported in part by a grant from American Cancer Society.

REFERENCES

1. Lang RA and Burgess AW: Autocrine growth factor and tumorigenic transformation. Immunol Today 11:244, 1990
2. Nagarkatti M, Hassuneh M, Seth A, Manikasundari K and Nagarkatti PS: Constitutive activation of IL-2 gene in the induction of spontaneous *in vitro* transformation and tumorigenicity of T cells. Proc Natl Acad Sci USA 91:7638, 1994
3. Nagarkatti M, Clary SR and Nagarkatti PS: Characterization of tumor-infiltrating CD4$^+$ cells as Th1 cells based on lymphokine secretion and functional properties. J of Immunol 144:4898, 1990
4. Seth A, Gote L, Nagarkatti M and Nagarkatti PS: T-cell-receptor-independent activation of cytolytic activity of cytotoxic T lymphocytes mediated through CD44 and gp90^{Mel-14}. Proc Natl Acad Sci USA 88:7877, 1991
5. Hammond DM, Nagarkatti PS, Gote LR, Seth A, Hassuneh MR and Nagarkatti M: Double-negative T cells from MRL-*Lpr/Lpr* mice mediate cytolytic activity when triggered through adhesion molecules and constitutively express perforin gene. J Exp Med 178:2225,1993
6. Gootenberg JA, Ruscetti FW, Mier GW, Gazdar A and Gallo RC: Human cutaneous T cell lymphoma and leukemia cell lines produce and respond to T cell growth factor. J Exp Med 154:1403, 1981

69

THE MYELOPROLIFERATIVE DISORDERS AND THEIR TREATMENT WITH INTERFERON ALFA

Richard T. Silver

INTRODUCTION

Under normal circumstances, the marrow progenitor cells respond selectively to differing specific stimuli. Examples include the rise in white blood cell count occurring after a bacterial infection or an elevation in the platelet count seen following hemorrhage. These reactions are self-limited, for the marrow reverts to its normal status once the stimulus subsides. Myeloproliferative disorders (MPD), on the other hand, comprise a group of diseases in which pluripotent stem cells proliferate more or less en masse, at least at the onset of the disease. In the hematology literature, opinions differ as to which diseases should be included in the MPD group, but based upon refinements in cytogenetics and molecular biology the diseases are chronic myeloid leukemia (CML), polycythemia vera (PV), essential thrombocythemia (ET) and agnogenic myeloid metaplasia (AMM). For example, although the prime abnormality in polycythemia vera (PV) is an increased red cell mass, these patients often have increases in the white blood cell count and platelet counts; patients with chronic myeloid leukemia (CML) may also have increased platelet counts. Common clinical features include an enlarged spleen and similar biochemical abnormalities including elevation in the serum values of uric acid, vitamin B-12 and B-12 binding protein. A distinction exists in the concentration of leukocyte alkaline phosphatase. It is absent in patients with CML, increased in PV and ET and variably increased, normal or decreased in AMM. Although cytogenetic abnormalities have been seen in all patients with the myeloproliferative diseases, the only consistent cytogenetic one occurs in CML, sharply distinguishing it from the other MPDs. There is increased reticulin fibers in the marrow of all the MPD as the disease progresses, and increased evidence of fibroblastic proliferation, probably a reactive phenomenon. Acute leukemia developing as an end result of any of the myeloproliferative disorders is extraordinary difficult to treat. In decreasing order of frequency, it occurs in CML, AMM, PV and ET. Remissions ranging about 8 months have been seen after intensive chemotherapy in the blast phase of CML, but less often in treatment of secondary acute leukemia of PV or AMM.

Megakaryocytes are central to the genesis of fibrosis in the MPD. They pervasively proliferate in all the MPDs. They persist in fibrotic marrows when other blood cells have

Molecular Biology of Hematopoiesis 5, edited by Abraham et al.
Plenum Press, New York, 1996

disappeared and they produce platelet-derived growth factor (PDGF), a stimulant of fibroblastic proliferation and factor 4, an inhibitor of collagenase. The understanding of the essential role of the megakaryocyte has broad implications with respect to newer treatments of this disease, particularly with interferon and anagrelide, which will be discussed.

HEMATOLOGIC CHARACTERISTICS

Chronic Myeloid Leukemia (CML)

CML develops with an insidious onset of symptoms and signs including splenomegaly and leucocytosis marked by an increase in neutrophils, eosinophils and basophils. The median age of onset is 50 years with no sex preference. The median survival approximates 50 months.

There are 3 phases in the disease, the first, a chronic phase lasting about 30 to 40 months, the second, a transitional phase called the accelerated phase and the terminal blast or acute phase in which the disease appears like acute leukemia of either the myeloid or lymphoid type.

Chronic Phase of CML. In the chronic phase, the white blood cell count approximates around 200,000/dL with myelocytes predominating. There may be slight anemia and increased platelets. Of unique importance is the presence of the Philadelphia chromosome occurring in approximately 95% of all CML patients. Until recently, no significant advances have been made in improving survival time in most patients with CML. Initially, the great majority of patients respond to treatment by many drugs and even radiotherapy. Although the quality of life in CML patients is thus improved, survival time is not significantly increased since the Philadelphia chromosome persists.

The Ph chromosome is formed by a balanced translocation of chromatin between chromosomes 9 and 22. Thus, overall cellular cytogenetic material is conserved. During this translocation, the c-abl oncogene is moved from chromosome 9 into the breakpoint cluster region (bcr) within the BCR gene of chromosome 22 resulting in a BCR-ABL gene which encodes an 8.5 kbase messenger RNA which translates into a 210kDA protein. This protein has tyrosine kinase activity. The latter is uniquely present in the leukemic cells of CML patients and is intimately involved in cell growth and regulation.

Currently, for conventional treatment hydroxyurea in doses of 5 to 15 mg/kgm is the treatment of choice. Alternatively, busulfan in a dose of approximately 0.1 mg/kg to a maximum daily dose of 6 mg may be employed. The doses of both drugs are lowered as the WBC recedes.

Polycythemia Vera (PV)

Polycythemia vera is characterized not only by proliferation of erythroid progenitors but also by early myeloid and megakaryocytic cells. An increasing number of cases are now being detected prior to development of symptoms and about half the patients are now seen in their 30's and 40's. As the disease progresses, symptoms develop that are related to hypervolemia and hyperviscosity, often accentuated by an increase in platelet count. Headaches, lightheadedness, vertigo, blurred vision and both arterial and venous thromboses are frequent. Spontaneous bruising, peptic ulcers and gastrointestinal hemorrhage are seen as the disease progresses. A most troublesome symptoms is pruritus, worse after a hot bath or shower. Secondary gout and uric acid kidney stones are relatively frequent. About 75% of the patients have an enlarged spleen. In addition to the increased red cell mass (for men, rbc

≥ 36 cc/kg; women ≥ 32 cc/kg), approximately 60% of patients have an elevated WBC at the time of diagnosis which involves neutrophils, basophils, and eosinophils. Usually the platelet count is elevated and platelet function is abnormal. Evidence of iron deficiency exists and may be severe. On marrow examination the iron stores are reduced or absent, and the marrow reticulin stain is most often increased in intensity.

As the disease progresses, and probably accentuated by phlebotomy, extramedullary hematopoiesis occurs often accompanied by myelofibrosis in the bone marrow. The end stage of PV is morphologically indistinguishable from that of the disease agnogenic myeloid metaplasia.

Essential Thrombocythemia (ET)

This disease is characterized by a platelet count of more than 1 million/dL due to proliferation of an abnormal clone of megakaryocytes unresponsive to control mechanisms normally governing platelet production. Reactive thrombocytosis owing to bleeding, cancer or infection must always be considered although in these conditions, the platelet count is rarely elevated to the degree seen in ET. All patients should have cytogenetic studies performed to exclude CML. The diagnosis of thrombocythemia is based on exclusion. For patients with borderline elevated RBC values, chromium-51 red blood cell mass study is mandatory to exclude p.vera. This disease affects young and middle aged men and women. Despite the elevated platelet count, recurrent bleeding and thromboses are paradoxically the cardinal features. The platelet abnormality cannot be quantified by any measurement of platelet dysfunction for despite a great deal of research, the pathophysiologic basis for excessive bleeding and thrombosis is not clearly understood. The result of *in vitro* platelet function tests are variable and show impaired or absence epinephrine induced platelet aggregation. In general, however, the higher the platelet count the greater is the likelihood of thrombosis or hemorrhage but there is no exact correlation. Reduction in the platelet count towards normal is associated with reversal of these findings. Although some patients, particularly younger ones, remain asymptomatic for long periods, approximately 50% of patients with ET are first seen as emergencies because of an episode of acute bleeding or vascular occlusion. In my experience, most younger patients have presented with thromboses rather than hemorrhage. Thus, patients with this disorder may have transient schemic attacks, strokes, and thrombosis of the splenic and/or hepatic veins and evidence of peripheral circulatory impairment characterized by digital cyanosis, erythremia and burning owing to microvascular thromboses. Clinical bleeding is manifested by easy bruising, epistaxis, unexplained hemorrhage and post operative hemorrhage. The majority of patients have splenomegaly. On rare occasions, the disease terminates in acute leukemia. It is impossible to distinguish on marrow biopsy the cellular phase of agnogenic myeloid metaplasia from cases of essential thrombocythemia. However, as more and more patients with essential thrombocythemia are seen, a substantial number seem to progress to myelofibrosis with myeloid metaplasia.

The peripheral smear shows clusters and clumps of megakaryocytes and giant platelets. Of particular clinical relevance is pseudohyperkalemia which may occur due to loss of potassium from large numbers of platelets during clotting of blood *in vitro*. This abnormality is not associated with any effect of increased serum potassium.

Agnogenic Myeloid Metaplasia (AMM)

Agnogenic myeloid metaplasia, AMM (also called idiopathic myelofibrosis with myeloid metaplasia), is characterized by a hypercellular bone marrow, extramedullary hematopoiesis, splenomegaly and a leukoerythroblastic peripheral blood picture. The disease can be defined as the causally unknown proliferation of marrow cells growing in organs or tissues that are not usually involved in blood cell formation. Extramedullary hematopoiesis is a term derived from the fact that this blood cell formation occurs outside the medullary or bone marrow cavity.

AMM occurs primarily in older age. Nearly two thirds of the cases occur between the ages of 50 and 70 years, about equally in men and in women. Most often, symptoms are related to anemia or to an enlarged spleen. Otosclerosis causes deafness in about 10 per cent of patients.

When the patient is first seen the spleen is almost always enlarged; this finding is so important that the diagnosis is suspect if splenomegaly is not detected. The liver is palpable in about two thirds of the patients. Although extramedullary hematopoiesis occurs frequently in the lymph nodes, it rarely accounts for significant nodal enlargement.

The hemoglobin ranges between 9 and 13 g/dL. The leukocyte count is elevated in about half of the patients, normal in one third, and low in the remainder. The peripheral blood smear shows a shift toward granulocytic immaturity, and a few myeloblasts and promyelocytes. Significant changes in the red cells include variation in size and shape and teardrop-shaped forms (teardrop forms are not unique to AMM). Fragments and clumps of megakaryocytes may be seen. The platelet count may be increased, normal, or low.

The bone marrow aspirate in almost every patient yields a dry tap even when the marrow biopsy is very cellular. Marrow biopsy shows panhypercellularity in the early stages of the disease and, most strikingly, increased numbers of morphologically atypical megakaryocytes. Special silver stains demonstrate an increase in the amount of reticulin fibers even before the classic increase in collagen tissue occurs. Obviously, the true morphologic diagnosis of myeloid metaplasia or extramedullary hematopoiesis rests upon the demonstration of this process in the spleen and/or the liver; however, rarely are the risks justified to make the diagnosis by closed needle biopsy.

Because of the problem of accurately determining the onset of the illness, survival time in AGM is difficult to state with assurance . As the disease progresses, the spleen gradually enlarges and so may the liver. Anemia becomes more severe and is complicated both by iron deficiency owing to bleeding and by relative folic acid deficiency. The high portal blood flow due to the enlarged spleen may cause "forward liver failure" and portal hypertension. Sclerosis of the hepatic vein and development of the Budd-Chiari syndrome have been recognized. Eventually the spleen occupies the entire abdomen. Splenectomy results in symptomatic improvement but shortly afterwards, the liver enlarges, hepatic failure and ascites develop. There is no known successful treatment and the patient dies.

TREATMENT OF THE MYELOPROLIFERATIVE DISORDERS

During the past decade, the development and clinical use of the interferons, has had considerable impact on the treatment of the myeloproliferative diseases.

Interferons. The fact that the interferons have a wide range of biologic activities including antiviral, antiproliferative, immunomodulatory and oncogene regulatory properties lead to their trial in the myeloproliferative disorders. There are three types of interferon, alpha, beta and gamma. Interferon beta is of more interest to neurologists; gamma interferon has been used by us and others in hematologic/oncologic diseases without much success. Interferon alpha has had the widest clinical experience. The common side effects of interferon can be troublesome, are dose related and are seen especially during the first 6 months of therapy. These include muscle aches and pain, fever, diarrhea, depression and somnolence. Acetaminophen or other non-steroidal drugs are helpful. Delayed side effects of IFN-α include nephrotic syndrome, Peripheral neuropathy, lupus-like syndrome and hyperthyroidism.

Treatment of CML

Because of the properties mentioned, interferon was first used therapeutically in CML by investigators at MD Anderson Hospital and then by us and others. The dose of rIFN-α was 5 million units/m^2/day which, for an average-sized person, is 5 to 10 million units per day. We can summarize the recent literature by stating that in CML, complete hematologic response occurs in about 70% of patients. Cytogenetic responses characterized by a decrease in the number of Philadelphia positive cells are seen in about 50% of patients. Of these, 25% are complete or major cytogenetic responses. There is a correlation between dose and response so that adequate doses of rIFN must be given. Patients treated early in the course of disease do better than those treated later. The fact that interferon improves survival compared with conventional therapy has been established unequivocally by the Italian Study Group in Chronic Myeloid Leukemia. In a study of 322 patients treated with either interferon-α-2a or hydroxyurea, total overall response and survival was significantly different in the interferon treated group. At 72 months, the survival was 85% for responders versus 50% for the non-responders. Significantly, durability of remission has been directly associated with the degree of cytogenetic remission.

Since 75% of patients with CML do not have durable remissions, alternative treatment modalities have been sought. There is abundant evidence to suggest that the combination of cytosine arabinoside (ara-C) and interferon may be synergistic. We have just completed a study of 91 patients with CML treated with CML and ara-C. Based upon preliminary experience, it would appear that compared to rIFN only, the clinical and cytogenetic response may be improved with the combination.

Treatment of PV

Because interferon clearly has myelosuppressive activity (first demonstrated in CML) it was considered for use in PV. Phlebotomy is the initial step in treating any newly diagnosed patient with PV. Since blood viscosity correlates with the elevation of the hematocrit, every effort should be made to reduce it to below 45% at which point the problems associated with hypervolemia and hyperviscosity decrease. Patients may require 2 to 5 phlebotomies a year although many will require more. Phlebotomy often increases the degree of thrombocytosis and thus the risk of thrombotic complications, and increases signs and symptoms of iron deficiency which may become marked.

Patients managed with phlebotomy-only may have an increase in severe and often fatal thrombotic complications, particularly during the first 2 to 4 years of treatment. Treatment with phlebotomy only does not stop the natural progression of PV which includes increasing myelofibrosis and splenomegaly. Treatment with marrow suppressives including radioactive phosphorus and alkylating agents such as chlorambucil have had serious consequences, notably a significant increase in leukemia and in malignancies of the gastrointestinal tract. Thus alternative myelosuppressive drugs have been sought. The most commonly used is hydroxyurea but this drug is not without its own problems since increasing evidence suggests it also is associated with leukemia, myelodysplasia and lymphoma in about 10% of cases.

Concurrently, the experimental data generated in the past few years justified the use of rIFN in PV: Platelet-derived growth factor (PDGF), a product of megakaryocytes, has been found to initiate proliferation of fibroblasts which may be involved in the development of secondary myelofibrosis. IFN-α antagonizes the action of PDGF and also inhibits the proliferation of early and late erythroid progenitor cells. Thus, for the first time a drug is available which provides a physiologic basis for treating PV without the risk of leukemia. Of 34 patients with active PV, IFN-α controlled and normalized red blood cell values within 6 to 12 months, eliminating the need for phlebotomy. No thrombohemorrhagic events occurred. Thrombocytosis and splenomegaly were reversed in the majority of patients. Pruritus and other constitutional symptoms abated. The dose of interferon was considerably less than that used in CML, approximately 3 million units subcutaneously 3 times a week. As the disease was controlled, the dose could be reduced in about half the patients.

For patients with elevated platelets who cannot undertake interferon therapy, the addition of anagrelide to phlebotomy can be considered. For elderly patients with PV, radioactive phosphorus can still be used for its activity is smooth and relatively predictable.

Treatment of ET

Until recently, it was thought that many patients with ET did not require treatment. However, as previously mentioned the development of thrombotic episodes in these patients is capricious. Recent evidence suggests that the use of one or more alkylating agents before or after hydroxyurea leads to a significant increase in acute leukemia in patients with ET. Low dose interferon is extremely useful and may prevent subsequent myelofibrosis. Continued therapy is required with interferon.

Treatment of AMM

The treatment of this disease is completely unsatisfactory and very limited progress has been made in the last 50 years. Chemotherapy, radiation therapy to the spleen, and glucocorticoids are generally not helpful. Patients with anemia sometimes respond to erythropoietin although this is often accompanied by enlargement of the spleen. Transplantation has been attempted, but this requires an allogeneic match, and moreover most patients with this disease are too old. Splenectomy is useful in very selective cases, but it is an operation of considerable magnitude. Usually the liver then enlarges and assumes the role of the major site of extramedullary hematopoiesis and liver failure subsequently occurs. Thus, the natural history of the disease has not changed. For these reasons, alternative treatments are mandatory. Currently, it is my belief that the early use of interferon may abort the progression of the fibrosis by interfering with megakaryocytic dysfunction. After fibrosis becomes established, rIFN-α may still be helpful because it may reduce spleen size and secondary hypersplenism, anemia and thrombocytopenia. For patients with established bony sclerosis, the use of stem cell rescue after bone marrow rongeur will be tested in the immediate future. This approach is based on the observation that curettage of the endosteal lining of the marrow may alter the microenvironment for facilitating stem cell implantation and regeneration.

ACKNOWLEDGMENT

This study was supported in part by an educational grant from Hoffman-La Roche, Inc., Nutley, New Jersey.

70

INHIBITION OF HEMATOPOIESIS BY IN VIVO ACTIVATED T CELLS SECRETING HIGH AMOUNTS OF INTERFERON-γ (IFN-γ) PLUS INTERLEUKIN 10 (IL-10)[*†]

E. M. Schneider,[‡] I. Lorenz, B. Harms, and R. E. Scharf

Institute of Hemostasis and Transfusion Medicine
Heinrich Heine University of Düsseldorf, Düsseldorf, Germany

ABSTRACT

T cell clones were isolated from patients showing suppressed hematopoiesis without any signs of acute or chronic viral infections. All clones were CD4-positive and were derived from the peripheral blood by limiting dilution in the presence of low concentrations of IL-2. Characteristically, the majority of these clones secreted 50 - 300 IU/ml of IFN-γ[#] and 100 to 2500 pg/ml of IL-10 per $5x10^5$ clone cells on day 3-5 following subculture. Culture supernatants displayed a weak suppressive activity on erythropoiesis using standard colony forming assays and ficoll-separated bone marrow cells. However, the addition of 1/10 of 20 Gy-irradiated clone T cells secreting both cytokines impaired hematopoiesis in vitro. Here, myeloid cell colony formation appeared to be more affected than erythroid burst formation.

Coculture of resting peripheral blood mononuclear cells (PBMC) with the irradiated T cell clones resulted in the induction of a T cell proliferative response in vitro on day 3 until day 6. The clone-induced T cells did not affect hematopoiesis. Thus, the clone T cells did not function as suppressor inducer cells. In contrast, the supernatant of these clone-stimulated PBMC harvested on day 3 and day 6 suppressed erythroid colony formation. Analysis of cytokines secreted by c.lone-stimulated PBMC revealed extremely high amounts of macrophage-inflammatory protein-1 α (MIP-1α) ranging from 7.5 to 20 ng/ml/$1x10^6$ PBMC

[*] This work has been dedicated to Andreas Sievers, Professor of Botany on the occasion of his 65th birthday.

[†] Supported in part by the Histiocytosis Association of America.

[‡] Correspondence to: E. Marion Schneider, Ph.D., Immunology Laboratory, Institut für Hämostaseologie und Transfusionsmedizin, Heinrich-Heine-Universität Düsseldorf, Geb. 12.43, Postfach 101007, D-40001 Düsseldorf, Germany.

[#] See abreviations section immediately preceding references.

seeded. IL-10 was negative and IFN-γ concentrations ranged between 8 and 44 IU/ml/ 1×10^6 PBMC seeded.

Proliferation assays were performed with either monocyte-depleted bone marrow mononuclear cells or with CD34-enriched and G-CSF cultured (300 U/ml) progenitors isolated from G-CSF-treated patients to test the effect of IL-10, IFN-γ and MIP-1α. MIP-1α counterregulated the stimulatory effect by IFN-γ on bone marrow cells. In progenitor assays, MIP-1α neutralized the IFN-γ mediated suppression ($p<0.01$).

These results demonstrate another aspect of T cell-mediated suppression at stages of hematopoietic development defined by their susceptibility to IFN-γ.

INTRODUCTION

Amongst multiple T cell-derived cytokines, IFN-γ appears to be one of the major effector molecules to suppress hematopoiesis.[1,2] Recently, however, the identification of defined differentiation stages amongst hematopoietic stem cell populations revealed a selective effect on primitive secondary colony forming cells but not on more mature hematopoietic cells expressing the stem cell antigen CD34 in addition to the proliferation marker CD38.[3]

In combination with other cytokines, stimulatory effects by IFN-γ have been also described by several authors.[4,5,6] Interferon-γ is a specific product of NK-cells and T-helper 1 (Th1) cells and is antagonized by the T-helper 2 (Th2) specific cytokine, IL-10.[7] This study reports on the hematopoietic effect of a number of T cell clones isolated from a patient with anemia but without any signs of acute infection who displayed a mixed Th1/Th2 type phenotype by secreting IFN-γ and IL-10 simultaneously and in high amounts. Coculture experiments of irradiated T cell clones co-incubated with resting allogeneic PBMC revealed the capacity to induce T cell proliferation as well as the secretion of high amounts of MIP-1α. MIP-1α could be shown to reverse the effects displayed by IFN-γ on hematopoietic cell proliferation.

Subsequent analyses of in vivo activated T cells isolated from other patients with dysregulated hematopoiesis indicate that in vivo activation of T cells with the same Th1/Th2 mixed cytokine secretion pattern can occur.

MATERIALS AND METHODS

Patients

Patients were selected due to impaired hemoglobin values and/or granulocytopenia and the simultaneous presence of CD69+, CD25+ and HLA-DR+ T lymphocytes in the peripheral blood designated as a lymphoproliferative syndrome.[8] One 46-year-old patient had anemia of unknown cause and gave rise to the isolation of the T cell clones described here. Another 4-month-old patient suffered from hepatosplenomegaly, anemia and granulocytopenia and a third patient had aplastic anemia. From all these patients, an uncloned T cell line was generated with a stable IFN–γ plus IL-10 secretion pattern.

T Cell Clones

T cell clones activated in vivo were isolated by limiting dilution on 30 Gy-irradiated lymphocyte feeder cells (pool of PBMC of 20 unrelated individuals) in medium containing 20 IU/ml of recombinant human interleukin 2 (rh-IL-2, a gift of Dr. Loeliger, Sandoz AG,

Basel, Switzerland) and 10% endotoxin-free fetal calf serum (FCS, Boehringer, Mannheim, Germany) in Iscove's modified Dulbecco's medium (IMDM). Fresh feeders ($1x10^5$ PBMC/ml) were added on every second subculture. Clones were characterized by surface marker staining using flow cytometry. Cytotoxicity assays were performed using sodium [^{51}Cr]-chromate-labeled K562 target cells. Released radioactivity was quantified in the supernatant using scintillation counting in a β-Plate (Pharmacia/LKB, Freiburg, Germany).

Inducer T Cell Assay

Induction of T cell proliferation was tested by stimulating $5x10^4$ peripheral blood mononuclear cells (PBMC) with 0.5 - $1x10^4$ 20 Gy-irradiated cloned T cells in round-bottomed microtiter plates in 200 μl of medium containing 5% heat-inactivated human AB-serum and IMDM. The proliferative response of PBMC responder cells was monitored by [^{3}H]-Thymidine incorporation on day 3 and 6. Moreover, microscopical evaluation of the stimulated PMBC on day 3 was performed in order to quantify the percentage of T cell blast formation.

Supernatants of PBMC stimulated with cloned T cells for either 3 or 6 days were harvested, dialyzed and added to erythroid colony assays.

Cytokine Secretion Pattern

Cytokines released during the constitutive phase of in vitro culture were tested in culture supernatants harvested on day 5 after each subculture. IFN-γ ELISA (Medgenix, Ratingen, Germany), IL-10 ELISA (Medgenix) and MIP-1α ELISA (R&D Systems purchased via Biermann, Bad Nauheim, Germany) were used.

Hematopoietic Assays

Hematopoietic progenitor assays were performed with adherence-depleted non-separated bone marrow cells, stimulated with 3 IU/ml of rhIL-3 (Genzyme, Cambridge, MA, USA), 100 U/ml rhGM-CSF (Amersham, Braunschweig, Germany), and with or without 1-4 IU/ml of erythropoietin (Behring-Werke, Marburg, Germany) in semisolid methylcellulose medium; myeloid (CFU-GM) and erythroid colony formation (BFU-e) were evaluated in the presence of either cloned T cells or the dialyzed culture supernatant of the clones harvested at a clone cell density of 5-$8x10^5$ cells/ml and on day 5 after each subculture.

Proliferation Assay with Hematopoietic Cells

Two types of hematopoietic cells were tested in proliferative assays: i) Ficoll-isolated bone marrow cells obtained from aspirates of healthy volunteers after depletion from monocytes by plastic adherence; T cell contents were less than 20 % as determined by FACS analysis, and ii) CD34+ cells isolated from a patient prophylactically treated with G-CSF before the development of sepsis (treated with 3 μg Neupogen/kg/day, Amgen, Munich, Germany) by labelling mononuclear cells with the anti-CD34 antibodies Birma K3 (DAKO, Hamburg, Germany), Qbend-10 (Immunotech, Hamburg, Germany) and positive enrichment with anti-mouse Ig-specific Dynabeads (Dynal, Hamburg, Germany) according to the manufacturers instruction. After isolation, the CD34+ cells were cultured with 300 U/ml of G-CSF in medium containing 10% FCS (endotoxin-free, Boehringer, Mannheim, Germany) and MEM-alpha medium (Gibco/BRL, Karlsruhe, Germany). For proliferation assays, the hematopoietic cells were incubated with growth factors in microtiterplates for 48 hours, labeled with 0.5 μCi of [^{3}H]-Thymidine (Amersham, Braunschweig, Germany) per well and harvested 16 hours later. At the time of functional testing, the CD34+ enriched and 14 days

Table 1. Cytokines were determined in culture supernatants of T cell clones derived from the peripheral blood of a patient with anemia, following 5 days after subculture. The suppression of BFU-e plated with 1×10^6 bone marrow cells is given in percent of control BFU-e grown in the presence of 1×10^5 PBMC of an allogeneic healthy donor (112 BFU-e/1×10^6 plated cells = 100 %)

Patient	IL10 pg/ml per 5×10^5 cells	IFN-γ IU/ml per 5×10^5 cells	2×10^4 clone cells added (%BFU-E)	4% clone derived spnt (%BFU-E)
parent line	<50	66	n.d.	26
clone				
2	203	<15	130	100
4	2000	282	25	90
8	438	141	90	80
11	1700	146	35	105
12	2461	23	80	n.d.
13	360	30	120	120
15	203	79	80	90
16	80	38	125	110
17	1652	43	120	80

n.d.: not determined.

cultured hematopoietic cells were homogenously positive for CD33 and HLA-DR. Of these cells, 35 % were CD13+ and 12 % were CD34+. These cells as well as the non-separated but monocyte-depleted bone marrow cells were stimulated with various growth factors of recombinant origin: rhIL-1α (produced in CHO cells, Boehringer, Mannheim, Germany), IL-4 (a gift by Klaus Rüde, Sandoz, Basel, Switzerland), rhIL-3 (produced in E. coli, Genzyme, Cambridge, MA, USA), rhG-CSF (Neupogen®, produced in E.coli, Amgen, Munich, Germany), rhGM-CSF (Imukin®, produced in yeast, Amersham, Braunschweig, Germany), rhIFN-γ (produced in E. coli, a gift from Thomae, Biberach, Germany) rhTNF-α (produced in E. coli, Boehringer, Mannheim, Germany), rhMIP-1α and rhIL-10 (produced in E.coli, PeproTech, Rocky Hill, NJ, USA), and rhIL-1RAP (a gift by Daniel Tracey, The Upjohn Company, Kalamazoo, MI, USA).

RESULTS

Table 1 summarizes the cytokine secretion pattern of the parent line and 9 T cell clones isolated by limiting dilution. During 3 to 5 days following the last subculture in medium containing IL-2, the cloned cells (concentration: 5×10^5 cells/ml) secreted significant amounts of IL-10 and IFN-γ except for clone # 2. All clones were CD4+ and some clones exhibited a weak NK-like cytotoxicity by lysing K562 as well as Jurkat target cells (data not shown).

The addition of 1×10^5 20 Gy-irradiated clone cells to an erythroid progenitor assay of 1×10^6 non-separated peripheral blood or bone marrow cells resulted in a variable inhibition. The supernatant of the cloned T cells containing IFN-γ and IL-10 displayed no significant inhibition on erythroid burst formation. In contrast, myeloid colony formation was more readily suppressed (data not shown).

It was further tested whether such Th1/Th2 hybrid T cells could play a role in the T cell network by coculturing 5×10^4 resting peripheral blood mononuclear cells (PBMC) of healthy donors with 5×10^3 of 20 Gy-irradiated clone T cells for 3 and 6 days.

Figure 1a shows the induction of T cell proliferation measured by [^{3}H]-Thymidine incorporation. Indeed, these T cell clones functioned similarly to antigen presenting cells

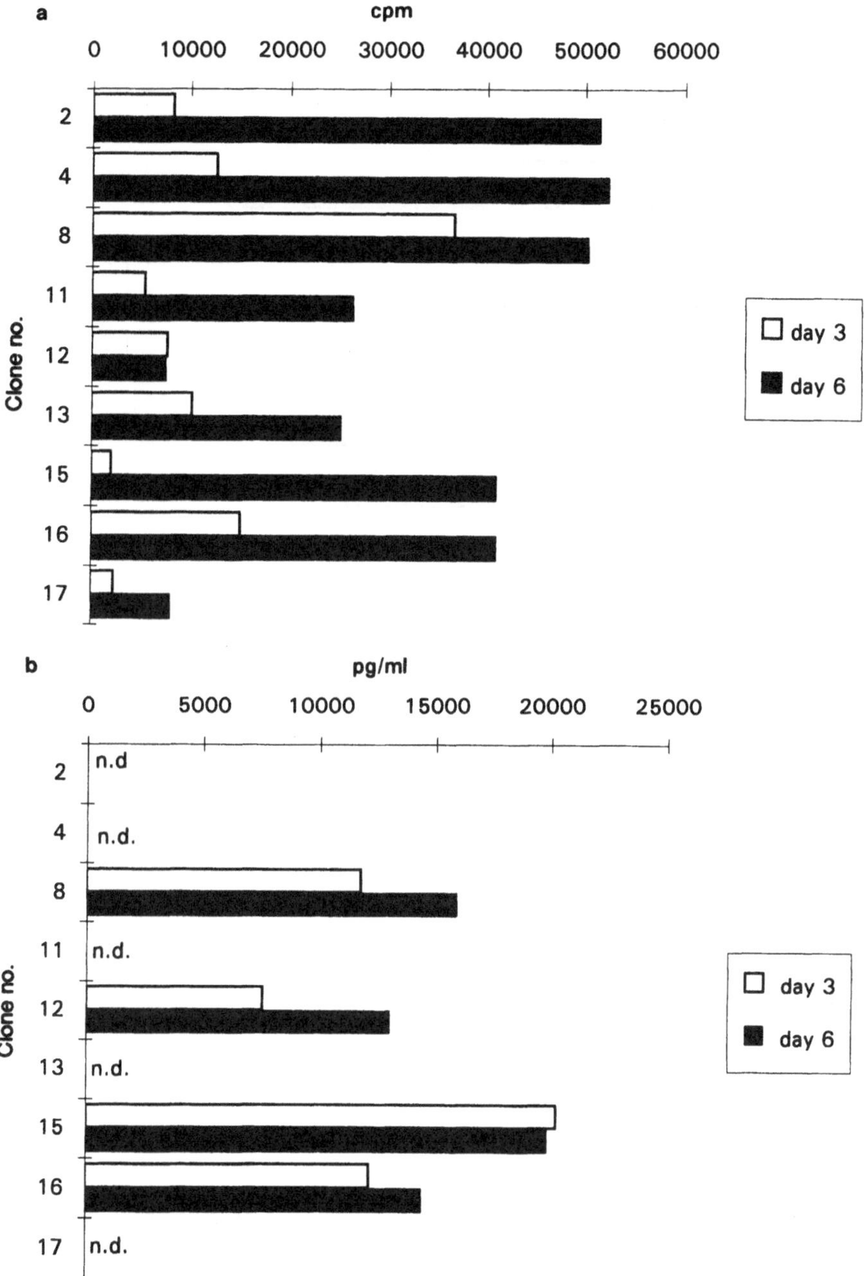

Figure 1. a) Proliferation of PBMC stimulated with clone cells. Five x 10^4 responder PBMC of a healthy unrelated donor were stimulated with 5 x10^3 20 Gy-irradiated cloned T cells in 200µl of 96 well round-bottomed microtiter plates. [^{3}H]-Thymidine incorporration was measured on day 3 and on day 6. **b)** MIP-1α secretion by PBMC stimulated with clone cells. Five x10^6 PBMC were stimulated with 5x10^5 cloned T cells in 2 ml medium, and MIP-1α was quantified in the supernatant by ELISA; n.d.: not determined.

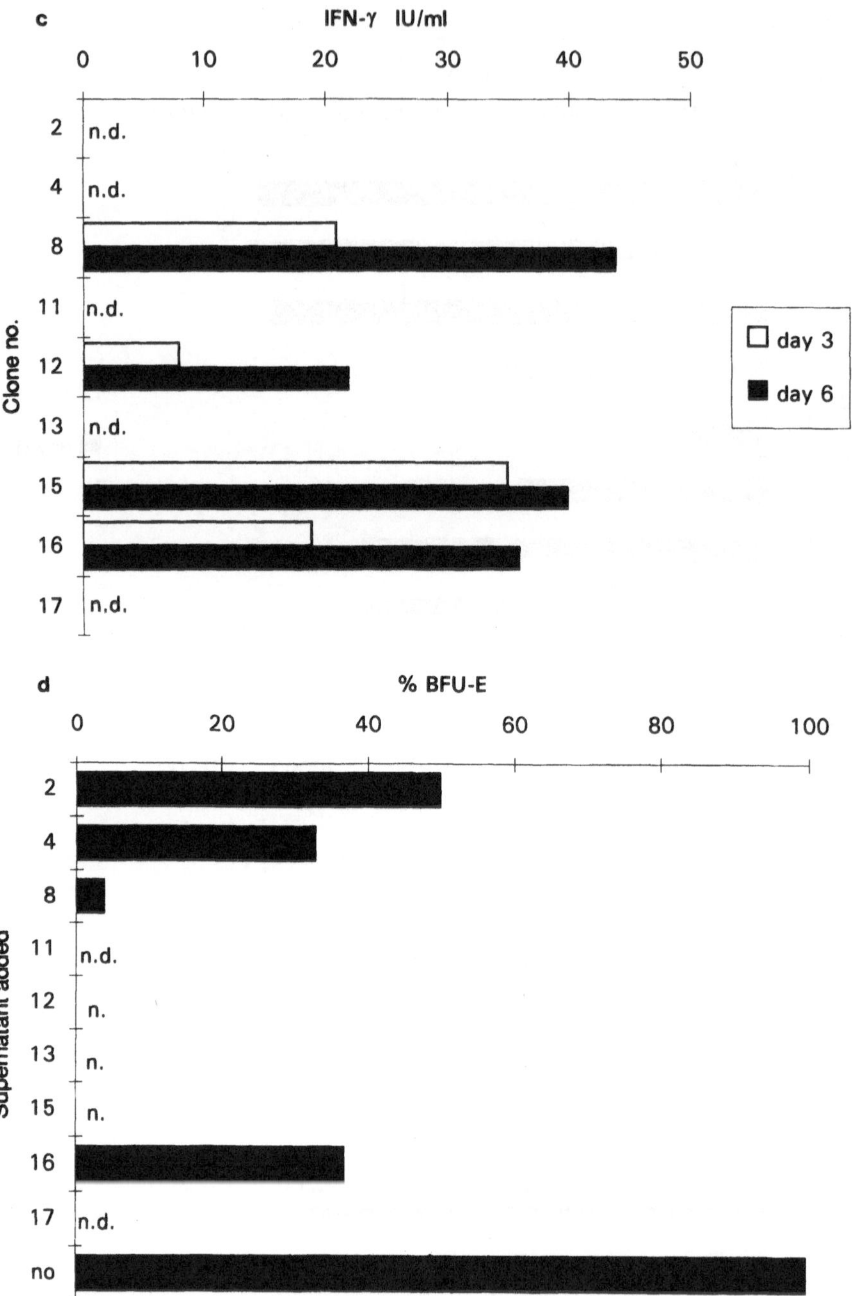

Figure 1. c) IFN-γ secretion by PBMC stimulated with clone cells. PBMC were stimulated with cloned T cells (as described in Fig. 1b) and secreted IFN-γ was measured in the supernatant by ELISA; n.d.: not determined. **d)** Influence of factors secreted by PBMC after 6 day stimulation with clone cells. PBMC were stimulated with cloned T cells as in Fig. 1b, and the supernatant harvested on day 6 was dialyzed and added to colony forming assays with non-adherent bone marrow cells. Two $x10^5$ non-adherent bone marrow cells in the absence of clone-stimulated supernatant gave 34 +/- 7 BFU-E. The residual BFU-E formed under the influence of factors secreted by PBMC after stimulation with cloned T cells are given in percent of the control culture; n.d.: not determined.

capable of rising an allogeneic response. Except for clones # 11 and 17, the induction of proliferation in PBMC was comparatively high amongst all the other clones.

We also harvested cytokines secreted during the induction phase. Accordingly, PBMC stimulated with the cloned T cells # 8, 12, 15 and 16 for 3 and 6 days in vitro, secreted high amounts of MIP-1α (Figure 1b), low amounts of IFN-γ (Figure 1c) and no detectable amounts of IL-10 (<50 pg/ml, data not shown).

Remarkably, only 4% (v/v) of the supernatant derived from clone-stimulated PBMC on day 6 were sufficient to suppress BFU-e formation potently when performed with allogeneic adherence-depleted bone marrow cells (Figure 1d). The inhibition of BFU-e caused by the MIP-1α containing supernatants was significant (96% inhibition) when derived from PBMC stimulated with clone # 8. This clone represents one of the clones with the highest level of induction of proliferation (Figure 1a).

In another set of experiments, we tested the influence of hematopoietically active cytokines on proliferative responses of bone marrow cells in order to compare the results with the response pattern obtained with CD34+-enriched and G-CSF-cultured progenitor cells isolated from the peripheral blood of G-CSF-treated patients.[9] Bone marrow cells and hematopoietic progenitors differed due to the effect by exogeneous IFN-γ. Significance was calculated according to Student's T-test from triplicate cultures.

Figure 2a demonstrates that bone marrow stem cells proliferate at low level but respond to IL-1α, IL-3, IFN-γ (each not significant), IL-4 ($p<0.05$), G-CSF ($p<0.01$) and GM-CSF ($p<0.001$). TNF-α appears to display a variable stimulatory effect, whereas MIP-1α, IL-10 and recombinant human IL-1 receptor antagonist protein (IL-1RAP) have no effect.

Figure 2b summarizes the results obtained with in vitro cultured progenitor cells isolated from the peripheral blood. The following factors, including IL-1α, IL-3 (each $p<0.005$) and IL-10 ($p<0.001$) had a weak but significant stimulatory effect, GM-CSF ($p<0.005$) and G-CSF ($p<0.001$) caused a strong and significant stimulation, whereas IFN-γ resulted in a significant inhibition of proliferation ($p<0.001$).

Figure 3 displays the modulation of the IFN-γ effects obtained by addition of MIP-1α, IL-10 or IL-1RAP. Thus, the stimulatory effect of IFN-γ on bone marrow cells appears to be neutralized by MIP-1α and, to a lesser extent, by IL-10 or IL-1RAP.

By contrast, the inhibition mediated by IFN-γ on the G-CSF cultured hematopoietic cells is neutralized by MIP-1α ($p<0.01$), IL-10 ($p<0.001$) and also by IL-1RAP ($p<0.05$).

DISCUSSION

The present study contributes another mechanism to control hematopoiesis by T cells, i.e. by a Th1/Th2 hybrid T cell type secreting simultaneously IFN-γ and IL-10. The functional significance of this T cell type is proven by anemia or neutropenia documented in patients, one of which gave rise to the T cell clones and uncloned lines described here. In addition to the cytokine secretion pattern which has been previously described by Nanda et al.[10] for minor antigen-specific suppressor T cells, the present T cell clones characteristically stimulate growth and cytokine release in resting peripheral blood mononuclear cells. In this setting, stimulation is as high as a proliferative response induced by allogeneic antigen presenting cells. However, a mixed lymphocyte response against alloantigens is not expected to reach such proliferation levels already on day 3. Importantly, the Th1/Th2 hybrid clones induce the secretion of high amounts of MIP-1α and maintain a fairly high secretion of IFN–γ from peripheral blood mononuclear cells.

MIP-1α has been characterized by multiple effects on the recruitment of resting T cells and macrophages and on the survival and maintenance of non-proliferative stem cells.[11]

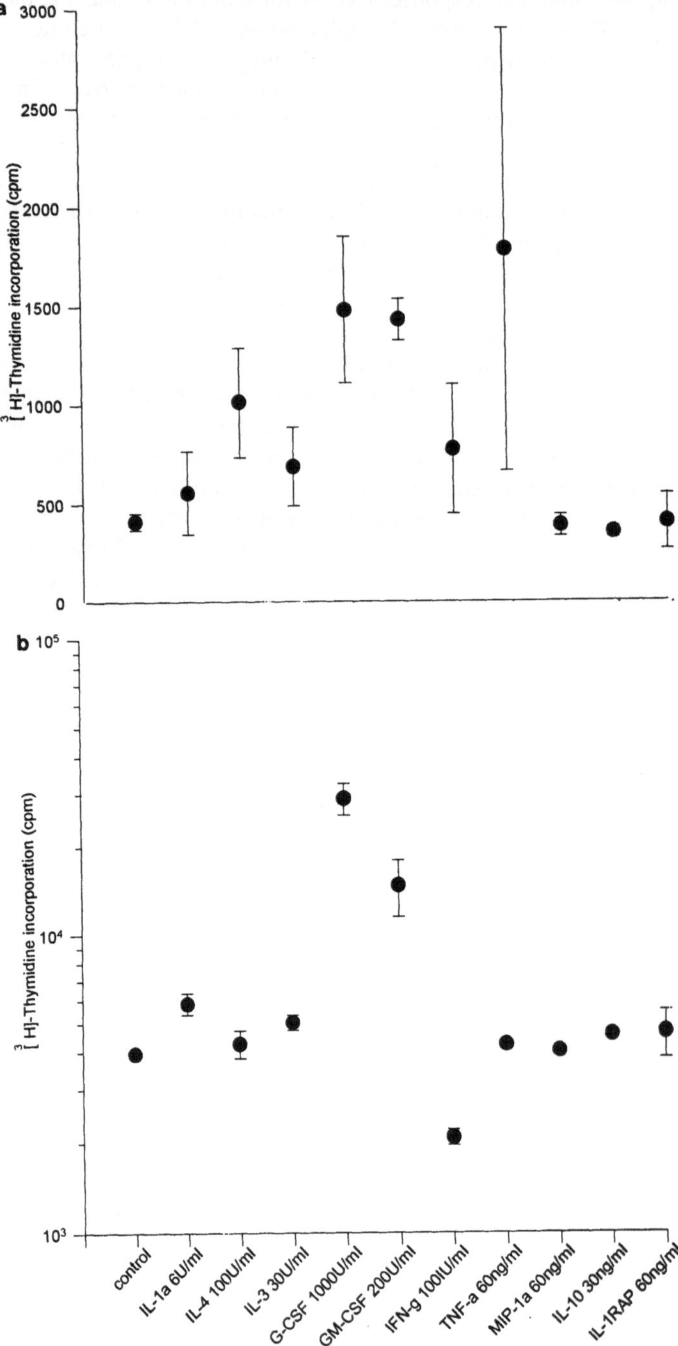

Figure 2. a) The proliferative response of adherence-depleted bone marrow cells is shown in the presence of recombinant growth factors given on the x-axis. Factors were added at the beginning of the culture and proliferation was measured on day 2. Results are given as mean±SE of triplicate cultures and one of two independent experiments. The stimulatory effects by IFN-γ and IL-3 are not significant, those by IL-4 ($p<0.05$), G-CSF ($p<0.01$) and GM-CSF ($p<0.01$) are significant. b) The proliferation of G-CSF mobilized and G-CSF cultured hematopoietic cells isolated from a patient are significantly inhibited by IFN-γ ($p<0.001$) and significantly stimulated by G-CSF ($p<0.001$), by IL-1, IL-3, and GM-CSF (each $p<0.005$).

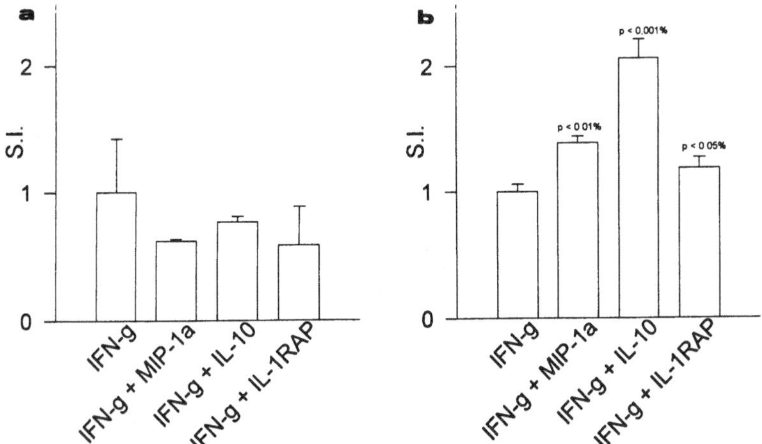

Figure 3. a. Modulation of the IFN-γ mediated stimulation on the proliferative response of adherence-depleted bone marrow cells by MIP-1α, by IL-10 and by IL-1RAP. Cytokine concentrations are as in Figure 2. Neutralization of the stimulatory IFN-γ effect by 60 ng of MIP-1 is visible but does not reach significance. b. The IFN-γ mediated inhibition of the proliferative response of myeloid progenitors (Fig. 2b) is modulated by MIP-1α ($p<0.01$), by IL-10 ($p<0.001$), by IL-1RAP ($p<0.05$).

Moreover, MIP-1α is apparently also able to suppress the proliferation of G-CSF-cultured hematopoietic cells (Figure 2d). Its potential relevance for dysregulated hematopoiesis as in aplastic anemias has been previously discussed.[12]

Importantly, addition of MIP-1α is also capable of neutralizing the stimulatory effect of IFN-γ on the less mature hematopoietic cells.

However, if MIP-1α is the main effector molecule induced by such in vivo activated, IFN-γ/IL-10 secreting T cells, it could also counteract the hematopoietically stimulative effects by IFN-γ to preserve more immature non-cycling stem cells. The positive effect by MIP-1α alone has been already demonstrated by its use in preserving long-term bone marrow culture initiating cells (LTBMIC) from CD34+/CD33- cells.[13] Moreover, MIP-1α can efficiently recruit T cells into the hematopoietic microenvironment by its chemotactic properties.[14] Thus, T cells secreting IL-10 plus IFN-γ simultaneously, may cause further T cell activation and induction of MIP-1α and also IFN-γ (Figure 2c) with the consequence of a reduced "output" of hematopoietic progenitor cells. Their cytokine phenotype appears to be resistant to inhibition by classical Th1 and Th2-type effectors. Thus, further studies will aim at the identification of counterregulatory immune responses capable of interfering with this functional phenotype in vivo and in vitro.

ABBREVIATIONS

BFU-E	Burst forming unit-erythrocyte
CFU-G	Colony forming units-granulocyte
CFU-GM	Colony forming units-granulocyte/macrophage
IL	Interleukin
IL-1RAP	Interleukin-1 receptor antagonist protein
IFN-γ	Interferon-gamma
MIP-1α	Macrophage inflammatory protein 1-alpha
n.d.	not determined
PBMC	Peripheral blood mononuclear cells

REFERENCES

1. Zoumbos, NC, Dieu JY, Young NS: Interferon is the suppressor of hematopoiesis generated by stimulated lymphocytes in vitro. J Immunol 133: 769, 1984
2. Broxmeyer HE, Lu L, Platzer E, Feit C, Juliano L, Rubin BY: Comparative analysis of the influences of human gamma, alpha and beta interferons on human multipotential (CFU-GEMM), erythroid (BFU-E) and granulocyte-macrophage (CFU-GM) progenitor cells. J Immunol 131: 1300, 1983
3. Snoeck H-W, van Bockstaele DR, Nys G, Lenjou M, Lardon F, Haenen L, Rodrigus I, Peetermans ME, Berneman ZN: Interferon-γ selectively inhibits very primitive $CD34^{2+}$ $CD38^-$ and not more mature $CD34^+CD38^+$ human hematopoietic progenitor cells. J Exp Med 180: 1177, 1995
4. Caux CI, Moreau S, Sealand S, Banchereau J: Interferon-γ enhances factor dependent proliferation of human CD34+ hematopoietic progenitor cells. Blood 79: 2628, 1992
5. Kawano Y, Takaue Y, Hirao A, Abe T, Saito S, Matsunaga K, Watanabe T, Hirose M, Ninomiya T, Kuroda Y, Yokobayshi A, Asano S: Synergistic effect of recombinant interferon-γ and interleukin-3 on the growth of immature hematopoietic progenitors. Blood 77: 2118, 1991
6. Brugger W, Möcklin W, Heimfeld S, Berenson RJ, Mertelsmann R, Kanz L: Ex vivo expansion of enriched peripheral blood CD34+ progenitor cells by stem cell factor, interleukin-1β (IL-1β), IL-6, IL-3, interferon-γ and erythropoietin. Blood 81: 2579, 1993
7. Fiorentino DF, Bond MA, Mosmann TR: Two types of mouse helper T cells IV. Th2 clones secrete a factor that inhibits cytokine production by Th1 clones. J Exp Med 170: 2081, 1989
8. Woessner S, Feliu E, Villamor N, Zarco MA, Domingo A, Milla F, Florensa L, Rozman M, Abella E, Soler J, Vallespi T, Irriguible MD, Sole F: Granular lymphocyte proliferative disorders: a multicenter study of 20 cases. Ann Hematol 68: 285, 1994
9. Weiβ M, Gross-Weege W, Harms B, Schneider EM: Filgrastim (rhG-CSF) related modulation of the inflammatory response in patients at risk of sepsis or with sepsis Cytokine 8:260-265, 1996
10. Nanda NK, Sercarz EE, Hsu DH, Kronenberg M: A unique pattern of lymphokine synthesis is a characteristic of certain antigen-specific suppressor T cell clones. Int Immunol 6: 731,1994
11. Avalos BR, Bartynski KJ, Elder PJ, Kotur MS, Burton WG, Wilkie NM: The active monomeric form of macrophage inflammatory protein-1 alpha interacts with high and low-affinity classes of receptors on human hematopoietic cells. Blood 84: 1790, 1994
12. Holmberg LA, Seidel K, Leisenring W, Torok-Storb B: Aplastic anemia: analysis of stromal cell function in long-term marrow cultures. Blood 84: 3685, 1994
13. Verfaille CM, Catanzarro PM, Li WN: Macrophage inflammatory protein 1 alpha, interleukin 3 and diffusible marrow stromal factors maintain human hematopoietic stem cells for at least eight weeks in vitro. J Exp Med 179: 643,1994
14. Taub DD, Conlon K, Lloyd AR, Oppenheim JJ, Kelvin DJ: Preferential migration of activated CD4+ and CD8+ T cells in response to MIP-1α and MIP-1β. Science 260: 355,1993

ROLE OF NITRIC OXIDE IN THE ACTIONS OF CYTOKINES ON HYPOTHALAMIC RELEASING HORMONE SECRETION

S. M. McCann,* S. Karanth, N. Belova, A. Kamat, K. Lyson, M. Gimeno, and V. Rettori

Department of Internal Medicine
The University of Texas Southwestern Medical Center at Dallas
5323 Harry Hines Blvd., Dallas, Texas 75235-8873

ABSTRACT

Interleukins act within the hypothalamus to cause the release of corticotropin-releasing hormone (CRH) and inhibit release of luteinizing hormone-releasing hormone (LHRH). Nitric oxide synthase (NOS), the enzyme that converts arginine into citrulline plus nitric oxide (NO), the latter a highly active free radical, occurs in a large number of neurons in the brain, including certain neurons in the hypothalamus. Our *in vivo* experiments in rats employing N^G-monomethyl-L-arginine (NMMA), an inhibitor of NOS and nitroprusside, a releaser of NO, have determined that interleukin-2 (IL-2) activates CRH release by acting on its receptors on cholinergic neurons which stimulate NOergic neurons by muscarinic receptors. The NO diffuses to the CRH neurons and activates them as revealed by *in vitro* experiments with hypothalamic explants. On the other hand, norepinephrine or glutamic acid activate LHRH release by stimulating NOergic neurons that in turn induce LHRH release by activation of guanylate cyclase and cyclooxygenase. IL-1α acts directly on the LHRH neuron to block its response to NO on the basis of *in vitro* experiments. *In vivo*, NO stimulates LHRH release which induces pulsatile LH but not FSH release on the one hand and induces mating behavior on the other. The IL-1-induced blockade of LHRH release by NO may account for the reduction in libido in infection. *In vivo* experiments employing injection of NMMA into the 3rd ventricle have revealed that NO mediates the prolactin-releasing action of interleukins. It also mediates pulsatile growth hormone (GH) release via stimulation of growth hormone-releasing hormone (GRH) release. IL-1 inhibits pulsatile GH release by blocking the response of the GRH neurons to NO. IL-1 also stimulates somatostatin release.

* To whom reprint requests should be addressed: S. M. McCann, Pennington Biomedical Research Center (LSU), 6400 Perkins Road, Baton Rouge, Louisiana 70808-4124. Phone: (504) 763-3042; Fax: (504) 763-3030.

Molecular Biology of Hematopoiesis 5, edited by Abraham et al.
Plenum Press, New York, 1996

Thus, the results implicate NO in control of release of various hypothalamic peptides, and indicate that interleukins have their action at various loci in these pathways.

INTRODUCTION

Previous work by ourselves and others has shown that the important actions of cytokines at both hypothalamic and pituitary to produce the pattern of pituitary hormone secretion observed in various infections. In general, the various cytokines cause the release of adrenocorticotropin hormone (ACTH) by stimulating the release of corticotropin-releasing factor (CRF), which passes to the portal vessels and then causes release of ACTH. Vasopressin may also play a role. Although it was originally controversial, it is now clear that there are also direct actions of cytokines to release ACTH by directly stimulating the corticotropes.

At the same time that ACTH release is stimulated, there is an inhibition of luteinizing hormone (LH) and to a lesser extent, follicle-stimulating hormone (FSH) release mediated by blockade of the release of LH-releasing hormone (LHRH). Here also there may be inhibitory actions at the pituitary. Prolactin is usually stimulated by cytokines by an action primarily on the hypothalamus, whereas, growth hormone (GH) and thyrotropin-stimulating hormone (TSH) is inhibited. It is our belief that the rapid responses to cytokines are largely mediated through the hypothalamus; however, slower responses are probably mediated at the pituitary level, and the actions at the pituitary in general amplify the hypothalamic response. Not only do the cytokines reach the pituitary and brain via the general circulation after the secretion from immune cells, but also they are almost certainly made in the brain and in the pituitary gland. In our own work, we demonstrated a interleukin-1α (IL-α) neuronal system with cell bodies in the region of the thermal sensitive neurons in the hypothalamic preoptic region and relatively short axons and dendrites, which may be involved in affecting temperature regulation and also the releasing hormone neurons. Other has shown the existence of an IL-1ß system and there is evidence also for production for cytokines by glial elements, and in the pituitary gland itself, it appears that folliculostellate cells produce IL-6 and probably other cytokines as well.

It has been known for some time that bacterial lipopolysaccharide or cytokines can induce by stimulating the messenger RNA synthesis, the synthesis of cytokines in macrophages and other immune cells, this has been termed the inducible nitric oxide synthase (NOS). After delay, large quantities of NO are produced by this inducible enzyme which is important in killing bacteria and viruses, and also adjacent cells. In the meantime, a constitutive form of the enzyme was discovered in mass? endothelium and then finally in the nervous system in neurons. This neuron NOS converts arginine into citrulline, plus NO, a highly active free radical in a large number of neurons in the brain including certain ones within the hypothalamus. There are large numbers of these neurons in the supraoptic and paraventricular nucleus and their axons project down to the median eminence and neural lobe of the pituitary gland. In addition, we have demonstrated the presence of a number of small neurons containing NOS in the arcuate median eminence region of the hypothalamus. Therefore, we began an evaluation of the role of NOergic neurons in control of the release of the various hypothalamic releasing and inhibiting hormones.

We have carried out both *in vivo* and *in vitro* studies in rats. In the *in vivo* studies, we microinjected the various compounds into the third ventricle (3V) of conscious freely moving rats, whereas with indwelling atrial catheters so that we can draw blood samples throughout the experiment. In the *in vitro* studies, we used either a piece of tissue consisting of the paraventricular region extending down to the median eminence, or a medial basal hypothalamic (MBH) piece including the arcuate median eminence region and carry out static incubation in Krebs-Ringer bicarbonate buffer (KRB).

In our initial studies *in vitro*, we demonstrated that N^G-monomethyl-L-arginine (NMMA), a competitive inhibitor of NOS would inhibit the release of CRF induced by IL-2 (10^{-13}M). Conversely, sodium nitroprusside (NP), a releaser of NO, caused a release of CRF. To determine how IL-2 was activating NO, we examined the effect of the inhibitor, NMMA on the CRF release induced by cholinergic stimulation with acetylcholine or carbachol and found that this would block the CRF release induced by these compounds. Furthermore, that atropine, the muscarinic receptor blocker could block the cholinergic stimulation of CRF release and the IL-2-induced stimulation of CRF release. Therefore, we concluded that IL-2 activates corticotropin-releasing hormone (CRH) release by action on putative receptors on cholinergic interneurons. These cholinergic interneurons, in term stimulate NOergic neurons by muscarinic receptors. The NO diffuses the CRH neurons and activates them. This causes a activation of soluble guanylate cyclase within the CRF neuron, and also a concurrent activation of cyclooxygenase and this then brings about the release of CRF. The pathway is visualized as follows, namely, the increase in cyclic guanosine monophosphate (cGMP) increases intracellular free calcium which then activates phospholipase A_2 (PLA_2) causing generation of arachidonate from membrane phospholipids. This is converted by the NO activated cyclooxygenase to prostaglandin E_2, (PGE_2) which then activates adenylate cyclase causing generation of cyclic adenosine monophosphate (cAMP), activation of protein kinase A, an extrusion of CRF secretory granules into the hypophyseal portal vessels.

In contrast to this, cytokines inhibit LHRH release, and in turn, LH release. In our initial *in vitro* experiments, we demonstrated that NO stimulates the release of LHRH just as it stimulates the release of CRH, and we demonstrated this by showing *in vitro* that NP, the releaser of NO, would stimulate LHRH release, whereas the release stimulated by norepinephrine (NE) was inhibited by NMMA, the blocker of the enzyme. NE acts by stimulating the release of NO from the NOergic neurons since it increases the release of NO measured by the labeled citrulline method. This method employs addition of labeled arginine, incubation of the tissue, then measurement of the conversion of arginine to citrulline after chromatography in counting the labeled citrulline in a *sc lation* counter. Since arginine is converted to equal molar quantities of NO and citrulline and NO disappears rapidly, whereas citrulline is relatively stable, this provides a convenient method to assay NOS activity. The NO released from the NOergic diffuses to the LHRH terminals in the median eminence arcuate region, and activates soluble guanylate cyclase leading to generation of cGMP. Indeed, we have measured the increase in cGMP induced by NP. The cGMP then plays a critical role in activating LHRH release. We hypothesized that it increases the intracellular free calcium, which then activates PLA_2 leading to conversion of membrane phospholipids into arachidonate. We have now shown that NO also activates cyclooxygenase. Since we can add labeled arachidonate and demonstrate that NP causes the formation of PGE_2. Furthermore, inhibitors of cyclooxygenase blocked the release of LHRH. The activated cyclooxygenase then converts the increased arachidonate formed into PGE_2, and we have demonstrated that indeed we get an increase in PGE_2 following treatment with NP. This PGE_2 activates adenylate cyclase leading to generation of cAMP, which activates protein kinase A and this leads to extrusion of LHRH secretory granules into the portal vessels and activation of LH release from the gonadotropes. This latter part of the pathway was studied earlier and has not been reexamined.

We found to our surprise that when we activated the LHRH terminals directly with NP, we could block the response to NP with IL-1α (10^{-11}M). We interpret this to mean that IL-1α acts directly on the LHRH neuron to block its response to NO. We confirmed these *in vitro* results employing castrate male rats with 3V cannulae, in which we showed that NO stimulates LHRH release which induces pulsatile LH, but FSH release since we could block

LH but FSH release by injecting NMMA into the 3V cannulae. Microinjection of 0.5 pmoles of IL-1α produced a result exactly similar to that with NMMA in that it inhibited pulsatile LH, but not FSH release. The response to IL-1α *in vitro* is very similar in that it prevents the release of PGE_2 and LHRH induced by NE. Consequently, we believe that we have established the mechanism of action of IL-1 to inhibit the LHRH directly.

In other experiments *in vivo*, we showed the important role of NO in inducing mating behavior in the rat. Indeed, in the ovariectomized estrogen-primed rat with a 3V cannulae, injection of progesterone into the cannulae will bring on mating as evidence by lordosis behavior within 30 to 45 min of the injection. This can be blocked by injection of NMMA into the 3V at the same time as the administration of the progesterone, whereas the NMMA has no effect. In addition, one can induce mating in the estrogen-primed animals with sodium NP to release NO. That the mating behavior is caused by the release of LHRH was demonstrated by injecting antisera against LHRH in the cannula. These were capable of blocking not only progesterone-induced mating, but also the mating induced by NP. Therefore, NO controls the LHRH release into the portal vessels which is involved in mediating pulsatile LH release, but also the LH controlling mating behavior by action on brain stem centers.

In fact, NO acts at all levels in the organism to induce reproduction acting also by release from terminals in the corpora cavernosa penis to induce penile erection, in the vagina to either contract or relax uterine smooth muscle and in the male by also a central action to induce penile erection as well as this direct peripheral effect. Thus IL-1 induce blockade of LHRH release by NO may account not only for the suppression of gonadotropin secretion in infection, but also for the reduction in libido in infection.

As indicated above, there is a prolactin-releasing action of IL-1α when injected into the 3V and this can also be blocked by the inhibitor of NOS, NMMA indicating that NO mediates the prolactin-releasing action of interleukins. The mechanism is not further elucidated, but may involve the release of oxytocin which is a prolactin-releasing factor since the oxytocinergic neurons are located in regions containing larger numbers of NOergic neurons.

NO also mediates pulsatile GH release via stimulation of GH-releasing hormone (GHRH) release, since NMMA injected into the 3V will also block pulsatile GH release. Other experiments have shown that IL-1α stimulates somatostatin release and inhibits GRH *in vitro*. The mechanism of stimulation of somatostatin release by GHRF is via ß endorphinergic activation of the NOergic neurons which then release NO that stimulates not only somatostatin release, but also somatostatin mRNA synthesis. Thus, IL-1 inhibits pulsatile GH release by blocking the response of the GRH neurons to NO, and by stimulating the release of the inhibitory hormone somatostatin.

In summary, NO has a potent effect to alter the release of various hypothalamic peptides and the cytokines interact at various steps in the activation of process to either stimulate or inhibit the release of the relevant releasing hormone. NO stimulates the release of CRF, LHRH, prolactin-releasing factors and GHRF as well as somatostatin, but has no effect on the putative FSHRF. IL-1 stimulates the cholinergic neurons to release NO which then brings about CRF release, but acts to inhibit the response of the LHRH terminals to NO, whereas IL-1 activates prolactin-releasing activity and inhibits NO-induced GRF release and activates NO-induced somatostatin release. Thus, the results implicate NO and control of release of various hypothalamic peptides indicate that interleukins have their action at various loci in these pathways.

ACKNOWLEDGMENT

This work was supported by NIH Grant DK43900.

REFERENCES

1. McCann SM, Karanth S, Kamat A, Dees WL, Lyson K, Gimeno M, Rettori V: Induction by cytokines of the pattern of pituitary hormone secretion in infection. Neuroimmunomodulation 1: 2-13, 1994.
2. Rettori V, Dees, WL, Hiney JK, Lyson K, McCann SM: An interleukin-1-alpha-like neuronal system in the preoptic-hypothalamic region and its induction by bacterial lipopolysaccharide in concentrations which alter pituitary hormone release. Neuroimmunomodulation 1: 251-258, 1994.
3. Breder CD, Dinarello CA, Saper CB: Interleukin-1 immunoreactive innervation of the human hypothalamus. Science 240: 321-324, 1988.
4. Spangelo BL, MacLeod, RM, Isakson PC: Production of interleukin-6 by anterior pituitary cells in vitro. Endocrinology 126: 582-586, 1990.
5. Lancaster JR Jr: Nitric oxide in cells. Am Sci 80: 248-259, 1992.
6. Moncada S, Palmer RMJ, Higgs EA: Nitric oxide: Physiology, pathophysiology, and pharmacology. Pharmacol Rev 43: 109-142, 1991.
7. Bredt DS, Snyder SH: Nitric oxide, a novel neuronal messenger. Neuron 8: 3-11, 1992.
8. Canteros G, Rettori V, Franchi A, Genaro A, Cebral E, Saletti A, Gimeno, M, McCann SM: Ethanol inhibits luteinizing hormone-releasing hormone (LHRH) secretion by blocking the response of LHRH neuronal terminals to nitric oxide. Proc Natl Acad Sci 92: 3416-3420, 1995.
9. Rettori V, Belova N, Dees WL, Nyberg CL, Gimeno M, McCann SM: Role of nitric oxide in the control of luteinizing hormone-releasing hormone release in vivo and in vitro. Proc Natl Acad Sci 90: 10130-10134, 1993.
10. Rettori V, Belova N, Kamat A, Lyson K, McCann SM: Blockade by interleukin-1 alpha of the nitricoxidergic control of luteinizing hormone-releasing hormone release *in vivo* and *in vitro*. Neuroimmunomodulation 1: 86-91, 1994.
11. Mani SK, Allen JMC, Rettori V, O'Malley BW, Clark JH, McCann SM: Nitric oxide mediates sexual behavior in female rats by stimulating LHRH release. Proc Natl Acad Sci 91: 6468-6472, 1994.
12. Rettori V. Belova N, Gimeno M, McCann SM: Inhibition of nitric oxide synthase in the hypothalamus blocks the increase in plasma prolactin induced by intraventricular injection of interleukin-1α in the rat. Neuroimmunomodulation 1: 116-120, 1994.
13. Rettori V, Belova N, Yu WH, Gimeno M, McCann SM: Role of nitric oxide in control of growth hormone release in the rat. Neuroimmunomodulation 1: 195-200, 1994.
14. Karanth S, Aguila MC, McCann SM: The influence of interleukin-2 on the release of somatostatin and growth hormone-releasing hormone by mediobasal hypothalamus. Neuroendocrinology 58: 185-190, 1993.
15. Aguila MC Growth hormone-releasing factor increases somatostatin release and mRNA levels in the rat periventricular nucleus via nitric oxide by activation of guanylate cyclase. Proc Natl Acad Sci USA 91: 782-786, 1994.

NEUROENDOCRINE CONTROL OF THE THYMUS

Role of Pituitary Peptides

Mireille Dardenne[1*] and Wilson Savino[2]

[1] CNRS URA 1461
Hôpital Necker, 75015 Paris, France
[2] Department of Immunology
The Oswaldo Cruz Foundation
Rio de Janeiro, Brazil

INTRODUCTION

The existence of a physiological immunoneuroendocrine network working in fine harmony, and clearly contributing to homeostasis, has now been demonstrated. In this context, nervous, endocrine, and immune systems communicate with each other, using common mediators and respective receptors.[1]

An interesting aspect of this network involves the interactions between the thymus and the pituitary. In the present review, we shall focus on recent data concerning the influence of pituitary hormones on the thymic microenvironment and more particularly its epithelial component.

Before presenting the recent data on this subject, we shall briefly discuss some general aspects of the microenvironmental compartment of the thymus and its involvment in intrathymic T cell differentiation.

INTRATHYMIC T CELL DIFFERENTIATION AND THE THYMIC MICROENVIRONMENT

It is presently well established that intrathymic events of T cell differentiation are largely driven by the so-called thymic microenvironment.[2] This non-lymphoid compartment of the organ is formed by different cell types; the most prominent is the epithelial reticulum, forming a tridimensional cellular network in which lymphocytes migrate and differentiate.

* Corresponding author : Mireille Dardenne, CNRS URA 1461, Université Paris V, Hôpital Necker, 161, rue de Sèvres, 75743 Paris Cedex 15, France. Tel: 33 (1) 44 49 53 92; Fax: 33 (1) 44 49 06 76.

Molecular Biology of Hematopoiesis 5, edited by Abraham et al.
Plenum Press, New York, 1996

Thymic epithelial cells (TEC) influence thymocyte differentiation by means of secretory products as well as cell-cell interactions. Among the latter are those involving class I and class II major histocompatibility complex gene products, largely expressed by TEC, and that interact with the T cell receptor in the context of CD8 and CD4 molecules respectively. These interactions are crucial for defining intrathymic selection of the T cell repertoire.[3]

Additionally, classical adhesion molecules expressed by TEC, namely ICAM-1 and LFA3, interact with their respective counter-receptors on thymocytes, namely LFA-1 and CD2 antigens.[4] More recently, extracellular matrix-mediated TEC/thymocyte interactions were also observed to be involved in thymocyte migration and differentiation.[5]

Several secretory products that were able to modulate T cell activities had already been demonstrated as being produced by TEC, for example, interleukins 1,3 and 6,[6,7] granulocyte-macrophage colony stimulating factor and transforming growth factor α.[8]

TEC also secrete a group of polypeptides termed thymic hormones, among which thymosin $\alpha 1$, thymopoietin and thymulin have well determined aminoacid sequences.[9]

Thymulin, originally described by our group,[10] is a nonapeptide that is coupled to zinc in an equimolecular ratio.[11] Such metallopeptidic configuration is actually the biologically active form of the hormone, and can be distinguished from the zinc-free peptide by means of specific monoclonal antibodies and nuclear magnetic resonance profiles.[12,13] This hormone is produced only by TEC,[14] apparently following a classical biosynthetic pathway for secretory polypeptides, in which zinc is probably coupled at the level of Golgi-derived secretory vesicles.[15]

Functionally, thymulin was shown to induce T cell differentiation markers and to modulate T cell activities such as those involving skin graft rejection.[9] Furthermore, we recently showed that thymulin-containing TEC-derived culture supernatants were able to induce thymocyte proliferation, an effect that was largely abrogated in the presence of anti-thymulin monoclonal antibodies.[16]

MECHANISMS CONTROLLING THYMULIN SECRETION

Although circulating levels of thymulin decrease with age, they are rather constant during a large period of life, suggesting the existence of regulatory mechanisms. In fact, we demonstrated both *in vivo* and *in vitro* that the amount of thymulin can itself up- or down-regulate its own production by TEC,[17,18] suggesting an autocrine feedback loop.

Nonetheless, extrinsic circuits, particularly those involving hormones and neuropeptides, also play a significant role in the maintenance of thymulin levels. We demonstrated that thyroid and steroid hormones modulate thymulin secretion both *in vivo* and *in vitro*.[19] Interestingly, we noted that not only thymulin secretion, but other aspects of TEC physiology, including cytokeratin expression and extracellular matrix production, can be modulated by these hormones.[19]

ROLE OF PITUITARY HORMONES IN THE THYMIC EPITHELIUM

The involvement of pituitary hormones in thymus physiology was initially evidenced by the precocious age-dependent decline of circulating thymulin levels observed in dwarf mice.[20] In fact, we recently confirmed that children having congenital GH deficiency also exhibit reduced serum thymulin levels, as already demonstrated by Mocheggiani et al.[21]

A further important aspect concerning the influence of pituitary hormones upon the thymus was raised by Kelley and co-workers who showed a reversal of the age-dependent thymic atrophy and of the accumulation of $CD4^-CD8^-$ thymocytes[22,23] in aging rats injected with the pituitary cell line GH3. It thus appeared worthwhile to address a number of questions concerning the putative influence of growth hormone (GH) and prolactin (PRL) on different aspects of TEC physiology.

Prolactin: A Pleiotropic Modulator of TEC Physiology

We first demonstrated that mice injected with PRL exhibit a progressive and dose-dependent increase in thymulin serum levels. This effect was also observed *in vitro* using murine and human TEC cultures.[24]

This was in keeping with the high thymulin serum levels evidenced in hyperprolactinemic patients bearing pituitary tumors.[25] It was also noteworthy that PRL could enhance thymulin production in normal aging mice as well as in dwarf animals. Additionally, PRL treatment modulated other aspects of TEC physiology such as cytokeratin expression, cell growth and extracellular matrix (ECM) production.

Effects of Growth Hormone on Thymic Epithelial Cells; Relationship to IGF-1

Similar to prolactin, GH can also up-regulate thymulin secretion. This was first demonstrated in aging dogs[26] and more recently extended to the rat and mouse models.[27] Moreover, we showed that acromegalic patients consistently present higher thymulin serum levels (compared to age-matched controls) that were significantly decreased after specific therapy. It is relevant to point out that the enhancing effects of GH upon thymulin secretion were also observed *in vitro*, using murine TEC lines as well as primary cultures of human TEC.[28]

An interesting aspect of the role of GH in the thymic epithelium is the probable mediation by insulin-like growth factor 1 (IGF-1). Not only did we show that IGF-1 can also enhance thymulin production by cultured TEC, but also that the effects of GH can be prevented by treatment with anti-IGF-1 or anti-IGF-1 receptor antibodies. Additionally, in acromegalic patients, a strict correlation between the serum levels of thymulin and IGF-1 was established.[28] Thus it is possible that GH triggers on thymic epithelial cells the production of IGF-1, which in turn stimulates thymulin secretion by an autocrine pathway.

Modulation of TEC/Thymocyte Interactions by Pituitary Hormones

Very recently, we demonstrated that pituitary hormones are also able to modulate, at least in vitro, direct TEC/thymocyte interactions. We first found that both PRL and GH enhance adhesion of freshly isolated thymocytes on cultured TEC. Similar effects were obtained when treating TEC cultures with IGF-1. Moreover, the GH-induced enhancement of TEC/thymocyte adhesion was prevented with anti-IGF-1 or anti-IGF-1 receptor antibodies. Importantly, the enhancement of TEC/thymocyte adhesion following PRL, GH or IGF-1 treatment is at least partially mediated by extracellular matrix molecules, since it can be abrogated if TEC are incubated with specific anti-ECM antibodies.[29]

As a second model for TEC/thymocyte interactions we used the so-called thymic nurse cells (TNC). These are large lympho-epithelial complexes in which one epithelial cell harbours 20-200 differentiating thymocytes.[30] When settled in culture, TNCs spontaneously release their thymocytes by an active cytoskeleton dependent mechanism.[31] In this respect, TNCs can be regarded as an *in vitro* model for TEC-driven thymocyte migration.

We observed that PRL, GH and IGF-1 were able to enhance thymocyte release from TNCs, as assessed by the relative proportions of lymphocyte-containing versus lymphocyte-free TNCs after 3 days of culture.

NEUROENDOCRINE CONTROL OF THYMOCYTE DIFFERENTIATION

Since hormones and neuropeptides can affect TEC functions related to thymocyte differentiation, it is apparent that the latter process is also under neuroendocrine control. However, besides the indirect influences mediated by the thymic microenvironment, direct effects of pituitary hormones have been reported. At least in some experimental conditions, PRL appears to be effective in stimulating thymocyte proliferation by enhancing IL-2 production and IL-2 receptor expression.[32]

In addition to the in vitro data showing direct effects of peptidic hormones upon thymocytes, a series of in vivo experiments evidenced important changes in thymocyte differentiation under neuroendocrine influence. For example, mice injected with anti-PRL antibodies exhibited changes in thymocyte subpopulations an effect which was also observed following treatment with the dopamine agonist bromocriptine. It was also shown that GH injections in aging mice increased total thymocyte numbers and the percentage of CD3-bearing cells,[33] in keeping with our data showing an enhanced concanavalin-A mitogenic response as well as IL-6 production by thymocytes from GH-treated aging animals.[27] Interestingly, similar findings were observed in animals treated with IGF 1.[34,35] In the same vein, it was shown that cyclosporin A-induced thymic atrophy was restored by in vivo treatment with recombinant GH of IGF-1.[36] Additionally, IGF-1 was able to induce repopulation of the atrophic thymus from diabetic rats.[37] Moreover, mouse substrains, selected for bearing high or low IGF-1 circulating levels, exhibited differential thymus developmental patterns that positively correlated with IGF-1 levels.[38]

The role of GH on thymus development is also stressed by the findings obtained with GH-deficient dwarf mice. In these animals, beside the precocious decline in thymulin serum values,[20] there is progressive thymic hypoplasia with decreased numbers of $CD4^+/CD8^+$ double-positive thymocytes. Such defects were largely restored by long-treatment with GH.[39,40]

EXPRESSION OF RECEPTORS FOR PROLACTIN AND GROWTH HORMONE ON THYMIC CELLS

In order to provide a further molecular basis for defining the influence of pituitary hormones on thymic epithelial cells, we also studied the presence of specific receptors for PRL and GH on these cells.

Binding studies revealed a specific GH-receptor in TEC with an affinity constant of 0.14-0.27nM. Such a receptor was present in both murine and human TEC cultures.[41] Since GH probably stimulates an autocrine circuit in the thymus involving IGF-1 receptor, we also investigated the expression of this molecule on TEC membranes. Preliminary results using an anti-IGF-1 receptor monoclonal antibody suggest the presence of this receptor on TEC (unpublished data). Lastly, we demonstrated by means of immunocytochemistry, immunoblotting and Northern blotting the presence of PRL receptors on TEC.[42] Importantly,

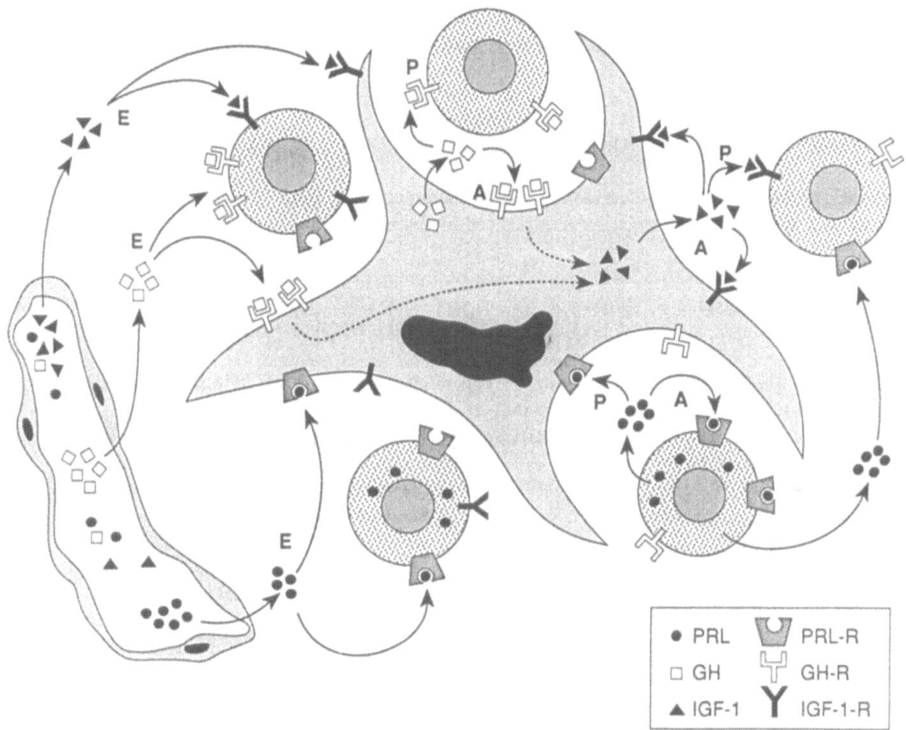

Figure 1. Proposed endocrine, paracrine and autocrine circuits for PRL, GH and IGF-1 action on thymic cells. In this model, PRL, secreted by pituitary cells or thymocytes, and GH, probably via IGF-1 secretion, would exert influences on thymic epithelial cells and thymocytes via specific receptors.

agonist concentrations of anti-PRL receptor antibodies mimicked PRL, in terms of enhancing *in vitro* both thymulin secretion and TEC proliferation.

In addition to the above data, we recently evidenced the expression of GH and PRL receptors on murine and human thymocytes, comprising both mature and immature CD4/CD8-defined subsets.[43,44] These findings, together with the evidence of messenger RNA for PRL in human thymocytes (but not TEC)[45] and for GH in human TEC[46] raise the hypothesis that within the thymus, endocrine, paracrine and autocrine PRL/GH circuits may co-exist, as proposed in figure 1.

CONCLUDING REMARKS

It seems clear from the above data that pituitary hormones, namely prolactin and growth hormone (the latter probably acting via an IGF-1 mediated circuit), are involved in the general neuroendocrine control of the thymus. In fact, by acting pleiotropically on the thymic epithelium, these hormones should ultimately intervene in thymocyte proliferation and differentiation. Nonetheless, it is also possible that these hormones exert direct effects on thymocytes. Moreover, other components of the thymic microenvironment, as for example dendritic cells, may also be under PRL/GH influence.

Lastly, a relevant question that has not so far been addressed is the possibility that, directly and/or indirectly, PRL and GH could modulate the differential expansion of the intrathymic T cell repertoire. This is plausible, since it has already been demonstrated for the *in vivo* influence of oestradiol on intrathymic T cell differentiation.[47]

REFERENCES

1. Blalock JE: Neuroimmunoendocrinology. Chem Immunol 52: 1, 1992
2. Van Ewijk W: T-cell differentiation is influenced by thymic micro-environments. Ann Rev Immunol 9, 591, 1991
3. Boyd RL, Hugo, P : Towards an integrated view of thymopoiesis. Immunol Today 12: 71, 191.
4. Nonayama S, Nakayama S, Shiohara T, Yata J: Only dull CD3+ thymocytes bind to thymic epithelial cells. The binding is elicited by both CD2/LFA-3 and LFA-1/ICAM-1 interactions. Eur J Immunol 30: 783, 1989
5. Savino W, Villa Verde DM, Lannes Vieira J: Extracellular matrix proteins in intrathymic T cell migration and differentiation ? Immunol Today 14: 158, 1993
6. Le PT, Tuck DT, Dinarello CA, Haynes BF, Singer KH: Thymic epithelial cells produce interleukin 1. J Immunol 138: 2520, 1988
7. Le PT, Lazorich S, Whichard LP, Yang, YC, Clarck, SC, Haynes BF, Singer KH: Human thymic epithelial cells produce IL-6, granulocyte-monocyte CSF and leukemia inhibitory factor. J Immunol 145: 3310, 1990
8. Le PT, Kurtzberg J, Brant SL, Nieldel JE, Haynes BH, Singer KH: Human thymic epithelial cells produce granulocyte and macrophage colony-stimulating factors. J Immunol 141: 1211, 1988.
9. Bach JF: Thymic hormones. Clin Immunol Allergy 3: 1, 1983.
10. Bach JF, Dardenne M, Pleau JM, Rosa J: Biochemical characterization of a serum thymic hormone. Nature 266: 55, 1977
11. Dardenne M, Pleau JM, Nabarra B, Lefrancier P, Derrien M, Choay J, Bach JF: Contribution of zinc and other metals to the biological activity of the serum thymic factor (FTS). Proc Natl Acad Sci USA 79, 5370, 1982
12. Dardenne M, Savino W, Berrih S, Bach JF: Evidence for a zinc dependent epitope in the molecule of thymulin (FTS-Zn). Proc Natl Acad Sci USA 33: 687, 1985
13. Cung MT, Marraud M, Lefrancier P, Dardenne M, Bach JF, Laussac JP: NMR study of a lymphocyte differentiating thymic factor. J Biol Chem 263: 5574, 1988
14. Savino W, Dardenne M, Papiernik M, Bach JF: Thymic hormone containing cells. Characterization and localization of serum thymic factor in young mouse thymus studied by monoclonal antibodies. J Exp Med. 156: 628, 1982
15. Savino W, Dardenne M: Thymic hormone containing cells. Characterization of colchicine, cytochalasin B and monensin on the secretion of thymulin by cultured human thymic epithelial cells. J Histochem Cytochem 34: 1719, 1986
16. Villa-Verde DMS, Defresne MP, Greimers R, Dardenne M, Savino W, Boniver J: Induction of thymocyte proliferation by supernatants from a mouse thymic epithelial cell line. Cell Immunol. 136: 113, 1991.
17. Savino W, Dardenne M, Bach JF: Thymic hormone containing cells. III. Evidence for a feed-back regulation of the secretion of the serum thymic factor (FTS) by thymic epithelial cells. Clin Exp Immunol 52: 7, 1983
18. Cohen D, Berrih S, Dardenne M, Bach JF: Feedback regulation of the secretion of a thymic hormone (thymulin) by human thymic epithelial cells in culture. Thymus 8: 109, 1986
19. Dardenne M, Savino W: Neuroendocrine circuits controlling the physiology of the thymic epithelium. Immunol Today 15: 518,1994
20. Pelletier M, Montplaisir S, Dardenne M, Bach JF: Thymic hormone activity and spontaneous auotimmunity in dwarf and their littermates. Immunology 30: 783, 1976
21. Mocchegiani E, Paolucci P, Balsamo A, Cacciari E, Fabris N: Influence of growth hormone on thymic endocrine activity in humans. Horm Res 33: 248, 1990.
22. Kelley KW, Brief S, Weatly HJ, Novakofski J, Bechtel PJ, Simon, J, Walker EB: GH3 pituitary adenoma cells can reverse thymic aging in rats. Proc Natl Acad Sci USA 83: 5663, 1986
23. Li YM, Brunke DL, Dantzer R, Kelley KW: Pituitary epithelial cell implants reverse the accumulation of CD4-CD8- lymphocytes in thymus glands of aged rats. Endocrinology 130: 2703, 1992
24. Dardenne M, Savino W, Gagnerault MC, Itoh T, Bach JF: Neuroendocrine control of thymic hormonal production. I. Prolactin stimulates in vivo and in vitro the production of thymulin by human and murine thymic epithelial cells. Endocrinology 125: 1251, 1989

25. Timsit J, Safieh B, Gagnerault MC, Lubetzki J, Savino W, Bach JF, Dardenne M: Increased levels of thymulin in patients with hyperprolactinemia and acromegaly. C R Acad Sci [III] 310: 7, 1990
26. Goff BL Roth JA, Arp LH, Incefy GS: Growth hormone treatment stimulates thymulin production in aged dogs. Clin Exp Immunol 68: 580, 1987
27. Goya RG, Gagnerault MC, Leite de Moraes MC, Savino W, Dardenne M: In vivo effects of growth hormone on thymus function in aging mice. Brain Behav Immunity 6: 341, 1992
28. Timsit J, Savino W, Safieh B, Chanson P, Gagnerault MC, Bach JF, Dardenne M: Growth hormone and insulin-like growth factor-I stimulate hormonal function and proliferation of thymic epithelial cells. J Clin Endocrinol Metab 75: 183, 1992
29. Mello-Coelho V, Villa-Verde DMS, Lannes-Vieira J, Dardenne M, Savino W: Pituitary hormones modulate extracellular matrix-mediated interactions between thymocytes and thymic epithelial cells (submitted)
30. Wekerle H, Ketelsen UP: Thymic nurse cells-Ia-bearing epithelium involved in T-lymphocyte differentiation ? Nature 283: 402, 1980
31. Andrews P, Boyd RL: The murine thymic nurse cell an isolated thymic micro-environment. Eur J Immunol 15: 36, 1985
32. Mukherjee P, Mastro AM, Hymer WC: Prolactin induction of interleukin-2 receptors on rat splenic lymphocytes. Endocrinology 126: 88, 1990
33. Rebar RW, Miyake A, Low TL, Goldstein AL: Thymosin stimulates secretion of luteinizing hormone-releasing factor. Science 214: 669, 1981
34. Clark R, Strasser J, McCabe S, Robbins K, Jardieu P: Insulin-like growth factor-1 stimulation of lymphopoiesis. J Clin Invest 92: 540, 1993
35. Dorup I, Flyvbjerg A: Effects of IGF-1 infusion on growth and muscle Na (+)-K+ pump concentration in K(+)-deficient rats. Am J Physiol 264:E810, 1993
36. Beschorner WE, Divic J, Pulido H, Yao X, Kenworthy P, Bruce G: Enhancement of thymic recovery after cyclosporine by recombinant human growth hormone and insulin-like growth factor I. Transplantation 52: 879, 1991
37. Binz K, Joller P, Froesch P, Binz H, Zapf J, Froesch ER: Repopulation of the atrophied thymus in diabetic rats by insulin-like growth factor I. Proc Natl. Acad. Sci. USA 87: 3690, 1990
38. Siddiqui RA, McCutcheon SN, Blair HT, Mackenzie DD, Morel PC, Breier BH, Gluckman PD: Growth allometry of organs, muscles and bones in mice from lines divergently selected on basis of plasma insulin-like growth factor-I. Growth Dev Aging 56: 53, 1992
39. Murphy WJ, Durum SK, Longo DL: Role of neuroendocrine hormones in murine T cell development. Growth hormone escents thymopoietic effects in vivo. J Immunol 149: 3851, 1992
40. Murphy WJ, Durum SK, Longo DL: Differential effects of growth hormone and prolactin on murine T cell development and function. J Exp. Med 178: 231, 1993
41. Ban E, Gagnerault MC, Jammes H, Postel-Vinay MC, Haour F, Dardenne M: Specific binding sites for growth hormone in cultured mouse thymic epithelial cells. Life Sciences 48: 2141, 1991
42. Dardenne M, Kelly PA, Bach JF, Savino W: Identification and functional activity of prolactin recetors in thymic epithelial cells. Proc Natl Acad Sci USA 88: 9700, 1991
43. Gagnerault MC, Touraine P, Savino W, Kelly PA, Dardenne M: Expression of prolactin receptors in murine lymphoid cells in normal and autoimmune situations.J Immunol 150: 5673, 1993.
44. Dardenne M, Leite de Moraes MC, Kelly PA, Gagnerault MC. Prolactin receptor expression in human hematopoietic tissues analysed by flow cytofluorometry. Endocrinology, 1994, 134: 2108-2114.
45. Pellegrini I, Lebrun JJ, Ali S, Kelly PA: Expression of prolactin and its receptor in human lymphoid cells. Mol. Endocrinol. 6: 1023, 1992.
46. Maggiano N, Piantelli M, Ricci R, Larocca LM, Capelli A, Ranelletti FO: Detection of growth hormone-producing cells in human thymus by immunohistochemistry and non-radioactive in situ hybridization. J Histochem Cytochem 42:1349, 1994
47. Screpanti I, Meco D, Morrone S, Gulino A, Mathieson BJ, Frati L: In vivo modulation of the distribution of thymocyte subsets: effects of estrogen on the expression of different T cell receptor V beta gene families in CD4-, CD8- thymocytes. Cell Immunol 134: 414, 1991

73

DEVELOPMENTAL AND EVOLUTIONARY ASPECTS OF THYMIC T CELL EDUCATION TO NEUROENDOCRINE SELF

Vincent Geenen,[2*] Abdellah Benhida,[1] Ouafae Kecha,[1] Imane Achour,[1] Eric Vandersmissen,[1] Yves Vanneste,[1] Béatrice Goxe,[1] and Henri Martens[1]

[1] Liège University Medical School
Department of Endocrinology and Molecular Medicine
Institute of Pathology CHU-B23
and
[2] Institute of Chemistry B6
B-4000 Sart Tilman, Belgium

ABSTRACT

Thymic epithelial cells, including nurse cells (TEC/TNC), from various species synthesize neuroendocrine-related precursors belonging to neurohypophysial, tachykinin and insulin hormone families. The thymic repertoire of neuroendocrine-related polypeptides transposes at the molecular level the paradoxical role of the thymus in both T-cell positive and negative selection. On one hand indeed, these precursors are a source of *signals* which interact with neuroendocrine-type receptors expressed by target pre-T cells according to the cryptocrine type of cell-to-cell signaling. On the other hand, the same precursors constitute a source of *self antigens* which are presented to pre-T cells by the thymic major histocompatibility complex (MHC) system. Basically, the model of thymic T-cell education to neuroendocrine self was established from the identification in TEC/TNC of immunoreactive (ir-) oxytocin (OT) as the self-antigen of the neurohypophysial family. Nevertheless, through the expression in TEC/TNC of ir-neurokinin A (NKA) and ir-insulin-like growth factor-II (IGF-II), it perfectly applies to the tachykinin and insulin hormone families.

* Send correspondence to: Vincent Geenen MD, PhD, Research Associate Professor of NFSR, Belgium, Institute of Pathology CHU-B23, Laboratory of Radio-Immunology and Neuroendocrin-Immunology, University of Liège, B-4000 Liège-Sart Tilman/Belgium. Telephone: (32) 41 662550; Fax: (32) 41 662977.

Molecular Biology of Hematopoiesis 5, edited by Abraham et al.
Plenum Press, New York, 1996

THE NEUROHYPOPHYSIAL PEPTIDE/GENE FAMILY

Oxytocin (OT) and vasopressin (VP) belong to a highly conserved family of peptides which is widely extended throughout the animal kingdom.[1] The name of this family comes from the fact that originally these peptides have been identified from the mammalian hypothalamo-neurohypophysial axis. In most of vertebrates, this family is divided into two lineages. The oxytocin (OT) lineage participates in the regulation of the reproductive process at different levels; the vasopressin (VP) lineage mainly intervenes in the control of water metabolism, as well as in the regulation of cardiovascular functions. All members of this family are characterized by a sequence of nine amino acids, with two residues cysteine in position 1 and 6. They are synthesized under the form of larger precursors which all possess a 10-kD neurophysin-like domain in their primary structure. Members of this family have been demonstrated in molluscs and insects, and the corresponding genes were shown to exhibit the same general structure as their mammalian homologs (for a recent review on the molecular biology of the neurohypophysial family, see reference 2).

Despite their high conservation during evolution, the physiological role of the neurophysins remains obscure. In the so-called higher vertebrates, the neurophysin domain of the neurohypophysial precursors binds and transports the active nonapeptides OT and VP along the axons of the hypothalamo-neurohypophysial neurons.[3] However, the expression of genes encoding neurohypophysial-related precursors have been recently established in coelenterates, molluscs and insects. The physiological significance of neurophysins remains unknown in those species, as well as in vertebrate species which do not possess a hypothalamo-neurohypophysial tract.

OT AS A CRYPTOCRINE SIGNAL IN INTRATHYMIC T-CELL DEVELOPMENT

Thymic epithelial and nurse cells (TEC/TNC) from various species synthesize neurohypophysial-related precursors, with a marked dominance of OT and its associated neurophysin.[4,5,6] Using a 3' RACE-PCR protocol, the two neurohypophysial OT and VP genes were demonstrated to be transcribed in the human and murine thymuses.[7] Since OT is the dominant peptide of the family expressed by TEC/TNC, the discrepancy of the data between the thymic cDNA and peptide levels strongly suggests the existence of post-transcriptional modifications. Amazingly, the intrathymic *OT* gene expression and OT synthesis are not correlated with a secretion of the nonapeptide or its neurophysin in the supernatant of primary cultures of human or murine TEC/TNC.[8] An elegant ultrastructural study further demonstrated that thymic ir-OT and neurophysin are expressed in TEC only (and not in immature T cells). Thymic ir-OT is not located in secretory granules, but is diffuse in the cytosol, clear vacuoles and the juxtamembranar space of murine TEC.[9] The model of cryptocrine cell-to-cell signaling has been advanced by Funder to describe the transmembrane exchanges of chemical informations between large nursing cells and immature elements differentiating at their contact.[10] On the basis of the above findings, we have advanced the hypothesis that thymic OT could intervene in a cryptocrine-type signaling between TEC and pre-T cells.

As an important argument supporting an effective cryptocrine signaling in the thymus, neurohypophysial receptors are expressed in the rat thymus, by rat thymocytes (pre-T cells), by a murine pre-T cell line (RL12-NP), as well as by murine cytotoxic T cells. A molecular maturation of the neurohypophysial reception system expressed by T cells seems to occur in parallel with T-cell differentiation since pre-T cells express neurohypophysial V1-type receptors whereas cytotoxic T cells express neurohypophysial receptors of the OT-type.[11] These receptors are functional and are able to transduce neurohypophysial signals according to the rules established in other cellular systems. The interaction between neurohypophysial signals and their pre-T cell receptors is followed by the phosphorylation of several focal adhesion kinases, and this event could play an important role in T-cell interactions with the thymic microenvironment which are important for their developmental programme.

OT AS THE SELF-ANTIGEN OF THE NEUROHYPOPHYSIAL HORMONE FAMILY

The thymus is now thought to be the main site associated with the presentation of the self molecular structure to immature T cells.[12] Some analogous features exist in the biochemical properties between the neurohypophysial peptides and the sequence of peptides naturally presented by MHC class I molecules.[13] To further investigate the hypothesis that OT and associated neurophysin could behave as the self antigens of the neurohypophysial family, human thymic stromal cell membranes were purified, solubilized, and run through an affinity-column prepared with a mAb directed against the monomorphic part of human MHC class I molecules. In the pool of membrane proteins retained by anti-MHC class I, a 55-kD protein was identified by its labeling with an antiserum against neurophysins. This protein could also be precipitated both by anti-neurophysin and anti-MHC class I antibodies.[14] Most plausibly, this 55-kD membrane protein is a hybrid protein with a neurophysin domain (10 kD), as well as a MHC class I heavy chain (45 kD)-related domain. The MHC-related domain is probably implicated in the membrane translocation of this chimeric/hybrid protein, while its neurophysin domain could bind thymic OT for presentation to pre-T cells. The precise biochemical mechanisms leading to the synthesis of this hybrid neurohypophysial/MHC class I protein remain to be further deciphered. Since the three exons of neuro-hypophysial genes are correctly transcribed in the thymus, the origin of this protein could reside at a post-transcriptional level (such as a *trans*-splicing-like event), or at a post-translational level (like the ATP-dependent covalent binding to ubiquitin of intracellular proteins targeted to proteolysis).

As another experimental argument for the role of thymic OT as the self antigen of the neurohypophysial family, we have investigated the effects of the immune recognition of neurohypophysial antigens on the cytokine profile secreted by human TEC in primary cultures. Only Abs directed to OT (but neither anti-VP Abs, nor different preparations of Igs) were able to stimulate the TEC production of IL6 and leukemia-inhibitory factor (LIF) as measured by specific EASIAs (Medgenix Diagnostics, Fleurus, Belgium) (manuscript submitted).

On the basis of these studies on thymic OT, a model (Fig. 1) has been constructed which proposes that neuroendocrine-related thymic peptides (X) mediate into two types of interactions with pre-T cells depending on their intervention as signals or as self antigens of their respective family.

NEUROENDOCRINE-RELATED SELF-ANTIGENS *VERSUS* AUTO-ANTIGENS

With regard to thymic T-cell education to neuroendocrine families, it now appears a kind of economical principle in the organization of the thymic peptide repertoire. TEC are

Thymic epithelial/nurse cell

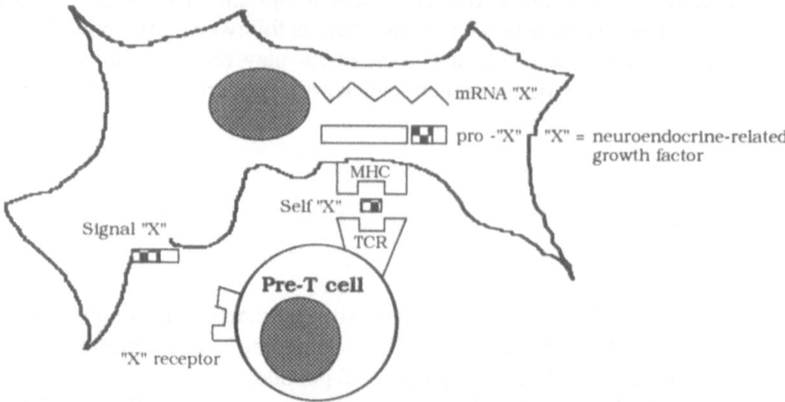

Cryptocrine signaling

Physiology
Accessory signal in T-cell differentiation/activation

Pathophysiology
Paraneoplastic syndromes (*signal* "X")
Oncogenesis ("X" *receptor*)

Pharmacology
Immunomodulation by "X" agonists/antagonists

T-cell self-education

Physiology
Central T-cell tolerance of "X"-related endocrine functions

Pathophysiology
Autoimmunity against "X"- related endocrine functions

Pharmacology
Blockade of "X"- specific autoimmunity

Figure 1. The dual role played by the thymic repertoire of neuroendocrine-related growth factors in T-cell life and death. Thymic polypeptide precursors are a source of membrane-targetted cryptocrine *signals* (signal "X") inducing accessory pathways in T-cell differentiation and activation following their interactions with cognate receptors (receptor "X") expressed by pre-T cells. The same precursors are also a source of neuroendocrine *self-antigens* (self "X") which are presented in relationship with the thymic MHC system. The presentation of neuroendocrine self-antigens could induce the deletion or the developmental arrest of self-reactive T cells bearing a randomly rearranged TCR against their related polypeptide family. These two distinct types of molecular interactions between TEC/TNC and immature T cells are proposed to mediate the dual physiological role of the thymus in T-cell selection. As already shown,[22] immunomodulatory effects may be expected from the pharmacological manipulation of these interactions, both in T-cell activation or in T-cell self-reprogramming.

not the site of expression of all members of one given family, but a representative member is dominantly expressed by TEC/TNC (Table I). From our studies, a difference also appears between neuroendocrine auto-antigens expressed by peripheral tissues tackled by an autoimmune response and their homologous self antigens expressed in thymic epithelium. This difference is important to take into account since autoantigens are known to activate autoreactive immune cells whereas, at least theoretically, self antigens should delete or anergize autoreactive cells. In the neurohypophysial family, while VP could behave as an autoantigen in some cases of idiopathic diabetes insipidus,[15] there is now good evidence that OT is the neurohypophysial self antigen. Since OT intervenes at different steps of the reproductive process, the higher tolerance of the OT lineage would indirectly contribute to the preservation of the species.

In the insulin family, many authors have shown that insulin is an autoantigen involved in the diabetogenic autoimmune process. Thymic epithelium is the site of a marked expression of IGF-II,[16] although IGF-II is not secreted by primary cultures of human or rat TEC.

Table 1. The organization of the thymic repertoire of neuroendocrine self peptide precursors

Neuroendocrine families	Physiological functions	Neuroendocrine self-antigens
Neurohypophysial family		
VP	Water metabolism	OT
OT	Reproduction	
Insulin family		
Insulin	Glucose metabolism	IGF-II
IGF-I	Growth	
IGF-II	Fetal development	
Tachykinins		
SP	Pain, sensory innervation	NKA[23]
NKA	Growth, development	
NKB	?	
Parathormones		
PTH	Calcium metabolism	PTH-rP[24]
PTH-rP	Fetal development	
Calcitonins		
CT	Calcium metabolism	
CGRP	Trophic factor, Fetal development	CGRP[25]
Natriuretic peptides	Sodium metabolism	ANP[26]

(Abbreviations: VP = vasopressin, antidiuretic hormone; OT = oxytocin; IGF = insulin-like growth factor; SP = substance P; NKA or NKB= neurokinin A or B; PTH = parathormone; PTH-rP = parathormone-related peptide; CT = calcitonin; CGRP = calcitonin gene-related peptide; ANP = atrial natriuretic peptide.)

In a recent study, we have identified and characterized IR-IGF-II in the membrane preparations from human thymuses (submitted). This result is surprising because pro-IGFs do not possess any transmembrane domain. In addition, our studies have shown that thymic IGF-II is a source of self peptides that could induce central T-cell tolerance of the whole insulin family.[16] We currently investigate the hypothesis that a molecular defect in thymic T-cell education to IGF-II or to IGF-II-derived self peptides could play a role in the physiopathology of insulin-dependent diabetes (IDDM). IGF-II is very homologous, but not identical to insulin and this biochemical difference could lead to completely opposite immune responses! In other words, while insulin has been shown to activate autoreactive cytotoxic T cells, IGF-II could be able of reprogramming their tolerant state. These differences in the biochemical identity and immunological responses elicited by insulin-related antigens could be fundamental for the design of an efficient and secure prevention of autoimmune IDDM.

A NOVEL ROLE FOR NEUROPHYSIN AND OTHER BINDING PROTEINS?

For a long time, the binding of neurohypophysial nonapeptides to their associated neurophysins for their axonal tranport has been a useful model for understanding the interactions between small peptides and larger proteins.[3,17,18] Interestingly, the residue tyrosine in position 2 of OT and VP was shown to play an important role in this binding. According to our data, even if MHC class I pathways are involved, thymic neurophysin seems to be the final step in the process of OT presentation by TEC to immature T cells. A very close functional analogy thus exists for neurophysins between, on one hand, the binding and transport of OT along the neurohypophysial axons until the nerve endings of the posterior pituitary and, on the other hand, the binding and presentation of the self antigen OT to immature T cells at the TEC membrane. The major implication of this new physiological role of neurophysin is that the central T-cell education to neurohypophysial self is *not* restricted by MHC class I alleles. Another selective advantage resides in the fact that, through thymic neurophysin, the *structure* of neurohypophysial-related peptides can be presented to pre-T cells. Because of their ring structure, a presentation by classic MHC class I alleles was indeed excluded.[13] The absence of a tight MHC allelic restriction in the process of central T-cell tolerance to neuroendocrine self principles should have important implications in the design of preventive strategies for autoimmune endocrine diseases (such as vaccination procedures based on neuroendocrine self antigens).

In the insulin-related peptide system, the role of binding and transport proteins is assumed by IGF-binding proteins (IGFBPs). Contrary to neurophysins, IGFBPs are not part of IGF precursors and are encoded by separate genes. Interestingly, some IGFBPs are anchored to cell membranes, but the hypothesis of a relationship with MHC has never been explored. The still hypothetical presentation by a membrane-anchored IGFBP could explain why IGF-II is detected in membranes but not in the supernatants of TEC cultures.

DEVELOPMENTAL AND EVOLUTIONARY ASPECTS[19]

Both in ontogeny and phylogeny, a continuum of interactions clearly appears between the neurohypophysial family and the Ig/MHC/TCR superfamily. At the cellular level in the thymus, TNCs were shown to constitute a crucial microenvironment where such interactions occur. This continuum of neuroendocrine-immune interactions culminates at the biochemical level with the identification in thymic membranes of a hybrid neurohypophysial/MHC class I-related protein. Though the post-transcriptional mechanisms leading to this protein remain to be further explored, this hybrid protein strongly argues for a common ancestral origin of the main families of molecules implicated in cell-to-cell signaling and recognition. It is also noteworthy that the diversification of both families seems to have occurred at about the same time than the emergence of early vertebrates. With regard to the Ig superfamily, its extreme diversification was catalyzed by the recombinases (RAG 1 and 2) recently identified.[20] Since the existence of neurohypophysial precursors has been established in molluscs and insects, it is tempting to hypothesize that some biochemical properties of this family[21] could have used as "structural guides" during further evolution of the Ig/MHC/TCR superfamily.

The immune system has primarily evolved to protect the integrity of "self" against aggression from "nonself" infectious invaders. Given the common peptide nature of most allo-, auto-, and self-antigens, the immune system has to be educated to recognize and to tolerate the molecular structure of "self". Even though peripheral tolerogenic pathways are increasingly established, the thymus is clearly recognized as playing the central role in allowing T cells to recognize self-antigens. Since differentiation of the whole T cell repertoire involves recombination at random of gene segments coding for the antigen receptor (TCR) chains, the emergence of self-reactive T cells may naturally follow this highly hazardous biological phenomenom. The thymus thus exerts a radical "anti-hazard" constraint by purging the immune system of self-reactive T cells which otherwise represent a serious threat for survival. In the same global perspective, pathological autoimmunity may be considered as the tribute paid by mammalian species for the higher complexity and efficiency of their immune defenses.

ACKNOWLEDGMENT

This work was supported by the special research fund of Liège University Medical School, by the National Fund of Scientific Research (grant n° 3.4562.90), by Télévie/FRSM (grants n° 7.4611.91 and 7.4548.93), by the Belgian University Foundation (Brussels), by the Association contre le Cancer (Belgium), by the Association pour la Recherche contre le Cancer (France), by the European Science Foundation, and by Juvenile Diabetes Foundation International. A part of this paper has been published in Journal of Molecular Medicine 75: 449-455 (1995).

REFERENCES

1. Acher R: Neurohypophysial peptide systems: Processing machinery, hydroosmotic regulation, adaptation and evolution. Regul Pept 45: 1, 1993.
2. Gainer H, Wray S: Cellular and molecular biology of oxytocin and vasopressin, in Knobil E, Neill JD (eds): The Physiology of Reproduction, Second Edition. New York, NY, Raven Press, 1994, p 1099.
3. Breslow E: Chemistry and biology of the neurophysins, Annu Rev Biochem 48: 251, 1979.
4. Geenen V, Legros JJ, Franchimont P, Baudrihaye M, Defresne MP, Boniver J: The neuroendocrine thymus: Coexistence of oxytocin and neurophysin in the human thymus, Science 232: 508, 1986.
5. Geenen V, Legros JJ, Franchimont P, Defresne MP, Boniver J, Ivell R, Richter D: The thymus as a neuroendocrine organ. Synthesis of vasopressin and oxytocin in human thymic epithelium, Ann NY Acad Sci 496: 56, 1987.
6. Geenen V, Defresne MP, Robert F, Legros JJ, Franchimont P, Boniver J: The neurohormonal thymic microenvironment: Immunocytochemical evidence that thymic nurse cells are neuroendocrine cells, Neuroendocrinology 47: 365, 1988.
7. Geenen V, Vandersmissen E, Martens H, Goxe B, Kecha O, Legros JJ, Lefèbvre PJ, Benhida A, Rentier-Delrue F, Martial J: Cellular and molecular aspects of thymic T-cell education to neurohypophysial principles, in Saito T, Yoshida S, Imura H (eds): Proceedings of the First Joint World Congress of Neurohypophysis and Vasopressin. Amsterdam, NL, Elsevier, 1995 (in press).
8. Geenen V, Robert F, Martens H, Benhida A, Degiovanni G, Defresne MP, Boniver J, Legros JJ, Martial J, Franchimont P: Biosynthesis and paracrine/cryptocrine actions of "self" neurohypophysial-related peptides in the thymus, Mol Cell Endocrinol 76: C27, 1991.
9. Wiemann M, Ehret G: Subcellular localization of immunoreactive oxytocin within thymic epithelial cells of the male mouse, Cell Tissue Res 273: 79, 1993.
10. Funder JW: Paracrine, cryptocrine, acrocrine, Mol Cell Endocrinol 70: C21, 1990.
11. Martens H, Robert F, Legros JJ, Geenen V, Franchimont P: Expression of functional neurohypophysial peptide receptors by murine immature and cytotoxic T-cell lines, Prog NeuroEndocrinImmunol 5: 31, 1992.
12. Kruisbeek A: Development of $\alpha\beta$ T cells, Curr Opin Immunol 5: 227, 1993.
13. Rammensee HG, Falk K, Rötzschke O: Peptides naturally presented by MHC class I molecules, Annu Rev Immunol 11: 213, 1993.
14. Geenen V, Vandersmissen E, Cormann-Goffin N, Martens H, Legros JJ, Degiovanni G, Benhida A, Martial JA, Franchimont P: Membrane translocation and relationship with MHC class I of a human thymic neurophysin-like protein, Thymus 22: 55, 1993.
15. Geenen V, Achour I, Robert F, Vandersmissen E, Sodoyez JC, Defresne MP, Boniver J, Lefebvre PJ, Franchimont P: Evidence that insulin-like growth factor II (IGF-II) is the dominant thymic peptide of the insulin superfamily, Thymus 21: 115, 1993.
16. Imura H, Nakao K, Shimatsu A, Ogawa Y, Sando T, Fujisawa I, Yamabe H: Lymphocytic infundibuloneurohypophysitis as a cause of central diabetes insipidus, N Engl J Med 329: 683, 1993.
17. Ando S, McPhie P, Chaiken, IM: Sequence redesign and the assembly mechanism of the oxytocin/bovine neurophysin I biosynthetic precursor, J Biol Chem 262: 12962, 1987.
18. Chen L, Rose JP, Breslow E, Yang D, Chang WR, Furey Jr WF, Sax M, Wang BC: Crystal structure of a bovine neurophysin II dipeptide complex at 2.8A determined from the single-wavelength anomalous scattering signal of an incorporated iodine atom, Proc Natl Acad Sci USA 88: 4240, 1991.
19. Geenen V, Wiemann M, Martens H: Thymus Gland: Neuroendocrin-Immunology, in Adelman G, Smith B (eds), Encyclopedia of Neuroscience, Second Edition. New York, NY, Elsevier, 1995: in press.

20. Bartl S, Baltimore D, Weissman IL: Molecular evolution of the vertebrate immune system, Proc Natl Acad Sci USA 91: 10769, 1994.
21. Capra JD, Cheng KW, Friesen HG, North WG, Walter R: Evolution of neurophysin proteins: the partial sequence of human neurophysin-I, FEBS Lett 46: 71, 1974.
22. Geenen V, Martens H, Robert F, Vrindts-Gevaert Y, De Groote D, Franchimont P: Immunomodulatory properties of cyclic hexapeptide oxytocin antagonists. Thymus 20: 217, 1992.
23. Ericsson A, Geenen V, Robert F, Legros JJ, Vrindts-Gevaert Y, Franchimont P, Brene S, Persson H: Expression of preprotachykinin A and neuropeptide-Y messenger RNA in the thymus. Mol Endocrinol 4: 1211, 1990.
24. Kramer S, Reynolds FH Jr, Castillo M, Valenzuela DM, Thorikay M, Sorvillo JM: Immunological identification and distribution of parathyroid hormone-like protein polypeptide in normal and malignant tissues. Endocrinology 128: 1927, 1991.
25. Bulloch K, McEwen BS, Diwa A, Baird S: Relationship between dehydro-epiandrosterone and calcitonin-gene related peptide in the mouse thymus. Am J Physiol 268: E168, 1995.
26. Vollmar AM, Schulz R: Atrial natriuretic peptide is synthesized in the human thymus. Endocrinology 126: 2277, 1990.

HEMATOPOIETIC RESCUE VIA α1-ADRENOCEPTORS ON BONE MARROW B CELL PRECURSORS AND ENDOGENOUS TRANSFORMING GROWTH FACTOR-β*

Mauro Togni and Georges J. M. Maestroni†

Center for Experimental Pathology
Istituto Cantonale di Patologia, 6604 Locarno, Switzerland

ABSTRACT

We demonstrated that adrenergic agents may affect hematopoiesis via high and low affinity α1-adrenergic receptors present on bone marrow cells.[1-3] Here we show that the high affinity, α1-adrenoceptor is present on $Mac1^-B220^+sIgM^-$ (pre-B) cells. Conversely, the low affinity α1-adrenoceptor seems to be present on $Mac1^+B220^-$ cells. Noradrenaline administration in mice rescued hematopoiesis from the toxic effect of carboplatin or X-rays sublethal irradiation. The protection was reflected by higher leukocyte and platelets counts as well as by increased bone marrow granulocyte/macrophage colony-forming units (GM-CFU). At its most effective dose (3 mg/ kg, s.c.), noradrenaline protected 77% of the mice injected i.v. with 200 mg/ kg of carboplatin (LD 100 : 170 mg/kg) but not mice which were lethally irradiated (900cGy). In vitro, noradrenaline (1 μM) rescued GM-CFU in unseparated bone marrow cells containing the adherent population expressing the high affinity α1-adrenoceptor, but not in non-adherent cells containing the low affinity receptor. Consistently, the hematopoietic rescue was counteracted by low concentrations (0.1 nM-10 nM) of the α1-antagonist prazosin. Anti-transforming growth factor-β (TGF–β) monoclonal antibodies prevented the hematopoietic rescue exerted by noradrenaline. This suggests that activation of α1-adrenoceptor on pre-B cell results in production of endogenous TGF-β. Our findings describe a novel mechanism of hematopoietic regulation and might find application in preventing the myeloablative effect of anti-cancer treatments.

* This study has been supported by the Fondazione Ticinese per la Ricerca contro il Cancro.

† Send correspondence to: Dr. Georges J. M. Maestroni, Center for Experimental Pathology, Istituto Cantonale di Patologia, 6604 Locarno, Switzerland. Tel. 091 756 26 71; Fax 091 756 26 90.

INTRODUCTION

The existence of multiple hematopoietic regulators such as cytokines and growth factors is well documented.[4] The multiplicity of hematopoietic regulators seems to reflect the need for a subtle physiological control of the complex cell mixtures required in certain situations.[4] If this pose several problems in our understanding of hematopoiesis, the situation is much worse from the clinical point of view. Single hematopoietic regulators are already used to counteract the bone marrow toxicity of cancer chemotherapy compounds or to enhance regeneration after bone marrow transplantation. However, such procedures remain problematic because of negative side effects and high costs.[5] An endogenous modulation of hematopoietic regulators is likely to present substantial advantages over exogenous administration and might also circumvent the need for a very difficult clinical testing of thousands of regulators combinations. We have recently shown that the neurohormone melatonin may rescue hematopoiesis from the toxic effect of cancer chemotherapy compounds via induction of an interleukin-4-like factor which in turn stimulates the endogenous production of colony-stimulating factors.[6, 7] In a related series of studies we have shown that chemical sympathectomy by 6-hydroxydopamine or administration of the α1-adrenergic antagonist prazosin enhance myelopoiesis and exert an inhibitory effect on lymphopoiesis.[1, 3] Further studies demonstrated that noradrenaline and/or other adrenergic agonists can inhibit the growth of GM-CFU in vitro and that this effect is exerted on adrenoceptors present on bone marrow cells.[2] Functional and pharmacological studies revealed the presence of two specific binding sites for ^{3}H-prazosin which differed in their affinity. Competition studies characterized the high affinity site as an α1B-adrenergic receptor.[2] The remaining site was of less clear characterization and the results obtained were compatible with a low affinity α1-adrenoceptor. Separation of bone marrow cells by counterflow centrifugal elutriation resulted in separation of the two adrenoceptors, with the α1B-adrenergic receptor being partially eluted in a lymphoid fraction containing no blasts and no assayable GM-CFU.[2]

This study is aimed at identifying the cell types bearing the α1-adrenergic receptors and at investigating the feasibility of an adrenergic modulation of the hematological toxicity of anti-cancer chemotherapy compounds or X-rays irradiation. Here we report that the high affinity, α1B-adrenergic receptor is present in loosely adherent, Mac1$^-$, B220$^+$, sIgM$^-$ cells while the low affinity α1 adrenoceptor is present in adherent and non-adherent Mac1$^+$B220$^-$ cells. Activation of the α1B-adrenergic receptor seems to induce the release of TGF-β which in turn rescues hematopoiesis from the toxic action of carboplatin or sublethal X rays exposure.

ADRENERGIC RECEPTORS IN ADHERENT VS NON-ADHERENT CELLS

We have demonstrated that counterflow centrifugal elutriation can separate the bone marrow cell population bearing the high affinity adrenoceptor.[2] However, the separation was only qualitative and the efficiency was poor. In the attempt to find a better method to separate such population we found that the simple procedure of adherence could be used to separate the α1-adrenoceptors. The high affinity adrenergic receptor was completely separated in the adherent and loosely adherent cell population.[8] In some experiments in which non-adherent cells were separated by repeated, vigorous washings, we got an indication that loosely adherent cells contains probably the population bearing such receptor. Conversely, the low affinity 3H-prazosin binding seems distributed both in adherent and non-adherent cells. The Kd and Bmax values were fully consistent with our previous results.[2] Taking advantage of

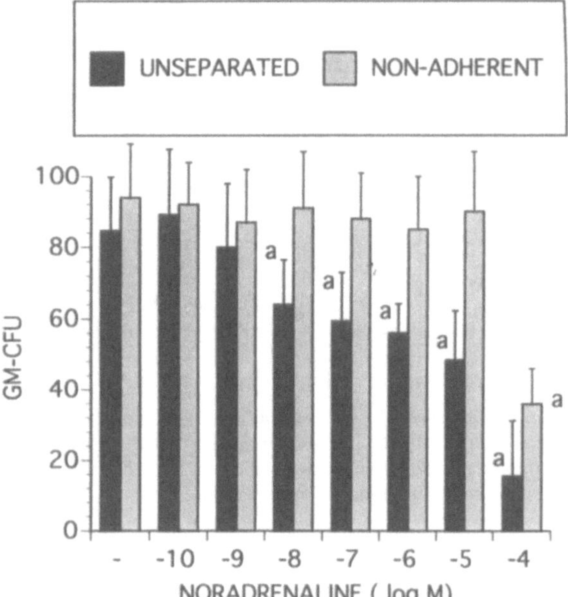

Figure 1. Noradrenergic inhibition of GM-CFU in unseparated vs non adherent bone marrow cells. NA was dissolved in α–MEM/ 20% horse serum and added at the reported concentrations in the cell suspension before plating. For each concentration the mean value from 5 experiments ±SD is shown. a: $p < 0.01$ (ANOVA on colony counts).

such different distribution, we investigated which adrenoceptor mediates the GM-CFU inhibition that occurs upon addition of noradrenaline in cultures of unseparated bone marrow cells.[2] The results shown in figure 1 indicate that when added in GM-CFU cultures of non-adherent bone marrow cells, noradrenaline was effective only at 0.1 mM, while in unseparated cells the effect was already significant at 10 nM. We did not investigate the noradrenaline sensitivity of adherent and loosely adherent cells because, as it has been reported,[9] the concentration of GM-CFU precursors in such population was too low. However, a reasonable interpretation of our results is that, at physiological concentrations, noradrenaline inhibits GM-CFU via the high affinity α1-adrenoceptor which is contained in adherent and loosely adherent but not in non-adherent bone marrow cells.

CHARACTERIZATION OF THE α1-ADRENOCEPTORS POSITIVE CELLS

We concentrated our efforts on adherent and loosely adherent cells which represented 10-13% of the whole bone marrow and contained all the high affinity and part of the low affinity 3H-prazosin binding. Flow cytometry of such population in 3 experiments revealed that the phenotypic markers most represented were Mac1 (49.1 ± 13.7 %) and B220 (38.1 ± 8.8%). Adherent and loosely adherent cells were therefore processed by magnetic cell sorting to deplete or enrich $Mac1^+$ or $B220^+$ cells. The depleted or enriched subsets were then evaluated for their capacity to bind 3H-prazosin. Table 1 shows the results of such experiments. The extent of enrichment or depletion was relatively incomplete because we had to limit the magnetic cell sorting to a single run in order to get enough cells for the binding studies. It should be noted, in fact, that the number of cells in the isotherm saturation studies in enriched fractions was as low as 250×10^3 per tube. In spite of this, Ligand analysis[10] gave values which were consistent with those obtained in the whole subset of adherent and loosely adherent cells or in unseparated cells.[2, 8] Taken together the results obtained indicate the presence of the high affinity α1-ad-

Table 1. Characterization of the $\alpha 1$–adrenoceptors positive adherent bone marrow cells

Adherent cells	% MAC1$^+$	% B220$^+$	Kd_{high} nM	$Bmax_{high}$ fM / 10^6cells	Kd_{low} nM	$Bmax_{low}$ fM / 10^6cells
Mac1 depleted	6 ± 4	63 ± 6.9	2 ± 1	14 ± 6.3	-	-
B220 depleted	67.9 ± 7.8	10.7 ± 10.6	-	-	46 ± 19	200 ± 80
MAC1 enriched	81 ± 4.3	12 ± 5	-	-	54 ± 12	189 ± 36
B220 enriched	13.4 ± 1.3	74.4 ± 14.1	1.5 ± 0.9	56.4 ± 5.4	-	-

The mean values of 5 experiments are shown. Variations are SD of percentage and SE of Kd and Bmax. The number of cells per tube in the 3H-prazosin saturation studies was of 250 x 10^3 in the enriched populations and 10^6 in the depleted ones. The Kd and Bmax values were obtained by analyzing together the isotherm saturation curves by Ligand.

renoceptor in Mac1$^-$B220$^+$ cells. Whereas the low affinity receptor appears to be present in Mac1$^+$B220$^-$ cells. Although we cannot exclude it, by our approach no population seems to bear both adrenoceptors. At this point, we studied whether the high affinity $\alpha 1$-adrenoceptor bearing cells are B-cell precursors (sIgM$^-$) or mature B cells (sIgM$^+$). We discarded a further MACS fractionation of B220$^+$ cells because of the poor recovery rate (see above). We considered that spleen mononuclear cells, which contain only mature B cells, bear the low affinity receptor alone (Kd: 65 nM; Bmax: 115 fM/10^6 mononuclear cells). Therefore we performed 3H-prazosin saturation studies in cells harvested from 28 days-long, IL7 cultures. Such cell population was constituted of 97 ± 1.5 % Mac1-, sIgM-, B220+ B precursors and expressed indeed the high affinity $\alpha 1$-adrenoceptor.[8]

HEMATOPOIETIC RESCUE

We reasoned that the noradrenaline-induced inhibition of GM-CFU might influence the hematopoietic toxicity of myeloablative anticancer treatments. To investigate this point, noradrenaline or saline were injected in mice either before or after X rays irradiation or carboplatin injection. The effect of such treatments on peripheral blood leukocytes, platelets and bone marrow GM-CFU is reported in table 2. Apparently, noradrenaline could rescue hematopoiesis from the toxic effect of X rays or carboplatin. When survival was evaluated, noradrenaline protected mice that were injected with a lethal dose of carboplatin. The dose-response curve of noradrenaline is bell-shaped with beginning of protection at 1 mg and decrease of rescue capacity over 3 mg / kg body weight (fig. 2). In contrast, noradrenaline did not protect lethally irradiated (900 cGy) mice. No difference in survival between noradrenaline and PBS treated animals was observed (data not shown).

EFFECT ON CARBOPLATIN TOXICITY IN VITRO

To establish whether noradrenaline can rescue GM-CFU via the high affinity $\alpha 1$-adrenoceptor, unseparated bone marrow cells were incubated with carboplatin in presence or

Table 2. Hematopoietic rescue by NA in mice treated with carboplatin or exposed to X-rays

	NA			PBS		
	L/µl (x 10³)	P/µl (x 10³)	GM-CFU / femur	L/µl (x 10³)	P/µl (x 10³)	GM-CFU / femur
TBI 300 cGy (6)	1.9 ± 0.11 [b]	26.5 ± 1.2 [b]	3634 ± 861 [b]	0.9 ± 0.2	18.1 ± 1.6	1659 ± 611
TBI 400 cGy (6)	1.3 ± 0.15 [b]	23.1 ± 0.9 [b]	4680 ± 979 [b]	0.7 ± 0.1	17.3 ± 0.6	1749 ± 627
TBI 500 cGy (6)	1.02 ± 0.2	19 ± 0.35	1029 ± 208 [c]	1.1 ± 0.3	17.8 ± 1.4	423 ± 249
CP (15)	5.8 ± 1.6 [a]	54.3 ± 21.9 [a]	3630 ± 1721 [a]	2.5 ± 0.7	26.1 ± 13	995 ± 535
CONTROL (7)	10.6 ± 2.1	232 ± 37	7965 ± 1356	9.6 ± 2.6	225 ± 27	8870 ± 1235

C57BL/6 mice were either treated with one single, lethal intravenous injection of carboplatin (200 mg / kg body weight) or exposed to 300-900 cGy (X rays) total body irradiation (TBI) using a linear accelerator (15 MV energy equivalent). NA was injected subcutaneously 1 h and immediately before. as well as 2 and 4 h after carboplatin. In the irradiation experiments, noradrenaline was injected 4 h. 2 h and 30 min before X rays exposure. Controls were injected with noradrenaline or saline according to the schedule used in the carboplatin experiments. The mean concentration ± the standard deviation of peripheral blood leukocytes (L), platelets (P) and marrow GM-CFU, 3 days after treating C57BL/6 mice with NA (or PBS) and carboplatin (CP) or X-rays radiation is shown. The values relative to carboplatin represent the mean of 3 experiments. a : $p < 0.001$; b: $p < 0.005$; c: $p < 0.05$.

absence of noradrenaline ± the α1-antagonist prazosin. After incubation, cells were plated in the GM-CFU assay. As shown in figure 3, prazosin was able to counteract the protective effect of noradrenaline already at low concentration (0.1 nM) which suggested the direct involvement of the high affinity α1-adrenoceptors present in pre-B cells. To further substantiate this point, non-adherent bone marrow cells which contain only the low affinity receptor, were incubated with carboplatin ± various concentrations of noradrenaline. In analogy with the effect in GM-CFU cultures only the highest concentration (0.1 mM) of noradrenaline could exert a weak protective effect.[8]

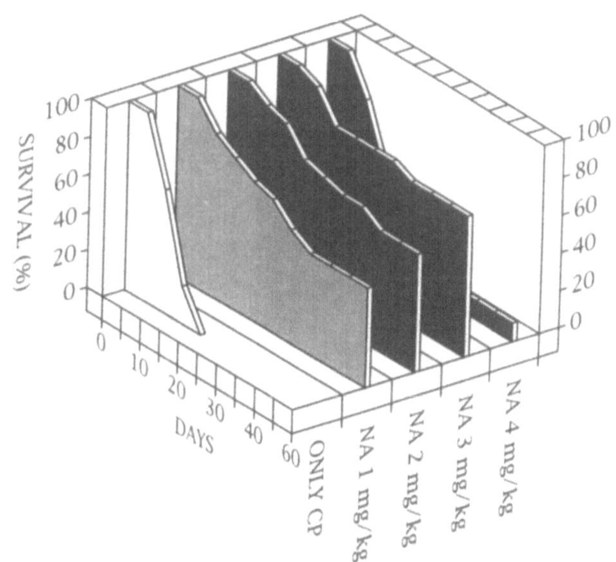

Figure 2. Survival of mice treated with carboplatin (CP) ± NA. The mice were treated with the reported NA doses and with CP according to the protocol described in the legend to table 2 and then observed daily for survival.

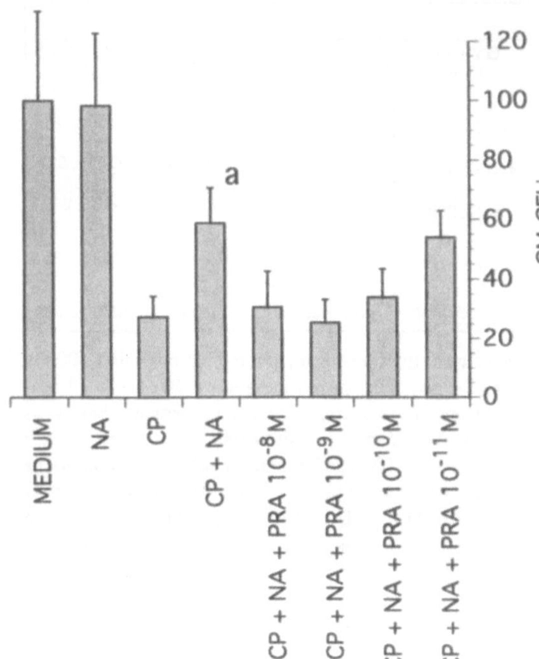

Figure 3. Rescue of GM-CFU in bone marrow cells incubated with carboplatin via α1-adrenoceptors. Unseparated bone marrow cells were incubated (37° C, 8 h) with carboplatin (CP, 25 μM), in presence or absence of NA (1μM) and with the reported concentration of the α1–adrenergic antagonist prazosin (PRA). Control cultures were incubated with tissue culture medium (MEDIUM) or NA alone. The values are the mean of 5 experiments ± SD. a: $p < 0.01$ vs CP; CP+NA+PRA 10^{-8} - 10^{-10} M.

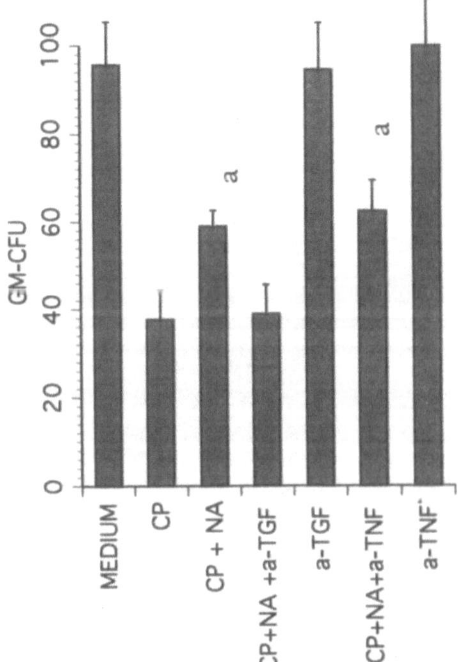

Figure 4. Effect of anti-TGF-β monoclonal antibody on the noradrenergic inhibition of GM-CFU. Unseparated bone marrow cells were assayed for GM-CFU in presence or in absence of NA (1 μM) and anti-mouse TGF-β (a-TGF), 10 μg/ml or anti-mouse TNF-α. (a-TNF), 25 μg /ml.Controls were cells alone, with anti-mouse TGF-β or anti-mouse TNF-α antibobies. The values represent the mean of 3 experiments ± SD. a: $p < 0.002$.

EFFECT OF ANTI-MOUSE TGF-β MONOCLONAL ANTIBODY ON HEMATOPOIETIC RESCUE INDUCED BY NORADRENALINE

Cells of the B lineage can produce inhibitory cytokines such as tumor necrosis factor-α (TNF–α) and TGF-β.[11] Boh cytokines have been reported to rescue hematopoiesis from myeloablative treatments,[12-13] TGF-β has been also proposed to be a physiological regulator of B cell growth and differentiation.[14] We, therefore, investigated whether anti-TNF–α or anti-TGF-β antibodies could neutralize the inhibitory effect exerted by noradrenaline on GM-CFU and, eventually, the noradrenaline-induced hematopoietic rescue. As shown in figures 4 and 5, anti-mouse TGF-β but not anti-TNF–α antibodies were indeed able to counteract the effect of noradrenaline both in GM-CFU cultures (Fig. 4) or in the rescue experiments (Fig.5). These results together with those shown in figure 3 suggest that activation of high affinity α1-adrenoceptors present in B cell precursors induces the production or activation of TGF-β which in turn mediates the hematopoietic inhibitory and rescue effects.

DISCUSSION

In this study, we show that high affinity α1-adrenoceptors are present in $Mac1^-$ $B220^+sIgM^-$bone marrow cells. The low affinity receptor seems located in $Mac1^+B220^-$ cells. This result is consistent with our previous studies performed in counterflow centrifugal elutriation fractions.[2]

Moreover, we show that noradrenaline can induce, either in vivo or in vitro, a biologically significant hematopoietic rescue from the toxic effect of X rays or cancer chemotherapy compounds. The in vitro data suggest that such rescue is mediated by the high affinity α1-adrenoceptor via TGF-β production or activation.

The noradrenaline stimulation of endogenous TGF-β seems relevant to our understanding of the complex mechanisms which regulate the blood forming system. TGF–β is a negative regulator of hematopoiesis[15, 16] and its involvement as product of α1-adrenergic

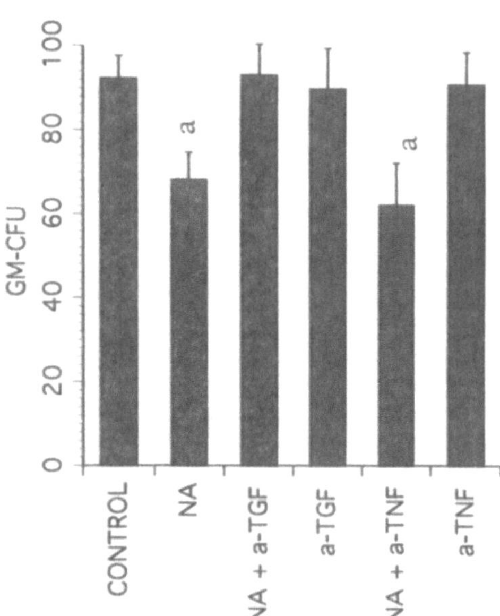

Figure 5. Effect of anti-TGF-β monoclonal antibody on the noradrenergic rescue of bone marrow cells. Unseparated bone marrow cells were incubated (37° C, 8 h) with carboplatin (25 μM) alone or in presence of NA (1 μM) ± anti-TGF-β (a-TGF,10 μg/ml), or ± anti-mouse TNF-α (a-TNF; 25 μg/ml) antibobies. and assayed for GM-CFU. Controls were cells incubated with culture medium (MEDIUM), anti-TGF-β or anti-TNF-α monoclonal antibodies alone. The values are from 3 experiments ± SD. a: $p < 0.001$.

activation may account for the granulocytic hyperplasia observed in mice treated chronically with the α1-adrenergic antagonist prazosin.[1-3] TGF–β has been also reported to regulate pre–B cell differentiation and proliferation.[14, 17] Our results suggest that pre–B cells may regulate their own proliferation and differentiation in an autocrine manner via TGF–β secretion. TGF–β seems, in turn, to be under control of sympathetic nerve fibers entering the bone marrow.[18] Our finding might have also a pathological relevance in that TGF–β has been reported to be an endogenous growth inhibitor of chronic lymphocytic leukemia B cells.[19] Beside hematopoiesis, TGF–β has been reported to influence a large number of other physiological phenomena. To remain whithin the bone marrow microenvironment, TGF–β has been shown to promote bone formation[20] and this might reflect an unsuspected sympathetic nervous influence on osteogenesis.

The hematopoietic rescue exerted by noradrenaline seems also interesting. Our findings are consistent with previous reports which show that administration of TGF–β may protect mice from the toxic effect of cancer chemotherapeutic agents.[21, 22] However, noradrenaline could rescue mice that were injected with a lethal dose of carboplatin but not lethally irradiated mice. The reason for such difference is not clear and should be matter of further studies.

In conclusion, our findings suggest that the sympathetic neurotransmitter noradrenaline participates in controlling hematopoiesis by inducing the release of TGF–β upon activation of high affinity α1-adrenoceptors present in pre–B cells. Such noradrenergic hematopoietic regulation provides a new conceptual and pharmacological approach for further investigations and relevant clinical applications.

ACKNOWLEDGMENTS

The authors thank Dr. G. Garavaglia, Ospedale S.Giovanni, Bellinzona for the irradiation facilities and Ms. E. Hertens and Ms. P. Galli for excellent technical assistance.

REFERENCES

1. Maestroni GJM, Conti A Pedrinis E: Effect of adrenergic agents on hematopoiesis after syngeneic bone marrow transplantation in mice. Blood 5: 1178, 1992.
2. Maestroni GJM Conti A: Modulation of hematopoiesis via alpha-1 adrenergic receptors on bone marrow cells. Exp. Hematol. 22: 314, 1994.
3. Maestroni GJM Conti A: Noradrenergic modulation of lymhohematopoiesis. Int. J. Immunopharmacol. 16: 117, 1994.
4. Metcalf D: Hematopoietic regulators: redundancy or subtlety ? Blood 82: 3515, 1993.
5. Nemunaitis J, Appelbaum FR, Singer K, Lilleby K, Wolff S, Greer JP, Bierman P, Resta D, Campion M, Levitt D, Zeigler Z, Rosenfeld C, Shadduck RK Buckner CD: Phase I trial with recombinant human interleukin-3 in patients with lymphoma undergoing autologous bone marrow transplantation. Blood 82: 3273, 1993.
6. Maestroni GJM, Covacci V Conti A: Hematopoietic rescue via T-cell-dependent, endogenous GM-CSF by the pineal neurohormone melatonin in tumor bearing mice. Cancer Res. , 54:2429, 1994.
7. Maestroni GJM, Conti A Lissoni P: Colony-stimulating activity and hematopoietic rescue from cancer chemothereapy compounds are induced by melatonin via endogenous interleukin 4. Cancer Res. 54:4740, 1994.
8. Togni M, Maestroni GJM: Hematopoietic rescue in mice via α1-adrenoceptors on bone marrow B cell prescursors. Int. J. Oncology 9:313, 1996.
9. Kiefer F, Wagner F Keller G: Fractionation of mouse bone marrow by adherence separates primitive hematopoietic stem cells from in vitro colony-forming cells and spleen colony-forming cells. Blood 78: 2577, 1991.

10. Munson PJ Rodbard D: LIGAND: A versatile computerized approach for the characterization of ligand binding systems. Anal. Biochem. 107: 220, 1980.
11. Callard RE (ed): Cytokines and B lymphocytes. London, Academic Press, 1990.
12. Dalmau SR, Freitas CS Tabak DG: Interleukin-1 and tumor necrosis factor-alpha as radio and chemoprotectors of bone marrow. Bone Marrow Transplant. 12: 551, 1993.
13. Pierce DF Coffey RJ: Therapeutic manipulation od cytokines : transforming growth factor beta-1 protects mice treated with lethal doses of cytarabine. Am. Surg. 60: 18, 1994.
14. Rehaman JA LeBien TW: Transforming growth factor-beta regulates normal human pre-B cell differentiation. Int. Immunol. 6: 315, 1994.
15. Ruscetti FW, Dubois C, Falk LA, Jacobsen SE, Sing G, Longo DL, Wiltrout RH Keller JR: In vivo and in vitro effects of TGF-b 1 normal and neoplastic haemopoiesis. Ciba Found. Symp. 1991 157: 212, 1991.
16. Dubois CM, Ruscetti FW, Stankova J Keller JR: Transforming growth factor-beta regulates c-kit message stability and cell-surface protein expression in hematopoietic progenitors. Blood 83: 3138, 1994.
17. Lee G, Namen AE, Gillis S, Ellingsworth LR Kincade PW: Normal B cell precursors responsive to recombinant murine IL-7 and inhibition of IL-7 activity by transforming growth factor-b. J. Immunol. 142: 3875, 1989.
18. Felten SY Felten DL: Innervation of lymphoid tissue, in Ader R, Felten DL Cohen N (eds.): Psychoneuroimmunology II. San Diego, Academic Press, 1991, p. 27–71.
19. Lotz M, Ranheim E Kipps TJ: Transforming growth factor beta as endogenous growth inhibitor of chronic lymphocytic leukemia B cells. J. Exp. Med. 179: 999, 1994.
20. Baylink DJ, Finkelman RD Mohan S: Growth factors to stimulate bone formation. J. Bone Miner. Res. 8: 1993.
21. Pierce DFJ Coffey RJ: Therapeutic manipulation of cytokines: transforming growth factor beta-1 protects mice treated with lethal doses of cytarabine. Am. Surg. 60: 18, 1994.
22. Grzegorzewski K, Ruscetti FW, Usui N, Damia G, Longo DL, Carlino JA, keller JR Wiltrout RH: Recombinant transforming growth factor beta 1 and beta 2 protect mice from acutely lethal doses of 5-fluorouracil and doxorubicin. J. Exp. Med. 180: 1047, 1994.

MELATONIN-INDUCED HEMATOPOIETIC-NEUROENDOCRINE CYTOKINES*

Georges J. M. Maestroni†

Center for Experimental Pathology
Istituto Cantonale di Patologia
6604 Locarno, Switzerland

ABSTRACT

We have reported that melatonin may rescue bone marrow cells from apoptosis induced either in vivo or in vitro by cancer chemotherapy compounds. The number of granulocyte-macrophage colony-forming units cultured with suboptimal concentrations of colony stimulating factor was higher in presence of melatonin both at physiological and pharmacological concentration. CD4+,Thy-1.2+cells depletion or addition of anti mouse interleukin-4 (IL-4) monoclonal antibodies prevented both effects of melatonin. We proposed that melatonin represents a neuroendocrine regulator of IL-4 production in bone marrow T-helper cells and that IL-4 stimulates adherent stromal cells to produce granulocyte/macrophage colony-stimulating factor. However, in further investigations we did not find any direct evidence of the ability of melatonin to stimulate IL4. We found that beside anti-IL4 antibodies also the specific opioid antagonist naltrexone neutralized the colony stimulating activity of melatonin. SDS-PAGE and blotting analysis of gel filtration fractions of supernatants from bone marrow cells cultures revealed that upon melatonin stimulation, T-helper cells release a 15 and 67 kDa opioid peptides which are recognized both by anti-common opioid sequence and anti-IL4 monoclonal antibodies. The term "neuroendocrine cytokines " seems thus to fit the properties of these IL4-like and opioid-like melatonin-induced peptides which might represent a new family of immunological and hematopoietic regulators.

INTRODUCTION

Our previous work has shown that melatonin can augment the immune response and correct immunodeficiency states which may follow acute stress, viral diseases, aging or drug

* This study was supported by Swiss Nationalfonds Grant 31-36128.92.

† Telephone: 091 756 26 71; Fax 091 756 26 90.

treatment.[1-7] Relevant to the studies hereunder exposed, we observed that melatonin was able to antagonize the effect of high dose cyclophosphamide on antibody production.[8] This finding has been then confirmed and extended to other immune parameters by other authors.[9,10] Such interesting effects of melatonin seem to depend on activated CD4+, T-cells which upon melatonin stimulation show an enhanced synthesis and/or release of opioid peptides and cytokines.[5-7,9-11] A large body of evidence indicates also that melatonin may inhibit carcinogenesis and tumor growth in a variety of experimental and clinical situations.[12] On the basis of our animal studies,[6] we have investigated the clinical effect of melatonin in association with low-dose interleukin-2 in cancer patients and found that this association represents a well tolerated strategy capable of determining an apparent control of tumor growth in patients with advanced solid neoplasms.[13-16] Taken together, these findings imply that the circadian information provided by melatonin contributes in maintaining a correct immune homeostasis.

Programmed cell death or apoptosis is a normal process by which cells are eliminated during embrionic development and in adult life. Disruption of this normal process can cause developmental abnormalities and facilitate cancer development.[17] Normal haemopoietic cells require certain viability and growth factors like CSFs and undergo apoptosis when these factors are withdrawn.[18] Programmed cell death can be induced by removal of CSFs, by cytotoxic therapeutic agents, or by the tumor suppressor gene wild-type p53.[19]

We investigated whether we could take advantage of the melatonin-T-helper cell connection to rescue bone marrow cells from apoptosis induced by cancer chemotherapy compounds. We found that melatonin may indeed protect bone marrow progenitors via a cytokines cascade involving colony stimulating factors.

HEMATOPOIETIC RESCUE VIA T-CELL-DEPENDENT, ENDOGENOUS CSF INDUCED BY MELATONIN

Melatonin was injected in 2 months old, female C57BL/6 mice which were transplanted with Lewis lung carcinoma and treated with the antitumor agents etoposide or cyclophosphamide. The results which are shown in table 1 demonstrate that melatonin can protect bone marrow functions from the toxic effect of cancer chemotherapy compounds without interfering with their anti-cancer action in vivo. Such effect was apparent and significant on leukocytes, platelets and marrow granulocyte/macrophage-colony forming units (GM-CFU).[20] Melatonin was able to antagonize the hematopoietic toxicity of cyclophosphamide especially when injected together with the drug. These results indicated that melatonin is able to protect the bone marrow in the course of cytotoxic anti-cancer treatments.[20]

To study whether melatonin acted directly on haemopoietic cells, different concentrations of melatonin (10^{-10} to 10^{-6} M) in presence of etoposide 10^{-5} M were incubated with bone marrow cells. Melatonin proved to rescue bone marrow cells from apoptosis induced by etoposide.[20] Table 2 shows that the protective effect exerted by melatonin in vitro was already evident at physiological concentration (10^{-9} M) and was reflected by an increased number of the lineage-committed myeloid precursors GM-CFU but not of the less differentiated, pluripotent spleen-colony forming units (S-CFU). These results suggested that melatonin can induce the production of endogenous CSFs. In particular, the in vivo effect on platelets and the rescue of GM-CFU but not of S-CFU candidated GM-CSF as possible mediator of melatonin.[18,19,21] Indeed, anti-GM-CSF monoclonal antibodies were capable of neutralizing the protective effect of melatonin and this suggests that melatonin exerted its effect by inducing bone marrow cells to produce a GM-CSF-like substance.[20] In bone

Table 1. Effect of melatonin and cancer chemotherapy agents in Lewis lung carcinoma-bearing mice

GROUPS (n)	LEUKOCYTES / µl	PLATELETS / µl (x10^3)	GM-CFU / FEMUR	No. of METASTASES
CONTROL (10)	10300 ± 1070	253 ± 69.5	8459 ± 1674	-
PBS (64)	17128 ± 3264 (c)	300 ± 79.6	10303 ± 2907	16.8 ± 6.7
ME (46)	12960 ± 2667	287 ± 74.1	9835 ± 4067	12.8 ± 5.2
CY160 (10)	5650 ± 980	161 ± 43.3	2471 ± 490	0
CY160 + ME (12)	11083 ± 2882 (a)	316 ± 59.7 (a)	5927 ± 1898 (b)	0
ET40 (14)	8943 ± 251	149 ± 23.8	4089 ± 1528	1.4 ± 1.4
ET40 + ME (14)	10912 ± 1352 (a)	316 ± 77 (a)	6459 ± 2826 (c)	1 ± 0.9

Cyclophosphamide (CY) and etoposide were injected at the doses indicated (mg / kg b.w.) in C57 BL/ 6 mice from day 8 through day 12 after LLC transplantation. Melatonin (ME) was injected from day 8 throughout the experiments. The values represent the mean ± the standard deviation. Leukocytes and platelets were evaluated 16 days after tumor transplantation, while the remaining parameters were measured at the end of the experiments (day 20-22). CONTROL: normal healthy mice; PBS: phosphate saline; a: $p < 0.001$; b: $p < 0.005$; c: $p < 0.01$.

marrow cell suspensions, the principal sources of GM-CSF are macrophages and T cells. Athymic nude mice are T-cell-deficient, a condition which is in part balanced by enhanced macrophage functions. When bone marrow cells were obtained from T-cell-deficient mice, melatonin did not exert any protection against etoposide toxicity.[20] This indicated that T lymphocytes are the target of melatonin in the bone marrow. Consistently, we found the presence of high affinity (Kd: 346 ± 24 pM) melatonin binding sites in bone marrow CD4+, T cells.[22]

Table 2. Effect of melatonin on survival of hematopoietic progenitor cells upon incubation with etoposide of murine bone marrow cells

INCUBATION	GM-CFU / 10 ^{5} CELLS	S-CFU / 10 ^{5} CELLS
CONTROL	75 ± 19	17.6 ± 2.6
ET (10 µM)	16.6 ± 4.7	6.5 ± 3.9
ET (10 µM) + ME (1 µM)	35.7 ± 9.5 (a)	6.8 ± 1.9
ME (1µM)	72.5 ± 16	12.4 ± 3.1 (b)

Bone marrow cells from C57BL/6 mice were incubated for 8 hours at 37°C, 5% CO2 either in culture medium alone (CONTROL) or with the reported concentrations of etoposide (ET) and melatonin (ME). After washing the cells were assayed for GM-CFU or S-CFU 12 in lethally irradiated mice. The values represent the mean of 4 experiments ± the standard deviation. a : $p < 0.001$ ET + ME vs ET; b: $p < 0.005$ ME vs CONTROL. (From Maestroni et al., Cancer Res. 54, 2429, 1994, with permission)

COLONY-STIMULATING ACTIVITY AND HEMATOPOIETIC RESCUE INDUCED BY MELATONIN ARE NEUTRALIZED BY ANTI-INTERLEUKIN-4 MONOCLONAL ANTIBODY

The melatonin-induced GM-CSF proved to rescue lineage committed GM-CFU. In contrast, more primitive, multipotent progenitor cells such as S-CFU, not only were not protected from the toxic effect of anticancer compounds but rather were decreased upon incubation with melatonin alone (Table 2). A T-cell cytokine which may exert a selective inhibitory effect on primitive progenitor cells is IL4.[23] In addition, IL4 has been also reported to stimulate macrophages and fibroblast cells to produce colony stimulating activity.[23, 24] To evaluate whether IL4 is involved in the ability of melatonin to rescue hematopoietic cells, we added rat anti mouse IL4 mAb to bone marrow cells that were incubated with etoposide in presence of melatonin. As a matter of fact, anti-IL-4 mAb neutralized the effect of melatonin both on apoptosis and number of GM-CFU.[25] As GM-CSF proved to be also involved[20] and IL4 has been reported to synergize with or stimulate GM-CSF production,[23,24,26] we investigated whether melatonin could substitute for or augment the effect of suboptimal concentrations of GM-CSF on GM-CFU. Starting from 0.1 nM, a concentration which approaches the lower physiological plasma level,[27] melatonin increased significantly the number of GM-CFU . However, melatonin was effective only in presence of GM-CSF.[25] As in the rescue experiment, anti-IL4 mAb neutralized the effect of melatonin. (fig. 1) Consistently with our previous findings, melatonin failed to show any action when added in GM-CFU cultures performed with bone marrow cells that were depleted from Thy^+, $CD4^+$ cells.[25]

Although IL4 gene expression has been found also in bone marrow stromal cell clones,[28] $CD4^+$, T cells are considered the major source of IL4.[23] Conversely, GM-CSF may be equally synthesized in both T cells and stromal cells. [21] A simple method to separate bone marrow stromal cells is by adherence.[29] Therefore, we investigated the effect of melatonin in adherent vs non-adherent bone marrow cells. We found that melatonin looses its activity when added in GM-CFU cultures performed either with adherent or non-adherent bone marrow cells. The activity reappeared when the two cell populations were mixed again.[25] As

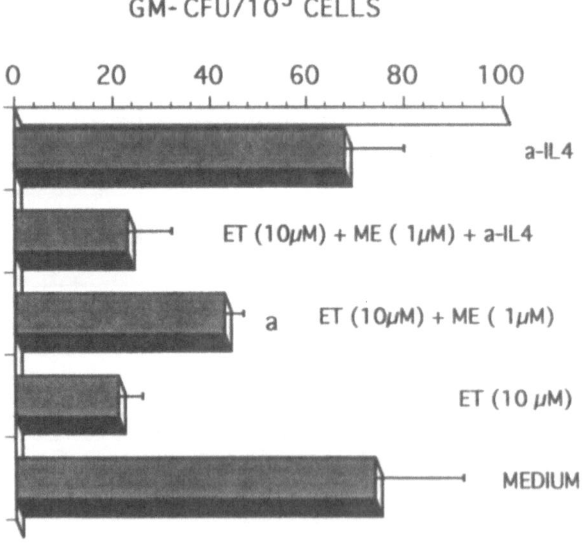

Figure 1. Effect of anti-IL4 mAb on melatonin-induced hematopoietic rescue. Bone marrow cells from C57BL/6 mice were incubated for 8 h at 37°C, 5% CO_2 in culture medium alone (MEDIUM) or in presence of anti-IL4 mAb, (a-IL4, 10 μg/ml), etoposide (ET, 10 μM) and/or melatonin (ME, 1 μM). The values represents the mean of three experiments ± SD. a: $p < 0.01$.

T lymphocytes are non-adherent cells, this result suggested that the colony stimulating activity was released by adherent cells upon stimulation of the melatonin-induced, T cell-derived IL4. To prove this hypothesis, we incubated non-adherent cells with melatonin in order to produce the IL4 activity in the supernatant. The supernatant was dialyzed to get rid of melatonin and then tested for its activity on GM-CFU in unseparated, non-adherent or adherent bone marrow cells. The results show that the supernatant was active on unseparated and adherent bone marrow cells but not on non-adherent cells.[25] As expected anti-IL4 mAb neutralized the supernatant activity.

IS INTERLEUKIN-4 REALLY INVOLVED IN THE MELATONIN HEMATOPOIETIC ACTION?

Taken together, our results suggested that melatonin was acting on T-helper cells inducing IL4 production. Therefore, we decided to investigate deeper the relative mechanism. To our surprise, ELISA measures of IL4 concentration in active supernatants of non-adherent bone marrow cells indicated quite clearly that melatonin does not affect IL4 secretion. We also set up a reverse transcripatse polymerase chain reaction assay to detect IL4 mRNA in non-adherent bone marrow cells incubated overnight with various concentrations of melatonin. Again, no evidence of a melatonin action on IL4 mRNA was obtained. At this point we were confronted with a rather puzzling situation. We thought that the ability of anti-IL4 mAb to neutralize melatonin could be due to crossreactivity with another substance. We knew that melatonin stimulates activated T-helper cells to release opioid peptides which exert anti-stress and immunoenhancing effects.[5, 30] Therefore we wondered whether such melatonin-induce-immuno-opioids (MIO) could mediate also the hematopoietic effect of melatonin. We investigated whether the specific opioid antagonist naltrexone could neutralize the colony stimulating activity of melatonin. Figure 2 shows that naltrexone could indeed abolish the effect of melatonin on GM-CFU. We then studied whether bone marrow T-helper cells produce MIO upon melatonin incubation. We found that, both activated spleen T-helper cells and non-adherent bone marrow cells which contain T-helper cells release 2 opioid peptides with an apparent molecular weight of 15 and 67 kDa.[31] To test the hypothesis that MIO were recognized by the anti-IL4 mAb, the peptides were partially purified by ion exchange chromatography and analyzed by polyacrylamide gel electrophoresis and immunoblotting. We found that anti-IL4 mAb recognized the same 15

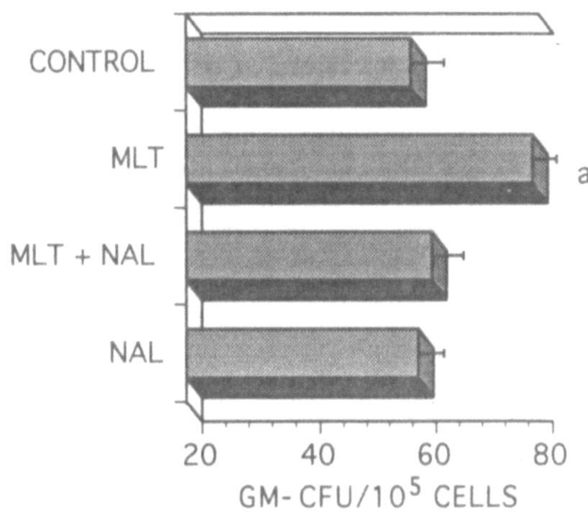

Figure 2. Melatonin neutralization by naltrexone in the GM-CFU assay. Bone marrow cells were incubated in presence of a suboptimal concentration of GM-CSF and with or without melatonin 10 nM. The specific opioid antagonist naltrexone (1 μM) was added at the beginning of the culture. The values represent the mean of 3 experiments± SD. a: $p < 0.01$.

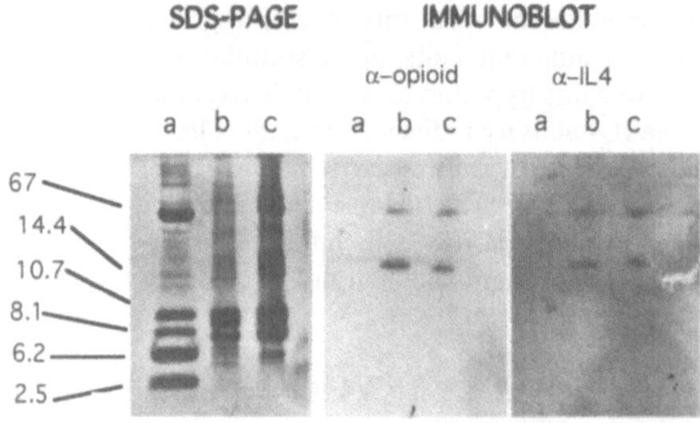

Figure 3. SDS-Polyacrylamide gel electrophoresis (SDS-PAGE) and immunoblotting. Supernatant of antigen-activated mononuclear spleen cells (b) or of non-adherent bone marrow cells (c) incubated overnight with melatonin (2 nM) were dialyzed, liophylized and analyzed. PAGE was run with SDS denatured material (1mg/ml,5 min, 100°C) in 20 % polyacrylamide gels. The Coomassie blue staining was compared with that of molecular weight markers (a). The gels were then blotted on 0.45 μm nitrocellulose membranes and incubated overnight with mouse mAb directed against the Tyr-Gly-Gly-Phe- opioid sequence (α–opioid) , or rat-anti mouse IL4 mAb (α-IL4) diluted in 0.05 M Tris, 0.5 M NaCl and 0.1% Tween-20 , pH 9. After washing, an alkaline phosphatase-conjugated second antibody directed against mouse or rat IgG was then added for 3 hours in 0.1M Tris, 0.1 M NaCl, 50 mM $MgCl_2$ for 3 hours. Staining was then performed by addition of nitro blue tetrazolium and 5-Br-4 Cl-3 indolylphosphate.

and 67 kDa peptides that were bound by the anti-opioid monoclonal antibody (fig.3). This anti-opioid antibody was specific for the free amino terminal Tyr-Gly-Gly-Phe opioid sequence (fig.3).

DISCUSSION

In this paper we demonstrate that melatonin influences hematopoiesis via a 15 and/or 67 KDa opioid peptides (MIO) which are released by bone marrow T-helper cells. The previously reported involvement of IL4 was suggested by a crossreactivity of anti-IL4 mAb with such opioid peptides. In addition, our previous findings indicate that melatonin stimulates bone marrow T-helper cells to release such opioid-IL4-like substances which in turn act on bone marrow stromal cells inducing the production of colony stimulating factors.[25] In fact, we found that the colony stimulating activity of melatonin could be inhibited by the opioid antagonist naltrexone and, most recently, we got evidence that the naltrexone-sensitive colony stimulating activity elutes in gel filtration fractions containing MIO (data not shown). Although in our previous work we demonstrated that anti-IL4 mAb neutralized also the hematopoietic rescue exerted by melatonin, we do not yet know whether MIO mediate the hematopoietic rescue too. In any case, our findings open interesting perspectives concerning the real nature of MIO and their potential pharmacological use. In regard to their identification, there is a reasonable possibility that MIO belong to the enkephalin-containing peptides family which has been shown to be released by T-helper type 2 cells.[23,32,33] The fact that the monoclonal antibodies we used in our blotting analysis recognize peptides with the common opioid sequence at their amino terminal (fig.3), agrees with the MIO action which is naltrexone sensitive,[5, 30] contrary to that reported for other known enkephalin containing peptides.[33]

The crossreactivity of anti-IL4 mAb with MIO and their IL4-like hematopoietic effects speak possibly for a new family of endogenous hematopoietic regulators. We would like to define such regulators as neuroendocrine cytokines, i.e. immuno-derived agents with both neuroendocrine and cytokine properties. Beside melatonin, these neuroendocrine cytokines might represent a key pharmacological tool which would regulate in an integrative and physiological fashion both the neuroendocrine and the immune-hematopoietic systems.

ACKNOWLEDGMENTS

The skillful technical assistance of Ms. Elisabeth Hertens and Paola Galli is gratefully acknowledged. I also thank Dr. Lelio Flamigni for carrying out part of the biochemical work.

REFERENCES

1. Maestroni GJM, Conti A Pierpaoli W: Role of the pineal gland in immunity. Circadian synthesis and release of melatonin modulates the antibody response and antagonize the imunosuppressive effect of corticosterone. J. Neuroimmunol. 13: 19, 1986.
2. Maestroni GJM, Conti A Pierpaoli W: Role of the pineal gland in immunity: II. Melatonin enhances the antibody response via an opiatergic mechanism. Clin. Exp. Immunol. 68: 384, 1987.
3. Maestroni GJM, Conti A Pierpaoli W: Role of the pineal gland in immunity. III. Melatonin antagonizes the immunosuppressive effect of acute stress via an opiatergic mechanism. Immunology 63: 465, 1988.
4. Maestroni GJM Conti A: Beta-endorphin and dynorphin mimic the circadian immunoenhancing and anti-stress effects of melatonin. Int. J. Immunopharmacol. 11: 333, 1989.
5. Maestroni GJM Conti A: The pineal neurohormone melatonin stimulates activated CD4+, Thy-1+ cells to release opioid agonist(s) with immunoenhancing and anti-stress properties. J. Neuroimmunol. 28: 167, 1990.
6. Maestroni GJM: The immunoneuroendocrine role of melatonin. J. Pineal Res. 14: 1, 1993.
7. Maestroni GJM, Conti A Reiter RJ (ed): Advances in Pineal research 7. London, John Libbey & Co., 1994.
8. Maestroni GJM Conti A: Melatonin in relation to the immune system, in Yu H-S Reiter RJ (eds.): Melatonin. Biosynthesis, Physiological effects and Clinical Applications. Boca Raton, CRC Press, 1993, p 290.
9. Pioli C, Caroleo MC, Nistico G Doria G: Melatonin increases antigen presentation and amplifies specific and non specific signals for T-cell proliferation. Int. J. Immunopharmacol. 15: 463, 1993.
10. Caroleo MC, Frasca D, Nistico G Doria G: Melatonin as immunomodulator in immunodeficient mice. Immunopharmacology 23: 81, 1992.
11. Del Gobbo V, Libri V, Villani N, Caliò R Nstico G: Pinealectomy inhibits interleukin-2 production and natural killer activity in mice. Int. J. Immunopharmacol. 11: 567, 1989.
12. Blask DE: Melatonin in oncology, in Yu H-S Reiter RJ (eds.): Melatonin. Biosynthesis, Physiological Effects, and Clinical Applications. Boca Raton, CRC Press, 1993, p 447.
13. Lissoni P, Barni S, Ardizzoia A, Brivio F, Tancini G, Conti A G.J.M. M: Immunological effects of a single evening subcutaneous injection of low-dose interleukin-2 in association with the pineal hormone melatonin in advanced cancer patients. J. Biol. Reg. Homeos. Agents 6: 132, 1992.
14. Lissoni P, Barni S, Rovelli F, Brivio F, Ardizzoia A, Tancini G, Conti A Maestroni GJM: Neuroimmunotherapy of advanced solid neoplasms with single evening subcutaneous injection of low-dose interleukin-2 and melatonin: preliminary results. Eur. J. Cancer 29A: 185, 1993.
15. Lissoni P, Barni S, Tancini G, Ardizzoia A, Rovelli F, Cazzaniga M, Brivio F, Piperno A, Aldeghi R, Fossati D, Characjeius D, Kothari L, Conti A Maestroni GJM: Immunotherapy with subcutaneous low-dose interleukin-2 and the pineal indole melatonin as a new effective therapy in advanced cancers of the digestive tract. B. J. Cancer 67: 1404, 1993.
16. Lissoni P, Barni S, Tancini G, Ardizzoia A, Ricci G, Aldeghi R, Brivio F, Tisi E, Rescaldani R, Quadro G Maestroni GJM: A randomised study with subcutaneous low-dose interleukin 2 alone vs interleukin 2 plus the pineal neurohormone melatonin in advanced solid neoplasms other than renal cancer and melanoma. Br. J. Cancer 69: 196, 1994.
17. Arends MJ Willie HA: Apoptosis: Mechanisms and roles in pathology. Inter. Rev. Exp. Pathol. 32: 223, 1991.

18. Williams GT, Smith CA, Spooncer E, Dexter TM Taylor DR: Haemopoietic colony stimulating factors promote cell survival by suppressing apoptosis. Nature 343: 76, 1990.
19. Sachs L Lotern J: Control of programmed cell death in normal and leukemic cells: New implications for therapy. Blood 82: 15, 1993.
20. Maestroni GJM, Covacci V Conti A: Hematopoietic rescue via T-cell-dependent, endogenous GM-CSF by the pineal neurohormone melatonin in tumor bearing mice. Cancer Res., 54:2429, 1994.
21. Garland JM: Colony stimulating factors, in Thomson A (eds.): The Cytokine Handbook. London, Academic Press, 1991, p 269.
22. Maestroni GJM: T helper 2 lymphocytes as a peripheral target of melatonin. J. Pineal Res. 18: 84, 1995.
23. Banchereau J: Interleukin-4, in Thomson A (eds.): The Cytokine Handbook. London, Academic Press, 1991, p 119.
24. Wieser M, Bonifer R, Oster W, Lindemann A, Mertelsmann R Hermann F: Interleukin-4 induces secretion of CSF for granulocytes and CSF for macrophages by peripheral blood monocytes. Blood 73: 1105, 1989.
25. Maestroni GJM, Conti A Lissoni P: Colony-stimulating activity and hematopoietic rescue from cancer chemothereapy compounds are induced by melatonin via endogenous interleukin 4. Cancer Res. 54:4740, 1994.
26. Peschel C, Paul WE, Ohara J Green I: Effect of B cell stimulatory factor/ interleukin 4 on hematopoietic progenitor cells. Blood 70: 254, 1987.
27. Yu H-S, Tsin ATC Reiter RJ: Melatonin: History, Biosynthesis, and Assay Methodology, in Yu H-S Reiter RJ (eds.): Melatonin. Biosynthesis, Physiological Effects, and Clinical Applications. Boca Raton, CRC Press, 1993, p 1.
28. Gutierrez-Ramos JC, Olsson C Palcios R: Interleukin (IL1 to IL7) gene expression in fetal live and bone marrow stromal clones: Cytokine-mediated positive and negative regulation. Exp. Hematol. 20: 986, 1992.
29. Sullivan AK, Claxton D, Shematek G Wang H: Cellular composition of rat bone marrow stroma. Antigen-defined subpopulations. Lab. Invest. 60: 667, 1989.
30. Maestroni GJM Conti A: Anti-stress role of the melatonin-immuno-opioid network. Evidence for a physiological mechanism involving T cell-derived, immunoreactive β-endorphin and met.enkephalin binding to thymic opioid receptors. Int. J. Neurosci. 61: 289, 1991.
31. Maestroni GJM, Flamigni L, Hertens E Conti A: Biochemical and functional characterization of melatonin-induced opioids in spleen and bone marrow T-helper cells. Neuroendocrinol. Lett., 17:145, 1995.
32. Kuis W, Villigher PM, Leser H-G Lotz M: Differential processing of proenkephalin-A by human peripheral blood monocytes and T lymphocytes. J. Clin. Invest. 88: 817, 1991.
33. Hiddinga HJ, Isaak DD Lewis RV: Enkephalin-containing peptides processed from proenkephalin significantly enhance the antibody-forming cell responses to antigens. J. Immunol. 152: 3748, 1994.

76

INNERVATION OF BONE MARROW BY TYROSINE HYDROXYLASE-IMMUNOREACTIVE NERVE FIBERS AND HEMOPOIESIS-MODULATING ACTIVITY OF A β-ADRENERGIC AGONIST IN MOUSE

David L. Felten,[1*] Kimberly Gibson-Berry,[1] and J. H. David Wu

[1] Department of Neurobiology and Anatomy
[2] Department of Chemical Engineering
University of Rochester
Rochester, New York 14267-0166

ABSTRACT

The bone marrow is a primary lymphoid tissue and the site of origin for all blood cells in adult mammals. The presence of sympathetic fibers in bone marrow has been noted previously with histofluorescence staining for catecholamines. In the present study, we demonstrate the presence of sympathetic fibers in murine bone marrow by immunostaining for tyrosine hydroxylase (TH), the rate-limiting enzyme in the synthesis of catecholamines. Dense TH-immunoreactive nerve bundles were observed to take a winding course along the larger arterioles and gave rise to branches that travel in parallel with smaller arterioles. Some nerve bundles of varicose fibers course between the marrow sinuses, the site of the hemopoietic parenchyma. The proximity of nerve fibers to the hemopoietic parenchyma suggests a direct influence of sympathetic neurotransmitters on the bone marrow microenvironment. At least one such direct influence was demonstrated by examining the effect of exogenous isoproterenol (ISO), a catecholamine and β-adrenergic receptor agonist, on hemopoiesis in a three-dimensional long-term bone marrow culture (LTBMC) system. ISO significantly stimulated hemopoietic cell output (up to a 3-fold increase) at two of the concentrations tested (10^{-7} and 10^{-5}M) under the experimental conditions. These results provide the anatomical basis for the nerve-hemopoiesis interactions and demonstrate the

* Corresponding author: David L. Felten, M. D. , Ph. D. , University of Rochester School of Medicine and Dentistry, 601 Elmwood Ave. Box 603, Rochester, New York 14642. Telephone: (716) 275-6539; Fax: (716) 442-8766; Email: dfelten@medinto. rochester.. edu.

Molecular Biology of Hematopoiesis 5, edited by Abraham et al.
Plenum Press, New York, 1996

direct hemopoiesis-modulating activity of catecholamines at the same order of magnitude as that achieved by hemopoietic growth factors such as IL-3.

INTRODUCTION

It is becoming evident that the nervous and hemopoietic systems are intimately associated [1-3]. Several criteria must be met in order to establish that cells of the hemopoietic system are targets for neurotransmission. These criteria include the presence of nerve terminals and release of neurotransmitters in close proximity to hemopoietic cells, the presence of functional receptors for neurotransmitters on those cells, and functional changes associated with nerve or neurotransmitter manipulation[4].

Hemopoiesis occurs in the space between the bone marrow sinuses, the vascular network of bone marrow. Within the hemopoietic cords, the marrow stromal cells form a three-dimensional scaffolding that supports the self renewal and differentiation of the hemopoietic stem cells into all lineages of blood cells through the production of extracellular matrices and the secretion of hemopoietic growth factors[5].

The presence of nerve fibers in bone marrow has been demonstrated using histological techniques that stain general nerve fibers [6-8]. Furthermore, fluorescence histochemistry revealed the presence of catecholamine-containing fibers in bone marrow [1,4,9], suggesting that at least some of these fibers are sympathetic noradrenergic (NA) nerves. In addition to the presence of nerve fibers in bone marrow, the involvement of neurotransmission in hemopoiesis is suggested by the presence of functional β-adrenergic receptors coupled to adenylate cyclase on erythroid progenitors and all mature cell types originating from the bone marrow [4,10-12]. Finally, catecholamines have been found to modulate erythropoiesis [13-15] and granulopoiesis [16-18].

Although catecholamine-containing nerve fibers have been demonstrated in bone marrow, it remains to be shown that these fibers contain tyrosine hydroxylase (TH). TH is the rate-limiting enzyme in the synthesis of norepinephrine (NE), the primary neurotransmitter in sympathetic fibers. In this work, we examined the presence of sympathetic nerve fibers in bone marrow using tyrosine hydroxylase as a marker. This method permits a permanent stained section with better background histology and secondary staining than is available with the use of histofluorescence [19]. In addition, we examined the direct effect of isoproterenol (ISO), a catecholamine and β-adrenergic receptor agonist, on hemopoiesis. To provide a bone marrow model mimicking the hemopoietic microenvironment *in vivo*, we used a novel three-dimensional long term bone marrow culture (LTBMC) system previously reported [20] for accessing the effect of ISO.

MATERIALS AND METHODS

Immunocytochemical Staining of Nerve Fibers

The bone marrow in femora and tibia of adult male BALB/c mice (4-6 weeks; Jackson laboratories; Bar Harbor, ME) was fixed *in situ* by whole body perfusion through cardiac puncture. The animals were first perfused with phophate buffered saline (PBS, 0.15 M, pH 7.2) containing 1% sodium nitrite and 0.1% heparin followed by 4% paraformaldehyde in PBS. The bones were removed, cleaned of excess tissue, and decalcified in 0.1 M sodium acetate containing 10% ethylenediaminetetraacetic acid (EDTA) for 2-3 weeks. The decalcified bones and marrow were allowed to equilibrate in 30% sucrose overnight followed by freezing on dry ice and storage at -70°C until further processing. Tissues were cut on a cryostat at 15 μm and mounted on slides followed by immunocytochemical staining. Sections were stained with a rabbit polyclonal antibody against rodent tyrosine hydroxylase (TH; Chemicon; Temecula, CA). A biotinylated goat-anti-rabbit secondary antibody (Jackson Immunoresearch; West Grove, PA) was used and complexes were visualized using the standard avidin-peroxidase diaminobenzidine nickel-enhanced immunocytochemical methodology (Vector Laboratories; Burlingame, CA.) [21].

Chemical Sympathectomy

To confirm that the TH-immunoreactive nerve fibers are noradrenergic (NA) in nature and capable of synthesizing norepinephrine (NE), the primary neurotransmitter in sympathetic neurons, animals were treated with the neurotoxin 6-hydroxydopamine hydrobromide (6-OHDA; Sigma; St. Louis, MO), followed by analysis of marrow NE content. Adult male BALB//c mice (n=6; 4-6 weeks) received a single intraperitoneal injection of 6-OHDA dissolved in sterile saline at a dosage of 150 mg per kg of body weight (BW). Control animals (n=6) received intraperitoneal injections of sterile saline at equivalent volumes per kg BW. The bone marrow NE contents from the treated animals were analyzed as described below.

Analysis of Bone Marrow Norepinephrine (NE) Content

The day after intraperitoneal injection of 6-OHDA, fresh bone marrow samples were obtained. Bone marrow in freshly isolated femora and tibia from 6-OHDA- and saline-treated mice were flushed out with 0.5 ml of 0.1 M perchloric acid ($HClO_4$; which prevents oxidation of NE) to which 0.025 μM 3,4-dihydroxybenzylamine was added as an internal standard. The marrow in perchloric acid acid was frozen on dry ice and stored at -70°C until further use. The thawed marrow samples were sonicated and the homogenates were centrifuged at 11,000 x g for 10 min. The supernatant from each sample was added to 1 ml of 0.15 M sodium phosphate buffer (pH 6.1) and 1 ml of 1.5 M Tris/EDTA (pH 8.6) and the cell pellets were saved and later analyzed for protein content. To the solutions containing supernatant and buffer, 50 mg acid washed alumina powder was added and the samples were shaken for 5 min. to absorb the NE onto the alumina powder. The supernatant from each sample was removed, alumina was washed twice with distilled water, and the NE was extracted from the alumina by adding 200 μl of 0.1 M $HClO_4$, followed by centrifugation and transfer of the extraction medium into a new tube.

The NE content in each sample was analyzed using a system consisting of a Waters model 510 solvent delivery system, a Waters 717 plus refrigerated automatic sample injector, a Biophase 5 μm C18 reverse phase column (Waters Corp; Milford MA) and a glassy carbon thin layer electrochemical detector (TL-5, Bioanalytical Systems; West Lafayette, IN). The detector potential was set at 0.85 V (vs. an Ag-AgCl reference electrode) using an LC-4C amperometric controller (Bioanalytical Systems). The mobile phase was 0.33 M citrate -0.67 M phosphate (pH 4.5) with sodium octyl sulfate (1.2-1.4mM) used as the ion-pairing agent. The signal from the detector was recorded and peak heights and areas determined using a Waters Millennium data and chromatography control station (NEC Image 466 PC running Millennium version 2.00 software; Waters). NE levels were expressed as pmole / mg protein found in the cell pellets as a means of standardization.

Long-Term Bone Marrow Culture (LTBMC)

Fresh bone marrow cells were collected from the femora and tibia of adult male BALB/c mice (4-6 weeks), pooled and resuspended in McCoy's 5A culture medium (Gibco;

Grand Island, NY) containing 12.5% fetal bovine and 12.5% horse serum, 50 U/ml penicillin and 50 mg/ml streptomycin, and 10^{-6}M hydrocortisone (Sigma). The long-term bone marrow culture was carried out suing a three-dimensional bone marrow bioreactor culture system as previously described [20] except that the medium chambers were omitted and that the reactor was operated in a batch mode instead of a continuous perfusion mode. The culture medium in the cell chamber was removed and replaced with fresh medium daily. To study the effects of a β-adrenergic agonist, the cultures were fed with culture medium containing various concentrations of isoproterenol (ISO; Research Biochemicals Inc; Natick, MA). Aliquots of cells were harvested weekly from the cultures, and cell numbers and viability were assessed by the trypan blue dye exclusion method.

RESULTS

Immunocytochemical Staining of TH-Immunoreactive Nerve Fibers

To examine the innervation of tyrosine hydroxylase (TH)-immunoreactive nerve fibers in murine bone marrow, frozen thin-sections were obtained after the tissue was fixed *in situ* by paraformaldehyde and the bone was decalcified as described in Materials and Methods. The sections were then stained with a polyclonal anti-TH antibody. A typical micrograph of the immunocytochemically stained bone marrow sections is shown in Figure 1. In this micrograph of murine femoral bone marrow, TH-immunoreactive fibers are clearly visible, demonstrating sympathetic innervation of bone marrow. Dense nerve bundles

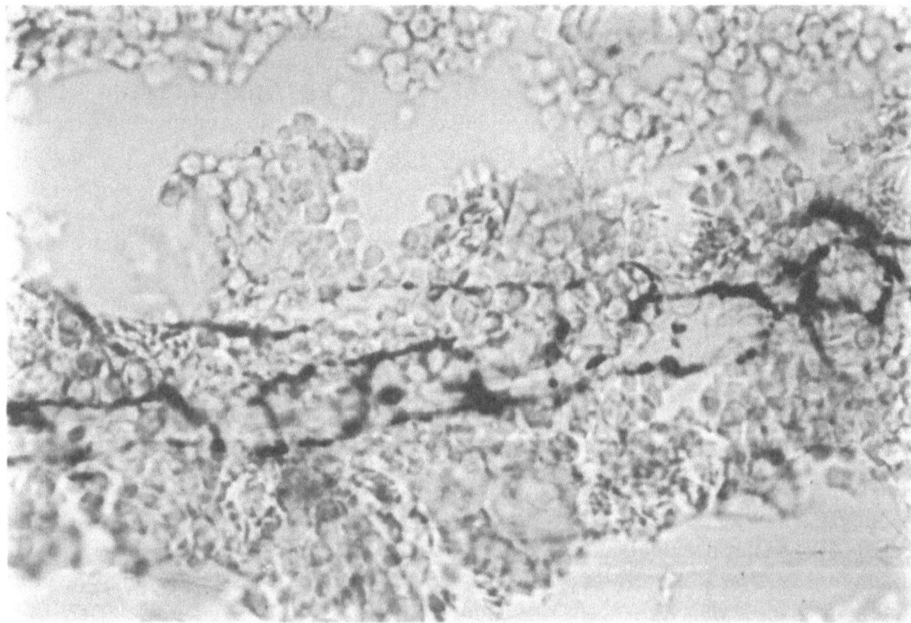

Figure 1. Micrograph of a murine bone marrow thin-section immunocytochemically stained with an antibody against tyrosine hydroxylase (TH) the rate-limiting enzyme in the synthesis of catecholamines and a marker for sympathetic neurons. Dense TH-positive nerve bundles are seen taking a winding course along a larger artery and sending branches that following smaller arterioles. These TH-immunoreactive fibers are indicative of the presence of catecholamine-synthesizing nerve fibers in murine bone marrow.

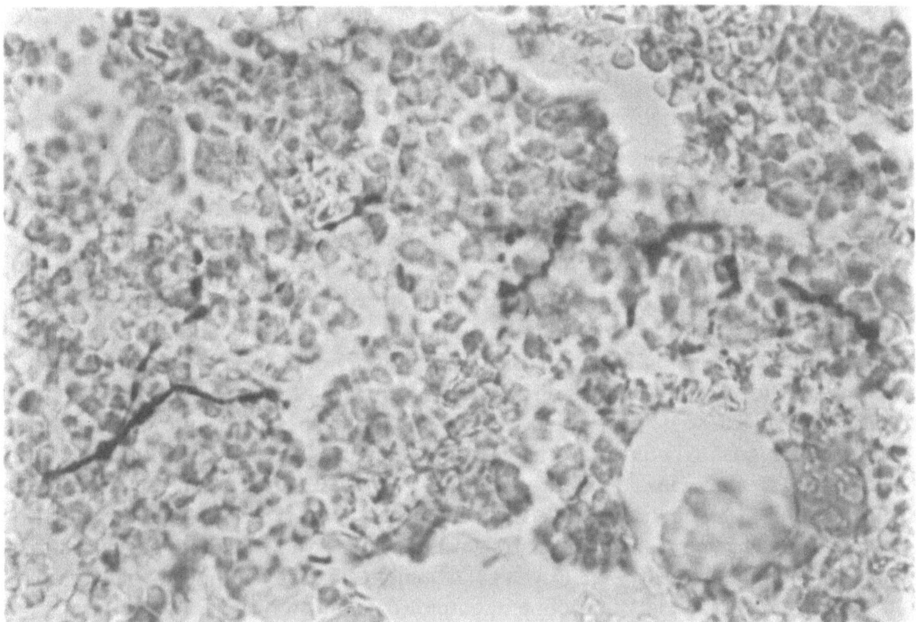

Figure 2. Micrograph of a murine femoral bone marrow thin-section, illustrating individual TH-immunoreactive fibers coursing through the hemopoietic parenchyma, the site of hemopoiesis in the marrow. Intimate contacts between bone marrow cells and these NA fibers, capable of releasing neurotransmitter at various points along the fiber, provide the anatomical basis for neural modulation of bone marrow function.

are observed to take a winding course along the larger arterioles as well as give rise to branches that travel in parallel with smaller arterioles (Figure 1). Some of these nerve bundles send individual varicose fibers into the marrow sinuses, the site of the hemopoietic parenchyma. Figure 2 shows typical TH-immunoreactive nerve fibers coursing along the hemopoietic compartment, intimately associated the bone marrow cells. Since TH is the rate-limiting enzyme of the synthetic pathway for catecholamines, these fibers presumably are capable of synthesizing and releasing the neurotransmitter, norepinephrine (NE). Thus, TH-positive nerve fibers are present along the vasculature as well as in the parenchyma of the bone marrow.

Chemical Sympathectomy

To confirm that the TH-immunoreactive nerve fibers are noradrenergic (NA) in nature and capable of synthesizing NE, the primary neurotransmitter in sympathetic neurons, marrow NE content was analyzed following treatment of mice with 6-hydroxydopamine (6-OHDA) which specifically destroys NA nerve fibers and eliminates fluorescence histochemical staining and TH immunocytochemical staining. NE content in the bone marrow from 6-OHDA-treated mice (n=6) was determined by HPLC analysis as described in Materials and Methods. In the marrow of these mice, the levels of NE were reduced to 20-25% of the levels observed in marrow from control animals (data not shown), indicating that the catecholamine nerve fibers in bone marrow are NA in nature and capable of synthesizing and releasing NE.

Effects of Isoproterenol on Hemopoiesis in LTBMC

Innervation by the TH-positive NA nerve fibers in bone marrow, particularly in the hemopoietic compartment (parenchyma), raises the likelihood that the catecholamines released by these fibers can have a direct modulating effect on the growth, differentiation, or function of bone marrow cells. To facilitate the examination of such modulating effect, we carried out long-term bone marrow culture (LTBMC) using the bone marrow harvested from BALB/c mice, the same strain used for the experiments described above, and the three-dimensional bone marrow culture system we previously reported [20]. Since the endogenous neurotransmitter (NE) has both α- and β-adrenergic activites, we chose isoprotererol (ISO), a β-adrenergic receptor agonist, as the first compound to be examined using this culture system.

Four days after the cultures were initiated, they were fed daily with medium containing various concentrations of ISO. Aliquots of cells were removed weekly for enumeration. The total cell outputs on week 3 are shown in Figure 3. At week 3, exposure to ISO resulted in a dose-dependent increase in cell output from the cultures when compared to controls. The total cell outputs were increased as much as two to three fold at the optimal concentration (10^{-7}M). This stimulatory effect of 10^{-7} M ISO on hemopoiesis was observed during the entire time course of the cultures, through 5 weeks, when the experiment was terminated. An enhancement of hemopoiesis was also observed with other concentrations of ISO, although prolonged exposure (5 weeks) to a high, supraphysiological concentration of ISO (10^{-5} M) eventually led to a decline in output (data not shown).

DISCUSSION

To establish cells of the hemopoietic system as targets of neurotransmission, several criteria must be met[4]. First, the anatomical basis must be present, including a documenta-

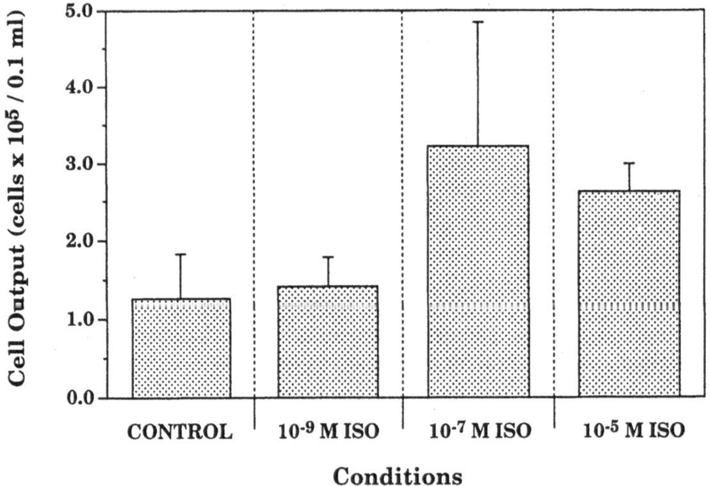

Figure 3. The effect of the β-adrenergic agonist, isoproterenol (ISO), on cell output in the three-dimensional long-term bone marrow culture (LTBMC). Aliquots of cells (100 μl) were harvested from the bioreactor on week 3 and assessed for numbers and viability as described in Materials and Methods. A dose-dependent increase in cell output was observed, with the greatest effect occurring at a concentration of 10^{-7} M ISO, providing evidence of sympathetic modulation of hemopoiesis. The data are expressed as the means (n=6) of individual cultures treated in the same manner. The vertical bars represent the SD of the mean for each group.

tion of nerve fibers in close proximity to effector cells combined with evidence for synthesis and release of the neurotransmitters for interaction with those cells. Second, the presence of specific and functional receptors for the neurotransmitters must be demonstrated. Finally, there must be evidence for functional effects, either by *in vivo* perturbation of the nervous system (e.g. neural stimulation and pharmacological blockade) or *in vitro* exposure to neurotransmitters.

Bone marrow has been found to contain myelinated and unmyelinated fibers nerve fibers that enter the bone marrow with the nutrient artery and follow the vasculature into the parenchyma [1,4, 6-8]. While the arterioles are heavily innervated, nerve profiles have also been seen in association with adventitial reticular cells, capillaries, vascular smooth muscle and hemopoietic elements. In an extensive ultrastructural study of nerves in murine marrow, the intimate association between nerve terminals and adventitial reticular cells, combined with the discovery of gap junctions between these stromal cells led the authors to propose the presence of a "functional neuro-reticular complex"[22]. Since it is believed that marrow stromal cells support the growth and differentiation of hemopoietic progenitors, this anatomical unit may be one path whereby neural signaling can alter hemopoiesis.

The identity of at least some of the nerve fibers in bone marrow was first suggested by the demonstration of neuronal degeneration following mechanical destruction of post-ganglionic sympathetic fibers supplying the limbs in dogs, cats and mice[6,23]. Catecholamine-containing fibers in rat marrow were subsequently documented with the use of fluorescence histochemistry[1,4,9]. However, the presence of catecholamine-containing nerve fibers does not prove the existence of neurons that synthesize and release NE as a neurotransmitter; it may only indicate a mechanism for uptake or sequestration. In this paper, we report the presence of nerve fibers in marrow that are immunoreactive for tyrosine hydroxylase, the rate-limiting enzyme in the synthesis of catecholamines. Furthermore, the NA nature of these fibers is demonstrated by the observation that NE content in the bone marrow was reduced dramatically following chemical sympathectomy by treatment with 6-OHDA. Together these results confirm the presence of sympathetic fibers in the bone marrow of mice, thereby fulfilling one of the criteria for the establishment of neurotransmission with hemopoietic cells as targets. The NA nerve fibers are associated with the vascular system in bone marrow, and also course along the hemopoietic compartment (Figure 2). NA sympathetic nerve fibers are believed to be capable of releasing NE at various sites along their path[24]. Therefore, the proximity of nerve fibers to the hemopoietic parenchyma raises the likelihood of direct effects of sympathetic neurotransmitters on the bone marrow microenvironment.

In addition to anatomical evidence, previous functional studies also suggest sympathetic modulation of hemopoiesis. Stimulation and ablation of the hypothalamus (which regulates sympathetic outflow as well as pituitary hormone secretion) was found to produce alterations in erythropoiesis, the direction of the effect being dependent upon which regions of the hypothalamus were manipulated. Some of these studies suggested that the mechanism of these effects were through alterations in erythropoietin secretion[25], while other studies suggested a direct effect of sympathetic fibers innervating the marrow[26,27]. A number of *in vitro* studies confirmed the idea of direct effects of catecholamines on erythropoiesis by demonstrating enhancement of erythropoietin-stimulated growth of erythroid colonies by β-adrenergic agonists *in vitro* [13-15]. These functional effects combined with data demonstrating the presence of adenylate cyclase-coupled β-adrenergic receptors on erythroid progenitors further strengthens the likelihood of sympathetic neural modulation of erythropoiesis [4,10-12].

Besides the apparent stimulatory effects of catecholamines on erythropoiesis, the sympathetic nervous system also has been implicated in the modulation of white blood cell production. Maestroni and colleagues demonstrated that chemical sympathectomy with the neurotoxin 6-OHDA produced an increase in granulopoiesis, particularly following bone marrow transplantation [16]. Daily subcutaneous injections of the α-adrenergic antagonist, prazosin, also enhanced the numbers of circulating granulocytes and bone marrow CFU-GM [17,18]. Since both of these manipulations result in a decrease in the noradrenergic contribution to the hemopoietic microenvironment, these results imply a tonic inhibitory effect of the sympathetic nervous system on granulopoiesis through α-adrenergic receptors. It is possible that catecholamines exert dual effects on hemopoiesis, a stimulatory action through β-adrenergic receptors and an inhibitory action through α receptors.

The studies described above were carried out suing chemical sympathectomy *in vivo* or colony-forming assay *in vitro*. In the present studies, we employed a unique and novel three-dimensional long term culture system for bone marrow to directly access the potential role of the sympathetic nervous system in modulating hemopoiesis. In this system, highly porous microcarriers are packed into a well and inoculated with fresh bone marrow cells[20]. This LTBMC system differs from the traditional flask-based LTMBC[28-30] in several ways: 1) the bone marrow cells in the culture system grow in a three-dimensional configuration, similar to that *in vivo;* 2) the cell output from the culture system at 37°C is virtually the same as that at 33°C; 3) in the absence of exogenous growth factors except those in the serum, the culture system produces lymphoid cells and all stages of committed cells including erythrocytes, granulocytes, macrophages, and megakaryocytes, indicating multilineal differentiation of the hemopoietic stem cells. Furthermore, cell clusters resembling erythroblastic islands are observed in the absence of exogenous erythropoietin (EPO). Aberrant morphological changes of the stromal cells, such as the flattening and fat cell formation commonly observed in the flask culture[29], are not found. The culture system therefore appears to provide a different microenvironment compared to that of the flask culture. The three-dimensional LTBMC is therefore a powerful alternative model for studying the hemopoietic process in general and for elucidating the potential modulation of bone marrow function by the sympathetic nervous system.

Using this system, we report an overall enhancement of hemopoiesis with addition of the β-adrenergic receptor agonist, ISO, to the culture medium. The total cell output was increased by two to three times at week 3 by 10^{-7} M ISO, and similar stimulatory effects on hemopoiesis were seen at other time points during cell culture (data not shown). In our experience with this three-dimensional bone marrow culture system, the degree of stimulation of the total cell output by a typical hemopoietic growth factor such as IL-3 is generally two to three fold. The effect of ISO in hemopoiesis *in vitro* is therefore comparable to that of a hemopoietic growth factor. Future studies will attempt to further determine lineage-specific effects and cytokine gene expression profiles as influenced by ISO. One question that remains to be answered is whether ISO exerts its effect directly on the hemopoietic cells, indirectly through the marrow stromal cells, or both. Although the presence of β-adrenergic receptors on hemopoietic cells has been demonstrated [10-12], it is a distinct possibility that the stomal cells also express receptors for catecholamines. In light of recent reports of modulation of cytokine production in marrow stromal cells by the neuropeptide, Substance P [31,32], it seems reasonable to anticipate a similar modulatory role for catecholamines.

In summary, in this work, we provided the anatomical basis for the potential neural-hemopoietic interactions by demonstrating the presence of TH-positive nerve fibers in bone marrow. Using a novel three-dimensional LTBMC system, we further demonstrated that ISO, an agonist for the endogenous neurotransmitter NE, has a profound direct influence on hemopoiesis. This LTBMC system is a promising hemopoiesis model for elucidating the mechanisms involved and for expanding the study to other neurotransmitters in bone marrow.

ACKNOWLEDGMENT

This work was financially supported by R37 MH42076 and an award from the Lucille P. Markey Charitable Trust. We appreciate the excellent technical assistance of John Housel and Charles Richardson in developing the immunocytochemical and neurochemical protocols. We also thank Suzanne Felten for her useful advise on technical matters and experimental design. Finally, we thank Sakis Mantalaris for his invaluable work in developing the three-dimensional LTBMC.

REFERENCES

1. Felten SY, Felten DL: Innervation of lymphoid tissue; in: Ader R, Felten DL, Cohen N (eds): Psychoneuroimmunology. San Diego, CA, Academic Press, 1992, vol2, pp 27-69.
2. Madden KS, Felten DL: Experimental basis for neural-immune interactions. Physiol Rev 1995; 75:77-106.
3. Madden KS, Sanders VM, Felten DL: Catecholamine influences and sympathetic neural modulation of immune responsiveness. Ann Reve Pharm Tox 1995; 35:417-448.
4. Felten DL, Felten SY, Bellinger DL, Carlson SL, Ackerman KD, Madden KS, Olshowka JA, Livnat S: Noradrenergic sympathetic neural interactions with the immune system: structure and function. Imm Reve 1987; 100:225-260.
5. Abboud CN, Litchman MA: Structure of the bone Marrow; in Beutler E, Litchman MA, Coller BS, Kipps TJ (eds): Williams Hematology. New York, McGraw-Hill, 1995, pp 25-38.
6. Takase B, Nomura S: Studies on the innervation of bone marrow. J Comp Neurol 1957; 108:421-443.
7. Calvo W: The innervation of the bone marrow in laboratory animals. Am J Anat 1968; 123:315-328.
8. Calvo W, Forteza-Vila J: On the development of bone marrow innervation in new born rats as studied with silver impregnation and electron microscopy. Amer J Anat 1969; 126:355-372.
9. DePace DM, Webber RH: Electrostimulation and morphologic study of the nerves to the bone marrow of the albino rat. Acta Anat 1975; 93:1-118.
10. Bilezikian JP: Dissociation of beta-adrenergic receptors from hormone responsiveness during maturation of the rat reticulocyte. Biochim Biophy Acta 1978; 542:263-273
11. Beckman B, Fisher JW: Changes in beta-2 receptor sensitivity with maturation of erythroid progenitors. Experientia 1979 35:1671-1672.
12. Setchenska MS, Arnstein HRV: Characteristics of the adenylate cyclase system of differentiating rabbit bone marrow erythroblasts. Biomed Biochim Acta 1983; 42:S192-S196.
13. Przala F, Gross DM, Dargon PA, Fisher JW: Effects of *in vitro* beta-adrenergic activation on rabbit bone marrow erythroid colony forming cells. Proc Soc Exp Biol Med 1977; 155:334-338.
14. Brown JE, Adamson JW: Modulation of *in vitro* erythropoiesis: The influence of β-adrenergic agonists on erythroid colony formation. J Clin Invest 1977; 60:70-77.
15. Mladenovic J. Adamson JW: Adrenergic modulation of erythropoiesis: *in vitro* studies of colony-forming cells in normal and polycythaemic man. Brit J Haematol 1984; 56:323-332.
16. Maestroni GJM, Conti A, Pedrinis E: Effect of adrenergic agents on hematopoiesis after syngeneic bone marrow transplantation in mice. Blood 1992; 80:1178-1182.
17. Maestoni GJM, Conti A: Modulation of hematopoiesis via α1-adrenergic receptors on bone marrow cells. Exp Hematol 1994; 22:313-320.
18. Maestroni GJM, Conti A: Noradrenergic modulation of lymphohematopoiesis. Int J Immunopharm 1994; 16:117-122.
19. Felten DL, Felten SY, Sladek JR, Notter MD, Carlson SL, Bellinger DL, Weigand SJ: Fluorescence histochemical techniques for catecholamines as tools in neurobiology. J Microscopy 1990; 157:271-283.
20. Wang T-Y, Brennan JK, Wu JHD: Multilineage hematopoiesis in a three-dimensional murine long-term bone marrow culture. Exp Hematol 1995; 23:26-32.
21. Ackerman KD, Felten SY, Bellinger DL, Felten DL: Noradrenergic sympathetic innervation of the spleen. III. Development of innervation in the rat spleen. J Neurosci Res 1987; 18:49-54.
22. Yamazaki K, Allen TD: Ultrastructural morphometric study of efferent nerve terminals on murine bone marrow stromal cells, and the recognition of a novel anatomical unit: the "neuro-reticular complex". Amer J Anat 1990; 187:261-276.
23. Kuntz A, Richins CA: Innervation of the bone marrow. J Compar Neuro 1945 83:213-222.

24. Shimizu N, Hori T, Nakane H: An interleukin-1β-induced noradrenaline release in the spleen is mediated by brain corticotropin-releasing factor: an *in vivo* microdialysis study in conscious rats. Brain Behav Immun 1994; 7:14-23.
25. Mirand EA, Grace JT, Johnston GS, Murphy GP: Effects of hypothalamic stimulation on the erythropoietic response in the rhesus monkey. Nature 1964 204:1163-1165.
26. Medado P, Izak G, Feldman S: The effect of electrical stimulation of the central nervous system on erythropoiesis in the rat. II. Localization of a specific brain structure capable of enhancing red cell production. J Lab Clin Med 1967 69:776-786.
27. Feldman S, Rachmilewitz EA, Izak G: The effect of central nervous system stimulation on erythropoiesis in rats with chronically implanted electrodes. J Lab Clin Med 1966; 67:713-725.
28. Dexter TM, Allen TD, Lajtha LG: Conditions controlling the proliferation of hemopoietic stem cells *in vitro.* J Cell Physiol 1977; 91:335.
29. Dexter TM, Moore MAS, Sheridan APC: Maintenance of hemopoietic stem cells and production of differentiated progeny in allogeneic and semi-allogeneic bone marrow chimeras *in vitro.* J Exp Med 1977; 145:1612.
30. Whitlock CA, Robertson D, Witte ON: Murine B cell lymphopoiesis in long term culture. J Immunol Methods 1984; 67:353.
31. Rameshwar P, Ganea D, Gascon P: In vitro stimulatory effect of Substance P on hemopoiesis. Blood 1993; 81:391-398.
32. Rameshwar P, Ganea D, Gascon P: Induction of IL-3 and granulocyte-macrophage colony-stimulating factor by Substance P in bone marrow cells is partially mediated through the release of IL-1 and IL-6. J Immunology 1994; 152:4044-4054.

IL-1, IL-6, AND TNF-α RELEASE IS DOWNREGULATED IN WHOLE BLOOD FROM SEPTIC PATIENTS

Jean-Pierre Kremer,[1*] Doraid Jarrar,[2] Ursula Steckholzer,[3] and Wolfgang Ertel[3]

[1] GSF-Forschungszentrum für Umwelt und Gesundheit
Institut für Experimentelle Hämatologie, Munich, Germany
[2] Chirurgische Klinik, Universitätsklinikum Grosshadern
Ludwig-Maximilians-Universität, Munich, Germany
[3] Klinik für Unfallchirurgie
Universitätsspital Zürich, Switzerland

ABSTRACT

Proinflammatory cytokines are important mediators during endotoxemia. In experimental models, injection of lipopolysaccharide (LPS) causes activation of macrophages with excessive secretion of TNF-α, IL-1β and IL-6; infusion of high doses of these mediators results in organ failure and death. Natural infection may be different, as it persists over days or even weeks, with a repeated challenge of macrophages through endotoxin. Little is known about the capacity of peripheral blood mononuclear cells (PBMCs) to release proinflammatory cytokines under those conditions. Therefore, as an *ex vivo* model of sepsis, the expression of proinflammatory cytokines after stimulation of whole blood with LPS was studied. A high dose of LPS (1 μg/ml) resulted in a maximum increase in secretion of TNF-α, IL-1β and IL-6 in controls, but a marked depression was observed in septic patients ($p<0.01$; 15 patients with severe sepsis versus 20 control patients without infection). This reduction persisted up to 10 days after diagnosis of sepsis. The release of TNF-α, IL-1β and IL-6 was markedly decreased in the septic group even when a lower and physiologically more relevant LPS concentration (1 ng/ml) was used. IL-1β mRNA was similar to controls, but a downregulation was observed in TNF-α and IL-6 transcript levels in PBMCs from blood of septic patients. This was at least in part due to a marked reduction in TNF and IL-6 mRNA half-life.

* To whom correspondence should be addressed: Dr. J.-P. Kremer, GSF-Institut für Experimentelle Hämatologie, Marchioninistrasse 25, D-81377 Munich, Germany. Tel.: 0049 89 7099-209; Fax: 0049 89 7099-225.

These results indicate that different mechanisms downregulate proinflammatory cytokine release in whole blood of septic patients. Although excessive secretion is known to be deleterious, low concentrations of these cytokines are involved in regulation of essential cellular and humoral immune functions. In this light, the reduced capacity to express and release adequate amounts of proinflammatory cytokines after exposure to endotoxin observed in whole blood PBMCs from septic patients may contribute to the development of immunodeficiency.

INTRODUCTION

Sepsis remains one of the major complications in hospitalized patients. Severe sepsis may lead to multiple organ dysfunction, to septic shock, and to death. It is mainly caused by bacterial endo- and exotoxins, the most frequent representative being LPS from gram-negative bacteria. Proinflammatory cytokines such as TNF-α, IL-1β and IL-6 have been shown to be important mediators during endotoxemia (1, 2, 3). In animal models, administration of high doses of these cytokines results in organ failure and death (4, 5, 6), whereas a passive immunization against inflammatory cytokines was able to protect from endotoxin induced septic shock (7, 8, 9).

Natural infection, however, may be different from the situation mimicked by animal models: it persists over days or even weeks, and monocytes/macrophages as the principal target cells are likely to have repeated contact with endotoxin. Little is known about the capacity of PBMCs obtained from patients with severe sepsis to release proinflammatory cytokines under these conditions. Therefore, as an *ex vivo* model of sepsis, we used a whole blood stimulation assay to study influence of natural infection on the capacity of PBMCs to express and release TNF-α, IL-1β and IL-6.

MATERIALS AND METHODS

Patients and Blood Collection

Patients included in this study fulfilled the criteria of sepsis syndrome (defined by fever or hypothermia, tachycardia, tachypnea, and clinical signs of altered organ perfusion), or septic shock (defined by hypotension in addition to diagnosis of sepsis syndrome). Among the 15 patients enrolled, 8 fulfilled the criteria of sepsis syndrome, and 7 of septic shock. In 6 cases, gram-negative bacteria were isolated, gram-positive in 3 cases, mixed populations in 5 cases and fungi in 1 case. Seven patients died from multiple-organ-failure during the observation period of 10 days. Blood from septic patients was collected on day of diagnosis (=d0) and on the following days until d10. Control patients had been admitted to the hospital for hernia repair or cholecystectomy (n=20). Their blood was collected once preoperatively.

Blood was drawn into heparinized syringes and stimulated with lipopolysaccharide (LPS; E. coli 055:B5, Difco, Detroit, MI) in sterile polypropylene tubes placed on a rotator in an incubator at 37°C and a 5% CO_2-atmosphere. We used either a dose of 1 μg LPS/mL, expected to induce maximum levels of proinflammatory cytokines, or a more physiological dose of 1 ng/mL, a concentration which has been detected in clinical infections (10,11). After 1, 2, 4, 8 and 24 h of incubation, blood samples were processed by centrifugation through Ficoll-Hypaque (Seromed, Berlin, Germany). The plasma was then stored at -70°C for cytokine assays, and PBMCs were lysed as described below for RNA analysis. The incubation lasting up to 24 h was not found to influence significantly the composition of cellular subpopulations in blood or the viability of PBMCs.

Determination of Cytokine in Plasma

TNF-α present in plasma was measured using WEHI 164 cytotoxicity assay (12; detection limit 0.1 U/mL). IL-6 was quantified by means of a bioassay using IL-6 dependent 7TD1 hybridoma (13). Specificity of both bioassays was demonstrated by the fact that an anti-TNF-α antibody or an anti-IL-6 antibody (both from Genzyme, Boston, MA) could completely abolish TNF-α or IL-6 bioactivity in the samples analyzed. IL-1β levels in plasma were measured by a specific ELISA (detection limit 15 pg/mL). Cytokine release data were normalized according to differential white blood cell counts and based on 1 x 10^6 monocytes/mL. They are presented as mean ± SEM. Significance was calculated with the unpaired Wilcoxon rank sum test with Bonferroni correction for multiple comparisons.

RNA Analysis

PBMCs from whole blood were lysed in guanidine thiocyanate. Total cellular RNA preparation, electrophoresis and Northern blotting was done according to standard methods (14, 15). Hybridization and autoradiography on x-ray film was done as previously described (16). The probes used for hybridization had been labeled with ^{32}P by Megaprime labeling (Amersham Life Science, Braunschweig, Germany). They were fragments of human TNF-α cDNA (0.8 kb *Eco*RI fragment kindly provided by Genentech Inc, San Francisco, CA), human IL-6 cDNA (0.44 kb *Ban*II-*Taq*I fragment, a generous gift from Toshio Hirano, Osaka University, Japan), and human IL-1β cDNA (1.5 kb *Pst*I fragment, generously provided by Genetics Institute, Cambridge, MA).The amount of RNA present on the blots was controlled by an additional hybridization to a murine 28S rRNA probe (kindly provided by I. Grummt, Deutsches Krebsforschungszentrum, Heidelberg, Germany).

Transcript stability of TNF-α and IL-6 mRNA was determined by stimulating whole blood with LPS (1 ng/mL) for 2 h, followed by inhibition of transcription with actinomycin D (5 μg/mL; Sigma, St. Louis, MO) and processing of samples at different time points

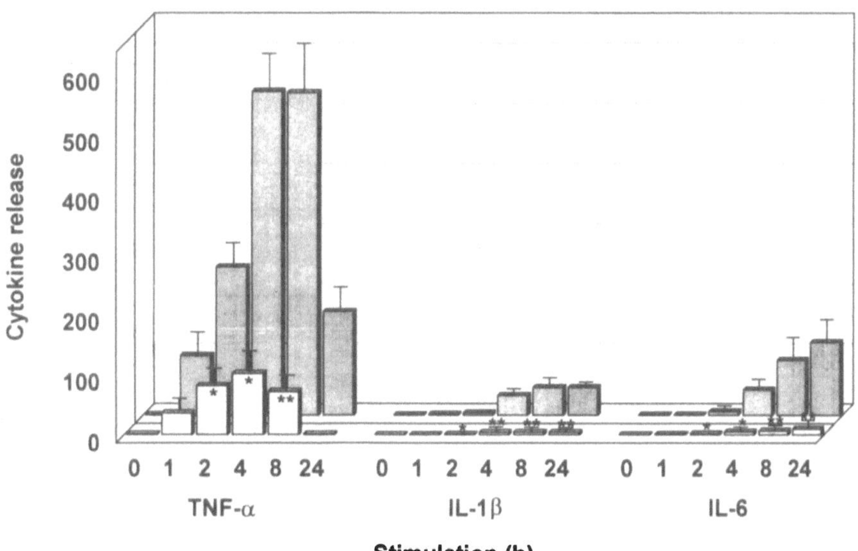

Figure 1. Kinetics of release of TNF-α (U/mL), IL-1β (ng/mL) and IL-6 (x10^3 U/mL) in septic patients. Whole blood obtained from 15 septic patients on day of diagnosis (open bars) and from 20 control patients (shaded bars) was stimulated with LPS (1 μg/mL) for up to 24 h as indicated. * P < .05, ** P < .01 (sepsis vs. control).

thereafter as described above. Northern blotting and hybridization was performed as indicated, and blots were then analyzed by means of the FUJI digital imaging system (FUJI imaging plates and Fujix BAS 1000 Bioimaging Analyzer from FUJI, Düsseldorf, Germany).

RESULTS AND DISCUSSION

We first investigated the response of whole blood PBMCs when activated by 1 μg LPS/mL, an endotoxin dose expected to elicite a maximal response (Ertel, unpublished observations). Release of TNF-α, Il-1β and IL-6 into whole blood over an observation period of 24 h is shown in Fig 1. Proinflammatory cytokine levels detected in plasma are markedly depressed in patients with severe sepsis when compared to controls. Studies performed with blood from both groups incubated in a similar manner but without LPS did not reveal any spontaneous cytokine release (data not shown).

We followed proinflammatory cytokine release over a period of 10 days after diagnosis of sepsis (Fig 2). As can be seen, the depression of protein release persisted over the whole observation period with no apparent tendency towards normalization.

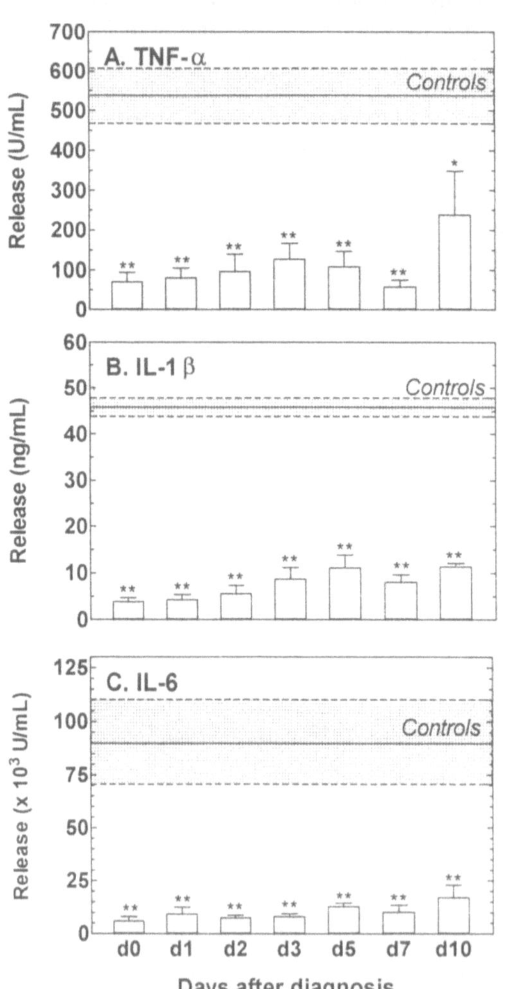

Figure 2. Release of proinflammatory cytokines on consecutive days of sepsis. Open bars show release of TNF-α, IL-1β and IL-6 per 1 x 10^6 monocytes into whole blood from septic patients obtained on consecutive days after diagnosis and stimulated with LPS (1 μg/mL) for 8 h (d0: n=15, patients surviving d10: n=8). Shaded aereas show normal values obtained from control patients (n=20). * P < .05, ** P < .01 (sepsis vs. control).

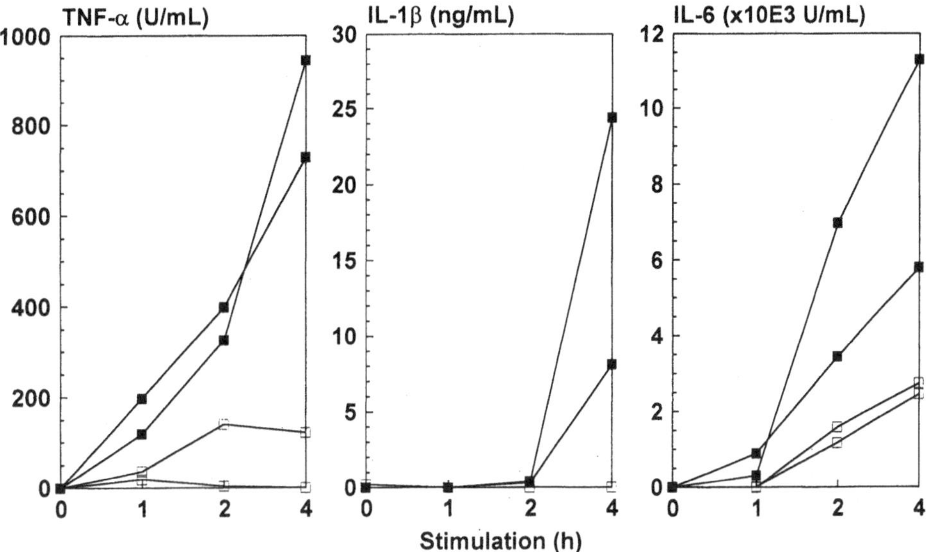

Figure 3. Proinflammatory cytokine release into blood stimulated by low dose LPS. Whole blood from 2 patients with sepsis (open symbols) and two control patients (closed symbols) was stimulated with LPS (1 ng/mL) for 0 to 4 h.

The depression of TNF-α, Il-1β, and IL-6 release into whole blood is not related to an unspecific monocyte/macrophage activation by extremely high doses of LPS, a mechanism suggested by Wright et al. (17). This is concluded from further experiments using 1 ng LPS/mL, a dose which has been detected in the plasma from septic patients (10,11). As shown in Fig 3, even at this clinically relevant dose, TNF-α, Il-1β, and IL-6 is severely depressed in septic patients when compared to controls.

We then investigated at which level the inhibition of proinflammatory cytokine release might occur. Northern blotting of PBMCs isolated from whole blood following stimulation with LPS and hybridization with the probes of interest resulted in detection of significantly reduced mRNA levels of TNF-α and IL-6 in septic patients, whereas transcript levels of IL-1β were near to normal (Fig 4). The depressed levels of TNF-α and IL-6 transcripts are at least in part due to a reduced stability of the mRNAs in question, as was shown by actinomycin D experiments followed by densitometric analysis of mRNA (Fig 5): half-life of TNF-α and IL-6 mRNA is clearly reduced versus controls.

These results indicate that in our *ex vivo* model of sepsis, PBMCs show a reduced responsiveness to an adequate endotoxin challenge. This depression was found to be a phenomenon persisting over the whole 10 days of observation and to occur at low and clinically relevant doses of LPS. Comparison of protein and RNA data show that in PBMCs of septic patients regulation of proinflammatory cytokines is affected at different levels. Whereas TNF-α and IL-6 was suppressed at the transcriptional level, IL-1β is likely to be regulated rather at the posttranscriptional level. This notion is supported by other studies showing that TNF and IL-1 expression is regulated independently after stimulation of PBMCs with LPS (6, 18, 19, 20).

The mechanisms of proinflammatory cytokine downregulation in sepsis remain unclear. A number of antiinflammatory mediators has been described, among them PGE_2, IL-4, IL-10, IL-13, and TGF-β (19-24). In addition, PGE_2 is elevated in plasma of septic patients (25), as is TGF-β (26).

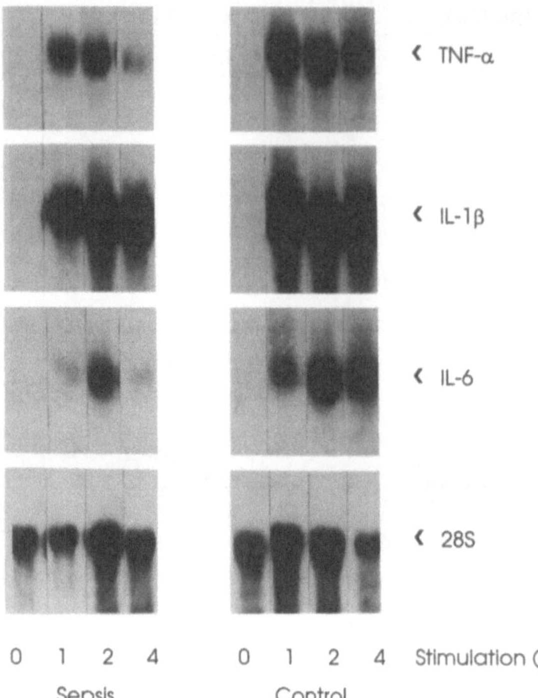

Figure 4. Expression of TNF-α, IL-1β and IL-6 mRNA. PBMCs were isolated from whole blood from a septic patient and from a control patient stimulated with LPS (1 ng/mL).

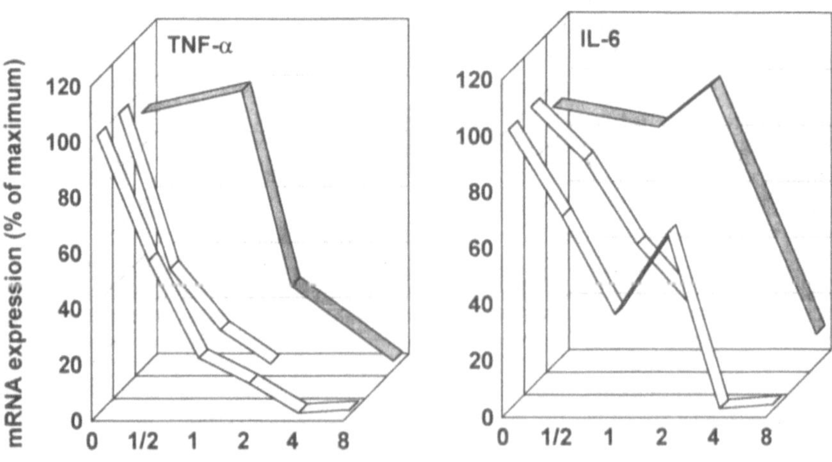

Figure 5. Stability of TNF-α and IL-6 mRNA. PBMCs from two septic patients (open lines) and a control patient (shaded line) were isolated from whole blood stimulated with 1 ng LPS/mL for 2 h followed by inhibition of transcription by addition of actinomycin D. Northern blotting of total RNA and hybridization was followed by quantitative autoradiography by means of FUJI digital imaging. TNF and IL-6 mRNA amounts were normalized based on 28S rRNA levels.

Our data are in line with the concept of endotoxin tolerance of monocytes as developed by Dinarello (6). It has been shown by other investigators that proinflammatory cytokine release of PBMCs isolated from normal volunteers who had received an endotoxin injection is depressed (27, 28). Our results confirm these observations and extend their relevance to the clinical situation of sepsis. At the same time, they show that one of the mechanisms responsible for this depression is a reduced half-life of TNF-α and IL-6 mRNA.

It is known that excessive secretion of proinflammatory cytokines is deleterious to the host and may result in organ failure and death (29, 30). Nevertheless, TNF-α, Il-1β, and Il-6 are essential for the regulation of cellular and humoral functions, and TNF-α and IL-1 have been shown to mediate a protective effect against infection in animal sepsis models (31 -33). In this light, sepsis leads to an imbalance in proinflammatory cytokine synthesis and release. Thus, the reduced capacity of LPS stimulated PBMCs to release adequate amounts of proinflammatory cytokines observed may contribute to the development of immunodeficiency during continuous sepsis or repeated septic episodes.

REFERENCES

1. Michie HR, Manogue KR, Spriggs DR, Revhaug A, O'Dwyer S, Dinarello CA, Cerami A, Wolff SM, Wilmore DW: Detection of cicrculating tumor necrosis factor after endotoxin administration. N Engl J Med 318:1481, 1988
2. Van Zee KJ, DeForge LE, Fischer E, Marano MA, Kenney JS, Remick DG, Lowry SF, Moldawer LL: IL-8 in septic shock, endotoxemia, and after IL-1 administration. J Immunol 146:3478, 1991
3. Martich DG, Danner RL, Ceska M, Suffredini AF: Detection of interleukin-8 and tumor necrosis factor in normal humans after intravenous endotoxin: The effect of antiinflammatory agents. J Exp Med 173:1021, 1991
4. Tracey KJ, Beutler B, Lowry SF, Merryweather J, Wolpe S, Milsark IW, Hariri RJ, Fahey III TJ, Zentella A, Albert JD, Shires GT, Cerami A: Shock and tissue injury induced by recombinant human cachectin. Science 234:470, 1986
5. Okusawa S, Gelfand A, Ikejima T, Connolly RJ, Dinarello CA: Interleukin 1 induces a shock-like state in rabbits. Synergism with tumor necrosis factor and the effect of cyclooxygenase inhibition. J Clin Invest 81:1162, 1988
6. Dinarello CA: Interleukin-1 and interleukin-1 antagonism. Blood 77:1627, 1991
7. Beutler B, Milsark IW, Cerami AC: Passive immunization against cachectin/tumor necrosis factor protects mice from lethal effect of endotoxin. Science 229:869, 1985
8. Silva AT, Bayston KF, Cohen J: Prophylactic and therapeutic effects of a monoclonal antibody to tumor necrosis factor-α in experimental gram-negative shock. J Infect Dis 162:421, 1990
9. Ohlsson K, Björk P, Bergenfeldt M, Hageman R, Thompson RC: Interleukin-1 receptor antagonist reduces mortality from endotoxin shock. Nature 348:550, 1990
10. Van Deventer SJH, Büller HR, ten Cate JW, Sturck A, Pauw W: Endodoxemia: An early predictor of septicemia in febrile patients. Lancet 1:605, 1988
11. Shenep JL, Flynn PM, Barrett FF, Stidham GL, Westenkirchner DF: Serial quantitation of endotoxemia and bacteremia during therapy for gram-negative bacterial sepsis. J Infect Dis 157:565, 1988
12. Espevik T, Nissen-Meyer J: A highly sensitive cell line, WEHI 164 clone 13, for measuring cytotoxic factor/tumor necrosis factor from human monocytes. J Immunol Methods 95:99, 1986
13. Van Snick J, Cayphas S, Vink A, Uyttenhove C, Coulie G, Rubira MR, Simpson RJ: Purification and NH_2-terminal amino acid sequence of a T-cell derived lymphokine with growth factor activity for B-cell hybridomas. Proc Natl Acad Sci USA 83:9679, 1989
14. Chomczynski P, Sacchi N: Single step method of RNA isolation by acid guanidinium thiocyanate-phenol-chloroform extraction. Anal Biochem 162:156, 1987
15. Sambrook J, Fritsch EF, Maniatis T: Molecular Cloning: A Laboratory Manual. Cold Spring Harbor, NY, Cold Spring Harbor Laboratory, 1989
16. Kremer JP, Reisbach G, Nerl C, Dörmer P: B-cell chronic lymphocytic leukemia cells express and release transforming growth factor-β. Brit J Haematol 80:480, 1992
17. Wright SD, Ramos RA, Tobias PS, Ulevitch RJ, Mathison JC: CD14, a receptor for comlexes of lipopolysaccharide (LPS) and LPS binding protein. Science 249:1431, 1990

18. Knudsen PJ, Dinarello CA, Strom TB: Prostaglandins posttranscriptionally inhibit monocyte expression of interleukin 1 activity by increasing intracellular cyclic adenosine monophosphate. J Immunol 137:3189, 1986
19. Schindler R, Clark BD, Dinarello CA: Dissociation between IL-1β mRNA and protein synthesis in human peripheral blood mononuclear cells. J Biol Chem 265:10232, 1990
20. Zuckerman SH, Evans GF, Butler LD: Endotoxin tolerance: Independent regulation of interleukin-1 and tumor necrosis factor expression. Infect Immun 59:2774, 1991
21. Esser R, Rhoades K, Mc Bride WH, Morton DL, Economou JS: IL-4 downregulates IL-1 and TNF gene expression in human monocytes. J Immunol 142:3857, 1989
22. De Waal Malefyt R, Abrams J, Bennett B, Figdor CG, de Vries JE: Interleukin 10 (IL-10) inhibits cytokine synthesis by human monocytes: An autoregulatory role of IL-10 produced by monocytes. J Exp Med 174:1209, 1991
23. Minty A, Chalon P, Derocq JM, Dumont X, Guillemot JC, Kagad M, Labit C, Leplatois P, Liauzun P, Miloux B, Minty C, Casellas P, Loison G, Lupker J, Shire D, Ferrara P, Caput D: Interleukin-13 is a new human lymphokine regulating inflammatory and immune responses. Nature 362:248, 1993
24. Bogdan C, Paik J, Vodovotz Y, Nathan C: Contrasting merchanisms for suppression of macrophage cytokine release by transforming growth factor-β and interleukin-10. J Biol Chem 267:23301, 1992
25. Grbic JT, Mannick JA, Gough DB, Rodrick ML: The role of prostaglandin E_2 in immune suppression following injury. Ann Surg 214:253, 1991
26. Ayala A, Knotts JK, Ertel W, Perrin MM, Morrison MH, Chaudry IH: Role of interleukin 6 and transforming growth factor-β in the induction of depressed splenocyte responses following sepsis. Arch Surg 128:89, 1993
27. Granowitz EV, Porat R, Mier JW, Orencole SF, Kaplanski G, Lynch EA, Ye K, Vannier E, Wolff SM, Dinarello CA: Intravenous endotoxin suppresses the cytokine response of peripheral blood mononuclear cells of healthy humans. J Immunol 151:1637, 1993
28. Rodrick ML, Moss NM, Grbic JT, Revhaug A, O'Dwyer ST, Michie HR, Gough DB, Dubravec D, Manson JM, Saporoschetz IB, Collins KH, Jordan AL, Wilmore DW, Mannick JA: Effects of in vivo endotoxin infusions on in vitro cellular immune responses in humans. J Clin Immunol 12:440, 1992
29. Rock CS, Lowry SF: Tumor necrosis factor-α. J Surg Res 51:434, 1991
30. Dinarello CA: The proinflammatory cytokines interleukin-1 and tumor necrosis factor and treatment of the septic shock syndrome. J Infect Dis 163:1177, 1991
31. Sheppard BC, Fraker DL, Norton JA: Prevention and treatment of endotoxin and sepsis lethality with recombinant human tumor necrosis factor. Surgery 106:156, 1989
32. Cross AS, Sadoff JC, Kelly N, Bernton E, Gemski P: Pretreatment with recombinant murine tumor necrosis factor α/cachectin and murine interleukin-1α protects mice from lethal bacterial infection. J Exp Med 169:2021, 1989
33. Czuprynski CJ, Brown JF: Recombinant murine interleukin-1 alpha enhancement of nonspecific antibacterial resistance. Infect Immun 55:2061, 1987

EXPRESSION OF TRANSFORMING GROWTH FACTOR-BETA TYPE II RECEPTORS IN THE CELLS OF THE HUMAN THYMIC MICROENVIRONMENT DURING ONTOGENESIS

Bela Bodey,[1*] Bela Bodey, Jr.,[1] and Frederick L. Hall[2]

[1] Department of Pathology
School of Medicine
University of Southern California
[2] Department of Orthopedic Surgery
Childrens Hospital of Los Angeles

ABSTRACT

Transforming Growth Factor Beta (TGF-beta), a multifunctional cytokine, has potent regulatory effects on a wide spectrum of cell types. In general, TGF-beta seems to be stimulatory for the growth of cells mesodermal/mesenchymal in their origin and inhibitory for cells epithelial/neuroectodermal in their origin. All three isoforms of the TGF-beta superfamily (beta-1, beta-2 beta-3) are homodimers with a molecular weight of approximately 25 kD. Three distinct receptors, referred to as receptor types I, II, and III (betaglycan), appear to mediate cellular responses. The molecular cloning of the human type II receptor enabled the development of specific antibodies suitable for immunochemical studies. In this study, we employed an indirect, alkaline phosphatase conjugated streptavidin-biotin immunocytochemical antigen detection technique on routine, formalin fixed and paraffin embedded tissue sections. We employed a rabbit polyclonal, affinity purified antibody developed in this laboratory and directed specifically against the extracellular domain of the human TGF-beta type II receptor (TGF-bIIR).

We observed fetal thymuses of varying ages (7.5-19 weeks) as well as postnatal tissues from 1 week after birth to 21 years of age. In prenatal tissues, only weak immunoreactivity was detected in cortical cells, however, we noticed the presence of TGF-bIIR expression located in the cortical region among the stem cells that had immigrated and were

[*] Corresponding author: Professor Bela Bodey, M.D., Ph.D. Division of Hematology/Oncology, 4650 Sunset Boulevard, Mailstop 54, Los Angeles, California 90054. Telephone: (213) 669-2205.

already committed to lymphopoietic differentiation. During ontogenesis, thymocytes and the reticulo-epithelial (RE) cells in the thymic cortex showed little to no expression of TGF-bIIR. In contrast, medullary I-lymphocytes, RE cells, and stromal cells exhibited strong immunoreactivity. From approximately 16 weeks in the thymic medulla and at the level of cortico-medullary junction (the location of interdigitating cells and dendritic cells) the cells demonstrated a strong presence of the TGF-bIIRs. The various types of hematopoietic stem cells located in the perilobular connective tissue and the cells of the RE network also contained TGF-bIIRs. A non-specific immunostaining was detected in the necrotic center of Hassall's bodies (HBs): the hypertrophied RE cells within the HBs expressed numerous TGF-bIIRs. In postnatal thymuses (1 week; 2 months; 1,2,3,5 and 21 years) TGF-Beta II receptor immunoreactivity was detected in virtually all cell types of the cortex, while a significant decrease in the immunoreactivity of the medullary regions was observed. These results demonstrate spacial and temporal differences in the expression of the TGF-B Type II receptor in the developing thymus, which is indicative of a functional role.

INTRODUCTION

It has been postulated that normal mammalian cells and tissues exist within a complex microenvironment of stimulatory and inhibitory polypeptide growth factors and cytokines that are produced by the cells themselves or by adjacent cells. Two different, autocrine and paracrine proteins have been termed transforming growth factors (TGFs).[1-2] TGF-alpha is structurally and functionally related to the epidermal growth factor (EGF), a growth stimulator. TGF-beta has emerged as the prototypical inhibitor of proliferation.[3] The term "transforming" is related to their discovery: they induced a reversible morphologic transformation of fibroblastic cells *in vitro* and stimulated colony formation in soft agar.[4-6] It has been also postulated that "autoinduction" represents a mechanism by which the cell-biological effects of a growth factor such as TGF-beta might be amplified.[7-9] TGF-beta was originally isolated from transformed cell-conditioned media that was able to stimulate anchorage independent growth of nontransformed fibroblasts.

The TGF-beta family of polypeptides includes a group of dimeric proteins, in which all nine cysteine residues are conserved, and distant polypeptides that share significant sequence identity and possess closely related cell-biological functions.[8] Three molecular isoforms of TGF-beta have been identified in mammals and have been designated as TGF-beta1, TGF-beta2 (also called BSC-1 cell growth inhibitor or polyergin), and TGF-beta3.[10-14] TGF-betas have been purified from kidney,[15] placenta,[16] and platelets.[17] These genes have been identified in humans on chromosomes 19, 1, and 14, respectively.[18-19] TGF-beta1 has been localized in hemopoietic cells of blood islands and capillaries and in the liver, commencing its hemopoietic role during early ontogenesis.[20] In mammals the sequence for TGF-beta1 is highly conserved: it is identical in man,[13] pig,[21] cow,[22] and monkey.[23] TGF-beta2 was first detected and isolated from a human glioblastoma derived cell line.[24-25] Presently primary glioblastomas are frequently associated with an immunosuppressed host.

All types of normal cells in physiologic conditions and a great percentage of neoplastically transformed cells express cell surface membrane receptors or binding proteins for TGF-beta.[26-30] Recently nine distinct TGF-beta binding proteins have been defined.[31] Three well glycosylated receptors of various molecular masses have been identified: *1)* 53 kD (type I); *2)* 70 kD (type II); and *3)* 280 kD (type III) employing photoaffinity and the method of cross-linking.[3,32] The type II receptor has been determined to act in a cooperative manner with the type I receptor in binding TGF-beta and signaling via a *de novo* formed heteromeric complex.[33] Loss of type I and/or type II receptors results in loss of cellular responsiveness to TGF-beta regulation.[34] The type II receptor has been cloned and a functional, cytoplasmic serine/theonine kinase domain has been identified.[35] The TGF-beta receptor (Endoglin) has received the CD105 designation at the 5th International Conference on Human Leukocyte Differential Antigens in 1993.

The molecular cloning of the TGF-beta type II receptor has enabled the development of specific immunochemical reagents to monitor its expression. In the present study, we examined a panel of prenatal and postnatal thymus tissues by immunocytochemical methods to characterize the developmental expression of the TGF-B type II receptor in the thymic cellular microenvironment.

MATERIAL AND METHODS

Tissue Handling and Storage

Exclusively formalin fixed, paraffin-embedded normal human tissues were used in this study! Fetal thymic tissue was obtained after spontaneaus and artificial abortions. The postnatal normal thymic tissues were obtained at the time of childhood open heart resections at the Childrens Hospital Los Angeles. All tissues were fixed in 10% neutral solution of formalin within 30-50 minutes following surgical removal and were embedded in paraffin following dehydration in ascending dilutions of ethyl-alcohol (50%, 70%, 80%, 95% and 100% twice). Routine histopathologic observation was carried on hematoxylin and eosin stained paraffin embedded tissue sections.[36]

Polyclonal Rabbit Antisera Employed in This Study

A newly developed, affinity purified polyclonal rabbit antisera directed against the N-terminal extracellular domain of the human Type II receptors of TGF-beta was employed in this study. The working concentration was determined with preliminary dilution experiments using antigen-positive and antigen-negative human normal postnatal tissues (cerebrum, cerebellum, lymph nodes, spleen, liver). The antibody was generally diluted 1:200 in purified mixture of proteins (Shandon, Pittsburgh, PA).

Immunocytochemical Antigen Detection Technique

The immunocytochemical antigen detection technique for formalin fixed and paraffin embedded tissues:

Prior to cutting, we carefully determined that formaldehyde fixation and routine paraffin embedding had not reacted or destroyed the antigenic epitope under investigation. 3-4 um thick paraffin embedded tissue sections were cut and mounted on clear uncoated slides. Sections were dewaxed in three changes (always fresh) of Xylene substitute (Shandon) and rehydrated in descending concentration of ethyl-alcohol (100%, 96% twice, 80%, 70%, 55%, 40%) and in the last step in three changes of ascending proportions of distilled water in TBS.

In our immunocytochemical research projects we never used the method of "antigen liberation" by single or combined enzymatic digestion prior to the primary antigen-antibody reaction. During this research project we employed a newly developed, so called antigen retrieval (AR) method,[37-38] but the nice staining properties without any AR made us confident to avoid the use of the AR method, except as part of the preliminary screening.

Staining Procedure

The immunocytochemical antigen detection technique for formalin fixed and wax embedded tissue sections was a highly sensitive, four to six step, indirect biotin-streptavidin (ABC) conjugated with alkaline phosphatase (AP) method,[39] following deparaffinization (in three changes of xylene substitute (Shandon)) for 20 to 30 minutes, and descending dilutions of alcohol 100%, 95%, 85%, 70% and 50% to TBS.

Blocking Step (1). To eliminate the endogeneous AP activity from the tissues, we used 1% glacial acetic acid mixed with the working buffer for 10 minutes. Use of levamisole solution is also described in our earlier observations. As we explained earlier,[39] GAA inhibition was preferred because of the possible presence of levamisole-resistant AP iso-enzyme.[40]

Blocking Step (2). The second blocking step was conducted with a purified mixture of proteins (Shandon) from various species for 5-10 minutes to block cross-reactive antigenic epitopes. Excess serum was removed from around the sections.

Incubation with the Primary Antibody. The tissue sections were incubated for 80-100 minutes with the primary (observed, detecting the desired antigenic epitope), affinity purified and biochemically characterized polyclonal rabbit anti-human polyclonal antisera directed against the type II receptors of transforming growth factor-beta.

Incubation with the Secondary Antibody. Incubation with biotinylated, whole goat anti-rabbit IgG molecule (IgG molecule diluted by ICN Biomedicals, Lisle, IL) for 20 minutes.

Streptavidin Conjugation. Incubation with AP conjugated streptavidin for 20 minutes.[41-44]

Alkaline Phosphatase – Color Development. Color visualisation of the primary antigen-antibody (Ag-Ab) reaction was accomplished with an alkaline phosphatase (AP) kit I (Vector Laboratories, Burlingame, CA 94010) which contains AS-TR with Tris-HCl buffer at pH 8.2. An alkaline phosphatase detection substrate, buffered with Tris-HCl (at pH 8.2), was added for 28-60 minutes to allow formation of a stable red precipitate from the primary Ag-Ab reaction product (various intensities of red precipitate). Sections were counterstained with a diluted solution of Gill's hematoxylin.

Morphologic Clearing in Xylene Substitute (Shandon). Short and long term clearing was used in two changes of xylene substitute (Shandon) during this study.

Human Tissue Controls

Every immunocytochemical staining needs controls for true evaluation. Early human postnatal tonsils and sections of decalcified un-fixed osteosarcomas or astrocytomas, all antigen positive tumors were employed as positive tissue controls.

Additional controls for all used tissues and MoABs included:

1. omission of the primary MoAB;
2. color developer only to scan for remaining endogenous AP activity.

Qualitative and Quantitative Evaluation

Qualitative and quantitative evaluation of the percent of antigen positive cells and the intensity of staining were conducted using a light microscope (Microphot, Nikon, Japan) counting 100-200 cells from each of five distinct areas in non-necrotic, non-hemorrhagic tissue. Artefacts were avoided, but, on the other hand, morphologically characteristic areas were sought out. The presence of numerous blood vessels and connective tissue elements or hemopoietic tissue (immigrated lympho- and non-lymphopoietic, already committed stem cells) required careful qualitative assessment of the staining intensity and the TGF-beta type II receptor antigen distribution. Major cell types of the thymic microenvironment bearing TGF-beta type II receptors were documented.

Qualitative and Quantitative Evaluation of the Antigen-Antibody (Ag-Ab) Reaction

Our Quantitative Evaluation is as Follows.[39] (++++) over 90% of the total cell number are positive; (+++) 50% to 90% of the total cell number are positive; (++) 10% to 50% of the total cell number are positive; (+) 1% to 10% of the total cell number are positive; (+/-) under 1% of the total cell number are positive; (-) negative.

Our Qualitative Evaluation of Staining Intensity is as Follows.[39] (A) very intense red staining; (B) intense red staining; (C) light red staining; (D) negative staining.

RESULTS

The comparative expression of the TGF-beta type II receptor (TGF-bIIR) in the different cell types of the thymic microenvironment and the qualitative evaluation (staining intensity) of the immunocytochemical reactions of this study are summarized in Table 1.

Fetal Thymus

The earliest stage of the thymic ontogenesis available for our observations was represented by a 7.5 embryonal week (2nd lunar month) thymus. A minimal expression of the TGF-bIIR was observed on subcapsular RE cells (the first ontogenetic appearence of thymic nurse cells). Minimal expression was also present on medullary thymocytes and on very few medullar RE cells. No TGF-beta IIR was found on the thymic stromal cells (dendritic cells; IDC). The IDC of the well formed cortico-medullary junction did not express the TGF-bIIRs. The immunocytochemical results observed at age of 16 fetal weeks (4th lunar month) were much different. All medullary thymic elements expressed the TGF-bIIR. The type II receptor was detected on over 90% of the medullary thymocytes, on between 50 and 90% of the medullary RE cells, and on between 10 and 50% of medullary stromal cells. Over 90% of IDC, located in the cortico-medullary junction also demonstrated presence of TGF-beta type IIR. Suprisingly over 90% of the RE cells that were located in the outer core of the Hassall's bodies expressed TGF-bIIR. The cortical area demonstrated minimal presence of TGF-bIIR. Expression was detected on the cortical thymocytes and the subcapsular large RE cells. During the 19th fetal week (end of the fifth lunar month), a very strong presence of the TGF-bIIR was observed on the thymic medullary elements. Over 90% of all cellular elements in the medulla expressed TGF-bIIR. The cortical distribution of the receptor was much lower ranging from sporadic staining of individual cells to approximately 10% of the cellular elements.

Table 1. TGF-beta type II receptor expression in the cells of the human thymus

AGES	CELLS OF THE THYMIC MICROENVIRONMENT							
	Cortical Thymocytes	*Subcapsular RE cells*	*Cortical RE cells*	*Cortical Stromal cells*	*Medullary Thymocytes*	*Medullary RE cells*	*Medullary Stromal cells*	*HB*
7.5 f.w.	-D			-D	+A	-D	-D	not present
16 f.w.	±A			-D	++++A	+++A	++A	++++A
18 f.w.	±A,B	±A	-D	-D	++++A	++++A	++++A	+++A,B
19 f.w.	±A,B	±A	±A	±A	++++A,B	++++A	++++A	+++A,B
1 p.n.w.	+++A	++A	+++A	+++A	+++A	+++A	+++A	++A,B
8 p.n.w.	+++A	++A	+++A	+++A	++A	++A	++A	++A,B
1 y.	+++A	++A	+++A	+++A	+++A	++A	++A	+A,B
2 y.	++++A	+++A	+++A	+++A	++A	++A,B	++A	±A
3 y.	++++A	+++A	+++A	+++A	++A	++A,B	++A	±A
5 y.	+++A	+++A	++++A	+++A	++A	++A	++A,B	+A,B
21 y.	±A	±A	±A	±A,B	±A	±A	±A	++A

Abbreviations: f.w. - fetal week
p.n.w. - postnatal week
y. - year

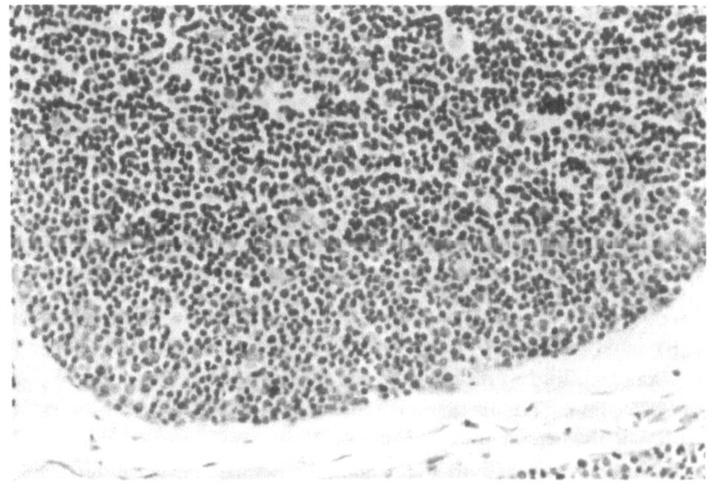

Figure 1A. Human embryonal thymus (7.5 weeks of ontogenesis). Paraffin section 3-5 um. Developing thymic tissue showing no separation of cortex and medulla. The immunologically immature thymocytes are dispersed throughout the thymus. In the subcapsular 2-3 cell layers on the RE cells TGF-beta type IIRs are present. Magnification: 100 x.

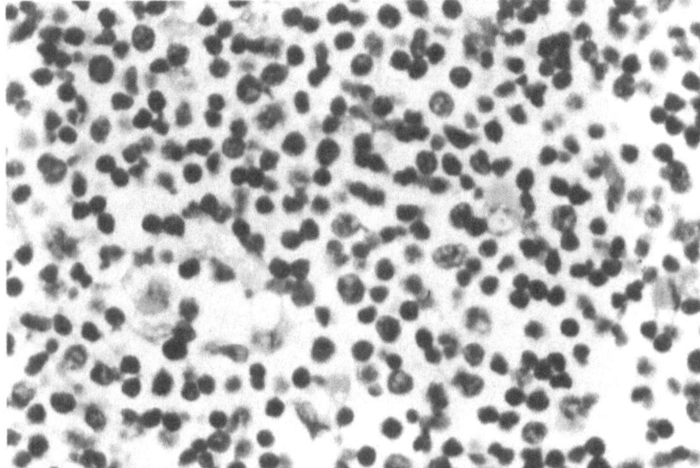

Figure 1B. The same thymus as Fig 1A. Paraffin section 3-5 um. Subcapsular thymic tissue (later the so-called Cortex Thymi). Some cells of the RE cellular network express TGF-beta type IIRs. Magnification: 1000 x.

Postnatal Thymus

The expression pattern of the TGF-bIIR in postnatal tissues was different when compared with the fetal period in that we observed an elevated expression of this receptor in the major cell types of the thymic microenvironment. The most important distinction was the presence of this receptor on all investigated cortical as well as medullary thymic cell types. By comparison, the medullary immunoreactivity was present but was less pronounced. The results were consistent until the 5th year of age. The only contrast was the lower

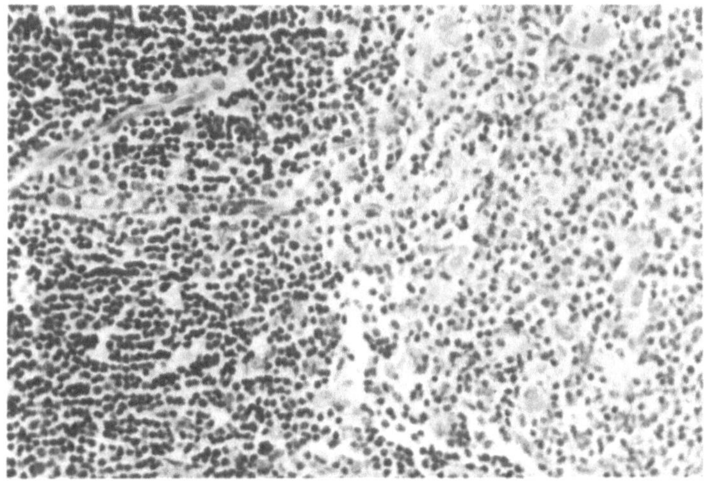

Figure 1C. Human prenatal thymus 16 weeks; (end of the 4th intrautrine month). Paraffin section 3-5 um. Morpho-physiologic division of the thymus into two functionally different subunits: cortex and medulla. The medullary RE cells express TGF-beta type IIRs. Magnification: 400 x.

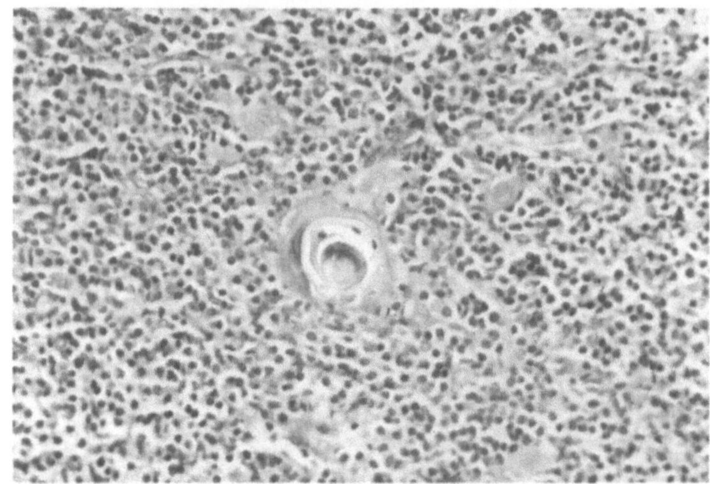

Figure 1D. Human prenatal thymus (16 weeks; end of the 4th lunar month). Paraffin section 3-5 um. Thymic medulla with well developed Hassall's body (HB). The cells of the RE network and the cells of the cortico-medullary junction express numerous TGF-beta type IIRs. Within the HBs, TGF-beta type IIR expression is detected only in the peripheral core, where the hypertrophied RE cells are located. Magnification: 400 x.

percentage of TGF-bIIR positive RE cells within the Hassall's bodies: 10 to 50% during the early postnatal period which further decreased to between 1 and 10% in the 1st year of age. We also observed a very involuted thymus at 21 years of age. Only various cell groups in the form of buds were present along with very large HBs. The TGF-bIIR expression was very low in all cell types. The receptor was detected only on 1 to 10% of all cells. Surprisingly, still 10 to 50% of the RE cells within the HBs demonstrated presence of the TGF-bIIR.

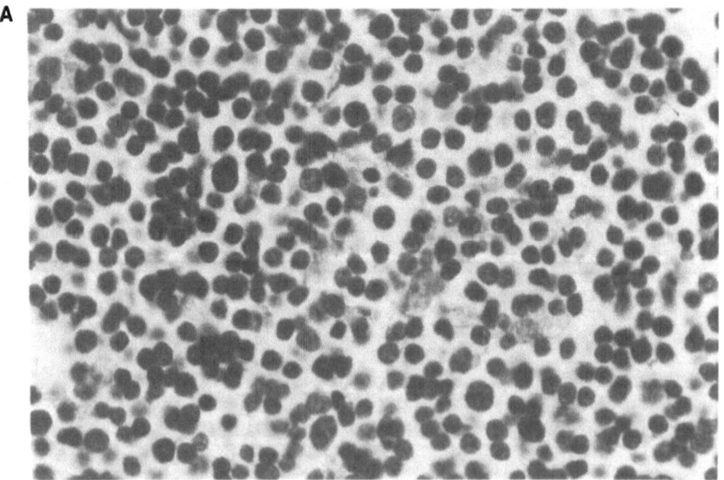

Figure 2A. Human postnatal thymus (7 days after birth). Paraffin section 3-5 um. The cells of the cortical region still contain only a minimal number of TGF-beta type IIRs. There are cortical RE cells expressing TGF-beta type IIRs in the deep cortex. Magnification: 1000 x.

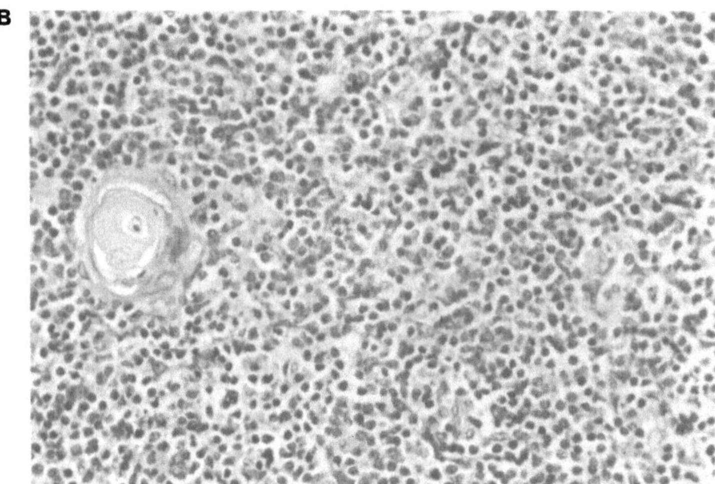

Figure 2B. Human postnatal thymus (2nd month of extrauterine development). Paraffin section 3-5 um. The RE and non-lymphatic cells of the cortico-medullary junction and the medullary region express numerous TGF-beta type II receptors. The immunologically mature medullary lymphocytes also contain TGF-beta type II receptors. A well developed H.B. is formed from hypertrophied RE cells (arrowheads). Magnification: 400 x.

DISCUSSION

Employing immunocytochemical antigen detection techniques, we were able to determine that TGF-beta receptors are expressed in a manner that is consistent with a role in the development and maturation of the thymus. The developmental pattern of receptor expression suggests that TGF-beta plays a role in intrathymic proliferation and differentia-

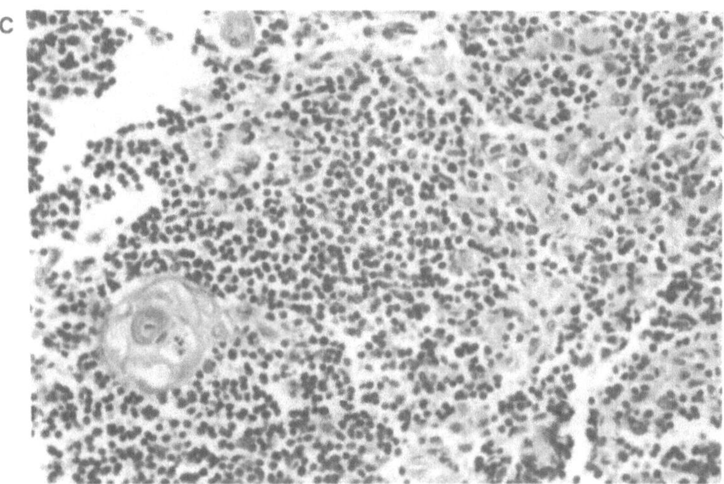

Figure 2C. Human postnatal thymus (3 years of age). Paraffin section 3-5 um. This microphotograph is very similar in characteristics to Fig. 2B. Hypertrophied RE cells within the HB (arrowheads), containing numerous TGF-beta type II receptors. The immunocytochemical reactivity demonstrates an active functional role of the HBs. Magnification: 400 x.

tion pathway of the immature cortical thymocytes in immunocompetent $CD4^+$ or $CD8^+$ T lymphocytes. TGF-beta type II receptors were also expressed by the interdigitating stromal cells (IDCs) and by dendritic cells located in the thymic cortico-medullary junction. This result suggests an active role of TGF-beta in intrathymic positive and negative selection of the immunologically matured two major subpopulations of T lymphocytes. The presence of TGF-beta type IIR in the thymic medulla suggests its involvement in regulating cell to cell interactions between the already mature T lymphocyte subsets and the cells of the RE network, prior to T cell migration to the periphery.

Among the earliest lymphatic hematopoietic cells to enter the CD4/CD8 intrathymic maturation pathway are $CD4^-$ CD810 precursor cells that differentiate into $CD4^+$ $CD8^+$ (double positive) thymocytes.[45] This maturation requires at least one cell division, and the progression through the cell cycle is specifically retarded within the thymic microenvironment by cell to cell contact with thymic RE cells which express transmembrane TGF-beta receptors. TGF-beta may also have has a specific regulatory effect on cell cycle progression and differentiation of $CD4^-$ $CD8^+$ lymphopoietic precursor cells. TGF-beta proteins do not appear to regulate the earlier step of intrathymic T lymphocyte maturation, that of generation of $CD4^-$ $CD8^+$ thymocytes from $CD4^-$ $CD8^-$ thymocytes.[46] These authors described TGF-beta type IIR expression on the cortical thymocytes and RE cells of an intact thymus. The subcapsular thymic nurse cells as a subpopulation of the cortical RE cellular network also contained TGF-beta type HRs. However, the study, did not observe TGF-beta type IIR expression during the fetal development of the thymus and FACS analysis was employed after *in vitro* cell culture of the ex vivo isolated thymocytes. Earlier work already described that TGF-beta indeed inhibits the proliferation of mature T lymphocytes in cell culture[46-47] and also *in vivo*.[48] Our present finding that TGF-Beta type II receptors are present on immigrating hematopoietic stem cells further suggests a functional role in early T-cell differentiation.

Certainly these factors also depress the proliferation and activation level of the cytotoxic T lymphocytes, the cytotoxicity of natural killer (NK) cells, lymphokine activated killer (LAK) cells and macrophages.[46,49-50] TGF-beta is a powerful chemoattractant for blood

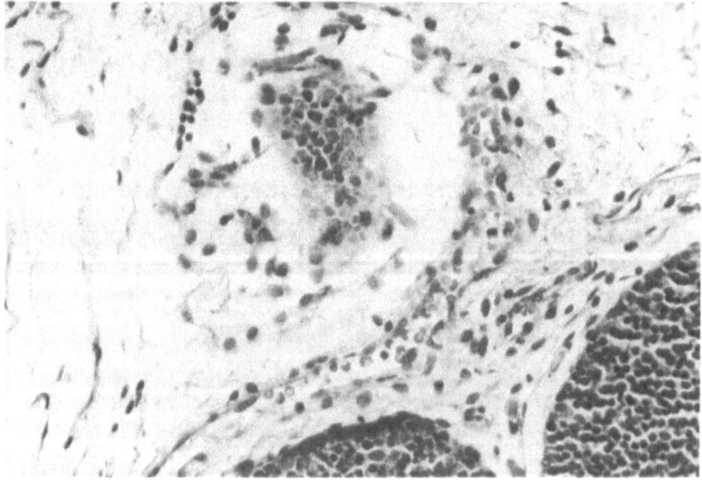

Figure 3A. Human prenatal thymus (16 weeks; end of the 4th lunar month). Paraffin section 3-5 um. The intralobular (INL) loose connective tissue is full of lymphopoietic and non-lymphatic hemopoietic stem cells. Erythro- and granulocytopoietic colonies are already formed at this time. Magnification: 1000 x.

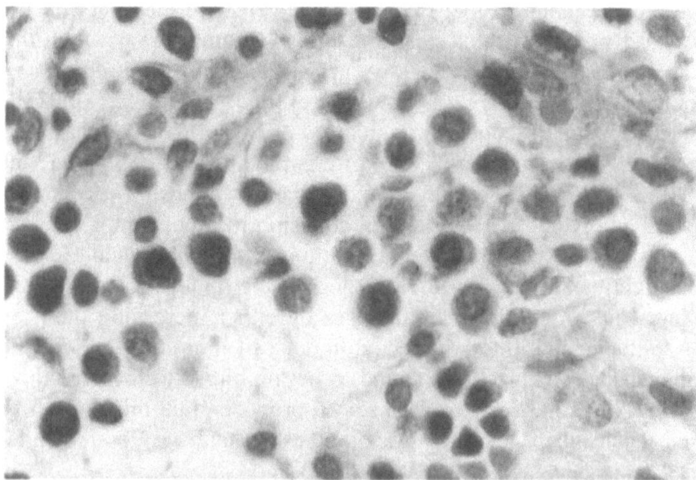

Figure 3B. Human prenatal thymus (16 weeks; end of the 4th lunar month). Paraffin section 3-5 um. Immunocytochemical reaction achieved with MoAB directed against TGF-beta type II receptors. Hemopoietic stem cells are forming non-lymphatic hemopoietic colonies in the thymic intralobular loose connective tissues. Presence of erythropoietic colony formation (small arrow) and organization of intrathymic granulocytopoiesis (large arrow). Magnification: 1000 x.

monocytes and also induces gene expression for various inflammatory mediators in monocytes.[51-53] In addition to its main effects on cell proliferation, TGF-beta also exhibits a remarkable diversity of functional activities. The physiological role of TGF-beta in nontransformed cells is not limited to the regulation of cell division, but may also have a stimulatory role on differentiation.

The effects of TGF-beta varies depending on the cell type and the microenvironmental growth conditions. TGF-beta is capable of influencing various differentiation processes: elevation of amino acid incorporation and glycolysis; activation of *c-sis* expression; production of effects similar to those produced by *ras* oncogene transfection; reversible conferring of the transformed immunophenotype; activity as an immunosuppressor; inbition of terminal differentiation of pre-adipocytes; hematopoiesis (lymphopoiesis); myogenesis (mesoderm induction); stimulation of collagen synthesis; acceleration of bone resorption; and promotion of wound healing. These facts suggest that the TGF-beta cytokine family may use various or several loci within the immune network to modulate and regulate immuno-inflammatory responses and cell-tissue repairs.[24]

TGF-beta1 influences the expression of several protooncogenes and their products such as *c-jun*, *jun B* and *c-fos* which are primarily controlled by the transcription factor AP1.[54-55] Increased expression of *c-fos* was registered almost immediately in T lymphocytes after a stimulation with mitogenic lectins (Con-A) or employing anti-CD3 or anti-CD2 monoclonal antibodies.[56-57] This may set the stage for the late steps of cell division. In keratinocytes and hematopoietic stem cells, TGF-beta1 reduces *c-myc* transcription which is required for completion of G_1 of the cell cycle.[54] Certainly, the immunological maturation pathway is complex and requires biological action from a number of other factors, such as action of various cytokines.

The TGF-beta growth factor family also plays a regulatory role in the immunology of malignant transformation, because it protects tumor cells from the host humoral immunological responses.[24,58] However, in neoplastic cells TGF-beta is excluded from tumor differentiation either due to a lower number of receptor sites, a deficient secondary messenger system, or the fact that TGF-beta responses are cell-type specific. The role of TGF-beta in tumor growth depends on such factors as the substratum upon which the tumor grows which may determine a positive or negative growth regulatory role for this multifunctional growth factor. Many neoplastic cells, however, produce high levels of TGF-beta and thus eventually lose response to its growth regulatory effects. Other neoplastic cells respond to the presence of TGF-beta in a completely different manner, as compared with their normal counterparts. TGF-beta also controls the expression of adhesion receptors presenting the possibility of its involvement in tumor cell adhesion to other tumor cells or host cells, as well as the adhesion molecule mediated migration of immunological effector cells through the blood vasculature. The level of invasiveness of the tumor is also affected by TGF-beta which stimulates the production of proteinase by either increasing the production of the enzyme or by reducing inhibitors. Directed tumor cell migration may also be affected by TGF-beta. Since TGF-beta has been found to be inhibitory to the growth of a number of human carcinoma cell lines in vitro, it has been proposed that this growth factor may find some place in human anti-cancer therapy, as well as immunosuppression and immunomodulation.

REFERENCES

1. Lehnert SA, Akhurst RJ: Embryonic expression pattern of TGF-beta type I RNA suggests both paracrine and autocrine mechanisms of action. *Development* 104:263, 1988.
2. Fox FE, Ford HC, Douglas R, Cherian S, Nowell PC: Evidence that TGF-beta can inhibit human T-lymphocyte proliferation through paracrine and autocrine mechanisms. *Cellular Immunol* 150:45, 1993.
3. Cheifetz S, Weatherbee JA, Tsang ML-S, Anderson JK, Mole JE, Lucas R, Massague J: The transforming growth factor-beta system, a complex pattern of cross-reactive ligands and receptors. *Cell* 48:409, 1987.
4. de Larco JE, Todaro GJ: Growth factors from murine sarcoma virus-transformed cells. *Proc Natl Acad Sci USA* 75:4001, 1978.
5. Todaro GJ, Fryling C, de Larco JE: Transforming growth factors produced by certain human tumor cells: polypeptides that interact with epidermal growth factor receptors. *Proc Natl Acad Sci USA* 77:5258, 1980.
6. Goustin AS, Leof EB, Shipley GD, Moses HL: Growth factors and cancer. *Cancer Res* 46:1015, 1986.
7. Van Obberghen-Schilling E, Roche NS, Flanders KC, Sporn MB, Roberts AB: Transforming growth factor beta 1 positively regulates its own expression in normal and transformed cells. *J Biol Chem* 263:7741, 1988.
8. Barnard JA, Beauchamp RD, Coffey RJ, Moses HL: Regulation of intestinal epithelial cell growth by transforming growth factor type beta. *Proc Natl Acad Sci USA* 86: 1578, 1989.
9. Bascom CC, Wolfshohl JR, Coffey RJ, Madisen L, Webb NR, Purchio AR, Derynck R, Moses HL: Complex regulation of transforming growth factor beta 1, beta 2, and beta 3 mRNA expression in mouse fibroblasts and keratinocytes by transforming growth factors beta 1 and beta 2. *Mol Cell Biol* 9:5508, 1989.
10. Derynck R, Jarrett JA, Chen EY, Eaton DH, Bell JR, Assoian RK, Roberts AB, Sporn MB, Goeddel DV: Human transforming growth factor-beta complementary DNA sequence and expression in normal and transformed cells. *Nature* 316:701, 1985.
11. Hanks SK, Armor R, Baldwin JH, Maldonado F, Spiess J, Holley RW: Amino acid sequence of the BSC-1 cell growth inhibitor (polyergin) deduced from the nucleotide sequence of the cDNA. *Proc Natl Acad Sci USA* 85:79, 1988.
12. Madisen L, Webb NR, Rose TM, Marquardt H, Ikeda T, Twardzik D, Seyedin S,Purchio AF: Transforming growth factor-beta 2: cDNA cloning and sequence analysis. *DNA* 7:1, 1988.
13. Ten Dijke P, Hansen P, Iwata KK, Peiler C, Foulkes JG: Identification of another member of the transforming growth factor type beta gene family. *Proc Natl Acad Sci USA* 85:4715, 1988.
14. Roberts AB, Sporn MB: The transforming growth factors beta. In: Sporn MB, Roberts AB (Eds.) Handbook of experimental pharmacology, Vol 95/I: Peptide growth factors and their receptors. Springer Verlag, Heidelberg, pp 419-472.

15. Roberts AB, Anzano MA, Meyers CA, Wideman J, Blacher R, Pan Y-CE, Stein S, Lehrman SR, Smith JM, Lamb LC, Sporn MB: Purification and properties of a type beta transforming growth factor from bovine kidney. *Biochem* 22:5692, 1983.
16. Frolik CA Dart LL, Meyers CA, Smith DM, Sporn MB: Purification and initial characterization of a type beta transforming growth factor from human placenta. *Proc Natl Acad Sci USA* 80:3676, 1983.
17. Assoian RK, Komoriya A, Meyers CA, Miller DM, Sporn MB: Transforming growth factor-beta in human platelets: identification of a major storage site, purification, and characterization. *J Biol Chem* 258:7155, 1983.
18. Fujii D, Brissenden JE, Derynck R, Francke U: Transforming growth factor beta gene maps to human chromosome 19 long arm and to mouse chromosome 7. *Somatic Cell Mol Genet* 12:281, 1986.
19. Barton DE, Foellmer BE, Du J, Tamm J, Derynck R, Francke U: Chromosomal mapping of genes for transforming growth factors beta 2 and beta 3 in man and mouse: dispersion of TGF-beta gene family. *Oncogene Res* 3:323, 1988.
20. Wilcox JN, Derynck R: Developmental expression of Transforming Growth Factors alpha and beta in mouse fetus. *Mol Cell Biology* 8:3415, 1988.
21. Derynck R, Rhee L: Sequence of the porcine transforming growth factor-beta precursor. *Nucleic Acids Res* 15:3187, 1987.
22. Van Obberghen-Schilling E, Kondaiah P, Ludwig RL, Sporn MB, Baker CC: Complementary deoxyribonucleic acid cloning of bovine transforming growth factor-beta 1. *Mol Endocrinol* 1:693, 1987.
23. Sharples K, Plowman GD, Rose TM, Twardzik DR, Purchio AF: Cloning and sequence analysis of simian transforming growth factor-beta cDNA. *DNA* 6:239, 1987.
24. Fontana A, Constam DB, Frei K, Malipiero U, Pfister HW: Modulation of the immune response by transforming growth factor beta. *Int Arch Allergy Immunol* 99:1, 1992.
25. Wrann M, Bodmer S, de Martin R, Siepl C, Hofer-Warbinek R, Fei K, Hofer E, Fontana A: T cell suppressor factor from human glioblastoma cells is a 12.5-kd protein closely related to transforming growth factor beta. *EMBO J* 6:1633, 1987.
26. Roberts AB, Frolik CA, Anzano MA, Sporn MB: Transforming growth factors from neoplastic and nonneoplastic tissues. *Fed Proc* 22:2621, 1983.
27. Frolik CA, Wakefield LM, Smith DM, Sporn MB: Characterization of a membrane receptor for transforming growth factor-beta in normal rat kidney fibroblasts. *J Biol Chem* 259:10995, 1984.
28. Tucker RF, Branum EL, Shipley GD, Ryan RJ, Moses HL: Specific binding to cultured cells of ^{125}I-labeled type beta transforming growth factor from human platelets. *Proc Natl Acad Sci USA* 81:6757, 1984.
29. Massague J: Type beta transforming growth factor from feline sarcoma virus-transformed cells. Isolation and biological properties. *J Biol Chem* 259:9756, 1984.
30. Segarini PR: TGF-beta receptors. Clinical applications of TGF-beta. Wiley, Chichester (CIBA Foundation Symposium # 157), pp. 29-50, 1991.
31. Massague J: Receptors for the TGF-beta family. *Cell* 69:1067, 1992.
32. Massague J, Like B: Cellular receptors for type beta transforming growth factor. Ligand binding and affinity labeling in human and rodent cell lines. *J Biol Chem* 260:2636, 1985.
33. Wrana JL, Attisano L, Carcomo J, Zentella A, Doody J, Laiho M, Wang X, Massague J: TGF-beta signals through a heteromeric protein kinase receptor complex. *Cell* 71:1003, 1992.
34. Laiho M, Weis FMB, Boyd FT, Ignotz RA, Massague J: Responsiveness of transforming growth factor beta (TGF-beta) restored by genetic complementation between cells defective in TGF-beta receptors I and II. *J Biol Chem* 266:9108, 1992.
35. Lin HY, Wang X-F, Ng-Eaton E, Weinberg RA, Lodish HF: Expression cloning of the TGF-beta type II receptor, a functional transmembrane serine/threonine kinase. *Cell* 68:775, 1992.
36. Partanen S: Immunohistochemically demonstrable pp60$^{c\text{-}src}$ in human breast cancer. *Oncology Reports* 1:603, 1994.
37. Shi S-R, Key ME, Kalra KL: Antigen retrieval in formalin fixed, paraffin-embedded tissues: an enhancement method for immunohistochemical staining based on microwave oven heating of tissue sections. *J Histochem Cytochem* 39:741, 1991.
38. Shi S-R, Cote C, Kalra KL, Taylor CR, Tandon AK: A technique for retrieving antigens in formalin-fixed, routinely acid-decalcified, celloidin-embedded human temporal bone sections for immunohistochemistry. *J Histochem Cytochem* 40:787, 1992.
39. Bodey B, Zeltzer PM, Saldivar V, Kemshead J: Immunophenotyping of childhood astrocytomas with a library of monoclonal antibodies. *Int J Cancer* 45:1079, 1990.
40. Yam LT, Janckila AJ, Epremian BE, Li C-Y: Diagnostic significance of levamisole-resistant alkaline phosphatase in cytochemistry and immunocytochemistry. *Am J Clin Pathol* 91:31, 1989.

41. Strasburger CJ, Amir-Zaltsman Y, Kohen F: The avidin-biotin reaction as an universal amplification system in immunoassays. *Prog Clin Biol Res* 285:79, 1988.
42. Wilchek M, Bayer EA: Introduction to avidin-biotin technology. *Methods Enzymol* 184:5, 1990.
43. Duhamel RC, Whitehead JS: Prevention of nonspecific binding of avidin. *Methods Enzymol* 184:201, 1990.
44. Diamandis EP, Christopoulos TK: The biotin- (strept)avidin system: principles and applications in biotechnology. *Clin Chem* 37:625, 1991.
45. Takahama Y, Letterio JJ, Suzuki H, Farr AG, Singer A: Early progression of thymocytes along the CD4/CD8 developmental pathway is regulated by a subset of thymic epithelial cells expressing growth factor beta. *J Exp Med* 179:1495, 1994.
46. Kehrl JH, Wakefield LM, Roberts AB, Jakowlew SB, Alvarez-Mon M, Derynck R, Sporn MB, Fauci AS: Production of transforming growth factor-beta by human T lymphocytes and its potential role in the regulation of T cell growth. *J Exp Med* 163:1037, 1986.
47. Wahl SM, Hunt DA, Bansal G, McCartney-Francis N, Ellingsworth L, Allen JB: Bacterial cell wall-induced immunosuppression. Role of transforming growth factor-beta. *J Exp Med* 168:1403, 1988.
48. Wahl SM, Hunt DA, Wong HL, Dougherty S, McCartney-Francis N, Wahl L, Elligsworth L, Schmidt JA, Hall G, Roberts AB, Sporn MB: Tranforming growth factor-beta is a potent immunosuppressive agent that inhibits IL-1-dependent lymphocyte proliferation. *J Immunol* 140:3026, 1988.
49. Kehrl JH, Roberts AB, Wakefield LM, Jakowlew SB, Sporn MB, Fauci AS: Transforming growth factor beta is an important immunomodulatory protein for human B-lymphocytes. *J Immunol* 137:3855, 1986.
50. Rook AH, Kehrl JH, Wakefield LM, Roberts AB, Sporn MB, Burlington DB, Lane HC, Fauchi FS: Effects of transforming growth factor beta on the functions of natural killer cells: depressed cytolytic activity and blunting of interferon responsiveness. *J Immunol* 136:3916, 1986.
51. Postlethwaite AE, Keski-Oja J, Moses HL, Kang AH: Stimulation of the chemotactic migration of human fibroblasts by transforming growth factor beta. *J Exp Med* 165:251, 1987.
52. Wiseman DM, Polverini PJ, Kamp DW, Leibovich SJ: Transforming growth factor-beta (TGF-beta) is a chemoattractant for monocytes and induces their expression of angiogenic activity. *Biochem Biophys Res Commun* 157:793, 1988.
53. Chantry D, Turner M, Abney E, Feldmann M, Modulation of cytokine production by transforming growth factor-beta. *J Immunol* 142:4295, 1989.
54. Chen R, Ebner R, Derynck R: Inactivation of the type II receptor reveals two receptor pathways for the diverse TGF-beta activities. *Science* 260:1335, 1993.
55. Okragly A, Balwit JM, Haak-Frendscho M: Transforming Growth Factor $beta_1$ (TGF-$beta_1$) a biological paradox. *Promega Notes* 47:10, 1994.
56. Reed JC, Alpers JD, Nowell PC, Hoover RG: Sequential expression of proto-oncogene during lectin-stimulated mitogenesis of normal human lymphocytes. *Proc Natl Acad Sci USA* 83:3982, 1986.
57. Nel AE, Taylor LK, Kumar GP, Gupta S, Wang SC-T, Williams K, Liao O, Swanson K, Landreth GE: Activation of a novel serine/threonine kinase that phosphorylates c-fos upon stimulation of T and B lymphocytes via antigen and cytokine receptors. *J Immunol* 152:4347, 1994.
58. Lee G, Larry R, Ellingsworth LR, Gillis S, Wall R, Kincade PW: Beta-transforming growth factors are potential regulators of B lymphopoiesis. *J Exp Med* 166:1290, 1987.

NEW DIRECTIONS IN IRON CHELATING THERAPY

Chaim Hershko[1*] and Gabriela Link[2]

[1] Department of Medicine
Shaare Zedek Medical Center Jerusalem
P O Box 3235, Israel
[2] Department of Human Nutrition and Metabolism
Hebrew University Hadassah Medical School

Although the iron chelating drug deferroxamine has been available for clinical use for over three decades, it has only gained acceptance as a useful therapeutic agent in the late seventies following the demonstration of its ability to deplete excess iron stores in thalassemic patients. Recognition of the central role of iron in the generation of toxic, oxygen-derived species through the Haber-Weiss reaction (1), documentation of the ability of deferoxamine (DF) to prevent the damage associated with free radical generation in reperfusion injury, and its ability to inhibit the proliferation of malignant cells and protozoa such as the malarial parasite by inactivation of the iron-dependent enzyme ribonucleotide reductase, resulted in a large number of studies exploring the novel therapeutic applications of iron chelating drugs. Some of the information in this field has been reviewed in previous publications (2). The purpose of the present chapter is to summarize the state of the art in the use of DF and other selective iron chelators in conditions unrelated to iron overload.

MODIFICATION OF REPERFUSION INJURY

The term reperfusion injury or "oxygen paradox" describes the aggravating of anoxic injury initiated during ischaemia by reexposure to normal oxygen concentrations, causing further damage to vital cellular membranes by lipid peroxidation (3). Because of the important role of iron in ·OH formation, DF has been introduced into free radical research as a specific in vitro and in vivo probe for iron-dependent radical reactions (4). DF has a very high, and specific affinity to $Fe3+$ and is very efficient in preventing its reduction to $Fe2+$ and the participation of iron in the Haber-Weiss reaction.

* Correspondence to: C Hershko, Department of Medicine, Shaare Zedek Medical Center, Jerusalem, Israel P O Box 3235. Fax: 972-2-6513946; Phone 972-2-6555111.

Molecular Biology of Hematopoiesis 5, edited by Abraham et al.
Plenum Press, New York, 1996

Early studies have shown that DF treatment limits *myocardial stunning* and increases the rate of recovery following reversible myocardial ischemia in experimental animals. Progress in recent years in the study of DF as a potentially useful agent for preventing myocardial reperfusion damage has been slow. Although the number of studies published in this field has continued to increase at a logarithmic rate, little new and original information has been generated within the last few years.

The role of low-molecular iron complexes in the pathogenesis of reperfusion injury has been addressed by a number of studies. Studies in isolated rat hearts have shown a striking increase in the concentration of low molecular iron generated within the cardiac tissue during hypoxia-reperfusion (5), while other studies have shown that the addition of micromolar concentrations of iron to the perfusion solution aggravates the damage caused by hypoxia-reperfusion. These observations provide a strong rationale for the use of iron chelators in this condition.

The importance of DF dosage and timing of administration has been studied by DeBoer et al (6). In line with previous observations, an optimal protective effect of DF was found when introduced prior to and throughout hypoxia-reperfusion. The beneficial effect of DF was limited to concentrations of 0.15 to 0.46 mM. at higher concentrations DF was toxic and aggravated, instead of preventing myocardial injury. This dose-related toxicity may explain the failure of DF added to the cardoplegic solution at nearly 2 mM concentrations, to improve clinical outcome in humans undergoing elective coronary bypass surgery. Other studies have explored the use of DF in cardioprotection following prolonged hypoxia such as 24 h cold storage or 72 h coronary artery ligation (7). The cardioprotective effect of DF under such conditions extends its potential of use to conditions such as the preservation of isolated perfused heart for transplantation or the containment of myocardial infarction after prolonged coronary occlusion.

Without exception, all recent studies have shown a beneficial effect of DF treatment on cardiac contractility, decreased infarct size and decreased CPK leakage. All of these observations imply that iron may play a significant role in the pathogenesis of myocardial reperfusion injury, and that DF treatment may result in a significant attenuation of myocardial dysfunction in this condition. The practical implications of these observations for the management of analogous clinical situations remain to be explored. Because of the importance of introducing the chelator *prior to* reperfusion damage, the most promising situations in clinical medicine where DF may be useful are elective surgical procedures in which DF treatment may be started before and throughout cardioplegia and hypothermic anoxia.

MODIFICATION OF ANTHRACYCLINE CARDIOTOXICITY

Anthracyclines are glyoside antibiotics that are very effective in the treatment of hematologic malignancies and solid tumours. However, the use of anthracycline antineoplastic drugs is limited by a dose-dependent risk of cumulative toxicity to the heart, with 20% of patients developing heart failure when the total dose exceeds 700 mg/m^2. Any measure preventing the cardiotoxicity of anthracyclines would therefore increase their curative effect by permitting dose escalation without increasing the risk of cardiac mortality (8).

Anthracyclines promote free radical production by redox cycling increasing the flow of electrons from NADH to molecular oxygen (9,10). The most likely target of free radical toxicity is cardiolipin, a major phospholipid component of the inner mitochondrial membrane (11). It has been shown in a number of studies that the antineoplastic effect of anthracyclines may be dissociated from their cardiotoxic effect. By employing cultured rat heart cells and mouse L-1210 leukemia cells, a dose- and time-dependent decrease in cellular

ATP and glutathione, and release of LDH was found in both myocytes and leukemic cells. However, at low anthracycline concentrations complete arrest of leukemic cell growth has been achieved without any indication of cardiotoxicity (12). Interventions aimed at reducing free radical formation and lipid peroxidation resulted in decreasing cardiac toxicity without interfering with tumour response. Tocopherol treatment in mice receiving adriamycin prevented both lipid peroxidation and cardiac toxicity without diminishing the response to therapy of P388 ascites tumour (13). Similarly, methylene blue treatment aimed at modifying intracellular redox balance and preventing the in vivo reduction of doxorubicin, resulted in a marked decrease in drug toxicity without interfering with the antineoplastic effect against L1210 ascites tumour cells in mice (14).

An important development in the application of iron chelating chemicals in cancer therapy has been the demonstration of the protective effect of the bispiperazonedione ICRF-187 against doxorubicin-induced cardiac toxicity in women with advanced breast cancer (15). ICRF-187 was originally synthesized by Creighton el al as a possible antitumor agent (16). However, it was soon recognized that ICRF-187 protects mice and other experimental animals against anthracycline cardiotoxicity (17,18) without interfering with their antitumoral effect (19). The mechanism of ICRF-187 cardioprotection is not entirely clear. ICRF-187 is hydrolyzed intracellularly into a bidentate chelator resembling EDTA. It has been proposed that this chelator interacts with intracellular iron and prevents the formation of harmful oxygen derivatives (15). Indeed, a marked decrease in adriamycin-induced free hydroxyl radical formation has been shown in isolated rat heart preparations treated by ICRF-187 (20).

Other investigators have studied the ability of known iron chelators such as DF, EDTA, or the 3-hydroxypyrid-4-one orally effective synthetic iron chelators to protect the heart from anthracycline toxicity in experimental animal systems (21). When studied in isolated mouse atria exposed *in vitro* to doxorubicin, DF at a concentration of 200 μM was the most effective of all chelators studied in protecting muscle contractility from the harmful effect of doxorubicin. However, the effect of DF was biphasic, with no protective effect at 500 μM. In a second series of studies by the same author, the *in vivo* effect of DF was studied in rats receiving weekly i.v. injections of doxorubicin 3 mg/kg to cumulative doses of 6, 9 and 12 mg/kg. DF was given at 100 mg/kg/d for 3 consecutive days prior to each weekly dose of doxorubicin. These *in vivo* studies have shown that DF was able to prevent the histopathological abnormalities and the decrease in heart function measured as rate-pressure-product induced by doxorubicin. Unfortunately, DF proved to be quite toxic in these non-iron-loaded animals and the mortality rate in DF and doxorubicin-treated animals was actually higher than those receiving doxorubicin only.

Our group has employed an *in vitro* system of beating heart cells in culture to examine the inter-relation between iron, iron chelation and anthracycline toxicity. Our previous studies in this experimental model have shown that *in vitro* iron loading of rat heart cells results in increased membrane lipid peroxidation and loss of thiolic membrane-protein enzyme activity, leading to abnormal contractility and electrophysiologic behavior, that the main targets of iron toxicity are the sarcolemmal and lysosomal membranes and, that *in vitro* iron chelation in iron-loaded heart cells is able to abolish the structural and functional abnormalities induced by iron (22-26).

We have employed two indicators of cellular damage: LDH release and cell contractility. Both of these indicators have shown a marked increase in doxorubicin toxicity by prior iron-loading of the cultured heart cells (27). This was not a simple additive effect of iron toxicity and anthracycline damage, since at the concentrations employed iron had only a minimal effect on LDH release and no effect at all on contractility whereas doxorubicin had only a minor effect on contractility. Thus, the marked increase in LDH release and abnormal contractility of iron-loaded heart cells after doxorubicin treatment are the expressions of a synergistic effect of anthracycline, and iron toxicity.

If iron aggravates anthracycline toxicity one could anticipate a protective effect of DF by depleting intra- and extracellular iron and by preventing its participation in free hydroxyl radical generation. Our results have shown that these expectations were only partly fulfilled. DF treatment of iron-loaded heart cells resulted in a marked decrease in anthracycline toxicity as judged both by LDH leakage and cell contractility. However, DF treatment of normal heart cells had no measurable protective effect against doxorubicin toxicity, irrespective of whether DF was administered prior to, or simultaneously with doxorubicin (27).

Our subsequent studies were aimed at identifying the organelle representing the main target of doxorubicin-iron interaction in myocardical toxicity. We found only a limited effect of doxorubicin on sarcolemmal thiolic enzyme inactivation and on lysosomal fragility, two of the most prominent targets of iron toxicity identified in our previous studies (28). In contrast, mitochondria proved to be not only extremely sensitive to anthracycline toxicity, but also an important target of the potentiation of anthracycline toxicity by iron. Mitochondrial respiration monitored by C^{14}-palmitate utilization was decreased to 33% of controls following 24h incubation of cultured heart cells with 1 μg/ml doxorubicin. This damage was almost doubled (18%) when iron-loading preceded anthracycline treatment. Conversely, removal of iron by deferoxamine resulted in an increase of palmitate utilization to 44% of controls.

Our data clearly indicate that iron overload aggravates anthracycline toxicity and that this interaction can be prevented by effective iron chelating therapy. Since DF itself is an inhibitor of cell proliferation via inactivation of ribonucleotide reductase (29,30), it is reasonable to assume that concurrent DF therapy may prevent toxicity while at the same time enhance the tumoricidal effect of anthracyclines. However, this assumption is far from being proven, as in some *in vitro* studies coadministration of DF and anthracycline prevented the tumor-killing effect of doxorubicin (31). In view of the extensive clinical experience accumulated within the last two decades in the use of iron chelators in medicine, such treatment may easily be applicable for patient management.

THE ANTIMALARIAL EFFECT OF IRON CHELATORS

A number of experimental and clinical studies indicate that iron deficiency may have an important inhibitory effect on the progression of malarial infection and, conversely, that iron repletion may result in the exacerbation of malaria (32). However, this hypothesis is not universally accepted, as other studies have been unable to show an adverse effect of iron administration on human malaria (33), and severe iron deficiency may interfere with the normal immune response thus aggravating, rather than inhibiting infection.

In view of the possible beneficial effects of iron depletion, DF has been studied as a potential antimalarial agent, DF inhibits the growth of *P. falciparum* cultures at concentrations above 20 μM (34). *In vivo* studies in rats infect with *P berghei*, mice with *P vinckei* and monkeys with *P falciparum* have shown that DF is able to suppress malaria if a continuous supply of the chelator is assured by frequent (8 hourly) subcutaneous injections (35,36), or by osmotic pumps.

Encouraged by these studies in experimental animals, several investigators have tested the antimalarial effect of DF in humans. Traore et al (37) have studied the effect of DF 0.5 g i.m. given twice daily for 3 days on the rate of clearance of parasitemia in patients with P. falciparum malaria who were also receiving chloroquine. Although parasitaemia appeared to decrease more rapidly in the 6 patients receiving DF and chloroquine than in the 3 controls treated by chloroquine only, the small number of patients, and the inclusion of chloroquine- resistant cases with resurgent malaria limit the value of this preliminary report. In another clinical study by Bunnag et al (38) 14 patients with symptomatic P. vivax and 14 with uncomplicated P. falciparum malaria received continuous i.v. DF 100 mg/kg for 72 hours. No other antimalarial treatment was given. In both groups DF reduced the parasitemia to zero within 57 to 106 hours. There was significant drug toxicity with transient visual blurring in 9 patients. Recrudescence was observed within the subsequent 3 weeks in all but 2 patients. A major weakness of this study was the absence of a control group.

Two controlled studies of DF in human malaria have been conducted by Gordeuk et al. In the first of these, the effect of DF therapy in partially immune adults with asymptomatic P. falciparum parasitemia has been tested (39). Twenty eight individuals were entered into a randomized, double-blind, crossover trial comparing DF, 100 mg/kg per day with placebo given by subcutaneous infusions via portable pumps. DF and placebo were each given continuously for 72 hours with the sequence of administration determined by random assignment. Compared with placebo, DF treatment was associated with an almost tenfold enhancement of the rate of parasite clearance during both the initial phase ($p=0.006$) and the crossover phase ($p=0.0001$). Mean (±SEM) steady state concentrations of DF were 7 ±1 μM at 36 hours and 8 ±1 μM at 72 hours. In the second randomized, double-blind, placebo-controlled trial by Gordeuk et al (40) the effect of DF 100 mg/kg/d given by intravenous infusion for 72 hours was studied in 83 children with cerebral (P.falciparum) malaria. All patients were receiving, in addition, standard therapy consisting of quinine and sulfadoxine-pyrimethamine. Among 50 patients with deep coma, median recovery time was decreased by DF from 68 to 24 hours ($p = 0.03$). The rate or parasite clearance was 2 times faster in the DF-treated group. However, there was no significant difference in rates of mortality.

Collectively, these studies leave no doubt as to the ability of DF to hasten recovery from malaria, presumably by inhibiting parasite growth in a similar fashion to its effect in experimental *in vitro* and *in vivo* systems. In cerebral malaria, an additional beneficial effect could be inhibition of oxidative brain damage by preventing the formation of toxic free radicals through the iron-driven Fenton reaction. However, as emphasized in several recent editorials (41), additional large-scale carefully controlled studies are needed, with particular emphasis on mortality and neurological sequelae, before DF could be recommended for the treatment of cerebral malaria. In spite of its effectiveness in suppressing malaria in experimental animals and in humans, it is unlikely that DF will be suitable for the management of clinical malaria because of its poor oral absorption, high price, and relatively slow rate of red cell penetration. In view of these considerations were and others (42,43) have studied other, orally effective iron chelating compounds, some of which have already been shown to be more effective iron chelators than DF. In a study of a group of phenolic iron chelators, we have found that N,N'-bis(o-hydroxybenzyl) ethylenediamine-N,N' - diacetic acid (HBED), the most powerful iron chelator in this group with an increased affinity to iron and increased lipid solubility, was also the most effective antimalarial agent (44). Consequently, we postulated that both affinity to iron and increased lipophilicity may contribute to the antimalarial activity of an iron chelator.

In order to explore the role of lipophilicity in antimalarial activity, we have examined the antimalarial effects of 3-hydroxypyrid-4-ones (42), a family of bidentate orally effective iron chelators. All 3-hydroxypyrid-4-ones have an identical stability constant for iron(IIII), but they may be made more, or less lipophilic by increasing or reducing the length of the R_2 substituent on the ring nitrogen. Of the hydroxypyridin-4-ones investigated in our studies), those with the highest lipid solubility proved to be the most efficient antimalarial compounds. Recent studies by Shanzer et al (45) employing a series of synthetic iron chelators have confirmed our conclusions that the antimalarial effect of iron chelators is determined by their lipophilicity as well as their affinity to iron.

Several recent studies have shown that DF is able to inhibit the proliferation *in vitro* and *in vivo* of *Leishmania donovani* (46), *Trypanosoma cruzi* (47), *Pneumocystis carinii* (48), and *Legionella pneumophila (*49). These intriguing observations on the antimicrobial effects of DF and other iron chelators lend new meaning to the term "Nutritional Immunity" (50) and open new channels for exploring the possibility of controlling infection by means of selective intracellular iron deprivation. Experimental models for studying the effect of iron chelators on other intracellular pathogens such as *Toxoplasm gondii*, *Chlamidia psittaci*, or *Mycobacterium tuberculosis* should be established. Packaging the chelator in liposomes or red cell ghosts, or manipulating their lipid solubility to improve their delivery to appropriate target organs such as the macrophage system may greatly improve their efficiency. In view of the short half-life and poor oral effectiveness of DF, it is unlikely that this drug will be suitable for clinical use as a practical antimicrobial agent. However, with the introduction of simple, orally effective new chelators, it is reasonable to expect that future research may lead to the identification of iron chelators with considerable usefulness in the control of infectious disease.

ACKNOWLEDGMENTS

Supported by grant RO1 DK48094-08 of the National Institute of Diabetes and Digestive and Kidney Diseases; and a grant by the joint Shaare Zedek - Bar Ilan University research foundation.

REFERENCES

1. Halliwell B, Gutteridge JMC: Oxygen, free radicals and iron in relation to biology and medicine: some problems and concepts. Arch Biochem Biophy 1986; 246:540-544.
2. Hershko C: Iron chelators. in JH Brock, JW Halliday, MJ Pippard & LW Powell eds *Iron Metabolism in Health and Disease* WB Saunders Ltd London 1994 pp 391—426
3. White BC, Krause GS, Aust SD, Eyster GE: (1985) Postischemic tissue injury by iron-mediated free radical- lipid peroxidation. Ann Emerg Med 1985; 14:804-809.
4. Gutteridge JMC, Richmond R, Halliwell B: Inhibition of the iron-catalysed formation of hydroxyl radicals from superoxide and of lipid peroxidation by desferrioxamine. Biochem J 1979; 184:469-472.
5. Voogd A, Sluiter W, van Eijk HG, Koster JF: Low molecular weight iron and the oxygen paradox in isolated rat hearts. J Clin Invest 1992; 90:2050-2055.
6. DeBoer DA, Clark RE: Iron chelation in myocardial preservation after ischemia-reperfusion injury: the importance of pretreatment and toxicity. Ann Thor Surg 1992; 53:412-418.
7. Chopra K, Singh M, Kaul N et al: Decrease of myocardial infarct size with desferrioxamine: possible role of oxygen free radicals in its ameliorative effect. Molec Cell Biochem 1992; 113:71-76.
8. Young RC, Ozols RF, Myers CE: The anthracycline antineoplastic drugs. New Engl J Med 1981; 305:139-153.
9. Singal PK, Deally CMR, Weinberg LE: Subcellular effects of adriamycin in the heart: a concise review. J Molec Cell Cardiol 1987; 19:817-828.
10. Olson RD, Mushlin PS: Doxorubicin cardiotoxicity: analysis of prevailing hypotheses. FASEB J 1990; 4:3076-3086.
11. Goormaghtigh E, Ruysschaert JM: Anthracycline glycoside-membrane interactions. Biochim Biophy Acta 1984; 779:271-288.
12. Singh Y, Ulrich L, Katz D et al: Structural requirements for antracycline-induced cardiotoxicity and antitumor effects. Toxicol Appl Pharm 1989; 100:9-23.
13. Myers CE, McGuire WP, Liss RH et al: Adriamycin: the role of lipid peroxidation in cardiac toxicity and tumor response. Science 1977; 197:165-167.
14. Hrushesky WJM, Olshefski R, Wood P et al: Modifying intracellular redox balance: an approach to improve therapeutic index. Lancet 1985; 1:565-567.

15. Speyer JL, Green MD, Kramer E et al: Protective effect of the bispiperazinedione ICRF-187 against doxurubicin-induced cardiac toxicity in women with advance breast cancer. New Engl J Med 1988; 319:745-752.
16. Creighton AM, Hellmann K & Whitecorss S: Antitumour activity in a series of bisDiketopiperazines. Nature 1969; 222:384-385.
17. Herman EH & Ferrans VJ: Reduction of chronic doxurubicin cardiotoxicity in dogs by pretreatment with (±)-bis(3,5-dioxopiperazinyl-l-yl) propane (ICRG-187). Cancer Res 1981; 41:3436-3440.
18. Yeung TK, Jaenke RS, Wilding D et al: The protective activity of ICRF-187 against doxorubicin-induced cardiotoxicity in the rat. Cancer Chemother Pharm 1992; 30:58-64.
19. Woodman RJ, Cysyk RL, Kline I et al: Enhancement of the effectiveness of daunorubicin (NSC-82151) or adriamycin (NSC-123127) against early mouse L1210 leukemia with ICRF-159 (NSC-129943). Cancer Chemother Rep 1975; 59:689-695.
20. Rajagopalan S, Politi PM, Sinha BK: Adriamycin-induced free radical formation in the perfused rat heart: Implications for cardiotoxicity. Cancer Res 1988; 48:4766-4769.
21. Voest EE: Iron chelation, oxygen radicals, and anthracyclines in the treatment of cancer. *Doctoral Thesis*. University of Utrecht. Kwiek BV Press, Utrecht, Holland 1993.
22. Link G, Pinson A, Hershko C: Heart cells in culture: a model of myocardial iron overload and chelation. J Lab Clin Med 1985; 106:147-153.
23. Link G, Pinson A, Kahane I, Hershko C: Iron loading modifies the fatty acid composition of cultured rat myocardial cells and liposomal vesicles: Effect of ascorbate and α-tocopherol on myocardial lipid peroxidation. J Lab Clin Med 1989; 114:243-249.
24. Link G, Athias P, Grynberg A, et al: Effect of iron loading on transmembrane potential, contraction and automaticity of rat ventricular muscle cells in culture. J Lab Clin Med 1989; 113:103-111.
25. Hershko C, Link G, Pinson A et al: Iron mobilization from myocardial cells by 3-hydroxypyridin-4-one chelators: Studies in rat heart cells in culture. Blood 1991; 77:2049-2053.
26. Link G, Pinson A & Hershko C: Iron-loading of cultured cardiac myocytes modifies sarcolemmal structure and increases lysosomal fragility. J Lab Clin Med 1993; 121:127-134.
27. Hershko C, Link G, Tzahor M et al: Anthracycline toxicity is potentiated by iron and inhibited by deferoxamine: Studies in rat heart cells in culture. J Lab Clin Med 1993; 122:245-251.
28. Link G, Pinson A, Hershko C: Ability of the orally effective iron chelators dimethyl- and diethyl-hydroxypyrid-4-one and of deferoxamine to restore sarcolemmal thiolic enzyme activity in iron-loaded heart cells. Blood 1994, 83:2692-2697.
29. Lederman HM, Cohen A, Lee JW, et al: Deferoxamine: a reversible S-phase inhibitor of human lymphocyte proliferation. Blood 1984; 66:748-753.
30. Hoffbrand AV, Ganeshaguru K, Hooton JWL, Tattersall MHN: Effect of iron deficiency and desferrioxamine on DNA synthesis in human cells. Brit J Haemat 1976; 33:517-526.
31. Doroshow JH: Prevention of doxorubicin-induced killing of MCF-7 human breast cancer cells by oxygen radical scavengers and iron chelating agents. Biochem Biophys Res Comm 1986; 135:330-335.
32. Murray MJ, Murray AB, Murray MB, et al: The adverse effect of iron depletion on the course of certain infections Brit Med J 1978; 2:1113-1115.
33. Harvey P, Heywood P, Nesheim MC, Habicht JP, Alperts M: Iron repletion and malaria. Fed Proc 1987; 46: 1161. (abstr).
34. Raventos-Suarez C, Pollack S, Nagel RL: Plasmodium falciparum: inhibition of in vitro growth by desferrioxamine. Amer J Trop Med Hyg 1982; 31:919-922.
35. Hershko C, Peto TEA: Deferoxamine -inhibition of malaria is independent of host iron status. J Exper Med 1988; 168:375-387.
36. Fritch G, Treumer J, Spira DT, Jung A: Plasmodium vinckei:Suppression of mouse infections with desferrioxamine B. Exper Path 1985; 60:171-174.
37. Traore O, Carnevale P, Kaptue-Noche L et al: Preliminary report on the use of desferrioxamine in the treatment of Plasmodium Falciparum malaria. Amer J Hem 1991; 37:206-208.
38. Bunnag D, Poltera AA, Viravan C et al: Plasmodicidal effect of desferrioxamine B in human vivax and faciparum malaria from Thailand. Acta Trop Basel 1992; 52:59-67.
39. Gordeuk VR, Thuma P, Brittenham GM et al: Effect of iron chelation therapy on recovery from deep coma in children with cerebral malaria. New Engl J Med 1992; 327:1473-1477.
41. Wyler DJ: Bark, weeds, and iron chelators - Drugs for malaria. New Engl J Med 1992; 327:1519-1521.
42. Hershko C, Theanacho EN, Spira DT et al: The effect of N-alkyl modification on the antimalarial activity of 3-hydroxypyrid-4-one oral iron chelators. Blood 1991; 77:637-643.
43. Heppner DG, Hallaway PE, Kontoghiorghes GJ, Eaton JW: Antimalarial properties of orally active iron chelators. Blood 1988; 72:358-363.

44. Yinnon AM, Theanacho EN, Grady RW et al: Antimalarial effect of HBED and other phenolic and catecholic iron chelators. Blood 1989; 74:2166-2171.
45. Shanzer A, Libman J, Lytton S, Glickstein H, Cabantchik ZI: Reversed siderophores act as antimalarial agents. Proc Nat Acad Sci 1991; 88:6585-6589.
46. Segovia M, Navarro A, Artero JM: The effect of liposome-entrapped desferrioxamine on Leishmania donovani in vitro. Ann Trop Med Parasitol 1989; 83:357-360.
47. Lalonde RG, Holbein BE: Role of iron in Trypanosoma cruzi infection in mice. J Clin Invest 1984; 73:470-476.
48. Clarkson AB, Saric S, Grady RW: Deferoxamine and eflornitine (DF-α-difluoromethylornithine) in a rat model of Pneumocystis carinii pneumonia. Antimic AG Chemoth 1990; 34:1833-1835.
49. Byrd TF, Horwitz MA: Interferon gamma-activated human monocytes downregulate transferrin receptors and inhibit the intracellular multiplication of Legionella pneumophila by limiting the availability of iron. J Clin Invest 1989; 83:1457-1465.
50. Kochan I: (1973) The role of iron in bacterial infections with special considerations of host-tubercle bacillus interaction. Curr Topics Microbiol Immunol 1973; 60:1-30.

MOLECULAR PATHOGENESIS OF HEMOCHROMATOSIS

C. Camaschella,[1*] A. Roetto,[1] L. Sbaiz,[1] P. Gasparini,[2] A. Totaro,[2] D. Girelli,[4] P. Fortina,[3] E. Rappaport,[3] S. Fargion,[5] and A. Piperno[6]

[1] Dipartimento di Scienze Biomediche e Oncologia Umana
Università di Torino, CNR CIOS, Torino
[2] IRCCS CSS San Giovanni Rotondo, Foggia, Italy
[3] Department of Pediatrics
University of Pennsylvania School of Medicine, Philadelphia, Pennsylvania
[4] Clinica Medica
Università di Verona, Verona, Italy
[5] Istituto di Medicina Interna
Università di Milano, Italy
[6] Istituto di Scienze Biomediche Ospedale S. Gerardo, Monza

MOLECULAR PATHOGENESIS OF HEMOCHROMATOSIS

Hereditary Hemochromatosis (HC) is the primary form of iron overload, due to a genetic defect of iron absorption, which is especially frequent in Caucasians. Affected subjects may develop in midlife liver cirrhosis, hearth failure and arrhytmias, diabetes and other endocrinopathies, skin pigmentation, arthropathies and increased susceptibility to hepatocellular carcinoma. Since excess iron may be effectively removed by phlebotomies, early diagnosis is important to prevent disease complications (1,2). HC is the most common autosomic recessive disorder in Northern-European populations (3). The frequency of homozygotes approximates 3 - 5 per 1000 (1-4). Clinical symptoms occur more frequently in males than in females. The latter, although genetically affected, may not express the disease during the fertile age due to physiological iron losses.

Simple criteria are available for patient diagnosis: transferrin saturation values greater than 62 % in males and of 50 % in females identify the affected genotype in more than 90 % of the affected patients. Increased serum ferritin is less accurate since it predicts only 71 % of the affected genotypes. Liver biopsy is required to confirm the diagnosis and

*Correspondence: Clara Camaschella, M.D., Dipartimento di Scienze Biomediche e, Oncologia Umana, Università di Torino, Ospedale S. Luigi, 10043 Orbassano, Torino, Italy. Phone: 39-11-9026610; Fax: 39-11-9038636.

Molecular Biology of Hematopoiesis 5, edited by Abraham et al.
Plenum Press, New York, 1996

to assess the liver iron concentration. A protocol for early hemochromatosis screening has been recently suggested (5).

Notwithstanding remarkable progresses in the knowledge of intracellular iron control (6-7), the biochemical defect of HC is still unknown. The disease is inherited in linkage with HLA-A (8-9), which is known to be located on the short arm of chromosome 6, providing the starting point for positional cloning of the gene. All the known proteins involved in iron metabolism have been ruled out as candidate proteins for the primary HC defect, since their corresponding genes have been mapped on chromosomes other than 6. Results of multipoint linkage analysis of highly informative markers of 6p in Italian and Australian HC families provided contrasting results on gene location: according to Italian data the most likely candidate area was centromeric to HLA-A (10). In Australian families the highest lod score values were obtained for D6S105 marker (11), which is approximately 3 cM telomeric to HLA-A. The existence of linkage disequilibrium between HLA-A3 serotype and the disease is well known (8-9). Several studies have tried to analyze linkage disequilibrium with other markers of 6p in order to restrict the HC candidate region. Data available at present indicate a strong disequilibrium both for markers close to HLA-A and at 3 cM on the telomeric side close to D6S105 (12-14). Differences in the degrees of disequilibrium observed probably reflect genetic differences between the populations studied. Therefore at present the candidate region is considered to extend for approximately 5 Megabases from HLA-B to telomeric to D6S105 (15). Only a part of this region is isolated in YACs, including the HLA-class I and a region extending 1000 Kb on its telomeric side (16,17). Cloning and characterization of the area close to D6S105 marker is in progress. Several cDNA have been produced from the cloned areas, especially from HLA class I region (18-20), but at present there is no evidence that one of these transcripts is a real candidate for HC. Very recently a few genes involved in iron transport, and linking copper metabolism to iron metabolism have been cloned in yeasts (21-22). It is expected that advances in this type of studies in humans will contribute possible HC candidates. Meanwhile molecular analysis in different populations has shown that a prevalent haplotype, HLA-A3-related, is present in several populations and accounts for 30 - 50 % of the chromosomes in the different populations studied (23-24). Most important, this haplotype is absent among normals, suggesting that it represents the ancestral haplotype of HC and carries a single ancient HC mutation. Preliminary data in Italian patients indicate that patients homozygotes for this haplotype express the most severe clinical disorder, in terms of disease complications and iron overload (Piperno et al, in preparation).

The isolation of the HC gene will allow to establish the molecular pathogenesis of the disorder, and will facilitate early diagnosis of at risk subjects and even population screening. It is also expected that the gene will provide the protein sequence and function to gain further insights into the complex topic of mechanisms of iron absorption and redistribution.

ACKNOWLEDGMENTS

This work was supported by MURST (Rome), Teleton Grant E32 (Rome) and A.I.R.C. (Milano).

REFERENCES

1. Simon M, Bourel M, Genetet B, Fauchet R: Idiopathic hemochromatosis: demonstration of recessive transmission and early detection by family HLA typing. N Engl J Med 297: 1017-1021, 1977.

2. Niederau C, Fischer R, Sonnenberg A, Stremmel W, Trampisch HJ, Strohmeyer G: Survival and causes of death in cirrhotic and non cirrhotic patients with primary hemochromatosis. N Engl J Med 313: 1256-1262, 1985.
3. Edwards CQ, Griffin LM, Goldgar D, Drummond C, Skolnick M, Kushner J: Prevalence of hemochromatosis among 11.065 presumably healthy blood donors. New Engl J Med 318: 1355-1362, 1988.
4. Dadone MM, Kushner JP, Edwards CQ, Bishop DT, Skolnick MH: Hereditary hemochromatosis: analysis of laboratory expression of the disease by genotype in 18 pedigrees. Am J Clin Pathol 78: 196-207, 1982.
5. Edwards CQ, Kushner JP: Current Concepts: Screening for Hemochromatosis. New Engl J Med 328: 1616-1620, 1993.
6. Casey JL, Koeller DM, Rami VC, Klausner RD, Harford JB: Iron regulation of transferrin receptor mRNA levels requires iron-responsive elements and a rapid turnover determinant in the 3' untranslated region of the mRNA. EMBO J 8: 3693-3699, 1989.
7. Rouault TA, Hentze MW, Caughman SW, Harford JB, Klausner RD: Binding of a cytosolic protein to the iron-responsive element of human ferritin messenger RNA. Science 241: 1207-1210, 1988.
8. Simon M, Bourel M, Fauchet R, Genetet B: Association of HLA A3 and HLA B14 antigens with idiopathic hemochromatosis. Gut 17: 332-334, 1976.
9. Simon M, Le Mignon L, Fauchet R, Yaouanq J, David V, Edan G, Bourel M: A study of 609 HLA haplotypes marking the hemochromatosis gene (1) mapping of the gene near the HLA-A locus and characters required to define a heterozygous population and (2) hypothesis concerning the underlying cause of hemochromatosis HLA association. Am J Hum Genet 41: 89-105, 1987.
10. Gasparini P, Borgato L, Piperno A, Girelli D, Olivieri O, Gottardi E., Roetto A, Dianzani I, Fargion S, Schinaia G, Cappellini MD, Gandini G, Pignatti PF, Fiorelli G, De Sandre G, Camaschella C: Linkage analysis of 6p21 polymorphic markers and the hereditary hemochromatosis: localization of the gene centromeric to HLA-F. Hum Mol Genet 5: 571-576, 1993.
11. Jazwinska EC, Lee SC, Webb SI, Halliday JW, Powell LW: Localization of the Hemochromatosis gene close to D6S105. Am J Hum Genet 53: 347-352, 1993.
12. Yaouanq J, Perichon M, Chorney M, Pontarotti P, Le Treut A, El Kahloun A, Mauvieux V, Blayau M, Jouanolle AM, Chauvel B, Moirand R, Nouel O, Le Gall JY, Feingold J, David V: Anonymous marker loci within 400 Kb of HLA-A generate haplotypes in linkage disequilibrium with the hemochromatosis gene (HFE). Am J Hum Genet 54: 252-263, 1994.
13. Worwood M, Raha-Chowdhury R, Dorak MT, Bowen DJ, Burnett AK: Alleles at D6S265 and D6S105 define a haemochromatosis specific genotype. Br J Haematol 86: 863-866, 1994.
14. Totaro A, Grifa A, Roetto A, Lunardi C, D'Agruma L, Sbaiz L, Zelante L, De Sandre G, Camaschella C, Gasparini P: New polymorphisms and markers in the HLA class I region: relevance to the Hereditary Hemochromatosis. Human Genet 95: 429-434, 1995
15. Camaschella C, Gasparini P: Hunting the hemochromatosis gene: progress and problems. Eur J Hum Genet 2: 141-147, 1994.
16. Abderrahim H, Sambucy JL, Iris F, Ougen P, Billault A, Chumakov IM, Dausset J, Cohen D, Le Paslier D: Cloning the human major histocompatibility complex in YACs. Genomics 23: 520-527, 1994.
17. Amadou C, Ribouchon MT, Mattei MG, Jenkins NA, Gilibert DJ, Copeland NG, Avoustin P, Pontarotti P: Localization of new genes and markers to the distal part of the human major histocompatibility complex (MHC) region and comparison with the mouse: new insights innto the evolution of mammalian genomes. Genomics 26: 9-20, 1995.
18. El Kahloun A, Chauvet B, Mauvieux V, Dorval I, Jouanoll AM, Gicquel I, Le Gall JY, David V: Localization of seven new genes around the HLA-A locus. Hum Mol Genet 2: 55-60, 1993.
19. Goei VL, Parimoo S, Capossela A, Chu TW, Gruen JR: Isolation of novel non-HLA-gene fragments from the hemochromatosis region (6p21.3) by cDNA hybridization selection. Am J Hum Genet 54: 244-251, 1994.
20. Wei H, Fan WF, Xu H, Parimoo S, Shukla H, Chaplin DD, Weissman SM Genes in one megabase of the HLA class I region. Proc Natl Acad Sci USA 90: 11870-11874, 1993.
21. Askwith C, Eide D, Van Ho A, Bernsard P, Li L, Davis-Kaplan S, Sipe DM, Kaplan J: The Fet3 gene of S. cerevisiae encodes a multicopper oxidase required for ferrous iron uptake. Cells 76: 403-410, 1994.
22. Yuan DS, Stearman R, Dancis A, Dunn T, Beeler T, Klausner RD: The Menkes/Wilson disease gene homologue in yeast provides copper to a ceruloplasmin-like oxidase required for iron uptake. Proc Natl Acad Sci USA 92: 2632-2636, 1995.
23. Jazwinska EC, Pyper WR, Burt MJ, Francis JL, Goldwurm S, Webb SI, Lee SC, Halliday JW, Powell LW: Haplotype analysis in Australian Hemochromatosis patients: evidence for a predominant ancestral haplotype exclusively associated with hemochromatosis. Am J Hum Genet 56: 428-433, 1995.

24. Worwood M, Gasparini P, Camaschella C: Report on the International Workshop on Molecular Genetics of Hemochromatosis held at Villa Feltrinelli Gargnano (Bs) Italy, 25th September 1994. J Med Genet 32: 320-323, 1995.

ANTIMALARIAL ACTION OF IRON CHELATORS

Appolinaire Tsafack,[1] Jacob Golenser,[2] and Z. Ioav Cabantchik[1*]

[1] Department of Biological Chemistry
Institute of Life Sciences
[2] Department of Parasitology
Hadassah Medical School
Hebrew University, Jerusalem, Israel 91904

ABSTRACT

Unlike mammalian cells, malaria parasites are uniquely sensitive to the action of iron chelators. By depriving cells of iron, natural and synthetic iron chelators and siderophores of the hydroxamate family, inhibit, among others, nucleic acid synthesis, arrest parasite growth and stop proliferation. Moreover, the chelator induced damage to parasites can often be irreversible, depending on the nature of the drug used and on the developmental stage of the parasite exposed to the drug. The uniqueness of parasite's susceptibility to iron chelators rests on the parasite's limited ability for mobilizing iron. However, the fact that most available iron chelators irreversibly affect parasites only at given stages of parasite development, limits their usefulness as prospective therapeutic agents. The aim is to develop iron chelators and treatments which lead to long lasting inhibitory effects on all stages of parasite development. This was accomplished with iron chelators which, in combination, outperformed the sum of single drug treatments. The synergistic properties of chelators herewith reported were obtained in malaria parasites grown in *Plasmodium falciparum* infected red blood cells. The studies were carried with desferrioxamine (DFO) in combination with various hydrophobic siderophores of the salicylaldehyde-isonicotinoyl or fluorobenzoyl hydrazone family and reversed siderophores.

INTRODUCTION

Plasmodia reside in erythrocytes during the major part of their life cycle. They grow and proliferate by acquiring nutrients from either the host cell or the external environment.

*Correspondence to: Prof. Z. Ioav Cabantchik, Department of Biological Chemistry, Institute of Life Sciences, Hebrew University, Jerusalem 91904, Israel. Telephone: 972-2-6585420; Fax: 972-2-6586974; E.mail: Ioav@vms.huji.ac.il.

Molecular Biology of Hematopoiesis 5, edited by Abraham et al.
Plenum Press, New York, 1996

This includes the mobilization of bioavailable iron and its integration into essential proteins [1]. The source of iron for parasite development seems to originate within the infected cell [2,3]. Support for that idea rests on hemoglobin's ability to release free iron in an artificial environment simulating that of the parasite food vacuole [4]. However, it remains to be shown whether such phenomena occur in the native system and whether they are of a regulated nature. This is particularly important, in terms of meeting parasite demands while avoiding excess supply of a potential toxic agent such as iron.

A most remarkable feature of malaria is that the plasmodia, despite living in a "sea of iron", showed a marked susceptibility to iron deprivation which can be induced by various types of iron chelators (reviewed in ref. 1). This property was therefore assessed as a possible chemotherapeutic target, which might be applicable to severe cases of the diesease. Present aims are to develop iron chelators with optimal performance as antimalarials and highest therapeutic index. Ideally, an iron chelator should preferentially affect infected cells at all stages of parasite growth and either spare host cells or impinge on them a minimal damage. In this work we analyze the biophysical properties which confer to chelators their access mode to intracellular parasites and their ability to arrest parasite proliferation. The studies were conducted *in vitro* with cultures of human erythrocytes infected with *Plasmodium falciparum.*

SELECTIVE PROPERTIES OF IRON CHELATORS AS ANTIMALARIALS

Iron chelators were shown to act on the intracellular iron pools associated with the parasite [1-3]. The antimalarial efficacy of the iron chelators is dependent upon both the chemical properties (*i.e.* lipophilicity), their iron binding capacity and their selectivity for the metal. This is depicted for various families of chelators (Table 1). Among these were derivatives of the hydrophilic desferrioxamine (DFOs) [3,5], a family of lipophilic siderophores referred to as reversed siderophores (RSFs) [6], as well as other permeable chelators such as salicylaldehyde isonicotinoyl hydrazone and naphthalene fluorobenzoyl hydrazone [7]. Structure-activity relationship studies for given series of chelators (*e.g.* RSFs), indicated

Table 1. The IC_{50} of RSFs and their combined effects with DFO

			IC_{50} (μM)	
Chelators	Partition coefficient	Speed of action	(-) DFO	(+) DFO
RSF.ileu	14	13±1	10±1	8±1
RSF.phe	>10	16±1	0.6±0.2	0.3±0.1
RSF.leum2	12.5	6±2	4±2	1.0±0.2
RSFm2	1.7	11±1	3±1	1.1±0.2
DFO	<1	<0.2	-	17±1

The combined action of RSF with DFO was assessed using a fixed concentration of 10μM of DFO with various concentrations of indicated RSFs (data adapted from refs 6,11,12 and unpublished ones). The effect was expressed as inhibition of nucleic acid synthesis. The experimental data for the combined effects of either RSF with DFO were statistically different to theoretical values for independent ($P<0.01$) and additive ($P<0.05$) modes of action when DRCs were analyzed using Poch approach to drug combination anlysis [13]. The various RSFs are named according to the nomenclature given elsewhere [6,12], the amino acid representing part of the hydroxamate arm and the m the number of C atoms in the tripode bound connector [6].

that the antimalarial activity was dictated by the product of lipophilicity and the iron(III) association constant [8].

DIFFERENTIAL EFFECTS OF CHELATORS ON PARASITES AND MAMMALIAN CELLS

Iron chelators of the hydroxamate family are essentially cytostatic to mammalian cells, that is they reduce cell proliferation in a reversible manner, since cell fully recover after drug removal [9]. However, *P. falciparum* parasites in culture differed markedly from mammalian cells in drug susceptibility and in the cells' capacity to recover from the treatment (Fig. 1). The mode of action of the chelators as antimalarials depended on their chemical properties and varied with the developmental stage of the parasite. These properties are depicted in the model in Fig. 1 for hydrophilic chelators such as desferrioxamine and hydrophobic chelators such as reversed siderophores (RSF). The model stipulates that parasite iron is primarily acquired from the trophozoite stage onward. RSFs, which are demonstrably permeant to all stages of parasite development, swiftly arrest growth at all stages but act irreversibly only on rings [8]. Thus, parasites recover from RSF treatment during the trophozoite-schizont stage because they can apparently meet the demands for iron upon drug removal. On the other hand, because of its poor permeation, DFO fails to act on rings but permeates slowly into mature parasites and affects them irreversibly because of its retention within cells [8,10]. The relatively poor efficacy of DFO stems from its slow permeation relative to parasite's progression through the life cycle. The depicted model suggested possible schemes for improving the antimalarial action of iron chelators by using combinations of DFO and RSFs [11,12]. Fast and slow permeating drugs, when used in combination, are expected to produce either additive or synergistic effects, depending on their stage specificity and drug action profile. These were recently assessed with DFO and either RSFs [11,12] or isonicotinoyl and fluorobenzoyl hydrazones [7].

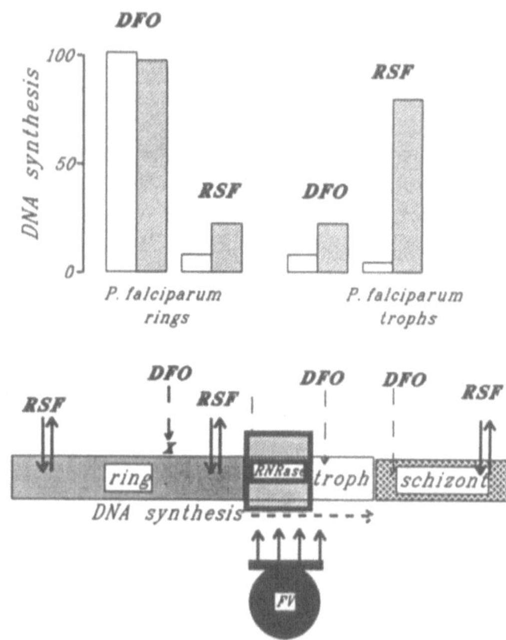

Figure 1. Mode of action of iron chelators as antimalarials: *Bottom*: The growth stages of intra-erythrocytic malaria parasites are depicted on a linear form. The differential speed and efficiency of DFO and RSFs antimalarial are indicated by the thickness of the arrows. The iron mobilization from degraded hemoglobin (upward pointing arrows from the parasite's food vacuole-FV), iron integration into iron-dependent enzymes such as ribonucleotide reeuctase (RNRase) and DNA synthesis start at the trophozoite stage (adapted from ref. 8). *Top*: Reversible (empty bars) and irreversible (slashed bars) effects of DFO (100 μM) and RSF (100 μM) administered to parasites at the indicated stages of growth (adapted from ref. 1).

ANTIMALARIAL PROPERTIES OF COMBINATION OF DRUGS

The application of pairs of drugs which act on different parasite growth stages is expected to produce additive effects on asynchronously growing cultures. The complementary effects were observed for RSFs and DFO when administered to mixed parasite cultures and after drug exposure and withdrawal [11,12]. However, on synchronously growing cultures, the pattern of action of drug combinations might be of a more complex nature, depending on the speed of action of the drug and the nature of the inhibition.

Of particular interest and therapeutic relevance is the possibility of designing drug combinations which produce synergistic effects, thus reducing the risk of drug toxicity to the host. For drugs acting on similar putative targets, such as iron chelators, one drug may potentiate the action of the other either by facilitating the others' drug entry or by rendering the cells more susceptible to it. DFO's restricted permeation ability to mature parasite forms has a dual effect: on the one hand it allows parasites to "escape" from its action by progressing through the life cycle. However, on the other hand, once it has reached the parasite, DFO is more likely to be retained intracellularly for a considerable period of time and produce a long lasting biological effect. Based on the proposed mode of drug action on parasites (Fig. 1), RSFs' fast action on trophozoites has immediate consequence: it slows down parasite progression through the cycle. In doing so, RSFs widen the window of exposure to DFO, allowing more drug to permeate into the mature stages of growth. In this fashion, RSFs apparently potentiate DFO effects even at concentrations in which DFO alone produces no detectable inhibitions. In fact,the profiles of inhibition of RSFs in the presence of DFO were higher than additive, as exemplified in Table 1 and analytically proven by mathematical analysis according to Poch [11-13].

An additional mode of drug potentiation was hypothesized on the basis of one impermeant drug (e.g. DFO at the ring stage) facilitating removal of iron from parasitized cell by another- permeating drug. The iron chelator SIH was chosen based on its ability to swiftly permeate across cell membranes and its propensity for donating chelated iron [14]. Having shown that Fe could be transferred from Fe-SIH to DFOs, we assessed whether the ring impermeant DFO or the ring- and trophozoite- impermeant hydroxyl-ethyl-starch-DFO (HES-DFO), potentiated SIH's action on rings. The antimalarial activity of SIH was assessed by exposing cultures of rings to the free drug or its iron complex for 40 hrs and estimating nucleic acid synthesis during the last 20 hrs. The SIH-Fe complex was not inhibitory to parasite growth, whereas SIH alone was found to effectively suppress *P. falciparum* growth in culture (IC_{50} 23±3 μM). When used in combination with constant (non-inhibitory) concentrations of DFOs, SIH antimalarial action on rings was markedly potentiated, as the above IC_{50} value was reduced to 5±1 μM and 10±1 μM with DFO and HES-DFO, respectively. These results conform with SIH's ability to mobilize iron into the medium from the intracellular compartment of infected cells [14] from *P. falciparum* infected erythrocytes [7]. The mobilized iron could be trapped extracellularly using impermeant chelators in the medium, such as DFO on ring stage or HES-DFO on both rings and trophozoites. Thus the impermeant DFOs acted as extracellular sinks of iron scavenged from infected cells by SIH, thereby potentiating its antimalarial action.

CONCLUSIONS

The goal of this work was to investigate the mode of action of iron chelators as antimalarials, with the aim of improving their therapeutic efficacy and index as antimalarial agents. This involved understanding how drugs reach their intracellular targets and how they ultimately interfer with the metabolism of the parasites. The working model of iron chelators as antimalarials described here allowed us: a. to assess those factors which limit the action of given chelators to particular parasite growth stages and those which determine their speed of action and b. to design regimens of treatment for covering the entire life cycle of the intra-erythrocytic parasite. The chosen regimens were aimed at obtaining synergistic action of drugs which could potentially act on similar targets, but whose permeation ability limited their speed of action to particular parasite growth stages. The combined action of DFO (used at subtoxic levels) with various hydrophobic iron chelators, was shown to improve the pharmacological profile of chelators on *in vitro* cultures of parasites. For the *in vivo* situation, the working model has to take into consideration pharmacokinetic factors, such as the mode and rate of drug administration, volume of distribution, drug clearance, etc.. These properties are being studied in animal models of malaria with the aid of slow drug release systems. The application of drugs entrapped in polymers, is anticipated to contribute significantly to improving the antimalarial performance of iron chelators.

ACKNOWLEDGMENTS

This work was supported in part by an NIH grant AI20342 and by the National Israel Research Fund administered by the Israel Academy of Sciences and Humanities.

REFERENCES

1. Cabantchik ZI. Iron chelators as antimalarials. The biochemical basis of selective cytotoxicity. Parasitology Today 1995;11:74-78.
2. Scott MD, Ranz A, Kuypers FA, Lubin BH, Meshnick SR: Parasite uptake of desferroxamine: a prerequisite for antimalarial activity. Brit J Haematol 1990;75:598-602.
3. Loyevsky M, Lytton SD, Mester B, Libman J, Shanzer A, Cabantchik ZI: The antimalarial action of desferal involves a direct access route to erythrocytic (*Plasmodium falciparum*) parasites. J Clin Invest 1993;91:218-224.
4. Gabay T and Ginsburg H. Hemoglobin denaturation and iron release in acidified red blood cell lysate. A possible source of iron for intra-erythrocytic malaria parasites. Exp Parasitol 1993;77:261-272.
5. Lytton SD, Cabantchik ZI, Libman J, Shanzer A. Reversed siderophores as antimalarial agents. II. Selective scavenging of Fe(III) from parasitized erythrocytes by a fluorescent derivative of desferal. Mol Pharmacol 1991;40:584-590.
6. Shanzer A, Libman J, Lytton SD, Glickstein H, Cabantchik ZI: Reversed siderophores act as antimalarial agents. Proc Natl Acad Sci USA 1991;88:6585-6589.
7. Tsafack A, Loyersky M, Ponka P, Cabantchik ZI; Mode of action of iron (III) chelators as antimalarials IV. Potentiation of desferal action by benzoyl and isonicotinoyl hydrazone derivatives. J Lab Clin Med 1996; 127:574-582.
8. Lytton SD, Mester B, Libman J, Shanzer A, Cabantchik ZI: Mode of action of iron (III) chelators as antimalarials: II. Evidence for differential effects on parasite iron-dependent nucleic acid synthesis. Blood 1994;84:910-915.
9. Glickstein H, Breuer W, Loyevsky M, Konijn AM, Libman J, Shanzer A, Cabantchik ZI: Differential cytotoxicity of iron chelators on malaria infected cells versus mammalian cells. Blood 1996; 87:4871-4878.
10. Whitehead S and Peto TEA. Stage-dependent effect of desferrioxamine on growth of *Plasmodium falciparum in vitro*. Blood 1990;76:1250-1255.
11. Golenser J, Tsafack A, Libman J, Shanzer A, Cabantchik ZI: Antimalarial action of hydroxamate-based iron chelators and potentiation of desferrioxamine action by reversed siderophores. Antimicrob Agents & Chemo 1995;39:61-65.
12. Tsafack A, Golenser J, Libman J, Shanzer A, Cabantchik ZI: Mode of action of iron(III) chelators as antimalarials. III. Overadditive effects in the combined action of hydroxamate-based agents on in vitro growth of *Plasmodium falciparum*. Mol Pharmacol 1995;47:403-409.

13. Poch G. *Combined Effects of Drugs and Toxic Agents. Modern Evaluation in Theory and Practice.* Springer-Verlag, New York (1993).
14. Ponka P, Grady RW, Wilczynska A and Schulman M. The effect of various chelating agents on the mobilization of iron from reticulocytes in the presence and absence of pyridoxal isonicotinoyl hydrazones. BBA 1984;802:477-489.

INHIBITION OF REPERFUSION INJURY BY IRON CHELATORS

J. J. M. Marx*

Department of Internal Medicine
University Hospital Utrecht
Utrecht, The Netherlands

ABSTRACT

If tissue is deprived from oxygen by ischaemia, cellular damage may occur during reperfusion. This reperfusion injury is associated with damage to membranes, release of intracellular enzymes and influx of calcium which finally will kill the affected cells. Strong evidence exists for the pivotal role of free radical formation in initiating cell damage after reoxygenation. The most toxic oxygen species appears to be the hydroxyl radical which is only formed in the presence of catalytic transition metals, especially iron and copper. Iron mediated tissue damage by free radicals can even occur in subjects with low or absent iron stores [1,2]. Free radical scavengers are able to ameliorate reperfusion damage. Iron chelators have been proven to be very effective in prevention of free radical mediated tissue injury. The generation of toxic oxygen species and the role of iron in microvascular damage after ischaemia and reperfusion is described, as well as the results of recent experimental studies showing that reperfusion injury can be prevented by iron chelators.

GENERATION OF TOXIC OXYGEN SPECIES

A free radical is a molecular species that contains one or more unpaired electrons (an electron which occupies one atomic or molecular orbital by itself) [3]. The most toxic oxygen metabolite is the hydroxyl radical, which is generated during transition metal catalyzed single electron reduction of oxygen. The hydroxyl radical has a half-life of 10^{-9} sec and a diffusion radius of only 2.3 nm [4]. The sequence of events leading to hydroxyl radical formation is summarized in figure 1, showing the most important molecular species which are formed from oxygen (O_2), each accepting one electron for transformation to the next

* Correspondence: Prof. Dr. J. J. M. Marx, Department of Internal Medicine, Room G02.228, University Hospital Utrecht, P.O. Box 85500, 3508 GA Utrecht, The Netherlands. Phone: *31-30-2507394; Fax: *31-30-2518328.

Molecular Biology of Hematopoiesis 5, edited by Abraham et al.
Plenum Press, New York, 1996

$O_2 + e \rightarrow O_2^{\cdot-}$ — *NADPH-oxidase system*, *Xanthine / Xanthine-oxidase*

$O_2^{\cdot-} + e + 2H^+ \rightarrow H_2O_2$

$H_2O_2 + e$

$$O_2^{\cdot-} + Fe^{3+} \rightarrow O_2 + Fe^{2+}$$
$$H_2O_2 + Fe^{2+} \rightarrow OH^- + HO^\cdot + Fe^{3+}$$
$$O_2^{\cdot-} + H_2O_2 \rightarrow O_2 + OH^- + HO^\cdot$$

(Iron catalyzed Haber-Weiss reaction)

$HO^\cdot + e + H^+ \rightarrow H_2O$ — *Tissue electron donors (lipids, proteins, DNA)*

Figure 1. Simplified model of the univalent reduction-steps of molecular oxygen. The first electron that generates the superoxide radical is derived by the NADPH-oxidase system of activated phagocytes, or by xanthine oxidase from hypoxanthine and xanthine, during reoxygenation after e period of anoxia. Both sources of the first electron are present during vascular reperfusion after a period of ischaemia. Formation of the hydoxyl radical ($HO^\cdot$) is only possible if a catalytically active transition metal is available.

species: superoxide ($O_2^{\cdot-}$), hydrogen peroxide (H_2O_2), the hydroxyl radical ($HO^\cdot$) and water (H_2O). The superoxide radical is constantly formed in all living cells, and in particular during activation of phagocytes or during reoxygenation following ischaemia. Once $HO^\cdot$ is formed, it will oxidize all substances in its direct environment, causing peroxidation of membrane lipids, proteins or DNA. The peroxidation of membrane lipids can lead to the development of gaps in the membrane through which calcium can enter the cell, increasing the intracellular concentration of free calcium [5].

As H_2O_2 is a relatively stable and non-toxic molecule, formation of $HO^\cdot$ is fully dependent upon the availability of catalytic trace elements such as iron, copper or cobalt as indicated in figure 1. The important role of iron in catalyzing the formation of hydroxyl radicals from $O_2^{\cdot-}$ and H_2O_2 was emphasized by Haber and Weiss [6]. Although it is well recognized that catalytic iron is essential for the formation of the hydroxyl radical, the source of iron is not always clear. Not only low molecular weight (LMW) iron complexes can present iron in a catalytically active form. Also iron from ferritin [7], lactoferrin [8], uteroferrin [9], haem [10], hemoglobin [11], myoglobin, cytochromes and from microorganisms [12] can catalyze formation of hydroxyl radicals. Many reducing agents can cause decompartmentalization of iron from ferritin [13]. Direct catalytic activity of iron in haem or hemoglobin, requires redox conditions and modification of prosthetic haem groups [14]. Under physiological conditions toxic activity of oxygen metabolites is attenuated by scavengers, such as superoxide dismutase (SOD), catalase, glutathione, and vitamins E, C and A.

REPERFUSION INJURY

The most important source of toxic amounts of hydroxyl radicals in the body is single electron reduction of oxygen, which is initiated by formation of superoxide [15]. The electron needed for formation of $O_2^{\cdot-}$ can be produced at the cellular level during reoxygenation after a period of anoxia. The situation in a physiological environment with vasculated tissue is more complicated, and injury is caused by the sequence of ischaemia and reperfusion (this is called "reperfusion injury") [16], which is associated with damage to membranes, release of intracellular enzymes and influx of calcium. Increase of intracellular calcium causes activation of Ca^{++}-dependent proteases, phospholipase, mitochondrial dysfunction, cytoskeletal disruption, and conversion of xanthine dehydrogenase to xanthine oxidase [17]. Cells in a permanent anoxic environment will loose their viability. If anoxia is only temporary, reoxygenation will result in functional recovery, depending on the period of anoxia or hypoxia. Depending on the duration of anoxia, however, damage will accelerate during reoxygenation and is the result of oxygen radical formation, mediated by xanthine oxidase (XO), as summarized in figure 2. XO is the rate limiting enzyme during degradation and terminal oxidation of nucleic acids. It is able to generate $O_2^{-\cdot}$ and H_2O_2 during the oxidation of hypoxanthine and xanthine. In non-ischemic conditions XO exists as the NAD^+ -dependent xanthine dehydrogenase (XDH), NAD^+ being the electron acceptor during oxidation of purines. XDH, however, is converted into XO during tissue ischaemia [18], proportional to the duration of ischemia. During anoxia the production of hypoxanthine from ATP is increased, and XO catalyzes upon reperfusion the oxidation of hypoxanthine, with formation of urate, $O_2^{-\cdot}$ and H_2O_2. At the same time the intracellular concentration of low molecular weight iron increases enormously, probably as a result of reductive release from ferritin, stimulated by acidification during ischaemia [19]. This means that all the ingredients for formation of the hydroxyl radical are present. Microvascular injury is further aggravated by activated neutrophils, because hypoxia and reoxygenation stimulate neutrophil adhesion to endothelium [20]. XO-derived oxidants, in particular $O_2^{-\cdot}$, initiate production of inflammatory mediators and increase expression of the preformed granule membrane protein 140 (GMP-140), and de novo synthesis of the endothelial leukocyte adhesion molecule 1

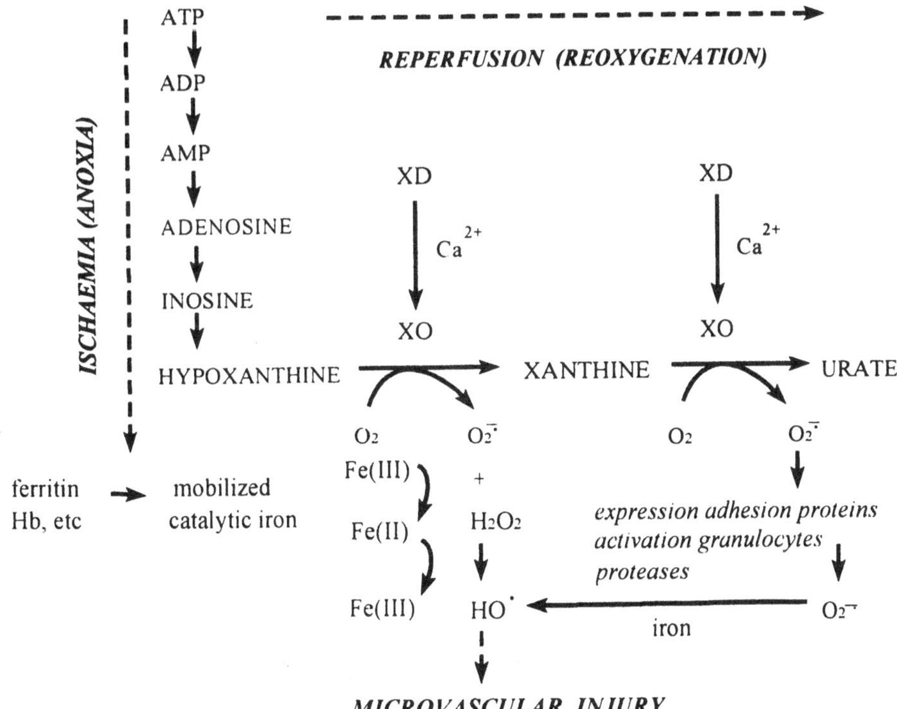

Figure 2. Schematic representation of microvascular injury, caused by reperfusion following a limited period of ischaemia.

(ELAM-1) [21]. Adhesion of granulocytes to endothelial cells can be enhanced in the presence of non-transferrin bound iron [22], which may be released from cells during hypoxia.

In a physiological system activated neutrophils, eosinophils, monocytes and macrophages are the most important sources of superoxide. Their NADPH-oxidase system, assembled during the respiratory burst, is responsible for the one electron shuttling through the plasma membrane, which results in the formation of superoxide at the cell surface or within a phagocytic vesicle [23]. Free radicals produced by phagocytes can become harmful if the generation is in excess of the need for killing of microorganisms, and when they are formed in the vicinity of healthy cells, e.g. at the surface of endothelial cells. Addition of iron to activated granulocytes and monocytes, or release of catalytic iron from adjacent cells, impairs their phagocytic function [24], and is associated with lipid peroxidation [25].

Tissues are capable of adaptive changes if short, sublethal periods of ischaemia are alternated with short periods of reperfusion, which may for instance reduce myocardial infarct size after a longer period of ischaemia [26]. Among other factors "ischemic preconditioning" is caused by increased antioxidant enzyme activity [27].

IRON CHELATORS AS SCAVENGERS OF TOXIC OXYGEN PRODUCTS

Cells are protected against oxidative damage by scavengers, including specific antioxidant enzymes (superoxide dismutase, catalase and gluthatione peroxidase), vitamins A, C and E, or molecules with a high affinity for iron (transferrin, lactoferrin, albumin), haem (haemopexin) or haem-containing proteins (haptoglobin) [28]. Some antioxidants also have prooxidant activity. Ascorbic acid for example is an excellent antioxidant, but also a strong reductant, able to reduce Fe^{3+} to Fe^{2+}.

The ability of iron to catalyze production of hydroxyl radicals depends on the characteristics of the chelating agent [29]. Some iron chelators protect against oxygen radical damage, while others promote toxicity. In aqueous solution iron ions are dissolved with six coordinated water molecules. The most effective iron chelators are multidentate compounds which are able to form coordinated covalent bonds by which all six water molecules are displaced. An example is deferoxamine (DFO, a hydrophilic, hexadentate iron chelator with a polyamine structure) with six oxygen donor atoms, which force iron into the ferric form. An important and highly effective modification of DFO is its conjugation with macromolecules, increasing the plasma half-life, and limiting its acute and chronic toxicity [30]. These drugs seem especially promising in treatment of reperfusion injury. As an alternative to DFO, which must be administered parenterally, many new orally active iron chelators have been developed [31], but only one of these, L1 (deferiprone) has been evaluated clinically [32].

Some iron chelators are toxic because they have prooxidant activity. Two examples of highly toxic iron chelators are nitrilotriacetate (NTA) [33] and EDTA [34]. Like deferoxamine, EDTA is a hexadentate ligand with six iron coordination sites. EDTA, however, is too small to completely encompass the iron ion. A seventh coordination site is induced by distortion of the usual coordination symmetry, which leaves iron in a catalytically active form. Iron-EDTA in a superoxide generating system catalyzes the formation of the hydroxyl radical [1]. At present only DFO, and its macromolecular derivatives, can be considered as safe and clinically effective iron chelators.

INHIBITION OF REPERFUSION INJURY BY IRON CHELATORS

After the early description of myocardial enzyme release during reoxygenation of the anoxic perfused rat hart [35], reperfusion injury was reported to be the cause of tissue damage in an unexpectedly wide variety of pathological conditions. Almost all known scavengers of toxic oxygen metabolites have been tested in experimental models of reperfusion injury to prevent tissue damage. For instance SOD, allopurinol (which inhibits XO), glutathione, and raffinose could improve preservation of organs for transplantation [36]. SOD and glutathione could effectively reduce microvascular leakage after tourniquet-induced ischaemia in striated muscle of hamsters [37], which may be an equivalent for tissue damage in pressure sores. Especially allopurinol is often used as an inhibitor of reperfusion injury, as it is a non-toxic drug, widely used for prevention of excess production of uric acid. The results from clinical studies, however, are disappointing.

A logical and probably more effective approach to prevent reoxygenation damage is use of iron chelators which should be able to take the sting out of the Haber-Weiss reaction by directly preventing the generation of hydroxyl radicals. It was demonstrated in rats that iron loading of the hearts enormously increased sensitivity to reoxygenation after a mild anoxic insult [38]. Indeed, the results of deferoxamine and some orally active iron chelators are promising. Some selected publications, in which reoxygenation or reperfusion injury was effectively prevented by DFO, are mentioned in table 1. In a number of studies hydroxyethyl starch (HES-)DFO was used. HES-DFO and also HES-FO had no effect on hemodynamics, while DFO often causes a marked reduction in systemic blood pressure [39]. In contrast to DFO, HES-DFO does not cause ocular toxicity [49]. The plasma-$T^1/_2$ of intravenously administered HES-DFO is more then tenfold increased compared to DFO [50]. It is disappointing that the evidence for oxygen metabolite scavengers and iron chelators in preventing reperfusion injury in man is still scarce. Because DFO is a registered and relatively safe drug, there is no good reason why DFO should not be assayed in patients who may suffer from reoxygenation damage. A problem may be the short plasma-$T^1/_2$, which can prohibit high plasma levels, and the hemodynamic side effects of DFO. Both problems can be overcome by using HES-DFO. Another interesting aspect of this drug is that HES-DFO, which does not enter cells, is just as effective as DFO. This means that catalytically active iron is chelated in the microcirculation, probably near the surface of endothelial cells. Results

Table 1. Experimental models of oxygen free radical mediated tissue damage, caused by anoxia followed by reoxygenation, in which iron chelators could prevent tissue injury

Pathological condition	Experimental Model	Chelator
Myocardial infarction [39]	coronary occlusion and reperfusion in dog hearts	DFO, HES-DFO
Myocardial ischaemia [40]	isolated perfused rabbit hearts	DFO
Cardiopulmonary bypass [41]	Bypass surgery in man	DFO
Postischemic skeletal muscle injury [42]	partial hindlimb ischaemia in rats	DFO
Lung transplant preservation [43]	canine single lung transplantation	DFO
Heart transplant preservation [44]	24-hour cold rat heart preservation and reperfusion	DFO
Heart-lung transplantation [45]	swine heart-lung blocks	HES-DFO
Renal hypoxia and reoxygenation [46]	cultured rat proximal tubular epithelial cells	DFO, HES-DFO
Hepatic ischaemic damage [47]	hemorrhagic shock in rats	HES-DFO
Ischaemic brain injury [48]	resuscitation following cardiac arrest in rats	HES-DFO
Retinal Ischaemia [49]	intraoccular pressure and vascular ligation of cat eyes	HES-DFO

of experimental studies using iron chelators are promising enough to perform clinical investigations in the near future.

REFERENCES

1. Voest EE, Vreugdenhil G, Marx JJM: Iron-chelating agents in non-iron overload conditions. Ann Intern Med 120:490, 1994
2. Hershko C: Control of disease by selective iron depletion: a novel therapeutic strategy utilizing iron chelators. Clin Haematol 7:965, 1994
3. Halliwell B, Gutteridge JMC: Free radicals in biology and medicine. 2nd ed. Oxford, Clarendon Press, 1989.
4. Roots R, Okada S: Estimation of life times and diffusion distances of radicals involved in X-ray-induced DNA strand breaks or killing of mammalian cell. Radiat Res 64:306, 1975
5. Shasby DM, Lind SE, Shasby SS, Goldsmith JC, Hunninghake GW: Reversible oxidant-induced increases in albumin transfer across cultured endothelium: alterations in cell shape and calcium homeostasis. Blood 65:605, 1985
6. Haber F, Weiss J: The catalytic decompensation of hydrogen peroxide by iron salts. Proc Roy Soc Lond (A) 147:332, 1934
7. Biemond P, Van Eijk HG, Swaak AJG, Koster JF: Iron mobilization from ferritin by superoxide derived from stimulated polymorphonuclear leukocytes. J Clin Invest 73:1576, 1984
8. Vercellotti GM, Van Asbeck BS, Jacob HS: Oxygen radical-induced erythrocyte hemolysis by neutrophils. Critical role of iron and lactoferrin. J Clin Invest 76:956, 1985
9. Sibille JC, Doi B, Aisen P: Hydroxyl radical formation and iron-binding proteins. Stimulation by the purple acid phosphatases. J Biol Chem 262:59, 1987
10. Balla G, Vercellotti GM, Muller-Eberhard U, Eaton J, Jacob HS: Exposure of endothelial cells to free heme potentiates damage mediated by granulocytes and toxic oxygen species. Lab Invest 64:648, 1991
11. Seibert AF, Taylor AE, Bass JB, Haynes J Jr: Hemoglobin potentiates oxidant injury in isolated rat lungs. Am J Physiol 260:H1980, 1991
12. Hoepelman IM, Bezemer WA, Vandenbroucke-Grauls CMJE, Marx JJM, Verhoef J: Bacterial iron enhances oxygen radical-mediated killing of Staphylococcus aureus by phagocytes. Inf Immunity 58:26, 1990
13. Reif DW: Ferritin as a source of iron for oxidative damage. Free Rad Biol Med 12:417, 1992
14. Ortiz de Montellano P: Free radical modification of prosthetic heme groups. Pharmac Ther 48:95, 1990
15. Liochev SI, Fridovich I: The role of $O_2^{\cdot-}$ In the production of $HO^{\cdot}$: in vitro and in vivo. Free Radic Biol Med 19:29, 1994
16. Granger DN, Rutili G, McCord JM: Superoxide radicals in feline intestinal ischemia. Gastroenterol 81:22, 1981
17. Paller MS, Greene EL: Role of calcium in reperfusion injury of the kidney. Ann NY Ac Sci 723:59, 1995
18. Parks DA, Williams TK, Bechman JS: Conversion of xanthine dehydrogenase to oxidase in ischemic rat intestine. A reevaluation. Am J Physiol 254:G768, 1988
19. Voogd A, Sluiter W, Van Eijk HG, Koster JF: Low molecular weight iron and the oxygen paradox in isolated rat hearts. J Clin Invest 90:2050, 1992
20. Kurose I, Granger DN: Evidence implicating xanthine oxidase and neutrophils in reperfusion-induced microvascular dysfunction. Ann NY Ac Sci 723:158, 1995
21. Palluy O, Morliere L, Gris JC, Bonne C, Modat G: Hypoxia/reoxygenation stimulates endothelium to promote neutrophil adhesion. Free Rad Biol Med 13:21, 1992
22. Hoepelman IM, Verhoef J, Marx JJM: Effects of nontransferrin-bound iron on the aggregation and adhesion of polymorphonuclear granulocytes (PMN). Ann N York Ac Sci 562:363, 1988
23. Cohen MS, Britigan BE, Hassett DJ, Rosen GM: Phagocytes, O_2 reduction, and hydroxyl radical. Rev Infect Dis 10:1088, 1988
24. Hoepelman IM, Jaarsma EY, Verhoef J, Marx JJM: Polynuclear iron complexes impair the function of polymorphonuclear granulocytes. Br J Haematol 68;385, 1988
25. Hoepelman IM, Bezemer WA, Van Doornmalen E, Verhoef J, Marx JJM: Lipid peroxidation of human granulocytes (PMN) and monocytes by iron complexes. Br J Haematol 72:584, 1989
26. Reimer KA, Vander Heide S, Jennings RB: Ischaemic preconditioning slows ischemic metabolism and limits myocardial infarct size. Ann NY Ac Sci 723:99, 1995

27. Das DK, Prasad MR, Tu R, Jones RM: Preconditioning of heart by repeated stunning. Adaptive modification of antioxidative defense system. J Cell Mol Biol 38:739, 1992
28. Krinsky NI: Mechanism of action of biological antioxidants. Proc Soc Exp Biol Med 200:248, 1992
29. Van Asbeck BS, Marx JJM, Struyvenberg A, Van Kats JH, Verhoef J: Effect of Iron(III) in the presence of various ligands on the phagocytic and metabolic activity of human polymorphonuclear leukocytes. J Immunol 132:851, 1984
30. Hedlund BE, Hallaway PE: High-dose systemic iron chelation attenuates reperfusion injury. Bioch Soc Transact 21:340, 1993
31. Porter JB, Huehns ER, Hider RC: The development of iron chelating drugs. Clin Haematol 2:257, 1989
32. Al-Refaie FN, Hoffbrand AV: Oral iron-chelating therapy: the L1 experience. Clin Haematol 7:941, 1994
33. Marx JJM: The use of ferric nitrilotriacetate and ferrozine in iron transport studies. Eur J Clin Invest 25:721, 1995
34. Marx JJM, Vreugdenhil G, Voest EE: Iron chelating agents are not useful in treating atherosclerosis. Ann Int Med 121:384, 1994
35. Hearse DJ, Humphry SM, Chain EB. Abrupt reoxygenation of the anoxic potassium-arrested perfused rat heart: a study of myocardial enzyme release. J Mol Cell Cardiol 5:395, 1973
36. Bulkley GB: Reactive oxygen metabolites and reperfusion injury: aberrant triggering of reticuloendothelial function. Lancet 433:934, 1994
37. Menger MD, Pelikan S, Steiner D, Messmer K: Microvascular ischemia-reperfusion injury in striated muscle: significance of "reflow paradox". Am J Physiol 263:H1901, 1992
38. Van der Kraaij AMM, Mostert LJ, van Eijk HG, Koster JF. Iron-load increases the susceptibility of rat hearts to oxygen reperfusion damage. Protection by the antioxidant (+)-cyanidanol-3 and deferoxamine. Circulation 78:442, 1988
39. Maruyama M, Pieper GM, Kalyanaramen B, Hallaway PE, Hedlund BE, Gross GJ: Effects of hydroxyethyl starch conjugated deferoxamine on myocardial functional recovery following coronary occlusion and reperfusion in dogs. J Cardiovasc Pharmacol 17:166, 1991
40. Williams RE, Zweier JL, Flaherty JT: Treatment with deferoxamine during ischemia improves functional and metabolic recovery and reduces reperfusion-induced oxygen radical generation in rabbit hearts. Circulation 83:1006, 1991
41. Menasche PO, Pasqueir C, Jaillon P, Piwnica A. Deferoxamine reduces neutrophil mediated free radical production during cardiopulmonary bypass in man. J Thorac Cardiovasc Surg 96:582, 1988
42. Fantini GA, Yoshioka T: Deferoxamine prevents lipid peroxidation and attenuates reoxygenation injury in postischemic skeletal muscle. Am J Physiol 264:H1953, 1993
43. Conte JV, Katz NM, Foegh ML, Wallace RB, Ramwell PW: Iron chelation therapy and lung transplantation effects of deferoxamine on lung preservation in canine single lung transplantation. J Thor Cardiovasc Surg 101:1024, 1991
44. Ely D, Dunphy G, Dollwet H, Richter H, Sellke F, Azodi M: Maintenance of left ventricular function (90%) after twenty-four-hour heart preservation with deferoxamine. Free Rad Biol Med 12:479, 1992
45. Qayumi AK, Jamieson WRE, Poostizadeh A, Germann E, Gillespie KD: Comparison of new iron chelating agents in the prevention of ischemia/reperfusion injury: a swine model of heart-lung transplantation. J Invest Surg 5:115, 1992
46. Paller MS, Hedlund BE: Extracellular iron chelators protect kidney cells from hypoxia/reoxygenation. Free Rad Biol Med 17:597, 1994
47. Bauer M, Feucht K, Ziegenfuss T, Marzi I: Attenuation of shock-induced hepatic microcirculatory disturbances by the use of a starch-deferoxamine conjugate for resuscitation. Crit Care Med 23:316, 1995
48. Rosenthal RE, Chanderbhan, Marshall G, Fiskum G: Prevention of post-ischemic brain lipid conjugated diene production and neurological injury by hydroxyethyl starch-conjugated deferoxamine. Free Rad Biol Med 12:29, 1992
49. Gehlbach P, Purple RL: Enhancement of retinal recovery by conjugated deferoxamine after ischemia-reperfusion. Invest Ophtal Vis Sci 35:669, 1994
50. Hallaway PE, Eaton JW, Panter SS, Hedlund BE. Modulation of deferoxamine toxicity and clearance by covalent attachment to biocompatible polymers. Proc Natl Acad Sci USA 86:10108, 1989

CLINICAL STUDIES OF IRON-CHELATING TREATMENT IN MALARIA

Victor R. Gordeuk,[1*] Godfrey Biemba,[2] and Philip E. Thuma[3]

[1] Department of Medicine
The George Washington University Medical Center
Washington, DC
[2] Macha Mission Hospital
Choma, Zambia
[3] Department of Pediatrics
Hershey Medical Center
Penn State University, Hershey, Pennsylvania

ABSTRACT

Clinical studies have demonstrated that iron chelation therapy has acitivity against human malaria. Desferrioxamine B clears parasitemia and symptoms in patients with uncomplicated falciparum and vivax malaria as a single agent, but recrudescence is common. In combination with quinine, desferrioxamine B enhances parasite clearance and speeds recovery from deep coma in children with cerebral malaria. At least two mechanisms have been proposed for the clinical activity of desferrioxamine B: (i) the inhibition of parasite growth by the withholding of iron and (ii) the protection of central nervous system tissue from free radical-mediated damage. Numerous in vitro studies have supported the first mechanism: the withholding of iron by iron chelators suppressed plasmodial growth in erythrocytes and hepatocytes. Studies of transferrin saturation in children with cerebral malaria have been consistent with the second mechanism. Prolonged coma was associated with abnormally high transferrin saturations, and the addition of iron chelation to the therapeutic regimen appeared to hasten recovery of full consciousness specifically in the children with high transferrin saturations.

INTRODUCTION

New chemotherapeutic strategies are needed to treat malaria, a parasitic infection which potentially affects 40% of the world's population [1]. Falciparum malaria, the most

* Address correspondence to: Victor R. Gordeuk, M. D., Suite 3-428, 2150 Pennsylvania Avenue NW, Washington, D. C., 20037.

severe form, is especially a threat to children in Africa and more than one million die each year from this condition [2]. With vector control of malaria unsatisfactory and with a reliably effective vaccine not yet available, chemotherapy remains the principal weapon in the fight against malaria. The emergence and rapid spread of multi-drug resistant *Plasmodium falciparum* strains makes the search for new agents of critical importance [3].

The pathogenicity of *P. falciparum* is related to its ability to reproduce at an extremely rapid rate. In the asexual erythrocytic stage of the parasite's life cycle, which is responsible for the clinical manifestations, a single merozoite invades the red blood cell, matures into a trophozoite and undergoes DNA replication to give rise to up to 32 daughter cells in just 48 hours. Iron is required for a number of parasite enzyme systems necessary for this explosive growth and proliferation [4-8]. The withholding of iron from the parasite by iron chelators could be expected to inhibit parasite growth, and studies in vitro [9-12] and in animal models [13-15] have shown that this is the case.

A number of studies have now been conducted examining the clinical effect of desferrioxamine B, the only iron chelator approved for use in humans, in malaria. This agent is a naturally occurring trihydroxamic acid derived from cultures of *Streptomyces pilosus* which has been used extensively for over two decades for iron chelation therapy in patients with iron loading anaemias [16]. Desferrioxamine B (molecular weight 657) complexes with iron in a 1:1 molar ratio to produce ferrioxamine with a molecular weight of 713. The drug must be given by continuous parenteral infusion to be effective. Administered in this manner, daily doses of up to 150 mg/kg are tolerated [16,17].

CLINICAL ANTIMALARIAL EFFECTS OF DESFERRIOXAMINE B

Desferrioxamine B as a Single Agent in Adults with Mild to Moderate Malaria

To determine if iron chelation has activity against human malaria, desferrioxamine B has been administered as a single agent in amounts of 100 mg/kg/day by continuous 72-hour subcutaneous infusions to 47 adult Zambian volunteers with asymptomatic infection with *P. falciparum* [18,19] and by contionous 72-hour intravenous infusions to 28 Thai subjects with symptomatic but uncomplicated falciparum or vivax malaria [20]. The results of these studies are summarized in Table 1. Desferrioxamine B was effective in enhancing the clearance of peripheral blood asexual parasites in five cohorts of subjects with *P. falciparum* infection and one cohort of subjects with *P. vivax* infection (Table 1). Inspection of the results shown in Table 1 suggests that the intravenous route of administration is more effective than the subcutaneous route. At the same dosage, Thai subjects with much higher parasite counts who received the chelator intravenously tended to have equivalent or faster parasite clearance then Zambian subjects with low parasite counts and subcutaneous administration. The likely explanation is the achievement of higher serum concentrations with intravenous compared to subcutaneous administration. When steady state serum concentrations of desferrioxamine B + ferrioxamine (the iron complex of desferrioxamine B) were measured in 26 Zambian subjects who received desferrioxamine B subcutaneously at a dose of 100 mg/kg per day for 72 hours [18], the mean steady state concentrations of 6.9 μM at 36 hours and 7.7 μM at 72 hours were at the lower end of the range of values for the ID_{50} of desferrioxamine B against *P. falciparum* as determined in vitro [9,12]. On the other hand, a previous study of desferrioxamine B levels in non-iron-loaded subjects who were given the chelator intravenously suggests that the Thais given desferrioxamine B intravenously at the

Table 1. Desferrioxamine B, 100 mg/kg per day, as a single agent in adults with mild to moderate malaria

Administration	N	Initial mean parasitemia (no./μL)[a]	Time to reduce parasitemia[b] 50%	90%	100%	P-value (comparison to placebo)	Fever clearance time[c]	Country [ref.]
A. Plasmodium falciparum infection								
Continuous s.c.	12	1,150	30 h	47 h	80 h[d]	0.005	—	Zambia [22]
Continuous s.c.	13	280	24 h	43 h	46 h[d]	0.0001	—	Zambia [22]
Continuous s.c.	16	540	28 h	60 h	104 h[d]	0.001	—	Zambia [23]
Continuous s.c.	6	320	24 h	61 h	> 72 h[d]	—	—	Zambia [23]
Continuous i.v.	14	13,540	9 h	25 h	48 h	—	62 h	Thailand[24]
B. *P. vivax* infection								
Continuous i.v.	14	19,380	17 h	39 h	72 h	—	55 h	Thailand [24]

s.c. (sub-cutaneous infusion for 3 days); i.v. (intra-venous infusion for 3 days).
[a]Geometric mean
[b]Extrapolated from plots of geometric mean parasite counts.
[c]Mean
[d]Time to mean value<20/μL, the limit of detection of the assay.

same dosage would be expected to have a mean steady state concentration of about 20 μM [21], a level with greater anti-parasitic effect based on in vitro studies.

Because the Zambian studies were conducted in a malaria transmission area, it was not possible to conclude whether recrudescence of parasitaemia occurred. The study in Thailand, in which subjects were treated in a location where malaria transmission does not occur, indicated that recrudescence is important [20]. Recrudescence was observed in all 14 Thai subjects with falciparum malaria averaging 10 days after the start of therapy, and in 12 of the 14 Thai subjects with vivax malaria averaging 15 days after the start of therapy. Desferrioxamine B has been associated with visual and auditory disturbances when used in the treatment of iron overload, especially with the long term administration of high doses [22-25]. Formal audiologic and visual evaluations were not performed in the present studies of desferrioxamine B as an anti-malarial agent, but the chelator was, in general, well tolerated. No side effects or toxicity could be attributed to desferrioxamine B in the Zambian studies [18,19], but about one-third of the subjects in the Thai study reported transient visual blurring [20].

Desferrioxamine B in Combination with Quinine

The possibility that iron chelation might have both an anti-parasitic effect and a protective effect for host tissues threatened by ischemia and hemorrhage sparked an interest in examining desferrioxamine B in addition to standard anti-malarials in the setting of complicated malaria. To determine if iron chelation added to standard quinine-based therapy enhances parasite clearance and speeds recovery of full consciousness in cerebral malaria, a prospective, randomized, double-blind, placebo-controlled trial of desferrioxamine B was conducted in 83 Zambian children treated with a standard, quinine-based regimen [26]. Entrance criteria included age less than 6 years, *P. falciparum* parasitaemia, normal cerebral spinal fluid and unarousable coma. Each child received quinine, 10 mg/kg, every eight hours for five days and a single dose of sulfadoxine/pyrimethamine, 25/2.5 mg/kg. In addition, either desferrioxamine B, 100 mg/kg, or placebo was given as a 72 hour intravenous infusion. Cox proportional hazards regression analysis was used to examine times to clear parasites and to recover full consciousness, and adjustment was made for significant variables at study entry.

Parasite clearance, measured in 69 subjects, was shorter in children who received desferrioxamine B compared to those who received placebo (p=0.01). The estimated relative rate of parasite clearance with desferrioxamine B was 2.0 times (95% confidence interval of 1.2 to 3.6) the rate with placebo. The time to recover full consciousness was examined after stratification according to the depth of coma at enrollment as suggested by Molyneux et al [27]. In 33 subjects with light coma (scored initially as 3 to 4), the time to regain full consciousness among children given desferrioxamine B did not differ significantly from those who received placebo. Among 50 children with deep coma (scored initially as 0 to 2), the rate of recovery of full consciousness with desferrioxamine B was 2.2 times the rate with placebo (95% confidence interval of 1.1 to 4.7). Estimated median recovery times were 68 hours with placebo (n=28) and 24 hours with desferrioxamine B (n=22). Mortality was 16.7% in 42 children receiving desferrioxamine B versus 22.0% among 41 given placebo (P=0.5), but this study included too few children to detect even a 50% reduction in mortality with statistical significance. None of the survivors treated with desferrioxamine B had neurologic squelae but two of the children given placebo (6%) had paresis at discharge. This study provided preliminary evidence that the addition of iron chelation therapy to standard, quinine-based therapy in children with cerebral malaria may improve the outcome. Whether this approach could reduce mortality and neurologic sequelae would require a larger study. The possible mechanisms for the apparently beneficial effect of desferrioxamine B in cerebral malaria are discussed in the next section.

POTENTIAL MECHANISMS OF ACTION OF IRON CHELATORS IN PATIENTS WITH MALARIA

At least two mechanisms might account for a beneficial effect of iron chelators in the treatment of malaria, including the inhibition of parasitaemia by the withholding of iron from the intra-erythrocytic trophozoite and the protection of ischemic tissue from peroxidant damage.

The Withholding of Iron from the Intra-Erythrocytic Parasite

Evidence from experiments in vitro suggests that desferrioxamine B exerts a stage specific, suppressive and lethal effect on the erythrocytic *P. falciparum* parasite by the withholding of iron [9,12,28-30]. The steady state desferrioxamine B concentrations of 7 to 8 micromolar, which we observed with subcutaneous administration in Zambian subjects with malaria, are within the range of levels that have produced inhibition of plasmodial growth in vitro and thus are consistent with an iron-withholding mechanism of action.

Protection from Peroxidant Tissue Damage in Cerebral Malaria

The obstruction of cerebral microvasculature by *P. falciparum*-infected erythrocytes, local ischaemia and microhaemorrhage are important in the pathogenesis of cerebral malaria; in addition, the release of free haemoglobin and iron probably contribute to the process [31,32]. Studies in vitro and in vivo suggest that the final pathway in ischemic and hemorrhagic injury to the brain and other organs is mediated by iron-generated free radicals that induce lipid peroxidant damage to cellular and sub-cellular membranes [33-36]. Free haemoglobin can serve as a biologic Fenton reagent to provide iron for the generation of the hydroxyl radical [35,36]. With these considerations in mind, it is possible that iron chelation therapy with desferrioxamine B could protect against damage to the central nervous system by inhibiting iron-induced peroxidant damage to cells and sub-cellular structures of the brain, and that the observed effect of the chelator in hastening recovery from deep coma in the trial described above [26] could be at least partially due to this mechanism.

While the inflammatory response to other severe infections is usually associated with cytokine-induced reductions in plasma iron levels [37,38], the presence of hemolysis and dyserythropoiesis which occur in children with severe malaria may raise transferrin saturation in some children with cerebral malaria [39]. Elevations in transferrin saturation would diminish the capability of transferrin to complex with non-protein bound iron that enters the circulation as a result of hemolysis or damage to tissues. Because 'free' iron or hemoglobin may foster the production of free radicals or the formation of more toxic iron-centered activated oxygen species from radicals of lesser potency such as superoxide [35,40], elevated transferrin saturations could permit the generation of these toxic molecules and consequent tissue damage. If free radical generation is important in the pathophysiology of cerebral malaria, more severe or prolonged central nervous system abnormalities might be expected in patients with elevated transferrin saturations and iron chelation therapy might be especially beneficial in this group of patients.

To determine how frequently elevated transferrin saturations develop in the setting of cerebral malaria and if they are associated with an adverse outcome, transferrin saturations at presentation were determined retrospectively in the children with cerebral malaria included in the trial described above [26,41]. More than one-third of children in both the placebo- and iron chelator-treated groups had transferrin saturations exceeding 43%, 3 S.D. above the expected mean for age. Among children given quinine and placebo, those with elevated transferrin saturations had a delayed estimated median time to recover full consciousness (68.2 hours) compared to those with saturations ≤43% (25.4 hours) ($P = 0.006$). The addition of iron chelation to quinine therapy in children with high saturations appeared to hasten recovery ($P = 0.046$). No significant relationship between transferrin saturation and parasite clearance was found. Thus, when other baseline variables are accounted for, elevated transferrin saturations seem to be associated with delayed recovery of consciousness among children with cerebral malaria given only standard therapy, and the addition of an iron chelator seems to hasten the rate of recovery of consciousness in children with elevated saturations.

These results are consistent with the hypothesis that iron-generated free radicals play a role in the pathogenesis of coma in cerebral malaria. Desferrioxamine has been shown to inhibit peroxidant damage to lung tissue in mice [40], to the myocardium in rabbits [42] and to central nervous system tissue in cats [36]. The finding that desferrioxamine B appeared to hasten recovery from coma specifically in children with high transferrin saturations is consistent with the possibility that iron chelation in cerebral malaria diminishes central nervous system damage by protecting against lipid peroxidation induced by iron-generated free radicals.

CONCLUSION

Iron chelation therapy appears to have clinical activity in both uncomplicated and severe malaria. The mechanisms for the effect of iron chelators in clinical malaria are probably complex, and may include the withholding of iron from the parasite as well as the protection of ischemic tissue from peroxidant injury in the patient. To date, all of the clinical studies of iron chelation in malaria have employed agents designed for long-term use to bind-up and remove iron from the body in patients with iron overload. It may now be appropriate to consider the design and clinical testing of iron chelating compounds which are safe for short-term use, orally effective and specifically targeted to bind up parasite-associated iron.

ACKNOWLEDGMENT

Supported in part by an FDA Orphan Products Development Grant (FD-R-000975) and by a grant from the Office of Minority Health to the Cell Biology and Metabolism Branch, National Institute of Child Health and Human Development.

REFERENCES

1. Gilles HM: Management of Severe and Complicated Malaria: A Practical Handbook. Geneva, World Health Organization, 1991.
2. World Health Organization: Malaria Control in Countries where Time-Limited Eradication is Impracticable at Present. Technical Report Series No. 537. Geneva, World Health Organization, 1974.
3. Clyde DF: Recent trends in the epidemiology and control of malaria. Epidemiol Rev 1987;9:219-243.
4. Bezkorovainy A: Biochemistry of Nonheme Iron. New York, Plenum, 1980.
5. Scheibel LW, Sherman IW: Metabolism and organellar function during various stages of the life cycle: Proteins, lipids, nucleic acids and vitamins; in Wernsdorfer W, McGregor I (eds): Malaria: Principles and Practice of Malariology. New York, Churchill Livingstone, 1988, pp 219-252.
6. Wrigglesworth JM, Baum H: The biochemical features of iron; in Jacobs A, Worwood M (eds): Iron in Biochemistry and Medicine II. San Diego, Academic, 1980, pp 29-86.
7. Scheibel LW: Plasmodial parasite biology: Carbohydrate metabolism and related organellar function during various stages of the life cycle; in Wernsdorfer W, McGregor I (eds): Malaria: Principles and Practice of Malariology. New York, Churchill Livingstone, 1988, pp 171-217.
8. Scheibel LW, Rodriguez S: Antimalarial activity of selected aromatic chelators v. localization of ^{59}Fe in *Plasmodium falciparum* in the presence of oxines. Prog Clin Biol Res 1989;313:119-149.
9. Raventos-Suarez C, Pollack S, Nagel RL: *Plasmodium falciparum:* Inhibition of in vitro growth by deferoxamine. Am J Trop Med Hyg 1982;31:919-912.
10. Fritsch G, Sawatzki G, Treumer J, Jung A, Spira DT: *Plasmodium falciparum:* Inhibition *in vitro* with lactoferrin, desferrithiocin, and desferricrocin. Exp Parasitol 1987;63:1-9.
11. Yinnon AM, Theanacho EN, Grady RW, Spira DT, Hershko C: Antimalarial effect of HBED and other phenolic and catecholic iron chelators. Blood 1989;74:2166-2171.
12. Whitehead S, Peto TEA: Stage-dependent effect of deferoxamine on growth of *Plasmodium falciparum in vitro*. Blood 1990;76:1250-1255.
13. Fritsch G, Treumer J, Spira DT, Jung A: Suppression of mouse infections with deferoxamine B. Exp Parasitol 1985;60:171-174.
14. Hershko C, Peto TEA: Deferoxamine inhibition of malaria is independent of host iron status. J Exp Med 1988;168:375-387.
15. Pollack S, Rossan RN, Davidson DE, Escajadillo A: Deferoxamine suppresses *Plasmodium falciparum* in *Aotus* monkeys. Proc Soc Exp Biol Med 1987;184:162-164.
16. Modell B, Berdoukas V: Desferrioxamine; in: The Clinical Approach to Thalassaemia. London, Grune and Stratton, 1984, pp 216-241.
17. Brittenham GM: Iron chelating agents; in: Current Therapy in Hematology Oncology 3. St. Louis, Mosby, 1987, pp 149-153.
18. Gordeuk VR, Thuma PE, Brittenham GM, Zulu S, Simwanza P, M'hango A, Fleisch G, Parry D: Iron chelation with deferoxamine B in adults with asymptomatic *Plasmodium falciparum* parasitemia. Blood 1992;79:308-312.
19. Gordeuk VR, Thuma PE, Brittenham GM, Biemba G, Zulu S, Simwanza G, Kalense P, M'hango A, Parry D, Poltera AA, Aikawa M. Iron chelation as a chemotherapeutic stragegy for falciparum malaria. Am J Trop Med Hyg 1993;48:193-197.
20. Bunnag D, Poltera AA, Viravan C, Looareesuwan S, Harinasuta KT, Schindlery C: Plasmodicidal effect of desferrioxamine B in human vivax and falciparum malaria from Thailand. Acta Tropica 1992;52:59-67.

21. Summers M, Jacobs A, Tudway D, Perera P, Ricketts C. Studies in desferrioxamine and ferrioxamine in normal and iron-loaded subjects. Brit J Haematol 1979:42:547-555.
22. Olivieri NF, Buncic JR, Chew E, Gallant T, Harrison RV, Keenan N, Logan W, Mitchell D, Ricci G, Skarf B, Taylor M, Freedman MH: Visual and auditory neurotoxicity in patients receiving subcutaneous deferoxamine infusions. N Engl J Med 1986;314:869-873.
23. Blake DR, Winyard P, Lunec J , Williams A, Good PA, Crewes SJ, Gutteridge JVC, Rowley D, Halliwell B, Cornish A, Hider RC: Cerebral and ocular toxicity induced by desferrioxamine. Q J Med 1985;56:345-355.
24. Polson RJ, Jawed A, Bomford A, Berry H, Williams R: Treatment of rheumatoid arthritis with desferrioxamine: Relation between stores of iron before treatment and side effects. Br Med J 1985;291:448.
25. Porter JB, Huehns ER: The toxic effects of desferrioxamine. Balliere's Clin Haematolo 1989;2:459-474.
26. Gordeuk VR, Thuma PE, Brittenham GM, McLaren CE, Parry D, Backenstose A, Biemba G, Msiska R, Holmes L, McKinley E, Vargas L, Gilkeson R, Poltera AA: Effect of iron chelation therapy on recovery from deep coma in children with cerebral malaria. New Engl J Med 1992; 327:1473-1477.
27. Molyneux ME, Taylor TE, Wirima JJ, Borgstein A: Clinical features and prognostic indicators in children with paediatric cerebral malaria: A study of 131 Malawian comatose children. Q J Med 1989; 71: 441-459.
28. Scott MD, Ranz A, Kuypers FA, Lubin BH, Meshnick SR: Parasite uptake of desferrioxamine: a prerequisite for antimalarial activity. Brit J Haematol 1990;75:598-602.
29. Loyevsky M, Lytton SD, Mester B, Libman J, Shanzer A, Cabantchik ZI: The antimalarial action of Desferal involves a direct access route to erythrocytic (*Plasmodium falciparum*) parasites. J Clin Invest 1993;91:218-224.
30. Atkinson CT, Bayne MT, Gordeuk VR, Brittenham GM, Aikawa M: Stage-specific ultrastructural effects of desferrioxamine on *Plasmodium falciparum* in vitro. Am J Trop Med Hyg 1991;45:593-601.
31. MacPherson GG, Warrell MJ, Looareesuwan S, Warrell DA: Human cerebral malaria: A quantitative ultrastructural analysis of parasitised erythrocyte sequestration. Am J Pathol 1985;119:385-401.
32. Aikawa M: Human cerebral malaria. Am J Trop Med Hyg 1988;39:3-10.
33. McCord JM: Oxygen-derived free radicals in postischemic tissue injury. N Engl J Med 1985;312:159-163.
34. Henson PM, Johnson RB: Tissue injury in inflammation: Oxidants, proteinases, and cationic proteins. J Clin Invest 1987;79:669-674.
35. Sadrzadeh SMH, Graf E, Panter SS, Hallaway PE, Eaton JW: Hemoglobin: A biologic Fenton reagent. J Biol Chem 1984;259:14354-14356.
36. Sadrzadeh SM, Anderson DK, Panter SS, Hallaway PE, Eaton JW: Hemoglobin potentiates central nervous system damage. J Clin Invest 1987;79:662-664.
37. Alvarez-Hernandez X, Liceaga J, McKay IC, Brock JH: Induction of hypoferremia and modulation of macrophage iron metabolism by tumor necrosis factor. Lab Invest 1989;61:319-322.
38. Dinarello CA, Cannon JG, Mier JW, Bernheim HA, Lo Preste G, Lynn DL, Love RN, Webb AC, Auron PE, Reuben RC, Rich A, Wolff SM, Putney SD: Multiple biological activities of human recombinant interleukin 1. J Clin Invest 1986;77:1734-1739.
39. Pootrakul P, Kitcharoen K, Yansukon P, Wasi P, Fucharoen S, Charoenlarp P, Brittenham G, Pippard MJ, Finch CA: The effect of erythroid hyperplasia on iron balance. Blood 1988;71:1124-1129.
40. Ward PA, Till GO, Kunkel R, Beauchamp C: Evidence for role of hydroxyl radical in complement and neutrophil-dependent injury. J Clin Invest 1983;72:369-371.
41. Gordeuk VR, Thuma PE, McLaren CE, Biemba G, Zulu S, Poltera AA, Askin JE, Brittenham GM: Transferrin saturation and recovery from coma in cerebral malaria. Blood 1995;85:in press.
42. Ambrosio G, Zweier JL, Jacobus WE, Weisfeldt ML, Flaherty JT: Improvement of postischemic myocardial function and metabolism induced by administration of deferoxamine at the time of reflow: The role of iron in the pathogenesis of reperfusion injury. Circulation 1987;76:906-915.

84

IRON REGULATION OF TRANSFERRIN RECEPTOR AND FERRITIN EXPRESSION IN DIFFERENTIATING FRIEND LEUKEMIA CELLS

Eliana Marina Coccia,[1] Emilia Stellacci,[1] Giovanna Marziali,[1] Roberto Orsatti,[1] Edvige Perrotti,[1] Nicoletta Del Russo,[1] Ugo Testa,[2] and Angela Battistini[1*]

[1] Laboratory of Virology and
[2] Laboratory of Hematology and Oncology
Istituto Superiore di Sanità, 00161 Rome, Italy

SUMMARY

Transferrin receptor (TfR) and ferritin expression has been investigated in Friend erythroleukemia cells (FLCs) induced to differentiate by dimethylsulfoxide (Me_2SO). In differentiating FLCs, administration of hemin increases the ferritin content approximately 20-25 fold; conversely, iron salts have only mild stimulatory effects on ferritin accumulation and iron chelators only slightly inhibit the stimulatory effect exerted by hemin. Moreover, in Me_2SO-induced FLC, the negative feedback reported in a variety of other cell types for the regulation of TfR expression by heme is not operative. We conclude that in FLCs induced to differentiate, hemin acts synergistically with the differentiation inducer increasing ferritin expression and is not able to down modulate the TfR number. These effects appear characteristic of differentiating erythroid cells since they are not observed in other cell types (i.e. fibroblastic cell lines).

INTRODUCTION

Cellular iron homeostasis is an essential feature of living organisms and is mainly maintained by the coordinate regulation of the expression of the transferrin receptor (TfR),

* Address correspondence to: Angela Battistini, PhD, Department of Virology, Istituto Superiore di Sanità, Viale Regina Elena, 299, 00161 Rome, Italy. Phone: (396) 49903266; Fax: (396) 4453369 - 49902082.

Molecular Biology of Hematopoiesis 5, edited by Abraham et al.
Plenum Press, New York, 1996

which mediates cellular iron uptake, and the iron storage protein ferritin, which sequesters iron not immediately utilized for cellular metabolism.

In cells requiring more iron, the levels of TfRs rise while the levels of intracellular ferritin fall and, conversely, when cells have adequate iron the opposite occurs. (For a review see 1).

The molecular mechanisms underlying this regulation have been studied extensively and are mostly exerted posttranscriptionally by specific mRNA-protein interactions between the iron regulatory protein (IRP) and iron responsive elements (IREs) contained in the 5' untranslated region (UTR) of ferritin and the 3' UTR of TfR mRNAs, respectively.[1,2] During depletion of the extracellular iron supply, the high affinity binding between the IRP and the IRE in the 5' UTR of the ferritin mRNA inhibits its translation by preventing the stable association of the small ribosomal subunit. At the same time, the interaction of IRP with five IREs in the 3' UTR of the TfR mRNA protects this transcript against targeted endonucleotidic degradation. When iron levels in the medium rise, the IRE binding is prevented, the TfR mRNA is not degraded and ferritin mRNA is actively translated.

This mechanism may be especially relevant for cellular physiology and has been observed in several cell lines such as fibroblasts,[3] leukemia cell lines[4] and mitogen activated T lymphocytes.[5] Conversely, in particular cell types such as macrophages,[6] which are physiologically involved in iron-storage, and in differentiating erythroid cells,[7] which must incorporate high levels of iron required to sustain hemoglobin synthesis, additional mechanisms seem to be operative. Hemoglobin (Hb) synthesis in red cells is the major iron utilization pathway in the human body and accounts for > 80% of systemic iron turnover. Therefore, iron acquisition and utilization are phenomena of central importance in differentiating red cells and must be tightly regulated. Friend leukemia cells (FLCs) provide an useful model for studies of erythroid differentiation at both the cellular and molecular levels. They are erythroid precursors blocked in their differentiation pathway at the proerythroblastic stage and can be induced to differentiate by treatment with dimethylsulfoxide (Me_2SO) and other chemical inducers such as hexamethylenbisacetamide (HMBA). (For a review see 8)

In this report we analyze TfR and ferritin expression in FLCs induced to differentiate. In contrast to the negative feedback observed in other cell types in which iron or hemin treatment induces a decline in TfR synthesis and a rise in ferritin expression, FLCs undergoing erythroid differentiation show a heme-induced up-regulation of both TfR and ferritin expression. Moreover, in contrast to what is observed in fibroblastic cell lines, only heme and not ferric ammonium citrate (FAC) is able to affect ferritin synthesis in FLCs.

MATERIALS AND METHODS

Cells and Treatments

FLC 745A cells were grown in RPMI 1640 supplemented with 5% fetal calf serum (FCS) and antibiotics. B_6 cells were grown in Dulbecco supplemented with 10% FCS and antibiotics. Cells were seeded at $5x10^5$ cells/ml and treated for the indicated times with 1.5% Me_2SO (stock solution undiluted at room temperature Merck), hemin (Sigma), and ferric ammonium citrate (FAC) (Sigma) at the indicated concentrations. Deferoxamine mesylate (Desferal; CIBA-Geigy, Geneva, Switzerland) was used at a final concentration of 100 μM.

Hemin stock solution was prepared by dissolving the powder in 0,5 M NaOH followed by buffering at pH 7,4 with tris (hydroxymethyl) aminomethane. The concentration was evaluated spectrophotometrically at 409 nm in pyridine solution ($\varepsilon = 0{,}163$).

Cells were counted daily with a Coulter counter or a hemocytometer. Cell mortality, evaluated by the trypan blue dye exclusion method, did not exceed 2%. The degree of differentiation was determined by evaluating the percentage of benzidine-positive (B^+) cells according to Orkin et al.[9]

Transferrin Binding Assay

Transferrin (Tf) binding to FLCs was carried out using [^{125}I]-Tf (DuPont, NEN) according to a method previously described.[7]

Total RNA Extraction and Analysis

FLCs (10^8) were treated as indicated and subsequently washed twice with phosphate-buffered saline, and total RNA was isolated by the guanidine-cesium chloride method.[10] RNA samples (10 μg) were analyzed after electrophoresis through denaturing agarose gels (1,2%) containing formaldehyde, transferred onto a Hybond-N nylon membrane (Amersham), and hybridized with random-primed ^{32}P-labeled (1,5 x 10^6 cpm/ml) human H-chain ferritin cDNA (ferritin minigene [Fer mg]),[11] TfR cDNA[12] and glyceraldehyde 3'-phosphate dehydrogenase (GaPDH) cDNA probes. The hybridization conditions were as follows: 50 mM sodium phosphate (pH 7,0)-50% formamide-5xSSC (1xSSC is 0,15 M NaCl plus 0,015 M sodium citrate), 4xDenhardt's solution-0,1% sodium dodecyl sulfate (SDS) and 200 μg of sonicated salmon sperm DNA per ml at 42°C for 20 h. Filters were washed twice in 1xSSC-0,1% SDS for 30 min at room temperature and twice in 0,1xSSC-0,1% for 30 min at 42°C and then exposed to Fuji X-ray film, using intensifying screens.

Evaluation of Ferritin Content

The level of ferritin in the cellular extracts of FLCs was evaluated by Western blot as previously described.[13] Briefly, proteins (50 to 100 μg) from postmitochondrial supernatant fraction were heated, electrophoresed on 10% SDS-polyacrylamide gel, and transferred onto nitrocellulose paper. Ferritin subunits were then shown by using antiferritin antibody (Boehringer Mannheim, Germany) and then [^{125}I]-labeled protein A (DuPont, NEN). A LKB 2202 UltroScan XL laser microdensitometer was used to quantify individual polypeptide bands. The values are expressed in arbitrary units.

RESULTS

Stimulation of Erythroid Differentiation and Modulation of Transferrin Receptor Expression by Me_2SO and Hemin

In FLC, the administration of both Me_2SO and hemin is associated with an increase of globin mRNA, globin chain synthesis and Hb accumulation.[14,15] Cells treated with increasing amounts of hemin displayed up to 40% of B^+cells at day 4 of culture at maximal concentrations of hemin (75μM). As expected, Me_2SO treatment elicited the induction of a higher number of B^+ cells (~60%); the addition of hemin to Me_2SO-treated cells induced a further rise of B^+ cells up to > 95% (Table 1). The analysis of the binding of Tf to its receptor indicates that the addition of hemin to undifferentiated or to Me_2SO-induced FLCs resulted in differential effects (see Table 1). In fact (a) in undifferentiated FLCs, hemin addition elicited a marked down-modulation of the Tf binding capacity, this phenomenon being apparent even with low hemin concentrations (5μM); (b) Me_2SO induced a significant enhancement of the level of Tf binding as compared to the values observed at corresponding

Table 1. Effect of heme addition on hemoglobin synthesis, transferrin receptor and ferritin expression in FLCs

Treatments	(^{125}I)-Tf bound cpm/400,000 cells	B^+ %
Control	3,328	1
Hemin (5 μM)	1,983	5
Hemin (10 μM)	1,823	6
Hemin (25 μM)	1,677	13.5
Hemin (50 μM)	1,443	23
Hemin (75 μM)	1,401	41
Me_2SO (1,5%)	7,633	64
Me_2SO (1,5%) + Hemin (5 μM)	10,173	93
Me_2SO (1,5%) + Hemin (10 μM)	10,173	92
Me_2SO (1,5%) + Hemin (25 μM)	9,430	89
Me_2SO (1,5%) + Hemin (50 μM)	7,436	89
Me_2SO (1,5%) + Hemin (75 μM)	7,076	95.5

The percentage of B+ cells ans [^{125}I]-transferrin binding were evaluated as described in the Materials and Methods after 4 days of culture.

days of culture in control cells, this phenomenon being particularly apparent at day 4 of culture when FLCs are terminally differentiated; and (c) in differentiated (Me_2SO-treated) FLCs, hemin addition elicited a slight, but significant, up-modulation of the Tf binding capacity, this phenomenon being particularly evident at low hemin concentrations.

Effect of Hemin on the TfR mRNA Steady-State Level in Control and Me_2SO-Induced FLCs

In order to clarify the molecular mechanism responsible for the differential effect of hemin on the binding of transferrin to its receptor in Me_2SO treated cells versus undifferentiated cells, TfR mRNA levels were analyzed by Northern blot. Untreated FLCs and FLCs treated with Me_2SO for 24 h and 72 h were incubated with increasing concentrations of hemin. Total RNA was extracted and analyzed by Northern blot. Densitometric quantitation of the autoradiographs expressed in arbitrary units after normalization with GaPDH is shown in Fig. 1.

In control cells, the addition of increasing amounts of hemin elicited the expected down regulation of TfR mRNA both at days 1 and 3 of culture. In Me_2SO-treated cells, a higher level of TfR mRNA with respect to control FLCs was observed, in agreement with ^{125}I Tf binding data. In differentiated FLCs (i.e. day 3 of Me_2SO treatment) hemin was not able to down modulate the TfR mRNA even at higher doses.

Finally in Me_2SO treated but not yet-differentiated cells (day 1 of culture) hemin induced a down modulation of the TfR mRNA similar to that observed in control cells incubated with hemin alone.

Ferritin Regulation by Hemin Treatment in Me_2SO-Induced FLCs

The analysis of the ferritin expression after hemin treatment of FLCs induced or not to differentiate with Me_2SO is shown in Fig. 2.

Dose-response experiments on the third day of culture showed that maximum ferritin accumulation is achieved at hemin concentrations corresponding to 50-75 μM both in control and in Me_2SO-treated FLCs. However, two differences observed between control and Me_2SO-treated cells in response to hemin seem relevant: (i) in control, but not in Me_2SO-

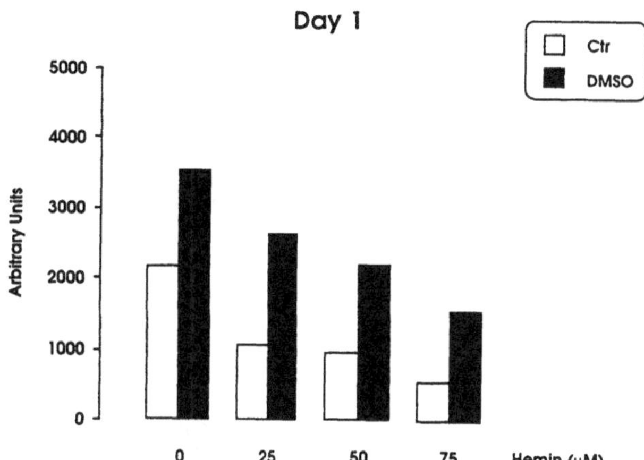

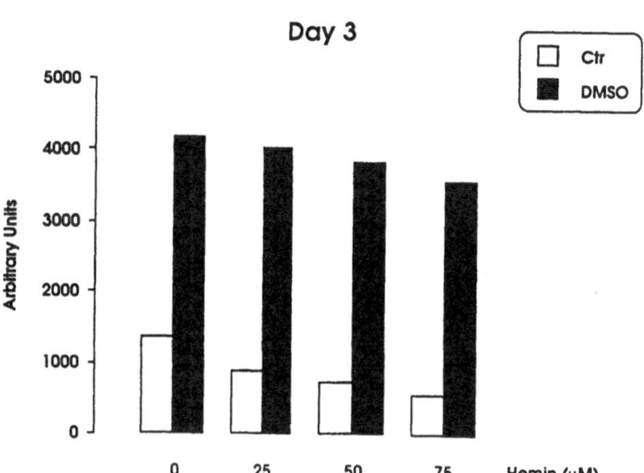

Figure 1. Dose-response effect of hemin treatment on TfR mRNA accumulation in control (open bars) and Me_2SO-induced (black bars) FLCs. After 1 and 3 days of culture, $5X10^7$ cells were collected and total RNA purified for Northern blot analysis as described in the Materials and Methods. The filters were hybridized with TfR and GaPDH cDNA probes.The level of TfR mRNA was evaluated by means of laser densitometry of auto-radiograms. The data represent the mean values observed in three separate experiments and are expressed in arbitrary units after normalization with respect to GaPDH.

treated cells, low hemin concentrations (5-10 μM) (not shown) induced a significant rise of ferritin content and (ii) at optimal hemin concentrations (50-75 μM) Me_2SO-treated cells showed a ferritin content significantly higher than control cells.

This synergistic effect on ferritin synthesis, exerted by hemin plus Me_2SO is clearly linked to the process of erythroid differentiation occurring in FLCs. In fact, control experiments (not shown) indicate that in the fibroblastic B6 cell line, which does not undergo erythroid differentiation, the combined treatment with Me_2SO and hemin does not increase the ferritin accumulation over the value obtained with hemin alone.

Accumultion of Ferritin mRNA and Protein in FLC after Treatment with Different Iron Compounds

Iron is known to enhance the biosynthesis of ferritin without alteration of the steady state level of ferritin mRNA.[16,17,18] However, we recently showed[19] that, in FLCs, 100 μM hemin induces an increase between 5- and 10-fold for both H- and L-chain ferritin mRNAs. Here we extended our observations to the effect of heme on ferritin H-chain expression in FLCs induced to differentiate. As shown in Fig. 3, after 24 h of incubation, Me_2SO induces

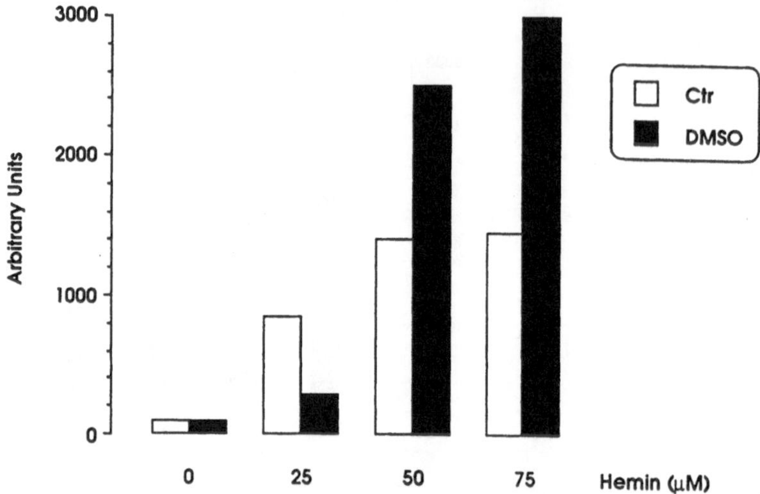

Figure 2. Dose-response effect of hemin treatment on ferritin accumulation. FLCs were treated with the indicated doses of hemin in the absence (open bars) or in the presence (black bars) of 1,5% Me_2SO. On the third day of culture, cells were harvested and processed for Western blot analysis as described in the Materials and Methods. The autoradiograms were scanned by an LKB 2202 UltroScan XL laser microdensitomer and the values, expressed in arbitrary units, are reported in the graph.

a slight (2-fold) increase in ferritin H-chain mRNA accumulation, whereas hemin per sè[19] induces a 5-fold increase. When the inducer is used in combination with 100 μM hemin, a maximal induction of ferritin mRNA is reached (10- to 15-fold depending upon the experiment).

Since we observed a synergistic effect of hemin and the differentiation inducers upon ferritin expression, the effect of another source of iron, i.e. FAC, was also analyzed. As shown in Figure 3 FAC (30 μg per ml) has only a slight effect (2-fold) on mRNA stimulation both in control and in Me_2SO-treated cells. The same results were obtained after 48-72 h of treatment (data not shown).

Iron administration is thought to regulate ferritin expression at the level of mRNA translation, therefore we compared the effect of chelatable iron and heme upon the accumulation of ferritin polypeptides in control and in Me_2SO-treated FLCs (Fig. 4). Cells were treated for 24 h with control medium (lane 1), with 30 μg per ml FAC (lane 2), with 100 μM hemin (lane 3), with 1,5% Me_2SO (lane 4), with 1,5% Me_2SO plus 30 μg per ml FAC (lane 5), and with 1,5% Me_2SO plus 100 μM hemin (lane 6). The ferritin content was determined by immunoblotting using a polyclonal antiferritin antibody. Hemin significantly stimulates ferritin expression both in differentiated and undifferentiated FLCs. On the contrary, in FAC-treated cells, ferritin content rises about 2-fold above control level and never reaches the values observed in hemin-treated cells.

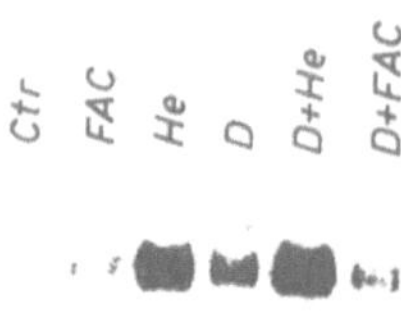

Figure 3. Analysis of ferritin H-chain mRNA steady-state level following iron perturbation. FLCs were treated for 24 h with control medium (ctr), 30 μg per ml FAC (FAC), 100 μM hemin (He), 1.5% Me_2SO (D), 1.5% Me_2SO+100 μM hemin (D+He) and 1.5% Me_2SO+30 μg per ml of FAC. (D+FAC). Total RNA was purified for Northern blot analysis as described in the Materials and Methods. Filters were hybridized with ^{32}P-labelled cDNA probe for ferritin H-chain.

Figure 4. Western blot analysis of ferritin in undifferentiated and differentiated FLCs after treatment with different iron compounds. The cells were treated for 24 h with control medium (lane 1), with 30 μg per ml FAC (lane 2), with 100 μM hemin (lane 3), with 1,5% Me_2SO (lane 4), with 1,5% Me_2SO plus 30 μg per ml FAC (lane 5), and with 1,5% Me_2SO plus 100 μM hemin (lane 6). After 24 h cells were harvested, centrifuged and processed for Western blot analysis as described in the Materials and Methods.

Effect of FAC and Desferal on Ferritin Expression in Differentiating FLCs

To further address the question of which biological form of intracellular iron, either heme or non heme-chelatable iron released from hemin through heme oxygenase activity is directly involved in the induction of ferritin expression in FLCs, noninduced cells or cells induced to differentiate by Me_2SO were treated with FAC and/or with 50 μM hemin plus a molar excess (100μM) of desferal. After the treatments, cells were harvested and both ferritin mRNA accumulation and ferritin biosynthesis were analyzed, respectively, by Northen blot and immunoprecipitation of metabolically labeled proteins. As shown in Table II, the effect of hemin on ferritin H-chain mRNA accumulation is only marginally affected by desferal treatment. Moreover, in Me_2SO + hemin treated cells, the effect of desferal is even weaker with respect to that observed in noninduced cells. In line with the mRNA data, the biosynthesis of ferritin is not significantly affected by the desferal treatment in either noninduced or in Me_2SO-induced cells. FAC (see also Figs. 3 and 4) has only a slight effect on ferritin mRNA and protein accumulation (2- and 4-fold increase over the control value). Instead, in Me_2SO-treated cells, virtually no effect on ferritin mRNA stimulation and only a 2-fold increase in protein accumulation over the value obtained for noninduced FAC-treated cells is observed.

These results suggest that, in cells of erythroid origin such as FLCs, the effect of heme on ferritin accumulation could be mediated by a direct mechanism independent from iron delivery after cleavage by heme-oxigenase.

DISCUSSION

In the present report we analyzed the regulation of ferritin and TfR expression by iron compounds in Friend leukemia cells induced to differentiate.

Iron is an essential nutrient for living cells, but it can exhibit significant toxicity because of its propensity to catalyze the formation of free radicals.[20] Therefore, cells have developed highly efficient mechanisms for acquiring, as well detoxifying, iron.

Cellular iron homeostatis is mainly maintained by the coordinate posttranscriptional regulation of the expression of TfR, which mediates iron uptake, and of the intracellular iron storage protein ferritin, whose major roles are those of iron storage and sequestration. There is increasing evidence that in some specialized cells such as macrophages, which are physiologically involved in iron storage,[6] and FLCs undergoing erythroid differentiation[7,13,19] the regulation of TfR and ferritin expression is mediated by additional and/or different mechanisms with respect to those described in fibroblasts or liver cells. In particular, in erythroid cells, heme synthesis accounts for the majority of total body iron turnover

Table 2. Effect of hemin and desferal on ferritin content in Me_2SO induced FLCs

	Ferritin H- chain content in arbitrary units	
Treatments	mRNA	Protein
Control	0.3 ± 0.15	0.2 ± 0.1
Hemin 50 μM	1.7 ± 0.8	3 ± 0.5
Hemin 50 μM + Desferal 100 μM	1.2 ± 0.5	1.7 ± 0.5
Me_2SO (1.5%)	0.6 ± 0.1	0.5 ± 0.15
Me_2SO (1.5%) + Hemin 50 μM	4 ± 0.7	6.5 ± 1
Me_2SO (1.5%) + Hemin 50 μM + Desferal 100 μM	3.8 ± 0.5	5 ± 1
FAC 50 μg/ml	0.8 ± 0.2	0.8 ± 0.1
Me_2SO (1.5%) + FAC 50 μg/ml	0.8 ± 0.15	2 ± 0.11

Ferritin mRNA and protein content were determined as described in the Materials and Methods. Results are expressed in arbitrary units in terms of mean values ± SE observed in three separate experiments.

and appears itself to play a regulatory role in erythroid cell metabolism.[21,22] This suggests that the regulation of iron acquisition and storage by erythroid cells should be coordinate with the control of iron utilization for heme synthesis.

Our results indicate that in FLCs differentiating to mature erythroblasts hemin addition is not able to down modulate the number of TfRs which instead slightly increases. Experiments in a variety of cell types[4,5,23,24] indicate that the expression of TfRs is modulated by the addition of iron and/or heme through negative feedback, via the intracellular iron and/or heme level. Particularly, experiments performed on Hela cells[25], human leukemic and erythroleukemic lines[4], mitogen-activated human T lymphocytes[5], cultured hepatocytes[26], and uninduced FLCs (this report) show a decrease in the number of TfRs upon the addition of hemin. On the other hand, recent data indicate that, in human primary, *in vitro* grown monocytes-macrophages, iron modulates TfR expression through a positive feedback mechanism.[6] It has also been reported that, in contrast to uninduced cells, iron appears to have less influence on TfR mRNA expression in hemoglobin synthesizing FLCs.[27] These observations are fully consistent with our data which indicate that there is a coordinate control of globin, heme synthesis and iron uptake in erythroid cells.

Our results also indicate that hemin plays a key role in the control of ferritin expression both in uninduced and Me_2SO-induced FLCs at the transcriptional and translational levels (Fig.5). This induction is more pronounced when hemin is added together with the inducer of differentiation.

This stimulatory effect of hemin on ferritin H expression in FLCs does not seem to be related to the iron donating capacity of heme, as suggested by two experimental evidences: (i) iron salts have only a slight stimulatory effect and (ii) iron chelators together with hemin only moderately inhibit the effect of heme on ferritin H expression.

The question of which biological form of intracellular iron, either heme or nonheme chelatable iron, is directly involved in the induction of ferritin mRNA translation (or of TfR mRNA degradation) has long been debated.[28,30] Recently, Eisestein et al.[31] investigated this point in Rat-2 fibroblasts and suggested that iron liberated from heme by heme oxygenase activates ferritin synthesis and that the heme molecule itself is not the agent which is primarily responsible for the induction of ferritin biosynthesis. *In vitro*, both hemin[30] and chelatable iron[28] modulate RNA-binding of IRP.

In FLCs the effect of hemin on ferritin accumulation seems, however, a direct one and not mediated by its capacity to donate iron to the chelatable iron pool through the action of heme oxigenase.

HIGH IRON SUPPLY

FIBROBLASTS DIFFERENTIATING RED CELLS

Transcriptional Regulation

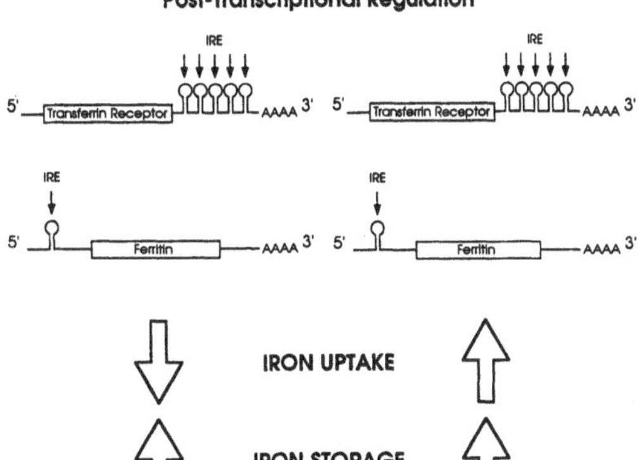

Figure 5. A model for the regulatory mechanisms of TfR and ferritin expression by heme (iron) in differentiating FLCs. In several cellular systems iron homeostasis is maintained by the coordinate post-transcriptional regulation of TfR and ferritin. In particular, high iron supply induces a destabilization of TfR mRNA and allows the ferritin mRNA translation. In contrast to this negative feedback, differentiating FLCs show a heme-induced up-regulation of TfR expression. Furthermore, in FLCs, hemin increases the ferritin H-chain mRNA levels through an enhanced transcription of the gene. We suggest that these effects could be characteristic of a few specialized cells such as those of erythroid origin.

The translational regulation of ferritin biosynthesis has been clearly demonstrated in a variety of *in vivo* and *in vitro* systems as mediated by interaction between the IRE and the IRP. The same elements modulate the stabilization of TfR mRNA in opposite way.

The picture observed in differentiating FLCs is particularly intriguing (Fig. 5). In fact, in this cellular system, an increase in heme availability leads to an increase in ferritin expression without the corresponding down-modulation of TfR binding. It has already been reported that in differentiated murine erythroleukemia cells, ferritin mRNA translation in the presence of desferal appears to be considerably less repressed then in other cell types,[32] while the expected increase in TfR mRNA level is not observed under the same experimental conditions.[33] These observations are fully consistent with our results and raise the question of whether additional and/or different molecular mechanisms regulate the coordinate TfR and ferritin expression in differentiated erythroid cells. In fact, while it has already been demonstrated that IRE/IRP interaction is necessary and sufficient for translational regulation of ferritin mRNA,[11,34-36] other elements in addition to the IREs are required for regulation of the TfR mRNA.[37] Moreover, a recent study showed that erythroid cells exhibit very high levels of TfR transcription as compared to other cell types.[38]

In conclusion our studies support the hypothesis that the increase in heme content occurring during the maturation of FLCs plays a key role not only in stimulating globin-chain synthesis, but also in mediating optimal expression of ferritin and TfRs at the protein level, which is in turn necessary to sustain an optimal synthesis of heme itself. The coordinate regulation of these events by intracellular heme suggests the involvement of unknown regulatory factors specific to a few specialized cells such as those of erythroid origin.

ACKNOWLEDGMENTS

We thank Ramona Ilari and Eleonora Benedetti for technical assistance and Giulia Pacetto and Deborah Wool for editorial assistance of the manuscript.

This work was supported by grants from the Associazione Italiana per la Ricerca sul Cancro, and Consiglio Nazionale delle Ricerche-Progetto Finalizzato ACRO.

REFERENCES

1. Klausner RD, Rouault TA, Harford JB: Regulating the fate of mRNA: The control of cellular iron metabolism. Cell 72:15, 1993
2. Kühn LC, Hentze MW: Coordination of cellular iron metabolism by post-transcriptional gene regulation. J Inorg Biochem 47:183, 1992
3. Ward JH, Kushner JP, and Kaplan J: Transferrin receptors of human fibroblasts. Analysis of receptor properties and regulation. Biochem J 208:19, 1982
4. Louache F, Testa U, Pelicci P, Thomopoulos P, Titeux M, and Rochant H: Regulation of transferrin receptors in human haematopoietic cell lines. J Biol Chem 259:11576, 1984
5. Pelosi E, Testa U, Louache F, Thomopoulos P, Salvo G, Samoggia P, Peschle C: Expression of transferrin receptors in phytohemagglutinin-stimulated human T-lymphocytes. Evidence for a three-step model. J Biol Chem 261:3036, 1986
6. Testa U, Petrini M, Quaranta MT, Pelosi-Testa E, Mastroberardino G, Camagna A, Boccoli G, Sargiacomo M, Isacchi G, Cozzi A, Arosio P, Peschle C: Iron upmodulates the expression of TfRs during monocyte-macrophage maturation. J Biol Chem 264:13181, 1989
7. Battistini A, Marziali G, Albertini R, Habetswallner D, Bulgarini D, Coccia EM, Fiorucci G, Romeo G, Orsatti R, Testa U, Affabris E, Peschle C, Rossi GB: Positive modulation of hemoglobin, heme, and transferrin receptor synthesis by murine interferon-alpha and -beta in differentiating Friend cells. Pivotal role of heme synthesis. J Biol Chem 266:528, 1991
8. Reuben RC, Rifkind RA, Marks PA: Chemically induced murine erythroleukemic differentiation. Biochim Biophys Acta 605:325, 1980
9. Orkin SH, Swan D, Leder P: Differential expression of alpha- and beta-globin genes during differentiation of cultured erythroleukemic cells. J Biol Chem 250:8753, 1975
10. Chirgwin JJ, Przybyla AE, MacDonald RJ, Rutter WJ: Isolation of biologically active ribonucleic acid from sources enriched in ribonuclease. Biochemistry 18:5294, 1970
11. Hentze MW, Rouault TA, Caughman SW, Dancis A, Harford JB, Klausner RD: A cis-acting element is necessary and sufficient for translational regulation of human ferritin expression in response to iron. Proc Natl Acad Sci USA 84:6730, 1987
12. Stearne PA, Pietersz GA, Goding JW: cDNA cloning of the murine TfR: Sequence of trans-membrane and adjacent regions. J. Immunol. 134:3474, 1985
13. Battistini A, Coccia EM, Bulgarini D, Scalzo S, Fiorucci G, Romeo G, Affabris E, Testa U, Rossi GB, Peschle C: Intracellular heme coordinately modulates globin chain synthesis, transferrin receptor number, and ferritin content in differentiating Friend erythroleukemia cells. Blood 78:2098, 1991
14. Ross J, Sautner D: Induction of globin mRNA accumulation by hemin in cultured erythroleukemic cells. Cell 8:513, 1976
15. London IM, Clemens MJ, Ranu RS, Levin DH, Cherbas LF, Ernst V: The role of hemin in the regulation of protein synthesis in erythroid cells. Fed Proc 35:2218, 1976
16. Aziz N, Munro HN: Both subunits of rat liver ferritin are regulated at a translational level by iron induction. Nucleic Acids Res 14:915, 1986
17. Zahringer J, Bahga BS, Munro HN. Novel mechanism for translational control in regulation of ferritin synthesis by iron. Proc Natl Acad Sci USA 73:857, 1976
18. Rouault TA, Hentze MW, Dancis A, Caughman W, Harford JB, Klausner RD: Influence of altered transcription on the translational control of human ferritin expression. Proc Natl Acad Sci USA 84:6335, 1987
19. Coccia EM, Profita V, Fiorucci G, Romeo G, Affabris E, Testa U, Hentze MW, Battistini A: Modulation of ferritin H-chain expression in Friend erythroleukemia cells: transcriptional and translational regulation by hemin. Mol Cell Biol 12:3015, 1992

20. Theil EC: Ferritin: structure, gene regulation, and cellular function in animals, plant, and microorganism. Annu Rev Biochem 56:289, 1987
21. London IM, Levin DH, Matts RL, Thomas NSB, Petryshyn R, Chen JJ: Regulation of protein synthesis, in Boyer PD (ed): The enzymes, Academic Press, New York, 1987
22. Ponka P, Schulman HM, Cox TM: Iron metabolism in relation to heme synthesis, in Dailey HA (ed): Biosynthesis of heme and chlorophyllis, McGraw-Hill, NY, 1990
23. Ward JH, Kushner JP, Kaplan J: TfRs of human fibroblasts. Analysis of receptor properties and regulation. Biochem J 28:19, 1982
24. Testa U, Louache F, Titeux M, Thomopoulos P, Rochant H: The iron chelating agent picolinic acid enhances TfR expression in human erythroleukemic lines. Br J Haematol 60:491, 1985
25. Ward JH, Jordan I, Kushner JP, Kaplan J: Regulation of Hela cells transferrin receptors. J Biol Chem 259:13235, 1984
26. Muller-Eberhard U, Liem H, Grasso J, Giffhorn-Katz S, De Falco M, Katz N: Increase in surface expression of transferrin receptors on cultured hepatocytes of adult rats in response to iron deficiency. J Biol Chem 263:14753, 1988
27. Chan YY, Seiser C, Schulman HM, Kühn LC, Ponka P: Regulation of transferrin receptor mRNA expression distinct regulatory features in erythroid cells. Eur J Biochem 220:683, 1994
28. Rogers J, Munro HN: Translation of ferritin light and heavy subunit mRNAs is regulated by intracellular chelatable iron levels in rat hepatoma cells. Proc Natl Acad Sci USA 84:2277, 1987
29. White K, Munro HN: Induction of ferritin subunit synthesis by iron is regulated at both the transcriptional and translational levels. J Biol Chem 263:8938, 1988
30. Lin JJ, Daniels-McQueen S, Gaffield L, Patino MM, Walden WE, Thach RE: Specificity of the induction of ferritin synthesis by hemin. Biochem Biophys Acta 1050:146, 1990
31. Eisenstein RS, Garcia-Mayol D, Pettingell W, Munro HN: Regulation of ferritin and heme oxygenase synthesis in rat fibroblasts by different forms of iron. Proc Natl Acad Sci USA 88:688, 1991
32. Coccia EM, Goossen B, Hentze MW: Characterization of iron-responsive regulatory elements localized in erythroid δ-aminolevulinic acid synthase mRNAs: Implications for the regulation of heme biosynthesis, in Abraham N et al. (eds): Molecular Biology of Haematopoiesis Andover, Intercept Ltd, 1992
33. Ponka P, Schulman HM, Cox TM: Iron metabolism in relation to haem biosynthesis, in Dailey H (ed): Biosynthesis of Heme and Chlorophyllis, Mc-Graw-Hill. New York, 1990
34. Aziz N, Munro HN: Iron regulates ferritin mRNA translation through a segment of its 5' untranslated region. Proc Natl Acad Sci USA 84:8478, 1987
35. Hentze MW, Caughman SW, Rouault TA, et al: Identification of the iron-responsive element for the translational regulation of human ferritin mRNA. Science 238:1570, 1987
36. Caughman SW, Hentze MW, Rouault TA, Harford JB, Klausner RD: The iron-responsive element is the single element responsible for iron-dependent translational regulation of ferritin biosynthesis. Evidence for function as the binding site for a translational repressor. J. Biol Chem 263:19048, 1988
37. Casey JL, Koeller DM, Ramin VC, Klausner RD, Harford JB: Iron regulation of transferrin receptor mRNA levels requires iron-responsive elements and a rapid turnover determinant in the 3' untranslated region of the mRNA. EMBO J 8:3693, 1989
38. Chan L-NL, Gerhardt EM: Transferrin receptor gene is hyperexpressed and transcriptionally regulated in differentiating erythroid cells. J Biol Chem 267: 8254, 1992

LACTOFERRIN*

A Nuclear Factor that Down-Modulates the Granulocyte-Macrophage Colony-Stimulating Factor Expression

Silvana Penco, Giovanna Bianchi-Scarrà, and Cecilia Garrè†

The Institute of Biology and Genetics
University of Genova, Genova, Italy

ABSTRACT

Lactoferrin (Lf) from human neutrophils has an inhibitory effect on granulocyte-macrophage colony-stimulating factor (GM-CSF) production via Interleukin-1 (IL-1). The nuclear localization of Lf and its ability to bind DNA, even in a specific way, suggest that it may be involved in the transcriptional regulation of GM-CSF. To check this possibility we used two cellular systems: 5637 cell line with constitutive high production of GM-CSF and IL-1β, and human embrional fibroblast (PEU cells), with low basal GM-CSF production, inducible to higher level by IL-1β. We transiently transfected 5637 and PEU cells with a Lf cDNA expression vector and we then analyzed the effect of transfected Lf on the GM-CSF and IL-1β production. Only a 20% of down-modulation of GM-CSF mRNA and protein was observed in 5637 cells; in PEU cells we observed an increase of the GM-CSF mRNA and protein in IL-1β induced cells, whereas Lf transfection was able to inhibit such an effect. Cotransfection of Lf expression vector together with a plasmid, containing 2 kb of human GM-CSF promoter fused to the CAT reporter gene, revealed that Lf was able to decrease the promoter activity in 5637 and in PEU cells. This effect was more evident in IL-1β stimulated cells.

INTRODUCTION

LACTOFERRIN (Lf) is an iron-binding glycoprotein synthetized by several glandular epithelia and polymorphonuclear leukocytes. The physiological function of Lf have not

* Supported by C.N.R. 93.2011.14 (to C.G.), and 93.01977.14 (to G. B-S.).

† Address correspondence to: Dr. Cecilia Garrè, Institute of Biology and Genetics, University of Genova, Viale Benedetto XV, 6, 16132, Genova, Italy. Phone: (+39)103538949, Fax: (+39)103538978 E mail: IBIG@IGECUNIV.CISI.UNIGE.IT.

Molecular Biology of Hematopoiesis 5, edited by Abraham et al.
Plenum Press, New York, 1996

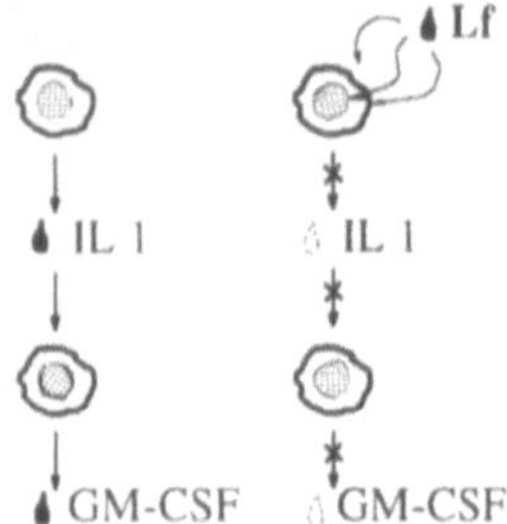

Figure 1. The scheme represents the model proposed according to the findings obtained by Broxmeyer and his group. IL-1β stimulates GM-CSF secretion. Lf is able to inhibit the secretion of IL-1β. This, in turn, results in the inhibition of GM-CSF production.

been precisely defined although several biological function have been attributed to Lf by in vitro studies. Our aim was to investigate the role of neutrophil Lf in myelopoiesis. Lf decreases the production of granulocyte-macrophage colony stimulating factor (GM-CSF). This effect seems to be secondary to the inhibited production of interleukin-1 (IL-1) (1), which is an inducer of GM-CSF expression in physiological and in pathological conditions such as acute myeloblastic leukemia (2, 3). It should be pointed out that these results are not in agreement with other previous reports (4).

We started our studies from the model proposed according to the findings obtained by Broxmeyer and his group (1). The production of IL-1β by monocytes stimulates GM-CSF secretion by fibroblasts. Lf is able to inhibit the production and/or the release of IL-1β by monocytes. This, in turn, results in the inhibition of GM-CSF production (Fig. 1).

The GM-CSF expression is a result of both transcriptional and post-transcriptional controls. Several studies identified cis-elements and related trans-acting factors able to regulate GM-CSF transcription (5). On the other hand, a conserved AU-rich motif in the 3' untranslated region as well as specific binding proteins involved in selective mRNA degradation were found (6, 7).

The molecular mechanism responsible for down-modulation of GM-CSF production is still to be clarified. Using mouse macrophages, Thorens and co-workers showed that Lf reduces the high level of GM-CSF mRNA induced by fetal calf serum (FCS) (8).

Our work was aimed to clarify the molecular mechanisms of GM-CSF down-regulation by Lf. Our hypothesis was based on the observations of the ability of Lf to interact with DNA, even in a specific way (9, 10). Lf-DNA interactions suggest the possible involvement of Lf as a nuclear factor in GM-CSF gene regulation.

MATERIALS AND METHODS

Cells

The following cells were used: bladder carcinoma cell-line 5637, which costitutively produces GM-CSF and IL-1β, and primary cultures of human embryonic lung fibroblasts (PEU cells), obtained from Istituto Zooprofilattico Sperimentale della Lombardia e dell'Emilia, Brescia, Italy, used in passages 10 to 25 in all experiments. Cells were maintained in Dulbecco's modified Eagle's medium (DMEM), supplemented with 10% FCS, 2 mM glutamine, and 0.05 mg/ml gentamycin.

Transient Transfections

5637 cell-line and PEU cells were plated at 70% confluence 24 h before transfection. Cells were transfected by the calcium phosphate coprecipitate method (11). Transfections were performed with 10 μg of the p91023-B expression vector containing Lf cDNA (pLf). Cells, 18 h later, were reincubated for a further 18 h, with or without 15U/ml IL-1β, before harvesting. In other experiments, cells were pLf co-transfected together with 6 μg of pPF2000 containing the promoter GM-CSF (position -2010 to +26) cloned in pBLCAT3 vector. Transfection experiments included 4 μg of pRSV-β-gal, as an internal standard. In control experiments an equimolar concentration of a control plasmid pBluescript (pBS) was used instead pLf. All experiments were performed in duplicate. Harvested cells were utilized for total RNA extraction and for cytoplasmatic extract preparation. Conditioned medium of transfected-cell was utilized for GM-CSF protein assay.

mRNA Assay

Total RNA was purified according to Chomczynski and Sacchi (12). Purified RNA was analyzed by Northern blot (13). Lf cDNA (2.0 kb EcoR1/TthIII fragment in p91023-B), GM-CSF cDNA, (0.8 kb Xho1 fragment in pXM) and IL-1β cDNA (1.3 kb Pst1 fragment in pSP64), and β-actin cDNA (1.8 kb HindIII fragment in pEMBL8); all cDNAs, labeled by random priming, were used as probes.

Lf In Situ Immunofluorescence

Cells, grown on glass coverslips, were transfected with pLf. Fixed cells were permeabilized with 0.1% Triton X100 for 5 min and blocked with 1% iron-saturated human transferrin (Sigma Co., St. Louis, Mo, U.S.A.) for 1 h at 25 °C. Incubation with mouse monoclonal anti-Lf antibody Lf.2B8 (Labometrics, Milano, Italy) and a fluorescent streptavidin-biotin detection system was performed. This antibody was highly specific and did not cross-react with human transferrin nor with bovine Lf. Fluorescent cells were visualized and photographed with Zeiss microscopy.

GM-CSF Protein Assay

Assay of GM-CSF protein in cellular conditioned medium was performed by a solid-phase immunoenzymetric assay (EASIA, Medgenix Diagnostic, Fleurus, Belgium). This assay is highly sensitive and allows detection of 3 pg/ml GM-CSF. No cross reaction was observed with other cytochines.

Chloramphenicol Acetyltransferase (CAT) Assay

Extracts from cell co-transfected with pLf and pPF2000 were normalized for β-galactosidase activity and CAT assay was carried out as described (14). CAT activity is expressed as a percentage of chloramphenicol conversion.

RESULTS

First of all we investigated the effect of transfected-Lf on the expression of GM-CSF by utilizing two cellular systems: 5637 cell line, which costitutively produces high levels of GM-CSF and IL-1β, and human embrional lung fibroblasts (PEU cells) that produce very low levels of GM-CSF, which can be induced to higher levels by IL-1β-treatment.

We started by transiently transfecting 5637 cells with a Lf cDNA expression vector (pLf). Lf in these cells, which do not contain endogenous Lf (data not shown), was detected by in situ immuno-fluorescence (Fig. 2a)

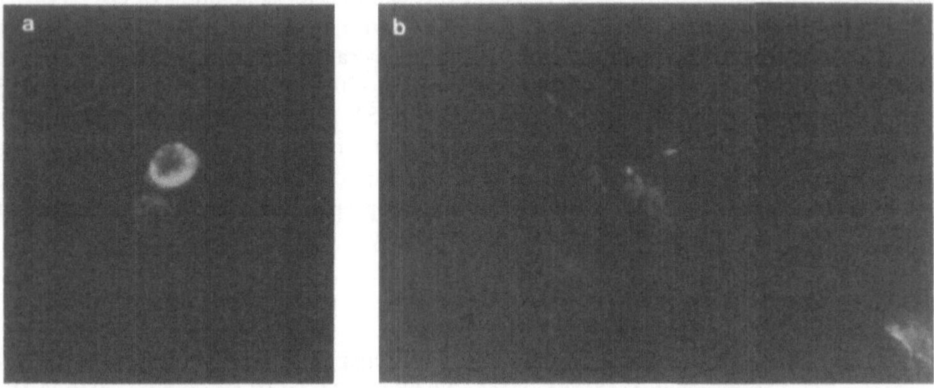

Figure 2. Immune-fluorescent detection of transfected Lf in 5637 (a) and PEU (b) cells. Cells transfected with pLf were visualized by immune-fluorescent staining.

We then analyzed the effect of transfected-Lf on the level of GM-CSF and IL-1β mRNAs. The levels of IL-1β and GM-CSF mRNA were not significantly modified by transfected Lf, in comparison with β-actin mRNA; we observed only a 20% reduction in GM-CSF mRNA level (Fig. 3). We also checked the GM-CSF protein level in conditioned medium of Lf-transfected cells; the level of GM-CSF protein was 20% lower than that of control cells (Fig. 5).

We performed the same experiments in PEU cells and we found that Lf-transfected (Fig. 2b) in IL-1β-induced cells was able to decrease mRNA significantly (Fig. 4). With regard to GM-CSF protein secreted in cell culture medium, we found values around 0.10 ng/ml, whereas IL-1β treatment increased such a values to 0.30 ng/ml; Lf transfected in IL-1β stimulated cells reduced the values to 0.15 ng/ml.

We then explored whether Lf could mediate a reduction in the transcriptional activity of the GM-CSF promoter. In 5637 and PEU cells, the Lf expression vector was co-transfected together with a plasmid containing 2 kb of human GM-CSF promoter fused to the CAT reporter gene. In PEU cells, CAT activity was about three fold increased in IL-1β-treated cells compared with untreated cells. Co-transfection with the Lf expression vector caused a

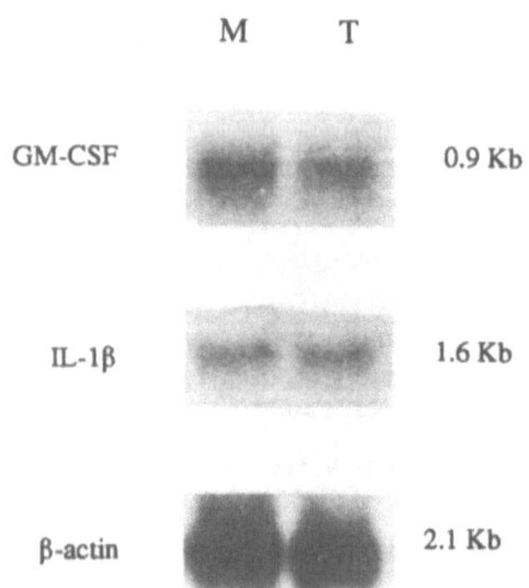

Figure 3. GM-CSF mRNA decrease induced by Lf in 5637 cells. Cells were transfected with either a control plasmid (M) or with pLf (T). RNA (10 μg) was subjected to Northern blot analysis; sequential hybridizations were performed with ^{32}P labeled GM-CSF, IL-1β and β-actin probes.

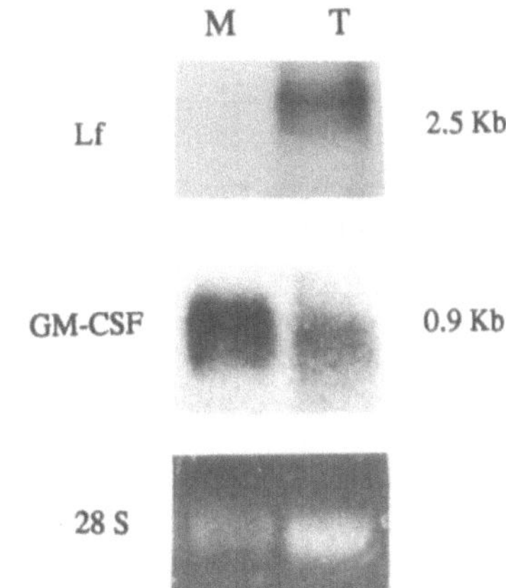

Figure 4. GM-CSF mRNA decrease induced by Lf in IL-1β-stimulated PEU cells. Cells, treated with 15 U/ml of IL-1β, were transfected with either a control plasmid (M) or with pLf (T). RNA was analyzed as indicated in Fig.3; sequential hybridizations were performed with ^{32}P labeled Lf and GM-CSF probes; ethidium bromide staining of 28 S rRNA is also shown.

six fold reduction in CAT activity of IL-1β-treated cells, while less than two fold was observed in untreated cells (Table 1). In 5637 cells, in which IL-1β is endogenously expressed, the CAT expression driven by the GM-CSF promoter was high both in IL-1β-treated and untreated cells and was about four fold higher compared with PEU untreated cells. Co-transfection with Lf plasmid caused a seven fold reduction in CAT activity in both IL-1β-treated and untreated cells (Table 1).

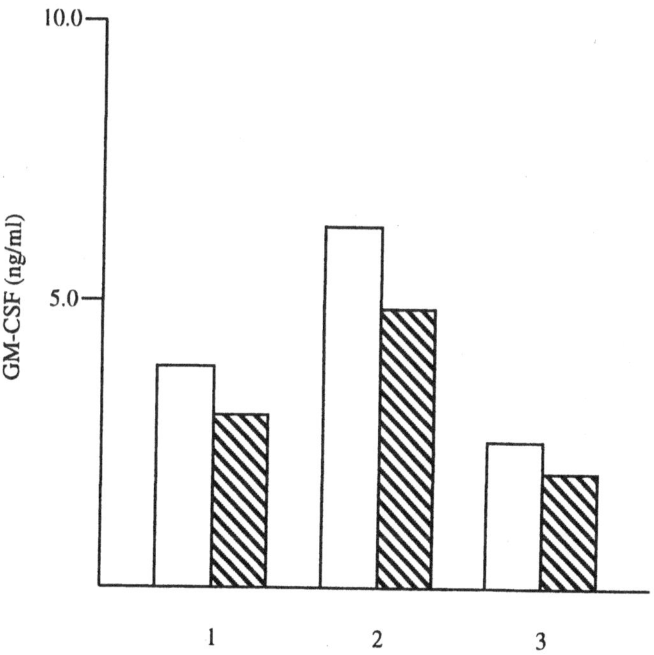

Figure 5. GM-CSF protein concentration in culture medium of Lf transfected or untrasfected 5637 cells. Cells were transfected with 10 μg of Lf or with the control plasmid pBS and GM-CSF protein assay was performed by EASIA method. Three different experiments were performed in duplicate. White bars represent Lf-untransfected cells and stripped bars Lf-transfected cells.

Table 1. Values of GM-CSF promoter activity induced by IL-1β and inhibited by Lf

	Relative CAT activity %			
IL-1β	–	–	+	+
pLf	–	+	–	+
PEU cells	100 ± 10.7	57.4 ± 1.1	286.9 ± 31.5	38.9 ± 6.1
5637 cells	390 ± 10.9	63.7 ± 28.5	427.4 ± 51.8	58.4 ± 16.9

Values represent mean ± SEM of CAT activities of four different experiments performed in duplicate. CAT activity of PEU Lf-untransfected and IL-1β-unstimulated-cells is expressed as 100%; all the other CAT activities are reported as relative %.

DISCUSSION

In co-transfection experiments, we observe a strong effect of Lf-treatment on exogenous GM-CSF promoter activity (seven fold reduction) to respect with a little effect of Lf-transfected on GM-CSF mRNA and protein level (20% reduction) in 5637 cells. Our results show that IL-1β stimulates GM-CSF promoter activity in PEU fibroblasts, and provide additional information about the mechanism by which IL-1 regulates GM-CSF expression (15, 16). Lf-transfection caused a significant reduction in GM-CSF mRNA and protein as well as in CAT activity in IL-1β-stimulated PEU cells.

How Lf acts on the GM-CSF promoter activity is not known. Lf may interact directly with DNA elements that are located within transcriptional control regions. Such a possibility is consistent with the ability of Lf to bind DNA, even in a sequence-specific way (9, 10).

Moreover Lf, very prone to bind other macromolecules (17, 18), might be engaged in interactions with factors that are involved in the regulation of GM-CSF transcription. We are now working to identify regions of the GM-CSF promoter responsive to the Lf inhibitory effect; cotrasfections of pLf together with different subclones of the full-length GM-CSF promoter fused to the CAT reporter gene, will be done in 5637 cells. Interestingly we observed that specific consensus sequences Lf bound (10), are present in the 2.0 kb of the GM-CSF promoter. We found that two of the three above mentioned sequences are present in more than one copy in our promoter; in particular these sequences are well conserved in base positions necessary for the Lf binding. Preliminary results indicate that Lf affect the CAT activity less strongly when we used some subclones with progressive deletion of the Lf bound sequences.

ACKNOWLEDGMENT

We thank Dr. O. Conneely (Baylor College of Medicine, Houston, Texas) for the gift of pLf, Dr. P. Fiorentini (Institute of Biology and Genetics, University of Genova, Genova, Italy) for the gift of pPF2000 and Dr S.C. Clark (Genetics Institute, Cambridge, U.S.A.) for the gifts of IL-1β and GM-CSF cDNAs.

REFERENCES

1. Zucali JR, Broxmeyer HE, Levy D, Morse C: Lactoferrin decreases monocyte-induced fibroblast production of myeloid colony-stimulating activity by suppressing monocyte release of interleukin-1. Blood 74: 1531, 1989.

2. Seelentag WK, Mermod JJ, Montesano R, Vassalli P: Additive effects of interleukin 1 and tumor necrosis factor-α on the accumulation of three granulocyte and macrophage colony-stimulating factor mRNAs in human endothelial cells. EMBO J. 6:2261, 1987
3. Rodriguez-Cimadevilla JC, Beauchemin V, Villeneuve L, Letendre F, Shaw A, Hoang T: Coordinate secrection of interleukin-1β and granulocyte-macrophage colony-stimulating factor by the blast cells of acute myeloblastic leukemia: role of interleukin-1 as an endogenous inducer. Blood 76: 1481, 1990
4. Sawatzski G, Rich IN: Lactoferrin stimulates colony stimulating factor production in vitro and in vivo. Blood cells 15: 371, 1989
5. Gasson JC: Molecular physiology of granulocyte-macrophage colony-stimulating factor. Blood 77: 1131, 1991
6. Shaw G, Kamen R: A conserved AU sequence from the 3' untranslated region of GM-CSF mRNA mediates selective mRNA degradation. Cell 46: 659, 1986
7. Bickell M, Iwai Y, Pluznik DH, Cohen R: Binding of sequence-specific proteins to the adenosine-plus uridine-rich sequences of the murine granulocyte-macrophage colony-stimulating factor mRNA.P.N.A.S 89:10001, 1992
8. Thorens B, Mermod J-J, Vassalli P: Phagocytosis and inflammatory stimuli induce GM-CSF mRNA in macrophages through posttrascriptional regulation. Cell 48: 671, 1987
9. Garrè C, Bianchi-Scarra' G, Sirito M, Musso M, Ravazzolo R: Lactoferrin binding sites and nuclear localization in K562(S). J Cell. Physiol. 153: 477, 1992
10. He J, Furmanski P: Sequence specificity and transcriptional activation in the binding of lactoferrin to DNA. Nature 373: 721, 1995 11. Chen C, Okayama H: High-efficiency trasformation of mammalian cells by plasmid DNA. Mol. Cell. Biol. 7: 2745, 1987
11. Chen C, Hokayama H: High-efficiency transformation of mammalian cells by plasmid DNA. Mol. Cell Biol. 7:2745, 1987
12. Chomczynski P, Sacchi N: Single-step method of RNA isolation by acid guanidium thiocyanate-phenol-chloroform extraction. Anal. Biochem. 162: 156, 1987
13. Rambaldi A, Young DC, Griffin JD: Expression of the M-CSF (CSF-1) gene by human monocytes. Blood 69: 1409, 1987
14. Gorman CM, Moffat Lf, Howard BH: Recombinant genomes express chloramphenicol acetyltransferase in mammalian cells. Mol. Cell. Biol. 2: 1044, 1982
15. Kaushansky K: Control of granulocyte-macrophage colony-stimulating factor production in normal endothelial cells by positive and negative regulatory elements. J. Immunol 143: 2525, 1989
16. Nimer SD, Gates MJ, Koeffler HP, Gasson JC: Multiple mechanisms control the expression of granulocyte-macrophage colony-stimulating factor by human fibroblasts. J. Immunol. 143: 2374, 1989
17. Lampreave F, Pineiro A, Brok JK, Castillo H, Sanchez L, Calvo M: Interaction of bovine lactoferrin with other proteins of milk whey. Int. J. Biol. Macromol. 12: 2, 1990
18. de Lillo A, Tejerina JM, Fierro JF: Interaction of calmodulin with lactoferrin. FEBS 298: 195, 1992

AUTHOR INDEX